Oxford Textbook of

Cancer Biology

Oxford Textbook of

Cancer Biology

EDITED BY

Francesco Pezzella

Nuffield Division of Clinical Laboratory Sciences,
Radcliffe Department of Medicine,
University of Oxford, Oxford, UK

Mahvash Tavassoli

Department Mucosal and Salivary Biology,
King's College London,
London, UK

David J. Kerr

Nuffield Division of Clinical Laboratory Sciences,
Radcliffe Department of Medicine,
University of Oxford, Oxford, UK;
Weill Cornell College of Medicine,
New York, USA

OXFORD
UNIVERSITY PRESS

OXFORD
UNIVERSITY PRESS

Great Clarendon Street, Oxford, OX2 6DP,
United Kingdom

Oxford University Press is a department of the University of Oxford.
It furthers the University's objective of excellence in research, scholarship,
and education by publishing worldwide. Oxford is a registered trade mark of
Oxford University Press in the UK and in certain other countries

First Edition published in 2019

Impression: 2

Published in the United States of America by Oxford University Press
198 Madison Avenue, New York, NY 10016, United States of America

British Library Cataloguing in Publication Data
Data available

Library of Congress Control Number: 2018960018

ISBN 978–0–19–877945–2

Printed and bound in Great Britain by
CPI Group (UK) Ltd, Croydon CR0 4YY

Preface

The textbook is dead. Long live the textbook! With increased output of rapidly published new data and availability of teaching material on the web, it has often been predicted that the textbook will become extinct. However, in our experience, it has also become increasingly difficult to find a comprehensive text which enables us to catch up with the current state of art in multiple fields, within a wider contextual framework. While the high number of research and review papers provide a continuous update on increasingly narrow and specialized topics in cancer biology, we think there will be always a need for concise, coherent descriptions of the fundamentals on areas like cell cycle or cell death. This is particularly important for students who require a platform of basic information before venturing more deeply into the literature. We have assembled a fantastic cast of authors, each of whom are outstanding in their field, and have attempted, when relevant, to make the translational link to the application of cancer biology for patient benefit. We must understand that without novel basic science and the generation of new knowledge, there cannot be sustainable innovations in cancer diagnosis and therapy. We have structured this book logically and trust that the inquisitive reader will select which chapters to explore in greater depth.

There is a difference between the textbooks of today and yesterday: before, publication was the terminus or end of the work for its authors; now, because of the integration between the printed book and online resources, this is no longer the case. This will allow us to annually review, revise, and update the chapters on the online version of the book to reflect recent developments in the field.

Finally, as this is a cancer textbook, we would like to remember our parents, relatives, friends and, of course, patients whose lives have been affected and in many cases, ended too soon by this disease. We hope this book is another small step forward in the right direction.

Acknowledgements

We would like to thank our friends Sandor Paku, Balazs Dome, and Andrew Reynolds for granting us permission to use the picture on the cover of the book, illustrating a non-angiogenic tumour growing in a mouse model.

We would also like to acknowledge the help and support by Oxford University Press staff that guided us through the process of creating this book: Andrea, Caroline, Janine, Sree, and Anya. We also would like to thank all the authors for their work and their willingness and commitment to write.

Contents

SECTION IV
Cancer microenvironment

SECTION V
Global vision of cancer

SECTION VI
The biology of cancer treatment

SECTION VII
Conclusions

Abbreviations

AID	activation-induced deaminase
AIDS	acquired immune deficiency syndrome
ALCL	anaplastic large cell lymphoma
ALK	anaplastic lymphoma kinase
ALK+DLBCL	anaplastic lymphoma kinase-positive diffuse large B cell lymphoma
ALO17	lymphoma oligomersation parter on chromosome 17
ALT	adult T-cell lymphoma
Alt-NHEJ	alternative non-homologous end-joining
AML	acute myeloid leukaemia
AMPK	adenosine-monophosphate-activated protein kinase
AP	alkaline phosphatase
AP1	Activator protein 1
APAAP	alkaline phosphatase-anti-alkaline phosphatase
APB	ALT-associated PML body
APE1	apurinic/apyrimidinic endonuclease 1
AT	ataxia telangiectasia
Atg	autophagy-related protein
ATIC	5-aminoimidazole-4-carboxamide ribonucleotide formyl transferase/IMP cyclohydrolase
ATM	ataxia telangiectasia mutated protein kinase
ATP	adenosine triphosphate
ATR	ataxia telangiectasia and rad-3-related protein kinase
ATRIP	ATR-interacting protein
B	biotin
β-CAT-	β-catenin
B-CLL	B-cell chronic lymphocytic leukaemia
BCL2	B-cell leukemia/lymphoma 2
BCL6	B-cell leukemia/lymphoma 6
BCR-ABL	breakpoint cluster region - Abelson
BER	base excision repair
BL	Burkitt lymphoma
BLM	bloom syndrome protein
BM	bone marrow
BRCA1, 2	breast cancer type 1 and 2
CAH IX	congenital adrenal hyperplasia IX
CAK	CDK-activating kinase
CAM	cell adhesion molecules
CARs	cysteinyl-tRNA synthetase
cART	combination antiretroviral therapy
CBP	CREB-binding protein
CDC25A, C	cell division cycle 25A and 25C
CDK	cyclin-dependent kinase
CDKI	cyclin-dependent kinase inhibitor
CHK1, 2	checkpoint protein kinase 1 and 2
CIN	chromosomal instability
CK7	cytokeratin 7
CK8/18	cytokeratin 8/18
CKI	CDK-inhibitory
CLTC	clathrin heavy chain
CMIP	CpG island methylator phenotype
c-RAF	RAF proto-oncogene serine/threonine-protein kinase
CRE	cyclic AMP response element
CRKL	CRK avian sarcoma virus CT-10 homologue-like
CS	Cockayne syndrome
CSC	cancer stem cell
CSR	class switch recombination
CtIP	CtBP-interacting protein
C3G-CRKL	protooncogene c-CRK
D-loop	displacement loop
DAPK	death-associated protein kinase
DDR	DNA damage response
Deptor	DEP domain-containing mTOR-interacting protein
DNA-PKcs	DNA-dependent protein kinase catalytic subunit
DNA Pol	DNA polymerase
DR5	death receptor 5
DRAM	damage-regulated autophagy modulator
DSB	double-strand break
E-cad	E-cadherin
E2F1	E2F transcription factor 1
EBV	Epstein–Barr virus
EGF	epidermal growth factor
EGFR	epidermal growth factor receptor
EME1	essential meiotic endonuclease 1
EML4	echinoderm microtubule-associated protein like 4
ER	endoplasmic reticulum
ER	oestrogen receptor
ERCC1, 5	excision repair cross-complementation group 1 and 5
ERK	extracellular-signal-regulated kinase
EXO1	exonuclease 1
FA	Fanconi anaemia
FANCA, B, C, D2, M, I, P	Fanconi anaemia complementation group A, B, C, D2, M, I, P
FGFR	fibroblast growth factor receptor
FITC	fluorescein isothiocyanate
FOXM1	forkhead box M1
FOXO3A	forkhead box protein O3
FOXP1	forkhead box protein 1
FOXP3	forkhead box protein 3

FRS2 fibroblast growth factor receptor substrate 2
GAB2 GRB2 associated binding protein
GADD45 growth arrest and DNA damage 45
GATA4 GATA-binding protein 4
GC germinal centre
GF growth factor
GFR growth factor receptors
GG-NER global genome nucleotide excision repair
GGR global genome repair
GRB2 growth factor-receptor-bound protein 2
HAV hepatitis A virus
HBV hepatitis B virus
HCV hepatitis C virus
HDAC histone deacetylase
HER2 human epidermal growth factor receptor 2
HIV Human Immunodeficiency Virus
HIF-1 hypoxia-inducible factor
HNSCC head and neck squamous cell carcinoma
HPV human papilloma virus
HR homologous recombination
HRP horseradish peroxidase
HSC haematopoietic stem cells
HTLV human T-lymphotropic retrovirus
HVS herpesvirus saimiri
IARC International Agency for Research on Cancer
ICL interstrand DNA crosslink
IDL insertion and deletion loop
IMT inflammatory myofibroblastic tumours
IR ionizing radiation
IR Insulin receptor
IRF-4 interferon regulatory factor 4
IRS-1 insulin receptor substrate 1
JAK3 Janus kinase 3
JNK-C Jun N-terminal kinase
KAP-1 KRAB-associated protein-1
KEGG *Kyoto Encyclopedia of Genes and Genomes*
KIF5B kinesin/family member 5B
KRAS Kirsten rat sarcoma viral oncogene homologue
KS Kaposi's sarcoma
KSHV Kaposi's sarcoma-associated herpesvirus
LC3 microtubule-associated protein 1A/1B-light chain 3
LFS Li-Fraumeni syndrome
LUCA last unknown common ancestor
MAP MUTYH-associated polyposis
MAPK mitogen-activated protein kinase
MCD multicentric Castleman disease
MCL mantle cell lymphoma
MCPV Merkel cell polyomavirus
MDC1 mediator of DNA-damage checkpoint 1
MDM2 mouse double minute 2 homologue
MIN microsatellite instability
MLH1 MutL homologue 1
MMR mismatch repair
MMP matrix metalloproteinase
MRN Mre11/Rad50/Nbs1 complex
MSH2 MutS homologue 2
MSN moesin
mTOR mammalian target of rapamycin
MYC MYC proto-oncogene
MYCN MYCN protooncogene, neuroblastoma derived
MYH9 non-muscle heavy chain
NAD nicotinamide adenine dinucleotide
NBS Nijmegen breakage syndrome
NER nucleotide excision repair
NF-kB nuclear factor kappa-light-chain-enhancer of activated B cells
NHEJ non-homologous end-joining
NPC nasopharyngeal carcinoma
NPM nucleophosmin
NPM-ALK nucleophosmin-anaplastic lymphoma kinase
NSCLC non-small cell lung cancer
p53BP1 p53 binding protein 1
p130CAS breast cancer anti-estrogen resistance 1
p130Cas breast cancer anti-oestrogen resistance protein 1
pAMPK phosphorylated 5' adenosine monophosphate-activated protein kinase
PAR poly-ADP-ribose
PARG poly (ADP-ribose) glycohydrolase
PARP poly (ADP-ribose) polymerase
PCAF P300/CBP-associated factor
PCNA proliferating cell nuclear antigen
PDGF platelet-derived growth factor
PI3K phosphoinositide 3-kinase
PKB protein kinase B
PLC-g phospholipase C-gamma
PLK1 polo kinase 1
PLWHA people living with HIV/AIDS
PML promyelocytic leukaemia
PR progesterone receptor
PTEN phosphatase and tensin homologue
p16 cyclin-dependent kinase inhibitor 2A
p53 TP53 or tumour protein
p56 phosphoglycerate kinase
P53R2 p53-inducible ribonucleotide reductase small subunit 2-like protein
RAD51, RAD52 radiation repair 51 and 52
RANBP2 Ran-binding protein 2
KRAS rat sarcoma viral oncogene homologue
Rb retinoblastoma protein
Redd1 regulated in development and DNA damage response 1
RISC RNA-induced silencing complex
RNA ribonucleic acid
ROS reactive oxygen species
RPA replication protein A
RSV rouse sarcoma virus
RTK receptor tyrosine kinase
SAC spindle assembly checkpoint
SASP senescence-associated secretory phenotype
SEC31L1 SEC31 homologue A
SH2 src homology 2
SHM somatic hypermutation
SHP2 protein-tyrosine phosphatase 1D or protein-tyrosine phosphatase 2C

SIRT1	sirtuin 1
SOS	Son of Sevenless
SRC	non-receptor protein kinase sarcoma
ssDNA	single-strand DNA
SSB	single-strand break
SSBR	single strand break repair
STAT5	signal transducer and activator of transcription 5
STORM	stochastic optical reconstruction microscopy
T-loop	telomere loop
T-SCE	telomere-sister chromatid exchange
T-stump	telomere stump
TAL-1	T-cell acute lyphocytic leukaemia protein 1
TC-NER	transcription-coupled nucleotide excision repair
TCP	tumour control probability
TCR	transcription-coupled repair
TEN	telomerase N-terminal
TFG	TRK-fused gene
TFIIH	transcription factor IIH
TLS	translesion DNA synthesis
TNFSF10	tumour necrosis factor superfamily member 10
TopBP1	topoisomerase (DNA) II binding protein 1
TPM	tropomyosin
TrkA	tropomyosin receptor kinase A
TSC2	tuberous sclerosis 2
TTD	trichothiodystrophy
ULK1	Unc-51-like kinase 1
VEGF	vascular endothelial growth factor
VEGFR	vascular endothelial growth factor receptor
VIM	vimentin
Wip1	wild-type P53-induced phosphatase 1
XLF	XRCC4-like factor
XP	xeroderma pigmentosum
XPA, B, C, D, F, G	xeroderma pigmentosum complementation group A, B, D, F, G
XRCC1, 4	X-ray cross-complementing protein 1 and 4
ZAP70	zeta-chain (TCR) associated protein kinase 70 kD

Contributors

Balkees Abderrahman, Department of Breast Medical Oncology, University of Texas, MD Anderson Cancer Center, Houston, USA

Margaret Ashcroft, Department of Medicine, University of Cambridge, Cambridge, UK

Nicholas Athanasou, Nuffield Department of Orthopaedics, Rheumatology, and Musculoskeletal Science, University of Oxford, Oxford, UK

Michael Baumann, German Cancer Research Center (DKFZ); and Department of Radiotherapy and Radiation Oncology, Faculty of Medicine and University Hospital Carl Gustav Carus, Technische Universität Dresden, Germany

Karim Bensaad, Department of Oncology, University of Oxford, Oxford, UK

Jessica Bullenkamp, Molecular and Clinical Sciences Research Institute, St. George's University London, London, UK

Giacomo Buscemi, Department of Biosciences, University of Milan, Milan, Italy

Simon Carr, Department of Oncology, University of Oxford, Oxford, UK

David N. Church, Cancer Genomics and Immunology Group and NIHR Comprehensive Biomedical Research Centre, The Wellcome Centre for Human Genetics, University of Oxford, Oxford, UK

Laura C. Collopy, Cancer Institute, Faculty of Medical Sciences, University College London, London, UK

Pedro Cutillas, Cell Signalling and Proteomics Group, Barts Cancer Institute (CRUK Centre), Queen Mary University of London, London, UK

Blossom Damania, Lineberger Comprehensive Cancer Center and Department of Microbiology and Immunology, School of Medicine, University of North Carolina, Chapel Hill, USA

Dirk P. Dittmer, Lineberger Comprehensive Cancer Center and Department of Microbiology and Immunology, School of Medicine, University of North Carolina, Chapel Hill, USA

Tom Donnem, Department of Oncology, University Hospital of North Norway and the Arctic University of Norway, Tromso, Norway

Anna Dubrovska, OncoRay-National Center for Radiation Research in Oncology, Faculty of Medicine and University Hospital Carl Gustav Carus, Technische Universität Dresden; and Helmholtz-Zentrum Dresden-Rossendorf, Institute of Radiooncology-OncoRay; Cancer Consortium (DKTK), partner site Dresden, and German Cancer Research Center (DKFZ), Germany

Nadège Gaborit, Institut de Recherche en Cancérologie de Montpellier, INSERM U1194, Université de Montpellier, Institut régional du Cancer de Montpellier, Montpellier, France

Kevin Gatter†, Nuffield Division of Clinical Laboratory Sciences, Radcliffe Department of Medicine, University of Oxford, Oxford, UK

Mark A. Glaire, Cancer Genomics and Immunology Group and NIHR Comprehensive Biomedical Research Centre, The Wellcome Centre for Human Genetics, University of Oxford, Oxford, UK

Betty Gration, MRC Molecular Haematology Unit, Radcliffe Department of Medicine, Weatherall Institute of Molecular Medicine, University of Oxford, Oxford, UK

Daniel Haller, Department of Medicine, Perelman School of Medicine, University of Pennsylvania, Philadelphia, USA

Adrian L. Harris, Department of Oncology, University of Oxford, Oxford, UK

Edward Hookway, Nuffield Department of Orthopaedics, Rheumatology and Musculoskeletal Sciences, University of Oxford, Oxford, UK

Jiangting Hu, Radcliffe Department of Medicine, University of Oxford, Oxford, UK

V. Craig Jordan, Department of Breast Medical Oncology, University of Texas MD Anderson Cancer Center, Houston, USA

Robert Kerbel, Biological Sciences Platform, Sunnybrook Research Institute, Department of Medical Biophysics, University of Toronto, Toronto, Canada

David J. Kerr, Nuffield Division of Clinical Laboratory Sciences, Radcliffe Department of Medicine, University of Oxford, Oxford, UK; and Weill Cornell College of Medicine, New York, USA

Benedikt M. Kessler, Target Discovery Institute, Nuffield Department of Medicine, University of Oxford, Oxford, UK

Mechthild Krause, Department of Radiotherapy and Radiation Oncology, Faculty of Medicine and University Hospital Carl Gustav Carus, Technische Universität Dresden; German Cancer Consortium (DKTK), partner site Dresden, and German Cancer Research Center (DKFZ); OncoRay – National Center for Radiation Research in Oncology, Faculty of Medicine and University Hospital Carl Gustav Carus, Technische Universität Dresden, Helmholtz-Zentrum Dresden - Rossendorf; Helmholtz-Zentrum Dresden - Rossendorf, Institute of Radiooncology – OncoRay, Dresden, Germany; National Center for Tumor Diseases (NCT), Partner Site Dresden; German Cancer Research Center (DKFZ); Faculty of Medicine and University Hospital Carl Gustav Carus, Technische Universität Dresden; and Helmholtz Association / Helmholtz-Zentrum Dresden - Rossendorf (HZDR), Germany

Nicholas La Thangue, Department of Oncology, University of Oxford, Oxford, UK

Andrew P. Mazar, Monopar Therapeutics, Wilmette, USA

Wilma Mesker, Department of Surgery, Leiden University Medical Center, Leiden, the Netherlands

Kingsley Micklem, Nuffield Division Clinical Laboratory Science, Radcliffe Department of Medicine, University of Oxford, Oxford, UK

Gwennaëlle C. Monnot, Ludwig Cancer Research Center, Department of Fundamental Oncology, Faculty of Biology and Medicine, University of Lausanne, Lausanne, Switzerland

Udo Oppermann, Nuffield Department of Orthopaedics, Rheumatology, and Musculoskeletal Science, University of Oxford, Oxford, UK

Francesco Pezzella, Nuffield Division of Clinical Laboratory Sciences, Radcliffe Department of Medicine, University of Oxford; and Cellular Pathology Clinical Service Unit, Oxford University Hospitals, Oxford, UK

David H. Phillips, Department of Analytical, Environmental and Forensic Sciences, School of Population Health and Environmental Sciences, King's College London, London, UK

Karen Pulford, Emeritus Reader in Immunodiagnostics, Nuffield Division of Clinical Laboratory Sciences, Radcliffe Department of Medicine, University of Oxford, Oxford, UK

† It is with regret we report the death of Kevin Gatter during the preparation of this textbook.

Chao-Nan Qian, Department of Nasopharyngeal Carcinoma, State Key Laboratory of Oncology South China, Sun Yat-Sen University Cancer Center, Guangzhou, China

Lynn Quek, MRC Molecular Haematology Unit, Radcliffe Department of Medicine, Weatherall Institute of Molecular Medicine, University of Oxford, Oxford, UK

Andrea Rasola, Department of Biomedical Sciences, University of Padova, Padova, Italy

Pedro Romero, Ludwig Cancer Research Center, Department of Fundamental Oncology, Faculty of Biology and Medicine, University of Lausanne, Lausanne, Switzerland

Almut Schulze, Department of Biochemistry and Molecular Biology, Biocenter, University of Würzburg, Würzburg, Germany

Connor Sweeney, MRC Molecular Haematology Unit, Radcliffe Department of Medicine, Weatherall Institute of Molecular Medicine, University of Oxford, Oxford, UK

Mahvash Tavassoli, Department Mucosal and Salivary Biology, King's College London, London, UK

Rob Tollenaar, Department of Surgery, Leiden University Medical Center, Leiden, the Netherlands

Kazunori Tomita, Cancer Institute, Faculty of Medical Sciences, University College London, London, UK

Andrey Ugolkov, Division of Hematology and Oncology, Feinberg School of Medicine, Northwestern University, Chicago, USA

Pieter-Jan van Dam, Faculty of Medicine and Health Sciences, University Antwerp, Antwerp, Belgium

Steven Van Laere, HistoGeneX NV, Antwerp, Belgium

Lieven Verbeke, Department of Information Technology, Ghent University, Ghent, Belgium

Jaap Verweij, Department of Medical Oncology, Erasmus University Medical Centre, Rotterdam, the Netherlands

Paresh Vyas, MRC Molecular Haematology Unit, Radcliffe Department of Medicine, Weatherall Institute of Molecular Medicine, University of Oxford, Oxford, UK

Herman Waldmann, Sir William Dunn School of Pathology, University of Oxford, Oxford, UK

Yan-Qun Xiang, Department of Nasopharyngeal Carcinoma, Sun Yat-Sen University Cancer Center, Guangzhou, China

Yosef Yarden, Department of Biological Regulation, Weizmann Institute of Science, Rehovot, Israel

SECTION I
The multicellular organism

1 The multicellular organism and cancer

Francesco Pezzella, David J. Kerr, and Mahvash Tavassoli

Introduction

Cancer has a lot to do with the way life has developed on our planet and the successful evolution of multicellular organisms. Cells are the smallest unit containing all the features necessary and sufficient to life, the viruses occupying a special place. In 1863, the German pathologist Rudolf Virchow introduced the concept of cellular pathology (Virchow, 1863) stating that diseases are due to the occurrence of a pathological process at cellular level. This is very much the case with cancer that is definitively a disease of a cell belonging to a multicellular organism.

A brief history of the cell: Eubacteria (Bacteria), Archaea, Eukaryotes, and the last unknown common ancestors

The defining moment for the appearance of the cell has been the formation of what we now call the cell membrane. This is a complex structure able to form vesicles allowing the segregation inside of genetic material (the Genotype) plus the molecular machinery (the Phenotype) needed for this new structure to grow and reproduce copies of itself, through the cell cycle.

Cells are divided into two taxons, Prokaryota and Eukaryota, a taxon being formed by organisms included in a particular entity (e.g. in a family or in a genus; (Thain and Hickman, 2004). This distinction is based on the structure and organization of the cell: in the Eukaryotes (Composite), cell membranes are present also inside the cells delimiting discrete internal structures such as, for example, nucleus and mitochondria, while no such division can be found in the prokaryotes (non-composite; Fig. 1.1). All the cells share a set of common features: they contain their genetic information, replicate throughout the cell cycle, their activity is governed through cell signalling and can produce energy through a metabolic apparatus. Approximately 200 gene families are common to the two taxons.

The introduction of genomic studies, as a tool to investigate the evolutionary correlations between organisms, has unveiled within the prokaryotes two distinct groups or domains, the Bacteria and the Archaea, as distant from one other as they are from the Eukaryotes. It has therefore been proposed that, above the division into animal Kingdoms, exists a division into three domains: the Eubacteria (or Bacteria), the Archaea, and the Eukaryotes (Woese et al., 1990), each domain comprising a variety of kingdoms (Fig. 1.2). Molecular studies have demonstrated that the two domains Bacteria and Archaea derive from the last unknown common ancestors (LUCA), while the Eukaryotes evolved from the Archaea (Fig. 1.2). LUCA is defined as the last organism preceding, in the evolutionary tree, the division into the two domains of Bacteria and Archaea and it is assumed to be the living organism from which all present living organisms descend. It is estimated that LUCA lived between some 3.5 to 3.8 billion years ago (Fig. 1.3).

The genetic division of cells into these three domains is reflected by their biological characteristics, some of which are summarized in Table 1.1. The mechanism of transcription, translation, and splicing in the Archaea is close to that of the Eukarya and both differ from the one found in the Prokaryota. Crucially, although the Archaea do not have a nucleus, they have histone proteins that bind to DNA double strand, compacting it into nucleosome-related structures, and Archaea RNA polymerases have the multisubunit complexity of Eukarya RNA polymerases. On the other side, the metabolism of the Archaea is more similar to Eubacteria than to Eukarya (Olsen and Woese, 1997). Despite the closer similarity in metabolic functions of Eubacteria to Archaea, there is one exception: the use of photosynthesis that can be found both in Eubacteria and Eukarya but is absent in Archaea. This is due to the fact that, although genetic evidences show that the Eukarya evolved from the Archaea, horizontal transfer of genes has happened between Eubacteria and Eukarya (Hedges, 2002).

Basic anatomy of the eukaryotic cell in Metazoa

In the cytological classification dividing the prokaryote from the eukaryotic cell the latter is distinguished as it is composed by several organelles, some possibly reminding a more primitive cell, which have learned to live in symbiosis. Each of these organelles contributes to specific need(s) of the eukaryotic cell. It is now believed that the crucial moment to the transition from a simpler cell to the more complex eukaryote was when different cells started to live inside others. Crucial to all this was the formation of the nucleus and the appearance of mitochondria. The main anatomical

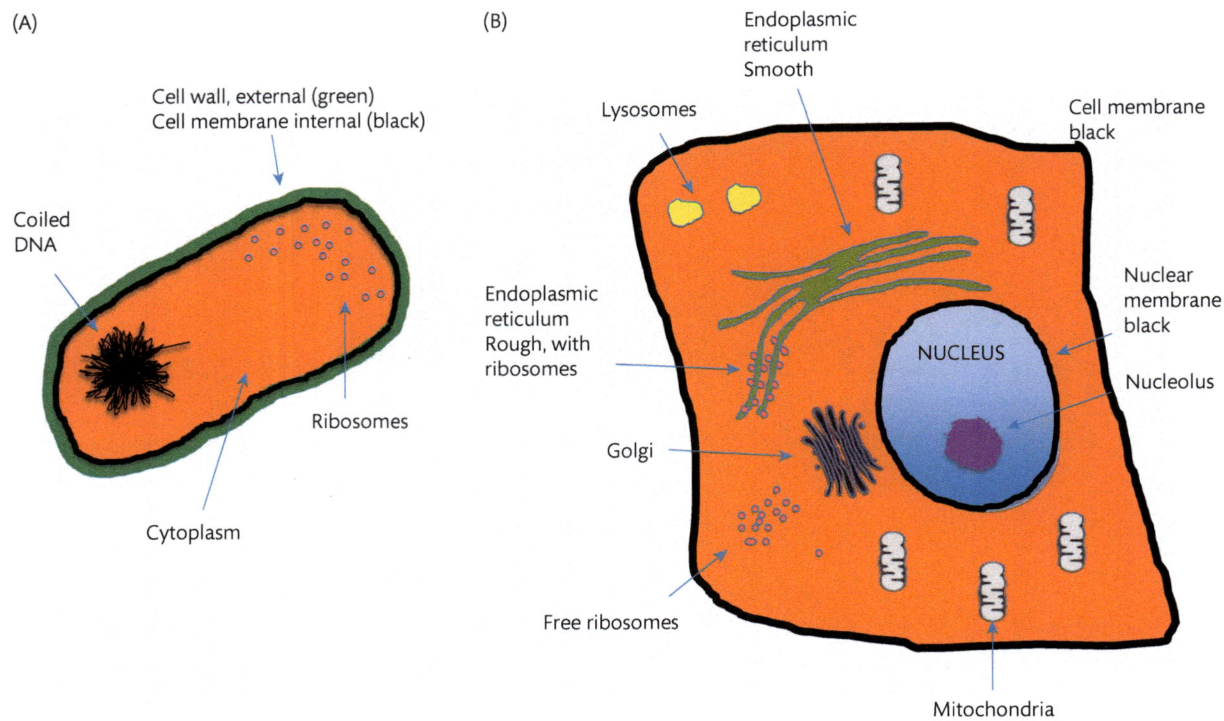

Fig. 1.1 The prokaryotic and the eukaryotic cells. (A) The prokaryotic cell is defined by the cell membrane. Inside the space delimitated by this membrane is the cytoplasm in which all the molecules are contained in one unique space. (B) The eukaryotic cell is also defined by the cell membrane, however the cell membrane is also present inside the cells where defines different organelles. The most prominent is the nucleus, in which the genetic material, the DNA, is segregated. Other cell membrane-defined organelles are the mitochondria, the Golgi apparatus, lysosomes, and the endoplasmic reticulum (ER). The former is divided into the ER rough, when ribosomes are attached to its membrane, and smooth, when ribosomes are not present.

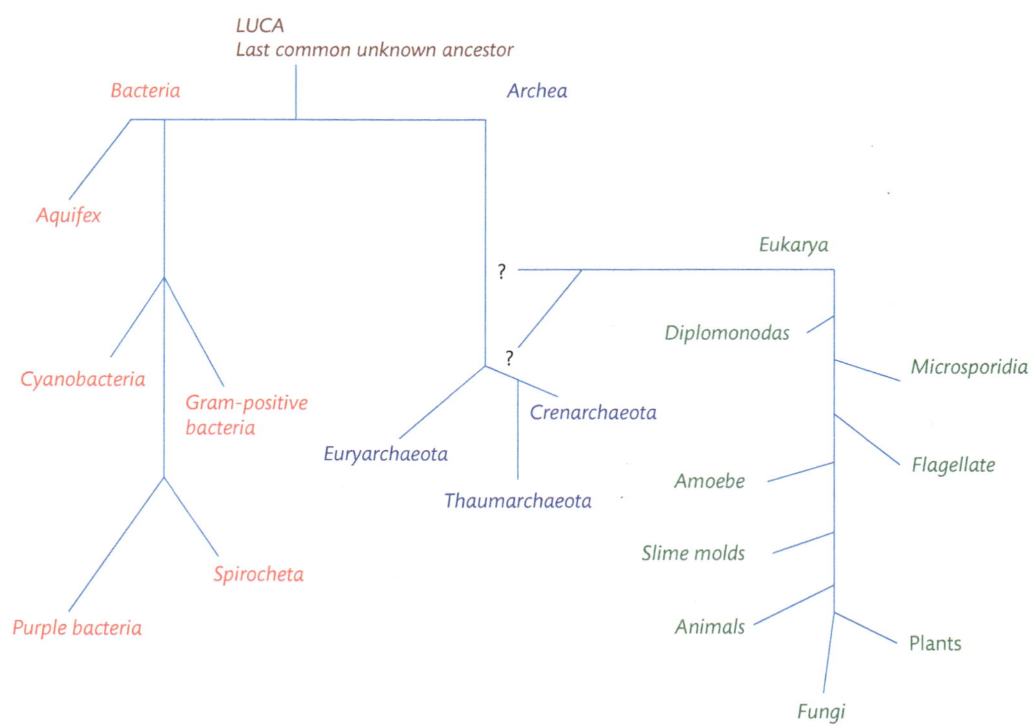

Fig. 1.2 The three domains: Bacteria, Archaea, and Eukaryota. The last unknown common ancestor (LUCA) evolved into the first two domains, Bacteria and Archaea, which, as far as the anatomical structure is concerned, are prokaryotic cells. Subsequently from the Archaea, the third domain evolved: the Eukaryota. Each of these three domains evolved into several kingdoms.
Adapted from https://commons.wikimedia.org/wiki/File:Phylogenic_Tree-en.svg © Conquistador / Wikimedia Commons / CC BY-SA 3.0

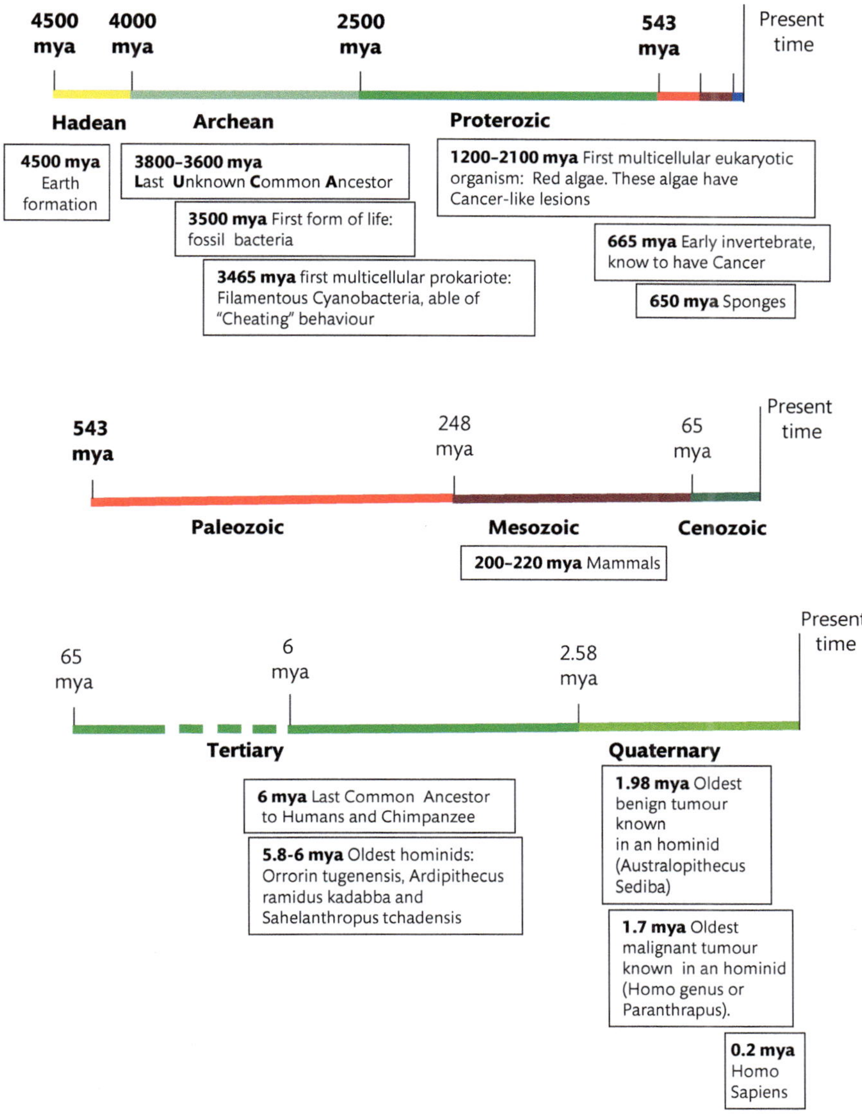

Fig. 1.3 Timeline. Mya, millions of years ago.

characteristics of the eukaryotic cells, when not dividing, are represented in Fig. 1.1.

The cell membrane

The cell membrane is a bilayer of phospholipids, each with a hydrophilic head and a hydrophobic tail. In aqueous environments, the phospholipids spontaneously organize themselves as a double layer with the hydrophobic tails inside and the head outside, so originating the cell bio membrane (Fig. 1.4). In a cell, the membranous network is divided into two main components: the cell surface membrane, the plasma membrane delimiting the actual cell and representing the border with the extracellular world, and those membranes delimiting the internal cellular compartments. In the plasma membrane within the scaffolding formed by the phospholipid bilayer are numerous different structures. These are made by proteins, 500 (five hundred) types of lipid molecules and 10,000 (ten thousand) proteins are involved in the making of the cell membrane. The main structures formed by intramembranous proteins are channels (e.g. ion pumps) and receptors (e.g. epidermal growth factor receptor). Some molecules however, like oxygen, can diffuse through the membrane without needing a specific pump.

The plasma membrane is a highly dynamic fluid structure and all the protein complexes are 'floating' wtihin it and also the very same lipid molecules are continuing moving within the membrane. Groups of lipids can also form units called 'rafts', which move among the other lipids. This dynamic nature of the plasma membrane was firstly described in 1972 as the fluid mosaic model (Singer and Nicolson, 1972; Edidin, 2003). The external cell membrane is in continuity with the internal membranes that not only defines the internal organelles of the cells, but also provides a framework for countless biochemical reactions and trafficking of molecules.

Table 1.1 Comparison of the main biological characteristics of Eubacteria, Archaea, and Eukaryota

	Eubacteria	Archaea	Eukaryota
Cell membrane	Yes	Yes	Yes
Transcription and translation	Yes	Yes	Yes
Signal transduction	Yes	Yes	Yes
Epigenetic change	Yes	Yes	Yes
Protein chaperons	Yes	Yes	Yes
Nucleus	No	No	Yes
Cytoskeleton	No	No	Yes
Organelles	No	No	Yes
DNA	Circular	Circular	Linear
Operons	Yes	Yes	No
Ribosome	70 s	70 s	80 s
Grow above 80°C	Yes	Yes	No
Number of genes	1,000–6,000	1,000–6,000	6,000–50,000
Operons	Yes	Yes	No
Multicellularity	No	No	Yes
Introns	No	Yes	Yes
Histone proteins	No	Yes	Yes
DNA-dependent RNA polymerase	Simple subunit	Complex subunit	Complex subunit
tRNA initiator	Formylmethionine	Methionine	Methionine
Transcription factors	Yes	No	Yes
Spore formation	Yes	No	No
Photosynthesis	Yes	No	Yes

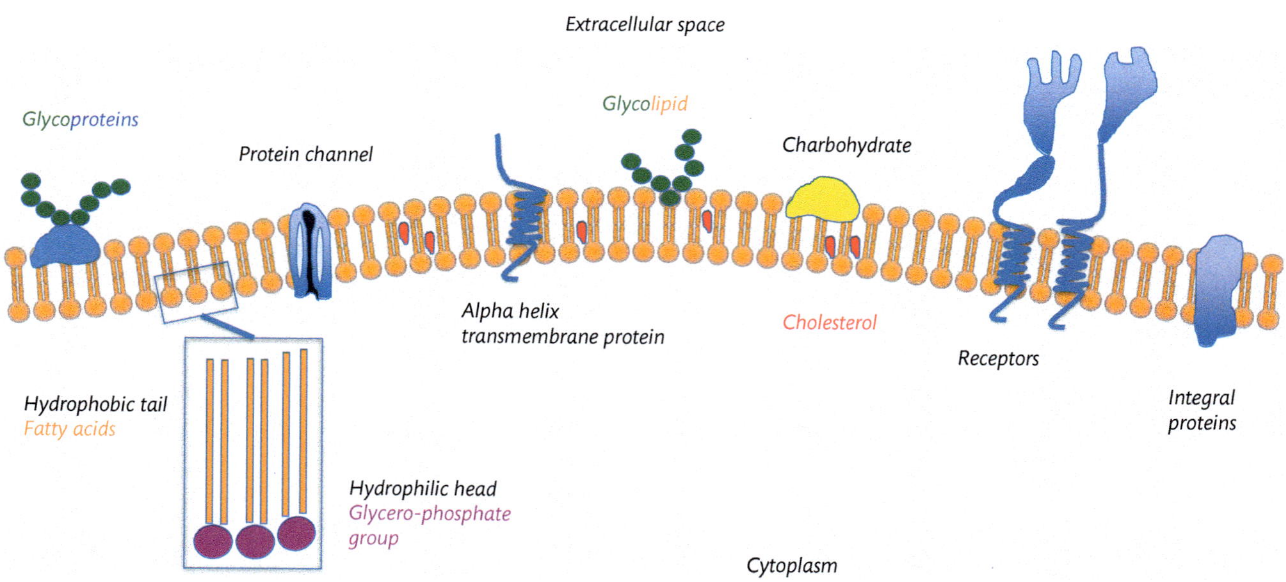

Fig. 1.4 The cell membrane. The cell membrane is made up by phospholipid molecules with a hydrophilic head (orange) and a hydrophobic tail (yellow). Within the membrane are several different structures that can 'float' across the membrane, which has fluid property. Transmembrane proteins span all thickness of the membrane and the main types are the protein channels, the integral protein, and the alpha-helix proteins. Glycoproteins and carbohydrates are present on the external surface.

The two major compartments inside the cells are the nucleus and the cytoplasm; the latter includes all the intracellular volume which is not nucleus.

The cytoplasm

The cytoplasm is occupied by cytosol, an aqueous medium rich in proteins and salts accounting for approximately 50% of the cytoplasm. The main site of protein synthesis and degradation and of intermediate metabolism, forms the cytosol that permeates all the organelles. The main reason for different organelles is the need for keeping apart different biochemical reactions. A single membrane delimits all the organelles, with the only exceptions being the mitochondria and the nucleus, which have an outer and one inner membrane.

Mitochondria are organelles formed by an external and one internal membrane filled with matrix. It is where the oxidative phosphorylation (i.e. respiration), occurs and where adenosine triphosphate (ATP) is produced. ATP is the source of energy for all cellular functions: such an energy is liberated when an ATP molecule is hydrolysed producing adenosine diphosphate (ADP), phosphate, and energy. The endoplasmic reticulum, or ER, forms the major cytoplasmic network, most of which is Rough ER where ribosomes are located and protein synthesis occurs. The remaining one is called the Smooth ER.

Another prominent function of the ER is lipid synthesis. The Golgi apparatus is another system of cisterns where simpler molecules are packaged into more complex ones: it is also where lysosomes are built. The lysosomes are spherical membrane vesicles containing a wide range of hydrolytic enzymes able to degrade many molecules and their purpose is to eliminate any damaged or unwanted molecules. Such molecules are transported to the lysosome by specialized vesicles called endosomes. Peroxisomes are instead involved in several metabolic and catabolic functions. The most important are catabolism of very long chain fatty acids into branched chain fatty acids, D-amino acids, and polyamines with reduction of reactive oxygen species. They are also the place where phospholipids are synthesized and the pentose phosphate pathway, critical for the energy metabolism, is located. Finally, free ribosomes are also present within the cytoplasmic matrix.

The nucleus

The nucleus is the largest organelle. A double bilayer membrane, the nuclear membrane or nuclear envelope, in which numerous pores (nuclear pore complexes) are present, thus allowing communication with the cytoplasm, which delimits it. Within the nucleus is the genome (with only the exception of mitochondrial DNA) and its transcriptional machinery. It can be divided into two main structures: the nuclear membrane and the nuclear interior (Lammerding, 2011).

Between the two layers of the nuclear membrane is the perinuclear space. Under the nuclear membrane is the nuclear lamina, mainly made up by laminin filaments. This membrane is perforated by the nuclear pore complexes which cause the inner and outer membranes to fuse. Nuclear pore are large complexes made of approximately 50 nucleoporins and regulate the trafficking between nucleus and cytoplasm. Nuclear pore complexes are not the only protein structure within the nuclear membrane, with some spanning the whole thickness (Lammerding, 2011).

When not dividing, approximately half of the nuclear volume is occupied by chromatin made of the unfolded DNA packed around histone proteins. There are two types of chromatin: heterochromatin, more packed and less transcriptionally active, and euchromatin, which is not so condensed and in which most of the transcription occurs (Lammerding, 2011). The aggregates of DNA and histones form structures called nucleosomes: packaged DNA from different chromosomes occupy distinct areas in the non-dividing nucleus. Nucleoli are discrete bodies formed by proteins and nucleic acids and are the production site of the ribosome. Cajal bodies are located in the proximity of the nucleoli and contain different formations: for example, the snurposome and spliceosome are involved in the processing of the recently transcribed mRNA. Finally, there is the nucleoskeleton, a protein scaffolding supporting the different nuclear components. All these structures are immersed in the nucleoplasm, a very protein rich aqueous medium equivalent to the cytosol.

The life of the single cell

All the unicellular organisms tend to grow without limitation according to the availability of resources. The life cycle of each of these organisms therefore coincides with the time required to duplicate (i.e. to complete the cell cycle), the cell cycle being a complex of events that brings one cell to divide into two. Prokaryotes do not age: their life cycle is very simple. They die when conditions became adverse and food is scanty, although this is not always the case: in determinate conditions, some cells can become 'dormant' and resume growth when the environment becomes permissive again. In most eukaryotic cells ageing does appear: their lifespan is regulated by an internal clock made up by telomeres and telomerase. The main physiological events in the life cycle of a single cell are reproduction through mitosis, response to damage, cell death, and movement. Cells divide through mitosis, a process in which the genetic code is duplicated and then the cells divide into equal new cells. As cells are exposed to external insults, either chemical or physical, repair mechanisms are present that can block the further division until the necessary corrections are made. Should the repairs fail, the cells can trigger their own death through apoptosis. Apoptosis does not only follow damage within a multicellular organism but can also be triggered at appropriate moments during the organism's development or life.

Multicellular organisms and the development of cancer

During the evolution of life, multicellularity has appeared independently at several different times, exploiting different strategies (Kaiser, 2001). There are therefore several mechanisms leading to the formation of multicellular organisms. For example, while plants relied on the formation of a rigid cell wall which brings different cells into one organism, the animal cells, which do not have cell walls, had to rely on membrane proteins called adhesion molecules to provide a mechanism allowing the cells to stick to each other (Bonner, 1998).

When confronted by an aggregate of cells, the first issue is how to differentiate a multicellular organism from a colony. The most commonly used, and the broadest criteria, is the existence of a spatial division of work in multicellular organisms, compared to the colony in which each unicellular member performs the same tasks. According to this definition, the oldest known unambiguous multicellular organisms belong to the Bacteria domain and are the filamentous cyanobacteria. These emerged, as suggested by fossil dating, 3,465 million years ago (mya), approximately 1,000 million years after the Earth's formation, which is estimated at 4,500 mya (**Fig. 1.3**).

Cyanobacteria were the first organisms to develop photosynthesis and release oxygen. However, these bacteria also rely on the enzyme nitrogenase to convert nitrogen gas into ammonia, necessary to build their proteins and other essential structural components when combined nitrogen (i.e. reactive molecules containing nitrogen), like nitrate, nitrite, ammonium, urea, and amino acids, are not available (**Fig. 1.5**). However, nitrogenase is irreversibly destroyed in the presence of oxygen and therefore cyanobacteria had to find a way to be able to perform oxygen-producing photosynthesis and, at the same time, to maintain nitrogenase function. This problem has been solved by multicellularity: formation of filaments, made up by a line of cyanobacteria in which two differently evolved cyanobacteria can be found. Those containing chlorophyll, performing photosynthesis, and releasing oxygen, are more numerous. These differentiated into a form called heterocysts, in which nitrogenase function is maintained in the absence of photosynthesis, as they do not contain chlorophyll (**Fig. 1.3**). Therefore, a simple prokaryotic multicellular organism containing two types of cells was formed (Bonner, 1998; Adams, 2000; Flores and Herrero, 2010).

The time of the emergence of eukaryotic multicellular organisms as assessed today is still a broad estimate, possibly sometime between 2,100 (Donoghue and Antcliffe, 2010) and 1,200 (Rokas, 2008) mya. Red algae are so far considered as the first eukaryotic multicellular organisms, appearing 1,200 mya (**Fig. 1.3**). The main strategies employed by eukaryotic cells to build a multicellular entity include: lack of cell separation after mitosis; mostly found in aquatic organisms; and aggregation of single cells prevalent in terrestrial creatures. A final fundamental characteristic in the classification of multicellular organisms is complexity. The easiest and most practical approach to 'measure' complexity is the number of cells making up the organism (Rokas, 2008).

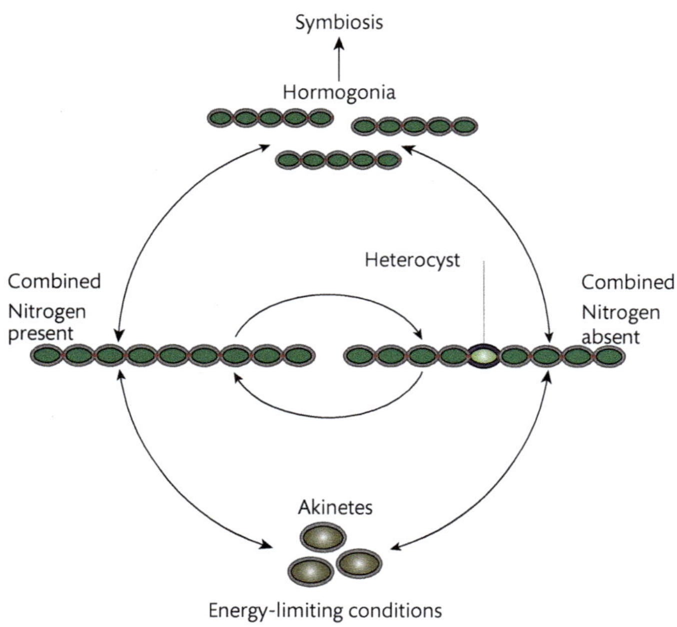

Fig. 1.5 The filamentous cyanobacteria. Cyanobacteria exists more commonly as vegetative form but can differentiate into three further forms: heterocyst, akinetes, and hormogonial cells. In absence of 'combined nitrogen', like nitrate, nitrite, ammonium, urea, and amino acids which easily react and combine with other molecules and can be used for protein production, cyanobacteria needs to 'fixate' the poorly reactive nitrogen and transform it into the more reactive ammonia according to the following reaction $N_2 + 8\,H^+ + 8\,e^- \rightarrow 2\,NH_3 + H_2$, catalysed by a nitrogenase enzyme. Vegetative cells cannot do that as they produce oxygen which is toxic for the nitrogenase enzyme. Therefore, some vegetative cells differentiate into heterocysts, cells that do not produce oxygen, but are able to fixate N_2 into ammonia in response to deprivation of combined nitrogen. Subsequently, the filamentous cyanobacteria acquires its new structure, characterized by a number of vegetative cells regularly interrupted by one heterocyst. When nutrients and energy are scanty some vegetative cells differentiate into akinetes, which can than start to proliferate again and produce vegetative cells when nutrients became available again. Vegetative cells from some filamentous cyanobacteria can also differentiate into hormogonia, which are than dispersed and can subsequently originate new filamentous cyanobacteria growing in symbiosis with plants.

Adapted by permission from Macmillan Publishers Ltd: Springer Nature, *Nature Reviews Microbiology*, 'Compartmentalized function through cell differentiation in filamentous cyanobacteria', Flores E and Herrero A.

Hallmarks of multicellularity

Multicellularity required the acquisition of functions not present or diversely utilized in single cell organisms. Up to seven hallmarks of multicellularity have been described (Rokas, 2008; Srivastava et al., 2010; Aktipis et al., 2015).

Regulation and control of the cell cycle

A strict control of proliferation is essential to the development and survival of a multicellular organism. For the organism to maintain itself, proliferation can occur only in well-defined circumstances and is regulated by a series of positive signals, inducing it, and suppressive signals, blocking it. Furthermore, these control rules are different from tissue to tissue (e.g. the neurons do not enter proliferation ever), while, on the other extreme, bone marrow stem cells are continuously proliferating to provide new blood cells, which have a very high turnover. To guarantee this strict control, redundant mechanisms are present.

Apoptosis (programmed cell death)

While unicellular organisms just proliferate, within multicellular organisms remodelling takes place, mostly during development when some embryonic structures are temporary and need to be eliminated as the fetus develops. Apoptosis is also required in adulthood (e.g. after immune stimulations, only some of the immune cells specifically responding will survive; the others, responding in a non-specific way, will undergo apoptosis and disappear). This is possible thanks to the appearance of apoptosis, or programmed cell death, which causes, when necessary, the death of selected cells according to the organism's blueprint.

The interaction with the extracellular environment

The extracellular matrix is essential for cells to maintain their physiological functions. Furthermore, it is where cells come into contact with the immune system. Multicellular entities need to protect themselves from the intrusion of external pathogenic organisms and, at the same time, to maintain tolerance against 'self' antigens.

Specialization of cell types and division of work

As discussed before, the need for specialized cells cooperating is the very reason for which multicellularity developed. Different tissues need to perform a huge variety of different tasks, allowing the multicellular organisms to development degrees of complexity well outside the reach of single cell living forms. Again, this requires a strict control as each different tissue needs to differentiate in a very precise way according to its designated function.

Resources transport and allocation

According to the work each cell needs to do, different resources are required. While smaller organisms can rely on diffusion, the larger ones have different approaches with some relying on body cavities providing the required transport system, while the most complex have developed a branching vascular network. This is of course the case in humans, where the vascular and the lymphatic systems carry out this function.

Cell-cell and cell-matrix adhesion

As already discussed, different strategies for creation of multicellular organisms have emerged across the history of life. The cardinal function, which allows multicellularity in Metazoa, is that of adhesion (i.e. the creating stable mechanical contacts between cell and cell or cell and extracellular matrix). This is obtained by a variety of molecules like intercellular junctions, adhesion molecules, and adaption of cytoskeleton proteins.

Signalling and gene regulation

To develop and maintain a multicellular organism it is fundamental to have an efficient signalling system, both between and inside cells. This system is responsible for having each cell acting in synchrony with the other according to the organism's blueprint. This is realized by a high regulation of gene transcription leading to the synthesis of proteins making up appropriate signalling pathways.

Hallmarks of multicellularity and cancer

Disruption of these functions has been found to be closely linked to the development of cancer (Hanahan and Weinberg, 2011; Aktipis et al., 2015).

Regulation and control of cell cycle

Uncontrolled proliferation due to either an excess of proliferative stimuli, classic oncogenes, or the loss of inhibitory functions, loss of tumour suppressor genes are covered in more detail in Chapters 11 and 13.

Apoptosis (programmed cell death)

Resistance to programmed cell death or to ageing leads to abnormal neoplastic cell accumulation, covered in Chapters 14 and 15.

The interaction with the extracellular environment

This includes avoiding immune destruction and cross-talk between tumours, their supportive stroma, and the immune system. Alteration of these functions leads to neoplastic cells to escape potential anticancer activity of the immune system (see Chapters 23 and 29).

Specialization of cell types and division of work

Cancers develop from specific tissues but the ability to reproduce the structure and specialization of the cells seen in the normal tissue is lost to a variable degree across different malignancies (see Chapter 20).

Resources transport and allocation

Disruption of the normal blood supply is followed by the establishment of a new relationship between the neoplastic cells and the blood vessels. Cancer needs a resource transport system, which can be achieved in a variety of ways (Chapter 22). Also, a metabolic reprogramming of the cancer cell follows these changes (see Chapters 16 and 18).

Cell-cell and cell-matrix adhesion

Disruption or inhibition of these functions leads to the formation of abnormal neoplastic organ-like structures and to metastatic dissemination throughout the body (see Chapter 19).

Signalling and gene regulation

The normal network maintaining coordination gets disrupted following alterations at different levels along the way causing a pathological resetting of behaviour (Chapters 9, 10, and 12).

Cancer as a disease of the multicellular organism

Cancer is a disease of multicellular organisms in which the emergence of changes in the DNA disrupts the instructions controlling the growth and physiology of some of the organism's cells. Eventually one cell acquires enough changes to be able to abnormally grow outside the organism blueprint into a clone (i.e. a group of cells deriving from the same ancestor cell) of neoplastic cells. It is therefore an 'information' disease caused by alteration in the information blueprint (i.e. the genome). As DNA codes such instructions, any damage can lead to two main effects on the single host organism. The first is that the damage has actually no effect, if the area of changed genetic code is silent or redundant or not active at the time in which the damage occurs. The second leads to a change in the instruction blueprint, which is followed by pathological events of various nature. If the event is cellular death, some type of disease other than cancer can occur (e.g. degenerative diseases). Cancer is one of the pathological situations that can follow a genetic injury. It is characterized by a cellular growth that follow a reset of growth instructions that varies from one type of cancer to another and, indeed, from patient to patient causing the cancer cells to replace and destroy the normal body structures in an apparently chaotic way.

The instruction for changes leading to cancer are fundamentally those governing the set of functions necessary to 'make' a multicellular organism. As the cancer-linked pathways and cellular functions are associated with the appearance of multicellular organisms, it is not surprising that cancer, characterized by invasion and metastasis, or cancer-like phenomenon, characterized by abnormal proliferation and differentiation of 'cheating' cells (Aktipis et al., 2015), have been found across the spectrum of multicellular organisms

(Schlumberger and Lucke, 1948; Scharrer and Lochhead, 1950; Leroi et al., 2003; Aktipis et al., 2015).

As shown in Fig. 1.6, 'cheating' of multicellular cooperation has been observed already in multicellular bacterial organisms, like overgrowing due to loss of proliferative inhibition. 'Cheating' describes 'the breaking of shared rules, including genetically encoded phenotypes or behaviour, that leads to a fitness advantage for the cheater' (Aktipis et al., 2015). Cheating has been described as early as bacteria multicellular organisms (Fig. 1.6), involves mostly the functions of proliferation and/or apoptosis, and leads to forming a 'mass' of cells. Cancer is instead defined as a primary mass causing metastases and is mostly occurring in Metazoan but not only, as some plants have cancer-like lesions. Actually, the simplest organism in which lesions appear, sharing many characters of what we call cancer, is red algae. In these plants, primary tumours due to loss of proliferation and apoptotic control occur. These lesions can ulcerate and propagate in a metastasis-like fashion.

Sponges are the oldest surviving metazoans and appeared approximately 650 million years ago (Srivastava et al., 2010) and already contain all the pathways characterizing both metazoans and cancer (Domazet-Loso and Tautz, 2010; Srivastava et al., 2010; Aktipis et al., 2015). Sponges do not have distinct organs but have specialized structures like pores, canals, ostia, chambers, and a rudimentary immune system. No cancer has been observed in sponges.

Cancer-like lesions and/or cheating, a clear distinction between the two being sometime difficult, has been seen in other early Metazoa such as hydra and corals, where fast-growing, destructive lesions with loss of architecture grow. Proper malignant tumours such as lymphoid, epithelial, neuronal, and those from gonad cells are instead commonly seen in protostomes, invertebrates, and occur in all the more complex types of Metazoa (Aktipis et al., 2015).

However it must be noted that as complexity, dimensions, and lifespan increases, not all the species are susceptible to cancer (Aktipis et al., 2015): the so-called Peto's paradox (Peto et al., 1975; Caulin and Maley, 2011; Roche et al., 2013).

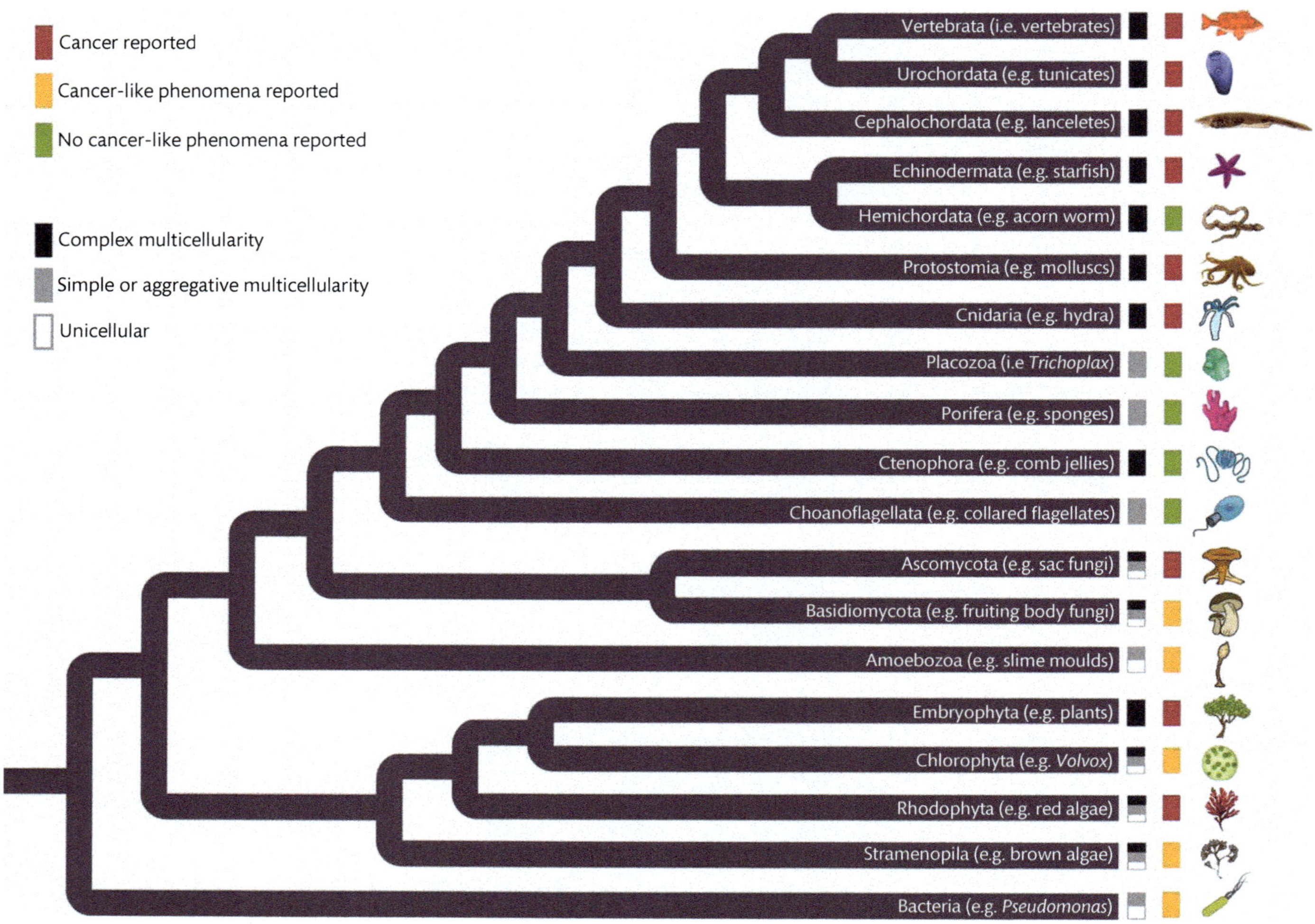

Fig. 1.6 Cancer across the tree of life. Black, grey, or white boxes at branch tip indicates the cellularity status as unicellular (white), aggregative multicellularity (grey), or complex multicellularity (black). Red, yellow, or green boxes represent whether a cancer phenotype (invasion or metastasis) was reported or cancer-like/'cheating'-type lesion was observed (abnormal proliferation or differentiation) such as callus or galls (yellow box). If no cancer or cancer-like/cheating lesions were reported, there is a green box.

The Peto paradox

Every time that a cell proliferates, there is risk of an error at the time when the DNA is copied. Different types of mistakes can happen, like single mutations, duplication, and/or redistribution of the genetic material among the daughter cells. Consequently, as Metazoa increased in complexity and size and the lifespan got longer and longer, the risk of cancer was expected to grow in direct proportion; the larger and long-lived the animal, the higher the number of mitosis occurring in its body, and therefore the higher the chance of DNA damage to occur. However, this turned out not to be the case as large dimensions and longer life do not necessarily means increased risk of cancer (Aktipis et al., 2015): this is the 'Peto's paradox', which get its name from a study published in 1995 by Richard Peto et al. (Peto et al., 1975). In this experiment a large cohort of mice of different ages were exposed to topical application of the carcinogen. The rate of appearance by epithelial tumours was related to the duration of exposure to the chemical but not to the mouse's age: it was the time of exposure to the carcinogen dictating the risk of developing cancer and not the age of the exposed mouse—and neither the span of survival after the exposure. This study demonstrated that, against the then current wisdom, increased lifespan per se can be irrelevant as far as increase in cancer risk is concerned. More broadly, 'Peto's paradox' is now a term to indicate a counter-intuitive event.

As far as cancer is concerned, two classic examples are those of the blue whales and the elephants. Blue whales are approximately six million times larger than mice; however, variation in cancer in non-laboratory animals varies in average for no more than a factor of two, independently of the mass. In the whale's case, the paradox is even more remarkable as only very rarely do they die of cancer (Leroi et al., 2003). Calabrese and Shibate (Calabrese and Shibata, 2010), using a mathematical model, have been able to correctly approximate the risk of colorectal cancers in humans: their equation resulted in an overall risk of approximately 2.5% by the age of 90 years while the risk actually observed is 5 (Caulin and Maley, 2011). However, using the same model they predicted a risk for the blue whales of 100% of getting this cancer by the age of 80 years while they can live beyond 100 years and cancer of any type is a rare event.

The question raised by Peto's paradox is therefore how some large long-lived animals manage to have such a low rate of cancer. According to natural selection, this is because the mechanisms giving protection from cancer must have been selected in order to allow large animals to exist and to have a long life span. However, the nature of these mechanisms remains unclear. While no studies are available for blue whales, some have been carried out on other animals and the main mechanisms involved in low susceptibility to cancer are concerned with telomerases replicative senescence and cell proliferation control, tumour suppressor activity, and genome stability.

One example is the elephant, which is notoriously a large mammal but can live in the wilderness to 60–70 years with a low cancer incidence. In this pachyderm, the protective mechanism is suggested to be redundancy of tumour suppressor function as both the African and the Asian species have 20 copies of the tumour suppressor *TP53* gene (Abegglen et al., 2015). Gain and loss of genes involved in DNA repair, cell-cycle regulation, and ageing could be responsible for the lack of malignancies in the bowhead whale, possibly the longest-lived mammal, which is estimated to live in excess of 200 years (Keane et al., 2015). The largest wealth of data is available for rodents. To this order belong animals with a great variability of longevity, body mass, and cancer incidences. Naked mole rats are the longest living, in excess of 30 years, while mice and rats live approximately 3 or 4 years. The difference in cancer incidence is striking: mice, among the smallest rodents, are prone to cancer and in some strains the incidence goes up to 95% while the larger naked and the blind mole rats are virtually cancer free.

Systemic studies on rodents have started to unravel the mechanisms behind these differences. Small but long-lived rodents appear to have cells which proliferate more slowly than small short-lived animals, while larger long-lived rodents are protected by shorter telomerases and therefore enhanced replicative senescence. Long-lived animals also show higher levels of expression of DNA repair genes, raising the hypothesis of a more efficient DNA repair activity (Gorbunova et al., 2014).

Resistance to cancer evolved several times independently as demonstrated by the comparison of two long-lived cancer-resistant rodents: the Naked mole rat (*Heterocephalus glaber*) and the blind mole rat (*Spalax ehrenbergi*). These two species are phylogenetically distant from each other. In naked mole rats, there are large levels of high molecular mass hyaluronan polysaccharides, five times longer than those of humans. These longer forms bind to CD44 triggering cell cycle arrest, while the low molecular mass hyaluronans promote cell cycle. When the *Has2* gene responsible for hyaluronan synthesis is knocked down or when the hyaluronoglucosaminidase 2 (*Hyal2*) gene, responsible for breaking down hyaluronan, is overexpressed, naked mole rat cells start forming tumours. Furthermore, in these rodents the 28S rRNA is cleaved in two, increasing the fidelity of translation. The mechanisms in the blind mole rat are different; one is the secretion of interferon by premalignant cells, which causes a massive necrosis in the surrounding tissue eliminating the premalignant cells. The second is again linked to hyaluronan, but this time the hyaluronan present is not able to block mitosis but is rather a powerful antioxidant (Gorbunova et al., 2014).

Conclusion

Cancer is a disease due to the malfunctioning of the biological functions necessary for cell growth and for the formation and maintenance of the multicellular organisms. Because of the complexity of large multicellular animals, this means that many different types of errors and damages can lead to what is known as cancer leading to malignant lesions with varied and complex biology and clinical behaviour. Cancer must be regarded from a practical point of view as many different diseases, each to be individually unravelled to fully understand it and produce an effective treatment.

TAKE-HOME MESSAGE

- Cancer is a disease due to the malfunctioning of the pathways necessary to the life of a multicellular organism.
- It is a disease of information, as genetic damage alters the information blueprint of the organism: the DNA.
- Evidence of cancer-like behaviour has been found even in the simplest prokaryotic multicellular organisms.
- To be prone to cancer is not inevitable for a multicellular organism as some are very resistant to the disease, as per the Peto paradox.

OPEN QUESTIONS

- Open questions about the cancer are numerous! They are scattered through the book.

FURTHER READING

Alberts, B., Johnson, A., Lewis, J., et al. (2015). *Molecular Biology of the Cell*, 6th edition. New York: Garland Science.

Allen, T. & Cowling, G. (2011). *The Cell: A Very Short Introduction*. Oxford: Oxford University Press.

Benton, M. J. (2008). *The History of Life: A Very Short Introduction*. Oxford: Oxford University Press.

Dawkins, R. & Wong, Y. (2016). *The Ancestor's Tale*, 2nd edition. London; Weidenfeld & Nicolson.

Diamond, J. C. (1998). *Because Cowards Get Cancer Too*. London: Vermilion.

Schiffman, J., Maley, C. C., Nunney, L., Hochberg, M., & Breen, M. (eds.) (2015). Theme issue 'Cancer across life: Peto's paradox and the promise of comparative oncology'. *Philosophical Transactions of the Royal Society B*, **370** (1673), DOI: 10.1098/rstb.2015.0198.

Weinberg, R. A. (1998). *One Renegade Cell*. New York: Basic Books.

REFERENCES

Abegglen, L. M., Caulin, A. F., Chan, A., et al. (2015). Potential mechanisms for cancer resistance in elephants and comparative cellular response to DNA damage in humans. *JAMA*, **314**, 1850–60.

Adams, D. G. (2000). Heterocyst formation in cyanobacteria. *Curr Opin Microbiol*, **3**, 618–24.

Aktipis, C. A., Boddy, A. M., Jansen, G., et al. (2015). Cancer across the tree of life: cooperation and cheating in multicellularity. *Philos Trans R Soc Lond B Biol Sci*, **370**, pii: 20140219.

Bonner, J. T. (1998). The origins of multicellularity. *Integrative Biology Issues News and Reviews*, **1**, 27–36.

Calabrese, P. & Shibata, D. (2010). A simple algebraic cancer equation: calculating how cancers may arise with normal mutation rates. *BMC Cancer*, **10**, 3.

Caulin, A. F. & Maley, C. C. (2011). Peto's paradox: evolution's prescription for cancer prevention. *Trends Ecol Evol*, **26**, 175–82.

Domazet-Loso, T. & Tautz, D. (2010). Phylostratigraphic tracking of cancer genes suggests a link to the emergence of multicellularity in metazoa. *BMC Biol*, **8**, 66.

Donoghue, P. C. & Antcliffe, J. B. (2010). Early life: origins of multicellularity. *Nature*, **466**, 41–2.

Edidin, M. (2003). Lipids on the frontier: a century of cell-membrane bilayers. *Nat Rev Mol Cell Biol*, **4**, 414–18.

Flores, E. & Herrero, A. (2010). Compartmentalized function through cell differentiation in filamentous cyanobacteria. *Nat Rev Microbiol*, **8**, 39–50.

Gorbunova, V., Seluanov, A., Zhang, Z., Gladyshev, V. N., & Vijg, J. (2014). Comparative genetics of longevity and cancer: insights from long-lived rodents. *Nat Rev Genet*, **15**, 531–40.

Hanahan, D. & Weinberg, R. A. (2011). Hallmarks of cancer: the next generation. *Cell*, **144**, 646–74.

Hedges, S. B. (2002). The origin and evolution of model organisms. *Nat Rev Genet*, **3**, 838–49.

Kaiser, D. (2001). Building a multicellular organism. *Annu Rev Genet*, **35**, 103–23.

Keane, M., Semeiks, J., Webb, A. E., et al. (2015). Insights into the evolution of longevity from the bowhead whale genome. *Cell Rep*, **10**, 112–22.

Lammerding, J. (2011). Mechanics of the nucleus. *Compr Physiol*, **1**, 783–807.

Leroi, A. M., Koufopanou, V., & Burt, A. (2003). Cancer selection. *Nat Rev Cancer*, **3**, 226–31.

Olsen, G. J. & Woese, C. R. (1997). Archaeal genomics: an overview. *Cell*, **89**, 991–4.

Peto, R., Roe, F. J., Lee, P. N., Levy, L., & Clack, J. (1975). Cancer and ageing in mice and men. *Br J Cancer*, **32**, 411–26.

Roche, B., Sprouffske, K., Hbid, H., Misse, D., & Thomas, F. (2013). Peto's paradox revisited: theoretical evolutionary dynamics of cancer in wild populations. *Evol Appl*, **6**, 109–16.

Rokas, A. (2008). The molecular origins of multicellular transitions. *Current Opinion in Genetics & Development*, **18**, 472–8.

Scharrer, B. & Lochhead, M. S. (1950). Tumors in the invertebrates: a review. *Cancer Res*, **10**, 403–19.

Schlumberger, H. G. & Lucke, B. H. (1948). Tumors of fishes, amphibians, and reptiles. *Cancer Res*, **8**, 657–753.

Singer, S. J. & Nicolson, G. L. (1972). The fluid mosaic model of the structure of cell membranes. *Science*, **175**, 720–31.

Srivastava, M., Simakov, O., Chapman, J., et al. (2010). The Amphimedon queenslandica genome and the evolution of animal complexity. *Nature*, **466**, 720–6.

Thain, M. & Hickman, M. (2004). *Dictionary of Biology*. London: Penguin Books.

Virchow, R. K. (1863). *Cellular Pathology as Based Upon Physiological and Pathological Histology*. Philadelphia, PA: J. B. Lippincott and Co.

Woese, C. R., Kandler, O., & Wheelis, M. L. (1990). Towards a natural system of organisms: proposal for the domains Archaea, Bacteria, and Eucarya. *Proc Natl Acad Sci U S A*, **87**, 4576–9.

2

DNA repair and genome integrity

Giacomo Buscemi

Introduction

DNA damage and repair studies started in the late 1930s by physicists' experiments on recovery of cells inadvertently exposed to long-wave light. Nowadays, it is estimated that mammalian cells suffer ~2 × 10^5/day DNA lesions induced by normal metabolism products or environmental agents (Barnes and Lindahl, 2004). This number is further increased by the genotoxic effects of air pollution, cigarette smoking, food additives, toxins, and nuclear plant disasters. In the early 1960s, the discovery that the carcinogenicity of polycyclic aromatic hydrocarbons, the classic components of tobacco smoke, is directly dependent on their ability to form DNA adducts provided an unambiguous link between tumorigenesis and chemical DNA modifications. Now, we are aware that most carcinogens operate by generating DNA damage and causing mutations. Similar data were also obtained about radiation-induced mutations through the analysis of atomic bomb survivors, although the evidence that X-ray exposure causes an increased risk of malignancies was already accepted in 1895, soon after their discovery. Moreover, the role of DNA repair in cancer was further supported by the disclosure, in the last 15 years, of rare syndromes harbouring mutations in DNA repair genes characterized by a predisposition to cancer, starting from xeroderma pigmentosum or Lynch syndrome.

The necessity of defects in DNA repair as an early or late event during cancer development is still debated. However, the consciousness that endogenous and exogenous DNA damage induces mutations potentially leading to carcinogenesis, and that efficient DNA repair mechanisms are required to protect organisms from cancer, are key concepts in cancer aetiology. More recently emerged the notion that general DNA damage response (DDR) mechanisms, more than repair pathways per se, are essential to prevent cancer. Indeed, the global cellular response to DDR is more complex than the simple activation of a specific DNA repair mechanism, since it is formed by network of pathways involving hundreds of proteins affecting cell cycle progression, cell survival, metabolism, and ageing. Alterations of DDR are essential to express some of those features that characterize a cancer cell, like uncontrolled replication or resistance to cell death. Notably, DDR signalling defects frequently have also a greater impact on chromosomal stability upon damage than DNA repair pathways, which mostly influence cell survival. Current models of tumorigenesis (Fig. 2.1) indicate that single or multiple initiating events, often caused by mutations, lead to hyper-replication and replicative stress. Replication stress promotes cancer development by inducing breaks, particularly at common fragile sites, specific genomic regions showing increased fragility when DNA replication is perturbed. These events, coupled with pre-existing genetic alterations or acquired mutations that downregulate DDR mechanisms, enable replicative immortality and resistance to cell death, finally enhancing the possibility of misrepaired lesions and genome instability (Fig. 2.1). These features are essential for the rapid adaptation of a cancer cell to its ever-changing microenvironment and for malignancy progression.

Molecular aspects of the DNA damage response

In the 1940s the DNA duplex, which contains essentially all genetic information, was initially perceived as a highly stable macromolecule. Therefore, it was a surprise to find that DNA is subjected to incessant damage. Spontaneous DNA alterations include deamination, hydrolysis, non-enzymatic methylation, and oxidation of DNA bases (Lindahl and Barnes, 2000). Some of them are generated indirectly by normal cellular processes: base oxidations, for example, are induced by reactive oxygen species (ROS), that are continuously produced in living cells as toxic by-products of oxygen metabolism. In addition, the frequency of DNA lesion is further increased by exogenous sources including ionizing (IR) and ultraviolet (UV) radiation, and various chemicals agents.

To prevent the accumulation of nuclear DNA lesions, all organisms have evolved a complex signalling cascade to repair damage and eliminate cells that are beyond repair. Named DDR (Ciccia and Elledge, 2010), this cascade involves, in eukaryotes, hundreds of proteins that control the outcome of DNA repair at different levels (Fig. 2.2). Indeed, if a cell suffering DNA damage survives and continues growing with a restored unaltered genome, it will depend on the ability of DDR:

- To tightly regulate the activity of a multitude of DNA repair enzymes and regulators, with the final goal to optimize repair;
- To recruit chromatin remodelling proteins around the injured region thus improving the access of repair factors;

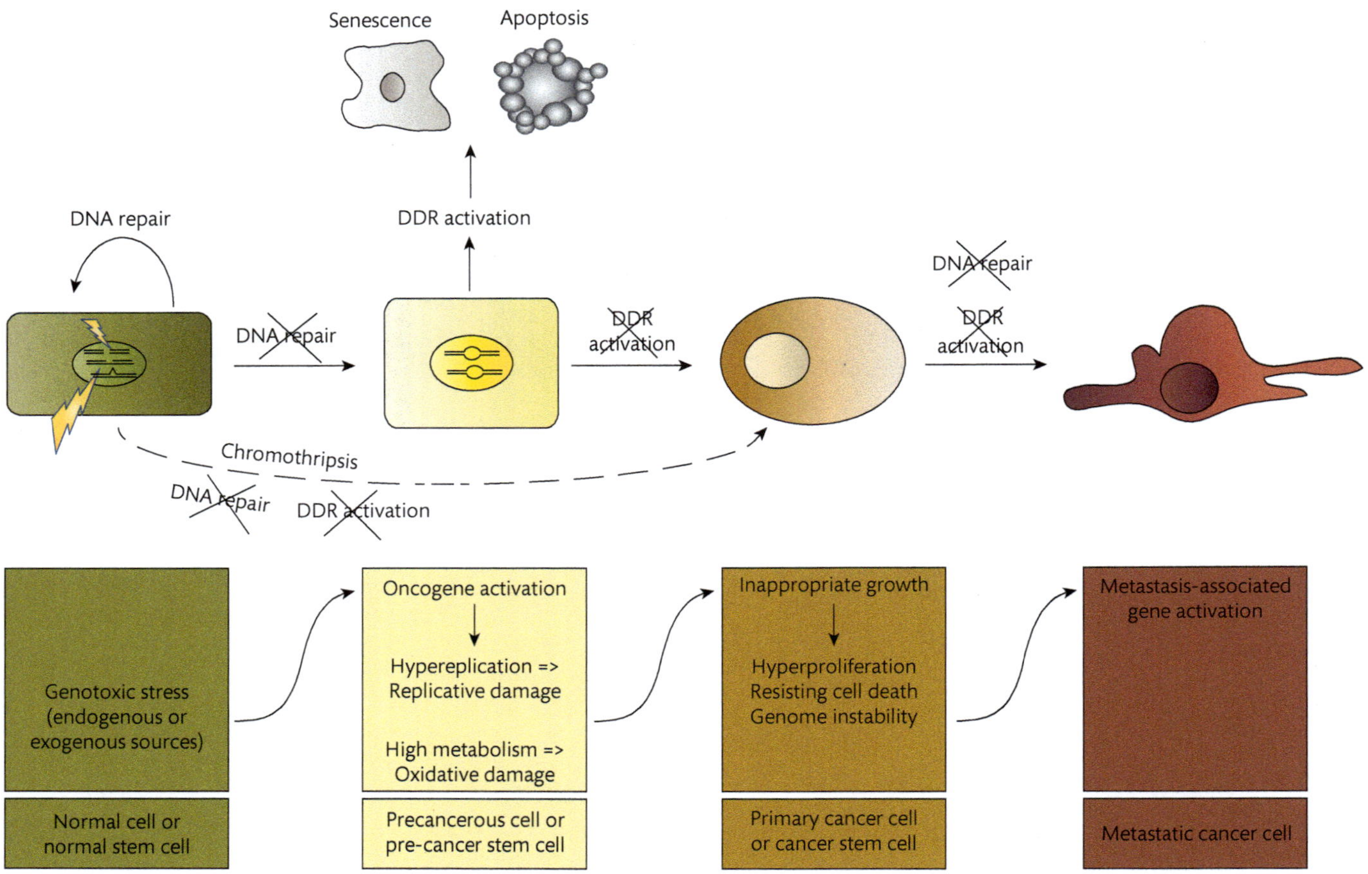

Fig. 2.1 *DDR activities on the route to cancer.* A schematic representation showing how alterations of DNA damage response (DDR) activity promote critical steps during carcinogenesis. Cells are under the constant assault of endogenous and exogenous sources of damage. A healthy cell has a plethora of DDR processes to protect DNA. Nonetheless, a mutation could occur by error or due to defects in DNA repair pathways. This may directly or indirectly result in oncogene activation, which leads to replicative and oxidative stress and damage. Normal activation of DDR processes will lead cells accumulating lesions to apoptosis or premature senescence. Defects in DDR will dismantle this barrier promoting progression from this precancerous state to inappropriate growth. The genome instability deriving from DNA repair and DDR defects could also fuel the activation of later stage of tumorigenesis. A more dramatic event of damage and defective repair, like chromothripsis, could accelerate this process.

- To transiently arrest cell cycle progression at checkpoints implementing time to restore the correct DNA structure;
- To act at different cellular level providing an opportune environment to enable DNA repair.

The main steps regulated by DDR in presence of DNA damage are: DNA lesion recognition; transduction of damage signal; DNA repair; and eventual activation of secondary activities like cell cycle arrest, apoptosis, and senescence (Fig. 2.2).

DNA lesion detection

DDR activation is induced when sensor proteins, which constantly control the DNA, find structural base distortions or breaks (Ciccia and Elledge, 2010). The sensitivity of the system is so high that a single base modification or misalignment is detected within billions of normal base pairs tightly packed inside chromatin. To detect the myriad of possible base alterations too many sensors would be necessary, so it was proposed that a limited number of proteins recognizes, more than the lesion itself, distortions of the double helix structure commonly produced by base alteration, mispairing, or DNA cross-link. DNA unwinding and free ends are instead the signal of DNA breaks presence.

Damages affecting only one of the two DNA strands are generally corrected by excision repair systems. Base excision repair (BER), nucleotide excision repair (NER) and mismatch repair (MMR) mechanisms (Fig. 2.3) all show steps in which the injury is cut out and the resulting gap refilled using, as template, the complementary DNA strand. Specifically, base modification due to oxidation, deamination, or alkylation are all recognized and excised by the same protein family: glycosylases of the BER system (Krokan and Bjoras, 2013). Differently, lesions distorting the helix, including UV-induced damage and bulky adducts, are fixed by global genome NER (GG-NER), or, if occurring in transcribed region, by a specialized NER system, named transcription coupled repair (TC-NER). Finally, base incorporation errors are mainly resolved by mismatch repair (MMR).

In particular, during NER the damage is generally detected by proteins of the xeroderma pigmentosum (XP) group (in particular XPE, C, and A, Fig. 2.3), known to be mutated in XP. However, the same lesion occurring in transcribed regions are detected through arrest of the transcriptional machinery and require proteins mutated in Cockayne syndrome (CS; Fig. 2.3). In the MMR pathway, the sensor of mismatches is the MSH2

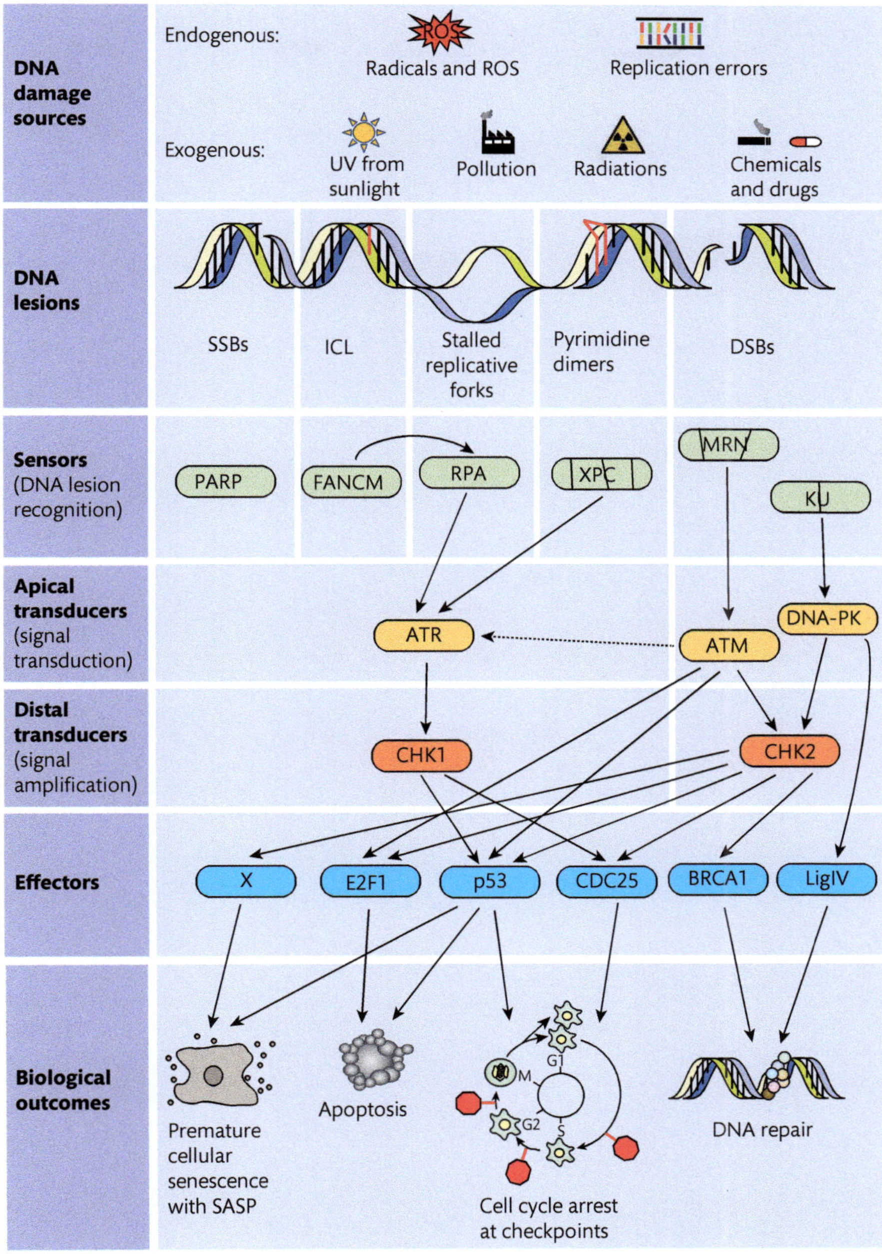

Fig. 2.2 *Schematic view of the essential DDR steps.* Schematic representation of the main steps of DDR activity in multicellular organism, in response to endogenous or exogenous nuclear DNA damage. The signalling cascade is essentially constituted by sensors, a limited number of apical and distal kinases and hundreds of effectors that activate in a fine-tuned way the correct biological response in relation to damage characteristics.

protein in a heterodimeric association with MSH3 or MSH6, thus forming the MutS complex.

Single-strand breaks (SSBs) directly produced by radiation and radicals or, in some cases, indirectly left during defective BER or NER, are recognized by the poly(ADP-ribose) polymerase (PARP) family of proteins (Caldecott, 2008) and repaired by SSB repair (SSBR) pathway. PARP1 and PARP2 activation and the subsequent synthesis of poly(ADP-ribose) (PAR) chains by these proteins occur within seconds at sites of damage (Fig. 2.4). The major substrates of DNA damage-induced poly(ADP-ribosyl)ation are PARP1 itself and histones surrounding DNA lesions. PAR structures constitute a platform to start the recruitment of DNA repair factors. Then PAR chains are rapidly degraded by PARG, an hydrolysing enzyme, thus providing a transient response that lasts for minutes only. This transient nature of the response is an essential feature of the DDR.

Interstrand DNA cross links (ICLs) covalently connect the two strands of DNA and constitute a dangerous bidirectional barrier to replication or transcription. Understanding the mechanism of ICLs repair is extremely important since agents causing this kind of lesions are widely used in cancer therapy. Indeed, nitrogen mustards and derivatives (melphalan, chlorambucil), psoralens, mitomycin C, platinum-based compounds like cisplatin, and nitrosoureas such as bis-chloroethylnitrosourea are clinically useful interstrand

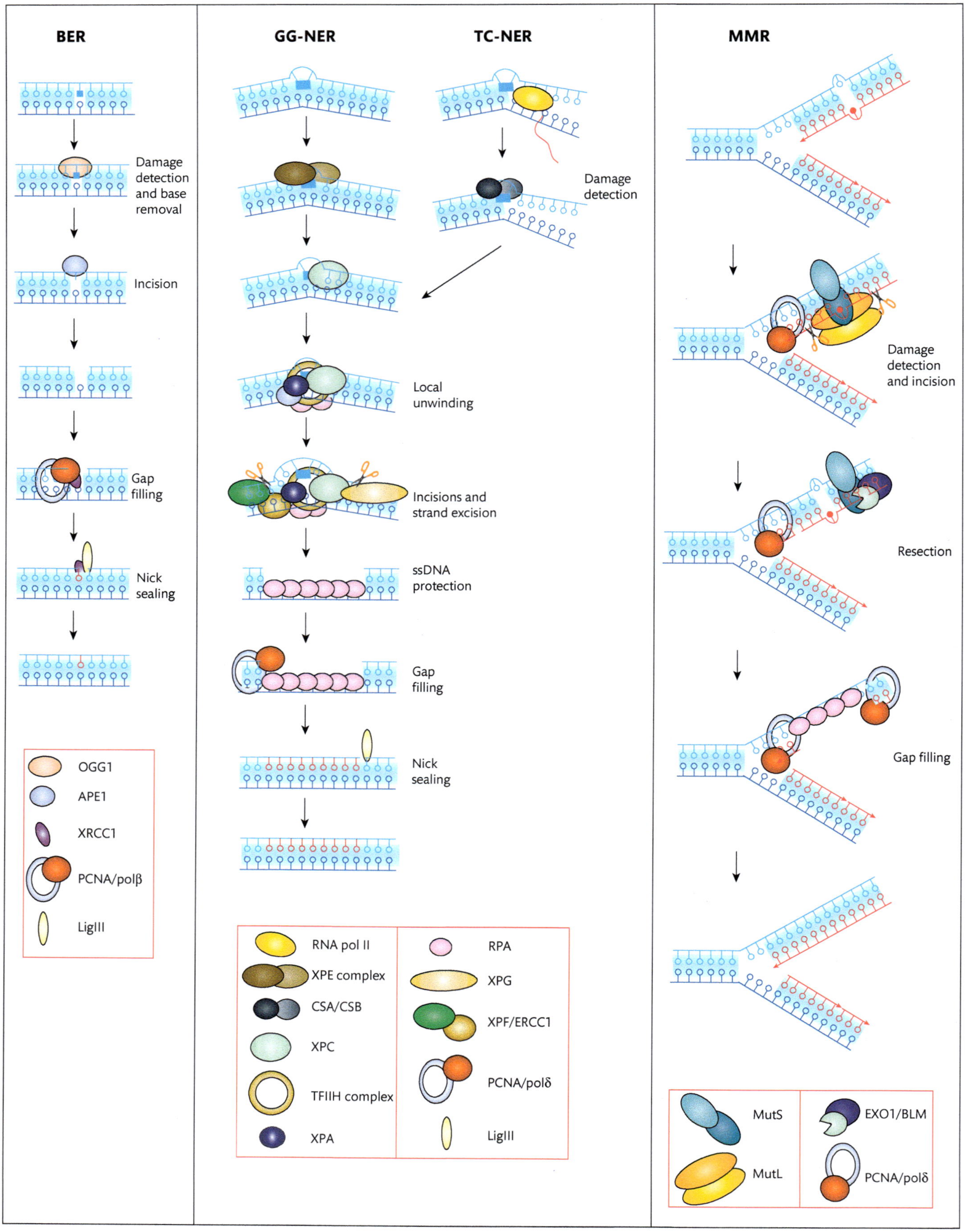

Fig. 2.3 *Excision repair systems.* Simplified schematic view of base excision repair (BER), global genome (GG-) and transcription coupled (TC-) nucleotide excision repair (-NER), and mismatch repair (MMR). (In red = neosynthesis.)

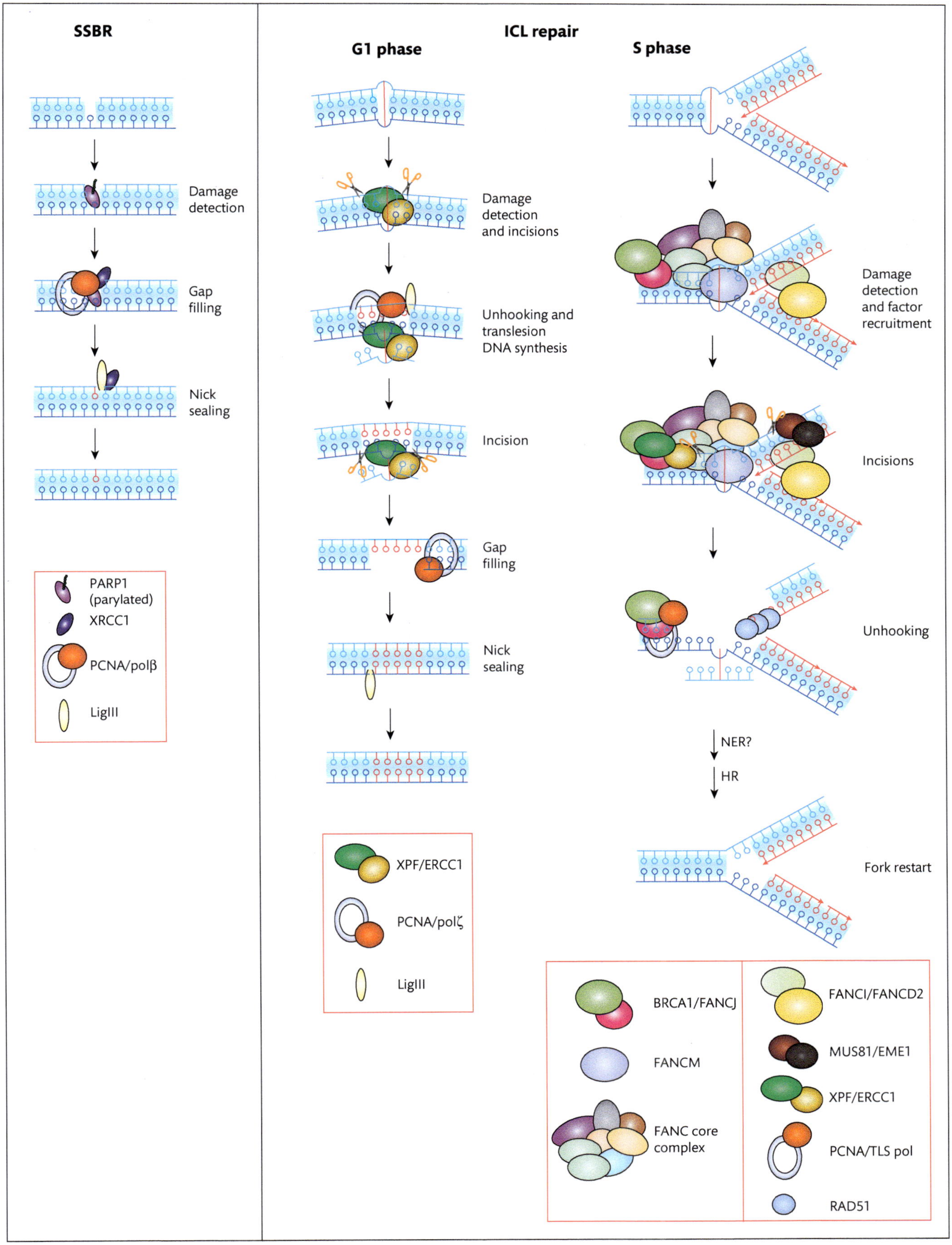

Fig. 2.4 *Single-strand break and interstrand cross-link repair systems.* Simplified schematic view of single-strand break repair (SSBR) and interstrand cross-link repair (ICL repair) during G1 or in the case that a replicative fork reaches an ICL during S-phase. (In red = neosynthesis.) The last step of ICL repair during S-phase creates a DSB and a hook, repaired, respectively, by homologous recombination and nucleotide excision repair.

cross-linking compounds (Deans and West, 2011). In addition, environmental agents (like furocoumarins from plants and nitrous acid in food) or cellular products (i.e. nitric oxide and lipid peroxidation by-products) are natural sources of ICLs. Central components of the ICL repair pathway are genes mutated in Fanconi anaemia (FA) syndrome. In particular, for damage detection FANCM shows DNA-binding activity (Fig. 2.4) and has been implicated in targeting the Fanconi core complex to DNA (Kim et al., 2008).

DSBs, although occurring infrequently (about 10 per cell per day), can lead to severe chromosomal rearrangements or loss of genetic material. For this reason human cells have evolved at least three partially independent sensors to detect these lesions: Ku70/Ku80, PARP, and the MRE11/RAD50/NBS1 (MRN) complex.

DSBs are rapidly surrounded by the Ku complex (formed by a Ku70/Ku80 heterodimer) toroidal structure (Fig. 2.5; see Mahaney et al., 2009). Ku complex loads and activates the catalytic subunit of DNA-PK (DNA-PKcs; Fig. 2.5) to initiate the main DSB repair pathway in human cells: the non-homologous end joining (NHEJ; see Ciccia and Elledge, 2010). In some cases the PARP1/2 complex, beside SSBs, could recognize DSBs and compete with Ku to promote alternative subpathways of NHEJ. DSBs can also be bound by the MRN complex, which preferentially promotes the preparation of DNA for the homologous recombination (HR) repair system (Fig. 2.5). On the whole, this competition between sensors is a first way towards the choice of an appropriate DSBs repair pathway.

In addition, RPA, an essential heterotrimeric complex (RPA1, RPA2, RPA3) that binds ssDNA during replication, can coat the 3' ssDNA tail deriving from DSB processing by resection, generating a platform for the activation of the apical kinases of DSBs signal transduction pathways.

DNA damage signal transduction

BER, NER, or MMR repair proteins are immediately recruited on DNA lesions by sensors to start their activity. These proteins are ready to use inside the cell, and if the damage is not widespread or difficult to fix, repair occurs mainly in a 'silent way', without the activation of additional responses.

However, particularly dangerous lesions (e.g. even a single difficult to repair DSB, or an extended amount of base modifications, like a diffuse presence of pyrimidine dimers due to prolonged UV exposure) activate an alarm signal transduction pathway.

This signalling activates secondary biological outcomes (Fig. 2.2) such as:

- the transient arrest of cell cycle at checkpoints during G1 or G2 phase or the slowdown of replicative fork progression during S-phase, to avoid that damaged cells could replicate or divide the genetic material before repair, thus enhancing the risk of genomic instability;
- the permanent arrest of cell cycle or the programmed suicide to exclude the propagation of an altered genetic information;
- the induction of extracellular signals with autocrine and paracrine effects.

The signal transduction (Ciccia and Elledge, 2010) starts from sensor proteins that attract to DNA lesions essentially three serine/threonine-protein kinases, namely ATM (ataxia telangiectasia mutated), ATR (ataxia telangiectasia and Rad3-related protein), and DNA-PKcs. These proteins belong to the phosphatidylinositol-3 kinase (PI3K) family and are the apical (initiating) kinases of the DDR cascade. These three large kinases enhance their basal activity by an autophosphorylation step, an event facilitated by sensors ability to recruit huge amounts of these proteins around the DNA lesion. Whereas ATM and DNA-PKcs are triggered primarily by DSBs (Shiloh and Ziv, 2013), ATR is activated by the presence of ssDNA regions directly caused by stalled replication forks, or indirectly induced by the processing of DSBs or UV lesions (Shiotani and Zou, 2009). Indeed, in the presence of diffused and/or clustered amount of UV-induced pyrimidine dimers, the NER activity produces sufficient ssDNA/RPA-coated DNA to induce ATR firing and signalling. On the contrary, with the same rare scattered lesions ssDNA is immediately refilled, thus preventing ATR activation.

At the same time, the apical kinases cooperate with two other classes of proteins: the mediators and the transducer kinases (Ciccia and Elledge, 2010). Mediators (MDC1, 53BP1, and BRCA1 for ATM; TopBP1 and claspin for ATR), by indirectly binding the lesions in an ATM/ATR-dependent way, contribute to further reinforce ATM and ATR activity facilitating the recruitment of specific targets (Lindsey-Boltz and Sancar, 2011; Shiloh and Ziv, 2013). A similar function is also carried out by histone modifications: the most widely known being phosphorylation of H2AX histone variant in serine139 (Rogakou et al., 1998), targeted by the apical kinases. Mediators and histone modifications spread up to megabases around the lesion and are detectable by immunofluorescence techniques as foci inside the nucleus, becoming invaluable markers to assess the presence of DNA damage.

The second class of proteins, the transducer kinases, transmits the DNA damage signal: CHK2 is the transducer for ATM (Matsuoka et al., 2000) and CHK1 for ATR (Kumagai et al., 2004). Apical and transducer kinases phosphorylate hundreds of effector proteins, which are the executors of DDR functions to induce the appropriate biological outcome (Ciccia and Elledge, 2010; Zannini et al., 2014).

The presence of two kinases in the DDR has the advantage of rapidly and strongly enhancing the initial signal, since a single molecule of ATM/ATR and CHK2/CHK1 can phosphorylate several targets. In addition, kinases could be rapidly shut down by phosphatases or degradation activities.

Furthermore, the DDR kinases phosphorylating their targets promote protein activation, physical interaction, and a cascade of post-translational modifications such as phosphorylation, ubiquitylation, sumoylation, and methylation (Lukas et al., 2011).

Recently, new levels of regulation have been discovered for DDR signalling. Particularly, microRNAs (miRNAs), ~21-nucleotide-long RNA regulators, have been found to control, at the post-transcriptional level, DDR gene expression (Wang and Taniguchi, 2013). ATM protein amount, for example, is controlled by miR-100 and miR101. On the other side, the expression of several miRNAs is altered in response to DNA damage. ATM itself can control miRNA expression at transcriptional or post-transcriptional levels but the roles for these molecules in DDR are still poorly understood. miRNAs activity is particularly intriguing since they can act as oncogenes or tumour suppressors with important roles during carcinogenesis.

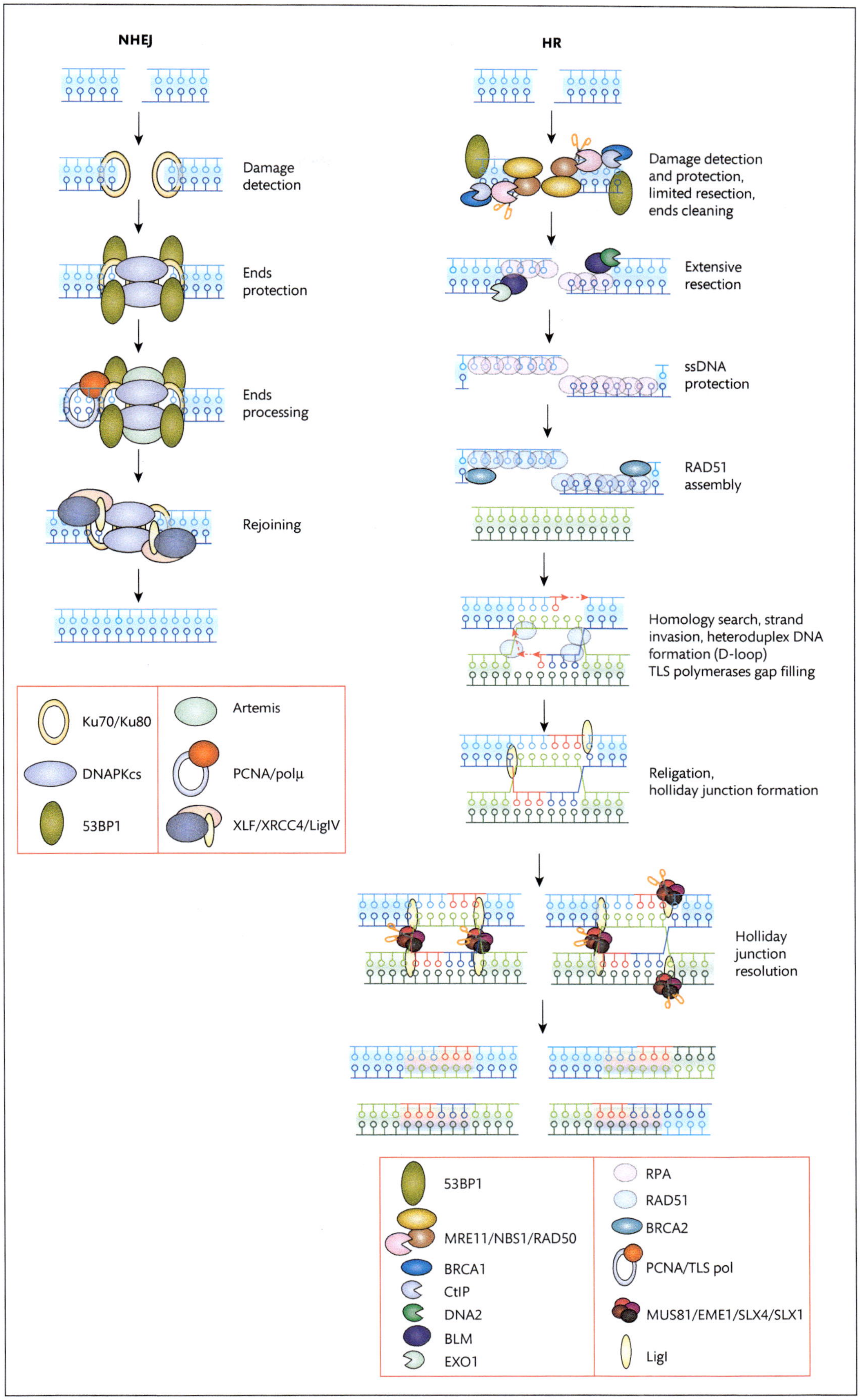

Fig. 2.5 *Double-strand break repair pathways.* Simplified schematic view of double-strand break repair by non-homologous end joining (NHEJ) or homologous recombination (HR). (Pale blue/blue and pale green/green DNA strands represent sister chromatids, neosynthesis is in red.) The final outcome of HR repair can significantly differ depending on D-loop and Holliday junction (HJ) formation, migration, and resolution. The activation of different HR subpathways could influence the presence and extension of crossover between chromatids. The MUS81/EME/SLX1/SLX4 complex is depicted but HJ resolution can be performed by GEN1 and HJ dissolution by BLM/TopIII/RMI1/RMI2 complex.

DNA repair

To correct limited base lesions, the core BER function requires essentially only four proteins (Krokan and Bjoras, 2013): a DNA glycosylase that removes the base; APE1 endonuclease that cuts the DNA backbone creating a nick; DNA polymerase (Pol) β that fills in the gap; and DNA ligase III that rejoins the chain (Fig. 2.3). XRCC1 has a relevant scaffold activity.

A collective feature of preferred NER substrates is that they are bulky and thus they thermodynamically destabilize the DNA duplex. The presence of sensors on this damage allows the formation of the preincision complex (Marteijn et al., 2014) formed by XPC, XPA, and transcription factor IIH (TFIIH; Fig. 2.3). TFIIH, which has a well-known role during transcription, consists of 10 subunits including the two helicases XPD and XPB. The helicase activity of TFIIH further opens the double helix around the lesion, while XPA binds chemically altered nucleotides. In this step RPA is also recruited and coats the undamaged strand. Successively, the structure-specific endonucleases XPF/ERCC1 and XPG are recruited and produce two cuts that eliminate the region of the DNA chain containing the lesion. The ssDNA gap created, covered by RPA, is ~30 nucleotides long and is refilled by Polδ, κ, or ε and associated factors. Then DNA ligase III or I seal the nick completing the process (Fig. 2.3).

The specificity of MMR is primarily for base–base mismatches and insertion/deletion mispairs that have escaped the proofreading activity of replication polymerases or have been produced during recombination. Since the parental strand carries the correct genetic sequence, the repair process must be directed to the nascent DNA strand containing the error. The presence of Okazaki fragments and/or the positioning of PCNA replication factor allows MMR to discriminate the parental and newly synthesized DNA strand (Kunkel and Erie, 2015). Degradation of the strand containing the error is performed by exonuclease 1 (EXO1) helped by the endonuclease activity of MutLα, which is enhanced by PCNA presence (Kunkel and Erie, 2015). Gap filling and sealing occurs using NER factors (Fig. 2.3).

Since ICLs engage both DNA strands, repair mechanisms involving a single round of excision followed by template resynthesis are not sufficient. Therefore, in G1 phase of the cell cycle the NER elements XPF and ERCC1, with the help of FANCP and the MMR factor MutSβ (Hashimoto et al., 2016) perform two rounds of incisions on both DNA strands (Fig. 2.4). Since the first incision creates a ssDNA gap, this is filled by translesion polymerases (like Polκ or ζ or REV1) before the excision of the opposite strand. If an ICL occurs during DNA replication or is produced in G1 or G2, but left unrepaired up to S-phase, the presence of an open replicative fork crashing against the lesion promotes a different NER pathway. In this case the ICL is processed with the participation of FA proteins (Hashimoto et al., 2016; Fig. 2.4). The FANCM sensor complex recruits the Fanconi core complex, consisting of seven FANC proteins (A, B, C, E, F, G, and L). This complex monoubiquitinates both FANCD2 and FANCI promoting the incisions of the ICL using structure-specific endonucleases such as XPF/ERCC1, MUS81/EME1, SLX4/SLX1 or FAN1. This first incision step is sufficient to introduce a DSB at replicative fork (Fig. 2.4), which is repaired by HR. The specificity of this pathway underlines another feature of the DDR: the correlation between cell cycle phase and lesion repair. This is particularly relevant for cancer studies in consideration of the hyper-replicative status of cancer cells and the high levels of damages during replication. A lesion occurring in S-phase is different from the same lesion in G1, for the presence of active replication forks that change the topological, structural, and protein-bound features of DNA. For these reasons S-phase cells can activate dedicated repair systems.

DSBs are mainly repaired by NHEJ and HR (Fig. 2.5). NHEJ is an easy rejoining activity that fuses the broken ends together:

- It can work during any phase of the cell cycle.
- It is fast, thus reducing the risk of managing highly recombinogenic free DNA ends.
- It could be error prone producing small deletions, particularly when it works on dirty DNA ends.

HR is a homology-directed repair system that utilizes as a template to direct the error-free repair the correct identical sequence present in the sister chromatid after DNA replication.

- It can work almost exclusively during or after S-phase, since, for steric reasons, it prefers to pair sister chromatids and not homologous chromosomes.
- It is slow, since based on many successive preparatory steps.
- It can produce loss of heterozygosity since it works by exchanging/copying between sister chromatids.
- It is a high-fidelity repair system, but since mammalian genomes are characterized by about 25% of repetitive sequence, it can fail during homology search, inducing rearrangements and loss of genetic material.

NHEJ is the dominant repair pathway in mammalian cells (Polo and Jackson, 2011), as it is estimated that HR is used to repair only 15–30% of DSBs.

NHEJ is initiated by DNA-PKcs that translocates with Ku to the DNA ends of the break (Fig. 2.5). The presence of DNA-PKcs molecules on opposing DSB termini promotes synapsis or tethering of the two DNA molecules. At the same time DNA ends cleaning (i.e. eliminate single-stranded overhangs or altered bases or even proteins blocked on damage) is supported by DNA polymerases (Polμ and λ), nucleases (Artemis, EXO1, WRN, CtIP, MRN complex), or BER enzymes. Extended DNA resection on the other side is limited by 53BP1 localization on the lesion. After accurate (but sometimes inappropriate) DNA ends processing, a ligation step is performed by DNA ligase IV in conjunction with its binding partners XRCC4 and XLF.

Differently, much of our current knowledge about the mechanism of eukaryotic HR derived from studies in budding yeast, where this pathway is more efficient (Ciccia and Elledge, 2010). The DNA ends of the DSB are initially processed through 5' to 3' end resection performed by the MRN complex together with CtIP, to generate molecules with 3'-single-stranded tails (Fig. 2.5). CtIP activity is enhanced by the presence of BRCA1 that has also the ability to hinder the 53BP1 antiresection function. The resected trait is successively extended by the combined action of EXO1 and/or DNA2 exonuclease with BLM (mutated in Bloom syndrome) helicase (Fig. 2.5). Then, ssDNA ends are coated by RPA, subsequently replaced by Rad51, an event facilitated by mediator proteins such as Rad52 and BRCA2 (Fig. 2.5). Rad51 plays a central role in the homology search and promotes the processes of strand invasion and heteroduplex formation. While DNA synthesis is carried out by DNA Polη, strand ligation creates a cross-shaped structure known as a Holliday junction (Fig. 2.5): This intermediate is resolved, alternatively, by the BLM/TopIIIα, GEN1 or SLX4/SLX1/MUS81/EME1 endonucleolitic complexes, a step that defines HR subpathways and the entity of crossover between chromatids. Finally, DNA ligase I performs the ligation step.

Current models suggest that the decision between NHEJ and HR is regulated essentially by a competition between pro- and antiresection factors presence and positioning: if resection is extensive the ssDNA tail will promote HR (Huertas, 2010). In turn resection is regulated by the combination of several events like chromatin status around the break, sensors binding competition, cell cycle phase, and exo- and endo-nuclease local activities. For example, resection is promoted by MRN/CtIP/BRCA1 and counteracted by 53BP1/Rif1. The competitive binding of these complexes is cell cycle-regulated by the S-phase specific cyclin dependent kinases (CDKs), that phosphorylate CtIP promoting BRCA1-CtIP binding and HR (Huertas and Jackson, 2009). However, 53BP1 recruitment and positioning is also strongly dependent on histone post-translational modifications (PTMs) and it has been recently hypothesized that actively transcribed genes, characterized by an open chromatin status and specific histone modifications, are preferentially repaired by HR (Aymard et al., 2014). This will help to preserve relevant genes from the mutation-prone NHEJ.

DNA structure condensation through histone and non-histone proteins is a barrier for an efficient repair. While in the past DNA repair studies were focused on DNA molecule integrity restoration, almost considered as a 'naked entity', more recently we started to consider the complexity of genome structure and epigenome as a target of DDR and as an important step during repair. The impact of chromatin structure on DNA repair was first described in the 'access-repair-restore' model (Smerdon, 1991). In the last few years several remodelling factors, histone chaperones, histone-modifying enzymes, and histone covalent modifications (including phosphorylation, acetylation, methylation, and ubiquitylation) have been identified as involved in opening chromatin structures, DNA repair enhancement, and pre-existing chromatin restoration (Polo and Almouzni, 2015). For example, in DSBs repair two steps have been extensively studied: chromatin PARylation performed by PARP proteins, which promotes the transient recruitment of chromatin-remodellers to DSBs; and the ATM/CHK2-dependent delocalization of KAP-1, a repressor protein that interacts with histone-modifying enzyme maintaining heterochromatin (Iyengar and Farnham, 2011).

Despite all these repair systems it is likely that some lesions will be misrepaired or left unrepaired. In this situation, tolerance mechanisms mitigate the interference of the persisting lesions with replication and transcription. The translesion DNA synthesis (TLS), for example, causes the bypass of base damage by the replication machinery, allowing normal DNA replication and gene expression downstream of the unrepaired damage (Sale et al., 2012). TLS represses cell cycle arrest and requires specialized low-fidelity DNA polymerases to permit replication, but nevertheless TLS introduces mutations into the newly synthesized DNA sequence.

Main additional biological outcomes of the DNA damage response: cell cycle arrest, premature senescence, apoptosis

Cell cycle arrest

By transiently arresting the cell cycle at checkpoints, the DDR provides the necessary time for the repair of a lesion before the critical phases of DNA replication and mitosis. DNA repair is tightly interconnected with cell cycle progression and unrepaired DNA lesions induce signalling pathways that arrest the cell cycle before DNA replication (G1/S arrest) or cell division (G2/M arrest) phases (Warmerdam and Kanaar, 2010). These mechanisms are relevant when damaged cells are cycling and not resting, or terminally differentiated. To arrest cell cycle progression ATM/CHK2 and ATR/CHK1 act, directly or indirectly, on the cyclin/Cdk complexes, the master regulators of cell cycle. Indeed, for example, they can target and inhibit the CDC25 family of phosphatases that are required to promote Cdk activity. The p53/p21 axis (p21 is a CDK inhibitor) is also important to prolong G1 and G2 cell cycle arrest. Cells suffering damages during S-phase can only slow down replication to avoid forks stalling and collapse.

Premature senescence

Normal diploid cells have the ability to proliferate in cell culture for a limited period of time, then they cease to divide and enter a state of cellular senescence. This phenomenon (replicative senescence; Ohtani et al., 2009) has been for a long time considered the result of an exhaustion of the proliferative lifespan. More recently, the senescence programme has been identified as a barrier to the replication of cells suffering a chronic damage derived from genotoxic agents' exposure (premature senescence) or from oncogene-induced hyperproliferation (oncogene-induced senescence, OIS; Gorgoulis and Halazonetis, 2010). In these conditions, senescence seems preferred to apoptosis for those cells characterized by essential structural functions. The molecular mechanisms characterizing senescence are mainly unknown, but there are evidences that crucial players are the p53/p21 pathway and p16, whose activities converge on the cell cycle regulator protein Rb (Munoz-Espin and Serrano, 2014).

The senescent phenotype is not limited to an intracellular signalling that arrest cell proliferation. Indeed, in the context of higher organisms, cells have evolved intricate mechanisms of intercellular communication that the DDR employs to trigger extracellular alarm signals. Senescent cells suffering chronic damages are metabolically active and can secrete cytokines, chemokines, and proteases (Rodier et al., 2009) on the whole named the senescence-associated secretory phenotype (SASP). The SASP can have both positive or negative effects, depending on the context (Tchkonia et al., 2013). For example, the SASP cytokines IL-6 and IL-8 can reinforce the growth arrest prompted by senescence as a helpful defence against cancer but can also induce epithelial-to-mesenchymal transitions, thus promoting carcinogenesis. Furthermore, SASP can alert nearby cells to potential danger and promote the immune clearance of the damaged cells. Conversely, it might cause local and systemic inflammation, disrupt tissue architecture, and stimulate the growth of nearby malignant cells. The molecular mechanisms that drive SASP are under investigation, but a role for CHK2, p53 and NF-κB has been ascertained (Rodier et al., 2009; Chien et al., 2011). SASP is a late response to the original injury, but there is also evidence that damaged cells can rapidly transmit a DDR-dependent stress signal to neighbouring healthy cells, thus causing the paracrine activation of a bystander DDR (Najafi et al., 2014). Bystander effects, particularly detected upon radiation-induced DNA damage, include a wide range of biological processes, such as secondary DNA damage, malignant transformation, chromosomal aberrations, cell death, apoptosis, and adaptive responses. These events implicate various clastogenic factors and signalling molecules, transmitted through gap junctions, as well as released outside cells. Moreover, also reactive oxygen/nitrogen species as well as cytokines are involved in mediating the bystander effect.

Autophagy

Autophagy (from the Greek, 'self-eating') is a catabolic process, tightly regulated and evolutionary conserved, in which damaged proteins and organelles are degraded in lysosomes, finally resulting in the release of amino acids and fatty acids that can be used again by the cell. Autophagy is triggered in response to various stress stimuli, including: nutrient and energy stresses, hypoxia, redox stress, and mitochondrial damage (Kaur and Debnath, 2015). Autophagy is a protective and pro-survival mechanism, but extensive autophagy may lead to cell death.

The best described pathway leading to autophagy is activated during starvation (Kaur and Debnath, 2015). In this pathway mTOR (mammalian target of rapamycin) plays a central role since the mTOR complex 1 (mTORC1) inhibits autophagy through the phosphorylation of ULK1 (Unc-51-like kinase 1) and Atg13. During starvation, mTORC1 inhibition leads to dephosphorylation of ULK1 and Atg13 and the formation of an active complex of Atg13, ULK1, and FIP200. This event, in combination with Vps34 (a PI3K) activation, mediated by Beclin-1 and other factors, starts autophagosome, a double membrane structure, maturation. Autophagosome specificity for targets and lysosomes are all events regulated by lipidated LC3 recruitment both at inner and outer membrane.

In principle, autophagy and DDR should be usefully interconnected. Indeed, sources of damage, like ROS or radiations, hit, even before nuclear DNA, cytoplasmic macromolecules, and organelles structure, that should be promptly removed. Furthermore, senescence should be positively regulated by autophagy, while apoptosis and autophagy seem alternative. Therefore, it is not surprising that increasing evidences suggest an interplay between DDR and autophagy, but up to now the presence of a direct correlation remains elusive.

The DDR could exert a control (positive or negative) on autophagy through transcriptional regulation (Czarny et al., 2015). Indeed, several ATM/ATR kinases targets can influence the expression of genes associated with autophagy:

- **a.** NF-κB upregulates Beclin-1
- **b.** p53 transcriptionally regulates adenosine-monophosphate-activated protein kinase (AMPK) subunits and activators, PTEN (an mTOR inactivator), DAPK (phosphorylates Beclin-1) and DRAM (as a role in a late step of autophagy)
- **c.** ΔNp63α transcribes ULK1, several Atg family genes, and Beclin-1
- **d.** Che-1 upregulates Reddl and Deptor (two mTOR inhibitors)

Additionally, the ATM interactor FOXO3a regulates transcription of autophagy-related genes, including LC3. On the other side autophagy specifically degrades DDR components like p62 (involved in Rad51 binding to DNA damage), HP1α (a protein promoting chromatin condensation, displaced from DNA breaks by DDR) and CHK1.

Furthermore, a prolonged DDR response might activate autophagy through energy consumption. Indeed, PARP-1 hyperactivation can cause adenosine triphosphate (ATP) depletion and the consequent adenosine monophosphate (AMP) increase (Huang and Shen, 2009), thus promoting the activation of AMPK, a well-known inhibitor of mTOR and therefore an autophagy promoter.

Interestingly, other connections between ATM and autophagy seem more related to ROS presence than to ROS-deriving DNA damage. For example, a cytoplasmic fraction of ATM, in response to elevated ROS, can induce autophagy (Alexander et al., 2010) while another ATM localized in mitochondria regulates mitophagy (autophagy of damaged mitochondria). Of note, the loss of one Beclin-1 allele in an ATM-*null* mouse induces a significant delay in the tumour-prone phenotype of these mutants reducing mitochondrial abnormalities more than improving the DDR function (Valentin-Vega and Kastan, 2012). Therefore, the absence of ATM could promote genome instability directly through defects in DNA repair but also indirectly through a dysfunctional mitochondrial clearance that enhances free ROS and DNA damage.

Recent data suggest that sirtuins, a family of NAD+-dependent protein deacetylases, may also play an important role in autophagic control of DDR. Indeed, SIRT1 can induce the formation of autophagosome or it can directly regulate autophagy by targeting mTOR and FOXO. At the same time SIRT1 interacts with many proteins which can be, directly or indirectly, involved in DDR, like p53 (Lin and Fang, 2013).

Finally, a cross talk between senescence and autophagy was recently described (Kang et al., 2015) since ATM and ATR can suppress the autophagic degradation of GATA4 transcription factor, promoting NF-κB transcription. NF-κB has a crucial role in SASP initiation and facilitate senescence induction. The accumulation of GATA4 in tissues of aged humans may contribute to inflammation in age related diseases including cancer.

Apoptosis

Cells with an irreparable damage activate suicide by apoptosis to prevent the replication and propagation of a modified, and thus potentially harmful genome. The induction of apoptosis proceeds through at least two main routes, the extrinsic and intrinsic pathways, and the activation of a series of cysteine-aspartic proteases, named caspases. Effector caspases cleave the inhibitor of the DNAse (iCAD) inducing nuclear DNA fragmentation and promote the degradation of kinases, DNA repair, and cytoskeletal proteins, contributing to the typical morphological alterations of apoptotic cells. To activate apoptosis the DNA damage signalling exploits primarily the p53 pathway, but ATM and CHK2 can also promote proapoptotic p53-independent pathways targeting, respectively, NF-κB and c-Abl, or p38 and E2F1 (Zannini et al., 2014).

The p53 network

The p53 protein is considered the 'master gatekeeper' protein and the 'guardian of the genome' in human cells. p53 is one of the most important and studied tumour suppressor, with more than 2000 articles per year in the last two decades. This is legitimated by the fact that the p53 gene (*TP53*) is mutated in around 50% of tumour cells, with the rate varying from 10–12% in leukaemia to 38–70% in lung cancers, and 43–60% in colon cancers (Murray-Zmijewski et al., 2008). Indeed, p53 knockout mice develop tumours with short latency and 100% penetrance.

This protein performs its function primarily as a transcription factor, controlling the expression of more than 100 target genes, responding to a great variety of stresses. Among p53 targets the most abundant are involved in DNA repair, cell cycle arrest, and apoptotic programme. Various PTMs, over 100 cofactors, and p53 cellular localization can contribute to determine when and what kind of proteins are produced.

Differently from other tumour suppressor genes, most *TP53* mutations in tumours are missense, predominantly affecting those residues that are located in the DNA-binding domain of the protein, causing the loss of its tumour suppressor function and, in some cases, the gain of novel oncogenic activities.

Treatment of normal cells with either genotoxic agents or non-genotoxic agents stresses results in the phosphorylation of p53 at about 20 serine and threonine residues throughout the protein, and acetylation at about a half-dozen lysines in the C-terminus (Fig. 2.6; see Appella and Anderson, 2001). Phosphorylation by ATM or ATR at serine15 is a key priming event for the phosphorylation of several other residues at the N-terminus (serine15 cluster). These events are specific to genotoxic stresses and principally promote p53 protein stabilization. Indeed, a protein so relevant for cell life is practically absent in unstressed conditions. In normally growing cells, nuclear p53 has low activity and a short half-life because it is complexed with the E3 ubiquitin protein ligase (MDM2) which causes p53 ubiquitination and degradation by the proteasome (Fig. 2.7). Only very low p53 levels and activities allow a normal growth. After DNA damage, serine15 phosphorylation promotes MDM2 displacement, allowing p53 accumulation in the nucleus where it can perform its function as transcription factor (Cheng and Chen, 2010). The stability of p53 protein is under the control of a negative feedback, since MDM2 is a target of p53 transcriptional activity. As a consequence, single cells exposed to DSBs inducing agents show p53 pulses of fixed amplitude, duration and period, and the mean number of pulses increases with the extent of DNA damage (Lev Bar-Or et al., 2000; Lahav et al., 2004). This effect combined with specific mRNA decay of p53-transcribed genes generates different profiles of protein expression (Porter et al., 2016). Consistently, altering p53 dynamics pharmacologically changes patterns and extension of target genes expression and, ultimately, cell fate (Purvis et al., 2012). Differently, UV radiation triggers a single p53 pulse with a dose-dependent amplitude and duration; this well correlates with the observation that IR and UV activate a different set of p53-dependent genes.

Another important interactor and repressor of p53 activity is HDMX. Normally this protein shuttles between the nucleus and the cytoplasm, but in response to DSBs, nuclear HDMX is phosphorylated by ATM and CHK2 and retained there, where it is degraded (Pereg et al., 2006).

However, p53 activity is not only regulated by the time of accumulation or the interaction with HDMX, but also by several PTMs and cofactors, that modulate the ability of p53 to bind specific sequences to the promoters of its target genes. Indeed, the ATM- and ATR-dependent DDR signalling induces directly or indirectly a multitude of different p53 posttranslational modifications that can determine an appropriate and proportionate response according to the type of damage and stress intensity. An example of this mechanism is serine46 phosphorylation by the ATM/ATR-activated HIPK2 kinase which drives p53 towards a 'killer' activity, stimulating the transcription of proapoptotic targets (D'Orazi et al., 2002). Furthermore, phosphorylation of the serine15 cluster also promotes the recruitment of acetyltransferases like p300, CBP e PCAF, that acetylate several C-terminal lysines on p53. Acetylation of p53, which is counteracted by deacetylases like SIRT1, regulates promoter specificity, driving damage response towards apoptosis.

As for p53 stability, p53 activity is under the control of a negative feedback loop. Wip1 phosphatase, a p53 transcription target, dephosphorylates and deactivates both ATM and p53 (Shreeram et al., 2006), while activates MDM2, thus enhancing p53 degradation.

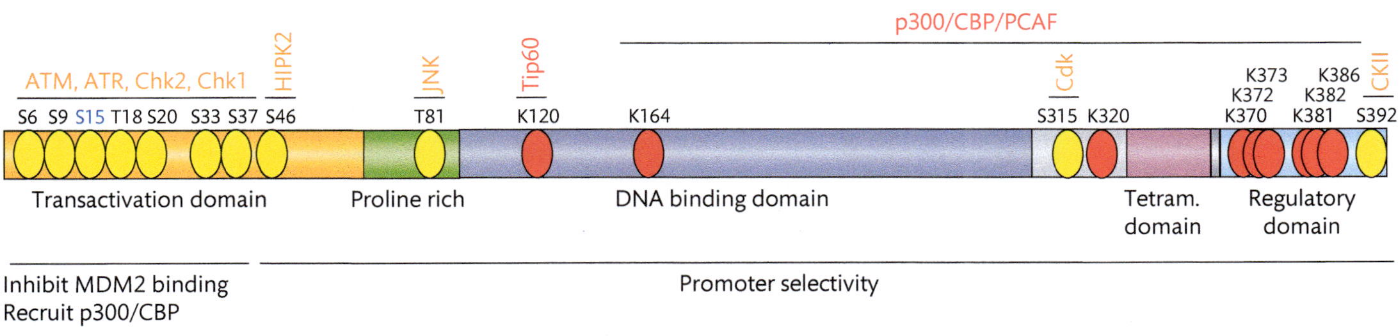

Fig. 2.6 *p53 protein structure and main post-translational modifications (PTMs) associated with the DNA damage response.* p53 protein domains and key post-translational modifications are depicted. The transactivation domain, proline-rich region, DNA-binding domain, tetramerization domain, and regulatory domain are shown. The most relevant serine (S) and threonine (T) residues phosphorylated after DNA damage are indicated (yellow ellipses) together with the protein kinases (orange letters) known to phosphorylate them. Acetylated tyrosines (K) are shown as red ellipses together with the acetylases (red letters) known to acetylate them. Up to serine37, the indicated phosphorylations are known to mainly influence p53 protein half-life acting on the interaction between p53 and MDM2 and to recruit acetylases. Starting from serine46 the indicated PTMs are more related to p53 promoter selectivity.

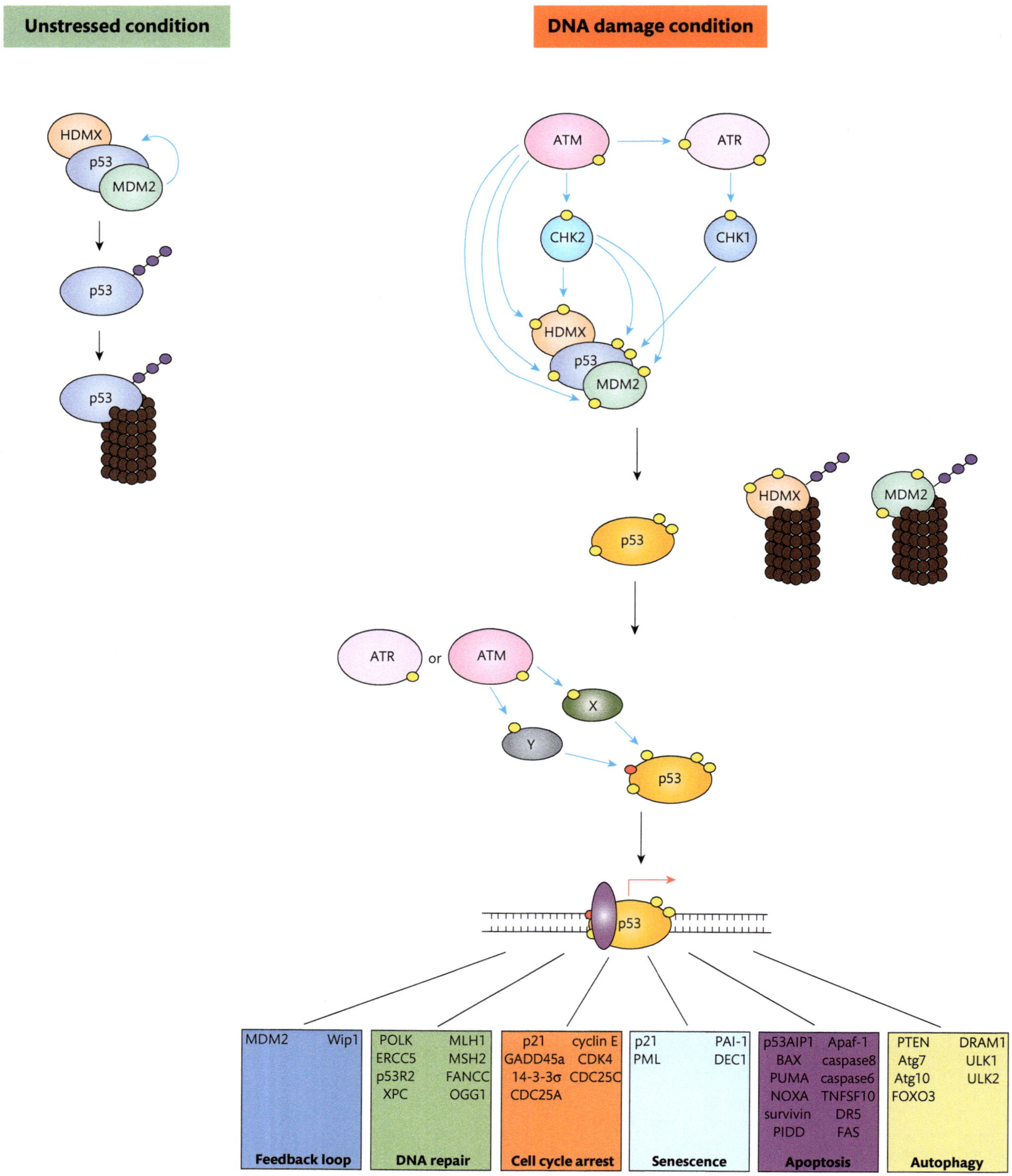

Fig. 2.7 *p53 protein regulation and main transcriptional targets associated with the DDR.* Under normal conditions, two major negative regulators—MDM2 and HDMX—bind to p53, repressing its activity, and inducing its degradation by a poly-ubiquitination (violet circles) and proteasome (brown)-dependent pathway. In response to various stress signals, including DNA damage, p53, MDM2, and HDMX are phosphorylated (yellow circles) by apical and transducer kinases (ATM, ATR, CHK2, CHK1), leading to MDM2 and HDMX degradation and p53 stabilization and activation. ATM or ATR can also directly or indirectly (through X and Y proteins) induce other post-translational modification that regulates p53 promoter selectivity (i.e. HIPK2 activation). Various interactors (purple ellipse) can also redirect p53 activity. The figure includes lists of representative p53 transcriptional target genes involved in processes that are important for p53 inactivation (feedback loop) and DDR functions (DNA repair, cell cycle arrest, senescence apoptosis, and autophagy).

On the whole, p53 transcription of prorepair, proarrest, and proapoptotic genes allows the cell to take the decision between transient or permanent cell cycle arrest and cell death (Batchelor et al., 2009). This response is dependent on cell stress, cell type, and tissue type. To exert these activities p53 has the ability to induce or repress the transcription of several targets in response to DNA lesions (Fig. 2.7).

To enhance repair p53 can promote transcription of gene-encoding proteins like ERCC5, FANCC, XPC, MLH1, MSH2, GADD45a, and Polκ. It can also enhance the transcription of p53R2, encoding a ribonucleotide reductase that can supply deoxyribonucleotides to DNA repair.

The *CDKN1A* gene, encoding the p21 protein, is one of the most studied p53 targets and is implicated in the induction of cell cycle arrest; p21 binds and inhibits the CDKs, thus preventing G1 and G2 progression. This protein can also start a programme of premature senescence that in this case could be considered as the evolution of a chronic cell cycle arrest.

In case of a damage beyond repair, p53 stimulates the extrinsic (or death receptor pathway) or the intrinsic (or mitochondrial pathway) apoptotic pathways depending on the DNA damage. In particular, p53 starts the intrinsic pathway through the transactivation of proteins such as BAX, BID, PUMA, and NOXA that permeabilize the mitochondrial membrane thus causing the release of proapoptotic factors. Furthermore, p53 can also transrepress apoptosis inhibitors like Bcl-2 and survivin. Moreover, p53 also initiates extrinsic apoptosis by transcribing genes encoding for proteins like cell death receptors (DR5, FAS) and cell death ligands (TNFSF10). Interestingly, p53 has also transcription-independent effects on the permeability of the mitochondria membrane by directly activating BAX or by antagonizing the antiapoptotic proteins BCL2 or BCL-XL.

Finally, p53 can also promote a favourable cellular situation to take under control the DNA damage stress, transcribing components of metabolism regulation, antioxidants and autophagic proteins of the Atg family (see 'Main additional biological outcomes of the DNA damage response: cell cycle arrest, premature senescence, autophagy, apoptosis' section).

A recent field of p53 activity investigation is related to the role of cancer stem cells (CSCs) in carcinogenesis and cancer treatments. CSCs are rare quiescent cells within the tumour that possess augmented tumorigenic potential and drug resistance. Alike normal stem cells, CSCs are able to self-renew and differentiate and they contribute to various aspects of tumorigenic process including tumour initiation, progression, invasiveness, and metastasis. CSCs may originate from normal stem cells that underwent genetic and epigenetic alterations, or alternatively by de-differentiation of progenitor or mature cells induced by specific signals from the microenvironment. Wild-type p53 serves as a potent barrier to CSCs formation regardless of their origin (Aloni-Grinstein et al., 2014), while mutant p53 proteins exhibit their oncogenic gain-of-function by facilitating the acquisition of CSCs phenotype.

The DNA damage response activity during normal cell physiology

The DDR is also involved in cellular functions characterized by alterations of DNA structure, but independent of the presence of DNA damage. In particular, DDR helps to maintain DNA integrity by participating in telomere length regulation (O'Sullivan and Karlseder, 2010), viral DNA processing (Turnell and Grand, 2012), and antigen receptor assembly by lymphocytes (Callen et al., 2007). The DDR is also implicated in mitosis (Heijink et al., 2013), meiosis (Richardson et al., 2004), and differentiation programmes (Nagaria et al., 2013) regulation.

DDR activities at telomeres, during meiosis, and in lymphocytes maturation explain why several human syndromes linked to DDR defects show progeroid features, infertility, immunodeficiency, predisposition to lymphoma or leukaemia. Neurodegeneration, another common symptom in DDR syndromes was hypothesized to derive from specific features of the central nervous system like an accelerated metabolism producing ROS and high levels of transcription that produce unstable structures like RNA/DNA hybrids and high amounts of ssDNA. The predisposition to damage, combined with the absence of cell replacement in case of apoptosis or senescence, could partially explain neurodegeneration.

Telomeres

Telomeres are chromosomal ends, structurally related to DNA breaks and for this reason under the control of the DDR. A protein complex, named shelterin, binds the repeated telomeric DNA sequence (de Lange, 2005) and protects normal telomere structures from exonucleases, damage sensors and repair proteins by forming a three-dimensional structure called telomere loop that hides chromosomal ends and prevents chromosome fusions. Moreover, shelterin proteins inhibit ATM, ATR, and CHK2 masking their activation domains (Karlseder et al., 2004; Buscemi et al., 2009). Telomere stress or shortening partially uncovers telomeres creating the opportunity for ATM and Chk2 to function, finally leading to DDR activation, permanent cell cycle arrest, and acquisition of senescence features, like cellular flattening and vacuolization (Kuilman et al., 2010).

Meiosis

During meiosis homologous chromosomes undertake a programmed sequence of compaction, synapsis, recombination, and segregation to split in half the genetic material. A characteristic of the meiotic process is the controlled production of a high number of DSBs, which are repaired specifically by HR. In this case and differently from what happens in somatic cells, HR prefers homologous chromosomes over sister chromatids as templates for repair, thus enabling the exchange of parental genetic information, an important source of offspring genetic diversity. As a consequence, alterations of HR mechanisms affect the genome integrity and the development of germ cells. Studies in the past decade have highlighted common and peculiar steps occurring during meiotic crossover and genomic DNA repair (Youds and Boulton, 2011).

Lymphocytes development and maturation

During the development and maturation of lymphocytes, the immune system is a site of intense DNA modifications. Both T and B cells use V(D)J recombination, to cut and rejoin segments of DNA that, when assembled, generates diversity encoding the N-terminal variable portion of the T-cell receptors or immunoglobulin. V(D)J is initiated by the cleavage step performed by recombination-activating genes 1 and 2 (Rag1 and Rag2) proteins and the formation of a DSB. This break, which occurs in G1, is repaired through NHEJ pathways (Malu et al., 2012) and, coherently, defects in several NHEJ factors (DNA-PKcs, Artemis, ATM, NMR complex) have been identified in humans and mice immunodeficient conditions.

Furthermore, upon antigen recognition in secondary lymphoid organs, mature B cells increase their repertoire by class switch recombination (CSR) and somatic hypermutation (SHM). CSR is a process of DNA rearrangement finally resulting in an orchestrated change from IgM to IgG, IgE, or IgA expression. To induce CSR, activation-induced deaminase (AID) activity hits two switch repeat regions, converting several dC bases to dU that are successively transformed in abasic sites, nicked by MMR activity, finally resulting in DSBs. The following process of intrachromosomal DNA rearrangement between broken ends, is characterized by synapsis and DNA repair processes mediated by DDR signalling and by error-prone NHEJ pathways (Xu et al., 2005). Rearranged Ig genes are further modified by point mutations introduced by the SHM process. As for CSR, AID deamination of DNA is the initiating step that produces a minimum of 800-fold more uracils in the Ig loci than elsewhere during cellular metabolism. Whether these bases are repaired by BER and MMR in a correct or mutagenic way depends on several factors (Xu et al., 2005). For example, if abasic sites deriving from dU processing are faithfully repaired by Polδ or mutagenically processed by Polη, or if they hit replicative forks and so on. The final frequency of mutation is 10^{-2}-10^{-3} mutations/bp, which is a million times higher than in the other part of the genome.

Defect in DNA damage response and cancer

Several years ago, the role of DDR in cancer was supported by the discovery that defective MMR and NER underlie somatic and hereditary colorectal and skin cancer. Nowadays, the DDR and DNA repair are considered key pathways in normal cellular physiology, since their defects are relevant for multiple pathological conditions, including ageing, neurodegeneration, and cancer. Research into the molecular bases of DNA repair, that started more than 50 years ago, has now translated into drugs that highly specifically inactivate key players of the DDR. Indeed, at the end of 2014, PARP inhibitors, the first molecularly targeted anticancer drugs exploiting the DNA repair defects present in *BRCA1*- or *BRCA2*-mutant cancers were FDA-approved.

Genome instability and the hypermutation phenotype

Cancer development is induced by the progressive accumulation of genomic changes that lead to the loss of tumour suppressor activities, oncogenes activation, and/or the generation of fusion genes with protumourigenic potential. These genomic alterations in turn promote changes that can originate waves of clonal expansion, ultimately facilitating the phenotypes of malignant cancer cells. Genomic instability is a hallmark of cancer, but it should be a consequence of tumour progression or an active process that drives tumour evolution. Therefore, it is not clear whether DDR alterations, that enhance genomic instability, should be an initiating step for carcinogenesis or not and if they fuel cancer development and ability to adapt. The mutator phenotype hypothesis, formulated 40 years ago, suggests that cancer occurrence is promoted by an increased frequency of stochastic mutations, initially supposed to derive from DNA polymerases defects and later extended to DNA repair impairment. On the other side, for high proliferating cells, like CSCs, exposed to a selective pressure, the basal frequency of replication errors could account for those single or few mutations, sufficient to induce tumour cell proliferation in *in vitro* model systems. The discovery by next generation sequencing approaches that tumours show high number of mutations (from 500 in acute myelogenous leukaemia to 100,000 in melanomas and glioblastomas; Loeb et al., 2016) supports the mutator hypothesis, although do not fully clarify if multiple mutations and a mutator phenotype can cause malignancy or are instead a consequence of it. The progressive accumulation of genomic changes in cancer is still debated. Indeed, the recurring presence in tumours of hundreds of clustered rearrangements on one or more chromosome(s), was recently suggested to derive from chromothripsis (Stephens et al., 2011), a single catastrophic event of multiple DSBs formation, possibly arising from telomere crisis. These chromosome fragments are then pieced together inaccurately by erroneous DNA repair. Although most cells experiencing chromothripsis *in vivo* would presumably die because of the initial DNA damage signal or of the severe genetic imbalances, those cells that will survive could already be on the way to cancer (Fig. 2.1). It is not clear at what stage of carcinogenesis chromothripsis can occur, but obviously cells bearing DDR defects are predisposed to these events.

Uncontrolled replication and cell survival/death unbalancing

Immunohistochemistry analysis of surgically resected specimens from patients with untreated breast, lung, colon, and bladder melanomas have shown a strong activation of the DDR (ATM, CHK2, p53) in dysplastic lesions, but not in normal tissues. However, these markers are less frequently present, or also absent, in more advanced or invasive lesions (Gorgoulis et al., 2005). The constitutive activation of the ATM/CHK2 response in preneoplastic cells, which precedes the occurrence of p53 mutations and induction of genomic instability, is apparently driven by cancer-promoting oncogenic stimuli (e.g. unscheduled expression of cyclin E, aberrant growth stimuli, Rb impairments), which deregulate the DNA replication machinery and generate large amounts of intermediates such ssDNA and DSBs (Fig. 2.1). In this respect, the DDR pathway would act as a tumour suppressor by creating a barrier to oncogene-driven uncontrolled proliferation through the induction of cellular senescence or apoptosis programmes, thereby limiting cancer progression. However, cells may eventually evade this constraint to deregulated DNA replication through a selection pressure for inactivating mutations of the DDR machinery (Fig. 2.1).

On the whole, the DDR is constituted by a limited number of apical and distal kinases (Fig. 2.2), which through thousands of PTMs regulate hundreds of effectors. In normal cells the combination of a subset of these PTMs create a 'barcode' that activate the correct biological response in relation to damage characteristics. Even small modifications of this code could change the fate of a cell allowing unwanted survival, and therefore for example cancer proliferation, or increased cell death, as demonstrated by neurodegenerative features in DDR-associated syndromes.

Sporadic mutations in the DNA damage response genes and cancer

Practically all cancers show alterations in one or more DDR components. Beside the well-known presence of p53 mutations, the apical kinase ATM has been found frequently mutated. Mutations in the ATM gene have a role in sporadic tumours, such as T-cell prolymphocytic leukaemia, B-CLL, and mantle cell lymphoma (MCL). ATM mutations were found in the early stages of B-CLLs and associate with a shorter treatment-free interval (Austen et al., 2005). ATM inactivation in MCL correlated with a high number of chromosomal translocations, suggesting that ATM deficiency may be an early event predisposing to chromosomal instability in these tumours. Moreover, aberrant methylation of the ATM proximal promoter, together with reduced ATM transcription was found in more than 70% of cases of sporadic breast cancer (Vo et al., 2004). CHK1 may be involved in the pathogenesis of a subset of aggressive diffuse large B-cell lymphomas (Tort et al., 2005), while CHK2 appears to function like a multiorgan tumour susceptibility gene.

Other DNA repair pathways have been detected as altered in cancer. MMR defects cause microsatellite instability (MIN) predisposing to colorectal and endometrial carcinoma. MMR deficiency, in some cases due to epigenetic silencing, has been found in 15–17% of all primary colorectal cancer, 30% of endometrial cancers and approximately 10% of ovarian cancers. Furthermore, chromosomal instability (CIN) is present in most sporadic solid tumours. Oncogenes activation and DNA replication stress with DSBs formation promote CIN continuously. Moreover, at late stages of cancer progression, chronic hypoxia, and/or cycles of hypoxia and reoxygenation might also favour DNA damage and genomic instability in cells with a deregulate DDR background.

Inherited mutations in the DNA damage response and cancer-prone syndromes

More than 30 syndromes have been described up to now, correlating with mutations in a DDR component (Table 2.1), and most of them showing a significant predisposition to develop cancer.

A growing family of NER diseases includes XP, Cockayne syndrome, cerebro-oculofacial skeletal syndrome, trichothiodystrophy (TTD), and other pathologies with mixed or milder symptoms. About 40 genes are directly involved in NER, but only about a dozen of them have been found to be deregulated in NER-related human diseases. The others represent either essential genes that would be lethal if mutated or that might induce such mild clinical effects that people bearing these defects fall into the population of 'sun-sensitive' individuals. Although repair absence in XP can be readily correlated to cancer through increased mutagenesis induced by UV light, the lack of cancer in CS despite their sun sensitivity is still an enigma. CS and XP patients have similar increased mutation rates in response to UV, but additional chromosome instability could derive from the role of XP proteins in both GG- and TC-NER.

Another paradigm for the relationship between cancer and DDR is the case of ATM: a growing body of evidence indicates that inherited ATM mutations confer increased susceptibility to cancer. In ataxia telangiectasia (AT) patients who have germline homozygous mutations of ATM, a well-established clinical feature is the increased risk of childhood malignancies, primarily lymphoid tumours. Moreover, there is a strong association between heterozygous mutations of ATM and tumour susceptibility, illustrated by the elevated rates of breast cancer among female carrier relatives of AT patients (estimated 3.9–5.5 times; Hall, 2005), that are essentially asymptomatic. ATM alterations can also be associated with increased risk of prostate and lung cancer (Kim et al., 2006). It is important to note that while AT is a rare disease, carriers of ATM are estimated to be 1% of the total population. Promising therapeutic approaches in these cases explore synthetic lethality with DDR inhibitors (i.e. for ATR or CHK1; see 'DNA damage response pathway components as targets for chemo- and radio-sensitization').

In accordance with the essential role of the MRE11/RAD50/NBS1 (MRN) complex for genome stability maintenance, mutations of its components can predispose to cancer. For example, Nijmegen breakage syndrome patients, characterized by biallelic mutations of NBS1, show cancer predisposition, particularly lymphomas and leukaemias. NBS1 heterozygous carriers of the founder mutation 657del5 (1/177 in the Central Europe Slavic population) may have increased risk of prostate cancer (Cybulski et al., 2004). In addition, also the risk of breast, prostate, and colorectal cancers, lymphoblastic leukaemia, and non-Hodgkin's lymphoma is also elevated in NBS1 carriers (di Masi and Antoccia, 2008).

Li-Fraumeni syndrome (LFS) is a rare autosomal dominant hereditary cancer syndrome characterized by germline mutations in the *TP53* gene (McBride et al., 2014). About 50% of the people carrying mutations in *TP53* gene will develop cancer by the age of 30 years, with a lifetime risk of up to 70% in men and almost 100% in women, the latter predominantly ascribable to breast cancer. In these conditions, breast cancer (27%) and sarcomas (25%), are the most common reported tumours, but the incidence of several forms of cancer is also increased. A fraction of 40–60% of tumours in LFS patients exhibits wild-type *TP53* loss of heterozygosity. Indeed, while mutation and loss of both *TP53* alleles is a quite frequent event in sporadic cancers, the tumours occurring in LFS patients often retain a structurally intact wild-type *TP53* allele. However, there is evidence that many *TP53* missense mutations can have a dominant-negative role and functionally inactivate wild-type *TP53* forms; in this case the selective pressure for loss of the wild-type allele in tumour is reduced. On the contrary, tumours from LFS patients with a null *TP53* allele constantly also exhibit the loss of the remaining wild-type allele.

DNA damage response pathway components as targets for chemo- and radio-sensitization

The knowledge that DDR was altered in cancer opened the opportunity to act on these pathways as a therapeutic approach, although the possibility to identify targeted treatments and predict therapeutic outcome is complicated by the high degree of complexity of the DDR. Nowadays, radiotherapy and chemotherapies are the most relevant cancer treatments beside surgery and they mainly act by generating DNA damage (Cheung-Ong et al., 2013). In part, this reflects the DDR impairment present in most cancer cells beside their ability to proliferate more rapidly than most normal cells, since S-phase is a particularly vulnerable time for DNA damage exposure. Indeed, reduced or absence of DDR factors is usually positively associated with therapeutic outcome, with the exception of defects in p53 and other proapoptotic proteins, which commonly increase therapy resistance.

Table 2.1 *Human syndromes associated with inherited DDR defects.* A selection of human diseases known as deriving from mutations of DDR components. Mutated genes and their functions in DDR are indicated. The predisposition to cancer is indicated (N = no predisposition). In some cases, cancer predisposition is uncertain due to the low number of known patients and/or their youthfulness

Disease	Mutated gene	Function	Cancer
Xeroderma pigmentosum (XP)	XPA-G, POLH	GG-NER	Squamous and basal cell carcinoma, melanoma
Cockayne syndrome (CS)	CSA, CSB	TC-NER	N
Trichothiodystrophy (TTD)	XPB, XPD, TTDA	TC-NER	N
Cerebro-oculo-facio-skeletal syndrome	XPD, XPG, CSB, ERCC1	TC-NER	N
XFE syndrome	XPF	NER, ICL repair	N
Spinocerebellar ataxia with axonal neuropathy (SCAN1)	TDP1	SSBR	N
Ataxia with oculomotor apraxia type 1 (AOA1)	APTX	SSBR	N
Ligase I syndrome	LIG1	SSBR, NER	N
MYH-associated polyposis	MYH	BER	Colorectal cancer
Hereditary non-polyposis colorectal cancer (HNPCC)	MSH2, MLH1, MSH6, others	MMR	Colorectal cancer, carcinomas
Fanconi anaemia (FA)	Fanconi anaemia compl. groups	ICL repair	AML, myelodysplasia, squamous cell carcinoma
Fanconi anaemia-like disorder	RAD51C	DNA repair	N
Bloom syndrome	BLM	DNA repair	Carcinomas, leukaemias, lymphomas
Werner syndrome	WRN	DNA repair	Sarcomas
Ataxia with oculomotor apraxia type 2 (AOA2)	SETX	DNA repair	N
Rothmund-Thompson syndrome	RECQL4	DSBs repair	Osteosarcoma, skin cancers
Jawad syndrome	CtIP	DSBs repair	N?
MSCZ syndrome	PNKP	NHEJ, SSBR	N
Immunodeficiency with microcephaly	XLF	NHEJ	N
Ligase IV syndrome	LIG4, XLF	NHEJ	ALL, lymphomas
Radiosensitive severe combined immunodef. (RS-SCID)	Artemis, XLF, DNA-PKcs	NHEJ	Lymphomas
Hereditary breast and ovarian cancer syndrome (HBOC)	Brca1, Brca2	HR	Breast and ovarian cancers
Ataxia Telangiectasia	ATM	DSBs signalling	Leukaemias, lymphomas, breast cancer
Ataxia Telangiectasia-like disorder	MRE11	DSBs signalling	Y?
Nijmegen breakage syndrome	NBS1	DSBs signalling	B-cell lymphoma
Nijmegen breakage-like disorder	RAD50	DSBs signalling	Y?
Riddle syndrome	RNF168	DSBs signalling	N?
Seckel syndrome	ATR, ATRIP, CtIP, SCKL3	Damage signalling	Y?
Autosomal dominant oropharyngeal cancer syndrome	ATR	Damage signalling	Y
Primary microcephaly 1	MCPH1	Damage signalling	N
Hutchinson-Gilford progeria syndrome (HGPS)	LMNA	Damage signalling	N
Li-Fraumeni syndrome	p53	Signalling	Brain and breast cancer, sarcomas

In the 1940s, studies of the victims of chemical warfare during World Wars I and II allowed the discovery of the first widely used cancer drugs; soldiers exposed to sulphur mustard gas were found to have depleted bone marrow and reduced lymph nodes. More stable nitrogen mustard compounds were demonstrated to cause tumour regressions in mice with transplanted lymphoid tumours (Gilman, 1963). Only a decade later, nitrogen mustards were found to directly alkylate DNA, leading to replication forks stalling resulting in cell death via apoptosis. In the 1960s to the 1970s, there was an increased interest in developing anticancer compounds that chemically react with DNA. Nowadays, ICL agents, such as mitomycin C, psoralens, and platinum compounds, are used in chemotherapy for the treatment of various cancers. Platinum compounds have been very successful in the treatment of solid tumours since cisplatin therapy can cure over 90% of all testicular cancer cases and also has good efficacy in the treatment of ovarian, bladder, head and neck, and cervical cancers (Kelland, 2007). Since cross-linking drugs are unselective, they are restricted by dose-limiting toxicities in the blood. It is predictable that NER defects could increase the efficacy of low doses of ICL agents. Indeed, cell lines derived from testicular cancer show reduced levels of ERCC1 and XPF proteins and impaired NER. Similarly, tumours characterized by *BRCA2* mutations could be more usefully treated with these drugs.

Other drugs that directly induce DNA lesions are radiomimetic agents, such as bleomycin. Indirect mechanisms characterize other classes of DNA-damaging agents used for chemotherapy like antimetabolites (5-fluorouracil, capecitabine, floxuridine, gemcitabine) and topoisomerase poisons (etoposide, camptothecin, doxorubicin). Antimetabolites mimicking normal cellular molecules interfere with DNA replication thus producing secondary DNA lesions. Topoisomerases are enzymes responsible for releasing the torsional strain of the double helix, through the transient production and relegation of nicks. Drugs that act on topoisomerases preventing the relegation step induce accumulation of DNA breaks.

However, in normal cells DDR pathways are often redundant and strongly related: one pathway can sometimes 'back-up' or rescue defects in another. The acquired deficiencies in DDR that can promote the transformation of a cell can render the same cell dependent on specific DDR cascades for its survival. Therefore, the inhibition of a DDR pathway using drugs should in some cases have a greater effect (synthetic lethality) on cancer than on normal tissues (Fig. 2.8). These drugs might be usefully tested as single therapy, taking advantage of endogenous DNA damage, or in combination with genotoxic agents (Fig. 2.8). An example is provided by drugs targeting PARP1, which is involved in SSBs repair. Blocking PARP1 catalytic activity results in the accumulation of SSBs, that might be transformed to DSBs particularly during replication. The correction of these DSBs, during S or the subsequent G2 phase, is based on HR (Chevanne et al., 2010). Therefore, tumours characterized by defects in HR components, such as *BRCA* mutations, and disruption of *FA* and *ATM* genes, are more likely sensitive to small molecules that inhibit PARP1 (Fig. 2.8). Trials are currently underway to test PARP1 inhibitors (like olaparib or veliparib) as adjuvant, neoadjuvant, and metastatic settings for the treatment of *BRCA*-defective breast cancer or ovarian cancer patients.

Similarly, a lethal interaction between inhibitors of ATR/CHK1 or ATM/CHK2 pathways and deficiency in other DDR players has been reported (Manic et al., 2015). Pharmacologic inactivation of CHK1-reduced cell growth in several cell lines depleted of *BRCA2*. Moreover, *ATM*- or p53-deficient cancer cells were selectively killed by an ATR or CHK1 inhibitor, since ATR/CHK1 pathway might back-up *ATM*/CHK2/p53 activity in presence of a severe damage (Fig. 2.8). On the other side, it is important to note that several DDR components show pro-survival or pro-cell death roles in relation to DNA damage amount and complexity. The inhibition of these enzymes could be beneficial or detrimental depending on the lesions

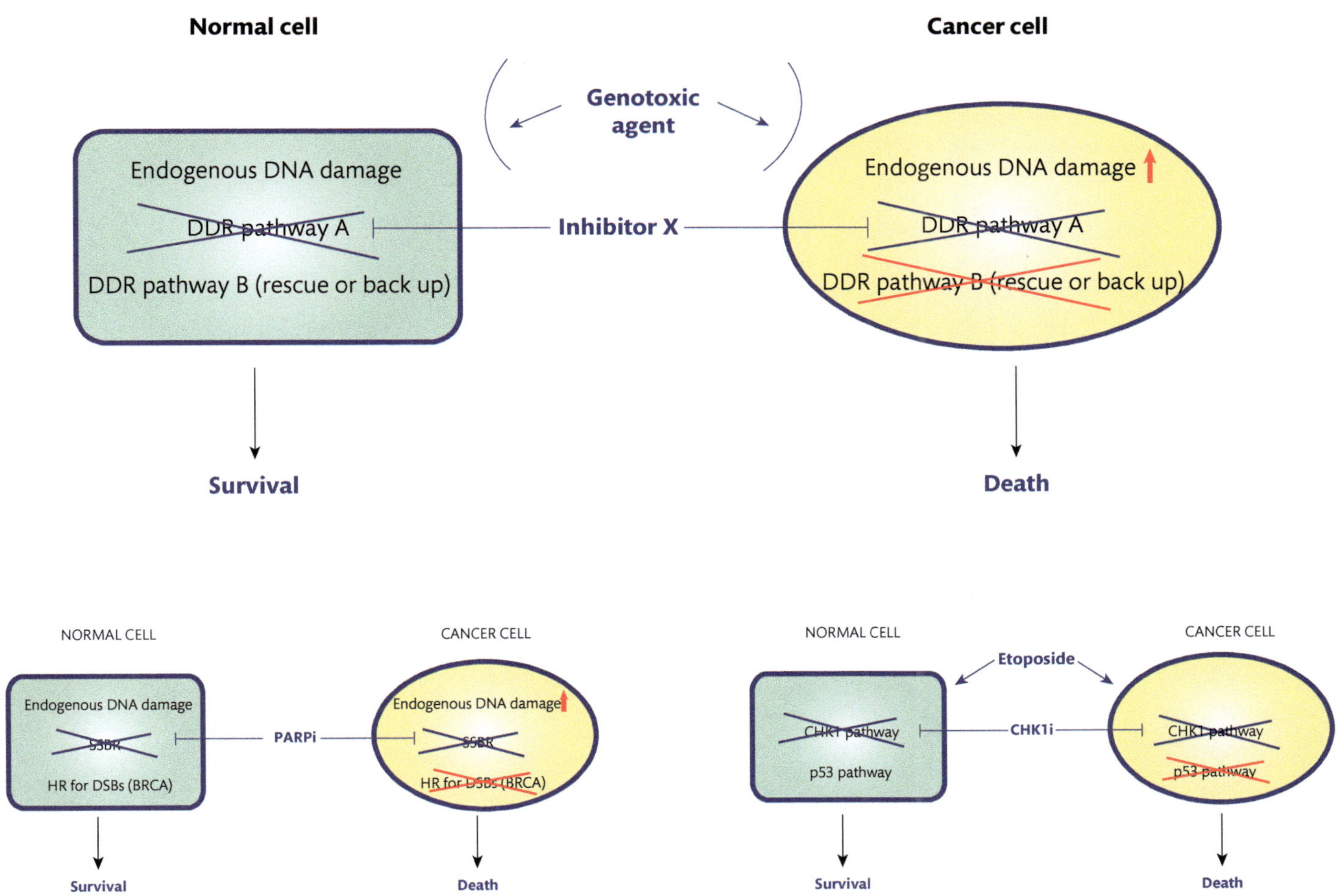

Fig. 2.8 *Synthetic lethality for targeted cancer therapy.* The upper panel shows a general scheme of synthetic lethality for treating cancer-exploiting defects in DDR. The therapy with DDR inhibitors could be employed as a single agent, taking advantage of the increased endogenous damage deriving from hyperproliferation (see also Fig. 2.1). Alternatively, inhibitors could synergize with the exposure to genotoxic chemotherapeutic agents. In the lower panels, two practical examples of ongoing studies with PARP inhibitors (PARPi) and CHK1 inhibitors (CHK1i) as single agents or in combination with DNA-damaging therapeutic drugs are shown.

produced by the combined genotoxic treatment, an effect that may differ from person to person. CHK2 is paradigmatic for this issue since it can activate preferentially cell cycle arrest and senescence but also apoptosis in relation to an elusive threshold of damage varying from cell to cell.

DNA repair could also provide a mechanism for the resistance to cancer therapy (Bouwman and Jonkers, 2012). It has been shown that glioma stem cells, that display a heightened DDR, are refractory to radiation treatment, thus potentially explaining why glioblastoma is difficult to cure. Therefore, DDR inhibition might enhance the efficacy of DNA-damaging therapies: various DDR-inhibitory drugs are in preclinical and clinical development to test this hypothesis. Furthermore, as DDR activation is present during oncogenesis, screening for DDR-markers could enhance the accuracy and sensitivity of cancer diagnosis and might allow effective detection of premalignant disease. In the long term, it might be even possible to develop drugs that transiently increase DDR activation to reduce cancer incidence due to genotoxic agent exposure. In this regard, it is important to note that mice engineered to have enhanced p53-dependent DNA damage responses are less tumour-prone than wild-type mice.

An important effort is ongoing to explore p53 regulatory mechanisms as a possible target for cancer therapy (Joerger and Fersht, 2016). For tumours that retain wild-type p53 but have defects in p53-regulatory pathways, such as overexpression or amplification of MDM2 and HDMX, the best method to restore p53 activity has been the inhibition of negative regulators of the p53 response. Indeed, most of the current efforts are focused on small-molecule drugs that block p53 interaction with MDM2 or HDMX, like nutlin and MI-219 (Zanjirband et al., 2016).

Many oncogenic mutations that inactivate p53 function disrupt the direct binding to specific DNA or prevent the proper folding of p53 DNA-binding domain. Many of these p53 mutants are temperature-sensitive, which has encouraged the attempts to discover molecules that can restore p53 activity by acting as p53 chaperones.

Around 8% of *TP53* mutations associated with cancer are nonsense mutations, and recent developments in the identification of small molecules and drugs that promote the read-through of nonsense codons could provide a novel method for treating tumours carrying this type of mutation.

Therefore, progresses in the diagnostic procedures for the identification of DDR differences between cancer and normal cells is a great promise for the intelligent targeting of DNA-damaging and DDR-inhibitor therapies for the individual patient.

TAKE-HOME MESSAGE

Nuclear DNA stability is essential to avoid that cells of a multicellular organism deviate from their correct fate. The DDR protects the genome stability providing time and resources for a high fidelity DNA repair. On the other side, DDR neutralizes those cells unable to completely repair DNA lesions. Even small defects in this complex system of pathways and their balancing could promote a large plethora of syndromes, among them cancer.

Essential characteristics of the DDR are:

- Fast to switch on and off.
- Highly sensitive to lesion amount and complexity with the ability to reveal and amplify the signal even of a single DNA lesion.
- Formed by highly interconnected pathways.
- Equipped with rescue and back-up systems.
- Cell cycle phase and cell-type specific.
- Constituted by sensors, a limited number of apical and distal kinases and hundreds of effectors and thousands of PTMs that in a 'barcode' way activate the correct biological response (survival versus death) in relation to damage characteristics.
- Strongly linked to any aspect of cell physiology.
- Recyclable for those physiological processes that require to alter and restore the DNA structure like meiosis or antibody maturation.

Defects in DDR components are essential events to promote cell transformation during tumorigenesis and to provide cancer cells the plasticity necessary for their adaptability. At the same time, DDR defects provide a vulnerable side to cancer therapy. Indeed, specific genotoxic agents and synthetic lethality approaches with DDR protein inhibitors are valuable tools for a personalized therapy.

OPEN QUESTIONS

- How we can use DDR activity assays to reveal early stages of carcinogenesis?
- How is the activity of the DDR in stem cells and how we can take advantage of their characteristics to treat cancer?
- How relevant is the role of non-coding RNAs in the DDR?
- It is possible to explore a personalized gene therapy for those hereditary syndromes deriving from mutations in DDR components?

FURTHER READING

Forment, J. V., Kaidi, A., Jackson, S. P. (2012). Chromothripsis and cancer: causes and consequences of chromosome shattering. *Nat Rev Cancer*, **12**(10), 663–70.

Hanahan, D., Weinberg, R. A. (2011). Hallmarks of cancer: the next generation. *Cell*, **144**(5), 646–74.

Jeggo, P. A., Pearl, L. H., Carr, A. M. (2016). DNA repair, genome stability and cancer: a historical perspective. *Nat Rev Cancer*, **16**(1), 35–42.

Loeb, L. A. (2016). Human cancers express a mutator phenotype: hypothesis, origin, and consequences. *Cancer Res*, **76**(8), 2057–9.

Roos, W. P., Thomas, A. D., Kaina, B. (2016). DNA damage and the balance between survival and death in cancer biology. *Nat Rev Cancer*, **16**(1), 20–33.

REFERENCES

Alexander, A., Cai, S. L., Kim, J., et al. (2010). ATM signals to TSC2 in the cytoplasm to regulate mTORC1 in response to ROS. *Proc Natl Acad Sci U S A*, **107**(9), 4153–8.

Aloni-Grinstein, R., Shetzer, Y., Kaufman, T., & Rotter, V. (2014). P53: the barrier to cancer stem cell formation. *FEBS Letters*, **588**(16), 2580–9.

Appella, E. & Anderson, C. W. (2001). Post-translational modifications and activation of p53 by genotoxic stresses. *Eur J Biochemistry*, **268**(10), 2764–72.

Austen, B., Powell, J. E., Alvi, A., et al. (2005). Mutations in the ATM gene lead to impaired overall and treatment-free survival that is independent of IGVH mutation status in patients with B-CLL. *Blood*, **106**(9), 3175–82.

Aymard, F., Bugler, B., Schmidt, C. K., et al. (2014). Transcriptionally active chromatin recruits homologous recombination at DNA double-strand breaks. *Nat Struct Mol Biol*, **21**(4), 366–74.

Barnes, D. E. & Lindahl, T. (2004). Repair and genetic consequences of endogenous DNA base damage in mammalian cells. *Annual Review of Genetics*, **38**, 445–76.

Batchelor, E., Loewer, A., & Lahav, G. (2009). The ups and downs of p53: understanding protein dynamics in single cells. *Nature Rev Cancer*, **9**(5), 371–7.

Bouwman, P. & Jonkers, J. (2012). The effects of deregulated DNA damage signalling on cancer chemotherapy response and resistance. *Nature Rev Cancer*, **12**(9), 587–98.

Buscemi, G., Zannini, L., Fontanella, E., Lecis, D., Lisanti, S., & Delia, D. (2009). The shelterin protein TRF2 inhibits Chk2 activity at telomeres in the absence of DNA damage. *Curr Biol*, **19**(10), 874–9.

Caldecott, K. W. (2008). Single-strand break repair and genetic disease. *Nat Rev Genet*, **9**(8), 619–31.

Callen, E., Nussenzweig, M. C., & Nussenzweig, A. (2007). Breaking down cell cycle checkpoints and DNA repair during antigen receptor gene assembly. *Oncogene*, **26**(56), 7759–64.

Cheng, Q. & Chen, J. (2010). Mechanism of p53 stabilization by ATM after DNA damage. *Cell cycle (Georgetown, Tex.)*, **9**(3), 472–8.

Cheung-Ong, K., Giaever, G., & Nislow, C. (2013). DNA-damaging agents in cancer chemotherapy: serendipity and chemical biology. *Chem Biol*, **20**(5), 648–59.

Chevanne, M., Zampieri, M., Caldini, R., et al. (2010). Inhibition of PARP activity by PJ-34 leads to growth impairment and cell death associated with aberrant mitotic pattern and nucleolar actin accumulation in M14 melanoma cell line. *J Cell Physiol*, **222**(2), 401–10.

Chien, Y., Scuoppo, C., Wang, X., et al. (2011). Control of the senescence-associated secretory phenotype by NF-kappaB promotes senescence and enhances chemosensitivity. *Genes Dev*, **25**(20), 2125–36.

Ciccia, A. & Elledge, S. J. (2010). The DNA damage response: making it safe to play with knives. *Mol Cell*, **40**(2), 179–204.

Cybulski, C., Gorski, B., Huzarski, T., et al. (2004). CHEK2 is a multiorgan cancer susceptibility gene. *Am J Human Genet*, **75**(6), 1131–5.

Czarny, P., Pawlowska, E., Bialkowska-Warzecha, J., Kaarniranta, K., & Blasiak, J. (2015). Autophagy in DNA damage response. *Int J Mol Sci*, **16**(2), 2641–62.

de Lange, T. (2005). Shelterin: the protein complex that shapes and safeguards human telomeres. *Genes Dev*, **19**(18), 2100–10.

Deans, A. J. & West, S. C. (2011). DNA interstrand crosslink repair and cancer. *Nat Rev Cancer*, **11**(7), 467–80.

di Masi, A. & Antoccia, A. (2008). NBS1 heterozygosity and cancer risk. *Current Genomics*, **9**(4), 275–81.

D'Orazi, G., Cecchinelli, B., Bruno, T., et al. (2002). Homeodomain-interacting protein kinase-2 phosphorylates p53 at Ser 46 and mediates apoptosis. *Nat Cell Biol*, **4**(1), 11–19.

Gilman, A. (1963). The initial clinical trial of nitrogen mustard. *Am J Surg*, **105**, 574–8.

Gorgoulis, V. G. & Halazonetis, T. D. (2010). Oncogene-induced senescence: the bright and dark side of the response. *Curr Opin Cell Biol*, **22**(6), 816–27.

Gorgoulis, V. G., Vassiliou, L. V., Karakaidos, P., et al. (2005). Activation of the DNA damage checkpoint and genomic instability in human precancerous lesions. *Nature*, **434**(7035), 907–13.

Hall, J. (2005). The ataxia-telangiectasia mutated gene and breast cancer: gene expression profiles and sequence variants. *Cancer Letters*, **227**(2), 105–14.

Hashimoto, S., Anai, H., & Hanada, K. (2016). Mechanisms of interstrand DNA crosslink repair and human disorders. *Genes Environ*, **38**(9), doi: 10.1186/s41021-016-0037-9. eCollection 2016.

Heijink, A. M., Krajewska, M., & van Vugt, M. A. (2013). The DNA damage response during mitosis. *Mutation Res*, **750**(1–2), 45–55.

Huang, Q. & Shen, H. M. (2009). To die or to live: the dual role of poly(ADP-ribose) polymerase-1 in autophagy and necrosis under oxidative stress and DNA damage. *Autophagy*, **5**(2), 273–6.

Huertas, P. (2010). DNA resection in eukaryotes: deciding how to fix the break. *Nat Struct Mol Biol*, **17**(1), 11–16.

Huertas, P. & Jackson, S. P. (2009). Human CtIP mediates cell cycle control of DNA end resection and double strand break repair. *J Biol Chem*, **284**(14), 9558–65.

Iyengar, S. & Farnham, P. J. (2011). KAP1 protein: an enigmatic master regulator of the genome. *J Biol Chem*, **286**(30), 26267–76.

Joerger, A. C. & Fersht, A. R. (2016). The p53 pathway: origins, inactivation in cancer, and emerging therapeutic approaches. *Annu Rev Biochem*, **85**, 375–404.

Kang, C., Xu, Q., Martin, T. D., et al. (2015). The DNA damage response induces inflammation and senescence by inhibiting autophagy of GATA4. *Science*, **349**(6255), aaa5612.

Karlseder, J., Hoke, K., Mirzoeva, O. K., et al. (2004). The telomeric protein TRF2 binds the ATM kinase and can inhibit the ATM-dependent DNA damage response. *PLoS Biol*, **2**(8), E240.

Kaur, J. & Debnath, J. (2015). Autophagy at the crossroads of catabolism and anabolism. *Nat Rev Mol Cell Biol*, **16**(8), 461–72.

Kelland, L. (2007). The resurgence of platinum-based cancer chemotherapy. *Nat Rev Cancer*, **7**(8), 573–84.

Kim, J. H., Kim, H., Lee, K. Y., et al. (2006). Genetic polymorphisms of ataxia telangiectasia mutated affect lung cancer risk. *Human Mol Genet*, **15**(7), 1181–6.

Kim, J. M., Kee, Y., Gurtan, A., & D'Andrea, A. D. (2008). Cell cycle-dependent chromatin loading of the Fanconi anemia core complex by FANCM/FAAP24. *Blood*, **111**(10), 5215–22.

Krokan, H. E. & Bjoras, M. (2013). Base excision repair. *Cold Spring Harb Perspect Biol*, **5**(4), a012583.

Kuilman, T., Michaloglou, C., Mooi, W. J., & Peeper, D. S. (2010). The essence of senescence. *Genes Dev*, **24**(22), 2463–79.

Kumagai, A., Kim, S. M., & Dunphy, W. G. (2004). Claspin and the activated form of ATR-ATRIP collaborate in the activation of Chk1. *J Biol Chem*, **279**(48), 49599–608.

Kunkel, T. A. & Erie, D. A. (2015). Eukaryotic mismatch repair in relation to DNA replication. *Annu Rev Genet*, **49**, 291–313.

Lahav, G., Rosenfeld, N., Sigal, A., et al. (2004). Dynamics of the p53-Mdm2 feedback loop in individual cells. *Nat Genet*, **36**(2), 147–50.

Lev Bar-Or, R., Maya, R., Segel, L. A., Alon, U., Levine, A. J., & Oren, M. (2000). Generation of oscillations by the p53-Mdm2 feedback loop: a theoretical and experimental study. *Proc Natl Acad Sci U S A*, **97**(21), 11250–5.

Lin, Z. & Fang, D. (2013). The roles of SIRT1 in cancer. *Genes Cancer*, **4**(3–4), 97–104.

Lindahl, T. & Barnes, D. E. (2000). Repair of endogenous DNA damage. *Cold Spring Harb Symp Quant Biol*, **65**, 127–33.

Lindsey-Boltz, L. A. & Sancar, A. (2011). Tethering DNA damage checkpoint mediator proteins topoisomerase II beta-binding protein 1 (TopBP1) and Claspin to DNA activates ataxia-telangiectasia mutated and RAD3-related (ATR) phosphorylation of checkpoint kinase 1 (Chk1). *J Biol Chem*, **286**(22), 19229–36.

Loeb, K. R., Cherian, S., Becker, P. S., et al. (2016). Comparative analysis of flow cytometry and morphology for the detection of acute myeloid leukaemia cells in cerebrospinal fluid. *Br J Haematol*, **172**(1), 134–6.

Lukas, J., Lukas, C., & Bartek, J. (2011). More than just a focus: the chromatin response to DNA damage and its role in genome integrity maintenance. *Nat Cell Biol*, **13**(10), 1161–9.

Mahaney, B. L., Meek, K., & Lees-Miller, S. P. (2009). Repair of ionizing radiation-induced DNA double-strand breaks by non-homologous end-joining. *Biochem J*, **417**(3), 639–50.

Malu, S., Malshetty, V., Francis, D., & Cortes, P. (2012). Role of non-homologous end joining in V(D)J recombination. *Immunol Res*, **54**(1–3), 233–46.

Manic, G., Obrist, F., Sistigu, A., & Vitale, I. (2015). Trial watch: targeting ATM-CHK2 and ATR-CHK1 pathways for anticancer therapy. *Mol Cell Oncol*, **2**(4), e1012976.

Marteijn, J. A., Lans, H., Vermeulen, W., & Hoeijmakers, J. H. (2014). Understanding nucleotide excision repair and its roles in cancer and ageing. *Nat Rev Mol Cell Biol*, **15**(7), 465–81.

Matsuoka, S., Rotman, G., Ogawa, A., Shiloh, Y., Tamai, K., & Elledge, S. J. (2000). Ataxia telangiectasia-mutated phosphorylates Chk2 *in vivo* and *in vitro*. *Proc Natl Acad Sci U S A*, **97**(19), 10389–94.

McBride, K. A., Ballinger, M. L., Killick, E., et al. (2014). Li-Fraumeni syndrome: cancer risk assessment and clinical management. *Nat Rev Clin Oncol*, **11**(5), 260–71.

Munoz-Espin, D. & Serrano, M. (2014). Cellular senescence: from physiology to pathology. *Nat Rev Mol Cell Biol*, **15**(7), 482–96.

Murray-Zmijewski, F., Slee, E. A., & Lu, X. (2008). A complex barcode underlies the heterogeneous response of p53 to stress. *Nat Rev Mol Cell Biol*, **9**(9), 702–12.

Nagaria, P., Robert, C., & Rassool, F. V. (2013). DNA double-strand break response in stem cells: mechanisms to maintain genomic integrity. *Biochim Biophys Acta*, **1830**(2), 345–53.

Najafi, M., Fardid, R., Hadadi, G., & Fardid, M. (2014). The mechanisms of radiation-induced bystander effect. *J Biomed Phys Eng*, **4**(4), 163–72.

Ohtani, N., Mann, D. J., & Hara, E. (2009). Cellular senescence: its role in tumor suppression and aging. *Cancer Sci*, **100**(5), 792–7.

O'Sullivan, R. J. & Karlseder, J. (2010). Telomeres: protecting chromosomes against genome instability. *Nat Rev Mol Cell Biol*, **11**(3), 171–81.

Pereg, Y., Lam, S., Teunisse, A., et al. (2006). Differential roles of ATM- and Chk2-mediated phosphorylations of Hdmx in response to DNA damage. *Mol Cell Biol*, **26**(18), 6819–31.

Polo, S. E. & Almouzni, G. (2015). Chromatin dynamics after DNA damage: The legacy of the access-repair-restore model. *DNA Repair*, **36**, 114–21.

Polo, S. E. & Jackson, S. P. (2011). Dynamics of DNA damage response proteins at DNA breaks: a focus on protein modifications. *Genes Dev*, **25**(5), 409–33.

Porter, J. R., Fisher, B. E., & Batchelor, E. (2016). p53 pulses diversify target gene expression dynamics in an mRNA half-life-dependent manner and delineate co-regulated target gene subnetworks. *Cell Systems*, **2**(4), 272–82.

Purvis, J. E., Karhohs, K. W., Mock, C., Batchelor, E., Loewer, A., & Lahav, G. (2012). P53 dynamics control cell fate. *Science (New York, N. Y.)*, **336**(6087), 1440–4.

Richardson, C., Horikoshi, N., & Pandita, T. K. (2004). The role of the DNA double-strand break response network in meiosis. *DNA Repair*, **3**(8–9), 1149–64.

Rodier, F., Coppe, J. P., Patil, C. K., et al. (2009). Persistent DNA damage signalling triggers senescence-associated inflammatory cytokine secretion. *Nat Cell Biol*, **11**(8), 973–9.

Rogakou, E. P., Pilch, D. R., Orr, A. H., Ivanova, V. S., & Bonner, W. M. (1998). DNA double-stranded breaks induce histone H2AX phosphorylation on serine 139. *J Bio Chem*, **273**(10), 5858–68.

Sale, J. E., Lehmann, A. R., & Woodgate, R. (2012). Y-family DNA polymerases and their role in tolerance of cellular DNA damage. *Nat Rev Mol Cell Biol*, **13**(3), 141–52.

Shiloh, Y. & Ziv, Y. (2013). The ATM protein kinase: regulating the cellular response to genotoxic stress, and more. *Nat Rev Mol Cell Biol*, **14**(4), 197–210.

Shiotani, B. & Zou, L. (2009). Single-stranded DNA orchestrates an ATM-to-ATR switch at DNA breaks. *Mol Cell*, **33**(5), 547–58.

Shreeram, S., Hee, W. K., Demidov, O. N., et al. (2006). Regulation of ATM/p53-dependent suppression of myc-induced lymphomas by Wip1 phosphatase. *J Exp Med*, **203**(13), 2793–9.

Smerdon, M. J. (1991). DNA repair and the role of chromatin structure. *Curr Opin Cell Biol*, **3**(3), 422–8.

Stephens, P. J., Greenman, C. D., Fu, B., et al. (2011). Massive genomic rearrangement acquired in a single catastrophic event during cancer development. *Cell*, **144**(1) 27–40.

Tchkonia, T., Zhu, Y., van Deursen, J., Campisi, J., & Kirkland, J. L. (2013). Cellular senescence and the senescent secretory phenotype: therapeutic opportunities. *J Clin Investig*, **123**(3), 966–72.

Tort, F., Hernandez, S., Bea, S., et al. (2005). Checkpoint kinase 1 (CHK1) protein and mRNA expression is downregulated in aggressive variants of human lymphoid neoplasms. *Leukemia*, **19**(1), 112–17.

Turnell, A. S. & Grand, R. J. (2012). DNA viruses and the cellular DNA-damage response. *J General Virology*, **93**(10), 2076–97.

Valentin-Vega, Y. A. & Kastan, M. B. (2012). A new role for ATM: regulating mitochondrial function and mitophagy. *Autophagy*, **8**(5), 840–1.

Vo, Q. N., Kim, W. J., Cvitanovic, L., Boudreau, D. A., Ginzinger, D. G., & Brown, K. D. (2004). The ATM gene is a target for epigenetic silencing in locally advanced breast cancer. *Oncogene*, **23**(58), 9432–7.

Wang, Y. & Taniguchi, T. (2013). MicroRNAs and DNA damage response: implications for cancer therapy. *Cell Cycle (Georgetown, Tex.)*, **12**(1), 32–42.

Warmerdam, D. O. & Kanaar, R. 2010. Dealing with DNA damage: relationships between checkpoint and repair pathways. *Mutation Res*, **704**(1–3), 2–11.

Xu, Z., Fulop, Z., Zhong, Y., Evinger, A. J., 3rd, Zan, H., & Casali, P. (2005). DNA lesions and repair in immunoglobulin class switch recombination and somatic hypermutation. *Ann N Y Acad Sci*, **1050**, 146–62.

Youds, J. L. & Boulton, S. J. (2011). The choice in meiosis—defining the factors that influence crossover or non-crossover formation. *J Cell Science*, **124**(Pt 4), 501–13.

Zanjirband, M., Edmondson, R. J., & Lunec, J. (2016). Pre-clinical efficacy and synergistic potential of the MDM2-p53 antagonists, Nutlin-3 and RG7388, as single agents and in combined treatment with cisplatin in ovarian cancer. *Oncotarget*, **7**(26), 40115–34.

Zannini, L., Delia, D., & Buscemi, G. (2014). CHK2 kinase in the DNA damage response and beyond. *J Mol Cell Biol*, **6**(6), 442–57.

3

Evolution and cancer

Tom Donnem, Kingsley Micklem, and Francesco Pezzella

Introduction

As defined by the *Oxford English Dictionary*, evolution is the process by which something develops gradually into a different form. It is therefore a concept that can be applied not only to biology but to many other fields, we can talk for example of evolution of ideas, of a political system or of the design of an object.

The idea that living organisms evolve is an old one: the Greek philosopher Anaximander, who lived in the sixth century BC, is considered as one of the most important precursors of the concept that organisms change and the more complex derive from simpler ones. For many centuries, the first and main biological question relevant to biological evolution was the classification of plants and animals into species. The big problem to be solved was to explain the origin of these species. The explanation of how species evolved into other species was required.

Eventually, at the same time, Darwin and Wallace proposed the theory that natural selection acting on variations within species was sufficient to drive evolution (Box 3.1). Darwin in his book *On the Origin of Species by Means of Natural Selection* formulated and demonstrated a hypothesis on the mechanism of the evolution of species using a wide variety of evidence.

Darwin theory established three conditions that need to be satisfied for evolution to happen:

- the occurrence of variation
- the possibility to inherit these variations
- the presence of selective pressure

Cancer is a disease in which evolutionary forces play a role at two levels. The first is the natural selection, within populations of organisms, of traits that are involved with cancer, either promoting it or

Box 3.1 The 1858 Darwin–Wallace paper

Alfred Russell Wallace (1823–1913) was from an impoverished family and never had any formal academic training. He left school at age of 14 and started to become interested in science by attending educational science evenings for working men. Trained as surveyor, he cultivated his blooming interest in natural science with extensive reading and research. He came up with the idea of earning a living by providing specimens to the rapidly growing group of naturalists working in Britain. Eventually he became one of the most important naturalists in the history of biology. While doing field work in the Far East he came to formulate, independently from Darwin, the hypothesis that the origin of new species was driven by natural selection. Wallace and Darwin by then had been in epistolary contact and actually Wallace was providing, on a professional basis, specimens for Darwin's work. At the beginning of 1858, Wallace wrote a 20-page assay entitled 'On the tendency of varieties to depart indefinitely from the original type' that he sent to Darwin, as he was very aware that Darwin was writing a large work on the subject. Darwin's reluctance to go public with his work has assumed throughout the years a legendary status. He was well aware of the enormous resonance his work would generate and that many reactions would not be positive. However, when on 18 June 1858 he received a parcel from Wallace containing the essay, things changed as he realized that it was not longer possible to avoid going public. Wallace had sent it to Darwin with the request to forward its contents to Lyell, the famous geologist. Although Darwin realized that the publication of Wallace's paper ahead of his own would undermine the originality of his own work, he still offered to help with the paper's publication. In order to give justice to both scientists, Lyell proposed that the Wallace paper should be made public at the same time as an equivalent essay from Darwin and Wallace promptly agreed. With just 24 hours to spare, Lyell managed to have the two contributions unified and inserted in the programme of the Linnaean Society meeting of 1 July 1858 as a paper entitled: 'On the Tendency of Species to form Varieties; and on the Perpetuation of Varieties and Species by Natural Means of Selection. By Charles Darwin, Esq., F.R.S., F.L.S., & F.G.S., and Alfred Wallace, Esq. Communicated by Sir Charles Lyell, F.R.S., F.L.S., and J. D. Hooker, Esq., M.D., V.P.R.S., F.L.S, &c.' The paper was read to a crowd of approximately 30 fellows by the Secretary of the Society. Despite all predictions, the reading went unnoticed with no public reaction whatsoever. Darwin and Wallace remained lifelong friends, each recognizing the other's merits, as witnessed by Darwin helping Wallace to obtain a state pension and Wallace being one of Darwin's coffin bearers to his final resting place (Desmond and Moore, 1992; van Wyhe, 2012).

suppressing it. This raises the question is why such an unfavourable character, the high incidence of cancer, has been preserved during human evolution. As discussed in Chapter 1, the risk of cancer is highly variable between species and is not linked to either dimension or life span of animal. Humans, unfortunately for us, are among the ones with the highest risk (Greaves, 2015). The second level concerns how evolutionary forces act on cancer cells. Cancer is a clonal disease (i.e. is made up by cells all deriving from a single ancestor cell; see Nowell, 1976); however, more or less rapidly, cancer lesions become heterogeneous, with different subclones deriving from the original cells. Some clones will progress, while others will disappear due to either evolutionary forces within the tumour or external to it (Vincent, 2010).

Human evolution and cancer

Evolutionary medicine tries to understand why the human body, during its evolution, has become subject to some diseases rather than others and asks the question why inherited susceptibility to disease exists. One fundamental issue is that natural selection is driven by characteristics leading to effective and successful reproduction and is far less affected by negative events happening after successful reproduction (Nesse, 2001). Why are human so affected by cancer is a classic case study of evolutionary medicine. As discussed in Chapter 1, cancer is widespread in the animal kingdom but with a very variable incidence and is not associated with body dimension or lifespan and it is a fact today that humans are among the most affected animals (Greaves, 2015).

Some evolutionary factors could have allowed cancer-traits to be preserved

The risk of cancer is linked to many factors (e.g. inherited genotype, chance genetic events, habitat, and lifestyle). One prevalent theory (Antolin, 2009; Greaves, 2015) is that in the last few thousand years the lifespan of humans has dramatically and very quickly increased not because of selection of more long-lived, less cancer vulnerable humans, but because of sudden advances mostly in availability of food and healthcare and environmental exposure to man-made carcinogens. Some evolutionary factors have been identified which possibly affecting the risk of cancer in humans (Greaves, 2007; Nesse and Stearns, 2008):

Adaptation is not perfect and has limits. There are limits to what selection can do: it can only select from the available phenotypes. These genetic changes occur continuously but they completely driven by chance. Therefore, the features present in an organism on which selective forces are applied are a random assembled series of characters far from being optimal or organized. A negative trait can be tolerated for many generations until circumstances change and its negative effect can be felt.

Mutation happens. The genetic code is continuously subject to events, either random or driven from the environment, causing mutations. Therefore, unfavourable characters are bound to emerge.

Natural selection does not plan for the future. Selection privileges reproduction 'now' and has 'no eyes to the future'. Our African ancestors had deeply pigmented skin to protect from tropical ultraviolet (UV) exposure from the sum. Migrants to higher latitudes in Europe some 50,000 years ago adapted to the reduced solar UV by reducing skin pigmentation to allow effective vitamin D synthesis. However, this adaptation among white-skinned people leads to a many-fold increased risk of skin cancer.

The only benchmark for natural selection is successful reproduction. Conventionally post-reproductive negative effects do not exert selection pressure and indeed natural selection does not maximize health except for what needed to generate a successful offspring. However, the implication that there may be kin and group selection suggests that there may be increased fitness due to post-reproductive longevity, for example, the 'grandmother effect' (i.e. that groups whose members are long-lived enough to help to raise the offspring of their own child, gain a selective advantage and, as a group, reproduce more successfully).

There are trade-offs between conflicting traits. The selection of a character which is potentially unfavourable under certain conditions can confer benefits in other situations. In this was a potentially unfavourable trait can persist in a population. A classic case is sickle cell anaemia. Subjects with homozygotic abnormal haemoglobin succumb quickly and fail to reproduce but heterozygotic subjects are more resistant to malaria, allowing it to persist in areas endemic for this infective disease. The work of Crespi and Summers (2006) described next is a cancer-related example.

Antagonistic co-evolution as a cause for 'trade-off' leading to cancer-promoting traits

A hypothesis trying to explain why trade-offs can happen, is that of antagonistic coevolution in which the presence of two sides with divergent interest (e.g. mother and fetus or male and female development) lead to positive selections of genes that, on one side, accommodate the divergent interest but on the other are associated in cancer. In many cases these genes are 'associated with' cancer rather than oncogenic, furthermore many of them are highly preserved among mammals but still the risk of cancer among mammals is highly variable. One example is that of the adhesion molecules E-cadherin and VE-cadherin which are expressed on placental villi enhancing their ability to 'invade' in an effective way the uterine mucosa supporting the placental function. However, these molecules are also present on malignancies enhancing their ability to invade (Crespi and Summers, 2006).

There are also cases in which a genuine cancer-causing trait appears to be positively selected. *BRCA2* inherited mutations and other genetic variations predispose to breast cancer. A polymorphism on axon 10 of *BRCA* has been found associated with increased disposition as homozygote patients have a 1.3-fold increased risk, but still it remains present in the general population without being eliminated. The incidence of such a polymorphism is lower in women compared to men, apparently because it impairs the female fetus vitality. Conversely the higher incidence in the male population is presumably involved in improving the male fetus vitality (Healey et al., 2000). In this case therefore the trade-off comes from a conflict between a negative role in women but a positive in males (Crespi and Summers, 2006).

Another unfavourable character selected in humans is that the relative size of breasts is larger than in other mammals and they develop to such a size independently from pregnancy. This is caused by activation of pathways causing a prolonged and premature breast development resulting in an increased mass of glandular tissue to hormones which put it at higher risk of cancer. The selection pressures which have led to the development of high cancer risk breasts are not clear: mate choice and extra deposit of fat reserves are among the possible reasons (Crespi and Summers, 2006).

Why are human cancer rates so high in humans compared with other animals?

We have earlier described a series of hypothesis explaining why cancer-favouring traits have been preserved in human populations. This raises the issue why humans have such high cancer rates compared with other animals, even closely-related species such as primates. Reasons are likely to be many and here are some examples.

As we discussed in Chapter 1, elephants are protected by having 20 copies of the tumour suppressor gene p53 and, in the wild, they live until they are approximately 60 or 70 years old. How was such a population selected? Like humans, elephants are able to reproduce since their early teens, however there is a difference in mating habits. One explanation is that human males reproduce at a relatively young age, like the female. However, among reproductively successful elephants, the males are usually older (typically 40 years older) and consequently stronger, out-competing younger males. This has possibly led to select the characters benefiting the health of an older animal population, which have evolved traits that protect against cancer.

Another long-lived mammal studied is the bowhead whale which can live for 200 years. Analysis of its genome reveal numerous genes and specific mutations positively selected for DNA repair, cell-cycle regulation and longevity compared with its close relative the minke whale which has a typical lifespan of 50 years (Keane et al., 2015).

Modern humans are also long-lived but this longevity is recently acquired, and it has been speculated that we are not adapted to our contemporary lifestyle (Greaves 2015). Of course, long life in itself will increase the incidence of cancer but there are other factors which play a part. Further, the lowering of infant mortality as well as other economic changes, have caused a reduction in number of offspring. Contraception, delayed pregnancy, and an ample diet result in modern females undergoing many oestrus cycles which contribute to proliferative stress for breast and ovary cells. Calorie intake alone may influence cancer risk, experimentally calorie restriction or genetic manipulation of insulin-like growth factor 1 in mice reduces cancer risk in mice although its importance in humans is not clear (Pollak, 2004). Certainly obesity is a risk factor in many cancers.

In addition, there are all the anthropogenic carcinogens in the modern environment from industrial chemicals, asbestos, and motor vehicle exhausts, to name a few. Most significantly, some social drugs (e.g. tobacco, alcohol, and in South Asia, areca nuts) are carcinogenic.

The conclusion, so far, is that the high incidence of cancer we see nowadays in humans can be partly explained in terms of evolutionary biology as a maladaptation to our modern lifespan. That we do not possess adaptive mechanisms to reduce the risk of cancer that are present in organisms adapted to longevity means that we are also uniquely susceptible to the man-made carcinogens in our environment.

Natural selection and the cancer cell

It is a common experience for clinicians treating cancer to observe changes in the disease characteristics as time and treatment go by. Despite originating from a single 'renegade' cell, tumours become quickly made up by a rather heterogeneous collection of cells, which belong to different subclones. As these subclones find themselves in a changeable environment, the body, where resources are limited, competition among different cells is inevitable with some emerging over the others as able to grow into large clones. These changes can be regarded as a case of biological evolution. The concept of Cellular Darwinism refers specifically to the application of the concept of 'evolution of the species by natural selection' to cell populations. Such cellular evolution is a mechanism in non-pathogenic situations such as the development of the immune response as well the cells of a neoplastic lesion.

Cancer heterogeneity and the intrinsic selective forces

The heterogeneity of the neoplastic population forming a tumour is the reason for which the laws of natural selection act on cancer: should a malignancy been made by exactly equal and genetically stable cells, there would be no competition and no antagonists, and therefore no clones evolving over others. Genotypic and phenotypic heterogeneity exists obviously between different tumours (intertumour heterogeneity) but it is also present within neoplastic lesions and between primary and metastatic lesions (intratumour heterogeneity) as was firstly formally described by Peter Nowell (Nowell, 1976).

There are four general causes of intratumour variation: genetic heterogeneity plus three non-genetic sources: epigenetic mutations, differentiation hierarchies, and stochastic mechanisms (Almendro et al., 2013). The natural forces acting on cancer and driving clone selection can be divided into intrinsic and extrinsic and these four causes of heterogeneity represent also the internal or *intrinsic* selective forces. These can be classified as selectively advantageous, selectively disadvantageous, and neutral, with only the two former providing selective pressures (Greaves and Maley, 2012).

Genetic heterogeneity is due to the occurrence of random mutations leading to increasing genomic instability (i.e. an increased tendency of accumulating changes in the genome; Shen, 2011). Point mutations can be due to different causes, both intrinsic and extrinsic: to defects in DNA repairs and mismatch repairs pathways or because of more extensive damage involving chromosomes (Burrell et al., 2013). Instability at chromosomal levels leads to another family of alterations in which chromosomal number and/or structure are altered (Burrell et al., 2013).

Epigenetic changes are defined as stable, heritable phenotypes that are not caused by a change in the underlying DNA sequence (Bird, 2002). The commonest and best understood type of epigenetic change is methylation of DNA promoter sequences which leads to transcription silencing. It has been demonstrated in cancer by studying the spectrum of MLH1 promoter methylation: within a single tumour different subclones were identified by the patterns of methylation of this promoter (Varley et al., 2009).

Differentiation hierarchies. The identification of cancer stem cells has led to the cancer stem cell model, which states that 'growth and progression of many cancers are driven by a small subpopulation of cancer stem cells'. Many cancers therefore are assumed to be organized in much the structure as normal tissues: the cancer stem cells undergo epigenetic changes analogous to the differentiation of normal cells, generating neoplastic cells which have lost the ability to generate further tumour but forming the bulk of the neoplastic mass (Shackleton et al., 2009) and still showing heterogeneity. However, such heterogeneity is due to epigenetic changes rather than further genetic changes. Some of these cells, although have lost the standard stem cells characteristics, can however still originate tumour growth. It is therefore possible that in many tumours 'common' cancer cells are accumulating changes and being able to grow and therefore to compete for selection (Shackleton et al., 2009; Almendro et al., 2013).

Stochastic non-genetic events cause genetically identical cells to have a variable behaviour in tissue cultures. This is due to the random variation in the biochemical process inside the cells (Almendro et al., 2013). Gene expression has been described as variable in genetically identical organisms, not only in those of single cell but also in multicellular organisms like the *Caenorhabditis elegans*, a mutation introduced into the skn-1 gene that produces cells with a variable number of transcripts leading to embryos with variable level of defects in the intestinal tube. This is due to the presence of redundant mechanisms of downstream gene expression control (Raj et al., 2010).

Clonal evolution of cancer cell populations: The Peter Nowell model

In 1976 Peter Nowell proposed a model of clonal development of cancer: from the original neoplastic cell, an increasingly heterogeneous population of cells develops as new genetic damage accumulates. Some of these cells will be more successful, expanding more than others (Nowell, 1976). Using the same evolutionary tree firstly drawn by Darwin (Fig. 3.1), Nowell illustrated the evolution of malignant tumours. He proposed that most neoplasms arise from a single cell. As further genetic changes accumulate, new clones appear. Non-viable clones die out. Expansion of viable clones leads to metastatic disease and/or treatment resistant disease. Nowell therefore formalized the concept that cancer cells are subject to natural selection according to the same laws governing the selection of species (Nowell, 1976).

Cytogenetics has demonstrated that some abnormalities are common to all the cells, indicating that they all have a common precursor (i.e. are clonal while other cytogenetic abnormalities are present only in a limited number of cells indicating that subclones have developed, some larger, some smaller).

Characterization of the isoenzyme glucose-6-phosphate dehydrogenase provided evidence that all the neoplastic cells in a tumour derives from a common ancestor. This enzyme is located on chromosome X and two isoenzymes exist which can be differentiated as they have a different motility in electrophoresis (Linder and Gartler, 1965). By this approach it has been possible to demonstrate that with cancer cells in tumours from women

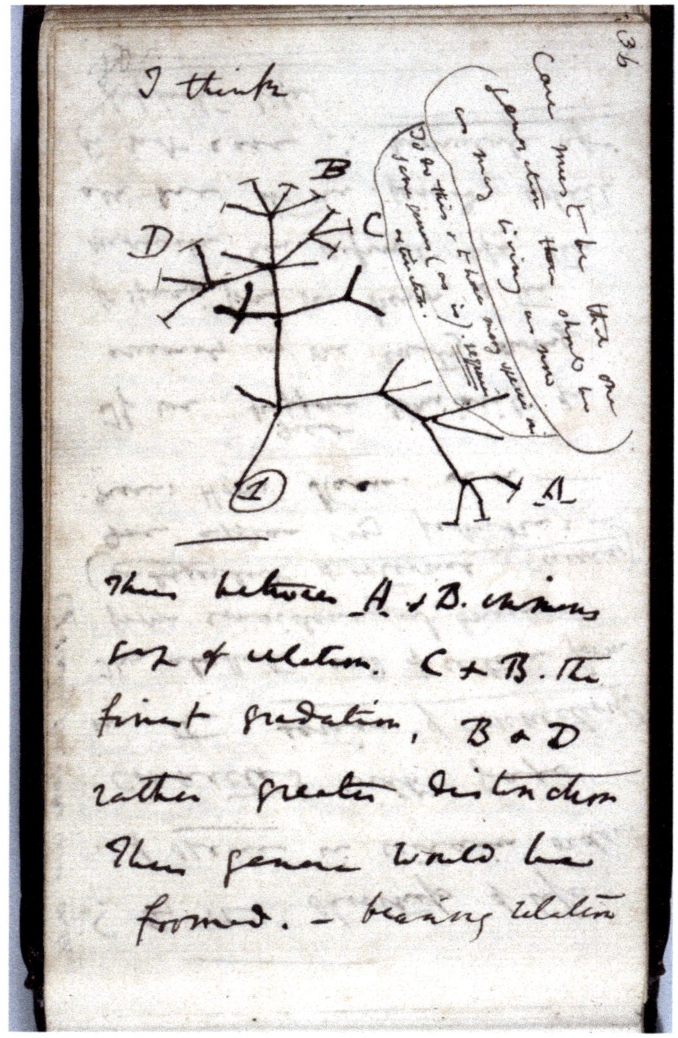

Fig. 3.1 The evolutionary tree designed by Darwin to represent the evolution of species accurately depicts the evolution of cancer clones within the neoplastic population. Charles Darwin Notebook B (1837–1838), p. 36: 'I think. Case must be that one generation then should be as many living as now. To do this and to have many species in same genus (as is) requires extinction. Thus between A and B immense gap of relation. C and B the finest gradation, B and D rather greater distinction. Thus genera would be formed—bearing relation.'
Reproduced by kind permission of the Syndics of Cambridge University Library (DAR121, p. 36).

with the two isoforms in their body, just one type of isoform is active.

Finally, it was noticed that in plasma cell malignancies, the multiple myelomas produce the same type of light chain as the immunoglobulin: either lambda or kappa chain.

As the original neoplastic clone acquires more genetic damages some cells will become non-vital, others will be not influenced and will keep growing as before, but some subclones will acquire an advantage and will expand. This is evidenced by the occurrence of extra chromosomal alterations; many variables are involved, like the degree of genetic instability, the tumour microenvironment, immune status and, last but not least, treatment received. This model is still valid and has been supported by the last 40 years of progress in the field of cancer biology.

The cancer ecosystem and the extrinsic selective forces including medical treatments

The extrinsic forces are ultimately those present in the tumour microenvironment forming the cancer ecosystem (Merlo et al., 2006). As the cellular microenvironment affects the cancer cell function, similarly the cancer cells affect the non-neoplastic cells of the microenvironment. The final selective pressure is therefore determined by the interaction of the intrinsic and extrinsic forces and therefore the environment they create for the cells to grow in constitutes an ecosystem (Greaves and Maley, 2012).

Selective forces from the ecosystem are numerous. Inflammation is one such factor: the presence or absence of it and the type of inflammatory cells are dictated by the biology of the tumour, on one side, and the status of the host immune system on the other. Eventually the resulting type of inflammatory infiltration will affect the tumour. Hormonal dependence is a common factor and tumour growth is driven by the levels of hormones present. Ability of the cancer cells to either induce angiogenesis or to exploit pre-existing vessels and the type of vascularization present can also results in which type of cells will be more successful with non-angiogenic tumours thriving in lung and liver while angiogenic tumours are more present in breast tissue or other subcutaneous areas.

This ecosystem is dramatically affected by chemotherapy, radiotherapy, and/or immunotherapy like humanized monoclonal antibodies. In the simplest scenario a subclone resistant to the incoming treatment is already present among the others and keeps thriving while the sensitive clones are eliminated. In more complex but very common situations, some of the genetic defects, although not providing resistance themselves, favour the development of the same. A classic example is that of chemotherapy which relies on inducing DNA damages: these drugs aim to either kill cancer cells by producing a catastrophic DNA breakdown or producing limited DNA damage which leads to slower growth as DNA repair mechanisms are activated. However, cells not fatally damaged but which have damaged and inactivated DNA repair genes will keep growing and accumulating mutations, and therefore increasing the chance that a completely resistant clone will develop (Almendro et al., 2013).

While it is well known that, following treatment, three main situations occurs, one in which the tumour is completely eradicated, one in which the tumour relapses later on after a complete remission and finally in one which the tumour fails to respond (Fig. 3.2) the underlying dynamic and mechanisms of what happens to the cancer cells and how the different clones behave is still poorly understood (Navin, 2014). Only a few models are available at the present time and they relate to different stages of the disease.

The evidence for selection after treatment has been demonstrated in studies on breast cancer patients receiving neo-adjuvant treatment before surgery. In this approach chemotherapy is given and then surgery performed when the lesion is reduced but still present. Both genetic alterations, detected using iFISH, chromosomal damage, and phenotypic variations in immunostaining with antibodies against CD44 and CD24 (two markers with a heterogeneous expression in breast cancer), were looked at. Genetic heterogeneity is present both before and after treatment while study of phenotypic heterogeneity showed mixed results: heterogeneity was maintained in some tumours but lost in others with an overall increase in slow growing CD24 positive cells (Almendro et al., 2014).

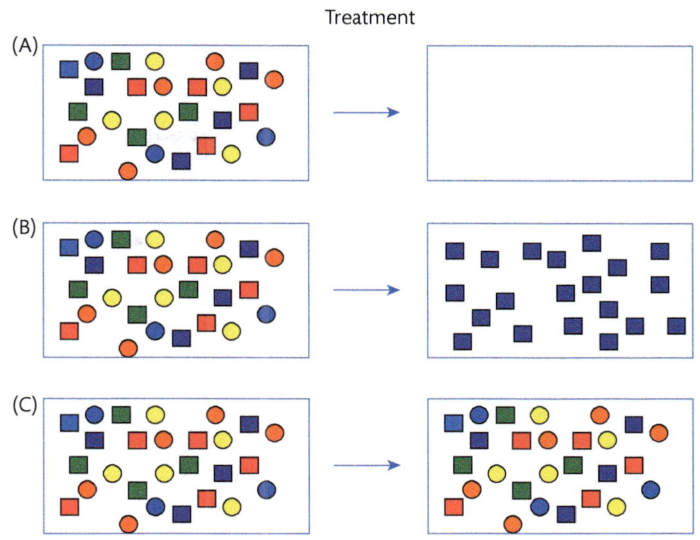

Fig. 3.2 Selective pressure from treatment. Some of the different possible types of outcome of the selective pressure applied by treatment on tumours. (A) The tumour has responded completely to treatment: all the neoplastic cells have died. (B) All but one clones responded to treatment: after a variable period of time, the resistant cells have grown enough to produce a new clinically appreciable lesion. (C) Complete lack of response: all the neoplastic cells remain alive.

Adapted with permission from *Cell Reports*, Volume 6, Issue 3, Navin EN, Preview Tumor Evolution in Response to Chemotherapy: Phenotype versus Genotype, pp. 417–19, Copyright © 2014 The Authors. Published by Elsevier Inc. under the terms of the Creative Commons license CC BY-NC-ND 4.0.

Some tumours respond completely, achieving clinical remission but then later relapse (i.e. a new lesion occurs). The evolution of the tumour in this situation has been investigated in acute myeloid leukaemia using next generation sequencing. In agreement with the first model (Nowell, 1976; Merlo et al., 2006) several clones are present, each bearing a set of genetic alterations, in the primary lesion. Treatment and remission follow but eventually the disease relapses. Less genetic heterogeneity is present in the post-treatment relapsing disease, consistent with the successful growth of a resistant small clone (Ding et al., 2012). Two main different patterns are found: in the first pattern the dominant clone in the presenting leukaemia gained further alterations as the patient was treated and evolved into relapsing disease. In the second pattern a minor subclone resisted treatment and, having gained further mutations, expanded after treatment.

Genetic drift versus natural selection

Genetic drift, known also as Sewall Wright effect, is a significant variation in the frequency of a given genotype because of the occurrence of a random event leading to the disappearance of particular genes, as individuals die or do not reproduce, because of this completely fortuitous even. Genetic drift is therefore usually observed in small population but can rarely occurs in larger ones (Fig. 3.3; see Thain and Hickman, 2004). The main effect of genetic drift on cancer is seen at the very early stages. Assuming that neoplastic transformation happens in a stem cell and considering that the number of neoplastic stem cells will be relatively small, a mutant is likely to go extinct many times by chance only, before being able to develop into a tumour. As exemplified by Merlo et al. (2006): in

Fig. 3.3 Genetic drift. Genetic drift is a change in population gene frequency resulting from causes operating randomly and not because of selection, immigration, or emigration (Thain and Hickman, 2004). Its effects are usually more evident in small populations; but 65 million years ago, a massive meteorite impacted the Earth at what is now the Gulf of Mexico. The event was of such a magnitude that it can be considered as causing the biggest genetic drift known so far: the extinction of terrestrial dinosaurs.

Adapted with permission from International Space Station 24 hours a day. © NASA

an effective population of stem cells of 10^6, if a stem cell acquires a mutation that gives it a 10% fitness advantage on the surrounding stem cells, there is still a 91% probability that this mutated stem cell will go extinct by genetic drift before it could escape extinction (i.e. reach 'fixation'). The chance of a cell being eliminated by genetic drift are therefore directly proportional to the total number of cells present and inversely correlated with the number of stem cells (Merlo et al., 2006).

Consequently, the variation in organization and number of the stem cells from which cancer can arise determines how much genetic drift will affect the chance of developing a tumour in different tissues (Rozhok and DeGregori, 2015).

An effective illustration of the impact of genetic drift has been illustrated by Rozhok and colleagues (Rozhok and DeGregori, 2015) by comparing the events in the intestine to those in the bone marrow. The intestinal mucosa is organized in criptae, each cripta contains at its bottom a limited number of stem cells (**Fig. 3.4**) while the marrow tissue is hosted in a large communal space in which the stem cells represent a significant population (**Fig. 3.4**). Therefore, a limited number of stem cells in each cripta makes elimination more likely to happen by the genetic drift of a mutated stem cell, than in the larger pool of stem cells in bone marrow.

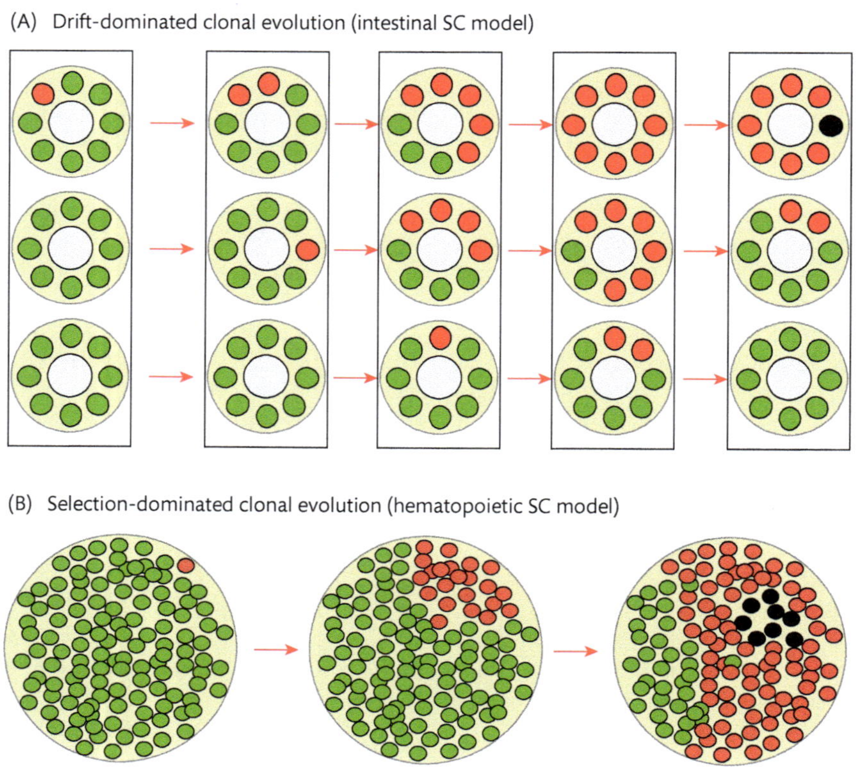

Fig. 3.4 Genetic drift: illustration of the effect of tissue architecture on sequential oncogenic mutation accumulation in stem cells pools. (A) Schematic representation of genetic drift in the intestine where each gland is separated from the other and has its own small pool of stem cells. Schematic section of intestinal epithelium with three crypts (normal cells are shown in green, mutated cells in red). In the upper line of glands, mutated cells become predominant owing to the drift and eventually a different neoplastic cell (shown in black) appears. In the second line of glands, only a few mutated cells manage to survive while in the bottom line the mutated cell, despite being 'more fit' than the normal, fails to develop and disappears. (B) In a very large pool of cells, the one cell carrying an advantageous mutation produces the dominant neoplastic clone as it responds to selective pressure. Eventually an evolved, more aggressive cell starts to grow (black cell).

Reproduced with permission from Andrii I. Rozhok, James DeGregori, 'Toward an evolutionary model of cancer: Considering the mechanisms that govern the fate of somatic mutations', *PNAS*, Jul 2015, 112 (29) 8914–21; DOI: 10.1073/pnas.1501713112

Conclusion

In recent years, the application of evolutionary biology methodologies to the study of cancer has allowed us to start to answer two big questions: why cancer-favouring traits are preserved in variable ways in animals, and how natural selection affects the growth of the neoplastic cells in the body. The persistence of traits which favour cancer is variable between species. *Homo sapiens* is one of the species at highest risk. The main hypothesis that has emerged so far is that until recently human lifespan was much shorter and developments in health and society, while increasing longevity, have resulted in humans being maladapted to such long life.

When a tumour develops, the cancer population is subject to evolutionary forces which drive its growth. These can be intrinsic to the cell (i.e. random genetic events which leads to the appearance in the neoplastic mass of different clones with different 'fitness'). Extrinsic forces instead include mainly the cellular microenvironment and the challenge of chemo or radiotherapy. The ability of some cells to be selected from either the microenvironment and/or what should be a therapeutic intervention, lead eventually to the cancer killing its host.

TAKE-HOME MESSAGE

- The study of the effect of natural selection of species evolution helps to understand why cancer-promoting traits have been positively selected in many species.
- Cancer cells are living organisms and as such evolve and are subject to natural selection.
- Evolutionary biology helps to explain how cancer grows and how interact with treatment.

OPEN QUESTIONS

- How can a better understanding of the relationship between cancer cells and their microenvironment be achieved to better understand the behaviour of neoplastic cells?
- The developments in molecular pathology are improving our understanding on one side but also highlighting unexpected and still not well explained behaviour, like tumours developing clones which have actually lost the genetic alteration believed to be the disease hallmark.
- Evolutionary studies using full genome sequencing will have to become more common to fully explain how cancer cell populations behave and respond to treatment.
- Can the study of the diversity of the cancer population become a tool to predict cancer behaviour?
- Increasing collaboration between biologists and mathematicians is needed.

FURTHER READING

Almendro, V., Marusyk, A., Polyak, K. (2013). Cellular heterogeneity and molecular evolution in cancer. *Annu Rev Pathol*, **8**, 277–302.

Darwin, C. (1859). *On the Origin of Species by Means of Natural Selection*. London: John Murray.

Darwin Online. Available at: http://darwin-online.org.uk/

Darwin, C. R. & Wallace, A. R. (1858). On the tendency of species to form varieties; and on the perpetuation of varieties and species by natural means of selection. [Read 1 July] Journal of the Proceedings of the Linnean Society of London. *Zoology*, **3**, 45–50. Available at: http://darwin-online.org.uk/content/frameset?itemID=F350&viewtype=text&pageseq=1

Greaves, M. & Maley, C. C. (2012). Clonal evolution in cancer. *Nature*, **481**, 306–13.

Greaves, M. (2000). *Cancer: The Evolutionary Legacy*. Oxford/New York: Oxford University Press.

Greaves, M. The Darwin Cancer Blog. Available at: https://thedarwincancerblog.com/author/bjcadmin1/

Lyons, S. (2011). *Evolution: The Basics*. London/New York, Routledge.

Merlo, M. F., Pepper, J. W., Reiuds, B. J., & Maley, C. C. (2006). Cancer as an evolutionary and ecological process. *Nat Rev Cancer*, **6**, 924–35.

Wallace Online. Available at: http://wallace-online.org/

REFERENCES

Almendro, V., Cheng, Y. K., Randles, A., et al. (2014). Inference of tumor evolution during chemotherapy by computational modeling and *in situ* analysis of genetic and phenotypic cellular diversity. *Cell Rep*, **6**, 514–27.

Almendro, V., Marusyk, A., & Polyak, K. (2013). Cellular heterogeneity and molecular evolution in cancer. *Annu Rev Pathol*, **8**, 277–302.

Antolin, M. F. (2009). Evolutionary biology of diseases and Darwinian medicine. In: Ruse, M. & Travis, J. (eds) *Evolution: The First Four Billion Years*. Harvard: Belknap Press Harvard University Press.

Bird, A. (2002). DNA methylation patterns and epigenetic memory. *Genes Dev*, **16**, 6–21.

Burrell, R. A., McGranahan, N., Bartek, J., & Swanton, C. (2013). The causes and consequences of genetic heterogeneity in cancer evolution. *Nature*, **501**, 338–45.

Crespi, B. J. & Summers, K. (2006). Positive selection in the evolution of cancer. *Biol Rev Camb Philos Soc*, **81**, 407–24.

Desmond, A. & Moore, J. (1992). *Darwin*, London: Penguin.

Ding, L., Ley, T. J., Larson, D. E., et al. (2012). Clonal evolution in relapsed acute myeloid leukaemia revealed by whole-genome sequencing. *Nature*, **481**, 506–10.

Greaves, M. (2007). Darwinian medicine: a case for cancer. *Nat Rev Cancer*, **7**, 213–21.

Greaves, M. (2015). Cancer's Darwinian dilemma: an evolutionary tale in three acts. *BMJ*, **351**, h6581.

Greaves, M. & Maley, C. C. (2012). Clonal evolution in cancer. *Nature*, **481**, 306–13.

Healey, C. S., Dunning, A. M., Teare, M. D., et al. (2000). A common variant in *BRCA2* is associated with both breast cancer risk and prenatal viability. *Nat Genet*, **26**, 362–4.

Keane, M., Semeiks, J., Webb, A. E., et al. (2015). Insights into the evolution of longevity from the bowhead whale genome. *Cell Rep*, **10**, 112–22.

Linder, D. & Gartler, S. M. (1965). Glucose-6-phosphate dehydrogenase mosaicism: utilization as a cell marker in the study of leiomyomas. *Science*, **150**, 67–9.

Merlo, L. M., Pepper, J. W., Reid, B. J., & Maley, C. C. (2006). Cancer as an evolutionary and ecological process. *Nat Rev Cancer*, **6**, 924–35.

Navin, N. E. (2014). Tumor evolution in response to chemotherapy: phenotype versus genotype. *Cell Rep*, **6**, 417–9.

Nesse, R. M. (2001). How is Darwinian medicine useful? *West J Med*, **174**, 358–60.

Nesse, R. M. & Stearns, S. C. (2008). The great opportunity: evolutionary applications to medicine and public health. *Evol Appl*, **1**, 28–48.

Nowell, P. C. (1976). The clonal evolution of tumor cell populations. *Science*, **194**, 23–8.

Pollak, M. N. (2004). Insulin-like growth factors and neoplasia. *Novartis Found Symp*, **262**, 84–98; discussion 98–107, 265–8.

Raj, A., Rifkin, S. A., Andersen, E., & Van Oudenaarden, A. (2010). Variability in gene expression underlies incomplete penetrance. *Nature*, **463**, 913–18.

Rozhok, A. I. & Degregori, J. (2015). Toward an evolutionary model of cancer: Considering the mechanisms that govern the fate of somatic mutations. *Proc Natl Acad Sci U S A*, **112**, 8914–21.

Shackleton, M., Quintana, E., Fearon, E. R., & Morrison, S. J. (2009). Heterogeneity in cancer: cancer stem cells versus clonal evolution. *Cell*, **138**, 822–9.

Shen, Z. (2011). Genomic instability and cancer: an introduction. *J Mol Cell Biol*, **3**, 1–3.

Thain, M. & Hickman, M. (2004). *The Penguin Dictionary of Biology*. London: Penguin Books.

Van Wyhe, J. (2012). *Alfred Russel Wallace. A Biographical Sketch* [Online]. Available at: http://wallace-online.org/Wallace-Bio-Sketch_John_van_Wyhe.html

Varley, K. E., Mutch, D. G., Edmonston, T. B., Goodfellow, P. J., & Mitra, R. D. (2009). Intra-tumor heterogeneity of MLH1 promoter methylation revealed by deep single molecule bisulfite sequencing. *Nucleic Acids Res*, **37**, 4603–12.

Vincent, M. D. (2010). The animal within: carcinogenesis and the clonal evolution of cancer cells are speciation events sensu stricto. *Evolution*, **64**, 1173–83.

SECTION II

The aetiology of cancer

4

Genetics and genetic instability in cancer

Mark A. Glaire and David N. Church

Background

Cancer is a disease of a disordered genome. Since Theodor Boveri made his pioneering observations of abnormally segregated chromosomes in tumour cells more than 100 years ago, our understanding of the genetic basis by which cells undergo neoplastic transformation has increased hugely (Cleveland and Don, 2009). We now know that the integrity of cellular DNA is under continuous threat as a consequence of errors made during DNA replication, and from mutagens both endogenous and exogenous (Fig. 4.1; Helleday et al., 2014). We know that eukaryotes have evolved multiple highly effective mechanisms to prevent and repair such mutations. Additionally, we know that failure of these cellular mechanisms, leading to genetic instability and eventually tumour development, occurs in a substantial fraction of human cancers.

The fidelity of DNA replication in humans is truly remarkable, with an error rate of approximately 1×10^{-9} per base replicated, or one mutation per genome duplication (Roach et al., 2010). This is due to a combination of the accurate base incorporation and exonuclease proofreading by the replicative DNA polymerases (Pols) ε and δ (Rayner et al., 2016), and post-replication surveillance by the mismatch repair apparatus (Jiricny, 2006). The importance of mismatch repair (MMR) to human cancer is well established. Evidence that defective polymerase proofreading contributes to malignancy has been forthcoming only recently, largely as a result of the large-scale sequencing of tumour genomes (TCGA, 2012). Similarly, advances in technology have also permitted a much more accurate assessment of the contribution of other mutational processes to genetic instability and human cancer. For example, while it has long been known that both endogenous and exogenous factors are capable of damaging DNA and inducing malignancy, the distinctive patterns of mutation caused by these insults, often referred to as a mutational scar, may provide a 'smoking gun' that gives an historical account of the mutagenic forces that contributed to the carcinogenic process (Helleday et al., 2014). The consequences of DNA alkylation, deamination, and double-stranded breaks caused by reactive oxygen species are all detectable by next generation sequencing. Similarly, DNA damaging agents such as UV radiation and chemicals like aflatoxin or benzo (α)pyrene all leave a distinct genomic 'scar'. Furthermore, defects in DNA repair, such as the deficiency in homologous recombination caused by mutations in *BRCA1* or *BRCA2*, often cause characteristic genomic alterations identifiable by next generation sequencing (NGS) approaches (Nik-Zainal et al., 2012).

In this chapter, we briefly review the mechanisms by which eukaryotes maintain genome stability and suppress mutagenesis. We highlight the insights that NGS technologies have provided into the molecular basis of mutagenesis in tumours. We summarize the key the drivers of genome instability in human cancer, with a particular focus on emerging causes, such as DNA polymerase exonuclease domain mutations. We review the emerging evidence that indicates that genomic instability is an important determinant of prognosis in tumours. Also, we highlight exciting and novel therapeutic strategies to target cancers with genomic instability, some of which have demonstrated encouraging antitumour activity in patients.

Mechanisms that maintain genome stability

Under normal circumstances, DNA replication is a highly accurate process, resulting in an error rate of less than one mutation every 1×10^{9}–1×10^{10} bases replicated (Roach et al., 2010). The bulk of DNA replication in eukaryotes is performed by the DNA polymerases Pol ε and Pol δ (Budd and Campbell, 1993). Studies in yeast and other model organisms indicate that Pol ε synthesizes the leading strand, and Pol δ duplicates the Okazaki fragments of the lagging strand following priming by Pol α (Fig. 4.2A) (Nick McElhinny et al., 2008). Both enzymes are heterodimers, the major subunits of which are encoded in humans by *POLE* and *POLD1*, respectively (Shevelev and Hubscher, 2002). These subunits contain both the polymerase catalytic domain, which adds bases to the primer strand, and an exonuclease domain, which proofreads the newly synthesized DNA strand and excises mispaired bases incorporated during replication. Loss of exonuclease proofreading function of either Pol ε or Pol δ results in an approximately 100-fold increase in mutation rate in model systems, and induces tumour development in mice (Simon et al., 1991).

The fidelity of DNA duplication is further increased by the actions of the mismatch repair (MMR) system (Fig. 4.2B), which surveys the newly replicated DNA for base mispairs and small insertions and deletion loops (IDLs) caused by polymerase slippage at repetitive DNA microsatellites (Jiricny, 2006). The MMR system in humans comprises several protein complexes formed from various

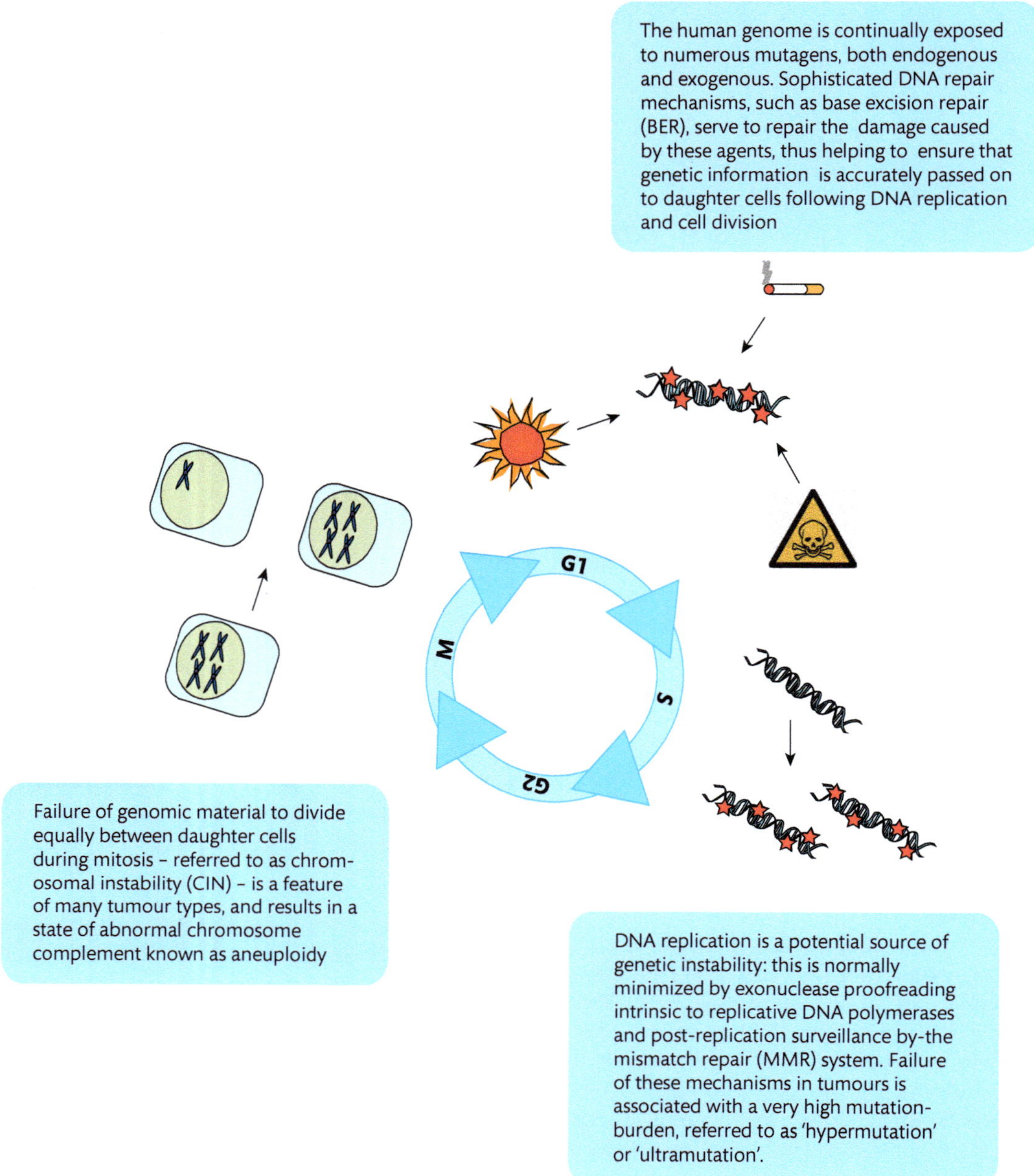

Fig. 4.1 Causes of genomic instability in cancers. Schematic highlighting important causes of genetic instability in human cancers. Further details are provided in the main text.

combinations of the main MMR proteins MLH1, MSH2, MSH3, MSH6, and PMS2. These heterodimers serve several complementary functions. Mismatch repair is initiated by the binding of MutS, a heterodimer that exists in two main forms, to the aberrant DNA region. MutSα, a heterodimer of MSH2 and MSH6, is responsible for the repair of base pair mismatches and small IDLs. MutSβ, which comprises MSH2 and MSH3, is responsible for the correction of larger IDLs. Recognition of the DNA lesion by MutS heterodimers leads to the recruitment of the secondary MutL protein complex, which comprises a heterodimer of MLH1 with either PMS2 (MutLα), PMS1 (MutLβ) or MLH3 (MutLγ). MutL forms a ternary complex with MutS and the DNA strand, which in turn recruits additional proteins including Exo1 and PCNA, which excise the mismatch or IDL. Resynthesis and religation of the DNA strand is subsequently performed by Pol δ and DNA ligase (Jiricny, 2006).

As noted earlier, DNA is also at risk of damage following replication by the products of normal cellular metabolism and exogenous mutagens. There are multiple mechanisms by which such DNA damage is recognized and repaired, which vary according to the type of lesion and the circumstances under which repair takes place. While a comprehensive discussion of these is beyond the scope of this chapter, those germane to our discussion are outlined in the following sections.

Relative to the potential for disruption of the human genome, the rarity of cancer is a testament to the evolutionarily conserved mechanisms that recognize and correct genomic damage. Should one, or

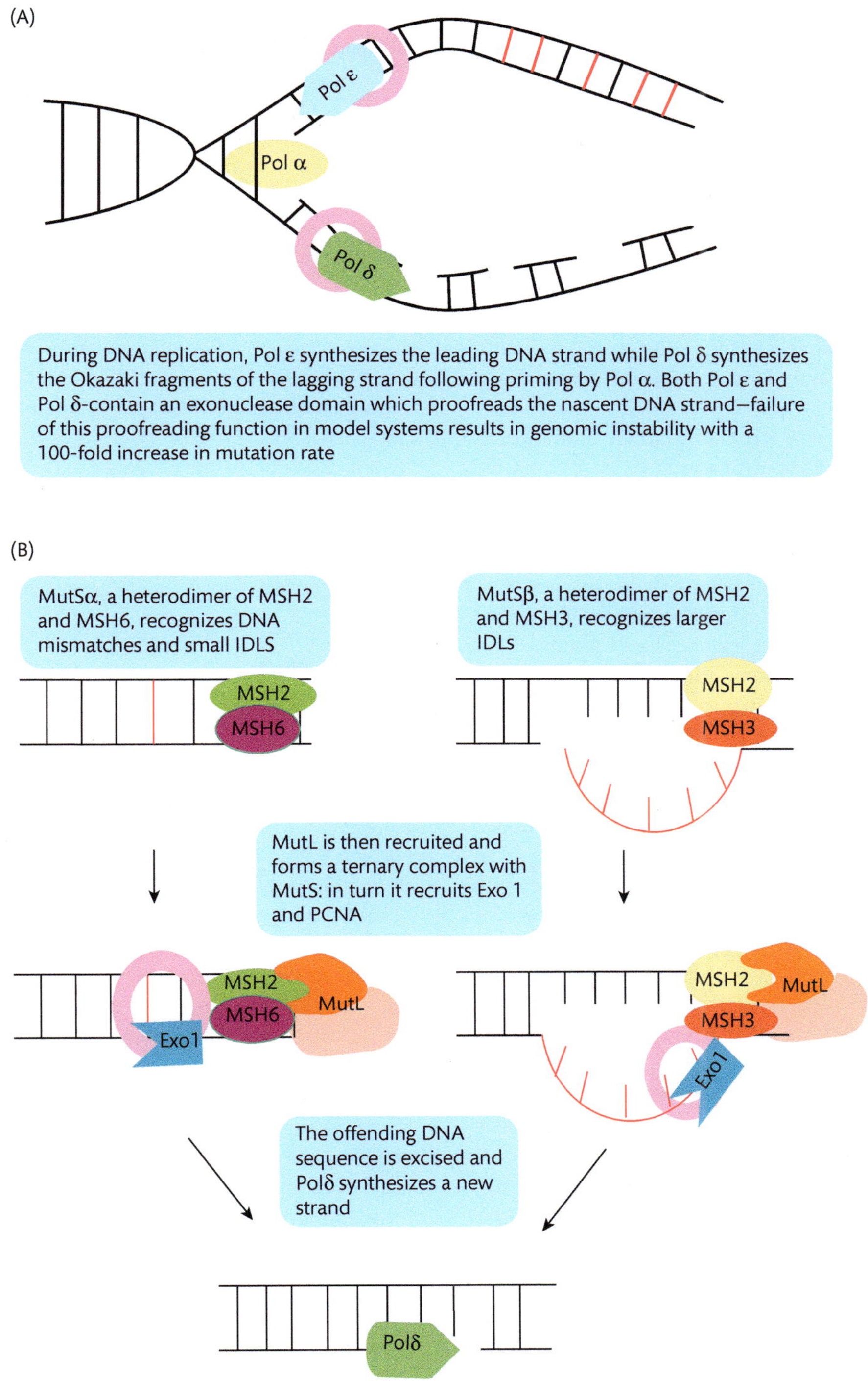

Fig. 4.2 Polymerase proofreading and DNA mismatch repair (MMR) protect against genomic instability. Under normal circumstances DNA replication is an exceptionally accurate process. (A) The bulk of DNA replication is performed by the DNA polymerases pol ε and pol δ, the major subunits of which contain the polymerase and exonuclease domains and are encoded by *POLE* and *POLD1* in humans. The essential contribution that polymerase proofreading makes to DNA replication is demonstrated by the increased mutation rate caused by its loss in yeast and mouse models. Germline mutations in the proofreading exonuclease domain of *POLE* and *POLD1* cause intestinal polyposis and predisposition to early-onset cancer, while somatic polymerase proofreading domain mutations appear limited to *POLE*, and cause a phenotype of ultramutation in sporadic cancers (mispaired bases shown in red in the schematic). (B) Post-replication surveillance by the DNA mismatch repair (MMR) system functions to correct mispaired bases and insertion-deletion loops (IDLs) introduced during DNA replication. The first stage in the repair process is recognition of the DNA aberration by a MutS heterodimer, the precise form of which depends on the type of DNA lesion. MutSα, composed of the proteins MSH2 and MSH6, recognizes repair base-base mismatches and small indels, while MutSβ, a heterodimer of MSH2 and MSH3, recognizes larger IDLs. Recognition of the DNA lesion by MutS heterodimers leads to the recruitment of the secondary MutL protein complex (a heterodimer of MLH1 with either PMS2, PMS1, or MLH3), and in turn to the recruitment of other proteins including PCNA and Exo1. The aberrant DNA lesion is then excised, and the resynthesis and relegation of the DNA strand completed by Pol δ and DNA ligase. Germline mutations in MMR genes cause Lynch syndrome, characterized by predisposition to multiple cancer types, while loss of MMR function in sporadic cancers is typically caused by hypermethylation of the MLH1 promoter with loss of gene expression. In both cases, MMR-deficient tumours are hypermutated, with an enrichment of frameshift mutations.

more, of these repair mechanisms fail, then integrity of the genome is at risk, and the process of carcinogenesis may begin.

Molecular insights into the cancer genome

The advent of NGS technologies has hugely advanced our understanding of cancer genetics. To date, most studies have used whole exome sequencing to focus on the ~1% of the genome that is protein coding, though the number of cancer genomes that have been analysed is increasing rapidly as sequencing costs fall. The latter approach has numerous advantages, including the ability to identify large-scale structural rearrangements—a key feature of several types of genomic instability.

Perhaps the best-characterized type of somatic mutation involves the substitution of a single base for another, a class commonly known as single nucleotide variants (SNVs) or point mutations. SNV should be distinguished from SNP (single nucleotide polymorphism): SNV is a stochastic difference in a base without any implication as far as its frequency and inheritance is concerned while a SNP is not random, is inherited and occurs in a section of the population (at least 1%). SNVs may be transitions, where the substitution involves replacement of a pyrimidine by another pyrimidine or replacement of a purine by another purine, or transversions, where a pyrimidine is replaced by a purine, or vice versa (Stratton et al., 2009). While the number of possible transitions is half the number of possible transversions, transitions account for the majority of mutations in most tumour types. However, these proportions may be reversed in cancers associated with particular mutagens (e.g. cigarette smoking) or other causes of genetic instability. When studying SNV mutations, a more detailed mutational profile can be derived by considering not just the affected base but also the bases immediately 5' and 3' to it: this generates 96 possible combinations (6 possible base substitutions multiplied by the 16 possible sequence contexts). In a seminal study published in 2013, Alexandrov and colleagues used this 96-channel variant signature to define 21 distinct mutational signatures across 30 different tumour types (Alexandrov et al., 2013). Crucially, they were able to correlate the presence of particular signatures with specific mutagens and other molecular defects. For example, Signature 7, which is characterized by a high frequency of C>T mutations at trinucleotides Cp<u>C</u>pN and Tp<u>C</u>pN, was primarily found in melanoma, corresponding to the distinct type of DNA damage caused by UV radiation exposure. Signature 4, which exhibits a transcriptional strand bias for C>A mutations, is found in adenocarcinomas, squamous and small cell carcinomas of the lung, squamous cell carcinomas of the head and some types of liver cancer. Cigarette smoking is a key aetiological factor in these cancers, and it is believed that Signature 4 arises from bulky DNA adducts generated by carcinogenic compounds in tobacco smoke and their removal by transcription coupled nucleotide excision repair (Alexandrov et al., 2013).

Another common type of mutation in cancers are insertions and deletions (indels). These occur as a consequence of a failure to recognize and correct small insertion and loops caused by slippage of the DNA replication machinery at repetitive DNA microsatellites. The presence of indels in these regions—a phenomenon known as microsatellite instability (MSI)—is most commonly caused by defects in the DNA mismatch repair apparatus and is discussed in detail next (see 'Aneuploidy and chromosomal instability'). The distinct signature of mutations this causes was identified by Alexandrov and colleagues as Signature 6 in their pivotal study (Alexandrov et al., 2013).

Alternatively, somatic genomic aberrations in cancer may occur at level of the chromosome—indeed the chromosome 9:22 translocation, or Philadelphia chromosome that causes chronic myeloid leukaemia was one of the first specific molecular defects to be described in cancer (Rowley, 1973). Other examples include deletions, inversions, and duplications. When large, such alterations may cause whole gene deletions, or the production of novel oncogenes through gene fusions. Several important causes of genomic instability, such as that caused by *BRCA1* or *BRCA2* mutations, cause chromosomal rearrangements, and more are likely to be discovered as the characterization of mutational processes is extended beyond the analysis of alterations in nucleotide sequence.

DNA sequence instability in cancer

Mismatch repair deficiency (Lynch syndrome, sporadic cancers)

Perhaps the best-characterized type of genomic instability at the nucleotide level is that which results from mismatch repair deficiency (MMR-D), which causes a phenotype of hypermutation with base pair mismatches and IDLs. Interestingly, while MMR-D promotes tumorigenesis in both familial cancers (Lynch syndrome; see Lynch et al., 2015) and sporadic malignancies, the molecular mechanisms responsible for this differ somewhat between the two (Jiricny, 2006).

Lynch syndrome (previously referred to as hereditary non-polyposis colorectal cancer or HNPCC), is an autosomal dominant condition that predisposes sufferers to various malignancies including colorectal, ovarian, endometrial, stomach, and prostate cancer. Individuals with Lynch syndrome inherit a deleterious mutation in one of the MMR genes, most commonly *MSH2*, but *MLH1*, *PMS2*, and *MSH6* may be affected. While heterozygous MMR mutations do not disrupt DNA repair function, the acquisition of a 'second hit' during adult life causes loss of MMR function, resulting in increased mutation rate and eventually, a sufficient complement of mutations to cause cancer. While there are several mechanisms by which the second hit may occur, by far the most common is loss of heterozygosity. Lynch syndrome causes approximately 3% of colorectal cancers and a similar fraction of endometrial cancers, which appear to be enriched among carriers of *MSH6* mutations. Most Lynch syndrome-associated tumours display microsatellite instability, though this is not universal, particularly among cases with *MSH6* variants.

As noted previously, defective mismatch repair is also common in sporadic cancers. Data from The Cancer Genome Atlas analyses and other studies indicate that approximately 15% of colorectal cancers (TCGA, 2012), 15–20% of endometrial cancers and a smaller but appreciable fraction of stomach and ovarian cancers display MSI as a consequence of deficient mismatch repair. MMR-D in sporadic tumours of all types is most commonly caused by hypermethylation of the *MLH1* promoter region, resulting in loss of gene expression, though biallelic somatic mutations account for a small fraction of cases.

Irrespective of its cause, at the genomic level, MMR deficiency results in a striking degree of genomic instability, with a 10–100-fold increase in tumour mutation burden, and a characteristic preponderance of indel mutations as noted earlier. The tendency to mutation at microsatellites results in a distinct pattern of driver mutations compared to mismatch repair proficient tumours. For example, in colorectal cancer, MMR-D tumours frequently display mutations in genes such as *TGFBR2* and *IGF2R* and only rarely harbour *TP53* mutations, in contrast to MMR-P tumours where the proportions are reversed (TCGA, 2012).

MMR-D tumours also show other notable molecular characteristics. They are typically diploid, without evidence of the chromosomal instability that characterizes most colorectal and some endometrial tumours. Sporadic MMR-D tumours also commonly display a CpG island methylator phenotype (CMIP), usually in combination with *BRAF* mutation (Weisenberger et al., 2006)—a feature that can be used to triage cases with MLH1 loss on immunohistochemistry for molecular testing and genetics referral (Domingo et al., 2004). Emerging evidence indicates that the hypermutation, and particularly the high number of frameshift mutations observed in these tumours may explain their tendency to display a strong cytotoxic immune response—a characteristic which could plausibly account for their favourable prognosis (see 'Genomic instability and prognosis'; and Llosa et al., 2015).

Polymerase proofreading domain mutations (sporadic ultramutated cancers)

As reviewed earlier in this chapter, the importance of MMR deficiency to cancer development has been appreciated for more than two decades. During the last few years, another important cause of tumour genomic instability has become apparent, with the discovery that mutations within the replicative DNA polymerases exonuclease domains perturb proofreading and cause a substantial increase in mutation burden in a subset of common cancers (Rayner et al., 2016).

The initial reports of these defects arose from two broadly contemporaneous, yet separate studies. The first of these was the analysis of human colon and rectal cancer reported by the TCGA, which performed whole exome sequencing as part of the large-scale molecular characterization of more than two hundred tumours. In addition to detecting the anticipated 15% of tumours with hypermutation as a consequence of mismatch repair deficiency, they also noted the existence of a small subset of cancers with an exceptionally high mutational load, and no evidence of defective mismatch repair. These 'ultramutated' cancers all harboured somatic mutations within the exonuclease domain of *POLE*, though detailed characterization of this subset was not performed (TCGA, 2012). In the second study, Palles and colleagues used linkage analysis and whole genome sequencing to identify heterozygous germline mutations within the exonuclease domains of *POLE* and the other replicative polymerase *POLD1* in patients with an unexplained history of familial intestinal polyposis and early-onset colorectal cancer (Palles et al., 2013). Interestingly, the *POLD1* variant was also shown to predispose to endometrial cancer. In addition to confirming pathogenicity of these mutations by modelling in yeast, the authors further analysed the TCGA sporadic colorectal cancers and showed that the somatic *POLE* variants were highly likely to be functional based on sequence alignments and structural modelling (Palles et al., 2013).

Subsequent studies have demonstrated that somatic mutations in the *POLE* exonuclease domain occur 1–2% of colorectal cancers (TCGA, 2012), 7–12% of endometrial cancers (Cancer Genome Atlas Research et al., 2013), and less commonly in other tumour types (Rayner et al., 2016). Interestingly, somatic *POLD1* mutations appear to be very uncommon. Somatic *POLE* exonuclease domain mutations are associated with a distinctive pattern of base substitution mutations, with strong bias for Tp<u>C</u>pT>Tp<u>A</u>pT changes, corresponding to Signature 10 in the analysis by Alexandrov and colleagues. These *POLE* variants tend to be recurrent substitutions at particular codons, most commonly P286R and V411L, and in several cases have been confirmed to disrupt exonuclease activity and cause a mutator phenotype in cell-free assays and yeast models. Interestingly, the increase in mutation rate caused by the P286R mutation substantially exceeds that which results from loss of polymerase proofreading alone, suggesting that this, and possibly other, somatic *POLE* mutations may exert other, as yet unknown effects on replication fidelity (Rayner et al., 2016).

Sporadic tumours with somatic *POLE* exonuclease domain mutations are notable for having the highest burden of mutation of any tumour type, suggesting that the degree of genetic instability caused by these variants may approach the maximum compatible with continued viability. The potential consequences of this and other aspects of these tumours interesting biology are discussed further in the subsequent sections.

APOBEC overexpression

In contrast to the previous two examples, where an increased mutation rate occurs as a result of a failure of cellular processes that serve to suppress errors, genomic instability can also arise as a result of dysregulation of mechanisms designed to *promote* mutagenesis. The cytidine deaminases are a family of enzymes that play important and highly evolutionarily conserved roles in pathogen defence by inducing DNA damage. One member of this family, known as activation-induced damage, functions to aid antibody diversification by causing DNA damage at immunoglobulin loci, particularly at cytosine residues flanked by a 5' purine (Di Noia and Neuberger, 2007). Another class of members, the APOBEC (apolipoprotein B mRNA editing, catalytic polypeptide) enzymes, serve vital roles in innate immunity. APOBEC3F and APOBEC3G provide protection against retroviruses: these enzymes attach themselves to viral particles, and cause deamination of cytosine residues during the first retroviral DNA strand synthesis. The resulting excess of uracil is toxic for the invading retrovirus, resulting in lethality (Harris and Liddament, 2004).

Work performed during the last decade has revealed that APOBEC overexpression is a major contributor to genetic instability in human cancers. Harris and colleagues first demonstrated that APOBEC activity mutates DNA by deaminating cytosine residues (Harris et al., 2002); Nik-Zainal and colleagues later correlated the overexpression of APOBECs with a specific mutational signature in breast cancers (Nik-Zainal et al., 2012). They defined a characteristic pattern of base substitutions in the genomes of 21 breast cancers which they attributed to the deaminase action of the APOBEC enzymes, predominantly APOBEC3A, APOBEC3B, and APOBEC1 (Nik-Zainal et al., 2012). This signature accounted for the majority of mutations in 10% of oestrogen receptor (ER) positive breast cancers and was characterized by a preponderance of

C>T, C>G, C>A substitutions at TpCpX trinucleotides. Further investigation suggested that APOBECs are responsible for kataegis—a phenomenon of hypermutation confined to localized genomic regions (Nik-Zainal et al., 2012). The evidence suggesting that APOBEC causes genomic instability is substantial: transgenic overexpression has been shown to induce cancer (Yamanaka et al., 1995), and a dose–response is observed in that the level of expression of APOBECs in cancers correlates with the prevalence of its mutational signature (Roberts et al., 2013). Building on this work, Alexandrov and colleagues attributed two signatures to the likely action of APOBECs: Signature 2, which is characterized by C>T and C>G mutations at TpCpN trinucleotides, and Signature 13, which is characterized by thymine preceding the mutated cytosine (TpCpN). In total, Signatures 2 and 13 have been detected in sixteen different types of malignancy including cervical, head and neck, stomach, pancreas, and haematological cancers such as ALL, CLL, and B-cell lymphoma (Alexandrov et al., 2013). What mechanisms could induce APOBEC-mediated mutagenesis? One possible link is viral infection, and it is noteworthy that both cervical and head and neck cancers are strongly associated with human papilloma virus (HPV) infection. To this end, it was shown that HPV infection is correlated with APOBEC-mediated mutagenesis in head and neck squamous cancers; where its mutational signature is often manifest as *PIK3CA* helical domain mutations, a relatively specific oncogenic event (Henderson et al., 2014). Similarly, HPV infection induces aberrant APOBEC3B expression in breast cancer, where it appears to be an early event in carcinogenesis (Ohba et al., 2014).

BRCA1/2 mutations and other defects in DNA repair

BRCA1 and BRCA2 are DNA repair enzymes that maintain chromosomal integrity by directing genomic repair in response to DNA double-strand breaks (DSBs). Following a DSB, the cell has several repair mechanisms available to correct the error, the most prominent of which are non-homologous end joining (NHEJ) and homologous DNA recombination (HR). The former is referred to as an error-prone process, as it may result in loss of DNA fidelity, whereas homologous recombination is considered error-free, as DNA exchange occurs between identical chromatids or homologous chromosomes (Venkitaraman, 2014).

BRCA1 and BRCA2 play essential, but differing roles in DSB repair by HR. In essence, BRCA1 acts early in the repair process by identifying DNA lesions and initiating repair by HR; BRCA2 appears to stabilize stalled replication forks and ensures HR repair by regulating the activity of RAD51. In response to DNA damage, BRCA1 colocalizes with the resection factor MRE11-RAD50-NBS1 and engages with the resection factor CtIP. BRCA2's main role is to facilitate loading of the RAD51 complex onto the resected ssDNA, which is essential for HR and ensures that DSB repair does not proceed by the deleterious single-strand annealing pathway. BRCA1 or BRCA2-deficient cells are specifically deficient in HR, whereas other DSB repair pathways remain intact (Moynahan et al., 1999; Moynahan et al., 2001). In addition to protecting against chromosomal structural aberrations, BRCA proteins also prevent numerical chromosome errors. Aneuploidy in BRCA-deficient cells can result from bypassing of the G_2 DNA damage checkpoint, abnormal spindle formation and erroneous cell division during cytokinesis (Venkitaraman, 2014).

Females who inherit germline heterozygous mutations in *BRCA1* have a 55–65% lifetime risk of developing breast cancer and a 40–60% risk of developing epithelial ovarian cancer. The corresponding risk for both diseases among *BRCA2* mutation carriers is somewhat lower at approximately 45% and 10–20%, respectively (Chen and Parmigiani, 2007). It is estimated that roughly 5% of breast cancers and 10% of epithelial ovarian cancers are caused by germline mutations in *BRCA1* or *BRCA2*. Consistent with the function of both genes in HR, tumours that arise as a result of their loss of function are characterized by genetic instability with a high frequency of genomic rearrangements. This is detectable as a distinct mutational signature of small deletions with rearrangements showing overlapping homology, which occurs as a consequence of error-prone repair by NHEJ following the failure of HR (Turner et al., 2004; Alexandrov et al., 2013). While this signature is evident in breast and ovarian cancers with germline *BRCA1* or *BRCA2* mutations, it is not specific to loss of function of these genes. Rather it reflects an underlying HR defect, which may alternatively be caused by genomic alterations in other HR genes, including *EMSY, PTEN, RAD51C, ATM*, or *ATR* (Lord and Ashworth, 2016). The presence of the molecular characteristics associated with defective HR in the absence of germline *BRCA1* or *BRCA2* mutations is referred to as *BRCA*ness, the identification of which may be important for therapeutic targeting, as discussed next (see 'Targeting genomic instability for therapy'; see also Turner et al., 2004).

Interestingly, given their importance in hereditary cancer, somatic mutations of *BRCA1* and *BRCA2* in sporadic tumours appear to be relatively rare, though they have been reported in epithelial ovarian cancer. However alternative mechanisms of inactivation have been reported, including silencing of *BRCA1* by promoter hypermethylation (present in 10–15% of sporadic breast cancers), by increased expression of ID4, a negative regulator of *BRCA1* (Beger et al., 2001) and by overexpression of miR-146a and mir-146b-5p (Garcia et al., 2011). Contrastingly there is little evidence to support hypermethylation silencing of *BRCA2*. Rather, it has been suggested the *BRCA2* loss in sporadic cancers may be the result of an amplification of the *ESMY* gene; the product of which is capable of abrogating activation of *BRCA2* (Hughes-Davies et al., 2003).

Nucleotide excision repair defects (XP)

Nucleotide excision repair (NER) is a particularly versatile DNA repair mechanism in that it is able to correct the damage caused by a large variety of insults. NER is activated in response to distortion of the DNA double helix and/or chemical alteration of the DNA and occurs in two contexts: global genome repair (GGR) or transcription coupled repair (TCR). The former functions by surveying the whole genome for helix distortions, whereas the latter acts to resolve DNA modifications that cause a pause in transcription (Nouspikel, 2009). GGR is initiated by the XPC complex in response to distortions in the double helix structure. Smaller lesions such as UV exposure induced cyclobutane pyrimidine dimers, which do not cause destabilization of the DNA duplex, may escape recognition by XPC, but are recognized by the DDB complex: a heterodimer of DDB1 and DDB2/XPE. The DDB complex is also part of a multisubunit E3 ubiquitin ligase that poly ubiquitinates both XPC and XPE, with contrasting effects. Ubiquitination increases XPC's affinity for DNA but causes degradation of XPE: this in effect returns control of DNA repair back to the XPC pathway (Nouspikel, 2009). Once the

DNA lesion is correctly identified, a 'denaturation bubble' is formed around the DNA lesion by the actions of the TFIIH complex and the offending lesion exposed and excised by the actions of the XPG nuclease (Mu et al., 1996) and the XPF-ERCC1 complex (O'Donovan et al., 1994). The resultant gap is then filled by DNA polymerases ε and δ (Popanda and Thielmann, 1992), and the nick-sealed by DNA ligase III and with a minor contribution from ligase I in replicating cells (Moser et al., 2007).

NER is the principal mechanism of repair of covalent DNA modifications such as the adducts caused by chemicals such as benzo(a) pyrene diol-epoxide and aflatoxins. It also repairs the damage caused by cyclobutane pyrimidines (CPDs) and pyrimidines-pyrimidone photoproducts ((6–4)PPs), both of which are induced by UV radiation, as well as the DNA cross-links caused by drugs such as cisplatin (Nouspikel, 2009).

Two diseases highlight the important and varying roles of NER system. Xeroderma pigmentosum (XP) is a hereditary condition, primarily characterized by its cutaneous features that include skin atrophy, pigmentation anomalies, and a substantially increased risk of squamous cell carcinoma of the skin (Kraemer et al., 1987). Sufferers are also at an increased risk of developing gastrointestinal and lung cancers, likely reflecting the important role of NER in protecting guarding against dietary and inhaled carcinogens. XP is caused by mutations in either *XPC* or *XPE*, resulting in defective GGR. In contrast, Cockayne syndrome, in which TCR function is lost because of germline mutations in ERCC genes (*ERCC6* or *ERCC8*) is predominantly a developmental disease with a lower risk of cancer. As both syndromes are associated with a similar magnitude of increase in mutational rate, the reasons for this discordance are unclear, though it has been suggested that loss of GGR also results in a degree of chromosomal instability that is necessary for carcinogenesis to occur (Cleaver, 2005).

Base excision repair defects (*MUTYH*)

Base excision repair (BER) is a highly evolutionarily conserved mechanism that repairs lesions caused by alkylation, deamination, and oxidative damage, which if left unrepaired have the potential to cause mutation. BER is initiated by specific DNA glycosylases that recognize DNA damage and remove the aberrant DNA base by severing the N-glycosylic bond attaching it to the sugar phosphate backbone. Following this, an AP endonuclease (APE1 in humans), generates a single-strand break (SSB) or 'nick' in the DNA at the site of the excised base, before incorporation of the missing base by DNA polymerase β and sealing of the nick by DNA ligase III. DNA damage that results in an SSB that with unligatable ends, may be repaired by long patch BER, in which a larger segment of DNA is resynthesized by the action of Pol δ (Robertson et al., 2009).

BER performs an essential role in genome maintenance. Pathogenic germline mutations in *MUTYH*, one of the 11 recognized human glycosylases, cause an autosomal recessive cancer predisposition syndrome known as *MUTYH*-associated polyposis (MAP). Patients with MAP have a substantially increased lifetime risk of developing CRC: 43% rising to 100% in the absence of surveillance (Kastrinos and Syngal, 2007). These individuals are also prone to development of polyps of the upper gastrointestinal tract and cutaneous tumours. Tumours in patients with MAP display a characteristic mutation spectrum that corresponds to the causative molecular defect, with a marked increase in G>T transversion mutations (Al-Tassan et al., 2002).

Structural variation in cancers

Aneuploidy and chromosomal instability

In addition to changes in nucleotide sequence, genomic instability may occur at a chromosomal level, either as aneuploidy, or chromosomal instability (CIN; **Fig. 4.3**). More than a century after Boveri proposed that cancer was caused by abnormal chromosome segregation, an abundance of data support central roles of both aneuploidy and CIN in human tumours. While aneuploidy and chromosomal instability are terms that are often used interchangeably they are not synonymous. Aneuploidy is the state of a cell having too few or too many chromosomes; thus it refers to its karyotype. Chromosomal instability refers to an increased rate of chromosomal gain or loss, and in the strictest sense of its definition refers to whole chromosome mis-segregation and excludes subchromosomal changes such as translocations, inversions, or deletions. CIN can lead to aneuploidy, but aneuploidy may occur in the absence of CIN, as illustrated by Down's syndrome—in which there is aneuploidy, but not CIN (Thompson et al., 2010).

Aneuploidy is generally caused by errors in the segregation of chromosomes during mitosis. This process is regulated by the spindle assembly checkpoint (SAC), which ensures that the chromosomes are correctly orientated on the microtubule spindle before the onset of anaphase. There are numerous genes that function in the SAC, which, if mutated, result in aneuploidy in model organisms (Michel et al., 2001). However, very few of these are mutated in human cancers, and the mechanisms responsible for aneuploidy in malignancy remain only partially understood. Indeed, while mutations in *STAG2*, a component of the cohesin complex and as such a plausible candidate gene, were postulated to cause aneuploidy in multiple cancers in one high-profile study, a subsequent analysis found no evidence for this in bladder cancers (Balbas-Martinez et al., 2013). Similarly, while several recurrently-mutated cancer genes, including *APC*, *TP53*, and *CTNNB1* have been implicated in aneuploidy, the evidence for this is far from conclusive.

While its cause may be opaque, it is clear that aneuploidy is a prevalent feature of many cancer types. In addition to generating insights into the mechanisms by which it occurs, future studies may also suggest novel therapeutic strategies against aneuploid tumours, discussed next (see 'Targeting genomic instability for therapy').

Interaction between genomic instability and other hallmarks of cancer

The previous sections have highlighted some of the distinct mechanisms that cause genetic instability in human cancer. However, far from occurring independently of the other pathological processes that drive malignancy, genetic instability is often a consequence of the other hallmarks of cancer. Here we briefly review some of those most germane to our discussion.

Genetic instability in cancers is frequently the result of epigenetic alterations. Perhaps the best described of these is the subset of colorectal tumours defined by the CIMP, which manifests as an

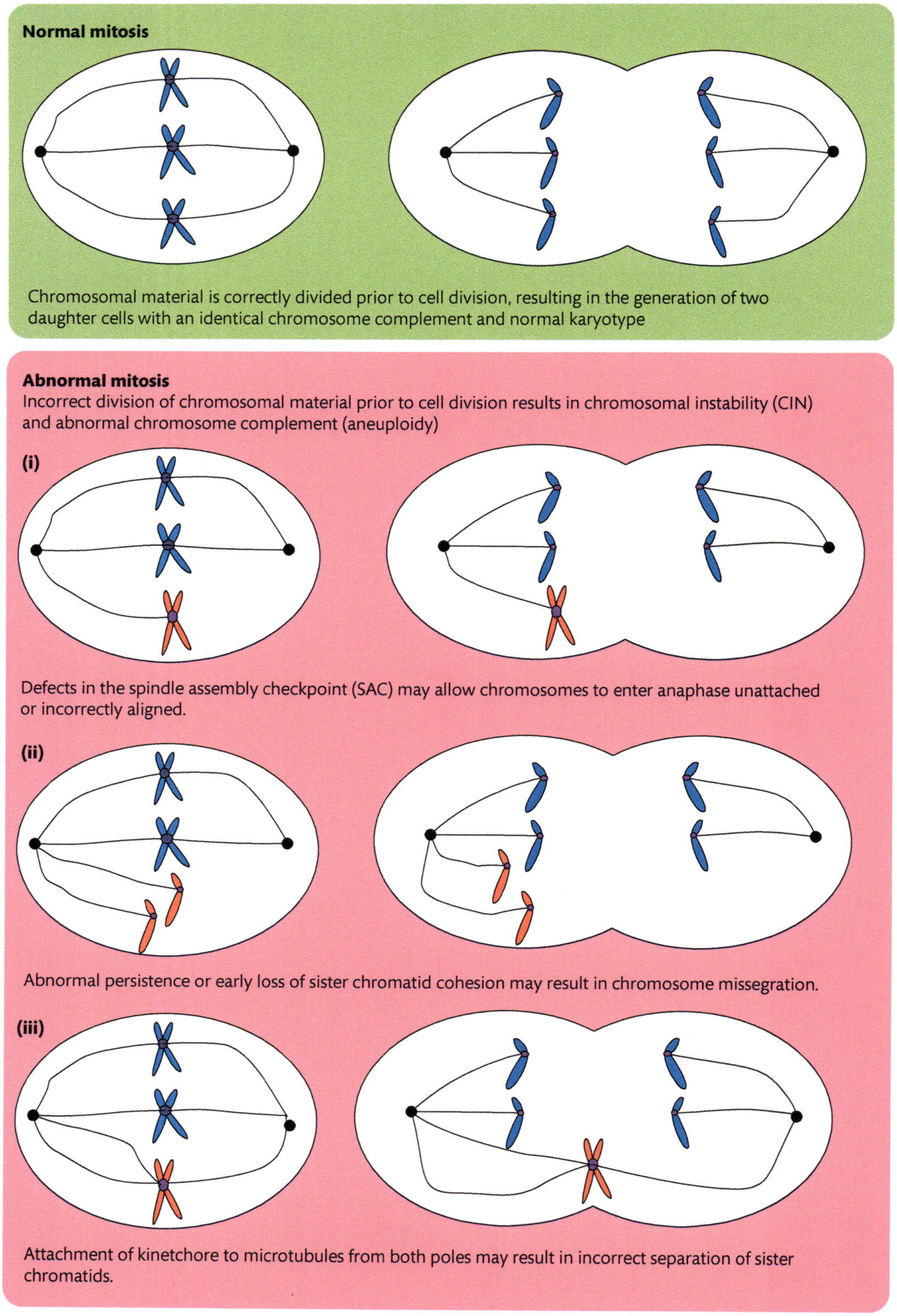

Fig. 4.3 Cartoon showing mechanisms of chromosomal instability (CIN). Normal mitosis (upper panel) is characterized by correct division of chromosomal material prior to cell division. Chromosomal instability refers to a change in the rate of chromosomal derangements and is often associated with aneuploidy, which refers to an aberrant chromosomal complement. Chromosomal instability is common in solid tumours and may result from defects in several processes (I–III in lower panels).

increased frequency of hypermethylation at CpG islands. CIMP colorectal cancers frequently display hypermethylation of the *MLH1* promoter, resulting in its silencing and a loss of mismatch repair function, as discussed previously in 'Mismatch repair deficiency (Lynch syndrome, sporadic cancers)'. There is consequently substantial overlap between the CIMP and MSI groups among colorectal cancers (Weisenberger et al., 2006).

Achieving replicative immortality is one of the hallmarks of cancer (Hanahan and Weinberg, 2011). One of the most important mechanisms that guards against this is the shortening of telomeres—specialized chromatin structures present at the end of chromosomes—that occurs with each cell division. Telomere shortening can be thought of as a proverbial 'biological clock', which if sufficient, results in cell cycle arrest and senescence, thus exerting a tumour suppressor effect. While cancer cells are often able to restore telomere stability, usually via reactivation of telomerase, the resultant telomeres are typically greatly shortened. These shortened telomeres are dysfunctional in that they are only able to partly prevent chromosomal degradation and recombination during cell division, resulting in chromosomal instability. Interesting data have suggested that activation of telomerase by itself is not sufficient to induce tumorigenesis, but may participate in disease progression by enabling the propagation of chromosomal instability (Shay and Wright, 2011).

Oncogene-associated replication stress refers to the phenomenon of DNA damage response activation in response to stalled replication forks. Importantly, studies indicated that fork stalling, and subsequent DSBs could occur as a consequence of oncogene activation in the absence of defects in DNA repair, suggesting that many of the common aberrations found in cancer could potentially cause genetic instability. This postulate has been supported by the observation that activating mutations in a number of key oncogenes, including *Ras*, *MYC*, and *CCNE1*, are capable of inducing DSBs and genomic instability (Gaillard et al., 2015).

Consequences of genomic instability in cancer

Genomic instability and prognosis

The preceding sections have summarized the abundant data that demonstrate that genomic instability is a central feature of many tumour types. As such, it is perhaps unsurprising that it has the potential to influence both tumour behaviour and clinical outcome. Accumulating data have confirmed that this is indeed the case, but intriguingly, have also indicated that its impact may depend on the extent to which it is perturbed, and the context in which it occurs.

Arguably the strongest evidence for and impact of genomic instability on prognosis has been obtained from the study of hypermutated mismatch repair deficient colorectal cancers. Both a large meta-analysis and the subsequent study of multiple large clinical trials have consistently demonstrated that in early stage (stage II/III) colorectal cancers with MMR-D have an improved prognosis, with a risk of recurrence that is approximately half to two-thirds that of mismatch repair proficient tumours (Popat, 2004). Investigation of the prognostic implications of other causes of increased SNV burden suggest they may also influence clinical outcome. In a recent study, Leonard and colleagues studied the relationship between overexpression of APOBEC3B and APOBEC3G enzymes and clinical outcome in high-grade serous ovarian carcinoma; they reported significantly better prognosis of patients with APOBEC3G upregulation (Leonard et al., 2016). Recent studies of the ultramutated subset of endometrial and colorectal cancers with *POLE* exonuclease domain mutations have also indicated improved clinical outcome of patients with these tumours (Meng et al., 2014). Provocatively in colorectal cancer, the prognosis of *POLE*-mutant tumours appears to be even better than that of MMR-D cases.

How do these forms of genetic instability influence prognosis? All result in hypermutation though this varies in severity, from a relatively modest increase in mutations with APOBEC3G overexpression, to a burden of more than 100 mutations per Mb in *POLE*-mutant cancers. However, in each of these cases, the hypermutation is often associated with an enhanced host cytotoxic T-cell immune response when compared to non-hypermutated tumours (van Gool et al., 2015). Plausibly, the larger number of mutations among these tumours results in a greater pool of mutated peptides with the ability to act as neoantigens—that is, novel-mutated peptides that are recognized as 'non-self' by the immune system. While the factors that determine antigenicity of these peptides are incompletely understood, predicted neoantigen burden has been shown to correlate with cytolytic immune activity across multiple tumour types, consistent with the speculation that this accounts for the apparently increased immunogenicity of these cancers (Rooney et al., 2015).

However, the association of hypermutation with enhanced immune response and better good prognosis is far from universal across cancers. While the 15–20% of endometrial cancers with mismatch repair deficiency also display an enhanced immune response, for reasons that are unclear they do not appear to have a better prognosis than mismatch repair proficient endometrial tumours (Caduff et al., 1996), in contrast to the situation in colorectal cancer. Furthermore, metastatic mismatch repair-deficient colorectal cancers have a very poor prognosis (Tran et al., 2011). While it is intriguing to speculate on the mechanisms that may account for this observation, they await explanation. While these examples illustrate the limitations of our knowledge of the relationship between hypermutation, immune response, and clinical outcome, it is nevertheless clear that the impact of genomic instability on prognosis must be considered in the context of the host immune system.

In addition to an effect on immunogenicity, genomic instability per se may alter tumour fitness and thus impact on prognosis. While mechanisms such as error catastrophe—a decrease in viability as a consequence of exceeding a mutation threshold—have not been proven to exist in human cancers, the upper limit of mutation observed among ultramutated *POLE*-mutant tumours suggests that this may apply. Similarly, studies in model organisms have demonstrated that while chromosomal instability is frequently observed in human cancers very high levels may decrease cellular fitness. Provocatively, correlative analysis of human cancers supports this hypothesis—while CIN overall is associated with a worse prognosis in colorectal cancer (Walther et al., 2008), very high levels of CIN correlate with better outcomes in multiple tumour types (**Fig. 4.4**; see also Birkbak et al., 2011).

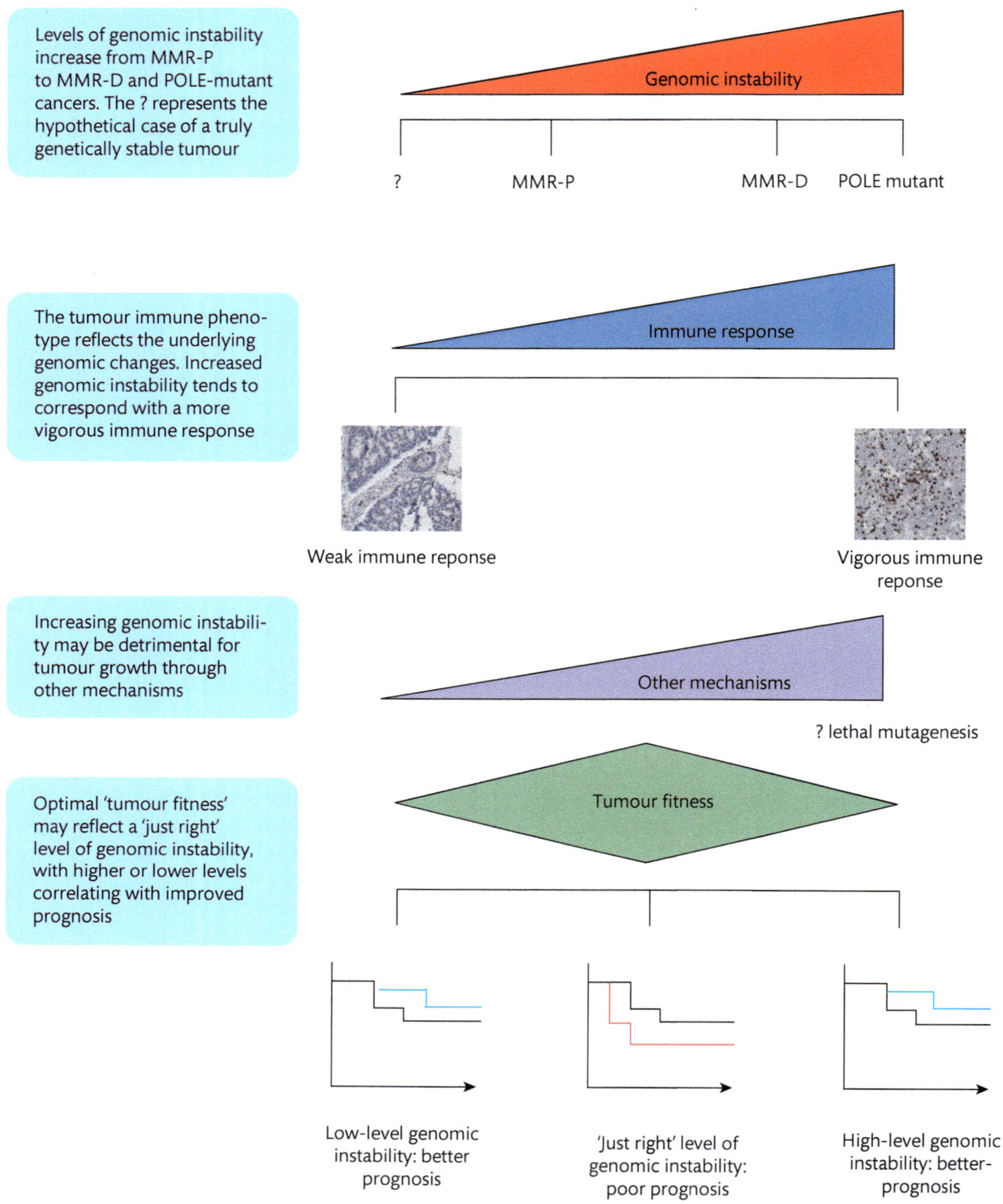

Fig. 4.4 Proposed relationship between genomic instability and cancer prognosis. Genomic instability is a hallmark of most cancers and may be a prerequisite for tumour initiation and progression. Genomic instability allows tumours to accumulate mutations beneficial for tumour growth, however high levels of genomic instability may reduce tumour fitness. Most of the evidence for this relates to the enhanced immune response against highly mutated mismatch repair deficient, and polymerase proofreading domain-mutant tumours, though other mechanisms, such as lethal mutagenesis, could potentially also reduce tumour fitness. Conceivably, a moderate, 'just right' level of genomic instability may be optimum for tumour fitness, resulting in a poor prognosis for patients with such tumours.

Targeting genomic instability for therapy

The frequency with which genetic instability occurs in cancers at the very least raises the possibility that it may confer a selective advantage during tumour development, and some have argued that a mutator phenotype is a defining characteristic of cancer. As discussed previously, in most cases tumour genetic instability is a consequence of a failure of a normal cellular failsafe that exists to protect against the deleterious consequences of mutation. While cancer cells harbouring such defects are clearly viable, a substantial and emerging body of evidence indicates that they may be particularly reliant on alternative pathways of DNA repair or other cellular processes in order to survive. Targeting these may result in selective lethality against genetically unstable tumour cells—a phenomenon known as synthetic lethality.

The best-characterized example of this it the use of PARP inhibitors against epithelial ovarian cancers with defective homologous recombination repair as a consequence of *BRCA1* or *BRCA2* mutations. The deficiency of HRR in these cells causes them to rely on alternative mechanisms to repair DNA damage, prominent among which is a pathway that relies on the enzyme polyadenylate (ADP ribose) polymerase (PARP). Several academic laboratories have developed inhibitors of PARP which vary both in their potency and selectivity. While these drugs exert little effect in cells with wild-type *BRCA* genes and normal HRR, in cells with deficient HRR they are able to promote apoptosis as a result of a failure of DNA repair (Farmer et al., 2005). Several clinical trials have shown that PARP inhibitors induce antitumour responses and improve survival in advanced ovarian cancer with *BRCA* mutations (Audeh et al., 2010). Strikingly, a common mechanism of resistance to these drugs is restoration of HRR by secondary mutations that restore *BRCA1* or *BRCA2* function and HRR, presumably enabling them to escape the synthetic lethality (Lord and Ashworth, 2016).

The possibility of synthetic lethality has also been explored in the context of mismatch repair deficiency. In an exciting preclinical study, Martin and colleagues performed a short interfering RNA (siRNA) screen to evaluate sensitivity of MMR-D cells to loss of components of the DNA replication and repair machinery. They found that MMR-D cells were particularly sensitive to loss of the repair polymerase Pol beta, leading to a model in which defective MMR leaves cells reliant on this enzyme for continued viability. This strategy has not yet been tested in the clinic (Martin et al., 2010).

An alternative therapeutic strategy is based on exploiting the elevated mutation rate in hypermutated tumours. The apparent upper limit to mutation burden evident from sequencing studies is consistent with the prediction that there is an upper limit to mutation rate in tumour cells, which if exceeded results in loss of viability due to the likelihood that the cell will acquire a deleterious mutation in an essential gene with each division. Consequently, tumours with an elevated mutation rate may be sensitive to agents that further increase this—a strategy referred to as lethal mutagenesis. While this has not been tested in tumours, it is believed to account for the activity of several types of antiviral agents against HIV (Loeb et al., 1999). Similarly, the data presented previously suggest that a similar phenomenon may operate for chromosomal instability. Therefore, targeting the spindle assembly checkpoint in CIN high tumours may further increase instability, resulting in lethality.

A note of caution is required, as inhibiting DNA repair and driving mutagenesis or chromosomal instability could have potentially serious consequences on normal cells. Thus such approaches are likely to be confined to patients with incurable malignancies until such time as they have been conclusively shown to be safe for normal cells. Nevertheless, the results of such clinical trials are eagerly awaited.

Conclusions and future directions

Cancer is a disease of disordered genomes, and it is generally accepted that many, if not most tumours display some form of genetic instability. Next generation sequencing technologies have provided an unprecedented ability to identify the causes and define the consequences of genetic instability in common cancers. This has revealed novel tumour subgroups with distinctive molecular and clinical characteristics, such as ultramutated colorectal and endometrial cancers caused by *POLE* exonuclease domain mutation. The extension of such approaches to study tumours with forms of genetic instability other than point mutations is likely to provide similar advances, as illustrated by the identification of a subgroup of pancreatic cancers that display evidence of HRR defects as a consequence of *BRCA* mutations. It is increasingly evident that many forms of genomic instability may influence prognosis, either favourably or unfavourably, and that the direction of effect may reflect the type and severity of instability. Novel therapeutic strategies against *BRCA1* or *BRCA2*-mutant tumours have demonstrated activity in clinical trials, and those against mismatch repair-deficient tumours have shown promise in preclinical models.

Theodor Boveri was undoubtedly far ahead of his time when he postulated that cancer is a disease of disordered chromosomes. We can only speculate on what Boveri might make of the multiple mechanisms that we now know to cause genetic instability in cancer, and on what future research will discover.

TAKE-HOME MESSAGE

- Cancer is a disease of an aberrant genome.
- Genomic integrity is maintained by several sophisticated repair systems; defects in these systems, either inherited or acquired, can lead to genomic instability and carcinogenesis.
- Both the type and degree of genomic instability can influence the clinical characteristics and prognosis of cancers.
- Targeting genomic instability may represent a promising novel therapeutic strategy against human tumours.

OPEN QUESTIONS

- Does the concept of inducing lethal mutagenesis represent a viable treatment strategy?
- How can assessment of genomic instability be integrated into clinical practice as a prognostic marker or a therapeutic target?
- In addition to the effects on the host immune response, by what other mechanisms does tumour genomic instability determine clinical outcome?
- Can lessons from the management of familial genomic instability syndromes can be applied to somatic tumours with these defects?

FURTHER READING

Lord, C. J., & Ashworth, A. (2012). The DNA damage response and cancer therapy. *Nature*, **481**, 287–94.

O'Connor, M. J. (2015). Targeting the DNA damage response in cancer. *Mol Cell*, **60**(4), 547–60.

Roberts, S. A., & Gordenin, D. A. (2014). Hypermutation in human cancer genomes: footprints and mechanisms. *Nature Reviews Cancer*, **14**, 786–800.

Tubbs, A., & Nussenzweig, A. (2017). Endogenous DNA damage as a source of genomic instability in cancer. *Cell*, **168**, 644–56.

REFERENCES

Al-Tassan, N., Chmiel, N. H., Maynard, J., et al. (2002). Inherited variants of MYH associated with somatic G:C-->T:A mutations in colorectal tumors. *Nat Genet*, **30**, 227–32.

Alexandrov, L. B., Nik-Zainal, S., Wedge, D. C., et al. (2013). Signatures of mutational processes in human cancer. *Nature*, **500**, 415–21.

Audeh, M. W., Carmichael, J., Penson, R. T., et al. (2010). Oral poly(ADP-ribose) polymerase inhibitor olaparib in patients with *BRCA1* or *BRCA2* mutations and recurrent ovarian cancer: a proof-of-concept trial. *Lancet*, **376**, 245–51.

Balbas-Martinez, C., Sagrera, A., Carrillo-De-Santa-Pau, E., et al. (2013). Recurrent inactivation of STAG2 in bladder cancer is not associated with aneuploidy. *Nat Genet*, **45**, 1464–9.

Beger, C., Pierce, L. N., Kruger, M., et al. (2001). Identification of Id4 as a regulator of *BRCA1* expression by using a ribozyme-library-based inverse genomics approach. *Proc Natl Acad Sci U S A*, **98**, 130–5.

Birkbak, N. J., Eklund, A. C., Li, Q., et al. (2011). Paradoxical relationship between chromosomal instability and survival outcome in cancer. *Cancer Res*, **71**, 3447–52.

Budd, M. E. & Campbell, J. L. (1993). DNA polymerases delta and epsilon are required for chromosomal replication in Saccharomyces cerevisiae. *Mol Cell Biol*, **13**, 496–505.

Caduff, R. F., Johnston, C. M., Svoboda-Newman, S. M., Poy, E. L., Merajver, S. D., & Frank, T. S. (1996). Clinical and pathological significance of microsatellite instability in sporadic endometrial carcinoma. *Am J Pathol*, **148**, 1671–8.

Cancer Genome Atlas Network (TCGA) (2012). Comprehensive molecular characterization of human colon and rectal cancer. *Nature*, **487**, 330–7.

Cancer Genome Atlas Research Network, Kandoth, C., Schultz, N., et al. (2013). Integrated genomic characterization of endometrial carcinoma. *Nature*, **497**, 67–73.

Chen, S. & Parmigiani, G. (2007). Meta-analysis of *BRCA1* and *BRCA2* penetrance. *J Clin Oncol*, **25**, 1329–33.

Cleaver, J. E. (2005). Cancer in xeroderma pigmentosum and related disorders of DNA repair. *Nat Rev Cancer*, **5**, 564–73.

Cleveland, A. J. H. & Don, W. (2009). Boveri revisited: chromosomal instability, aneuploidy and tumorigenesis. *Nature Reviews Molecular Cell Biology*, **10**, 478–87.

Di Noia, J. M. & Neuberger, M. S. (2007). Molecular mechanisms of antibody somatic hypermutation. *Annu Rev Biochem*, **76**, 1–22.

Domingo, E., Laiho, P., Ollikainen, M., et al. 2004. BRAF screening as a low-cost effective strategy for simplifying HNPCC genetic testing. *J Med Genet*, **41**, 664–8.

Farmer, H., McCabe, N., Lord, C. J., et al. (2005). Targeting the DNA repair defect in BRCA mutant cells as a therapeutic strategy. *Nature*, **434**, 917–21.

Gaillard, H., Garcia-Muse, T., & Aguilera, A. (2015). Replication stress and cancer. *Nat Rev Cancer*, **15**, 276–89.

Garcia, A. I., Buisson, M., Bertrand, P., et al. (2011). Down-regulation of *BRCA1* expression by miR-146a and miR-146b-5p in triple negative sporadic breast cancers. *EMBO Mol Med*, **3**, 279–90.

Hanahan, D. & Weinberg, R. A. (2011). Hallmarks of cancer: the next generation. *Cell*, **144**(5), 646–74.

Harris, R. S. & Liddament, M. T. (2004). Retroviral restriction by APOBEC proteins. *Nat Rev Immunol*, **4**, 868–77.

Harris, R. S., Petersen-Mahrt, S. K., & Neuberger, M. S. (2002). RNA editing enzyme APOBEC1 and some of its homologs can act as DNA mutators. *Mol Cell*, **10**, 1247–53.

Helleday, T., Eshtad, S., & Nik-Zainal, S. (2014). Mechanisms underlying mutational signatures in human cancers. *Nat Rev Genet*, **15**, 585–98.

Henderson, S., Chakravarthy, A., Su, X., Boshoff, C., & Fenton, T. R. (2014). APOBEC-mediated cytosine deamination links PIK3CA helical domain mutations to human papillomavirus-driven tumor development. *Cell Rep*, **7**, 1833–41.

Hughes-Davies, L., Huntsman, D., Ruas, M., et al. (2003). EMSY links the BRCA2 pathway to sporadic breast and ovarian cancer. *Cell*, **115**, 523–35.

Jiricny, J. (2006). The multifaceted mismatch-repair system. *Nat Rev Mol Cell Biol*, **7**, 335–46.

Kastrinos, F. & Syngal, S. (2007). Recently identified colon cancer predispositions: MYH and MSH6 mutations. *Semin Oncol*, **34**, 418–24.

Kraemer, K. H., Lee, M. M., & Scotto, J. (1987). Xeroderma pigmentosum. Cutaneous, ocular, and neurologic abnormalities in 830 published cases. *Arch Dermatol*, **123**, 241–50.

Leonard, B., Starrett, G. J., Maurer, M. J., et al. (2016). APOBEC3G expression correlates with T cell infiltration and improved clinical outcomes in high-grade serous ovarian carcinoma. *Clin Cancer Res*, **22**(18), 4746–55.

Llosa, N. J., Cruise, M., Tam, A., et al. (2015). The vigorous immune microenvironment of microsatellite instable colon cancer is balanced by multiple counter-inhibitory checkpoints. *Cancer Discov*, **5**, 43–51.

Loeb, L. A., Essigmann, J. M., Kazazi, F., Zhang, J., Rose, K. D., & Mullins, J. I. (1999). Lethal mutagenesis of HIV with mutagenic nucleoside analogs. *Proc Natl Acad Sci U S A*, **96**(4), 1492–7.

Lord, C. J. & Ashworth, A. (2016). *BRCA*ness revisited. *Nat Rev Cancer*, **16**, 110–20.

Lynch, H. T., Snyder, C. L., Shaw, T. G., Heinen, C. D., & Hitchins, M. P. (2015). Milestones of Lynch syndrome: 1895–2015. *Nat Rev Cancer*, **15**, 181–94.

Martin, S. A., McCabe, N., Mullarkey, M., et al. (2010). DNA polymerases as potential therapeutic targets for cancers deficient in the DNA mismatch repair proteins MSH2 or MLH1. *Cancer Cell*, **17**, 235–48.

Meng, B., Hoang, L. N., McIntyre, J. B., et al. (2014). POLE exonuclease domain mutation predicts long progression-free survival in grade 3 endometrioid carcinoma of the endometrium. *Gynecol Oncol*, **134**, 15–19.

Michel, L. S., Liberal, V., Chatterjee, A., et al. (2001). MAD2 haplo-insufficiency causes premature anaphase and chromosome instability in mammalian cells. *Nature*, **409**, 355–9.

Moser, J., Kool, H., Giakzidis, I., Caldecott, K., Mullenders, L. H., & Fousteri, M. I. (2007). Sealing of chromosomal DNA nicks during nucleotide excision repair requires XRCC1 and DNA ligase III alpha in a cell-cycle-specific manner. *Mol Cell*, **27**, 311–23.

Moynahan, M. E., Chiu, J. W., Koller, B. H., & Jasin, M. (1999). Brca1 controls homology-directed DNA repair. *Mol Cell*, **4**, 511–18.

Moynahan, M. E., Pierce, A. J., & Jasin, M. (2001). *BRCA2* is required for homology-directed repair of chromosomal breaks. *Mol Cell*, **7**, 263–72.

Mu, D., Hsu, D. S., & Sancar, A. (1996). Reaction mechanism of human DNA repair excision nuclease. *J Biol Chem*, **271**, 8285–94.

Nick McElhinny, S. A., Gordenin, D. A., Stith, C. M., Burgers, P. M., & Kunkel, T. A. (2008). Division of labor at the eukaryotic replication fork. *Mol Cell*, **30**, 137–44.

Nik-Zainal, S., Alexandrov, L. B., Wedge, D. C., et al. (2012). Mutational processes molding the genomes of 21 breast cancers. *Cell*, **149**, 979–93.

Nouspikel, T. (2009). DNA repair in mammalian cells: nucleotide excision repair: variations on versatility. *Cell Mol Life Sci*, **66**, 994–1009.

O'Donovan, A., Davies, A. A., Moggs, J. G., West, S. C., & Wood, R. D. (1994). XPG endonuclease makes the 3' incision in human DNA nucleotide excision repair. *Nature*, **371**, 432–5.

Ohba, K., Ichiyama, K., Yajima, M., et al. (2014). *In vivo* and *in vitro* studies suggest a possible involvement of HPV infection in the early stage of breast carcinogenesis via APOBEC3B induction. *PLoS One*, **9**, e97787.

Palles, C., Cazier, J.-B., Howarth, K. M., et al. (2013). Germline mutations affecting the proofreading domains of POLE and POLD1 predispose to colorectal adenomas and carcinomas. *Nat Genet*, **45**, 136–44.

Popanda, O. & Thielmann, H. W. (1992). The function of DNA polymerases in DNA repair synthesis of ultraviolet-irradiated human fibroblasts. *Biochim Biophys Acta*, **1129**, 155–60.

Popat, S. (2004). Systematic review of microsatellite instability and colorectal cancer prognosis. *J Clin Oncol*, **23**, 609–18.

Rayner, E., Van Gool, I. C., Palles, C., et al. (2016). A panoply of errors: polymerase proofreading domain mutations in cancer. *Nat Rev Cancer*, **16**, 71–81.

Roach, J. C., Glusman, G., Smit, A. F., et al. (2010). Analysis of genetic inheritance in a family quartet by whole-genome sequencing. *Science*, **328**, 636–9.

Roberts, S. A., Lawrence, M. S., Klimczak, L. J., et al. (2013). An APOBEC cytidine deaminase mutagenesis pattern is widespread in human cancers. *Nat Genet*, **45**, 970–6.

Robertson, A. B., Klungland, A., Rognes, T., & Leiros, I. (2009). DNA repair in mammalian cells: base excision repair: the long and short of it. *Cell Mol Life Sci*, **66**, 981–93.

Rooney, M. S., Shukla, S. A., Wu, C. J., Getz, G., & Hacohen, N. (2015). Molecular and genetic properties of tumors associated with local immune cytolytic activity. *Cell*, **160**, 48–61.

Rowley, J. D. (1973). Letter: a new consistent chromosomal abnormality in chronic myelogenous leukaemia identified by quinacrine fluorescence and Giemsa staining. *Nature*, **243**, 290–3.

Shay, J. W. & Wright, W. E. (2011). Role of telomeres and telomerase in cancer. *Semin Cancer Biol*, **21**, 349–53.

Shevelev, I. V. & Hubscher, U. (2002). The 3' 5' exonucleases. *Nat Rev Mol Cell Biol*, **3**, 364–76.

Simon, M., Giot, L., & Faye, G. (1991). The 3' to 5' exonuclease activity located in the DNA polymerase delta subunit of Saccharomyces cerevisiae is required for accurate replication. *EMBO J*, **10**, 2165–70.

Stratton, M. R., Campbell, P. J., & Futreal, P. A. (2009). The cancer genome. *Nature*, **458**, 719–24.

Thompson, S. L., Bakhoum, S. F., & Compton, D. A. (2010). Mechanisms of chromosomal instability. *Curr Biol*, **20**, R285–95.

Tran, B., Kopetz, S., Tie, J., et al. (2011). Impact of *BRAF* mutation and microsatellite instability on the pattern of metastatic spread and prognosis in metastatic colorectal cancer. *Cancer*, **117**, 4623–32.

Turner, N., Tutt, A., & Ashworth, A. (2004). Hallmarks of 'BRCAness' in sporadic cancers. *Nat Rev Cancer*, **4**, 814–19.

Van Gool, I. C., Eggink, F. A., Freeman-Mills, L., et al. (2015). POLE proofreading mutations elicit an antitumor immune response in endometrial cancer. *Clin Cancer Res*, **21**, 3347–55.

Venkitaraman, A. R. (2014). Cancer suppression by the chromosome custodians, *BRCA1* and *BRCA2*. *Science*, **343**, 1470–5.

Walther, A., Houlston, R., & Tomlinson, I. (2008). Association between chromosomal instability and prognosis in colorectal cancer: a meta-analysis. *Gut*, **57**, 941–50.

Weisenberger, D. J., Siegmund, K. D., Campan, M., et al. (2006). CpG island methylator phenotype underlies sporadic microsatellite instability and is tightly associated with BRAF mutation in colorectal cancer. *Nat Genet*, **38**, 787–93.

Yamanaka, S., Balestra, M. E., Ferrell, L. D., et al. (1995). Apolipoprotein B mRNA-editing protein induces hepatocellular carcinoma and dysplasia in transgenic animals. *Proc Natl Acad Sci U S A*, **92**(18), 8483–7.

5

Epigenetics

Edward Hookway, Nicholas Athanasou, and Udo Oppermann

Introduction to epigenetics in cancer biology

The classical paradigm in cancer biology has been that a change in DNA leads to an alteration of the transcribed mRNA resulting in a change of function of a protein that contributes to the development of the cancer phenotype. The underlying change in the DNA can include point mutations that lead to either a gain of function in an oncogene, such as the V600E mutation in *BRAF* causing increased kinase activity in some melanomas (Kumar et al., 2004); deletions of DNA leading to the loss of tumour suppressor genes such as *TP53*; or chromosomal translocations resulting in the expression of fusion proteins that drive cancer development as in Ewing Sarcoma (Delattre et al., 1992).

Dysregulated expression of a normal protein without any underlying DNA mutation is implicated in cancer development and maintenance of cancer. A cell that inappropriately does not express a functioning tumour suppressor is equivalent to a cell expressing a mutant tumour suppressor which has lost its activity: in both examples the cells are lacking a functional protein.

Epigenetics is a broad term that encompasses many diverse processes that are important within the cell for regulating the level of expression of genes and controlling the phenotype of the cell. The term derives from the Greek 'epi' meaning 'above' or 'over' and signifies the position of epigenetics being 'above the genome' and in a position to control gene expression. A consensus definition of epigenetics developed at Cold Harbour Springs is a 'stable, heritable phenotype that is not caused by a change in the underlying DNA sequence' (Berger et al., 2009).

In this chapter we will explore the underlying molecular and structural biology behind epigenetic processes and discuss the implications of these mechanisms in cancer biology.

Epigenetics and development

With the exception of gamete cells produced by meiosis and immune cells that have undergone somatic recombination, every cell in the body contains the same DNA sequence from the moment of conception until death. Despite containing the same DNA 'blueprint', an individual eukaryotic organism displays many different cell types with different transcriptional profiles over its lifespan. Recent advances in sequencing technology have allowed the transcriptomes of individual cells to be accurately recorded and this has demonstrated that at any given time only a small proportion of the genome is being actively transcribed (Djebali and Davis, 2012). Therefore, it is clear that there exist mechanisms which tightly control gene expression both in a cell-type specific and a temporal manner.

Many insights into this process of gene expression control have been gained from the study of embryogenesis and developmental biology. All cells are derived from the zygote and within five days the undifferentiated totipotent cells of the morula give rise to distinct cell populations in the blastocyst of the trophoblast and pluripotent cells of the inner cell mass and trophoblast (De Paepe et al., 2014). As embryogenesis progresses, cells becoming increasingly differentiated and specialized, losing their ability to differentiated into any cell type. Multipotent stem cells, such as the mesenchymal stem cells, can differentiate into a limited number of related cell types (Kokabu et al., 2016), whereas terminally differentiated cells have lost the ability to change into any other cell type under normal physiological conditions.

The developmental biologist C. H. Waddington developed the metaphor of the 'epigenetic landscape' to describe the process of cell differentiation (Waddington, 1942). He imagined a hill with many valleys, representing different epigenetic changes, with a ball at the top representing an undifferentiated cell in its totipotent state. As the ball rolls down the hill into a valley, it loses the ability to cross over into other valleys, indicating the increasing specialization and differentiation of the cell, until it finally comes to rest at the bottom in a terminally differentiated state. Each differentiated state is intricately associated with a different 'landscape', as the nature of the landscape prevents the ball from transitioning into a different valley, just as a cell that has terminally differentiated into a hepatocyte does not spontaneously become a neuron.

One feature associated with the development of cancer is the loss of cellular differentiation where cancer cells lose phenotypic features associated with their cell of origin. Using Waddington's analogy, the epigenetic landscape has changed so the ball is able to escape from its terminally differentiated state at the bottom of a valley. The change in the epigenetic landscape can be one of the causes of the development of the cancer, as in the case of NUT-midline carcinoma (see the 'Histone acetylation' section) or can be secondary to other driver mutations within the cell. The ability to target epigenetic processes, to change the transcriptomic profile of a cancer cell and cause inappropriately silenced tumour suppressor genes to become re-expressed or decrease the expression of oncogenes, offers a mechanism for treating cancer cells by targeting the underlying molecular mechanisms that have led to cancer development instead of relying on cytotoxic chemotherapy (Azad et al., 2013).

Molecular mechanisms of epigenetic regulation

The diverse range of epigenetic mechanisms can be split into two broad categories: firstly, mechanisms involving modification to the structure of DNA and chromatin and, secondly, mechanisms involving the expression of regulatory RNA sequences. It is important to appreciate that any individual epigenetic modification cannot truly be considered in isolation as there exists a complex interplay between the different mechanisms (Vaissière et al., 2008; Murr, 2010). While some of the interactions have been documented, it is likely that in this emerging field of cancer biology there are further interactions that have yet to be discovered, let alone fully understand.

DNA structure and organization

Many mechanisms of epigenetic regulation are dependent on the structure of DNA and how it is organized within the cell nucleus. The image of DNA made famous by Watson and Crick is of a double helix formed from two antiparallel strands of DNA (Watson and Crick, 1953). Each strand contains a backbone formed from alternating a phosphate group with a deoxyribose sugar and also attached to the deoxyribose is one of the four nitrogen-containing bases: adenine (A), thymine (T), guanine (G), and cytosine (C). Between the two strands of DNA, bases form non-covalent hydrogen bonds with their partner such that adenine binds thymine and guanine binds cytosine: this is referred to as Chargaff's rules (Chargaff et al., 1952). Each turn of the helix comprises 10 base pairs with the distance between each pair being 3.4 Angstrom (equal to 0.34 nm or 3.4×10^{-10}m). While this may seem at first glance to be a very small distance, when it is considered that the length of the human genome is approximately 3 billion base pairs long and somatic cells contain two copies of the genome, this results in a total length of DNA per cell of several metres. An estimate of the number of human cells in the 'average' human body of 40 trillion cells (4×10^{13}; Bianconi et al., 2013) therefore gives a total length of DNA equivalent to the distance from the sun to Pluto and back again. Clearly a mechanism is needed for efficiently packing this amount of DNA in a cell nucleus.

Double-stranded DNA interacts with proteins and RNA within the cell nucleus to form chromatin, a more condensed and structured molecule that allows the long lengths of DNA be packaged within the cell. One of the core components of chromatin are the histone proteins. There are five major families of histone protein and these are highly conserved through evolution. Heterodimers are formed between H2A and H2B, and H3 and H4, and then two each of these dimers join together to form an octamer consisting of two copies each of H2A, H2B, H3, and H4. This histone octamer forms a core region, around which 147 base pairs of DNA wrap around 1.67 times forming a nucleosome. In-between adjacent beads on the string is a section of linker DNA approximately 80 bp long. The wrapping of DNA around the core of histone proteins has been referred to as having a 'bead on a string' appearance when viewed under an electron microscope with a diameter of approximately 10 nm.

Histone 1 does not form part of the core complex of histones around which DNA wraps but interacts with the DNA wrapped around the nucleosome and in the linker region to help stabilize the structure and form an even more compact structure where multiple histones wrap around each other to form a 30 nm-wide fibre (Orthaus et al., 2009). During cell division, further protein interaction occurs leading to even further condensation of the chromatin, resulting in the appearance of chromosomes as seen during metaphase.

The packaging of DNA and the structure of chromatin has important implications for gene expression. When the chromatin is in the more open beads on a string configuration, referred to as euchromatin, transcription factors and RNA polymerases can bind to the DNA and therefore euchromatin is normally observed in areas of the genome that are actively transcribed. Chromatin that is in the more densely packed 30 nm fibre conformation, referred to as heterochromatin, is less able to bind the protein machinery necessary for transcription and is therefore associated with areas not associated with active gene transcription (Fig. 5.1). Areas of heterochromatin can either be constitutive or facultative (Saksouk et al., 2015).

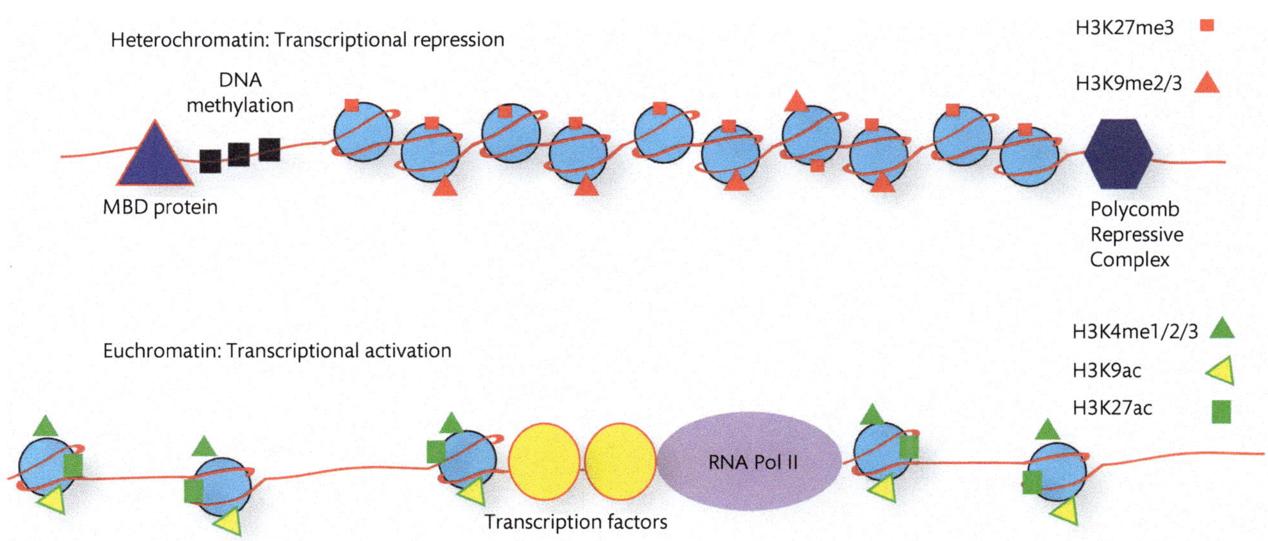

Fig. 5.1 Heterochromatin and euchromatin: post-translational modification of histone tails leads to changes in the packing of histones (blue) and the accessibility of DNA to transcriptional machinery. In heterochromatin, which is associated with transcriptional inhibition, histones are tightly packed, preventing transcription. Heterochromatin is associated with repressive histone marks such as H3K27me3 and H3K9me2/3, DNA methylation, and the binding of methyl-CpG-binding proteins (MBD) and the polycomb repressive complex. In contrast, euchromatin has an open structure allowing binding of transcriptional machinery such as DNA polymerases and transcription factors and is associated with activating histone marks such as H3K4me3 and H3K27ac.

Constitutive heterochromatin is associated with areas of the genome that are not usually transcribed such as regions around centromeres and telomere or highly repetitive areas such as microsatellites or tandem repeats. Facultative heterochromatin is found in areas of the genome where transcription of the underlying DNA occurs in a cell-type specific manner, meaning it may be transcribed at some points in development in certain cell types but not in others.

Modulating the structure of chromatin from the open euchromatin to closed heterochromatin therefore offers a mechanism of controlling the level of gene expression and an explanation for gene silencing: a cell may express the necessary transcription factors for the expression of a particular gene but if it is located in an area of heterochromatin it may not be actively transcribed.

Histone modification and the 'histone code'

Post-translation modification of histone proteins is one of the major mechanisms of epigenetic regulation (Kouzarides, 2007; Patel and Wang, 2013). Each of the histone proteins that participates in the formation of the hetero-octamer contains a globular C-terminal domain about which the DNA wraps itself but also contain an N-terminal tail that protrudes outwards from the core. Each of the N-terminal tails contains multiple amino acids that can be subjected to a range of post-translational modifications and enzymes have been identified capable of methylation (Zhang and Reinberg, 2001), acetylation (Sterner and Berger, 2000), phosphorylation (Nowak and Corces, 2004), sumoylation (Nathan et al., 2006), poly-ADP ribosylation (Hassa et al., 2006), deamination (Cuthbert et al., 2004), and ubiquitination (Gutiérrez et al., 2012; Fig. 5.2).

To add to the complexity, it is possible for multiple different post-translational modifications to occur on any given histone tail and some modifications, such a methylation can occur up to three times on a lysine residue. The multiple permutations of different modifications which can act to either enhance or repress transcription has given rise to what is referred to as the 'histone code' (Jenuwein and Allis, 2001). It is also important to remember that there are multiple interactions between different types of epigenetic modification: repressive histone marks are often associated with DNA methylation, which also act to repress transcription (Rose and Klose, 2014).

An example of the importance of histone tails in tumour biology has been demonstrated in chondroblastoma, a rare tumour that occurs in the epiphysis of long bones. A mutation was found in 95% of cases coding for the same residue in the tail of a subtype of histone 3 (Behjati et al., 2013). When referring to histone modification, it is conventional to refer to the histone number first, the single letter amino acid code of the residue being modified second, the position of the residue third, and then finally any modification. The mutation resulted in the change on histone 3 lysine 36 (H3K36), which can normally be post-translationally modified by methylation. While it is unclear of the exact mechanistic implications of this mutation, its near universal prevalence in a rare tumour type suggests it must have important downstream implications.

Proteins which are involved in the 'histone code' can be conveniently divided into three categories: 'writers', which are responsible for adding marks to histones; 'erasers', responsible for removing marks; and 'readers' that recognize the presence of modification and help recruit other proteins to that location.

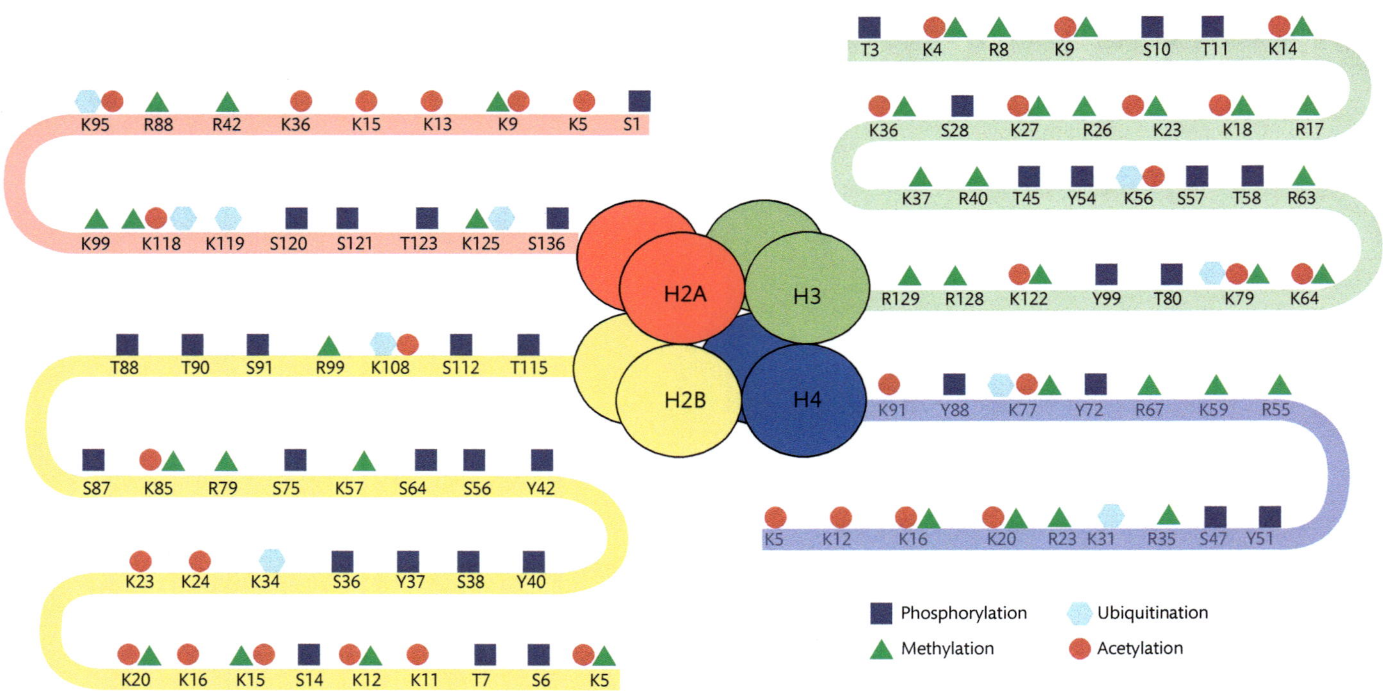

Fig. 5.2 Histone modifications: histone tails can undergo many post-translation modifications including phosphorylation, ubiquitination, methylation, and acetylation, the common sites of which are shown. It is possible for an individual histone to carry several different modifications increasing the complexity of the 'histone code'. In addition to the modifications shown, many other modifications have been observed including citrullination, succinylation, hydroxylation, and formylation.

Histone acetylation

One of the earliest post-translational modification of histones discovered was acetylation (Allfrey and Mirsky, 1964). Acetyl marks are added to the positively charged lysine residues in histone tails by histone acetyltransferases (HATs) using acetyl-coenzyme A as a substrate. The negatively charged phosphate backbone of DNA normally interacts with the positively charged lysine residues. Acetylation removes the positively charged group from the lysine, reducing the strength of the interaction between the DNA and the nucleosome and therefore makes the DNA more accessible for transcriptional machinery to bind. Histone acetylation is therefore commonly associated with areas of active gene transcription (Sterner and Berger, 2000).

There are more than 30 enzymes in humans with histone acetyltransferase activity and they can be categorized either on the basis of their subcellular localization or overall protein structure and function of the catalytic domain (Wapenaar and Dekker, 2016). Located in the nucleus, Type A HATs are responsible for interaction with chromatin whereas Type B HATs are located within the cytoplasm and interact with newly synthesized histone molecules before they become integrated into chromatin (Lee and Workman, 2007). Residues in all four of the core histones can be acetylated but transcriptional activation is particularly associated with acetylation of H3K9 and H3K14 (Karmodiya et al., 2012). Many HATs can be considered to be promiscuous in that they have a wide range of targets which will be acetylated both within histones and in other proteins (Friedmann and Marmorstein, 2013). It may therefore be more appropriate to refer to these proteins as 'lysine acetyltransferases'.

Histone deacetylases (HDACs) are responsible for the removal of acetyl marks from lysine residues on histones and a range of proteins (Delcuve et al., 2012). There are approximately 20 HDACs found in humans and can be separated into those which use zinc as a cofactor, the so-called HDACs, and those that use NAD+, the sirtuins (Carafa et al., 2016). Deacetylation of histones is associated with an increased interaction between DNA and the histone and is therefore associated with transcriptional repression (Sterner and Berger, 2000). An example of the interaction between different epigenetic mechanisms is the interplay between DNA methylation and histone deacetylation, both repressive marks, via the methyl-CpG binding domain (MBD) protein family (discussed in 'DNA base modification') which are able to recruit HDACs (Saito and Ishikawa, 2002).

A number of HDAC inhibitors (HDACi) have been licensed for clinical use against a range of cancers. Cutaneous T-cell lymphoma was the first cancer for which an HDACi, vorinostat, was licensed (Mann et al., 2007) and the HDACi panobinostat is now licensed for multiple myeloma in Europe (Laubach et al., 2015). Clinical trials are underway in a range of haematological malignancies and solid tumours (West and Johnstone, 2014).

Bromodomains are 'readers' of acetylated lysine residues and found in over 40 proteins in humans including HATs and HDACs (Filippakopoulos and Knapp, 2014). An important example in cancer biology of a bromodomain-containing protein is involved in the rare NUT-midline carcinoma (NMC), an aggressive cancer with poor prognosis. In NMC, a chromosomal translocation occurs creating a fusion protein between the nuclear protein in testis (NUT) gene on chromosome 15 with the bromodomain-containing protein 4 (BRD4) on chromosome 19 (French et al., 2004; Fig. 5.3). The fusion protein is oncogenic and normally found in isolation from other mutations. BRD4 contains two bromodomains and can interact with acetylated lysine in the promoter region of genes, recruiting positive-transcription elongation factor b (P-TEFb) and leading to transcriptional activation. BRD4 also binds extensively to enhancer and super enhancer regions of DNA, which can lead to the upregulation of other oncogenic genes. The use of a small molecule inhibitor JQ1, which disrupts the ability of BRD4 and other members of the BET bromodomain family from binding to acetylated lysines, has proven very effective in animal models at inhibiting the growth of cancer cells containing the NUT-BRD4 fusion (Filippakopoulos et al., 2010). Inhibition of BRD4 has also been shown to decrease the expression of MYC which is often overexpressed in cancers, leading to the possibility that inhibitors of bromodomain-containing proteins may be more generally useful in the treatment of cancers (Grayson et al., 2014).

BRD4 also plays an important role in epigenetics due to its function as a 'mitotic bookmark' (Dey et al., 2000). During mitosis, many proteins normally associated with chromatin are removed but BRD4 remains bound to H4K5ac around the site of genes that were being actively transcribed immediately prior to mitosis (Zhao et al., 2011). BRD4 also recruits transcription machinery to these bookmarked sites after mitosis by binding positive translation elongation factor b (P-TEFb), which is required for phosphorylation of RNA polymerase and activation of transcription (Jang et al., 2005). Recently, it has been demonstrated that in addition to having the function of an epigenetic reader of acetylated lysine residues, BRD4 also contains histone acetyltransferase activity (Devaiah et al., 2016).

Histone lysine methylation

In addition to acetylation, lysine residues in histone tails can also be methylated. Lysine methyltransferases (KMT) are enzymes that add methyl marks to the ε-nitrogen of lysine residues using S-adenosyl methionine (SAM) as the methyl donor. Lysine demethylases (KDM) remove the methyl marks from lysine residues. Methylated lysines can exist in three different methylation states which have been associated with different functions: *mono*-methyl (me1), *di*-methyl (me2), and *tri*-methyl (me3).

Histone lysine methylation plays an important role in development and the function of many of the proteins involved were first elucidated by studying embryogenesis and development in the fruit fly Drosophila (Black et al., 2012). Proteins belonging to the trithorax group are associated with activation gene transcription (Schuettengruber et al., 2011) whereas proteins in the Polycomb group (PcG) are associated with transcriptional repression (Steffen and Ringrose, 2014).

PcG proteins join together to form multimeric polycomb repressive complex 1 and 2 (PRC1/2) that have different actions but work together to cause gene silencing. PRC2 contains either *enhancer of Zeste 1 or 2* (EZH1/2) a lysine methyltransferase that is specifically able to add methyl groups to H3K27, forming the repressive histone mark H3K27me3. EZH2 overexpression has been detected in a wide range of malignancies where it causes dysregulation of H3K27me3 and therefore alters gene expression (Yamaguchi and Hung, 2014). Mutations in two other proteins that form part of PRC2, SUZ12, and EED result in deficiency of H3K27me3 and are associated with malignant peripheral nerve sheath tumours (Lee et al., 2014). PRC1 is a multimeric protein complex that can be composed of several different subunits (Schwartz and Pirrotta, 2013). The chromobox

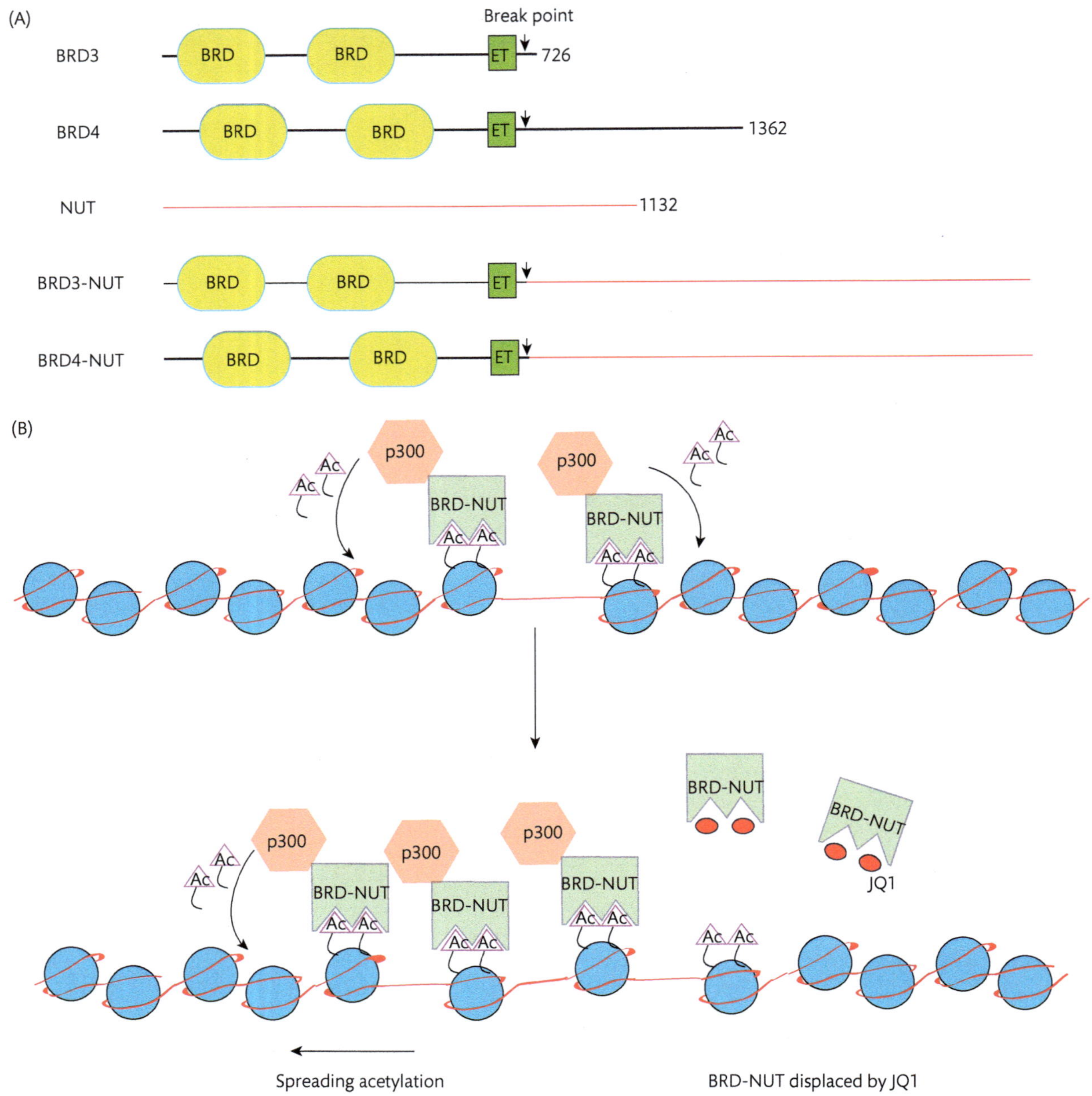

Fig. 5.3 NUT midline carcinoma: (A) schematic representation of BRD3, BRD4 and NUT protein demonstrating the bromodomains (BRD) and extraterminal (ET) domains and the breakpoint for formation of the fusion transcript. (B) Spreading acetylation caused by the BRD-NUT fusion proteins. Bromodomains recognize and bind to acetylated histones. The histone acetyltransferase p300 is recruited through an interaction with the NUT portion of the fusion protein, resulting in increased histone acetylation and therefore more binding of BRD-NUT protein and spreading of acetylation. This process can be inhibited by the small molecule inhibitor JQ1 by interfering with the binding of the bromodomains in the BRD-NUT fusion to acetylated histones.

proteins (CBX) represent a group of evolutionary conserved proteins that are associated with heterochromatin and hence transcriptional repression and function as a reader domain for H3K27me3 and the repressive mark H3K9me3. Several closely related CBX proteins (CBX2/5/7/8) can be part of the PRC1 complex, and changes in the expression of CBX proteins has been shown to play an important role in differentiation (Ren et al., 2008). The role of different CBX proteins in cancer is an active area of research. CBX7 downregulation has been associated with poor prognosis in various cancers, including anaplastic thyroid cancer (Pallante et al., 2008), but overexpression of CBX7 has been documented in lymphomas (Scott et al., 2007).

Also in the PRC1 complex is ring finger protein 1A or 1B (RING1A/1B), which acts to ubiquitinylate H2AK119, a modification that leads to further compaction of chromatin and therefore transcriptional repression (Wang et al., 2004; Cao et al., 2005).

The method by which polycomb complexes are targeted to the genome in humans remains under active study. The canonical model involves PRC2 initiating the repression by methylating H3K27 at an unmethylated CpG island with subsequent recruitment of PRC1 via its chromobox to maintain the repression by ubiquitinylating H2AK119 (Di Croce and Helin, 2013). In *Drosophilla*, PRC2 is targeted to PRC response elements in DNA but due to the larger complexity of PRCs in humans, the mechanism is not so clear (Frey et al., 2016).

One mechanism by which PRC2 can be recruited is via interacting with non-coding RNAs. An example of ncRNA recruiting PRC2 occurs in X-chromosome inactivation. X-inactivation specific transcript (XIST) is transcribed from the X-inactivation centre of the chromosome that is to be silenced producing an RNA that is capped, spliced, and polyadenylated but not translated. XIST diffuses down the length of the X chromosome where it interacts with the SUZ12 component of PRC2, recruiting the complex and causing H3K27me3 to be formed via the action of EZH1/2 (Hernández-Muñoz et al., 2005).

A further example of ncRNA being implicated in the recruitment of PRC2 is that of HOTAIR, a 2 kb non-coding RNA transcribed from chromosome 12 that has been shown to be involved in transcriptional silencing of genes on other chromosomes through the recruitment of PRC2 (Gupta et al., 2010). HOTAIR has been shown to be overexpressed in a wide range of solid tumours including gliomas, urothelial tumours, and breast cancers (Wu et al., 2014).

An alternative explanation for the recruitment of PRC to DNA differs from the canonical explanation by having a modified PRC1 responsible for the initial event. PRC1 containing KDM2B recognizes unmethylated CpG island and using the RIN1A/B subunit of PRC1 causes ubiquitination of H2AK119. PRC2 is then recruited to this site of the ubiquitination through the action of AE binding protein 2 (AEBP2), which interacts with several members of the PRC2 family leading to trimethylation of H3K27 (Wu et al., 2013).

The importance of H3K27 in controlling gene expression is observed in glioblastoma multiforme where the mutation from a lysine to a methionine at position 27 in H3F3A is common (Schwartzentruber et al., 2012; Sturm et al., 2012). This mutation has been associated with decreased levels of H3K27me3 globally and aberrant gene expression (Chan et al., 2013; Lewis et al., 2013). The same mutation has subsequently been identified in a range of central nervous system tumours including high-grade astrocytomas (Shankar et al., 2016) and midline gliomas (Aihara et al., 2014).

Counteracting the polycomb group proteins are members of the trithorax group associated with activation of gene expression via histone modification by methylation of H3K4, H3K36, and ATP-dependent chromatin remodelling. H3K4me3 marks are found around the transcription start site of genes that are being actively transcribed. Proteins of the trithorax group form multisubunit complexes comprising proteins with a range of functions including 'writer', 'eraser', and 'reader' domains and these are largely conserved in eukaryotic organisms (Schuettengruber et al., 2011). The trithorax complex contains a protein with methyltransferase activity of either the SET1-family or mixed lineage leukaemia (MLL) family in complex with ASH2-like histone lysine methyltransferase complex subunit (ASH2L), WD repeat domain 5 (WDR5), retinoblastoma binding protein 5 (RBBP5), and DPY-30. A key component of the trithorax group of proteins is the MLL family of genes that contain methyltransferase activity with the ability to mono-, di-, and trimethylated H3K4. MLL genes are often involved in chromosomal translocations in haematological malignancies (DiMartino and Cleary, 1999). ASH2L is known to modulate the methyltransferase activity towards H3K4 of the trithorax complex and has been found to be overexpressed in certain oestrogen receptor positive breast cancers (Qi et al., 2014). WDR5 is important in the formation of multimeric protein complexes and its overexpression in bladder cancers has been associated with increased H3K4me3 (Chen et al., 2015). Retinoblastoma (Rb) binding protein 5 (RBBP5) binds Rb, a key regulator of cell cycle progression. Overexpression of DPY-30 has been implicated in increasing proliferation in gastric cancer cells (Lee et al., 2015).

Other than the core components of the trithorax complex, additional proteins have been noted to bind to certain forms of the complex in different cell types. For example, the histone demethylase UTX which is able to demethylate the repressive H3K27me3 mark has been found in complexes with MLL2/3 (Schuettengruber et al., 2011). This complex will therefore have the dual activity of increasing the permissive mark H3K4me3 and reducing H3K27me3. DOT1L is a histone methyltransferase that is responsible for methylation of H3K79 and interacts with MLL-AF9 fusion proteins associated with the development of leukaemia (Nguyen et al., 2011).

Some gene loci are marked by both H3K27me3 and H3K4me3—so-called bivalent or poised promoters. Bivalent promoters are often found associated with genes during development that are required to be rapidly expressed when required but remain repressed until then (Voigt et al., 2013). It has been suggested that loss of the repressive H3K27me3 mark from bivalent promoters may be associated with activation of oncogenes and the initiation or propagation of cancer (Martinez-Garcia and Licht, 2010).

Histone methyl marks can be removed by enzymes of the histone demethylase family (KDM). Demethylation occurs via two distinct mechanism. The first involves a flavin adenine dinucleotide (FAD)-dependent amine oxygenase in the KMD1 family able to move mono- and dimethyl but not trimethyl groups from H3K4 and H3K9 (Zheng et al., 2015). The second, removing the trimethyl modification, occurs via the jumonji-domain containing (JmjC) proteins that use alpha-ketoglutarate and iron as cofactors for the demethylation reaction (Johansson et al., 2014; **Fig. 5.4**). Like the lysine methyltransferases, the demethylase family is important during embryogenesis and development and mutations and deletions in these genes leads to developmentally abnormal phenotypes (Nottke et al., 2009). Possibly due to the fact that lysine methylation is associated with both activation of gene expression, for example H3K4me3, and the repression of expression with marks such as H3K27me3, both over- and underexpression of various KDMs have been associated with a cancer phenotype (Rotili and Mai 2011; Thinnes et al. 2014). It is likely that the phenotypic effect of aberrant expression or mutation in the KDMs is highly dependent on the cell type of origin and the presence of any other mutations present.

DNA base modification

According to our definition, an epigenetic modification is one that is stable and heritable but does not depend on a change to the underlying DNA sequence consisting of the four nitrogen-containing bases (adenine (A), thymine (T), guanine (G) and cytosine (C)) attached to a deoxyribose-phosphate backbone. A chemical modification to a base which causes a change in the interaction of proteins involved with transcription but does not change how the base is read and

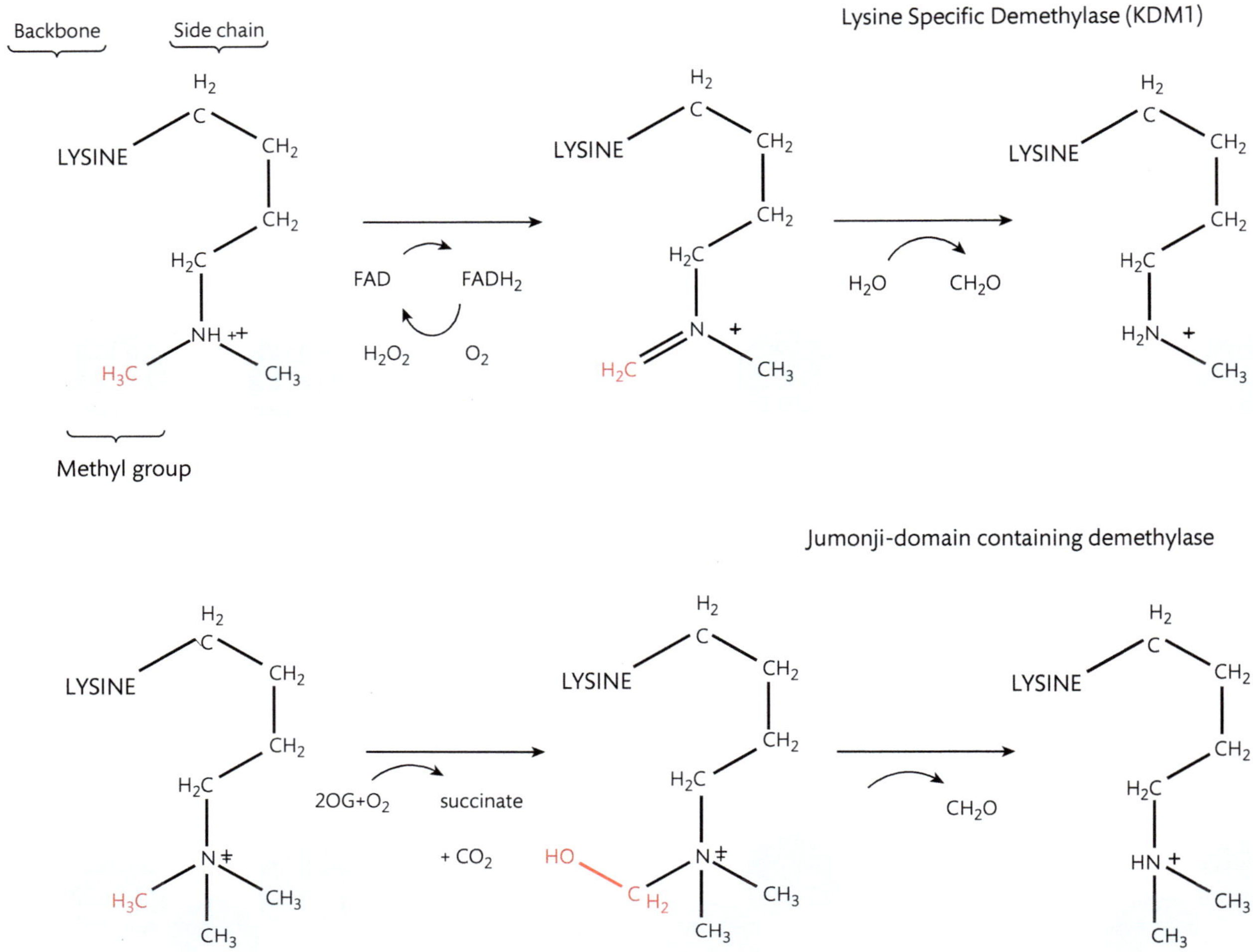

Fig. 5.4 Mechanism of demethylases: lysine-specific demethylase (KDM1) demethylates H3K4me1 and H3K4me2 by forming an imine intermediate using FAD as a cofactor that subsequently undergoes non-enzymatic hydrolysis releasing formaldehyde. The requirement to form an imine intermediate prevents the demethylation of trimethylated lysine residues. Jumonji-domain containing demethylases use a different mechanism requiring 2-oxoglutarate (2OG) and iron (Fe^{2+}) as cofactors, leading to the formation of an unstable hydroxymethyl that spontaneously is released as formaldehyde. Importantly, jumonji-domain containing demethylases can demethylate mono-, di-, and trimethylated lysine residues.

copied during DNA replication is therefore able to exert an epigenetic effect. In prokaryotes, modification to both adenine and cytosine has been recorded (Sánchez-Romero et al., 2015) whereas in eukaryotic organisms only modifications to cytosine within DNA have been observed.

The major form of modification to DNA observed is the addition of a methyl group to cytosine resulting in 5-methylcytosine (5-mC). When found in promoter regions of a gene, 5-mC is associated with silencing of that gene by inhibiting transcription (Doerfler, 1983; Bird, 2002; Suzuki and Bird, 2008). This inhibition of transcription has an important role in cancer biology as, if the promoter of a tumour suppressor gene is hypermethylated, there will be a lack of functional protein produced (Laird and Jaenisch, 1996; Baylin and Herman, 2000; Esteller, 2007). As described in the general introduction to epigenetics, a wild type protein that is not expressed by a cell is in many ways equivalent to an expressed mutant protein that has lost the ability to perform its normal tumour suppressor role: in both situations there is the lack of a functioning protein within the cell. Hypermethylation of tumour suppressor promoter regions is a common mechanism found in many different cancer types including haematological malignancies, carcinomas, and sarcomas.

There are several related mechanisms by which DNA methylation in the promoter of a gene leads to decrease transcription. Part of the inhibitor effect on transcription may be explained by the methyl group of the 5-mC projecting into the major groove of the DNA helix interfering with the binding of transcription factors and other proteins necessary for transcription (Dantas Machado et al., 2015). More important in mediating the effect of methylated DNA on gene silencing is the interaction between 5-mC and the methyl-CpG binding proteins (MBD), proteins which selectively bind to methylated DNA and have low affinity to DNA that is not methylated (Du et al., 2015). The presence of proteins bound to the 5-mC helps to physically prevent the binding of transcription factors that will reduce the level of transcription. MBDs consist of a family of proteins that in addition to containing domains that allow recognition of methylated DNA are able to affect chromatin structure, another important epigenetic mechanism discussed in the section on histone modification. Some MBDs contain domains that allow them to directly alter chromatin structure, such as SETDB1 and SETDB2, which contain a histone methyltransferase domain that increase H3K9 trimethyl level associated with transcriptional repression and heterochromatin formation. Other MBDs, such as MBD1-4 and MeCP2, interact with histone deacetylases (HDACs) that also results in the formation of heterochromatin associated with decreased transcription.

5-methylcytosine is formed by enzymes of the DNA 5-cytosine methyltransferase family (DNMT) using S-adenosyl methionine (SAM) as the methyl donor. The activity of the DNMTs is not equal for all cytosine molecules within DNA and has a preference for a cytosine molecule immediately upstream of a guanine (Jeltsch, 2006), commonly referred to as a CpG dinucleotide where the 'p' represents the phosphate group of the DNA backbone. The convention of referring to a CpG dinucleotide is to distinguish it from a cytosine and guanine base pairs on opposite strands of DNA that are connected via hydrogen bonding.

The CpG dinucleotide occurs at approximately one-quarter of the frequency in mammalian genomes than would be expected given their overall GC content however, compared to other eukaryotic organisms, mammals have a significantly higher rate of methylation at CpG nucleotides (Han et al., 2008). A suggested reason for the lower than expected lever of CpG is that when a methylated cytosine spontaneously deaminates it forms a thymine and while there exist an enzyme that can detect a thymine/guanine mismatch, thymine DNA glycosylase (TDG), the enzyme is not completely efficient. Over evolutionary time, this results in the transition of 5-methylcytosine to thymine, leading to a C:G to A:T mutation. MBD4, mentioned previously as a methylated DNA-binding protein, also contains glycosylase activity that is able to repair deamination from 5-methylcytosine to thymine and therefore plays an important role in preventing mutations at the site of CpG (Yoon et al., 2003). MBD4 itself is subject to epigenetic regulation and is found to be transcriptionally silenced as an early event in many colorectal adenocarcinomas (Neri and Lucci-Cordisco, 2009). MBD4 mutations also occur in colorectal cancers with microsatellite instability and defects in mismatch repair (Bader et al., 1999).

The distribution of CpG across the vertebrate genome is uneven, with large areas between genes containing very few CpGs followed by a much higher density found in the promoter regions of many genes—referred to as 'CpG islands' (Varriale and Bernardi 2010). CpG islands are regions at least 200 base pairs long which contain greater than 50% GC content and an observed to expected CpG ratio greater than 60%. In contrast to promoters with a high level of DNA methylation that are associated with transcriptional silencing, promoters of actively expressed genes contain a larger number of unmethylated CpGs. The increased frequency of CpG islands in gene promoters may be explained firstly by the decreased deamination rate of unmethylated cytosine to uracil which occurs compared to that of 5mC to thymine and secondly by the improved detection and repair of U: G by base excision repair appears compared to that of T:G. Evolutionary pressure may be an additional explanation of the presence of a CpG islands close to the transcription start site where such a pressure does to exist in intergenic areas to a mutation following deamination of 5mC to uracil has little impact.

There are three catalytically active members of the DNMT family: DNMT1, DNMT3a, and DNMT3b (Subramaniam et al., 2014). The enzyme originally named DNMT2 was found not to methylate cytosines in DNA but instead to methylate a specific cytosine in the tRNA of aspartic acid, so it was renamed tRNA Cytosine-5 methyltransferase (TRDMT1; see Goll et al., 2006). DNMT1 is important in maintaining the methylation pattern when DNA is replicated. As DNA polymerase replicates the complementary strand, the appropriate bases are faithfully replicated but the newly synthesized strand of DNA will not contain any methylated bases. DNMT1 has a high activity for methylating the cytosine in CpG dinucleotides when the double-stranded DNA is hemimethylated—that is, one strand contains a methylated cytosine and the other strand does not as occurs immediately after DNA replication (Jeltsch, 2006). DNMT1 interacts with Proliferating Cell Nuclear Antigen (PCNA) and is localized at the replication fork allowing the methylation pattern to be propagated (Iida et al., 2002). The activity of DNMT1 explains how DNA methylation is an epigenetic mark that can be inherited from parent to daughter cells.

The hemimethylated status of newly replicated DNA is used as a way of the organism detecting which strand is newly synthesized and is important for the correct functioning of the mismatch repair (MMR) system in prokaryotes (Fukui and Fukui, 2010). Although the MMR system is highly evolutionarily conserved, the presence or absence of methylated bases does not appear to be used as a detection mechanism for newly synthesized DNA in eukaryotes yet DNMT1 appears to play a crucial role in this pathway. The interaction of DNMT1 with PCNA brings it into contact with many other proteins of the MMR system and studies where DNMT1 has been knocked-down have indicated that this results in increased microsatellite instability, a hallmark of deficient MMR that is not related to DNA methylation (Loughery et al., 2011).

Compared to the activity at hemimethylated sites, the activity of DNMT1 for *de novo* methylation, adding a methyl group to a cytosine when neither strand contains an existing methyl group is much lower under physiological conditions (Bird, 1999). Overexpression of DNMT1 though has been shown to induce *de novo* promoter methylation and aberrant regulation of DNMT1 in cancer cells may play an important role in gene silencing (Etoh et al., 2004).

Members of the DNMT3 family (DNMT3a, DNMT3b, and DNMT3L, a catalytically inactive accessory protein) are associated with de novo methylation and in contrast to DNMT1, are able to methylate cytosine when the opposite strand is not methylated with the same efficiency as they can methylated cytosine in hemimethylated DNA (Bird, 1999). While DNMT3s show a preference for methylation at CpG sites they are also able to methylate cytosines in other contexts including CpA dinucleotides at very low levels. DNMT3 are important for methylating areas of repetitive DNA which may represent transposable elements and the methylation and silencing of these regions are important for maintaining genomic stability. DNMT3s therefore plays an important role in establishing the methylation pattern within the cell.

In a further example of interaction between different mechanisms of epigenetic control, DNTM3a is known to interact with proteins associated with repressive chromatin marks. DNMT3a interacts via its PHD domain with unmethylated histone 3 lysine 4, a mark associated with gene repression (Otani et al., 2009). EZH2, the catalytically active histone methyltransferase subunit of PRC2, is also associated with repression of transcription and interacts with DNMT3 linking the repressive marks of DNA methylation and histone modifications (Viré et al., 2006).

Genome editing studies using a human embryonic stem cell line have shown that removal of both DNMT3a and DNMT3b either singularly or together is not fatal in these cells and does not diminish their pluripotency. Silencing of DNMT1 however was fatal in these cells (Liao et al., 2015). This is in contrast to studies in mice where double knockout of DNMT3a and b was embryological lethal (Okano et al., 1999).

DNMT3a appears to be particularly important in the haematological malignancies, especially acute myeloid leukaemia (AML), in which up to 22% of cases have a mutation in the gene (Ley et al., 2010). In mouse haematopoietic stem cells (HSC), mutations in DNMT3a have been associated with an advantage in self-renewal. As HSCs survive for many decades in human, mutations in DNMT3a may be an early event which when coupled to an advantage in self-renewal can lead to an increasing prevalence of HSCs containing the mutation over a lifetime. Extensive genomic sequencing studies of thousands of people without haematological malignancies have shown that over time HSCs become increasingly clonal with 5–10% of 70-year-olds relying on only one clone for almost all haematopoiesis (Genovese et al., 2014; Jaiswal et al., 2014; Xie et al., 2014). These same sequencing studies have demonstrated the prevalence of mutations in DNMT3a in haematopoietic stem cells. The exact mechanism by which the mutations in DNMT3a leads to haematological malignancy is not clear and, in many cases, secondary mutations are also found. However, these studies indicate the importance that DNMT3 has in maintaining genomic stability and preventing malignant development (Yang et al., 2015).

DNA methylation also plays an important role in other epigenetic mechanisms implicated in cancer biology such as X-chromosome inactivation and genetic imprinting. Genetic imprinting is an epigenetic process by which the parental origin of a particular allele is able to be transmitted to the daughter cells. For imprinted genes, normally only those inherited from a particular parent are expressed but can lead to disease for several reasons. One situation in which imprinting can lead to disease is when there is a mutation in a gene that is expressed while its unmated partner allele is silenced. Errors in meiosis can result in a daughter cell receiving two copies of a gene from the same parent instead of the usual situation in which one allele is received from each, a situation referred to as uniparental disomy. If a gene normally expresses the maternal copy of the gene but both copies have come from the father this will result in disease.

An example of imprinting in cancer biology is the GTP-binding protein Di-Ras3 (DIRAS3, also known as ARHI and NOEY2). The *DIRAS3* gene is maternally imprinted meaning that only the parental allele is expressed (Yu et al., 1999). DIRAS3 is found normally expressed in breast and ovarian tissue but is absent in cancerous tissue (Yu et al., 2003). It is a tumour suppressor gene as the loss of function of a gene associated with cancer development. Unlike most tumour suppressor genes which require 'two hits' to deactivate them, because of the two copies contained within the cell, DIRAS3 only requires 'one hit' due to the silencing of the imprinted maternal gene. DIRAS3 is a further example of the interplay between different mechanisms of epigenetic regulation. The 'hit' observed in preventing gene function is often due to promoter methylation by DNMT of the unimprinted paternal allele leading to gene silencing. Furthermore, in certain breast cancers it has been shown that the microRNA miR-221 is upregulated and by binding to the three prime untranslated regions of the DIRAS3 mRNA it can lead to degradation and therefore decrease in protein expression levels (Li et al., 2013).

Demethylation of cytosine bases is the reverse process of methylation, and the balance between the two is important for regulating the methylation pattern within the cell. Demethylation can happen either actively or passively. As during each round of DNA replication the methylation pattern has to be re-established on the newly synthesized strand by the action of DNMT1, inhibiting the process of maintenance methylation will result in a global decrease in the methylation level of the cells over multiple cell divisions. Two drugs are licensed for use that work by inhibiting methylation of the newly synthesized strand of DNA, 5-azacitidine, and 5-azadeoxycitidine, which are also used in the treatment of myelodysplastic syndrome and other haematological malignancies such as AML and chronic myeloid leukaemia (Gros et al., 2012). Both are cytidine analogues containing a nitrogen in the 5 position of the nitrogenous ring, the position to which DNMTs would normally attach the methyl group. After activation by intracellular kinases, the trinucleotide can be incorporated into DNA in place of a dCTP, where the hemimethylated CpG is recognized by DNMT1 and the process of methylation begins with the enzyme forming a covalent adduct to the base. The presence of the nitrogen in the 5 position of the nitrogenous ring means that the DNTM1 is unable to complete the methylation reaction and so becomes permanently attached to the base which is subsequently removed by DNA repair machinery leading to degradation of DNMT1 and depletion of intracellular levels.

DNA methylation can also be an active process catalysed by enzymes of the ten-eleven translocation (TET) family of enzymes, originally discovered as a fusion partner to MLL in acute myeloid leukaemia (Ono et al., 2002). An interesting observation is that MLL, also known as lysine methyltransferase 2A (KMT2A) is responsible for histone 3 lysine 4 methylation, a mark of transcriptional activation. There are three members of the TET family in humans all containing an oxygenase domain preceded by a cysteine rich area and TET1 and TET3 also contain a CXXC domain, important for binding to methylated CpG dinucleotides. The TET family of proteins results in the oxidation of 5-methylcytosine to 5-hydroxymethylcytosine (5-hmC) using molecular oxygen as a substrate and requiring alpha-ketoglutarate as a cofactor. Further oxidation using the same mechanism is possible from 5hmC to 5-formylcytosine (5fC) and then to 5-carboxylcytosine (5caC; Fig. 5.5). Following the activation oxidation from 5mC to 5hmC using the TET enzymes, passive dilution of the methylated cytosine mark occurs as subsequent rounds of DNA replication occur without the methyl mark being replace by DNMT as it does not recognize the 5hmC as a mark to indicate that the newly synthesized strand should be methylated. Like 5hmC, 5fC and 5caC also undergo passive dilution following DNA synthesis; it can also be actively repaired by removal of the oxidized base using TDG followed by repair using the base excision repair system (Hashimoto et al., 2012).

The ability of TET family enzymes to demethylate cytosine is important in early embryogenesis for establishing the methylation pattern of the new organism and consolidating the methylation states of the maternal and paternal genomes (Zhao and Chen, 2013). TET3 rapidly oxidizes the 5mC to 5hmC in the paternal genome but appears to be excluded from the maternal genome and certain areas of the paternal genome that remain methylated as part

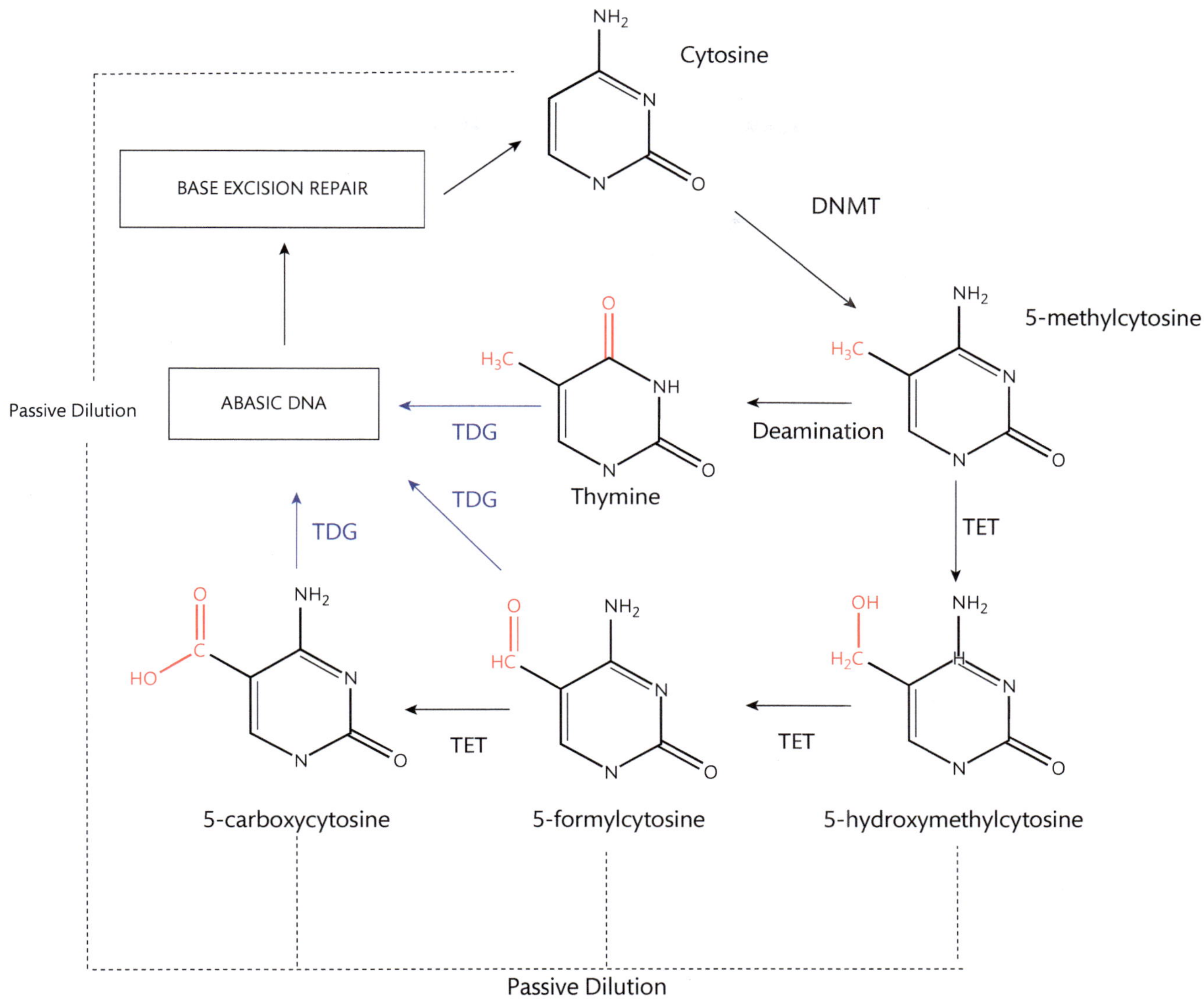

Fig. 5.5 DNA methylation: Cytosine residues can undergo methylation by DNA methyltransferase (DNMT) to form 5-methylcytosine (5-mC). 5-mC can undergo spontaneous deamination to form thymine, leading to a G:T mismatch in base pairing. This mismatch is corrected by removal of the thymine base by thymine DNA glycosylase (TDG) forming a basic DNA that is then repaired by the base excision repair machinery of the cell. Demethylation can occur due to oxidation by the TET enzymes. 5-mC is oxidized first to 5-hydroxymethylcytosine (5-hmC), then 5-formylcytosine (5-fC) and finally 5-carboxylcytosine (5-caC). 5-hmC, 5-fC, and 5-caC are poorly recognized by DNMT so passive dilution of the mark occurs following DNA replication and cell division. 5-fC and 5-caC are also substrates of TDG and undergo base excision repair to restore the base to an unmodified cytosine.

of genomic imprinting. The maternal genome is rich in the histone mark H3K9me2 to which additional proteins such as development pluripotency-associated 3 (DPPA3, also known as STELLA or PGC7) are recruited, excluding TET3 and therefore preventing the active demethylation by oxidation (Nakashima et al., 2013). Areas of the paternal genome that are imprinted appear to use the same mechanisms in preventing demethylation.

In addition to the role of TET1 in association with MLL in the fusion gene associated with AML, TET2 is overexpressed in a wide range of both solid tumours and haematological malignancies (Delhommeau et al., 2009). TET2 mutations have been found in a number of cancers resulting in increased levels of 5mC and decreased levels of 5hmC. It has been observed that in *de novo* AML, mutations in TET2 and isocitrate dehydrogenase 1 or 2 (IDH1/IDH2) appears to be mutually exclusive (Figueroa et al., 2010). IDH is an enzyme of the tricarboxylic acid cycle that catalyses the oxidative decarboxylation of isocitrate to alpha-ketoglutarate, the cofactor required for TET2 protein function. Mutations in IDH can lead to the production of alpha-hydroxyglutarate which accumulates within the cell and competes with the alpha-ketoglutarate for the active site of TET2. Therefore, mutation in either IDH or TET2 results in dysfunctional demethylation within the cell that contributes to the cancer phenotype.

Non-coding RNAs

A further method of epigenetic control is through the expression of non-coding RNAs (ncRNA). Non-coding RNAs are transcribed from the genome and often undergo post-transcriptional processing such as polyadenylation or splicing, but do not result in a protein product being formed. Studies by the ENCODE consortium have suggested that many areas of the genome not traditionally associated with protein-coding are transcribed at very low rates, but this does not necessarily imply that all of these transcripts have a function (Djebali and Davis, 2012). Several examples of non-coding RNAs have been described in the section on histone lysine modifications in relation to the role they play in the recruitment of polycomb complex to DNA during X-inactivation. In the context of epigenetics, 'non-coding RNAs' is used to refer to those RNA that have a regulatory effect on other genes especially the long non-coding RNAs (lncRNA) which are over 200 nt long (Kung et al., 2013) and short non-coding RNAs, such as microRNAs (miRNA) and piwi-interacting RNAs (piRNAs), but it must be remembered that ribosomal RNA and tRNAs transcribed at high levels in the cell also do not get translated to form a protein product.

MicroRNAs (miRNA) are 22 bp RNAs that can alter gene expression by binding to the three prime untranslated regions of mRNAs leading to their destabilization and degradation (He and Hannon, 2004). Primary miRNAs are transcribed from DNA by RNA polymerase 2 and are processed within the nucleus by the Microprocessor complex that includes DROSHA, a ribonuclease enzyme and DiGeorge critical region 8 (DGCR8, also known as pasha) to form a stem-loop structured pre-miRNA of 70–80 bp in length (Denli et al., 2004). The processed pre-miRNA is exported to the cytoplasm where it is processed by DICER1 which removes the loop leaving a mature double-stranded miRNA (Hutvágner et al., 2001). One strand of the miRNA is then attached to an Argonaute protein to form the miRNA-induced silencing complex (miRISC) which is able to bind via complimentary base pairing with the 3'UTR of the target protein. Binding of the miRISC complex directs the mRNA:miRNA complex to the processing bodies (P-body), where the mRNA is degraded preventing further translation. Therefore, increased levels of a particular miRNA than targets a given mRNA will lead to decreased levels of the mRNA and therefore reduce expression of the protein (**Fig. 5.6**).

Disruption of the miRNA system has been observed in many cancers, either by aberrant transcription of the primary miRNA (Jansson and Lund, 2012) or dysregulation of the processing system (Palanichamy and Rao, 2014). This may involve either over- or underexpression of proteins, leading to a change in the level of mature miRNA and therefore altered gene expression.

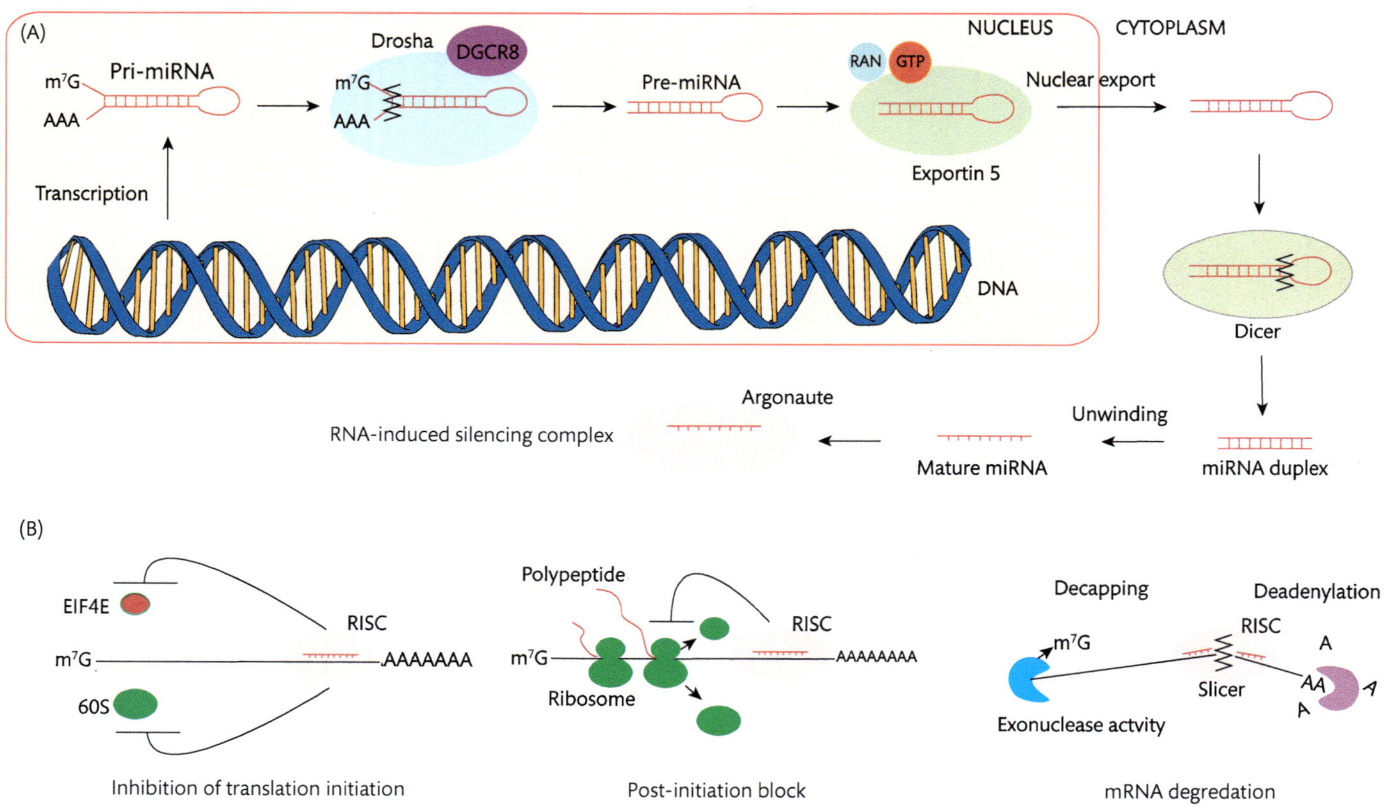

Fig. 5.6 MicroRNA processing: (A) miRNA genes are transcribed by RNA polymerase 2 or 3 and the produced transcript, referred to as pri-miRNA, forms a double-stranded hairpin structure and contains a 5′ cap and 3′ polyadenylation. The pri-miRNA is recognized by DiGeorge Syndrome Critical Region 8 (DRCG8, also known as 'pasha') and processed by Drosha to remove the 5′ cap and poly A tail, forming a pre-miRNA. The pre-miRNA is exported to the cytoplasm by exportin 5 in complex with RAN-GTP. In the cytoplasm, the pre-miRNA is further processed by Dicer to remove the loop structure leaving a miRNA duplex. Following unwinding of the duplex, one strand is incorporated into the RNA-induced silencing complex (RISC) in association with proteins of the Argonaute family. (B) miRNA associated with RISC can modify protein expression by interfering with the initiation of translation via inhibiting binding of either the 60S ribosomal subunit of EIF4E. Polypeptide elongation can also be inhibited by causing ribosomes to disengage prematurely from the mRNA transcript. miRNA-RISC can also lead to mRNA degradation by interacting with proteins that result in decapping and de-adenylation promoting degradation of the transcript through exonucleases. The interaction of miRNA and RNA forming a double-stranded RNA is the substrate for the Argonaute protein Slicer that results in mRNA cleavage.

Piwi (in *Drosophilla* P-element induced wimpy testis) RNAs (piRNA) are longer than miRNAs at approximately 30 bp long and were originally discovered as important in development in *Drosophilla* (Juliano et al., 2011). Expression of piRNAs in cancerous tissues appears to differ from that of healthy tissue of the same origin (Martinez et al., 2015). The exact role and mechanism of action of piRNAs is a topic of active research.

Summary and conclusions

In this chapter an overview of epigenetics and its relation to cancer biology has been described. The link between epigenetics and development has been emphasized with the correlation between the epigenetic landscape of the cell and the resulting phenotype that results as a sum of the expression genes within the cells. The complexity of epigenetic regulation with many interactions between different mechanisms of epigenetic control is clear, and the difficulty in understanding these interactions due to our incomplete knowledge of the underlying biology explained.

A deeper understanding of epigenetics offers the prospect of a greater understanding of gene regulation in cancer and the possibility of novel-treatments targeting the specific epigenetic processes that have become deranged within an individual cancer type.

TAKE-HOME MESSAGE

- The epigenetic 'landscape' of many cancers is deranged leading to aberrant gene expression.
- The interaction between different epigenetics mechanisms is complex and incompletely understood.
- Inappropriate silencing of a non-mutated gene may be functionally equivalent to a loss of function genetic mutation.
- Targeting epigenetic processes is a valid therapeutic strategy.

OPEN QUESTIONS

- How can we unravel the complexity of epigenetic interactions and the histone code?
- Do we understand the implication of specific epigenetic changes in different cancer types and the relationship to other cancer mutations?
- What do we know of the relationship between epigenetic aberrations and patient outcomes?
- How can novel therapeutics that selectively target specific epigenetic processes be developed?

FURTHER READING

Feinberg, A. P., Koldobskiy, M. A., & Göndör, A. (2016). Epigenetic modulators, modifiers and mediators in cancer aetiology and progression. *Nat Rev Genet*, **17**(5), 284–99.

Huarte, M. (2015). The emerging role of lncRNAs in cancer. *Nat Med*, **21** (11), 1253–61.

Iacobuzio-Donahue, C. A. (2009). Epigenetic changes in cancer. *Annu Rev Pathol*, **4**(1), 229–49.

Johansson, C., Tumber, A., Che. K., Cain, P., et al. (2014). The roles of jumonji-type oxygenases in human disease. *Epigenomics*, **6**(1), 89–120.

Jones, P. A. (2012). Functions of DNA methylation: islands, start sites, gene bodies and beyond. *Nat Rev* Genet, **13**(7), 484–92.

Ohnishi, K., Semi, K., Yamamoto, T., et al. (2014). Premature termination of reprogramming *in vivo* leads to cancer development through altered epigenetic regulation. *Cell*, **156**(4), 663–77.

REFERENCES

Aihara, K., Mukasa, A., Gotoh, K., et al. (2014). H3F3A K27M mutations in thalamic gliomas from young adult patients. *Neuro-oncology*, **16**(1), 140–6.

Allfrey, V. G. & Mirsky, A. E. (1964). Structural modifications of histones and their possible role in the regulation of RNA synthesis. *Science (New York, N.Y.)*, **144**(3618), 559.

Azad, N., Zahnow, C. A., Rudin, C. M., & Baylin, S. B. (2013). The future of epigenetic therapy in solid tumours—lessons from the past. *Nat Rev Clin Oncol*, **10**(5), 256–66.

Bader, S., Walker, M., Hendrich, B., et al. (1999). Somatic frameshift mutations in the MBD4 gene of sporadic colon cancers with mismatch repair deficiency. *Oncogene*, **18**(56), 8044–7.

Baylin, S. & Herman, J. (2000). DNA hypermethylation in tumorigenesis: epigenetics joins genetics. *Trends Genet*, **16**(4), 168–74.

Behjati, S., Tarpey, P. S., Presneau, N., et al. (2013). Distinct H3F3A and H3F3B driver mutations define chondroblastoma and giant cell tumor of bone. *Nat Genet*, **45**(12), 1479–82.

Berger, S. L., Kouzarides, T., Shiekhattar, R., & Shilatifard, A. (2009). An operational definition of epigenetics. *Genes Dev*, **23**(7), 781–3.

Bianconi, E., Piovesan A, Facchin F, et al. (2013). An estimation of the number of cells in the human body. *Ann Hum Biol*, **40**(6), 463–71.

Bird, A. (1999). DNA methylation *de novo*. *Science (New York, N.Y.)*, **286**(5448), 2287–8.

Bird, A. (2002). DNA methylation patterns and epigenetic memory. *Genes Dev*, **16**(1), 6–21.

Black, J. C., Van Rechem, C., & Whetstine, J. R. (2012). Histone lysine methylation dynamics: establishment, regulation, and biological impact. *Mol Cell*, **48**(4), 491–507.

Cao, R., Tsukada, Y.-I., & Zhang, Y. (2005). Role of Bmi-1 and Ring1A in H2A ubiquitylation and Hox gene silencing. *Mol Cell*, **20**(6), 845–54.

Carafa, V., Rotili, D., Forgione, M., et al. (2016). Sirtuin functions and modulation: from chemistry to the clinic. *Clin Epigenetics*, **8**, 61.

Chan, K.-M., Fang, D., Gan, H., et al. (2013). The histone H3.3K27M mutation in pediatric glioma reprograms H3K27 methylation and gene expression. *Genes Dev*, **27**(9), 985–90.

Chargaff, E., Lipshitz, R., & Green, C. (1952). Composition of the desoxypentose nucleic acids of four genera of sea-urchin. *J Biol Chem*, **195**(1), 155–60.

Chen, X., Xie, W., Gu, P., et al. (2015). Upregulated WDR5 promotes proliferation, self-renewal and chemoresistance in bladder cancer via mediating H3K4 trimethylation. *Sci Rep*, **5**, 8293.

Di Croce, L. & Helin, K. (2013). Transcriptional regulation by Polycomb group proteins. *Nat Struct Mol Biol*, **20**(10), 1147–55.

Cuthbert, G. L., Daujat, S., Snowden, A. W., et al. (2004). Histone deimination antagonizes arginine methylation. *Cell*, **118**(5), 545–53.

Dantas Machado, A. C., Zhou, T., Rao, S., et al. (2015). Evolving insights on how cytosine methylation affects protein-DNA binding. *Brief Funct Genomics*, **14**(1), 61–73.

Delattre, O., Zucman, J., Plougastel, B., et al. (1992). Gene fusion with an ETS DNA-binding domain caused by chromosome translocation in human tumours. *Nature*, **359**(6391), 162–5.

Delcuve, G. P., Khan, D. H., Davie, J. R., et al. (2012). Roles of histone deacetylases in epigenetic regulation: emerging paradigms from studies with inhibitors. *Clin Epigenetics*, **4**(1), 5.

Delhommeau, F., Dupont, S., Della Valle, V., et al. (2009). Mutation in TET2 in myeloid cancers. *N Engl J Med*, **360**(22), 2289–301.

Denli, A. M. Tops, B. B., Plasterk, R. H., Ketting, R. F., & Hannon, G. J. (2004). Processing of primary microRNAs by the microprocessor complex. *Nature*, **432**(7014), 231–5.

Devaiah, B. N., Case-Borden, C., Gegonne, A., et al. (2016). BRD4 is a histone acetyltransferase that evicts nucleosomes from chromatin. *Nat Struct Mol Biol*, **23**(6), 540–8.

Dey, A., Ellenberg, J., Farina, A., et al. (2000). A bromodomain protein, MCAP, associates with mitotic chromosomes and affects G(2)-to-M transition. *Mol Cell Biol*, **20**(17), 6537–49.

DiMartino, J. F. & Cleary, M. L. (1999). MLL rearrangements in haematological malignancies: lessons from clinical and biological studies. *Br J Haematol*, **106**(3), 614–26.

Djebali, S. & Davis, C. A. (2012). Landscape of transcription in human cells. *Nature*, **489**(7414), 101–8.

Doerfler, W. (1983). DNA methylation and gene activity. *Annu Rev Biochem*, **52**, 93–124.

Du, Q., Luu, P. L., Stirzaker, C., Clark, S. J., et al. (2015). Methyl-CpG-binding domain proteins: readers of the epigenome. *Epigenomics*, **7**(6), 1051–73.

Esteller, M. (2007). Cancer epigenomics: DNA methylomes and histone-modification maps. *Nat Rev Genet*, **8**(4), 286–98.

Etoh, T., Kanai, Y., Ushijima, S., et al. (2004). Increased DNA methyltransferase 1 (DNMT1) protein expression correlates significantly with poorer tumor differentiation and frequent DNA hypermethylation of multiple cpG islands in gastric cancers. *Am J Pathol*, **164**(2), 689–99.

Figueroa, M. E., Abdel-Wahab, O., Lu, C., et al. (2010). Leukemic IDH1 and IDH2 mutations result in a hypermethylation phenotype, disrupt TET2 function, and impair hematopoietic differentiation. *Cancer Cell*, **18**(6), 553–67.

Filippakopoulos, P., Qi, J., Picaud, S., et al. (2010). Selective inhibition of BET bromodomains. *Nature*, **468**(7327), 1067–73.

Filippakopoulos, P. & Knapp, S. (2014). Targeting bromodomains: epigenetic readers of lysine acetylation. *Nat Rev Drug Discov*, **13**(5), 337–56.

French, C. A., Kutok, J. L., Faquin, W. C., et al. (2004). Midline carcinoma of children and young adults with NUT rearrangement. *J Clin Oncol*, **22**(20), 4135–9.

Frey, F., Sheahan, T., Finkl, K., et al. (2016). Molecular basis of PRC1 targeting to Polycomb response elements by PhoRC. *Genes Dev*, **30**(9), 1116–27.

Friedmann, D. R. & Marmorstein, R. (2013). Structure and mechanism of non-histone protein acetyltransferase enzymes. *FEBS J*, **280**(22), 5570–81.

Fukui, K. & Fukui, K. (2010). DNA mismatch repair in eukaryotes and bacteria. *Journal of Nucleic Acids*, **2010**, 1–16.

Genovese, G., Kähler, A. K., Handsaker, R. E., et al. (2014). Clonal hematopoiesis and blood-cancer risk inferred from blood DNA sequence. *N Engl J Med*, **371**(26), 2477–87.

Goll, M. G., Kirpekar, F., Maggert, K. A., et al. (2006). Methylation of tRNAAsp by the DNA methyltransferase homolog Dnmt2. *Science (New York, N.Y.)*, **311**(5759), 395–8.

Grayson, A. R., Walsh, E. M., Cameron, M. J., et al. (2014). MYC, a downstream target of BRD-NUT, is necessary and sufficient for the blockade of differentiation in NUT midline carcinoma. *Oncogene*, **33**(13), 1736–42.

Gros, C., Fahy, J., Halby, L., et al. (2012). DNA methylation inhibitors in cancer: recent and future approaches. *Biochimie*, **94**(11), 2280–96.

Gupta, R. A., Shah, N., Wang, K. C., et al. (2010). Long non-coding RNA HOTAIR reprograms chromatin state to promote cancer metastasis. *Nature*, **464**(7291), 1071–6.

Gutiérrez, L., Oktaba, K., Scheuermann, J. C., Gambetta, M. C., Ly-Hartig, N., & Müller, J. (2012). The role of the histone H2A ubiquitinase Sce in polycomb repression. *Development (Cambridge, England)*, **139**(1), 117–27.

Han, L., Su, B., Wen-Hsiung, L., et al. (2008). CpG island density and its correlations with genomic features in mammalian genomes. *Genome Biol*, **9**(5), R79.

Hashimoto, H., Hong, S., Bhagwat, A. S., Zhang, X., & Cheng, X. (2012). Excision of 5-hydroxymethyluracil and 5-carboxylcytosine by the thymine DNA glycosylase domain: its structural basis and implications for active DNA demethylation. *Nucleic Acids Res*, **40**(20), 10203–14.

Hassa, P. O., Haenni, S. S., Elser, M., & Hottiger, M. O. (2006). Nuclear ADP-ribosylation reactions in mammalian cells: where are we today and where are we going? *Microbiol Mol Biol Rev*, **70**(3), 789–829.

He, L. & Hannon, G. J. (2004). MicroRNAs: small RNAs with a big role in gene regulation. *Nat Rev Genet*, **5**(7), 522–31.

Hernández-Muñoz, I. et al. (2005). Stable X chromosome inactivation involves the PRC1 Polycomb complex and requires histone MACROH2A1 and the CULLIN3/SPOP ubiquitin E3 ligase. *Proc Natl Acad Sci U S A*, **102**(21), 7635–40.

Hutvágner, G., McLachlan, J., Pasquinelli, A. E., Bálint, E., Tusch, T., & Zamore, P. D. (2001). A cellular function for the RNA-interference enzyme Dicer in the maturation of the let-7 small temporal RNA. *Science (New York, N.Y.)*, **293**(5531), 834–8.

Iida, T., Suetake, I., Tajima, S., et al. (2002). PCNA clamp facilitates action of DNA cytosine methyltransferase 1 on hemimethylated DNA. *Genes to Cells*, **7**(10), 997–1007.

Jaiswal, S., Fontanillas, P., Flannick, J., et al. (2014). Age-related clonal hematopoiesis associated with adverse outcomes. *N Engl J Med*, **371**(26), 2488–98.

Jang, M. K., Mochizuki, K., Zhou, M., Jeong, H. S., Brady, J. N., & Ozato, K. (2005). The bromodomain protein Brd4 is a positive regulatory component of P-TEFb and stimulates RNA polymerase II-dependent transcription. *Mol Cell*, **19**(4), 523–34.

Jansson, M. D. & Lund, A. H. (2012). MicroRNA and cancer. *Mol Oncol*, **6**(6), 590–610.

Jeltsch, A. (2006). On the enzymatic properties of Dnmt1: specificity, processivity, mechanism of linear diffusion and allosteric regulation of the enzyme. *Epigenetics*, **1**(2), 63–6.

Jenuwein, T. & Allis, C. D. (2001). Translating the histone code. *Science (New York, N.Y.)*, **293**(5532), 1074–80.

Johansson, C., Tumber, A., Che, K., et al. (2014). The roles of Jumonji-type oxygenases in human disease. *Epigenomics*, **6**(1), 89–120.

Juliano, C., Wang, J. & Lin, H. (2011). Uniting germline and stem cells: the function of Piwi proteins and the piRNA pathway in diverse organisms. *Annu Rev Genet*, **45**, 447–69.

Karmodiya, K., Krebs, A. R, Oulad-Abdelghani, M., Kimura, H., & Tora, L. (2012). H3K9 and H3K14 acetylation co-occur at many gene regulatory elements, while H3K14ac marks a subset of inactive inducible promoters in mouse embryonic stem cells. *BMC Genomics*, **13**(1), 424.

Kokabu, S., Lowery, J. W., & Jimi, E. (2016). Cell fate and differentiation of bone marrow mesenchymal stem cells. *Stem Cells Int*, **2016**, 3753581.

Kouzarides, T. (2007). Chromatin modifications and their function. *Cell*, **128**(4), 693–705.

Kumar, R., Angelini, S., Snellman, E., & Hemminki, K. (2004). BRAF mutations are common somatic events in melanocytic nevi. *J Invest Dermatol*, **122**(2), 342–8.

Kung, J. T. Y., Colognori, D., & Lee, J. T. (2013). Long noncoding RNAs: past, present, and future. *Genetics*, **193**(3), 651–69.

Laird, P. & Jaenisch, R. (1996). The role of DNA methylation in cancer genetics and epigenetics. *Annu Rev Genet*, **30**, 441–64.

Laubach, J. P., Moreau, P., San-Miguel, J. F., & Richardson, P. G. (2015). Panobinostat for the treatment of multiple myeloma. *Clin Cancer Res*, **21**(21), 4767–73.

Lee, K. K. & Workman, J. L. (2007). Histone acetyltransferase complexes: one size doesn't fit all. *Nature Rev Mol Cell Biol*, **8**(4), 284–95.

Lee, W., Teckie, S., Wiesner, T., et al. (2014). PRC2 is recurrently inactivated through EED or SUZ12 loss in malignant peripheral nerve sheath tumors. *Nat Genet*, **46**(11), 1227–32.

Lee, Y. J., Han, M. E., Baek, S. J., Kim, S. Y., Oh, S. O. (2015). Roles of DPY30 in the proliferation and motility of gastric cancer cells. *PloS One*, **10**(7), e0131863.

Lewis, P. W., Müller, M. M., Koletsky, M. S., et al. (2013). Inhibition of PRC2 activity by a gain-of-function H3 mutation found in pediatric glioblastoma. *Science (New York, N.Y.)*, **340**(6134), 857–61.

Ley, T. J., Ding, L., Walter, M. J., et al. (2010). DNMT3A mutations in acute myeloid leukemia. *N Engl J Med*, **363**(25), 2424–33.

Li, Y., Liu, M., Zhang, Y. et al. (2013). Effects of ARHI on breast cancer cell biological behavior regulated by microRNA-221. *Tumour Biol*, **34**(6), 3545–54.

Liao, J., Karnik, R., Gu, H., et al. (2015). Targeted disruption of DNMT1, DNMT3A and DNMT3B in human embryonic stem cells. *Nat Genet*, **47**(5), 469–78.

Loughery, J. E. P. Dunne, P. D., O'Neill, K. M., Meehan, R. R., McDaid, J. R., Walsh, C. P. (2011). DNMT1 deficiency triggers mismatch repair defects in human cells through depletion of repair protein levels in a process involving the DNA damage response. *Hum Mol Genet*, **20**(16), 3241–55.

Mann, B. S. Johnson, J. R., Cohen, M. H., Justice, R., Pazdur, R. (2007). FDA approval summary: vorinostat for treatment of advanced primary cutaneous T-cell lymphoma. *Oncologist*, **12**(10), 1247–52.

Martinez-Garcia, E. & Licht, J. D. (2010). Deregulation of H3K27 methylation in cancer. *Nat Genet*, **42**(2), 100–1.

Martinez, V. D., Vucic, E. A., Thu, K. L., et al. (2015). Unique somatic and malignant expression patterns implicate PIWI-interacting RNAs in cancer-type specific biology. *Scientific Reports*, **5**, 10423.

Murr, R. (2010). Interplay between different epigenetic modifications and mechanisms. *Adv Genet*, **70**, 101–41.

Nakashima, H. Kimura, T., Kaga, Y., et al. (2013). Effects of dppa3 on DNA methylation dynamics during primordial germ cell development in mice. *Biol Reprod*, **88**(5), 125.

Nathan, D., Ingvarsdottir, K., Sterner, D. E., et al. (2006). Histone sumoylation is a negative regulator in Saccharomyces cerevisiae and shows dynamic interplay with positive-acting histone modifications. *Genes Dev*, **20**(8), 966–76.

Neri, G. & Lucci-Cordisco, E. (2009). Silent beginning: early silencing of the *MED1/MBD4* gene in colorectal tumorigenesis. *Cancer Biol Ther*, **8**(2), 192–3.

Nguyen, A. T., Taranova, O., He, J., & Zhang, Y. et al. (2011). DOT1L, the H3K79 methyltransferase, is required for MLL-AF9-mediated leukemogenesis. *Blood*, **117**(25), 6912–22.

Nottke, A., Colaiácovo, M. P., & Shi, Y. (2009). Developmental roles of the histone lysine demethylases. *Development (Cambridge, England)*, **136**(6), 879–89.

Nowak, S. J. & Corces, V. G. (2004). Phosphorylation of histone H3: a balancing act between chromosome condensation and transcriptional activation. *Trends Genet*, **20**(4), 214–20.

Okano, M., Bell, D. W., Haber, D. A., & Li, E. (1999). DNA methyltransferases Dnmt3a and Dnmt3b are essential for *de novo* methylation and mammalian development. *Cell*, **99**(3), 247–57.

Ono, R., Taki, T., Taketani, T., Taniwaki, M., Kobayashi, H., & Hayashi, Y. (2002). LCX, leukemia-associated protein with a CXXC domain, is fused to MLL in acute myeloid leukemia with trilineage dysplasia having t(10;11)(q22;q23). *Cancer Res*, **62**(14), 4075–80.

Orthaus, S., Klement, K., Happel, N., Hoischen, C., & Diekmann, S. (2009). Linker histone H1 is present in centromeric chromatin of living human cells next to inner kinetochore proteins. *Nucleic Acids Res*, **37**(10), 3391–406.

Otani, J., Nankumo, T., Arita, K., Inamoto, S., Ariyoshi, M., & Shirakawa, M. (2009). Structural basis for recognition of H3K4 methylation status by the DNA methyltransferase 3A ATRX-DNMT3-DNMT3L domain. *EMBO Rep*, **10**(11), 1235–41.

De Paepe, C., Krivega, M., Cauffman, G., Geens, M., Van de Velde, H. (2014). Totipotency and lineage segregation in the human embryo. *Mol Hum Rep*, **20**(7), 599–618.

Palanichamy, J. K. & Rao, D. S. (2014). miRNA dysregulation in cancer: towards a mechanistic understanding. *Frontiers Genet*, **5**, 54.

Pallante, P., Federico, A., Berlingieri, M. T., et al. (2008). Loss of the CBX7 gene expression correlates with a highly malignant phenotype in thyroid cancer. *Cancer Res*, **68**(16), 6770–8.

Patel, D. J. & Wang, Z. (2013). Readout of epigenetic modifications. *Annu Rev Biochem*, **82**, 81–118.

Qi, J., Huo, L., Zhu, Y. T., & Zhu, Y.-J. (2014). Absent, small or homeotic 2-like protein (ASH2L) enhances the transcription of the estrogen receptor α gene through GATA-binding protein 3 (GATA3). *J Biol Chem*, **289**(45), 31373–81.

Ren, X., Vincenz, C., & Kerppola, T. K. (2008). Changes in the distributions and dynamics of polycomb repressive complexes during embryonic stem cell differentiation. *Mol Cell Biol*, **28**(9), 2884–95.

Rose, N. R. & Klose, R. J. (2014). Understanding the relationship between DNA methylation and histone lysine methylation. *Biochim Biophys Acta*, **1839**(12), 1362–72.

Rotili, D. & Mai, A. (2011). Targeting histone demethylases: a new avenue for the fight against cancer. *Genes Cancer*, **2**(6), 663–79.

Saito, M. & Ishikawa, F. (2002). The mCpG-binding domain of human MBD3 does not bind to mCpG but interacts with NuRD/Mi2 components HDAC1 and MTA2. *J Biol Chem*, **277**(38), 35434–9.

Saksouk, N., Simboeck, E., Déjardin, J. (2015). Constitutive heterochromatin formation and transcription in mammals. *Epigenetics Chromatin*, **8**(1), 3.

Sánchez-Romero, M. A., Cota, I., & Casadesús, J. (2015). DNA methylation in bacteria: from the methyl group to the methylome. *Curr Opin Microbiol*, **25**, 9–16.

Schuettengruber, B., Martinez, A. M., Iovino, N., & Cavalli, G. (2011). Trithorax group proteins: switching genes on and keeping them active. *Nat Rev Mol Cell Biol*, **12**(12), 799–814.

Schwartz, Y. B. & Pirrotta, V. (2013). A new world of polycombs: unexpected partnerships and emerging functions. *Nat Rev Genet*, **14**(12), 853–64.

Schwartzentruber, J., Korshunov, A., Liu, X. Y., et al. (2012). Driver mutations in histone H3.3 and chromatin remodelling genes in paediatric glioblastoma. *Nature*, **482**(7384), 226–31.

Scott, C. L., Gil, J., Hernando, E., et al. (2007). Role of the chromobox protein CBX7 in lymphomagenesis. *Proc Natl Acad Sci U S A*, **104**(13), 5389–94.

Shankar, G. M., Lelic, N., Gill, C. M., et al. (2016). BRAF alteration status and the histone H3F3A gene K27M mutation segregate spinal cord astrocytoma histology. *Acta Neuropathol*, **131**(1), 147–50.

Steffen, P. A. & Ringrose, L. (2014). What are memories made of? How Polycomb and Trithorax proteins mediate epigenetic memory. *Nat Rev Mol Cell Biol*, **15**(5), 340–56.

Sterner, D. E. & Berger, S. L. (2000). Acetylation of histones and transcription-related factors. *Microbiol Mol Biol Rev*, **64**(2), 435–59.

Sturm, D., Witt, H., Hovestadt, V., et al. (2012). Hotspot mutations in H3F3A and IDH1 define distinct epigenetic and biological subgroups of glioblastoma. *Cancer Cell*, **22**(4), 425–37.

Subramaniam, D., Thombre, R., Dhar, A., & Anant, S. (2014). DNA methyltransferases: a novel target for prevention and therapy. *Front Oncol*, **4**, 80.

Suzuki, M. & Bird, A. (2008). DNA methylation landscapes: provocative insights from epigenomics. *Nat Rev Genet*, **9**(6), 465–76.

Thinnes, C. C., England, K. S., Kawamura, A., Chowdhury, R., Schofield, C. J., & Hopkinson, R. J. (2014). Targeting histone lysine demethylases—progress, challenges, and the future. *Biochim Biophys Acta*, **1839**(12), 1416–32.

Vaissière, T., Sawan, C., & Herceg, Z. (2008). Epigenetic interplay between histone modifications and DNA methylation in gene silencing. *Mutat Res*, **659**(1), 40–8.

Varriale, A. & Bernardi, G. (2010). Distribution of DNA methylation, CpGs, and CpG islands in human isochores. *Genomics*, **95**(1), 25–8.

Viré, E., Brenner, C., Deplus, R., et al. (2006). The polycomb group protein EZH2 directly controls DNA methylation. *Nature*, **439**(7078), 871–4.

Voigt, P., Tee, W.-W., & Reinberg, D. (2013). A double take on bivalent promoters. *Genes Dev*, **27**(12), 1318–38.

Waddington, C. (1942). The epigenotype. *Endeavour*, **1**, 18–19.

Wang, H., Wang, L., Erdjument-Bromage, H., et al. (2004). Role of histone H2A ubiquitination in polycomb silencing. *Nature*, **431**(7010), 873–8.

Wapenaar, H. & Dekker, F. J. (2016). Histone acetyltransferases: challenges in targeting bi-substrate enzymes. *Clin Epigen*, **8**, 59.

Watson, J. D. & Crick, F. H. C. (1953). Molecular structure of nucleic aids: a structure for deoxyribose nucleic acid. *Nature*, **171**(4356), 737–8.

West, A. C. & Johnstone, R. W. (2014). New and emerging HDAC inhibitors for cancer treatment. *J Clin Invest*, **124**(1), 30–9.

Wu, X., Johansen, J. V., & Helin, K. (2013). Fbxl10/Kdm2b recruits polycomb repressive complex 1 to CpG islands and regulates H2A ubiquitylation. *Mol Cell*, **49**(6), 1134–46.

Wu, Y., Zhang, L., Wang, Y., et al. (2014). Long noncoding RNA HOTAIR involvement in cancer. *Tumour Biol*, **35**(10), 9531–8.

Xie, M., Lu, C., Wang, J., et al. (2014). Age-related mutations associated with clonal hematopoietic expansion and malignancies. *Nat Med*, **20**(12), 1472–8.

Yamaguchi, H. & Hung, M.-C. (2014). Regulation and role of EZH2 in cancer. *Cancer Res Treat*, **46**(3), 209–22.

Yang, L., Rau, R., & Goodell, M. A. (2015). DNMT3A in haematological malignancies. *Nat Rev Cancer*, **15**(3), 152–65.

Yoon, J.-H., Iwai, S., O'Connor, T. R., & Pfeifer, G. P. (2003). Human thymine DNA glycosylase (TDG) and methyl-CpG-binding protein 4 (MBD4) excise thymine glycol (Tg) from a Tg:G mispair. *Nucleic Acids Res*, **31**(18), 5399–404.

Yu, Y., Fujii, S., Yuan, J. et al. (2003). Epigenetic regulation of ARHI in breast and ovarian cancer cells. *Ann N Y Acad Sci*, **983**, 268–77.

Yu, Y., Xu, F., Peng, H., et al. (1999). *NOEY2* (ARHI), an imprinted putative tumor suppressor gene in ovarian and breast carcinomas. *Proc Natl Acad Sci U S A*, **96**(1), 214–9.

Zhang, Y. & Reinberg, D. (2001). Transcription regulation by histone methylation: interplay between different covalent modifications of the core histone tails. *Genes Dev*, **15**(18), 2343–60.

Zhao, H. & Chen, T. (2013). Tet family of 5-methylcytosine dioxygenases in mammalian development. *J Human Genet*, **58**(7), 421–7.

Zhao, R., Nakamura, T., Fu, Y., Lazar, Z., & Spector, D. L. (2011). Gene bookmarking accelerates the kinetics of post-mitotic transcriptional re-activation. *Nature Cell Biol*, **13**(11), 1295–304.

Zheng, Y.-C., Ma, J., Wang, Z. et al. (2015). A systematic review of histone lysine-specific demethylase 1 and its inhibitors. *Med Res Rev*, **35**(5), 1032–71.

6

Viral carcinogenesis

An overview

Dirk P. Dittmer and Blossom Damania

Introduction

Approximately 30% of human cancers are caused by infectious agents such as viruses, helicobacter, and helminths (IARC, 2012; Table 6.1). For these cancers epidemiological studies have established sufficient evidence to connect virus exposure and cancer development, thus fulfilling the postulate of association. Viral infection is the genetic driver of tumour development. For these cancers, viruses deliver additional genetic information in the form of viral proteins, and sometimes also viral micro ribonucleic acids (RNAs), to initiate transformation. Virus infection constitutes the first step in the stepwise progression model of tumorigenesis.

Progressive transformation can be approximated in culture models, such as primary rodent fibroblasts. These cells have a limited lifespan (the so-called Hayflick limit; Hayflick, 1965) that can be expanded by forced expression of viral or human oncogenes. This process is called immortalization and involves activation of telomerase and inactivation of tumour suppressor genes such as *TP53* and the *RB* family members PRB1/p105, p107, and PRB2/p130. Transformation is a distinct, often but not always, second step, whereupon cells acquire the ability to grow in low serum, independent of solid support, and where growth is no longer inhibited by neighbouring cells. Typically transformed cells form tumours in immunodeficient mice. Human mature B and T cells also can be transformed by tumour viruses in culture, in a growth-factor dependent manner (growth factors for mature B cells are IL6 and IL4, and for T cells IL2). Viral oncogenes are classified as immortalizing or transforming, or both. Oncogenic viruses either immortalize or transform cells in culture, thus fulfilling the criteria for oncogenic agents.

Viruses are classified on the basis of their replication cycle and the manner in which their genetic information is maintained, either in the form of deoxyribonucleic acid (DNA) or ribonucleic acid (RNA) (Baltimore, 1971). For instance, human papilloma virus (HPV) is a DNA virus, but it does not encode its own polymerase. Epstein–Barr virus (EBV) is a DNA virus that encodes its own DNA-dependent DNA polymerase. Both replicate in the nucleus. Human immunodeficiency virus (HIV) and hepatitis C virus (HCV) are RNA viruses. Whereas HCV encodes an RNA-dependent RNA polymerase and replicates in the cytoplasm, HIV first generates a DNA intermediate by reverse transcription and integration into the host DNA and then uses the host RNA polymerase to generate the virion RNA.

Bacteria, such as *Helicobacter pylori*, are autonomous life forms and can exist both inside and outside the cell. They do not experimentally transform cells in culture, cause mutation, or directly induce DNA replication. Rather, they affect the host by indirect mechanisms, such as inducing inflammation, producing carcinogenic substances, inducing angiogenesis, or modulating the immune system. These properties are considered the hallmarks of cancer (Hanahan and Weinberg, 2011) and if present over a long period of time can aid the survival and expansion of neoplastic cells.

Finally, helminths are multicellular organisms; they are parasites that infect humans and animals. Among the helminths *Schistosoma haematobium* is associated with bladder cancer, mostly in Egypt. *Opisthorchis viverrini* and *Clonorchis sinesis* are causally associated with cholangiocarcinoma, which is prevalent throughout Southeast Asia. The presumed mechanisms are again indirect and include chronic inflammation, which then leads to hyperplasia of the biliary epithelium. This creates an increased pool of proliferating cells susceptible to secondary mutation and a microenvironment susceptible to the expansion of abnormal cells. There is a second aspect of helminth infection that is worth mentioning: helminths severely skew the human immune response, which is normally balanced between cell-directed and antibody-mediated responses. Helminths polarize T-cell responses towards dealing with parasite infections (Th2 and IgE isotypes). This may have several consequences. As the immune system is occupied with fighting the parasite, intestinal bacteria, certain viruses, and perhaps even nascent cancer cells experience a less than optimal immune response.

Most viral infections cause immediate disease (e.g. Ebola haemorrhagic fever, influenza, measles, and mumps). In these cases, it is easy to establish an association that conforms to the postulates of the German physician Robert Koch (1843–1910) and that may be summarized as:

- The infective agent must be present in every organism suffering from the disease in question.
- The infective organism occurs in no other disease or in normal organisms.
- It must be possible to isolate and grow the organism from the infected subject and must be able to induce the same disease when injected in a healthy organism.

Table 6.1 Infectious agents known to cause cancer

Agent	Cancer	Key references and dates
Helicobacter pylori	Gastric	(Warren and Marshall, 1983)
Schistosoma haematobium	Bladder	(Mustacchi and Shimkin, 1958)
Opisthorchis viverrini	Cholangiocarcinoma	(Parkin et al., 1993)
Clonorchis sinesis	Cholangiocarcinoma	
HPV	Cervical cancer	(Durst et al., 1983)
	Anal cancer	(Palefsky et al., 1991)
	Head and neck cancers	(Gillison et al., 2000)
		(Ang et al., 2010)
	HPV vaccine	(Kirnbauer et al., 1993)
EBV	Burkitt lymphoma	(Henle et al., 1967)
		(Burkitt, 1962)
		(Epstein and Barr, 1964)
	NPC	(zur Hausen et al., 1970)
		(Nonoyama and Pagano, 1973)
	Hodgkin's disease	(Pallesen et al., 1991)
		(Herbst et al., 1991)
KSHV	Kaposi's sarcoma	(Chang et al., 1994)
	PEL	(Cesarman et al., 1995)
	Multicentric Castleman's disease	(Soulier et al., 1995)
Marek's disease virus	No cancer in humans	
	Lymphoma in chicken	(Jakowski et al., 1974)
SV40	No cancer in humans	
	Rat fibroblasts (p53)	(Linzer and Levine, 1979)
Murine polyomavirus	No cancer in humans	
	Rat fibroblasts (src)	(Courtneidge and Smith, 1983)
Merkel cell polyomavirus	Merkel cell carcinoma	(Feng et al., 2008)
Adenovirus	No cancer in humans	
	Rat fibroblasts (p53)	(Sarnow et al., 1982)
HTLV-I/II	T-cell lymphoma	(Poiesz et al., 1981)
ALV	No cancer in humans	
	Rat fibroblasts (src)	(Wang et al., 1976)
MoLV	No cancer in humans	
	Rat fibroblasts (myc)	(Stehelin et al., 1976)
HBV	Liver cancer	(Millman et al., 1969)
		(Dane et al., 1970)
HCV	Liver cancer	(Choo et al., 1989)

ALV, avian leukosis virus; EBV, Epstein–Barr virus; HBV, hepatitis B virus; HPV, human papilloma virus; HTLV-I/II, human T-lymphotropic retrovirus; KSHV, Kaposi's sarcoma-associated herpesvirus; MoLV, Moloney leukaemia virus; NPC, nasopharyngeal cancer; PEL, primary effusion lymphoma; SV40, simian virus 40.

For these agents, their roles in causing the disease hold also when more updated criteria are used (Fredricks and Relman, 1996).

By contrast human cancer represents a delayed phenotype presenting decades after exposure. The first experimental demonstration of an infectious cause of cancer is attributed to Ellerman and Bang working on leukaemia in chicken in 1908, and to Francis Peyton Rous in 1910 working on sarcoma in chicken. Rous' sarcoma is caused by an oncogenic retrovirus, which carries the activated human oncogene v-SRC. The discovery of oncogenic viruses in chicken is no coincidence. Chickens were and are of immense commercial value and thus diseases in chicken are of concern and subject to intense scrutiny by veterinary pathologists. Chickens are also housed in large numbers and thus there exists the opportunity to observe rare events, such as cancer. In fact, an oncogenic herpesvirus, Marek's disease virus, causes lymphoma in chicken and is a feared poultry pathogen (Jakowski et al., 1974).

The first indications that a DNA virus could cause cancer were obtained by Richard Shope and Francis Peyton Rous in the 1930s. They noticed warts in wild rabbits. Warts or papillomas are benign hyperplasias. In humans, genital warts are caused by variants of HPV that are not associated with cancer. Shope and Rous showed that cell-free extract made from these lesions could induce new lesions when inoculated into other rabbits. Moreover, they found that the transmitted agent induced frank carcinomas. This is due to the particular molecular mechanism caused by papillomaviruses and polyomaviruses, discussed next.

Merkel cell polyomavirus and animal polyomaviruses

Merkel cell polyomavirus (MCV) is a 'small' DNA tumour virus with a circular genome of approximately 5,000 bp and is found in histologically confirmed tumour cells (Feng et al., 2008). MCV is epidemiologically associated with Merkel cell carcinoma and individual viral proteins affecting key tumour suppressor pathways, for instance, *RB* signalling. MCV encodes a set of structural proteins (VP1, VP2) that make up the viral capsid and it encodes regulatory proteins, which are called small, large, and middle T antigen because of their sequence similarity and functional homology to regulatory proteins of other polyomaviridae. Other human polyomaviruses are the BK virus (named by the patient's initials in whom the virus was first isolated) and the John Cunningham (JC) virus. BK and JC virus are not associated with human cancer.

Two animal polyomaviruses derived from the simian virus 40 (SV40) of primates and polyomavirus of mice are transforming in animals. The T antigens carry the transforming functions for these viruses, and the SV40 T antigen was the means by which the human tumour suppressor protein TB53 was discovered (Linzer and Levine, 1979). The same T antigen was also found to interact with the product of the retinoblastoma gene (*RB*), another tumour suppressor protein. For murine polyomavirus, which is a tumour virus of mice, the transforming function maps to small T antigen, which engages *c-SRC* (the mammalian homologue of *v-SRC*, which is the transforming gene of Rous sarcoma virus in chicken; see Courtneidge and Smith, 1983). By contrast, the transforming activity of SV40 maps to regions in the large T antigen. SV40 T binds to and inactivates the human tumour suppressor proteins RB and p53 (Fig. 6.1). SV40 induces tumours in newborn (i.e. weakly immunodeficient) hamsters. The virus also transforms primary human cells in culture but is not associated with tumours in the human population.

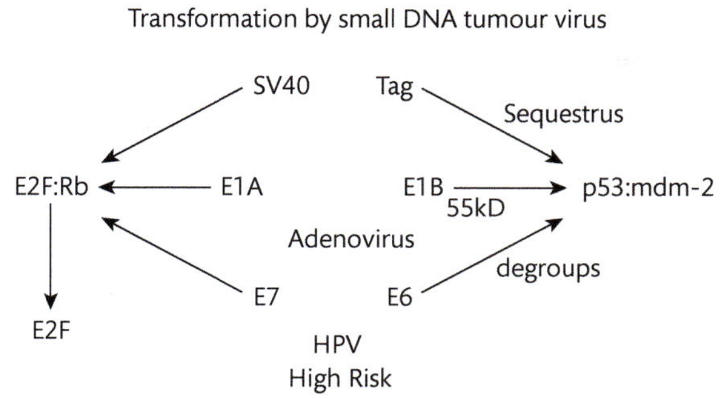

Fig. 6.1 Transformation principle for the small DNA tumour viruses SV40, adenovirus, and high-risk strains (principally 16, 18) of human papillomavirus. Only the high-risk papilloma viruses cause cancer in humans, but all readily transform rat fibroblasts in culture.

Rb, retinoblastoma protein; E2F, a transcription factor; Tag, large transforming (T) protein of SV40; E1A, E1B, early proteins of adenovirus; p53, tumour suppressor protein of 55 kD size; mdm-2, mouse-double-minute protein 2 (the human equivalent is hdm-2).

Human papilloma viruses

Like the polyomaviruses, the papilloma viruses are also small, double-stranded DNA viruses. They carry two oncogenes, E6 and E7. E6 binds to and inactivates TB53 (Scheffner et al., 1990), while E7 binds to and inactivates RB and its family members, p110, and others. These viruses infect the basal layer of epithelial cells in the skin. Here they stay dormant as circular plasmids. As the basal layer cells, which can be considered stem cells, differentiate into the upper layers of stratified epithelium, papilloma viruses begin to replicate. Like all small DNA viruses, papilloma viruses do not encode their own DNA polymerase. They depend on induction of cell polymerases (and supporting DNA synthesis enzymes) for replication. Inactivation of RB by E7 liberates the transcription factor E2F, which drives the cell into S-phase. E7 also affects other cell cycle regulators, such as cyclin A and E; it inactivates p21 and p27, which are negative regulators of cyclin dependent kinases, and it may induce other pro-growth pathways as well. Degradation of TP53 via ubiquitinylation by E6:E6-associated protein (E6-AP) prevents the sensing of improper S-phase entry and apoptosis. A consequence of this is the upregulation of the cell cycle inhibitor, p16INK4. Detection of p16INK4 has emerged as a clinically useful surrogate biomarker for cervical cancer and HPV head and neck cancer, more so than the detection of the E6 or E7 oncoproteins themselves.

In 9 out of 10 infected cells, reactivation of HPV leads to viral replication and virion formation, virus egress, and eventual destruction of the infected cell as expression of E6 and E7 is induced as part of the complete replication cycle. In a small fraction, however, the circular HPV genome integrates into the host genome of the still multipotent basal epithelial cell, and this integration event disrupts the regulatory region or open reading frame of the E2 transcriptional suppressor. E2 negatively regulates E6 and E7 transcription during dormancy. In such an abortively infected cell, E6 and E7 are transcribed under conditions where virion formation is not possible and improper replication (under conditions of inactive TP53) ensues. This leads to the rapid accumulation of secondary genetic changes, some of which lead to transformation and tumour formation. In head and neck cancer, mutation of *TP53* is inversely correlated with the presence of HPV (Cancer Genome Atlas, 2015). Almost all HPV-associated cancers show highly elevated levels of p16INK4. There are over 100 different HPV types. Only a fraction (the 'high-risk' types 16, 18, 31, 33, 35, 39, 45, 51, 52, 56, 58, 59, 68, 73) are associated with cancer, predominantly of the cervix. E6 and E7 oncoproteins of 'high-risk', but not 'low-risk' HPVs are able to immortalize primary human keratinocytes and, by virtue of specific variants in their amino acid sequence, efficiently inactive RB and TP53.

HPV and human hepatitis B virus (HBV) infection can be prevented by vaccination, which reduces the likelihood of persistently infected cells. HPV-associated cervical cancer is the most common cancer in women worldwide, particularly in low- and middle-income countries (Parkin et al., 2014). HPV vaccination has significantly reduced the incidence of high-grade cervical abnormalities and cancer in those countries where the vaccine is available and where the HPV types that are part of the vaccine drive the cancer. A new, nonavalent vaccine that protects against nine different HPV types is bound to increase the utility of this prevention strategy (Joura et al., 2015).

Hepatitis B virus

Notwithstanding geographic variation in infection rates, about a quarter of all cases of hepatocellular carcinoma (liver cancer) are associated with HBV infection. A quarter of cases are associated with HCV infection and the remainder have varying aetiologies. It is confusing but important to keep in mind that in molecular terms the different hepatitis viruses function very differently. HBV is a partially double-stranded DNA virus. HCV is a positive-strand RNA virus in the same family as dengue, Zika, and West Nile viruses. Other viruses also target the liver but are not associated with cancer. Hepatitis A virus (HAV) is a picornavirus and there are also hepatitis D, E, and G viruses (reviewed in Knipe et al., 2013).

It has been difficult to pinpoint any single HBV protein as driving transformation. Chronic inflammation as a result of persistent low-level replication in the liver and/or chronic induction of DNA damage by replication and integration into the host chromosome have emerged as the most plausible scenarios for HBV-associated tumorigenesis (Fig. 6.2). Integrated HBV genomes are observed in the majority of HBV-associated hepatocellular carcinomas. The integration appears random, but repeated integration events have been observed in the vicinity of growth-promoting genes, such a *MYC*, telomerase, cyclin A, fragile sites, and others. HBV also encodes viral proteins that modulate cell signalling for the benefit of the virus. The HBV surface protein S modulates membrane-proximal kinases, such as protein kinase C. The HBV transactivator X (HBX) has pleiotropic effects on multiple cellular signalling pathways associated with DNA replication, DNA damage, and cell division. HBX reportedly interacts with p53, with the nuclear export factor Crm1, the cytoplasmic Raf-1 kinase, and is also believed to modulate epigenetic modifying enzymes (Ringelhan and Protzer, 2015). Which of these interactions is the most important one is unclear; perhaps all contribute to some degree or at some point in liver carcinogenesis.

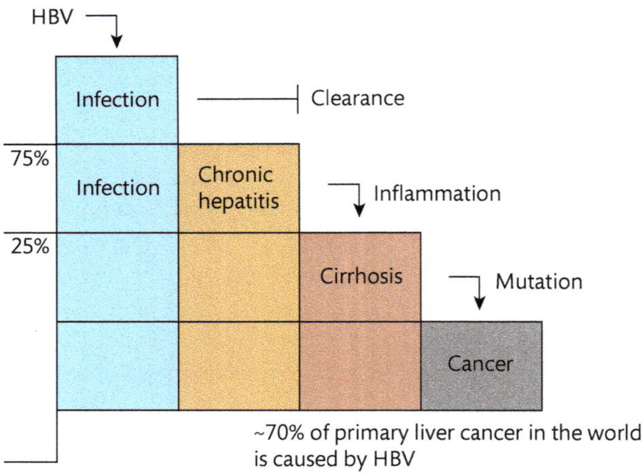

Fig. 6.2 Transformation principle for HBV. Around 75% of individuals exposed to HBV clear the infection (top row), while 25% experience chronic persistence and over time develop chronic hepatitis, an inflammation of the liver (second row). Chronic hepatitis can lead to cirrhosis, which provides an enlarged pool of replicating, activated liver cells (third row from the top), a fraction of which sustains mutations in their genome and develop into fully transformed cancer cells (bottom row).

Hepatitis C virus

HCV infection is now curable with a 12-week course of antiviral drugs (Sulkowski et al., 2014). These regimens do not prevent re-infection and cases of resistance have been reported, but clearance of HCV prevents progression to liver cancer. HCV is a single-stranded RNA virus of the flavivirus family. It replicates via an RNA-dependent RNA polymerase in the cytoplasm of infected cells. HCV only infects liver cells and is exquisitely sensitive to interferon-alpha (IFN). IFN-alpha plus ribavirin was the standard of care prior to the introduction of viral protease NS3/4a (e.g. telaprevir or boceprevir) and viral polymerase NS5b-targeting drugs (e.g. daclatasvir or sofosbuvir). For a long time, it was impossible to grow HCV in culture. Now cell-culture adapted versions of one particular isolate (JHF-1) replicate robustly in IFN-alpha deficient Huh-7 cells and genetic engineering uncovered additional host restrictions that pave the way for robust replication of many HCV strains in culture (Lindenbach et al., 2005; Wakita et al., 2005; Saeed et al., 2015).

Retroviruses and cancer

From a mechanistic standpoint we can distinguish multiple modes of retrovirus-induced oncogenesis:

a. By inducing immunosuppression, which leads to an increase in cancers caused by other viruses or cancers in which host immune surveillance is a major controlling factor. In HIV-infected patients, there is a high incidence of Kaposi's sarcoma (KS) and pleural effusion lymphoma, both caused by Kaposi's sarcoma-associated herpesvirus (KSHV). EBV-associated central nervous system lymphoma, HPV-associated anal cancer, and MCV-induced Merkel cell carcinoma are also seen at much higher rates in HIV-infected persons than in the general population.
b. Retroviral integration-driven lymphomas, which are rare, but have been observed secondary to early gene therapy trials (Hacein-Bey-Abina et al., 2008). Note that in the case of gene therapy, the viral vector inoculum exceeds that of natural infection by orders of magnitude and that the therapy is often associated with temporary immunodeficiency.
c. Viral-driven lymphomas caused by an oncogene-carrying virus, for example, human T-lymphotropic retrovirus (HTLV-1), leading to adult T-cell lymphoma (ATL). Here the oncogene is of purely viral origin and no homologue is found in the human genome.
d. Viral oncogene-driven cancers, wherein the viral oncogene represents an activated form of a human oncogene acquired by 'molecular piracy'. These led to the discovery of many dominant-acting oncogenes, such as v-myc (avian myelocytomatosis virus), v-src (Rouse sarcoma virus), Abl (Abelson murine leukaemia virus), v-erb-A and v-erb-B/EGFr (Avian erythroblastosis virus), and ras (Harvey sarcoma virus and Kirsten sarcoma virus). Of note, none of these examples apply to human cancers.

Human immunodeficiency virus

HIV is a single strand RNA virus, which depends on reverse transcription into a DNA intermediate and subsequent random integration into the host genome for its replication cycle. It is not found in histologically confirmed tumour cells. Neither HIV-1 nor HIV-2 transform cells in culture or cause tumours in animals.

Thus, HIV is not able to directly transform a normal cell into a neoplastic one; however, individual HIV proteins induce cellular changes that are associated with cancer. For instance, the HIV tat protein induces angiogenesis in experimental models (Albini et al., 1995). This coincidence is expected as many of the same molecular pathways that facilitate virus proliferation also facilitate host cell proliferation (i.e. hyperplasia). Long-term, but not primary, HIV infection also induces an environment that is conducive to the development of cancers in humans. This includes depletion of CD4 cells and thus a weakening of immune-mediated tumour surveillance (Fig. 6.3). As a result, several cancers are overrepresented in persons living with HIV/AIDS. These include the viral-associated cancers KS and central nervous system lymphoma, and anal cancer, as well as a smattering of non-viral-associated cancers, such as lung cancer. The latter show greater incidence rates in persons living with HIV and who also have been exposed to years of antiretroviral drugs (Robbins et al., 2015). There is, as of now, no evidence that the molecular biology of cancers that develop in individuals experiencing HIV-induced immune suppression differs from the molecular biology of cancers that develop in individuals experiencing chemically induced immune suppression (e.g. in the context of organ transplantation).

Human T-lymphotropic retrovirus

HTLV-I is a complex retrovirus, like HIV. It is associated with ATL. The geographic distribution of HTLV prevalence parallels the distribution of ATL and is highest in Japan and the Caribbean. HTLV-I expresses the oncogene Tax (Fig. 6.4). Tax binds to the cyclic adenosine monophosphate (AMP) response element (CRE) and forms a complex with CRE binding protein (CREB/ATF1), its accessory factor p300, and CREB binding protein CBP. Activation of CRE by

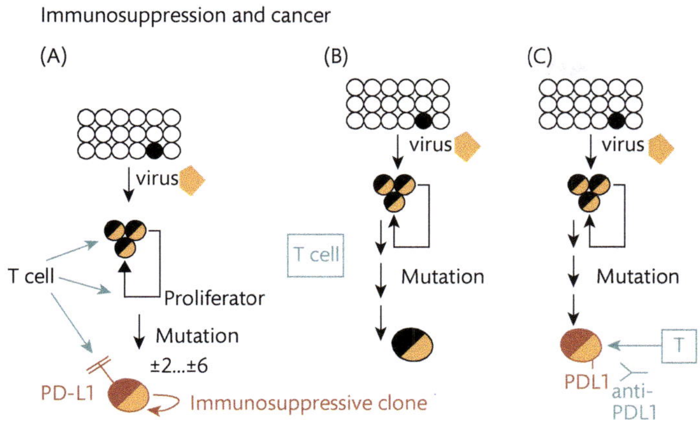

Fig. 6.3 Immune suppression and cancer. Cancers, in particular cancers with infectious aetiology are more frequent in long-term immunocompromised individuals. It does not matter if the immune deficiency is the result of congenital defects, chemical treatment, such as during chemotherapy or organ transplantation, or viral infections, such as by HIV. Mutations and viral infections happen all the time. Mutations happen at a constant frequency per cell lifetime. In humans between one to six mutations are required to convert a normal cell into a cancer cell; fewer if the cell is infected by a tumour virus. (A) T cells constantly survey all cells in the body and have the capacity to detect and kill virally infected and mutated cells. Hence, some of the molecular changes ('hallmarks of cancer') selected for its escape from T-cell surveillance, such as the upregulation of PD-L1. PD-L1 inhibits T-cell killing by engaging the inhibitor receptor PD1 on T cells. (B) During periods of immune suppression, the T cells are inactive. This has two effects; first, the number of cells carrying a mutation or a virus infection increases, as T cells no longer recognize and kill these abnormal cells. This accelerates tumour development. Second, in the absence of T cells, a selection for immune-suppressive tumour subclones (e.g. by upregulation of PD1) is no longer needed. Hence, tumours that develop during immune suppression, such as Merkel cell carcinoma, are highly immunogenic when the immune system is restored. (C) Adding anti-PD-L1, or comparable agents, which disrupt T-cell inhibitory receptors facilitates tumour cell recognition and killing by T cells.

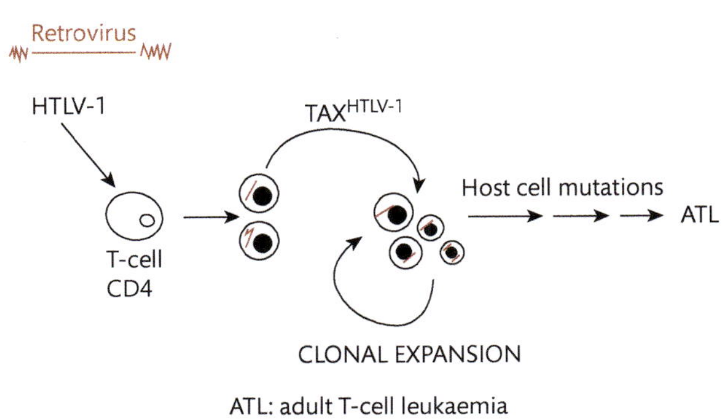

Fig. 6.4 Transformation principle for HTLV-1. Infection of CD4 T cells by the retrovirus HTLV-1 leads to clonal expansion driven by the viral Tax protein (and potentially other viral factors). This leads to clonal expansion of the infected cell, even in the absence of an antigenic stimulus. Clonal expansion results in an increased pool of activated cells susceptible to mutation and eventually in the emergence of ATL.

Tax is required for viral replication and Tax is required for transformation in culture. At the same time Tax:CREB:p300:CBP also modulates host genes with CRE elements and Tax reportedly also interacts with NFκB. These events reprogram the cell and are also believed to allow for increased rates of DNA damage and chromosomal instability (Poiesz et al., 1981). Since carcinogenesis is the result of accumulating oncogenic mutations, increasing the rates and tolerance for mutations leads to Tax-dependent, albeit indirect, transformation.

Oncogenic herpesviruses: EBV and KSHV

Herpesviruses have evolved with their host during most of speciation. Tumour-associated herpesviruses have been identified in birds (Marek's disease virus of chickens), amphibians (Luke herpesvirus of frogs; Naegele et al., 1974), dolphins and mammals, including human and non-human primates. Marek's disease virus is interesting because here transforming activity maps to a cluster of viral micro RNAs (miRNAs), rather than a protein (Zhao et al., 2011). Analogous to molecular piracy of oncoproteins, molecular piracy of miRNAs serves to target proteins of interest to viral replication.

KSHV carries an orthologue of human miR-155 (Gottwein et al., 2007; Skalsky et al., 2007), whereas EBV has evolved to aberrantly induce mir-155 in infected lymphocytes. Forced expression of either the viral or human miR-155 causes transformation in appropriate model systems.

KSHV and EBV both establish dormancy in B cells (Fig. 6.5). Analogous to the small DNA viruses, their genomes persist as extrachromosomal plasmids in infected cells. Because these herpesviruses express viral proteins that ensure synchronous genome replication between virus and host, as well as equal segregation during cell division, the viral genome is never diluted out and is not dependent on intermediate bursts of replication to be maintained in dividing lymphocytes. This feature distinguishes herpesviruses from small DNA tumour viruses and allows for extended, lifelong persistence, which is termed latency.

Herpesviruses encode their own DNA polymerase and associated DNA synthesis enzymes and therefore they do not depend on the cellular replication machinery. As matter of fact, most herpesviruses replicate productively in growth-arrested fibroblast monolayers or even postmitotic neurons. They do not transform fibroblasts, but EBV and the squirrel monkey homologue of KSHV called herpesvirus saimiri (HVS) transform mature, human lymphocytes in culture, the same lineage where these viruses establish latency. One could envision that the 'motivation' to evolve pro-growth capabilities for these viruses has been to ensure the preferential survival of latently infected lymphocytes. In the case of EBV and KSHV, infected B cells are in competition with uninfected B cells. The infected lymphoid cells are able to proliferate more, escape negative selection and terminal differentiation, and/or are able to respond to suboptimal levels of antigenic stimuli ahead of uninfected cells. As a result, more and more virus-infected cells accumulate. In such a scenario, viral mechanisms that target B-cell receptor and growth-factor signalling evolved. For instance, KSHV encodes a homologue of the B-cell growth factor IL-6. Human or viral IL6 overexpression

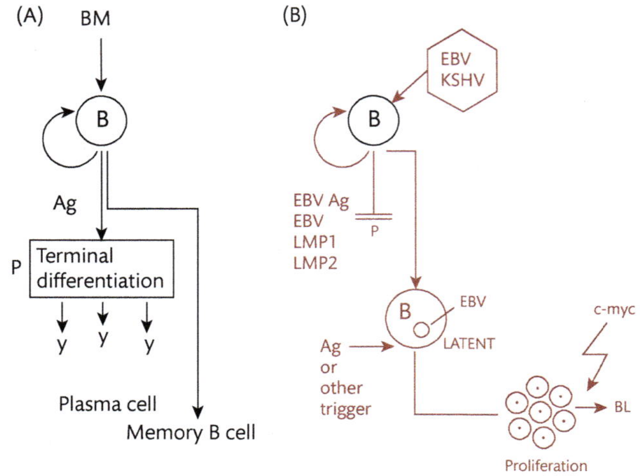

Fig. 6.5 Transformation principle for large DNA tumour viruses EBV and KSHV. (A) In normal B-cell development, cells originate from the bone marrow (BM) and mature in the spleen and lymph nodes, where they encounter antigen (ag). This leads to activation, proliferation, and affinity maturation in the germinal centre. Eventually B cells leave the germinal centre and differentiate into either plasma cells, which secrete IgG (IgM, IgE, or IgA), or memory B cells, which carry IgM/IgG on the cell surface. (B) EBV and KSHV infect B cells, and in the case of EBV induce rapid proliferation through the expression of LMP1 and LMP2 proteins, mimicking antigen exposure. These viruses block terminal differentiation and cell death and latently infected cells emerge from the germinal centre (GC). In the case of EBV, these are memory B cells. Upon Ag encounter or upon other triggers of normal B-cell differentiation, the virus either replicates and/or leads to preferential proliferation of virally infected B cells. Over time the fraction of infected cells increases and with it the probability to sustain additional mutations in human proto-oncogenes, foremost among them c-myc. Translocation of c-Myc leads to overexpression and Burkitt lymphoma (BL).

causes lymphoma in transgenic mice. Neutralization of IL6 (siltuximab or tocilizumab) in human patients reduces B-cell hyperplasia. KSHV, EBV, and HVS carry transforming oncogenes that localize to the cell membrane: K1 and K15 in the case of KSHV; LMP1 and LMP2A in the case of EBV; and STP and TIP in the case of HVS (Damania, 2004). The KSHV K1 protein can substitute for STP in HVS-mediated T-cell transformation and also extends the lifespan of primary endothelial cells in culture. LMP1 is a viral homologue of CD40 and thus able to mimic CD40L activation in LMP1-expressing cells (Uchida et al., 1999). The EBV LMP2 protein activates TRAF6 and ultimately NF-κB and the KSHV vFLIP protein also constitutively activates NF-κB in latently infected cells. KSHV encodes a D-type cyclin homologue that can activate CDK4/6 and is resistant to inhibition by p21, whereas EBV potently induces human cyclin-D upon primary infection.

Since herpesviruses have 'the luxury' of large genomes, they evolved to interfere with cellular signalling at multiple junctions. In the context of an immune cell, like a B cell, signalling drives response to antigen and specific effector functions. Yet, the same pathways, including TP53-associated signalling circuits contribute to oncogenesis in abnormal scenarios or at specific lymphocyte differentiation states. In addition to lymphoma, EBV and KSHV are also associated with non-lymphoid cancers: nasopharyngeal carcinoma (NPC) and a subset of upper gastric cancers in the case of EBV; and KS, an endothelial lineage cancer, in the case of KSHV.

TAKE-HOME MESSAGE

- Cancer of viral aetiology account for up to a quarter of all cancers worldwide. They occur disproportionally in low and middle-income countries.
- The study of tumour viruses per se led to fundamental insights into cancer biology (e.g. the discovery of the myc and ras oncogenes and of the TP53 tumour suppressor gene).
- Viral cancers are unique, since in many cases vaccination and screening can prevent cancer progression (e.g. human papilloma virus 16/18-associated cervical cancer).
- In other cases, virus-directed therapy offers additional avenues of treatment (e.g. hepatitis C virus-associated liver cancer).

OPEN QUESTIONS

- **HIV and cancer:** Being infected with HIV can be considered a chronic disease, at least for those with reliable access to combination antiretroviral therapy (cART). The leading causes of mortality and morbidity in people living with HIV/AIDS (PLWHA) in the United States are cancer and heart disease. For instance, KS remains the leading cancer in PLWHA in the United States even in patients on cART. One-third of KS cases in the United States now occur in patients with no detectable HIV (Krown et al., 2008). The damage to the immune system and likely the cancer microenvironment occurs during primary infection and latency; removal or cure of HIV afterwards is unlikely to alter cancer risk. There exists a dramatically increased risk for many cancers in PLWHA and cancer prevalence is expected to increase as the cohort of PLWHA ages. There remain many unanswered questions: we do not understand whether cancers that develop in the context of HIV infection are different from those that develop in its absence, and we do not know if PLWHA and cancer respond to the same therapies, or how anti-HIV and anticancer drugs interact. Many drugs used to treat cancer in HIV-negative individuals are contraindicated in combination with cART. Often clinical trials for cancer exclude participation of HIV-positive individuals. This deprives PLWHA of many advanced treatments. Understanding how anticancer treatments interact with cART represents an urgent research goal.
- **Virus-associated cancer in low- and middle-income countries:** By 2030, the majority of cancers will occur in developing nations (Thun et al., 2010). Presently the leading cause of cancer mortality and morbidity in the developing world are cancers of viral origin: KS, cervical cancer, liver cancer, and Burkitt lymphoma (Parkin et al., 2014). These can be prevented with screening and vaccination, and their burden can be relieved with adequate therapy.

FURTHER READING

Flint, S. J., Enquist, L. W., Racaniello, V. R., et al. (2003). *Principles of Virology: Molecular Biology, Pathogenesis, and Control of Animal Viruses*, 2nd edition. Washington, DC: ASM Press.

This Week in Virology (TWIV). Available at: http://www.microbe.tv/twiv/

Yarchoan, R. (2014). *Cancers in People with HIV and AIDS: Progress and Challenges*. New York: Springer.

REFERENCES

Albini, A., Barillari, G., Benelli, R., Gallo, R. C., & Ensoli, B. (1995). Angiogenic properties of human immunodeficiency virus type 1 Tat protein. *Proc Natl Acad Sci U S A*, **92**, 4838–42.

Ang, K. K., Harris, J., Wheeler, R., et al. (2010). Human papillomavirus and survival of patients with oropharyngeal cancer. *N Engl J Med*, **363**, 24–35.

Baltimore, D. (1971). Expression of animal virus genomes. *Bacteriol Rev*, **35**, 235–41.

Burkitt, D. (1962). A children's cancer dependent on climatic factors. *Nature*, **194**, 232–4.

Cancer Genome Atlas Network (2015). Comprehensive genomic characterization of head and neck squamous cell carcinomas. *Nature*, **517**, 576–82.

Cesarman, E., Chang, Y., Moore, P. S., Said, J. W., & Knowles, D. M. (1995). Kaposi's sarcoma-associated herpesvirus-like DNA sequences in AIDS-related body-cavity-based lymphomas. *N Engl J Med*, **332**, 1186–91.

Chang, Y., Cesarman, E., Pessin, M. S., et al. (1994). Identification of herpesvirus-like DNA sequences in AIDS-associated Kaposi's sarcoma. *Science*, **266**, 1865–9.

Choo, Q. L., Kuo, G., Weiner, A. J., Overby, L. R., Bradley, D. W., & Houghton, M. (1989). Isolation of a cDNA clone derived from a blood-borne non-A, non-B viral hepatitis genome. *Science*, **244**, 359–62.

Courtneidge, S. A. & Smith, A. E. (1983). Polyoma virus transforming protein associates with the product of the c-rc cellular gene. *Nature*, **303**, 435–9.

Damania, B. (2004). Oncogenic gamma-herpesviruses: comparison of viral proteins involved in tumorigenesis. *Nat Rev Microbiol*, **2**, 656–68.

Dane, D. S., Cameron, C. H., & Briggs, M. (1970). Virus-like particles in serum of patients with Australia-antigen-associated hepatitis. *Lancet*, **1**, 695–8.

Durst, M., Gissmann, L., Ikenberg, H., & Zur Hausen, H. (1983). A papillomavirus DNA from a cervical carcinoma and its prevalence in cancer biopsy samples from different geographic regions. *Proc Natl Acad Sci U S A*, **80**, 3812–15.

Epstein, M. A. & Barr, Y. M. (1964). Cultivation *in vitro* of human lymphoblasts from Burkitt's malignant lymphoma. *Lancet*, **1**, 252–3.

Feng, H., Shuda, M., Chang, Y., & Moore, P. S. (2008). Clonal integration of a polyomavirus in human Merkel cell carcinoma. *Science*, **319**, 1096–100.

Fredricks, D. N. & Relman, D. A. (1996). Sequence-based identification of microbial pathogens: a reconsideration of Koch's postulates. *Clin Microbiol Rev*, **9**, 18–33.

Gillison, M. L., Koch, W. M., Capone, R. B., et al. (2000). Evidence for a causal association between human papillomavirus and a subset of head and neck cancers. *J Natl Cancer Inst*, **92**, 709–20.

Gottwein, E., Mukherjee, N., Sachse, C., et al. (2007). A viral microRNA functions as an orthologue of cellular miR-155. *Nature*, **450**, 1096–9.

Hacein-Bey-Abina, S., Garrigue, A., Wang, G. P., et al. (2008). Insertional oncogenesis in 4 patients after retrovirus-mediated gene therapy of SCID-X1. *J Clin Invest*, **118**, 3132–42.

Hanahan, D. & Weinberg, R. A. (2011). Hallmarks of cancer: the next generation. *Cell*, **144**, 646–74.

Hayflick, L. (1965). The limited *in vitro* lifetime of human diploid cell strains. *Exp Cell Res*, **37**, 614–36.

Henle, W., Diehl, V., Kohn, G., Zur Hausen, H., & Henle, G. (1967). Herpes-type virus and chromosome marker in normal leukocytes after growth with irradiated Burkitt cells. *Science*, **157**, 1064–5.

Herbst, H., Dallenbach, F., Hummel, M., et al. (1991). Epstein-Barr virus latent membrane protein expression in Hodgkin and Reed-Sternberg cells. *Proc Natl Acad Sci U S A*, **88**, 4766–70.

IARC (2012). *A review of human carcinogens. Part B: Biological agents/ IARC Working Group on the Evaluation of Carcinogenic Risks to Humans* (2009: Lyon, France). Geneva, Switzerland: WHO Press.

Jakowski, R. M., Fredrickson, T. N., Schierman, L. W., & McBride, R. A. (1974). A transplantable lymphoma induced with Marek's disease virus. *J Natl Cancer Inst*, **53**, 783–9.

Joura, E. A., Giuliano, A. R., Iversen, O. E., et al. (2015). A 9-valent HPV vaccine against infection and intraepithelial neoplasia in women. *N Engl J Med*, **372**, 711–23.

Kirnbauer, R., Taub, J., Greenstone, H., et al. (1993). Efficient self-assembly of human papillomavirus type 16 L1 and L1-L2 into virus-like particles. *J Virol*, **67**, 6929–36.

Knipe, D. M. & Howley, P. M. (2013). *Fields Virology*, 6th edition. Philadelphia, PA: Wolters Kluwer/Lippincott Williams & Wilkins Health.

Krown, S. E., Lee, J. Y., Dittmer, D. P., & Consortium, A. M. (2008). More on HIV-associated Kaposi's sarcoma. *N Engl J Med*, **358**, 535–6; author reply 536.

Lindenbach, B. D., Evans, M. J., Syder, A. J., et al. (2005). Complete replication of hepatitis C virus in cell culture. *Science*, **309**, 623–6.

Linzer, D. I. & Levine, A. J. (1979). Characterization of a 54K dalton cellular SV40 tumor antigen present in SV40-transformed cells and uninfected embryonal carcinoma cells. *Cell*, **17**, 43–52.

Millman, I., Zavatone, V., Gerstley, B. J., & Blumberg, B. S. (1969). Australia antigen detected in the nuclei of liver cells of patients with viral hepatitis by the fluorescent antibody technic. *Nature*, **222**, 181–4.

Mustacchi, P. & Shimkin, M. B. (1958). Cancer of the bladder and infestation with Schistosoma hematobium. *J Natl Cancer Inst*, **20**, 825–42.

Naegele, R. F., Granoff, A., & Darlington, R. W. (1974). The presence of the Lucke herpesvirus genome in induced tadpole tumors and its oncogenicity: Koch-Henle postulates fulfilled. *Proc Natl Acad Sci U S A*, **71**, 830–4.

Nonoyama, M. & Pagano, J. S. (1973). Homology between Epstein-Barr virus DNA and viral DNA from Burkitt's lymphoma and nasopharyngeal carcinoma determined by DNA-DNA reassociation kinetics. *Nature*, **242**, 44–7.

Palefsky, J. M., Holly, E. A., Gonzales, J., Berline, J., Ahn, D. K., & Greenspan, J. S. (1991). Detection of human papillomavirus DNA in anal intraepithelial neoplasia and anal cancer. *Cancer Res*, **51**, 1014–19.

Pallesen, G., Sandvej, K., Hamilton-Dutoit, S. J., Rowe, M., & Young, L. S. (1991). Activation of Epstein-Barr virus replication in Hodgkin and Reed-Sternberg cells. *Blood*, **78**, 1162–5.

Parkin, D. M., Bray, F., Ferlay, J., & Jemal, A. (2014). Cancer in Africa 2012. *Cancer Epidemiol Biomarkers Prev*, **23**, 953–66.

Parkin, D. M., Ohshima, H., Srivatanakul, P., & Vatanasapt, V. (1993). Cholangiocarcinoma: epidemiology, mechanisms of carcinogenesis and prevention. *Cancer Epidemiol Biomarkers Prev*, **2**, 537–44.

Poiesz, B. J., Ruscetti, F. W., Reitz, M. S., Kalyanaraman, V. S., & Gallo, R. C. (1981). Isolation of a new type C retrovirus (HTLV) in primary uncultured cells of a patient with Sezary T-cell leukaemia. *Nature*, **294**, 268–71.

Ringelhan, M. & Protzer, U. (2015). Oncogenic potential of hepatitis B virus encoded proteins. *Curr Opin Virol*, **14**, 109–15.

Robbins, H. A., Pfeiffer, R. M., Shiels, M. S., Li, J., Hall, H. I., & Engels, E. A. (2015). Excess cancers among HIV-infected people in the United States. *J Natl Cancer Inst*, **107**, pii: dju503.

Saeed, M., Andreo, U., Chung, H. Y., et al. (2015). *SEC14L2* enables pan-genotype HCV replication in cell culture. *Nature*, **524**, 471–5.

Sarnow, P., Ho, Y. S., Williams, J., & Levine, A. J. (1982). Adenovirus E1b-58kd tumor antigen and SV40 large tumor antigen are physically associated with the same 54 kd cellular protein in transformed cells. *Cell*, **28**, 387–94.

Scheffner, M., Werness, B. A., Huibregtse, J. M., Levine, A. J., & Howley, P. M. (1990). The E6 oncoprotein encoded by human papillomavirus types 16 and 18 promotes the degradation of p53. *Cell*, **63**, 1129–36.

Skalsky, R. L., Samols, M. A., Plaisance, K. B., et al. (2007). Kaposi's sarcoma-associated herpesvirus encodes an ortholog of miR-155. *J Virol*, **81**, 12836–45.

Soulier, J., Grollet, L., Oksenhendler, E., et al. (1995). Kaposi's sarcoma-associated herpesvirus-like DNA sequences in multicentric Castleman's disease. *Blood*, **86**, 1276–80.

Stehelin, D., Varmus, H. E., Bishop, J. M., & Vogt, P. K. (1976). DNA related to the transforming gene(s) of avian sarcoma viruses is present in normal avian DNA. *Nature*, **260**, 170–3.

Sulkowski, M. S., Gardiner, D. F., Rodriguez-Torres, M., et al. (2014). Daclatasvir plus sofosbuvir for previously treated or untreated chronic HCV infection. *N Engl J Med*, **370**, 211–21.

Thun, M. J., Delancey, J. O., Center, M. M., Jemal, A., & Ward, E. M. (2010). The global burden of cancer: priorities for prevention. *Carcinogenesis*, **31**, 100–10.

Uchida, J., Yasui, T., Takaoka-Shichijo, Y., et al. (1999). Mimicry of CD40 signals by Epstein-Barr virus LMP1 in B lymphocyte responses. *Science*, **286**, 300–3.

Wakita, T., Pietschmann, T., Kato, T., et al. (2005). Production of infectious hepatitis C virus in tissue culture from a cloned viral genome. *Nat Med*, **11**, 791–6.

Wang, L., Galehouse, D., Mellon, P., Duesberg, P., Mason, W. S., & Vogt, P. K. (1976). Mapping oligonucleotides of Rous sarcoma virus RNA that segregate with polymerase and group-specific antigen markers in recombinants. *Proc Natl Acad Sci U S A*, **73**, 3952–6.

Warren, J. & Marshall, B. (1983). Unidentified curved bacilli on gastric epithelium in active chronic gastritis. *Lancet*, **1**, 1273–5.

Zhao, Y., Xu, H., Yao, Y., et al. (2011). Critical role of the virus-encoded microRNA-155 ortholog in the induction of Marek's disease lymphomas. *PLoS Pathog*, **7**, e1001305.

Zur Hausen, H., Schulte-Holthausen, H., Klein, G., et al. (1970). EBV DNA in biopsies of Burkitt tumours and anaplastic carcinomas of the nasopharynx. *Nature*, **228**, 1056–8.

7
Chemical carcinogens

David H. Phillips

Introduction to chemical carcinogens

The term 'carcinogen' was probably first used in 1853 by Sir James Paget in his *Lectures on Surgical Pathology* '. . . is there one material for cancer, one carcinogen, which, like an organic radical, may form different yet closely allied compounds, in its combinations with the various substances provided by different bloods, or different parts?'

While the principle remains the same, we now know that there are many carcinogens, defined as substances or agents that have the property of increasing tumour incidence in an exposed species.

Many substances and agents are carcinogenic. They can be chemical (synthetic or naturally occurring), physical (ionizing and non-ionizing radiation, mineral fibres), or biological (viruses, bacteria, parasitic organisms). This chapter focuses on chemical carcinogens.

Identifying and studying chemical carcinogens can serve several scientific and public health benefits. There is the potential to shed light on the causes of human cancer and thereby to initiate prevention strategies. Identifying the potential for carcinogenesis among the many thousands of new chemicals synthesized each year provides a means for safety testing and the elimination from the market of harmful products. Understanding how chemical carcinogens exert their biological effects furthers understanding of mechanisms of carcinogenesis and offers opportunities for monitoring human exposure to carcinogens, identifying early effects and approaches towards cancer prevention or abrogation.

Milestones in chemical carcinogenesis

The first observations of an association between chemical exposure and cancer were made in the late eighteenth century. First John Hill, a London physician, noted in 1761 that a number of cases of nasal cancer were apparently associated with excessive use of snuff; then Percival Pott, a London surgeon, wrote in 1775 that the high incidence of cancer of the scrotum among chimney sweeps 'seems to derive its origin from lodgement of soot in the rugae of the scrotum'. As the Industrial Revolution gathered pace through the nineteenth century, it became apparent that employment in industries where exposure to other fossil fuel products, including paraffin refining, shale oil, and coal tar industries, resulted in high incidences of skin cancer. Observations on increased incidences of bladder cancer were also made among workers in the dye industry. However, but it was not until the twentieth century that any of these substances were shown experimentally to be carcinogenic. This was achieved in 1915 by applying the materials repeatedly to the inner surface of the ears of rabbits and, in subsequent experiments, to the skin of mice. These successes then made it possible to elucidate the nature of the carcinogenic agents in these complex mixtures, and it was established in the following years that the principal components were confined to high-boiling fractions that were free of nitrogen, arsenic, and sulphur, and had characteristics of complex aromatic hydrocarbons. The production of synthetic carcinogenic tars by heating acetylene and isoprene in a hydrogen atmosphere strengthened the case, as did the observation that the fluorescence spectra of carcinogenic extracts of tars had features similar to the fluorescence spectrum of benz[a]anthracene, a polycyclic aromatic hydrocarbon (PAH). Kennaway and colleagues in London therefore undertook the synthesis of many PAHs and tested them for carcinogenicity. In 1930 they reported that dibenz[a,h]anthracene induced skin tumours in mice; this was the first demonstration of the carcinogenicity of a pure chemical. Then in 1933 they isolated benzo[a]pyrene from coal tar and demonstrated that it was also carcinogenic (Phillips, 1983).

Also, in the 1930s there were reports of the carcinogenicity in experimental animals of several aromatic amines, products, or by-products of the synthetic dye industry. By the early 1960s, chemicals with a broad range of structures, including many other PAHs, aromatic amines, oestrone, metal salts, nitrogen mustards, and N-nitroso compounds had been demonstrated to be carcinogenic, plus some naturally occurring chemicals, including pyrrolidine alkaloids (found in many plant species) and aflatoxin B_1, a mycotoxin produced by *Aspergillus flavus* (Miller and Miller, 1979).

Further studies on the carcinogenicity of substances on the skin of mice demonstrated two phases of cancer development (Miller and Miller, 1979). Application of single doses of PAHs (initiation) followed by repeated doses of croton oil (promotion) produced tumours with high frequency; these were mostly papillomas. Croton oil alone did not induce tumours, and interruption of the repeated dosing regimen also prevented tumour formation, although tumours subsequently appeared if the treatment was resumed. From these observations it was deduced that initiation, involving only a single treatment, was irreversible, but that promotion appeared to be a protracted and, at least in its early phases, reversible process.

Later a third phase, progression, was described to account for later events whereby some papillomas developed into malignant and invasive carcinomas, sometimes after further treatment with a carcinogen. Compounds that are initiators can also be complete carcinogens, whereby they can induce tumours alone if given in high enough dose, or repeatedly; for example, the PAHs. An example of a pure initiator, at least on mouse skin, is ethyl carbamate.

A similar multistage system was developed in experiments inducing liver tumours in rats. Feeding the animals 2-acetylaminofluorene for a limited period, followed by phenobarbital for many months led to a higher incidence of liver tumours than in rats fed 2-acetylaminofluorene alone, while rats fed only phenobarbital did not develop tumours at all; in this model it can been seen that 2-acetylaminofluorene is an initiator (and also a complete carcinogen), and phenobarbital is a promoter (Miller and Miller, 1979).

Metabolism and metabolic activation

From the earliest studies on chemical carcinogens it became apparent that compounds of different chemical classes, with diverse structures and chemical properties, could be carcinogenic. Despite this, initial efforts at finding a unifying theory to account for carcinogenic properties were founded on the assumption that the answer lay in there being some structural or electronic property of the carcinogens that distinguished them from non-carcinogens. Extensive theoretical studies and calculations were made, for example, on the PAHs, to define the characteristics that defined carcinogenicity, but ultimately these attempts all failed (Phillips, 1983).

Meanwhile other investigators had been studying the metabolic fate of carcinogens and several of these were found to have the property of binding strongly to proteins and also to DNA. As many carcinogens were chemically somewhat inert, the suspicion was that the binding was due to metabolites, rather than the parent compounds. Another clue to the conundrum came from the observation that a metabolite of a carcinogen was more potently carcinogenic than its parent compound. This strongly suggested that metabolic activation of some carcinogens might be required for them to exert their carcinogenic effects (Miller and Miller, 1979).

James and Elizabeth Miller proposed that what structurally diverse carcinogens had in common was their metabolism to electrophilic species, which they termed ultimate carcinogens (Miller, 1970; Miller, 1978; Miller and Miller, 1981). Metabolites with high(er) carcinogenic potency were termed proximate carcinogens (i.e. intermediates in the pathway to the ultimately reactive species). The initial hypothesis that proteins were the critical target for carcinogenesis gave way to the current understanding that DNA is in fact the true target, when it was shown in 1964 that the extent of binding of a series of PAHs to DNA in mouse skin, but not to RNA or protein, correlated with their carcinogenic potencies (Brookes and Lawley, 1964). Subsequent experiments that showed that many carcinogens were also mutagens in bacteria when tested in the presence of subcellular liver fractions that metabolize the compounds to their ultimately reactive forms (Ames et al., 1973) confirmed that DNA is the critical target for carcinogenesis and that cancer development is driven by mutation. It thus became clear that the property of covalent binding to DNA, originally found for direct-acting mutagenic carcinogens such as alkylating agents, including nitrogen mustard (Lawley, 1989), was also shared by the somewhat chemically inert environmental carcinogens (for example, PAHs), but that the latter required metabolic activation first.

Why do cells convert relatively inert chemicals into reactive intermediates that do them harm? The phenomenon can be

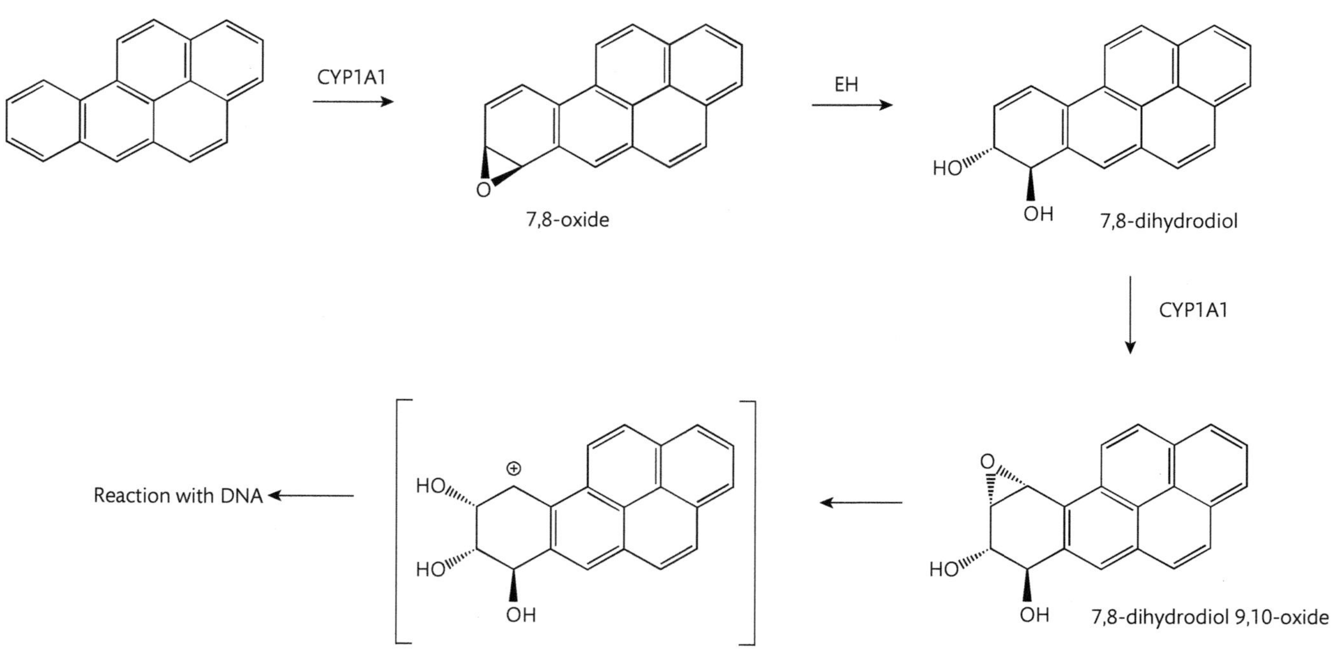

Fig. 7.1 Metabolic activation of benzo[a]pyrene.

Adapted with permission from Phillips, D. H. 'The formation of DNA adducts', pp. 338–350 in Alison M. R (Ed.), *The Cancer Handbook*, London, UK, Copyright © 2007 John Wiley and Sons, Inc. All Rights Reserved.

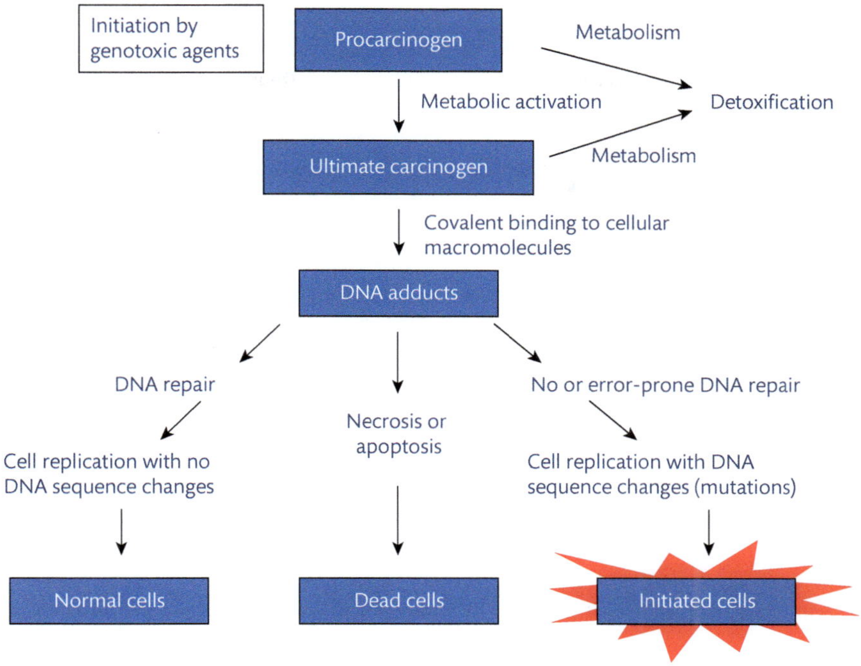

Fig. 7.2 Schematic showing how carcinogens are activated to cause DNA damage, and the cellular processes that ensue.

considered to be an aberration of a detoxication process that has evolved to render lipophilic compounds more hydrophilic (water-soluble) and thus more readily removed from the cell and excreted from the organism. The process can require several steps involving a number of enzymes and non-enzymatic events, any number of which can generate reactive intermediates. In most instances, rapid further metabolism of these intermediates means there is no harm to the cell. It is when such intermediates persist inadvertently for any length of time that their presence can result in damage to cellular macromolecules, including DNA, thereby initiating the deleterious effects of mutagenicity and carcinogenicity. Consider, as an example, the highly lipophilic carcinogen benzo[a]pyrene. Multiple positions in the molecule undergo oxidative metabolism to phenols, dihydrodiols, and diones. The pathway that leads to activation of benzo[a]pyrene (Fig. 7.1) involves an initial epoxidation at the 7,8-positions, catalysed by cytochrome P450, followed by a rapid conversion of the reactive intermediate 7,8-epoxide to the 7,8-dihydrodiol by epoxide hydrolase. Further epoxidation then occurs at the 9,10-double bond, to form a diol-epoxide, which is chemically reactive (Sims et al., 1974). This turns out to be a poor substrate for further metabolism by epoxide hydrolase and glutathione transferases, with the result that it can persist long enough in cells to react with cellular macromolecules, including DNA, and exert its mutagenic and carcinogenic effects.

For many xenobiotics, metabolism (detoxication) involves an initial oxidation step (phase I metabolism), which is commonly carried out by one of the members of the cytochrome P450 superfamily (Rendic and Guengerich, 2012) followed by conjugation to an acetate, sulfate, or glucuronide (Phase II metabolism), by *N*-acetyltransferases, sulfotransferases, or glucuronosyltransferases, respectively. For some carcinogens, however, the instability of these conjugates can lead to their dissociation, resulting in metabolic activation by generating electrophiles that can react with DNA, thereby exerting their carcinogenic effects.

A variety of nucleophilic sites in DNA can be modified by the electrophilic species generated by metabolic activation of carcinogens (see 'Classes of chemical carcinogens'). Most of the covalent binding occurs on the purine and pyrimidine bases, although some agents also react with the phosphodiester backbone to form phosphotriesters. The most commonly modified base is guanine, followed by adenine, with the pyrimidines cytosine and thymine less liable to modification. However, the sites of modification vary from agent to agent and it depends on the mechanism of the substitution reaction as to which base, and which position on the base, is the preferred site of reaction (Lawley, 1984).

Figure 7.2 demonstrates the principles of carcinogen activation outlined here, together with the possible consequences of carcinogen-induced damage—it can be repaired by one of several cellular DNA repair mechanisms (in which case the genome is restored to its undamaged state); it can cause cell death, or it can lead to mutation as a result of an error-prone process or as a result of miscoding of DNA during replication, leading to a daughter cell possessing a sequence alteration from that of the parent cell. Such mutations are the initiating events in carcinogenesis.

Classes of chemical carcinogens

An extensive and comprehensive series of authoritative publications list the many chemicals, lifestyle practices, and biological agents that are known to cause or are suspected of causing human cancer. Since 1971 the International Agency for Research on Cancer (IARC) has published over 100 monographs (IARC

Monographs, n.d.) identifying, of the 990 agents evaluated to date, 118 that are known to be carcinogenic to humans (Group 1), 80 that are probably carcinogenic to humans (Group 2A) and 289 that are possibly carcinogenic to humans (Group 2B). The evaluations are based on epidemiological evidence of human exposure and associated cancer, evidence from animal experiments and, where available, evidence from *in vitro* experiments providing insights into the mechanism(s) of action of the agents. Classification depends on whether these various categories of evidence are considered to be sufficient, limited, or inadequate. Therefore it is important to note that the classifications are based on the strength of the evidence and not on the potency of the carcinogen; they are designed to provide qualitative, rather than quantitative, judgements on carcinogenicity and identify cancer hazards, rather than assess risks.

Examples of carcinogens are shown in **Table 7.1**, together with their reactive metabolites and sites of modification of DNA. PAHs, which are metabolically activated to diol-epoxides, modify the exocyclic amino groups of guanine and adenine, while aromatic amines (e.g. 2-acetylaminofluorene, AAF) and heterocyclic amines (e.g. PhIP), activated through their amino groups, show highest affinity for the C8 position of guanine, as do nitroaromatic polycyclic hydrocarbons (nitro-PAHs; e.g. 1-nitropyrene), when activated through nitro reduction. Aflatoxin B_1, a naturally occurring mycotoxin, preferentially modifies the N7 position of guanine. This adduct carries an electronic charge rendering it unstable; it may undergo hydrolysis to yield aflatoxin dihydrodiol (the DNA returning to an unmodified state), it may lead to depurination to yield an aflatoxin-base adduct and an apurinic site in DNA, or it may undergo a ring opening of the guanine residue to yield a stable adduct. A consequence of the first and second options is that metabolites and modified bases of aflatoxin are detectable in the urine of exposure individuals, as is described later in this chapter. Adducts formed by alkyl halides (e.g. vinyl chloride) involve

Table 7.1 Some representative carcinogens, their active metabolites, and sites of modification of DNA

Carcinogen	Major active metabolite	Sites of modification of DNA
Benzo[*a*]pyrene (BP) 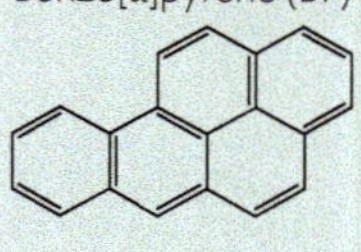	BP 7,8-diol 9,10-epoxide 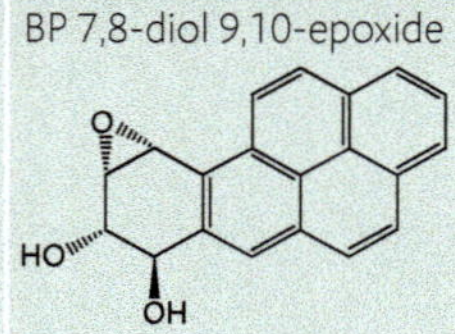	N^2-guanine, N^6-adenine
Benzo[*c*]phenanthrene (BcPh) 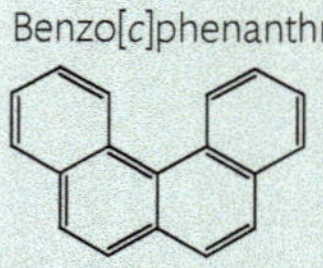	BcPh 4,3-diol 2,1-epoxide 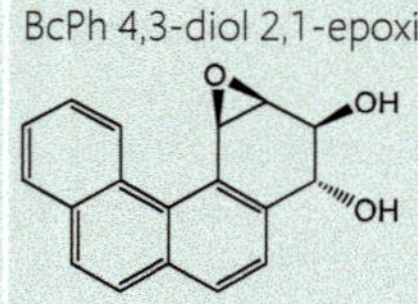	N^6-adenine, N^2-guanine
Aflatoxin B_1 (AfB_1) 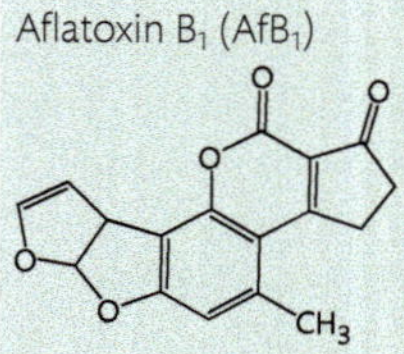	AfB_1 8,9-epoxide 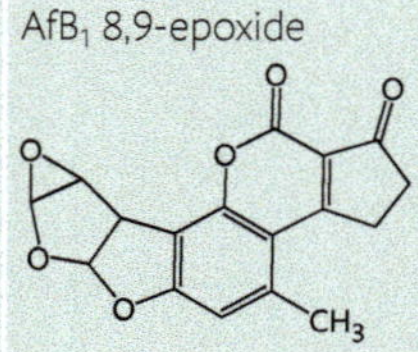	N7-guanine
Tamoxifen 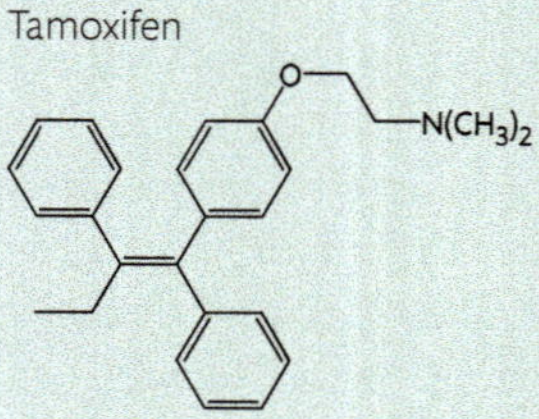	α-hydroxytamoxifen sulfate 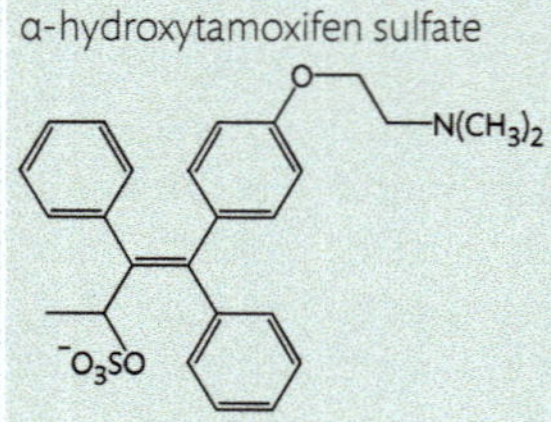	N^2-guanine, N^6-adenine
2-Acetylaminofluorene (AAF) 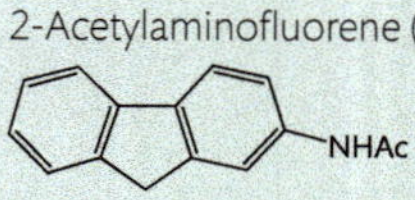	*N*-acetoxy-AAF 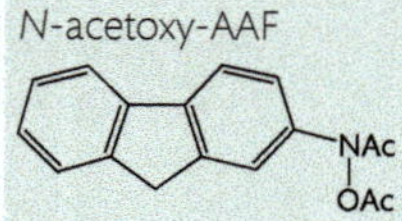	C8-guanine, N^2-guanine
Vinyl chloride 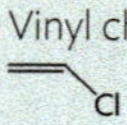	Chloroethylene oxide 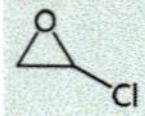	3,N^4-cytosine, 1,N^6-adenine, 3,N^2-guanine

(continued)

Table 7.1 Continued

Carcinogen	Major active metabolite	Sites of modification of DNA
2-Amino-1-methyl-6-phenylimidazo[4,5-*b*]pyridine (PhIP) 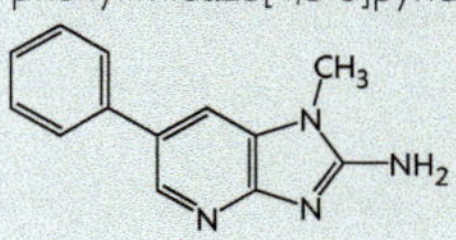	N-Acetoxy-PhIP 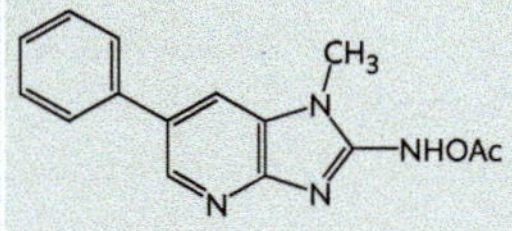	C8-guanine
1-Nitropyrene (1-NP) 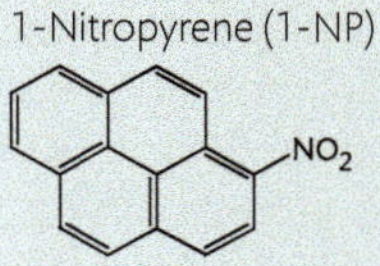	N-Hydroxy-1-aminopyrene 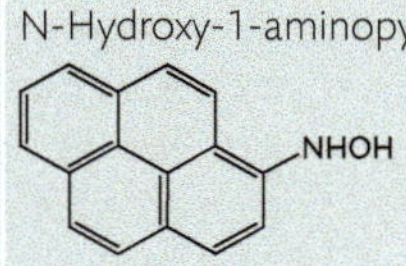	C8-guanine
3-Nitrobenzanthrone 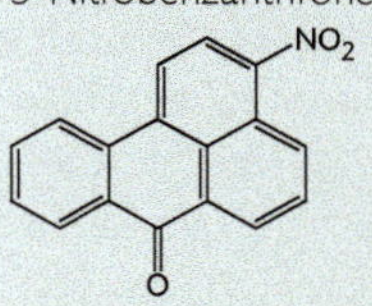	N-Hydroxy-3-aminobenzanthrone 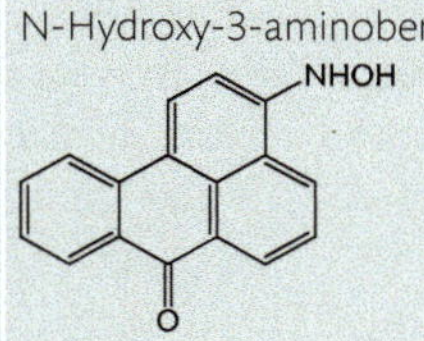	N^2-guanine, C8-guanine, N^6-adenine
Aristolochic acid I (*R = OCH3*) & Aristolochic acid II (*R = H*) 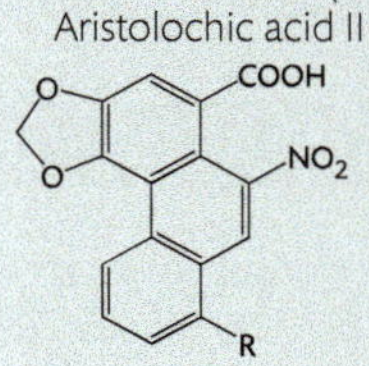	N-Hydroxyaristolactam 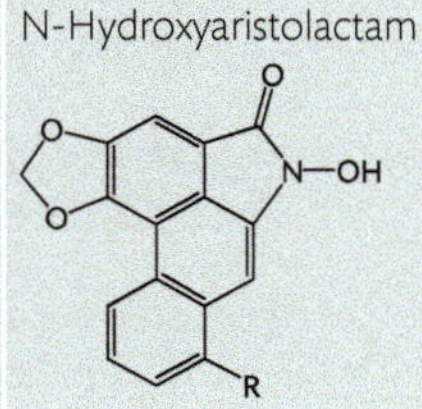	N^6-adenine, N^2-guanine

Adapted with permission from Phillips, D. H. 'The formation of DNA adducts', pp. 338–350 in Alison M. R (Ed.), *The Cancer Handbook*, London, UK,

formation of etheno adducts, with bonding to two positions in the base, forming a new ring structure.

Different carcinogens induce tumour formation in different organs in rodents. Thus, aflatoxin B_1 induces liver tumours in rats, fish, ducks, and monkeys, but not mice, following oral administration; in mice intraperitoneal administration caused lung tumours (Eaton and Gallagher, 1994). Tamoxifen induces liver tumours in rats, but not in mice (Phillips, 2001), AAF is carcinogenic in many species (rat, mouse, hamster, rabbit) and in many organs, including liver, bladder, mammary gland, ear duct and gastrointestinal tract (Verna et al., 1996). PAHs, such as benzo[a]pyrene, induce skin tumours in mice following topical application, and in lung, spleen, and forestomach (not liver) following oral administration (Hakura et al., 1998).

Human exposure to carcinogens is widespread. There are many chemicals that can cause cancer in occupational settings, such as the PAHs formed during combustion of fossil fuels that are known to increase cancer risk in workers involved in power generation, iron foundries, aluminium production, oil refining and other circumstances involving use or combustion of fossil fuel products (e.g. coal tar). These same chemicals are widespread environmental pollutants, contributing to the risk of cancer from air pollution, both indoor and outdoor (see 'Carcinogens that cause human cancer—exemplars'). Other industries historically associated with cancer risk and with exposure to known carcinogens are the rubber and dye industries (aromatic amines), welding (nickel and chromium fumes) and furniture making (inhalation of wood dust). Food is also an important potential source of human exposure. Food may contain carcinogens that are naturally occurring contaminants, such as fungal metabolites (aflatoxins), plant toxicants (aristolochic acid, pyrrolidine alkaloids), and others that may be formed during preservation (PAHs in smoked food; N-nitroso compounds in smoked and salt-preserved foods) or during the cooking process (PAHs when cooking over an open flame; heterocyclic amines formed during the cooking of protein-rich foods at high temperature; acrylamide formed during baking of carbohydrate-rich food). Many anticancer drugs are carcinogenic, posing a risk of secondary cancers to patients receiving the drugs as treatment for primary disease, but also a risk to medical personnel handling and administering them.

Mode of action

The characterization of tumour development consisting of initiation, promotion, and progression stages, which arose from the early classical experiments on tumour induction in rodents (see 'Milestones in chemical carcinogenesis'), still provides a basis for understanding some of the events in carcinogenesis. The irreversible process of initiation is now understood to emanate from the mutation of critical genes important for maintaining correct cellular function, such as proto-oncogenes, tumour suppressor genes, and DNA mismatch repair genes. The mutations may arise from interaction of mutagenic carcinogens with DNA and consequent misreplication during cell division, which will occur if the DNA damage is not repaired first, or if it does not lead to death of the cell through necrosis or apoptosis (see Fig. 7.2). Tumour promotion is a protracted, irreversible process that undoubtedly involves multiple events. In considering the differences between the properties of tumour cells and normal cells, a number of phenotypic characteristics that a cell needs to acquire to become malignant have been defined (Hanahan and Weinberg, 2000; Hanahan and Weinberg, 2011). These are self-sufficiency in growth signals, insensitivity to anti-growth signals, limitless replicative potential (i.e. immortality), evasion of apoptosis, sustained angiogenesis, reprogramming of energy metabolism, avoiding immune destruction, and tissue invasion and metastasis. In addition, underlying characteristics acquired include genome instability and mutation, and tumour promoting inflammation. However, this more complex and detailed description is not incompatible with the simpler initiation-promotion-progression model. Genome instability provides the driving force for acquiring new phenotypes, and mutations in 'caretaker' genes like *TP53* can accelerate the rate at which other phenotypic changes are acquired.

A current useful classification of carcinogenic modes of action (MOA) is to consider carcinogens to be either genotoxic, involving damage to DNA, or non-genotoxic. A genotoxic MOA can be *defined* as usually involving damage to DNA; a non-genotoxic MOA is less easy to define, apart from it *not* involving damage to DNA, and it can encompass a variety of mechanisms, including inducing hyperplasia, mitosis, and activating transcription factors through receptor-mediated perturbations.

A genotoxic MOA is characterized by:

- Positive results in one of more short-term tests for mutagenicity
- Usually carcinogenic in more than one species of test animal
- Carcinogenic to both sexes
- Exhibiting a dose response

A non-genotoxic MOA is characterized by:

- Negative results in all short-term tests for mutagenicity/genotoxicity
- Often carcinogenic in one sex of one species
- Carcinogenic at high dose only, with evidence or suggestion of a threshold

These properties should be regarded as indicative of a particular MOA, rather than proof of it, and further evidence should always be sought, such as the occurrence or absence of DNA damage. Even for some genotoxic agents, there may be mechanistic evidence for a threshold of effect. It may be that a genotoxic carcinogen will only cause DNA damage once a detoxification pathway is overloaded. An example is paracetamol, where DNA damage occurs once cellular glutathione has been depleted. Nevertheless, in the absence of *mechanistic* evidence to infer a threshold, it is prudent to assume that there is no threshold for mutagenicity.

There are also some exceptions to DNA being the target of genotoxic carcinogens. Examples include benomyl, carbendazim, and thiophanate-methyl, agents that induce aneugenicity by tubulin inhibition (i.e. interference with microtubule formation during mitosis). Other examples are topoisomerase inhibitors and antimetabolites (such as methotrexate).

Some carcinogens may have more than one MOA. Thus, tamoxifen is a genotoxic carcinogen in rat liver, but may be carcinogenic by a predominantly hormonal, non-genotoxic mechanism in human endometrium (Phillips, 2001).

As the majority of IARC Group 1 carcinogens appear to have a genotoxic mode of action, and non-genotoxic rodent carcinogens often exhibit a threshold, being active only at exposure levels much higher than would be encountered by humans, exposure to genotoxic carcinogens should be as low as reasonably achievable (the ALARA principle), while for non-genotoxic compounds allowable levels of exposure may be calculated based on risk assessment. Nevertheless, the fact that certain chronic inflammatory conditions and those associated with oxidative stress, including obesity, increase human cancer risk indicates that so-called promotion effects (to use the historical terminology) can have an impact on human cancer risk (Hussain et al., 2003) even if, mechanistically, they can be considered to exhibit a threshold of effect. It therefore remains important to understand MOA of rodent carcinogens so that the risk to humans can be assessed accurately.

Testing for carcinogenicity

Use of laboratory animals, most commonly rats and mice, to test chemicals for carcinogenicity has long been an integral component of safety testing and it is still widely employed today. Recent legislation, for example, an EU Directive on testing of cosmetics (European Commission, 2018) has banned the use of animals for testing some products. For those classes of products that still require animal testing, such as pharmaceuticals, the gold standard test is the long-term bioassay, involving administration of the test compound to both sexes of rats and/or mice in lifetime studies (generally for 2 years). Dose selection for such studies is controversial, as it often requires testing to the maximum tolerated dose, which can be many orders of magnitude higher than plausible human exposure levels. Such doses have been justified on the principle that genotoxic carcinogens do not generally have a threshold of effect, the response curve being linear without evidence of a threshold. Guidelines for the conduct of animal bioassays have been published by several regulatory and advisory organizations, including the Organisation for Economic Co-operation and Development (OECD, n.d.), the US Environmental Protection Agency (US EPA, n.d.), the US Food and Drug Administration (US FDA, n.d.), and the International Conference on Harmonisation of Technical Requirements for Registration of Pharmaceuticals for Human Use (ICH, n.d.).

Such long-term assays are time-consuming, costly, and raise ethical issues on animal welfare. There has been a constant search for alternatives, including the development of transgenic mice with high sensitivity to carcinogenesis that may form the basis of bioassays of shorter duration than the full 2 years. Transgenic animals can also be used for *in vivo* mutagenicity assays. The reader is referred to the websites of the organizations listed here for further details on these.

Bioassay responses that are purported to be rodent specific (and therefore, it can be argued, not relevant to humans) include:

- Male rat renal tubular adenomas associated with α2u-globulin nephropathy
- Forestomach papillomas and carcinomas associated with chronic forestomach hyperplasia
- Lung carcinomas with inhaled particulate burdens that may overwhelm clearance mechanisms
- Thyroid follicular adenomas associated with excessive hormone stimulation

Some rodent carcinogens have induced tumours only in tissues that have no direct human equivalent, such as the Zymbal gland, preputial/clitoral gland, Harderian gland, and forestomach. There are conflicting views on their relevance and the risk to humans. On the one hand it may be argued that tumour formation in these rodent-specific tissues will not have relevance for human risk, while on the other hand it has been argued that carcinogenesis in *any* epithelial tissue may reflect the general potential for epithelial tissue tumour responses in humans.

To test a compound for genotoxicity generally requires a combination of tests. These assess effects on three major endpoints of genetic damage: first, gene mutation (point mutations or deletions/insertions that affect single or blocks of genes); second, clastogenicity (structural chromosome changes); and third, aneuploidy (numerical chromosome aberrations). It is widely recognized that no single assay can detect all genotoxic substances, partly because no single assay can cover all three of these mutagenic processes (although some may cover more than one) and partly because no single test has been found that is fully predictive of animal bioassay results. All tests have less than 100% sensitivity:

$$\left(\begin{array}{c}\text{number of carcinogens found positive} \\ \div \text{number of carcinogens tested}\end{array}\right) \times 100$$

and less than 100% specificity:

$$\left(\begin{array}{c}\text{number of non-carcinogens found negative} \\ \div \text{number of non-carcinogens tested}\end{array}\right) \times 100.$$

The identification of mutagens and genotoxic carcinogens therefore requires selecting appropriate *in vitro* and *in vivo* tests (Eastmond et al., 2009). Developing a test strategy involves an initial assessment of the chemical's structure and class, its physicochemical properties, expected pathways of metabolism, and possible relationship to known genotoxic substances. Routes of exposure and bioavailability are also considered. These assessments can be supported by *in silico* analysis whereby use can be made of computational software to predict the likelihood that a chemical structure will have genotoxic properties on the basis of comparison with databases of known mutagenic structures.

The recommended test batteries of the organizations listed here generally have the following features in common:

1. A test for gene mutation in bacteria
2. An *in vitro* test for cytogenetic evaluation of chromosomal damage using mammalian cells; *or* an *in vitro* mouse lymphoma Tk assay
3. An *in vivo* test for chromosomal damage using rodent haematopoietic cells (only when animal experiments are permitted)

Transgenic mice containing one of a number of reporter genes can also be used to monitor for mutagenicity in multiple tissues rather than just in the bone marrow (Lambert et al., 2005). Several indicator assays may be conducted and used to provide supporting evidence. These detect genotoxicity—DNA damage—but not mutagenicity; in other words, they provide evidence of the occurrence (or lack of occurrence) of *pre*-mutagenic events. They include the single cell gel electrophoresis assay (the comet assay) which can detect DNA damage in single cells due to strand breaks, alkali-labile lesions, and DNA repair-induced breaks (Tice et al., 2000), and assays that detect the formation of DNA adducts (Phillips et al., 2000).

Monitoring human exposure to carcinogens

Because many carcinogens exert their biological effects through the formation of DNA adducts, the presence of such adducts in human tissues is therefore a tool for molecular epidemiological studies of cancer (Wild and Pisani, 1998). DNA adducts can be detected in human tissues by a variety of methods. These include immunoassays using antibodies raised against carcinogen-modified DNA, mass spectrometry, fluorescence spectroscopy, and ^{32}P-post labelling (Poirier et al., 2000). There is a large body of evidence to suggest DNA adducts are useful markers of carcinogen exposure, providing an integrated measurement of carcinogen uptake, metabolic activation, and delivery to the target cellular macromolecule in target tissues. For example, smoking-related adducts have been detected in many tissues of tobacco smokers (Phillips and Venitt, 2012), giving biological plausibility to the role of tobacco carcinogens in initiating cancer in many different human organs. For other studies, particularly those involving healthy individuals with known or suspected carcinogen exposure, the use of white blood cells as a readily obtainable source of DNA for such studies is commonplace. In a number of industrial settings, where exposure to PAHs is associated with increased risk of cancer, such as iron foundries, coke ovens, aluminium production plants, and graphite electrode manufacture, elevated levels of DNA adducts relative to unexposed controls have been detected (Phillips, 2005). *Environmental* exposure to carcinogens, for example, to PAHs from burning of fossil fuels, has also been monitored by means of DNA adduct detection in a variety of highly exposed populations (Phillips, 2005). Detection of DNA adducts by mass spectrometry has, until recently, been limited to detecting single characterized adducts. However recent studies have taken a more untargeted approach, either to detect all possible as-yet-unidentified adducts in a tissue, for example, human lung (Kanaly et al., 2006), or to detect known multiple adducts simultaneously (Monien et al., 2015).

In a number of prospective studies, the role of DNA adducts as predictors of cancer risk has been validated. In a study of men at risk of liver cancer from aflatoxin B_1 exposure in China, it was found that the levels of an aflatoxin-guanine adduct in urine samples were significantly higher in those subjects that subsequently developed hepatocellular cancer than those who did not (Qian et al., 1994). In another study, adduct levels in the white blood cells of smokers who later got lung cancer were significantly higher than those who did not (Tang et al., 2001).

Carcinogens that cause human cancer—exemplars

Cancer incidences can vary widely around the world, and when populations migrate their cancer incidence changes over time to match that of the host population. This tells us that cancer risk is primarily governed by environmental and lifestyle factors, with genetics playing a lesser role, at least in the majority of cases. Known risk factors currently account for about 45% of new cases. Occupational cancers are estimated to account for about 4% of all cases. What follows is a description of six of the major chemical causes of cancer in humans.

Arsenic

It has been known for many decades that arsenic and its compounds, when applied to the skin medicinally, or among miners and smelting workers exposed by inhalation, are carcinogenic. It is now very clear that environmental exposure to arsenic via drinking water poses a high risk of cancer to many people around the world, resulting in cancer of the skin, lung, and urinary bladder, and possibly of the kidney (Lubin et al., 2007). Regions of the world in which arsenic levels in ground water and drinking water are high include Asia (most of Bangladesh and parts of Mongolia, Thailand, India, Taiwan, and China), South America (parts of Chile, Argentina, and Bolivia), North America (parts of Mexico, Ontario, British Columbia, Arizona, Nevada, and California), Europe (parts of Germany, Hungary, Romania, and Greece), and Africa (parts of Ghana). The number of people at risk can be numbered in the tens of millions, particularly the residents of Bangladesh and West Bengal.

How arsenic exerts its carcinogenic effects is not fully understood. It has genotoxic properties, inducing chromosomal aberrations and aneuploidy (but not point mutations) in human cells and several test systems. Several different mechanisms have been suggested, including diminished DNA repair, altered DNA methylation, enhanced cell proliferation, suppression of p53-mediated pathways and induction of oxidative stress (Beyersmann and Hartwig, 2008).

Tobacco smoke

Although tobacco smoking is declining slowly in the developed world, it is still widespread and at the same time it is increasing rapidly in the developing world. It is causally linked to many cancers: lung, bladder, renal pelvis, oral cavity, oropharynx, hypopharynx, larynx, oesophagus, pancreas, ureter, liver, stomach, uterine cervix, tongue, nasal cavity and paranasal sinuses, nasopharynx, bone marrow (leukaemia), colorectum, and ovary. Involuntary smoking (inhalation of second-hand smoke) is also a risk of lung cancer for non-smokers. Millions of deaths worldwide are caused by smoking, with around 50% of lifelong smokers dying prematurely from a smoking-related disease, and around 15% of all cancer deaths worldwide (30% in Western countries), including around 90% of lung cancer, resulting from the habit.

Tobacco smoke is a complex mixture of more than 4,000 identified chemicals, of which at least 60 are carcinogenic (Hecht, 2003). These belong to a variety of different classes, as shown in Table 7.2 along with representative examples. Most of these are genotoxic, and indeed 'smoking-related' DNA adducts have been detected in many tissues of smokers (Phillips and Venitt, 2012); also, mutations detected in the tumours of smokers are often characteristically different from those in the tumours of non-smokers (see 'Mutational signatures'). Nevertheless, it is not yet clearly understood which components of tobacco smoke are responsible for its carcinogenicity, whether they differ according to the tumour site (e.g. between organs directly exposed to tobacco smoke, such as the respiratory tract, and those indirectly exposed, such as the bladder), and what other non-genotoxic compounds and processes contribute to the overall carcinogenic process.

Aflatoxins

Aflatoxins are a group of toxic chemicals produced mainly by two species of fungus, *Aspergillus flavus* and *Aspergillus parasiticus*, that contaminate crops, mainly maize and groundnuts, in areas of the world with hot and humid climates. Human exposure is associated with a high risk of hepatocellular carcinoma, one of the most frequent and fatal cancers worldwide, and the disease is particularly common in areas of high exposure such as China, Taiwan, Korea, and sub-Saharan Africa. Aflatoxin B_1 (Table 7.1) is the most carcinogenic and the most studied member of the family (see 'Classes of chemical carcinogens'), forming DNA adducts that readily depurinate are excreted in the urine, making this a suitable biological matrix for biomonitoring human exposure (see 'Monitoring human exposure to carcinogens'). Liver tumours from patients in high-risk areas who are exposed to aflatoxins bear a distinctive *TP53* mutation spectrum characteristic of mutations induced by aflatoxin B_1 (Hussain et al., 2007). When exposure to aflatoxin B_1 occurs in concert with hepatitis B infection, for example, in areas of China and Africa, these two factors appear to act synergistically such that the liver cancer risk from aflatoxin + hepatitis is the multiplicative product of the risks from each agent individually (Groopman et al., 2008). Thus, interventions to vaccinate against hepatitis B, and improved methods of grain storage to curtail mould growth and aflatoxin contamination are potentially realistic ways of preventing, or

Table 7.2 Classes of carcinogens in tobacco smoke, with representative examples

Class	Example	Class	Example
Polycyclic aromatic hydrocarbons	Benzo[a]pyrene	Phenolic compounds	Catechol
Heterocyclic hydrocarbons	Furan	Volatile hydrocarbons	Benzene
N-Nitrosamines	NNK	Miscellaneous organic compounds	Acrylamide, hydrazine, urethane
Aromatic amines	2-toluidine	Metals and metal compounds	Arsenic, beryllium, nickel, chromium, cadmium, cobalt, lead
N-Heterocyclic amines	PhIP	Radioisotope	Polonium-210
Aldehydes	Formaldehyde		

at least reducing, liver cancer incidence in these high-risk regions of the developing world.

Polycyclic aromatic hydrocarbons

As has already mentioned, the formation of complex mixtures of PAHs from the incomplete combustion of fossil fuels means that occupational exposure to high levels of them has been widespread, particularly in heavy industry, with elevated risks of cancer a consequence. These include lung cancer from inhalation, but also cutaneous cancer from chronic exposure to tars and mineral oils. There is also widespread occurrence of PAHs in the atmosphere, as a result of coal- and oil-powered power stations, industrial emissions, and automobile emissions (petrol and diesel). Air quality in cities is now a major concern for human health, with PAHs thought to be a major contributor to increased risk of cancer (Ravindra et al., 2001). For smokers, tobacco smoke is a significant source of exposure to PAHs, and there is good evidence for the involvement of PAHs in the mutagenic processes leading to tumour formation (see 'Mutational signatures'). For non-smokers, a major source of exposure to PAHs is the diet (see 'Classes of chemical carcinogens') and they are a possible cause of oesophageal squamous carcinoma in high-risk regions of China, South America, and Iran (Roth et al., 1998; Kamangar et al., 2005; Hakami et al., 2008).

Alcohol

Alcohol consumption is associated with an increased risk of a number of cancers. For some there is a linear relationship between consumption levels and risk, while for others increased incidences are associated only with higher levels of consumption. A recent UK report (COC Secretariat, 2015) estimated that between 4% and 6% of all new cancer cases in the United Kingdom are caused by alcohol consumption. Even at low levels of intake, less than 1.5 units per day (1 unit equates to 10 ml of pure alcohol), there is an increased risk of cancer of the oral cavity, pharynx, oesophagus and, in women, breast. At levels of intake above 1.5 units per day, there is increased risk of larynx and colorectal cancers, while at high levels (above ~6 units/day) drinkers are at increased risk of liver and pancreatic cancer.

It is not fully understood by what mechanism alcohol is carcinogenic, but the different dose-effect relationships in different tissues imply that multiple mechanisms may be involved. A possible genotoxic mechanism is through metabolism of ethanol to acetaldehyde, which is cytotoxic, mutagenic and clastogenic, and carcinogenic in experimental animals, but other potential ethanol-related mechanisms include generating reactive oxygen species, effects on hormone levels and on folate metabolism, and inducing cirrhosis (in the liver).

Aristolochic acid

Aristolochic acids are naturally occurring compounds found in plants of the *Aristolochia* genus. Aristolochic acid I (see Table 7.1) is genotoxic and induces renal tumours in experimental animals, but its role in human cancers is a relatively recent discovery. In the early 1990s several clients at a Belgian slimming clinic were prescribed a herbal concoction that inadvertently contained high levels of aristolochic acid. Many of them rapidly developed renal failure followed by urothelial cancer. Analysis of their tissue DNA showed the presence of aristolochic acid-DNA adducts, strongly implicating the drug as the causative agent (Arlt et al., 2002). Furthermore, mutations in the *TP53* gene of the tumours were found to be AT-TA transversions, characteristic of those induced by aristolochic acid in experimental systems (Nedelko et al., 2009) and unlike those normally found in urothelial cancers; as a result, the condition has become known as Aristolochic Acid Nephropathy (AAN). It was then noted that the pathology of AAN was similar to another chronic renal disease, Balkan endemic nephropathy (BEN) and its associated malignancies. The cause of BEN has now been ascribed to the occurrence of *Aristolochia clematitis* growing wild in the Balkan region and to its seeds contaminating harvested wheat grain used to make bread. BEN patients have also been found to have AA-DNA adducts in their urothelial tumours and, again, AT-TA transversions in *TP53* are the characteristic mutations.

Following these discoveries, attention turned to the Far East, where widespread use of traditional Chinese medicine has involved many formulations containing *Aristolochia* species and where rates of urothelial cancer are the highest in the world. Sequencing of the exomes, or whole genomes, of urothelial cancers from Taiwan has revealed a high frequency of AT-TA transversion mutations, further implicating aristolochic acid as a major causative agent in the aetiology of these diseases (Hoang et al., 2013; Poon et al., 2013). Most recently, an analysis of renal tumours from Romania (but from a non-BEN region) unexpectedly showed the same mutational signature (see 'Mutational signatures'), and aristolochic acid-DNA adducts were also detected in the tissues at levels similar to those reported in Asian and BEN populations (Turesky et al., 2016). Although the source of exposure is not clear, this finding reveals a silent epidemic beyond the geographical regions previously identified as being at risk, and with a previously unsuspected causative agent.

Mutational signatures

Cancers are caused by somatic mutations. The mutations in a cancer cell have accumulated over an individual's lifetime and are the result of multiple mutational processes that include both endogenous and exogenous mutagens and/or modifications to DNA repair and replication. Each process confers a distinct pattern, termed a 'mutational signature'. The most commonly mutated gene in human cancers is the tumour suppressor gene *TP53*. A database of *TP53* mutations identified in human tumours has catalogued over 30,000 mutations (WHO, 2018), from which it has been possible to identify correlations between the signatures and exposure to environmental carcinogens in some cases by reproducing the signatures experimentally; that is, by exposing cells in culture to the same agents and detecting the same patterns of *TP53* mutations in transformed clones. Thus C to T and CC to TT mutations commonly found in *TP53* in squamous carcinomas of the head and neck and in skin melanomas are associated with UV radiation exposure; G to T mutations in smokers' lung cancer are associated with exposure to PAHs, including benzo[a]pyrene; A to T mutations in urothelial tumours are characteristic of exposure to aristolochic acid; and the pattern of G to T mutations seen in some liver cancers is characteristic of exposure to aflatoxin B_1 (Liu et al., 2005; Nedelko et al., 2009).

With the advent of routine exome and whole genome sequencing, it has become apparent that human tumours commonly contain many thousands of mutations. In addition to driver mutations,

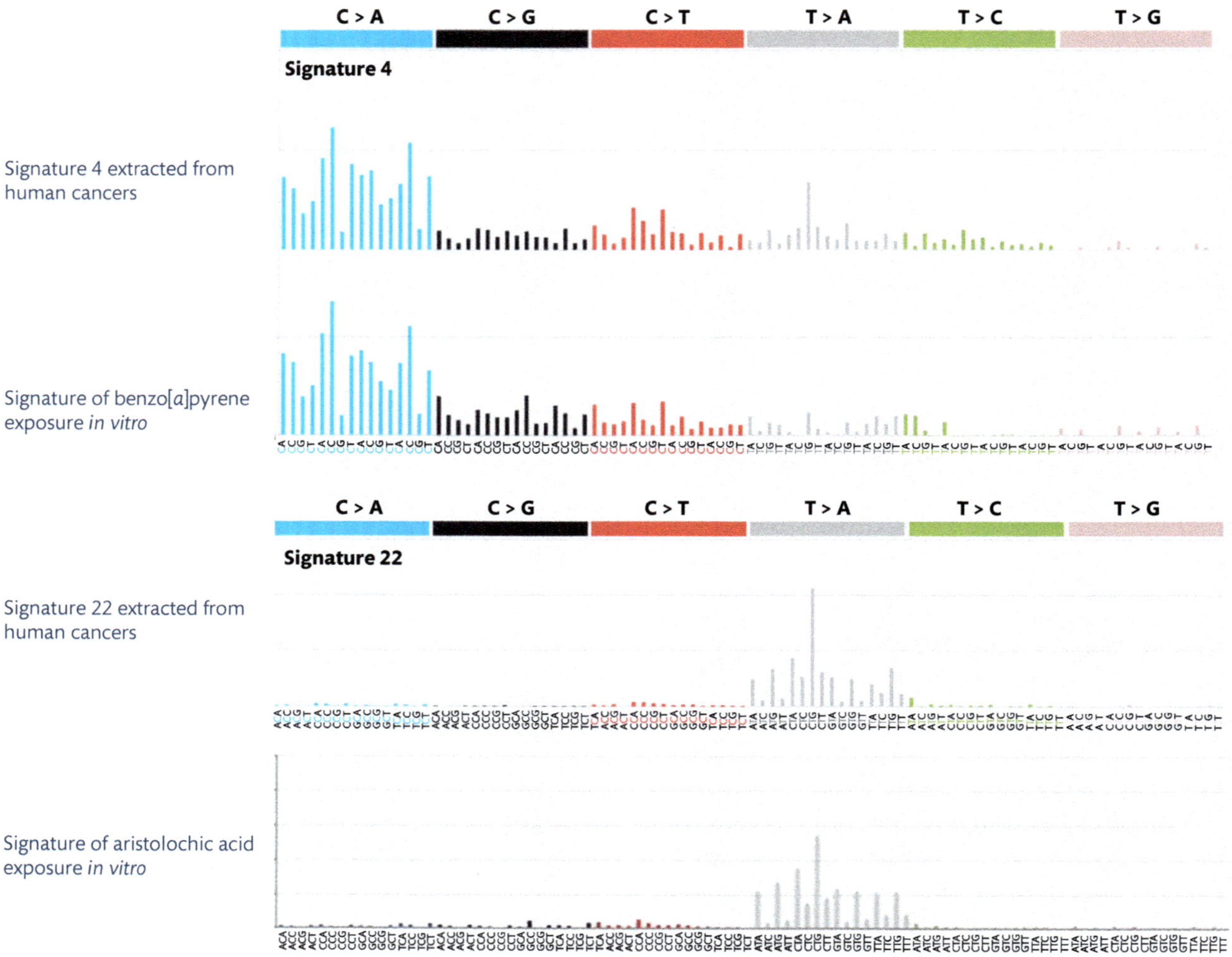

Fig. 7.3 Mutational signatures found in human tumours and induced in carcinogen-treated cells in culture. Each vertical bar represents the relative frequency of a particular base substitution. For each of six possible colour-coded base substitutions, there are 16 different possible sequence contexts, with four possible bases 3′ to the substitution and four possible bases 5′ to it, making 96 possibilities in all.

Adapted with permission from COSMIC, the Catalogue of Somatic Mutations in Cancer, Wellcome Trust Sanger Institute, Genome Research Limited, UK, Copyright © Wellcome Trust Sanger Institute. Available from http://cancer.sanger.ac.uk/cosmic/signatures. Source: data from Serena Nik-Zainal et al. 'The genome as a record of environmental exposure', *Mutagenesis*, Volume 30, Issue 6, 1 November 2015, pp. 763–770, https://doi.org/10.1093/mutage/gev073. Copyright © 2018 Oxford University Press. Published under the terms of the Creative Commons license CC BY 3.0.

such as those in genes like *TP53*, the signatures consist predominantly of passenger mutations. Analysis of some 12,000 cancer genomes has catalogued around 8 million mutations that have revealed 30 distinct base substitution signatures (COSMIC, n.d.). The signatures consist of the six possible base substitutions, each with 16 possible variations when the sequence context takes into account the four possible bases 5' to the mutation and the four possible bases 3' to it. Examples of these 96-component signatures are shown in **Figure 7.3**. It can be seen how one signature, which is prevalent in lung tumours of smokers, bears a close resemblance to the signature generated by exposing cells in culture to benzo[a]pyrene (a carcinogenic constituent of tobacco smoke; see earlier; Nik-Zainal et al., 2015). Another signature, associated with AAN (see 'Carcinogens that cause human cancer—exemplars'), closely matches the signature induced in cultured cells *in vitro* when treated with aristolochic acid I (**Fig. 7.3**).

The proof of principle demonstrated in these recent studies, revealing patterns of mutation across the whole genome previously observed in single gene analyses, will lead to greater precision of characterization of mutational signatures than could be achieved from analysis of a single gene (which, in the case of *TP53*, is mutated in only about 50% of human tumours). With many cancers suspected of being the result of as-yet-unidentified environmental causes or agents, the whole genome sequencing approach of identifying carcinogen-induced mutation signatures has the prospect of shedding new light on the aetiology of many cancers, as well as providing mechanistic insight into the carcinogenic process.

TAKE-HOME MESSAGE

- Most cases of human cancer are the consequence of environmental and lifestyle factors.
- Carcinogens are present in air, food, and water; some are naturally occurring, while others derive from human activity.

- Most chemical carcinogens share the property of being metabolized to reactive species that can damage DNA.
- Mutation is a consequence of DNA damage and tumours contain many mutations.
- Understanding the origin of mutations in tumours will shed new light on the environmental causes of cancer.

OPEN QUESTIONS

- How can the carcinogenicity of a chemical be predicted reliably?
- What determines the organ specificity of a carcinogen?
- What is a carcinogen's mode of action?
- How can human exposure to carcinogens best be monitored?
- What can mutational signatures in tumours tell us about the causative agents of human cancer?
- How can understanding of the causes of cancer be translated into effective prevention?

FURTHER READING

Gatehouse, D. (2007). Short-term testing for genotoxicity. In: Allison, M. R. (ed.) *The Cancer Handbook,* 2nd edition. Chichester: Wiley, pp. 388–400.

Gooderham, N. J. & Carmichael, P. L. (2007). Mechanisms of chemical carcinogenesis. In: Allison, M. R. (ed.) *The Cancer Handbook,* 2nd edition. Chichester, Wiley: pp. 322–37.

Hecht, S. S. (2007). Tobacco use and cancer. In: Allison, M. R. (ed.), *The Cancer Handbook,* 2nd edition. Chichester, Wiley, pp. 429–42.

Luch, A. (ed.) (2005). *The Carcinogenic Effects of Polycyclic Aromatic Hydrocarbons.* London: Imperial College Press.

Maronpot, R. R. (2007). Cancer bioassays. In: Allison, M. R. (ed.) *The Cancer Handbook,* 2nd edition. Chichester: Wiley, pp. 401–9.

Parry, J. M. & Parry, E. M. (eds) (2012). *Genetic Toxicology: Principles and Methods,* New York: Human Press.

Phillips, D. H. (2008). Biomarkers of exposure: adducts. In: Wild, C., Vineis, P., Garte, S. (eds) *Molecular Epidemiology of Chronic Diseases.* Chichester: Wiley, pp. 111–25.

Roberts, R. A. (2007). Non-genotoxic causes of cancer. In: Allison, M. R. (ed.) *The Cancer Handbook,* 2nd edition. Chichester: Wiley, pp. 359–71.

Sugimura, T., Wakabayashi, K., & Nagao, M. (2007). Dietary genotoxins and cancer. In: Allison, M. R. (ed.) *The Cancer Handbook,* 2nd edition. Chichester: Wiley, pp. 420–8.

Vineis, P. (2007). Molecular epidemiology of cancer and the use of biomarkers. In: Allison, M. R. (ed.) *The Cancer Handbook,* 2nd edition. Chichester: Wiley, pp. 410–19.

REFERENCES

Ames, B. N., Durston, W. E., Yamasaki, E., & Lee, F. D. (1973). Carcinogens are mutagens: a simple test system combining liver homogenates for activation and bacteria for detection. *Proc Natl Acad Sci U S A,* **70**(8), 2281–5.

Arlt, V. M., Stiborova, M., & Schmeiser, H. H. (2002). Aristolochic acid as a probable human cancer hazard in herbal remedies: a review. *Mutagenesis,* **17**(4), 265–77.

Beyersmann, D. & Hartwig, A. (2008). Carcinogenic metal compounds: recent insight into molecular and cellular mechanisms. *Arch Toxicol,* **82**(8), 493–512.

Brookes, P. & Lawley, P. D. (1964). Evidence for the binding of polynuclear aromatic hydrocarbons to the nucleic acids of mouse skin: relation between carcinogenic power of hydrocarbons and their binding to DNA. *Nature,* **202**, 781–4.

COC Secretariat (2015). Statement on consumption of alcoholic beverages and risk of cancer. Available at: https://assets.publishing.service.gov.uk/government/uploads/system/uploads/attachment_data/file/490584/COC_2015_S2__Alcohol_and_Cancer_statement_Final_version.pdf

COSMIC (n.d.). Signatures of mutational processes in human cancer. Available at: https://cancer.sanger.ac.uk/cosmic/signatures

Eastmond, D. A., Hartwig, A., Anderson, D., et al. (2009). Mutagenicity testing for chemical risk assessment: update of the WHO/IPCS Harmonized Scheme. *Mutagenesis,* **24**(4), 341–9.

Eaton, D. L. & Gallagher, E. P. (1994). Mechanisms of aflatoxin carcinogenesis. *Annu Rev Pharmacol Toxicol,* **34**, 135–72.

European Commission (2018). Cosmetics. Available at: https://ec.europa.eu/growth/sectors/cosmetics_en

Groopman, J. D., Kensler, T. W., & Wild, C. P. (2008). Aflatoxin, hepatitis B virus and liver cancer: a paradigm for molecular epidemiology. In: Wild, C., Vineis, P., & Garte, S. (eds) *Molecular Epidemiology of Chronic Diseases.* Chichester: Wiley, pp. 323–42.

Hakami, R., Mohtadinia, J., Etemadi, A., et al. (2008). Dietary intake of benzo(a)pyrene and risk of esophageal cancer in North of Iran. *Nutr Cancer,* **60**(2), 216–21.

Hakura, A., Tsutsui, Y., Sonoda, J., et al. (1998). Comparison between *in vivo* mutagenicity and carcinogenicity in multiple organs by benzo[a]pyrene in the lacZ transgenic mouse (muta mouse). *Mutat Res,* **398**(1–2), 123–30.

Hanahan, D. & Weinberg, R. A. (2000). The hallmarks of cancer. *Cell,* **100**(1), 57–70.

Hanahan, D. & Weinberg, R. A. (2011). Hallmarks of cancer: the next generation. *Cell,* **144**(5), 646–74.

Hecht, S. S. (2003). Tobacco carcinogens, their biomarkers and tobacco-induced cancer. *s,* **3**(10), 733–44.

Hoang, M. L., Chen, C. H., Sidorenko, V. S., et al. (2013). Mutational signature of aristolochic acid exposure as revealed by whole-exome sequencing. *Sci Transl Med,* **5**(197), 197ra102.

Hussain, S. P., Hofseth, L. J., & Harris, C. C. (2003). Radical causes of cancer. *Nat Rev Cancer,* **3**(4), 276–85.

Hussain, S. P., Schwank, J., Staib, F., Wang, X. W., & Harris, C. C. (2007). TP53 mutations and hepatocellular carcinoma: insights into the etiology and pathogenesis of liver cancer. *Oncogene,* **26**(15), 2166–76.

IARC Monographs (n.d.). IARC Monographs on the Evaluation of Carcinogenic Risks to Humans. Available at: https://monographs.iarc.fr/

International Conference on Harmonisation of Technical Requirements for Registration of Pharmaceuticals for Human Use (ICH) (n.d.). Available at: http://www.ich.org/home.html

Kamangar, F., Strickland, P. T., Pourshams, A., et al. (2005). High exposure to polycyclic aromatic hydrocarbons may contribute to high risk of esophageal cancer in northeastern Iran. *Anticancer Res,* **25**(1B), 425–8.

Kanaly, R. A., Hanaoka, T., Sugimura, H., Toda, H., Matsui, S., & Matsuda, T. (2006). Development of the adductome approach to detect DNA damage in humans. *Antioxid Redox Signal,* **8**(5–6), 993–1001.

Lambert, I. B., Singer, T. M., Boucher, S. E., & Douglas, G. R. (2005). Detailed review of transgenic rodent mutation assays. *Mutat Res,* **590**(1–3), 1–280.

Lawley, P. D. (1984). Carcinogenesis by alkylating agents. In: Searle, C. E. (ed.) *Chemical Carcinogens*, 2nd edition. Washington DC: American Chemical Society (ACS Monograph 182), pp. 325–84.

Lawley, P. D. (1989). Mutagens as carcinogens: development of current concepts. *Mutat Res*, **213**(1), 3–25.

Liu, Z., Muehlbauer, K. R., Schmeiser, H. H., Hergenhahn, M., Belharazem, D., & Hollstein, M. C. (2005). p53 mutations in benzo(a)pyrene-exposed human p53 knock-in murine fibroblasts correlate with p53 mutations in human lung tumors. *Cancer Res*, **65**(7), 2583–7.

Lubin, J. H., Beane Freeman, L. E., & Cantor, K. P. (2007). Inorganic arsenic in drinking water: an evolving public health concern. *J Natl Cancer Inst*, **99**(12), 906–7.

Miller, E. C. (1978). Some current perspectives on chemical carcinogenesis in humans and experimental animals: presidential address. *Cancer Res*, **38**(6), 1479–96.

Miller, E. C. & Miller, J. A. (1979). Milestones in chemical carcinogenesis. *Semin Oncol*, **6**(4), 445–60.

Miller, E. C. & Miller, J. A. (1981). Searches for ultimate chemical carcinogens and their reactions with cellular macromolecules. *Cancer*, **47**(10), 2327–45.

Miller, J. A. (1970). Carcinogenesis by chemicals: an overview--G. H. A. Clowes memorial lecture. *Cancer Res*, **30**(3), 559–76.

Monien, B. H., Schumacher, F., Herrmann, K., Glatt, H., Turesky, R. J., & Chesne, C. (2015). Simultaneous detection of multiple DNA adducts in human lung samples by isotope-dilution UPLC-MS/MS. *Anal Chem*, **87**(1), 641–8.

Nedelko, T., Arlt, V. M., Phillips, D. H., & Hollstein, M. (2009). TP53 mutation signature supports involvement of aristolochic acid in the aetiology of endemic nephropathy-associated tumours. *Int J Cancer*, **124**(4), 987–90.

Nik-Zainal, S., Kucab, J. E., Morganella, S., et al. (2015). The genome as a record of environmental exposure. *Mutagenesis*, **30**(6), 763–70.

OECD (n.d.). OECD Guidelines for the Testing of Chemicals. Available at: http://www.oecd.org/chemicalsafety/testing/oecdguidelinesfortthetestingofchemicals.htm

Paget, J. (1853). Lectures on surgical pathology. London: Longman. [quoted by Haddow, A., & Kon, G. A. R. (1947). Chemistry of carcinogenic compounds. *Br Med Bull*, **4**, 314–26].

Phillips, D. H. (1983). Fifty years of benzo(a)pyrene. *Nature*, **303**(5917), 468–72.

Phillips, D. H. (2001). Understanding the genotoxicity of tamoxifen. *Carcinogenesis*, **22**(6), 839–49.

Phillips, D. H. (2005). DNA adducts as markers of exposure and risk. *Mutat Res*, **577**(1–2), 284–92.

Phillips, D. H. (2007). The formation of DNA adducts. In: Allison, M. R. (ed.) *The Cancer Handbook*. Chichester: Wiley, pp. 338–50.

Phillips, D. H., Farmer, P. B., Beland, F. A., et al. (2000). Methods of DNA adduct determination and their application to testing compounds for genotoxicity. *Environ Mol Mutagen*, **35**(3), 222–33.

Phillips, D. H. & Venitt, S. (2012). DNA and protein adducts in human tissues resulting from exposure to tobacco smoke. *Int J Cancer*, **131**(12), 2733–53.

Poirier, M. C., Santella, R. M., & Weston, A. (2000). Carcinogen macromolecular adducts and their measurement. *Carcinogenesis*, **21**(3), 353–9.

Poon, S. L., Pang, S. T., McPherson, J. R., et al. (2013). Genome-wide mutational signatures of aristolochic acid and its application as a screening tool. *Sci Transl Med*, **5**(197), 197ra101.

Pott, P. (1775) Chirurgical observations. London: Hawes, Clarke and Collins. Reprinted in *Natl Cancer Inst. Monograph*, **10**, 7–13 (1963).

Qian, G. S., Ross, R. K., Yu, M. C., et al. (1994). A follow-up study of urinary markers of aflatoxin exposure and liver cancer risk in Shanghai, People's Republic of China. *Cancer Epidemiol Biomarkers Prev*, **3**(1), 3–10.

Ravindra, Mittal, A. K., & Van Grieken, R. (2001). Health risk assessment of urban suspended particulate matter with special reference to polycyclic aromatic hydrocarbons: a review. *Rev Environ Health*, **16**(3), 169–89.

Rendic, S. & Guengerich, F. P. (2012). Contributions of human enzymes in carcinogen metabolism. *Chem Res Toxicol*, **25**(7), 1316–83.

Roth, M. J., Strickland, K. L., Wang, G. Q., Rothman, N., Greenberg, A., & Dawsey, S. M. (1998). High levels of carcinogenic polycyclic aromatic hydrocarbons present within food from Linxian, China may contribute to that region's high incidence of oesophageal cancer. *Eur J Cancer*, **34**(5), 757–8.

Sims, P., Grover, P. L., Swaisland, A., Pal, K., & Hewer, A. (1974). Metabolic activation of benzo(a)pyrene proceeds by a diol-epoxide. *Nature*, **252**(5481), 326–8.

Tang, D., Phillips, D. H., Stampfer, M., et al. (2001). Association between carcinogen-DNA adducts in white blood cells and lung cancer risk in the physicians health study. *Cancer Res*, **61**(18), 6708–12.

Tice, R. R., Agurell, E., Anderson, D., et al. (2000). Single cell gel/comet assay: guidelines for *in vitro* and *in vivo* genetic toxicology testing. *Environ Mol Mutagen*, **35**(3), 206–21.

Turesky, R. J., Yun, B. H., Brennan, P., et al. (2016). Aristolochic acid exposure in Romania and implications for renal cell carcinoma. *Br J Cancer*, **114**(1), 76–80.

US Environmental Protection Agency (US EPA). (n.d.). Test Guidelines for Pesticides and Toxic Substances. Available at: https://www.epa.gov/test-guidelines-pesticides-and-toxic-substances

US Food and Drug Administration (US FDA). (n.d.). Guidance, Compliance, & Regulatory Information. Available at: https://www.fda.gov/Drugs/GuidanceComplianceRegulatoryInformation/

Verna, L., Whysner, J., & Williams, G. M. (1996). 2-Acetylaminofluorene mechanistic data and risk assessment: DNA reactivity, enhanced cell proliferation and tumor initiation. *Pharmacol Ther*, **71**(1–2), 83–105.

Wild, C. P. & Pisani, P. (1998). Carcinogen DNA and protein adducts as biomarkers of human exposure in environmental cancer epidemiology. *Cancer Detect Prev*, **22**(4), 273–83.

World Health Organization (WHO). (2015). IARC TP53 Database. Available at: http://p53.iarc.fr/

8

Radiation as a carcinogen

Yan-Qun Xiang and Chao-Nan Qian

Introduction

It is generally accepted that radiation can induce cancer. Electromagnetic radiation (EMR) is radiant energy released by certain electromagnetic processes that travels at the speed of light. EMR is all around us in the world. According to wavelengths, the EMR spectrum can be divided into low, median, and high frequencies (**Fig. 8.1**). Lower radio frequencies are classified as Group 2B carcinogens, which are considered possibly carcinogenic by the World Health Organization (WHO). This class includes possible carcinogens that have weaker effects, similar to that of coffee. Epidemiological studies have never been conclusive in this field, including the relationship

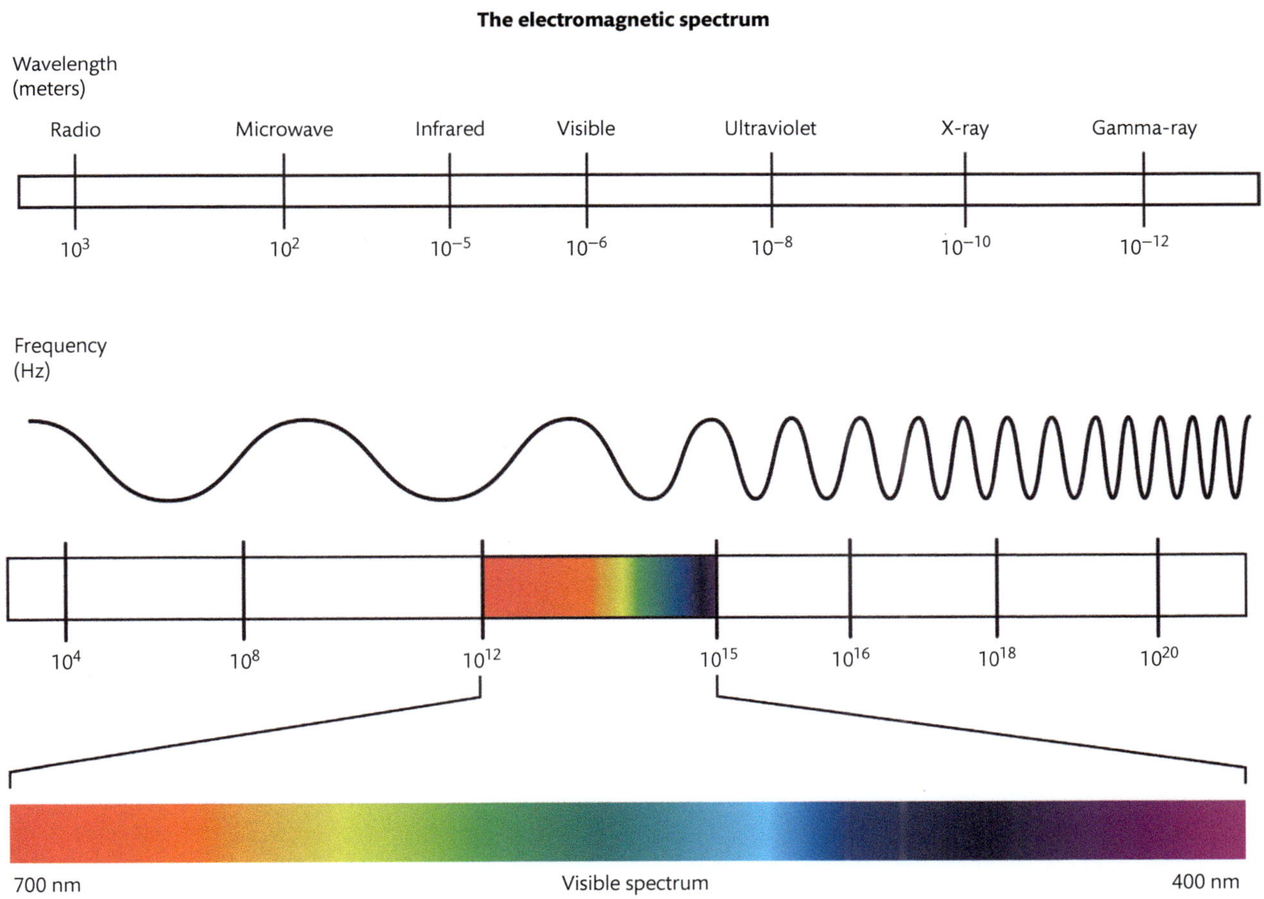

Fig. 8.1 The electromagnetic spectrum. The wavy line shows the relationship between frequency (ν) and wavelength (λ). One complete oscillation of the field (named the period) is the distance between two adjacent peaks. The wavelength of a complete cycle is related to the frequency by the speed of light, c ($\lambda = c/\nu$). Ultraviolet (UV) radiation (i.e. 'beyond' violet light in the sense of being of shorter wavelength and hence higher energy) occupies the wavelength bands from 400 to 315 nm (UVA), 315 to 280 nm (UVB) and 280 to 100 nm (UVC). 1 nm (nanometre) = 10-9 m.

between mobile phone use and brain cancer development. Recently, a French national study called CERENAT evaluated the use of mobile phones and the risk of brain tumours finding that there is an increased risk of meningioma from mobile phone use, and an increased risk of glioma with the use of mobile phones for a decade or longer. These researchers suggest that radiofrequency fields should be classified in Group 2A because they are 'probable' human carcinogens based on the criteria used by the International Agency for Research on Cancer (Lyon, France; see Morgan et al., 2015).

The higher frequencies of EMR (e.g. UV, X-ray, gamma radiation, alpha particles, beta particles, and neutrons) are referred to as ionizing radiation due to the ability of these radiations to produce ions and free radicals in materials, resulting in direct or indirect damage of biological molecules (e.g. DNA) in the cells. Box 8.1 shows the chemistry of the radiation-induced damaged. According to the WHO, all UV frequencies are classified as Group 1 carcinogens. Ultraviolet radiation from sun exposure is the primary cause of skin cancer. Photon radiation (including X-ray and gamma ray), alpha-particle emitters, beta particles, and neutrons are forms of invisible electromagnetic radiation with short wavelengths, which is important because they can damage living tissues by causing mutations in DNA and lead to carcinogenesis, which is the main topic of this chapter (Fig. 8.2). Sievert (Sv, J/Kg) is the unit used to evaluate the dose of radiation exposure (Table 8.1), and 1 Sv is equivalent to 1 Gy of X- or gamma radiation. Natural radioactivity comes from radioactive elements in the earth, food, water, and space and is unavoidable for human beings. This basic level of radiation equals approximately 3 mSv. Radiation exposure in diagnostic medical procedures, such as X-rays, computed tomography (CT), and nuclear scans, is seen in clinical practice. The effective doses of a conventional chest X-ray, CT scan, and bone scan (^{99m}Tc) are approximately 0.06 mSv, 8 mSv, and 4.4 mSv, respectively. Although each single medical exposure is not generally thought of as harmful compared to the natural radiation dose, multiple scans are usually conducted for a single patient, which should be taken into consideration.

Marie Curie, who was given the Nobel Prize twice 'in recognition of her services to the advancement of chemistry by the discovery of the elements radium and polonium, by the isolation of radium and the study of the nature and compounds of this remarkable element' died of aplastic anaemia, which was believed to be a consequence from her long-term exposure to radiation. The first major case of accidental exposure to radiation occurred in the 1920s following World War I as a result of the commercial use of radium at the United States (US) Radium Corporation to produce *Undark*, a high-tech paint that allowed America's infantrymen to read their wristwatches and instrument panels at night. They also marketed the pigment for non-military products, such as house numbers, pistol sights, light switch plates, and glowing eyes for toy dolls. The 'Radium Girls' sued US Radium Corporation for their health deterioration, which gained subsequent media attention of the hazards associated with radium and radiation.

The most famous exposure to artificial radiation came from the atomic bombs dropped on Hiroshima and Nagasaki in 1945. The acute effects of the atomic bombings killed 90,000–120,000 people out of a population of approximately 330,000 in Hiroshima and 60,000–80,000 out of a population of approximately 250,000 in Nagasaki. The after-effects of the bombs caused approximately 1,900 cancer deaths (Yamamoto et al., 2014). An epidemiology study showed that 46% of leukaemia deaths and 11% of solid cancer deaths among the bomb survivors were related to radiation from the bombs.

The worst nuclear accident was the explosion of a reactor at the Chernobyl power plant in 1986. Although the radiation was 400 times greater than that of the atomic bombs, the Scientific Committee on Problems of the Environment (Scope) suggested that the Chernobyl incident cannot be directly compared to number of atmospheric tests of nuclear weapons. A reason is that the isotopes released at Chernobyl have longer half-lives and larger pollution effects than those released by the detonation of atomic bombs. By 2005, the Chernobyl accident had caused 125,000 man-Sv to the population of Ukraine, Belarus, and Russia, and 115,000 man-Sv to most of the more distant European countries. Thyroid cancer among children increased in Belarus, Ukraine, and Russia.

Box 8.1 The chemistry of radiation-induced damage: The interaction between radiations and body water

All three ionization radiations have in common the property to displace electrons from atoms, giving the atoms positive electric charge. Radiations interact with any type of molecule, but as water is the most common in the human body, by probability, it is water radiation will mostly interact with, producing three intermediate forms: hydroxyl radicals, hydrogen peroxide, and superoxide radicals.

When a radiation interacts with a molecule of water, hydroxyl radicals (*OH) are the first to be produced:

$$H_2O \rightarrow H^+ + e^- + {}^*OH.$$

Another reaction leading the formation of hydroxyl radicals, in presence of water and iron, is the Fenton reaction:

$$H_2O_2 + Fe_2^+ \rightarrow OH^- + {}^*OH + Fe_3^+$$

hydroxyl radicals are extremely harmful as react very quickly with any possible molecule by snatching an electron.

The second product is hydrogen peroxide (H_2O_2) and this is the proposed chain of reactions leading to its formation:

First a molecule of water is ionized:

$$H_2O \rightarrow H_2O^+ + e^-$$

than the free electron travels at a distance and react with a second water molecule:

$$e^- + H_2O \rightarrow H + OH^-$$

while H_2O+, less mobile, splits into:

$$H_2O^+ \rightarrow H^+ + OH$$

and finally:

$$OH + OH \rightarrow H_2O_2.$$

In reacting with other molecules H_2O_2 can cause damage both by gaining or releasing an electron but is less reactive and harmful than hydroxyl radicals.

The third molecule is the superoxide radical (O_2*–)

$$O_2{}^{*-} + Fe^{3+} \rightarrow O_2 + Fe^{2+}$$

Not very dangerous by itself, the superoxide radical is harmful as it produces Fe_2+, which fuels the Fenton reaction, which results in highly toxic hydroxyl radicals.

Source: data from Nick Lane, *Oxygen: The molecule that made the world*, Oxford University Press, Oxford, Copyright © 2003; and Robert J. Shaleck, *The formation of hydrogen peroxide in water by ionizing radiation*, The Rice Institute, Houston, Texas, Copyright © 1953, https://scholarship.rice.edu/bitstream/handle/1911/18443/3079878.Pdf?sequence=1&isAllowed=y

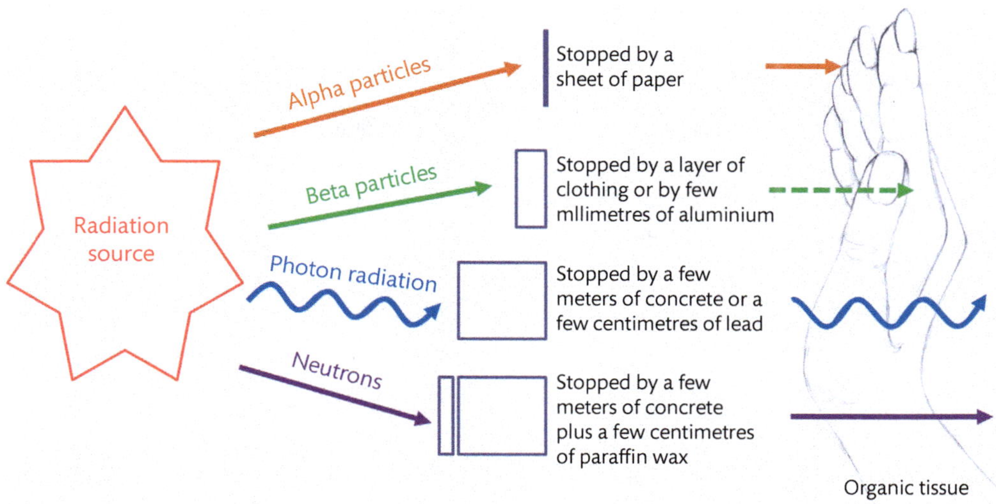

Fig. 8.2 Common radiations related to increased cancer risk. Alpha particles consist of two protons and two neutrons bound together into a particle identical to a helium nucleus. They are generally produced in the process of alpha decay but may also be produced in other ways. A beta particle, sometimes called beta ray, is a high-energy, high-speed electron or positron emitted in the radioactive decay. Photons are currently best explained by quantum mechanics and exhibit wave–particle duality, exhibiting properties of both waves and particles. Alpha particles, beta particles, and photons are directly ionizing radiations. Neutron radiation is indirectly ionizing radiation, because neutrons have no charge and do not directly ionize atoms in the way that charged particles do. However, neutron absorption results in gamma emission and the gamma ray (photon) is ionizing.

Radiocesium-137, which is able to pass easily from soil to grass and hence accumulate in sheep, was deposited on certain upland areas of the United Kingdom (UK). In 2009, sheep farmed in some areas of the UK were still subject to inspection, thereby prohibiting some sheep from entering the human food chain.

The recent nuclear accident at the Tokyo Electric Power Company Fukushima Daiichi Nuclear Power Stations occurred on 11 March 2011. This event was caused by an earthquake and resulted in a large amount of radionuclide released into the atmosphere. The estimated total release of iodine-131 (I-131), caesium-134 (Cs-134), and cesium-137 (Cs-137) was 120, 9.0 and 8.8 Peta Becquerels (PBq), respectively (PBq is the unit used to describe the radioactivity of a radiation source). Because the incidence of childhood thyroid cancer increased in those residing near the site following the Chernobyl accident, thyroid screening of all children in the Fukushima Prefecture was started. To date, screening of more than 280,000 children has resulted in the diagnosis of thyroid cancer in 90 children (approximate incidence, 313 per million; see Nagataki and Takamura, 2014; Rhodes, 2014).

Various parameters, including transients, harmonic content, resonance conditions, peak values and time above thresholds, and average levels, as well as tissue type and animal type, may be relevant in carcinogenesis from ionizing radiation. However, it is not known which of these parameters or what combination of parameters are critical risk factors for carcinogenesis. Additionally, there is no generally accepted mechanism or model to predict the incidence of radiation carcinogenesis.

Mechanisms of radiation transformation: Bystander and abscopal effects measured in the cellular system

Radiation-induced genome instability

Radiation-induced genome instability has been documented to play an important role in radiation transformation, both in directly exposed cells and in distant 'bystander' cells.

The direct mechanism for radiation-induced genomic instability has been speculated as a consequence of hampered DNA repairment in the susceptible cells. Relied upon for its direct effect on DNA damage, ionizing radiation is an effective and commonly used treatment for about two-thirds of human malignancies. DNA double-strand breaks generated by ionizing radiation are the most lethal form of damage to the cells and are mainly repaired via either homologous recombination or non-homologous end-joining pathways in mammalian cells. The benefit of radiotherapy can be gained from the less effective DNA repair in tumour cells versus highly effective DNA repair process in normal cells. However, in the mice with normal cells possessing a non-synonymous single nucleotide polymorphism in *CTIP* gene, Q418P, homologous recombination ability of the cells is reduced, and 9.7% of the mice develop acute myeloid leukaemia after whole-body exposure to 3 Gy of X-rays (Patel et al., 2016).

One indirect mechanism of genome instability is related to metabolic change by ionizing radiation. Oxidative stress contributes to

Table 8.1 Radiation measurements in common units and International System of Units (SI) units

	Radioactivity	Absorbed dose	Dose equivalent	Exposure
Common units	Curie (Ci)	Rad	Rem	Roentgen (R)
SI units	Becquerel (Bq)	Gray (Gy)	Sievert (Sv)	Coulomb/Kilogram (C/kg)

radiation responses, which are mainly produced due to reactive oxygen species (Ros) generation by the electron transport chain of the mitochondria, and by the cytoplasmic nicotinamide adenine dinucleotide phosphate oxidases. Persistent ionizing radiation disrupts metabolic processes and persists long after radiation exposure, which explains the delayed effects of ionizing radiation exposure. In a single cell, the consequences of oxidative stress depend on the interaction between the nucleus and the cellular population of several hundred or thousands of mitochondria that are genetically heterogeneous.

High intramitochondrial Ros

High intramitochondrial Ros levels also damage the mitochondrial (MT) DNA and thereby affect the epigenetic control mechanisms of the nuclear DNA (i.e. chromosomal DNA) by decreasing the activity of methyltransferases and thus causing global DNA hypomethylation (Spitz and Hauer-Jensen, 2014; Szumiel, 2015). Additionally, other epigenetic phenomena, including histone modifications and small RNA-mediated silencing, are also reported to contribute to the maintenance of IR-mediated effects (Ilnytskyy and Kovalchuk, 2011). These changes are transmitted to the progeny of the irradiated cells. Additionally, the production of reactive oxygen by ionizing radiation has the potential to damage single- or double-stranded DNA directly. With the accumulation of DNA damage, gene mutations will develop eventually.

The commonly observed radiation-induced acute myeloid leukaemia (rAML) has been studied for decades to elucidate the molecular mechanisms associated with multistage carcinogenesis. These radiation-induced cellular programmes include apoptosis, senescence, and impaired self-renewal within the haematopoietic stem cell pool. In the context of sporadic DNA damage to a cell, these cellular responses act in concert to restore tissue function and prevent selection for adaptive oncogenic mutations. A specific interstitial deletion of chromosome 2 has been found in a high proportion of rAML and is recognized as the initiating event of radiation-induced acute myeloid leukaemia. Olme et al. (2013) created a model with a reporter gene to confirm the deletion of chromosome 2, which leads to the loss of *SFPI*, a gene essential for haematopoietic development. Its product, the transcription factor PU.1, acts as a tumour suppressor in this model. Due to its loss, radiation-induced acute myeloid leukaemia develops. Denissova et al. (2011) reported that the frequency of loss of heterozygous mutants in mESCS can be induced several hundredfold by exposure to 5–10 Gy of X-rays. This induction is 50–100-fold higher than the induction reported for mouse adult or embryonic fibroblasts. The primary mechanism underlying the elevated loss of heterozygosity after irradiation is mitotic recombination, which is important to understanding the biological effects of ionizing radiation on early development and carcinogenesis. Gene expression analysis of mice mammary glands after exposure to 2 Gy of whole-body γ radiation showed that the mRNA levels of a total of 737 genes were significantly increased, by more than 2-fold of the control (Datta et al., 2012). More genes (493 genes; 67%) are upregulated than downregulated (244 genes; 33%). Upregulation of breast cancer-related canonical signalling pathways, including stathmin1 and human epidermal growth factor receptor-2 (HER-2) pathways are also observed. Multiple genes with tumour suppressor functions (GPRC5A, ELF1, NAB2, Sema4D, ACPP, MAP2 and RUNX1) persistently remain downregulated in response to radiation exposure. The related study has implications for breast cancer initiation and progression after radiation exposure.

Cell adhesion molecules are expressed at higher levels in malignantly transformed breast epithelial cells relative to levels in non-malignant cells (Calaf et al., 2013). However, reduced levels of adhesion molecules observed in the mouse xenograft-derived tumour 2 cell line compared to the pre-tumorigenic Alpha5 cell line suggest that the altered expression levels of adhesion molecules depend on the tumour tissue microenvironment. Expression levels of important cell adhesion molecules, such as α-catenin, β-catenin, γ-catenin, E-cadherin, and integrin, were found to be higher at the protein level in the Alpha5 and tumour cell lines relative to these levels in the non-tumorigenic Mcf-10F, oestrogen, and Alpha3 cell lines. Consistent with these results, a cDNA expression analysis revealed elevated levels of genes involved in cell adhesion (E-cadherin, integrin β6 and desmocollin3 [DSC3]) in the Alpha5 and tumour cell lines relative to the levels in the Mcf-10F, oestrogen, and Alpha3 cell lines. Evidence is increasing, however, that radiation exposure can also modify extrinsic factors by disturbing cells in the microenvironment. In a lung cancer susceptible mouse model, K-ras is spontaneously activated after radiation exposure in the later stages of the carcinogenic process, and dose fractionation creates a more permissive environment for the progression of lung cancer. These results strongly support the concept that radiation exposure can enhance cancer progression through the disruption of inflammatory responses.

Radiation-induced bystander effects (RIBEs) are the induction of biological non-targeted effects in cells that are not directly hit by radiation or by free radicals produced by ionization events (Hatzi et al., 2015). Although RIBEs have been demonstrated using various biological endpoints, the mechanism(s) of this phenomenon still remain unclear. The results of *in vitro* RIBEs and the evidence of non-targeted effects in various *in vivo* systems have been controversial (Mancuso et al., 2013), probably due to the neglection of cellular motility in the experiments, by which cancer cells or stem cells can travel to a distant organ in a given time. Patched1 (Ptch1) (+/-) mice were used to detect abscopal tumorigenesis of ionizing radiation (1, 2, or 3 Gy of X-rays) in shielded cerebellum, and it was concluded that the interplay between radiation dose and exposed tissue volume plays a critical role in non-targeted effects occurring in mouse central nerve system under conditions relevant to humans (Fleenor et al., 2015). These findings are consistent with the hypothesis that transformed cells or cancer stem cells might be able to travel to a distant organ in a given time.

Dose, dose rate, and carcinogenesis

Figure 8.3 is a radiation dose chart to summarize how different dosages of radiation can influence human health.

In the early 1960s, studies were performed at the TNO-Institutes for Health Research on the acute effects of high-dose total body irradiation (TBI) with X-rays and fission neutrons in Rhesus monkeys and the protective effect of autologous bone marrow transplantation (BMT). The surviving animals of this study were kept in order to investigate late radiation effects (i.e. tumorigenesis). The group of long-term surviving monkeys comprised nine

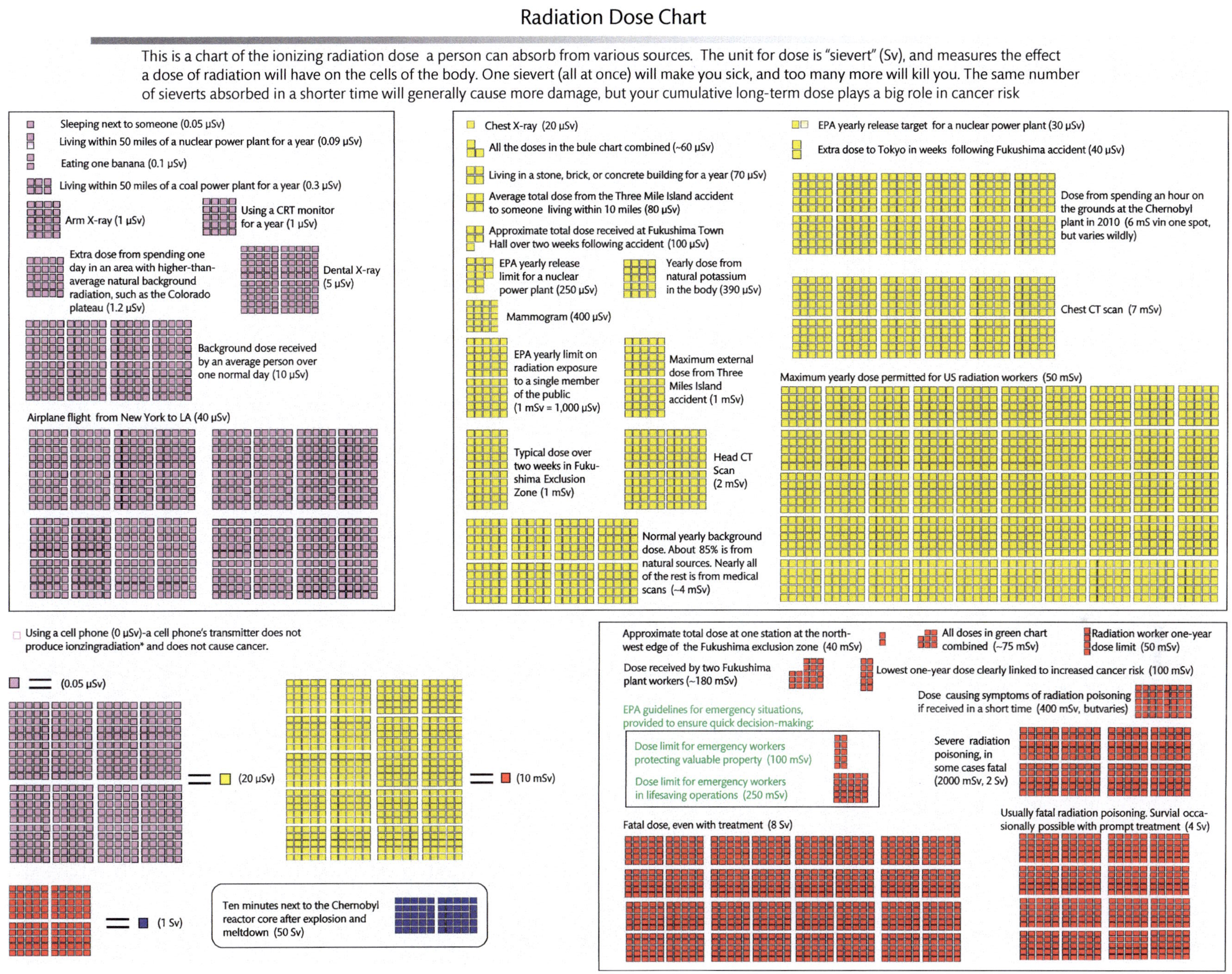

Fig. 8.3 Radiation dose chart to summarize how different dosages of radiation can influence human health.
Adapted from Randall Munroe, Radiation Dose Chart, https://xkcd.com/radiation/

neutron-irradiated animals (average total body dose 3 Gy, range 2.3–4.4 Gy) and 20 X-irradiated monkeys (average total body dose 7.1 Gy, range 2.8–8.6 Gy). At post-irradiation intervals of 4–21 years, an appreciable number of malignant tumours were observed. In the neutron-irradiated group, eight of nine animals died with one or more malignant tumours. In the X-irradiated group, this fraction was 10 of 20. The tumours in the control group, in seven of 21 animals, appeared at a much older age compared with those in the irradiated cohorts (Hollander et al., 2003). This study showed that various types of ionizing radiation produce different cancer risks regardless of radiation dose/dose rate. Sparsely ionizing radiation, such as gamma ray, generally produces linear or upwardly curving dose responses at low doses, either by fractionation or low dose rate, and results in a decreased biological effect. The risk decreases when the dose rate is reduced for a given dose. Densely ionizing radiation at medium and high linear energy transfer (LET, which describes how much energy an ionizing particle transfers to the material traversed per unit distance. It is identical to the retarding force acting on a charged ionizing particle travelling through the matter), such as neutrons, produces downwardly curving dose responses, in which the risk initially grows with the dose but eventually stabilizes or decreases. When the dose rate is reduced, under certain circumstances, there is an increase in carcinogenesis or other biological effects, which is called the inverse dose rate effect (Hill et al., 1982). There have been many reports on the dose-rate dependence of various high LET radiations. The dose-rate effect of 252-californium neutrons was investigated using confluent cultures of mouse m5S cells. The relative biological effectiveness (RBE) of neutrons for oncogenic transformation increased from 3.3 to 5.1 when the dose-rate was reduced from 1.8 to 0.12 cGy/min. Similarly, neutron RBE values for the HPRT-mutation were 4.9 and 7.4 at dose rates of 1.8 and 0.12 cGy/min, respectively (Komatsu et al., 1993).

Studies have shown that the inverse dose-rate effect is significant for a very limited range of LETs, from approximately 30–130 keV/microns. Compared with acute exposure to densely ionizing radiation, the biological effect of protracted exposure is sometimes, but not always, enhanced for transformational end points. Furthermore, the dose-rate modifying factors are independent of the dose and dose-rate transformation in experiments with fission neutrons (Balcer-Kubiczek et al., 1994). No increased transformation frequencies are observed when the alpha-particle dose is protracted over several hours (Hieber et al., 1987). The thyroid gland is sensitive to carcinogenesis due to ionizing radiation in human populations. Chromosomal rearrangements involving the Ret gene are prevalent in thyroid papillary carcinomas from patients with radiation history. After exposure to a range of doses of gamma radiation, a dose-dependent induction of RET/PTC rearrangements was observed in human thyroid cells after exposure to 0.1–10 Gy gamma radiation (Caudill et al., 2005).

With respect to low-dose and dose-rate studies, a joint study of nuclear workers of 15 countries focused on radiation risks of leukaemia in people with protracted external gamma exposure. The overall average cumulative bone marrow dose (per person) was approximately 0.02 Sv, which is 10 times less than the lifespan study (LSS) average dose. The study included 0.4 million nuclear workers with 5.2 million person-years (PY) under observation and 196 cases of leukaemia deaths and resulted in an ERR estimate of 1.93 Sv-1; the results were not significant, however. It is estimated that 1–2% of deaths from cancer among workers in this cohort may be attributable to radiation. These data did not show any evidence of a time-since-exposure effect for low-dose and low-dose-rate radiation (Cardis et al., 2005).

There is considerable controversy as to whether DNA damage induced by low doses and low dose rates of ionizing radiation is treated by cellular defence mechanisms in ways similar to that induced at high doses and high-dose rates and whether downstream delayed effects may be caused by low doses compared to moderate and high doses. This issue constitutes the major challenge in the linear no-threshold model currently used for radiological risk estimates. Among the various DNA lesions induced by ionizing radiation, DNA double-strand breaks (DSBS) are considered to be the most important due to their potential to cause cell death, mutagenesis, and carcinogenesis. One study examined the accumulation of DNA DSBS in mouse blood leucocytes and splenocytes after long-term, chronic low-dose γ-irradiation *in vivo* and how this exposure may alter cell sensitivity to acute high-dose irradiation. After an initial slight increase in the level of DNA DSBS at 40 days of exposure compared to controls, there was a subsequent drop after either 80 or 120 days of exposure. Interestingly, the level of DNA breaks after both 80 and 120 days of exposure was lower than that in the controls. This result indirectly indicated that low-level ionizing radiation *in vivo* may trigger inducible repair of both endogenous and exogenous DNA DSBS and that there is a dose threshold for this inducible defence mechanism, below which it does not occur. The dose–response for DNA DSBS at very low doses and dose rates is not linear (Osipov et al., 2013). However, adaptation against neoplastic transformation may be induced by exposure to very low-dose-rate low-LET radiation (Azzam et al., 1996). Doses of less than 100 mGy delivered at very low dose rates in the range 1–4 mGy/day can induce an adaptive response against neoplastic transformation *in vitro*. If the dose rate drops below approximately 1 mGy/day, this suppression is apparently lost, suggesting a possible dose-rate-dependent threshold in this process (Elmore et al., 2008).

Cellular studies

Cell transformation experiments are used to evaluate radiation carcinogenesis *in vitro*. In response to radiation, a small proportion of the normal tissue cells can morphologically change and develop into cells with malignant phenotypes. Since the 1970s, several cell lines, including the NIH/BALB/3T3 cell line, the Syrian hamster embryo cell culture system, and the mouse embryo C3H10T1/2 cell line, have been developed and widely used to study cell transformation by irradiation (Pollock and Todaro, 1968).

These cell models have provided good models to study the relationship between radiation carcinogenesis and irradiation dose and/or dose rate and have provided the evidence to quantify the incidence of radiation carcinogenesis *in vitro*. For a given radiation dose, the yield of foci per dish, which are clones of transformed cells, is approximately the same over orders of magnitude in the number of cells plated.

This finding indicated that there are at least two steps involved in the transformation: (1) radiation initially induces events in a large fraction of cells; (2) rare events, such as mutation, only happen in a few descendants of the irradiated cells. There is a typical correlation plot regarding cell survival, as measured by transformants per irradiated cell and transformants per surviving cell after exposure to low LET ionizing radiation. In the range of the shoulder of the survival curve, the number of transformed cells increases with the dose. However, in the high-dose range, the number decreases along with the dosage increment because of the cell killing effect of high-dose irradiation. After 4Gy, the transformants per surviving cells plateaus and remains constant up to approximately 12 Gy (Hiatt et al., 1977). The effectiveness of high LET is greater for transformants than for surviving cells, with more cells killed by high LET radiation. Single, split, and fractionated doses may have different efficacies in cell transformation depending on the growth phase of the cells. One study was conducted to determine the effects of single, split, and fractionated doses of 100 rad of X-irradiation on the morphological transformation of grown BALB/c 3T3 clone A31-11 mouse fibroblasts, and the results showed that a four-fraction 100-rad dose resulted in reduced transformants compared to a single dose of 100 rad when plateau phase cultures were utilized. However, the level did not decrease in low density cultures in which the cells proliferated during radiation exposure (Lurie and Kennedy, 1985). Additionally, adding chemical carcinogenic factors can enhance the effectiveness of transformation. MCF10A is an immortalized non-tumorigenic breast epithelia cell line. After exposure to radiation and cigarettes, they exhibited a fibroblastic phenotype. A gene expression profiling showed that the major changes were associated with tissue remodelling, metabolism, and altered cell membrane protein levels, whereas a more ambiguous outcome was observed for genes involved in inflammation and signalling events. In combination with tobacco smoke carcinogens, radiation may result in a possible initiation of cellular processes that give rise to breast cancer cells (Botlagunta et al., 2010).

In addition to the morphological changes of cells *in vitro* after irradiation, the transformed cells can grow into colonies in petri dishes, and form tumour in a syngeneic animal. Typically, the morphologically transformed cells cannot immediately grow into tumours when injected into syngeneic mice. They must be grown for many generations before they can grow into tumours, which is consistent with the long latent period of radiation carcinogenesis.

Animal studies

Experimental animals can provide valuable information on radiation carcinogenesis, although the results of animal studies are not always consistent with human studies. Radiation carcinogenicity experiments have only been conducted in mice, rats, hamsters, and dogs, in which they were exposed to alpha, beta, gamma, neutron, or X-ray radiation. These experiments are very expensive and time-consuming. Various animals and tissues have different sensitivities to irradiation-induced tumorigenesis. Additionally, various radiation types have different carcinogenic potentials. In most studies, experimental animals are used to investigate the relationships between radiation dose and the incidence of carcinogenesis and lifespan shortening.

The tumour types that have been induced in animal models by radiation include leukaemia, lymphoma, bone tumour, mammary gland tumours, ovarian tumours, lung tumours, and skin tumours. In mice, lymphosarcoma is a common type of radiation-induced leukaemia, and the other forms of leukaemia, such as granulocytic leukaemia, are observed less often in mice. In animal experiments, the incidence of lymphomas in mice exposed to whole-body radiation early in life increases up to 40% along with the increment of the radiation dosage (Upton, 1961).

Compared with a single dose exposure, protracted intermittent irradiation is considered more leukaemogenic. Germ cells are highly radiosensitive for tumorigenesis. Ovarian cancer can be produced by a low dose equal to or greater than 32 rad. However, the incidence of ovarian cancer tends to decrease at very high doses (Upton, 1961).

Lifespan can be shortened by exposure to radiation, as shown in experimental animals. Most experiments have shown a clear linear relationship between radiation dose and life span shortening in mice within the dose range of 50–800 cGy (Upton, 1961). These experiments investigated life span shortening under the conditions of single dose, daily exposure, weekly exposure, and continuous exposure, and the results confirmed that life span shortening only depended on total dose (Maisin et al., 1978; Thomson et al., 1981a; Thomson et al., 1981b). After a full post-mortem examination, the results showed that life span shortening was almost always due to the induction of cancer by radiation. An opinion believes that the reason of lifespan shortening is because of the deteriorate effect of radiation-induced carcinogenesis, that means, the cause of cancer and death are advanced by radiation, and the age of animal at irradiation may be an important factor in determining the rate of cancer progression.

The dose–response relationship regarding the incidence of radiation carcinogenesis has been debated for decades. The linear no-threshold (LNT) hypothesis currently serves as the basic theory and assumes that at low levels of exposure, radiation carcinogenesis is directly proportional to the dose, and regardless of how low the dose is, the excess cancer risk is increased without a threshold (**Fig. 8.4**; see Maisin et al., 1988; Duport et al., 2012).

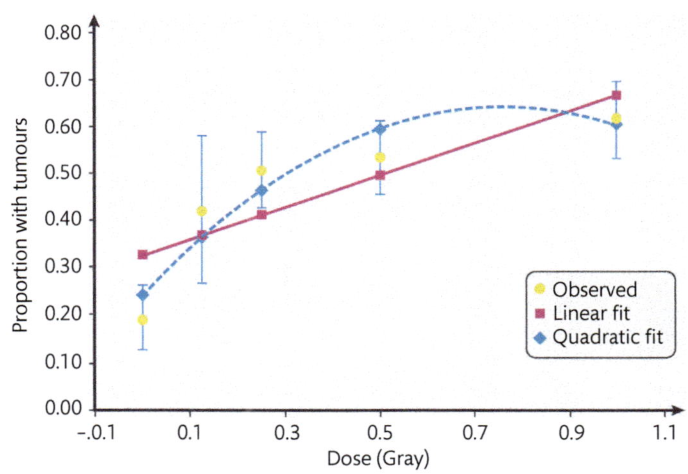

Fig. 8.4 Incidence rates for all cancers in male C57BL mice exposed to fission neutrons at an age of 21 day (Maisin et al., 1996). The data set ID is 458. The 90% confidence intervals for binomial parameters (cancer incidence rates) were computed using the Clopper–Pearson method.

However, different opinions regarding threshold dose and radiation hormesis have been presented. Radiation hormesis is the term for generally favourable biological responses to low-dose exposures to radiation by stimulating the activation of repair mechanisms, which is an adaptive response to overcompensate to a disruption in homeostasis (Crump et al., 2012).

Interestingly, pre-exposure to low 'priming' doses of radiation can result in increased longevity in rodents and insects, enhanced growth of multiple plant species, increased embryo production in rainbow trout, augmentation of immune responses in rabbits, mice, and *in vitro* human peripheral blood lymphocytes, and enhanced repair of cytogenetic damage (Upton, 2001). However, it remains uncertain whether these effects can lead to hormetic effects in cancer.

Experimental animals are the best models for conducting investigations into dose–response relationships based on their homogeneous biological backgrounds and the convenience of controlling the experimental conditions. However, according to a compiling analysis including 800 datasets from 262 experiments, with 87,982 exposed and 37,111 unexposed (control) animals, only 36% of the datasets showed a positive dose-effect relationship. These studies included animal types such as mouse, rat, dog, and hamster, and radiation types such as alpha, beta, gamma, neutron, and X-ray radiation. Although all the aforementioned radiation types have been proven to be carcinogenic, the probability of cancer per unit dose of radiation varies with some parameters, such as animal species, age and gender of animals, mode of radiation administration, dose, dose rate, type and volume of tissue irradiated, and type of radiation (Duport et al., 2012; Nagataki and Takamura, 2014). In fact, there is limited power for detecting hormesis based on the analysed animal database. Recently, a study also reported no observable effects of high- or low-dose radiation exposure on prostate cancer development using an autochthonous mouse model of prostate cancer (Lawrence et al., 2013).

Human effects

The most direct evidence of radiation-related cancer risk in humans is derived from epidemiological studies of populations exposed to ionizing radiation, some of which provide precious information on cancer risks at low doses. Most populations were exposed to radiation either routinely or during accidents. Another population with substantial risk of high exposure to radiation is the cohort with occupational exposure, which commonly occurs during nuclear fuel recycling, military activities, and medical applications. In an epidemiological study, radiation risk (RR) is defined as the increase in the number of cancer deaths over the expected number of deaths in an unirradiated population. Excess absolute risk (EAR) is the increased number of cancers per 10^4 person-year after exposure to 1 Sv. Excess relative risk (ERR) is the increase in cancers above that expected in an unirradiated population expressed as a fraction of the level in the unirradiated population.

Typically, accidental exposure and therapeutic exposure could occur with both low LET and high LET radiation. In accidental exposures, such as the atomic bombs at Hiroshima and Nagasaki and the explosion of the Chernobyl power plant, survivors received acute single doses. Leukaemia has the highest risk attributable to ionizing radiation with a minimum latent period of approximately 2 years. Among the excess radiation-induced cancer deaths of survivors of the atomic bombs in Hiroshima and Nagasaki that occurred between 1950 and 1987, more than 20% were due to leukaemia, with a total of 2.8 million PY. The average bone marrow dose of the survivors was approximately 0.2 Gy, which was between 0 and 4 Gy for 290 cases of leukaemia. EAR was estimated to be 2.7 cases per 10^4 PY Sv on average, decreasing by 6.5% annually, and the corresponding ERR was 3.9 per Sv, which is larger than the ERR estimates for any of the solid tumours (Preston et al., 1994; Preston et al., 2007). A special emphasis was placed on the health status of approximately 190,000 (6,000 were women) Chernobyl emergency workers. Of them, 78% have had individual dose assessments (Pitkevitch et al., 1997). The average dose of external whole-body gamma radiation is approximately 0.1 Gy, with a mean time of approximately 2 months. In the period from 1986 to 1997 (7–10 years since exposure), a statistically significant radiation risk of leukaemia incidence was observed, with an average ERR of 4.98 Gy^{-1}, and RR estimates in dose intervals were significant only for doses above 150 mGy. However, from 1998 to 2007, the radiation risks of leukaemia (excluding chronic lymphocytic leukaemia, CLL) were insignificant. Additionally, an increase in thyroid cancer rates was observed following the Chernobyl accident. After the Fukushima accident, the risk of developing thyroid cancer increased by 70%, and that of breast cancer increased by 6% for females exposed as infants. Additionally, the risk of developing leukaemia is 7% higher than average for males exposed as infants (Rhodes, 2014).

In therapeutic exposure, patients receive high total doses of radiation to tumour targets and relatively low total doses to the organ at risk, and these medical low levels are usually much higher than the level of natural background radiation. The incidence of second malignant tumours is a clinically observed adverse late effect of radiation therapy, especially in organs close to the treatment site (Kamran et al., 2016).

The risk of secondary malignant tumours is associated with age, hormonal influences, chemotherapy use, environmental influences, genetic predisposition, infection, and immunosuppression. Radiotherapy is an independent risk factor in the incidence of second tumours in patients with cervical cancer, head and neck cancer, non-Hodgkin lymphoma, and breast cancer. The common secondary cancers include cancers of the oesophagus, stomach, colon, rectum and anus, lung and mediastinum, bone and soft tissue, uterus, bladder, and kidneys; leukaemia; myeloma; and neoplasms of the bone and soft tissue, thyroid, central nervous system, skin, stomach, head and neck, liver and biliary tract, and the lungs and mediastinum, including both the irradiated organs and abscopal organs.

Typically, tissues that receive medium to high doses (>2.5 Gy) are apt to develop secondary malignancy. The reported 5- and 15-year probability to develop a histopathologically independent second tumour in or near the irradiated first tumour site (i.e. after intermediate or high radiation doses) is 0.5% and 2.2%, respectively. The secondary tumour latency ranges from one to over four decades, with a median of approximately 7 years. At 40 years after treatment, survivors of Hodgkin's lymphoma are at increased risk of secondary cancers, with a cumulative incidence reaching 48.5% (Schaapveld et al., 2015). Nasopharyngeal carcinoma is common in south China and Southeast Asia, which is curable by radiotherapy (Chen et al.,

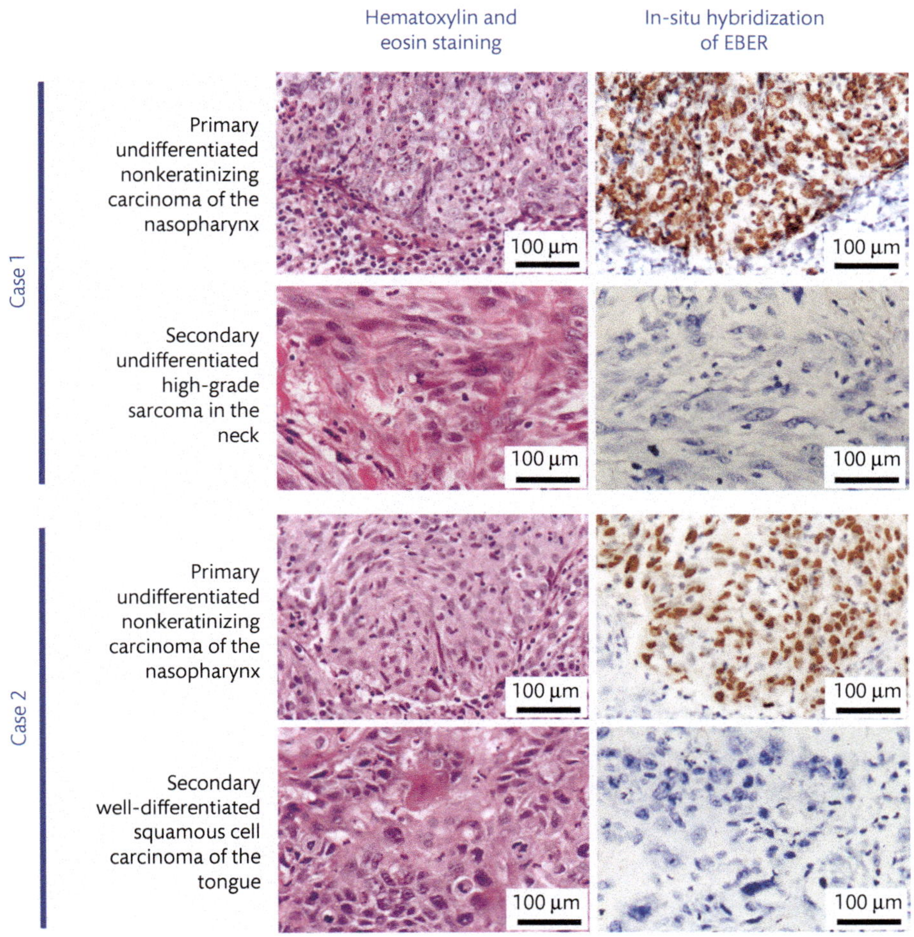

Fig. 8.5 Most nasopharyngeal carcinoma (NPC) cells in the patients from endemic areas harbour Epstein–Barr virus (EBV). Epstein-Barr encoding region *in situ* hybridization (brown in colour) is the methodology for the detection of the EBV in tissue sections. Presented here are tissue sections from primary NPCs and secondary radiation-induced tumours in the same patients. Notably, the secondary tumour cells do not harbour EBV, which is the typical characteristic of NPC cells. Case 1: A 35-year-old female with primary undifferentiated NPC underwent radiotherapy to the nasopharynx and the neck area for 68 Gy and 54 Gy, respectively. Thirty-two months later, a secondary undifferentiated high-grade sarcoma was developed in her right neck area. Case 2: A 38-year-old male with primary undifferentiated NPC underwent radiotherapy to the nasopharynx and the neck area for 70 Gy and 60 Gy, respectively. Fifty-five months later, a secondary well-differentiated squamous cell carcinoma had developed in his tongue.

2015; Wang et al., 2016). The most common secondary malignancies after radiotherapy for nasopharyngeal carcinoma include squamous cell carcinoma and sarcoma, which usually appear after 5 years of treatment (**Fig. 8.5**). Because of the long latency of radiation carcinogenesis, a follow-up period beyond 5 years is mandatory to identify radiogenic secondary malignancies. Although therapeutic exposure is often well characterized with respect to the dose and dose rate, possible confounders such as accompanying chemical treatment must be considered. Additionally, patients undergoing cancer therapy are usually in an age group that excludes the early years of life, which is of special importance based on the assumed greater sensitivity of children to radiation (Welte et al., 2010). Screening mammography offers the possibility of discovering malignant diseases at an early stage and provides a glandular dose of 2.5 mGy from a two-view per breast image. One study reviewed novel radiobiological data in a population of 100,000 individuals with or without screening mammography. The results showed that in women aged 50–74 years, the ratio of the induced incidence rate over the baseline incidence rate is approximately 1.6% for biennial screening. However, this number may be closer to approximately 0.7% if considering the dose and dose rate effectiveness factor values. This value suggests that individuals who are known to be carriers of risk-increasing genetic variations and/or have an inherited disposition of breast cancer and/or hyper-susceptibility to ionizing radiation should avoid ionizing radiation as much as possible and should receive ultrasound or magnetic resonance imaging (Pauwels et al., 2016). However, unselected populations of all ages are therefore of particular interest, and routine exposure with low-dose LET radiation is more concerning for people (**Fig. 8.6**). Radon is an important source of radiation exposure in the general public. It is estimated that in the United States, approximately 10–12% of lung cancer deaths are attributed to radon. A study on the oncogenic transformation incidence caused by radon daughter He4 ions with or without concurrent exposure to cigarette smoke condensate showed that He4 ions have an additive effect in oncogenic transformation potential in C3H 10T1/2 cells (Piao and Hei, 1993).

Site of cancer	Type of exposure																		
	Atomic bombs, Chernobyl				Radiotherapy and diadnosis										Occupation				
	Atomic bombs survivors	Inhabitants of Marshall island	Participants in weapons tests	Chernobyl	Ankylosing spondylitis (X-ray)	Ankylosing spondylitis (Radium)	Benign gynaecological disorders	Benign breast disease	Flurosocopy of the chest	Tinea capitis	Thymus	Thorolrast	Thyroid (I)	X-rays in vitro	Radium-dial painters	Radiologists	Underground miners	Nuclear workers	Aircraft personnel
Leukaemia	■		○	□	■		□	○		□	○	■	○	■		■		○	
Thyroid	■	□		■						■	■			○					
Breast	■					□		■	■		□				□				○
Lung	■				■							○					■		
Bone						■						○			■				
Stomach	■				□												○		
Oesophagus	□				□				□										
Bladder	■						□												
Lymphoma	□				□		○					○		□		□			
Central nervous system	○				■					□			○	□		□			○
Uterus	○						□												
Liver	■											■					○		
Skin										■	□					■	○		
Salivary gland	□									□	□								
Kidney					○	○													
Colon	■				■		○												
Small intestine															○				

■ Statistically significant correlation □ Strongly suspected but in some studies no significant correlation ○ Some correlation found but not significant

Fig. 8.6 Populations who received heavy exposure to ionizing radiation and were followed up for cancer and other long-term effects on health.
Reproduced by permission from Springer Nature, *Der Radiologe*, Health risks of ionizing radiation, Volume 38, Issue 9, pp. 719–25, Schneider G and Burkart W. Copyright © 1998.

Cosmic radiation is substantial for flight crew. Several epidemiologic or related studies of cancer incidence and mortality have reported that flight attendants have increased risks of female breast cancer, and there is higher melanoma incidence in both pilots and cabin crew. Occasionally, but not consistently, excesses of other cancers have been observed. Although the real causes of these excess cancer risks are not known, there is concern that they may be related to occupational exposures to ionizing radiation of cosmic origin. SES-adjusted combined RRs were elevated (>1.2) among male pilots for mortality from melanoma and brain cancer and for incidence of prostate and brain cancer. Among female flight attendants, increases were seen in the incidence of all cancers, especially melanoma and breast cancer (Sigurdson and Ron, 2004).

Nielsen et al. (Nielsen et al., 2014) examined numerous lung samples from a plutonium-exposed worker and human control using immunohistochemistry and quantitative reverse transcriptase-polymerase chain reaction (RT-PCR). The examination showed interstitial fibrosis in peripheral regions of the lung and significant modifications in the expression of Fas ligand (FASLG), B-cell lymphoma 2 (BCL2), and Caspase 3 (CASP3). Their study suggests that FASLG, BCL2, CASP3, and apoptosis play a role in the inflammatory responses following prolonged plutonium exposure. Ricarte-Filho and colleagues demonstrated that chromosomal rearrangements are the oncogenic 'drivers' in most post-Chernobyl carcinomas and that they often lead to unscheduled activation of the MAPK signalling pathway (Ricarte-Filho et al., 2013; Santoro and Carlomagno, 2013). These findings represent a major step forward in our understanding of radiation-induced carcinogenesis in humans and suggest various hypotheses about the mechanisms underlying the formation and selection of gene rearrangements during cancer cell evolution.

TAKE-HOME MESSAGE

- Ionizing radiations can generate ions and free radicals, resulting in damage of biological large molecules including DNA and therefore being able to induce cancer.
- Ionizing radiations induce cancer at low dose, and accumulative doses can increase the risk of cancer incidence.
- The most common types of cancers induced by radiation are leukaemia and thyroid cancer.
- The most hazardous source of radiation to general public health is natural background radiation, followed by medical radiation.
- Multiple approaches can be applied for minimizing the hazardous effects of medical radiations.

OPEN QUESTIONS

- How many ways can ionizing radiations induce genomic instability in the cells and therefore induce carcinogenesis?
- Why are bone marrow and the thyroid gland more susceptible to radiation-induced malignancies after the patient's exposure to nuclear disasters?
- How can we effectively suppress cancer formation in the individuals exposed to radiations?

FURTHER READING

Foray, N., Bourguignon, M., Hamada, N. (2016). Individual response to ionizing radiation. *Mutat Res*, **770**(Pt B), 369–86.

Hesketh, R. (2013). *Introduction to Cancer Biology*. Cambridge University Press 2013.

Ivanov, V. K., Tsyb, A. F., Khait, S. E., et al. (2012). Leukemia incidence in the Russian cohort of Chernobyl emergency workers. *Radiat Environ Biophys*, **51**(2), 143–9.

Lane, N. (2003). *Oxygen: The Molecule That Made the World*. Oxford University Press.

Shaleck, R. J. (1953). The formation of hydrogen peroxide in water by ionizing radiation. Available at: https://scholarship.rice.edu/bitstream/handle/1911/18443/3079878.Pdf?sequence=1&isAllowed=y

REFERENCES

Azzam, E. I., De Toledo, S. M., Raaphorst, G. P., & Mitchel, R. E. (1996). Low-dose ionizing radiation decreases the frequency of neoplastic transformation to a level below the spontaneous rate in C3H 10T1/2 cells. *Radiat Res*, **146**, 369–73.

Balcer-Kubiczek, E. K., Harrison, G. H., Torres, B. A., & McCready, W. A. (1994). Application of the constant exposure time technique to transformation experiments with fission neutrons: failure to demonstrate dose-rate dependence. *Int J Radiat Biol*, **65**, 559–69.

Botlagunta, M., Winnard, P. T., JR., & Raman, V. (2010). Neoplastic transformation of breast epithelial cells by genotoxic stress. *Bmc Cancer*, **10**, 343.

Calaf, G. M., Roy, D., Narayan, G., & Balajee, A. S. (2013). Differential expression of cell adhesion molecules in an ionizing radiation-induced breast cancer model system. *Oncol Rep*, **30**, 285–91.

Cardis, E., Vrijheid, M., Blettner, M., et al. (2005). Risk of cancer after low doses of ionising radiation: retrospective cohort study in 15 countries. *BMJ*, **331**, 77.

Caudill, C. M., Zhu, Z., Ciampi, R., Stringer, J. R., & Nikiforov, Y. E. (2005). Dose-dependent generation of RET/PTC in human thyroid cells after *in vitro* exposure to gamma radiation: a model of carcinogenic chromosomal rearrangement induced by ionizing radiation. *J Clin Endocrinol Metab*, **90**, 2364–9.

Chen, Y. P., Zhao, B. C., Chen, C., et al. (2015). Pretreatment platelet count improves the prognostic performance of the TNM staging system and aids in planning therapeutic regimens for nasopharyngeal carcinoma: a single-institutional study of 2,626 patients. *Chin J Cancer*, **34**, 137–46.

Crump, K. S., Duport, P., Jiang, H., Shilnikova, N. S., Krewski, D., & Zielinski, J. M. (2012). A meta-analysis of evidence for hormesis in animal radiation carcinogenesis, including a discussion of potential pitfalls in statistical analyses to detect hormesis. *J Toxicol Environ Health B Crit Rev*, **15**, 210–31.

Datta, K., Hyduke, D. R., Suman, S., Moon, B. H., Johnson, M. D., & Fornace, A. J., Jr. (2012). Exposure to ionizing radiation induced persistent gene expression changes in mouse mammary gland. *Radiat Oncol*, **7**, 205.

Denissova, N. G., Tereshchenko, I. V., Cui, E., Stambrook, P. J., Shao, C., & Tischfield, J. A. (2011). Ionizing radiation is a potent inducer of mitotic recombination in mouse embryonic stem cells. *Mutat Res*, **715**, 1–6.

Duport, P., Jiang, H., Shilnikova, N. S., Krewski, D., & Zielinski, J. M. (2012). Database of radiogenic cancer in experimental animals exposed to low doses of ionizing radiation. *J Toxicol Environ Health B Crit Rev*, **15**, 186–209.

Elmore, E., Lao, X. Y., Kapadia, R., Giedzinski, E., Limoli, C., & Redpath, J. L. (2008). Low doses of very low-dose-rate low-LET radiation suppress radiation-induced neoplastic transformation *in vitro* and induce an adaptive response. *Radiat Res*, **169**, 311–18.

Fleenor, C. J., Higa, K., Weil, M. M., & Degregori, J. (2015). Evolved cellular mechanisms to respond to genotoxic insults: implications for radiation-induced hematologic malignancies. *Radiat Res*, **184**, 341–51.

Hatzi, V. I., Laskaratou, D. A., Mavragani, I. V., et al. (2015). Non-targeted radiation effects *in vivo*: a critical glance of the future in radiobiology. *Cancer Lett*, **356**, 34–42.

Hiatt, H. H., W. J., Winston, J. A. (1977). *Origins of Human Cancer.* Cold Spring Harbor, NY: Cold Spring Harbor Laboratory.

Hieber, L., Ponsel, G., Roos, H., Fenn, S., Fromke, E., & Kellerer, A. M. (1987). Absence of a dose-rate effect in the transformation of C3H 10T1/2 cells by alpha-particles. *Int J Radiat Biol Relat Stud Phys Chem Med*, **52**, 859–69.

Hill, C. K., Buonaguro, F. M., Myers, C. P., Han, A., & Elkind, M. M. (1982). Fission-spectrum neutrons at reduced dose rates enhance neoplastic transformation. *Nature*, **298**, 67–9.

Hollander, C. F., Zurcher, C., & Broerse, J. J. (2003). Tumorigenesis in high-dose total body irradiated rhesus monkeys—a lifespan study. *Toxicol Pathol*, **31**, 209–13.

Ilnytskyy, Y. & Kovalchuk, O. (2011). Non-targeted radiation effects-an epigenetic connection. *Mutat Res*, **714**, 113–25.

Kamran, S. C., Berrington De Gonzalez, A., Ng, A., Haas-Kogan, D., & Viswanathan, A. N. (2016). Therapeutic radiation and the potential risk of second malignancies. *Cancer*, **122**(12), 1809–21.

Komatsu, K., Sawada, S., Takeoka, S., Kodama, S., & Okumura, Y. (1993). Dose-rate effects of neutrons and gamma-rays on the induction of mutation and oncogenic transformation in plateau-phase mouse m5S cells. *Int J Radiat Biol*, **63**, 469–74.

Lawrence, M. D., Ormsby, R. J., Blyth, B. J., et al. (2013). Lack of high-dose radiation mediated prostate cancer promotion and low-dose radiation adaptive response in the Tramp mouse model. *Radiat Res*, **180**, 376–88.

Lurie, A. G. & Kennedy, A. R. (1985). Single, split and fractionated dose X-radiation-induced malignant transformation in A31-11 mouse BALB/3T3 cells. *Cancer Lett*, **29**, 169–76.

Maisin, J. R., Decleve, A., Gerber, G. B., Mattelin, G., & Lambiet-Collier, M. (1978). Chemical protection against the long-term effects of a single whole-body exposure of mice to ionizing radiation II. Causes of death. *Radiat Res*, **74**, 415–35.

Maisin, J. R., Wambersie, A., Gerber, G. B., et al. (1988). Life-shortening and disease incidence in C57Bl mice after single and fractionated gamma and high-energy neutron exposure. *Radiat Res*, **113**, 300–17.

Mancuso, M., Giardullo, P., Leonardi, S., et al. (2013). Dose and spatial effects in long-distance radiation signaling *in vivo*: implications for abscopal tumorigenesis. *Int J Radiat Oncol Biol Phys*, **85**, 813–19.

Morgan, L. L., Miller, A. B., Sasco, A., & Davis, D. L. (2015). Mobile phone radiation causes brain tumors and should be classified as a probable human carcinogen (2A) (review). *Int J Oncol*, **46**, 1865–71.

Nagataki, S. & Takamura, N. (2014). A review of the Fukushima nuclear reactor accident: radiation effects on the thyroid and strategies for prevention. *Curr Opin Endocrinol Diabetes Obes*, **21**, 384–93.

Nielsen, C. E., Wang, X., Robinson, R. J., et al. (2014). Carcinogenic and inflammatory effects of plutonium-nitrate retention in an exposed nuclear worker and beagle dogs. *Int J Radiat Biol*, **90**, 60–70.

Olme, C. H., Finnon, R., Brown, N., Kabacik, S., Bouffler, S. D., & Badie, C. (2013). Live cell detection of chromosome 2 deletion and Sfpi1/PU1 loss in radiation-induced mouse acute myeloid leukaemia. *Leuk Res*, **37**, 1374–82.

Osipov, A. N., Buleeva, G., Arkhangelskaya, E., & Klokov, D. (2013). *In vivo* gamma-irradiation low dose threshold for suppression of DNA double strand breaks below the spontaneous level in mouse blood and spleen cells. *Mutat Res*, **756**, 141–5.

Patel, A., Anderson, J., Kraft, D., et al. (2016). The influence of the CTIP polymorphism, Q418P, on homologous recombination and predisposition to radiation-induced tumorigenesis (mainly rAML) in mice. *Radiat Res*, **186**, 638–49.

Pauwels, E. K., Foray, N., & Bourguignon, M. H. (2016). Breast cancer induced by X-ray mammography screening? A review based on recent understanding of low-dose radiobiology. *Med Princ Pract*, **25**, 101–9.

Piao, C. Q. & Hei, T. K. (1993). The biological effectiveness of radon daughter alpha particles. I. Radon, cigarette smoke and oncogenic transformation. *Carcinogenesis*, **14**, 497–501.

Pitkevitch, V. A., Ivanov, V. K., Tsyb, A. F., Maksyoutov, M. A., Matiash, V. A., & Shchukina, N. V. (1997). Exposure levels for persons involved in recovery operations after the Chernobyl accident. Statistical analysis based on the data of the Russian National Medical and Dosimetric Registry (Rnmdr). *Radiat Environ Biophys*, **36**, 149–60.

Pollock, E. J. & Todaro, G. J. (1968). Radiation enhancement of SV40 transformation in 3T3 and human cells. *Nature*, **219**, 520–1.

Preston, D. L., Kusumi, S., Tomonaga, M., et al. (1994). Cancer incidence in atomic bomb survivors. Part III. Leukemia, lymphoma and multiple myeloma, 1950–1987. *Radiat Res*, **137**, S68–97.

Preston, D. L., Ron, E., Tokuoka, S., et al. (2007). Solid cancer incidence in atomic bomb survivors: 1958–1998. *Radiat Res*, **168**, 1–64.

Rhodes, C. J. (2014). The Fukushima Daiichi nuclear accident. *Sci Prog*, **97**, 72–86.

Ricarte-Filho, J. C., Li, S., Garcia-Rendueles, M. E., et al. (2013). Identification of kinase fusion oncogenes in post-Chernobyl radiation-induced thyroid cancers. *J Clin Invest*, **123**, 4935–44.

Santoro, M. & Carlomagno, F. (2013). Oncogenic rearrangements driving ionizing radiation-associated human cancer. *J Clin Invest*, **123**, 4566–8.

Schaapveld, M., Aleman, B. M., Van Eggermond, A. M., et al. (2015). Second cancer risk up to 40 years after treatment for Hodgkin's lymphoma. *N Engl J Med*, **373**, 2499–511.

Sigurdson, A. J. & Ron, E. (2004). Cosmic radiation exposure and cancer risk among flight crew. *Cancer Invest*, **22**, 743–61.

Spitz, D. R. & Hauer-Jensen, M. (2014). Ionizing radiation-induced responses: where free radical chemistry meets redox biology and medicine. *Antioxid Redox Signal*, **20**, 1407–9.

Szumiel, I. (2015). Ionizing radiation-induced oxidative stress, epigenetic changes and genomic instability: the pivotal role of mitochondria. *Int J Radiat Biol*, **91**, 1–12.

Thomson, J. F., Williamson, F. S., Grahn, D., & Ainsworth, E. J. (1981a). Life shortening in mice exposed to fission neutrons and gamma rays I. Single and short-term fractionated exposures. *Radiat Res*, **86**, 559–72.

Thomson, J. F., Williamson, F. S., Grahn, D., & Ainsworth, E. J. (1981b). Life shortening in mice exposed to fission neutrons and gamma rays II. Duration-of-life and long-term fractionated exposures. *Radiat Res*, **86**, 573–9.

Upton, A. C. (1961). The dose–response relation in radiation-induced cancer. *Cancer Res*, **21**, 717–29.

Upton, A. C. (2001). Radiation hormesis: data and interpretations. *Crit Rev Toxicol*, **31**, 681–95.

Wang, H. Y., Chang, Y. L., To, K. F., et al. (2016). A new prognostic histopathologic classification of nasopharyngeal carcinoma. *Chin J Cancer*, **35**, 41.

Welte, B., Suhr, P., Bottke, D., et al. (2010). Second malignancies in high-dose areas of previous tumor radiotherapy. *Strahlenther Onkol*, **186**, 174–9.

Yamamoto, M., Serizawa, T., Shuto, T., et al. (2014). Stereotactic radiosurgery for patients with multiple brain metastases (JLGK0901): a multi-institutional prospective observational study. *Lancet Oncol*, **15**, 387–95.

SECTION III
How the cancer cell works

9

Growth factors and associated signalling pathways in tumour progression and in cancer treatment

Nadège Gaborit and Yosef Yarden

Introduction

In 1948, Elmer Bueker and Victor Hamburger made an interesting observation: when implanting a fragment of a mouse connective tissue tumour (sarcoma) into the body wall of a 3-day-old chick embryo, they noticed that the malignant tissue was penetrated by fibres of murine sensory origin. This observation led their collaborator, Rita Levi-Montalcini, to the discovery of the first growth factor (GF), the nerve growth factor (NGF). Using sensory ganglia of chick embryos that were exposed to fragments or cell-free extracts of the tumour tissue, she concluded that the tumour released a diffusible factor able to promote neurite outgrowth. The *in vitro* assay established by Montalcini allowed the young biochemist who joined her, Stanley Cohen, to isolate a 26-kDa polypeptide, NGF, from extracts of either mouse tumours (Cohen et al., 1954), snake venoms, or the mouse salivary gland. Employing extracts of the salivary gland and another assay, namely precocious opening of the eyes of newborn mice injected with fractions of the extract, Cohen and Savage later isolated and determined the primary structure of another GF, the epidermal growth factor (EGF), a 53-amino acid-long polypeptide (Savage et al., 1972). Montalcini and Cohen shared in 1986 the Nobel Prize for the discovery of the first GFs, an accomplishment that established a new field in biology and opened wide the door for many medical applications, including prognosis and therapy of cancer.

In hindsight, the discovery of the first GFs in a tumour reflects the ability of most solid and liquid tumours, such as sarcomas, carcinomas, and myelomas, to produce and shed a multitude of GFs (see Fig. 9.1). The frequent secretion of GFs by tumours is called autocrine regulation (Sporn and Todaro, 1980). It differs from the more physiological modes found in healthy tissues and organs: exocrine (secretion into the extracellular space) and endocrine (secretion into the circulation). It is worthwhile considering the analogy between tumours and other storehouses of GFs, such as damaged skin and other wounds (Schafer and Werner, 2008). Shortly after a skin injury and vascular damage, local blood clotting and inflammation are set in motion. The clot consists mainly of cross-linked fibrin and fibronectin, and it serves as a reservoir of various cytokines and GFs, which are required for the healing process. The clot also acts as a matrix for different cell types that are attracted to the wound from the adjacent unaffected tissues or from the circulation (e.g. lymphocytes and macrophages; see Fig. 9.2). Importantly, malignant tumours may develop at sites of chronic injury, and tissue injury has an important role to play in the pathogenesis of malignant diseases, with chronic inflammation being an important risk factor. Like in wounds, tumours often deposit an extracellular matrix and attract to their stroma lymphoid and other cell types. Along with remarkable similarities between wound repair and cancer, the long tissue remodelling phase that follows the initial steps of wound repair is replaced by an escalating phase of mutation-driven altered metabolism and enhanced invasive growth of malignant tumours. These phenotypic differences were originally described by Rudolf Virchow (1821–1902), a German pathologist who argued that tumours resemble wounds that never heal.

This chapter reviews the major families of GFs and the respective roles they play in tumour progression, including increased tumour mass, recruitment of blood vessels, enhanced invasiveness, and resistance to cytotoxic treatments. We also highlight the shared biochemical mechanisms of GF action on target cancer cells, from binding with cell surface receptors harbouring an intrinsic tyrosine-specific kinase activity, to the ultimate mechanisms controlling gene expression. We conclude by listing the ever-expanding repertoire of anticancer therapies aimed at intercepting GFs and their downstream signalling pathways.

The cyclic activation-inactivation process of GF signalling identifies multiple targets for oncogenic mutations

GF binding with the respective receptor instigates a well-orchestrated series of biochemical events aimed at pulsed activation of the target cell, followed by inactivation and regain of the resting state (Citri and

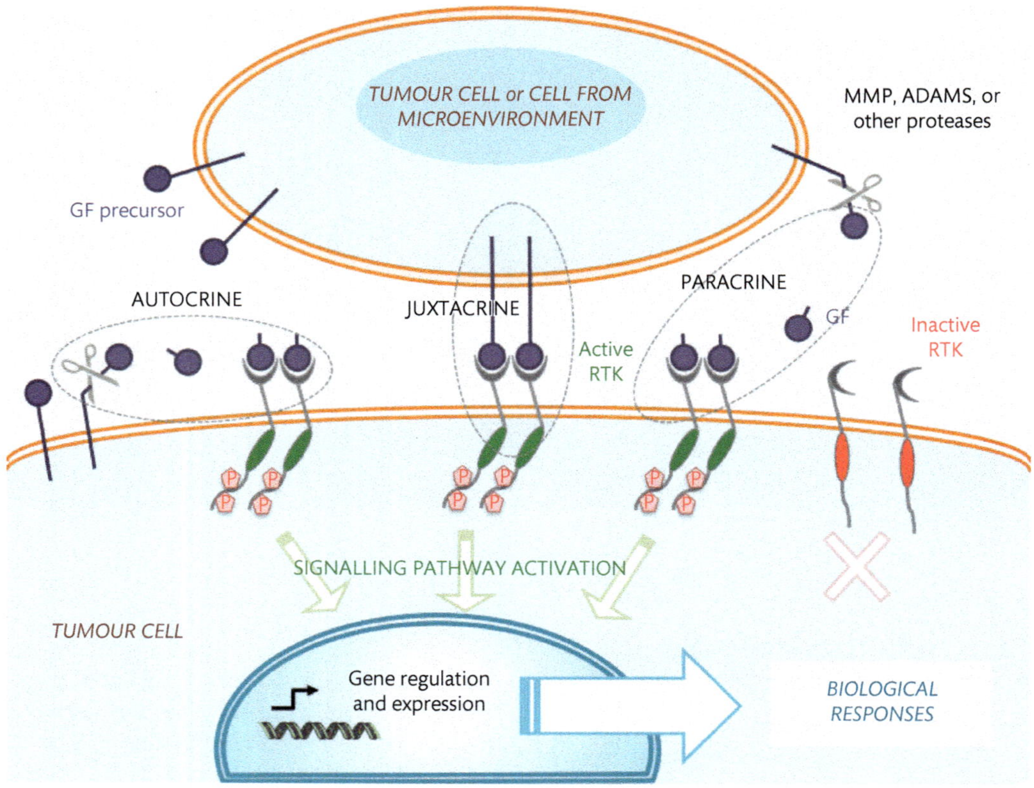

Fig. 9.1 A simplified view of growth factor shedding and modes of action. Shown schematically is the interface between two cells within a tumour. Although some growth factors (GFs) might perform endocrine functions (i.e. their delivery to target cells is mediated by the body's circulatory system), most GFs are short-range messengers, which act in a paracrine manner under normal physiological conditions. Cancer cells might adopt an autocrine mode, namely: cells respond to the factors they secrete. The precursors of several GFs are transmembrane proteins that undergo cleavage by extracellular proteases like matrix metaloproteinases (MMPs). Nevertheless, when still immobilized at the cell surface and un-cleaved, the pro-molecules might bind receptor tyrosine kinases (RTKs) expressed on neighbouring cells (juxtacrine mode of action). Independent of the exact mode of action, downstream signalling requires receptor dimerization and auto-phosphorylation (shown by encircled P letters).

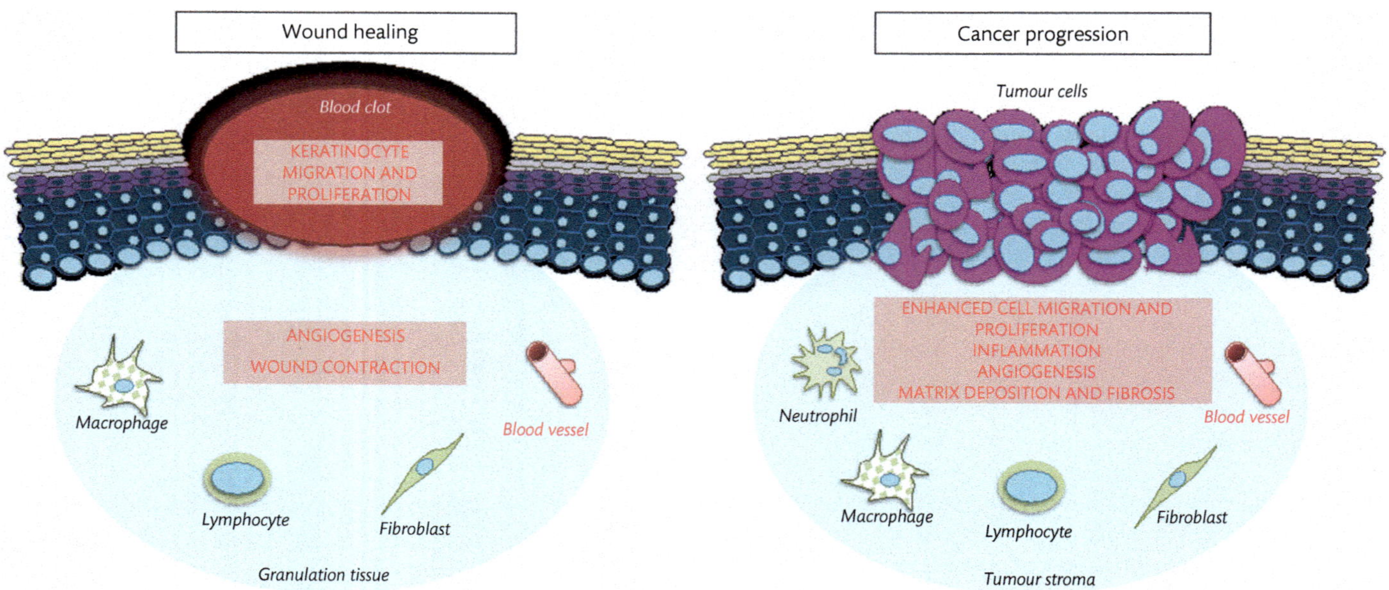

Fig. 9.2 Comparison between wound healing and cancer progression. A skin wound is schematically compared to a skin tumour, while emphasizing interactions with the granulation layer, a vascularized tissue that replaces the fibrin clot in a skin wound, and tumour's stroma. Both tissues recruit diverse cell types, which secrete multiple growth factors, and both deposit extracellular matrix. Note that recruited immune cells secrete GFs like PDGF, HB-EGF, interleukins and TGF-beta, while connective tissues secrete activin, HGF and EGF-like ligands. These GFs are essential for enhanced cell migration and proliferation, along with angiogenesis, local inflammation and occasionally also fibrosis.

Source: data from Antsiferova M and Werner S, 'The bright and the dark sides of activin in wound healing and cancer', *Journal of Cell Science*, Volume 125, pp. 3929–37,

Yarden, 2006; Lemmon and Schlessinger, 2010). Diverse oncogenic mutations interfere with this cycle by either enhancing the emanating signals or by prolonging the active state, as we describe next.

Ligand-induced receptor dimerization

This review concentrates on GFs that bind with surface receptors harbouring intrinsic tyrosine kinase activity (receptor tyrosine kinases, RTKs). Altogether, 58 RTKs are encoded by the human genome. RTKs like the receptor for the insulin-like growth factor 1 (IGF1) might exist as homodimers prior to ligand binding. Yet, both predimerized receptors and monomeric receptors, like KIT and EGFR, undergo conformational changes upon ligand binding, and these position the receptors in a way that enables auto-phosphorylation of tyrosine residues located at the cytoplasmic domains. Note that each receptor phosphorylates the other receptor within a dimer (i.e. transphosphorylation). The best understood conformational changes were uncovered by means of crystallization of portions of EGFR (Burgess et al., 2003) and they are presented in Figure 9.3. Yet, ligands that are dimeric, such as the platelet-derived growth factors (PDGFs), or trimeric (e.g. NGF), might induce oligomerization of their receptors by means of their own oligomerization (i.e. ligand-mediated oligomerization). Certain oncogenic mutations might promote constitutive (ligand-independent) dimers of RTKs. For example, intracellular juxtamembrane mutations in EGFR activate the receptor, apparently by promoting dimers. Likewise, variant III EGFR, which is highly expressed in brain tumours (glioblastomas), encodes a receptor lacking the EGF-binding domain and the arm. This constitutively active receptor is probably driven by receptor aggregation. The 1986 nuclear reactor accident in Chernobyl and the subsequent development of papillary thyroid carcinomas in children uncovered ligand-independent dimerization of another RTK, RET. This is due to fusion of the kinase domain and unrelated coiled-coil regions with intrinsic dimerization potentials. Another RTK, MET (a.k.a. HGFR, hepatocyte growth factor receptor) is similarly activated following a genomic rearrangement that generates a hybrid protein containing the tyrosine kinase sequences fused to dimerization-inducing leucine zipper motifs. HER2 (also called ErbB2 and NEU), a ligand-less receptor similar to EGFR, is overexpressed in breast, gastric, and other tumours, and this biases formation of active heterodimers containing ErbB2/HER2 and other ErbB/HER family members (Garrett et al., 2003).

Kinase activation and downstream signalling

In the resting state, the tyrosine kinase domains (TKDs) of RTKs are inhibited by intramolecular interactions imposed by the kinase's activation loop, which occludes the active site (e.g. in the insulin receptor) or the TKD flanking regions, either the carboxyl-terminal or the juxtamembrane domain. Upon formation of receptor dimers, these modes of auto-inhibition are relieved, primarily by means of transphosphorylation of critical tyrosine residues that negate auto-inhibition. Subsequently, most RTKs undergo stimulatory transphosphorylation within the activation loop. A different mechanism underlays stimulation of EGFR: an asymmetric dimer in which the C-lobe (carboxyl region) of one TKD (called the 'Activator') makes contacts with the N-lobe of the second TKD (called the 'Receiver'). Subsequent conformational changes in the N-lobe of the Receiver disrupt auto-inhibitory interactions such that the Receiver can adopt the characteristic active configuration without activation loop phosphorylation (Kovacs et al., 2015). Apparently, the ensuing RTK auto-phosphorylation events follow a precise order that both activates the kinase and establishes binding sites for cytoplasmic proteins containing phosphotyrosine-binding domains (e.g. SH2 and PTB domains). The recruited proteins include both adaptors, such as GRB2, IRS, FRS, and GAB1, or enzymes, such as phospholipase

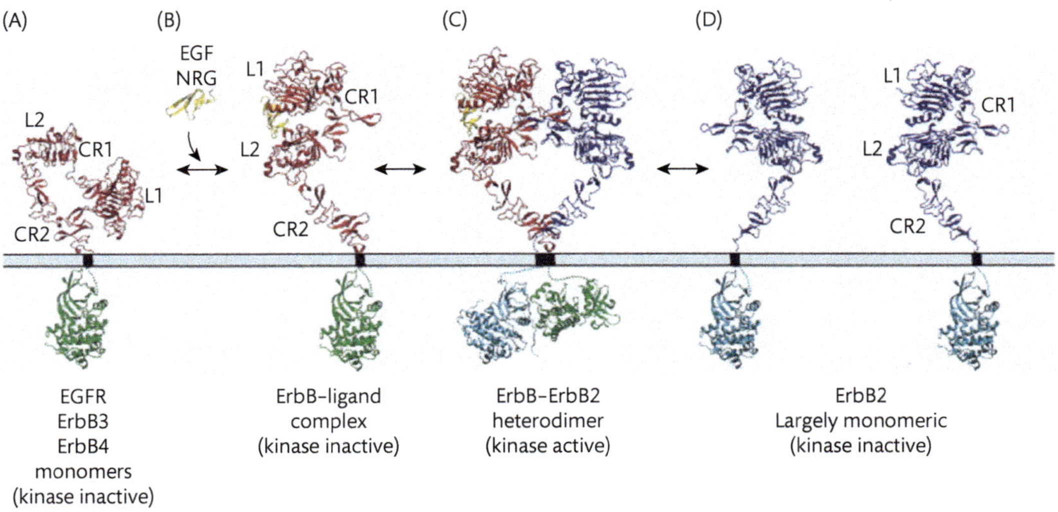

Fig. 9.3 Schematic representation of the process leading to formation of heterodimers involving ErbB2/HER2. A prototypical tethered structure of an ErbB/HER monomer is shown (A). Dimerization is prevented by binding of the dimerization arm, located in cysteine-rich domain 1 (CR1), with CR2, but ligand binding to domains L1 and L2 is allowed. Once a ligand (e.g. EGF or NRG) binds, the extended conformation of the receptor is stabilized, and the dimerization arm is exposed (B). This configuration is poised for dimer formation (C). Although both homodimers and heterodimers may form, ErbB2/HER2-containing heterodimers appear to be more stable. Within dimers, transphosphorylation of intrinsic tyrosine residues of the intracellular domain is rapidly induced. This step establishes phosphotyrosine-centred binding sites for adaptors or cytoplasmic enzymes involved in signal transduction. Note that unlike other monomers, monomeric forms of ErbB2/HER2 are constitutively in the extended configuration (D).

Source: data from Burgess AW et al., 'An open-and-shut case? Recent insights into the activation of EGF/ErbB receptors', *Molecular Cell*, Volume 12, Issue 3, pp. 541–52,

Box 9.1 Intracellular signalling pathways activated by growth factors

Once occupied by a GF and fully auto-phosphorylated, the stimulated RTK becomes the origin of several independent signal-generating pathways. Because every receptor's phosphotyrosine is flanked by unique amino acid sequences, each able to recruit a different set of phosphotyrosine-binding SH2 and PTB domains of various effector proteins (either enzymes or adaptors), instant and simultaneous firing takes place. Some of the major downstream pathways are described next.

The RAF-MEK-ERK pathway

The principle of instant firing initiated by an active dimer of RTKs was originally established when characterizing the linear cascade of protein kinases, the RAF-MEK-ERK pathway, which became a prototypic route (Plotnikov et al., 2011). Like in some other signalling pathways, a phosphotyrosine-binding adaptor (e.g. GRB2 or SHC), physically associates with the active RTK, along with a GTP-exchange factor (called SOS) specific to RAS proteins. Consequently, SOS replaces RAS-bound GDP molecules with RAS-GTP molecules, thereby activates RAS family members. One effector group of GTP-RAS is the RAF family of protein kinases (also called mitogen-activated protein kinase kinase kinase; MAPKKK). Active RAF proteins phosphorylate a serine residue located in the activation loop of a downstream kinase, MEK (MAPKK), which in turn phosphorylates both a threonine and a tyrosine residue within the activation loop of ERK. Unlike the narrow specificity of MEK towards ERKs, the latter can phosphorylate multiple proteins, including MAPK-activated protein kinases, the 90-kilodalton ribosomal S6 kinase (RSK), which regulates cell proliferation and survival, HDAC6, which regulates histone acetylation, and EIF4EBP2, which regulates translation rates. Once phosphorylated, ERK translocates into the nucleus, where it activates several transcription factors, most notable are members of the ETS family.

The phospholipase C pathway

Unlike the physical complexes and transphosphorylations, which are typical of the RAF-MEK-ERK pathway, the phospholipase C-gamma pathway generates second messengers. The SH2 domain of PLC-gamma directly engages active RTKs, and this entails phosphorylation and catalytic activation of PLC (Yang et al., 2013). As a result, a burst of PtdIns(4,5)P2 hydrolysis is one of the early events that follow the binding of many GFs (e.g. EGF, FGF and PDGF) with their respective receptors. PtdIns(4,5)P2 hydrolysis gives rise to two second messengers: diacylglycerol, which stimulates specific members of the protein kinase C family, and inositol triphosphate (IP3). Through binding to specific receptors located in the endoplasmic reticulum, IP3 stimulates release of calcium ions acting as second messengers and initiators of calcium/calmodulin-activated protein kinases.

The PI3K/AKT pathway

The regulatory p85 subunit of the phosphoinositol 3-kinase (PI3K) directly interacts with several RTKs (Cantley, 2002). Alternative engagement of PI3K is enabled, among other mechanisms, by loading of RAS with GTP, recruitment of CBL, or engagement of the GRB2-GAB-p85 complex. Once activated, the p110 catalytic subunit phosphorylates the 3′ carbon of the sugar ring of inositol lipids, primarily PtdIns(4,5)P2, to generate PtdIns(3,4,5)P3, as well as other second messengers. Incorporation of PtdIns(3,4,5)P3 into the inner leaflet of the plasma membrane establishes binding sites for PH domain containing proteins, such as the kinases AKT and phosphoinositide-dependent kinase 1 (PDK1). Phosphorylation of AKT proteins by PDK1 and additional kinases (e.g. PDK2, integrin-linked kinase, and mechanistic target of rapamycin complex, mTORC) permits subsequent activation of several downstream kinases, which inhibit apoptosis, promote cell migration or regulate cell cycle progression. In parallel, AKTs mediate transcription factor activation (e.g. CREB) or inhibition (e.g. FOXO family members).

The STAT signalling pathway

Several RTKs recruit and directly phosphorylate transcription factors of the STAT family. Alternatively, STAT activation may be the result of RTK-induced secretion of interleukin 6 or other cytokines able to stimulate the p130-JAK pathway. Once tyrosine phosphorylated, STATs undergo dimerization and translocation to the nucleus, where they form complexes at promoters of specific target genes. Activated STATs increase tumour cell proliferation, survival, and invasion, while suppressing antitumour immunity (Yu et al., 2009).

C-gamma and the phosphatidylinositol 3-kinase (PI3K), which generate lipid second messengers from phosphoinositol precursors (Box 9.1). Because each receptor recruits a specific combination of multiple signalling enzymes and adaptor proteins, different GFs generate specific intracellular signals, which translate to simultaneous effects on the cytoplasm, cytoskeleton, and gene expression programmes (Pawson, 2004). Additional variation at the level of signal processing is due to cross-talk between RTKs and other signalling pathways, which are concisely described in Box 9.2. Importantly, some of the enzymes and the adaptors recruited to active receptors are targets for oncogenic (driver) mutations frequently detected in tumours. For example, mutant forms of PI3K are prevalent in many types of carcinoma, and RAS proteins, which are activated once the GRB2 or SHC adaptors are engaged, display various mutations in tumours of the pancreas, colon, skin, and lung. Such mutations maintain constantly active signalling pathways, thereby gain accelerated cell proliferation and/or evasion of cell death.

Signal termination by means of receptor endocytosis, degradation, or recycling

Prior to the availability of anti-EGFR antibodies, Graham Carpenter and Stanley Cohen used a radioactive derivative of EGF, which they incubated with living fibroblasts, and observed internalization of the radioactive molecule, followed by intracellular degradation that was sensitive to inhibitors of lysosomal enzymes (Carpenter and Cohen, 1976). Later studies resolved a highly robust mechanism of ubiquitination-dependent rapid receptor endocytosis, which is modular and involves multiple components (Zwang and Yarden, 2009; Haglund and Dikic, 2012; see Fig. 9.4). Receptor endocytosis serves several functions: this process rapidly depletes the extracellular pool of the GF, as well as targets active receptors to degradation (or recycling), thereby terminates the active state. Upon stimulation with a GF, RTKs like EGFR exit a slow recycling pathway that involves several membrane coat adaptors, such as adaptor protein 2 (AP2). This constitutive trafficking is overwhelmed by kinase-dependent tagging of active receptors with monomeric or dimeric ubiquitin serving as entry tickets to late endosomes and to degradation (Fig. 9.4). Receptor ubiquitinylation is carried out by E3 ubiquitin ligases, such as CBL family members. Once recruited to active receptors, CBL establishes multiple ubiquitin-centred attachment sites for proteins containing a ubiquitin-binding domain, like EPS15 and Epsin. CBL mediates both monoubiquitinylation and lysine-63-linked diubiquitinylation at multiple sites of EGFR. These covalent modifications are required at several steps along the endocytic

Box 9.2 Cross-talk between RTKs and parallel pathways of signal transduction

Because the number of RTKs exceeds the number of distinct downstream signalling routes, their effect partly overlaps and it can be found similarity among the receptor-proximal components for the downstream pathways. Thus, the identity of an extracellular signal depends not only on the exact combination of signalling routes, but also on the kinetics of activation/inactivation and the cross-talk with parallel and antagonizing routes (Kholodenko et al., 2010). As a result, GF signalling is conceived as a complex system featuring a layered configuration and partly redundant modules. Thus, reverse phosphorylation, along with feedback and feed-forward loops confer to GF signalling remarkable robustness. The ultimate manifestation of this robustness is exemplified by the ability of RTK systems to re-wire signals in the face of target-specific drugs (e.g. kinase inhibitors and antibodies), thereby evolve resistance in clinical settings (Brand et al., 2011; Mancini et al., 2015). Next, we describe several signalling pathways, which are parallel to the generic RTK pathway, and maintain mutual cross-talk with the growth factor activated routes.

The TGF-beta pathway

Transforming growth factor beta (TGF-beta) is a secreted polypeptide, which is deposited in an inactive form, at abundant quantities in the extracellular matrix. TGF-beta is implicated in two cancer-related processes, namely EMT and the inflammatory cascade associated with cancer progression. TGF-beta receptors (Type I and Type II) dimerize to form an active serine- and threonine-specific protein kinase, which autophosphorylates, recruits, and activates SMAD proteins. SMADs open up to expose a dimerization surface when they are phosphorylated. Similar to STATs, SMADs translocate to the nucleus, where they induce transcription. Interestingly, the type I receptor can recruit SHC and activate the MAPK pathway. In addition, an anchoring protein (called SARA, for SMAD anchor for receptor activation) helps to recruit SMAD2 or SMAD3 to the activated type I receptor by binding to the receptor, to the SMAD, and to inositol phospholipid molecules in the plasma membrane.

The Notch pathway

Cell–cell communication playing critical roles in tumorigenesis, metastasis, and stem cell survival, is mediated by the Notch pathway. In addition, the Notch pathway maintains cross-talk with RTKs, such as ErbB/HER proteins. Within the cross-talk, ErbB/HER receptors exploit Notch as a compensatory pathway. The compensatory Notch pathway maintains RTK-induced downstream signals transmitted to routes like ERK and PI3K, to allow cancer cells to survive molecular targeted therapies. There are four Notch paralogs (Notch-1, -2, -3, -4) and five Notch-ligands (Jagged-1 and -2, and Delta-Like-1, -3, and -4), which bind to and activate the Notch receptor when two neighbouring cells are in close proximity to each other. The Notch receptor is a single-pass transmembrane receptor, which is composed of two non-covalently held chains. The intracellular domain of Notch contains the active portion of the Notch receptor, namely the Notch Intracellular Domain (NICD). Notch activation requires a series of proteolytic events mediated by ADAM family proteases, as well as by the γ-secretase complex. By means of a pair of nuclear localization sequences, the NICD is directed to the nucleus and initiates transcriptional activation of target genes. Within the nucleus, NICD binds to CBF-1, which is already bound to DNA, thereby enabling the release of negative co-regulatory proteins (e.g. CtBP1) and the recruitment (to CBF-1) of co-activating proteins (e.g. MAML-1), which establishes the Notch transcriptional activating complex (see also Chapter 24 Cancer and vessels).

The WNT pathway

This pathway is generally dissected into three subpathways: canonical, non-canonical planar cell polarity pathway, and non-canonical WNT/calcium pathway. There are multiple WNT ligands able to engage a cognate receptor complex, consisting of a serpentine receptor of the Frizzled family and a member of the LDL receptor family, Lrp5/6. Beta-catenin serves as a main target of the WNT pathway. This cytoplasmic protein is highly unstable; in the absence of WNT ligands, beta-catenin is targeted to degradation by a so-termed destruction complex, composed of the tumor suppressor adenomatous polyposis coli, the scaffolding protein AXIN, and two kinases CK1-alpha (casein kinase 1-alpha) and GSK-3-beta (glycogen synthase kinase 3-beta). The two kinases sequentially phosphorylate a set of conserved serine and threonine residues in the amino terminus of beta-catenin. The resulting phosphorylated residues recruit the beta-TrCP-containing E3 ubiquitin ligase, which targets beta-catenin for proteosomal degradation. Receptor occupancy inhibits the kinase activity of the destruction complex and, as a consequence, beta-catenin accumulates and travels into the nucleus, where it engages the N-terminus of DNA-binding proteins of the TCF/LEF family. Thus, the canonical pathway translates a WNT signal into the transient transcription of a TCF/LEF target gene programme.

The JAK-STAT signalling pathway

The JAK pathway has been found constitutively activated in several malignancies (e.g. prostate cancer, sarcomas and lymphomas), leading to induction of tumour cell proliferation and apoptosis inhibition. In addition, mutations of *JAK2* have been reported in the majority of patients with myeloproliferative neoplasms, such as thrombocythemia, myelofibrosis, and polycythemia vera. The JAK family include, in addition to JAK1 and JAK2, JAK3 and the non-receptor tyrosine-protein kinase 2 (TYK2). JAK proteins harbour binding sites for activated cytokine receptors. Ligand binding to cytokine receptors induces cross-phosphorylation of associated JAK kinases. These in turn can phosphorylate, among many other pathways, also the STAT family transcription factors, which form parallel dimers upon tyrosine phosphorylation, and can subsequently move into the nucleus. In healthy cells, JAK-STAT signalling is under tight control by the protein inhibitor of activated STATs (PIAS), which negatively regulates STAT signalling by interfering with the DNA-binding activity of STATs. Another inhibitory mechanism entails the suppressors of cytokine signalling, which interfere with the activity of JAK kinases, competing with STATs for binding or by enhancing proteasomal degradation.

pathway. In addition to CBL, two groups of GTPases play major roles in the initial and later steps of RTK internalization: dynamin, which promotes fission of vesicle stalks, and RAB family members acting as compartment-specific GTP-regulated switches. CBL continues to accompany internalized RTKs when they enter early endosomes and later transfer into multivesicular bodies (MVBs), which are formed by invagination of the limiting endosomal membrane to generate intraluminal vesicles. MVBs then fuse with pre-existing lysosomes or mature into lysosomes. Importantly, decelerated endocytosis and enhanced recycling of certain RTKs has emerged as a hallmark of several types of tumours (Mellman and Yarden, 2013). For instance, mutant forms of p53 enhance recycling of EGFR and MET by controlling the RAB-coupling protein. Similarly, transforming forms of RAS proteins determine the rate of clathrin-dependent endocytosis and vesicle transport by regulating RAB5 and RIN1, which binds with EGFR and controls actin fibres. At the receptor level, mutant forms of MET and EGFR that lack the respective CBL-specific binding sites are stable, undergo no ubiquitinylation, and exhibit strong tumorigenesis. Lastly, mutations and aberrant expression of components of the endocytic machinery, such as ACK1, endophilin, cortactin, and cool-1, skew the fate of RTKs towards sustained signalling and accelerated cell proliferation.

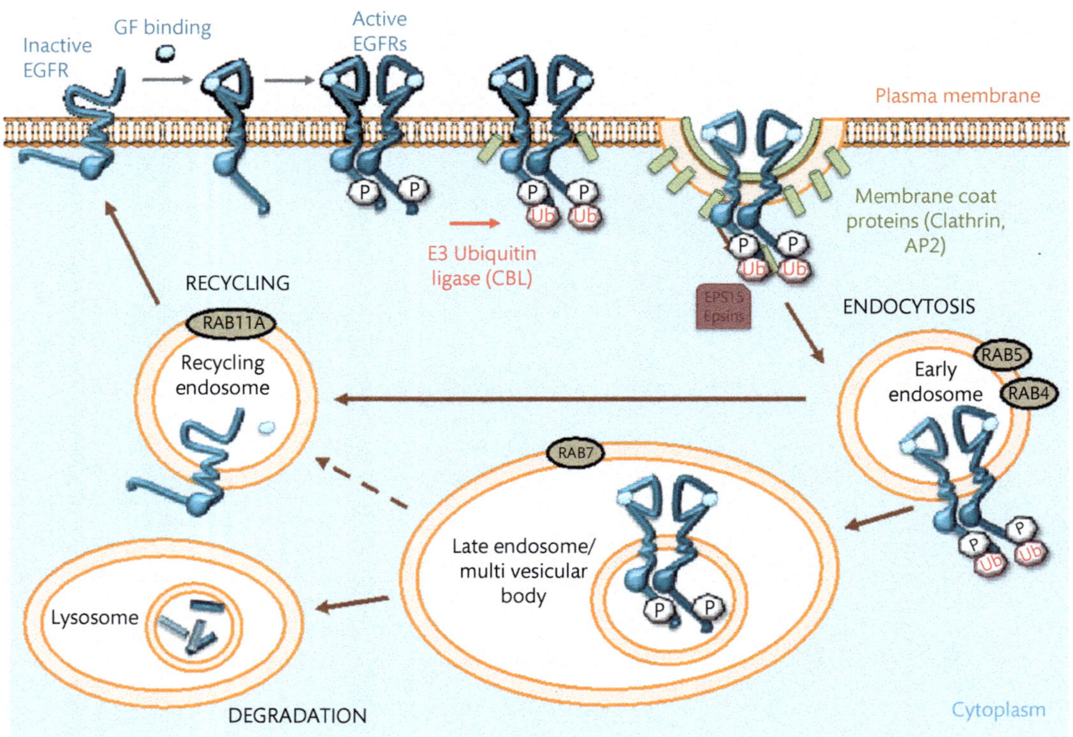

Fig. 9.4 Sorting of active RTKs for intracellular degradation. Inactive forms of RTKs, like EGFR, undergo constitutive but slow endocytosis. Following ligand binding and receptor auto-phosphorylation, EGFR binds with several cytoplasmic proteins, including CBL family ubiquitin ligases. Consequently, EGFRs undergo rapid endocytosis, primarily via clathrin-coated areas of the plasma membrane. This step requires ubiquitin-binding proteins, such as epsin family members, membrane coats, and small GTP-binding proteins of the RAB family. Receptors routed to early endosomes may be sorted for recycling, via a RAB11-controlled pathway, or they may be routed into intraluminal vesicles of late endosomes and the multivesicular body (MVB). This late sorting event depends on the receptor's ubiquitins and another set of ubiquitin binders.

Growth factor families and their cell surface receptors

The RTK family of mammals comprises 20 subfamilies, each containing several individual receptors sharing architectural features and a similar size (Fig. 9.5). Accordingly, the ligand GFs able to bind with each subfamily of receptors share structural features, which reflect high affinity binding of each GF family to the respective group of RTKs. Because grouping and multiplicity of RTKs hardly exist in invertebrate versions of the RTK family, they are attributable to duplications of ancient genomes or single chromosomes. In line with this, several RTK genes are clustered in specific chromosomes (e.g. the q arm of chromosome 5) and the respective GF genes might also be closely positioned. What follows is a description of several GF families frequently implicated in tumour progression.

Neuregulins and the epidermal growth factor (EGF) family

The precursors of this family of 11 members are expressed at the cell surface as transmembrane proteins. Once cleaved by proteases like matrix metalloproteinases (MMPs), the soluble ligands bind with and activate specific members of the EGF-receptor group (called ErbB or HER family) through a conserved 50–60 amino acid long domain containing six cysteine residues, and three disulphide bonds. The EGFR family includes four members, EGFR (also called HER1, or ErbB1), HER2 (NEU or ErbB2), HER3 (ErbB3), and HER4 (ErbB4). These tyrosine kinase receptors contain an extracellular part of four subdomains (from I to IV), a single transmembrane domain and an intracellular domain carrying the tyrosine kinase enzymatic function. It is worthwhile noting that the third member of the family, ErbB3/HER3, presents an impaired tyrosine kinase domain, whereas the second member, ErbB2/HER2, binds with no known GF. Both HER3 and HER2 are activated through ligand dependent hetero-dimerization, which enables their phosphorylation in trans. The 11 GFs of the family include ligands specific to EGFR, namely EGF, transforming growth factor alpha (TGF-alpha), amphiregulin, and epigen, along with three ligands able to bind with both EGFR and HER4/ErbB4 (heparin binding EGF-like growth factor; HB-EGF, epiregulin and betacellulin). In addition, several GFs bind with both ErbB3/HER3 and ErbB4/HER4, namely both neuregulin 1 (NRG-1, also called heregulin) and NRG-2, and two ligands, NRG-3 and NRG-4, are specific to ErbB4/HER4 only. In addition to cancer, the ErbB family is involved in non-malignant pathologies, such as in schizophrenia (NRG1 and ErbB4/HER4; see Buonanno, 2010), autoimmune demyelination of oligodendrocytes (Cannella et al., 1998) and psoriatic epidermal hyperplasia (Law et al., 2007). Enhanced expression or proteolytic release of specific EGF-like GFs have been associated with poor prognosis of cancer patients (Hynes and MacDonald, 2009). For example, release of free amphiregulin, HB-EGF, and TGF-alpha may occur spontaneously or through up-regulation of specific proteinases, such as MMP1,

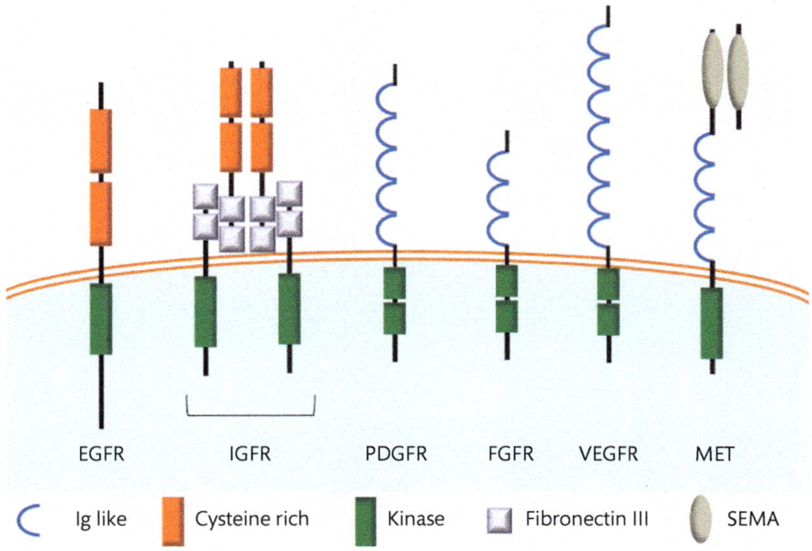

Fig. 9.5 Schematic structures of representative receptor tyrosine kinases. The plasma membrane is shown as a double line (in orange) and the homologous tyrosine kinase domains are in green. Representative receptors are shown, along with conserved structural motifs.

ADAM17 (a.k.a. TACE), and ADAM12, to promote tumorigenesis (Rosenthal et al., 1986). In addition, multiple mutations affect the four ErbB/HER proteins in human tumours, including frequent *ErbB4/HER4* mutations found in melanomas (Prickett et al., 2009). The gene encoding for EGFR is one of the most frequently mutated in non-haematopoietic cancers (see **Fig. 9.6**). Overexpression of EGFR, with or without gene amplification, has been reported in a large spectrum of tumours, including glioblastoma. In 2004, several groups reported activating mutations within the kinase domain of EGFR; these mutations often associate with patient response to kinase inhibitors like erlotinib (Lynch et al., 2004; Paez et al., 2004; Pao et al., 2004; Sordella et al., 2004). Similar to EGFR, amplification of the *ErbB2/HER2* gene characterizes 12–20% of breast (Slamon et al., 1989) and gastric cancer, along with smaller fractions of lung,

	ALL	AML	CLL	DLBCL	MM	BLCA	RCCC	ESCA	STAD	PAAD	COREAD	HC	SCLC	LUAD	LUSC	NSCLC	MB	LGG	GBM	NB	PA	HNSC	CM	UCEC	OV	BRCA	PRAD	THCA
EGFR		1	1				1	1	1		1			4	1			2	5			2		1		1	1	
ERBB2														2												1		
ERBB3						5					3												1	3				
FGFR1																					2							
FGFR2			1			1		1	3					1	1							1		4			1	
FGFR3			1			3									1								1					
MET						2	1	1	1					2	1			1	1	1				1				
PDGFRA																			2									
BMPR2						2	1				1		1	1								1	1	1	1	1		
TGFBR2						2				3	2			1	1				1			1	2	1	1	1		
	Blood and plasma cells							gastrointestinal					Lung				Brain							Hormono-dependent site				

Key	
1	0–3%
2	3–5%
3	5–10%
4	10–20%
5	>20%

Fig. 9.6 Frequency distribution of kinase mutations corresponding to RTKs in cancer types, for which the corresponding kinase is identified as a driver by genomic studies (based on (Fleuren et al., 2016)). Listed are the major RTKs known to be frequently mutated in human tumours. The prevalence of mutations is indicated according to the colour key. Note that two members of the transforming growth factor beta family are listed for reference. These are BMPR2 and TGFBR2. The following abbreviations were used: AML, acute myeloid leukaemia; ALL, acute lymphocytic leukaemia; CLL, chronic lymphocytic leukaemia; DLBCL, diffuse large B cell lymphoma; BLCA, bladder carcinoma; RCCC, renal clear cell carcinoma; STAD, stomach adenocarcinoma; PAAD, pancreatic ductal adenocarcinoma; COREAD, colorectal adenocarcinoma; HC, hepatocellular carcinoma; SCLC, small-cell lung cancer; LUAD, lung adenocarcinoma; LUSC, lung squamous cell carcinoma; NSCLC, non-small-cell lung cancer; CM, cutaneous melanoma; UCEC, uterine corpus endometrioid carcinoma; OV, serous ovarian adenocarcinoma; BRCA, breast carcinoma; PRAD, prostate adenocarcinoma; THCA, thyroid carcinoma; ESCA, oesophageal carcinoma; HNSC, head and neck squamous cell carcinoma; MM, multiple myeloma; LGG, lower grade glioma; GBM, glioblastoma; MB, medulloblastoma; NB, neuroblastoma; PA, pilocytic astrocytoma.

Source: data from Fleuren ED et al., 'The kinome 'at large' in cancer', *Nature Review Cancer*, Volume 16, pp. 83-98, Copyright © 2016, Rights Managed by Nature Publishing Group.

ovary, and colorectal tumours. Initial studies showed that administration of an anti-HER2 antibody inhibited NEU-transformed murine fibroblasts (Drebin et al., 1985), which led to the development of a humanized antibody, called trastuzumab. This antibody demonstrated efficacy in *HER2* amplified metastatic breast cancer (Cobleigh et al., 1999; Vogel et al., 2002; see also Table 9.1). Later clinical trials showed that trastuzumab, in combination with chemotherapy, can confer improved survival in patients with *HER2*-amplified gastric or gastro-oesophageal junction cancer (Bang et al., 2010). Unlike trastuzumab, which is specific to HER2, the small molecule GW572016/lapatinib was shown to inhibit both EGFR and HER2. A clinical trial involving patients with metastatic

Table 9.1 List of clinically approved drugs targeting RTKs or their ligand growth factors

Protein target	Name (trade name)	Year of first approval	Company	Condition	Remarks
EGFR	Gefitinib (Iressa)	2003	AstraZeneca	Non-small cell lung cancer (NSCLC)	Resistance conferred by the T790M mutation and additional mechanisms
EGFR	Erlotinib (Tarceva)	2004	OSI	Non-small cell lung cancer (NSCLC) and pancreatic cancer	Resistance conferred by the T790M mutation and additional mechanisms
EGFR	Afatinib (Gilotrif)	2013	Boehringer	Non-small cell lung cancer (NSCLC)	Irreversible inhibitor of EGFR and HER2
EGFR	Osimertinib (Tagrisso)	2015	AstraZeneca	Non-small cell lung cancer (NSCLC)	Irreversible inhibitor of EGFR, specific to T790M-EGFR
EGFR	Cetuximab (Erbitux)	2004	ImClone	Colorectal cancer, squamous cell cancer of head and neck and small cell lung cancer (SCLC)	Applied in combination with chemotherapy. KRAS mutations confer resistance of colorectal tumours
EGFR	Panitumumab (Vectibix)	2006	Amgen	Colorectal cancer in patients with wild-type KRAS	Applied in combination with chemotherapy. KRAS mutations confer resistance of colorectal tumours
EGFR	Necitumumab (Portrazza)	2015	Eli Lilly	Metastatic squamous non-small cell lung cancer	Applied in combination with two chemotherapy drugs
EGFR&HER2	Lapatinib (Tykerb)	2007	SmithKline	Breast cancer	ER activation might confer resistance
HER2	Trastuzumab (Herceptin)	1998	Genentech	Breast cancer, gastric and gastroesophageal junction cancer	Indicated in combination with chemotherapy for patients with tumours overexpressing HER2
HER2	Trastuzumab emtansine (Kadcyla)	2013	Genentech	Advanced breast cancer	An antibody-drug conjugate
HER2	Pertuzumab (Perjeta)	2012	Genentech	Breast cancer	Inhibits heterodimer formation by HER2; applied in combination with trastuzumab and chemotherapy
VEGF-A	Bevacizumab (Avastin)	2004	Genentech	Ovarian, fallopian tube, and cervical cancer; colon, rectum, lung, glioblastoma, and renal cell cancer	Administered in combination with chemotherapy
VEGF-A, VEGF-B, and PGF	Aflibercept (Zaltrap)	2012	Regeneron	Colorectal cancer	A recombinant decoy, combined with chemotherapy
VEGFR-2	Ramucirumab (Cyeramza)		Eli Lilly	CRC, NSCLC, gastric and oesophageal junction cancer	Combined with chemotherapy
VEGFR1/2/3	Axitinib (Inlyta)	2012	Pfizer	Renal cell and pancreatic cancer	Increased glucose metabolism
VEGFR, FGFR, PDGFR	Nintedanib (Ofev)	2014	Boehringer Ingelheim	Idiopathic pulmonary fibrosis and non-small cell lung cancer (NSCLC)	Might inhibit fibrotic and angiogenic processes
RET, MET, VEGFR1/2/3, KIT, TrkB, Flt3, Axl, Tie2	Cabozantinib (Cometriq & Cabometyx)	2012	Exelixis	Metastatic medullary thyroid cancer, prostate cancer, and renal cell cancer	MET activation might confer resistance
ALK, IGF-1R, InsR, ROS1	Ceritinib (Zykadia)	2014	Novartis	ALK-positive NSCLC, after crizotinib resistance	ALK mutations might confer resistance
ALK, Met, and Ros	Crizotinib (Xalkori)	2011	Pfizer	ALK-positive NSCLC	ALK mutations might confer resistance
VEGFRs, FGFRs, PDGFR, KIT, RET	Lenvatinib (Lenvima)	2015	Easai	Thyroid cancer	MET activation might confer resistance
VEGFR1/2/3, PDGFR, FGFR1/3, KIT, and FMS	Pazopanib (Votrient)	2009	GSK	Renal cell cancer and soft tissue sarcoma	Angiogenic switch, EMT

(continued)

Table 9.1 Continued

Protein target	Name (trade name)	Year of first approval	Company	Condition	Remarks
BCR-Abl, VEGFR, PDGFR, FGFR, Eph, Src, KIT, RET, Tie2, and Flt3	Ponatinib (Iclusig)	2012	Ariad	CML, Ph chromosome positive ALL	BCR-ABL mutations confer resistance
VEGFR1/2/3, BCR-Abl, B-RAF, KIT, PDGFR, RET, FGFR1/2, TIE2, and Eph2A	Regorafenib (Stivarga)	2012	Bayer	Colorectal cancer, gastrointestinal stromal tumours (GIST)	An analogue of sorafenib
VEGFR1/2/3, PDGFR, RAF, KIT, Flt3, and RET	Sorafenib (Nexavar)	2005	Bayer	Hepatocellular carcinoma and renal cell cancer	EMT might confer resistance
PDGFR, VEGFR1/2/3, KIT, Flt3, CSF-1R, and RET	Sunitinib (Sutent)	2006	Pfizer	Renal cell cancer, gastrointestinal stromal tumours (GIST), pancreatic neuroendocrine tumours	KIT mutations might confer resistance
VEGFRs, EGFR, RET, Brk, Tie2, EphRs, and Src	Vandetanib (Caprelsa)	2011	IPR Pharms	Medullary thyroid cancer	RET mutations (in vitro)

Drugs approved as of mid-2016 are listed. Note that the suffix -ib refers to kinase inhibitors whereas the suffix ab refers to monoclonal antibodies.

malignancies overexpressing *HER2* proved that the addition of lapatinib to chemotherapy was superior to chemotherapy alone in patients with HER2 positive metastatic breast cancer (Geyer et al., 2006), which led to the approval of lapatinib in combination with chemotherapy in this group of patients (Ryan et al., 2008). Despite the very low catalytic activity of HER3, somatic HER3 mutations exist in ~11% of colon and gastric cancers (Jaiswal et al., 2013), and when tested *in vitro*, the mutants were able to transform colonic and breast epithelial cells in a ligand-independent, but HER2-dependent manner. Moreover, relatively high expression levels of HER3 might associate with shorter survival of patients with breast, colorectal, melanoma, pancreatic, head and neck, and ovarian cancer (Ocana et al., 2013).

The insulin-like growth factor family

Insulin is an essential hormone secreted by beta cells of the pancreatic islets and controls body's glucose homeostasis, as well as carbohydrate metabolism. Insulin signals through binding to the insulin receptor (IR), made of two glycosylated chains: the alpha chain is fully extracellular and the beta chain, a transmembrane unit, harbours the kinase domain. IRs exist as two splice isoforms, IRA and IRB. Insulin binding to IR's alpha chain leads to kinase activation. Unlike insulin, the insulin-like growth factor (IGF) family—IGF1 and IGF2—are secreted by various tissues, although the liver is their main source. IGFs are endowed with characteristics of both GFs and hormones, and function in autocrine, paracrine, and endocrine modes. Unlike IR, the IGF1 receptor (called IGF1R) has only one isoform, but IGF2R is a non-RTK molecule. When dimerized, the holoreceptors IRA, IRB and IGF1R form six functional receptors: IRA/IRA, IRB/IRB, IRA/IRB, IGF1R/IRA, IGF1R/IRB and IGF1R/IGF1R. While insulin signals through binding to IRA or IRB, IGF1 signals through IGF1R, whereas IGF2 is able to signal through both IGF1R and IRA. In addition, IGF signalling is regulated by IGF-binding proteins (IGFBP-1 through -7), which trap the GFs and inactivate them prior to signal transduction (Pollak, 2012). IGF1 signals through two main pathways (see **Box 9.1** and **Fig. 9.7**). First, by recruiting the complex SHC/GRB2/SOS it activates the RAS/RAF/MAPK pathway, leading to cell proliferation. Second, by associating with GRB2/GAB and the adaptor called IRS it activates the PI3K/AKT pathway, a key mechanism of apoptosis resistance. IGF1 signalling plays a key role in embryonic development, cell survival, cell proliferation, ageing, and cellular adhesion. Multiple lines of evidence link IGF and insulin signalling to cancer progression. Firstly, IGF1R is essential for the transforming ability of several oncogenes (Sell et al., 1994). Secondly, it appears that secretion of insulin and IGF1 is relevant to cancer prognosis. For example, plasma levels of IGF1 correlate with risk of prostate and breast cancer. However, although IGF1R mutations were detected in breast tumours and in melanoma, the overall incidence is very low (Tognon and Sorensen, 2012). Attempts to intercept IGF1 signalling have included anti-IGF1R antibodies and specific kinase inhibitors. However, despite initial promising effects in phase II clinical trials, further tests showed that efficacy was limited to small cohorts of patients with chemotherapy-resistant solid tumours. Clinical failures have been attributed to the ability of IGF2 to bypass IGF1R inhibition by activating IRA (Janssen and Varewijck, 2014).

The hepatocyte growth factor family

The biology of the hepatocyte growth factor (HGF), which is also called scatter factor, has been reviewed (Trusolino et al., 2010). HGF has first been described as a humoral mediator of liver regeneration, much before the realization that it binds with the RTK called MET (or HGFR). MET is a transmembrane receptor of 190 kDa, mainly expressed on epithelial cells. A related RTK, called RON (Ronsin et al., 1993), binds with the macrophage stimulating protein. This chapter will focus on HGF and MET. HGF is synthesized primarily by mesenchymal cells and localizes to the plasma membrane as an inactive precursor, pro-HGF. The pro-protein is organized in two chains, an alpha chain displaying high affinity binding towards MET, and a beta chain containing a serine protease homology domain. Three serine proteinases have been implicated in pro-HGF processing: HGF activator, the type II transmembrane enzyme called matriptase, and hepsin. Two inhibitors are known to regulate HGF activation, the HGF activator inhibitor 1 and 2 (HAI1 and HAI2). MET includes two subunits: an entirely extracellular alpha chain, and a transmembrane beta chain carrying the kinase domain. HGF binding, through its serine proteinase homology domain, to the Sema domain of MET, leads to MET dimerization and transphosphorylation. Several substrate proteins, such as GRB2, SHC, PI3K, PLC-gamma and GAB1 are then recruited and trigger

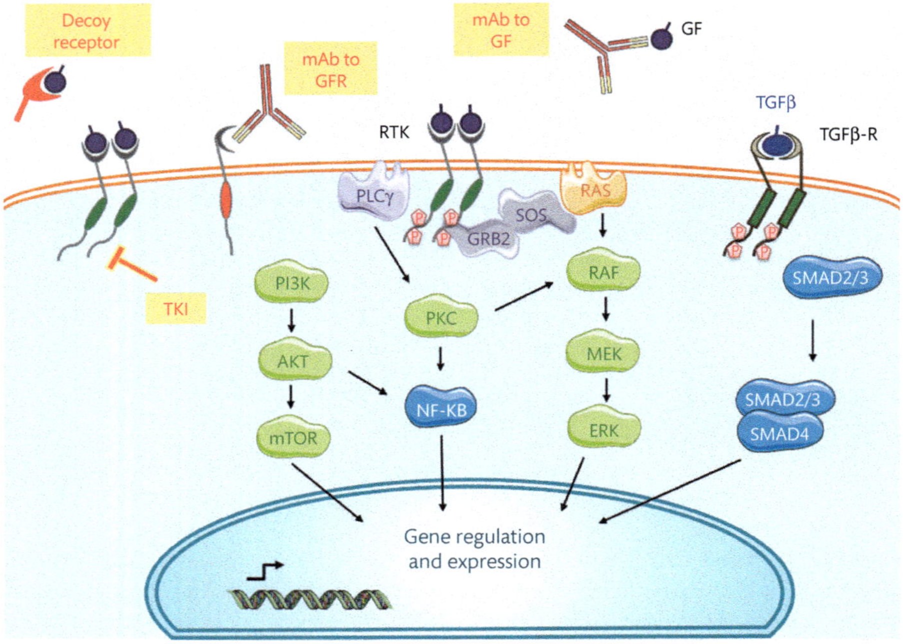

Fig. 9.7 The major intracellular signalling pathways engaged by ligand-activated RTKs and potential pharmacological interceptors. The plasma membrane and the nuclear membrane are shown in red and blue, respectively. Some linear routes are schematically presented downstream to a prototypic RTK. Also shown is the receptor for the transforming growth factor (TGF) beta and its downstream SMAD pathway. Four strategies of pharmacological interception of RTK signals are shown: a decoy (soluble) receptor, a monoclonal antibody (mAb) specific to a growth factor, an antibody to the respective RTK and a tyrosine kinase inhibitor (TKI) specific to a growth factor receptor. See Box 9.1 for details and abbreviations.

multiple signalling pathways, including PI3K/AKT, RAS-ERK, the RAC1-CDC42 (cell division control protein 42), RAP1/FAK and the WNT/βeta-catenin pathways. Accordingly, MET signalling is involved in many biological responses: cell survival, proliferation, epithelial-mesenchymal transition (EMT), migration/invasion and formation of branching tubules, cytoskeleton regulation, and metabolism. Because of this wide contribution to homeostasis and development, the link between aberrant HGF/MET signalling and pathology (e.g. cancer, diabetes, and autism) has been well studied. Aberrant HGF/MET signalling occurs in cancer through multiple mechanisms, such as genetic abnormalities (receptor activating mutations and amplification of *MET* or *HGF*) and receptor degradation defects. Importantly, germ line and somatic mutations in the tyrosine kinase domain of MET were found in papillary renal carcinomas (Schmidt et al., 1997). In support of MET's involvement in metastasis, transcripts encoding mutant forms of MET were found to be enriched in lymph node metastases, but they were barely expressed in primary hepatocellular carcinomas (Park et al., 1999). In addition, MET is involved in compensatory responses to anti-EGFR drugs (e.g. erlotinib and cetuximab) and consequent acquisition of patient resistance (Engelman et al., 2007). Interestingly, several clinical trials that applied antibodies directed to MET or HGF (i.e. onartuzumab and rilotumumab, respectively), or used specific tyrosine kinase inhibitors (TKIs), reported relatively moderate effects when tested on advanced stage solid tumours.

Vascular endothelial growth factors

The vascular endothelial growth factor (VEGF) family consists of five glycosylated GFs and several variants, which are the products of alternative mRNA splicing: VEGF-A (called VEGF), VEGF-B, VEGF-C, VEGF-D and PGF (placenta growth factor). VEGFs regulate both vasculogenesis and angiogenesis by binding with the respective RTKs, co-receptors like neuropilins (NPs) and proteoglycans (Kowanetz and Ferrara, 2006). Several VEGFRs exist: VEGFR-1 (FLT-1), VEGFR-2 (FLK-1 or KDR) and VEGFR-3 (FLT-4). While VEGF-A binds to both VEGFR-1 and VEGFR-2, VEGF-B, and PGF exclusively bind with VEGFR-1. VEGFR-1 and VEGFR-2 are expressed in vascular endothelial cells, as well as in monocytes, macrophages, and haematopoietic stem cells. Importantly, in addition to endothelial and other normal cells, solid tumours might also express functional VEGFR-1 and VEGFR-2, as well as the co-receptors NP1 and NP2 (Wu et al., 2006), but VEGFR-3 is largely restricted to lymphatic endothelial cells. VEGFR-2 seems to mediate most known cellular responses to VEGF and engages more intracellular signalling intermediates than VEGFR-1 (Waltenberger et al., 1994). VEGFRs control their target cells in several ways: while cell survival and proliferation are instigated through activation of the AKT and MAPK pathways, and cell migration is controlled by active Src family members interacting with the focal adhesion kinase (FAK), cell proliferation might be stimulated by the protein kinase C pathway. Several studies uncovered VEGF effects, which are independent of vascular processes. They include immune suppression, an ability to recruit bone marrow progenitors and exert direct effects on cancer cell survival and invasion. Thus, the combination of vascular effects and direct actions on cancer cells might underlay homing of tumour cells to premetastatic permissive niches for incoming tumour cells (Kaplan et al., 2005). Blocking angiogenesis is one of the most successful endeavours of molecular targeted cancer therapy (see Table 9.1). In addition, anti-VEGF drugs were found to be effective in eye diseases, such as age-dependent macular degeneration. Altogether,

12 anti-VEGF agents have been approved so far for clinical application. They fall into three classes: kinase inhibitors, antibodies, and a single VEGF decoy receptor (VEGF-Trap).

The platelet-derived growth factor family

Five dimeric isoforms of PDGFs exist: PDGF-AA, PDGF-BB, PDGF-CC, PDGF-DD, and the heterodimer PDGF-AB (Li et al., 2000; Bergsten et al., 2001). These isoforms differentially bind with the two receptors: PDGFR-alpha binds the A-, B-, and C-chains, whereas PDGFR-beta binds the B- and D-chains. Unlike PDGF-AA and PDGF-BB, which are synthesized as precursor molecules that undergo cleavage during synthesis, PDGF-CC and PDGF-DD are secreted as latent forms. The two PDGFRs are single transmembrane RTKs containing five immunoglobulin domains in the extracellular region and homologous TKDs, which are divided into two parts by a non-catalytic region of approximately 100 amino acids. The receptors are activated by ligand-induced dimerization and transphosphorylation within symmetric dimers. Up to 11 receptor's tyrosine residues undergo auto-phosphorylation and they recruit about 10 different families of phosphotyrosine-binding proteins, including Src, STAT5, GRB2, SHC, NCK, RAS-GAP, PI3K, SHP2, and PLC-gamma. Notably, PDGFR-beta recruits RAS-GAP (Which negates RAS activation), unlike PDGFR-alpha, hence PDGF-AA activates ERKs more rapidly and efficiently compared to PDGF-BB (Jurek et al., 2011). In general, PDGFs are secreted by epithelial and endothelial cells and they act in a paracrine manner on nearby mesenchymal cells (e.g. fibroblasts, pericytes, and smooth muscle cells). Along with critical functions of PDGFR-alpha and PDGF-AA in early mesenchymal embryonic derivatives, and similarly critical functions of PDGFR-beta (and PDGF-B) in recruitment of pericytes and smooth muscle cells to blood vessels, both ligands and receptors have roles to play in repair of skin wounds. Under pathological conditions, PDGFR-beta counteracts oedema formation by controlling interstitial fluid formation, and excessive activity of both receptors leads to development of fibrosis in lung, liver, kidney, and other organs. PDGF-AA has been implicated in EMT and increased invasiveness in tumour models, whereas PDGFR-alpha is mutated in approximately 5% of gastrointestinal stromal tumours (GIST). In addition, this receptor is amplified in subsets of glioblastoma, oligodendroglioma, and sarcoma. A fusion protein comprising portions of the TEL transcription factor and PDGFR-beta (cytoplasmic domain) was found in myelomonocytic leukaemia. Another fusion protein comprising segments of PDGF-B and collagen 1A1 was identified in a rare skin tumour (dermatofibrosarcoma), suggesting an autocrine mechanism. The increased intestinal fluid pressure characteristic to solid tumours has been attributed to PDGFR-beta expressed on the surface of stromal cells. Reducing this pressure might increase uptake of chemotherapeutic agents, but so far, no drugs targeting specifically PDGF signalling have been approved for clinical use.

The fibroblast growth factor family

The human fibroblast growth factor (FGF) family includes 22 members, but only 18 act as FGF receptor (FGFR) ligands. The FGFs are secreted glycoproteins, which are normally sequestered by heparin sulphate proteoglycan of the extracellular matrix. Most members of the family, including the extensively described acidic (FGF1) and basic (FGF2) molecules, function as autocrine or paracrine GFs. Yet, the non-canonical FGFs, 19, 21, and 23, diffuse from their source of production into the circulation to act in an endocrine mode. In addition, some FGFs might function within the nucleus. Signal transduction networks activated by FGFs regulate cell proliferation, differentiation, and survival during embryonic development and wound repair, as well as in tissue homeostasis (Turner and Grose, 2010). Once released from the extracellular matrix by heparanases or specific FGF-binding proteins, FGFs are trapped at the cell surface by klotho and other proteins. This interaction stabilizes the binding of FGF to FGF-receptors, leading to ternary complex formation. There are four highly conserved RTKs, called FGFR1, 2, 3, and 4, able to bind diverse FGFs. Another molecule, called FGFR5 (or FGFR1), is able to bind with FGFs but it carries no tyrosine kinase domain and might inhibit FGF signalling. Transphosphorylation of the kinase domain and intracellular tail of FGFRs allows phosphorylation of adaptor proteins, such as the FGFR substrate 2 (FRS2). Phosphorylated FRS2 recruits GRB2, which might activate the PI3K-AKT pathway or the RAS-RAF-ERK pathway. Additional pathways, such as signal transducer and activator of transcription (STAT) and PLC-gamma, have also been reported as downstream targets of FGFRs. Aberrant FGF signalling might promote cancer development by driving cancer cell proliferation or survival, and by supporting tumour neoangiogenesis. Analysis of 4,853 solid tumours detected aberrant FGFRs in 7% of tumours, mostly gene amplification (66%) and mutations (26%). Among the reported aberrations, 49% occur in FGFR1, 19% in FGFR2 and 19% in FGFR3. In this study, every analysed type of cancer showed FGFR aberrations, including urothelial carcinoma (32%), breast carcinoma (18%), endometrial adenocarcinoma (13%), squamous lung carcinoma (13%) and ovarian carcinoma (9%; see Helsten et al., 2016). To date several anti-FGF signalling strategies have been developed for treatment of solid tumours. They include FGFR-specific TKIs and pan-FGFR inhibitors, monoclonal antibodies to FGFR3 and FGF ligand traps (Hallinan et al., 2016).

The myriad of roles played by growth factors in tumour progression

Solid tumours, such as cancer of the colon and breast, are rarely initiated by familial (inherited) mutations. Instead, extrinsic mutators, either physical, chemical, or biological (e.g. viruses), along with intrinsic factors, such as sustained replicative stress, expose genetic instability early during tumour formation (Cahill et al., 1999). Subsequent clonal outgrowth and tumour progression depend on additional mutations altering oncogenes and tumour-suppressor genes. The facilitators of this process, which recurs once secondary and tertiary tumours outgrow, are multiple GFs produced by tumour cells or by the tissue microenvironment (see Fig. 9.8). The GFs execute several vital functions: they attract blood and lymph vessels, confer invasiveness across tissue barriers, permit colonization of distant organs and prevent death of tumour cells exposed to chemotherapeutic agents or to ionizing radiation. Next, we review the multiple roles played by GFs during cancer progression.

GFs act as mitogens for cancer cells

By supporting clonal expansion, GFs permit fixation of oncogenic mutations, as well as increase the pool of initiated cells susceptible

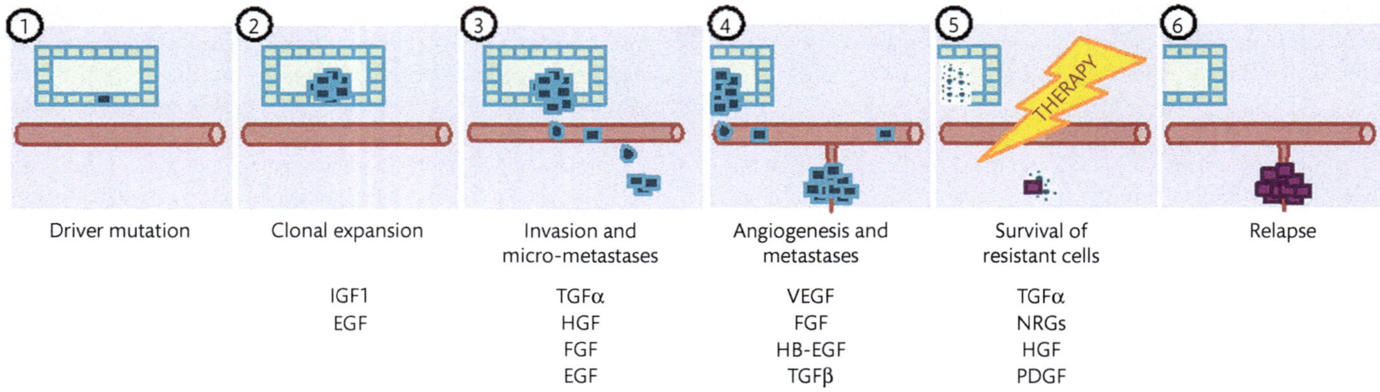

Fig. 9.8 Involvement of growth factors in the stepwise progression of solid tumours. The schematic picture shows an epithelial tissue (rectangles), situated next to a blood vessel (red tubes), and indicates some of the GFs involved in each step. The malignant process is instigated by the emergence of a driver mutation (Step 1), followed by an intraepithelial clonal expansion, which may be supported by GFs like IGF1 and EGF/NRG family members (Step 2). Tumour cell invasion (Step 3) entails migration of individual cells and depends on the ability of cells to invade through tissue barriers, as well as survive in secondary sites, where micro-metastases might form. In Step 4 either blood vessels are recruited to the primary and secondary tumour sites (angiogenic tumours) or pre-existing vessels are co-opted (non-angiogenic tumours) to supply nutrients and remove toxic metabolites. Steps 5 and 6 might be quite critical: specific GFs may drive survival of tumour cells under cytotoxic therapy (i.e. chemotherapeutic agents and irradiation) and promote relapses.

to additional mutations. Along with the ability of cancer cells to synthesize GFs to which they are responsive (Sporn and Todaro, 1980), which differentiates them from most normal cells, several distinct mechanisms may lead to constitutive pathway activation in tumours. They include RTK overexpression or activating mutations conferring ligand-independent signalling, as well as mutations affecting downstream mediators that similarly confer growth autonomy (Blume-Jensen and Hunter, 2001). Perhaps the most fundamental function of GFs involves their ability to sustain cell growth and division cycles. Proliferating cells repeatedly undergo four successive phases: in G1 (Gap 1) they commit to mitosis, in S they replicate their chromosomes, in G2 they prepare for mitosis, and in M (mitosis) they divide. Non-dividing cells arrest in the quiescent state (G0), which they exit in response to a GF. A group of cyclin-dependent kinases (CDKs) and their regulatory subunits, called cyclins, controls the cell cycle. Whereas cyclin D/Cdk4–6 and cyclin E/Cdk2 promote the G1/S transition, the activation of cyclin A/Cdk2 ensures progression in S and G2, and cyclin B/Cdk1 permits progression into mitosis. By elevating expression of MYC and D-type cyclins, many GFs and cytokines allow G1 phase progression. Conversely, therapeutic mAbs specific to RTKs like HER2 might act by up-regulating the CDK inhibitor $p27^{kip1}$. Two prominent growth suppressors, p53 and Rb, govern the decisions of cells to proliferate or, alternatively, activate senescence or apoptotic programmes. Hence, loss of p53 or Rb associates with tumorigenesis. The Rb protein integrates GF signals primarily by regulating E2F, a transcription factor involved in cell cycle regulation. In its hypophosphorylated form, Rb binds and inactivates the DNA-binding and transactivating functions of E2F. Stimulation of cells with GFs leads to CDK activation, Rb hyperphosphorylation, and subsequent dissociation from E2F. In turn, free E2F proteins can induce transcription of several genes involved in cell cycle entry, including expression of various cytokines and GFs (Polager and Ginsberg, 2009). Similar to Rb, p53 relays GF signals and permits irreversible crossing of a restriction point, called R, which commits to cell cycle progression. The underlying mechanism involves a two-pulse process: the first pulse induces essential metabolic enzymes and activates p53-dependent restraining processes, while the second pulse eliminates, via the PI3K/AKT pathway, the suppressive action of p53 and enables S phase entry (Zwang et al., 2012). Thus, in cells lacking p53 a single pulse of GFs may be sufficient for cell cycle progression, whereas normal cells require the second, p53-mediated pulse.

GFs induce angiogenesis

Tumour growth beyond a critical diameter of a few millimetres is limited due to lack of nutrient and oxygen supply, as well as because routes to evacuate metabolic waste are unavailable. Thus, the generation of new vessels is considered an early limiting step (Folkman, 2007), and an integral hallmark of cancer (Hanahan and Weinberg, 2011). In adults, new vessels are formed either by sprouting of mature endothelial cells (angiogenesis) or by bone marrow-derived endothelial progenitor cells (EPCs), which home to foci of angiogenesis (vasculogenesis; Asahara et al., 1997). Although much attention has been focused on the VEGF family, especially VEGF-A acting through VEGF receptor 2 (VEGFR-2), all three members of the VEGF-receptor family, along with oncogenes like *RAS* and *MYC* and GFs like FGFs and TGF-beta, also play important roles in both vasculogenesis and angiogenesis. Furthermore, hypoxia, GFs like EGF, steroid hormones and chemokines strongly induce secretion of VEGFs by activating either the hypoxia-inducing factor or transcription factors of the MAPK-dependent ETS family. It is worthwhile noting that transitory angiogenesis takes place during embryonic development and in adults, for example within the process that heals skin wounds. In contrast, during tumour progression, an 'angiogenic switch' is almost always activated and stays on. However, the chronically activated angiogenesis of tumours is typically aberrant. The vessels are enlarged, distorted, and leaky. Mobilization of circulating endothelial progenitor cells to tumours emerges as a critical step in cancer progression (Lyden et al., 2001). EPCs regulate the angiogenic switch through paracrine secretion of proangiogenic GFs, as well as by direct luminal incorporation into sprouting nascent vessels. For example, the tumour stroma promotes neoangiogenesis by recruiting EPCs, an effect mediated in part by the ability to secrete stromal cell-derived factor 1 (SDF-1; see Orimo et al., 2005). Further, it has been reported that cancer therapy, including high dose chemotherapy, could induce mobilization of EPCs to the viable rim of tumours (Shaked et al., 2006). Curiously, along with highly vascularized tumours (e.g. renal cancer), which are accessible to systemic drug delivery, some tumours, including several highly aggressive types like pancreatic ductal adenocarcinomas, are hypovascularized. Stromal 'deserts' might be due to antiangiogenic factors, such as thrombospondin 1 (TSP-1), fragments of plasmin (angiostatin) and type 1 collagen. The balance between endogenous inhibitors and stimulators of angiogenesis may thus control the angiogenic switch of solid tumours. Hence, anticancer therapies that make use of angiogenesis blockers, such as the anti-VEGF antibody called bevacizumab, are efficacious under certain clinical conditions (Ferrara and Kerbel, 2005).

GFs promote invasive growth and support metastasis of cancer cells

Unlike the well-established scheme of the cell cycle, which is relevant to all cells of metazoans and involves well-characterized molecular complexes and checkpoints, no universal scheme describes the transition of polarized, densely packed cells and tissues into collections of motile cells, which might colonize distant organs. Nevertheless, several features of this highly complex and varied hallmark of cancer are emerging and they all ascribe pivotal roles to both GFs and the tumour microenvironment (Hanahan and Weinberg, 2011). For example, tumour metastasis is licensed only when the basement membrane decorating secretory ducts undergoes dissolution, and accordingly GFs play critical roles in basement membrane disruption, penetration by cancer cells into the vascular or lymphatic systems (intravasation), as well as their departure from the bloodstream (extravasation) and subsequent colonization of distant organs (Yilmaz and Christofori, 2009). Several molecular switches underlay this sequence of phenotypic events as we describe next.

Early molecular switches

The actin cytoskeleton, along with the vesicular transport system, is essential for sustained motility. Actin-rich membrane protrusions, such as lamellipodia and filopodia, as well as stress fibres, generate and apply mechanical forces (Ridley, 2011), while spatiotemporally regulated cycles of vesicle exocytosis and endocytosis recycle into the leading edge the large amounts of plasma membrane needed for crawling and turnover of integrin-based and other types of adhesion sites. By anchoring several actin-binding proteins to the plasma membrane (e.g. cofilin and gelsolin), phosphoinositides play critical roles in growth factor-induced motility and the invasion-metastasis cascade. For example, EGF-induced cleavage of phosphoinositol 4,5 bisphosphate (PtdIns(4,5)P2) by PLC-gamma is essential for the induction of motility (Chen et al., 1994). Several substrates of GF-activated ERK are similarly essential. They include the WAVE2 regulatory complex (which stimulates the ARP2/3 actin nucleator), calpain, a protease specific to adhesion-related proteins, the myosin light chain kinase, involved in actin polymerization, and paxillin, which recruits the FAK to adhesion sites. In a similar way, the PI3K pathway controls migration directionality and speed, by engaging a large spectrum of effector proteins (Kolsch et al., 2008), such as the RHO family of small GTPases (RAC, RHO, and CDC42). Importantly, the early switches enabling migration and invasion of cancer cells might be driven not only by autocrine loops; paracrine mechanisms engaging stromal myeloid, endothelial, and mesenchymal cells likely complement the autocrine mechanisms. For example, intravasation of mammary cancer cells that secrete the colony-stimulating factor 1 (CSF-1) is enhanced by attraction of macrophages to tumours and by local secretion of macrophage-derived GFs, including EGFR ligands (Goswami et al., 2005). Likewise, the hypoxic and inflammatory conditions that occur during tumour progression increase secretion of TGF-beta by macrophages. TGF-beta-mediated induction of angiopoietin-like 4 (ANGPTl4) in breast cancer cells enables retention and possibly colonization of lungs by cancer cells (Padua et al., 2008). Another stromal source of pro-metastasis cytokines are cancer-associated fibroblasts (Ostman and Augsten, 2009).

Late, transcription-dependent molecular switches

The process commonly known as the EMT involves loss of epithelial markers like E-cadherin, MUC1, syndecan, and laminin-1, concomitantly with gain of mesenchymal markers, such as N-cadherin, vimentin, fibronectin and several transcription factors, including ETS-1 and snail (Kalluri and Weinberg, 2009; Thiery et al., 2009). The roles for GF-induced EMT and other molecular switches are exemplified by the disassembly of junctional complexes such as adherens junctions. These junctions are comprised primarily of the transmembrane protein epithelial cadherin (E-cadherin), which connects to the actin cytoskeleton via catenins (Takaishi et al., 1997). The loss of E-cadherin expression is considered a hallmark of EMT, and in some cases it is coupled to up-regulation of another calcium-dependent cell adhesion molecule, N-cadherin (Hazan et al., 2000). Mutations and downregulation of E-cadherin are frequently observed in human cancer. Likewise, GF-induced degradation of E-cadherin underlays disassembly of adherens junctions, which precedes loss of contact-mediated arrest of cell growth and cell migration (Fujita et al., 2002). Another switch involves the tensin family, which comprises four members, all localized to the cytoplasmic tails of integrins at focal adhesions. Unlike tensins 1, 2, and 3, tensin 4 (also called COOH-terminus tensin-like molecule; CTEN) harbours no N-terminal actin-binding domain that is present in the other tensin proteins (Lo, 2004). CTEN is up-regulated and associates with poor prognosis in breast cancer, thymomas, gastric cancers, and lung cancer. Importantly, stimulation of mammary cells with EGF is followed by transcriptional up-regulation of CTEN, concomitant with downregulation of tensin 3. This reciprocal switch enables CTEN to displace tensins from the cytoplasmic tail of integrins, thereby disassemble focal adhesions and promote cell migration (Mouneimne and Brugge, 2007).

GF-induced escape from apoptosis initiated by cytotoxic therapies

Intrinsic resistance to death signals like TRAIL is a well-understood hallmark of cancer (Hanahan and Weinberg, 2000). Some tumour types display very long (>20 years) dormancy, despite repeated cycles of cytotoxic therapies and, eventually, they evolve metastatic outgrowths. This reflects survival and adaptation of disseminated tumour cells to their new environments. In the clinical setting of patients repeatedly treated with radiation and cytotoxic drugs, survival of micrometastases can be acquired through the action of GFs and by loss of a proapoptotic regulator (e.g. expression of mutant forms of *p53*). The antiapoptotic survival signals generated by the PI3K–AKT and the RAS-ERK pathways are often involved in weakening therapy-induced apoptosis in human tumours (Cantley, 2002; see Fig. 9.7). GFs like IGF-1, HGF, and several EGF-like ligands have been implicated in evasion from apoptosis. This might explain the ability of anti-EGFR antibodies to sensitize tumours to chemotherapy both in animal models (Aboud-Pirak et al., 1988) and in colorectal patients treated with the anticancer antibodies cetuximab or panitumumab. Interestingly, unoccupied RTKs, such as insulin and IGF-I receptors, might exert proapoptotic signals, whereas the ligand-occupied receptors elicit antiapoptotic signals (Boucher et al., 2010). Apoptosis is an energy-dependent process that involves activation of caspases, a group of cysteine proteases, through a complex cascade linking the initiating stimulus to effectors of cell death. Apoptosis is mediated either by an intrinsic pathway, which engages mitochondria, or by an extrinsic pathway involving death-promoting ligands like FasL or the tumour necrosis factor. The intrinsic mechanism entails increased mitochondrial permeability, which is regulated by BCL family members and p53, release of cytochrome C and recruitment of SMACs (small mitochondria-derived activators of caspases), which inactivate the inhibitors of apoptosis (IAPs), a group of caspase repressors. Accordingly, antiapoptotic actions of GFs are relayed either by means of regulating BAD, a BCL family member (the intrinsic pathway), or by repressing the receptors for death-promoting ligands (the extrinsic pathway).

Clinically approved cancer drugs and experimental strategies targeting growth factor signalling

Commensurate with their roles in cancer initiation and progression, pharmacological interception of GFs, the cognate receptors and the respective downstream signalling pathways has emerged as an effective way to inhibit specific types of tumours. Importantly, not only driver mutations directly activating RTKs, or mutated, constitutively activated downstream signalling components, might serve as appropriate drug targets; the respective wild-type forms might serve as suitable targets in case they are essential for survival of cancer cells (so called 'non-oncogene addiction'). For example, small molecule inhibitors specific to mutant forms of EGFR can inhibit certain lung tumours and anti-EGFR antibodies are able to arrest some colorectal tumours, although the colorectal receptor harbours no oncogenic mutation. As described next and listed in Table 9.1, four types of pharmacological interceptors of GF signalling have so far (as of mid-2016) been approved for clinical applications.

Antibody-drug conjugates (ADCs)

Covalent complexes of antireceptor antibodies and specific drugs or toxins are typically internalized and transported to intracellular organelles, where the payload is released and interferes with various cellular processes, leading to cell death (Sievers and Senter, 2013; Hamilton, 2015). Although several experimental anti-RTK ADCs have been designed, the only licensed conjugate is the HER2-targeting trastuzumab emtansine (T-DM1). Through a stable linker, trastuzumab (an anti-HER2 antibody) is conjugated to the microtubule-inhibitory agent DM1 (a derivative of maitansine; see Junttila et al., 2011, Lewis Phillips et al., 2008). Clinical trials that compared T-DM1 and lapatinib (a HER2/EGFR-specific kinase inhibitor) demonstrated significantly improved overall survival of T-DM1-treated patients with HER2-positive metastatic breast cancer (Verma et al., 2012), which led to the approval of T-DM1 for metastatic breast cancer.

Recombinant soluble forms of RTKs

Decoy receptors are designed to retain high affinity binding of all, or only some ligands of the respective receptor. For example, a decoy combining the extracellular form of EGFR and ErbB4/HER4 can bind all 11 GFs of the EGF family (Lindzen et al., 2012). So far, ziv-aflibercept, a recombinant fusion protein comprising the VEGF-binding portions of VEGFR and the Fc portion of human immunoglobulin G, is the only clinically approved decoy. Although this molecule binds VEGF-A, VEGF-B, and PGF, it seems that its tumour-inhibitory activity is limited to colorectal cancer.

Tyrosine kinase inhibitors

Typically, TKIs are ATP mimetic drugs that form reversible (hydrogen) bonds with the ATP-binding site of specific kinases. So far, only two approved inhibitors, afatinib and ibrutinib, form covalent (irreversible) bonds with a cysteine proximal to the ATP-binding site of the target protein kinases. The reversible inhibitors fall into several groups: the majority are type I inhibitors, which are ATP-competitive drugs recognizing the active kinase conformation. By contrast, type II inhibitors recognize the inactive conformation of the kinase. With a few exceptions most of the approved drugs inhibit tyrosine-specific kinases, rather than serine/threonine-specific kinases (Levitzki, 2013), and while most compounds are considered mono-specific, some inhibit with similar potency more than one kinase. The first approved RTK-specific TKIs targeted mutant forms of EGFR in non-small lung cancer. Two mutations, the L858R point mutation and the exon 19 deletion (del746-750), represent the vast majority of the activating EGFR mutations, and they possess higher affinity towards the inhibitors, gefitinib and erlotinib, compared to the wild-type receptor (Carey et al., 2006; Yun et al., 2007). Rearrangement of the *ALK* gene provides a second targeting agent in patients with advanced non-small cell lung cancer (NSCLC). *EML4* (echinoderm microtubule-associated protein-like 4)-*ALK* is the predominant *ALK* fusion in NSCLC, but additional fusions exist (Rikova et al., 2007; Soda et al., 2007). Crizotinib, a type I small compound, achieved remarkable clinical effects when tested in selected cohorts of patients with *ALK* rearrangements, which led to the 2011 accelerated approval of crizotinib for treatment of *ALK*-positive advanced NSCLC. Lapatinib, a highly selective, small molecule inhibitor of both ErbB2/HER2 and EGFR (Rusnak et al., 2001), is currently the only clinically approved TKI for the treatment of advanced stage HER2-positive breast cancer. Notably, lapatinib is a type II ATP-competitive inhibitor that binds with the inactive conformations of its targets. Although a significant advancement in the treatment of breast cancer, the clinical efficacy of lapatinib has been limited by the evolvement of acquired resistance (Geyer et al., 2006; Johnston et al., 2008). In contrast to other kinase inhibitors (e.g. erlotinib and crizotinib), for which intrinsic target alterations, mainly mutations within the ATP-binding pocket, represent a major escape route, ErbB2/HER2 mutations do not play a major role in lapatinib resistance. Instead, several lines of evidence implicate de-repression and/or activation of compensatory,

ErbB2/HER2-independent survival pathways (Garrett and Arteaga, 2011). Altogether, more than 15 clinically approved TKIs are specific, at least to some extent, to RTK molecules. Many of those are multitarget inhibitors, such as pazopanib, which inhibits VEGFRs, along with KIT (the stem cell factor receptor) and several other RTKs. Pazopanib has been approved for the treatment of renal cell cancer and soft tissue sarcoma.

Antibodies intercepting growth factor signalling

Immunotherapy entails administration of recombinant monoclonal antibodies (mAbs) that are specific to a particular antigen, such as a GF or the respective receptor (Scott et al., 2012; Carvalho et al., 2016). Originally, therapeutic mAbs of murine origin were generated using the hybridoma technology (Köhler and Milstein, 1975). However, due to short half-life in serum, immunogenicity in humans, and ineffective recruitment of human immune effector responses, the murine antibodies are commonly humanized. This entails grafting the entire murine variable regions, or only the murine complementary-determining regions, into a human immunoglobulin G backbone, to create a chimeric or a humanized antibody, respectively. Cetuximab, a chimeric antibody, and panitumumab a fully human mAb, are EGFR-specific mAbs that block ligand binding and prevent activation of downstream signalling cascades (Mendelsohn and Baselga, 2006). Both mAbs have been approved for treatment of metastatic colorectal cancer, in combination with chemotherapy. Cetuximab is also approved for treatment of head and neck tumours, in combination with radiotherapy. Trastuzumab and pertuzumab are humanized mAbs that target HER2. Trastuzumab is used, with chemotherapy, for patients with ErbB2/HER2-overexpressing metastatic breast cancer, gastric or gastroesophageal junction cancer. Pertuzumab, another anti-HER2 mAb, is approved, in combination with trastuzumab and chemotherapy, for the treatment of patients with HER2-positive breast cancer. Although anti-GF antibodies are active in animal models, so far only one mAb, bevacizumab, has been approved. This anti-VEGF-A antibody is used in combination with chemotherapy for the treatment of patients with recurrent epithelial ovarian, fallopian tube, or cervical cancer, and for the treatment of colorectal cancer. In combination with chemotherapy, bevacizumab is also used for treatment of patients with lung cancer.

TAKE-HOME MESSAGE

- Along with stepwise accumulation of oncogenic mutations, cancer progression entails diverse actions of GFs, including their ability to drive cell proliferation, cell migration, invasion through tissue barriers, recruitment of blood vessels, and evasion from cell death induced by cytotoxic treatments.
- GFs are compact, highly stable molecules that bind transmembrane receptors harbouring an intrinsic tyrosine kinase activity.
- The receptors form dimers following their binding with highly specific growth factors. Within dimers, receptor transphosphorylation on tyrosine residues instigates intracellular signals, which translate to simultaneous activation of cytoplasmic, nuclear, and cytoskeletal processes involved in context-dependent biological outcomes.
- Oncogenic mutations might elevate growth factor secretion, mimic ligand binding, constitutively activate the kinase function of growth factor receptors, or switch intracellular effectors to their ON state.
- Twenty groups of growth factors and a corresponding number of receptor subtypes perform distinct, or partly overlapping functions, in embryonic development, wound repair and other physiological processes, and these are reflected in their actions in tumour progression.
- Several drugs, especially kinase inhibitors and monoclonal antibodies able to intercept growth factor signalling, have been approved for treatment of patients with different cancers.

OPEN QUESTIONS

- Which roles are played by non-coding RNAs in growth factor signalling and in tumour progression?
- How does integration of multiple GF signals, along with diverse intracellular signals, occur at the cytoplasmic and genomic levels?
- Which routes mediate cross-talk between growth factors and cells of the immune and vascular systems within tumour's stroma?
- Is there a unified, perhaps cytoskeleton-based, model of the metastatic process and the underlying molecular switches?

FURTHER READING

Hynes, N. E. & Watson, C. J. (2010). Mammary gland growth factors: roles in normal development and in cancer. *Cold Spring Harbor Perspective Biol*, **2**, a003186.

Weinberg, R. A. (2014). *The Biology of Cancer.* New York: Garland Science.

Wheeler, D. I. & Yarden, Y. (eds) (2015). *Receptor Tyrosine Kinases: Family and Subfamilies*. New York: Humana Press.

Wheeler, D. I. & Yarden, Y. (eds) (2015). *Receptor Tyrosine Kinases: Structure, Functions and Role in Human Disease*. New York: Humana Press.

Witsch, E., Sela, M., & Yarden, Y. (2010). Roles for growth factors in cancer progression. *Physiology (Bethesda)*, **25**, 85–101.

REFERENCES

Aboud-Pirak, E., Hurwitz, E., Pirak, M. E., et al. (1988). Efficacy of antibodies to epidermal growth factor receptor against KB carcinoma *in vitro* and in nude mice. *J Natl Cancer Inst*, **80**, 1605–11.

Asahara, T., Murohara, T., Sullivan, A., et al. (1997). Isolation of putative progenitor endothelial cells for angiogenesis. *Science*, **275**, 964–7.

Bang, Y. J., Van Cutsem, E., Feyereislova, A., et al. (2010). Trastuzumab in combination with chemotherapy versus chemotherapy alone for treatment of HER2-positive advanced gastric or gastro-oesophageal junction cancer (ToGA): a phase 3, open-label, randomised controlled trial. *Lancet*, **376**, 687–97.

Bergsten, E., Uutela, M., LI, X., et al. (2001). PDGF-D is a specific, protease-activated ligand for the PDGF beta-receptor. *Nature Cell Biol*, **3**, 512–16.

Blume-Jensen, P. & Hunter, T. (2001). Oncogenic kinase signalling. *Nature*, **411**(6835), 355–65.

Boucher, J., Macotela, Y., Bezy, O., Mori, M. A., Kriauciunas, K., & Kahn, C. R. (2010). A kinase-independent role for unoccupied insulin and IGF-1 receptors in the control of apoptosis. *Sci Signal*, **3**, ra87.

Brand, T. M., Iida, M., & Wheeler, D. L. (2011). Molecular mechanisms of resistance to the EGFR monoclonal antibody cetuximab. *Cancer Biol Ther*, **11**, 777–92.

Buonanno, A. (2010). The neuregulin signaling pathway and schizophrenia: from genes to synapses and neural circuits. *Brain Res Bull*, **83**, 122–31.

Burgess, A. W., Cho, H. S., Eigenbrot, C., et al. (2003). An open-and-shut case? Recent insights into the activation of EGF/ErbB receptors. *Mol Cell*, **12**, 541–52.

Cahill, D. P., Kinzler, K. W., Vogelstein, B., & Lengauer, C. (1999). Genetic instability and Darwinian selection in tumours. *Trends Cell Biol*, **9**, M57–60.

Cannella, B., Hoban, C. J., Gao, Y. L., et al. (1998). The neuregulin, glial growth factor 2, diminishes autoimmune demyelination and enhances remyelination in a chronic relapsing model for multiple sclerosis. *Proc Natl Acad Sci U S A*, **95**, 10100–5.

Cantley, L. C. (2002). The phosphoinositide 3-kinase pathway. *Science*, **296**, 1655–7.

Carey, K. D., Garton, A. J., Romero, M. S., et al. (2006). Kinetic analysis of epidermal growth factor receptor somatic mutant proteins shows increased sensitivity to the epidermal growth factor receptor tyrosine kinase inhibitor, erlotinib. *Cancer Res*, **66**, 8163–71.

Carpenter, G. & Cohen, S. (1976). 125I-labeled human epidermal growth factor. Binding, internalization, and degradation in human fibroblasts. *J Cell Biol*, **71**, 159–71.

Carvalho, S., Levi-Schaffer, F., Sela, M., & Yarden, Y. (2016). Immunotherapy of cancer: from monoclonal to oligoclonal cocktails of anti-cancer antibodies: IUPHAR Review 18. *Br J Pharmacol*, **173**, 1407–24.

Chen, P., Xie, H., Sekar, M. C., Gupta, K., & Wells, A. (1994). Epidermal growth factor receptor-mediated cell motility: phospholipase C activity is required, but mitogen-activated protein kinase activity is not sufficient for induced cell movement. *J Cell Biol*, **127**, 847–57.

Citri, A. & Yarden, Y. (2006). EGF-ErbB signalling: towards the systems level. *Nat Rev Mol Cell Biol*, **7**, 505–16.

Cobleigh, M. A., Vogel, C. L., Tripathy, D., et al. (1999). Multinational study of the efficacy and safety of humanized anti-HER2 monoclonal antibody in women who have HER2-overexpressing metastatic breast cancer that has progressed after chemotherapy for metastatic disease. *J Clin Oncol*, **17**, 2639–48.

Cohen, S., Levi-Montalcini, R., & Hamburger, V. (1954). A nerve growth-stimulating factor isolated from sarcom as 37 and 180. *Proc Natl Acad Sci U S A*, **40**, 1014–18.

Drebin, J. A., Link, V. C., Stern, D. F., Weinberg, R. A., & Greene, M. I. (1985). Down-modulation of an oncogene protein product and reversion of the transformed phenotype by monoclonal antibodies. *Cell*, **41**, 697–706.

Engelman, J. A., Zejnullahu, K., Mitsudomi, T., et al. (2007). MET amplification leads to gefitinib resistance in lung cancer by activating ERBB3 signaling. *Science*, **316**, 1039–43.

Ferrara, N. & Kerbel, R. S. (2005). Angiogenesis as a therapeutic target. *Nature*, **438**, 967–74.

Fleuren, E. D., Zhang, L., WU, J. & Daly, R. J. 2016. The kinome 'at large' in cancer. *Nat Rev Cancer*, **16**, 83–98.

Folkman, J. (2007). Angiogenesis: an organizing principle for drug discovery? *Nat Rev Drug Discov*, **6**, 273–86.

Fujita, Y., Krause, G., Scheffner, M., et al. (2002). Hakai, a c-CBL-like protein, ubiquitinates and induces endocytosis of the E-cadherin complex. *Nat Cell Biol*, **4**, 222–31.

Garrett, J. T. & Arteaga, C. L. (2011). Resistance to HER2-directed antibodies and tyrosine kinase inhibitors: mechanisms and clinical implications. *Cancer Biol Ther*, **11**, 793–800.

Garrett, T. P., Mckern, N. M., Lou, M., et al. (2003). The crystal structure of a truncated ErbB2 ectodomain reveals an active conformation, poised to interact with other ErbB receptors. *Mol Cell*, **11**, 495–505.

Geyer, C. E., Forster, J., Lindquist, D., et al. (2006). Lapatinib plus capecitabine for HER2-positive advanced breast cancer. *N Engl J Med*, **355**, 2733–43.

Goswami, S., Sahai, E., Wyckoff, J. B., et al. (2005). Macrophages promote the invasion of breast carcinoma cells via a colony-stimulating factor-1/epidermal growth factor paracrine loop. *Cancer Res*, **65**, 5278–83.

Haglund, K. & Dikic, I. (2012). The role of ubiquitylation in receptor endocytosis and endosomal sorting. *J Cell Sci*, **125**, 265–75.

Hallinan, N., Finn, S., Cuffe, S., Rafee, S., O'Byrne, K., & Gately, K. (2016). Targeting the fibroblast growth factor receptor family in cancer. *Cancer Treat Rev*, **46**, 51–62.

Hamilton, G. S. (2015). Antibody-drug conjugates for cancer therapy: The technological and regulatory challenges of developing drug-biologic hybrids. *Biologicals*, **43**, 318–32.

Hanahan, D. & Weinberg, R. A. (2000). The hallmarks of cancer. *Cell*, **100**, 57–70.

Hanahan, D. & Weinberg, R. A. (2011). Hallmarks of cancer: the next generation. *Cell*, **144**, 646–74.

Hazan, R. B., Phillips, G. R., Qiao, R. F., Norton, L., & Aaronson, S. A. (2000). Exogenous expression of N-cadherin in breast cancer cells induces cell migration, invasion, and metastasis. *J Cell Biol*, **148**, 779–90.

Helsten, T., Elkin, S., Arthur, E., Tomson, B. N., Carter, J., & Kurzrock, R. (2016). The FGFR landscape in cancer: analysis of 4,853 tumors by next-generation sequencing. *Clin Cancer Res*, **22**, 259–67.

Hynes, N. E. & Macdonald, G. (2009). ErbB receptors and signaling pathways in cancer. *Curr Opin Cell Biol*, **21**, 177–84.

Jaiswal, B. S., Kljavin, N. M., Stawiski, E. W., et al. (2013). Oncogenic ERBB3 mutations in human cancers. *Cancer Cell*, **23**, 603–17.

Janssen, J. A. & Varewijck, A. J. (2014). IGF-IR targeted therapy: past, present and future. *Front Endocrinol (Lausanne)*, **5**, 224.

Johnston, S., Trudeau, M., Kaufman, B., et al. (2008). Phase II study of predictive biomarker profiles for response targeting human epidermal growth factor receptor 2 (HER-2) in advanced inflammatory breast cancer with lapatinib monotherapy. *J Clin Oncol*, **26**, 1066–72.

Junttila, T. T., LI, G., Parsons, K., Phillips, G. L., & Sliwkowski, M. X. (2011). Trastuzumab-DM1 (T-DM1) retains all the mechanisms of action of trastuzumab and efficiently inhibits growth of lapatinib insensitive breast cancer. *Breast Cancer Res Treat*, **128**, 347–56.

Jurek, A., Heldin, C.-H., & Lennartsson, J. (2011). Platelet-derived growth factor-induced signaling pathways interconnect to regulate the temporal pattern of Erk1/2 phosphorylation. *Cell Signal*, **23**, 280–7.

Kalluri, R. & Weinberg, R. A. (2009). The basics of epithelial-mesenchymal transition. *J Clin Invest*, **119**, 1420–8.

Kaplan, R. N., Riba, R. D., Zacharoulis, S., et al. (2005). VEGFR1-positive haematopoietic bone marrow progenitors initiate the pre-metastatic niche. *Nature*, **438**, 820–7.

Kholodenko, B. N., Hancock, J. F., & Kolch, W. (2010). Signalling ballet in space and time. *Nat Rev Mol Cell Biol*, **11**, 414–26.

Köhler, G. & Milstein, C. (1975). Continuous cultures of fused cells secreting antibody of predefined specificity. *Nature*, **256**, 495–7.

Kolsch, V., Charest, P. G., & Firtel, R. A. (2008). The regulation of cell motility and chemotaxis by phospholipid signaling. *J Cell Sci*, **121**, 551–9.

Kovacs, E., Zorn, J. A., Huang, Y., Barros, T., & Kuriyan, J. (2015). A structural perspective on the regulation of the epidermal growth factor receptor. *Annu Rev Biochem*, **84**, 739–64.

Kowanetz, M. & Ferrara, N. (2006). Vascular endothelial growth factor signaling pathways: therapeutic perspective. *Clin Cancer Res*, **12**, 5018–22.

Law, A. J., Kleinman, J. E., Weinberger, D. R., & Weickert, C. S. (2007). Disease-associated intronic variants in the ErbB4 gene are related to altered ErbB4 splice-variant expression in the brain in schizophrenia. *Hum Mol Genet*, **16**, 129–41.

Lemmon, M. A. & Schlessinger, J. (2010). Cell signaling by receptor tyrosine kinases. *Cell*, **141**, 1117–34.

Levitzki, A. (2013). Tyrosine kinase inhibitors: views of selectivity, sensitivity, and clinical performance. *Annu Rev Pharmacol Toxicol*, **53**, 161–85.

Lewis Phillips, G. D., Li, G., Dugger, D. L., et al. (2008). Targeting HER2-positive breast cancer with trastuzumab-DM1, an antibody-cytotoxic drug conjugate. *Cancer Res*, **68**, 9280–90.

Li, X., Ponten, A., Aase, K., et al. (2000). PDGF-C is a new protease-activated ligand for the PDGF alpha-receptor. *Nat Cell Biol*, **2**, 302–9.

Lindzen, M., Carvalho, S., Starr, A., et al. (2012). A recombinant decoy comprising EGFR and ErbB-4 inhibits tumor growth and metastasis. *Oncogene*, **31**, 3505–15.

Lo, S. H. (2004). Tensin. *Int J Biochem Cell Biol*, **36**, 31–4.

Lyden, D., Hattori, K., Dias, S., et al. (2001). Impaired recruitment of bone-marrow-derived endothelial and hematopoietic precursor cells blocks tumor angiogenesis and growth. *Nat Med*, **7**, 1194–201.

Lynch, T. J., Bell, D. W., Sordella, R., et al. (2004). Activating mutations in the epidermal growth factor receptor underlying responsiveness of non-small-cell lung cancer to gefitinib. *N Engl J Med*, **350**, 2129–39.

Mancini, M., Gaborit, N., Lindzen, M., et al. (2015). Combining three antibodies nullifies feedback-mediated resistance to erlotinib in lung cancer. *Sci Signal*, **8**, ra53.

Mellman, I. & Yarden, Y. (2013). Endocytosis and cancer. *Cold Spring Harb Perspect Biol*, **5**, a016949.

Mendelsohn, J. & Baselga, J. (2006). Epidermal growth factor receptor targeting in cancer. *Semin Oncol*, **33**, 369–85.

Mouneimne, G. & Brugge, J. S. (2007). Tensins: a new switch in cell migration. *Dev Cell*, **13**, 317–19.

Ocana, A., Vera-Badillo, F., Seruga, B., Templeton, A., Pandiella, A., & Amir, E. (2013). HER3 overexpression and survival in solid tumors: a meta-analysis. *J Natl Cancer Inst*, **105**, 266–73.

Orimo, A., Gupta, P. B., Sgroi, D. C., et al. (2005). Stromal fibroblasts present in invasive human breast carcinomas promote tumor growth and angiogenesis through elevated SDF-1/CXCL12 secretion. *Cell*, **121**, 335–48.

Ostman, A. & Augsten, M. (2009). Cancer-associated fibroblasts and tumor growth--bystanders turning into key players. *Curr Opin Genet Dev*, **19**, 67–73.

Padua, D., Zhang, X. H., Wang, Q., et al. (2008). TGF-beta primes breast tumors for lung metastasis seeding through angiopoietin-like 4. *Cell*, **133**, 66–77.

Paez, J. G., Janne, P. A., Lee, J. C., et al. (2004). EGFR mutations in lung cancer: correlation with clinical response to gefitinib therapy. *Science*, **304**, 1497–500.

Pao, W., Miller, V., Zakowski, M., et al. (2004). EGF receptor gene mutations are common in lung cancers from 'never smokers' and are associated with sensitivity of tumors to gefitinib and erlotinib. *Proc Natl Acad Sci U S A*, **101**, 13306–11.

Park, W. S., Dong, S. M., Kim, S. Y., et al. (1999). Somatic mutations in the kinase domain of the MET/hepatocyte growth factor receptor gene in childhood hepatocellular carcinomas. *Cancer Res*, **59**, 307–10.

Pawson, T. (2004). Specificity in signal transduction: from phosphotyrosine-SH2 domain interactions to complex cellular systems. *Cell*, **116**, 191–203.

Plotnikov, A., Zehorai, E., Procaccia, S., & Seger, R. (2011). The MAPK cascades: signaling components, nuclear roles and mechanisms of nuclear translocation. *Biochim Biophys Acta*, **1813**, 1619–33.

Polager, S. & Ginsberg, D. (2009). p53 and E2f: partners in life and death. *Nat Rev Cancer*, **9**, 738–48.

Pollak, M. (2012). The insulin and insulin-like growth factor receptor family in neoplasia: an update. *Nat Rev Cancer*, **12**, 159–69.

Prickett, T. D., Agrawal, N. S., Wei, X., et al. (2009). Analysis of the tyrosine kinome in melanoma reveals recurrent mutations in ERBB4. *Nat Genet*, **41**, 1127–32.

Ridley, A. J. (2011). Life at the leading edge. *Cell*, **145**, 1012–22.

Rikova, K., Guo, A., Zeng, Q., et al. (2007). Global survey of phosphotyrosine signaling identifies oncogenic kinases in lung cancer. *Cell*, **131**, 1190–203.

Ronsin, C., Muscatelli, F., Mattei, M. G., & Breathnach, R. (1993). A novel putative receptor protein tyrosine kinase of the met family. *Oncogene*, **8**, 1195–202.

Rosenthal, A., Lindquist, P. B., Bringman, T. S., Goeddel, D. V., & Derynck, R. (1986). Expression in rat fibroblasts of a human transforming growth factor-alpha cDNA results in transformation. *Cell*, **46**, 301–9.

Rusnak, D. W., Affleck, K., Cockerill, S. G., et al. (2001). The characterization of novel, dual ErbB-2/EGFR, tyrosine kinase inhibitors: potential therapy for cancer. *Cancer Res*, **61**, 7196–203.

Ryan, Q., Ibrahim, A., Cohen, M. H., et al. (2008). FDA drug approval summary: lapatinib in combination with capecitabine for previously treated metastatic breast cancer that overexpresses HER-2. *Oncologist*, **13**, 1114–19.

Savage, C. R., JR., Inagami, T., & Cohen, S. (1972). The primary structure of epidermal growth factor. *J Biol Chem*, **247**, 7612–21.

Schafer, M. & Werner, S. (2008). Cancer as an overhealing wound: an old hypothesis revisited. *Nat Rev Mol Cell Biol*, **9**, 628–38.

Schmidt, L., Duh, F. M., Chen, F., et al. (1997). Germline and somatic mutations in the tyrosine kinase domain of the MET proto-oncogene in papillary renal carcinomas. *Nat Genet*, **16**, 68–73.

Scott, A. M., Wolchok, J. D., & Old, L. J. (2012). Antibody therapy of cancer. *Nat Rev Cancer*, **12**, 278–87.

Sell, C., Dumenil, G., Deveaud, C., et al. (1994). Effect of a null mutation of the insulin-like growth factor I receptor gene on growth and transformation of mouse embryo fibroblasts. *Mol Cell Biol*, **14**, 3604–12.

Shaked, Y., Ciarrocchi, A., Franco, M., et al. (2006). Therapy-induced acute recruitment of circulating endothelial progenitor cells to tumors. *Science*, **313**, 1785–7.

Sievers, E. L. & Senter, P. D. (2013). Antibody-drug conjugates in cancer therapy. *Annu Rev Med*, **64**, 15–29.

Slamon, D. J., Godolphin, W., Jones, L. A., et al. (1989). Studies of the HER-2/neu proto-oncogene in human breast and ovarian cancer. *Science*, **244**, 707–12.

Soda, M., Choi, Y. L., Enomoto, M., et al. (2007). Identification of the transforming EML4-ALK fusion gene in non-small-cell lung cancer. *Nature*, **448**, 561–6.

Sordella, R., Bell, D. W., Haber, D. A., & Settleman, J. (2004). Gefitinib-sensitizing EGFR mutations in lung cancer activate anti-apoptotic pathways. *Science*, **305**, 1163–7.

Sporn, M. B. & Todaro, G. J. (1980). Autocrine secretion and malignant transformation of cells. *N Engl J Med*, **303**, 878–80.

Takaishi, K., Sasaki, T., Kotani, H., Nishioka, H., & Takai, Y. (1997). Regulation of cell–cell adhesion by rac and rho small G proteins in MDCK cells. *J Cell Biol*, **139**, 1047–59.

Thiery, J. P., Acloque, H., Huang, R. Y., & Nieto, M. A. (2009). Epithelial-mesenchymal transitions in development and disease. *Cell*, **139**, 871–90.

Tognon, C. E. & Sorensen, P. H. (2012). Targeting the insulin-like growth factor 1 receptor (IGF1R) signaling pathway for cancer therapy. *Expert Opin Ther Targets*, **16**, 33–48.

Trusolino, L., Bertotti, A., & Comoglio, P. M. (2010). MET signalling: principles and functions in development, organ regeneration and cancer. *Nat Rev Mol Cell Biol*, **11**, 834–48.

Turner, N. & Grose, R. (2010). Fibroblast growth factor signalling: from development to cancer. *Nat Rev Cancer*, **10**, 116–29.

Verma, S., Miles, D., Gianni, L., et al. (2012). Trastuzumab emtansine for HER2-positive advanced breast cancer. *N Engl J Med*, **367**, 1783–91.

Vogel, C. L., Cobleigh, M. A., Tripathy, D., et al. (2002). Efficacy and safety of trastuzumab as a single agent in first-line treatment of HER2-overexpressing metastatic breast cancer. *J Clin Oncol*, **20**, 719–26.

Waltenberger, J., Claesson-Welsh, L., Siegbahn, A., Shibuya, M., & Heldin, C. H. (1994). Different signal transduction properties of KDR and Flt1, two receptors for vascular endothelial growth factor. *J Biol Chem*, **269**, 26988–95.

Wu, Y., Hooper, A. T., Zhong, Z., et al. (2006). The vascular endothelial growth factor receptor (VEGFR-1) supports growth and survival of human breast carcinoma. *Int J Cancer*, **119**, 1519–29.

Yang, Y. R., Follo, M. Y., Cocco, L., & Suh, P.-G. (2013). The physiological roles of primary phospholipase C. *Adv Biol Reg*, **53**, 232–41.

Yilmaz, M. & Christofori, G. (2009). EMT, the cytoskeleton, and cancer cell invasion. *Cancer Metastasis Rev*, **28**, 15–33.

Yu, H., Pardoll, D., & Jove, R. (2009). STATs in cancer inflammation and immunity: a leading role for STAT3. *Nat Rev Cancer*, **9**, 798–809.

Yun, C. H., Boggon, T. J., Li, Y., et al. (2007). Structures of lung cancer-derived EGFR mutants and inhibitor complexes: mechanism of activation and insights into differential inhibitor sensitivity. *Cancer Cell*, **11**, 217–27.

Zwang, Y., Oren, M., & Yarden, Y. (2012). Consistency test of the cell cycle: roles for p53 and EGR1. *Cancer Res*, **72**, 1051–4.

Zwang, Y. & Yarden, Y. (2009). Systems biology of growth factor-induced receptor endocytosis. *Traffic*, **10**, 349–63.

10

Hormones and cancer

Balkees Abderrahman and V. Craig Jordan

Historical background

An understanding of the hormonal control of breast and prostate cancer has evolved over the past 100 years (Jordan, 2009). Patient care went from physicians and surgeons documenting the apparently random success or failure of endocrine ablation, to targeted therapeutics aimed at blocking sex-steroid binding to the oestrogen receptor (ER) or androgen receptor (AR) in breast and prostate cancer, respectively. It is recognized that the ER target in breast cancer is the most important target in cancer, and the development of antioestrogenic therapeutics has saved more lives than any other strategic approach in cancer treatment (Sledge et al., 2014). Since prostate cancer is the most prevalent cancer in men, and the AR is present in the majority of tumours, one can double the success rate in the community by targeting sex steroids to save lives.

In the first half of the twentieth century, endocrine ablation (oophorectomy, adrenalectomy, and hypophysectomy) became standards of care that produced approximately 30% response rates in breast cancer. Unfortunately, these improvements are transient and last for no more than a few years. In a meta-analysis of seven randomized, placebo-controlled studies, metastatic prostate cancer patients responded to gonadectomy (orchiectomy) with a 33% response rate (Dijkman et al., 1995). Paradoxically, breast cancer in women, more than 5 years after menopause, or prostate cancer in men, both responded to high-dose oestrogen therapy in the absence of ablative surgery. Despite the fact that this paradox had no biological rationale, high-dose oestrogen remained the standard of care with endocrine ablation for 30 years from the 1940s.

The major question in the second half of the twentieth century, was whether it was possible to predict which breast cancer patients would not respond to endocrine ablation or high-dose oestrogen therapy. The primary goal was to avoid hospitalization and morbidity in patients, who would not have a responsive tumour. The discovery of the ER in the laboratory, and the knowledge that the ER was necessary to produce target site-specific effects in oestrogen-target tissues (uterus, vagina, and the pituitary gland) (Jensen and Jacobson, 1962), was subsequently translated to breast cancer tissue. A range of ER levels were noted from zero to high. This led to the correlation of the presence of the ER in a breast tumour to a high response rate of metastatic breast cancer (MBC) to endocrine ablation or high-dose oestrogen. Patients whose tumours had very low or no ER, were unlikely to respond to either endocrine ablation or high-dose oestrogen therapy. These studies established the ER assay used clinically today. The application in medicine became the identifications of patients, who would receive the new targeted antioestrogenic therapies, which were to emerge through translational research in the 1970s. Not only did the ER assay predict responsiveness for MBC, but also, the subsequent use of antioestrogenic strategies as adjuvant therapies, depended on the ER status of the primary tumour. By contrast, AR assays are not utilized to guide prostate cancer therapeutics.

During the second half of the twentieth century, the development of the non-steroidal antioestrogen, tamoxifen (Jordan, 2003), and its subsequent ubiquitous use for the treatment of ER-positive MBC, as an adjuvant therapy for all stages of breast cancer after surgery, the treatment of indolent lesions of epithelial origin (IDLE), and as a chemopreventive agent in high-risk populations of women, revolutionized the way cancer specific-targeted therapies would be developed in the future. Tamoxifen is a competitive inhibitor of oestrogen binding at the tumour ER. However, tamoxifen can also be a partial agonist and not a 'pure antioestrogen' in other cells and tissues in a woman's body. There is a low, but significant incidence, of oestrogen-like side effects in postmenopausal patients, such as endometrial cancer and thromboembolic events. However, this is only relevant for postmenopausal patients during long-term adjuvant therapy. The development of specific agents to block the aromatization enzyme (CYP19) for use in postmenopausal patients with ER-positive breast cancer, is an improvement over current tamoxifen treatment, with a decrease in oestrogen-like side effects and decreases in recurrence rates and mortality. It appears that no oestrogen synthesis is preferable to the blockade of estrogen action at the ER by tamoxifen, but this mantra may not necessarily be the way forward for future therapeutic advances.

Comparing and contrasting oestrogen and androgen receptors and their signal transduction pathways

The ER has two subtypes (Thomas and Gustafsson, 2011): ERα which was discovered in the late 1950s, and ERβ identified in 1996. ERα is a product of the gene ESR1, and ERβ is a product of ESR2 on a different chromosome (Thomas and Gustafsson, 2011). The ER has several structural and functional domains (**Fig. 10.1**; Thomas and

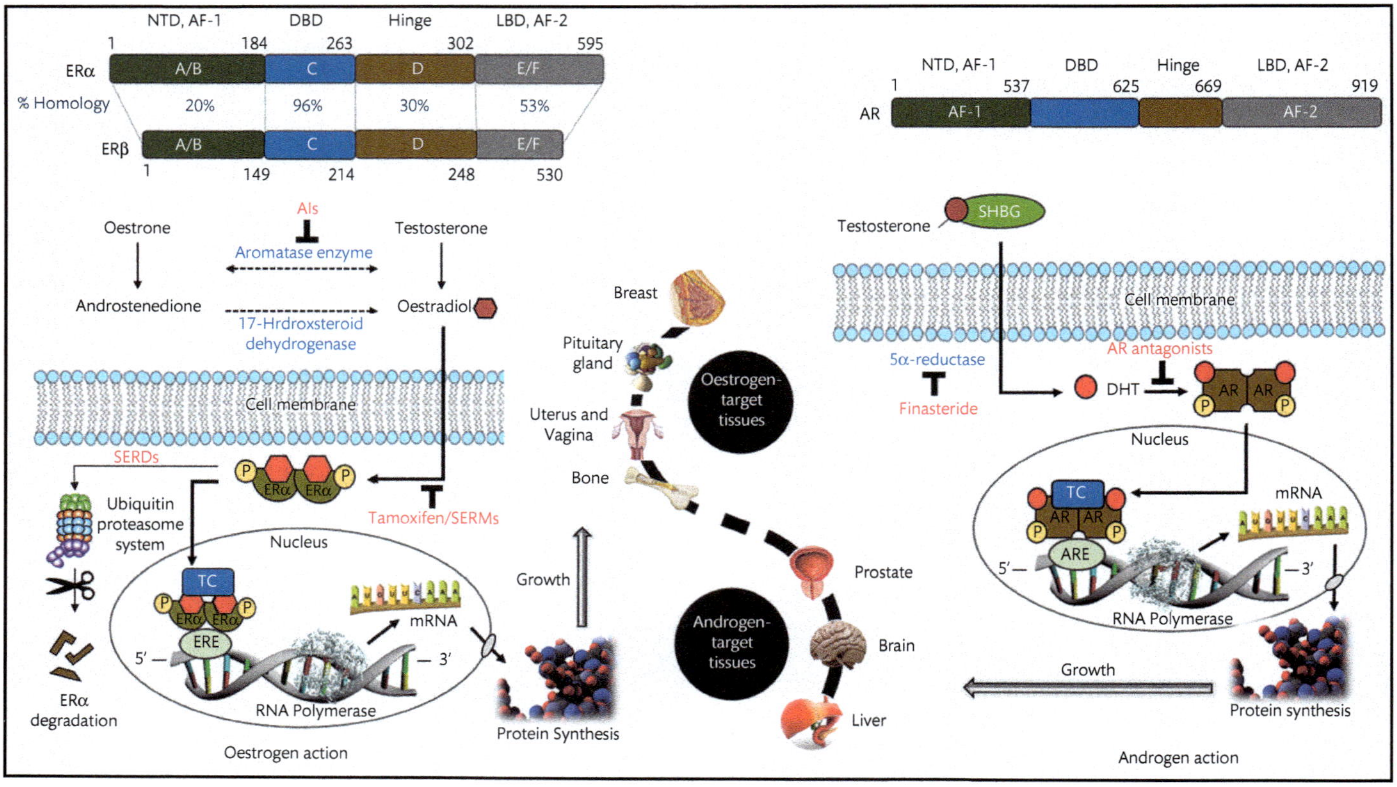

Fig. 10.1 Schematic diagram of the ER and AR structural and functional domains, their corresponding signal transduction pathways, commonly used therapeutic agents, and target tissues. The DBD is most conserved among ERα and ERβ with a homology of 96%. Oestrone is converted to androstenedione and testosterone is converted to oestradiol through the aromatase enzyme system, which is inhibited by AIs, androstenedione in turn is converted to oestradiol through 17-hydroxysteroid dehydrogenase. Tamoxifen/SERMs competitively inhibit oestrogen binding to the ER, while SERDs destroy the ER through the ubiquitin proteasome system. Oestrogen binding to the ER initiates a cascade of events throughout the ER signal transduction pathway. Similarly, the binding of dihydrotestosterone (DHT) to the AR, after its synthesized from testosterone by 5α-reductase, initiates a cascade of events throughout the AR signal transduction pathway. Finasteride inhibits 5α-reductase, and AR antagonists competitively inhibit the binding of androgens to the AR.

Gustafsson, 2011): the amino-terminal A/B region contains a transactivation domain (AF1), which is pivotal to the transcriptional activity of ERα through a ligand-independent function, and a coregulatory domain responsible for coactivators and corepressors recruitment. ERβ is truncated and lacks AF-1. The C region represents the DNA-binding domain (DBD), which is the most conserved region among ERα and ERβ. This region is crucial for binding to specific oestrogen response elements (EREs) in the proximal promoter region or at distal regulatory elements of ERE. The D region or the hinge region, is part of the ligand-dependent activating domain and the nuclear localization signal. The regions E and F contain the ligand-binding domain (LBD), a coregulatory binding surface, the dimerization domain, the second nuclear localization signal, and the activation function 2 (AF2). Both AF-1 and AF-2 act together at the promoter region for a full oestrogen-like activity in ERα, but ERβ does not have an active AF1 site. As there is no AF-1 in ERβ, heterodimerizations of ERα and ERβ cause antioestrogenic effects. The amino acids Leu384 and Met421 in the LBD regions of ERα are replaced by Met336 and Ile373 in ERβ, respectively. This similarity of the LBD of ERα and ERβ had created problems for targeting ligands to a specific ER. Human ERα and ERβ isoforms are expressed differently in malignant tissues. This naturally impacts cancer biology. Both receptors, however, exert opposite effects on cellular proliferation and apoptosis.

Isoform ERα-36 (Wang and Yin, 2015), also known as the 'dwarf or truncated ER', lacks both trans-activation domains. ERα-36 maintains a 'non-genomic' signalling pathway through mitogen-activated protein kinase, and is resistant to tamoxifen treatment.

In basic terms, once the ligand binds to the ER in the cytoplasm, ER dissociates from heat shock proteins (HSPs), dimerizes, gets phosphorylated and relocates within nucleus (Thomas and Gustafsson, 2011). The ligand:ER complex then binds to a gene specific ERE and recruits corresponding coactivators or corepressors. This, in turn, initiates or inhibits the cascade of transcription and translation.

Androgen synthesis is finely regulated by the hypothalamic–pituitary–gonadal axis. Upon the stimulation of the hypothalamus, luteinizing hormone releasing hormone (LHRH) is produced. This hormone works in a pulsatile fashion to stimulate the release of LH by the anterior pituitary. This, in turn, induces the synthesis of androgen at the testicular level. Moreover, LHRH, also stimulates the production of adrenocorticotropic hormone by the anterior pituitary, which augments overall androgen production, but in the adrenal gland. Testosterone is metabolically converted to dihydrotestosterone by 5α reductase (Fig. 10.1), which then binds to the AR, causing the dissociation of corresponding HSPs, and subsequent dimerization and phosphorylation of the AR. The AR has three distinctive functional domains: the N-terminal domain, the DBD and the LBD (Wong et al., 2014). In the nucleus, the androgen:AR complex (Edwards and Bartlett, 2005) binds to androgen response elements/genes, including *TMPRSS2:ERG* and prostate-specific antigen. This, in turn, recruits the DNA transcriptional machinery to initiate gene transcription (Edwards and Bartlett, 2005). Although the AR is the essential mediator to regulate normal growth, it also participates in promoting the oncogenic fusion genes, especially E-twenty-six family (Wang et al., 2009). Mechanisms of maintained AR signalling in castration-resistant prostate cancer (CRPC) have been identified (Attard et al., 2011; Ryan and Tindall, 2011), and include: increased AR signalling whether it was increased AR expression or gene amplification, point mutations in LBD, expression of active AR splice variants, cross talk with other pathways, presence of residual androgens, and changes in coregulators proteins.

A unique transcription factor named forkhead-box A1 (FOXA1) plays a critical role in chromatin remodelling and decompaction (Yang and Yu, 2015). This, in turn, allows the genomic access by the nuclear hormone receptors such as the AR and ER. The complex of FOXA1:AR remains in equilibrium states in the nucleus and defines the prostatic AR binding profile. In prostate cancer, this equilibrium is disturbed with FOXA1 and/or AR de-regulation (Yang and Yu, 2015). A recent meta-analysis (Shou et al., 2016) showed that higher levels of FOXA1 expression is associated with a better prognosis in breast cancer.

Translational research to create treatment strategies

Despite the fact that Charles Huggins was awarded the Nobel Prize in 1966, for his laboratory and clinical work defining the treatment strategy of androgen deprivation or high-dose oestrogen therapy to suppress pituitary gland function, basic research on prostate cancer has tended to lag behind translational research of breast cancer, by a decade or two. Indeed, the major stimulus for translational cancer research was the National Cancer Act signed into law by President Nixon in 1971. The plan was to take laboratory discoveries to aid patient survival as rapidly as possible, through the National Cancer Institute and a network of new comprehensive cancer centres.

The rapid progress with breast cancer treatment was because strategies were being developed with the non-steroidal antioestrogen, tamoxifen, which would revolutionize breast cancer patient care. The advances in cancer research have depended upon the selection of appropriate models for human disease, but in the early 1970s, few were generally available. Cell culture models for ER-positive breast cancer, were confined only to the MCF7 cell line (Levenson and Jordan, 1997), and this remained the status quo with two or three additions (T47D, ZR75) over the next decade. Although there were numerous high incidence strains of mice, which developed hormone-responsive mammary cancer, these model systems had been used historically to demonstrate that the mouse mammary tumour virus (Bittner's milk factor), was the agent responsible for transferring mother to offspring mammary cancer. However, the viral theory as the cause of breast cancer fell into disrepute, and as a result, research on mouse mammary tumorigenesis decreased. Nevertheless, mouse models that could be genetically engineered for different oncogenes became the standard to test and validate susceptibility to different cancers. The mouse made a comeback in the 1980s, with the new technology of molecular biology.

The dimethylbenzanthracene-induced rat mammary carcinoma model (DMBA), developed by Huggins (Huggins et al., 1961), was the standard laboratory model used to study hormones and cancer during the 1960s and 1970s. Despite obvious limitations (non-metastatic disease), Huggins deciphered the precise chain of events necessary for the single dose of a mammary carcinogen, to produce mammary cancer in all female rats. The protocol required the administration of 10 mg of DMBA, dissolved in 2 ml of arachis oil, but mammary tumorigenesis would only occur with administration between 50 and 65 days of age. Tumorigenesis does not occur if animals are ovariectomized at the time of carcinogen administration. By contrast, progesterone administration increases tumorigenesis. The mammary tumours are 90% ER-positive and respond to ovariectomy with tumour regression.

During the 1970s, translational breast cancer research advanced rapidly, primarily because well-defined laboratory treatment strategies were subsequently implemented as clinical trials. Three strategies were described (Jordan, 2008): aim at the ER target present in the breast tumour, deploy long-term adjuvant tamoxifen therapy as the most effective way to treat breast cancer clinically, and exploit the potential of tamoxifen in preventing breast cancer in women at high risk, or whats known as chemoprevention. Each of these strategies, defined using the DMBA-induced rat mammary carcinoma model, were successfully translated to clinical care. It is now established, through the Oxford Overview of Adjuvant Clinical Trials (EBCTCG, 1998), that tamoxifen is only successful in reducing recurrences and preventing deaths, if the primary tumour is ER-positive. Longer adjuvant tamoxifen therapy (5–10 years), has proven to be superior compared to shorter adjuvant tamoxifen therapy (1–2 years). Tamoxifen is the first medicine to reduce the risk of breast cancer, and indeed, any cancer in high-risk populations.

Prostate cancer has followed a similar treatment strategy, essentially exploiting the clues and principles established for the treatment of breast cancer. Androgen deprivation can still be achieved by gonadectomy; however, high-dose oestrogen treatment is now replaced by the use of a sustained release of an LHRH superagonist. This suppresses the release of gonadotropins, which in, turn suppresses androgen synthesis in the testes. Antiandrogens that bind to and block the AR, have been refined and improved over the past three decades, based on the experiences with the modulation of ER. The next generation 'antiandrogenic' blocking agents, abiraterone and enzalutamide, have significantly prolonged survival in patients with CRPC (Wong et al., 2014). Abiraterone acetate is considered a first-in-class inhibitor of cytochrome P450c17, which is responsible for androgen synthesis at the testicular and extragonadal level (Ryan et al., 2013). The combination of abiraterone acetate and prednisone, is used for CRPC after exposure to docetaxel (de Bono et al., 2011). Abiraterone also improves overall survival and delays the initiation of chemotherapy in metastatic CRPC (Ryan et al., 2013).

Mechanism of action of selective oestrogen receptor modulators

The discovery of synthetic non-steroidal antioestrogens, created an important opportunity in drug discovery (Lerner and Jordan,

1990). In the 1960s, these compounds were classified as post-coital contraceptives in laboratory animals, and the opportunity to create occasional, 'morning after pills' for women, was investigated in clinical trial. In contrast to findings in laboratory models, the non-steroidal antioestrogens, were found to induce ovulation in subfertile women, and the compound clomiphene (Clomid, a mixture of *cis/trans*-geometric isomers), was marketed for the induction of ovulation. Similarly, ICI46,474 (brand name Nolvadex, scientific name tamoxifen, the pure antioestrogenic *trans*-isomer) was marketed in some countries for the same use. Subsequently, tamoxifen was rigorously investigated as a long-term therapy for the adjuvant treatment of ER-positive breast cancer, and subsequently was evaluated as a chemopreventive agent in high-risk women (Jordan, 2008). These broad applications demanded the study of toxicology and antioestrogenic mechanisms.

In the mid-1980s, laboratory studies demonstrated that tamoxifen, and its high affinity metabolite 4-hydroxytamoxifen, is bound in oestrogen-target tissues around the animal's body (**Fig. 10.2**), but the tamoxifen:ER complex was perceived in different target tissues, either as an oestrogen or as an antioestrogen (Jordan and Robinson, 1987). The same ligand:ER complex was being interpreted differently in different target tissues. Tamoxifen was oestrogen-like in maintaining bone density, and lowering circulating cholesterol, whereas, the same complex was antioestrogenic in animal mammary cancers. These laboratory data extrapolated to patients with breast cancer, which demonstrated that long-term adjuvant tamoxifen therapy may also provide healthcare benefits, by retarding the development of osteoporosis and coronary heart disease. Most importantly, laboratory studies showed that immune deficient mice bitransplanted with an ER-positive breast cancer in one axilla, and an ER-positive endometrial cancer in the other, had a tamoxifen-controlled oestrogen-stimulated growth of the breast cancer, but a stimulated growth of the endometrial cancer (Gottardis et al., 1988). These data of the oestrogenic and antioestrogenic targeted effects in different human cancers, quickly extrapolated to clinical practice to involve gynaecological monitoring for patients receiving tamoxifen. This change in the practice of healthcare, not only provided important toxicological monitoring of patient populations in the 1990s during the multiple chemopreventive trials, but also was important in monitoring patients taking tamoxifen as long-term adjuvant therapy.

Overall, the laboratory and clinical work with tamoxifen on target-site specificity, along with the evaluation of the pharmacology of raloxifene not stimulating the rodent uterus, and being less oestrogenic in the human uterus, led to the concept that SERMs could be developed to prevent osteoporosis or coronary heart disease, while preventing breast cancer at the same time (Lerner and Jordan, 1990). Tamoxifen is FDA approved for the treatment and prevention of breast cancer in high-risk premenopausal and postmenopausal women, whereas, raloxifene (see **Fig. 10.2**) is the first SERM approved for the treatment and prevention of osteoporosis. Raloxifene also reduces the risk of breast

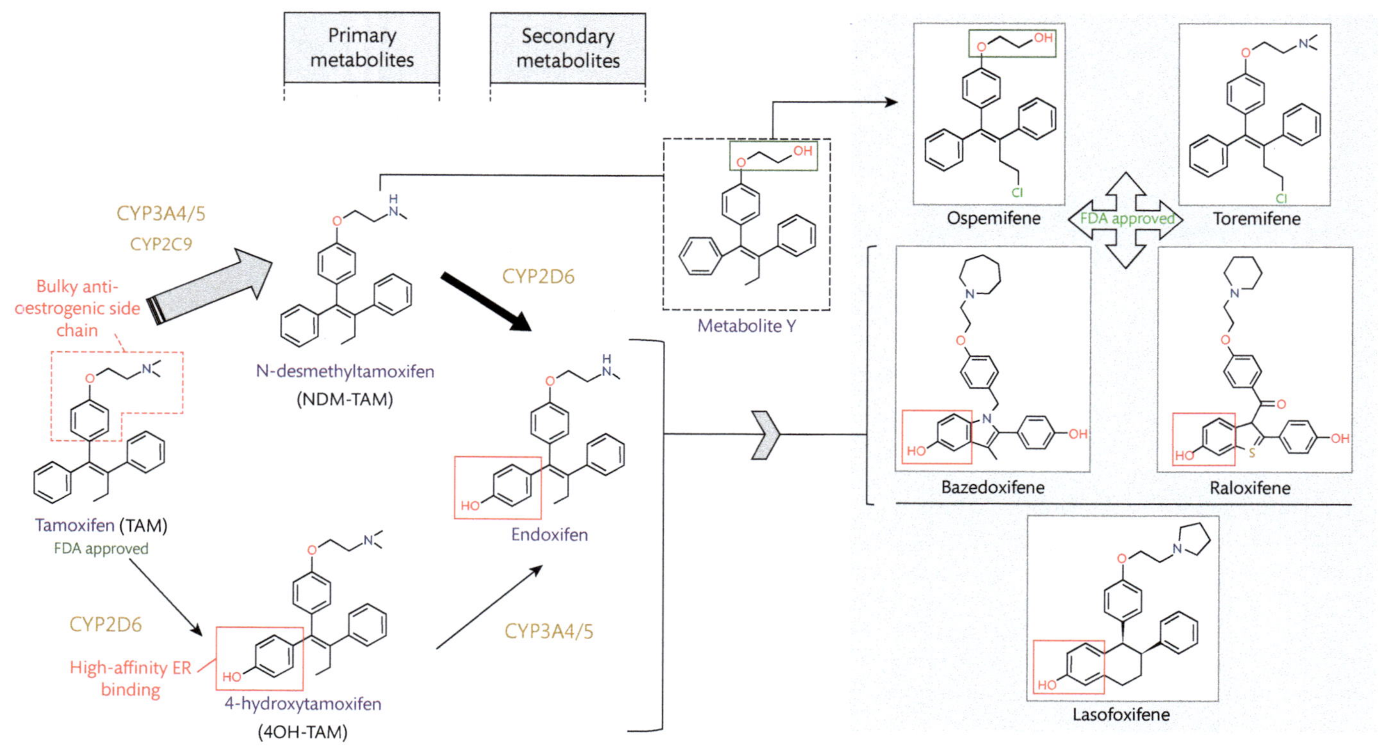

Fig. 10.2 Schematic representation of tamoxifen metabolism and the currently available SERMs. Tamoxifen undergoes oxidative metabolism by CYP2D6 to 4OH-TAM, and mainly to NDM-TAM by CYP3A4/5. The compound 4OH-TAM has a lower plasma concertation in comparison to NDM-TAM, which means that the primary route of tamoxifen metabolism is through N-demethylation to other metabolites. Endoxifen in turn is synthesized from the metabolism of two parent drugs: hydroxylation of NDM-TAM by CYP2D6, and from 4OH-TAM by CYP3A4/5. The compound NDM-TAM has an intermediate molecule known as metabolite Y which is metabolized to ospemifene. This diagrams also shows the five FDA-approved SERMs, and the currently pending SERM for approval, lasofoxifene.

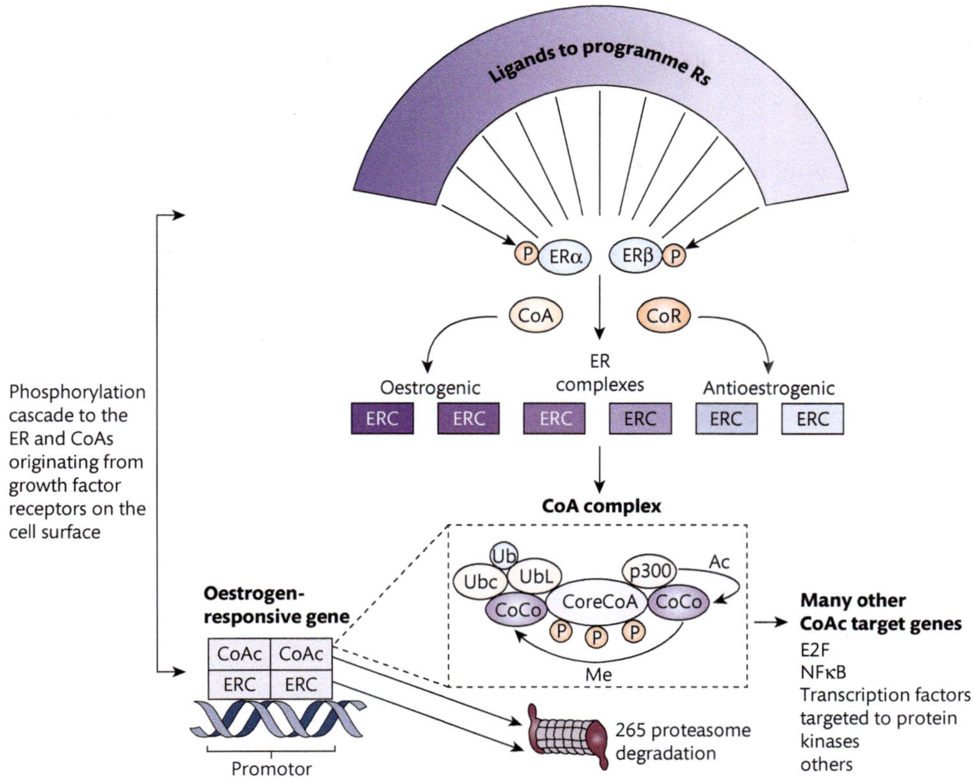

Fig. 10.3 The shape of the ligands that bind to the oestrogen receptors (ERs) α and β programmes the complex to become an oestrogenic or antioestrogenic signal. The context of the ER complex (ERC) can influence the expression of the response through the numbers of corepressors (CoR) or coactivators (CoA). In simple terms, a site with few CoAs or high levels of CoRs might be a dominant antioestrogenic site. However, the expression of oestrogenic action is not simply the binding of the receptor complex to the promoter of the oestrogen-responsive gene, but a dynamic process of CoA complex assembly and destruction (Lonard and O'Malley, 2006). A core CoA, for example, steroid receptor coactivator protein 3 (SRC3), and the ERC are influenced by phosphorylation cascades that phosphorylate target sites on both complexes. The core CoA then assembles an activated multiprotein complex containing specific co-co-activators (CoCo) that might include p300, each of which has a specific enzymatic activity to be activated later. The CoA complex (CoAc) binds to the ERC at the oestrogen-responsive gene promoter to switch on transcription. The CoCo proteins then perform methylation (Me) or acetylation (Ac) to activate dissociation of the complex. Simultaneously, ubiquitinylation by the bound ubiquitin-conjugating enzyme (Ubc) targets ubiquitin ligase (UbL) destruction of protein members of the complex through the 26S proteasome. The ERs are also ubiquitylated and destroyed in the 26S proteasome. Therefore, a regimented cycle of assembly, activation and destruction occurs on the basis of the preprogrammed ER complex (Lonard and O'Malley, 2006). However, the coactivator, specifically SRC3, has ubiquitous action and can further modulate or amplify the ligand-activated trigger through many modulating genes (O'Malley, 2006) that can consolidate and increase the stimulatory response of the ERC in a tissue. Therefore, the target tissue is programmed to express a spectrum of responses between full oestrogen action and antioestrogen action on the basis of the shape of the ligand and the sophistication of the tissue-modulating network. NFkB, nuclear factor kB.

cancer in postmenopausal high-risk patients, and does not increase the risk for endometrial cancer.

There are now three additional FDA-approved SERMs (**Fig. 10.2**): a bazedoxifene/oestrogen combination is used to ameliorate menopausal symptoms, and bazedoxifene is used for the treatment of osteoporosis, ospemifene is used to ameliorate dyspareunia (painful sexual intercourse), and toremifene is used in advanced MBC and is being evaluated for the prevention of prostate cancer.

A new SERM investigated extensively in clinical trials, but not yet approved, is lasofoxifene (**Fig. 10.2**). The molecule is a miracle of medicinal chemistry, as it is effective at 1/100th (0.5 mg/daily) the dose of raloxifene (60 mg/daily), for the treatment of osteoporosis, but at the same time, decreases the incidence of breast cancer, strokes, coronary heart disease, and without any increases in endometrial cancer risk.

Given the important clinical use of SERMs, the molecular mechanism of action is summarized in **Figure 10.3** (Jordan, 2007).

Mechanism of action of selective oestrogen receptor downregulators

The discovery that the steroidal antioestrogen ICI164,384 (Wakeling and Bowler, 1987) could bind to the ER, but produces no oestrogen-like effects in any oestrogen-target tissues in laboratory animals, raised the possibility that new 'pure antioestrogens' could be of a practical value for therapeutics. The clinically available 'pure antioestrogen', fulvestrant (Wakeling et al., 1991), binds to the ER and perturbs the shape of the complex to identify it as a foreign protein in the cell. The complex then becomes the target for ubiquitination and destruction by 26S proteasome. The clinical use of fulvestrant is established for the treatment of MBC, but the medicine is inconvenient for administration at 500 mg IM monthly to postmenopausal patients for extended adjuvant therapy.

The novel mechanism of action of fulvestrant encouraged research for orally active agents (McDonnell et al., 2015), for long-term adjuvant therapy following breast surgery. An early breakthrough was the discovery of GW5638 (McDonnell et al., 2015), a non-steroidal compound with a unique acrylic acid side chain, which is metabolically 4-hydroxylated to GW7604, in the same way tamoxifen is metabolically activated to 4-hydroxytamoxifen and endoxifen (Fig. 10.2). Today, multiple putative orally active SERDs are undergoing clinical trials (Abderrahman and Jordan, 2016). Although GW5638 is classified as a SERD because it destroys the ER complex, the molecule is actually classified as a SERM as well, in the sense that it lowers the cholesterol and maintains bone density in ovariectomized rats. Clearly, new orally active SERDs must be evaluated thoroughly to establish that they are not SERMs as well, like GW5638. Ironically, a compound that destroys the ER in breast cancer, but enhances patients' healthcare by maintaining bone density and gives protection from coronary heart disease, may actually be medically advantageous. These agents could be classified as 'Super-SERDs'.

Aromatase inhibitors

Although compounds (e.g. aminoglutethimide), had been used clinically to treat MBC, by blocking constitutive oestrogen synthesis in postmenopausal women (Santen et al., 2009), these compounds were not specific for the CYP19 aromatase enzyme system in the breast, and also interfered with glucocorticoid synthesis. In the early 1970s, during the development of tamoxifen, the compound 4-hydroxyandrostenedione, was discovered to be a specific and irreversible inhibitor of the CYP19 aromatase enzyme system in breast cancer cells (Jordan and Brodie, 2007). Today, there are three established aromatase inhibitors (AIs), available for the treatment of MBC, or as long-term adjuvant therapy in postmenopausal women (Goss et al., 2016). Anastrozole and letrozole are competitive inhibitors of steroid aromatization at the CYP19 enzyme, whereas, exemestane is a non-competitive inhibitor by irreversibly binding in the active site of CYP19. Precise details of the mechanism of action of these compounds are described by Santen and coworkers (Santen et al., 2009). The overview of worldwide randomized clinical trials, comparing tamoxifen with AIs, demonstrate the superiority of AIs over tamoxifen, to prevent recurrence from breast cancer in postmenopausal patients (Early Breast Cancer Trialists' Collaborative Group et al., 2015). It is noted that there is a reduced incidence of thromboembolic events and endometrial cancer, compared to tamoxifen, when AIs are used as adjuvant therapy in postmenopausal patients. However, AIs increase the risk of osteoporosis. Different AIs have been used successfully for the treatment of IDLE, and in chemoprevention to reduce the incidence of breast cancer in high-risk postmenopausal women.

Acquired resistance to antihormone therapy

Molecular mechanisms and mutations in the oestrogen and androgen receptor

Laboratory studies demonstrate that either tamoxifen or oestrogen deprivation increase the concentration of both the tamoxifen:ER complex, or the unoccupied ER, during AI treatment (Pink and Jordan, 1996). The turnover of the ER complex is increased by oestrogen binding to the ER, through the activation of the ubiquitin proteasome system, to fine tune the stimulatory signal of the ER complex within the cell. By contrast, tamoxifen binding to the ER decreases the turnover of the antioestrogenic ER complex, which accumulates in the cell in both therapeutic situations in breast cancer. As excess of ER production increases the probability for resistance mechanisms developing with the most important signal transduction pathway in cancer (Fig. 10.1). This aids cancer cell survival by trial and error.

Laboratory studies in the 1980s, first identified the mutation (T877A) in the AR of LNCaP prostate cancer cells (Umekita et al., 1996; Chuu et al., 2011), which had a putative role in resistance to antiandrogenic drugs. Subsequent studies with androgen resistant prostate cancer cell lines, and human tumour material, have described an extensive range of mutations (several hundred; Gottlieb et al., 2012, Grasso et al., 2012; Watson et al., 2015), which putatively play a role in the acquired resistance to androgen deprivation therapy.

Early studies, using site-directed mutagenesis in the 1990s, demonstrated the potential role of mutations in the ER, which could alter the actions of tamoxifen, or cause super sensitivity to oestrogen, or impact the un-liganded ER. Little progress occurred to demonstrate the clinical relevance in primary tumours. In the laboratory, a unique mutation was noted in the ER, which was obtained from an experimentally produced acquired resistance to tamoxifen (Wolf and Jordan, 1994). This mutation asp351tyr, was the first to show super-sensitivity to the oestrogen-like effect of tamoxifen, but more importantly, was the first to demonstrate the conversion of the non-oestrogenic SERM, raloxifene, from an antioestrogen to an oestrogen, at an oestrogen-target gene (Levenson and Jordan, 1998). Crystallization of the ER with SERMs showed how, and provided an initial insight into the role of asp351 in the SERM action of both raloxifene (Brzozowski et al., 1997), or 4-hydroxytamoxifen (Shiau et al., 1998). Today, drug resistance with AIs is attributed to amino acid (aa) asp351 tyr.

Multiple studies comparing the mutation frequency in primary tumours, and recurrent breast cancer primarily after AI therapy, have now documented a range of different mutations of low frequency, but with '2 hot spots' at amino acids Y537S and D538G (Fig. 10.4). The data on mutations found clinically, has been integrated with the known experimental molecular pharmacology on the structure-function relationships of SERMs, and how they interact through their antioestrogenic side chain with the pivotal aa asp351 (Jordan et al., 2015). Additionally, the molecular modelling of ER mutations at Y537S and D538G, following the failure of AI therapy, demonstrate that the ligand-binding domain of the ER, is able to close without a ligand through its interaction with asp351, to activate a cascade of oestrogen-like actions in breast cancer cells (Toy et al., 2013). The amino acid, asp 351, in fact, turns out to be the critical anchor aa in the majority of mutations that occur in the ER after AIs (Jordan, 2015). As a result of these findings and

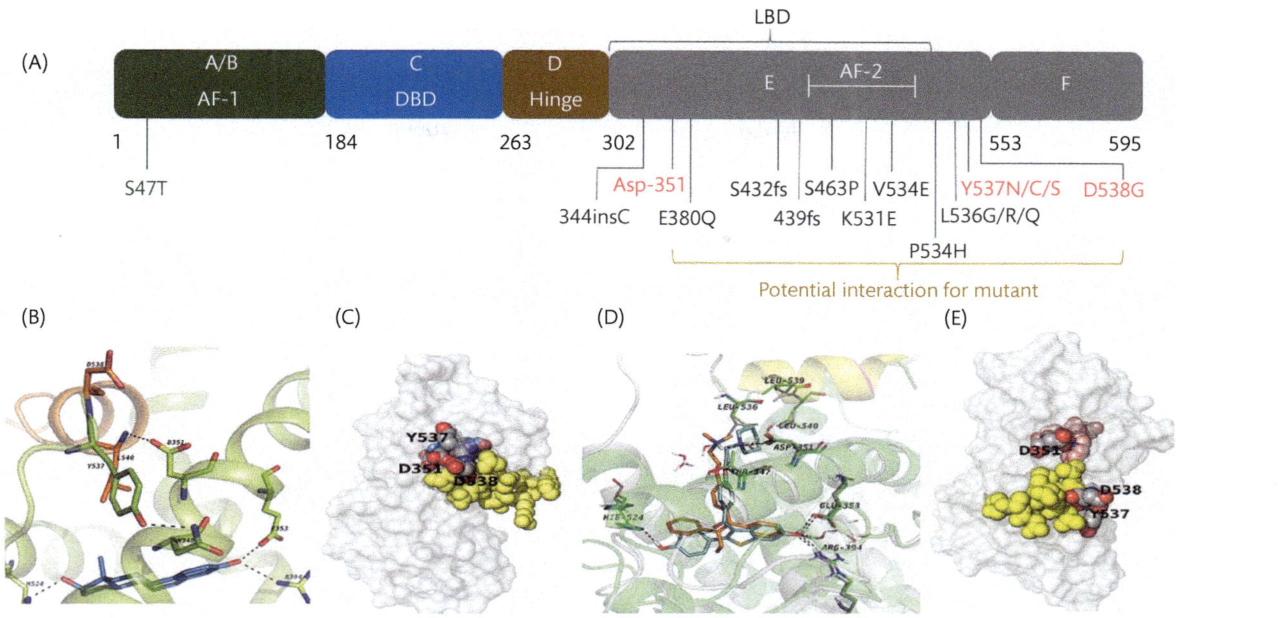

Fig. 10.4 Mutations and molecular interactions of the oestradiol (E_2)–oestrogen receptor (ER) complex. (A) Schematic representation of the wild-type human ER cDNA. The position initially known for the natural single-point mutations such as Asp351Tyr is indicated. The activating function (AF)–2 region and various mutant receptors generated by random chemical or site-directed mutagenesis are shown that either cause loss of AF-2 activity (i.e. 537, 538) or cause an increase in oestrogenic activity if the receptor is unliganded or liganded with an antioestrogen (other mutations). The orange line connecting 537, and 538 to the anchor Asp351 illustrates the current finding of Toy et al. (2013) that D538G interacts and closes the empty ER pocket. The most common and important mutations in the LBD are highlighted in red (B) The interaction of E_2 (blue) in the ligand-binding domain (LBD) with relevant amino acids and the associated amino acids in the vicinity from helix 12 (Brzozowski et al., 1997). (C) A space filled model from the top of the E_2 LBD showing the closed helix 12 (yellow) securing E_2 within. Three amino acids of relevance are indicated Asp(D)351 on the surface of the LBD complex and Tyr(Y)537 and Asp(D)538. (D). The selective ER modulators 4-hydroxytamoxifen (4-OHT) and raloxifene (Ral) secured within the LBD of the ER by the same two amino acids, Glu353 and Arg394, via a phenolic hydroxyl on both 4-OHT and Ral, as noted with the three phenolic hydroxyl on ring A of E_2 (B). (E) A space filler model from the top of the Ral LBD showing helix 12 pushed back (yellow) and the piperidine ring of Ral-neutralizing Asp(D)351. The amino acids Asp(D)538 and Tyr(Y)537 are far away from potential interactions to influence the position of helix 12 closure.

Adapted from Jordan VC et al., 'Estrogen Receptor Mutations Found in Breast Cancer Metastases Integrated With the Molecular Pharmacology of Selective ER Modulators', *Journal of National Cancer Institute*, 2015, Volume 1-7, Issue 6, by permission of Oxford University Press. DOI:10.1093/jnci/djv075,

the need to prevent the overproduction of unoccupied ER, it is reasoned that SERDs would be valuable therapeutic agents to replace AIs and tamoxifen. The promiscuous pool of excess ER would not create acquired resistance through mutational trial and error.

Evolution of acquired resistance in breast and prostate cancer

Antihormone resistance in ER-positive breast cancer segregates into two forms: intrinsic resistance and acquired resistance. Intrinsic resistance is defined as an ER-positive tumour, which is initially unable to respond to antihormone therapy. The ER is usually present in low concentrations in the tumour, but the pivotal role for the ER signal transduction pathway in tumour growth, has now been subverted by growth factor receptors human epidermal growth factor receptor 2 (HER2)/insulin-like growth factor 1 receptor (IGF-1R) and other growth factors at the cell membrane. It is reasoned that the growth factor receptors have a dominant role to play in tumour cell survival, and as a result, agents such as trastuzumab and lapatinib have achieved outstanding success in the control of HER2-positive amplified disease. Experimentally, the ER concentrations can be increased by blocking the HER2 receptor pathway and the converse is true. The current therapeutic goal is to resurrect the ER signal transduction pathway and reintroduce effective antioestrogenic therapies. Much fundamental work on the use of combinations of antihormone therapies, and inhibitors of growth factor receptors, have been completed using athymic models of ER-positive human breast cancer cells, transfected with the HER2 gene (Osborne and Schiff, 2011). However, questions must be asked to decipher the role of acquired resistance to antihormone therapy during adjuvant therapy, and address the question of how tamoxifen, a competitive inhibitor of oestrogen action, has the ability (as does an AI) to decrease mortality after the cessation of therapy.

As stated previously, the finding that continuous tamoxifen administration could produce acquired resistance to tamoxifen within a 1–2-year period, can be viewed as a model that replicates antihormone treatment of MBC. However, acquired resistance with tamoxifen is unique, as the tumours grow because of tamoxifen, not despite of tamoxifen. Tumours do not lose the ER and respond in laboratory models to either tamoxifen or oestrogen to initiate growth. This laboratory finding translated to clinical care with the use of either fulvestrant or an AI to treat MBC with acquired tamoxifen resistance (Howell et al., 2002). Recent laboratory studies have described the molecular mechanisms of cancer cell growth, in response to either tamoxifen or oestradiol (Fan et al., 2014).

If tamoxifen was only a competitive inhibitor of oestrogen action, then tumour recurrences would occur because of a woman's own

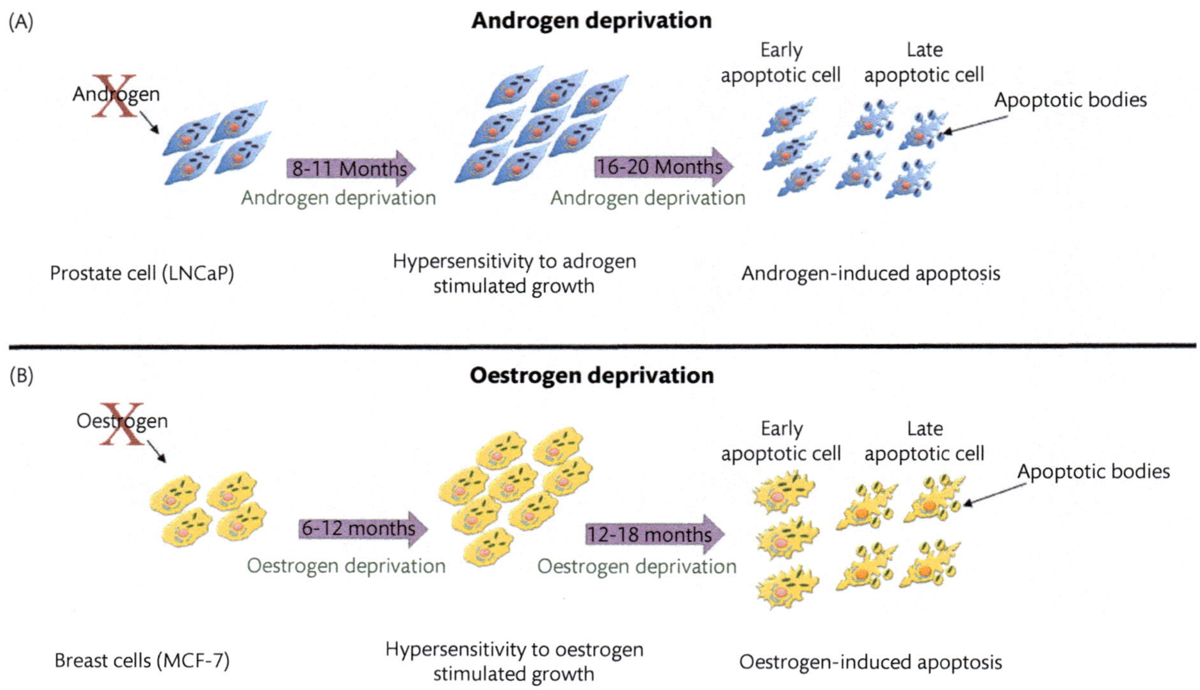

Fig. 10.5 The parallel evolution of hormone deprivation in prostate cancer and breast cancer cells *in vitro*. A: mimics the clinical relapsed androgen-ablation resistant in prostate cancer cells. The androgen sensitive prostate cancer cells (LNCaP) become hypersensitive to androgen when they are cultured in androgen-depleted condition *in vitro* for 8–11 months. After androgen deprivation for a longer period (16–20 months), androgen induces apoptosis. B: represents the oestrogen-deprived breast cancer cells (MCF-7) become hypersensitive to oestrogen when they are cultured in media with lower levels of oestrogen for 6–12 months. After oestrogen deprivation for a longer period (12–18 months), oestrogen starts to induce apoptosis.
Adapted with permission from Jordan VC et al., 'Sex Steroid Induced Apoptosis as a Rational Strategy to Treat Anti-hormone Resistant Breast and Prostate Cancer', *Discovery Medicine*, Volume 21, Number 117, pp. 411–27, Copyright © 2016 Discovery Medicine. All Rights Reserved.

oestrogen that would be stimulating tumour growth. It does not; this is a paradox, and laboratory studies provide intriguing clues.

This new principle is illustrated in the laboratory experiments of the retransplantation of early acquired resistance, into successive generations of tamoxifen treated athymic animals for up to 5 years (Yao et al., 2000). During the first few years of tamoxifen treatment, tumours grow with either tamoxifen treatment or if tamoxifen is stopped, physiological oestrogen will cause tumour growth. Obviously, this is why short-term adjuvant therapy is less effective than long-term adjuvant therapy. In contrast, long-term tamoxifen exposure for 3–5 years is defined by only tamoxifen-stimulated growth, but once the tamoxifen stops, physiologic oestrogen causes small tumours to regress and disappear. This cytotoxic effect of oestrogen would clearly be important clinically, at the end of 5 years or more of adjuvant therapy. Vulnerable micrometastases would be destroyed by a woman's own oestrogen, following the completion of therapy.

A similar phenomenon has been noted with oestrogen deprivation of MCF7 breast cancer cells in *vitro*. Cell populations advance through an initial phase of hypersensitivity to oestrogen following a year of oestrogen deprivation. This hypersensitivity is responsible for cell replication at very low concentrations of oestrogen and may in fact, be an early mechanism of acquired resistance to AIs. Further oestrogen deprivation, results in the selection of a cell population that grows spontaneously without oestrogen, but now physiological oestrogen causes apoptosis. The same evolution of acquired resistance under androgen deprivation occurs in prostate cancer cells under laboratory conditions (Jordan et al., 2016).

Acquired resistance to antiandrogens and androgen deprivation, have been documented in LNCaP cells that are AR-positive (Umekita et al., 1996; Chuu et al., 2011). Androgen deprivation again evolves through a period of androgen hypersensitivity within a year. As with the MCF7 breast cancer cells (**Fig. 10.5**), the LNCaP prostate cancer cells, now require only low concentrations of androgen to replicate. Further androgen deprivation for roughly a year, exposes their vulnerability to androgen-induced apoptosis. Transplantation of LNCaP cells, which are androgen-deprived and resistant prostate cancer cells, demonstrates spontaneous growth and tumour regression with testosterone administration. Testosterone requires conversion to dihydrotestosterone, since tumour regression is blocked by finasteride, an inhibitor of 5α-reductase (Umekita et al., 1996). The clinical importance of this laboratory observation is illustrated in a recent study (Schweizer et al., 2015), demonstrating 50% response rates in prostate cancer patients, who are resistant to antiandrogens, but respond to testosterone.

Sex-steroid induced apoptosis following long-term antihormonal therapy in breast and prostate cancer

The Liao group adapted androgen-deprived cell lines (Umekita et al., 1996), which developed into a broad series of mechanistic studies of androgen-induced apoptosis. Studies by Chuu and coworkers (Chuu et al., 2011), used LNCaP prostate cancer cells, and antiandrogen resistant prostate cancer cells ARCaP, which like oestrogen-deprived

MCF-7 breast cancer cells, are ER rich and have an AR rich phenotype. The authors (Chuu et al., 2011) conclude that androgens produce a G_1 cell cycle blockade by regulating c-myc, Skp2, and $p27^{kip}$ by the AR.

The MCF7 breast cancer cell line, following oestrogen deprivation for several years, responds to physiologic oestrogen with oestrogen-induced apoptosis. There have been several recent reviews on the mechanisms of oestrogen-induced apoptosis (Fan et al., 2015b; Jordan, 2015), but it is important to describe this clinically important and unique biology in cancer cells. The majority of laboratory studies have been completed using MCF-7 cells that do not have a mutant p53. However, the mechanism of apoptosis does not depend on a G_1 blockade with oestrogen and is independent of p53. Breast cancer cells T47D, with a mutant p53 and high PKCα, have been used for laboratory studies. The cells T47D, stably transfected with PKCα, grow spontaneously in athymic mice, but regress rapidly with physiologic oestrogen (Jordan, 2015). Death mechanisms, other than the traditional p53 decision network, require consideration in the new biology of oestrogen-induced apoptosis. To aid our understanding, some of the components of the overall process involving a cellular stress response (the unfolded protein response (UPR)), and a subsequent stress-induced cell death (mitochondria apoptosis) will be described, as each is a player in a much bigger and sophisticated machinery credited with the decision-making of apoptosis over survival.

The oestrogen-ER complex is the key trigger that initiates the series of events to commit the cell to apoptotic death (Fan et al., 2015b; Jordan, 2015). Unlike cytotoxic chemotherapy-induced cell death, which occurs rapidly within 24 hours in AI-resistant breast cancer cells, oestrogen-induced apoptosis requires at least 36 hours of cellular replications and protein synthesis (Fan et al., 2015a), before the AI-resistant cells commit irreversibly to apoptosis. This unusual subcellular biology has been investigated in detail to decipher the precise role of the ligand: ER complex (Jordan, 2015).

The original classification of a natural or synthetic oestrogen used the Allen Doisy ovariectomized mouse vaginal cornification assay. The daily s.c. administration of compounds, either caused a dose-related increase in the proportions of mice having a cornification of vaginal epithelium, or not. It was a yes/no answer based on a triggered oestrogen-mediated response. However, the recent classification of planar or angular oestrogens, using a transforming growth factor alpha (TGFα) gene target, changed our understanding of the gene regulation of oestrogens at a target site (Jordan et al., 2001). This knowledge created an important insight into the early events that occur with the oestrogenic ligand:ER complex. The shape of the oestrogenic ligand is very important for the modulation of oestrogen-induced apoptosis. A planar ligand such as oestradiol fits precisely within the LBD, and coactivators bind to the oestrogen:ER complex to activate oestrogen-responsive genes involved in protein synthesis and cellular replication. By contrast, an angular oestrogen binds to ER to produce a 'pseudo-antioestrogenic ER complex', which retards oestrogen action. Oestradiol commits the vulnerable oestrogen-deprived breast cancer cells, first to several days of cellular replication, but then UPR occurs in the endoplasmic reticulum, which is detected through the activation

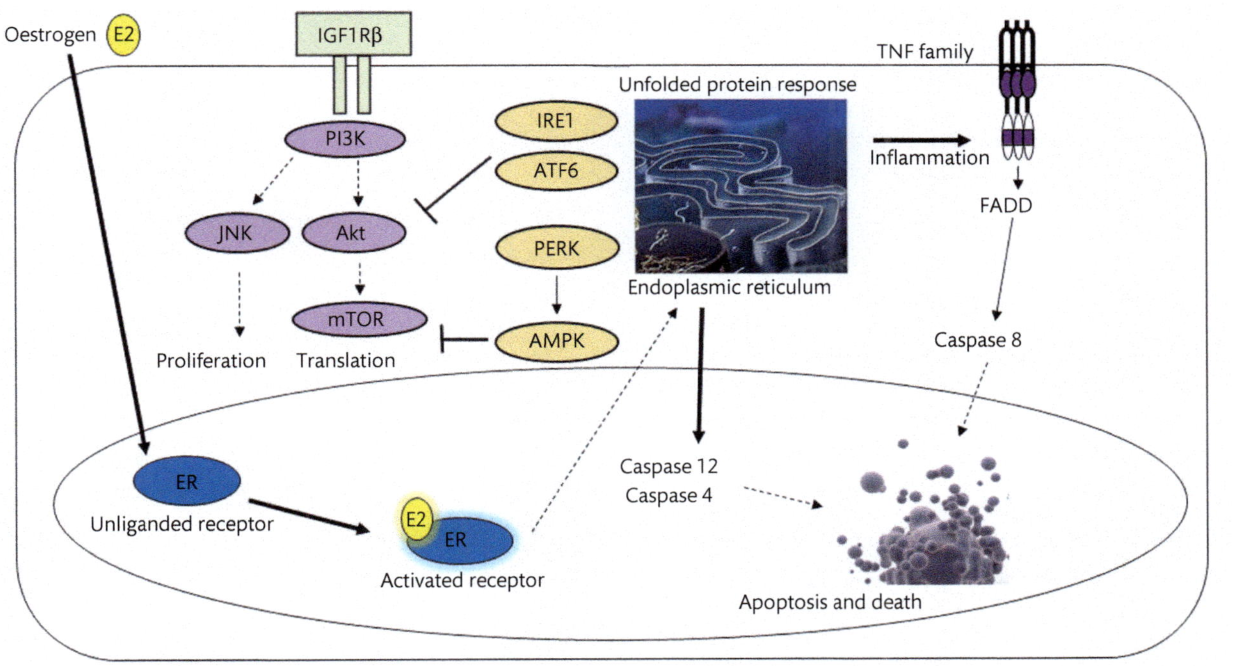

Fig. 10.6 Endoplasmic reticulum is a joint regulatory site to integrally modulate growth or apoptosis associated pathways. In the stages of E_2 triggering apoptosis E_2 activates IGF-1R/PI3K and its downstream signals, Akt and JNK, to promote cell growth. Simultaneously, E_2 activates nuclear ER to cause endoplasmic reticulum stress which activates a set of signalling pathways including three sensors (PERK, IRE1α, and ATF6), inflammatory responses, and adenosine monophosphate (AMP)-activated protein kinase (AMPK). AMPK and Akt converge on mTOR with opposing regulatory effects to coordinate bioenergetics and cell viability. IRE1α and ATF6 are involved in the degradation of Akt. The biological result in long-term oestrogen-deprived (LTED) ER-positive breast cancer cells is initial cell growth followed by events that lead to oestrogen-induced apoptosis.

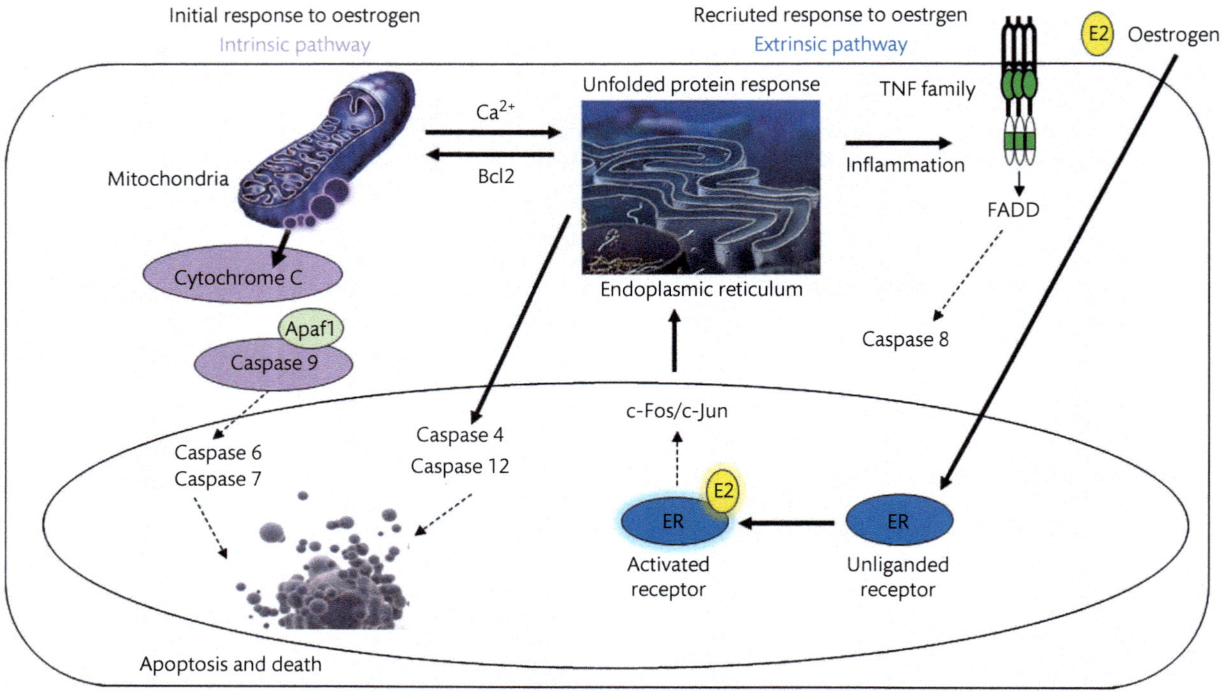

Fig. 10.7 Mechanisms of oestrogen-induced apoptosis. E_2 activates nuclear ER to activate multiple nuclear transcriptional factors including AP-1 family members. c-Fos/AP-1 associates with endoplasmic reticulum to activate unfolded protein response (UPR) with the function to maintain the homeostasis in the endoplasmic reticulum. Overaccumulation of unfolded protein will cause the endoplasmic reticulum stress which activates intrinsic and then extrinsic apoptosis pathways to cause cell death.

of the protein kinase regulated by RNA-like endoplasmic reticulum kinase (PERK) sensor pathway (**Fig. 10.6**). This sensor system triggers apoptosis through the intrinsic pathway that is located in the mitochondria. As a result, numerous caspases are activated, but then the extrinsic pathway, which is regulated through the death receptor (Fas/FasL), is recruited for the complete execution of cells (**Fig. 10.7**).

Numerous angular oestrogens based on the simple synthetic molecule triphenylethylene, have been investigated as oestrogenic triggers for UPR stress and subsequent apoptosis induction (Obiorah et al., 2014). The angular oestrogen has a poor fit in the LBD and the dysfunctional complex binds fewer coactivators. This fact of molecular pharmacology causes a delay in oestrogen-induced apoptosis in AI-resistant breast cancer cells. Taking this model for the activation of the oestrogen-ER complex to the extreme, a bulky antioestrogenic side chain (**Fig. 10.2**) as the hallmark of non-steroidal antioestrogens, will bind in the LBD but helix 12 but cannot close. The strategically placed N atom within the bulky antioestrogenic side chain neutralizes and shields aa asp351 (**Fig. 10.4**), so that the activation of the complex is not possible and the SERM (**Fig. 10.2**) is classified as an antioestrogen in breast cancer tissue (**Fig. 10.3**).

The non-steroidal synthetic, triphenylethylene derivatives, were originally used by Sir Alexander Haddow, as the first successful chemical therapy to treat any cancer. Nevertheless, this use of oestrogen therapy to treat MBC, was only successful when used at least 5 years following menopause (Jordan, 2009). Whereby, the breast cancer cells have evolved in the tumour to an oestrogen-deprived state. Today, this mimics the '5 to 10-year' adjuvant therapy period of oestrogen deprivation with tamoxifen or AIs, which is used currently in clinical practice. This is the shared rule of sex-steroid deprivation in both breast and prostate cancer.

New therapeutic approaches to delay antihormone-resistant breast cancer

The current success of the translational strategy of long-term (5-10 years) adjuvant antioestrogenic therapy for breast cancer, has sparked a multitude of innovations for improving response rates beyond tamoxifen or AIs alone. These approaches can be divided into three major strategies that will be considered based on their priority, and each will be evaluated for therapeutic success.

a. The subversion of acquired resistance to antihormonal therapies. The conventional clinical model to test the therapeutic success of new combination therapies in breast cancer, utilizes the clinical model of MBC that compares an AI, as a standard of care, to a candidate medicine predicted to subvert antihormone resistance mechanisms. Currently, positive results have occurred with either a cyclin-dependent kinase 4/6 (CDK4/6) inhibitor (e.g. palbociclib), or a mammalian target of rapamycin (mTOR) inhibitor (e.g. everolimus), combined with either an AI or fulvestrant. Promising that as it may, to improve adjuvant therapy with 'antioestrogenic' monotherapy, the patient cost is approximately more than \$10,000 dollars per month with palbociclib, and more than \$11,000 dollars per month with everolimus. There is also roughly a 50% incidence of grade 3 and 4 side effects including neutropenia, which makes the implementation of CDK4/6 inhibitors as an adjuvant therapy a challenge. The major factor for success with long-term antioestrogenic adjuvant monotherapy is adherence. Nonetheless, this drops down by 50% by year 5. As a result, the treatments benefits are not harnessed and lives are lost. Clearly, this percentage will be higher with severe side effects from combination therapy.

b. The exploitation of the knowledge of the drug resistance and survival venerability mechanisms seen with long-term adjuvant antioestrogenic therapies including the selection of vulnerable cell populations to oestrogen-induced apoptosis under long-term antioestrogenic adjuvant therapy. The Study of Letrozole Extension (SOLE), has completed patient recruitment to address the question whether five years of intermittent treatment with 3-month annual drug holidays, is superior to continuous therapy. The trial is based upon the idea (Jordan, 2014) that cessation of AI treatment will result in a woman's own oestrogen killing vulnerable populations of occult oestrogen-deprived breast cancer cells. Unfortunately, a 3-month annual drug holiday for 4 years might not kill cancer efficiently and instead oestrogen alone or in combination with other drugs might need to be administered for a few weeks. A proposed alternative to convert the 30% response rate to oestrogen therapy to a response rate that is much higher or a 100%, is the discovery and development of a 'combination cocktail' of inhibitors of cancer cell survival or promoters of cancer growth inhibition or apoptosis. Short-term treatments with the cocktail alone or in combination with oestrogen for a few weeks, at the end of planned adjuvant therapy, should be considered and pursued (Jordan et al., 2016).

c. Discovery and development of 'Super-SERDs' as a new innovative adjuvant therapy. The current plan is to replace AIs and tamoxifen with a new orally active SERD. The primary goal, by the pharmaceutical industry, is to build on the success of the AIs that creates an oestrogen free-state for women with breast cancer. The SERD will destroy the tumour ER and prevent autostimulatory mutations. Progress has been achieved with numerous compounds in clinical trials and recently reviewed (Abderrahman and Jordan, 2016). However, the application of a 'pure antioestrogen' for a decade may be inappropriate. A 'multifunctional medicine' would benefit women's health overall. A compound that has SERM properties in switching on and switching off oestrogen-target sites around a postmenopausal woman's body, to improve cardiac health, bone health and control the growth of occult breast and endometrial cancer (Lerner and Jordan, 1990), and has SERD proprieties in degrading the ER, and therefore, preventing future ER mutations in tumour cells, should be considered. The experimental compound GW5638 accomplishes these tasks in animal studies, but most importantly, destroys the ER. The development of a 'Super-SERD' could control numerous diseases simultaneously while destroying the ER in breast cancer. The innovations would not only extend the responsiveness of antioestrogenic adjuvant therapy in breast cancer by retarding acquired resistance, but also improve women's health during that treatment programme. As a result, adherence will become a desirable commitment by the patient to their future health.

Acknowledgement

This work was supported by the National Institutes of Health NIH MD Anderson's Cancer Center support grant (CA016672), the Susan G. Komen for the Cure Foundation (SAC100009), and the Cancer Prevention Research Institute of Texas (CPRIT) for the STARs and STARs Plus Awards. VC J thanks the benefactors of the Dallas/Ft Worth Living Legend Chair of Cancer Research for their generous support.

TAKE-HOME MESSAGE

Therapeutics which target the ER for the treatment and prevention of breast cancer, has proved to be the most successful treatment in cancer medicine. Millions of patients have benefited by either having their lives saved or extended. Similarly, target therapies for the AR in prostate cancer have improved the prognosis of hundreds of thousands of men. Strategies to continue exploiting the ER or AR target have merit, as these nuclear steroid receptors are conserved during the development of acquired resistance. Nevertheless, the fact that benefit only accrues during long-term adjuvant therapy (5–10 years) or during chemoprevention, with a decrease in ER-positive disease, presents cancer medicine with a challenge. Adherence is essential to extract benefits from treatments and hence save lives. The challenge is to subvert the development of acquired resistance to antihormone therapy with the addition of other targeted treatments to the long-term adjuvant treatment protocol. Current clinical trials demonstrate that both mTOR inhibitors and CDK4/6 inhibitors have potential to treat endocrine resistant MBC, but the high incidence of serious or unpleasant side effects, may not allow their translation to long-term adjuvant therapy.

The development of a SERD that replicates the pharmacology of the pure antioestrogen fulvestrant, might not be the optimal strategy for long-term adjuvant antihormone therapy. What is required, in part, is a SERD that destroys the ER that accumulates in the breast cancer cells during long-term AI or tamoxifen adjuvant therapy. This could avoid one mechanism of acquired resistance (i.e. mutations of the ER). However, it may be a strategic advantage to develop the same compound to have SERM properties for long-term adjuvant therapy that supports women's health, by preventing osteoporosis, and coronary heart disease, while improving the cancer therapeutics by reducing endometrial cancer, contralateral breast cancer, and breast cancer recurrences caused by acquired resistance. This new medicine design that destroys the tumour ER, but is proven to improve women's health at the same time, may be viewed as a 'Super-SERD'. Regrettably, healthcare costs have become prohibitive globally to provide optimal cancer patient care. New solutions to leverage our understanding of the death mechanisms that occur following long-term antihormone therapy (e.g. UPR and sex-steroid induced apoptosis), has the potential to decrease sex-steroid-dependent cancer mortality, both cheaply and effectively worldwide.

OPEN QUESTIONS

- Is it possible to develop new 'Super-SERDs' as long-term adjuvant therapies, thereby improving adherence?
- Can our knowledge of UPR and sex-steroid induced apoptosis in breast and prostate cancer, be deployed to maintain patients disease-free for their natural life, thereby preventing the fracture of the family?

FURTHER READING

Jordan, V. C. (2003). Tamoxifen: a most unlikely pioneering medicine. *Nat Rev Drug Discov*, **2**, 205–13.

Jordan, V. C. (2009). A century of deciphering the control mechanisms of sex steroid action in breast and prostate cancer: the origins of targeted therapy and chemoprevention. *Cancer Res*, **69**, 1243–54.

Jordan, V. C. (2015). The new biology of estrogen-induced apoptosis applied to treat and prevent breast cancer. *Endocr Relat Cancer*, **22**, R1–31.

Mcdonnell, D. P., Wardell, S. E., & Norris, J. D. (2015). Oral selective estrogen receptor downregulators (SERDs), a breakthrough endocrine therapy for breast cancer. *J Med Chem*, **58**, 4883–7.

Santen, R. J., Brodie, H., Simpson, E. R., Siiteri, P. K., & Brodie, A. (2009). History of aromatase: saga of an important biological mediator and therapeutic target. *Endocr Rev*, **30**, 343–75.

Shou, J., Lai, Y., XU, J., & Huang, J. (2016). Prognostic value of FOXA1 in breast cancer: a systematic review and meta-analysis. *Breast*, **27**, 35–43.

Thomas, C. & Gustafsson, J. A. (2011). The different roles of ER subtypes in cancer biology and therapy. *Nat Rev Cancer*, **11**, 597–608.

Toy, W., Shen, Y., Won, H., et al. (2013). ESR1 ligand-binding domain mutations in hormone-resistant breast cancer. *Nat Genet*, **45**, 1439–45.

Watson, P. A., Arora, V. K., & Sawyers, C. L. (2015). Emerging mechanisms of resistance to androgen receptor inhibitors in prostate cancer. *Nat Rev Cancer*, **15**, 701–11.

Yang, Y. A. & Yu, J. (2015). Current perspectives on FOXA1 regulation of androgen receptor signaling and prostate cancer. *Genes Dis*, **2**, 144–51.

REFERENCES

Abderrahman, B. & Jordan, V. C. (2016). Improving long-term adjuvant anti-oestrogenic therapy for breast cancer. *Clinical Pharmacist*, **8**, 6.

Attard, G., Richards, J., & De Bono, J. S. (2011). New strategies in metastatic prostate cancer: targeting the androgen receptor signaling pathway. *Clin Cancer Res*, **17**, 1649–57.

Brzozowski, A. M., Pike, A. C., Dauter, Z., et al. (1997). Molecular basis of agonism and antagonism in the oestrogen receptor. *Nature*, **389**, 753–8.

Chuu, C. P., Kokontis, J. M., Hiipakka, R. A., et al. (2011). Androgens as therapy for androgen receptor-positive castration-resistant prostate cancer. *J Biomed Sci*, **18**, 63.

De Bono, J. S., Logothetis, C. J., Molina, A., et al. (2011). Abiraterone and increased survival in metastatic prostate cancer. *N Engl J Med*, **364**, 1995–2005.

Dijkman, G. A., Fernandez Del Moral, P., Debruyne, F. M., & Janknegt, R. A. (1995). Improved subjective responses to orchiectomy plus nilutamide (anandron) in comparison to orchiectomy plus placebo in metastatic prostate cancer. International Anandron Study Group. *Eur Urol*, **27**, 196–201.

Early Breast Cancer Trialists' Collaborative Group, G., Dowsett, M., Forbes, J. F., et al. (2015). Aromatase inhibitors versus tamoxifen in early breast cancer: patient-level meta-analysis of the randomised trials. *Lancet*, **386**, 1341–52.

EBCTCG (1998). Tamoxifen for early breast cancer: an overview of the randomised trials. Early Breast Cancer Trialists' Collaborative Group. *Lancet*, **351**, 1451–67.

Edwards, J. & Bartlett, J. M. (2005). The androgen receptor and signal-transduction pathways in hormone-refractory prostate cancer. Part 1: modifications to the androgen receptor. *BJU Int*, **95**, 1320–6.

Fan, P., Agboke, F. A., Cunliffe, H. E., Ramos, P., & Jordan, V. C. (2014). A molecular model for the mechanism of acquired tamoxifen resistance in breast cancer. *Eur J Cancer*, **50**, 2866–76.

Fan, P., Cunliffe, H. E., Maximov, P. Y., et al. (2015a). Integration of downstream signals of insulin-like growth factor-1 receptor by endoplasmic reticulum stress for estrogen-induced growth or apoptosis in breast cancer cells. *Mol Cancer Res*, **13**, 1367–76.

Fan, P., Maximov, P. Y., Curpan, R. F., Abderrahman, B., & Jordan, V. C. (2015b). The molecular, cellular and clinical consequences of targeting the estrogen receptor following estrogen deprivation therapy. *Mol Cell Endocrinol*, **418** Pt 3, 245–63.

Goss, P. E., Ingle, J. N., Pritchard, K. I., et al. (2016). Extending aromatase-inhibitor adjuvant therapy to 10 years. *N Engl J Med*, **375**, 209–19.

Gottardis, M. M., Robinson, S. P., Satyaswaroop, P. G., & Jordan, V. C. (1988). Contrasting actions of tamoxifen on endometrial and breast tumor growth in the athymic mouse. *Cancer Res*, **48**, 812–15.

Gottlieb, B., Beitel, L. K., Nadarajah, A., Paliouras, M., & Trifiro, M. (2012). The androgen receptor gene mutations database: 2012 update. *Hum Mutat*, **33**, 887–94.

Grasso, C. S., Wu, Y. M., Robinson, D. R., et al. (2012). The mutational landscape of lethal castration-resistant prostate cancer. *Nature*, **487**, 239–43.

Howell, A., Robertson, J. F., Quaresma Albano, J., et al. (2002). Fulvestrant, formerly ICI 182,780, is as effective as anastrozole in postmenopausal women with advanced breast cancer progressing after prior endocrine treatment. *J Clin Oncol*, **20**, 3396–403.

Huggins, C., Grand, L. C., & Brillantes, F. P. (1961). Mammary cancer induced by a single feeding of polynuclear hydrocarbons, and its suppression. *Nature*, **189**, 204–7.

Jensen, E. V. & Jacobson, H. I. (1962). Basic guides to the mechanism of estrogen action. *Recent Progress in Hormone Research*, **18**, 387–414.

Jordan, V. C. (2003). Tamoxifen: a most unlikely pioneering medicine. *Nat Rev Drug Discov*, **2**, 205–13.

Jordan, V. C. (2007). Chemoprevention of breast cancer with selective oestrogen-receptor modulators. *Nat Rev Cancer*, **7**, 46–53.

Jordan, V. C. (2008). Tamoxifen: catalyst for the change to targeted therapy. *Eur J Cancer*, **44**, 30–8.

Jordan, V. C. (2009). A century of deciphering the control mechanisms of sex steroid action in breast and prostate cancer: the origins of targeted therapy and chemoprevention. *Cancer Res*, **69**, 1243–54.

Jordan, V. C. (2014). Linking estrogen-induced apoptosis with decreases in mortality following long-term adjuvant tamoxifen therapy. *J Natl Cancer Inst*, **106**, dju296.

Jordan, V. C. (2015). The new biology of estrogen-induced apoptosis applied to treat and prevent breast cancer. *Endocr Relat Cancer*, **22**, R1–31.

Jordan, V. C. & Brodie, A. M. (2007). Development and evolution of therapies targeted to the estrogen receptor for the treatment and prevention of breast cancer. *Steroids*, **72**, 7–25.

Jordan, V. C., Curpan, R., & Maximov, P. Y. (2015). Estrogen receptor mutations found in breast cancer metastases integrated with the molecular pharmacology of selective ER modulators. *J Natl Cancer Inst*, **107**, djv075.

Jordan, V. C., Fan, P., Abderrahman, B., et al. (2016). Sex steroid induced apoptosis as a rational strategy to treat anti-hormone resistant breast and prostate cancer. *Discov Med*, **21**, 411–27.

Jordan, V. C. & Robinson, S. P. (1987). Species-specific pharmacology of antiestrogens: role of metabolism. *Fed Proc*, **46**, 1870–4.

Jordan, V. C., Schafer, J. M., Levenson, A. S., et al. (2001). Molecular classification of estrogens. *Cancer Res*, **61**, 6619–23.

Lerner, L. J. & Jordan, V. C. (1990). Development of antiestrogens and their use in breast cancer: eighth Cain memorial award lecture. *Cancer Res*, **50**, 4177–89.

Levenson, A. S. & Jordan, V. C. (1997). MCF-7: the first hormone-responsive breast cancer cell line. *Cancer Res*, **57**, 3071–8.

Levenson, A. S. & Jordan, V. C. (1998). The key to the antiestrogenic mechanism of raloxifene is amino acid 351 (aspartate) in the estrogen receptor. *Cancer Res*, **58**, 1872–5.

Lonard, D. M. & O'Malley, B. W. (2006). The expanding cosmos of nuclear receptor coactivators. *Cell*, **125**, 411–14.

Mcdonnell, D. P., Wardell, S. E., & Norris, J. D. (2015). Oral selective estrogen receptor downregulators (SERDs), a breakthrough endocrine therapy for breast cancer. *J Med Chem*, **58**, 4883–7.

O'Malley, B. W. (2006). Molecular biology. Little molecules with big goals. *Science*, **313**, 1749–50.

Obiorah, I., Sengupta, S., Curpan, R., & Jordan, V. C. (2014). Defining the conformation of the estrogen receptor complex that controls estrogen-induced apoptosis in breast cancer. *Mol Pharmacol*, **85**, 789–99.

Osborne, C. K. & Schiff, R. (2011). Mechanisms of endocrine resistance in breast cancer. *Annu Rev Med*, **62**, 233–47.

Pink, J. J. & Jordan, V. C. (1996). Models of estrogen receptor regulation by estrogens and antiestrogens in breast cancer cell lines. *Cancer Res*, **56**, 2321–30.

Ryan, C. J., Molina, A., & Griffin, T. (2013). Abiraterone in metastatic prostate cancer. *N Engl J Med*, **368**, 1458–9.

Ryan, C. J. & Tindall, D. J. (2011). Androgen receptor rediscovered: the new biology and targeting the androgen receptor therapeutically. *J Clin Oncol*, **29**, 3651–8.

Santen, R. J., Brodie, H., Simpson, E. R., Siiteri, P. K., & Brodie, A. (2009). History of aromatase: saga of an important biological mediator and therapeutic target. *Endocr Rev*, **30**, 343–75.

Schweizer, M. T., Antonarakis, E. S., Wang, H., et al. (2015). Effect of bipolar androgen therapy for asymptomatic men with castration-resistant prostate cancer: results from a pilot clinical study. *Sci Transl Med*, **7**, 269ra2.

Shiau, A. K., Barstad, D., Loria, P. M., et al. (1998). The structural basis of estrogen receptor/coactivator recognition and the antagonism of this interaction by tamoxifen. *Cell*, **95**, 927–37.

Shou, J., Lai, Y., Xu, J., & Huang, J. (2016). Prognostic value of FOXA1 in breast cancer: a systematic review and meta-analysis. *Breast*, **27**, 35–43.

Sledge, G. W., Mamounas, E. P., Hortobagyi, G. N., Burstein, H. J., Goodwin, P. J., & Wolff, A. C. (2014). Past, present, and future challenges in breast cancer treatment. *J Clin Oncol*, **32**, 1979–86.

Thomas, C. & Gustafsson, J. A. (2011). The different roles of ER subtypes in cancer biology and therapy. *Nat Rev Cancer*, **11**, 597–608.

Toy, W., Shen, Y., Won, H., et al. (2013). ESR1 ligand-binding domain mutations in hormone-resistant breast cancer. *Nat Genet*, **45**, 1439–45.

Umekita, Y., Hiipakka, R. A., Kokontis, J. M., & Liao, S. (1996). Human prostate tumor growth in athymic mice: inhibition by androgens and stimulation by finasteride. *Proc Natl Acad Sci U S A*, **93**, 11802–7.

Wakeling, A. E. & Bowler, J. (1987). Steroidal pure antioestrogens. *J Endocrinol*, **112**, R7–10.

Wakeling, A. E., Dukes, M., & Bowler, J. (1991). A potent specific pure antiestrogen with clinical potential. *Cancer Res*, **51**, 3867–73.

Wang, Q., Li, W., Zhang, Y., et al. (2009). Androgen receptor regulates a distinct transcription program in androgen-independent prostate cancer. *Cell*, **138**, 245–56.

Wang, Z. Y. & Yin, L. (2015). Estrogen receptor alpha-36 (ER-alpha36): a new player in human breast cancer. *Mol Cell Endocrinol*, **418** Pt 3, 193–206.

Watson, P. A., Arora, V. K., & Sawyers, C. L. (2015). Emerging mechanisms of resistance to androgen receptor inhibitors in prostate cancer. *Nat Rev Cancer*, **15**, 701–11.

Wolf, D. M. & Jordan, V. C. (1994). The estrogen receptor from a tamoxifen stimulated Mcf-7 tumor variant contains a point mutation in the ligand binding domain. *Breast Cancer Res Treat*, **31**, 129–38.

Wong, Y. N., Ferraldeschi, R., Attard, G., & De Bono, J. (2014). Evolution of androgen receptor targeted therapy for advanced prostate cancer. *Nat Rev Clin Oncol*, **11**, 365–76.

Yang, Y. A. & Yu, J. (2015). Current perspectives on Foxa1 regulation of androgen receptor signaling and prostate cancer. *Genes Dis*, **2**, 144–51.

Yao, K., Lee, E. S., Bentrem, D. J., et al. (2000). Antitumor action of physiological estradiol on tamoxifen-stimulated breast tumors grown in athymic mice. *Clin Cancer Res*, **6**, 2028–36.

11

Oncogenesis and tumour suppression

Mahvash Tavassoli and Francesco Pezzella

Oncogenes and tumour suppressors: A short history

Oncogenes were discovered in 1909 when Peyton Rouse at the Rockefeller Institute discovered a virus, subsequently named the Rouse sarcoma virus (RSV), that causes sarcomas in chickens (Rous, 1911). However, how this virus could cause cancer remained a mystery until 1970 when Stephen Martin at the University of California, Berkeley, isolated a temperature-sensitive mutant of RSV that he demonstrated did not transform cells at the non-permissive temperature although these cells can grow normally, indicating the existence of a viral gene that is necessary for cell transformation but dispensable for replication (Martin, 1970). This observation proved that there was at least one gene that was clearly responsible for eliciting the neoplastic transformation and the genes belonging to this class became known as oncogenes. Later studies revealed that RSV has four genes arrayed along its RNA genome, three of which are responsible for viral reproduction; the fourth is responsible for cellular transformation and is not required for viral replication. This finding led to the discovery that this gene might actually be acquired from a normal cell through an accident of nature. Indeed, that proved to be the case and it led to the hypothesis that during its evolution, the virus that became known as RSV acquired a cellular gene and inserted it into its own genome, creating the viral oncogene SRC.

In 1977 and 1979, genetics and biochemical analysis of the genomes of avian acute leukaemia viruses MC29 and avian erythroblastosis virus led to the discovery of specific sequences unrelated to replicative genes or to the prototypic *src* oncogene (Duesberg et al., 1977; Bister and Duesberg, 1979). These novel viral oncogenes were later shown to be derived from cellular oncogenes, which today are known as major drivers of cancer: *Myc* and the *ERBB/EGFR* gene, respectively. These findings led to a larger hypothesis: if a proto-oncogene can create a viral oncogene why could not the same sort of process occur within the cell without the intervention of a virus. Subsequently, many other retroviral oncogenes were identified and almost in every instance they were found to be acquired from the normal mammal cells. Cellular genes that gave rise to the viral oncogenes became known as proto-oncogenes. Proto-oncogene is referred to the normal wild type form of the gene. Many proto-oncogenes are essential genes involved in fundamental processes in normal cells, like growth, metabolism, or differentiation, and when become affected by genetic abnormalities they can induce cancer. To date, several hundred proto-oncogenes have been implicated in human cancers. The consequence is exaggerated activity of the gene, again a function which is genetically dominant, meaning only one copy of the gene needs to acquire this change to induce its effect. Oncogenes and proto-oncogenes will remain a main focus in biology, biochemistry, and medicine.

In 1965, Henry Harris and John Watkins reported the successful fusion of one human cell with a murine one. This opened up to the possibility to conduct several studies to understand how the cell works but also lead Henry Harris to investigate an issue that dominated the cancer biology community. At the time it was assumed that changes leading to cancer would be dominant (i.e. an oncogenic damage in one allele would be enough to lead toward malignant transformation, despite the presence of a second, normal copy of the gene). Harris however suspected that this may well not be always the case. He calculated that, if this was true, we should observe far more cancers than we actually do. He formulated the hypothesis that there must be also recessive genetic alterations (i.e. one in which if one copy is damaged, the other, normal copy will maintain the cell phenotype as normal rather than malignant). He defined a phenotypically 'malignant' cell as one which, when injected into an animal, would growth into a cancer, while phenotypically 'normal' cells as those that would fail to do so. By fusing normal and malignant cells, he was able to demonstrate that some of these hybrid cells failed to produce the cancer phenotypes of the malignant line used in the fusion. Therefore, there was in the 'normal cell' genetic material that could 'overrun' the malignant phenotype (Harris, 1970). This genetic material was eventually discovered to be the genes we call tumour suppressors.

The oncogenes

The term 'oncogene' was first introduced by R. Huebner and G. Todaro in 1969 (Huebner and Todaro, 1969). In their virogene/oncogene hypothesis they postulated that cancer is the result of spontaneous and/or induced derepression of an endogenous specific viral oncogene(s). According to their hypothesis this oncogene is vertically transmitted in vertebrates as part of the genome in a covert form, the 'provirus', and radiation, chemical carcinogens, and the natural ageing process can activate its expression. Today the term 'oncogene' is one of the most important keywords of human cancer research literature according to PubMed database records (http://www.ncbi.nih.gov).

The term *oncogene* nowadays refers to a gene that encodes for a protein (oncoprotein) that under certain circumstances can transform a normal cell into a cancer cell. An oncogene is mainly derived from a normal cellular gene (the proto-oncogene) which has become abnormally 'activated'. Activation of a proto-oncogene can occur via various mechanisms. Three main forms of abnormalities were discovered; the first is known as gene amplification where the DNA at specific locus on a chromosome replicates many times over, sometimes giving rise to a set of small outside chromosomes, called the double minute chromosomes, which can reinsert themselves into a chromosome to create what is called a homogeneously staining region. This was first observed with *Myc* oncogene in human cancer cells and the consequence is an overproduction of the gene product. The second abnormality involved translocation of the sort first described for the Philadelphia chromosome in CML. Translocation of *Myc* gene was again the first to be seen at molecular level and this translocation alters the control of *Myc* expression and leads to overproduction of Myc product. The third type is single-point mutations which were discovered in another proto-oncogene known as *Ras*. In tumour cells this point mutation can convert the protein from a control state to an uncontrolled state.

Oncogenes encode proteins that control many cellular processes such as driving cellular proliferation, inhibition of programmed cell death (apoptosis), altered metabolism, and angiogenesis. Proteins encoded by oncogenes can be subdivided in different functional classes including transcription factors, chromatin remodellers, growth factors, and growth factor receptors as well as signal transducers and apoptosis regulators. In recent years, non-coding RNAs such as microRNAs and long non-coding RNAs have also been shown to have oncogenic properties.

Transcription factors

Transcription factors are among the most intensively studied oncogenes. Aberrant activity of a transcription factor can result in dysregulated expression of a diverse set of its target genes and this can ultimately contribute to malignant transformation and oncogenesis (e.g. by affecting cellular proliferation and differentiation). Transcription factors are often members of multigene families that share common structural domains. Many transcription factors interact with other partner proteins to become functional. For example, the Fos transcription protein dimerizes with the Jun transcription factor to form the AP1 transcription factor; AP1 complex increases the expression of several genes that control cell division (Shaulian and Karin, 2001). Activation of genes with a role in transcription can occur by chromosomal translocation and is commonly associated with lymphoid cancers, as well as in some solid tumours (Croce, 1987; Tomlins et al., 2005). Genes with a role in transcription can become activated by chromosomal translocation in both leukaemia and solid tumours (Croce, 1987; Tomlins et al., 2005). An example of this is the fusion of EWS gene with one of several partner genes, in Ewing's sarcoma. The EWS protein is an RNA-binding molecule with transcriptional activity, its translocation to other genes results in aberrant transcriptional activity of the fused proteins inducing oncogenic function. ETS family of transcription regulators, is present throughout the body and is involved in a wide variety of functions including the regulation of cellular differentiation, cell cycle control, cell migration, cell proliferation, apoptosis and angiogenesis. The members of the family have been implicated in the development of different tissues as well as cancer progression. Translocation of TMPRSS2 to ERG and ETV1 are common gene fusions found in prostate cancer (Tomlins et al., 2005). TMPRSS2, an androgen-regulated member of the type II transmembrane serine protease family is preferentially expressed in prostate tissues and is regulated by androgen-responsive elements (AREs) in a promoter. Androgen stimulation induces the overexpression of ERG in a TMPRSS2-ERG-positive cancers. (Cerveira et al., 2006).

A prototypic (proto-)oncogenic transcription factor is MYC (v-myc avian myelocytomatosis viral oncogene homologue), also known as c-Myc. Myc is also a very potent oncogene, a transcription factor which binds as a heterodimer with another protein partner Max. When bond to Max, it binds to histone acetyl transferase and acetylate histone and promotes transcription. Myc targets about 15% of the genome, a broad activating gene (Amati et al., 1992; Amati et al., 1993). This transcription factor is normally involved in a wide range of cellular features including ribosome, nucleotide, and mitochondrial biogenesis as well as RNA polymerase activity, RNA processing, RNA stability, and differentiation. As a consequence, MYC plays a crucial role in cell growth and proliferation (Lin et al., 2012). C-Myc (8q24) belongs to the MYC oncogene family which also includes N-MYC (2p24.1) and L-MYC (1p34.3). MYC was originally identified more than 30 years ago as the cellular homologue of the retroviral *v-myc* oncogene of avian acute leukaemia virus MC29 in chicken (myelocytomatosis virus 29) (Kalkat et al., 2017). It was later shown to act as a bona fide oncogene in humans, since chromosomal translocations which juxtapose MYC to immunoglobulin enhancers result in deregulated expression of MYC and contribute to the development of B-cell Burkitt lymphomas. Also, MYC can cooperate with activated *Ras* oncogene to transform normal cells. MYC, however, is not only driving tumour initiation and progression, but is also essential for tumour maintenance. Deregulation of MYC is a frequent event and is detected in more than 50% of all human cancers (Beroukhim et al., 2010). Importantly, the protein-coding region of MYC is rarely mutated; instead, regulatory aberrations of MYC are common which result in unconstrained high levels of gene expression.

Normally, the expression and activity of the MYC protein itself is tightly regulated at the transcriptional and post-transcriptional levels. Transcriptional control of MYC is exerted by a variety of transcription factors such as CNBP and FBP, but also by non-B DNA structures including single-stranded bubbles, G-quadruplexes, and Z-DNA (Michelotti et al., 1995). Post-translational modification of the MYC protein itself by kinases or acetyltransferases and other interacting proteins has been described. The MYC protein is ubiquitinated and degraded, and has a half-life of about 15–20 min (Farrell and Sears, 2014). Furthermore, MYC activity and protein stability can be affected by microRNA and long non-coding RNAs (LncRNA; see Mei and Wu, 2016). In non-transformed cells, the tumour suppressors p19/ARF and TP53 play important roles in the protection against aberrant MYC activity; in transformed cells, MYC activation is often accompanied by TP53 and/or ARF inactivation (Dang et al., 2005).

Chromatin remodellers

Mammalian cells package DNA into higher order structures, commonly referred to as chromatin. Expression of genes is not only based on the general transcription machinery and specific transcription factors, but also largely depend on proteins capable of modifying the chromatin architecture. These unique proteins are called chromatin remodelling proteins or remodellers (Peterson and Workman, 2000). Modifications in the degree of compaction of chromatin play a critical role in the control of gene expression, replication, and repair, and of chromosome segregation. Deregulation of chromatin leads to altered gene activation and/or inappropriate gene silencing. Recent studies suggest that gene translocations leading to fusions of transcription factors promote oncogenesis by altering chromatin structures. Two kinds of enzymes remodel chromatin: ATP-dependent enzymes that move the positions of nucleosomes, the repeating subunits of the histones in chromatin around which DNA winds; and enzymes that modify the N-terminal tails of histones. The pattern of histone modification constitutes an epigenetic code that determines the interaction between nucleosomes and chromatin-associated proteins. These interactions, in turn, determine the structure of chromatin and its transcriptional capacity (Jenuwein and Allis, 2001). In acute lymphocytic leukaemia and acute myelogenous leukaemia, the *ALL1* (also known as *MLL*) gene can fuse with one of more than 50 genes. The majority of these proteins are components of the human transcription complexes TFIID (including TBP), SWI/SNF, NuRD, hSNF2H, and Sin3A. Others are involved in histone methylation and RNA processing (Nakamura et al., 2002). The entire complex remodels, acetylates, deacetylates, and methylates nucleosomes and free histones. The fusion of ALL1 with one of more than 50 proteins results in the formation of the chimeric proteins that underlie acute lymphoblastic leukaemia and acute myelogenous leukaemia. ALL1 (MLL) fusion proteins deregulate homeobox genes which encode transcriptions factors, the *EPHA7* gene which encodes a receptor tyrosine kinase and microRNA sequences such as *miR191*.

Growth factors

Growth factors (GFs) are compact polypeptides, which bind to transmembrane receptors harbouring kinase activity, to stimulate specific combinations of intracellular signalling pathways, such as the mitogen-activated protein kinase (MAPK), the phosphatidylinositol 3-kinase (PI3K), phospholipase C-γ, and transcription factors like the signal transducers and activators of transcription (STATs) or SMAD proteins. These modules of cellular activation and the respective GFs are co-opted in several phases of tumour progression (**Fig. 11.1**).

Constitutive activation of a growth factor gene can contribute to malignant transformation and major regulators of all subsequent steps of tumour progression, namely clonal expansion, invasion across tissue barriers, angiogenesis, and colonization of distant niches. The role of GFs in cancer emerged from studies performed on nerve growth factor in 1950 (Cohen et al., 1954). The link between growth factors and oncogenesis was discovered when Waterfield and colleagues reported in 1983 that the transforming gene of the simian sarcoma virus is structurally related to platelet-derived growth factor (PDGF; Waterfield et al., 1983). Subsequently in 1984, another link between GFs and cancer was discovered; partial sequencing of the EGF-receptor (EGFR/ErbB-1) revealed homology to another oncogene, the erbB gene of the avian erythroblastosis virus (Downward et al., 1984).

The molecular cloning of EGFR by Ullrich and colleagues significantly helped the understanding of the intracellular mechanisms of GF action (Ullrich et al., 1984). Identification of oncogenic activation in receptors uncovered cancer-associated activating mutations that impinge on these pathways to relieve, in part, the reliance of tumours on growth factors. On the other hand, growth factors are frequently involved in evolvement of resistance to therapeutic regimens, which extends the roles for polypeptide factors to very late phases of tumour progression and offers opportunities for cancer therapy (Witsch et al., 2010).

GFs play important roles in various process of cancer formation, metastasis, and resistance to therapy (for a review, see Witsch et al., 2010). Examples include factors involved in growth and clonal expansion (endothelial growth factor (EGF) family), basement membrane breakdown and invasive growth (HGF and FGFs), evasion from apoptosis (IGF-1, as well as several EGF-like ligands), vasculogenesis and angiogenesis (VEGFs, FGFs, and TGF-β), tumour progression (neuregulins and the EGF family, IGF1 and IGF2), intravasation, extravasation, and dissemination (PDGF).

PDGF is a classic GF involved in cancer; it consists of α and β chains and is released from platelets during coagulation (Heldin and Westermark, 1999). It can induce the proliferation of various cell types and stimulate fibroblasts to participate in wound healing. The *sis* oncogene of the simian sarcoma virus is structurally similar to the gene for the β chain of PDGF. Overexpression of PDGF induces the *in vitro* transformation of fibroblasts containing PDGF receptors. An antibody against PDGF-β, or its receptor and small molecules, block the receptor inhibit growth of the transformed fibroblasts.

Another class of growth factors with direct roles in cancer is the WNT family of secreted glycoproteins which inhibit phosphorylation of β-catenin, which is involved in cell–cell adhesion and the activation of several signal-transduction pathways. In familial adenomatous polyposis, inactivating mutations of *APC* block the degradation of β-catenin by inhibiting its phosphorylation. As a result, free β-catenin in the cytoplasm translocates to the nucleus, where it activates genes involved in cell proliferation and invasion (Croce, 2008).

Growth factor receptors

Many growth factors (GFs) bind and activate transmembrane glycoproteins of the receptor tyrosine kinase (RTK) family. All RTKs contain an extracellular ligand-binding domain, a single transmembrane domain, and an intracellular part that contains a tyrosine kinase domain and several regulatory tyrosines, which are modified through auto- or transphosphorylation. Upon binding to their respective receptors, GFs drive the formation of receptor dimers, leading to the activation of the intrinsic tyrosine kinase domain. Subsequent phosphorylation of specific tyrosines enables the recruitment of various signalling adaptors containing Src homology 2 (SH2) and phosphotyrosine-binding (PTB) domains (Katz et al., 2007). Several RTKs have been implicated in the development and progression of neoplastic diseases. Genetic, epigenetic, and somatic changes deregulate the expression of growth factor receptors (GFRs), leading to cancer initiation and progression (Heldin, 1995). A classic example of GFR with important roles in cancer development and drug resistance is the epidermal growth factor receptor (EGFR), a transmembrane protein with tyrosine kinase activity. A deletion of the ligand-binding domain of EGFR causes constitutive activation of the receptor in the absence of the ligand. The activated receptors deregulate signalling in several pathways. EGFR overexpression without

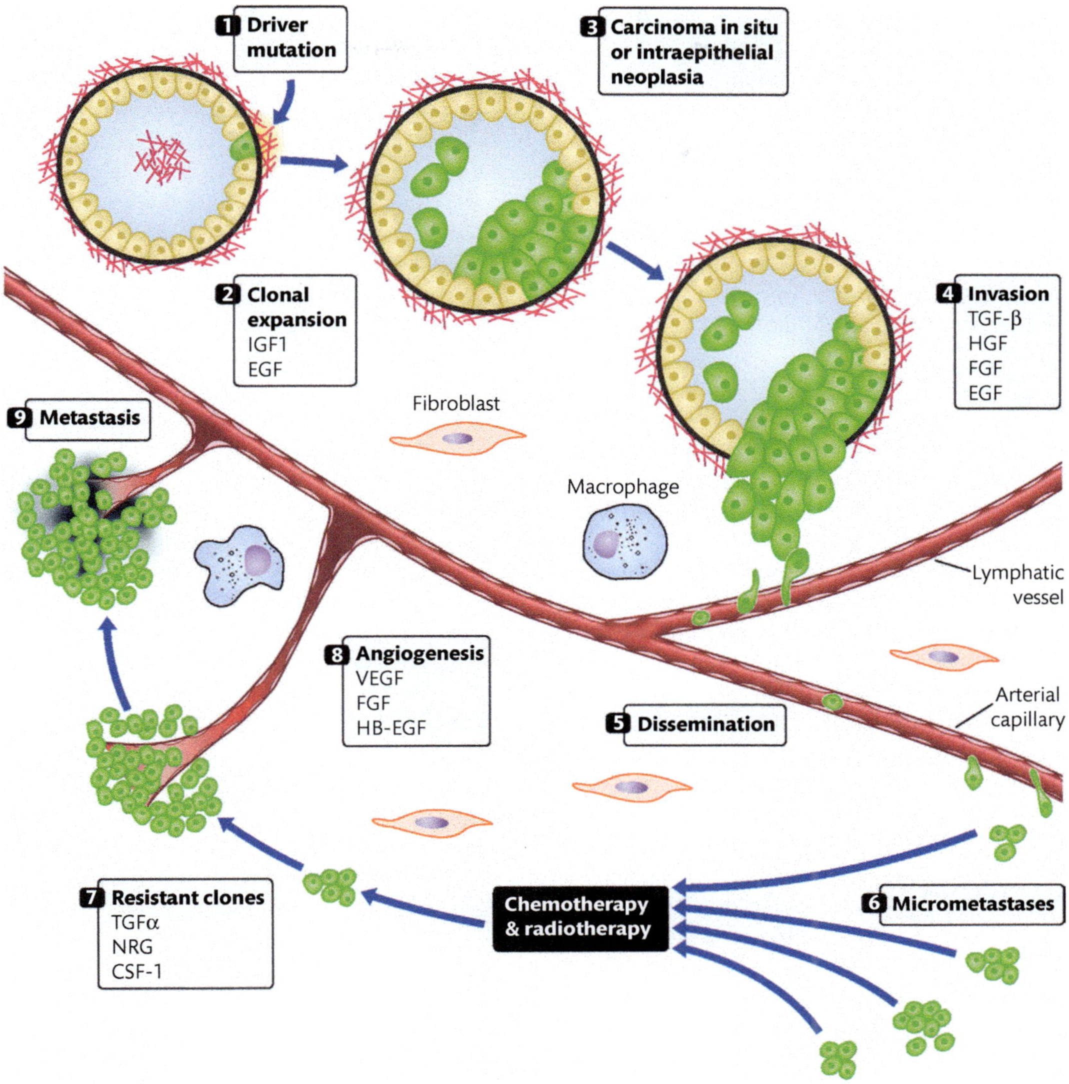

Fig. 11.1 Growth factors (GFs) play roles in different steps of malignant progression. The process is instigated by a somatic mutation, which confers considerable survival and growth advantages to the initiated cell. GFs like EGF and IGF1 support the consequent expansion of mutation-bearing clones, often leading to intraluminal lesions, such as carcinoma *in situ* or intraepithelial neoplasia, which are surrounded by the basal membrane. Invasion refers to the migration and penetration by cancer cells into neighbouring tissues. This process involves loss of epithelial polarity, *acquisition* of a motile, mesenchymal-like phenotype, and secretion of proteases. Both oncogenes and tumour suppressors, along with a large group of GFs, control this critical phase of tumour development. Cancer cells enter (extravasation) and exit (intravasation) lymphatic and blood vessels to disseminate and metastasize to distant organs. Extra- and intravasation entail the supporting functions of macrophages, platelets, and endothelial cells. The resulting micrometastases usually display sensitivity to chemotherapy and radiotherapy. However, the acquisition of new mutations and the ability of cancer cells to produce GFs (autocrine loops) propel the outgrowth of resistant clones. Angiogenesis is essential for the establishment of secondary tumours larger than 1 ml. Both sprouting of existing vessels and recruitment of bone marrow-derived endothelial progenitor cells are stimulated by GFs secreted by tumour and stromal cells. In the final phase, relatively large metastases populate a distinct set of target organs. Note that a latency period of several years may precede this final phase. (CSF-1, colony stimulating factor 1; EGF, epidermal growth factor; FGF, fibroblasts growth factor; HB-EGF, heparin-binding EGF; NRG, neuregulin; TGF, transforming growth factor; VEGF, vascular endothelial growth factor.)

any genetic alteration has also been implicated in many types of solid cancers (Normanno et al., 2006). Activating mutations within the kinase domains of other signalling receptors such as HER2/ neu and KIT has been detected in lung and breast cancer and gastrointestinal stromal tumours. In recent years multiple drugs targeting activated receptors have been developed. There are two classes of clinically active agents: monoclonal antibody against the extracellular domain of the receptor such as cetuximab which targets EGFR, second class is the competitive small molecule inhibitors of the tyrosine kinase activity, such as erlotinib and gefitinib specific to EGFR and afatinib a pan-HER inhibitor (Joensuu et al., 2001; Joensuu and Dimitrijevic, 2001; Marquez-Medina et al., 2015).

Another important growth factor involved in cancers is the vascular endothelial growth factor (VEGF), which stimulates angiogenesis in a variety of cancers. VEGF interacts with three receptor tyrosine kinases: VEGFR1 (FLT1), VEGFR2 (FLK1-KDR), and VEGFR3 (FLT4). Several inhibitors of VEGF and of the VEGFRs have been developed; bevacizumab a monoclonal anti-VEGF antibody, and SU5412, a small molecule, bind and block the activity of the receptor tyrosine kinases of VEGFR1 and VEGFR2 as well PDGF and KIT receptors. Imatinib, a successful drug in the treatment of BCR-Abl positive, chronic myelocytic leukaemia (CML), also inhibits signalling through PDGF and KIT receptor kinases; imatinib has shown to be effective in Gastrointestinal stromal tumours that carry activating mutations of KIT (Tamborini et al., 2004).

Signal transducers

Interaction of receptor tyrosine kinases with their respective ligand induces receptors autophosphorylation on the tyrosine residues within the intracellular part of the molecules. Autphosphorylation of proteins causes changes in their function and/or enzymatic activity resulting in specific biological responses. Phosphorylated receptors interact with multiple cytoplasmic proteins containing domains known as SRC Homology 2 Domain (SH2) and Phospho Tyrosine Binding (PTB) domain. Proteins containing SH2 and PTB are effectors and regulators of intracellular signalling pathways. Over 100 different human proteins downstream of receptor tyrosine kinases contain SH2 domains. Some of these proteins possess enzymatic activity, whereas others function as adapter proteins bridging the activated receptors to downstream signalling proteins (Pawson, 2007; Pawson and Warner, 2007). Many oncogenes encode members of signal-transduction pathways. There are two classes of signalling molecules, non- receptor protein kinases and guanosine- triphosphate– binding proteins (Lahiry et al., 2010). The non-receptor protein kinases include both tyrosine kinases (e.g. SYK, SRC, ABL, LCK) and serine and threonine kinases (e.g. AKT, RAF1, MOS, and PIM1). Mutations in these proteins lead to constitutive kinase activity resulting in the formation of oncogenes, associated with the development of both leukaemia and solid tumours. An important example is PI3K and some of its downstream targets, such as AKT and SGK, which are critical to tyrosine kinase signalling and can be mutated in cancer cells (Yuan and Cantley, 2008). Many of the genes commonly mutated in cancer encode components or targets of the PI3K-Akt and Ras-ERK pathways. Ordinarily these pathways are transiently activated in response to growth factor or cytokine signalling and ligand occupancy of integrin adhesion receptors, but genetic alterations can lead to constitutive signalling even in the absence of growth factors.

The Ras gene is a typical example of an oncogene (Shih et al., 1980), one of the first proto-oncogenes discovered at a single-point mutation in the coding sequence changing a guanosine to thymidine, changing glycine to valine. Glycine is the smallest amino acid and valine is very large, resulting in a change in protein conformation and function, triggering Ras protein to become hyperactive (Cooper, 1982; Fernandez-Medarde and Santos, 2011). Ras is a GTPase, a protein that binds a guanosine nucleotide and hydrolyses GTP to GDP the inactive state. This is helped by GAP, which is a negative activator of Ras and the activation is reinstated by GEF.

In the 1980s it was discovered that 30% of solid tumours show mutations in the Ras gene; some regions are more sensitive, and replacement of glycine with other amino acids was shown to affect Ras function, switching off-Ras with on-Ras. The human genome contains three Ras genes, H-RAS (Harvey sarcoma virus-associated oncogene) on chromosome 11, N-RAS on chromosome 1, and K-RAS (Kirsten sarcoma virus) on chromosome 12. These genes encode for a total of 4 ~21 kDa Ras proteins, since the K-Ras gene encodes two splice isoforms, K-RAS-4B, which is the major isoform, and K-RAS-4A.

The Ras oncogenes were first discovered in the rat genome in 1981 as cause of rat sarcomas, which is also were their name originates from (RAt Sarcomas). Subsequently, they were also detected in the mouse and human genomes. Activating Ras mutations are major drivers of tumour initiation and maintenance and are found in more than 30% of all human cancers. Interestingly, H-Ras is the least frequently mutated Ras isoform in human cancers (4%), whereas K-Ras is the predominantly mutated isoform (85%), followed by N-Ras (11%). Additionally, the Ras protein affected by mutation differs depending on the cancer type. In pancreatic ductal, lung (mostly non-small-cell lung cancer) and colonic carcinoma K-Ras mutations are most common, while bladder and head and neck squamous cell carcinoma mainly show H-Ras mutations; in lymphoid malignancies and cutaneous melanoma, N-Ras mutations are primarily found. These non-random cancer-type-specific mutational profiles of Ras gene isoforms suggest tissue-distinct roles for Ras in driving oncogenesis.

Other oncogenes with multiple function in driving cell division, promoting cell survival, promoting cell motility, promoting invasion/spread are cyclins, cyclin-dependent kinases (CDK), cytoskeletal components, cell adhesion molecules (CAM), matrix metalloproteases (MMP), antiapoptotic proteins (Bcl2 family) telomerase. Some of these have been discussed in other parts of this book.

Apoptosis regulators

Evasion of cell death by apoptosis is one of the hallmarks of cancer. *BCL-2* was the first antideath gene discovered, a milestone with far-reaching implications for tumour biology. Multiple members of the human Bcl-2 family of apoptosis-regulating proteins have been identified, including six antiapoptotic, three structurally similar proapoptotic proteins, and several structurally diverse proapoptotic interacting proteins that operate as upstream agonists or antagonists (Yip and Reed, 2008). Bcl-2-family proteins regulate all major types of cell death, including apoptosis, necrosis, and autophagy, thus operating as nodal points at the convergence of multiple pathways with broad relevance to oncology. The *BCL2* gene, which is involved not only in the initiation of almost all follicular lymphomas and some diffuse large B-cell lymphomas but also in solid tumours (Pezzella et al., 1993), encodes a cytoplasmic protein that localizes to mitochondria and increases cell survival by inhibiting apoptosis (Tsujimoto et al., 1984). The *BCL2* family members BCL-XL and BCL2 inhibit apoptosis and are upregulated in many cancers.

There are two main pathways to programmed cell death, or apoptosis. First is the intrinsic or mitochondrial pathway and is triggered by proteins that contain the BCL2 homology 3 domain (BH3); this domain inactivates BCL2 and BCL-XL (which normally inhibit apoptosis) and thereby activates the caspases that induce apoptosis. Several studies have recently shed light onto the role of pro- and antiapoptotic Bcl2 family members in tumour-pathogenesis and in mediating the effects of classical as well as novel frontline anticancer agents, allowing the development of more efficient and more precisely targeted treatment regimens. Most excitingly, recent progress in our understanding of how Bcl2-like proteins maintain or perturb mitochondrial integrity has finally enabled the development of rational-design based anticancer therapies that directly target Bcl2 regulated events at the level of mitochondria (Frenzel et al., 2009). Drugs that mimic BH3 domain and can bind to BCL-XL or BCL2 (peptides or small organic molecules that bind in a groove of these proteins) are in development (see Chapter 14). This approach has attracted considerable attention because many tumours overexpress BCL2 or related proteins. The second is the death-receptor pathway which is activated by the binding of Fas ligand, TRAIL, and tumour necrosis factor α, to their corresponding (death) receptors on the cell surface. Activation of death receptors activates caspases that cause cell death (Wu, 2009).

The tumour suppressor genes

Genetic and epigenetic basis of tumour suppression

Loss of heterozygosis and the two-hits model

Tumour suppressor genes are known for their roles in inhibiting cell growth and their antitumour effect. According to the Harris model (Harris, 1970; Harris, 1987), growth suppressor genes should be recessive and therefore both copies need to be inactivated in order to induce an effect. This follows the well-established principle that according to the classic genetic laws, in a diploid organism a mutation is 'dominant' if a heterozygous lead to an abnormal phenotype, but 'recessive' if the 'normal' phenotype is maintained (Fig. 11.2). Dominant mutations are usually 'gain of function', while recessive types are 'loss of function' (Berger and Pandolfi, 2011).

In 1971 Knudson described the 'two hits' hypothesis (Knudson, 1971). Knudson investigated a series of 48 paediatric retinoblastomas: 23 bilateral, and 25 unilateral. By combining his data with the general epidemiological data available he showed that the bilateral and/or multiple tumours were likely to be due to an inherited form, in which the child is born carrying already one copy of the affected gene; subsequent damage occurring on the other normal copy is necessary in order to develop a tumour. Instead, patients developing unilateral tumour are more likely to have a sporadic form in which both alterations are somatic and not inherited (Knudson, 1971; Paige, 2003). Molecular biology has confirmed this hypothesis when the retinoblastoma (Rb) gene was discovered 15 years later in 1986 (Friend et al., 1986). Inactivating mutations of the genes, blocking the production of the active protein, were found on both alleles of the genes in retinoblastoma cells, while only one mutated allele was found in the germline of patients with the hereditary form. This process involves two steps: the first is called loss of heterozygosis, that is, when the activity of one of the two allele is

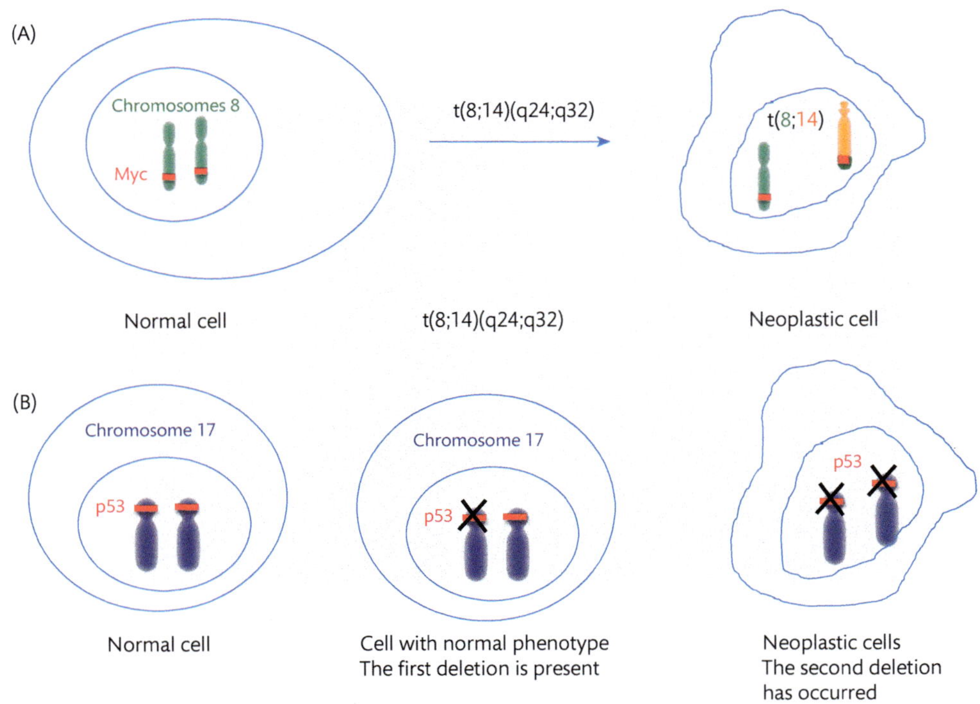

Fig. 11.2 Dominant and recessive changes. (A) An example of dominant (oncogenic) genetic alteration. The oncogene Myc is situated on the chromosome 8, band q24. In the classic t(8; 14) (q24; q32), the commonest observed in Burkitt's lymphoma, the fragment of the long arm of chromosome 8, containing the *Myc* gene, is translocated to the long arm of the chromosome 14 and goes under the influence of the heavy chain immunoglobulin promoter which lead to increases transcription. This event is by itself oncogenic. (B) An example of recessive (tumour suppressor) genetic alteration. The tumour suppressor gene p53 is located on chromosome 17p13.1. After the first deletion has occurred, the cell is still phenotypically normal and behaves like any normal cell. A second deletion, with complete loss of p53 activity, is necessary for the cell to acquire a neoplastic phenotype and behaviour.

lost due to inactivating damage; the second is the inactivation of the remaining allele.

Haploinsufficiency

More recently, inactivation of just one allele of a suppressor gene has been described to cause some degree of suppression and function loss. Such a gene is not considered to behave as classic tumour suppressor gene, as the two-hits model described earlier does not apply and a 'one hit' event can cause some degree of loss of function (Paige, 2003). Several mechanisms have been proposed to explain these 'exceptions' to the 'two hits' model. Haploinsufficiency (single (haplo) insufficient) is a condition that arises when the normal phenotype requires the protein product of both alleles to suppress growth, as one copy of the gene alone is insufficient and reduction of 50% of gene function results in an abnormal phenotype. It has also emerged that some classic tumour suppressor genes can behave in a haploinsufficient fashion in some conditions (Berger et al., 2011; Berger and Pandolfi, 2011). P53 is an example: it can behave in a fashion consistent with the two-hits hypothesis but in some situations the loss of just one allele can cause an abnormal phenotype (Berger and Pandolfi, 2011). Another situation is one described as obligate haploinsufficiency: the PTEN gene is a classic example (Berger and Pandolfi, 2011). While inactivation or loss of just one allele causes increased proliferation (loss of suppression), loss of both alleles triggers senescence and cellular death (complete suppression; Chen et al., 2005).

Genetic alterations causing loss of function

All the genetic alterations are divided into germline (i.e. present in the fertilized egg), or somatic, where they occur in any of cells of the organism. One group of alterations leading to loss of heterozygosis are gross chromosomal abnormalities; these were the first type to be identified (Solomon et al., 1991). They can be summarized into three main types represented in Fig. 11.3: (1) localized, in which limited areas of chromosome are deleted; (2) extensive, in which wide areas of chromosome are lost, these type of damage can be easily seen at cytogenetic level; and, (3) complete loss of one chromosome (Thiagalingam et al., 2002).

More subtle genetic damage occurs at the level of gene sequence and generally manifests as mutations. These include:

Single-base substitution. According to where it occurs it can lead to different outcomes. (a) As some amino acids are coded from alternative codons, a single mutation can change the codon but still the amino acid sequence remains the same. This type is called silent mutation and is not known to cause cancer. (b) Mutation can change the amino acid codon leading to a change in the final protein sequence; the consequences are variable, but it can activate an oncogene or

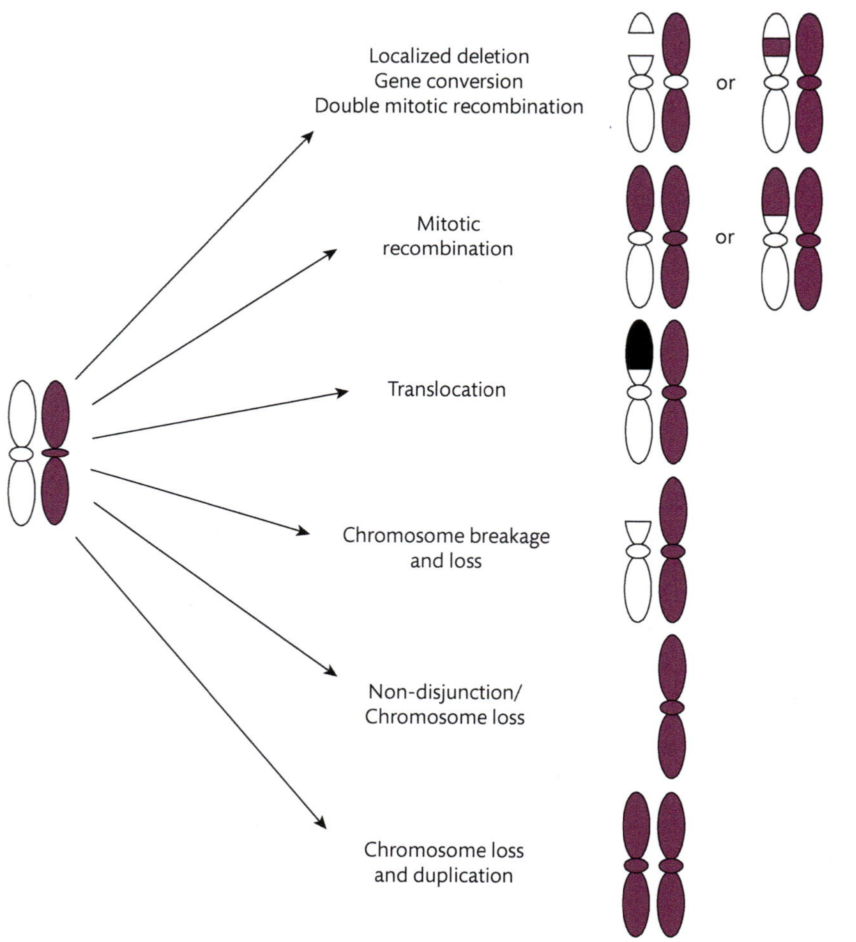

Fig. 11.3 Main types of chromosomal abnormalities detected by cytogenetics.

inactivate a suppressor protein. (c) A coding codon can be changed to a non-coding STOP codon resulting in a truncated protein, or no protein product.

Insertion and deletion. If one or more bases are inserted or deleted from a sequence, the result is a frame shift and, as one amino acid is coded by three bases, this alters all downstream coding or, if as a result of the shift a STOP codon is generated, this alteration will lead to a truncated protein or no protein at all.

Epigenetic alterations causing loss of function

Epigenetic regulation was discussed in Chapter 5. The main epigenetic mechanisms involved in the silencing suppressor genes are hypermethylation, histone modifications, and chromatin remodelling.

Hypermethylation of the gene promoter leads to silencing. Therefore, when it occurs in suppressor gene promoters, it leads to silencing of the transcript with consequent loss of function (Kaufman-Szymczyk et al., 2015).

A second epigenetic mechanism affecting suppressor genes is deacetylation by histone deacetylase (HDAC): acetyl groups are removed from lysine residues. As a result lysine residues, which are positively charged, become available to interact with the negatively charged DNA, and than the histones condensate around the DNA blocking transcription (Kaufman-Szymczyk et al., 2015). Examples of tumour suppressors belonging to this group are *BRCA2, CHD5*, and MEN1. Finally, senescence-associated heterochromatin foci are specialized domains of heterochromatin which are involved in repressing proliferative genes inducing senescence, producing a tumour suppressor effect. Any alteration in the assembly and function of these heterochromatin complexes are therefore likely to lead to loss of tumour suppression activity (Adams, 2007a; Adams, 2007b).

Gatekeeper and caretaker

Suppressor genes have been classified according to their function as gatekeeper (the classic suppressor genes) and caretaker. Gatekeepers regulate the number of cells by inhibiting cell growth and proliferation or promoting cell death (Kinzler and Vogelstein, 1997), while caretaker genes are involved in any activity contributing to maintenance of the integrity of the genome. Their inactivation leads to genetic instability. Only a few gatekeepers are activated in each cell type, like Rb in the retina and VHL in kidney. Loss of function by two hits of a gatekeeper causes tumour growth, while, the two-hits loss of function of a caretaker leads to increased genetic instability and alterations (in gatekeepers, suppressor genes, or in oncogenes). Occurrence of these further damages are necessary in order to develop a neoplasia (Kinzler and Vogelstein, 1997; see Fig. 11.4).

Gatekeepers: Blocking the cell cycle and inducing apoptosis

The cell cycle is tightly regulated, and any alteration in the action of oncogenic and suppressor genes can have severe consequences, including neoplastic growth. The two main suppressor genes whose loss of function deregulate cell cycle are those coding for p53 and Rb proteins. P53 induces arrest in G1 and G2 following DNA damage, while Rb blocks the cell in G1 when the proliferating stimulus is over. Their role in regulating proliferation has been discussed in Chapter 13 on cell cycle control and the role of p53 on apoptosis in Chapter 14 on cell death. Here we discuss the actual alterations leading to their loss of function (Sherr and McCormick, 2002).

The suppressive action of Rb and p53

Even a glance at a basic summary of how Rb and p53 exert their suppressive action (Fig. 11.5) reveals the complexity and high degree of cross-talk with cellular machinery. Rb prevents cell proliferation by forming a complex with E2F transcription factors inducing arrest in G1. If a growth factor activates its receptor, on one side c-Myc is induced, and on the other activation of the Ras MAPK pathway leads to the inhibition of Rb-E2F complex formation, removing the G1 arrest and leaving E2F available to cooperate with c-Myc to induce cellular proliferation. When the growth factor stimulation is withdrawn, levels of cyclin D1 fall, and Rb-E2F complex is formed again and G1 inhibition follows. Rb therefore is pivotal in arresting the cell cycle once the growth factor stimulus is halted. Nuclear levels of p53 are usually low and not involved in regulating growth under physiologic conditions; however, the active form of p53 stabilizes and accumulates following DNA damage or expression of abnormally high levels of p14Arf. As a result, accumulated p53 causes arrest in G1 or G2/

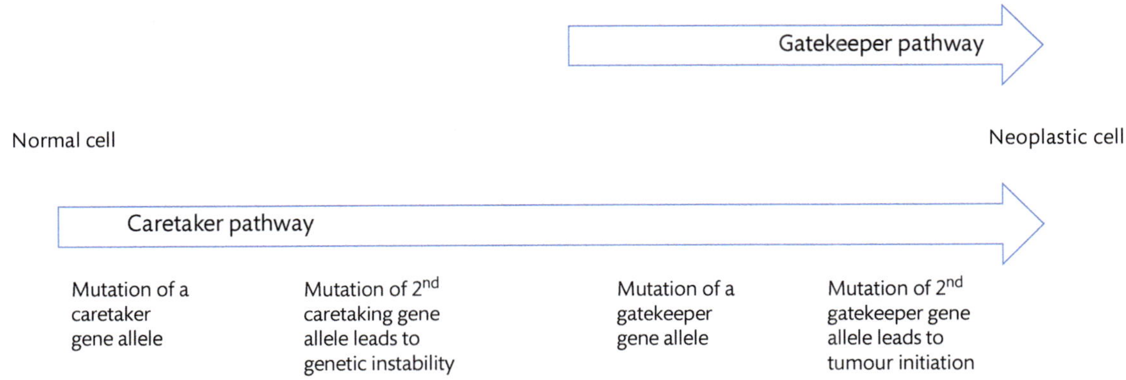

Fig. 11.4 Gatekeepers and caretakers. As two successive hits inactivate a caretaker gene, its function is lost. However, this does not induce a neoplastic transformation, but increases genetic instability causing an increased number of damaged genetic lesions and therefore increases the chance that two further successive hits inactivate a gatekeeper gene, leading eventually to neoplastic transformation.

Adapted by permission from Macmillan Publishers Ltd: Springer Nature, *Nature*, 'Gatekeepers and caretakers', Kenneth W. Kinzler and Bert Vogelstein, Volume 386, Issue 6627, Copyright © 1997 Springer Nature.

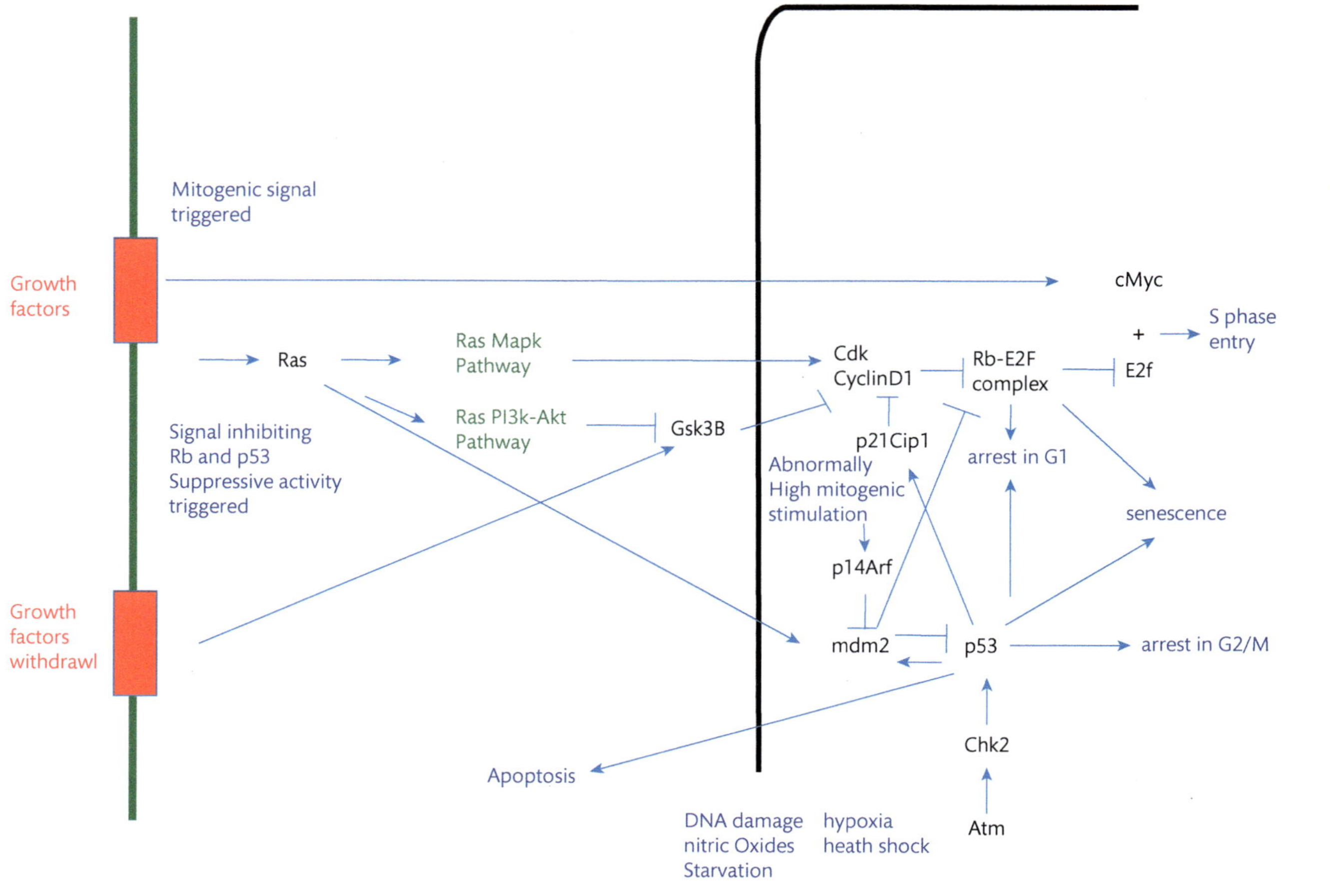

Fig. 11.5 P53 and Rb involvement in cell cycle control. Cyclin-dependent kinases (CDKs) are key regulatory enzymes regulating the progression through the phases of the cell cycle by modulating the activity of key substrates. Cyclin-CDK inhibitors (CKIs), such as p21Cip1, are negative regulators of CDKs. Transcription factor E2F and its regulator Rb are downstream targets of CDKs and Rb provides a negative loop by inhibiting proliferation. Any inactivation of this suppressor gene therefore leads to abnormal cell growth. DNA damage can occur due to either replication mistakes during the mitotic process or external agents (e.g. radiations or nitric oxide accumulation). In this case checkpoint kinase ATM phosphorylates and activates Chk2, which in turn directly phosphorylates and activates the tumour suppressor p53. P53 activation leads to cell cycle arrest and, if repair fails, induction apoptosis of the damaged cell. P53 loss of function therefore leads to growth of cells carrying DNA damage.

Source: data from Sherr CJ and McCormick F, 'The RB and p53 pathways in cancer,' *Cancer Cell*, Volume 2, pp. 103–12, Copyright © 2002 Cell Press; KEGG (Kyoto Encyclopedia of Genes and Genomes), *Cell cycle – Homo sapiens (human)*, Copyright © Kaneshisa Laboratories. Available from http://www.genome.jp/kegg-bin/show_pathway?map=hsa04110&show_description=show; and KEGG (Kyoto Encyclopedia of Genes and Genomes), *p53 signaling pathway – Homo sapiens (human)*, Copyright © Kaneshisa Laboratories. Available from http://www.genome.jp/kegg-bin/show_pathway?hsa04115

M and, if the insult persists, triggers cell death. P53 is also stabilized through p21Cip, indirectly inducing further G1 arrest.

Mechanisms of Rb gene inactivation

As discussed earlier, the inheritance of an inacivated suppressor gene leads to high likelihood of a second damage and therefore of developing a tumour. In the case of Rb which is the archetypal suppressor gene (see Chapter 13), this is the case with retinoblastoma tumours. This is a paediatric malignancy that can present in three forms: familial, when an already inactivated allele is received from one of the parents; sporadic bilateral, when the first hit happens in the parental germ cells or during embryonal development; and finally, sporadic unilateral, when both the first and second hit occurs after birth, the latter two forms being rarer. Furthermore, Rb has been found to also be involved in many other different types of sporadic cancers.

Germline genetic damage in retinoblastoma (Lohmann, 1999) includes large deletion and rearrangement detectable by cytogenetics involving the 13q14.2 locus where the gene is located; this type of genetic damage is found in up to 5% of the patients. Smaller deletions have been detected by Southern blotting technique in a further 10% of cases of familiar or bilateral tumours. Single-base substitution and small length mutation account for the remaining genetic lesions. The single-base substitution observed are nonsense mutations leading to a truncated transcript, so no protein or functional protein can be formed. Missense mutations leading to amino acid substitution occurs mostly in the pocket region, which is crucial to Rb function and is the region where the Human Papilloma Virus E7 oncoprotein binds (Lee et al., 1998). Epigenetic alterations (i.e. hypermethylation of CpG rich islands), are found in approximately 10% of patients (Lohmann, 1999; Singh et al., 2016). Small length mutations, including both deletion and insertions, are present in a minority of cases; the length of the deleted/inserted sequence can range between minus 39 to plus 55 base pairs. According to their location they can produce a termination codon or disrupt splicing. Less common, more complex alterations containing both insertion and deletion have also been described (Lohmann, 1999). Similar types of Rb damage have been reported in almost all types of tumours (Dyson, 2016).

Finally, in many cervical and head and neck carcinomas Rb is inactivated by the human papilloma virus (HPV). When HPV infects the epithelial cells, and links to the pocket region of the Rb protein binds to E7 oncoprotein with consequent inactivation of the protein (Dyson et al., 1993).

Emerging roles for the Rb gene

Two other related genes, *RBL2/p130* and *RBL1/p107*, are part of the Rb family. Recent data are starting to reveal that the Rb family is involved in a broader variety of functions than believed so far. Rb loss has been found to be associated also with increased chromosomal instability and maintenance of heterochromatin in telomeres and centromeres (caretaker functions). All three Rb proteins have been discovered also to be involved in inducing senescence by cross-talk with p53. Rb is also involved in regulation of apoptosis, although the full picture is yet to be understood. Finally, this gene family is also involved in differentiation and angiogenesis (Indovina et al., 2013).

Mechanisms of p53 gene inactivation.

Similar to Rb, P53 is also associated with familial and sporadic tumours and is widely recognized as the most commonly altered gene in human cancer. Familiar tumours due to p53 mutations are grouped together as Li-Fraumeni syndrome in which patients start to develop several types of malignancies at an early age. Li-Fraumeni syndrome was described, as clinical entity, in 1969 (Li and Fraumeni, 1969) but it wasn't until 1990 that the causative role of germline p53 mutations was firmly established (Malkin et al., 1990; Srivastava et al., 1990).

More than 250 different germline mutations have been described, mostly in the exons 5 to 8. Mutation hot spots are present in the region coding for the DNA-binding region of the protein or in the region coding for regulating transcription of target genes (Kamihara et al., 2014; Valdez et al., 2017). Mutations in non-coding regions have also been detected but their significance remains unknown (Valdez et al., 2017). The spectrum of somatic mutations is similar, although within the germline there is a higher incidence of mutations at the mutation-prone CpG residues (Kamihara et al., 2014).

The most common types of mutation, both germline and somatic are (in decreasing order): missense (73%), nonsense (9%), splice site (8%), frame shift (6%), large deletions (0.7%), intronic deletions (0.5%), and others (2%); (Valdez et al., 2017). Larger genetic damages have also been described by cytogenetic as the chromosome 17, where p53 resides at 17p13.1, is affected by a variety of alterations like complete loss of the chromosome, translocations involving p13, deletions of p13, and derivative chromosome 17 (Mitelman, 1991).

Emerging roles for the p53 gene

As our knowledge of cell biology expands, p53 also appears to be involved in a variety of previously unknown functions (Kastenhuber and Lowe, 2017). P53 is involved in the control of DNA methylation, and therefore epigenetic regulation and lack of p53 lead to abnormalities in embryonal stem cells differentiation through altered epigenetic regulation (Laptenko and Prives, 2017). Other cellular functions in which p53 is emerging to be involved are senescence, autophagy responding to starvation and reduction of protein synthesis, modulation of glycolysis, and oxidative phosphorylation. Some of these newly discovered functions could mean that, in some situations, p53 could actually promote cancer (e.g. increasing pentose phosphate pathway activity which can promote alternative pathways promoting growth or, in some other cases, by inhibiting autophagy; see Vousden and Prives, 2009).

Caretakers *BRCA1* (BReast CAncer1) and *BRCA2* (BReast CAncer2): Tumour suppression and genomic instability

Two caretaker genes are *BRCA1* and *BRCA2*. Approximately 10% of women with breast cancer have a family history of this disease; *BRCA1* and *BRCA2* are the genes identified as the responsible factors. As discussed in Chapter 4, 'Genetics and genetic instability in cancer', *BRCA1* acts early on in the repair process by identifying DNA lesions and initiating repair by homologous recombination (HR), while *BRCA2* appears to stabilize stalled replication forks and ensures HR repair by regulating the activity of *RAD51* (Boulton, 2006). The *BRCA1* gene is located on chromosome 17q21.3 and the encoded protein is 1,863 amino acids long (Savage and Harkin, 2015) with a shorter (1,399 amino acid) form called *BRCA1*-IRIS (Foulkes and Shuen, 2013). The protein forms an obligate heterodimer with *BARD1*, which has E3 ubiquitin ligase activity (Wu et al., 2008). *BRCA2* is on chromosome 13q12.3 and encodes a protein made up of 3,418 amino acids. It is prevalently present as a homodimer with apparently no enzymatic activity (Fradet-Turcotte et al., 2016). *BRCA1* has an E3 ubiquitin ligase function while *BRCA2* is part of the homologous recombinant machinery (Boulton, 2006). Both the heterodimer *BRCA1:BARD1* and the homodimer *BRCA2:BRCA2* form supercomplexes according to the specific task (**Fig. 11.6**).

As discussed in Chapter 4 (on genomic instability), *BRCA1* and *2* are essential for double strand breaks (DSBs) repair by homologous DNA recombination (HR). *BRCA1* however is also involved in cell cycle arrest (Wu et al., 2008; Savage and Harkin, 2015). Supercomplexes (b) and (c) (**Fig. 11.6**) are required to activate the ATR/Chk1 pathway which induces intra-S-phase arrest (Savage and Harkin, 2015). Supercomplex (c) also activates Chk1, which then induces cell cycle arrest at the G2/M checkpoint, while supercomplexes (e) promotes G2/M checkpoint arrest in a Chk1 independent way (Wu et al., 2008).

Both *BRCA1* and *BRCA2*, as components of the Fanconi anaemia pathway, are also involved in mitophagy and the clearance of damaged mitochondria, thereby also preserving genome stability indirectly as this function is distinct from the DNA repair activity (Sumpter et al., 2016). They are involved as *BRCA1:FANCS* and *BRCA2:FANCD1* complexes (Sumpter et al., 2016). Their inactivation, as does the inactivation of other components of the Fanconi ANemia Complementation group C (FANCC), leads to inhibition of mitophagy and increased sensitivity to cytokine-mediated cell death (Lord and Ashworth, 2016). Furthermore, there is increased accumulation of mitochondrial ROS and mitochondrial damage (Sumpter et al., 2016).

BRCA1 and *BRCA2* in cancer

As caretaker genes, inactivation of *BRCA1* and/or *BRCA2* does not directly causes cell growth but increases genomic instability, making more likely the chance of successive oncogenic or suppressor inactivating lesions occurring (**Fig. 11.4**). Germline mutations

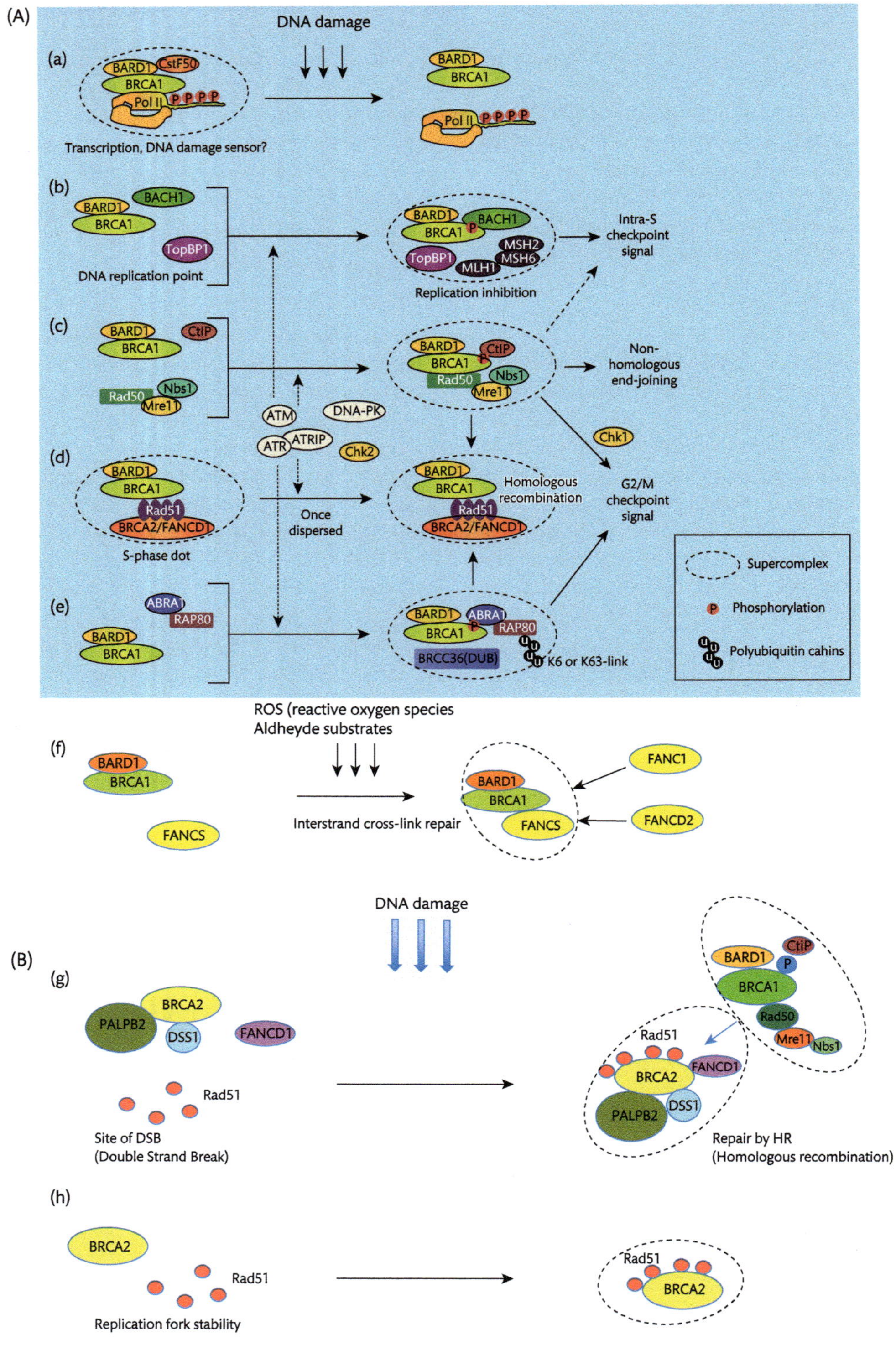

Fig. 11.6 *BRCA1* and *BRCA2* supercomplexes. (A) *BRCA1* supercomplex. *BRCA1* is present as a dimer with *BARD1*. According to the role played at different stages of DNA repair, it is involved in different supercomplexes. (a) These supercomplexes act both as DNA damage sensor and as transcription activator. Following some DNA damage, *BRCA1* dissociates from Pol II. (b) This supercomplex interacts with TopBP1 in an ATM dependent manner after DNA damage and activate the S-phase checkpoint. (c) Formation of this supercomplex depends on ATM- and Chk2-mediated phosphorylation. As far as DNA repair is concerned bridges the two DNA ends of the DSB, an intermediate of both non-homologous end joining (NHEJ) and homologous recombination (HR) It also activates the G2/M checkpoint. (d) After DNA damage this supercomplex localize to damaged DNA sites to provide homologous recombination repair of DSB. (e) This ubiquitinated supercomplex is involved both in DNA repair and G2/M checkpoint activation. (f) This supercomplex is involved in interstrand cross-link repair. (B) *BRCA2* supercomplexes (g) This *BRCA2* supercomplex is joined by the *BRCA1* supercomplex (b) at the site of homologous recombination. (h) *BRCA* and Rad51 localize at replication fork stability site.

(A) Adapted from Wu W et al. 'The ubiquitin E3 ligase activity of *BRCA1* and its biological functions', *Cell Division*, 3:1, Copyright © Wu et al; licensee BioMed Central Ltd. 2008, under the terms of the Creative Commons Attribution License (http://creativecommons.org/licenses/by/2.0), which permits unrestricted use, distribution, and reproduction in any medium, provided the original work is properly cited. Source: data from Fradet-Turcotte et al. '*BRCA2* functions: from DNA repair to replication fork stabilization', *Endocrine-Related Cancer*, Volume 23, Issue 10, Copyright © 2016; and Sumpter et al. 'Fanconi Anemia Proteins Function in Mitophagy and Immunity', *Cell*, Volume 165, Issue 4, pp. 867–81, Copyright © 2016 Elsevier Inc. (B) Source: data from Wu W et al. 'The ubiquitin E3 ligase activity of *BRCA1* and its biological functions,' *Cell Division*, 3:1, Copyright © Wu et al; licensee BioMed Central Ltd. 2008; Fradet-Turcotte et al. '*BRCA2* functions: from DNA repair to replication fork stabilization,' *Endocrine-Related Cancer*, Volume 23, Issue 10, pp. T1-T17, Copyright © 2016; and Boulton SJ, 'Cellular functions of the BRCA tumour-suppressor proteins,' *Biochemical Society Transactions*, Volume 34, Part 5, pp. 633–45, Copyright © 2016.

of both genes are associated with increased risk of breast, ovarian and, less frequently, other cancers types. Somatic mutations are also present in a wide spectrum of malignancies. Impairment of HR is the main consequence of inactivation of *BRCA1* or *BRCA2*. As one of these genes is inactive, *RAD51* fails to locate at the DNA break point (Lord and Ashworth, 2016). *BRCA1* has also transcriptional activities and is involved in chromatin remodelling: its inactivation therefore can lead to tumour formation also through this alternative pathway (Lord and Ashworth, 2016).

However, *BRCA*s also have some gatekeeper activity: deficient cells increase their ability to go through the cell cycle (Rodriguez et al., 2012). Following DNA damage, the activation of check points blocks the cell cycle progression and therefore the proliferation of damaged cells. DNA damage induces ATM-mediated phosphorylation of *BRCA1*, which in turn induces increased levels of p21 and GADDA45, which activate the S and G2/M checkpoints arresting progression through the cell cycle (Yoshida and Miki, 2004). *BRCA2* has been instead demonstrated to be involved, in complex with the PALB2 protein, in inducing arrest at the G2/M checkpoint, following DNA damage. In absence of either *BRCA2* or *PALB2, PLK1* and AURORA A phosphorylation is increased, promoting cell cycle progression as the AURORA A/BORA/PLK1 pathway is no longer inhibited (Menzel et al., 2011).

*BRCA*s and treatment

The unravelling of the *BRCA* pathways has led to a major pharmacological interest in targeted drug development. One of the many families involved in response to DNA damage is that of poly (ADP-ribose) polymerases (PARPs). This family is a key one in single-strand break (SSB) repair. Loss of PARPs activity in a normal cell can be substituted: the unrepaired SSBs are converted to DSBs and the BRCA pathway can proceed to the repair. However, cells that are deficient in *BRCA1* or *2* activity are, as previously discussed, less efficient in repairing DSB, therefore the inactivation of PARPs leads to a massive increases of DNA damage, eventually triggering cell death (Do and Chen, 2013). This concept has been called 'synthetic lethality' and has led to the development of a class of drugs, the PARP inhibitors, which are specifically used to treat tumours with either *BRCA1* or *BRCA2* inactivation (Bryant et al., 2005; Farmer et al., 2005).

Non-coding mRNAs

Less than 2% of the human genome DNA consists of protein-coding genes. This implies that the transcripts are mostly non-coding. Non-coding RNAs are functional RNA molecules that are not translated into protein and represent approximately 80% of the transcribed genome (Inamura, 2017). The main types of non-coding RNAs are summarized in Table 11.1. The two types more relevant to oncogenesis and tumour suppression are the small non-coding molecules microRNA (miRNA), and the long non-coding RNAs (lncRNA).

MicroRNA

MicroRNA (mRNA) is one of the three types of short non-coding RNAs. They are approximately 18–24 bases, most frequently 23 bases, and regulate their target genes. They are evolutionarily highly conserved in eukaryotic organisms to regulate and refine gene expression at the post-transcriptional level (Bartel, 2004). Over 2,500 mature miRNAs have been identified to date (miRBase.org), with each miRNA influencing the expression of hundreds of genes and a single messenger RNA (mRNA) being targeted by multiple miRNAs (Lewis et al., 2005). It is estimated that over one-third of human genes are conserved targets of miRNAs. They are involved in a wide variety of critical biological processes including cell cycle regulation, differentiation, development, metabolism, and death. MiRNAs have been described as micromanagers of protein output within a complex interactive network, fine-tuning the expression of many genes, with alteration of a single miRNA likely to result in subtle phenotypic consequences.

Table 11.1 Non-coding RNAs

Housekeeper RNAs		
rRNA	Ribosomal RNA	Ribosome catalyse protein synthesis from mRNA
tRNA	Transfer RNA	Transport amino acids and allows their inclusion in nascent proteins
Small non-coding RNAs <200 bases		
miRNA	MicroRNA	Sequence imperfectly complementary to those coding for proteins. Link to mRNA blocking transcription and inducing degradation
shRNA	Short hairpin RNA	Sequence perfectly complementary to those coding for proteins
tiRNA	Transcription initiation RNA	Involved in transcription initiation
piRNA or piwiRNA	Piwi RNA	Associate with Piwi proteins in germ cells. Mostly nuclear
snRNAs	Small nuclear spliceosomal RNA	Involved in splicing
snoRNAs	Small nucleolar RNAs	Involved in methylation
hnRNAs	Heterogenous nuclear RNAs	Contains repetitive retrotransposed sequences Involved in network regulation
lncRNA >= 200 bases	**Long non-coding RNAs**	Epigenetic regulation. Involved in development and differentiation
clncRNA	cytoplasmic lncRNA	
nlncRNA	nuclear lncRNA	

Source: data from Springer Nature, *Nature Review Genetics*, 'The rise of regulatory RNA', Morris, KV and Mattick JS, Volume 15, Issue 6, pp. 423–37, DOI:10.1038/nrg3722.

MiRNAs were first discovered in 1993 by Lee et al. (Lee et al., 1993) during their study of the nematode *Caenorhabditis elegans* gene *lin-4*. They found that this gene did not encode a protein but produced a small RNA. This small RNA repressed the expression of its target LIN-14 through base pairing with the 3'-untranslated region (3'-UTR) of the *lin-14* mRNA. However, the significance of this discovery was only realized when the second miRNA, let-7, was characterized in 2000, and subsequently found to be conserved in many species. Since the discovery of miRNAs, there has been huge interest in their role in tumorigenesis, and their potential as biomarkers and possible therapeutic targets.

Biogenesis of miRNAs

Thirty per cent (30%) of miRNA genes are located in intergenic regions, or in their own transcription units, and therefore transcribed as independent units. The rest are located within the introns or exons of non-coding or protein-coding genes (Lagos-Quintana et al., 2001; Rodriguez et al., 2004). These miRNAs are usually transcribed with the host gene, suggesting that they may be regulated together and derive from a common transcript. MiRNAs can be organized as individual genes or located close to each other and transcribed as clusters, with common function and sequence. MiRNAs are transcribed by RNA polymerase II into an initial miRNA precursor called pri-miRNA, which are several kilobases in length and fold into hairpin structures containing imperfectly base-paired stems. The pri-miRNA is processed in the nucleus by the ribonucleases Drosha and DiGeorge syndrome critical region 8 (DGCR8), which forms the microprocessor complex, to become a precursor of about 70–100 nucleotides, called pre-miRNA (Bartel, 2009). Pri-miRNA is exported to the cytoplasm by exportin 5, where it undergoes further processing by another ribonuclease, Dicer, into a mature 18–24 nucleotide double-strand miRNA. Only one strand (the 'guide' strand) is incorporated into a large protein complex called RNA-induced silencing complex (RISC), which is constituted by Argonaute family protein members, while the other strand (the 'passenger' strand or miRNA*) is removed and degraded. The mature miRNA will recognize the complementary sequences in the target mRNA and guide the miRNA-RISC complex to cleave the mRNA or inhibit protein translation (Fig. 11.7).

The main function of miRNAs is to inhibit protein synthesis, either by inhibition of translation or mRNA degradation by base pairing with their mRNA targets. The specificity depends on the degree of base pairing between the nucleotide positions 2 to 8 of the 5'-UTR of the mature miRNA, which is known as the seed region, with the 3'-UTR of the target mRNA.

MiRNAs with the same 2 to 8 nucleotide sequences belong to the same family. In plants most miRNAs base pair to mRNA with nearly perfect complementarity, resulting in mRNA degradation. This is rare in animals, where imperfect miRNA-mRNA base pairing leads to post-translational inhibition, including inhibition of translation

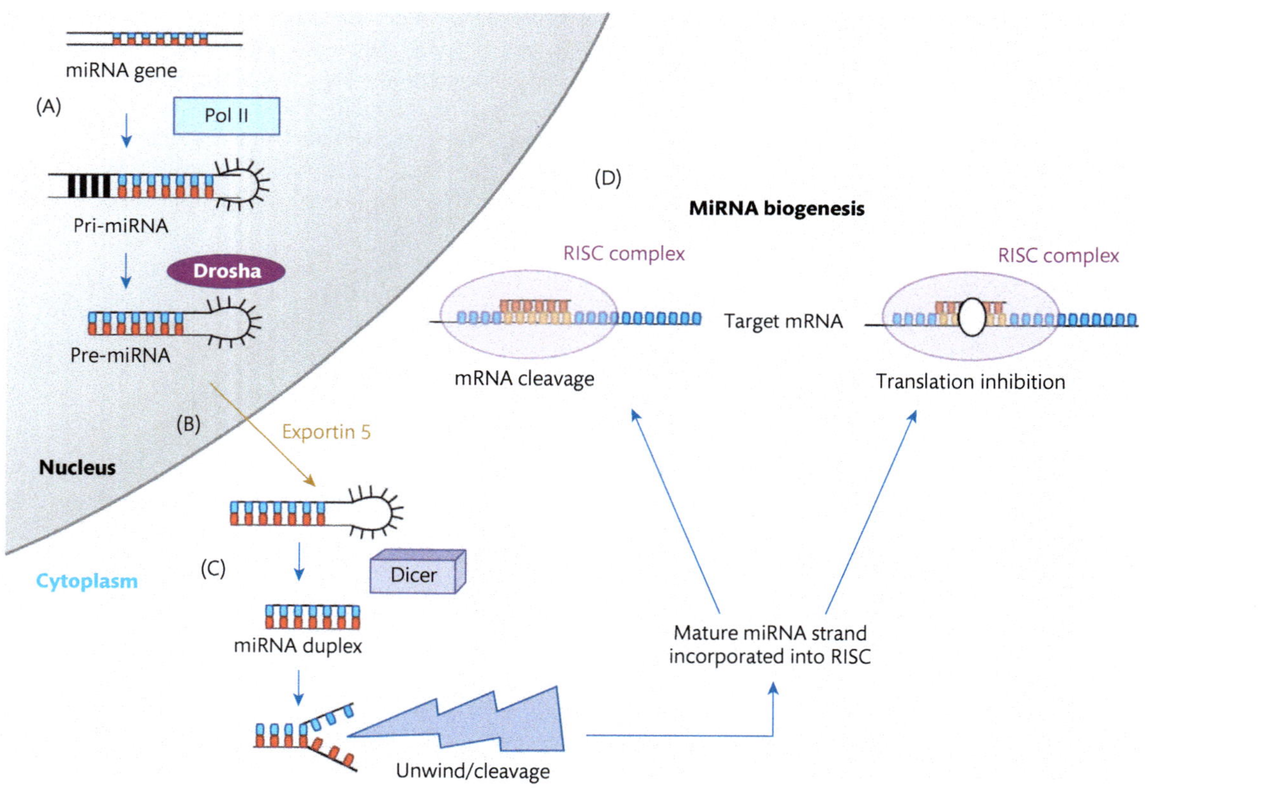

Fig. 11.7 MiRNA biogenesis. (A) MiRNA transcription by RNA polymerase II into pri-miRNA, which is cleaved by the enzyme Drosha to form a hairpin precursor pri-miRNA. (B) Pri-miRNA is exported to the cytoplasm by exportin 5. (C) Pri-miRNA is processed in the cytoplasm by the enzyme Dicer to form a transient miRNA duplex. One strand is incorporated into the protein complex RISC. (D) The mature miRNA leads the RISC to degrade the mRNA or induce translational repression, depending on the degree of complementarity between the miRNA and mRNA target.

Reproduced with permission from Garzon R, Fabbri M, Cimmino A, et al., 'MicroRNA expression and function in cancer', *Trends Mol. Med.* 12: 580–87.

initiation or post-initiation, mRNA destabilization and decay, or co-translational protein degradation. MiRNA biogenesis and function in animals are complex and do not always follow this pathway. They have been reported to target the 5'-UTR and coding regions of target mRNAs, with association of miRNAs with 5'-UTR target sites sometimes resulting in activation, rather than repression of translation. Some miRNAs mature by mRNA splicing with or without exonuclease trimming of strands, independent of the enzymes Drosha and Dicer, giving rise to miRtrons (Berezikov et al., 2007). Furthermore, in addition to gene silencing activity, miRNAs have been reported to also have decoy activity that interferes with the function of regulatory proteins, acting as molecular decoys for RNA-binding proteins (Eiring et al., 2010).

MicroRNAs in cancer

The first report of the role of miRNAs in cancer was published in 2002, when it was demonstrated that in more than 65% of cases of chronic lymphocytic leukaemia, a small region in chromosome 13q14 was deleted. This region did not contain a protein-coding tumour suppressor gene but two microRNAs, miR-15 and miR-16 (Calin et al., 2002). Since then altered miRNA expression has been reported in almost all types of cancer (Lu et al., 2005). MiRNA deregulation can be caused by a variety of mechanisms, including genomic deletion, mutation, or amplification; single nucleotide polymorphisms (Ryan et al., 2010); and epigenetic instability such as promoter hypermethylation and histone deacetylation (Chuang and Jones, 2007). Transcription factors, known to be deregulated in cancer, can also influence the regulation of miRNA transcription. For example, MYC, a well-established oncogenic transcription factor, has been shown to promote the transcription of the miR-17–92 cluster of oncogenic miRNAs. In addition, miRNA expression is under tight regulatory control and alterations to the enzymes in their biogenesis pathway can also affect tumorigenesis (Aguda et al., 2008).

MiRNAs are directly implicated in the tumorigenesis of cancer by acting as oncogenes or tumour suppressor genes. As miRNA's function is to inhibit the expression of a protein, an oncogenic miRNA (i.e. one whose overexpression induces cancer) in one that blocks production of a tumour suppressor protein. A tumour suppressor miRNA (i.e. one those loss lead to cancer) is one blocking the production of an oncogene.

They have been found to be involved in a variety of pathways in cancer development and progression, such as proliferation, apoptosis (le Sage et al., 2007), angiogenesis (Hua et al., 2006), maintenance of the cancer stem cell and metastasis (Takebe et al., 2011). In some cases the same miRNA can act as an oncogene in one cell type and a tumour suppressor gene in another, by acting on different targets or by functioning under differing transcriptional regulation. For example, miR-7 in head and neck cancer cell lines was shown to regulate EGFR expression and AKT activity and therefore had a tumour suppressor function. On the other hand miR-7, along with miR-21, has been associated with keratinization of oral tumours and poor prognosis, suggesting a possible oncogenic role (Jung et al., 2012).

Improved understanding of the role of miRNAs in both physiological and pathological processes has revealed their huge clinical potential, especially as they have been found to be stable even after many years in formalin-fixed paraffin-embedded (FFPE) samples and in body fluids (Chen et al., 2008; Hall et al., 2012). The biogenesis of miRNAs is tightly regulated and very sensitive to endogenous and exogenous stimuli, resulting in differing expression profiles between cell types and cell conditions. Global miRNA gene profiling of normal and tumour tissues has been carried out to assess whether miRNA signatures could be used as clinical biomarkers of diagnosis, classification, prognosis, and treatment response prediction. For example, miRNA expression profiling can differentiate between normal and diseased tissue and identify tissues of origin accurately (Lu et al., 2005). They have also been shown to be able to discriminate different subtypes of a particular cancer, such as basal and luminal breast cancer subtypes (Sempere et al., 2007). MiRNA profiling has revealed patterns of expression associated with disease outcome, such as miR-155 overexpression and let-7a downregulation in lung cancer, which predicts for poor disease outcome (Yanaihara et al., 2006), supporting the potential of miRNAs as prognostic biomarkers.

A more important potential role for miRNA signatures is the ability to predict the response to specific drugs or radiotherapy. For example, high miR-21 expression has been shown to predict poor response to adjuvant chemotherapy in pancreatic and colon cancer (Schetter et al., 2008) and radio resistance in lung cancer (Liu et al., 2013).

MiRNAs represent potential therapeutic targets for the diseases in which they are deregulated. MiRNAs that are overexpressed can be targeted using a class of anti-miRNA antisense oligonucleotides called antagomirs. In tumours where miRNAs are downregulated and function as tumour suppressor genes, miRNAs can be replaced using synthetic miRNAs, which mimic the expression of the protective miRNAs, or genes coding for downregulated miRNAs could be inserted into viral constructs. However, their development has the challenges of delivery and retention, safety, and potential off-target effects (Garzon et al., 2009; 2010). MicroRNAs are increasingly being integrated into clinical trials but mainly looking at their expression signatures as biomarkers for diagnosis, prognosis, or treatment response. However, there are currently two clinical studies, one phase 2 trial of miravirsen, a miR-122 inhibitor for hepatitis C infection (Lanford et al., 2010), and one phase 1 study to assess the safety of a miR-34 mimic in patients with primary liver cancer or liver metastases (Bouchie, 2013).

Long non-coding RNAs (lncRNA)

A lncRNA is an RNA molecule which is, by arbitrary definition, larger than 200 nucleotides, its transcription is mediated by RNA polymerase II and is not translated into protein. They are divided according to intracellular location as cytoplasmic and nuclear (Fatica and Bozzoni, 2014) and classified according to their genomic location relative to protein-coding transcripts: sense, antisense, bidirectional, intronic, and intergenic transcripts (also known as lincRNAs) have been reported (Salehi et al., 2017). LncRNAs are capped, polyadenylated, and spliced similar to protein-coding mRNAs and are regulated by transcription factors. Transcripts are often poorly conserved, unstable, and may be present in only a few copies. LncRNAs exert regulatory functions by interacting with mRNA, DNA, and proteins and can act as transcriptional regulators, molecular scaffolds, miRNA sponges, protein decoys, and reservoirs of small ncRNAs (Fatica and Bozzoni, 2014; Gao and Wei, 2017). The activity of lncRNAs is greatly affected by single nucleotide polymorphisms (Gao and Wei, 2017).

Long non-coding RNAs with oncogenic activity

Several lncRNA have been associated with tumorigenesis in recent years. These lncRNAs include HOX transcript antisense RNA (HOTAIR), metastasis-associated lung adenocarcinoma transcript 1 (MALAT1), also known as NEAT2 and H19. LncRNAs are believed to hold promise as biomarkers for diagnosis and prognosis since their highly tissue- and cell type-specific expression pattern could help in further classifying different subclasses of tumours. Owing to the tissue-specific expression patterns and site-specific action of lncRNAs, drugs targeting lncRNAs might also be more selective than conventional drugs.

HOTAIR is suggested to act as a pro-tumorigenic lncRNA since high expression levels have been linked to higher metastatic potential, poor survival, and cancer stage in a variety of cancers such as breast and colorectal cancer. Additionally, it has been linked to therapy resistance. HOTAIR is a 2.2 kb long intergenic non-coding RNA (lincRNA) which is transcribed from the HOXC cluster in an antisense manner. It acts as a molecular scaffold to link and target two histone modification complexes, PRC2 and LSD1, which consequently cause epigenetic gene silencing and repression of transcription in trans across 40 kb of the HOXD locus on chromosome 2. The effect of HOTAIR on PRC2 activity can be exploited when developing novel therapeutics. Synthetic oligonucleotide antagonists specifically blocking the lncRNA/PRC2 interaction could derepress PRC2 gene targets.

Another lncRNA with oncogenic properties is the 8.7 kb long non-coding transcript MALAT1, which was shown to be highly expressed in non-small cell lung cancer. MALAT1 is a nuclear lncRNA and has been linked to the regulation of alternative mRNA splicing and the modulation of the epigenetic machinery. High expression of MALAT1 has been associated with metastasis and poor prognosis first in lung cancer and later also in other types of cancer, such as liver, breast, and colon.

H19 is a 2.3 kb lncRNA which has been described to exhibit oncogenic and tumour suppressive properties depending on tumour and tissue type. H19 is a maternally imprinted gene and is found in an imprinted region of chromosome 11p15.5 near the insulin-like growth factor 2 (*IGF2*) gene. H19 can exert its oncogenic properties by various mechanisms. First of all, the H19 transcript can act as a decoy for let-7 microRNAs, playing a major role in development, cancer, and metabolism. Additionally, association of H19 with TP53 has been shown to induce (partial) inactivation of TP53. Interestingly, the plasmid BC-819 (DTA-H19) has been developed to make use of the tumour-specific expression of the H19 lncRNA. This plasmid contains the gene for a subunit of diphtheria toxin under the regulation of the H19 promoter. Intratumoural injection results in the reduction of tumour size in human trials in a broad range of carcinomas due to the induction the expression of high levels of diphtheria toxin as a result of high expression of H19 lncRNA, specifically in the tumour. This therapeutic strategy highlights one of the benefits of lncRNA-based approaches. The highly specific expression of some lncRNAs in tumour cells can allow the specific targeting of the therapeutic agent and can thereby help to reduce the risk of affecting normal tissues during treatment (Peng et al., 2017; Weidle et al., 2017).

Long non-coding RNAs with suppressor activity

Some lncRNAs are instead physiologically involved in maintenance of tumour suppression and therefore, their impairment promotes neoplastic growth (Huarte and Rinn, 2010). One of these lncRNAs is the intergenic lncRNA-p21 whose transcription is induced by p53. This transcript represses a variety of genes which antagonize the repressor activity of p53, inactivity of lncRNA-p21 therefore reduces the ability of p53 to induce apoptosis (Huarte and Rinn, 2010). LncRNA GAS5 (Growth Arrest Specific-5) is normally induced by starvation: its role is to antagonize glucocorticoid ligation to DNA and therefore reduce metabolism. Loss of GAS5 transcription has been found in cancer cells, perhaps allowing the cell to growth under starvation (Huarte and Rinn, 2010; Kunej et al., 2014).

LncCCND1, in a negative regulator loop, is induced by CyclinD1, then binds the FUS (fused in sarcoma protein); the LncCCND1:FUS complex then inhibits CyclinD1, slowing proliferation (Kunej et al., 2014).

Genes and mRNAs that can have oncogenic and suppressor activity

Increasingly, in recent years, genes have been described that can, according to the context, act either as oncogenes or as suppressor genes. One such example is p27Kip1 (**Fig. 11.8**), a member of the cyclin-dependent kinase inhibitor (CDKI) family (Lloyd et al., 1999). P27Kip1 blocks a substrate interaction domain on cyclins and prevent ATP linkage on CDK kinases exercising an antiproliferative activity (Besson et al., 2007).

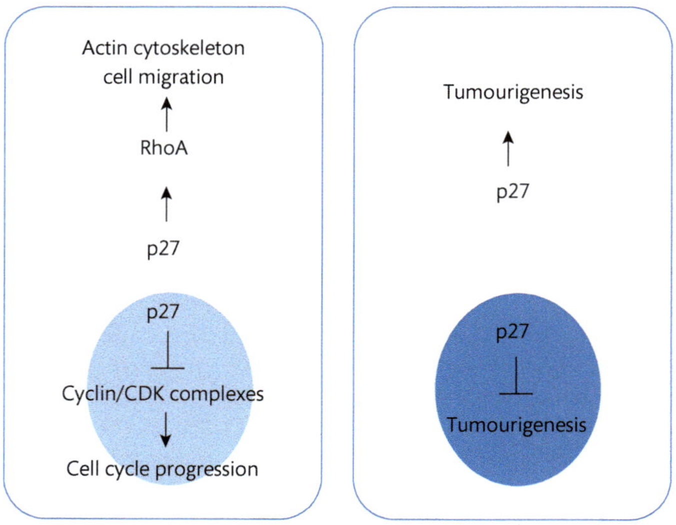

Fig. 11.8 Genes and miRNA with double activity: oncogenic and suppressive. p27Kip1 has suppressive and oncogenic activities. In the nucleus, p27 inhibits cell proliferation and therefore acts a suppressor of tumorigenesis. In the cytoplasm it promotes cell migration and therefore is oncogenic.

Tumours with lower p27 expression are more aggressive. p27Kip1 germline-inactivating mutations have been found in families affected by multiple endocrine neoplasia, while p27Kip1 mice develop pituitary adenomas and organomegaly (Sicinski et al., 2007). However, several factors were not consistent with its suppressive nature. The first was that p27Kip1+/− mice had an excess of mammary and prostate tumours compared to the mice p27Kip1−/− raising the suspect of an oncogenic activity of the residual allele in p27Kip1+/− (Muraoka et al., 2002; Sicinski et al., 2007). The second was the observation that patients with high levels of p27Kip1 in the cytoplasm of the neoplastic cells, rather than in the nucleus, have a poorer outcome (Sicinski et al., 2007; Svoronos et al., 2016).

Further studies have demonstrated that p27Kip1 has also other functions which are exerted in the cytoplasm, the two most important being regulation of actin cytoskeleton and promotion of cell motility (Besson et al., 2007). In a murine model it has been demonstrated that indeed p27Kip1 has an CDK-independent oncogenic activity in the cytoplasm. Mice expressing a p27Kip1 protein in which the CDK regulating activity is blocked, but the cytoplasmic function is preserved, have an incidence of a variety of tumours higher of both wild type animals and p27Kip1 knocked down (Besson et al., 2007).

Another example is the Notch gene and its signalling pathway, which has been implicated in the regulation of diverse functions in the hematopoietic system and other tissues. Notch is a binary cell-fate determinant, and its hyperactivation has been implicated as oncogenic in several cancers including breast cancer and T-cell acute lymphoblastic leukaemia (T-ALL). Recently, several studies also unravelled tumour suppressor roles for Notch signalling in different tissues, including tissues where it was before recognized as an oncogene in specific lineages (Nicolas et al., 2003). The Notch signalling pathway is a regulator of self-renewal and differentiation in several tissues and cell types. There is growing evidence that components of the same oncogenic pathway in lymphocytes may act as growth suppressors in myeloid as well as in epithelial malignancies such as head and neck cancers (Agrawal et al., 2011). It therefore appears that Notch signalling oncogenic or tumour suppressor abilities are highly context dependent (Lobry et al., 2014).

A similar finding is emerging also as far as miRNA is concerned. Several miRNA have been found to exert either effects according to the situation. The basic reason for this double behaviour is in the fact that each miRNA, those length his between 18 and 22 nucleotides, can target several different mRNAs (Svoronos et al., 2016). An example is miRNA-125b: predominantly oncogenic in haematological malignancies, is usually as suppressor in solid tumours (Shaham et al., 2012, Sun et al., 2013). This is due to the different sets of targets present in different tumours: miRNA-152b can silence many different genes, some oncogenic, some suppressors. So, it is likely that haematopoietic tumours express mostly suppressor genes targeted by miRNA152: their silencing has therefore an oncogenic effect. In solid tumours instead, most of the target genes are likely to be oncogenic and their silencing produces a net suppressor effect (Svoronos et al., 2016).

TAKE-HOME MESSAGE

- Oncogenes are translated into oncoprotein and when their function is abnormally increased they induce neoplastic transformation.
- Just one copy of an oncogene needs to be abnormal: dominant behaviour
- Tumour suppressor genes inhibit cell growth and have antitumour effects.
- According to the classic 'two hit' model, growth suppressor genes are recessive and therefore inactivation of both copies is required to induce an effect.
- It is now known that this is not the case: for some genes with suppressor activity, it is enough to have a partially diminished activity to produce pathological effects
- Non-coding mRNA can be oncogenic or suppressive.
- It is increasingly impossible to classify some genes and non-coding RNAs as 'oncogene' or 'tumour suppressor' as, according to the situation, they can behave in both ways!

OPEN QUESTIONS

- The full role of many genes we once thought were well characterized, is actually far from being completely understood.
- Increasingly non-coding RNAs are emerging as key players in oncogenesis but very little is known about their actual function.
- Epigenetic code, alongside genetic, is certainly very important in causing alterations leading to cancer. Much still needs to be learnt.
- More and more data are being collected about abnormalities in cancer cells; and mathematical models will be necessary to make sense of it all.

FURTHER READING

Carey, N. (2012). *The Epigenetics Revolution: How Modern Biology is Rewriting Our Understanding of Genetics, Disease and Inheritance.* London: Icon Books.

Harris, H. (1987). *The Balance of Improbabilities.* Oxford: Oxford University Press.

International Agency for Research on Cancer (IARC). IARC TP53 database. (Release R18, April 2016). Available at: http://p53.iarc.fr

Skloot, R. (2010). *The Immortal Life of Henrietta Lacks.* London: Pan Books.

Weinberg, R. A. (1997). *Racing to the Beginning of the Road: The Search for the Origin of Cancer.* London: Bantam Press.

REFERENCES

Adams, P. D. (2007a). Remodeling chromatin for senescence. *Aging Cell*, **6**, 425–7.

Adams, P. D. (2007b). Remodeling of chromatin structure in senescent cells and its potential impact on tumor suppression and aging. *Gene*, **397**, 84–93.

Agrawal, N., Frederick, M. J., Pickering, C. R., et al. (2011). Exome sequencing of head and neck squamous cell carcinoma reveals inactivating mutations in NOTCH1. *Science*, **333**, 1154–7.

Aguda, B. D., Kim, Y., Piper-Hunter, M. G., Friedman, A., & Marsh, C. B. (2008). MicroRNA regulation of a cancer network: consequences of the feedback loops involving miR-17–92, E2F, and Myc. *Proc Natl Acad Sci U S A*, **105**, 19678–83.

Amati, B., Brooks, M. W., Levy, N., Littlewood, T. D., Evan, G. I., & Land, H. (1993). Oncogenic activity of the c-Myc protein requires dimerization with Max. *Cell*, **72**, 233–45.

Amati, B., Dalton, S., Brooks, M. W., Littlewood, T. D., Evan, G. I., & Land, H. (1992). Transcriptional activation by the human c-Myc oncoprotein in yeast requires interaction with Max. *Nature*, **359**, 423–6.

Bartel, D. P. (2004). MicroRNAs: genomics, biogenesis, mechanism, and function. *Cell*, **116**, 281–97.

Bartel, D. P. (2009). MicroRNAs: target recognition and regulatory functions. *Cell*, **136**, 215–33.

Berezikov, E., Chung, W. J., Willis, J., Cuppen, E., & Lai, E. C. (2007). Mammalian mirtron genes. *Mol Cell*, **28**, 328–36.

Berger, A. H., Knudson, A. G., & Pandolfi, P. P. (2011). A continuum model for tumour suppression. *Nature*, **476**, 163–9.

Berger, A. H. & Pandolfi, P. P. (2011). Haplo-insufficiency: a driving force in cancer. *J Pathol*, **223**, 137–46.

Beroukhim, R., Mermel, C. H., Porter, D., et al. (2010). The landscape of somatic copy-number alteration across human cancers. *Nature*, **463**, 899–905.

Besson, A., Hwang, H. C., Cicero, S., et al. (2007). Discovery of an oncogenic activity in p27Kip1 that causes stem cell expansion and a multiple tumor phenotype. *Genes Dev*, **21**, 1731–46.

Bister, K. & Duesberg, P. H. (1979). Structure and specific sequences of avian erythroblastosis virus RNA: evidence for multiple classes of transforming genes among avian tumor viruses. *Proc Natl Acad Sci U S A*, **76**, 5023–7.

Bouchie, A. (2013). First microRNA mimic enters clinic. *Nat Biotechnol*, **31**, 577.

Boulton, S. J. (2006). Cellular functions of the BRCA tumour-suppressor proteins. *Biochem Soc Trans*, **34**, 633–45.

Bryant, H. E., Schultz, N., Thomas, H. D., et al. (2005). Specific killing of *BRCA2*-deficient tumours with inhibitors of poly(ADP-ribose) polymerase. *Nature*, **434**, 913–17.

Calin, G. A., Dumitru, C. D., Shimizu, M., et al. (2002). Frequent deletions and down-regulation of micro-RNA genes miR15 and miR16 at 13q14 in chronic lymphocytic leukemia. *Proc Natl Acad Sci U S A*, **99**, 15524–9.

Cerveira, N., Ribeiro, F. R., Peixoto, A., et al. (2006). TMPRSS2-ERG gene fusion causing ERG overexpression precedes chromosome copy number changes in prostate carcinomas and paired HGPIN lesions. *Neoplasia*, **8**(10), 826–32.

Chen, X., Ba, Y., Ma, L., Cai, X., et al. (2008). Characterization of microRNAs in serum: a novel class of biomarkers for diagnosis of cancer and other diseases. *Cell Res*, **18**, 997–1006.

Chen, Z., Trotman, L. C., Shaffer, D., et al. (2005). Crucial role of p53-dependent cellular senescence in suppression of Pten-deficient tumorigenesis. *Nature*, **436**, 725–30.

Chuang, J. C. & Jones, P. A. (2007). Epigenetics and microRNAs. *Pediatr Res*, **61**, 24R–29R.

Cohen, S., Levi-Montalcini, R., & Hamburger, V. (1954). A nerve growth-stimulating factor isolated from sarcoma as 37 and 180. *Proc Natl Acad Sci U S A*, **40**, 1014–8.

Cooper, G. M. (1982). Cellular transforming genes. *Science*, **217**, 801–6.

Croce, C. M. (1987). Role of chromosome translocations in human neoplasia. *Cell*, **49**, 155–6.

Croce, C. M. (2008). Oncogenes and cancer. *N Engl J Med*, **358**, 502–11.

Dang, C. V., O'Donnell, K. A., & Juopperi, T. (2005). The great MYC escape in tumorigenesis. *Cancer Cell*, **8**, 177–8.

Do, K. & Chen, A. P. (2013). Molecular pathways: targeting PARP in cancer treatment. *Clin Cancer Res*, **19**, 977–84.

Downward, J., Yarden, Y., Mayes, E., et al. (1984). Close similarity of epidermal growth factor receptor and v-erb-B oncogene protein sequences. *Nature*, **307**, 521–7.

Duesberg, P. H., Bister, K., & Vogt, P. K. (1977). The RNA of avian acute leukemia virus MC29. *Proc Natl Acad Sci U S A*, **74**, 4320–4.

Dyson, N., Dembski, M., Fattaey, A., Ngwu, C., Ewen, M., & Helin, K. (1993). Analysis of p107-associated proteins: p107 associates with a form of E2F that differs from pRB-associated E2F-1. *J Virol*, **67**, 7641–7.

Dyson, N. J. (2016). RB1: a prototype tumor suppressor and an enigma. *Genes Dev*, **30**, 1492–502.

Eiring, A. M., Harb, J. G., Neviani, P., et al. (2010). miR-328 functions as an RNA decoy to modulate hnRNP E2 regulation of mRNA translation in leukemic blasts. *Cell*, **140**, 652–65.

Farmer, H., McCabe, N., Lord, C. J., et al. (2005). Targeting the DNA repair defect in *BRCA* mutant cells as a therapeutic strategy. *Nature*, **434**, 917–21.

Farrell, A. S. & Sears, R. C. (2014). MYC degradation. *Cold Spring Harb Perspect Med*, **4**, pii: a014365.

Fatica, A. & Bozzoni, I. (2014). Long non-coding RNAs: new players in cell differentiation and development. *Nat Rev Genet*, **15**, 7–21.

Fernandez-Medarde, A. & Santos, E. (2011). Ras in cancer and developmental diseases. *Genes Cancer*, **2**, 344–58.

Foulkes, W. D. & Shuen, A. Y. (2013). In brief: *BRCA1* and *BRCA2*. *J Pathol*, **230**, 347–9.

Fradet-Turcotte, A., Sitz, J., Grapton, D., & Orthwein, A. (2016). *BRCA2* functions: from DNA repair to replication fork stabilization. *Endocr Relat Cancer*, **23**, T1–T17.

Frenzel, A., Grespi, F., Chmelewskij, W., & Villunger, A. (2009). Bcl2 family proteins in carcinogenesis and the treatment of cancer. *Apoptosis*, **14**, 584–96.

Friend, S. H., Bernards, R., Rogelj, S., et al. (1986). A human DNA segment with properties of the gene that predisposes to retinoblastoma and osteosarcoma. *Nature*, **323**, 643–6.

Gao, P. & Wei, G. H. (2017). Genomic insight into the role of lncRNA in cancer susceptibility. *Int J Mol Sci*, **18**, 1239.

Garzon, R., Calin, G. A., & Croce, C. M. (2009). MicroRNAs in Cancer. *Annu Rev Med*, **60**, 167–79.

Garzon, R., Marcucci, G., & Croce, C. M. (2010). Targeting microRNAs in cancer: rationale, strategies and challenges. *Nat Rev Drug Discov*, **9**, 775–89.

Hall, J. S., Taylor, J., Valentine, H. R., et al. (2012). Enhanced stability of microRNA expression facilitates classification of FFPE tumour samples exhibiting near total mRNA degradation. *Br J Cancer*, **107**, 684–94.

Harris, H. (1970). *Cell Fusion: The Dunham lectures*. Oxford: Clarendon Press.

Harris, H. (1987). *The Balance of Improbabilities*. Oxford: Oxford University Press.

Heldin, C. H. (1995). Dimerization of cell surface receptors in signal transduction. *Cell*, **80**, 213–23.

Heldin, C. H. & Westermark, B. (1999). Mechanism of action and *in vivo* role of platelet-derived growth factor. *Physiol Rev*, **79**, 1283–316.

Hua, Z., Lv, Q., Ye, W., et al. (2006). MiRNA-directed regulation of VEGF and other angiogenic factors under hypoxia. *PLoS One*, **1**, e116.

Huarte, M. & Rinn, J. L. (2010). Large non-coding RNAs: missing links in cancer? *Hum Mol Genet*, **19**, R152–61.

Huebner, R. J. & Todaro, G. J. (1969). Oncogenes of RNA tumor viruses as determinants of cancer. *Proc Natl Acad Sci U S A*, **64**, 1087–94.

Inamura, K. (2017). Major tumor suppressor and oncogenic non-coding RNAs: clinical relevance in lung cancer. *Cells*, **6**, pii: E12.

Indovina, P., Marcelli, E., Casini, N., Rizzo, V., & Giordano, A. (2013). Emerging roles of RB family: new defense mechanisms against tumor progression. *J Cell Physiol*, **228**, 525–35.

Jenuwein, T. & Allis, C. D. (2001). Translating the histone code. *Science*, **293**, 1074–80.

Joensuu, H. & Dimitrijevic, S. (2001). Tyrosine kinase inhibitor imatinib (STI571) as an anticancer agent for solid tumours. *Ann Med*, **33**, 451–5.

Joensuu, H., Roberts, P. J., Sarlomo-Rikala, M., et al. (2001). Effect of the tyrosine kinase inhibitor STI571 in a patient with a metastatic gastrointestinal stromal tumor. *N Engl J Med*, **344**, 1052–6.

Jung, H. M., Phillips, B. L., Patel, R. S., et al. (2012). Keratinization-associated miR-7 and miR-21 regulate tumor suppressor reversion-inducing cysteine-rich protein with kazal motifs (RECK) in oral cancer. *J Biol Chem*, **287**, 29261–72.

Kalkat, M., De Melo, J., Hickman, K. A., et al. (2017). MYC deregulation in primary human cancers. *Genes (Basel)*, **8**(6), pii: E151.

Kamihara, J., Rana, H. Q., & Garber, J. E. (2014). Germline TP53 mutations and the changing landscape of Li-Fraumeni syndrome. *Hum Mutat*, **35**, 654–62.

Kastenhuber, E. R. & Lowe, S. W. (2017). Putting p53 in Context. *Cell*, **170**, 1062–78.

Katz, M., Amit, I., & Yarden, Y. (2007). Regulation of MAPKs by growth factors and receptor tyrosine kinases. *Biochim Biophys Acta*, **1773**, 1161–76.

Kaufman-Szymczyk, A., Majewski, G., Lubecka-Pietruszewska, K., & Fabianowska-Majewska, K. (2015). The role of sulforaphane in epigenetic mechanisms, including interdependence between histone modification and DNA methylation. *Int J Mol Sci*, **16**, 29732–43.

Kinzler, K. W. & Vogelstein, B. (1997). Cancer-susceptibility genes. Gatekeepers and caretakers. *Nature*, **386**, 761, 763.

Knudson, A. G., Jr. (1971). Mutation and cancer: statistical study of retinoblastoma. *Proc Natl Acad Sci U S A*, **68**, 820–3.

Kunej, T., Obsteter, J., Pogacar, Z., Horvat, S., & Calin, G. A. (2014). The decalog of long non-coding RNA involvement in cancer diagnosis and monitoring. *Crit Rev Clin Lab Sci*, **51**, 344–57.

Lagos-Quintana, M., Rauhut, R., Lendeckel, W., & Tuschl, T. (2001). Identification of novel genes coding for small expressed RNAs. *Science*, **294**, 853–8.

Lahiry, P., Torkamani, A., Schork, N. J., & Hegele, R. A. (2010). Kinase mutations in human disease: interpreting genotype–phenotype relationships. *Nat Rev Genet*, **11**, 60–74.

Lanford, R. E., Hildebrandt-Eriksen, E. S., Petri, A., et al. (2010). Therapeutic silencing of microRNA-122 in primates with chronic hepatitis C virus infection. *Science*, **327**, 198–201.

Laptenko, O. & Prives, C. (2017). p53: master of life, death, and the epigenome. *Genes Dev*, **31**, 955–6.

Le Sage, C., Nagel, R., Egan, D. A., et al. (2007). Regulation of the p27(Kip1) tumor suppressor by miR-221 and miR-222 promotes cancer cell proliferation. *EMBO J*, **26**, 3699–708.

Lee, J. O., Russo, A. A., & Pavletich, N. P. (1998). Structure of the retinoblastoma tumour-suppressor pocket domain bound to a peptide from HPV E7. *Nature*, **391**, 859–65.

Lee, R. C., Feinbaum, R. L., & Ambros, V. (1993). The C. elegans heterochronic gene lin-4 encodes small RNAs with antisense complementarity to lin-14. *Cell*, **75**, 843–54.

Lewis, B. P., Burge, C. B., & Bartel, D. P. (2005). Conserved seed pairing, often flanked by adenosines, indicates that thousands of human genes are microRNA targets. *Cell*, **120**, 15–20.

Li, F. P. & Fraumeni, J. F., Jr. (1969). Soft-tissue sarcomas, breast cancer, and other neoplasms. A familial syndrome? *Ann Intern Med*, **71**, 747–52.

Lin, C. Y., Loven, J., Rahl, P. B., et al. (2012). Transcriptional amplification in tumor cells with elevated c-Myc. *Cell*, **151**, 56–67.

Liu, Z. L., Wang, H., Liu, J., & Wang, Z. X. (2013). MicroRNA-21 (miR-21) expression promotes growth, metastasis, and chemo- or radioresistance in non-small cell lung cancer cells by targeting PTEN. *Mol Cell Biochem*, **372**, 35–45.

Lloyd, R. V., Erickson, L. A., Jin, L., et al. (1999). p27kip1: a multifunctional cyclin-dependent kinase inhibitor with prognostic significance in human cancers. *Am J Pathol*, **154**, 313–23.

Lobry, C., Oh, P., Mansour, M. R., Look, A. T., & Aifantis, I. (2014). Notch signaling: switching an oncogene to a tumor suppressor. *Blood*, **123**, 2451–9.

Lohmann, D. R. (1999). *RB1* gene mutations in retinoblastoma. *Hum Mutat*, **14**, 283–8.

Lord, C. J. & Ashworth, A. (2016). *BRCA*ness revisited. *Nat Rev Cancer*, **16**, 110–20.

Lu, J., Getz, G., Miska, E. A., et al. (2005). MicroRNA expression profiles classify human cancers. *Nature*, **435**, 834–8.

Malkin, D., Li, F. P., Strong, L. C., et al. (1990). Germ-line p53 mutations in a familial syndrome of breast cancer, sarcomas, and other neoplasms. *Science*, **250**, 1233–8.

Martin, G. S. (1970). Rous sarcoma virus: a function required for the maintenance of the transformed state. *Nature*, **227**, 1021–3.

Marquez-Medina, D., & Popat, S. (2015). Afatinib: a second-generation EGF receptor and ErbB tyrosine kinase inhibitor for the treatment of advanced non-small-cell lung cancer. *Future Oncol*, **11**(18), 2525–40.

Mei, Y. & Wu, M. (2016). Noncoding RNAs Regulating p53 and c-Myc Signaling. *Adv Exp Med Biol*, **927**, 337–65.

Menzel, T., Nahse-Kumpf, V., Kousholt, A. N., et al. (2011). A genetic screen identifies *BRCA2* and *PALB2* as key regulators of G2 checkpoint maintenance. *EMBO Rep*, **12**, 705–12.

Michelotti, E. F., Tomonaga, T., Krutzsch, H., & Levens, D. (1995). Cellular nucleic acid binding protein regulates the CT element of the human c-Myc protooncogene. *J Biol Chem*, **270**, 9494–9.

Mitelman, F. (1991). *Catalog of Chromosome Aberrations in Cancer.* New York: Wiley-Liss.

Morris, K. V. & Mattick, J. S. (2014). The rise of regulatory RNA. *Nat Rev Genet*, **15**, 423–37.

Muraoka, R. S., Lenferink, A. E., Law, B., et al. (2002). ErbB2/Neu-induced, cyclin D1-dependent transformation is accelerated in p27-haploinsufficient mammary epithelial cells but impaired in p27-null cells. *Mol Cell Biol*, **22**, 2204–19.

Nakamura, T., Mori, T., Tada, S., et al. (2002). ALL-1 is a histone methyltransferase that assembles a supercomplex of proteins involved in transcriptional regulation. *Mol Cell*, **10**, 1119–28.

Nicolas, M., Wolfer, A., Raj, K., et al. (2003). Notch1 functions as a tumor suppressor in mouse skin. *Nat Genet*, **33**, 416–21.

Normanno, N., De Luca, A., Bianco, C., et al. (2006). Epidermal growth factor receptor (EGFR) signaling in cancer. *Gene*, **366**, 2–16.

Paige, A. J. (2003). Redefining tumour suppressor genes: exceptions to the two-hit hypothesis. *Cell Mol Life Sci*, **60**, 2147–63.

Pawson, T. (2007). Dynamic control of signaling by modular adaptor proteins. *Curr Opin Cell Biol*, **19**, 112–16.

Pawson, T. & Warner, N. (2007). Oncogenic re-wiring of cellular signaling pathways. *Oncogene*, **26**, 1268–75.

Peng, W. X., Koirala, P., & Mo, Y. Y. (2017). LncRNA-mediated regulation of cell signaling in cancer. *Oncogene*, **36**(41), 5661–7.

Peterson, C. L. & Workman, J. L. (2000). Promoter targeting and chromatin remodeling by the SWI/SNF complex. *Curr Opin Genet Dev*, **10**, 187–92.

Pezzella, F., Turley, H., Kuzu, I., et al. (1993). bcl-2 protein in non-small-cell lung carcinoma. *N Engl J Med*, **329**, 690–4.

Rodriguez, A., Griffiths-Jones, S., Ashurst, J. L., & Bradley, A. (2004). Identification of mammalian microRNA host genes and transcription units. *Genome Res*, **14**, 1902–10.

Rodriguez, A., Sosa, D., Torres, L., Molina, B., Frias, S., & Mendoza, L. (2012). A Boolean network model of the *FA/BRCA* pathway. *Bioinformatics*, **28**, 858–66.

Rous, P. (1911). A sarcoma of the fowl transmissible by an agent separable from the tumor cells. *J Exp Med*, **13**, 397–411.

Ryan, B. M., Robles, A. I., & Harris, C. C. (2010). Genetic variation in microRNA networks: the implications for cancer research. *Nat Rev Cancer*, **10**, 389–402.

Salehi, S., Taheri, M. N., Azarpira, N., Zare, A., & Behzad-Behbahani, A. (2017). State of the art technologies to explore long non-coding RNAs in cancer. *J Cell Mol Med*, **21**(12), 3120–40.

Savage, K. I. & Harkin, D. P. (2015). *BRCA1*, a 'complex' protein involved in the maintenance of genomic stability. *FEBS J*, **282**, 630–46.

Schetter, A. J., Leung, S. Y., Sohn, J. J., et al. (2008). MicroRNA expression profiles associated with prognosis and therapeutic outcome in colon adenocarcinoma. *JAMA*, **299**, 425–36.

Sempere, L. F., Christensen, M., Silahtaroglu, A., et al. (2007). Altered MicroRNA expression confined to specific epithelial cell subpopulations in breast cancer. *Cancer Res*, **67**, 11612–20.

Shaham, L., Binder, V., Gefen, N., Borkhardt, A., & Izraeli, S. (2012). MiR-125 in normal and malignant hematopoiesis. *Leukemia*, **26**, 2011–18.

Shaulian, E. & Karin, M. (2001). AP-1 in cell proliferation and survival. *Oncogene*, **20**, 2390–400.

Sherr, C. J. & McCormick, F. (2002). The RB and p53 pathways in cancer. *Cancer Cell*, **2**, 103–12.

Shih, T. Y., Papageorge, A. G., Stokes, P. E., Weeks, M. O., & Scolnick, E. M. (1980). Guanine nucleotide-binding and autophosphorylating activities associated with the p21src protein of Harvey murine sarcoma virus. *Nature*, **287**, 686–91.

Sicinski, P., Zacharek, S., & Kim, C. (2007). Duality of p27Kip1 function in tumorigenesis. *Genes Dev*, **21**, 1703–6.

Singh, U., Malik, M. A., Goswami, S., Shukla, S., & Kaur, J. (2016). Epigenetic regulation of human retinoblastoma. *Tumour Biol*, **37**, 14427–41.

Solomon, E., Borrow, J., & Goddard, A. D. (1991). Chromosome aberrations and cancer. *Science*, **254**, 1153–60.

Srivastava, S., Zou, Z. Q., Pirollo, K., Blattner, W., & Chang, E. H. (1990). Germ-line transmission of a mutated *p53* gene in a cancer-prone family with Li-Fraumeni syndrome. *Nature*, **348**, 747–9.

Sumpter, R., Jr., Sirasanagandla, S., Fernandez, A. F., et al. (2016). Fanconi anemia proteins function in mitophagy and immunity. *Cell*, **165**, 867–81.

Sun, Y. M., Lin, K. Y., & Chen, Y. Q. (2013). Diverse functions of miR-125 family in different cell contexts. *J Hematol Oncol*, **6**, 6.

Svoronos, A. A., Engelman, D. M., & Slack, F. J. (2016). OncomiR or tumor suppressor? The duplicity of microRNAs in cancer. *Cancer Res*, **76**, 3666–70.

Takebe, N., Harris, P. J., Warren, R. Q., & Ivy, S. P. (2011). Targeting cancer stem cells by inhibiting Wnt, Notch, and Hedgehog pathways. *Nat Rev Clin Oncol*, **8**, 97–106.

Tamborini, E., Bonadiman, L., Greco, A., et al. (2004). A new mutation in the KIT ATP pocket causes acquired resistance to imatinib in a gastrointestinal stromal tumor patient. *Gastroenterology*, **127**, 294–9.

Thiagalingam, S., Foy, R. L., Cheng, K. H., Lee, H. J., Thiagalingam, A., & Ponte, J. F. (2002). Loss of heterozygosity as a predictor to map tumor suppressor genes in cancer: molecular basis of its occurrence. *Curr Opin Oncol*, **14**, 65–72.

Tomlins, S. A., Rhodes, D. R., Perner, S., et al. (2005). Recurrent fusion of TMPRSS2 and ETS transcription factor genes in prostate cancer. *Science*, **310**, 644–8.

Tsujimoto, Y., Yunis, J., Onorato-Showe, L., Erikson, J., Nowell, P. C., & Croce, C. M. (1984). Molecular cloning of the chromosomal breakpoint of B-cell lymphomas and leukemias with the t(11;14) chromosome translocation. *Science*, **224**, 1403–6.

Ullrich, A., Coussens, L., Hayflick, J. S., et al. (1984). Human epidermal growth factor receptor cDNA sequence and aberrant expression of the amplified gene in A431 epidermoid carcinoma cells. *Nature*, **309**, 418–25.

Valdez, J. M., Nichols, K. E., & Kesserwan, C. (2017). Li-Fraumeni syndrome: a paradigm for the understanding of hereditary cancer predisposition. *Br J Haematol*, **176**, 539–52.

Vousden, K. H. & Prives, C. (2009). Blinded by the light: the growing complexity of p53. *Cell*, **137**, 413–31.

Waterfield, M. D., Scrace, G. T., Whittle, N., et al. (1983). Platelet-derived growth factor is structurally related to the putative transforming protein p28sis of simian sarcoma virus. *Nature*, **304**, 35–9.

Weidle, U. H., Birzele, F., Kollmorgen, G., & Ruger, R. (2017). Long non-coding RNAs and their role in metastasis. *Cancer Genomics Proteomics*, **14**, 143–60.

Witsch, E., Sela, M., & Yarden, Y. (2010). Roles for growth factors in cancer progression. *Physiology (Bethesda)*, **25**, 85–101.

Wu, G. S. (2009). TRAIL as a target in anti-cancer therapy. *Cancer Lett*, **285**, 1–5.

Wu, W., Koike, A., Takeshita, T., & Ohta, T. (2008). The ubiquitin E3 ligase activity of *BRCA1* and its biological functions. *Cell Div*, **3**, 1.

Yanaihara, N., Caplen, N., Bowman, E., et al. (2006). Unique microRNA molecular profiles in lung cancer diagnosis and prognosis. *Cancer Cell*, **9**, 189–98.

Yip, K. W. & Reed, J. C. (2008). Bcl-2 family proteins and cancer. *Oncogene*, **27**, 6398–406.

Yoshida, K. & Miki, Y. (2004). Role of *BRCA1* and *BRCA2* as regulators of DNA repair, transcription, and cell cycle in response to DNA damage. *Cancer Sci*, **95**, 866–71.

Yuan, T. L. & Cantley, L. C. (2008). PI3K pathway alterations in cancer: variations on a theme. *Oncogene*, **27**, 5497–510.

12

The signalling pathways in cancer

Jiangting Hu and Francesco Pezzella

Introduction

In the living cells there are many different pathways regulating their functions. This chapter will deal with the signalling network which allows the cell to 'communicate' with the outside world by receiving signals, process them, make appropriate changes accordingly and release feedback signals: a process defined by KEGG (*Kyoto Encyclopedia of Genes and Genomes*) as 'Environmental Information Processing'. Some of these pathways have been discussed in other chapters and they are listed in Table 12.1. One first issue is how to represent pathways: the standard method currently employed uses two-dimensional cartoons in which icons represent the different components (e.g. proteins, lipids, glycoprotein) while lines indicate the direction of the reaction and the type of reaction (e.g. production of a new protein, degradation, modification of a moiety, etc), lines with an arrow symbolize activation while line with a transversal bar at the end means inhibition (Iber and Fengos, 2012). While this approach is effective in mapping the basic features of a network, it also presents a highly simplified picture (Iber and Fengos, 2012). At the basis of the complexity of the intracellular pathways and their modelling are some basic features: the number of components, the number of interaction among them, the number of parameters involved and finally the fact that the observed interactions can be non-linear, that is, the changes in the output are not necessarily proportional to the changes in the input (Rangamani and Iyengar, 2008). The increasing use of high throughput techniques is providing us with increasing detailed 'snapshot' of the situation of the pathways in a given cell at a certain time. From these data maps built in the traditional way, showed in Figure 12.1, provide a very simplified picture. To resolve this problem increasing attention is devoted to develop mathematical models, which should lead to a more complete representation of the pathways' status and behaviour in a cell. In this chapter, however, we will have to rely on the classic two-dimensional graphic representation, which is currently standard in biology: our message is just to keep in mind all these issues while relying on the simplified representations currently used.

Table 12.1 Signalling pathways presented in other chapters

Pathways	Chapter
Notch pathway (dual address)	Cancer and blood vessels
VEGF pathway	Cancer and blood vessels
Apoptosis intrinsic pathway	Cancer cell death
Apoptosis extrinsic pathway	Cancer cell death
Energy metabolism	Cancer metabolism
Response to hypoxia	Oxygen and cancer: the response to hypoxia
Cell cycle control	Cell cycle control

General principles of cell signalling

Signalling pathways in metazoa are a sophisticated network used by cells to communicate with the outer environment namely other cells and extracellular matrix (ECM). They are made up by many different types of molecules which work together to control one or more cell functions in response to a given stimulus. After the first molecule in a pathway receives a signal, it activates another molecule(s) downstream. This process is repeated until the last molecule in the chain is activated and the cell function is carried out: a basic illustration of how intracellular pathways are organized and represented is shown in Figure 12.1.

Type of signals

In an animal cell, there are hundreds of different signalling molecules, including proteins, small peptides, amino acids, nucleotides, steroids, retinoids, fatty acid derivatives, and even dissolved gases, such as nitric oxide (Table 12.2). These molecules can be divided into two major types: the chemical signals secreted by exocytosis, the majority, and the plasma-membrane-bound molecules acting by direct contact between the two cells.

There are five major forms of signalling in the human body:

Endocrine. This is the most common type and involves the signalling cells, called endocrine cells, releasing their products, called hormones, into the bloodstream. For example, the islet of the pancreas constitutes an endocrine gland and produces the hormone insulin, which regulates the uptake of glucose in cells all over the body. Other examples of hormones include testosterone, oestrogen, progesterone, and gonadotropins.

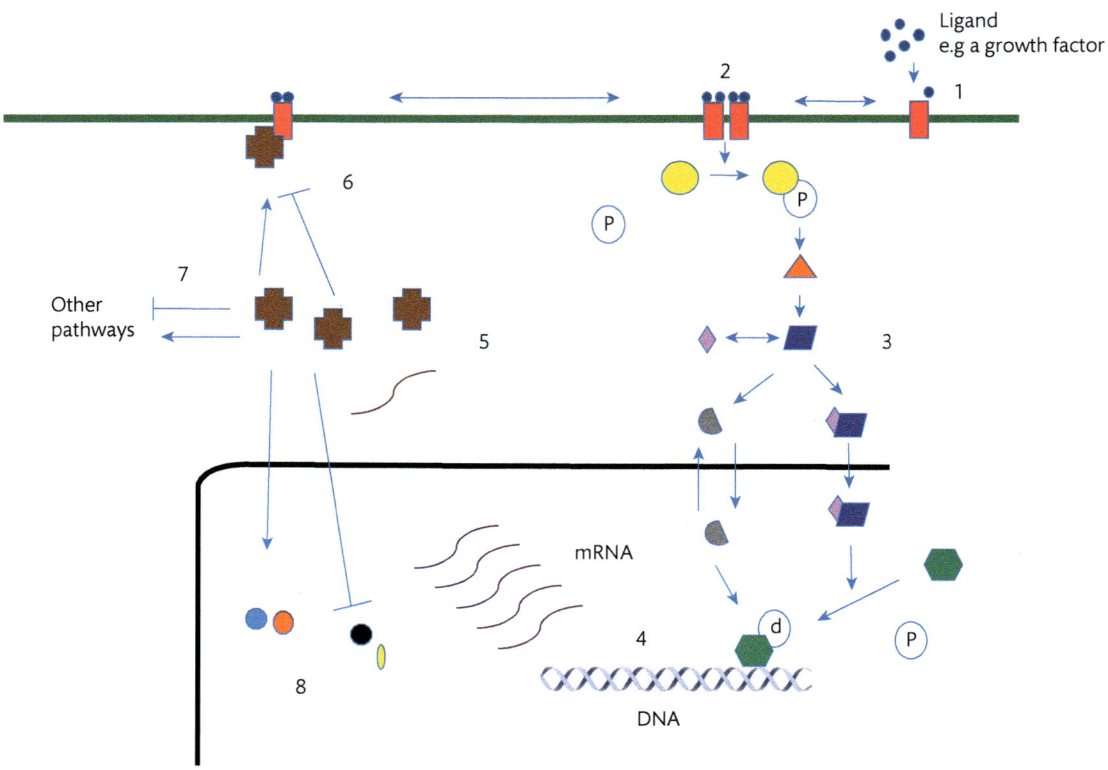

Fig. 12.1 Basic representation of a pathway. The extracellular ligand binds to its receptor (1) Linking to the receptor causes its activation, for example, by assembling in a dimer (2) being able to phosphorylate its intracellular substrate which in turn activate the intracytoplasmic component of the pathway (3). The last cytoplasmic component, once activated, enter the nucleus, or activate a nuclear receptor. Eventually a transcription factor is phosphorylated, becomes active and triggers transcription of its target gene (4). The mRNA moves to the cytoplasm where the protein is produced (5). The newly formed protein can have different effects: it can provide a positive or negative feedback on its own pathway; or (6), on other pathways (7) or on the transcription of other genes (8). Arrow indicates activation, line with a bar at the end indicates inhibition.

Table 12.2 Summary of molecular signals classification by mechanism of action

Molecules that bind to cell surface receptors				Molecules that bind to intracellular receptors
Through a kinase or phosphorylation cascade	Through second messenger cAMP	Through second messenger cGMP	Through second messenger Ca+ or PIs (phosphatidylinositol) or both	
Angiopoietin Chorionic somatomammotropin Colony stimulating factors (CSFs) Epidermal growth factor (EGF) Ephrins Erythropoietin (EPO) Fibroblast growth factor (FGF) Differentiation factor-9 (GDF-9) Hepatocyte growth factor (HGF) Insulin Insulin-like growth factor I & II (IGF I&II) Integrin Interleukins (IL1, 2, 3, 4, 5, 6, &7) Migration-stimulating factor (MSF) Neuregulins Neurotrophins Platelet-derived growth factor Prolactin Transforming growth factor (TGF) α and β Vascular endothelial growth factor (VEGF)	A2 and β-adrenergic catecholamines Calcitonin Corticotropin-releasing hormone Glucagon Lipotropin Melanocyte-stimulating hormone Somatostatin	Atrial natriuretic factor Nitric oxide	Acetylcholine(muscarinic) Adrenergic catecholamines Angiotensin II Antidiuretic hormone(vasopressin) Cholecystokinin Gastrin Gonadotropin-releasing hormone Oxytocin Substance P Thyrotropin-releasing hormone	Androgens Calcitriol Oestrogens Glucocorticoids Progestins Retinoic acid Thyroid hormones (T3 and T4)

Paracrine. This type of signalling molecules are released from cells in the immediate extracellular space and affect only the cells in the surrounding area by diffusing through the extracellular fluid. Many of the cells that are involved in inflammation, or that regulate cell proliferation utilize this type of signalling. For example, cancer cells sometimes enhance their own survival or proliferation in this way. Examples of signalling molecules that often function in a paracrine fashion include transforming growth factor-β (TGF-β) and fibroblast growth factors (FGFs).

Synaptic. Nerve cells (neurons) are specialized cells formed by a body and an axon: this is a very long prolongation of the cell. Once stimulated, the neuron sends an electric impulse along the axon membrane. When an impulse reaches the end of the axon, vesicles containing signalling molecules (neurotransmitters) fuse with the membrane and release the neurotransmitters into the synaptic space. These neurotransmitters are detected by receptors on the postsynaptic membrane cell, which may be another neuron or an effectors cell like a muscle cell. Examples of neurotransmitters include acetylcholine, serotonin, and histamine.

Juxtacrine. This is often referred to as a contact-dependent signalling. It does not involve the release of secreted molecules and only occurs over short distances. The cells involved make direct physical contact through signal molecules found in their cellular membrane. This type of signalling is extremely important during embryonic development and cell-fate determination (Zimmerman et al., 1993). The Notch pathway is an example and mediates juxtacrine signalling between adjacent cells. Notch receptors are single transmembrane proteins and they bind to specific ligands (e.g. Delta and Serrate ligands) on the membrane of adjacent cells. Ligand binding results in proteolytic cleavage of the Notch receptor, which releases an intracellular domain that is translocated to the nucleus where it regulates gene expression (see Chapter 22).

Exosome. Exosomes are small vesicles (up to 150 nano metre in diameter) formed by the cell membrane and released externally. They can subsequently fuse with the membrane of other cells. In this way their contents can be transferred from cell to cell (Smythies et al., 2014; Hessvik and Llorente, 2017). They were firstly described in 1987 (Johnstone et al., 1987) they were, initially dismissed as artefact of no interest (Johnstone, 2005). It has been only after some time that their importance has been eventually established (Johnstone, 2005; Smythies et al., 2014; Hessvik and Llorente, 2017). Exosomes represent a very original way in which cells can exchange 'messages'. These vesicles contain a large variety of molecules those inclusions in the exosome is not subject to a tight control like in the other forms of communication. They contain not only the like of growth factors, transcription factors, or hormones but also nucleic acids, most notably miRNA (Yang et al., 2017) and DNA (mitochondrial, single stranded, transposable elements and large (>10 kb) double strands; see Kahlert et al., 2014). Once transferred in new cells these factors and nucleic acids can therefore affect the host cell activity. The main difference from the other communication systems is that the active molecules transferred are not meant to be secreted: a classic example being that of miRNA. Cancer cells can produce exosomes containing miRNA produced by them and transfer it in the nearby normal stromal cell, altering therefore its characteristics, that is, transferring a character acquired by the cancer cell to a normal cell, and of course the vice versa also happens (Yang et al., 2017).

The cells receive the extracellular signals

Cells employ two principle tools to transduce an extracellular signal into an intracellular one. The first tool is the transmembrane receptor, in which the external ligand binds on its receptor located on the cell membrane. Each receptor is divided into three components: the extracellular domain, the transmembrane domain and the intracellular (cytosolic) domain. When the ligand binds to the extracellular domain, it induces either dimerization or rotation that cause the conformational change of the transmembrane domain, which eventually affects the spatial arrangement of the intracellular domain so that become able to interact with its intracellular substrate. One example is the insulin receptor (Tatulian, 2015). The second method is when the extracellular signalling molecules are sufficiently small and/or hydrophobic (i.e. lipophilic) to diffuse across the plasma membrane: they will bind to the cytosolic or nuclear-localized receptor to trigger the downstream signalling (Wijmans and Baker, 1995).

Cell surface transmembrane receptors

According to their structure and the signal transduction mechanism, cell surface transmembrane receptor proteins can be mainly divided into five known classes:

1. ion-channel linked receptors, also known as ligand-gated ion channels, transmitter-gated ion channels, or ionotropic receptors. They belong to a family of homologous, multipass transmembrane proteins involved in controlling, opening and closing of the ion channels allowing Na^+, K^+, Ca^{2+}, and/or Cl^- into and out of cells. A very common place to find ligand-gated channels is in electrically excitable cells like neurons that require a rapid and specific reaction to the stimulus. The signalling is mediated by a small number of chemical proteins called neurotransmitters (ligand) that are released through the nerve terminals when a presynaptic neuron is excited. If the ligand binds to an allosteric site of channel receptors located on the surface of postsynaptic neuron, then the change of the conformation of entire protein causes the channel to open, allowing the ions to flow. The flow of ions inside the target cells changes the electrical properties of the membrane converting the extracellular ligand signal into an electrical signal causing a response in the postsynaptic cell. The ligand-gated ion channels are different from the other two types: the voltage-gated ion channels and stretch-activated ion channels. The voltage-gated ion channels rely on the difference in membrane potential while the stretch-activated ion channels depend on the deformation of the cell membrane, or the cell membrane stretching (a mechano-signal).

2. G-protein-linked cell surface receptors, also known as G-protein coupled receptors (GPCRs). G-protein-linked receptors are only found in eukaryotes and form the largest family of the cell surface receptors. There are approximately 700–800 types of different GPCRs in human and each one has a specific function. They are unique transmembrane receptors estimated to be targeted by 30–50% of all modern medicinal drugs. GPCRs respond to a wide-range of extracellular signal molecules, including light-sensitive compounds, pheromones, hormones, and neurotransmitters. They can regulate immune system, growth, sense of smell, vision, taste, behaviour, mood, and many more unknown functions. Despite the chemical and functional diversities of the signal molecules that bind to them, all GPCRs have a similar structure that is they have seven transmembrane alpha helices, which is the most important characteristic. GPCRs interact with a specialized family of proteins called small GTP-binding protein (G proteins; Matozaki et al., 2000).

3. Enzyme-linked cell surface receptors, also known as catalytic receptors, like the ion-channel linked receptors and the GPCRs, the enzyme-linked cell surface receptors have two important domains: an extracellular ligand binding domain and an intracellular domain, which either have an intrinsic enzyme activity or associate directly with an enzyme. Each catalytic receptor has only one transmembrane helix. Up to now six known families of enzyme-linked cell surface receptors have been identified. They are:

a. Receptor guanylate cyclase, which catalyse the production of cyclic GMP in the cytosol;
b. Receptor tyrosine kinase, which have tyrosine amino acids attach to the intracellular catalytic sections involved in the phosphorylation of a small set of intracellular proteins. Most of the growth factor families belong to receptor tyrosine kinase, including epidermal growth factor receptor family: EGFR (ErbB-1), HER2/neu (ErbB-2), HER3 (ErbB-3), and HER4 (ErbB-4); Fibroblast growth factor receptor (FGFR) family: comprised of 23 members to become the largest family of growth factor; vascular endothelial growth factor receptor family; Ephrin receptor family and RET receptor family;
c. Tyrosine-kinase-associated receptors, which associate with proteins that have tyrosine kinase activity;
d. Receptor like tyrosine kinases, which are a group of enzymes able to catalyse the removal of a phosphate group attached to a tyrosine residue, using a cysteinyl-phosphate enzyme intermediate;
e. Receptor serine/threonine kinase, which specifically phosphorylate serine or threonine residues on sets of intracellular signalling proteins;
f. Histidine-kinase-associated receptors, which activate a 'two-component' signalling pathway in which the kinase phosphorylates itself on histidine and then immediately transfers the phosphate to a second intracellular signalling protein.

Among these types of enzyme-linked cell surface receptors families, the most wildly recognized and most common enzyme-linked receptors are the receptor tyrosine kinase (RTKs). RTKs have been shown not only to be the key regulators of normal cellular processes but also to have a critical role in the development and progression of many types of cancer. These receptors are involved in the regulation of cell growth, proliferation, differentiation and survival. RTKs have the ability to activate intracellular protein by phosphorylation, triggering the downstream signalling. RTKs act in pairs. When a signal molecule binds to an RTK, this one and the neighbouring RTK associate with each other forming a cross-linked dimer. Each RTK in the dimer phosphorylates the tyrosines on the other RTK. This process is called cross-phosphorylation. Once cross-phosphorylated, the intracellular enzymatic domain of the RTKs serve as docking-platform for different intracellular proteins that have SH2 domain to bind to the phosphorylated tyrosines. This process can cause multiple different SH2-containing proteins to bind at the same time to any of these phosphorylated domanis thus allowing the activation of multiple intracellular signalling at the same time.

Another type of receptors are the non-receptor tyrosine kinase (nRTKs). Non-receptor tyrosine kinases (nRTKs) do not have intrinsic enzymatic activity and therefore recruits intracellular tyrosine kinases (e.g. the Jak tyrosine kinase in the JAK-STAT pathway). Their function, again, is to regulate the protein (enzyme) functions by transferring a phosphate group from adenosine triphosphate (ATP) to the tyrosine residues of the protein. Based on the similarities in kinase domain structures, nRTKs to date can be divided into ten families sharing high degree of homology in the catalytic Src Homology 1 (SH1), p-Tyr binding Src Homology 2 (SH2), and protein–protein interaction Src Homology 3 (SH3) domains: Src family, Fak family, Csk family, Tec family, Abl family, Syk family, Jak family, Fes family, Frk family, and Ack family.

RTKs and nRTKs are involved in many signalling pathways in normal and cancer cells and can regulate cell differentiation, cell division, cell adhesion, stress responses, embryonic development, cell growth, ion transport, extracellular signalling, and other cellular processes. We will have detailed elaboration in this chapter.

4. Integrins. Integrins belong to the large family of transmembrane receptors for adhesion molecules of ECM and are a necessary component of all metazoans (Barczyk et al., 2010). Its heterodimeric extracellular parts contain α (Ca^{2+}-binding domain) and β (cysteine-rich domains) subunits, which are non-covalently bonded. The extracellular domains engage either ECM components or counter receptors of the adjacent cells while the cytoplasmic domains coordinate the assembly of cytoskeletal proteins and signalling complexes, thus integrin serve to bridge the two compartments, namely the ECM and the intracellular actin filamentous cytoskeleton, across the plasma membrane. Many ECM ligands and cell surface adhesion proteins, namely Fibronectin (FNs), arginine-glycine-aspartic acid (RGD), intercellular cell adhesion molecule, inactivated complement component C3b (iC3b), mucosal addressin cell adhesion molecule 1 (MAdCAM-1), platelet endothelial cell adhesion molecule 1(PECAM-1), vascular cell adhesion molecule (VCAM-1), epidermal growth factor (EGF), latency associated peptide transforming growth factor β (LAP-TGF-β) and milk fat globule EGF factor 8 (MFG-E8) bind to multiple integrin receptors. Integrins also exert an extensive cross talk with many growth factor receptors such as Fibroblast Growth Factor Receptor (FGFR) resulting in the formation of more complex protein association. The integrin-ligand complex then undergoes conformational changes leading to the rearrangement of intracellular part of integrin molecule and causes subsequent interactions with intracellular signalling pathways. This transduction is signalling from ECM to the cell (outside-in signalling), therefore to regulate the processes as protein phosphorylation, proliferation, differentiation, and apoptosis, for example, activation cytoplasmatic TPK (FAK) and serine/threonine kinases, activation of small GTPases, induction of calcium transport, or changes of phospholipid synthesis. On the other hand, conformational alterations also influence the interactions between cytoplasmatic domains of integrin, which will cause the changing of affinity of ligand binding domain (this is signalling from the cell to extracellular space (inside-out signalling; see Barczyk et al., 2010).

Another characteristic of the integrins is the possibility to be activated through mechanical rather than biochemical signals. Cells are anchored to the ECM through a network of molecules connecting the ECM components (e.g. collagen to the cytoskeleton). The integrins are pivotal in turning a mechanical signal, e.g. increased or decreased ECM rigidity or intracellular changes due to increased or decreased motility, into a biochemical signal (Roca-Cusachs et al., 2012). The first crucial step is the sensing by the integrins of a mechanical force which induces a conformational change of the integrin complexes on the cell membrane (**Fig. 12.2A**; Hynes, 2002; Ross et al., 2013). This modification, according to the type of integrin involved and the context than triggers a biochemical response. A model of the response to change of rigidity in the ECM is reported in **Figure 12.2B**.

5. Toll-like receptors (TLRs). TLRs are transmembrane proteins mainly expressed on the surface of sentinel cells such as macrophages and dendritic cells, that detect antigens derived from micro-organisms and act as the principal sensors of infection in mammals. In humans, there are 11 TLRs, TLR1 to TLR11. They were first discovered in 1997 by Ruslan Medzhitov and Charles Janeway and form a specific family of proteins associated with innate immune response

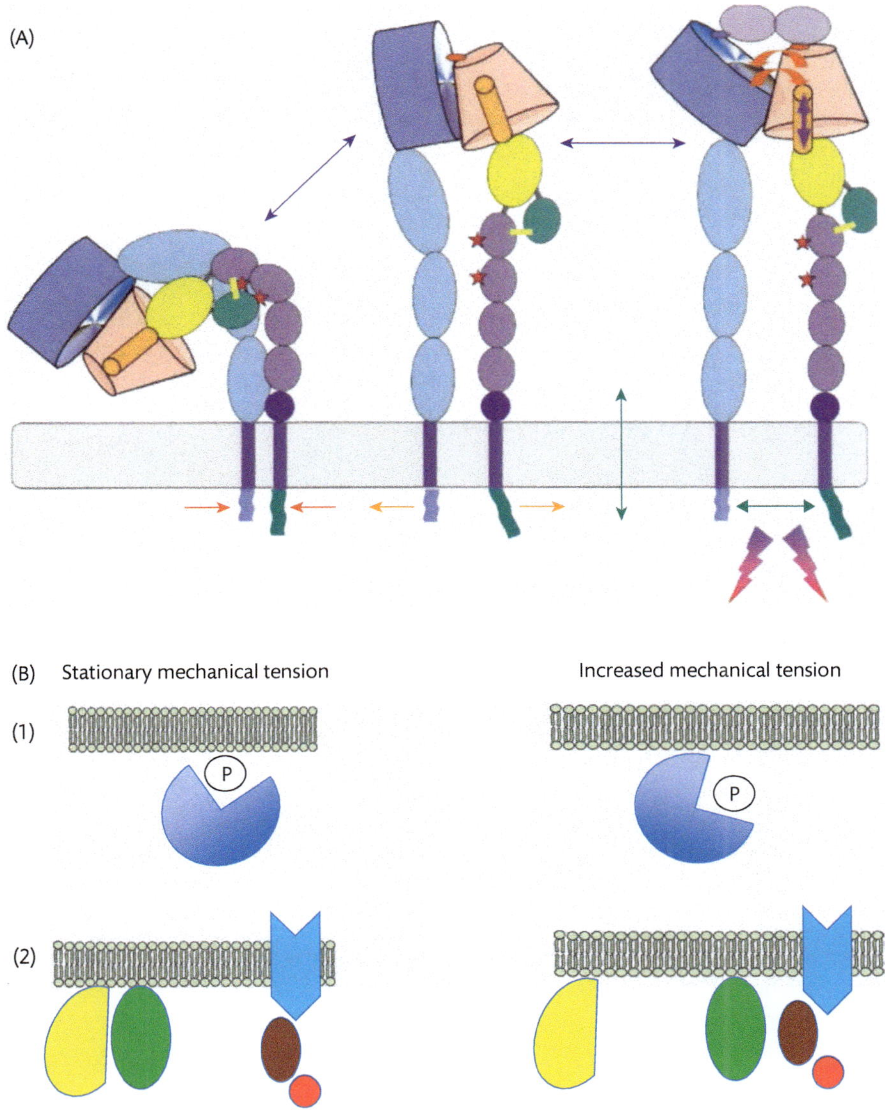

Fig. 12.2 Mechanosignalling. (A) Conformational changes resulting from bidirectional signalling by integrins. Integrin in its bent form which is believed to be inactive. The occurrence of a mechanical stress leads to the separation of the intracellular domains and to the straightening of the molecule. As a consequence, the orientation of the propeller (dark blue), the adjacent I domain (pink) and the hybrid domain (yellow) changes in a synchronized way. The hybrid domain (yellow), in turn, is linked to the I-EGF domains (purple) via the PSI domain (green), which is disulphide-bonded (yellow line) to the first I-EGF domain. Straightening and separation of the intracellular domains exposes activation epitopes in the I-EGF domains (red stars) and in the PSI domain. Separation of the cytoplasmic domains results also in conformational changes leading to binding of cytoplasmic proteins and therefore triggering signalling (lightning). All these changes are reversible and can operate in both directions, allowing both outside-in and inside-out signalling. (B) A model of activation following change of rigidity sensing. (1) A Phosphate group segregated, becomes available when a conformational change follows the occurrence a mechanical tension and can therefore activate an available substrate. (2) A tyrosine kinase (green) is moved away from coupling protein and can interact with a previously not available target

in mammalians. In cancer they are involved mostly with the cancer-associated inflammation and ECM response (Rakoff-Nahoum and Medzhitov, 2009).

Intracellular receptors

The intracellular receptors as the name implies, are located inside the cell, either in the cytoplasms or in the nucleus. Their ligands enter the cell according to the solution-diffusion path (Wijmans and Baker, 1995). The steroid hormones, thyroid hormones, lipophilic vitamins such as retinoids, vitamin D, and small molecules such as nitric oxide and hydrogen peroxide all bind to intracellular receptors. The steroid hormones can diffuse through the cell membrane into cytoplasm, in which they bind to steroid hormone receptors to form ligand-receptor complex. The forming of the ligand-receptor complex leads to activation of the receptor's nuclear localization signal activity that was originally concealed by binding to HSP90 in the absence of ligand. The ligand-receptor complex rapidly translocates to the nucleus where the receptor acts as a transcription factor binding to the DNA regulating in this way mRNA transcription. The thyroid hormones receptors, on the other hand, either located in mitochondria or inside the nucleus binding to the DNA. Once thyroid hormones enter into the cytoplasm, they can either bind to the receptor in the mitochondria which will cause the change of cell metabolic function or in the nucleus that cause the change of transcriptional activity of the receptor. A typical example relevant to cancer is that of oestrogen (Carroll, 2016). Gasotransmitters play a crucial role in all the cells and includes nitric oxide (NO), carbon monoxide (CO), and hydrogen sulphide (H_2S). Nitric oxide is the best understood so far (Hirst and Robson, 2011). NO activates soluble guanylyl cyclase, an enzyme that catalyses the production of cGMP from GTP. The cGMP produced has the primary effect of activating cGMP-dependent protein kinase, which has several effects including smooth muscle cell relaxation. This mechanism of vasodilatation is manipulated by a number of important drugs that either mimic NO (such as nitroglycerin) or prevent degradation of cGMP (Hirst and Robson, 2011).

Intercellular direct communications

Cells which are in direct contact with each other can also communicate through the direct connection, also known as the cell junction. There are three major types of cell junction in vertebrates:

Gap junctions (GJs). A gap junction is a narrow cell-to-cell channel that directly connects the cytoplasm of two adjacent cells allowing the exchange ions and low molecular weight molecules (Nielsen et al., 2012). The formation of the channel follows the direct contact between opposing plasma membranes and it can be open or closed. In cells from vertebrate, each gap junction is composed of a pair of channel proteins called connexons. Each connexon is composed of six connexins. In addition to form a pore in the centre of the assembled channel, these connexins can operate as hemichannels allowing exchanges of ions and small molecules between the cytoplasm and the extracellular space. Overall, most hydrophilic molecules including those small endogenous, xenobiotic chemicals that are less than 1–1.2 kD may easily pass through the channel (Nielsen et al., 2012). Large molecules or lipid molecules, which require carrier macromolecules for transport, would not be transferred through GJs. Gap junction play as a key role in tissue development and maintenance in multicellular organisms by homeostatic control of cells within tissues (Herve and Derangeon, 2013). Such mediate functions include 1: electrical coupling of cells, like in the heart muscle. 2: rapid exchanging of critical ionic electrical signals (e.g. Ca^{2+} ions) and the distributions of critical metabolites (e.g. cAMP); 3: cell-to-cell propagation of smaller signalling molecules. 4: exchange of virtually all soluble second messengers, amino acids, nucleotides plus glucose and its metabolites, allowing in this way the nourishing of sick or deprived cells by healthy neighbouring cells (reciprocity; Herve and Derangeon, 2013).

Anchoring junctions. These junctions make cell to adhere through proteins that are connected to the cell cytoskeleton. The main purpose of an anchoring junction is to hold cells together. Anchoring junctions are made up of three different proteins: cadherins, which form homodimers with each other, in the membranes of adjacent cells, α-catenin and β-catenin. Anchoring junctions have two forms. The first one, called a Desmosome, adheres cells with intermediate filaments in the cytoplasm. It also bonds with keratin. Desmosomes are composed of two cadherins: desmoglein and desmocollin. The second form, called an adherens junction, adheres the cells with microfilaments in the cytoskeleton. It is composed of a pair of E-cadherins. Desmosomes are abundant in tissues subject to great mechanical stress, such as muscles and skin. Anchoring junctions also transduce signals in and out of cells (Ferreira et al., 2015).

Tight junctions. Tight junctions are multiprotein structure which bound cells, mostly epithelial and endothelial, very closely together to form a barrier. Tight junctions are formed by a series of integral proteins connections between epithelial cells that are so strong that not even ions can passively move through cell layers. The tight junction proteins run along the plasma membranes like stitches, leaving some intercellular space between the junctions (Runkle and Mu, 2013). The only way that materials can enter the cell is via active transport. Its tightness depends on what molecules need to be blocked, and it can get weaker if there is an ATP deficiency. Despite this, tight junctions do not allow particles to transfer between the cells, for example, in digestive tract tight junction keep digestive enzymes and microorganisms in the intestine from leaking into the bloodstream (Liang and Weber, 2014).

The logistic system of intracellular signal transduction

Once a signal is received its must be spread through the intracellular network. This process is called intracellular signal transduction and is conducted according to two principles:

1. The cell must use protein-complexes, which either associated with the membranes or the cytoskeleton. The allosteric protein complex is formed when active and dissociated when inactive;
2. The cell responds in a very specific and selective manner according to the biophysical characteristics of the participating molecules themselves, as well as the specific character of the cell.

There are two principle intracellular signalling mechanisms: by phosphorylation and by GTP-binding protein. Both mechanisms share a common feature that is a signalling protein is activated by the addition of a phosphate group and inactivated by the removal of the same group.

The phosphorylation process is mediated by enzyme-linked cell surface receptors, predominantly by two types of protein kinase: receptor tyrosine kinase and serine/threonine kinase. Receptor tyrosine kinase (RTKs) is responsible for catalysing the transfer of a phosphoryl group from a nucleoside triphosphate donor, such as ATP to tyrosine residues in proteins. RTKs function in transmembrane signalling, whereas cytoplasmic tyrosine kinases are in signal transduction across the cytoplasm towards the nucleus, where the pathway proteins are phosphorylated at tyrosine residues during this process. Serine/threonine kinase, on the other hand, is a group of enzymes that catalyses the phosphorylation of serine or threonine residues in proteins, with ATP or other nucleotides as phosphate donors. Serine/threonine kinase enzymes participate in different pathway such as MAPK and TGF-beta. There are many different kinases that phosphorylate different targets in a cell. One example is the MAPK signalling following binding of EGF to the EGFR. EGF and EGFR form a complex undergoing conformational change activating the non-receptor kinase Raf. Active Raf triggers a phosphorylation cascades through MEK and ERKs ultimately activating variety of target molecules including transcription factors like c-Myc.

A second important intracellular signalling mechanism is through small GTP-binding protein (G proteins; Matozaki et al., 2000). The discovery of a family of GTP-binding proteins which couple the receptors (GPCRs) to specific cellular effectors was one of the major steps in the understanding of how the hormonal and sensory transduction works in eukaryotic cells. The absolute requirement of GTP for hormonal stimulation of adenylate cyclase was the initial observation which led to the purification of the G proteins. There are three types of G proteins associated with GPCRs: Gs, Gi, and Go, according to their function. However, all three G proteins have a common structure, which has three subunits, α, β, and γ (Hurowitz et al., 2000; Takai et al., 2001). The α and γ subunits have lipid anchors to attach to the cell membrane. In the inactive GDP-bound state, the α subunit bound with GDP. When a signalling molecule binds to GPCRs on the outer surface of the cell, the GPCRs undergo a conformational change that trigger the α subunit exchange of GDP for GTP, which will ultimately cause the trimer to dissociate into two activated components, an α subunit and a βγ dimer. These two effectors subunits will then regulate their target proteins, for example, enzymes and ion channels, namely second messengers, to relay the signalling messages (Johnston and Siderovski, 2007). The second messengers include Ca^{2+} ions, cyclic AMP (cAMP), a derivative of ATP and inositol phosphates (made from phospholipids) passing along the signals in pathways such as cAMP signal pathway and phosphatidylinositol signal pathway. When the extracellular signalling ligand is removed, the GTPase function of the activated α subunit will allow GTP to undergo hydrolysis reverting to GDP. Once this happens, the α subunit re-associates with a βγ dimer to re-form an inactive G protein, reversing the activation process (Matozaki et al., 2000; Johnston and Siderovski, 2007).

Targeted cell response

Given that target cell response is a multistep process of signal relay with the protein complex involvement and signalling amplification occurs at each step, one individual cell manages to display specific responses to so many different extracellular signals and one particular signal molecule can trigger different response according to the different physiological characteristics of the cells. Usually we can describe target cell response at two levels: at the molecular level, the responses of a signalling pathways frequently connect the cell surface to the nucleus involving transcription factors that lead to changes in gene expression; at the phenotypic level, cells undergo a process of changes in morphology and behaviour including appearance, growth, migration, identity, and metabolism reflecting the molecular changes. One of the key challenges in cell biology is to determine how a cell integrates all of this signalling information in order to make decisions. Many cells, for example, require a specific combination of extracellular survival factors to allow the cell to survive; when deprived of these signals, the cell undergo programmed cell death. For a normal cell, cell proliferation depends on a combination of signals that promote both cell division and survival, as well as signals that stimulate cell growth. However, tumour cells proliferation is usually independent from external stimuli. For example, the MAPK pathways, involving a series of protein kinase cascades play a critical role in regulation of cell proliferation, can be activated independently of growth factors stimulation.

Regulation of intercellular and intracellular signalling

As we discussed earlier, there are two basic mechanisms for intracellular signal transduction: protein phosphorylation and activation cascade of intracellular second messengers. For an effective and robust intracellular signalling, the molecules involved often form a complex at an activated receptor site followed by the signal relays from protein to protein along a signalling pathway.

The first line of self-regulation comes from the high affinity and specificity of the interactions between signalling molecules and their correct partners compared to the relatively low affinity of the interactions between inappropriate partners to avoid the unwanted cross-talk noises (Sharabi et al., 2013). For example, receptor protein kinases often contain additional docking sites that promote a high affinity and specific interaction with the downstream signalling molecules. Once the receptor activated by the extracellular signal molecules, the activated receptor then phosphorylates either itself at multiple sites including the docking site or the specific phospholipids in the adjacent plasma membrane, which can be served as docking sites.

Another effective strategy is the formation of scaffold of proteins, which bring together groups of interacting signalling proteins (scaffolding proteins) into signalling complexes by a scaffolding protein often before a signal has been received to ensure that they interact only with each other and not with inappropriate partners (Nussinov et al., 2013). "Scaffolding protein" is defined as a protein able to interact with several different members of a pathway forming multiprotein complexes. The assembly of effective signalling complex also depends on the proximity (i.e. various highly conserved, small interaction domains) of many intracellular signalling proteins. Interaction domains enable signalling proteins to bind to one another in multiple specific combinations so that the proteins can form linear or branching chains or even three-dimensional networks, which will determine the route followed by the signalling pathway.

The use of modular interaction domains is another way to facilitates signalling transmission. One example, linked to the EGF-EGFRs interaction, is that the activated receptor phosphorylates itself on tyrosines and EGFR2 (Erb2) may pair with another member of the ErbB receptor family (Erb1, Erb3, Erb4) to create an activated heterodimer (Yarden and Sliwkowski, 2001). The Shc PTB domain of adaptor protein Grb2 then interacts with the activated EGF receptor with high affinity and next, Grb2 uses one of its two SH3 domains to bind to a proline-rich region of a protein called Sos, which relays the signal downstream by acting as a GEF to activate a monomeric regulatory GTPase called Ras eventually resulting in the activation of Ras-signalling pathway.

The regulation of intracellular signalling is also controlled by signal amplification and the speed of the response in the signal transduction. It is achieved by four general mechanisms:

1. Specificity: signal molecule fits binding site on its complementary receptor, other signals do not fit;
2. Amplification: when enzymes activate enzymes, the number of affected molecules increases geometrically in a cascade-like manner;
3. Adaptation/desensitization: receptor activation triggers a feedback circuit that shuts off the receptor or removes it from the cell surface;
4. Integration/cross talk: when two pathways connected directly or in-directly it follows a combinational signal that triggers a different response from that the response triggered by each pathway alone (Vert and Chory, 2011).

Autocrine signalling

Apart from the cells signallings we discussed so far in this chapter, which are the signalling from one type of cells to another or between the different cells of same type, a cell can also secret signalling molecules that can bind back to its own receptors. This is called autocrine signalling.

Such a type of signalling is commonly seen in immune cells such as monocytes. Autocrine signalling is an important checkpoint in immune cell activation and allows immune cells to adjust their functional responses to the extracellular cues (Junger, 2011).

The major cancer-related signalling pathways

Once the receptors have activated their first substrate, as described earlier, a signalling pathway is activated. Because of their fundamental role in transmitting signals that regulate all the cell functions, these signalling pathways are widely involved in neoplastic growth. Some of these pathways are described in other chapters dealing with their roles (e.g. cell death or metabolism) (Table 12.1). Here we described the one which form the bulk of signalling involved, according to the KEGG definition, with 'environmental information processing' (i.e. transmitting to the cell the signals from the micro environment so that the cell can adjust its activity accordingly). In cancer either the signals reaching the cell or the pathways themselves can be altered, that result in a dysregulated signalling which can escape the normal control mechanisms. A simplified overall map of the signalling pathways more relevant to cancer is in Figure 12.3. This map is drawn according to the criteria normally used and discussed at the beginning of the chapter. However, we can speculate that the cells signalling representation in Figure 12.4 is perhaps closer to the real thing. In any case, it is immediately evident that in both maps the interconnections existing among all pathways.

The Ras Network

The Ras protein family, composed by N, H, and K Ras, control a large signalling network (Fig. 12.5) widely involved in cancer. Three main pathways have been identified: Ras Raf Erk (MAP kinases), Ras PI3K AKT/Pkb, and Ras Ral Gef. Upstream to the Ras proteins is a family of RTKs mostly binding to growth factors.

Raf Erk (MAP kinases)

Generalities. The mitogenic activated protein kinases pathway (Fig. 12.5A) is highly conserved and regulates mainly proliferation and differentiation. It also controls, with the PI3K pathway, the Retinoblastoma suppressor gene activity.

Mechanisms. Once activated, the Ras proteins interact with the proteins of the Raf family (A-Raf, B-Raf, and C-Raf) which in turn activated Mek and then the Erk scaffolding proteins (i.e. Raf1, Mek 1,2, Mp1, and Erk). 'Scaffolding protein' is defined as a protein able to interact with several different members of a pathway forming multiprotein complexes. The 'Erk scaffolding system' eventually activates the intranuclear targets (Elk1, cMyc, CyclinD1) acting on proliferation and differentiation (McKay and Morrison, 2007). According to the type of cells either a positive or a negative loop can be triggered leading to further proliferations or self limitation (Rauch et al., 2016).

Involvement in cancer. Persistent high levels of ligands in the extracellular environment, mostly growth factors, and activating mutations, mostly in the H-, K-, and N-Ras, B-Raf, C-Raf, or *Mek1/2* genes, are the main alterations found in cancer which lead to abnormal signalling of this pathway. This set of mutated proteins is becoming an attractive target for numerous drugs (Rauch et al., 2016).

PI3K AKT/Pkb

Generalities. The phosphatidylinositol 3'–kinase (PI3K)-AKT pathway (Fig. 12.5B) is one of the most crucial in cell physiology and commonly involved in cancer as it is at the centre of a large network regulating several vital functions of the cell (Martini et al., 2014). It is activated by several types of cellular stimuli or external toxic insults and regulates transcription of genes and translation into proteins, cell cycle, and cell survival. It is under control, through the Ras family, of RTK or, through the cAMP pathway of the GPCR.

Mechanisms. PI3K enzymes are divide into three classes: I, II, and III. Class I are dimers formed by a regulatory and a catalytic (p1100) subunits which, in normal cells, are activated by either growth factors, through RTKs receptors, GPCRs and Ras proteins.

Once activated, Ras stimulates class Ia and Ib PI3K isoforms respectively. PI3K catalyses the production of phosphatidylinositol-3,4,5-triphosphate (PIP3) at the cell membrane. PIP3 in turn serves as a second messenger that helps to activate AKT. This regulation is also mediated by Pdk1, a positive regulator, and Pten that has an inhibitory effect. Once active, AKT phosphorylates more than 200 substrates (Spangle et al., 2017) controlling key cellular processes by phosphorylating substrates involved in cell survival, motility, growth, glycolysis, cell cycle, and DNA repair (Martini et al., 2014; Courtnay et al., 2015). It also interacts with mTOR pathway which in turn controls cell survival, motility, microtubules lipids, autophagy, and protein synthesis (Martini et al., 2014).

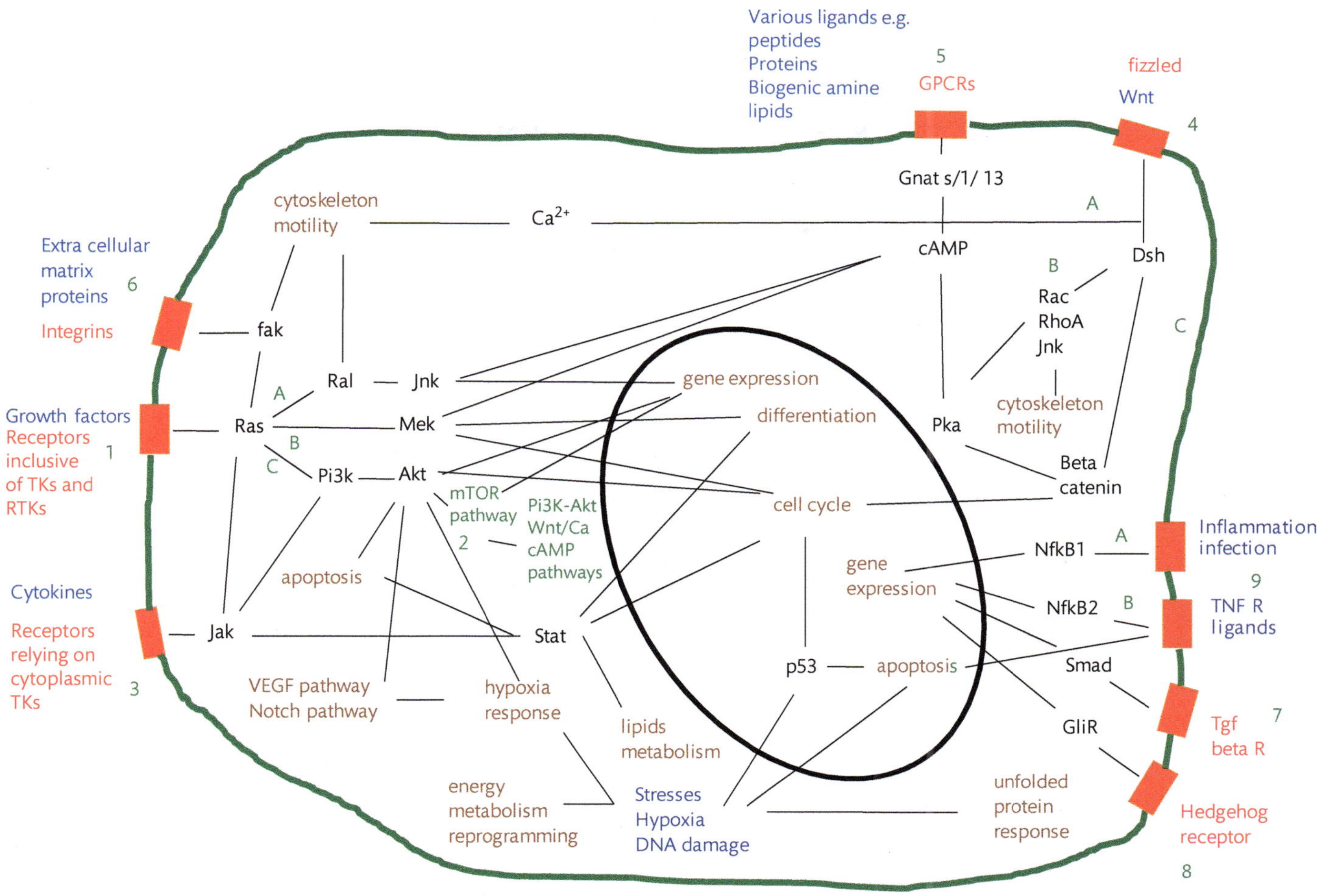

Fig. 12.3 An overview of the signalling pathways discussed in this chapter. Ligands or other activating functions in blue and receptors in red. Pathways and activated functions in green. **1:** Ras network (A: Ral pathway, B: MAPK pathway, C: PI3K/AKT pathway). **2:** mTOR pathway. **3:** Jack-Stat pathway. **4:** WNT pathways (A: Ca^{2+} pathway. B: Plain Cell Polarity Pathway C: Canonical Pathway). **5**: cAMP pathway. **6:** Focal adhesion pathway. **7:** Tgf-beta pathway. **8:** Hedgehog pathway. **9:** NfKb pathway (A: canonical, B: non-canonical). Note the cross talking between all the pathways.

Source: data from KEGG (Kyoto Encyclopedia of Genes and Genomes), *Pathways in cancer—Homo sapiens (human)*, Copyright © Kanehisa Laboratories. Available from http://www.genome.jp/kegg-bin/show_pathway?map=hsa05200&show_description=show

Class II PI3K are monomers mainly involved in regulating vesicular traffic including the transport on the cell membrane of the glucose transporter Glut4. PI3K-C2alpha is the member of this class for which we have more information and, alongside Glut4, is also involved in endocytosis and inducing sprouting angiogenesis. Another member of this class, PI3K-C2beta, less well investigated, appears to be involved instead in cell migration and inhibition of apoptosis. Both class I and class II are downstream to RTKs and GPCRs. Finally, Class III PI3K are involved in vesicular trafficking and nutrient sensing (Martini et al., 2014).

PI3K/AKT pathway also exercises its physiological regulatory roles through epigenetic mechanisms as AKT phosphorylates substrates that regulate chromatin conformation. Because of this epigenetic modification, the activity of these substrates is altered, making the DNA available, or not for transcription (Spangle et al., 2017).

Involvement in cancer. This pathway has a unique role in cancer, not only because of the variety of cellular functions affected but also because, unlike other pathways, every major hub of this pathway can be altered in cancer (Martini et al., 2014; Courtnay et al., 2015).

Class Ia PI3K includes four catalytic subunits (p110 alpha, beta, gamma, and delta) and one regulatory subunit p85, which are involved by genetic damage. Mutations are found in p110 alpha, beta, delta and p85. Amplification is instead found in p110 alpha, beta, gamma, and delta. Pten is a suppressor gene as it physiologically blocks AKT activation, and like all suppressor genes it can be also altered in hereditary tumours. AKT three members (1, 2 and 3) can be affected by mutation (AKT3) and amplification (AKT1 and AKT2).

As the Pik3 pathway is also involved in epigenetic regulation, alterations to its activity can lead to epigenetic changes which can promote oncogenesis. An example is the ability of AKT to reduce DNA replication-associated DNA methylation, increasing transcriptional activity. AKT stabilizes, through phosphorylations, DNA methyltransferase 1 (DNMT1), which in turn prevents DNA methylation by other enzymes. As a result, transcription increases. By phosphorylating the histone Ezh2, AKT diminishes the affinity of this histone for chromatin thus reducing its transcription-repressing activity. In other mechanisms, the histone acetyltransferase activity and the substrate affinity of p300/CBP is enhanced with

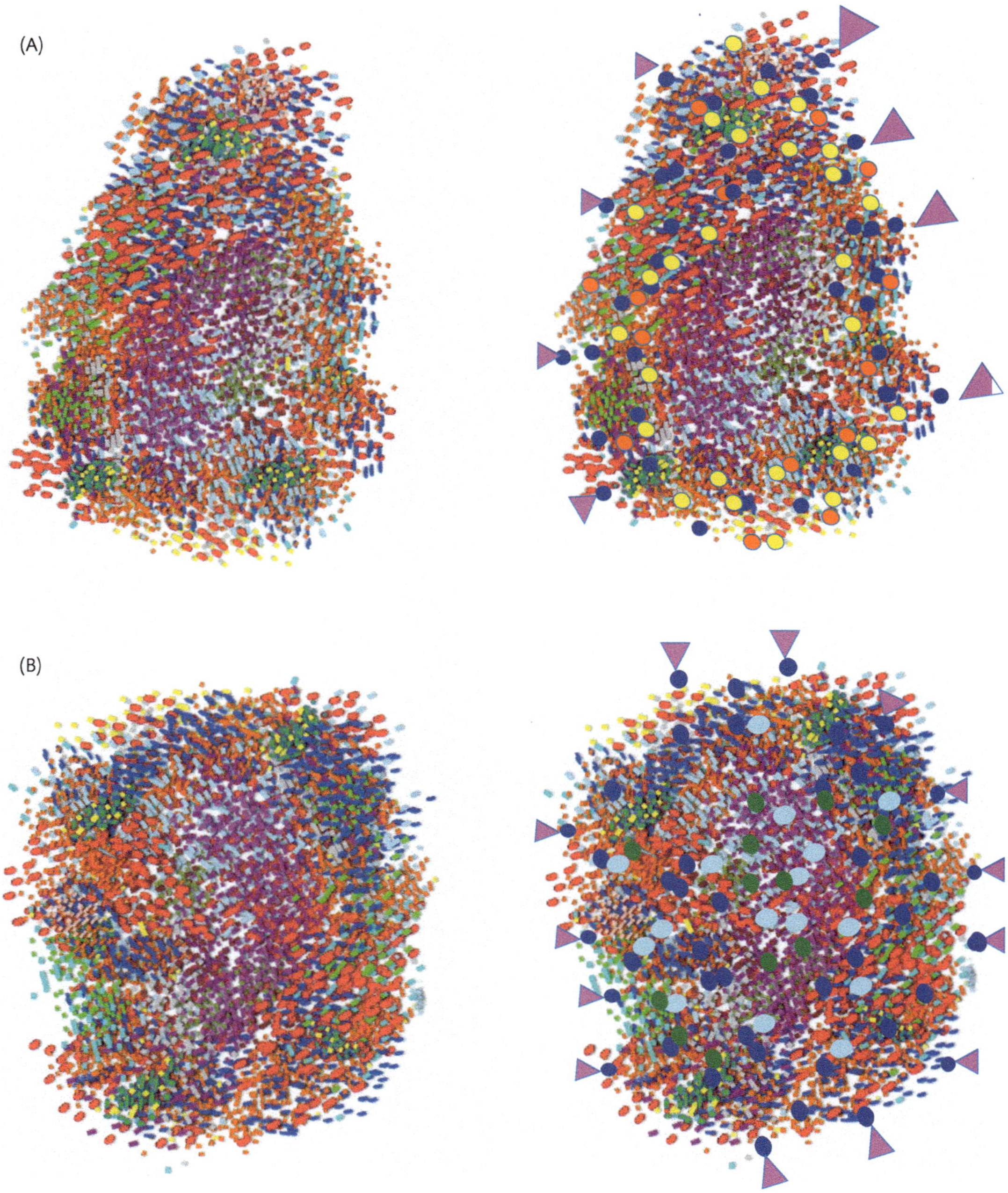

Fig. 12.4 What pathways look like in a cell. The two-dimensional diagram provides the 'anatomical scheme' of each pathway but the real situation is more complex. Inside the cells, the component of the pathways is scattered and, according to the type of cells, cross talk can produce different effects. The components of the pathways are represented by a coloured dot. Members of one pathway are of the same colour. (A) Cell A is stimulated with the purple growth factor which activates the blue pathway. In this particular cell, the activation from the blue pathways extend mostly to the yellow and orange pathways in the cytoplasm (e.g. causing increase in motility). (B) Cell B is also stimulated with the purple growth factor. This time the activation of the blue pathway extends mostly to the light blue and green pathways which transmit the signal to the nucleus (e.g. triggering transcription of a group of genes).

AKT phosphorylation, resulting in increased acetylation of H3K56 and other lysines, leading to transcriptional activation (Spangle et al., 2017).

Ral (Ras-like small GTPases)

Generalities. This is a small collateral pathway (Fig. 12.5C) which is divided into two: through RalA is involved in control of motility, glycolysis, and interact with the MAPK pathway. It is also connected with the Jack/Stat pathway to activate Jak. RalB controls endocytosis and apoptosis through direct binding to Exo84 and inducing downstream activation of ULK1 and Beclin1-VPS34 complex. (Feig et al., 1996; Bodemann et al., 2011). Both branches affect transcription, mitophagy, and autophagy.

Mechanisms. This is a small collateral pathway (Fig.12.5C), which is regulated by a family of guanine nucleotide exchange factors (Rgl). Rgl proteins are regulated by their interaction with the

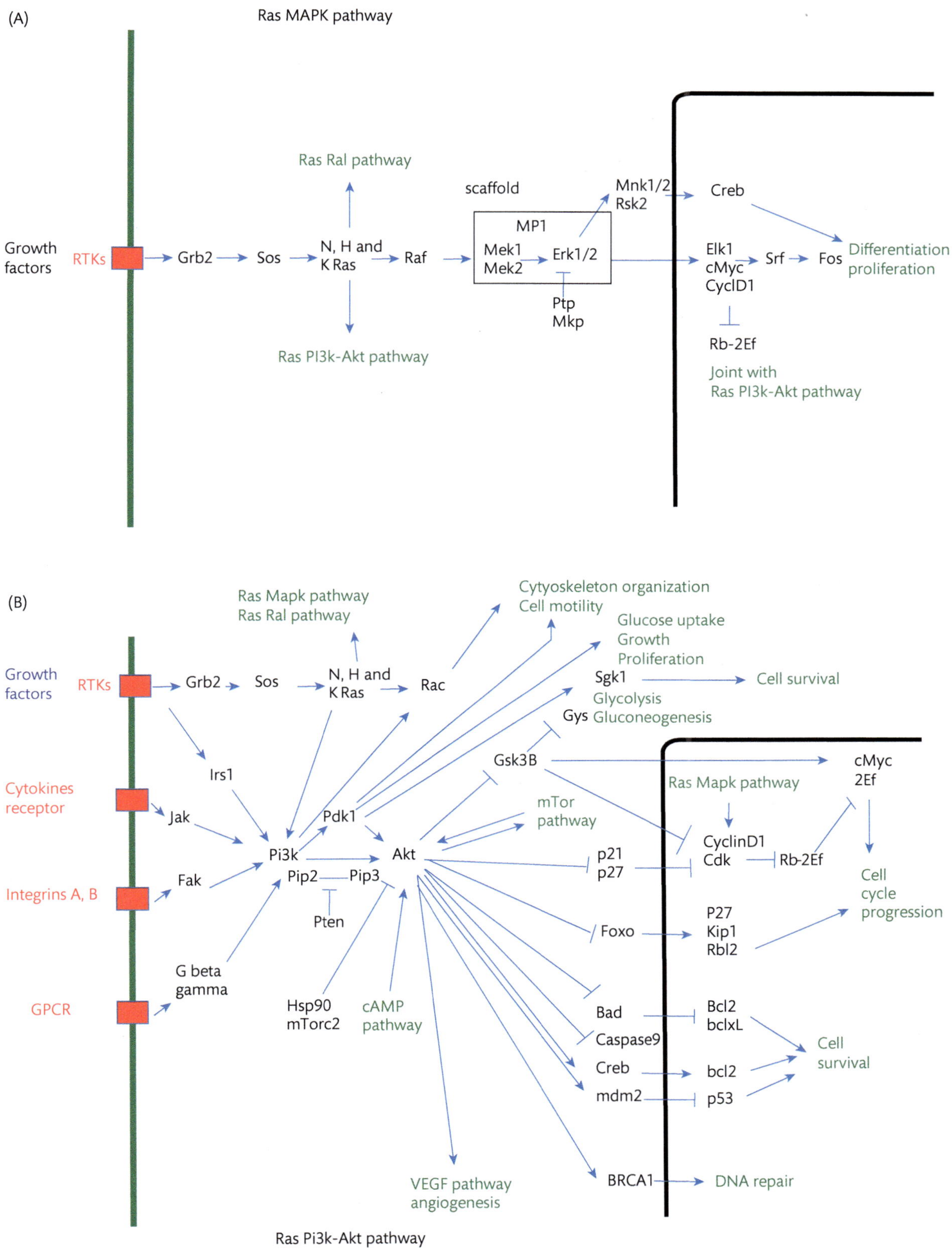

Fig. 12.5 The Ras network. It includes three pathways: the MAPK pathway, the PIk3/AKT pathway, and the Ral pathway. (A) The MAPK pathway. It is mostly concerned with proliferation and differentiation. (B) The PI3K/AKT pathway. It is centred on the PI3K-AKT axis. It is an highly complex one with involvement in many different functions. (C) The Ral pathway. Mostly involved in regulating cell polarity, cell survival, motility, and energy metabolism

(A) and (B) source: data from KEGG (Kyoto Encyclopedia of Genes and Genomes), *Pathways in cancer—Homo sapiens (human)*, Copyright © Kanehisa Laboratories, available from http://www.kegg.jp/pathway/hsa05200; KEGG (Kyoto Encyclopedia of Genes and Genomes), *MAPK signaling pathway—Homo sapiens (human)*, Copyright © Kanehisa Laboratories, available from http://www.kegg.jp/kegg-bin/show_pathway?hsa04010; KEGG (Kyoto Encyclopedia of Genes and Genomes), *Ras signaling pathway—Homo sapiens (human)*, Copyright © Kanehisa Laboratories, available from http://www.genome.jp/keggbin/show_pathway?map=hsa04014&show_description=show; KEGG (Kyoto Encyclopedia of Genes and Genomes), *KEGG Pathways in cancer—Homo sapiens (human)*, Copyright © Kanehisa Laboratories, available from http://www.kegg.jp/pathway/hsa05200; and Sherr CJ and McCormick F, 'The RB and p53 pathways in cancer', *Cancer Cell*, Volume 2, pp. 103–12, Copyright © 2002 Cell Press. (C) Source: data from KEGG (Kyoto Encyclopedia of Genes and Genomes), *KEGG Pathways in cancer—Homo sapiens (human)*, Copyright © Kanehisa Laboratories, available from http://www.kegg.jp/pathway/hsa05200; KEGG (Kyoto Encyclopedia of Genes and Genomes), *MAPK signaling pathway—Homo sapiens (human)*, Copyright © Kanehisa Laboratories, available from http://www.kegg.jp/kegg-bin/show_pathway?hsa04010; KEGG (Kyoto Encyclopedia of Genes and Genomes), *Ras signaling pathway—Homo sapiens (human)*, Copyright © Kanehisa Laboratories, available from http://www.genome.jp/keggbin/show_pathway?map=hsa04014&show_description=show; and Kashatus DF, 'Ral GTPases in tumorigenesis: emerging from the shadows', Experimental Cell Research, Volume 319, Issue 15, pp. 2337–42, Copyright © 2013 Elsevier Inc. All rights reserved.

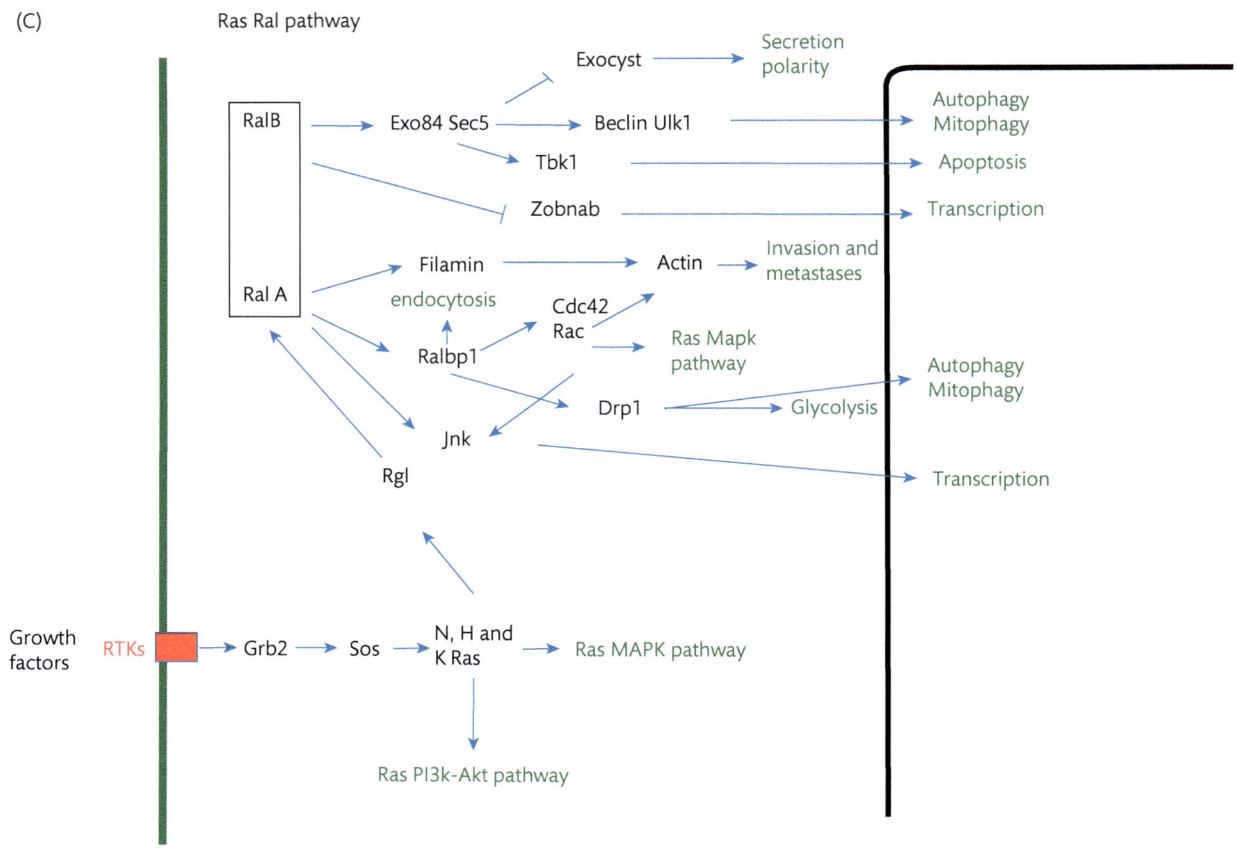

Fig. 12.5 Continued

effector's region of GTP-bound Ras (Kashatus, 2013). In this way, any oncogenic mutation of the Ras family can activate this pathway (Kashatus, 2013).

Involvement in cancer. Ral is very much involved in oncogenesis, although our knowledge is still sketchy. Ralbp1 is a large protein with several binding domains and is a hub involved in regulation of endocytosis, motility, autophagy, and metabolism regulation. Glycolysis is regulated by delivering phopso-Drp1 to mitochondria. Exo84 and Sec5 are other crucial hubs involved in transcription (Kashatus, 2013; Gentry et al., 2014).

The mTOR pathway

Generalities. mTOR (mammalian Target Of Rapamycin) pathway is a hub nested between other pathways (Fig. 12.3) and at its core are two super complexes (Fig. 12.6A): the mTorc1 and the mTorc2 (Fig. 12.6A) which have serine/threonine protein kinase activity. It owes its unusual name to the fact that this pathway was discovered to be the target of Rapamycin, an antifungal drug (Brown et al., 1994). Both super complexes are mainly controlled by the PI3K pathway (Fig. 12.6A), through the Tsc1/2-Tcb1D7 intersection mTorc1 and directly, mTorc2. However, the Tsc1/2-Tcb1D7 intersection mTorc1 is also regulated by the Wnt, the Mapk, the AMPK, and the hypoxia pathways. When active mTorc1 regulates, through transcription of target genes, a series of functions, including cell proliferation and macromolecules expression. It also inhibits mTorc2. mTorc 2 instead acts as a loop and sends regulatory signals back to the PI3K, the Wnt, and the cAMP pathways, eventually affecting cell survival, glucose metabolism, and motility.

Mechanisms. Activation of PI3K and Mapk is a key event to induce proliferation, however it is essential that a cell does not proliferate if certain conditions, like availability of nutrients, adequate energy metabolism, and molecular building blocks, are not available (Yu and Cui, 2016; Kim et al., 2017). mTor is a hub which provides coordination between proliferative signals and the presence of adequate intracellular environmental conditions (Saxton and Sabatini, 2017).

mTorc1 is a crucial hub which coordinates the presence of a proliferative signal with the availability of necessary features like ATP, oxygen, glucose, and also amino acids (amino acid sensors linked to mTorc1 have been recently discovered; see Wolfson and Sabatini, 2017). In a normal cell, through the mTorc1, the proliferating signal can be overrun and deactivated if the conditions are not right, as exemplified in Figure 12.6B and C. On the left (Fig. 12.6B) a proliferative signal reaches a cell in favourable conditions. Only a modest stimulation of the mTorc1 inhibitors Tsc1/2 is achieved through the AMPK pathway, leaving most of the mTorc1 active. The growth factor stimulation is therefore barely contrasted and a good level of proliferation is reached. However, if conditions are suboptimal (Fig. 12.6C) the lower ATP/AMP ratio, the hypoxia and/or the suboptimal amount of nutrients will increase the amount of active Tsc1/2 proteins, which will inactivate most of the mTorc1.

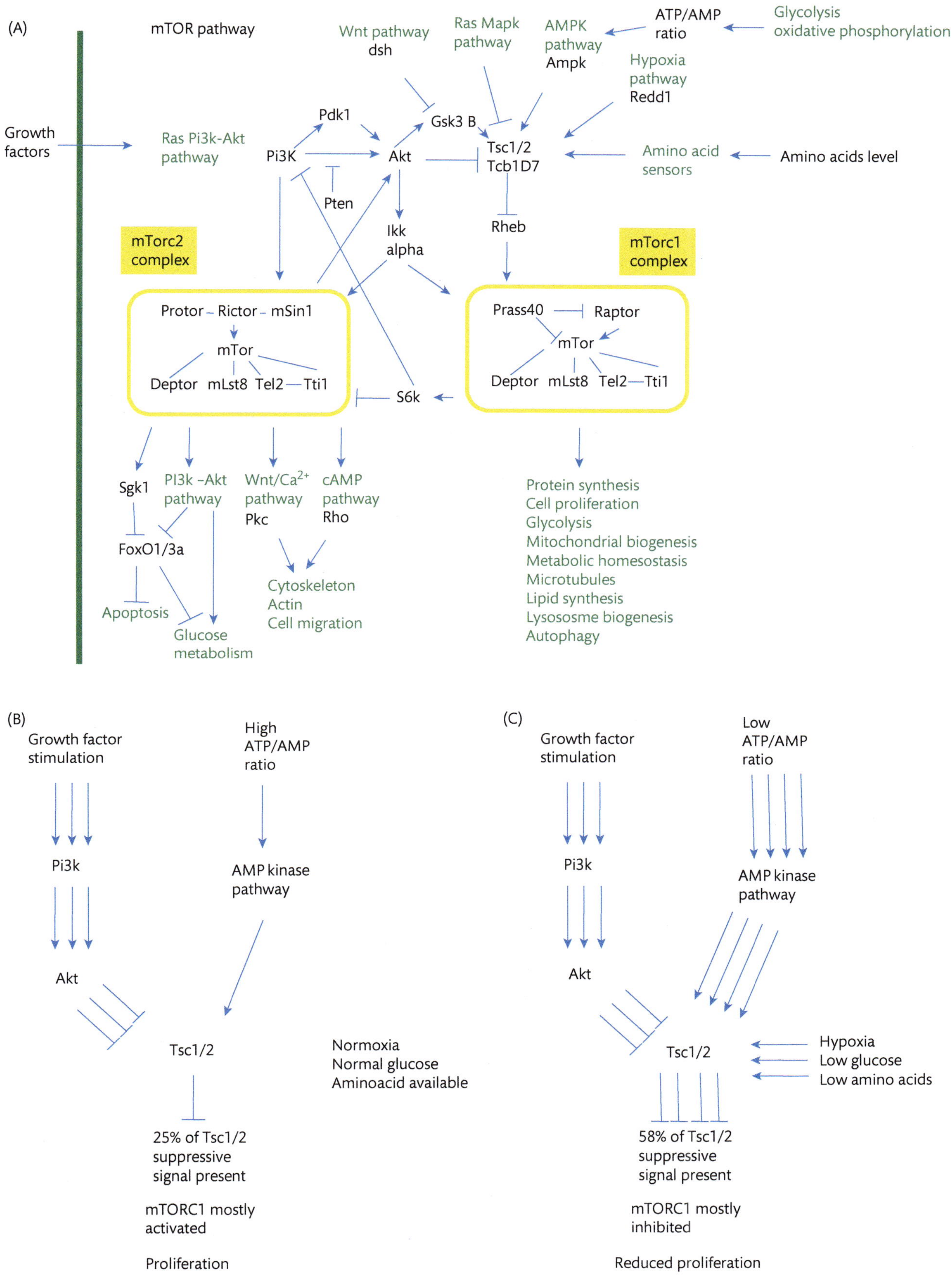

Fig. 12.6 The mTOR pathway. The mTOR pathway is nested between other pathways. Its main function is to coordinate proliferating activity with availability of resources. It is based on two complexes, termed mTOR complex 1 (mTorc1) and 2 (mTorc2).

Source: data from KEGG (Kyoto Encyclopedia of Genes and Genomes), *KEGG Pathways in cancer—Homo sapiens (human)*, Copyright © Kanehisa Laboratories, available from http://www.kegg.jp/pathway/hsa05200; KEGG (Kyoto Encyclopedia of Genes and Genomes), *mTOR signaling pathway—Homo sapiens (human)*, Copyright © Kanehisa Laboratories, available from http://www.genome.jp/kegg-bin/show_pathway?map=hsa04150&show_description=show; Kim LC et al. 'mTORC1 and mTORC2 in cancer and the tumor microenvironment', *Oncogene*, Volume 36, Issue 16, pp. 2191–201, Copyright © 2016 Springer Nature; and Saxton RA and Sabatini DM, 'mTOR Signaling in Growth, Metabolism, and Disease', *Cell*, Volume 168, Issue 6, pp. 960–76, Copyright © 2017 Elsevier Inc.

Consequently, most of the proliferative signal will be blocked and therefore, although the growth factor stimulation is still present as for the cell in optimal condition (Fig. 12.6), less proliferation will be achieved (Yu and Cui, 2016; Kim et al., 2017). If intracellular conditions are adequate and mTorc1 is activated, it will not only allow proliferation but will also stimulate mitochondria and glycolysis to maintain optimal ATP concentration, protein synthesis, and lipid synthesis to provide the building blocks for mitosis (Saxton and Sabatini, 2017).

Less is known about mTorc2. One function is to contribute to proliferating signals control acting as a loop on PI3K/AKT and contributing to phosphorylating members of the ACG family of protein kinases. Through the Sgk1 genes and, again the PI3K/AKT pathway, it is also involved in controlling apoptosis and glucose metabolism. Finally, through the regulation of Wnt/Ca^{2+} and cAPM pathways promotes cell motility and cytoskeleton organization (Yu and Cui, 2016; Kim et al., 2017; Saxton and Sabatini, 2017).

Involvement in cancer. The mTOR pathway is widely hyperactive in malignancies (Ersahin et al., 2015). Tsc1 and Tsc2 are tumour suppressor genes discovered because of their loss causes tuberous sclerosis, a congenital genetic condition in which an excess of benign tumours develop. As a consequences, mTorc1 is excessively activated (Kim et al., 2017). It is not known why only benign tumours are produced. One hypothesis is that in presence of a single alteration causing only mTorc1 activation, the benign tumours are self-limiting. Possibly because mTorc1, while promoting cell proliferation downstream, also inhibits PI3K activation through a feedback loop (Kim et al., 2017). Two other suppressors, p53 and Lkb1, are also negative regulators of mTorc1 and their inactivation leads to abnormal mTOR activity (Saxton and Sabatini, 2017). Finally activating mutations of *mTORC1* gene also occurs (Saxton and Sabatini, 2017).

mTorc2 is also involved in cancer as stimulate the PI3K/AKT axis. Amplification of the Rictor, one of the members of the mTorc2 super-complex, has been detected in several cancers and is associated with the activation of mTorc2 (Saxton and Sabatini, 2017).

Activation of any of the two mTOR components can also follow genetic damage on the upstream pathways, all heavily involved in oncogenesis. In this instance mTOR activity is very important as it allows adjustment of all the nutritional and metabolic requirements for the neoplastic cells to grow (Kim et al., 2017; Saxton and Sabatini, 2017). Selective targeting of mTOR is therefore very much under investigation, even in absence of direct genetic damage to the mTOR complex.

The JAK-STAT pathway

Generalities. The JAnus Kinase (JAK)/signal transducers and activators of transcription (STAT) pathway controls development and homeostasis: a canonical and a non-canonical pathway have been described (Fig. 12.7). In human, like in other mammals, the canonical pathway is the main signalling mechanism activated by several different ligands like cytokines, hormones, and growth factors. Following the binding of cytokines to their cognate receptor, proteins belonging to the STAT family are activated by the JAK family of non-receptor tyrosine kinases. It regulates apoptosis, cell cycle, lipid metabolism, and differentiation. The non-canonical pathway instead provide control of the heterochromatin by epigenetic mechanisms (Li, 2008). The JAK-STAT pathway is defined as a 'dual address' pathway (i.e. a pathway in which one of the components is translocated from cytoplasm to nucleus). The other dual address pathways are the Wnt canonical pathway, the TGF-beta pathway, the hedgehog pathway (discussed later in this chapter) and the Notch pathway (discussed in Chapter 22).

Mechanism. In the canonical pathway (Li, 2008), once the receptor is activated by the ligand, the JAK protein induced phosphorylation of STAT leading to formation of phosphorylated STAT dimer. As a dimer, STAT translocates into the nucleus induces transcription of target genes. In addition to the activation of STATs, JAK activation also affects the Ras Mapk pathway through PI3K, the PI3K/AKT and the mTOR pathways. The non-canonical pathway instead affects STAT's role in a different context. Alongside being in the cytoplasm as a signal transmitter, STATs are also present in the heterochromatin where they link to the heterochromatin protein 1 (HP1) in order to maintain heterochromatin stability and transcription repression. Upon phosphorylation by JAK, STATs are removed and HP1 is also displaced, leaving the DNA available for transcription (Li, 2008).

Involvement in cancer. The main pathological alteration found in cancer is the persistent phosphorylation of STATs. This can be achieved in different ways like presence of abnormal fusion proteins like TEL-JAK, which has the JAK portion abnormally active or presence in the extracellular spaces of abnormal levels of ligands (e.g. EGF or interleukin-6). Finally, abnormal high levels of non-receptors tyrosine kinases, like Abl, can keep the pathway over active (Bromberg, 2002).

The WNT pathway

Generalities. This pathway (Fig. 12.8) includes the canonical WNT pathway and non-canonical (defined as 'beta-catenin independent') branches: the Wnt/Pcp (planar cell polarity) and the Wnt/Ca^{2+} are the two best described. The established WNT is directly involved in cell cycle and proliferation and is now recognized as fundamental in the regulation of stem cells (Reya and Clevers, 2005). The non-canonical pathway Ca^{2+} controls cell fate in developmental processes and, through NfKb, transcription while both the Ca^{2+} and PCP regulates motility and migration, for example, migration of neural crest cells during development and invasion (Zhan et al., 2017). It should be however kept in mind that the canonical and non-canonical pathways are highly integrated and affect the activity of each other.

Mechanisms. Humans have 19 WNT proteins (ligand) and 10 frizzled receptors involved with this pathway. The Wnt canonical is mainly stimulated by Wnt16, which links to a dimmer formed by a frizzled receptor and the Lrp5/6 (LDL receptor related protein 5 and 6). In absence of stimulation, beta-catenin is linked to a group of other proteins (Gsk beta, Ckl alpha, annexin, and Apc) and this complex is ubiquitinated and degraded. Under activation, beta-catenin is stabilized and transferred in the nucleus where it promotes the transcription of Wnt-associated genes promoting proliferation (Corda and Sala, 2017; Zhan et al., 2017).

The non-canonical Wnt/Pcp pathway (Xiao et al., 2017), activates two GTPases (RhoA and Rac) leading to the activation of two further kinases (Jnk and Rock). Jnk, shared with the Ras Ral pathway,

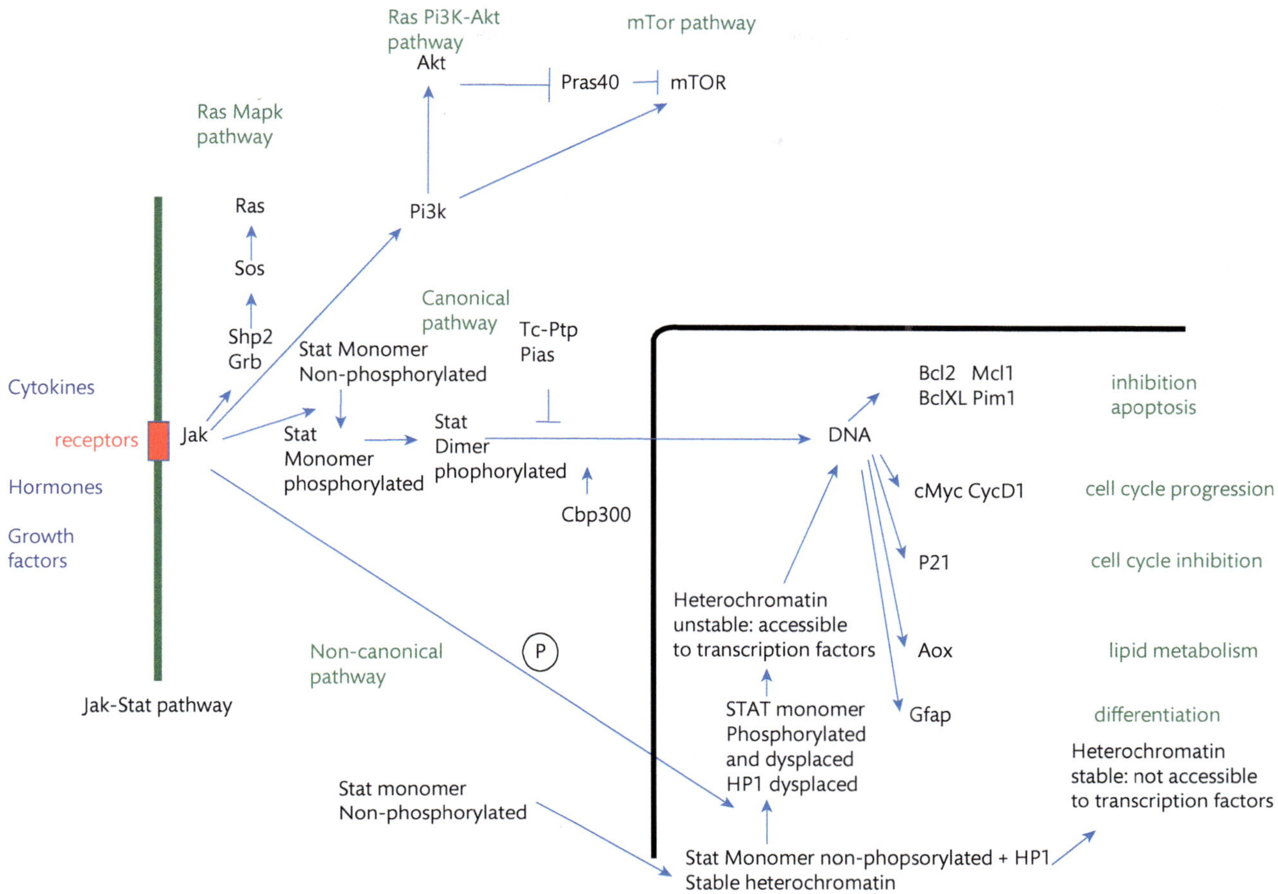

Fig. 12.7 JAK-STAT pathway. In the canonical pathway Jak phosphorylates the Stat monomer which move on to form a Statdimer. This latter pathway moves into the nucleus and act as transcription factor. The non-canonical pathway instead is involved in epigenetic regulation of transcription: non-phosphorylated Stat monomer links to DNA and maintain the heterochromatin stable. Following phosphorylation by Jak, Stat is displaced and the heterochromatin becomes accessible to transcription factors.

Source: data from KEGG (Kyoto Encyclopedia of Genes and Genomes), *KEGG Pathways in cancer—Homo sapiens (human)*, Copyright © Kanehisa Laboratories, available from http://www.kegg.jp/pathway/hsa05200; KEGG (Kyoto Encyclopedia of Genes and Genomes), *Jak-STAT signaling pathway—Homo sapiens (human)*, Copyright © Kanehisa Laboratories, available from http://www.genome.jp/kegg-bin/show_pathway?map=hsa04630&show_description=show; and Li WX, 'Canonical and non-canonical JAK-STAT signaling', *Trends in Cell Biology*, Volume 18, Issue 11, pp. 545–51, Copyright © 2018 Elsevier Inc.

induces transcription through the Jun transcription factors family. The Rock2 kinase instead, a key regulator of actin and cell polarity, phosphorylates myosin light chain, as a consequence, more actin binds to myosin and contractility increases (Riento and Ridley, 2003). Finally, the Wnt/Ca^{2+} pathway (Xiao et al., 2017) increases intracellular Ca^{2+} signalling activating a group of Ca^{2+}-dependent kinases. The first is the family of calcineurin (CaN) serin/threonine protein phosphatizes which targets the family of NFATC (nuclear factor of activated T-cells) which regulates cell fate and motility. The other two Ca^{2+} dependent kinases, CaMkII and Pkc, trigger transcription through NfkB and contribute to cell motility by activating cdc42 which induces the formation of podosomes (actin-rich protrusions), which enhances motility and invasion. Finally, CaMkII and Pkc have also an inhibitor effect on beta-catenin through the Talk1, Nlk cascade (Corda and Sala, 2017; Xiao et al., 2017).

Involvement in cancer: Canonical pathway

Mutations are a common event in the Wnt pathway in a wide spectrum of cancers (Zhan et al., 2017). Colorectal adenocarcinomas are the one more commonly hosting damage canonical pathway: this its a reflection of the key role that this pathway has in the biology of gut epithelial cells (Reya and Clevers, 2005). It is also commonly activated in breast, leukaemias, and melanoma (Zhan et al., 2017). The Apc tumour suppressor gene was identified as responsible for the familial adenomatous polyposis in which one Apc allele is missing in the germline. Subsequently, it has been found mutated and inactivated also in 80% of sporadic colorectal cancers and, less frequently, in other malignancies (Sedgwick and D'Souza-Schorey, 2016). When its function is lost, beta-catenin remains active. In a smaller number of tumours, it is another beta-catenin inactivator that is mutated, the axin suppressor gene. More rarely mutations in beta-catenin make it resistant to destruction by removing the N-terminal Ser/Thr destruction motif (Reya and Clevers, 2005; Sedgwick and D'Souza-Schorey, 2016; Zhan et al., 2017).

Involvement in cancer: Non-canonical pathway

Wnt5a overexpression, reported in melanoma and gastric cancer, results in increased cell migration and metastases as activates the Wnt/Ca^{2+} but not the canonical pathway (Corda and Sala, 2017). There are several members of the Wnt/PCP pathway abnormally

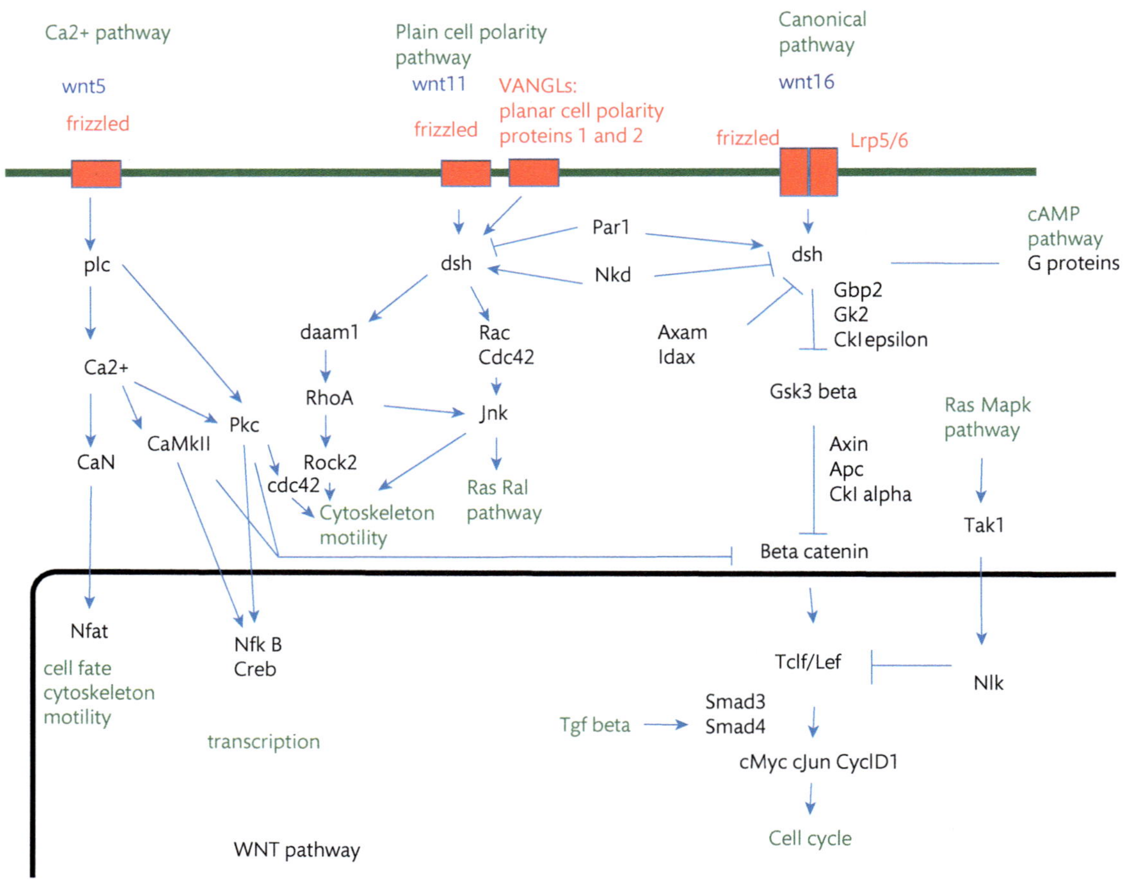

Fig. 12.8 The WNT pathways. There are three pathways: the WNT canonical (defined as beta-catenin dependent) and two non-canonical pathways, the planar cell polarity (PCP) Wnt/Pcp and the Wnt/Ca^{2+}. In the WNT canonical pathway, the binding of Wnt to its receptor leads to the stabilization of beta-catenin through inhibition of the specific degradation complex. Beta-catenin is then free to enter the nucleus and activate Wnt-regulated genes. The Wnt/Pcp pathway through activation of RhoA and Junk induces motility alongside the Wnt/Ca^{+} pathway.

Source: data from KEGG (Kyoto Encyclopedia of Genes and Genomes), *KEGG Pathways in cancer—Homo sapiens (human)*, Copyright © Kanehisa Laboratories. Available from http://www.kegg.jp/pathway/hsa05200; KEGG (Kyoto Encyclopedia of Genes and Genomes), *Wnt signaling pathway—Homo sapiens (human)*, Copyright © Kanehisa Laboratories. Available from http://www.kegg.jp/kegg-bin/show_pathway?hsa04310+5530;

Niehrs C. 2012. 'The complex world of WNT receptor signaling', *Nature Review Molecular Cell Biology*, Volume 13, Issue 12, pp. 767–79, Copyright © 2012 Springer Nature; and Sedgwick AE and D'Souza-Schorey C, 'Wnt Signaling in Cell Motility and Invasion: Drawing Parallels between Development and Cancer', *Cancers (Basel)*, Volume 8, Issue 9, pii: E80, Copyright © 2016 by the authors; licensee MDPI, Basel, Switzerland.

expressed in different types of cancer. VANGL1 (planar cell polarity protein 1), DSH and the frizzled receptor FZD6 are necessary for cell motility and FZD6 is found amplified in tumours (Sedgwick and D'Souza-Schorey, 2016; Corda and Sala, 2017). The VANGL2 (planar cell polarity protein 2), leads instead to increased proliferation through the other branch, activating the Jnk and Ras Ral pathways. Frizzled FZD6 overexpression is also involved in malignant transformation of B lymphoid cells (Corda and Sala, 2017).

The cAMP/GPCR pathway

Generalities. cAMP is one of the most common and universal second messengers, and its formation from ATP is induced by adenylyl cyclase (AC). Formation of cAMP from ATP is at the core of this pathway. Adenyl cyclase is downstream to and is activated by the family of GPCRs, the largest family of receptors in humans (Bar-Shavit et al., 2016) with a large number of ligands, including hormones and neurotransmitters. This pathway is involved in motility and apoptosis, but it has also a close cross talk with PI3K, Wnt canonical, and Wnt/Pcp pathways, which extend the influence of the cAPM network (Fig. 12.9)

Mechanisms. The GPCRs make the largest family of receptors present on cell membrane. Four main groups have been described: Group A, rhodopsin-like; Group B, secretin-like; Group C is comprised of metabotropic glutamate/pheromone receptors; and Group D, frizzled receptors, serving as receptors in the WNT pathway (Bar-Shavit et al., 2016). Once a GPCR binds to its ligand, changes conformation and activate adenyl cyclases, inducing generation of cAMP from ATP. cAMP acts primarily by activating PKA, cAMP gated ions, and Epac (Fig. 12.9; Godinho et al., 2015).

Involvement in cancer. Because of the key role of this pathway played in a number of critical functions, it is increasingly reported to be involved in cancer. A number of GPCRs are overexpressed leading, through cAMP accumulation, to pathological stimulation of the Wnt canonical and the PI3K pathway (Fig. 12.9). Alterations

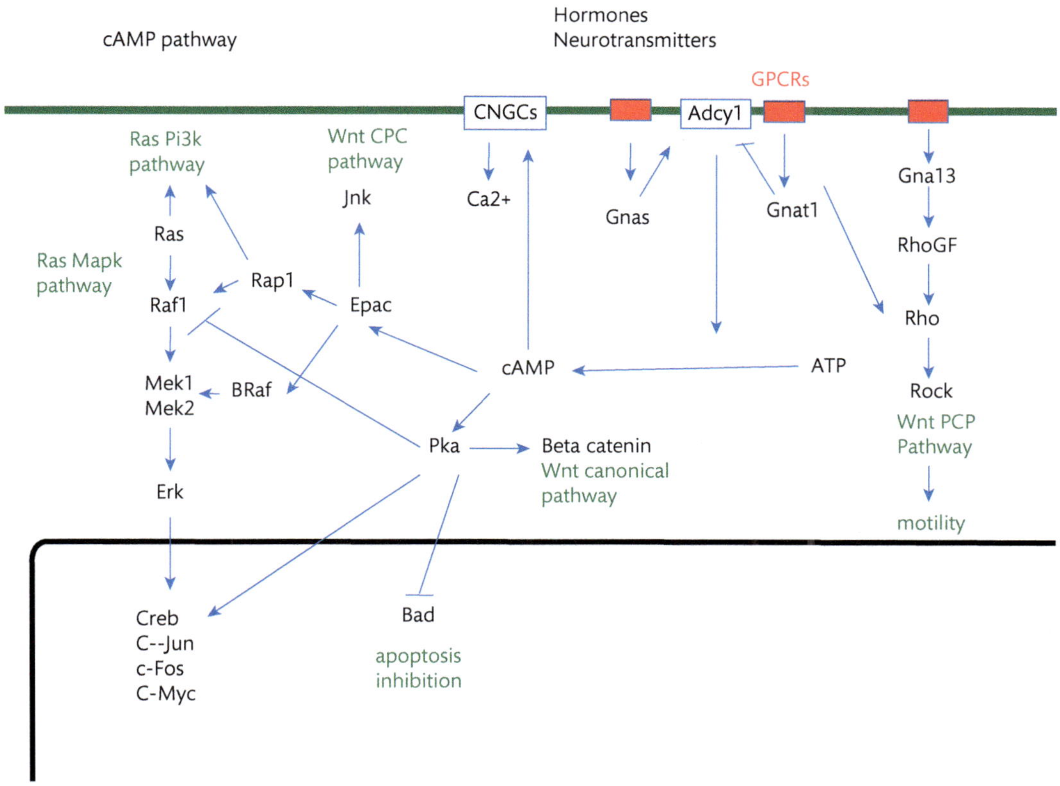

Fig. 12.9 The cAMP pathway. cAMP is one of the most common messengers. In this pathway, its formation is promoted by adenylyl cyclase (AC) activation, which follows the binding of ligands to that of G protein-coupled receptors (GPCRs). GPCRs make the largest family of receptors in humans and bind to a very broad spectrum of ligands including hormones and neurotransmitters. This pathway, alongside cell death regulation, acts mostly by interacting with other pathways.

Source: data from KEGG (Kyoto Encyclopedia of Genes and Genomes), *cAMP signaling pathway—Homo sapiens (human)*, Copyright © Kanehisa Laboratories. Available from http://www.genome.jp/kegg-bin/show_pathway?map=hsa04024&show_description=show

can also follow the occurrence of mutations in the GPCRs (in approximately 20% of human tumours). Another consequence can be the transactivation of the RTK family, mostly regulating the RAS network. The opposite also happens in which abnormal signalling from the RTKs can reach the cAPM pathway (Nogues et al., 2017).

The focal adhesion kinase pathway

Generalities. Focal adhesion kinase is the main pathway involved in cell–ECM interaction. Adhesion of ECM is an active process in which a continues exchange of signals occurs between the cell and the ECM. The main receptors involved are the integrin A and B which in turn activates the Fak (Focal adhesion kinase gene), the key hub for this pathway. This pathway affects cell motility and adhesion through actin regulation and cytoskeleton organisation, creating the structural links between cytoskeleton and integrin receptors to give the cell an adhesive or motile phenotype according to the physiological requests. It can also influence the cell cycle and cell differentiation through its interactions with the Mapk, PI3K, and Wnt canonical pathways (Fig. 12.10). It is therefore a pivotal pathway to coordinate adhesion, motility, and proliferation (Zhao and Guan, 2009).

Mechanisms. When activated by ligands present in the ECM, the integrins trigger the autophosphorylation of Fak. As a consequence, a binding site is freed on Fak, leading to the formation of a Fak-Src complex. A positive loop, inducing more phosphorylation on Fak, is also triggered. The Fak-Src complex leads to re-organization of actin cytoskeleton according to the adhesion signals received. Actin cytoskeleton arrangement allows the cell to adjust its adhesion strength, shape, and motility status. Cross talk with the Ras network assures the coordination between growth stimuli and adhesion (Zhao and Guan, 2009). As discussed earlier on in this chapter, it can also be activated by mechanic signals like increased rigidity of the ECM: this example is illustrated in Fig. 12.2.

Involvement in cancer. Central to the involvement of this pathway in cancer is the Fak activation in absence of physiological signals, in which increased levels of mRNA and proteins were found in a large number of different cancers although the molecular basis remains largely unknown (Zhao and Guan, 2009). Fak amplification has been found but only in a minority of cases. P53 and NfkB have been found to be able to, respectively, repress and induces Fak transcription, suggesting a role for these two genes in its abnormal expression (Zhao and Guan, 2009). Abnormal Fak activation has therefore pathological consequences in cancer spreading and, through the PI3K, cancer proliferation (Zhao and Guan, 2009; Tai et al., 2015). Another role in cancer, very much characteristic of this pathway, is the influence on the cancer-associated stroma.

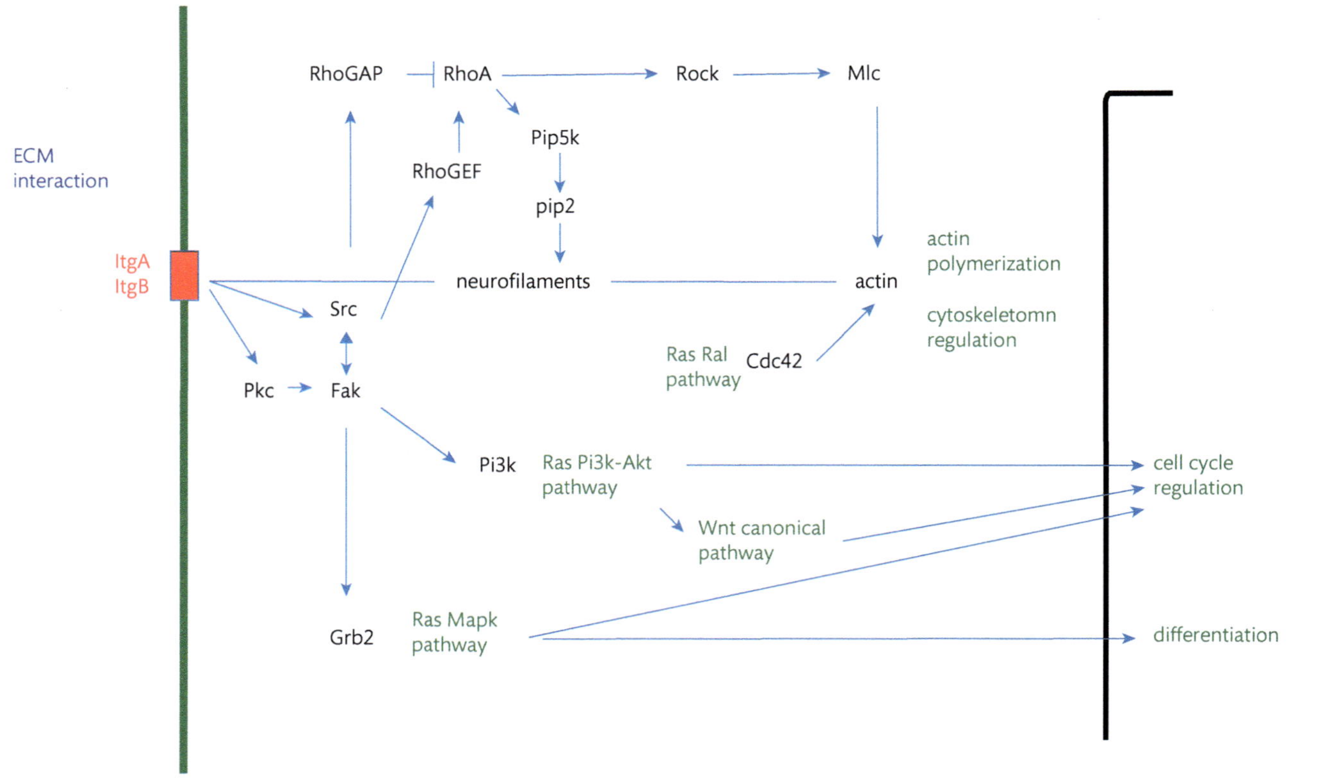

Fig. 12.10 Focal adhesion pathway. This pathway mediates the interaction between the cell and the extracellular matrix and coordinates proliferation and motility with the adhesion status of the cell.

Source: data from KEGG (Kyoto Encyclopedia of Genes and Genomes), *KEGG Pathways in cancer—Homo sapiens (human)*, Copyright © Kanehisa Laboratories. Available from http://www.kegg.jp/pathway/hsa05200; and KEGG (Kyoto Encyclopedia of Genes and Genomes), *Focal adhesion—Homo sapiens (human)*, Copyright © Kanehisa Laboratories. Available from http://www.genome.jp/kegg-bin/show_pathway?hsa04510

As this pathway is on one side affected by external signals but on the other side, it also transmits signals from the cell to the stroma, which can allow the cells get abnormal messages from the cancer associate stroma and also allow the stroma to modify cancer cells behaviour. An example is how tumour-derived lysyl oxidase-like 2 (LOXL2) eventually affect the cancer cell Fak. Excessive levels of LOXL2 from the cancer cell links to the normal fibroblasts (Tai et al., 2015) where activates their Fak and the PI3K-AKT pathways. These activated fibroblasts cause an excessive collagen deposition in the ECM, increasing the ECM rigidity and therefore activating, through mechanosigalling, beta integrins receptor on the surface of the cancer cell resulting in FAK activation leading to increased PIk3/AKT signalling (Tai et al., 2015; Wu and Zhu, 2015).

The other dual address pathways: NFKb, TGF-beta, and hedgehog

The NFKB pathway

Nuclear factor-kappa B (NF-kappa B) is a family of five transcription factors (NfKb1, NfKb2, Rel, RelA, RelB) that act as dimers as illustrated in Figure 12.11. This pathway regulates mainly genes involved in immunity, inflammation, and cell survival. The canonical pathway of which, defined as one in which the NfKB complex is phosphorylated through a multimeric complex formed by Ikk alpha, Ikk beta, and Ikk gamma (Nemo) while in the non-canonical pathway, phosphorylation is provided by an Ikk alpha dimmer (Fig. 12.11; Madonna et al., 2012; Mitchell et al., 2016). The canonical pathway is activated by TNF-alpha, Il1, lipopolysaccharides, and by antigen presentation on B cell, and T-cell receptors. It is also activated by viral infections. The non-canonical pathway is instead activated by TNF ligands.

Increased NfKb activity in cancer can be due to activating mutations, mostly in haematological malignancies, or by increased levels of activating ligands from tumour associated stroma (Hoesel and Schmid, 2013). In tumours NfKb is the main pathway allowing the cross talk between cancer cells and inflammation. Its effect is context dependent: active presence of inflammation can lead to NfKb activation in tumours cells while can also favour tumour growth by inducing immunosuppression. As far as action on the neoplastic cell is concerned, its activation can lead to transcription of antiapoptotic genes such as Flip and Bcl2, but this mechanism is rarely involved in tumours. It can also promote proliferation: a positive feedback loop has been described in Kras-induced tumours which involves activation of Erk, Erk–NF-kB–Timp1–CD63–FAK, and eventually Erk again. It can also promote metastatic spread when activated by the NfKb pathway, as described in the next section (Xia et al., 2014). Finally, it is also involved in metabolic reprogramming as it can induce transcription of Glut3, a glucose transporter, maintaining glycolysis. It also regulates cytochrome c oxidase (SCO_2), a subunit of the mitochondrial respiratory complex (Xia et al., 2014).

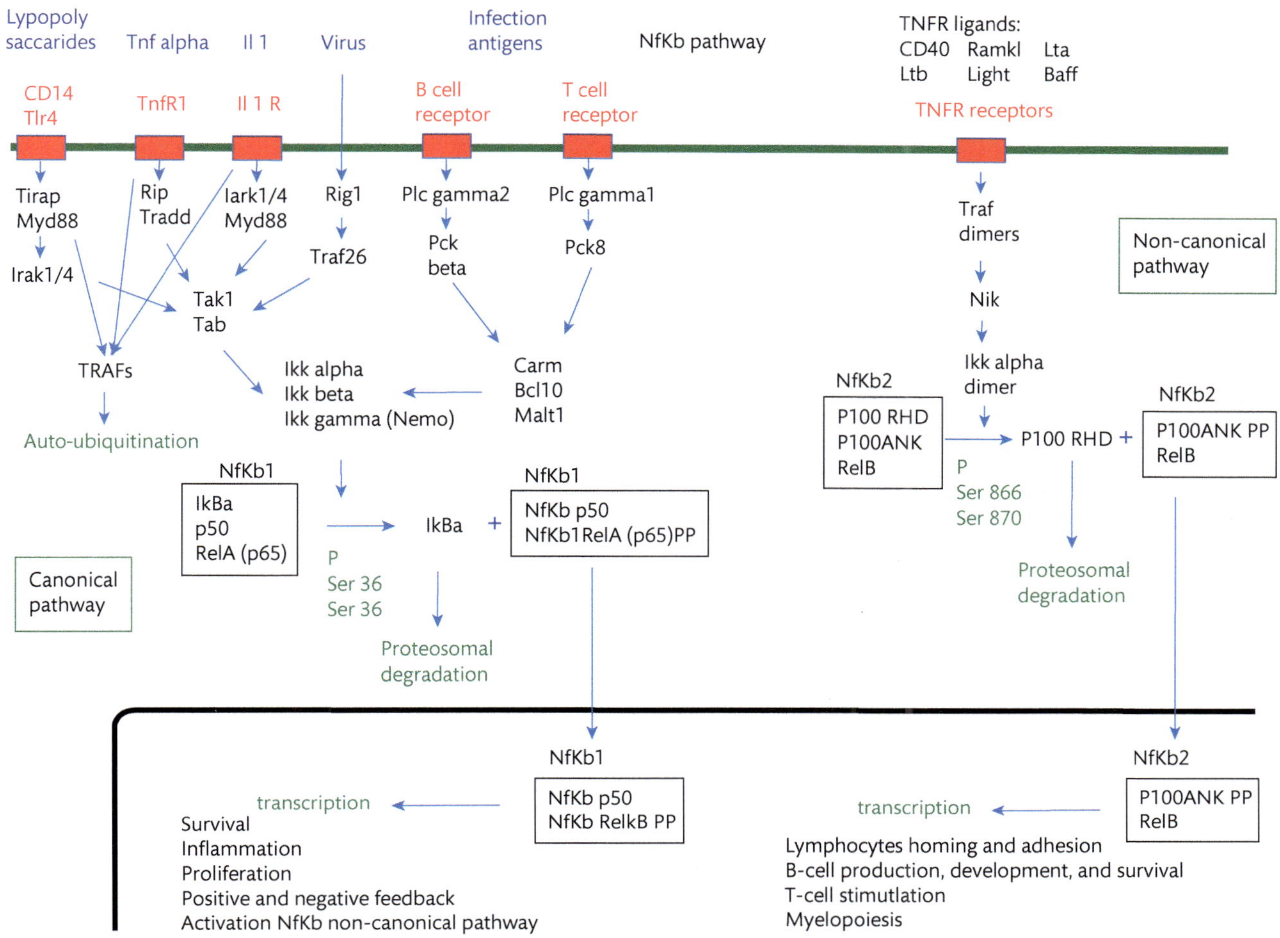

Fig. 12.11 The NF-kappa B pathway. NfKb is a family of transcription factors that work as dimmers. They regulate genes involved in immunity, inflammation, and cell survival. The canonical pathway is relying on IKK-mediated IkappaB-alpha phosphorylation on Ser32 and 36. The canonical pathway can also be activated by viruses. The non-canonical rely on phosphorylation of IkappaB-alpha on Tyr42 or on Ser residues in IkappaB-alpha PEST domain.

Source: data from KEGG (Kyoto Encyclopedia of Genes and Genomes), *NF-kappa B signaling pathway*, Copyright © Kanehisa Laboratories, available from http://www.genome.jp/kegg-bin/show_pathway?map=hsa04064&show_description=show

The TGF-beta pathway

The tumour growth factor pathway (**Fig. 12.12**) is among one of the oldest. As an organism grew more and more complex, the number of the TGF-beta superfamily members has increased accordingly during evolution (Ayyaz et al., 2017). This superfamily includes now the TGF-beta family, the bone morphogenic proteins family, the Activins family, and the Nodal family (Derynck and Zhang, 2003). This progressive expansion during evolution is due to the fact that this pathway is the main one involved in morphogenesis and affects numerous aspects of biology, like embryonic development (where migration is one of the main events), regeneration of tissues, stress response, haemopoiesis, neurogenesis and immunity (Ayyaz et al., 2017).

The canonical, Smad-dependent pathway, is activated when a ligand induces formation of a complex made up by TGF-beta R1 and TGF-beta R2. This complex then induces phosphorylation of Smad2/3 which in turn forms a complex with Smad4, such a complex moves to the nucleus and there, according to the context, it can act a corepressor or co-activator of transcription (**Fig. 12.12**) (Ayyaz et al., 2017; Zhang et al., 2017). In the non-canonical (Smad-independent) pathway, binding of a ligand to TNFR1/2 leads to the direct activation of the Mapk pathway and, through Tak1, of the NfKb pathway (**Fig. 12.12**; Derynck and Zhang, 2003).

Because of its central role in so many different biological processes, TGF-beta is deeply involved in cancer. The pathway can have either a tumour suppressor or an oncogenic effect according to the context. It is then involved in metastatic disease, mainly when activated by TGF-beta with consequent induction of migration.

Suppressor role

Deletion or mutations inactivating the affected molecule have been reported both in receptors and signal transduction molecules in a large variety of tumours (Huang and Blobe, 2016). In normal epithelial and haemopoietic cells, TGF-beta promotes the expression of the CDK-Is p15INK4B, p21CIP1, and p27KIP1, leading to inhibition of the cyclin–CDK complexes and to cell cycle arrest in G1 (Huang and Blobe, 2016). A second mechanism of suppression is related to apoptosis which TGF-beta can trigger through a variety of canonical

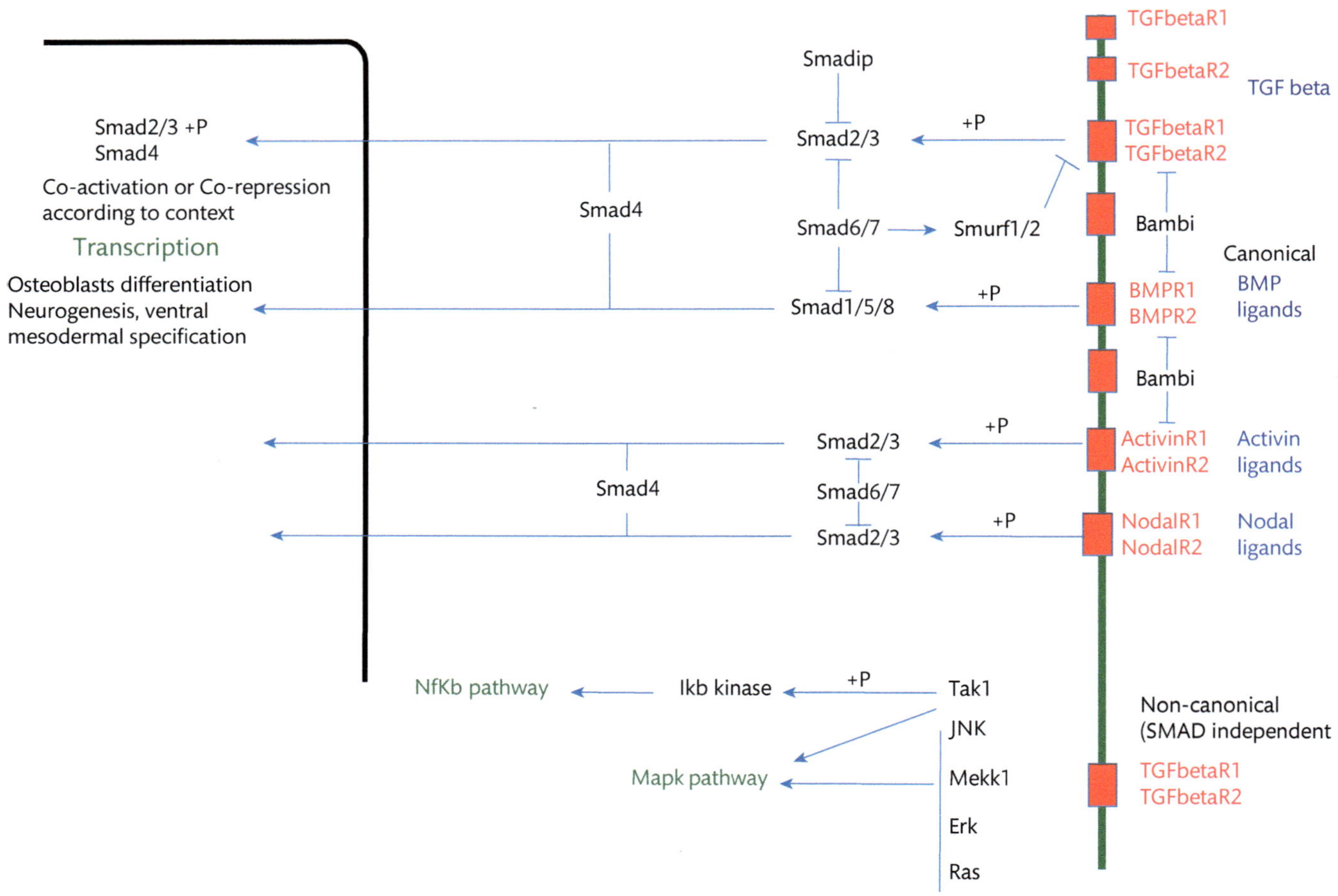

Fig. 12.12 The transforming growth factor-beta pathway. It includes a canonical (defined as Smad dependent) and non-canonical (Smad-independent) branch. Smads activation leads to transcription of a large variety of genes involved in several functions in embryogenesis, cell motility and differentiation, while the non-canonical activate NfKb and MAPK pathways.

Source: data from KEGG (Kyoto Encyclopedia of Genes and Genomes), *TGF-beta signaling pathway*, Copyright © Kanehisa Laboratories. Available from http://www.genome.jp/kegg-bin/show_pathway?map=hsa04350&show_description=show; and Derynck R and Zhang YE, 'Smad-dependent and Smad-independent pathways in TGF-beta family signaling', *Nature*, Volume 425, Issue 6958, pp. 577–84, Copyright © 2003 Springer Nature.

and non-canonical pathways (Huang and Blobe, 2016). These two suppressor mechanisms are mostly relevant to cancer early phases.

Oncogenic role

In more advanced tumours higher levels of TGF-beta pathway ligands are also very common but correlate with metastatic spread and poor outcome. One first mechanism is the induction of epithelial mesenchymal transition, by inducing expression of Snail, Slug, and Twist, with associated degradation of the ECM through metalloproteases. It can sustain angiogenesis through VEGF up-regulation and inhibits tumour surveillance, e.g. by inhibiting T-cytotoxic activity. (Huang and Blobe, 2016).

The hedgehog pathway

The hedgehog (Hh) signalling pathway (Fig. 12.13) plays a role in the control of cell proliferation, tissue patterning, stem cell maintenance, and development. Its insufficient activation leads to malformation in the embryo while hyperactivation is presented in many tumours. Because of its very important role in development, as the TGF-beta pathway, it is highly conserved through evolution (Gorojankina, 2016). Three hedgehog proteins have been identified so far in mammals: sonic hedgehog (SHH), indian hedgehog (IHH), and desert hedgehog (DHH; Briscoe and Therond, 2013).

That this pathway can be involved in cancer was discovered by investigating a congenital condition, Gorlin syndrome, in which, alongside craniofacial and skeletal malformations due to a mutation of a HH receptor Ptc, an increased incidence of basal cell carcinoma was presented. Mutations were subsequently discovered to occur also in Smo and Sufu (McMillan and Matsui, 2012). Although hedgehog pathway is involved in many types of cancers, these mutations are almost limited to basal cell carcinoma and medulloblastoma. In other tumours autocrine and paracrine stimulations are more common (McMillan and Matsui, 2012). These stimulations contribute to cancer by increasing proliferation, through cMyc and CyclinD1, and cell growth through IGF2. Recently, it has also been observed that hedgehog pathway can increase glycolysis (Briscoe and Therond, 2013).

TAKE-HOME MESSAGE

- The definition of a signal route as a pathway, is after all an arbitrary one, as there are extensive connections among all of them.

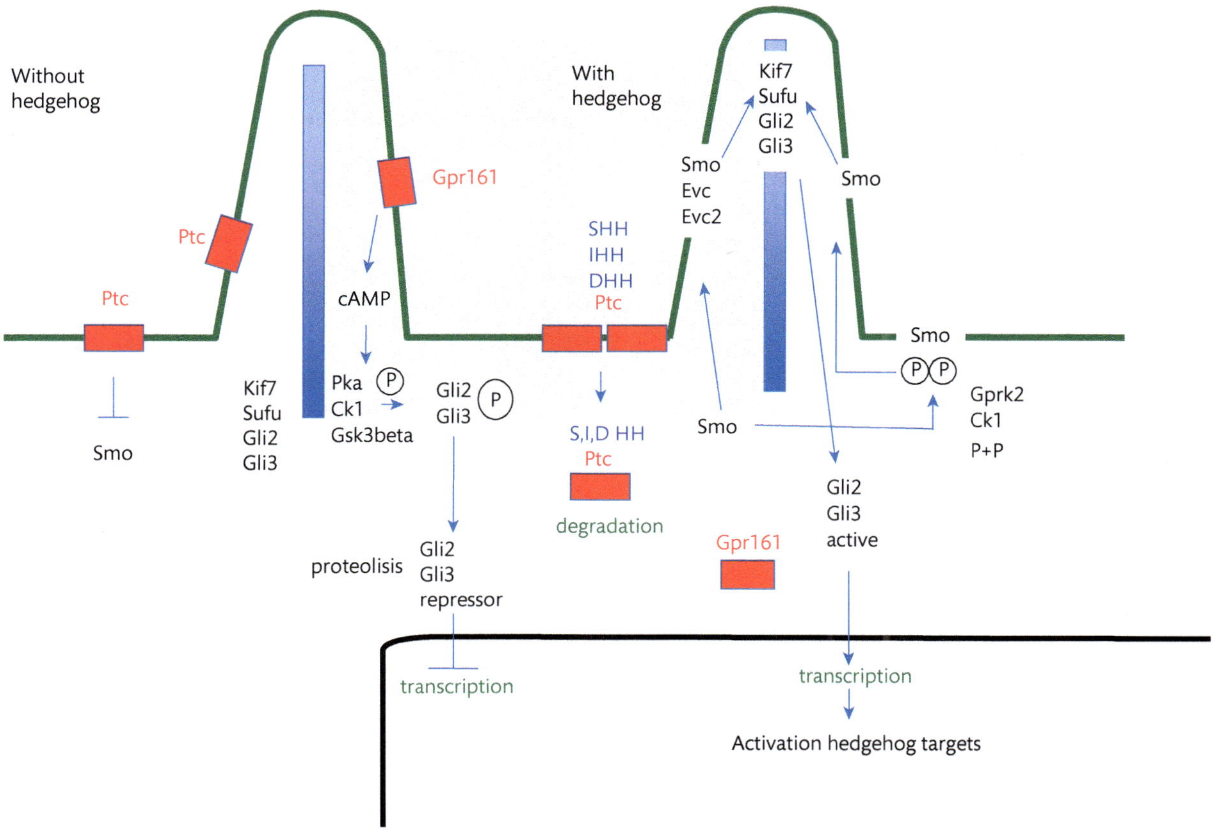

Fig. 12.13 The hedgehog (Hh) signalling pathway. Binding of one of the hedgehog ligands (SHH, IHH, DHH) leads to degradation of the Ptc receptor. In this way, Smo remains stable leading to the translocation on the tip of the cilium of the Kif7, Sufu, Gli2, Gli3 complex. The latter release active Gli2 and Gli3, which induce transcription.

Adapted with permission from KEGG (Kyoto Encyclopedia of Genes and Genomes), *Hedgehog signaling pathway*, Copyright © Kanehisa Laboratories. Available from http://www.genome.jp/kegg-bin/show_pathway?map=hsa04340&show_description=show. Source: data from Briscoe J and Therond PP, 'The mechanisms of Hedgehog signalling and its roles in development and disease', *Nature Reviews Molecular Cell Biology*, Volume 14, Issue 7, pp. 416–29, Copyright © 2013 Springer Nature.

- We have achieved a good understanding of their anatomy and basic physiology.
- We know what the dominant effects of the activation of a single pathway are, but we still do not have the tools to fully appreciate the complete effect of the interaction between multiple pathways.

OPEN QUESTIONS

- As discussed in the chapter on system biology (Chapter 26), the main question is now how to study the effect of one protein activation not just on the immediate downstream pathway, but on the entire cell.

FURTHER READING

Alberts, B., Johnson, A., Lewis, J., et al (eds) (2015). *Molecular Biology of the Cell*, 6th edition. New York: Garland Science.

Hanckock, J. T. (2017). *Cell Signaling*, 4th edition. Oxford: Oxford University Press.

Marks, F., Kingmuller, U., & Muller-Decker, K. (2017). *Cellular Signaling Processing*, 2nd edition. New York: Garland Science, Taylor and Francis Group.

Nagasaki, M., Saito, A. Doi, A., Matsuno, H., & Miyano S. (2017). *Foundations of Systems Biology: Using Cell Illustrator and Pathway Databases (Computational Biology)*. London: Springer.

Wagener, C. Stocking, C. Muller, O. (2016). *Cancer Signaling: From Molecular Biology to Targeted Therapy*. Weinheim: Wiley-VCH.

REFERENCES

Ayyaz, A., Attisano, L., & Wrana, J. L. (2017). Recent advances in understanding contextual TGF-beta signaling. *F1000Res*, **6**, 749.

Bar-Shavit, R., Maoz, M., Kancharla, A., et al. (2016). G protein-coupled receptors in cancer. *Int J Mol Sci*, **17**, 1320.

Barczyk, M., Carracedo, S., & Gullberg, D. (2010). Integrins. *Cell Tissue Res*, **339**, 269–80.

Briscoe, J., & Therond, P. P. (2013). The mechanisms of hedgehog signalling and its roles in development and disease. *Nat Rev Mol Cell Biol*, **14**, 416–29.

Bromberg, J. (2002). Stat proteins and oncogenesis. *J Clin Invest*, **109**, 1139–42.

Brown, E. J., Albers, M. W., Shin, T. B., et al. (1994). A mammalian protein targeted by G1-arresting rapamycin-receptor complex. *Nature*, **369**, 756–8.

Carroll, J. S. (2016). Mechanisms of oestrogen receptor (ER) gene regulation in breast cancer. *Eur J Endocrinol*, **175**, R41–9.

Corda, G., & Sala, A. (2017). Non-canonical WNT/PCP signalling in cancer: Fzd6 takes centre stage. *Oncogenesis*, **6**, e364.

Courtnay, R., Ngo, D. C., Malik, N., Ververis, K., Tortorella, S. M., & Karagiannis, T. C. (2015). Cancer metabolism and the Warburg effect: the role of HIF-1 and PI3K. *Mol Biol Rep*, **42**, 841–51.

Derynck, R., & Zhang, Y. E. (2003). Smad-dependent and Smad-independent pathways in TGF-beta family signalling. *Nature*, **425**, 577–84.

Ersahin, T., Tuncbag, N., & Cetin-Atalay, R. (2015). The PI3K/AKT/mTOR interactive pathway. *Mol Biosyst*, **11**, 1946–54.

Feig, L. A., Urano, T., & Cantor, S. (1996). Evidence for a Ras/Ral signaling cascade. *Trends Biochem Sci*, **21**, 438–41.

Ferreira, A. R., Felgueiras, J., & Fardilha, M. (2015). Signaling pathways in anchoring junctions of epithelial cells: cell-to-cell and cell-to-extracellular matrix interactions. *J Recept Signal Transduct Res*, **35**, 67–75.

Gentry, L. R., Martin, T. D., Reiner, D. J., & Der, C. J. (2014). Ral small GTPase signaling and oncogenesis: more than just 15 minutes of fame. *Biochim Biophys Acta*, **1843**, 2976–88.

Godinho, R. O., Duarte, T., & Pacini, E. S. (2015). New perspectives in signaling mediated by receptors coupled to stimulatory G protein: the emerging significance of cAmp efflux and extracellular cAMP-adenosine pathway. *Front Pharmacol*, **6**, 58.

Gorojankina, T. (2016). Hedgehog signaling pathway: a novel model and molecular mechanisms of signal transduction. *Cell Mol Life Sci*, **73**, 1317–32.

Herve, J. C. & Derangeon, M. (2013). Gap-junction-mediated cell-to-cell communication. *Cell Tissue Res*, **352**, 21–31.

Hessvik, N. P. & Llorente, A. (2017). Current knowledge on exosome biogenesis and release. *Cell Mol Life Sci*, **75**(2), 193–208.

Hirst, D. G. & Robson, T. (2011). Nitric oxide physiology and pathology. *Methods Mol Biol*, **704**, 1–13.

Hoesel, B. & Schmid, J. A. (2013). The complexity of NF-kappaB signaling in inflammation and cancer. *Mol Cancer*, **12**, 86.

Huang, J. J. & Blobe, G. C. (2016). Dichotomous roles of TGF-beta in human cancer. *Biochem Soc Trans*, **44**, 1441–54.

Hurowitz, E. H., Melnyk, J. M., Chen, Y. J., Kouros-Mehr, H., Simon, M. I., & Shizuya, H. (2000). Genomic characterization of the human heterotrimeric G protein alpha, beta, and gamma subunit genes. *DNA Res*, **7**, 111–20.

Hynes, R. O. (2002). Integrins: bidirectional, allosteric signaling machines. *Cell*, **110**, 673–87.

Iber, D. & Fengos, G. (2012). Predictive models for cellular signaling networks. *Methods Mol Biol*, **880**, 1–22.

Johnston, C. A. & Siderovski, D. P. (2007). Receptor-mediated activation of heterotrimeric G-proteins: current structural insights. *Mol Pharmacol*, **72**, 219–30.

Johnstone, R. M. (2005). Revisiting the road to the discovery of exosomes. *Blood Cells Mol Dis*, **34**, 214–19.

Johnstone, R. M., Adam, M., Hammond, J. R., Orr, L., & Turbide, C. (1987). Vesicle formation during reticulocyte maturation. Association of plasma membrane activities with released vesicles (exosomes). *J Biol Chem*, **262**, 9412–20.

Junger, W. G. (2011). Immune cell regulation by autocrine purinergic signalling. *Nat Rev Immunol*, **11**, 201–12.

Kahlert, C., Melo, S. A., Protopopov, A., et al. (2014). Identification of double-stranded genomic DNA spanning all chromosomes with mutated KRAS and p53 DNA in the serum exosomes of patients with pancreatic cancer. *J Biol Chem*, **289**, 3869–75.

Kashatus, D. F. (2013). Ral GTPases in tumorigenesis: emerging from the shadows. *Exp Cell Res*, **319**, 2337–42.

Kim, L. C., Cook, R. S., & Chen, J. (2017). mTORC1 and mTORC2 in cancer and the tumor microenvironment. *Oncogene*, **36**, 2191–201.

Li, W. X. (2008). Canonical and non-canonical JAK-STAT signaling. *Trends Cell Biol*, **18**, 545–51.

Liang, G. H. & Weber, C. R. (2014). Molecular aspects of tight junction barrier function. *Curr Opin Pharmacol*, **19**, 84–9.

Madonna, G., Ullman, C. D., Gentilcore, G., Palmieri, G., & Ascierto, P. A. (2012). NF-kappaB as potential target in the treatment of melanoma. *J Transl Med*, **10**, 53.

Martini, M., DE Santis, M. C., Braccini, L., Gulluni, F., & Hirsch, E. (2014). PI3K/AKT signaling pathway and cancer: an updated review. *Ann Med*, **46**, 372–83.

Matozaki, T., Nakanishi, H., & Takai, Y. (2000). Small G-protein networks: their crosstalk and signal cascades. *Cell Signal*, **12**, 515–24.

McKay, M. M. & Morrison, D. K. (2007). Integrating signals from RTKs to ERK/MAPK. *Oncogene*, **26**, 3113–21.

McMillan, R. & Matsui, W. (2012). Molecular pathways: the hedgehog signaling pathway in cancer. *Clin Cancer Res*, **18**, 4883–8.

Mitchell, S., Vargas, J., & Hoffmann, A. (2016). Signaling via the NFkappaB system. *Wiley Interdiscip Rev Syst Biol Med*, **8**, 227–41.

Niehrs, C. (2012). The complex world of WNT receptor signalling. *Nat Rev Mol Cell Biol*, **13**, 767–79.

Nielsen, M. S., Axelsen, L. N., Sorgen, P. L., Verma, V., Delmar, M., & Holstein-Rathlou, N. H. (2012). Gap junctions. *Compr Physiol*, **2**, 1981–2035.

Nogues, L., Palacios-Garcia, J., Reglero, C., et al. (2017). G protein-coupled receptor kinases (GRKs) in tumorigenesis and cancer progression: GPCR regulators and signaling hubs. *Semin Cancer Biol*, **48**, 78–90.

Nussinov, R., Ma, B., & Tsai, C. J. (2013). A broad view of scaffolding suggests that scaffolding proteins can actively control regulation and signaling of multienzyme complexes through allostery. *Biochim Biophys Acta*, **1834**, 820–9.

Rakoff-Nahoum, S. & Medzhitov, R. (2009). Toll-like receptors and cancer. *Nat Rev Cancer*, **9**, 57–63.

Rangamani, P. & Iyengar, R. (2008). Modelling cellular signalling systems. *Essays Biochem*, **45**, 83–94.

Rauch, N., Rukhlenko, O. S., Kolch, W., & Kholodenko, B. N. (2016). MAPK kinase signalling dynamics regulate cell fate decisions and drug resistance. *Curr Opin Struct Biol*, **41**, 151–8.

Reya, T. & Clevers, H. (2005). Wnt signalling in stem cells and cancer. *Nature*, **434**, 843–50.

Riento, K. & Ridley, A. J. (2003). Rocks: multifunctional kinases in cell behaviour. *Nat Rev Mol Cell Biol*, **4**, 446–56.

Roca-Cusachs, P., Iskratsch, T., & Sheetz, M. P. (2012). Finding the weakest link: exploring integrin-mediated mechanical molecular pathways. *J Cell Sci*, **125**, 3025–38.

Ross, T. D., Coon, B. G., Yun, S., et al. (2013). Integrins in mechanotransduction. *Curr Opin Cell Biol*, **25**, 613–18.

Runkle, E. A. & Mu, D. (2013). Tight junction proteins: from barrier to tumorigenesis. *Cancer Lett*, **337**, 41–8.

Saxton, R. A. & Sabatini, D. M. (2017). mTOR signaling in growth, metabolism, and disease. *Cell*, **168**, 960–76.

Sedgwick, A. E. & D'Souza-Schorey, C. (2016). Wnt signaling in cell motility and invasion: drawing parallels between development and cancer. *Cancers (Basel)*, **8**(9), pii: E80.

Sharabi, O., Shirian, J., & Shifman, J. M. (2013). Predicting affinity- and specificity-enhancing mutations at protein–protein interfaces. *Biochem Soc Trans*, **41**, 1166–9.

Sherr, C. J. & McCormick, F. (2002). The RB and p53 pathways in cancer. *Cancer Cell*, **2**, 103–12.

Smythies, J., Edelstein, L., & Ramachandran, V. (2014). Molecular mechanisms for the inheritance of acquired characteristics-exosomes, microRNA shuttling, fear and stress: Lamarck resurrected? *Front Genet*, **5**, 133.

Spangle, J. M., Roberts, T. M., & Zhao, J. J. (2017). The emerging role of PI3K/AKT-mediated epigenetic regulation in cancer. *Biochim Biophys Acta*, **1868**, 123–31.

Tai, Y. L., Chen, L. C., & Shen, T. L. (2015). Emerging roles of focal adhesion kinase in cancer. *Biomed Res Int*, **2015**, 690690.

Takai, Y., Sasaki, T., & Matozaki, T. (2001). Small GTP-binding proteins. *Physiol Rev*, **81**, 153–208.

Tatulian, S. A. (2015). Structural dynamics of insulin receptor and transmembrane signaling. *Biochemistry*, **54**, 5523–32.

Vert, G. & Chory, J. (2011). Crosstalk in cellular signaling: background noise or the real thing? *Dev Cell*, **21**, 985–91.

Wijmans, J. G. & Baker, R. W. (1995). The solution-diffusion model: a review. *J Mem Sci*, **107**(1–2), 1–21.

Wolfson, R. L. & Sabatini, D. M. (2017). The dawn of the age of amino acid sensors for the mTorc1 pathway. *Cell Metab*, **26**, 301–9.

Wu, L. & Zhu, Y. (2015). The function and mechanisms of action of LOXL2 in cancer (Review). *Int J Mol Med*, **36**, 1200–4.

Xia, Y., Shen, S., & Verma, I. M. (2014). NF-kappaB, an active player in human cancers. *Cancer Immunol Res*, **2**, 823–30.

Xiao, Q., Chen, Z., Jin, X., Mao, R., & Chen, Z. (2017). The many postures of noncanonical Wnt signaling in development and diseases. *Biomed Pharmacother*, **93**, 359–69.

Yang, F., Ning, Z., Ma, L., et al. (2017). Exosomal miRNAs and miRNA dysregulation in cancer-associated fibroblasts. *Mol Cancer*, **16**, 148.

Yarden, Y. & Sliwkowski, M. X. (2001). Untangling the ErbB signalling network. *Nat Rev Mol Cell Biol*, **2**, 127–37.

Yu, J. S. & Cui, W. (2016). Proliferation, survival and metabolism: the role of PI3K/AKT/mTOR signalling in pluripotency and cell fate determination. *Development*, **143**, 3050–60.

Zhan, T., Rindtorff, N., & Boutros, M. (2017). Wnt signaling in cancer. *Oncogene*, **36**, 1461–73.

Zhang, Y., Alexander, P. B., & Wang, X. F. (2017). TGF-beta family signaling in the control of cell proliferation and survival. *Cold Spring Harb Perspect Biol*, **9**, pii: a022145.

Zhao, J. & Guan, J. L. (2009). Signal transduction by focal adhesion kinase in cancer. *Cancer Metastasis Rev*, **28**, 35–49.

Zimmerman, G. A., Lorant, D. E., Mcintyre, T. M., & Prescott, S. M. (1993). Juxtacrine intercellular signaling: another way to do it. *Am J Respir Cell Mol Biol*, **9**, 573–7.

13 Cell cycle control

Simon Carr and Nicholas La Thangue

Introduction

The cell cycle can be divided into several discrete stages in which different events take place. Duplication of the cell's DNA occurs during the S phase, while the segregation of the genetic material to opposite poles of the cell, a process known as mitosis, occurs during the M phase. These two phases are preceded respectively by gap phase 1 (G1) and gap phase 2 (G2) periods of the cycle in which the cell can prepare for either S phase or M Phase. To complete cell division, a new cell membrane will form between the separated chromosomes to generate two daughter cells. Only a minority of cells in the human body are actively proliferating at any one time and are usually restricted to areas of tissues capable of self-renewal, such as the skin epithelium or haematopoietic system. Some cells will temporarily enter a quiescent state known as G0 in response to external stimuli while others may irreversibly enter into a terminally differentiated state, such as cardiomyocytes and neurons. Eventually some cells stop proliferating and enter a senescent state in response to internal or external insults (Morgan, 2007).

Mitosis itself can be subdivided into additional phases: (a) prophase, in which the chromosomes start to condense within the nucleus and the cell begins to prepare the structural components of the mitotic spindle required to efficiently separate the chromosomes; (b) prometaphase, in which the nuclear envelope begins to break down and the mitotic spindle attaches to chromosomes; (c) metaphase, in which the chromosomes become aligned at the mid-zone of the cell; (d) anaphase, in which the chromosomes begin to separate and move towards opposite poles of the cell; and (e) telophase, in which the nuclear envelope reforms around the separated chromosomes to generate two distinct nuclei (Morgan, 2007). At the end of telophase, a process known as cytokinesis will begin, in which a new plasma membrane forms between the two nuclei to generate distinct daughter cells (D'Avino et al., 2015; Srivastava et al., 2016).

The gap phases, G1 and G2, provide additional time in which protein synthesis and cell growth can occur, as a successful cell division also requires cytoplasmic contents, such as organelles, to be duplicated in addition to the cell's genetic material. The gap phases also represent a time in which several cell signalling pathways operate, to ensure environmental conditions are appropriate for the cell to progress into S phase or M phase. For example, the signalling pathways that operate in G1 are involved in executing the decision to begin another cell cycle round, while during G2, signalling events control the onset of mitosis and help regulate efficient chromosome segregation (Neganova and Lako, 2008; Bertoli et al., 2013a; Rhind and Russell, 2015; Wieser and Pines, 2015).

A group of specific signalling pathways constitute what is known as the cell cycle checkpoints, and they ensure that later cell cycle events only occur upon completion of the proceeding steps. These checkpoints therefore act as a quality control mechanism and can halt the cell cycle when things go wrong, or conditions become unfavourable (e.g. when the DNA in cells become damaged). The cell cycle contains three major checkpoints: the first, the G1/S checkpoint, occurs at the transition between G1 and S phase, a point in the cell cycle when it is important to determine if conditions are appropriate for cell proliferation (Neganova and Lako, 2008; Bertoli et al., 2013a); after this step the cell is committed to another cell cycle round and DNA replication will commence. The second occurs at the entry to mitosis and is known as the G2/M transition checkpoint (Rhind and Russell, 2015; Wieser and Pines, 2015). Assembly of the mitotic spindle is tightly regulated at this stage, and by metaphase the condensed chromosomes will become attached to the spindle microtubules that originate from each pole of the cell. The third checkpoint known as the metaphase-to-anaphase transition (spindle assembly checkpoint) controls the committing to chromosome separation and completion of mitosis (Rhind and Russell, 2015; Wieser and Pines, 2015). The importance of cell cycle checkpoints in cell division is highlighted by the high frequency of mutations found in their constituent components during cancer.

History of the cell cycle

Forty years ago, researchers were well aware that cells in the body could undergo cellular division, but the mechanisms regulating this process were still largely unknown. Many scientists were actively trying to understand the basics of the various stages of the cell cycle, including DNA replication and chromosome segregation, and it was thought that without this underlying knowledge of such processes, determining the control circuitry that regulated the cell cycle would be impossible. However, a number of researchers had been pursuing the question of cell cycle regulation with the help of a variety of model organisms. All eukaryotic cells employ similar machinery to duplicate themselves, so it was possible to exploit the experimental advantages offered by some organisms to gain a comprehensive view of cell cycle control. For example, single-celled budding and fission yeast are among the simplest eukaryotes to manipulate experimentally, since they proliferate rapidly, are easy to culture, and their genomes are fully defined (Morgan, 2007). The fertilized eggs of some animals, particularly those of the frog *Xenopus laevis*, also represent a powerful tool since they are relatively large in size and divide rapidly (Morgan, 2007).

In 2001, the Nobel Prize in physiology and medicine was awarded to Leland Hartwell, Paul Nurse, and Tim Hunt, who each made key discoveries in the field of cell cycle regulation. Hartwell used the genetic manipulation of budding yeast to identify a number of genes—cell division cycle, or *cdc* genes—whose function was required to progress beyond specific points in the cell cycle (Hartwell et al., 1970). He also used these mutant yeast strains as a tool to block the cell cycle at specific stages to determine the interdependence of cell cycle events (i.e. DNA replication is required for cell division to occur; see Hartwell, 1971). Indeed, a few years later several *cdc* genes had been identified and the 'start' signal which begins each cell cycle had been proposed and would later be identified as a unified signal across all organisms (Hartwell et al., 1974). Similar work was also being performed in fission yeast by Nurse, who had been isolating mutants in which the normal controls that regulate cell division had been lost. Nurse's laboratory dissected the control of cell division by examining the signalling events that regulate the activity of one of the mitotic protein kinases (Russell and Nurse, 1986, 1987; Gould and Nurse, 1991), and most importantly, they also demonstrated that this mechanism of control was conserved across multiple organisms (Lee and Nurse, 1987).

Meanwhile, biochemical approaches undertaken by Hunt in the embryos of sea urchins identified several proteins that accumulated to high levels during mitosis, only to rapidly decline before the eggs divided (Evans et al., 1983). These proteins hence became known as the cyclins and were observed in multiple organisms where their production and degradation help govern the timing of many key cell cycle events discussed in this chapter. Together with future work, this research highlighted the importance of the cyclins and the enzymes they regulated in determining the temporal control of multiple steps in the cell cycle (Murray and Kirschner, 1989; Murray et al., 1989).

Later, Weinert and Hartwell postulated the idea that the cell cycle constituted a series of checkpoints, which acted to halt cell cycle progression unless a number of conditions had been met. Yeasts were known to arrest in G2 if they sustained damage to their DNA, but a mutant was identified that failed to arrest when exposed to radiation. These mutant cells continued to divide and a large proportion of them subsequently died, indicating that a checkpoint to signal cell cycle arrest and DNA repair had been compromised (Weinert and Hartwell, 1988). This observation initiated a new field of research that identified several of these 'stop' proteins, functioning in numerous checkpoints, and such proteins were shown to be important not only in yeast but in human cells as well. Some of these checkpoint proteins are described in more detail throughout the rest of this chapter.

Effector proteins in cell cycle control: the kinases and ubiquitin ligases

A number of proteins are involved in cell cycle checkpoints, but perhaps the most important are two families of enzymes known as the cyclin-dependent kinases (CDKs) and the ubiquitin ligases.

The cyclin-dependent kinases

The CDKs target the addition of phosphate groups onto a large number of other proteins involved in processes as diverse as DNA synthesis, DNA repair, transcription, and cell division (Lim and Kaldis, 2013; Malumbres, 2014). Phosphorylation events on target proteins result in the modulation of their function, either to enhance or inhibit downstream activity, or to promote additional protein–protein interactions. There exists 20 different CDK enzymes in human cells (CDK1-CDK20), and each has a discrete set of target proteins, though there is some redundancy in terms of their substrates. However, all are important, one way or another, in integrating extracellular and intracellular signals that modulate gene transcription or events important to cell division. It is therefore important that their activity be tightly regulated, and this is primarily achieved via the association of the CDK with a regulatory subunit known as the cyclin, whose protein levels oscillate during different cell cycle stages (Malumbres, 2014; see Fig. 13.1). This ensures that there is temporal regulation to the formation of cyclin-CDK complexes, which in turn leads to discrete sets of phosphorylation events occurring during each cell cycle phase. A number of other kinases are also involved in the profound reorganizations that take place during mitosis. Indeed, a large proportion of the kinases so far identified in eukaryotes have been implicated in mitotic control. Again, these mitotic kinases are regulated both temporally and spatially to ensure efficient cell cycle progression and include members of the Aurora, Polo-like, and NEK families (Ma and Poon, 2011).

The Ubiquitin ligases

Although cell division is driven by multiple phosphorylation events, cells also use a more robust mechanism to regulate the cell cycle, involving ubiquitin-dependent destruction of targets. This removal of proteins requires the covalent attachment of ubiquitin chains to substrates, a modification that triggers proteolysis of the target via the proteasome. Ubiquitin is a small, highly conserved protein, and attachment of ubiquitin chains to substrates involves an enzymatic cascade carried out by large protein complexes known as ubiquitin ligases, of which two are primarily involved in the destruction of cell cycle regulators, namely the SCF complex and the APC/C (anaphase-promoting complex/cyclosome; see Teixeira and Reed, 2013; Bassermann et al., 2014). These two ubiquitin ligase complexes mediate many events during, before, and after mitosis when the cell prepares itself for re-entry into the next G1 phase, and recognition of their multitude of substrates is mediated by adaptor proteins that provide specificity and flexibility to the ubiquitin-proteasome system. The APC/C controls progression through mitosis and into G1 by ubiquitylating cell cycle regulators such as the mitotic cyclins, mitotic spindle components, and the DNA replication machinery. The SCF complex, on the other hand, plays a central role in the G1-S transition and entry into mitosis, by ubiquitylating G1 and S phase

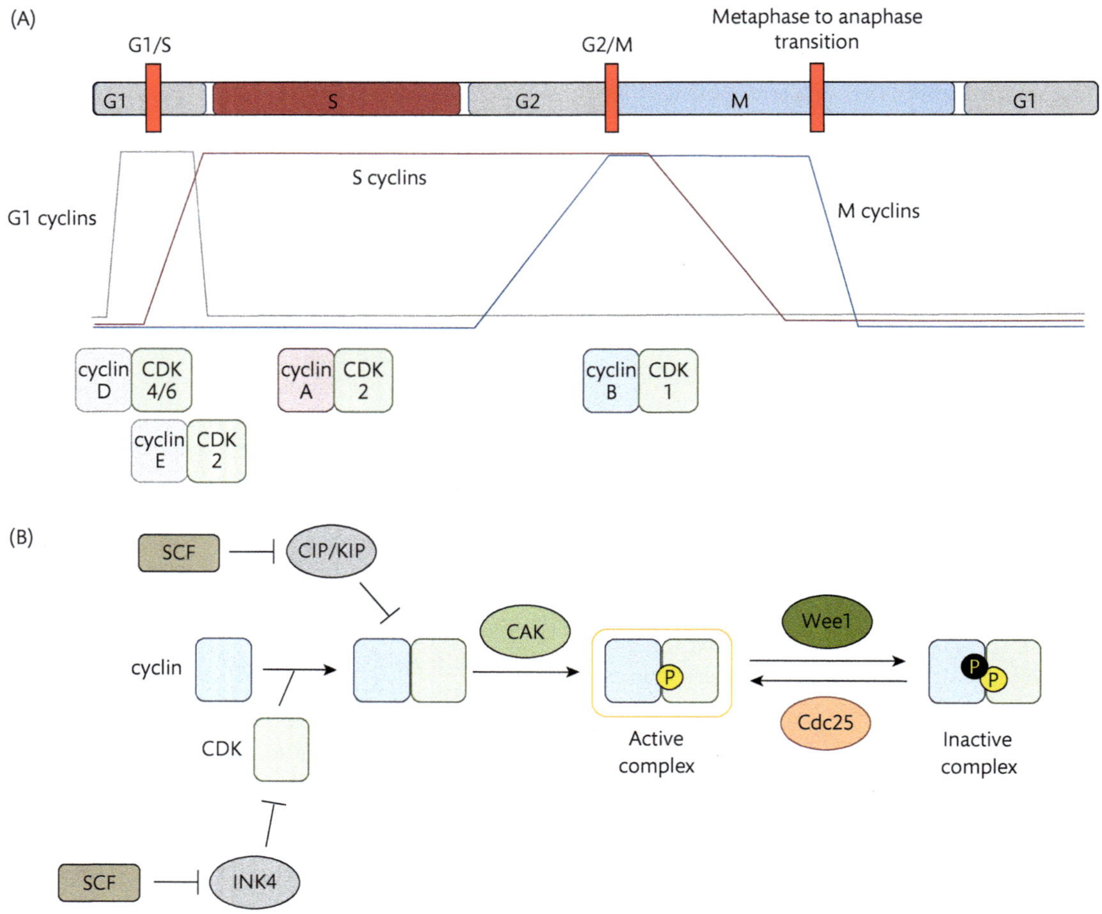

Fig. 13.1 Cyclin/CDK complexes. (A) Schematic representation of the cell cycle to indicate the oscillation in expression levels of G1, S, and M phase cyclins, and how this correlates with the three major checkpoints in the cell at the G1/S, G2/M, and metaphase-to-anaphase transition. Each cyclin has a distinct set of CDK binding partners, examples of which are illustrated here. The major G1 cyclin-CDKs are cyclin D-CDK4, cyclin D-CDK6, and cyclin E-CDK2. S phase cyclin-CDKs include cyclin A-CDK2, and the predominant M phase cyclin-CDK is cyclin B-CDK1. (B) Cyclin-CDK activity is antagonized by the CKIs, including members of the INK4 and CIP/KIP families, which function to prevent cell cycle progression during unfavourable conditions. As cells progress towards S phase, the activity of these CKIs is countered by their destruction, mediated by the SCF ubiquitin ligase complex. Full cyclin-CDK activity also requires the phosphorylation of CDKs by the enzyme CAK, and the removal of inhibitory phosphorylation events by the Cdc25 family of phosphatases. However, this inhibitory phosphorylation mark can be recreated by the Wee1 kinase in response to cellular stress. Inhibitory phosphorylation is shown in black, while activatory phosphorylation is shown in yellow.

cyclins and mitotic inhibitors (Teixeira and Reed, 2013; Bassermann et al., 2014).

Cell cycle commitment: The G1 to S phase transition

During G1, there is a wave of transcription events that effectively commit the cell to transit into S phase and commence DNA replication. Such changes in gene expression are regulated by specific CDK activities, which themselves result from the expression of discrete cyclin partners as the cell progresses towards the S phase. Cell cycle-regulated transcription is therefore both driven by and a driving force for cell cycle progression. After cells enter the S phase, this G1-S transcriptional wave is rapidly switched off; the mechanisms determining these events are discussed next.

The E2F transcription factor family and the pocket proteins

The majority of transcriptional events that occur at the G1-S transition are driven by a family of transcription factors known as the E2Fs, which act in concert with their dimerization partners, the DP proteins, to drive the expression of genes required for DNA replication and S phase entry (Polager and Ginsberg, 2008). The E2F family can broadly be divided into two subgroups based on their tendency to activate (E2F1, -2, -3A) or repress transcription (E2F3B, -4, -5, -6, -7, -8; see Fig. 13.2), though in reality there is some overlap between activators and repressors dependent on the cellular context. The importance of the E2F family for cell cycle control is highlighted by the fact that their misregulation is often found in cancer, when cells proliferate in an uncontrolled fashion (Polager and Ginsberg, 2008; Chen et al., 2009). However, E2F-driven transcription is usually restricted to the G1-S transition by the activity of the pocket proteins (pRB, p107, p130; see Giacinti and Giordano,

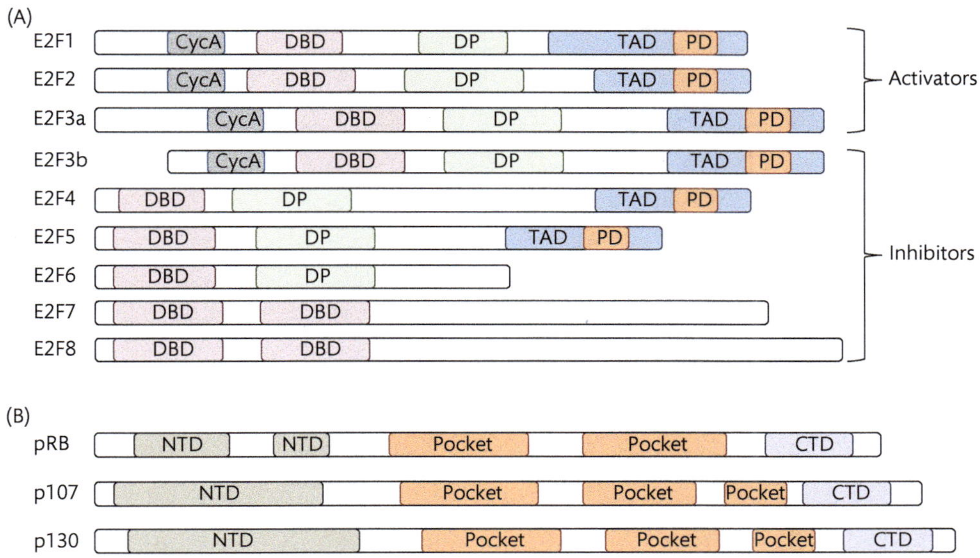

Fig. 13.2 The E2F transcription factors and the pocket protein family. (A) The E2F transcription factor family is responsible for regulating the expression of several genes required for cell cycle progression. It can be divided into two subgroups based on their ability to activate or repress transcription. (B) E2F-driven transcription is also negatively regulated by the activity of the pocket proteins, which bind to the transactivation domain of the E2Fs to inhibit their function or act as corepressors. CycA, cyclin A binding domain; DBD, DNA binding domain; DP, DP dimerization partner domain; TAD, transactivation domain; PD, pocket protein binding domain; NTD, N-terminal domain; CTD, C-terminal domain.

2006; Sun et al., 2007; Fig. 13.2). pRB can bind to the activator E2Fs and inhibit their function, while p107 and p130 act as corepressors for the inhibitor E2Fs. Inhibition by pRB is mediated primarily by direct binding of the pocket protein to the domain of E2F involved in transcriptional activity, although other mechanisms involving the recruitment of corepressors and chromatin remodelling factors also contribute. Once again, mutations affecting the pocket proteins are incredibly common in most human cancers, and cells lacking these proteins have severe defects in their ability to exit or halt the cell cycle in response to DNA damage (Burkhart and Sage, 2008; see also Box 13.1).

Transcriptional activation at the G1-S transition

The classic paradigm for transcriptional activation in G1 states that, initially, the activator E2Fs are bound and inhibited by pRB, while the inhibitor E2Fs like E2F4 and -5 associate with p107 and p130 on gene promoters to repress transcription (Neganova and Lako, 2008; Bertoli et al., 2013a; see Fig. 13.3). As cells progress through mid-G1 in response to pro-proliferative signals, early cyclin-CDK complexes (cyclin D-CDK4/6) will phosphorylate the pocket proteins, causing them to dissociate from the E2Fs which can then unleash their transcriptional activity (Connell-Crowley et al., 1997). When no longer linked to pRB, its target E2Fs can start promoting transcription.

Instead, when no longer bound by p107 and p130, E2F4 and -5 shuttle out of the nucleus into the cytoplasm, allowing the activator E2Fs, no longer inhibited by pRB, to replace them at the promoters of E2F-regulated genes. However, once again the story is not so simple, as in some cell types the ablation of all three activator E2Fs does not prevent normal proliferation (Chong et al., 2009). In addition, cells entering G1 from a state of quiescence (G0) are regulated in a different way to cells cycling into G1 from the previous M phase.

While in both scenarios E2F transcription is inhibited by the pocket proteins, in G0 this is mediated mostly by p130 and E2F4, whereas in cycling cells p107 and E2F4 predominate (Bertoli et al., 2013a). This simply seems to reflect the relative protein levels of the pocket proteins in different cell cycle stages: p130 is highly expressed in quiescent cells and its levels rapidly decline during proliferation, while p107 is barely detectable in quiescent cells but expressed more

Box 13.1 The retinoblastoma protein

The retinoblastoma protein (pRB) protein can be regarded as the archetypal tumour suppressor and was first identified in the retinoblastoma neoplasm of the eye, after which it derives its name. Characterization of the RB gene and identification of RB mutations in retinoblastoma patients confirmed that the protein indeed acts as a tumour suppressor, where it has an essential role in regulating cell cycle progression by blocking S phase entry. This repression is mediated by the ability of pRB to directly interact with the E2F family of transcription factors at the promoters of multiple genes, whose expression is required to drive cells from G1 into S phase. The pRB-E2F interaction effectively represses E2F-dependent transcription, though pRB is also able to recruit other protein factors that can modify histone residues or remodel chromatin structure at E2F promoters, which additionally contributes to pRB's function as a transcriptional repressor. Loss of normal pRB function is a common characteristic of tumour cells and leads to cell cycle deregulation. Indeed, retinoblastoma, osteosarcoma, and small-cell lung cancers display direct mutation or deletion of the RB gene itself, while the majority of other human tumours display functional inactivation of pRB via the altered expression or activity of pRB's many upstream regulators. pRB can also function in other cell cycle-related processes such as differentiation, where it interacts with a number of lineage-specific transcription factors and is necessary for the completion of the differentiation programme in erythroid, muscle, bone, and pancreatic development. It is now also widely accepted that pRB functions in a number of other cell signalling pathways related to cell cycle progression and cell growth, since loss of the RB gene impacts not only on G1-S phase progression, but on additional cellular processes such as mitotic progression, DNA repair, autophagy, apoptosis, and metabolism.

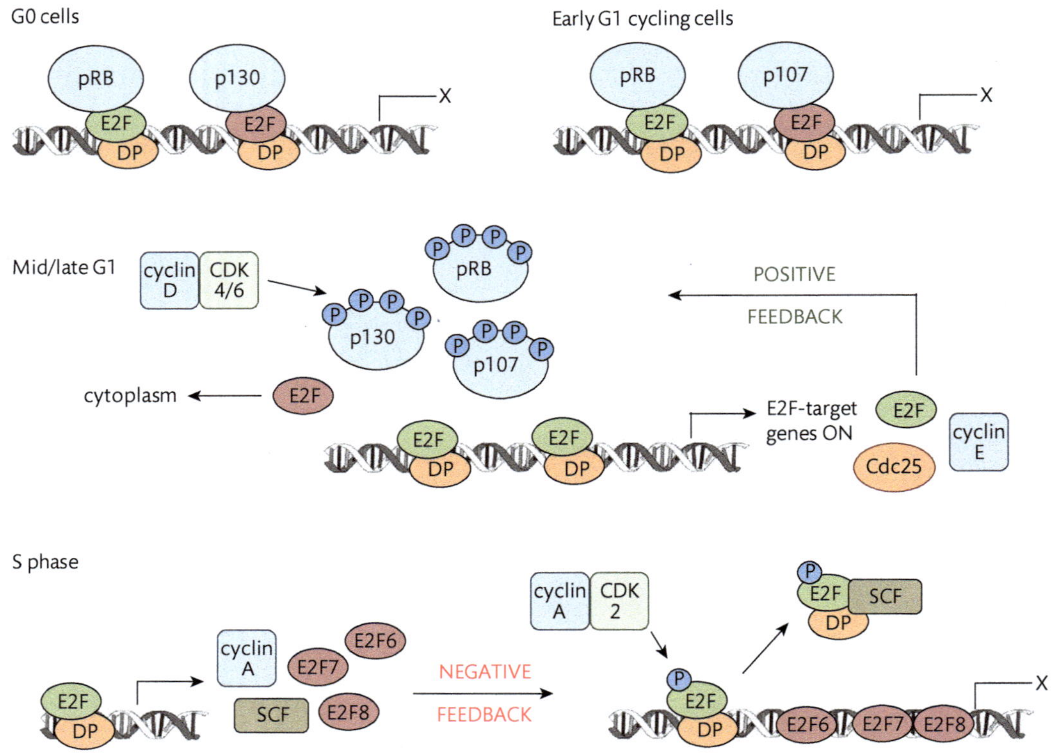

Fig. 13.3 Control of G1-S transcription. In quiescent cells (G0) and cycling (G1) cells, the activator E2Fs (shown in green) are inhibited by pRB binding, while p130 and p107 act as corepressors for the inhibitor E2Fs respectively (shown in red). As cells progress through G1 in response to proliferative signals, cyclin D-CDK4/6 complexes will target the pocket proteins for phosphorylation, causing them to dissociate from the E2Fs. The inhibitor E2Fs migrate out of the nucleus in the absence of their pocket protein partners and are replaced by the activator E2Fs. As transcription of early E2F targets proceeds, a positive feedback loop is established which commits the cell to cell cycle entry. This results from the formation of additional cyclin E-CDK2 complexes, which further phosphorylate and inactivate the pocket proteins, while at the same time, Cdc25A expression results in increased cyclin-CDK activity. The E2Fs drive expression of a plethora of genes required for S phase entry and DNA replication. After cells enter S phase, E2F transcription is inactivated by a negative feedback loop involving cyclin A-CDK2 mediated phosphorylation and dissociation of E2F1. E2F1 can also be degraded via the activity of the SCF ubiquitin ligase complex. The activator E2Fs are thus replaced at gene promoters by the inhibitor E2Fs, whose expression is also E2F-dependent.

highly in cycling cells. pRB is also more highly expressed in cycling cells but can be found at E2F-responsive promoters in quiescent, senescent, and differentiating cells as well.

As G1 transcriptional events proceed, a positive feedback loop promotes the cell's commitment to cell cycle entry. The point at which this commitment is reached is known as the 'restriction point', and after this period cells will continue to progress through the cell cycle independent of environmental signals (Neganova and Lako, 2008; Bertoli et al., 2013a). The inactivation of pocket proteins by cyclin-CDK complex-driven phosphorylation is essential to initiate this positive feedback loop, since it permits the expression of early E2F-target genes such as *cyclin E* and *CDC25A* (Duronio and O'Farrell, 1995; Vigo et al., 1999). This results in the formation of additional cyclin E-CDK complexes which can further target phosphorylation of pocket proteins (Fig. 13.3). G1 cyclins therefore result in further cyclin expression and a rapid increase in cyclin-CDK activity, which promotes cell cycle progression and the timely activation of a plethora of genes required for S-phase entry and DNA replication. Other positive feedback loops also exist, since the E2Fs are capable of driving their own transcription, leading to a rapid increase in activator E2F levels as cells progress through G1 (Wong et al., 2011).

Other methods of cyclin-CDK control

Working against the activity of the CDKs are the CDK-inhibitory proteins (CKIs) from the INK4 and CIP/KIP families (p15, p16, p18, p19, and p21, p27, p57, respectively), which can associate with cyclin-CDK complexes and inhibit their activity (Lim and Kaldis, 2013; Malumbres, 2014). The INK4 proteins act solely against the cyclin D-dependent kinases (CDK4 and CDK6), binding to the free CDK and inhibiting their activity and association with their cyclin partner, while the CIP/KIP members act against a broader spectrum of cyclin-CDK complexes once they have formed (Fig. 13.1B). However, all prevent the phosphorylation of pocket proteins by CDKs, and consequently induce G1 arrest. Since CKIs respond to conditions such as DNA damage or mitogen withdrawal, they therefore function to prevent cell cycle progression during periods of unfavourable conditions. As cells progress towards S phase, the activity of these CKIs needs to be countered, and this is primarily mediated by the targeted destruction of CKIs by the SCF ubiquitin ligase complex (Teixeira and Reed, 2013; Bassermann et al., 2014).

In addition to regulation by the CKIs, cyclin-CDK activity is also modulated by reversible phosphorylation events, which can function in both an activation and inhibitory fashion (Fig. 13.1B). For example, a protein complex known as the CDK-activating kinase (CAK) catalyses the phosphorylation of CDK subunits, and this event is required for full CDK activity (Malumbres, 2014). However, optimum CDK activity also requires the removal of inhibitory phosphorylation events, and this is mediated by a family of phosphatases known as the Cdc25 proteins. For example, Cdc25A promotes entry into the S phase by dephosphorylating the inhibitory marks on cyclin E/CDK2 and cyclin A/CDK2 complexes. Since *CDC25A* is an E2F target gene, its expression will increase as cells progress towards the G1-S boundary (Fig. 13.3). This is another example of positive feedback, since Cdc25A will increase cyclin-CDK complex activity, leading to pocket protein inactivation and further E2F-dependent transcription (Neganova and Lako, 2008; Wong et al., 2011; Bertoli et al., 2013a).

Role for other transcription factors: c-Myc

Entry into the cell cycle only occurs in response to appropriate growth signals from the extracellular environment, mediated by mitogenic signalling via receptors present at the cell's plasma membrane. Mitogenic signalling activates a number of transcription factors that drive the expression of genes required for cell cycle progression and cell growth. c-Myc is one such transcription factor, that is strongly implicated in cell cycle processes such as proliferation, differentiation, and cell death (Neganova and Lako, 2008; Bretones et al., 2015). Its importance in these pathways is highlighted by its high mutation rate in a variety of human cancers, and its expression is closely correlated with cell growth in response to mitogenic signals (Vita and Henriksson, 2006). A classic example is the t(8;14) translocation in Burkitt's lymphoma, in which c-Myc on chromosome 8 is translocated near an immunoglobulin (Ig) locus on chromosome 14, and becomes controlled by the Ig enhancer. This results in abnormally high levels of c-Myc protein expression (Janz, 2006).

c-Myc is not expressed in quiescent cells but is rapidly induced as cells become exposed to growth factors; while when cells exit the cell cycle during differentiation, c-Myc becomes downregulated. c-Myc can promote the expression of several genes including the cyclins, CDKs, CAK, CDC25, and the E2Fs, though c-Myc-target genes are also involved in protein synthesis and energetic metabolism (Neganova and Lako, 2008; Bretones et al., 2015). Since the cell also needs to increase in size during the cell cycle and synthesize components required for DNA replication, it is clear c-Myc has other important roles in cycling cells. It is also important to note that c-Myc can function as a transcriptional repressor in some circumstances; for example, it is able to inhibit the expression of CKI genes such as *p21* and *p27* (Claassen and Hann, 2000; Yang et al., 2001). This will further contribute to a cellular environment that favours proliferation.

Switching off G1-S transcription once DNA replication begins

After G1-S transcriptional events have driven cells into S phase, cells subsequently inactivate transcription of this set of genes. This control is mediated by the accumulation of transcriptional repressors during S phase and a negative feedback loop that targets the G1 transcriptional activators (Wong et al., 2011; Bertoli et al., 2013a). For example, during the S phase the E2Fs drive expression of cyclin A, which results in the formation of cyclin A-CDK2 complexes. Cyclin A-CDK2 is able to associate with and target E2F1 for phosphorylation, which then promotes the dissociation of the transcription factor from its promoters to inactivate E2F target genes (Fig. 13.3; Krek et al., 1994; Bertoli et al., 2013a). Furthermore, cyclin E and cyclin A-CDK2 complexes can phosphorylate the CKI p27, which promotes its association with and subsequent degradation by the SCF ubiquitin ligase complex (Montagnoli et al., 1999). This ensures cyclin A-CDK2 activity remains high during S phase. Since cyclin E, cyclin A, and SCF complex components are all E2F gene targets, this further contributes to the negative feedback loop affecting E2F1. The SCF complex has also been proposed to directly target E2F1 for degradation during S phase and G2, which further contributes to the inactivation of G1-S transcription events once DNA replication has begun (Teixeira and Reed, 2013; Bassermann et al., 2014).

E2F-target gene inactivation is also believed to result in part by the reactivation of the inhibitor E2Fs during S phase. The genes encoding these inhibitors are E2F targets themselves, so their protein levels accumulate at the G1-S transition just like the activator E2Fs. Unlike E2F4 and –5, E2F6, –7, and –8 can act as transcriptional repressors without the requirement of pocket proteins (which remain inactive during S phase due to high cyclin-CDK activity). They are therefore free to repress transcription during the S phase by replacing the activator E2Fs at target gene promoters (Westendorp et al., 2012; Bertoli et al., 2013b; Fig. 13.3).

DNA replication and S phase

DNA synthesis is tightly regulated to ensure the genetic information within the cell is copied only once per cell cycle. This prevents the abnormal gain or loss of genetic information which could disrupt cellular function and lead to diseases such as cancer, and is achieved by the use of a replication licensing system that coordinates cell cycle progression with DNA replication (Li and Jin, 2010; Masai et al., 2010; Diffley, 2011; Williams and Stoeber, 2012; see Fig. 13.4). The DNA licensing machinery consists of a large protein complex that forms on the DNA at origins of replication and helps to unwind the DNA helix. This begins at the end of the previous M phase and during early G1, when E2F transcriptional activity drives the expression of proteins such as the ORCs, Cdc6, Cdt1, and the MCM proteins, which then assemble into the prereplicative complex (pre-RC) to 'licence' DNA for replication in S phase.

When cells progress beyond the G1-S transition, firing of the licensed replication origins occurs in response to various signals mediated by cyclin-CDK complexes, and another kinase known as Cdc7 (Diffley, 2011; Williams and Stoeber, 2012). For example, Cdc7 targets phosphorylation of the MCM proteins, leading to their activation. The MCM proteins function as a replicative helicase which is loaded onto the DNA by Cdc6 and Cdt1, and once activated they unwind the DNA helix at origins of replication to form a template for the recruitment of the DNA synthesis machinery. Once cells have entered the S phase, inactivation of the licensing factors is critical for preventing re-initiation of DNA replication from origins that have already fired, and this is achieved primarily through the inactivation of pre-RC components by phosphorylation and ubiquitylation events (Li and Jin, 2010; Diffley, 2011; Truong and Wu, 2011). Cdc6 and ORC are targeted by cyclin-CDK complexes, and their

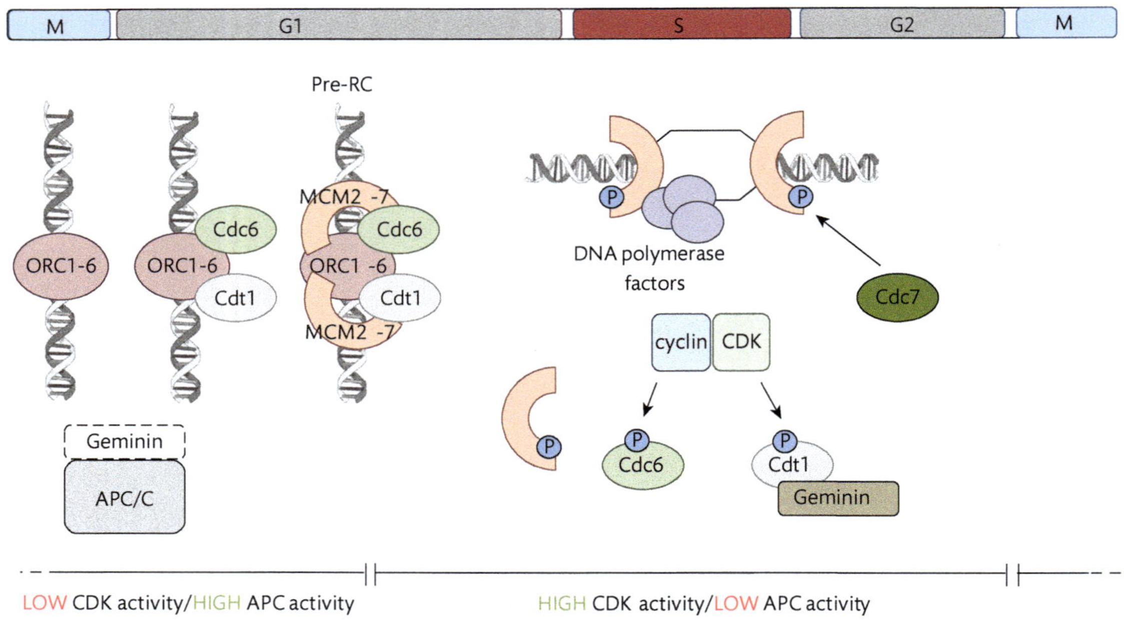

Fig. 13.4 Initiation of DNA replication. Cell cycle progression and DNA replication are coordinated to ensure the cell's genetic material is only replicated once per cell cycle. At the end of the previous M phase and throughout early G1, the expression of E2F-target genes such as ORC, MCM, Cdc6, and Cdt1 permits the formation of the prereplicative complex at origins of replication on the DNA. As cells progress into S phase, Cdc7 phosphorylates and activates the MCM helicase, which unwinds the DNA and permits the recruitment of the DNA polymerase machinery. Meanwhile, cyclin-CDK activity targets other pre-RC components to prevent reinitiation of DNA replication. This results in the translocation of Cdc6 out of the nucleus, while promoting Cdt1 degradation. Any remaining Cdt1 is inactivated by binding of the inhibitor geminin, which is highly expressed during S, G2, and M phases. During late M phase and early G1, the activity of the APC/C ubiquitin ligase complex promotes the destruction of geminin, while also mediating the degradation of S and M phase cyclins. This releases Cdt1 and also ensures CDK activity remains low, removing Cdc6 from inhibition and permitting MCM loading once more.

phosphorylation induces translocation of the proteins out of the nucleus and prevents pre-RC formation. The loading factor Cdt1 can also be targeted by phosphorylation, and this induces its subsequent ubiquitin-dependent degradation as cells progress through S phase and G2. Any residual Cdt1 is inactivated by the association of an inhibitor protein known as geminin, which is expressed at high levels during S, G2, and M phases. This association blocks the recruitment of the MCM proteins and hence prevents pre-RC formation. During late M phase and early G1, geminin itself becomes a target for degradation by the APC/C ubiquitin ligase complex, which then releases Cdt1 from inhibition (Teixeira and Reed, 2013; Bassermann et al., 2014). This results in a cyclical pattern of MCM helicase binding to replication origins, as it is displaced from DNA during the S phase and rebinds in late M phase and early G1.

When cells exist in a non-proliferative state, such as during differentiation, quiescence, and senescence, DNA replication needs to be switched off and once again this is mediated by regulation of the pre-RC licensing components. As cells exit the cell cycle, replication origins are converted to an unlicensed state by the down-regulation of Cdc6, Cdt1, and MCMs. Conversely, in highly proliferative cancer cells from many tumour types, the upregulation of licensing proteins is evident, since uncontrolled cell division is a common hallmark of cancer (Williams and Stoeber, 2012). This is generally a result of mutations in genes upstream of the licensing machinery, including the cyclins and the pRB-E2F pathway, which cause deregulation of licensing components and permit inappropriate entry into S phase, which can result in genetic errors and genomic instability.

Entering mitosis: The G2/M phase transition

After DNA replication is complete, there is a requirement to equally segregate the genetic material into two new daughter cells, each possessing an identical complement of chromosomes. To achieve this, cells first condense and arrange their chromosomes across the centre of the cell, to facilitate efficient separation, at which point the nuclear envelope breaks down and cells reorganize their microtubules to generate the mitotic spindle. This spindle will become attached to each of the chromosomes and is involved in segregating the genetic material to each cell pole. Once this separation is complete, the actin cytoskeleton is then engaged to divide the cell into two daughters by cytokinesis. This dramatic reorganization of cellular components during mitosis must be coordinated both spatially and temporally and is regulated once more by the activity of various cell cycle control components such as kinases and ubiquitin ligase complexes.

Mitotic kinases and ubiquitin ligases determine commitment to mitosis

Much like the restriction point during G1 when cells commit to cell cycle entry, once a cell is committed to mitosis there is no going back, and only cell death can prevent division. This switch is primarily regulated by the activity of the cyclin B-CDK1 complex, which forms as cells progress through G2 (Ma and Poon, 2011; Wieser and Pines, 2015). Cyclin B-CDK1 complexes target a very large number of substrates, including structural proteins and a number of other mitotic regulators. It is involved in restructuring the microtubule and actin cytoskeletons, promoting nuclear envelope breakdown, altering chromosome architecture, and regulating the timing of anaphase and cytokinesis. Cyclin B-CDK1 activity is often coordinated with the other mitotic kinases, most notably the Plk and Aurora kinases.

However, formation of the cyclin B-CDK1 complex alone is not always sufficient to drive entry into mitosis, as initially this complex remains inactive due to the presence of inhibitory phosphorylation events on the CDK1 subunit (Malumbres, 2014; Wieser and Pines, 2015; **Fig. 13.5**). These phosphorylation marks are created by a highly conserved family of protein kinases known as Wee1 and are antagonized by the Cdc25 family of phosphatases that act to remove the inhibitory phosphate groups and rapidly activate CDK1. This intricate method of CDK1 phospho-control permits several signals from a number of pathways to converge and regulate the timing of mitosis. For example, cellular stress or damage sustained during DNA replication activates checkpoint pathways that stabilize Wee1 but inhibit Cdc25, tipping the balance in favour of CDK1 inhibition (Wieser and Pines, 2015), while the activity of other important mitotic cyclin-CDK complexes (namely cyclin A-CDK), can help to promote mitosis by activating transcription of various mitotic regulators (Feruno et al., 1999). In some cells, the activity of cyclin B-CDK1 itself can drive a positive feedback loop, since both Wee1 and Cdc25 can be phosphorylated by CDK1 (Wieser and Pines, 2015). This inhibits Wee1 while promoting Cdc25 function, and together this ensures that once cyclin B-CDK1 activity reaches a certain threshold, activation rapidly proceeds to completion.

Other mitotic kinases such as Plk1 also strongly influence mitotic timing, though how it performs this regulation is not entirely clear to date (Lens et al., 2010; Bruinsma et al., 2012; Wieser and Pines, 2015). What is known, however, is that many CDK1 substrates are also targeted by Plk1, since the mitotic Plks have a conserved protein domain that allows them to bind to a phosphorylated substrate. Thus, for many targets, the action of cyclin B-CDK1 and Plk1 is coordinated by CDK1 phosphorylation creating a binding site for Plk1. Such coordination is exemplified when Plk1 works in concert with cyclin B-CDK1 to promote the association of Wee1 with the SCF ubiquitin ligase complex, which subsequently promotes the inactivation and degradation of Wee1 (**Fig. 13.5**; Watanabe et al., 2004). Plk1 activity is also an important determinant of cyclin B subcellular localization. B-type cyclins possess both a nuclear import signal and a nuclear export signal and as a result they constantly shuttle between the nucleus and the cytoplasm, though the bulk of cyclin B exists in the cytoplasm during G2. Plk1 activity during the onset of mitosis acts to prevent nuclear export, aiding in the nuclear accumulation of cyclin B-CDK1 during prophase (Toyoshima-Morimoto et al., 2002).

During mitosis, the activity of the Aurora kinases is also tightly regulated, since Aurora A kinase is intricately involved in the formation of the mitotic spindle. It is recruited by a number of cofactors to the microtubules themselves, where it phosphorylates proteins that function to nucleate microtubules from their organizing centres at each cell pole (Goldenson and Crispino, 2015). The Aurora B kinase also has a crucial role in correcting erroneous microtubule

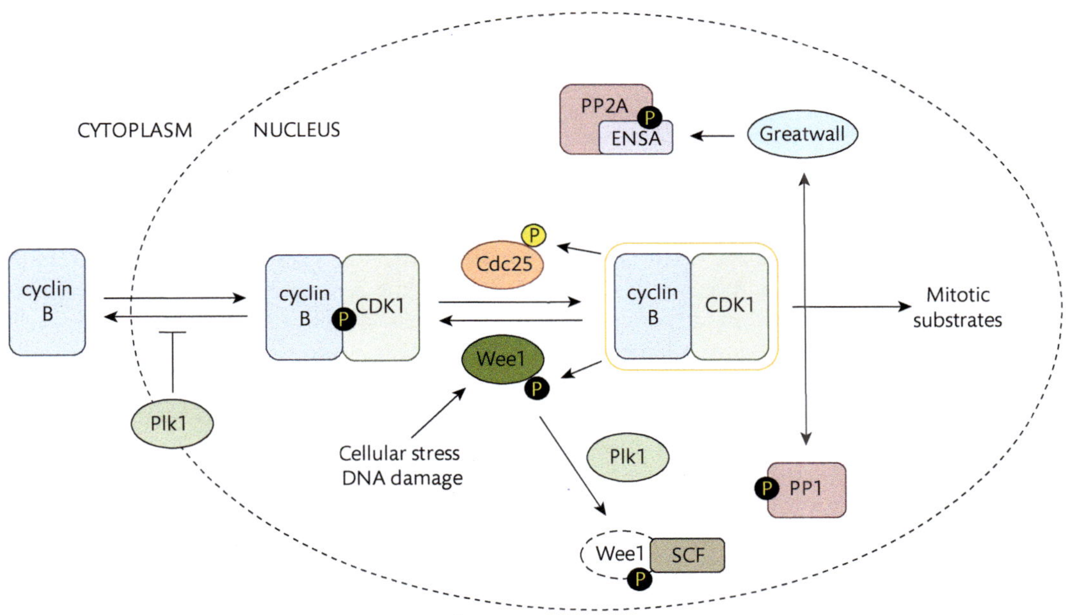

Fig. 13.5 Regulating cyclin B-CDK1 activity in early M phase. During G2, the bulk of cyclin B exists in the cytoplasm, though it can constantly shuttle in and out of the nucleus. Plk1 activity during prophase acts to prevent nuclear export, permitting cyclin B-CDK1 complexes to form. However, like all cyclin-CDK complexes, cyclin B-CDK1 is held in an inactive state by an inhibitory phosphorylation event created by the Wee1 family of kinases and opposed by the Cdc25 family of phosphatases. Cellular stress or DNA damage act to stabilize Wee1, tipping the balance in favour of CDK inactivation, while the activity of cyclin-CDK complexes (including cyclin B-CDK1 itself) set up a positive feedback loop that ensures activation proceeds to completion. This is mediated by the phosphorylation of Cdc25 and Wee1, which acts to promote the function of the first while inhibiting the function of the later. Plk1 acts in concert with cyclin B-CDK1 to phosphorylate many of the same downstream substrates, for example, it can also target Wee1 and promotes its association with SCF and subsequent degradation. The activity of phosphatases also needs to be regulated at this time, since they antagonize the function of cyclin B-CDK1. PP1 is directly inhibited by cyclin B-CDK1-mediated phosphorylation, while PP2A is indirectly inhibited via Greatwall kinase activity. This enzyme creates phosphorylated ENSA, a protein that acts as a competitive substrate inhibitor for PP2A. Inhibitory phosphorylations are shown in black, while activatory phosphorylation is shown in yellow.

attachments to the chromosomes once they are aligned along the centre of the cell, and therefore acts to ensure their correct separation. Once cells begin the anaphase stage, Aurora B moves to the microtubules, where it acts to help the coordination of cytokinesis (Lens et al., 2010; Goldenson and Crispino, 2015; Wieser and Pines, 2015).

Protein phosphatases also need to be controlled

While the phosphorylation of cellular substrates by the mitotic kinases is required to commit cells for entry into mitosis, activation of cyclin B-CDK1 alone in an experimental cell system is often not sufficient to drive these events (Mochida et al., 2009). This is because of the activity of phosphatase family members, enzymes that act to remove phosphorylation marks on proteins and hence antagonize the activity of the mitotic kinases. These phosphatases need to be inactivated at the onset of mitosis, to ensure that substrates are efficiently phosphorylated by cyclin B-CDK1 and the Plk and Aurora kinases. Like the ubiquitin ligases, phosphatases consist of large protein complexes that include the phosphatase catalytic subunit and other cofactors that act to target the complex to specific substrates (Wurzenberger and Gerlich, 2011; Mochida and Hunt, 2012). There are only a few protein phosphatase catalytic subunits in the cell, and they generally have broad substrate specificity, but the targeting subunits help to ensure recruitment of phosphatases only occurs in the correct place at the correct time.

The PP2A phosphatase family is directly involved in antagonizing phosphorylation events mediated by cyclin B-CDK1, but it becomes inactivated as cells enter mitosis by a negative feedback loop involving the cyclin-CDKs themselves. As the mitotic cyclin-CDK complexes form during G2 and M phase, they phosphorylate and activate another kinase known as Greatwall, which subsequently targets the small abundant ENSA protein that acts as a competitive substrate inhibitor for the PP2A complex (Wurzenberger and Gerlich, 2011; Mochida and Hunt, 2012). When bound to this inhibitor, PP2A can no longer target the removal of phosphate groups from other mitotic kinase targets, and the activity of cyclin B-CDK1 is no longer antagonized (Fig. 13.5). The ENSA inhibitor does slowly become dephosphorylated by PP2A however, which permits reactivation of PP2A at the end of mitosis when cyclin-CDK and Greatwall kinase activity begin to decline. Another highly abundant phosphatase in the cell, known as PP1, can also target a number of mitotic regulators for dephosphorylation, but its activity is also inhibited as cells enter mitosis (Wurzenberger and Gerlich, 2011; Mochida and Hunt, 2012). When cyclin B-CDK1 complexes form during M phase, the PP1 catalytic subunit becomes phosphorylated at an inhibitory site and is no longer able to catalyse the removal of phosphate groups from its substrates. Once again, this represents an example of the mitotic kinases ensuring the rapid activation of their targets by regulating the activity of the enzymes that act to antagonize them.

Chromosome cohesion and condensation

After DNA replication has been completed in S phase, each chromosome exists as two intertwined sister chromatids that are held together by a protein complex known as cohesin (Brooker and Berkowitz, 2014; Losada, 2014). This complex is highly conserved and forms a ring-shaped structure around the DNA fibres to entrap them. This chromosome organization is required for efficient housekeeping of the cell's genetic material and is essential for chromosome segregation and other processes such as DNA repair. At the onset of mitosis, most of the cohesin is released from chromatin to allow separation and segregation of the two sisters during anaphase, and this process requires the action of the mitotic kinases CDK1, Aurora B, and Plk1 (Losada, 2014). However, a small proportion of cohesin does remain enriched around the central portion of the chromosomes (the centromere), and this is protected from dissociation by the activity of PP2A and a protein known as shugoshin (Liu et al., 2013). The cohesin located at the centrosome permits alignment of the chromosomes across the mid-zone of the cell during metaphase, when the mitotic spindle becomes attached to the sister chromatids and begins to properly orientate the chromosomes. Cohesion between sisters, mediated by the centromeric cohesin, is important to prevent premature separation of the sister chromatids during this step. Once cells enter anaphase, the remaining cohesin complex is destroyed by the activity of an enzyme known as separase, and this allows the sister chromatids to finally segregate (Hirano, 2015).

As cells prepare for mitosis, the replicated DNA also needs to be condensed by progressive winding and folding of chromatin fibres, as trying to separate a tangled mass of DNA could lead to chromosome breakage and be disastrous for the cell. This process is mediated by a second multisubunit protein complex known as condensin, that shares some structural components with cohesin, and acts with other enzymes to supercoil the DNA (Thadani et al., 2012; Hirano, 2015). In higher eukaryotes, condensin is able to associate with the chromosomes during mitosis, once the nuclear envelope has broken down, and numerous aspects of its function are known to be regulated by the cell cycle control machinery. For example, mitotic cyclin-CDK and Plk1-dependent phosphorylation of condensin stimulates its supercoiling activity, while phosphorylation by Aurora B kinase promotes its association with chromatin and is required for maximal chromatin compaction in anaphase (Thadani et al., 2012).

Centrosome duplication and the mitotic spindle

Efficient separation of sister chromatids is dependent on the formation of the mitotic spindle, which consists of hundreds of proteins including the spindle microtubules themselves. These microtubules are focused at organizing centres present at each pole of the cell, known as the centrosomes. Each centrosome is composed of a large protein complex that include important cell cycle regulators and signalling molecules, as well as proteins that help to organize and nucleate microtubules. In G1, cells possess a single centrosome, but like DNA, during S phase, centrosomes duplicate once, and only once, per cycle (Agircan et al., 2014; Wang et al., 2014). These centrosomes remain tightly linked together until late in G2, when they separate to opposite poles of the cell. At the onset of mitosis, the centrosomes undergo a process of 'maturation', when they gain the ability to nucleate and anchor microtubules, and this process requires the concerted action of multiple mitotic enzymes including CDKs, Plks, NEKs, and Aurora kinases. Mitotic spindle assembly is therefore coupled to the centrosome cycle, which itself is tightly linked to the cell cycle via its regulation by mitotic kinases (Agircan et al., 2014; Wang et al., 2014). Defects in centrosome function often lead to abnormal mitotic spindle formation, chromosome segregation, and genetic instability (Cosenza and Krämer, 2016).

Centrosome duplication is primarily regulated by the activity of Plk4, which is recruited to the centrosome and phosphorylates a number of structural components to initiate centrosome biosynthesis (Wang et al., 2014). However, once centrosome duplication is complete, Plk4 protein is rapidly degraded by the action of the SCF ubiquitin ligase complex, which acts to prevent reduplication of centrosomes (Rogers et al., 2009). The centrosomes initially remain tightly bound together, but late in G2 several of the involved linker proteins become phosphorylated by members of the NEK family of kinases, themselves regulated by the action of Plk1. Such phosphorylation events cause these linker proteins to dissociate and this promotes the separation of centrosomes (Wang et al., 2014). The centrosomes are effectively 'pushed' apart to opposite poles of the cell by the activity of microtubules and their associated motor proteins, and this complex process is also regulated by phosphorylation events mediated by Plk1, the NEKs, CDK1, and Aurora A kinase. Aurora kinase activity is particularly important for spindle formation, as it phosphorylates centrosomal components involved in stabilizing spindle microtubules (Agircan et al., 2014; Wang et al., 2014).

Sister chromatid separation: Metaphase-to-anaphase transition checkpoint

After the chromosomes have been condensed and arranged along the mid-zone of the cell by the activity of the mitotic spindle, late mitotic events will drive the segregation of the sister chromatids to each cell pole. Once there, the mitotic spindle is broken down and the separated sister chromatids are enveloped once more into daughter nuclei. The decision to begin sister chromatid separation is known as the metaphase-to-anaphase transition and represents one of the main checkpoints in the cell: the spindle assembly checkpoint.

Once cells have entered mitosis, their progression is mainly regulated by the activity of the ubiquitin ligase complex APC/C, which has a number of targets including the mitotic kinases and proteins that help govern sister chromatid cohesion (Teixeira and Reed, 2013; Bassermann et al., 2014; Sivakumar and Gorbsky, 2015). Indeed, APC/C activity is one of the reasons why cyclin protein levels oscillate during the cell cycle, as it can target the S and M phase cyclin partners of CDK complexes. Without their binding partners, this effectively switches off the activity of the CDK. The APC/C therefore causes a shift in the balance between kinase and phosphatase activities, to promote an environment where target proteins become dephosphorylated (Fig. 13.6). Since phosphorylation events drive many of the steps required for mitotic entry, the removal of phosphate groups is one of the major signals that marks the end of mitosis (Wieser and Pines, 2015).

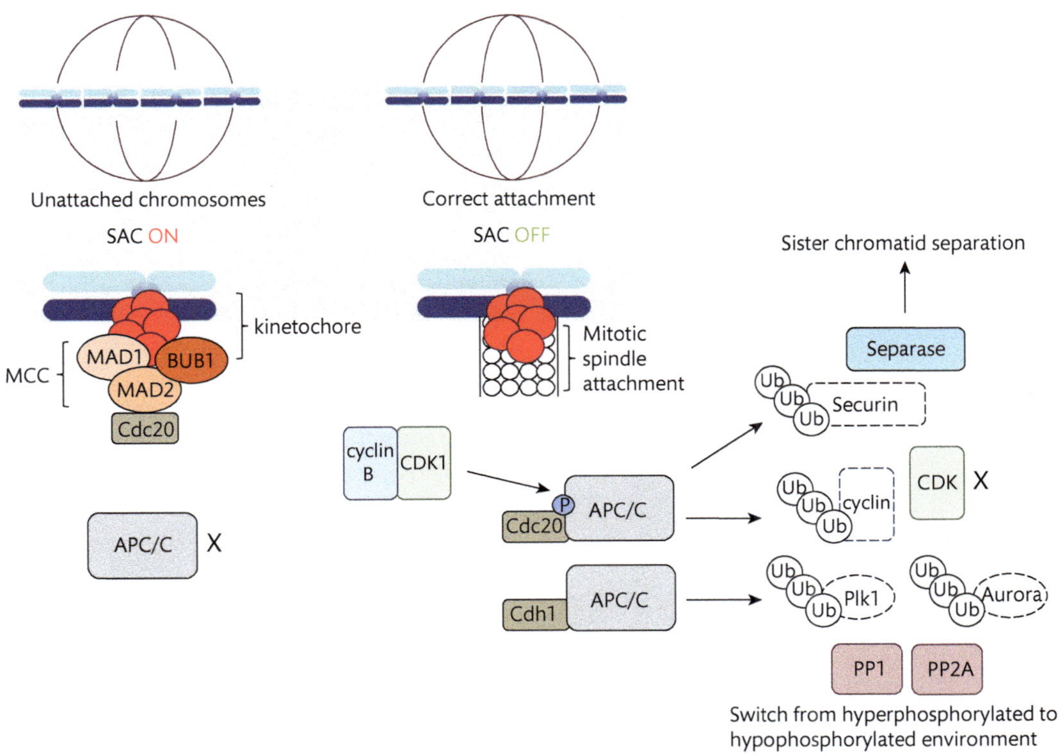

Fig. 13.6 APC/C activity regulates late mitotic events. Progression through mitosis is mainly regulated by the activity of the APC/C, which has a number of targets including mitotic kinases and proteins that govern sister chromatid cohesion. The APC/C complex is targeted to different substrates via the association of coactivator subunits (Cdc20 and Cdh1). By destroying and inactivating the mitotic kinases, the APC/C permits a shift in the balance between kinase and phosphatase activity, reverting the hyperphosphorylated cellular environment to a hypophosphorylated one. Such changes are essential for signalling the completion of mitosis and permitting the cell cycle control machinery to reset to a G1 state. APC/C activity is elegantly regulated by the spindle assembly checkpoint (SAC), which can detect the correct attachment of chromosomes to the mitotic spindle. In the absence of spindle attachment, a number of checkpoint proteins are recruited to the kinetochore to form the mitotic checkpoint complex (MCC), which prevents the activation of the APC/C by sequestering the Cdc20 coactivator. At metaphase, when spindle attachment is complete, the SAC is rapidly switched off, permitting the association of APC/C and Cdc20. This interaction is further enhanced by a phosphorylation event mediated by cyclin B-CDK1, which ensures efficient activation of the APC/C occurs at this time.

While the APC/C can recognize a plethora of different protein substrates, its association with these targets only occurs during discrete stages of mitosis or the G1 phase of the subsequent cell cycle. This regulation, in part, is mediated by the interaction of coactivator proteins with the APC/C, which are involved in recognizing specific amino acid motifs in target substrates (Teixeira and Reed, 2013; Sivakumar and Gorbsky, 2015). The two main coactivator subunits of import to APC/C-mediated mitotic events are Cdc20 and Cdh1, with Cdc20 being associated with the APC/C during mitosis and being replaced by Cdh1 when mitosis approaches completion (Fig. 13.6).

Regulation of APC/C activity by the metaphase-to-anaphase transition (spindle assembly) checkpoint

Given the importance of the APC/C in driving late mitotic events, its own activity is tightly regulated in a temporal fashion, so that it only becomes active once sister chromatid separation is appropriate. Once again, this regulation is mediated mainly by the activity of cyclin B-CDK1, which can phosphorylate the APC/C and enhance the binding of its coactivator Cdc20 (Shteinberg et al., 1999; Sivakumar and Gorbsky, 2015; Fig. 13.6). However, this in itself would appear to create a problem for the cell, since the APC/C would then target the destruction of cyclin B which would result in the inactivation of CDK1. This would severely restrict the time available to the cell to remain in mitosis, arrange its chromosomes, and generate the mitotic spindle, as all of these processes require the activity of CDK1 to remain high. This problem is resolved by the spindle assembly checkpoint (or SAC), that is able to detect the correct attachment of chromosomes to the mitotic spindle and will hold APC/C activity in check until this requirement is fulfilled (Foley and Kapoor, 2013; Musacchio, 2015). While the full details of how the SAC mediates this block are unknown, some of the details have been elucidated.

Attachment of the mitotic spindle to the chromosomes occurs at a protein structure present at the centromere known as the kinetochore, and in the absence of spindle attachment this region will recruit a number of checkpoint proteins that catalyse the assembly of an APC/C inhibitor called the mitotic checkpoint complex (or MCC). The recruitment of MAD1, MAD2, and BUB1 to unattached kinetochores seems to be the minimal requirement to form the MCC, which then closely associates with Cdc20 and prevents this coactivator subunit from associating with the rest of the APC/C (Fig. 13.6). The MCC also seems to be able to interact directly with the APC/C, though how this mechanism contributes to APC/C silencing is not yet understood (Foley and Kapoor, 2013; Musacchio, 2015).

The SAC produces a very strong inhibitory signal, capable of preventing APC/C activity even when only a few kinetochores remain unattached. At metaphase, when all the chromosomes are connected to the mitotic spindle, the SAC is rapidly switched off and APC/C activity will cause the destruction of mitotic cyclins and proteins that mediate sister chromatid cohesion (Fig. 13.6). One such target is an inhibitor protein known as securin, that associates with the separase enzyme and prevents it from inducing sister chromatid separation (Foley and Kapoor, 2013; Musacchio, 2015). The destruction of securin by the APC/C therefore releases separase activity, which can remove the remaining centromeric cohesin complex that holds sister chromatids together and permits their segregation during anaphase.

Phosphatases drive exit from mitosis

After sister chromatid separation, the cell is effectively committed to exit mitosis, since cyclin B has been targeted for destruction by the APC/C and CDK1 activity has declined. The cell now needs to restructure its nucleus and reshape its subcellular architecture to permit cytokinesis and return the two daughter cells to their original G1 state. These steps are in part driven by continued APC/C activity, now under the regulation of its other coactivator subunit Cdh1. Mitotic kinases including Plk and Aurora are the predominant targets, since these enzymes are not subject to degradation by the Cdc20-containing APC/C complex (Teixeira and Reed, 2013; Sivakumar and Gorbsky, 2015; Fig. 13.6). However, mitotic exit is primarily driven by the activity of phosphatases such as PP2A and PP1, which function to remove the manifold phosphorylation events introduced by the mitotic kinases. The cell therefore switches from a global hyperphosphorylated state during mitosis, to a hypophosphorylated state at the end of mitosis, and this shift in the balance between kinase and phosphatase activity defines the switch between early and late mitotic events (Wurzenberger and Gerlich, 2011; Wieser and Pines, 2015).

During mitosis, PP1 activity is held in check by inhibitory phosphorylation events mediated by cyclin B-CDK1, but as CDK1 activity declines PP1 can autocatalyse the removal of this inhibitory phosphate. PP1 is subsequently targeted to a number of substrates during anaphase where it helps to decondense chromosomes and induce the reformation of the nuclear envelope. PP2A also becomes active towards its substrates during late mitosis, since reduced CDK1 activity causes a subsequent decline in Greatwall kinase's ability to generate the phosphorylated competitive inhibitor ENSA. This shifts the balance towards PP2A dephosphorylating its own inhibitor, at which point it becomes free to target the removal of phosphate groups from other substrates (Wurzenberger and Gerlich, 2011; Mochida and Hunt, 2012). Again, these events are important in reshaping the cell and resetting the cell cycle control system in preparation for the next G1, or alternative cell fates such as G0 (quiescence) and differentiation.

Nuclear envelope reformation and cytokinesis

The mitotic kinases phosphorylate various substrates during mitosis, many of which include proteins present in the nuclear envelope, such as the lamins or components of the nuclear pore complex (NPC). Their phosphorylation causes their disassembly, a step which is important for nuclear envelope breakdown during prometaphase. It is therefore unsurprising that dephosphorylation of these proteins is associated with the reformation of the nucleus in telophase (Schellhaus et al., 2016). For example, cyclin B-CDK1 can target NPC components and nuclear membrane proteins, and this disrupts their association with chromatin and contributes to nuclear envelope disassembly, while the activity of PP1 late in mitosis acts to promote membrane recruitment to chromatin once more (Wurzenberger and Gerlich, 2011). The depolymerization of the nuclear lamina and its reassembly at the end of telophase also involves mitotic phosphorylation/dephosphorylation events mediated by CDK1 and PP1 (Peter et al., 1990; Steen et al., 2000).

Reforming of the nuclear envelope itself is thought to involve membrane structures derived from the cell's endoplasmic reticulum (ER), although the precise mechanism by which this membrane structure re-envelopes the decondensing chromosomes is still debated (Schellhaus et al., 2016). However, in late anaphase, ER derived membranes start approaching the chromatin, and this is believed to result from the chromatin binding affinity of nuclear envelope proteins associated with these membranes.

Cytokinesis represents the final step of cell division when the daughter cells become physically separated. It begins immediately after chromosome segregation, when a cleavage furrow forms at the equator and subsequently bisects the mother cell. This process is finely regulated to ensure equal distribution of cellular contents, and mistakes during cytokinesis can cause polyploidy and have been linked to various diseases including cancer (D'Avino et al., 2015; Srivastava et al., 2016). After anaphase, the mitotic spindle is completely reorganized into an array of antiparallel microtubules known as the central spindle, and many cytoskeletal proteins and signalling molecules cooperate to organize the orientation of this structure. CDK1 activity, for example, helps to prevent the interaction of these components prior to anaphase, when the mitotic spindle is still required, though their association is later promoted by the activity of Aurora B kinase. At the same time, actomyosin filaments begin to assemble at the cleavage furrow to form an annular structure known as the 'contractile ring'. The contraction of these filaments is the major driving force that triggers daughter cell separation and is intricately regulated by a series of signalling events involving proteins that stimulate actin polymerization and myosin contractility (D'Avino et al., 2015; Srivastava et al., 2016).

The cell cycle response to DNA damage

Cells are continuously being exposed to DNA damage caused by both endogenous and exogenous sources. A wide variety of DNA lesions can form, but in general all can result in genetic mutations that alter normal cellular function, therefore threatening the survival of the organism. In order to maintain the integrity of their genome, cells possess a signalling network that is able to both sense the type of damage sustained and repair it where appropriate, and this is known as the DNA damage response (DDR; Heijink et al., 2013; Sirbu and Cortez, 2013; Beishline and Azizkhan-Clifford, 2014). Defects in the DDR result in unrepaired damage which can lead to chromosomal alterations, a common cause of tumorigenesis. DNA repair can occur through a number of mechanisms, dependent on the type of damage sustained, but in order to provide cells with the time they require to perform this repair, cell cycle arrest is imposed immediately after detection of the damage. For example, damage sustained during G1 will result in cell cycle arrest prior to S phase entry to prevent replication of damaged DNA, while an intra-S phase checkpoint will halt ongoing DNA replication in the presence of damage.

A further checkpoint in G2 will prevent entry into mitosis if damaged DNA is detected, as this could cause incorrect chromosome segregation. In multicellular organisms, if the DNA damage is extensive, cells can also enter a state of permanent cell cycle arrest or be removed entirely from the organism by programmed cell death (apoptosis; Ouyang et al., 2012). While each of these DDR pathways is distinct, their signalling networks involve common 'sensor' proteins and downstream effector molecules (Maréchal and Zou, 2013) (Fig. 13.7). For example, DNA double-strand breaks will be recognized by a protein complex that recruits the ATM sensor kinase, while stalled replication forks will activate another sensor kinase known as ATR. These then target the activation of other 'effector' kinases, which mediate phosphorylation of hundreds of downstream targets involved in the establishment of cell cycle arrest. In particular, ATM and ATR activity lead to Cdc25 phosphorylation, resulting in the sequestration of Cdc25 in the cytoplasm or promoting its degradation by ubiquitin-mediated proteolysis (Sanchez et al., 1997; Mailand et al., 2002). As a result, they are unable to activate their cyclin-CDK substrates, establishing a block at S phase entry, or arresting S phase progression and the G2/M transition. DDR-mediated cell cycle control is not just restricted to CDK regulation however, since other kinases such as Wee1, Plk, and Aurora are also targeted for inhibition to enforce cell cycle arrest (Heijink et al., 2013; Sirbu and Cortez, 2013; Beishline and Azizkhan-Clifford, 2014). Interestingly, DNA damage sustained during mitosis does not induce arrest, even though the damage is sensed by the cell. This is because cells are programmed to prioritize mitotic progression over a full-scale DDR—prolonged mitosis is not beneficial as it can give rise to DNA breaks itself. The DDR is silenced during this cell cycle stage, though this does result in mitotic cells being more sensitive to DNA damage than interphase cells. In general, damage sustained during mitosis will be resolved in the next cell cycle by the G1 DNA damage checkpoint pathway (Heijink et al., 2013).

In addition to this direct cell cycle block, a transcription pathway is activated by DNA damage to enforce cell cycle arrest and induce the expression of genes required for DNA repair, and this involves the targeting of transcription factors such as p53 and E2F (Polager and Ginsberg, 2008; Biswas and Johnson, 2012; Meek, 2015; see Fig. 13.7). p53 levels in the cell are usually kept low by the activity of its negative regulator MDM2, which acts as a ubiquitin ligase to mediate p53 destruction. ATM-mediated phosphorylation of p53 and MDM2 result in an increase in p53 levels, which can then drive the transcription of genes such as p21, *GADD45*, *BAX*, and *PUMA* (Dai and Gu, 2010). p21 is a CKI which mediates cell cycle arrest in G1, while *GADD45* downregulates CDK1 activity at the G2/M transition. *BAX* and *PUMA* are proteins that help regulate the induction of apoptosis in response to significant levels of DNA damage.

Certain E2F family members are also known to be DNA damage responsive, in particular E2F1 whose protein levels increase in response to a variety of damage stimuli. This stabilization of E2F1 in part results from its phosphorylation mediated by ATM and ATR (Lin et al., 2001; Stevens et al., 2003). E2F-mediated G1-S transcriptional events are usually silenced during S phase by the activity of inhibitor E2Fs, but ATR signalling can also result in the phosphorylation and promoter disassociation of inhibitors such as E2F6 (Smolka et al., 2012; Bertoli et al., 2013b). These events result in the transcriptional stimulation of E2F-responsive genes involved in DNA repair or apoptosis, such as P73 and CASP7. It is important to note that p53, the pRB-E2F pathway, and other DDR/DNA repair components are heavily modified by post-translational modifications in addition to phosphorylation, and these events help to

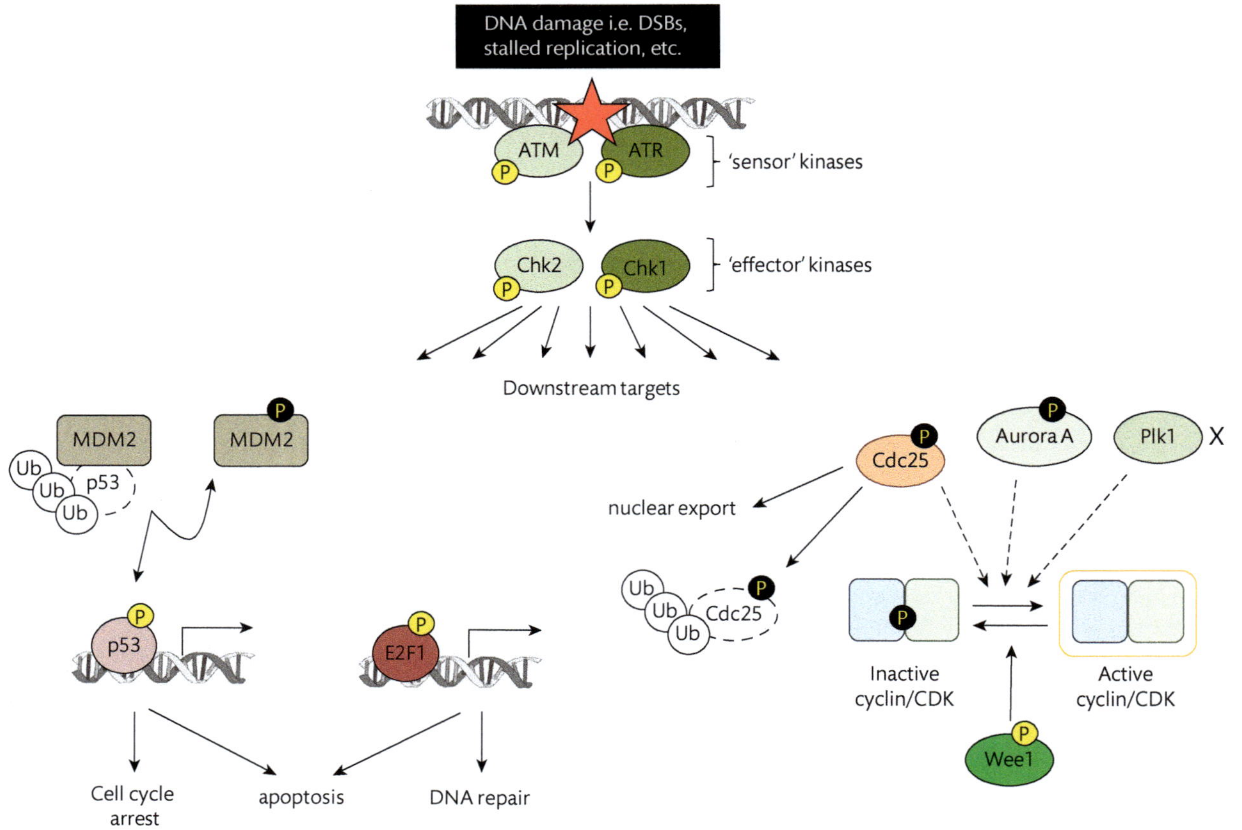

Fig. 13.7 The DNA damage response and cell cycle control. In response to DNA damage the cell will impose a cell cycle arrest to provide time for the damage to be repaired, or if the damage is extensive, the cell will be removed from the population via apoptosis. While each DDR pathway is distinct to the type of damage sustained, they involve the recruitment of common 'sensor' kinases (namely ATM and ATR), which then target downstream 'effector' kinases (Chk2 and Chk1), that mediate the phosphorylation of a plethora of substrates in the cell. For example, the Cdc25 family of phosphatases get phosphorylated, which induces their sequestration to the cytoplasm or degradation via ubiquitin-mediated proteolysis. As a result, they are unable to activate cyclin-CDK complexes. Other targets include the Wee1 kinases, which promote the formation of inhibitory phosphorylation events on cyclin-CDK complexes. Other mitotic kinases, including Aurora A and Plk1, are also directly or indirectly inhibited during the DDR, with consequent downstream impact on CDK1 activity. A transcriptional pathway is also activated in response to damage, via transcription factors such as p53 and E2F1. Phosphorylation of these factors (or their inhibitors in the case of MDM2), lead to protein stabilization and enhanced transcription of genes required for cell cycle arrest, DNA repair, or apoptosis. Inhibitory phosphorylation is shown in black, while activatory phosphorylation is shown in yellow.

modulate the precise cellular outcome in response to a variety of stresses (Dai and Gu, 2010; Munro et al., 2012; Dantuma and van Attikum, 2016).

Cancer and cell cycle deregulation

Gene mutations, cell cycle, and cancer progression

Although cancer is a broad grouping of diseases, all tumours share some fundamental hallmarks, one of which is the ability to proliferate in an uncontrolled fashion while avoiding signals that suppress growth or induce cell death. Under normal conditions, cells will only enter the cell cycle in response to external cues that indicate the environment is appropriate, and most cells within an organism are not cycling at all but exist in a terminally differentiated state. However, during the development of cancer, mutations in genes that regulate these processes permit the cell to circumvent the checks and controls that would usually restrain cell proliferation (Hanahan and Weinberg, 2011). The development of a tumour is therefore driven by the selection of new genetic traits that confer a competitive advantage over healthy cells. Given the complexity of the cell cycle control system, multiple mutations are usually required, as ablation of several key pathways is needed before a cell can proliferate uncontrollably (Hanahan and Weinberg, 2011).

Tumour growth begins with an abnormal increase in cell number following mutations that permit cancer cells to divide at an inappropriate rate. For example, genetic alterations in the mitogenic signalling pathway permit cells to grow in the absence of growth factors, while other mutations may inhibit the cellular response to signals that halt cell growth (Hanahan and Weinberg, 2011). Additionally, tumour cells may carry alterations that permit them to escape from apoptosis, thus avoiding cell death (Kelly and Strasser, 2011).

Oncogenes and tumour suppressors

Most genes involved in cell cycle control can be classified into one of two categories: the proto-oncogenes, which positively regulate progression through the cell cycle; and the tumour suppressors, that function to halt it (Morgan, 2007). Both groups of genes are frequently mutated in cancers. Mutations that affect the positive regulators and convert them to their oncogenic form usually involve gene amplifications and chromosome rearrangements, which can result from mis-segregation of chromosomes during anaphase or mistakes in DNA repair (Gordon et al., 2012). They result in protein overexpression or generate hyperactive proteins that drive proliferation. Mutations in tumour suppressor genes are often caused by point mutations or deletions of entire chromosomal segments, and lead to a loss of protein function with consequent cell cycle deregulation (Morgan, 2007).

Since tumour cells share the same underlying properties in terms of their proliferative behaviour, and many cell types use the same regulatory proteins to control their cell cycles, mutations in some proto-oncogenes and tumour suppressors are associated with many cancer types originating from a number of tissues. For example, mutation of the tumour suppressors p53 and pRB are associated with almost all types of human cancer (Sherr and McCormick, 2002). However, some mutations are tissue specific in their effects, and only promote cancer in one cell type. This is particularly true for familial cancers, where despite a mutation being shared by all somatic cells in the body, cancer will only develop in a small subset of tissues (Garber and Offit, 2005). The causes of this tissue specificity are still not always fully understood.

Next, we highlight some examples of proto-oncogenes and tumour suppressors that function in discrete cell cycle stages and describe how their mutation contributes to specific types of cancer.

Genetic alterations in mitogenic signalling

As stated earlier, most cells in the human body do not proliferate unless instructed to do so, while cancer cells can divide and grow in the absence of these signals. Defects in the mitogenic signalling pathway are usually the root cause of this uncontrolled proliferation, since they regulate the early activation of genes required for G1-S phase progression (see Section III of this book). Oncogenic mutations have been found at almost every step in these signalling pathways, including mutation of the genes encoding the mitogens and their receptors. For example, PDGF and the PDGF receptor are frequently expressed in several tumours, and their expression correlates with tumour growth, invasiveness, and poor clinical outcome (Cao, 2013). Many downstream signalling molecules also suffer from gain-of-function mutations, causing them to be hyperactive or always in an 'on' state. For example, the RAS gene encodes a protein that regulates several signalling networks and is found to be mutated in around 25% of all human cancers (Hobbs et al., 2016). Mitogen-responsive transcription factors such as Myc are also frequent targets for oncogenic mutation and are overexpressed in many cancers. A perfect example of this occurs with Burkitt's lymphoma, where a chromosome translocation leads to the production of MYC at abnormally high levels, which result in the proliferation of lymphoid cells (Janz, 2006).

Defective G1-S phase gene regulation

Cancer cells also usually carry mutations that impact on the expression of genes at the G1-S phase boundary, particularly those regulated by the E2F transcription factors. These genes are usually regulated by the activity of pRB, whose own function is controlled by the phosphorylation events driven by cyclin-CDK complexes. In almost all human cancers described, mutations will disrupt some arm of the pRB-E2F pathway. Gene amplifications of cyclin D and CDK4 are seen in some tumours and lead to the overproduction of these enzymes, though more commonly, tumour cells will lose CKI genes such as p16 and p27, which usually function to restrain the activity of early and late G1-regulated CDKs (Ortega et al., 2002; Malumbres and Barbacid, 2009). In addition, the gene encoding for pRB itself is lost or defective in many tumour types. Indeed, pRB was first linked with a familial disease that leads to the retinal cancer retinoblastoma (Garber and Offit, 2005).

Cancer and defects in cell cycle exit

Cell proliferation is not just governed by mitogens, but also by factors that govern cell survival, cell death, and differentiation. In general, tumour cells grow much more rapidly when they acquire mutations that permit them to circumvent their requirements on these factors. Mutation or deficiency of the gene *PTEN*, a phosphatase that counteracts one of the most critical growth signalling pathways in the cell, is prevalent in many types of human cancer and is associated with advanced tumour stage (Zhang and Yu, 2010), while antiapoptotic proteins such as BCL-2 and BCL-X_L are overproduced in certain cancers (Yip and Reed, 2008). Transcription factors that drive the expression of proapoptotic genes are also commonly mutated in many cancers, with p53 being a prime example, permitting cells to proliferate in the face of DNA damage or other cellular stresses (Rivlin et al., 2011). Tumour cells also acquire mutations that prevent differentiation, retaining the cell in a proliferative state while normal cells cease to divide. In general, genes like MYC and pRB also have roles in regulating differentiation in certain cell types, so loss of normal protein function by mutation prevents differentiation while also promoting proliferation.

Defects in the DNA damage response

There are several examples where there is a defined link between the dysfunction of DNA damage response components and cancer. Loss-of-function mutations in the DNA repair genes MSH2 and MLH1 are associated with the familial disease hereditary non-polyposis colorectal cancer, while familial forms of breast, ovarian, and pancreatic cancer have been associated with mutations in other DNA repair genes such as BRCA1, ATM, and RAD51 (Lord and Ashworth, 2012). A number of other rare syndromes exist where the underlying genetic cause is a defect in DNA repair, and these conditions include ataxia telangiectasia, Bloom's syndrome, and Nijmegen breakage syndrome, which all cause extreme radiosensitivity in patients (Lord and Ashworth, 2012). It is also important to remember that dysfunctional DNA damage response components will also prevent the cell cycle arrest or apoptosis that usually results from DNA damage. Thus, mutations that affect DNA repair and DNA damage responsive proteins will contribute towards genetic instability, since additional chromosomal damage will occur as the cell attempts to replicate or segregate damaged chromosomes (Aguilera and García-Muse, 2013). Indeed, many cancer cells are known to undergo chromosomal rearrangements that cause loss, amplification, or exchanges of chromosome segments, and these rearrangements are thought to result in part from mutations in DNA repair components (Thompson et al., 2010).

Mitotic defects: Centrosome duplication and the spindle microtubules

The mutation of various other genes involved in mitotic checkpoints and processes can also contribute to the genetic instability observed in many cancer types. For example, impairment of components of the SAC, important for preventing anaphase onset until all metaphase chromosomes attach correctly to the bipolar spindle, can significantly increases the likelihood of chromosome mis-segregations. Indeed, some aneuploid tumour cell lines display decreased expression of SAC components such as MAD2, while others have mutations in genes like BUB1, resulting in an inability to maintain mitotic arrest in response to spindle poisons (Thompson et al., 2010). Mouse models heterozygous for mutations in SAC genes like the MADs and BUBs also display elevated chromosome mis-segregation rates, with a corresponding increased tumour incidence. However, in human tumours mutations in SAC genes are very rare, though patients suffering from mosaic variegated aneuploidy do display mutations in a BUB-related kinase. This disease is associated with premature chromatid separation in lymphocytes and aneuploidy in many tissues, believed to be linked with the early onset of childhood cancer in these patients (Thompson et al., 2010).

Inefficient separation of sister chromatids can be observed during anaphase in many human tumours, and this mitotic defect results from a disruption to the normal spindle geometry in cells, which causes chromosome mis-segregation via the erroneous attachment of the mitotic spindle to sister chromatid kinetochores. Such conditions can arise in tumour cells possessing multiple centrosomes, as a result of deregulation of the centrosome duplication cycle or as a by-product of cells that fail to undergo correct cytokinesis (Thompson et al., 2010; Cosenza and Krämer, 2016). For example, breast and colorectal cancers frequently overexpress the Aurora A protein, which results in centrosome accumulation and spindle multipolarity (Thompson et al., 2010). Chromosomes therefore mis-segregate and aneuploidy is the result.

Summary and future directions

The cell cycle constitutes a complex series of cell signalling events that ensure cellular processes such as DNA replication and cell division are regulated in response to external cues and occur in a temporally regulated fashion. A huge number of genes, encoding both tumour suppressors and proto-oncogenes, act to regulate multiple steps in this cycle, and collectively they constitute the cell cycle control machinery. Many of these regulators can become mutated in cancer as a result of genetic mutations acquired spontaneously or in response to external stresses, and when enough of these genes fail to function correctly the outcome is a deregulated cell cycle that can result in tumour development. The challenge facing both scientists and clinicians today is how to use our developing understanding of cancer cell biology to generate better methods for treating the disease. Cancers themselves are a very heterogeneous collection of diseases, but available techniques allow us to characterize patients' tumours at the molecular level to discover groups of genes and proteins that are frequently mutated or deregulated in a particular cancer subtype. As our understanding of how these genes influence cell cycle control expands, we gain further insight into how mutations contribute to tumour growth and invasiveness. By designing precisely targeted drugs to exploit the differences between cancer and normal cells, we should be able to inhibit the inappropriate activity of mutated proteins, or indeed restore wild type protein function to cells which lack it. Indeed, many of the enzymes mentioned throughout this chapter represent novel targets for small molecule inhibitor design, with the ultimate goal being to generate a compound that can attack the tumour cells while having minimal off-target effects to healthy cells that do not possess the tumour-associated mutations. Having detailed molecular and genetic information on each patient's cancer will also enable us to stratify patients based on the distinct sets of mutations they possess. Indeed, such information will allow clinicians to correlate a cancer's genetic makeup with different prognoses and responses to individual types of therapy.

TAKE-HOME MESSAGE

- The cell cycle is divided into discrete phases in which different cellular events takes place. A number of cell cycle checkpoints exist that ensure later stages only occur upon completion of the proceeding step.
- Numerous proteins are involved in cell cycle checkpoints, but two broad families of enzymes are particularly important: the CDKs and the ubiquitin ligases.
- During the G1/S transition, CDK activity drives phosphorylation of the pocket proteins. This releases E2F transcription factor activity, which induces the expression of genes required for DNA replication and S phase progression.
- DNA replication during S phase is tightly regulated to ensure the genetic information within the cell is copied only once per cell cycle.
- The activity of mitotic kinases, particularly cyclin B-CDK1, is the predominant mechanism governing the entry into mitosis.
- Centrosome duplication and mitotic spindle formation is tightly linked to the cell cycle via its regulation by the mitotic kinases.
- Sister chromatid separation at the metaphase/anaphase transition is regulated by the spindle assembly checkpoint, which governs the activity of the APC/C ubiquitin ligase complex.
- Exit from mitosis is mediated primarily by the activity of phosphatases, and the APC/C-mediated destruction of the mitotic kinases.
- Cells have a complex signalling network that is capable of both sensing and repairing DNA damage. This DNA damage response coordinates DNA repair with cell cycle arrest.
- All cancer cells share a fundamental hallmark: the ability to proliferate in an uncontrolled fashion. During tumour development, cells acquire mutations in cell cycle checkpoint genes, which permits the circumvention of controls that would usually restrain cell proliferation.

OPEN QUESTIONS

- Why, in some genes, does a given mutation dysregulate cell cycle control only in some tissues but not in others?
- Many different factors can lead to deregulation of the cell cycle. It is now necessary, in each type of tumour if not in each patient, to map the specific alterations in that neoplasia deregulating proliferation. The aim is to select the most specific treatment causing the least side effects.
- Our understanding of the relationship between the main components of cell cycle regulation is still not complete (e.g. the role of p21 seems to be not only that of a repressor but in some conditions, could instead promote proliferation).

- Recent observations suggest that other factors like lipid molecules can affect the cell cycle; how this can affect cancer cells is still unknown.

FURTHER READING

Bertoli, C., Skotheim, J. M., & de Bruin, R. A. M. (2013a). Control of cell cycle transcription during G1 and S phases. *Nat Rev Mol Cell Biol*, **14**, 518–28.

Giacinti, C. & Giordano, A. (2006). RB and cell cycle progression. *Oncogene*, **25**, 5220–7.

Hanahan, D. & Weinberg, R. A. (2011). Hallmarks of cancer: the next generation. *Cell*, **144**, 646–74.

Heijink, A. M., Krajewska, M., & van Vugt, M. A. (2013). The DNA damage response during mitosis. *Mutat Res*, **750**, 45–55.

Ma, H. T. & Poon, R. Y. (2011). How protein kinases co-ordinate mitosis in animal cells. *Biochem J*, **435**, 17–31.

Meek, D. W. (2015). Regulation of the p53 response and its relationship to cancer. *Biochem J*, **469**, 325–46.

Rhind, N. & Russell, P. (2015). Signalling pathways that regulate cell division. *Cold Spring Harb Perspect Biol*, **4**, a005942.

Teixeira, L. K. & Reed, S. I. (2013). Ubiquitin ligases and cell cycle control. *Annu Rev Biochem*, **82**, 387–414.

Truong, L. N. & Wu, X. (2011). Prevention of DNA re-replication in eukaryotic cells. *J Mol Cell Biol*, **3**, 13–22.

Wieser, S. & Pines, J. (2015). The biochemistry of mitosis. *Cold Spring Harb Perspect Biol*, **7**, a015776.

Wurzenberger, C. & Gerlich, D. W. (2011). Phosphatases: providing safe passage through mitotic exit. *Nat Rev Mol Cell Biol*, **12**, 469–82.

REFERENCES

Agircan, F. G., Schiebel, E., & Mardin, B. R. (2014). Separate to operate: control of centrosome positioning and separation. *Philos Trans R Soc Lond B Biol Sci*, **369**, 1650.

Aguilera, A. & García-Muse, T. (2013). Causes of genome instability. *Annu Rev Genet*, **47**, 1–32.

Bassermann, F., Eichner, R., & Pagano, M. (2014). The ubiquitin proteasome system—implications for cell cycle control and the targeted treatment of cancer. *Biochim Biophys Acta*, **1843**, 150–62.

Bertoli, C., Skotheim, J. M., & de Bruin, R. A. M. (2013a). Control of cell cycle transcription during G1 and S phases. *Nat Rev Mol Cell Biol*, **14**, 518–28.

Bertoli, C., Klier S., McGowan, C., Wittenberg, C., & de Bruin, R. A. (2013b). Chk1 inhibits E2F6 repressor function in response to replication stress to maintain cell-cycle transcription. *Curr Biol*, **23**, 1629–37.

Beishline K. & Azizkhan-Clifford, J. (2014). Interplay between the cell cycle and double-strand break response in mammalian cells. *Methods Mol Biol*, **1170**, 41–59.

Biswas, A. K. & Johnson, D. G. (2012). Transcriptional and nontranscriptional functions of E2F1 in response to DNA damage. *Cancer Res*, **72**, 13–17.

Bretones, G., Delgado, M. D., & León, J. (2015). Myc and cell cycle control. *Biochim Biophys Acta*, **1849**, 506–16.

Brooker, A. S. & Berkowitz, K. M. (2014). The roles of cohesins in mitosis, meiosis, and human health and disease. *Methods Mol Biol*, **1170**, 229–66.

Bruinsma, W., Raaijmakers, J. A., & Medema, R. H. (2012). Switching Polo-like kinase-1 on and off in time and space. *Trends Biochem Sci*, **37**, 534–42.

Burkhart, D. L. & Sage, J. (2008). Cellular mechanisms of tumour suppression by the retinoblastoma gene. *Nat Rev Cancer*, **8**, 671–82.

Cao, Y. (2013). Multifarious functions of PDGFs and PDGFRs in tumor growth and metastasis. *Trends Mol Med*, **19**, 460–73.

Chen, H. Z., Tsai, S. Y., & Leone, G. (2009). Emerging roles of E2Fs in cancer: an exit from cell cycle control. *Nat Rev Cancer*, **9**, 785–97.

Chong, J. L, Wenzel, P. L., Sáenz-Robles, M. T., et al. (2009). E2f1-3 switch from activators in progenitor cells to repressors in differentiation cells. *Nature*, **462**, 930–4.

Claassen, G. F. & Hann, S. R. (2000). A role for transcriptional repression of p21CIP1 by c-Myc in overcoming transforming growth factor beta-induced cell-cycle arrest. *Proc Natl Acad Sci U S A*, **97**, 9498–503.

Connell-Crowley, L., Harper, J. W., & Goodrich, D. W. (1997). Cyclin D1/Cdk4 regulates retinoblastoma protein-mediated cell cycle arrest by site-specific phosphorylation. *Mol Biol Cell*, **8**, 287–301.

Cosenza, M. R. & Krämer, A. (2016). Centrosome amplification, chromosomal instability and cancer: mechanistic, clinical and therapeutic issues. *Chromosome Res*, **24**, 105–26.

Dai, C. & Gu, W. (2010). p53 post-translational modification: deregulated in tumorigenesis. *Trends Mol Med*, **16**, 528–36.

Dantuma, N. P. & van Attikum, H. (2016). Spatiotemporal regulation of posttranslational modifications in the DNA damage response. *EMBO J*, **35**, 6–23.

D'Avino, P. P., Giansanti, M. G., & Petronczki, M. (2015). Cytokinesis in animal cells. *Cold Spring Harb Perspect Biol*, **7**, a015834.

Diffley, J. F. (2011). Quality control in the initiation of eukaryotic DNA replication. *Philos Trans R Soc Lond B Biol Sci*, **366**, 3545–53.

Duronio, R. J. & O'Farrell, P. H. (1995). Development control of the G1 to S transition in Drosophila. Cyclin E is a limiting downstream target of E2F. *Genes Dev*, **9**, 1456–68.

Evans, T., Rosenthal, E. T., Youngblom, J., Distel, D., & Hunt, T. (1983). Cyclin: a protein specified by maternal mRNA in sea urchin eggs that is destroyed at each cleavage division. *Cell*, **33**, 389–96.

Feruno, N., den Elzen, N., & Pines, J. (1999). Human cyclin A is required for mitosis until mid-prophase. *J Cell Biol*, **147**, 295–306.

Foley, E. A. & Kapoor, T. M. (2013). Microtubule attachment and spindle assembly checkpoint signalling at the kinetochore. *Nat Rev Mol Cell Biol*, **14**, 25–37.

Garber, J. E. & Offit, K. (2005). Hereditary cancer predisposition syndromes. *J Clin Oncol*, **23**, 276–92.

Giacinti, C. & Giordano, A. (2006). RB and cell cycle progression. *Oncogene*, **25**, 5220–7.

Goldenson, B. & Crispino, J. D. (2015). The aurora kinases in cell cycle and leukemia. *Oncogene*, **34**, 537–45.

Gordon, D. J., Resio, B., & Pellman, D. (2012). Causes and consequences of aneuploidy in cancer. *Nat Rev Genet*, **13**, 189–203.

Gould, K. L. & Nurse, P. (1991). Tyrosine phosphorylation of the fission yeast cdc2 + protein kinase regulates entry into mitosis. *Nature*, **342**, 39–45.

Hanahan, D. & Weinberg, R. A. (2011). Hallmarks of cancer: the next generation. *Cell*, **144**, 646–74.

Hartwell, L. H., Culotti, J., & Reid, B. (1970). Genetic control of the cell-division cycle in yeast I. Detection of mutants. *Proc Natl Acad Sci U S A*, **66**, 352–9.

Hartwell, L. H. (1971). Genetic control of the cell-division cycle in yeast II. Genes controlling DNA replication and its initiation. *J Mol Biol*, **14**, 183–94.

Hartwell, L. H., Culotti, J., Pringle, J. R., & Reid, B. J. (1974). Genetic control of the cell division cycle in yeast. *Science*, **11**, 46–51.

Heijink, A. M., Krajewska, M., & van Vugt, M. A. (2013). The DNA damage response during mitosis. *Mutat Res*, **750**, 45–55.

Hirano, T. (2015). Chromosome dynamics during mitosis. *Cold Spring Harb Perspect Biol*, **7**, a015792.

Hobbs, G. A., Der, C. J., & Rossman, K. L. (2016). RAS isoforms and mutations in cancer at a glance. *J Cell Sci*, **129**, 1287–92.

Janz, S. (2006). Myc translocations in B cell and plasma cell neoplasms. *DNA Repair*, **5**, 1213–24.

Kelly, G. & Strasser, A. (2011). The essential role of evasion from cell death in cancer. *Adv Cancer Res*, **111**, 39–96.

Krek, W., Ewen, M. E., Shirodkar, S., Arany, Z., Kaelin, Jr. W. G., & Livingston, D. M. (1994). Negative regulation of the growth-promoting transcription factor E2F-1 by a stably bound cyclin A-dependent protein kinase. *Cell*, **1**, 161–72.

Lee, M. G. & Nurse, P. (1987). Complementation used to clone a human homologue of the fission yeast cell cycle control gene cdc2. *Nature*, **327**, 31–5.

Lens, S. M., Voest, E. E., & Medema, R. H. (2010). Shared and separate functions of polo-like kinases and aurora kinases in cancer. *Nat Rev Cancer*, **10**, 825–41.

Li, C. & Jin, J. (2010). DNA replication licensing control and rereplication prevention. *Protein Cell*, **1**, 227–36.

Lim, S. & Kaldis, P (2013). CDKs, cyclins and CKIs: roles beyond cell cycle regulation. *Development*, **140**, 3079–93.

Lin, W. C., Kashanchi, F., Radonovich, M. F., Shiekhattar, R., & Brady, J. N. (2001). Selective induction of E2F1 in response to DNA damage, mediated by ATM-dependent phosphorylation. *Genes Dev*, **15**, 1833–44.

Liu, H., Rankin, S., & Yu, H. (2013). Phosphorylation-enabled binding of SGO1-PP2A to cohesion protects sororin and centromeric cohesion during mitosis. *Nat Cell Biol*, **15**, 40–9.

Lord, C. J. & Ashworth, A. (2012). The DNA damage response and cancer therapy. *Nature*, **481**, 287–94.

Losada, A. (2014). Cohesin in cancer: chromosome segregation and beyond. *Nat Rev Cancer*, **14**, 389–93.

Ma, H. T. & Poon, R. Y. (2011). How protein kinases co-ordinate mitosis in animal cells. *Biochem J*, **435**, 17–31.

Mailand, N., Podtelejnikov, A. V., Groth, A., Mann, M., Bartek, J., & Lukas, J. (2002). Regulation of G(2)/M events by Cdc25A through phosphorylation-dependent modulation of its stability. *EMBO J*, **21**, 5911–20.

Malumbres, M. (2014). Cyclin-dependent kinases. *Genome Biology*, **15**, 122–31.

Malumbres, M. & Barbacid, M. (2009). Cell cycle, CDKs and cancer: a changing paradigm. *Nat Rev Cancer*, **9**, 153–66.

Maréchal, A. & Zou, L. (2013). DNA damage sensing by the ATM and ATR kinases. *Cold Spring Harb Perspect Biol*, **5**, a012716.

Masai, H., Matsumoto, S., You, Z., Yoshizawa-Sugata, N., & Oda, M. (2010). Eukaryotic chromosome DNA replication: where, when, how? *Annu Rev Biochem*, **79**, 89–130.

Meek, D. W. (2015). Regulation of the p53 response and its relationship to cancer. *Biochem J*, **469**, 325–46.

Mochida, S., Ikeo, S., Gannon, J., & Hunt, T. (2009). Regulated activity of PP2A-B55 delta is crucial for controlling entry into and exit from mitosis in Xenopus egg extracts. *EMBO J*, **28**, 2777–85.

Mochida, S. & Hunt, T. (2012). Protein phosphatases and their regulation in the control of mitosis. *EMBO Rep*, **13**, 197–203.

Montagnoli, A., Fiore, F., Eytan, E., et al. (1999). Ubiquitination of p27 is regulated by CDK-dependent phosphorylation and trimeric complex formation. *Genes Dev*, **13**, 1181–9.

Morgan, D. O. (2007). *The Cell Cycle: Principles of Control*. London: New Science Press.

Munro, S., Carr, S. M., & La Thangue, N. B. (2012). Diversity within the pRb pathway: is there a code of conduct? *Oncogene*, **31**, 4343–52.

Murray, A. W. & Kirschner, M. W. (1989). Cyclin synthesis drives the early embryonic cell cycle. *Nature*, **339**, 275–80.

Murray, A. W., Solomon, M. J., & Kirschner, M. W. (1989). The role of cyclin synthesis and degradation in the control of maturation promoting factor activity. *Nature*, **339**, 280–6.

Musacchio, A. (2015). The molecular biology of spindle assembly checkpoint signalling dynamics. *Curr Biol*, **25**, R1002–18.

Neganova, I. & Lako, M. (2008). G1 to S phase cell cycle transition in somatic and embryonic stem cells. *J Anat*, **213**, 30–44.

Ortega, S., Malumbres, M., & Barbacid, M. (2002). Cyclin D-dependent kinases, INK4 inhibitors and cancer. *Biochim Biophys Acta*, **1602**, 73–87.

Ouyang, L., Shi, Z., Zhao, S., et al. (2012). Programmed cell death pathways in cancer: a review of apoptosis, autophagy and programmed necrosis. *Cell Prolif*, **45**, 487–98.

Polager, S. & Ginsberg, D. (2008). E2F—at the crossroads of life and death. *Trends Cell Biol*, **11**, 528–35.

Peter, M., Nakagawa, J., Doree, M., Labbe, J. C., & Nigg, E. A. (1990). *In vitro* disassembly of the nuclear lamina and M phase-specific phosphorylation of lamins by Cdc2 kinase. *Cell*, **61**, 591–602.

Rhind, N. & Russell, P. (2015). Signalling pathways that regulate cell division. *Cold Spring Harb Perspect Biol*, **4**, a005942.

Rivlin, N., Brosh, R., Oren, M., & Rotter, V. (2011). Mutations in the p53 tumor suppressor gene: important milestones at the various steps of tumorigenesis. *Genes Cancer*, **2**, 466–74.

Rogers, G. C., Rusan, N. M., Roberts, D. M., Peifer, M., & Rogers, S. L. (2009). The SCF SLIMB ubiquitin ligase regulates Olk4/Sak levels to block centriole reduplication. *J Cell Biol*, **184**, 225–39.

Russell, P. & Nurse, P. (1986). CDC25 + functions as an inducer in the mitotic control of fission yeast. *Cell*, **45**, 145–53.

Russell, P. & Nurse, P. (1987). Negative regulation of mitosis by WEE1+, a gene encoding a protein kinase homolog. *Cell*, **49**, 559–67.

Sanchez, Y., Wong, C., Thoma, R. S., et al. (1997). Conservation of the Chk1 checkpoint pathway in mammals: linkage of DNA damage to CDK regulation through Cdc25. *Science*, **277**(5331), 1497–501.

Schellhaus, A. K., De Magistris, P., & Antonin, W. (2016). Nuclear reformation at the end of mitosis. *J Mol Biol*, **428**, 1962–85.

Sherr, C. J. & McCormick, F. (2002). The RB and p53 pathways in cancer. *Cancer Cell*, **2**, 103–12.

Shteinberg, M., Protopopov, Y., Listovsky, T., Brandeis, M., & Hershko, A. (1999). Phosphorylation of the cyclosome is required for its stimulation by Fizzy/cdc20. *Biochem Biophys Res Commun*, **260**, 193–8.

Sirbu, B. M. & Cortez, D. (2013). DNA damage response: three levels of DNA repair regulation. *Cold Spring Harb Perspect Biol*, **5**, a012724.

Sivakumar, S. & Gorbsky, G. J. (2015). Spatiotemporal regulation of the anaphase-promoting complex in mitosis. *Nat Rev Mol Cell Biol*, **16**, 82–94.

Smolka, M. B., Bastos de Oliveira, F. M., Harris, M. R., & de Bruin, R. A. (2012). The checkpoint transcriptional response: make sure to turn it off once you are satisfied. *Cell Cycle*, **11**, 3166–74.

Srivastava, V., Iglesias, P. A., & Robinson, D. N. (2016). Cytokinesis: robust cell shape regulation. *Semin Cell Dev Biol*, **53**, 39–44.

Steen, R. L., Martins, S. B., Tasken, K., & Collas, P. (2000). Recruitment of protein phosphatase 1 to the nuclear envelope by A-kinase anchoring protein AKAP149 is a prerequisite for nuclear lamina assembly. *J Cell Biol*, **150**, 1251–62.

Stevens, C., Smith, L., & La Thangue, N. B. (2003). Chk2 activates E2F-1 in response to DNA damage. *Nat Cell Biol*, **5**, 401–9.

Sun, A., Bagella, L., Tutton, S., Romano, G., & Giordano, A. (2007). From G0 to S phase: a view of the roles played by the retinoblastoma (Rb) family members in the Rb-E2F pathway. *J Cell Biochem*, **102**, 1400–4.

Teixeira, L. K. & Reed, S. I. (2013). Ubiquitin ligases and cell cycle control. *Annu Rev Biochem*, **82**, 387–414.

Thadani, R., Uhlmann, F., & Heeger, S. (2012). Condensin, chromatin crossbarring and chromosome condensation. *Curr Biol*, **22**, R1012–21.

Thompson, S. L., Bakhoum, S. F., & Compton, D. A. (2010). Mechanisms of chromosomal instability. *Curr Biol*, **20**, R285–95.

Toyoshima-Morimoto, F., Taniguchi, E., & Nishida, E. (2002). Plk1 promotes nuclear translocation of human Cdc25C during prophase. *EMBO Rep*, **3**, 341–8.

Truong, L. N. & Wu, X. (2011). Prevention of DNA re-replication in eukaryotic cells. *J Mol Cell Biol*, **3**, 13–22.

Vigo, E., Müller, H., Prosperini, E., et al. (1999). CDC25A phosphatase is a target of E2F and is required for the efficient E2F-induced S phase. *Mol Cell Biol*, **19**, 6379–95.

Vita, M. & Henriksson, M. (2006). The Myc oncoprotein as a therapeutic target for human cancer. *Semin Cancer Biol*, **16**, 318–30.

Wang, G., Jiang, Q., & Zhang, C. (2014). The role of mitotic kinases in coupling the centrosome cycle with the assembly of the mitotic spindle. *J Cell Sci*, **127**, 4111–22.

Watanabe, N., Arai, H., Nishihara, Y., et al. (2004). M-phase kinases induce phospho-dependent ubiquitination of somatic WEE1 by SCFbeta-TrCP. *Proc Natl Acad Sci U S A*, **101**, 4419–24.

Weinert, T. A., & Hartwell, L. H. (1988). The RAD9 gene controls the cell cycle response to DNA damage in *Saccharomyces cerevisia. Science*, **241**, 317–22.

Westendorp, B., Mokry, M., Groot Koerkamp, M. J. A., Holstege, F. C. P., Cuppen, E., & de Bruin, A. (2012). E2F7 represses a network of oscillating cell cycle genes to control S-phase progression. *Nucleic Acids Res*, **40**, 3511–23.

Wieser, S. & Pines, J. (2015). The biochemistry of mitosis. *Cold Spring Harb Perspect Biol*, **7**, a015776.

Williams, G. H. & Stoeber, K. (2012). The cell cycle and cancer. *J Pathol*, **226**, 352–64.

Wong, J. V., Dong, P., Nevins, J. R., Mathey-Prevot, B., & You, L. (2011). Network calisthenics: Control of E2F dynamics in cell cycle entry. *Cell Cycle*, **10**, 3086–94.

Wurzenberger, C. & Gerlich, D. W. (2011). Phosphatases: providing safe passage through mitotic exit. *Nat Rev Mol Cell Biol*, **12**, 469–82.

Yang, W., Shan, J., Wu, M., et al. (2001). Repression of transcription of the p27 (Kip1) cyclin-dependent kinase inhibitor gene by c-Myc. *Oncogene*, **20**, 1688–702.

Yip, K. W. & Reed, J. C. (2008). Bcl-2 family proteins and cancer. *Oncogene*, **27**, 6398–406.

Zhang, S. & Yu, D. (2010). PI(3)king apart PTEN's role in cancer. *Clin Cancer Res*, **16**, 4325–30.

14

Cancer and cell death

Jessica Bullenkamp and Mahvash Tavassoli

Summary

Programmed cell death is an ordered and orchestrated cellular process that occurs in physiological and pathological conditions. Regulated programmed cell death is one of the mechanisms by which multicellular organisms limit the growth and replication of cells. Cell death is essential to control cell growth and tissue homeostasis; it occurs in normal tissues to allow the removal of damaged cells or to maintain a constant number of cells in regenerating tissues and plays an important part in embryogenesis. In an average human adult 50–70 billion cells undergo programmed cell death every day. When cell death goes awry it can play a pivotal role in the pathogenesis of many diseases such as degenerative diseases. The abundance of literature suggests that defects in programmed cell death play a crucial role in carcinogenesis. The genetic alterations in the cancer cell not only lead to increased cellular proliferation, as discussed in other chapters of this book, but also lead to loss of programmed cell death increasing tumour growth.

Despite being part of the problem, cell death plays an important role in the treatment of cancer as it is an important target of many treatment strategies. The mechanisms of programmed cell death are complex and involve many pathways. Defects in programmed cell death can occur at any point along these pathways, leading to malignant transformation of the affected cells, tumour metastasis, and resistance to anticancer drugs. Unquestionably, apoptosis is the best-characterized and the most evolutionary conserved form of programmed cell death. Apoptosis is a process that requires activation of caspase proteases and in broad terms is regulated by two main mechanisms: the extrinsic pathway involving death receptors, or the intrinsic pathway comprising mitochondria and activated by stress signals. However, the actual interchange between different components of the apoptotic machinery and cross talk with other, less studied forms of cell death, has not been fully elucidated and is still under intense investigation. Recently, many studies have demonstrated the existence of several other forms of programmed cell death including autophagy, necroptosis, and pyroptosis. In this chapter we discuss several regulated forms of cell death; we outline what we know about their mechanisms and how we can exploit this knowledge to reactivate programmed cell death in tumour cells for the treatment of cancer.

Introduction

The number of cells in a multicellular organism is regulated not only by the rate of cell division but also by the rate of cells undergoing programmed cell death. Physiologically, programmed cell death is essential for the removal of unwanted or potentially dangerous cells and serves as a natural barrier for the unrestricted proliferation of malignant cells. The paradigmatic role in eliminating unwanted cells is that played during embryogenesis when remodelling is essential. In contrast, in the adult organism evasion of programmed cell death represents one of the major acquired features of cancer cells and allows for the expansion of cancer cells, promoting tumour development (Hanahan and Weinberg, 2000). In addition, resistance to cell death induction represents an obstacle for anticancer treatments attempting to kill malignant cells. It is therefore important to understand how cancer cells evade cell death induction to develop novel therapeutic agents.

Based on morphological features and distinct intracellular signalling processes we can distinguish at least three types of cell death, including apoptosis, autophagy, and necrosis. This chapter will introduce these different types of cell death in more detail, describing characteristic features, regulatory pathways, and common alterations in cancer cells.

Apoptosis

Apoptosis, the main form of programmed cell death, is a physiological process that is essential for the development and life of multicellular organisms. Among others, apoptosis is required for the maintenance of tissue homeostasis in adult organisms, the proper function of the immune system as well as a defence mechanism against external damaging agents.

The term 'apoptosis', named after the ancient Greek word for 'falling off', was first introduced by Kerr et al. in 1972 who defined morphological features of programmed cell death as opposed to necrosis (Kerr et al., 1972). Cells undergoing apoptosis display a characteristic morphology that is clearly distinct from that of healthy or necrotic cells. It is generally characterized by cell shrinkage and chromatin condensation (pyknosis) during early stages of apoptosis and membrane blebbing and nuclear fragmentation (karyorrhexis)

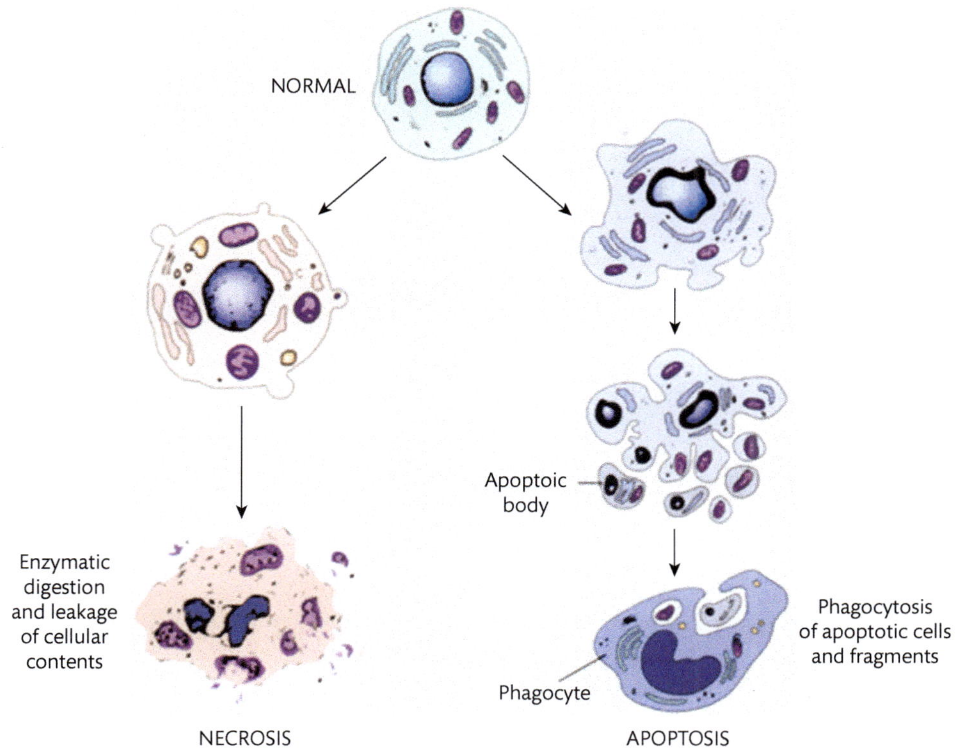

Fig. 14.1 Necrosis is characterized by swelling of the cell and organelles and loss of plasma membrane integrity, which results in leakage of cellular contents into the environment. In contrast, apoptosis is associated with shrinkage of cells, chromatin condensation (pyknosis), and membrane blebbing, while the plasma membrane stays intact. Cells undergoing apoptosis then disintegrate into apoptotic bodies which are engulfed by phagocytes.
Reproduced with permission from Kumar et al. (Eds.), *Robins and Cotran Pathologic Basis of Disease, Seventh Edition*, Saunders an imprint of Elsevier Inc., Copyright © 2004. Adapted with permission from Walker NI et al., 'Patterns of cell death', *Methods and Achievements in Experimental Pathology*, Volume 13, pp. 18–54, Copyright © 1988, S Karger AG, Basel.

at later stages. Eventually, the cell disintegrates into apoptotic bodies which are ultimately engulfed and digested by phagocytic cells due to exposure of phosphatidylserine on the cell surface (Fig. 14.1).

Apoptotic pathways

During apoptosis cells carry out a series of tightly regulated intracellular events. All apoptotic stimuli result in the activation of caspases, specific cysteine proteases that reside as inactive cytosolic zymogens, or procaspases, and require activation by proteolytic processing (Earnshaw et al., 1999). The human caspase family comprises 11 known members, at least seven of which are involved in apoptosis. The initiator phase of apoptosis is marked by the recognition of an apoptotic signal and activation of initiator caspases (caspase–2, –8, –9, and –10). These in turn activate effector caspases (caspase–3, –6, and –7) that specifically cleave cellular substrates during the effector phase leading to disintegration of the cell (Luthi and Martin, 2007). Activation of effector caspases is generally thought to represent a point of no return, inevitably followed by cell death. Some recent findings have started to challenge this assumption, suggesting that apoptosis can be reversed in a process named 'anastasis' from the Greek for 'rising to life' (Tang et al., 2012). The authors showed that dying cells having characteristic signs of apoptosis such as caspase-3 activation were able to regain normal morphology and proliferation once the apoptotic signal was removed. This might allow cancer cells to escape cell death induction e.g. during chemotherapy and potentially promote oncogenic transformation.

There are broadly two main apoptotic signalling pathways that are triggered by different apoptotic signals and employ different initiator caspases and regulatory components. Both pathways ultimately merge into activation of the same effector caspases, predominantly caspase-3, and induction of apoptosis (Fig. 14.2).

Intrinsic apoptotic pathway

The intrinsic, or mitochondrial, pathway is triggered by internal stress stimuli such as DNA damage, growth factor deprivation, or endoplasmic reticulum stress. Activation of the intrinsic pathway by these signals results in mitochondrial outer membrane permeabilization (MOMP), leading to the release of several proapoptotic factors from the mitochondrial intermembrane space into the cytoplasm, including cytochrome c and Smac/DIABLO (van Gurp et al., 2003). Cytochrome c then associates with apoptotic protease-activating factor (Apaf-1) and initiator caspase-9 to form the heptameric apoptosome in an ATP-dependent fashion (Hu et al., 1999). This results in activation of caspase-9, which subsequently cleaves caspase-3 and other effector caspases (Fig. 14.2).

Induction of MOMP is tightly regulated through interactions between members of the B-cell lymphoma 2 (Bcl-2) protein family, a family of over 20 proteins (Cory and Adams, 2002). Bcl-2 family proteins share homology in at least one out of four conserved Bcl-2 homology (BH) domains and can be grouped into proapoptotic and antiapoptotic proteins (Fig. 14.3). The proapoptotic proteins

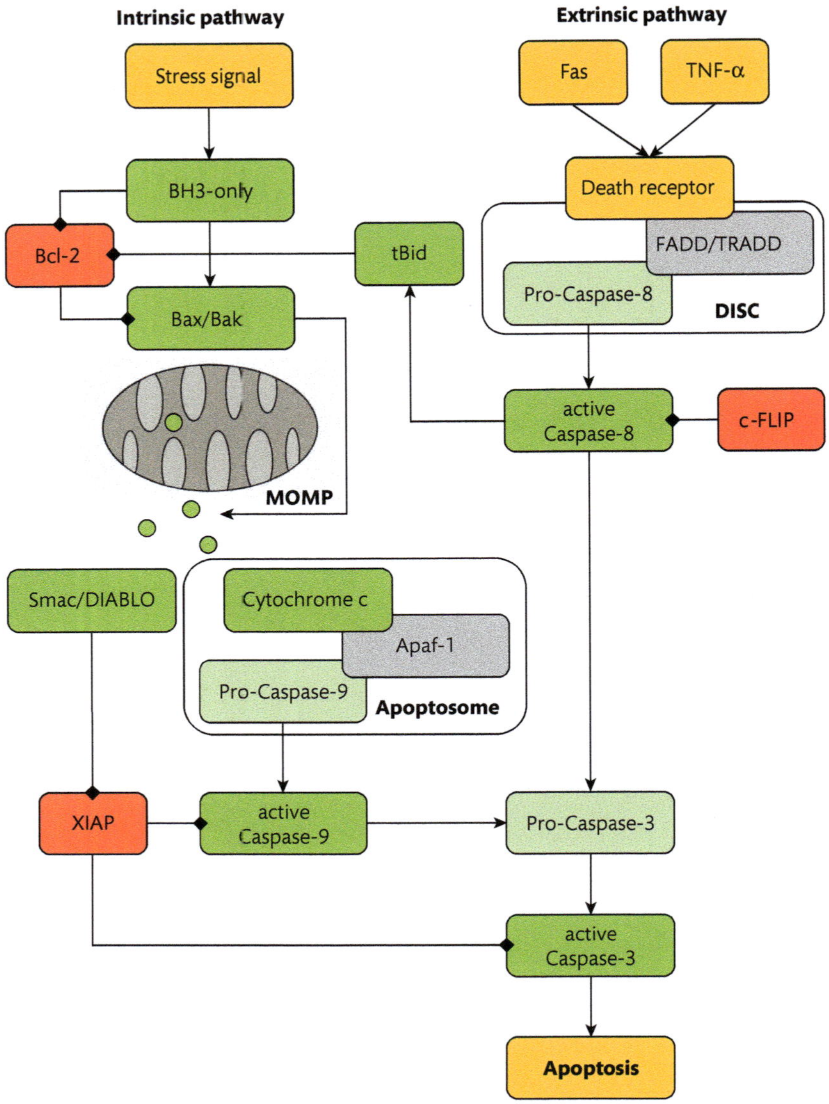

Fig. 14.2 The intrinsic pathway of apoptosis is activated by stress signals such as DNA damage or growth factor withdrawal and regulated by pro- and antiapoptotic members of the Bcl-2 protein family. It results in permeabilization of the outer mitochondrial membrane (MOMP) and the release of cytochrome c (cyt c) into the cytoplasm which triggers formation of the apoptosome and activation of initiator caspase-9. Engagement of death receptors results in the recruitment of adaptor proteins and formation of the DISC in which procaspase-8 is activated. Caspase-8 directly activates caspase-3 or mediates generation of tBid to trigger the intrinsic pathway in an amplification loop. Both pathways ultimately converge in the activation of effector caspases-3 or -7 leading to apoptosis. IAPs prevent the activation of caspases-9, -3, and -7 but can be inhibited by Smac released from the mitochondria while c-FLIP blocks caspase-8 activation.

Source: data from Strasser A. 'The role of BH3-only proteins in the immune system', *Nature reviews Immunology*, Volume 5, Issue 3, pp. 189–200, Copyright © 2005 Macmillan Publishers Limited, part of Springer Nature.

Bax and Bak are required for induction of MOMP but are kept in check by binding to antiapoptotic family members such as Bcl-2 or Bcl-x_L. Stress signals such as DNA damage induce activation of proapoptotic BH3-only proteins such as p53-upregulated modulator of apoptosis (PUMA), Bim, or Noxa through transcriptional or post-transcriptional regulation. Accumulation of BH3-only proteins overcomes the inhibitory effect of antiapoptotic proteins, allowing Bax and Bak to oligomerize in the outer mitochondrial membrane, resulting in the formation of pores and MOMP. The precise mechanism by which the BH3-only proteins induce activation of Bax and Bak is still under debate and involves both direct interaction of BH3-only proteins with Bax/Bak as well as indirect effects through binding to antiapoptotic Bcl-2 family members (Chipuk and Green, 2008).

Extrinsic apoptotic pathway

The extrinsic pathway is initiated following engagement of death receptors of the tumour necrosis factor (TNF) receptor family such as TNF-R1 (TNF receptor 1), TRAIL-R1/2 (TNF-related apoptosis-inducing ligand receptor) or CD95/Fas by their corresponding ligands (Walczak, 2013). Upon ligand binding death receptors undergo oligomerization and a conformational change exposing the receptor's death domain on the cytoplasmic side of the plasma membrane. This facilitates the recruitment of adaptor proteins

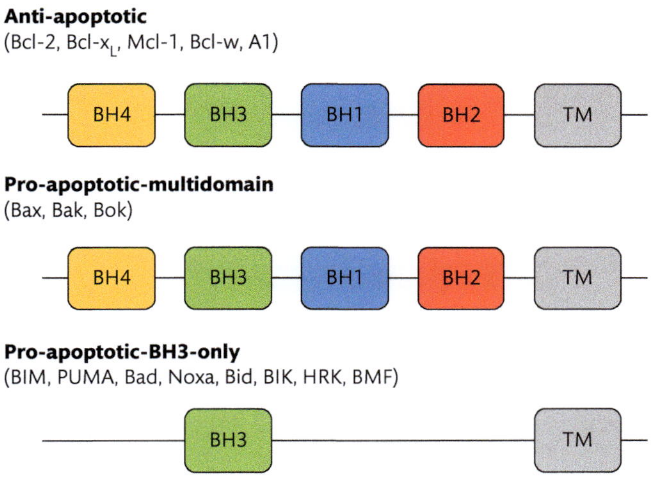

Fig. 14.3 Members of the Bcl-2 protein family can either prevent apoptosis (top) or promote it (bottom) and are classified into three different subsets. All Bcl-2 family proteins share a common BH3 domain, while multimeric proteins such as Bcl-2, Bax or Bak also contain several other Bcl-2-homology (BH) domains. In addition, many Bcl-2 family members contain a transmembrane domain (TM) which facilitates insertion into the outer mitochondrial membrane.

Source: data from Strasser A. 'The role of BH3-only proteins in the immune system', *Nature reviews Immunology*, Volume 5, Issue 3, pp. 189–200, Copyright © 2005 Macmillan Publishers Limited, part of Springer Nature.

such as Fas-associated protein with death domain (FADD) or TNF receptor-associated protein with death domain (TRADD), containing a death domain and a death-effector domain to the receptor and formation of a multiprotein signalling complex (Chinnaiyan et al., 1995). Procaspase-8 associates with the death effector domain of FADD or TRADD to form the death-inducing signalling complex (DISC), triggering proteolytic activation of procaspase-8 (Peter and Krammer, 2003). Activated caspase-8 either directly cleaves effector caspases (type I pathway) or truncates the BH3-only protein Bid to initiate the intrinsic apoptotic pathway (type II pathway; see Fig. 14.2, and Li et al., 1998).

Other regulators of apoptosis

Inhibitor of apoptosis (IAP) proteins function as negative regulators of initiator and effector caspases during apoptotic signalling. Recent evidence also indicates a role of IAPs in other cellular processes, such as activation of nuclear factor kappa B (NF-κB), innate immune signalling, cell migration, and survival (Gyrd-Hansen and Meier, 2010). So far eight mammalian IAPs have been identified, including XIAP (X-linked inhibitor of apoptosis), cIAP-1, cIAP-2, and survivin. They contain three baculovirus IAP repeat domains as well as a C-terminal *RING* (really interesting new gene) finger domain—a zinc finger type structural domain contained in many proteins that function as ubiquitin ligases and transfer ubiquitin from ubiquitin-conjugating enzymes to substrate proteins (Sun et al., 1999). XIAP blocks the activation of several caspases, for example, through binding to catalytic pockets (caspase-3, caspase-7), preventing dimerization (caspase-9) or ubiquitination and subsequent proteasomal degradation (caspase-3; see Riedl et al., 2001; Shiozaki et al., 2003). During apoptosis Smac/DIABLO is released from the mitochondrial intermembrane space into the cytoplasm where it binds to IAPs, thereby relieving the IAP-mediated inhibition of initiator and effector caspases (Chai et al., 2000). The antiapoptotic mechanism of survivin is not entirely understood; however, it has been suggested that it interacts with and stabilizes XIAP or disrupts the binding of Smac/DIABLO to IAPs (Ceballos-Cancino et al., 2007).

Cellular FLICE-inhibitory protein (c-FLIP) is an inactive caspase-8 homologue that interferes with death receptor signalling by competitive binding at the DISC (Krueger et al., 2001). The short splice variant c-FLIP$_S$ contains only the two DEDs and inhibits caspase-8 activation at the DISC, while the long form c-FLIP$_L$ shares significant homology with caspase-8 but is catalytically inactive. Depending on its expression levels c-FLIP$_L$ can exert either a pro or antiapoptotic effect on caspase-8 activation (Chang et al., 2002).

p53 and apoptosis

As discussed elsewhere in this book, p53 is one of the most studied proteins and genes, with over 80,000 entries recorded in PubMed. p53 is the most frequently mutated gene in cancer in both haematological and solid tumours and has since become known as the 'guardian of the genome' (Levine et al., 1991; Muller and Vousden, 2013). It is activated in response to a variety of stimuli such as DNA damage, oncogene activation, leading to growth suppression through induction of cell cycle arrest or apoptosis. In cancers, mutations in the TP53 gene alter its response pathway allowing the cell to escape from growth suppression, accumulate genomic instability, and lead to the development of many different types of cancer as a result.

The primary function of p53 is as a transcription factor; typically, p53 binds to target genes that contain a consensus p53 binding site. Recent genome-wide analysis has identified up to thousands potential p53 target genes that regulate multiple cellular processes, including cell cycle arrest, DNA repair, apoptosis, metabolism, hypoxia, autophagy, mRNA translation, and feedback mechanisms (Fischer, 2017). The importance of p53 in the intrinsic apoptosis pathway has been known for a long time. It directly induces expression of several proapoptotic Bcl-2 family members, including Bax, PUMA, and Noxa, as well as Apaf-1, a component of the apoptosome. Mutation of p53 therefore impairs the upregulation of many proapoptotic factors in response to different stress stimuli such as DNA damage and thereby prevents induction of apoptosis (Fig. 14.4).

Apoptosis and cancer

During tumour development cancer cells are subject to constant stress, including oncogenic stress, genomic instability, and hypoxia. To avoid induction of the intrinsic pathway of apoptosis in response to these stimuli, cancer cells have evolved various mechanisms to evade apoptosis (Hanahan and Weinberg, 2000). These can include the up- or downregulation of pro or antiapoptotic proteins, respectively, and modification of the stability and function of regulatory proteins through post-translational modifications (Fernald and Kurokawa, 2013).

The proapoptotic protein Bcl-2 was originally identified in B-cell follicular lymphoma where the t(14;18) chromosomal translocation results in overexpression of Bcl-2, decreasing the susceptibility of cells to apoptosis (Vaux et al., 1988). The resulting overexpression of Bcl-2 was shown to promote malignant cell survival through

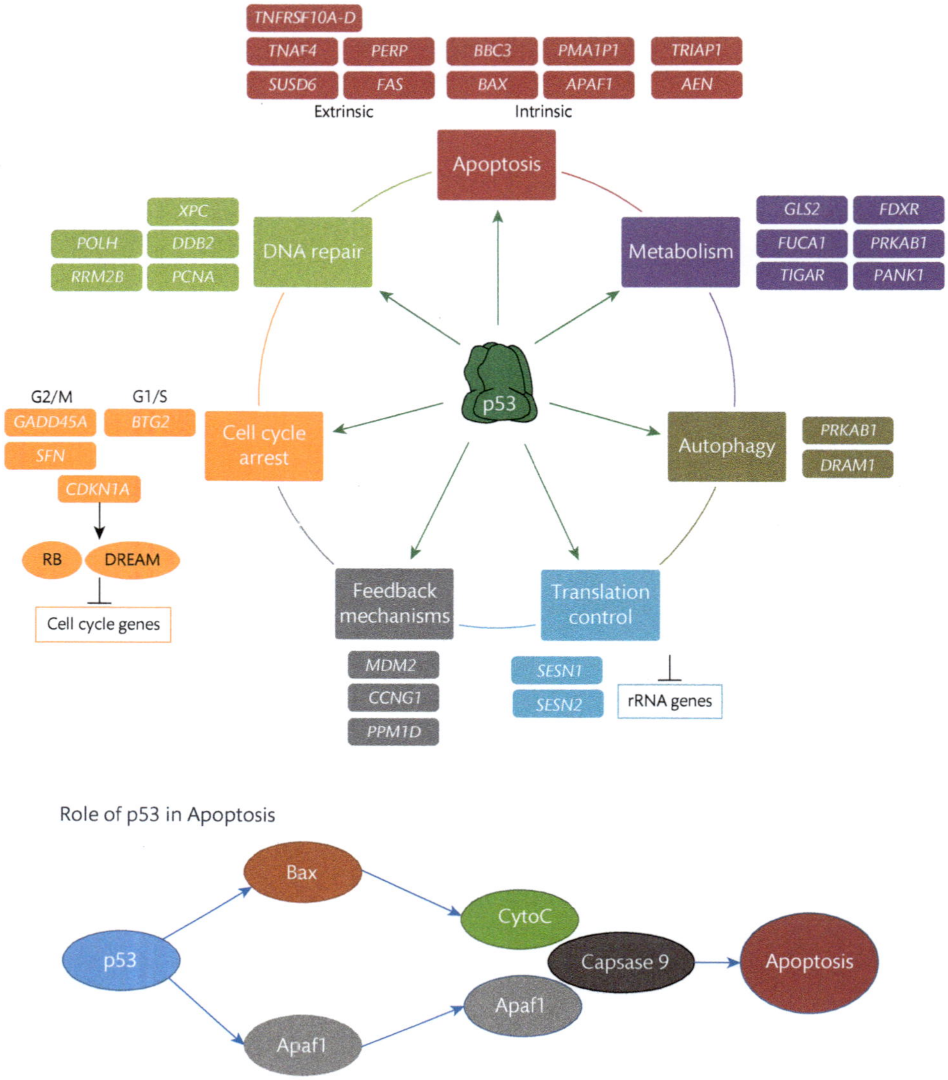

Fig. 14.4 p53 directly activates target genes that mediate various functions. Proteins encoded by p53 target genes function in multiple processes that include, but are not limited to, cell cycle arrest, DNA repair, apoptosis, metabolism, autophagy, translation control, and feedback mechanisms.

attenuation of apoptosis. It is estimated that approximately 50% of human tumours show dysregulated expression of Bcl-2. Similarly, overexpression of other antiapoptotic Bcl-2 family members (Bcl-x_L, Mcl-1) has been reported in various types of cancer, including multiple myeloma, CLL, ALL, AML, myelodysplastic syndrome as well as solid cancers such as breast or lung cancer and melanoma (Beroukhim et al., 2010; Ashkenazi et al., 2017). On the other hand, induction of antiapoptotic BH3-only proteins in response to cellular stress signals is often impaired, for example, due to mutations in the p53 tumour suppressor protein, the main transcription factor for many proapoptotic genes such as Bax, PUMA, Noxa and Apaf-1 (Oda et al., 2000; Yu and Zhang, 2003). In addition, frameshift mutations in the gene encoding Bax resulting in decreased expression levels have been identified in colon cancer (Rampino et al., 1997) while methylation of the Apaf-1 promoter results in functional inactivation in many melanoma cells (Soengas et al., 2001). Induction of apoptosis through the extrinsic pathway can additionally be impaired due to overexpression of FLIP, blocking activation of caspase-8, or decreased expression of death receptors.

Another strategy of cancer cells to evade apoptosis induction is the hyperactivation of kinase pathways such as Akt or Erk. These can negatively regulate the expression and stability of proapoptotic proteins such as Bim or Bad through suppression of transcription factors or direct phosphorylation that promotes proteasomal degradation or inhibits proapoptotic activity (Datta et al., 1997; Ley et al., 2003). On the other hand, constitutive activation of the transcription factor NF-κB in cancer cells not only promotes inflammation and angiogenesis but also activates a large number of antiapoptotic genes such as XIAP or TRAF that interfere with both the intrinsic and extrinsic pathways of apoptosis (Van Antwerp et al., 1998).

Expression of the antiapoptotic protein survivin is controlled by different tumour suppressors like p53 or oncogenic factors such as or NF-κB and Akt (Guha and Altieri, 2009). Consequently, while survivin is rarely present in adult normal tissues, it is overexpressed in most human tumours and higher levels of survivin have been associated with recurrence, therapy resistance and poor prognosis (Kanwar et al., 2013).

In summary, cancer cells demonstrate an imbalance between the expression of pro-and antiapoptotic proteins and often fail to activate proapoptotic proteins in response to apoptotic stimuli including DNA damage or growth factor withdrawal. This acquired resistance to apoptosis supports their survival despite excessive proliferation and also contributes to chemotherapy resistance.

Apoptosis and cancer therapy

The principal aim of cancer therapy is the complete removal of malignant cells without damaging the rest of the body. While surgical excision alone can be used to remove a primary tumour, combination with a systemic adjunct treatment is often required. Current conventional treatment strategies such as chemotherapy or radiotherapy are unspecific and inevitably produce unwanted side effects. In addition, dysregulation of apoptosis impairs the response of tumour cells to cytotoxic agents, resulting in persistence of cancer. It is therefore essential to develop more specific and more effective therapies to improve the care of cancer patients.

Novel anticancer agents have been designed to induce apoptosis on their own or to enhance the apoptotic effect of other agents by directly targeting regulators of apoptosis. As described earlier, dysregulation of antiapoptotic Bcl-2 family proteins has been reported in different types of cancers. Accordingly, selective inhibition of specific antiapoptotic Bcl-2 family proteins represents an important therapeutic opportunity and several small molecular inhibitors have progressed into phase I/II clinical trials (Scarfo and Ghia, 2013; Ashkenazi et al., 2017). These include small molecule inhibitors of Bcl-2 like Obatoclax which has undergone phase I/II clinical trials for several haematological malignancies but has since not been tested further (Joudeh and Claxton, 2012).

BH3 mimetics, such as ABT-737 or its orally available derivative Navitoclax (ABT-263), mimic the inhibitory function of BH3-only proteins on the prosurvival proteins Bcl-2, Bcl-x_L, and Bcl-w, thereby facilitating the induction of apoptosis (Billard, 2013). Navitoclax was the first BH3 mimetic to show efficacy in cancer patients, in particular B cell malignancies, either as a monotherapy or in combination with other chemotherapeutic agents (Wilson et al., 2010). However, its use in patients is limited due to thrombocytopenia resulting from inhibition of Bcl-x_L limiting the life span of blood platelets. This prompted the development of the highly selective Bcl-2 inhibitor venetoclax (ABT-199) which has a remarkable single agent activity and is now approved in the United States for the treatment of patients with CLL (Roberts et al., 2016; Stilgenbauer et al., 2016). Table 14.1 summarizes a selection of Bcl-2 family inhibitors currently undergoing clinical trials, including some selective inhibitors of Bcl-x_L and Mcl-1 (Ashkenazi et al., 2017).

Another approach to facilitate apoptosis induction is antagonizing the function of IAPs using antisense oligonucleotides or small molecule inhibitors that are currently in clinical trials. Survivin in particular has drawn much attention due to its dysregulated expression in most cancers compared to normal tissues. However, most anticancer agents designed to prevent survivin expression have shown only a limited response in clinical trials, including antisense oligonucleotides such as LY2181308 and small molecule inhibitors

Table 14.1 Targeting apoptosis pathways for cancer therapy

Therapeutic compound	Mechanism	Cellular target
Obatoclax (GX15-070)	Small molecule inhibitor	Bcl-2
ABT-737	Small molecule inhibitor	Bcl-2, Bcl-x_L, Bcl-w
Navitoclax (ABT-263)	Small molecule inhibitor	Bcl-2, Bcl-x_L, Bcl-w
BM-1197	Small molecule inhibitor	Bcl-2, Bcl-x_L
S44563	Small molecule inhibitor	Bcl-2, Bcl-x_L
BCL2-32	Small molecule inhibitor	Bcl-2, Bcl-x_L
AZD4320	Small molecule inhibitor	Bcl-2, Bcl-x_L
Venetoclax (ABT-199)	Small molecule inhibitor	Bcl-2
S55746 (BCL201)	Small molecule inhibitor	Bcl-2
WEHI-539	Small molecule inhibitor	Bcl-x_L
A-1155463	Small molecule inhibitor	Bcl-x_L
A-1331852	Small molecule inhibitor	Bcl-x_L
UMI-77	Small molecule inhibitor	Mcl-1
Compound 9	Small molecule inhibitor	Mcl-1
A-1210477	Small molecule inhibitor	Mcl-1
Compound 34	Small molecule inhibitor	Mcl-1
S63845	Small molecule inhibitor	Mcl-1
AMG176	Small molecule inhibitor	Mcl-1
Genasense/Oblimersen	Antisense oligonucleotide	Bcl-2
Smac mimetics	XIAP inhibition	XIAP
AEG35156	Antisense oligonucleotide	XIAP
Nutlin	Reactivation p53	p53/HDM2
HDAC inhibitors	Mitochondrial pathway	HDAC
Bortezomib	Mitochondrial pathway	Proteasome
TRAIL	Death receptor pathway	TRAIL-R1/2
Apoptin	Tumour-specific apoptosis	n/a
HAMLET	Tumour-specific apoptosis	n/a
mda-7	Tumour-specific apoptosis	n/a
adenoviral E4orf4	Tumour-specific apoptosis	n/a

Adapted with permission from Springer Nature, *Nature Review Drug Discovery*, 'From basic apoptosis discoveries to advanced selective BCL-2 family inhibitors', Ashkenazi A et al., Volume 16, Issue 4, pp. 273–84, Copyright © 2017 Springer Nature.

like YM155 (Peery et al., 2017). Due to its distinct expression in tumour tissues survivin might also have potential as a vaccine in cancer immunotherapy. A peptide mimetic of survivin, SVN53-67/ M57 (SurVaxM), was shown to induce a cytotoxic T-cell response and xenograft rejection in mouse models. It was subsequently showed promising activity in a small clinical study and its efficacy currently tested in phase I/II clinical trials for glioblastoma and multiple myeloma (Fenstermaker et al., 2016).

Alternatively, Smac mimetics inhibit IAPs similarly to the endogenous Smac/DIABLO protein and have shown potential in combination with other apoptosis-inducing agents such as TRAIL (TNF-related apoptosis-inducing ligand; see Chen and Huerta, 2009; Lu et al., 2011). Lastly, apoptosis in cancer cells can be promoted by re-introduction of wild type p53 into p53-mutant tumours using recombinant adenoviral vectors or using small molecule inhibitors such as nutlin that block MDM2-mediated p53 degradation (Fujiwara et al., 2000; Shangary and Wang, 2009).

In summary, understanding the complex role of the Bcl-2 family in cancers has enabled the development of effective apoptosis-inducing cancer therapeutics with the promise of refining cancer treatment in specific types of malignancies. However, despite advances in drug design there remains a need for more targeted therapies that will specifically affect tumour cells.

TRAIL is a member of the TNF family that was initially reported to induce apoptosis in a range of human cancer cell lines without affecting the viability of normal cells (Walczak et al., 1999). Several agents that target TRAIL receptors, including recombinant TRAIL and monoclonal antibodies, have shown *in vivo* antitumour activity and have been evaluated in phase I/II clinical trials for advanced tumours with promising results (Stuckey and Shah, 2013). However, despite its tumour-selective potential many cancers remain resistant to TRAIL, limiting the clinical application as a monotherapy (Newsom-Davis et al., 2009). This has led to the testing of TRAIL in combination with other therapeutic agents, aiming to sensitize cancer cells to TRAIL-mediated apoptosis or to augment the cytotoxic effect of TRAIL (Hellwig and Rehm, 2012). These include chemotherapeutic agents, proteasome inhibitors, histone deacetylase inhibitors or SMAC mimetics which have all shown promising effects *in vitro* and partially *in vivo* as a basis for future preclinical investigations. In addition to apoptosis induction, TRAIL has also been shown to trigger other cellular signalling pathways including NF-κB, mitogen-activated protein kinases (MAPK), c-Jun N-terminal kinase, and Akt, which might counteract its cytotoxic effect and represent a concern for the use of TRAIL in resistant cancers (Falschlehner et al., 2007).

In addition, recent studies have revealed a growing group of cellular or viral proteins that have the ability to induce tumour-specific cell death. These include HAMLET (human alpha-lactalbumin made lethal to tumour cells), mda-7/IL-24, adenoviral E4orf4, parvovirus-H1 NS1, or apoptin. Some of these are already in clinical development with a potential as novel tumour-selective agents while others are still in early stages of preclinical research (Grimm and Noteborn, 2010).

Necrosis/Necroptosis

For many years researchers distinguished two types of cell death: programmed cell death or apoptosis, and necrosis; the latter was thought to be an accidental form of cell death.

Necrosis

Necrosis is often associated with physical stresses to the cell such as trauma, toxins, nutrient deprivation, or osmotic shock. In contrast to morphological changes in apoptosis, induction of necrosis is characterized by swelling and rounding of cells and cell lysis due to rupture of the plasma membrane and release of cellular components into the environment, which often provokes an inflammatory response.

The lack of nutrients and oxygen in the centre of expanding tumours often induces necrosis, generating a core of necrotic tumour tissue. While this might represent an important mechanism to limit tumour growth it also generates an inflammatory environment which can in turn promote tumour progression.

However, the view of necrosis as an uncontrolled, accidental form of cell death has changed following the discovery of a tightly regulated type of cell death resembling necrosis, referred to as necroptosis.

Special forms of Necrosis

In addition, several other regulated forms of necrosis are emerging; including NETosis, ferroptosis, parthanatos, pyronecrosis, and pyroptosis (Tait et al., 2014; Vanden Berghe et al., 2014). Pyroptosis (from Greek for 'fire' and 'falling') refers to a caspase-dependent form of programmed cell death that requires the activity of caspase-1 or caspase-11. It is highly inflammatory, resulting in the release of proinflammatory cytokines, and possibly is part of an immune response against infection with intracellular pathogens.

In contrast, pyronecrosis, which occurs in response to different pathogens, is caspase-independent and requires components of the inflammasome. As the name suggests, ferroptosis is dependent on intracellular iron metabolism and involves reactive oxygen species (ROS) and lipid peroxidation (Xie et al., 2016). Dysregulation of ferroptosis has been implicated in various pathological processes including neurodegenerative diseases, kidney failure, and cancer. NETosis refers to the release of neutrophil extracellular traps (NETs) composed of chromatin and histones to immobilize bacteria as part of the innate immune response. Formation of NETs often results in rupture of the plasma membrane and neutrophil death which is considered a regulated form of necrosis. Lastly, parthanatos is a form of programmed cell death that is triggered by accumulation of PARP (poly(ADP-ribose) polymerase) proteins. While activation of PARP is required for cellular homeostasis (e.g. following DNA damage or ROS), excessive accumulation of PARP induces parthanatos through activation and nuclear translocation of apoptosis-inducing factor. Parthanatos is characterized by large-scale DNA fragmentation, but in contrast to apoptosis does not involve activity of caspases and is not associated with cell swelling as opposed to unregulated necrosis (Fatokun et al., 2014).

Necroptosis

Necroptosis was initially described as a caspase-independent but tightly regulated form of cell death resembling necrosis in response to TNF stimulation. Binding of TNF-α to its receptor TNF-R1 can have multiple outcomes, including survival, apoptosis, or necrosis—a process that depends on the formation of distinct intracellular signalling complexes (Fig. 14.5; see also Vandenabeele et al., 2010).

Mechanism of necroptosis

Ligation of TNF-R1 triggers the formation of the so-called complex I containing various proteins including the adapter protein TRADD, receptor-interacting protein (RIP) kinase RIP1, cIAPs, and TNF receptor-associated factor 2 (TRAF2). Polyubiquitination of RIP1, mediated by cIAPs, results in activation of NF-κB and MAPK signalling which generally promotes gene activation and cell survival.

Under certain conditions, involving the activity of deubiquitinases such as CYLD or lack of cIAP activity, complex I becomes unstable, resulting in the assembly of a second cytosolic complex, resembling a cytosolic DISC (Micheau and Tschopp, 2003). This so-called complex IIa generally contains the adapter proteins TRADD and FADD, as well as procaspase-8 and the RIP kinases RIP1 and RIP3 and normally triggers activation of caspase-8 to initiate the extrinsic

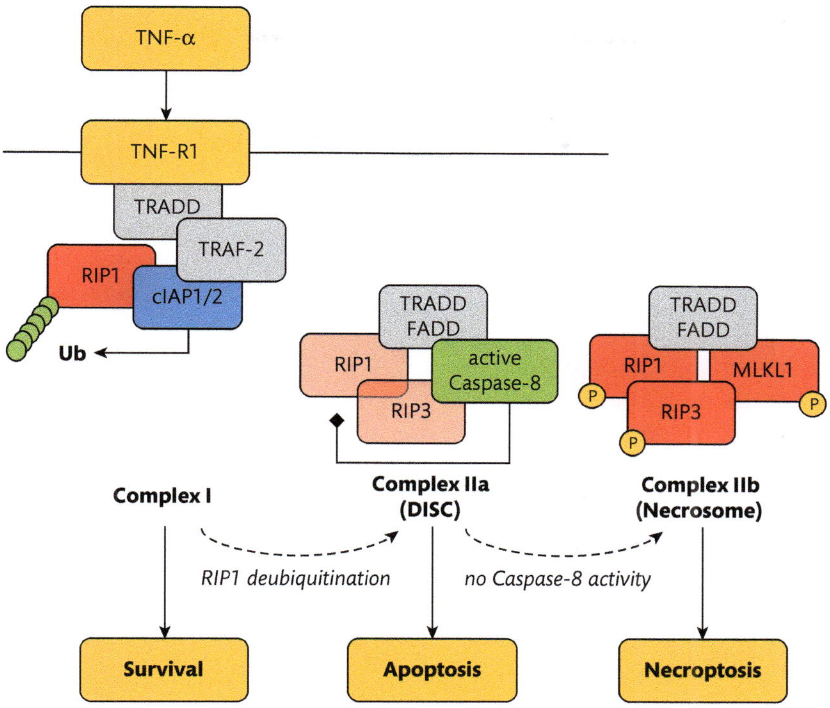

Fig. 14.5 Binding of TNF-a to its receptor (TNF-R1) induces formation of a protein complex containing the adapter protein TRADD, TRAF-2 and RIP-1, and cIAP1/2. This so-called complex I activates MAPK and NF-kB signalling and thereby mediates survival signals. Deubiquitination of RIP1 results in destabilization of complex I and formation of the cytosolic complex IIa or DISC (death-inducing signalling complex), which contains the adapter proteins (TRADD or FADD), procaspase-8, RIP1, and RIP3. This triggers activation of initiator caspase-8 which activates apoptotic signalling and inactivates RIP1 and RIP3. In the absence of caspase-8 activity, RIP1 and RIP3 are stabilized, phosphorylated, and activated, and formation of the so-called necrosome (complex IIb) occurs. This initiates necroptotic cell death, mediated by RIP1/3 and MLKL1.

Source: data from Han et al., 'Programmed necrosis: backup to and competitor with apoptosis in the immune system', *Nature Immunology*, Volume 12, Issue 12, pp. 1143–9, Copyright © 2011 Macmillan Publishers Limited, part of Springer Nature.

pathway of apoptosis. Active caspase-8 cleaves and thereby inactivates RIP kinases, preventing the induction of necroptosis (Feng et al., 2007).

However, suppression of apoptotic signalling, for example by pharmacological caspase inhibition or absence of caspase-8 (e.g. due to genetic depletion), results in stabilization of RIP1 and RIP3 (Declercq et al., 2009). This triggers phosphorylation and activation of RIP1 and RIP3 within this necrosome complex (complex IIb) to initiate necroptotic cell death (Cho et al., 2009; Declercq et al., 2009). How exactly necroptosis is executed still remains largely unclear, but it can involve several molecular mechanisms including MLKL1-mediated permeabilization of cellular and organelle membranes, generation of ROS, and induction of autophagy.

Necroptosis in cancer

There are currently no biomarkers of necroptosis and we still lack sufficient evidence to establish a clear role of necroptosis in cancer. Somatic mutations and downregulation of key mediators of necroptosis such as RIP1, RIP3, or MLKL1 have been detected in cancer cells, indicating a potential tumour suppressor function of necroptosis. However, elevated levels of the same proteins were demonstrated in other types of cancers, suggesting that further research is required to elucidate the contribution of necroptosis to tumour progression (Lalaoui and Brumatti, 2017).

Induction of necroptosis might also represent a promising cancer therapeutic approach to bypass resistance to apoptosis (Su et al., 2016). Cancer cells that are resistant to apoptosis-inducing agents such as chemotherapy (e.g. due to overexpression of Bcl-2 or mutations in p53) may be sensitive to compounds that induce necroptosis. Many types of cancer cells have been shown to undergo necroptosis *in vitro* in response to classic necroptosis inducers, natural compounds (e.g. shikonin) or chemotherapy drugs (e.g. DNA-damaging agents). In addition, as opposed to apoptosis, necroptosis can be classified as 'immunogenic', promoting the recruitment of immune effector cells and potentially inducing an antitumour immune response. However, further understanding of molecular events during necroptosis is required before exploiting this pathway for anticancer therapy.

Autophagy

Autophagy, literally translated as 'self-eating', refers to a catabolic process during which cells sequester organelles to recycle cellular materials such as sugars nucleotides, amino acids, and fatty acids (Rabinowitz and White, 2010). Physiologically, basal levels of autophagy ensure the continuous removal of damaged or old organelles; this is required for tissue homeostasis. Moreover, autophagy can be upregulated in response to various stimuli such as nutrient starvation, pathogen recognition, genotoxic stress, or hypoxia. In particular, in the context of metabolic stress and starvation, the degradation products of autophagy provide essential metabolites and energy to maintain cell survival.

Execution and control of autophagy

Like apoptosis autophagy is a tightly coordinated cellular programme. Initially a double membrane phagosome is formed that engulfs contents of the cytoplasm and matures into an autophagosome (Lamb et al., 2013). The origin of the membranous structures forming the autophagosome remains unclear and they may be derived from several sources, including the plasma membrane, endoplasmic reticulum, mitochondria, or Golgi apparatus. The autophagosome then fuses with lysosomes to form the autolysosome, where their cargo is eventually degraded by lysosomal hydrolases (Fig. 14.6).

Autophagosome formation is regulated by at least 18 proteins in mammalian cells, many of them homologues of the Atg (autophagy-related genes) family that were first described in yeast cells. One of the main regulators of autophagy, Beclin-1 (Atg6) was originally identified as a Bcl-2 interacting protein and is inhibited by Bcl-2, indicating cross talk between autophagy and apoptotic pathways (Pattingre et al., 2005). However, the precise molecular mechanisms underlying the assembly and maturation of autophagosomes will not be discussed here (He and Klionsky, 2009).

Many stimuli that induce autophagy such as nutrient depletion or endoplasmic reticulum stress trigger signalling pathways that eventually control the activity of the serine/threonine kinase mTOR (mammalian target of rapamycin). Under normal conditions, active mTOR, present in two distinct complexes (mTORC1 and mTORC2), suppresses autophagy through phosphorylation of ULK1/2 while also promoting cell growth and protein synthesis. Depletion of nutrients triggers activation of AMPK (AMP-activated protein kinase) which inhibits mTOR to upregulate autophagy (Laplante and Sabatini, 2012).

Role of autophagy in cancer

Autophagy is essential for the maintenance of tissue homeostasis both under physiological conditions and in response to different forms of stress. It is therefore not surprising that autophagy plays an important role in tumours, although its function can be both tumour-suppressive and tumour-promoting depending on the context such as genetic background and tumour stage (Fig. 14.7; Galluzzi et al., 2015).

Tumour-suppressive role of autophagy

Data from various murine models suggest that autophagy suppresses malignant transformation. For example, knockout of Beclin-1 results in spontaneous tumour formation (Qu et al., 2003). This tumour-suppressive function of autophagy might be based on several potential mechanisms. Intact autophagy prevents the generation of ROS by removing dysfunctional mitochondria and is required for cellular responses to genotoxic stress. Defective autophagy on the other hand interferes with oncogene-induced senescence, allowing the accumulation of malignant cells (Grasso and Vaccaro, 2014). Moreover, autophagy is implicated at different steps of innate and adaptive immune responses. Defective autophagy allows malignant cells to persist due to failure of the immune system to recognize

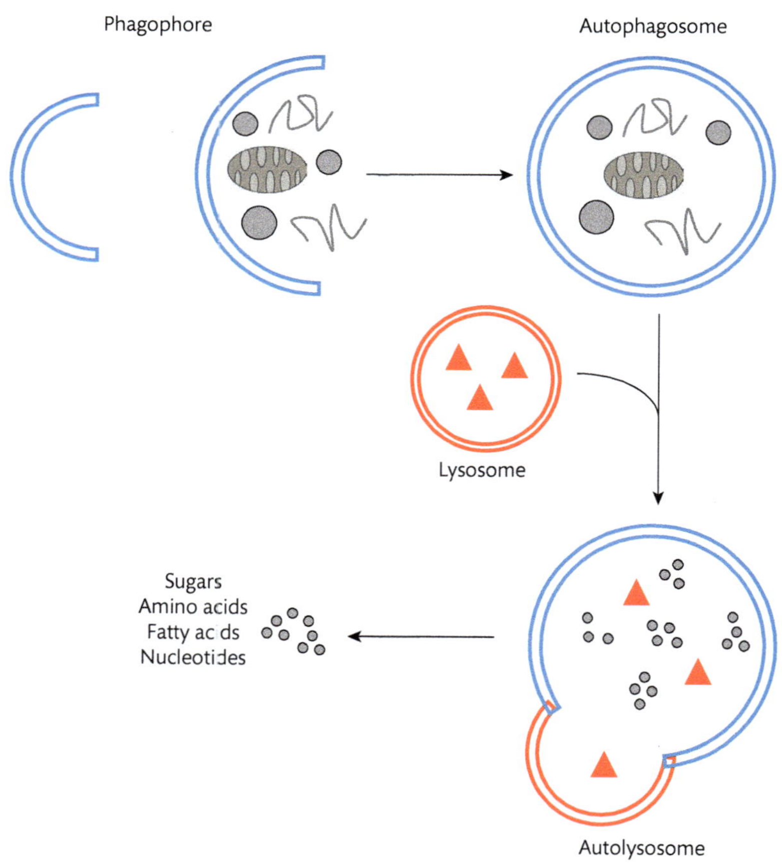

Fig. 14.6 Autophagy starts with the formation of a phagophore—a double membrane structure that engulfs cytoplasmic materials such as organelles. IT elongates and closes to form the autophagosome, which subsequently fuses with lysosomes. In this autolysosome lysosomal hydrolases mediate the degradation of cargo into products such as sugars, amino acids, or nucleotides which are transported back into the cytosol for recycling.

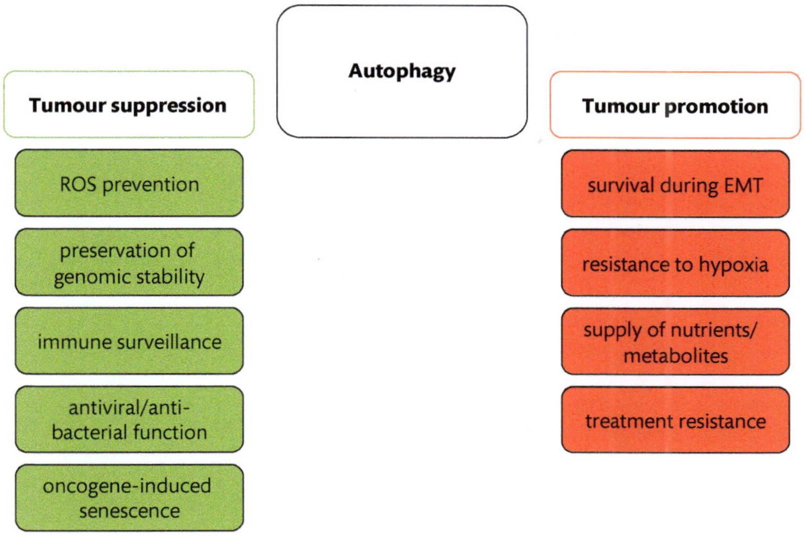

Fig. 14.7 Autophagy can have either a tumour-suppressive or tumour-promoting role, depending on the context. Mechanisms by which autophagy suppresses tumour formation include prevention of reactive oxygen species (ROS) and general preservation of genomic stability. Moreover, autophagy has a role in innate or adaptive immune responses and is important for the activation of oncogene-induced senescence. On the other hand, autophagy can promote tumour growth by improving tumour cell survival during hypoxia or nutrient deprivation as well as during epithelial-mesenchymal transition (EMT). In addition, autophagy can reduce the sensitivity of cancer cells to cell death during treatment.

and eliminate potentially dangerous cells at an early stage of malignant transformation (Zhong et al., 2016). In addition, autophagy is an important effector mechanism in the immune response to viral or bacterial infections, including several potentially carcinogenic pathogens such as human papilloma virus or *Helicobacter pylori* (Deretic et al., 2013).

Tumour-promoting role of autophagy

However, once a tumour has been established it seems to promote tumour progression rather than limit spread of the cancer. Rapidly proliferating cancer cells have a high energy demand which cannot be met by reduced nutrient availability in the changing tumour microenvironment. In this context, ongoing autophagy may confer a survival advantage by ensuring a continuous supply of nutrients and metabolites. In addition, induction of autophagy in response to cellular stresses such as hypoxia and DNA damage allows cancer to survive rather than undergo apoptosis (Kroemer et al., 2010). Overall, intact autophagy enables cancer cells to survive in adverse conditions such as a nutrient-deprived microenvironment in growing tumours. Moreover, autophagy has recently been suggested to maintain cancer cell survival during epithelial-mesenchymal transition (EMT) and thereby facilitate cancer spread and metastasis (Mowers et al., 2017).

Autophagy and cancer therapy

Due to the dual role of autophagy during malignant transformation and tumour growth, the use of autophagy-inducing agents or autophagy inhibitors should be carefully evaluated. Rapamycin and other mTOR inhibitors can trigger autophagy but have not yet demonstrated any clinical efficacy. On the other hand, pharmacological inhibitors of autophagy such as the antimalaria drug chloroquine might represent a novel therapeutic approach to target advanced human tumours that rely on autophagic flux to maintain their survival (Janku et al., 2011).

Conclusion

Programmed cell death plays pivotal roles in tumour progression, cancer therapeutics, and resistance of tumours to therapy. In this chapter we have highlighted different forms of programmed cell death, discussed their mechanisms and possible roles in cancer. Although the processes regulating apoptosis have been extensively studied, the importance of non-apoptotic cell death *in vivo* and in cancer remains largely unknown. Better understanding of how cells kill themselves might allow us to therapeutically manipulate it to treat cancers. However, the increased understanding of several key mechanisms involved in mediating programmed cell death and tumour resistance to therapy has resulted in a dramatic rise in therapeutic approaches to promote tumour-selective cell death. There are already a large number of new drugs and treatment strategies that are designed to enhance apoptosis. Most of these novel therapeutics are still in preclinical development but others have already entered clinical trials including inhibitors of the Bcl-2 family of proteins and IAP inhibitors. However, these agents are non-selective and can harm normal tissues; therefore, critical tests must be passed before they can be used safely in human subjects. Further understanding of the basic mechanisms of execution of cell death might provide opportunities to better manipulate cell death in a therapeutic context.

TAKE-HOME MESSAGE

- There are different types of cell death: the main ones are apoptosis, necrosis, and autophagy.
- Apoptosis (also called programmed cell death) is essential in embryos for development and in adults to maintain homeostasis.
- Loss of cell ability to undergo apoptotic cells death is an oncogenic mechanism.

- Cell death pathways are now explored as potential targets for anticancer drug aiming to induce death of neoplastic cells.

OPEN QUESTIONS

- Although apoptosis is the better understood so far among the different types of cellular death, the interchange between different components of the apoptotic machinery is still poorly understood.
- The significance of non-apoptotic cell death in physiological conditions and in cancer remains largely unknown and the underlying mechanisms poorly understood.
- Many compounds able to increase apoptosis are available but several are non-selective and can harm normal tissues. It is a future challenge to produce or identify compound that are effective but only in neoplastic cells they can be used safely in human subjects.
- The cross talk among the different forms of cell death needs to be elucidated.

FURTHER READING

Chen, D. J. & Huerta, S. (2009). Smac mimetics as new cancer therapeutics. *Anticancer Drugs*, **20**(8), 646–58.

Cory, S. & Adams, J. M. (2002). The Bcl2 family: regulators of the cellular life-or-death switch. *Nat Rev Cancer*, **2**(9), 647–56.

Fernald, K. & Kurokawa, M. (2013). Evading apoptosis in cancer. *Trends Cell Biol*, **23**(12), 620–33.

Fischer, M. (2017). Census and evaluation of p53 target genes. *Oncogene*, **36**, 3943–56.

Grimm, S. & Noteborn, M. (2010). Anticancer genes: inducers of tumour-specific cell death signalling. *Trends Mol Med*, **16**(2), 88–96.

Stuckey, D. W. & Shah, K. (2013). TRAIL on trial: preclinical advances in cancer therapy. *Trends Mol Med*, **19**(11), 685–94.

Walczak, H. (2013). Death receptor-ligand systems in cancer, cell death, and inflammation. *Cold Spring Harb Perspect Biol*, **5**(5), a008698.

REFERENCES

Ashkenazi, A., Fairbrother, W. J., Leverson, J. D., & Souers, A. J. (2017). From basic apoptosis discoveries to advanced selective BCL-2 family inhibitors. *Nat Rev Drug Discov*, **16**(4), 273–84.

Beroukhim, R., Mermel, C. H., Porter, D. et al. (2010). The landscape of somatic copy-number alteration across human cancers. *Nature*, **463**(7283), 899–905.

Billard, C. (2013). BH3 mimetics: status of the field and new developments. *Mol Cancer Ther*, **12**(9), 1691–700.

Ceballos-Cancino, G., Espinosa, M., Maldonado, V., Melendez-Zajgla, J. (2007). Regulation of mitochondrial Smac/DIABLO-selective release by survivin. *Oncogene*, **26**(54), 7569–75.

Chai, J., Du, C., Wu, J. W., Kyin, S., Wang, X., & Shi, Y. (2000). Structural and biochemical basis of apoptotic activation by Smac/DIABLO. *Nature*, **406**(6798), 855–62.

Chang, D. W., Xing, Z., Pan, Y., et al. (2002). c-FLIP(L) is a dual function regulator for caspase-8 activation and CD95-mediated apoptosis. *EMBO J*, **21**(14), 3704–14.

Chen, D. J. & Huerta, S. (2009). Smac mimetics as new cancer therapeutics. *Anticancer Drugs*, **20**(8), 646–58.

Chinnaiyan, A. M., O'Rourke, K., Tewari, M., & Dixit, V. M. (1995). FADD, a novel death domain-containing protein, interacts with the death domain of Fas and initiates apoptosis. *Cell*, **81**(4), 505–12.

Chipuk, J. E. & Green, D. R. (2008). How do BCL-2 proteins induce mitochondrial outer membrane permeabilization? *Trends Cell Biol*, **18**(4), 157–64.

Cho, Y. S., Challa, S., Moquin, D., et al. (2009). Phosphorylation-driven assembly of the RIP1-RIP3 complex regulates programmed necrosis and virus-induced inflammation. *Cell*, **137**(6), 1112–23.

Cory, S. & Adams, J. M. (2002). The Bcl2 family: regulators of the cellular life-or-death switch. *Nat Rev Cancer*, **2**(9), 647–56.

Datta, S. R., Dudek, H., Tao, X., et al. (1997). Akt phosphorylation of BAD couples survival signals to the cell-intrinsic death machinery. *Cell*, **91**(2), 231–41.

Declercq, W., Vanden Berghe, T., & Vandenabeele, P. (2009). RIP kinases at the crossroads of cell death and survival. Cell **138**(2), 229–32.

Deretic, V., Saitoh, T., Akira, S. (2013). Autophagy in infection, inflammation and immunity. *Nat Rev Immunol*, **13**(10), 722–37.

Earnshaw, W. C., Martins, L. M., Kaufmann, S. H. (1999). Mammalian caspases: structure, activation, substrates, and functions during apoptosis. *Annu Rev Biochem*, **68**, 383–424.

Falschlehner, C., Emmerich, C. H., Gerlach, B., & Walczak, H. (2007). TRAIL signalling: decisions between life and death. *Int J Biochem Cell Biol*, **39**(7–8), 1462–75.

Fatokun, A. A., Dawson, V. L., Dawson, T. M. (2014). Parthanatos: mitochondrial-linked mechanisms and therapeutic opportunities. *Br J Pharmacol*, **171**(8), 2000–16.

Feng, S., Yang, Y., Mei, Y., et al. (2007). Cleavage of RIP3 inactivates its caspase-independent apoptosis pathway by removal of kinase domain. *Cell Signal*, **19**(10), 2056–67.

Fenstermaker, R. A., Ciesielski, M. J., Qiu, J., et al. (2016). Clinical study of a survivin long peptide vaccine (SurVaxM) in patients with recurrent malignant glioma. *Cancer Immunol Immunother*, **65**(11), 1339–52.

Fernald, K. & Kurokawa, M. (2013). Evading apoptosis in cancer. *Trends Cell Biol*, **23**(12), 620–33.

Fischer, M. (2017). Census and evaluation of p53 target genes. *Oncogene*, **36**(28), 3943–56.

Fujiwara, T., Kataoka, M., Tanaka, N. (2000). Adenovirus-mediated p53 gene therapy for human cancer. *Mol Urol*, **4**(2), 51–4.

Galluzzi, L., Galluzzi, L., Pietrocola, F., Bravo-San Pedro, J. M., et al. (2015). Autophagy in malignant transformation and cancer progression. *EMBO J*, **34**(7), 856–80.

Grasso, D. & Vaccaro, M. I. (2014). Macroautophagy and the oncogene-induced senescence. *Front Endocrinol (Lausanne)*, **5**, 157.

Grimm, S. & Noteborn, M. (2010). Anticancer genes: inducers of tumour-specific cell death signalling. *Trends Mol Med*, **16**(2), 88–96.

Guha, M. & Altieri, D. C. (2009). Survivin as a global target of intrinsic tumor suppression networks. *Cell Cycle*, **8**(17), 2708–10.

Gyrd-Hansen, M. & Meier, P. (2010). IAPs: from caspase inhibitors to modulators of NF-kappaB, inflammation and cancer. *Nat Rev Cancer*, **10**(8), 561–74.

Hanahan, D. & Weinberg, R. A. (2000). The hallmarks of cancer. *Cell*, **100**(1), 57–70.

He, C. & Klionsky, D. J. (2009). Regulation mechanisms and signaling pathways of autophagy. *Annu Rev Genet*, **43**, 67–93.

Hellwig, C. T. & Rehm, M. (2012). TRAIL signaling and synergy mechanisms used in TRAIL-based combination therapies. *Mol Cancer Ther*, **11**(1), 3–13.

Hu, Y., Benedict, M. A., Ding, L., & Núñez, G. (1999). Role of cytochrome c and dATP/ATP hydrolysis in Apaf-1-mediated caspase-9 activation and apoptosis. *EMBO J*, **18**(13), 3586–95.

Janku, F., McConkey, D. J., Hong, D. S., & Kurzrock, R. (2011). Autophagy as a target for anticancer therapy. *Nat Rev Clin Oncol*, **8**(9), 528–39.

Joudeh, J. & Claxton, D. (2012). Obatoclax mesylate: pharmacology and potential for therapy of hematological neoplasms. *Expert Opin Investig Drugs*, **21**(3), 363–73.

Kanwar, J. R., Kamalapuram, S. K., & Kanwar, R. K. (2013). Survivin signaling in clinical oncology: a multifaceted dragon. *Med Res Rev*, **33**(4), 765–89.

Kerr, J. F., Wyllie, A. H., & Currie, A. R. (1972). Apoptosis: a basic biological phenomenon with wide-ranging implications in tissue kinetics. *Br J Cancer*, **26**(4), 239–57.

Kroemer, G., Mariño, G., & Levine, B. (2010). Autophagy and the integrated stress response. *Mol Cell*, **40**(2), 280–93.

Krueger, A., Schmitz, I., Baumann, S., Krammer, P. H., & Kirchhoff, S. (2001). Cellular FLICE-inhibitory protein splice variants inhibit different steps of caspase-8 activation at the CD95 death-inducing signaling complex. *J Biol Chem*, **276**(23), 20633–40.

Lalaoui, N. & Brumatti, G. (2017). Relevance of necroptosis in cancer. *Immunol Cell Biol*, **95**(2), 137–45.

Lamb, C. A., Yoshimori, T., & Tooze, S. A. (2013). The autophagosome: origins unknown, biogenesis complex. *Nat Rev Mol Cell Biol*, **14**(12), 759–74.

Laplante, M. & Sabatini, D. M. (2012). mTOR signaling in growth control and disease. *Cell*, **149**(2), 274–93.

Levine, A. J., Momand, J., & Finlay, C. A. (1991). The p53 tumour suppressor gene. *Nature*, **351**(6326), 453–6.

Ley, R., Balmanno, K., Hadfield, K., Weston, C., & Cook, S. J. (2003). Activation of the ERK1/2 signaling pathway promotes phosphorylation and proteasome-dependent degradation of the BH3-only protein, Bim. *J Biol Chem*, **278**(21), 18811–6.

Li, H., Zhu, H., Xu, C. J., & Yuan, J. (1998). Cleavage of BID by caspase 8 mediates the mitochondrial damage in the Fas pathway of apoptosis. *Cell*, **94**(4), 491–501.

Lu, J., McEachern, D., Sun, H., et al. (2011). Therapeutic potential and molecular mechanism of a novel, potent, nonpeptide, Smac mimetic SM-164 in combination with TRAIL for cancer treatment. *Mol Cancer Ther*, **10**(5), 902–14.

Luthi, A. U. & Martin, S. J. (2007). The CASBAH: a searchable database of caspase substrates. *Cell Death Differ*, **14**(4), 641–50.

Micheau, O. & Tschopp, J. (2003). Induction of TNF receptor I-mediated apoptosis via two sequential signaling complexes. *Cell*, **114**(2), 181–90.

Mowers, E. E., Sharifi, M. N., & Macleod, K. F. (2017). Autophagy in cancer metastasis. *Oncogene*, **36**(12), 1619–30.

Muller, P. A. & Vousden, K. H. (2013). p53 mutations in cancer. *Nat Cell Biol*, **15**(1), 2–8.

Newsom-Davis, T., Prieske, S., & Walczak, H. (2009). Is TRAIL the holy grail of cancer therapy? *Apoptosis*, **14**(4), 607–23.

Oda, E., Ohki, R., Murasawa, H., et al. (2000). Noxa, a BH3-only member of the Bcl-2 family and candidate mediator of p53-induced apoptosis. *Science*, **288**(5468), 1053–8.

Pattingre, S., Tassa, A., Qu, X., et al. (2005). Bcl-2 antiapoptotic proteins inhibit Beclin 1-dependent autophagy. *Cell*, **122**(6), 927–39.

Peery, R. C., Liu, J. Y., & Zhang, J. T. (2017). Targeting survivin for therapeutic discovery: past, present, and future promises. *Drug Discov Today*, **22**(10), 1466–77.

Peter, M. E. & Krammer, P. H. (2003). The CD95(APO-1/Fas) DISC and beyond. *Cell Death Differ*, **10**(1), 26–35.

Qu, X., Yu, J., Bhagat, G., et al. (2003). Promotion of tumorigenesis by heterozygous disruption of the beclin 1 autophagy gene. *J Clin Invest*, **112**(12), 1809–20.

Rabinowitz, J. D. & White, E. (2010). Autophagy and metabolism. *Science*, **330**(6009), 1344–8.

Rampino, N., Yamamoto, H., Ionov, Y., et al. (1997). Somatic frameshift mutations in the BAX gene in colon cancers of the microsatellite mutator phenotype. *Science*, **275**(5302), 967–9.

Riedl, S. J., Renatus, M., Schwarzenbacher, R., et al. (2001). Structural basis for the inhibition of caspase-3 by XIAP. *Cell*, **104**(5), 791–800.

Roberts, A. W., Davids, M. S., Pagel, J. M., et al. (2016). Targeting BCL2 with venetoclax in relapsed chronic lymphocytic leukemia. *N Engl J Med*, **374**(4), 311–22.

Scarfo, L. & Ghia, P. (2013). Reprogramming cell death: BCL2 family inhibition in hematological malignancies. *Immunol Lett*, **155**(1–2), 36–9.

Shangary, S. & Wang, S. (2009). Small-molecule inhibitors of the MDM2-p53 protein-protein interaction to reactivate p53 function: a novel approach for cancer therapy. *Annu Rev Pharmacol Toxicol*, **49**, 223–41.

Shiozaki, E. N., Chai, J., Rigotti, D. J., et al. (2003). Mechanism of XIAP-mediated inhibition of caspase-9. *Mol Cell*, **11**(2), 519–27.

Soengas, M. S., Capodieci, P., Polsky, D., et al. (2001). Inactivation of the apoptosis effector Apaf-1 in malignant melanoma. *Nature*, **409**(6817), 207–11.

Stilgenbauer, S., Eichhorst, B., Schetelig, J., et al. (2016). Venetoclax in relapsed or refractory chronic lymphocytic leukaemia with 17p deletion: a multicentre, open-label, phase 2 study. *Lancet Oncol*, **17**(6), 768–78.

Stuckey, D. W. & Shah, K. (2013). TRAIL on trial: preclinical advances in cancer therapy. *Trends Mol Med*, **19**(11), 685–94.

Su, Z., Yang, Z., Xie, L., DeWitt, J. P., & Chen, Y. (2016). Cancer therapy in the necroptosis era. *Cell Death Differ*, **23**(5), 748–56.

Sun, C., Cai, M., Gunasekera, A. H., et al. (1999). NMR structure and mutagenesis of the inhibitor-of-apoptosis protein XIAP. *Nature*, **401**(6755), 818–22.

Tait, S. W., Ichim, G., Green, D. R., et al. (2014). Die another way--non-apoptotic mechanisms of cell death. *J Cell Sci*, **127**(Pt 10), 2135–44.

Tang, H. L., Tang, H. M., Mak, K. H., et al. (2012). Cell survival, DNA damage, and oncogenic transformation after a transient and reversible apoptotic response. *Mol Biol Cell*, **23**(12), 2240–52.

Van Antwerp, D. J., Martin, S. J., Verma, I. M., & Green, D. R. (1998). Inhibition of TNF-induced apoptosis by NF-kappa B. *Trends Cell Biol*, **8**(3), 107–11.

van Gurp, M., Festjens, N., van Loo, G., Saelens, X., & Vandenabeele, P. (2003). Mitochondrial intermembrane proteins in cell death. *Biochem Biophys Res Commun*, **304**(3), 487–97.

Vanden Berghe, T., Linkermann, A., Jouan-Lanhouet, S., Walczak, H., & Vandenabeele, P. (2014). Regulated necrosis: the expanding network of non-apoptotic cell death pathways. *Nat Rev Mol Cell Biol*, **15**(2), 135–47.

Vandenabeele, P., Galluzzi, L., Vanden Berghe, T., & Kroemer, G. (2010). Molecular mechanisms of necroptosis: an ordered cellular explosion. *Nat Rev Mol Cell Biol*, **11**(10), 700–14.

Vaux, D. L., Cory, S., & Adams, J. M. (1988). Bcl-2 gene promotes haemopoietic cell survival and cooperates with c-myc to immortalize pre-B cells. *Nature*, **335**(6189), 440–2.

Walczak, H. (2013). Death receptor-ligand systems in cancer, cell death, and inflammation. *Cold Spring Harb Perspect Biol*, **5**(5), a008698.

Walczak, H., Miller, R. E., Ariail, K., et al. (1999). Tumoricidal activity of tumor necrosis factor-related apoptosis-inducing ligand *in vivo*. *Nat Med*, **5**(2), 157–63.

Wilson, W. H., O'Connor, O. A., Czuczman, M. S., et al. (2010). Navitoclax, a targeted high-affinity inhibitor of BCL-2, in lymphoid malignancies: a phase 1 dose-escalation study of safety, pharmacokinetics, pharmacodynamics, and antitumour activity. *Lancet Oncol*, **11**(12), 1149–59.

Xie, Y., Hou, W., Song, X., et al. (2016). Ferroptosis: process and function. *Cell Death Differ*, **23**(3), 369–79.

Yu, J. & Zhang, L. (2003). No PUMA, no death: implications for p53-dependent apoptosis. *Cancer Cell*, **4**(4), 248–9.

Zhong, Z., Sanchez-Lopez, E., & Karin, M. (2016). Autophagy, inflammation, and immunity: a troika governing cancer and its treatment. *Cell*, **166**(2), 288–98.

15

Telomerase and immortalization

Laura Collopy and Kazunori Tomita

Introduction: Chromosomal timer and immortality

Although most cells in our body age, eventually stopping the regeneration of tissues, cancer cells are able to escape this ageing programme and propagate indefinitely. Cells are programmed to have only a limited number of cell divisions. This ageing timer is located at the ends of chromosomes, called telomeres. Telomeres comprise repeats of a short, non-coding DNA sequence. Cells lose some of these telomere repeats at every chromosome replication and cell division. As telomeres are essential for chromosome integrity, cells stop dividing before the repeats are completely exhausted (**Fig. 15.1A**). To maintain lineage, telomere length is *reset* every generation. Germ and stem cells and unicellular eukaryotes like yeast express the telomere addition enzyme *telomerase* to counteract telomere erosion, therefore allowing reproduction of future generations to continue. Expression of telomerase is tightly controlled in differentiated cells. Most cancer cells appear to reactivate this enzyme to escape the fate of cellular ageing. In this chapter, we will navigate through the dawn of telomeres and the concept of cellular ageing, current knowledge of telomere structure and function, and finally how cancer cells manipulate telomerase to maintain their immortality.

History of cellular senescence and telomeres

Hayflick limit and senescence

Normal somatic cells have a limited proliferative potential and are not able to go on dividing indefinitely. This was demonstrated in the 1960s by Leonard Hayflick, who showed cells were capable of approximately 50 divisions before they underwent a growth arrest, termed *replicative senescence* (later called the Hayflick limit; see Hayflick and Moorhead, 1961; Harley et al., 1990). Senescence is irreversible, and although senescent cells are metabolically viable and can be maintained in culture, they have different properties form normal cells and, most importantly, are unable to divide.

Before the concept of cellular ageing was born, August Weismann speculated that 'death takes place because worn out tissue cannot forever renew itself and because a capacity for increase by means of cell division is not everlasting but finite' (Weismann, 1891). This limit is dictated by telomeres; in healthy individuals the Hayflick limit marks when telomeres have reached a critical length. In this way, telomeres restrict the proliferative lifespan of a cell, which ensures that DNA replication is stopped before chromosomal DNA can be lost or damaged. This concept was proven by immortalization of differentiated cells after ectopic expression of telomerase, which can maintain length of telomeres and proliferation capacity (Bodnar et al., 1998).

Discovering telomeres

In 1938, Hermann Muller examined how chromosomes responded to breakage using X-rays, a discovery that won him the 1946 Nobel Prize in Physiology or Medicine. He irradiated fruit flies to cause double-stranded DNA breaks. Muller observed that, when damaged, their chromosomes were prone to rearrangements and fusions. However, he noted that the original terminal ends of chromosomes were stable and exempt from these repair mechanisms. Muller termed these 'telomeres', from the Greek *telos* (end) and *meros* (component; Muller, 1938).

Muller initially predicted that a gene necessary for survival was located at the ends of the chromosomes in flies, thus exempting them from the rearrangement events caused by X-ray damage. However, it was elucidated that telomeres are protective of chromosomal DNA due to their intricate structure and relationship with several protein complexes. Deciphering the sequence and structure of telomeres was initially challenging, as most eukaryotic chromosomes are very long. However, the ciliated protozoan *Tetrahymena thermophila* has relatively short (<100 kb) palindromic minichromosomes. Elizabeth Blackburn discovered that these terminate in between 20 and 70 hexanucleotide repeats with a G-rich single-stranded overhang at the 3' end (Blackburn and Gall, 1978). After this finding, Blackburn and Jack Szostak identified that the telomere DNA sequence itself can protect chromosome ends (Szostak and Blackburn, 1982). This discovery established the concept of telomeres and, with the following discovery of telomerase, led to their Nobel prize in physiology or medicine in 2009. Thus, rather than containing a crucial gene located at chromosome terminal ends, the telomeric structure serves to protect DNA causing the effect observed earlier by Muller.

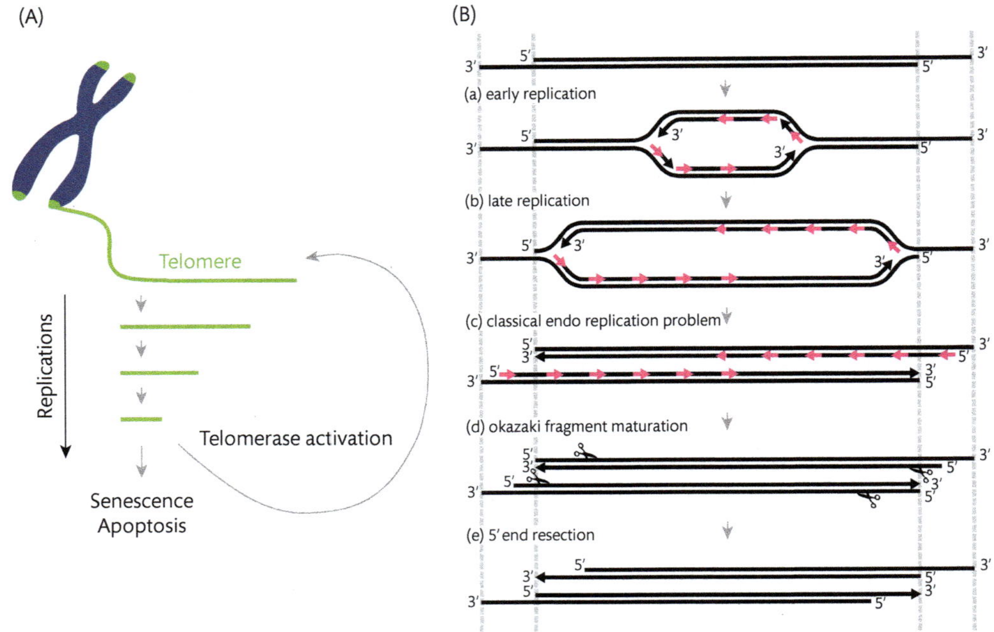

Fig. 15.1 Replicative senescence model. (A) Telomere erosion and cellular senescence. Telomeres can be found at the tip of chromosome arms seen at mitosis (metaphase chromosome spread). Telomere repeats are reduced with every chromosomal replication. Shortened telomeres elicit DNA damage and senescence checkpoints and cell cycle arrest. Cells decide whether to remain senescent or die through apoptosis. If telomerase replenishes telomeres, the cell cycle continues. (B) End replication problems: Schematic representation for chromosome replication. Prereplication chromosome is shown as the double-stranded DNA molecule on top. The 3′ ends of the single-strand DNA is protruding at the chromosome ends. Grey dotted lines define the position for original chromosome ends. (a) Early replication: DNA replication is initiated from an internal site of the chromosome. Unwound single-strands are replicated by polymerase towards the 5′ end direction, initiated by RNA primer (pink arrows). (b) Late replication: replication forks move towards chromosome ends. One strand can be synthesized by one polymerase (leading strand synthesis), and the other strand needs a number of RNA primers as polymerase can synthesize only in the 5′ to 3′ direction (lagging strand synthesis). (c) Telomere replication: chromosome ends are generally synthesized last (late S-phase). The leading strand ends (5′ end) can be fully replicated. However, 3′ end of lagging strand cannot be synthesized beyond an RNA primer (classical replication problem or lagging strand problem). (d) Okazaki fragment maturation. All RNA oligos are degraded. The gap between DNA fragments at the lagging strand (Okazaki fragment) is filled. (e) 5′ end resection: since lagging strand ends are blunt, 5′ ends are digested to generate the 3′ end. Not only leading strand ends but also lagging strand ends are thought to be resected to produce enough length of the overhang in mammals. Resulting sister chromatids are shorter than mother strands (leading strand problem).

The end replication problem and proliferative potential

After the double helical structure of DNA had been elucidated, Alexey Olovnikov described his 'Theory of Marginotomy' to suggest how it is replicated during the cell cycle. While waiting for a subway train, he imagined the track as a strand of DNA and the approaching train as a polymerase enzyme. He identified this 'end replication problem'; that there had to be a dead zone at the end of the DNA strand beneath the enzyme which could not be replicated, analogous to a region of track beneath the train. He therefore suggested that there must be DNA at the ends of chromosomes, which is lost through each round of cell division (Olovnikov, 1996). At this time, the nature of telomeric DNA was unknown and his theory was not widely accepted.

Telomeres provide a solution to the *end replication problem*. As DNA replicates, the daughter strands are extended by DNA polymerase only in the 5' to 3' direction. As a result, the DNA cannot be copied in the region where the RNA primers are laid down at the 5' end of each parent strand (Fig. 15.1B); the lagging strand problem. The protective structure of the telomere also elucidated another issue; the leading strand problem (Lingner et al., 1995). The polymerase on the leading strand can fill towards the 5' end of telomere, however this loses the 3' overhang, which is required for the protective telomere configuration in many organisms (see next). This issue is solved by the resection of the 5' end of the template telomere. Thus, each round of cell division results in the loss of 50–200 base pairs from both leading and lagging strand ends of the chromosome (Fig. 15.1B). Telomeres protect chromosomal DNA by shortening themselves. Consequently, telomeres act as a source of expendable DNA and get progressively shorter with age.

What this means is that telomere length in most somatic tissues is established from birth and will progressively shorten during each round of cell division throughout an individual's lifetime. This therefore poses the question, *Why do somatic cells have a replicative lifespan?* Although it may seem that the continued expression of telomerase could be the key to preventing ageing, replicative senescence exists as an antitumour mechanism to prevent the development of cancer.

Telomere structure and function

T-loops

The structure of telomeres is important to ensure chromosomal protection. Telomeric DNA is largely double-stranded, however, the ends are processed so that they terminate with a single-stranded, G-rich overhang, 130–210 bases in length in humans. This then undergoes a change in conformation, folding back to form a lariat-like structure called a t-loop (telomeric loop, Fig. 15.2A). The single-stranded portion embeds within the double-stranded region by homologous recombination, forming a d-loop (displacement loop). This effectively seals the chromosome ends for protection (Griffith et al., 1999). The t-loop configuration of the telomere end was first revealed *in vitro* and later observed in cells using the stochastic optical reconstruction microscopy (STORM) method (Doksani et al., 2013).

During his early experiments Muller noticed that telomeres are exempt from DNA damage repair mechanisms that are engaged in the presence of chromosome breakages. The t-loop structure of telomeres appeared to be a solution to this, as the DNA end is sheltered.

The telomeric DNA binding protein complex (shelterin) and DNA damage response machineries

Telomeres interact with a specialized protein complex called shelterin. The shelterin complex is made up of six family members: TRF1, TRF2, RAP1, TIN2, TPP1 and POT1 (Fig. 15.2B). TRF1 and TRF2 are the double-strand telomeric DNA binding proteins that recruit TIN2. TIN2 associates with TPP1. POT1 forms heterodimer with TPP1 and directly binds to the telomere single-strand DNA at the 3' overhang as well as at the d-loop within the t-loop structure. Collectively, they coat telomeric repeats (both single and double-stranded DNA) and protect chromosome ends (Fig. 15.2C). TRF2 is required for formation of t-loops by binding to the double-stranded portion and helping to remodel the terminal ends (Griffith et al., 1999; Doksani et al., 2013). Together with the shelterin complex, telomeres establish a heterochromatin structure and pack the ends of chromosomes.

As telomeric DNA can resemble a site of DNA breakage, telomeres face the risk of being recognized as damaged DNA. Failure to interact with shelterin and form t-loops can elicit DNA damage response pathways, leading to activation of senescent signalling or apoptosis. Furthermore, lack of telomeric protection can cause chromosome fusions or aberrant recombinations, leading to genetic instability and aneuploidy (van Steensel et al., 1998; Denchi and de Lange, 2007), a key step in the development of cancer.

In addition to t-loops, the shelterin complex interacts with DNA damage response factors and controls their functions so as not to induce their downstream pathways (Palm and de Lange, 2008). TRF2 directly binds to the kinase ataxia telangiectasia mutated (ATM) and a DNA repair complex called MRN (Mre11, Rad50, and Nbs1). By doing so, ATM and MRN fail to activate a p53 and H2AX (to give γH2AX), which induce cell cycle arrest and the DNA double-stranded break repair pathways (Karlseder et al., 1999; Denchi and de Lange, 2007; Feuerhahn et al., 2015). Loss of TRF2 results in non-homologous end-joining (NHEJ), where chromosomes undergo inappropriate end-to-end fusion events (van Steensel et al., 1998). The shelterin protein RAP1 binds to TRF2 and also plays a role in inhibiting NHEJ at the telomeres (Sarthy et al., 2009).

Like TRF2, the shelterin protein POT1 also has a role in repressing DNA damage response signalling (Denchi and de Lange, 2007). POT1 binding to the single-stranded telomeric DNA blocks the binding of replication protein A (RPA), which recruits another DNA damage checkpoint kinase, called ataxia telangiectasia and Rad3-related (ATR). Importantly, recruitment of POT1 to the

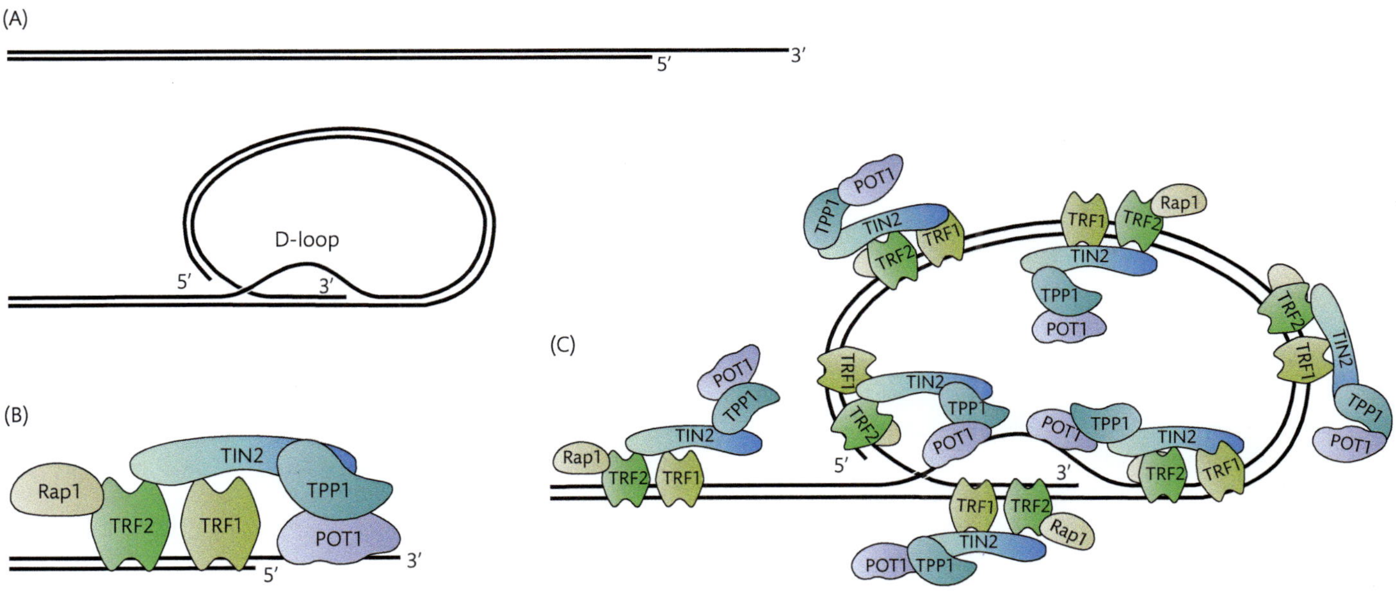

Fig. 15.2 Telomere structure. (A) T-loop formation. Telomere ends protrude at the 3′ end with a G-rich single-stranded tail (Top). The G-tail strand invades its own double-stranded telomere and replaces the G-strand to form a so-called d-loop. (B) The shelterin complex. Core telomeric protein components are TRF1, TRF2, RAP1, TIN2, TPP1, and POT1. TRF1 and TRF2 bind to the double-stranded telomeric DNA. RAP1 binds to TRF2. POT1 binds to the single-stranded G-rich telomeric DNA and forms a heterodimer with TPP1. TIN2 binds to TPP1, TRF1, and TRF2 to link the shelterin complex. (C) Model for shelterin mediated t-loop formation. TRF2 stabilizes the d-loop structure by holding the translocated overhang; Pot1 binds the G-strand within the d-loop. TIN2 is crucial for trafficking of POT1-TPP1 into the nucleus and TRF1-TIN2 interaction is required for recruitment for the TIN2 complex. TRF2 localization is also stabilized by interaction with TIN2. Thus, all proteins are likely to form the complex at the telomere.

telomere requires the TRF2 analogue TRF1 and other shelterin components TIN2 and TPP1, highlighting the importance of shelterin formation. However, TRF1 does not seem to impact on end protection. It appears to be required for replication of the repetitive G-rich strand. At the lagging strand synthesis, TRF1 binds and recruits the DNA helicases as well as TIN2-TPP1-POT1 to resolve the secondary structure G-quadruplex for replication (Sfeir et al., 2009). Thus, whereas each telomere protein has distinct role, the formation of shelterin is crucial for overall function of the telomere.

Telomerase and telomere length homeostasis

The telomerase complex

Telomerase is a specialized enzyme complex responsible for the lengthening of telomeres through the addition of hexanucleotide repeats. Its existence was initially postulated because there is heterogeneity in telomere length within and between cells. Therefore, it was predicted that this may arise due to the *de novo* addition of telomeric repeats to the chromosome ends. A series of experiments confirmed that a specialized ribonucleoprotein complex was responsible for the addition of telomeric repeats, and this was named 'telomerase' (Greider and Blackburn, 1985).

Telomerase requires two components; the telomerase reverse transcriptase (TERT) and the telomerase RNA component, TERC (or telomerase RNA, TR), which provides telomeric sequence as a template for reverse transcription (Fig. 15.3A and 17.3B). TERC is part of a group of RNAs called H/ACA. All H/ACA RNAs group with the four proteins: dyskerin, GAR1, NHP2, and NOP10, giving an H/ACA ribonucleoprotein complex, which stabilizes TERC (Fig. 15.3C). The 5' end forms a *pseudoknot domain,* which contains the telomeric template sequence CAAUCCCAAUC. These RNA binding proteins ensure the localization of telomerase to small organelles called Cajal bodies. Here, TERC binds with TERT to form the mature telomerase complex. In addition, the 3' end of TERC interacts with a protein called TCAB1 (Fig. 15.3C). TCAB1 associates only if TERT is bound to TERC and facilitates the trafficking of telomerase to the telomeres (Venteicher et al., 2009).

Telomerase activity and the telomeric repeat amplification protocol assay

The telomerase reaction proceeds by sequential addition of deoxynucleotide triphosphates onto the 3' end of the telomere. Telomerase is a *processive* enzyme. This means that each time it binds to the telomere, several telomeric repeats are added before it dissociates (Fig. 15.3D; see Greider, 1991; Hammond and Cech, 1997). Following the addition of one repeat, telomerase translocates and realigns at the 3' end, maximizing the efficiency of telomere extension. Structural modelling of telomerase predicts that the N-terminal region of TERT, TEN-domain (Fig. 15.3B), drives processivity by rotating telomeric DNA along as new telomeric repeats are synthesized (Steczkiewicz et al., 2011). The telomerase complex therefore extends telomeres to ensure maximum efficiency.

To examine the level of telomerase activity in different cell types, Jerry Shay and Woodring Wright's group developed the telomeric repeat amplification protocol assay (Kim et al., 1994). In this method, labelled single-stranded telomeric DNA seeds are mixed with cell lysates and elongation of the telomeric sequence is detected. Since only telomerase is able to extend the telomere seed, extension equates to the presence of telomerase in the tissue. This method was used to quantify telomerase activity in several cell types. They demonstrated that telomerase is not active in most somatic cells and it is in fact terminated during embryonic development. It does, however, remain active in male germ cells, lymphocytes, and stem cell populations, which is consistent with the high rate of proliferation in these tissues (Kim et al., 1994; Cong et al., 2002). Nevertheless, the majority of cancers are able to maintain their telomeres by reactivating telomerase expression (Kim et al., 1994). Thus, this method became a standard for diagnosis of the presence of cancer cells.

Telomerase *in vivo*

Telomerase localizes to the Cajal bodies for most of the cell cycle and possibly undergoes post-translational modifications and associates with chaperones and telomerase subunits necessary for its recruitment to the telomere (Hug and Lingner, 2006; Schmidt and Cech, 2015). Interaction of telomerase with the telomeric DNA requires the telomerase RNA template to hybridize with the tip of the telomere G-rich overhang (Fig. 15.3D). The reverse transcriptase and the C-terminal domains form the catalytic core of the molecule (Fig. 15.3B), necessary for performing the telomere lengthening reaction (Schmidt and Cech, 2015). The TEN-domain of TERT supports formation of telomerase RNA and telomeric DNA hybrid duplex for telomerase processivity. However, telomerase first needs to associate with the shelterin components to gain access to the telomere overhang. The shelterin complex has an important relationship with telomerase and, in addition to protecting telomeres, mediates telomerase access in order to control telomere length.

In normal cells where telomerase is expressed, telomere length is tightly regulated and kept within a species-specific and cell-specific range. Unlike *in vitro* situations, the association of telomerase with a telomeric DNA template is tightly controlled by the telomeric DNA binding proteins and the cell cycle. Regulation of telomerase action and the telomere length has been extensively investigated in the yeast model organisms *Saccharomyces cerevisiae* and *Schizosaccharomyces pombe*. Notably, two screenings using *S. cerevisiae* (so-called 'telomere silencing screening' and 'ever shorter telomeres screening'), first identified genes encoding telomerase subunits and a telomere DNA binding protein that control telomerase recruitment to telomeres (Singer and Gottschling, 1994; Lendvay et al., 1996). This discovery led to identification of telomerase encoding genes in *S. pombe* and humans (Nakamura et al., 1997). Elegant genetics assays using *S. cerevisiae* further elucidated that the shorter telomeres that accommodate fewer copies of the telomeric DNA binding protein Rap1 are preferentially extended by telomerase (Marcand et al., 1997; Teixeira et al., 2004). This selection mechanism allows many telomeres to be maintained by a few telomerase units, providing semi-conservative telomere maintenance. The *S. pombe* system determined that the shelterin component Taz1 (the TRF1/2 orthologue) restricts recruitment of telomerase to S-phase (Dehe et al., 2012). The cell cycle kinases and DNA replication factors overcome the inhibitory effect of telomeres, allowing telomerase activation at S-phase. These observations were confirmed in humans (Liu et al., 2002; Zhao et al., 2009; Xi et al., 2015).

It is proposed that telomere structure changes throughout the cell cycle to either restrict or allow the access of telomerase. Telomeres

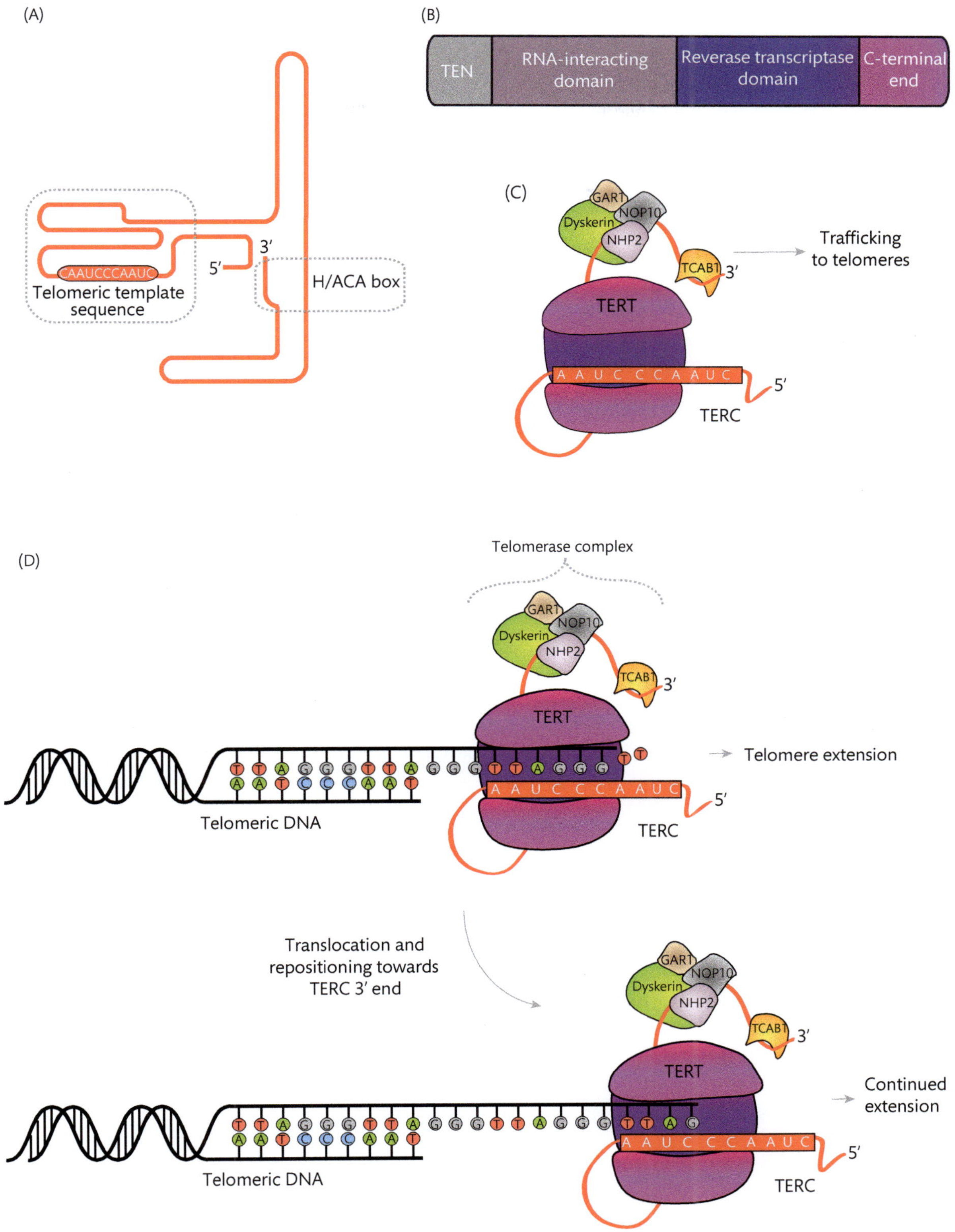

Fig. 15.3 Telomerase structure and action. (A) Structure of TERC RNA. Human TERC is 451 bp in length and folds to form four structural domains. TERC binds to TERT through the pseudoknot and CR4-CR5 domains. The H/ACA domain is required for stability and appropriate localization of the molecule. CR = conserved region. (B) Schematic of the TERT protein and key domains. The N-terminal domains signal the trafficking of TERT to the Cajal body where the mature telomerase complex can form. The TEN and C-terminal end domains are thought to be important in the recruitment of telomerase to the telomere. TEN = TERT N-terminal. (C) Schematic of the telomerase complex. The telomerase complex consists of the reverse transcriptase TERT which uses the RNA template provided by TERC to add telomere repeats. Through its 3′ H/ACA motif, TERC binds to dyskerin which is complexed with NOP10, NHP2, and GAR1. TCAB1 interacts with the mature telomerase complex to facilitate trafficking to the telomeres. (D) Extension of telomeres and telomerase processivity. Following synthesis of a telomeric repeat, telomerase translocates at the 5′ end of the telomere and realigns at the 3′ end.

therefore oscillate between a *non-extendable* state (e.g. when telomeres are folded into t-loops) and an *extendable* state (e.g. when telomeres have unfolded and are straight at the ends). Therefore, proteins like TRF2, which mediate the formation of t-loops, promote a non-extendable state, and negatively regulate telomerase activity. Hug and Lingner have also proposed a third state called the *extending state,* where telomerase is bound to the telomere and is actively lengthening it (Hug and Lingner, 2006). These mechanisms therefore help to ensure telomere length stays within an appropriate range.

Shelterin and telomerase control

The long tail of the G-rich single-stranded telomere can form a G-quadruplex that inhibits access of telomerase. POT1-TPP1 prevents formation of such secondary structure, and therefore promotes continuous extension of telomeric repeat by telomerase (Wang et al., 2007; Latrick and Cech, 2010). In addition, TPP1 itself associates with and recruits telomerase to the telomere. Without TPP1, telomerase is unable to bind at the telomeres and stalls in the Cajal bodies (Nandakumar et al., 2012; Zhong et al., 2012). However, recent work using *S. pombe* elucidated the presence of telomerase activation step after the recruitment (Armstrong et al., 2014). This highlights that the regulation of telomerase activity occurs at many stages and is highly intricate with much still to be deciphered (Schmidt and Cech, 2015).

TRF1 and TRF2 both bind to the double-stranded telomeric DNA and negatively regulate telomerase activity (Bianchi et al., 1997; Griffith et al., 1998). In addition, the role of TRF2 in forming t-loops and promoting a *non-extendable* state limits telomerase activity (Feuerhahn et al., 2015). Like yeast models, in humans longer telomeres have more TRF1 and TRF2 bound and this therefore has a greater inhibitory effect on telomerase making it preferentially target shorter telomeres with less TRF1 and TRF2 (Ancelin et al., 2002). Nevertheless, TRF1 and TRF2 are required for association of POT1-TPP1 *via* shelterin formation with TIN2. The shelterin complex therefore has an intricate relationship with telomeres and telomerase and disturbing the expression of shelterin proteins greatly impacts on telomere length which directly affects the cell cycle and senescence signalling (de Lange, 2005).

Telomere maintenance in cancer

Telomere protection and cancer

Why is telomerase so tightly regulated in normal cells? To answer this question, it must be considered that, in ~85% of cancers, telomerase has become reactivated (Table 15.1). Since telomere protection is essential for all dividing cells, tumour growth is restricted by the size of their telomeres. Loss of shelterin by critically shortened telomeres activates DNA damage response machineries that induce apoptosis or cell death *via* mitotic catastrophe. Cancers that gain telomere maintenance ability allows for the continued development of the tumour. Therefore, the regulation of telomerase is tightly controlled to avoid carcinogenesis (Shay and Bacchetti, 1997).

In 1998, Shay and Wright's group demonstrated that when TERT was ectopically expressed in human cells, they could continue to divide far beyond the normal number of population doublings (Bodnar et al., 1998). This showed that telomerase expression alone was sufficient to evade senescence. However, whereas these cells are immortal and maintain long telomeres, reactivation of telomerase alone is not sufficient to cause malignancy in patients and further cancerous mutations must occur.

Most tumours require 4–6 critical mutations in order to become malignant, unless a mutation dramatically changes transcriptional landscapes. Such mutations are unlikely to occur during the life of cell cycle, and senescence will likely occur in premalignant cells before they develop into a cancerous tumour (Shay and Wright, 2011). However, it is thought that short telomeres promote genetic instability, which in turn drives malignant transformation. Disruption of the senescence checkpoints such as p53 and retinoblastoma (pRb) can extend a number of cell divisions and accelerate telomere shortening (Fig. 15.4A). Critically shortened telomeres

Table 15.1 Telomerase activity in malignant tissue

Type of cancer	Telomerase-positive samples	Samples tested	% positive
Head, neck, and lung	239	314	76
Gastric carcinoma	72	85	85
Colorectal carcinoma	123	138	89
Pancreatic carcinoma	41	43	95
Hepatocellular carcinoma	149	173	86
Breast carcinoma	335	383	87
Female reproductive tract	52	56	93
Male reproductive tract	55	61	90
Kidney and urinary tract	273	306	89
Neural tissue	191	247	77
Skin	94	102	92
Haematological tissue	139	188	74

Source: data from Shay JW and Bacchetti S, 'A survey of telomerase activity in human cancer', *European Journal of Cancer*, Volume 33, Issue 5, pp. 787–91, Copyright © 1997 Published by Elsevier Ltd.
https://www.sciencedirect.com/science/article/pii/S0959804997000622

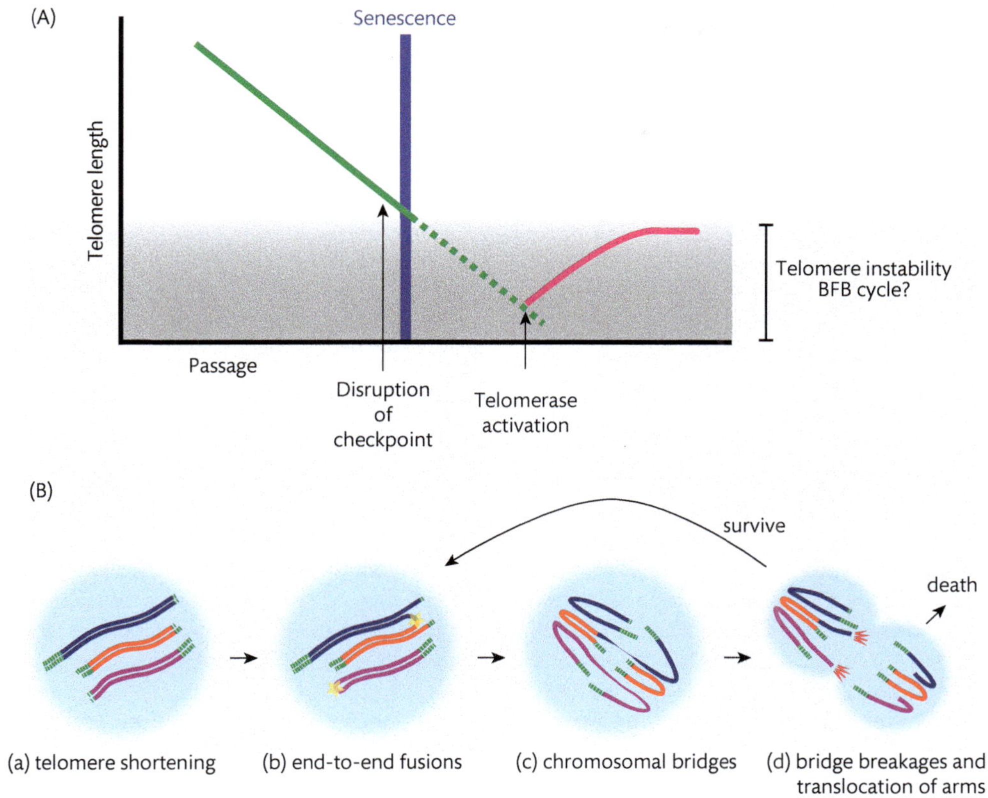

Fig. 15.4 Telomere crisis and breakage-fusion-bridge cycle. (A) Graph representing telomere crisis and immortalization. Y-axis and X-axis indicate length of telomeres and passage of cells, respectively. Telomeres are progressively shortened through generations (green line). Short telomeres are recognized by the senescence/apoptosis checkpoints to stop further divisions (blue column). Cells that escape this cell cycle arrest system further lose telomere repeats (green dotted line). Critically shortened telomeres elicit activation of DNA damage repair machineries that fuse chromosome ends (telomere crisis). Chromosome fusions and breakages lead to genomic instability (grey area). Cells gained ability to maintain chromosome ends (by activation of telomerase) dominate and establish their lineage. Most of telomerase-activated cancer cells maintain their telomeres relatively short, and some of them (t-stumps) are the minimum length required for telomere protection, retaining risk of telomere-induced chromosome instability. (B) Chromosome end-to-end fusions and bridge-fusion-breakage cycle. (a) Critically shortened telomeres lose ability to protect the chromosome ends. (b) Loss of telomere function allows fusion of chromosome ends. (c) Chromosome segregation at mitosis can be stalled by catenation of two chromosomes. (d) Nucleases and cytokinesis cut chromosome bridges to complete mitosis. This cause translocation of chromosomes and generation of new chromosome ends. A cell fuses the broken ends and moves to next round of cell cycle. A cell that has lost a chromosome arm is likely to die.

elicit DNA damage repair pathways and induce chromosome end fusions and anaphase bridges. This so-called 'breakage-fusion-bridge' cycle causes chromosome translocation and aneuploidy and greatly increases the incidence of mutations, hallmarks of cancer (McClintock, 1952; **Fig. 15.4B**). This selection process can lead to the activation of telomerase, which stabilizes telomeres and allows continued cell division. Telomere instability is therefore a critical step in driving oncogenesis (Shay and Wright, 2011). However, the mechanism of telomerase regulation in cancer cells remains to be established.

It is known that even in cancer cells telomerase is active at the telomere only during S-phase of the cell cycle (Zhao et al., 2009). At this time, only a small subset of telomeres is extended, suggesting processivity of cancer telomerase is low. Visualization of endogenous TERT determined that only 5–7% of the telomeres are associated with TERT at the given time of S-phase (Xi et al., 2015). Intriguingly, many cancer cells maintain their telomeres very short by telomerase to retain both telomere protection and chromosome instability. Some chromosomes in cancer cells have critically short telomeres, termed *t-stumps* (Xu and Blackburn, 2007). T-stumps are rarely found in normal telomerase-expressing cells (e.g. germ cells). They are still able to bind to the shelterin proteins TRF1 and TRF2, suggesting that despite their size they still have a protective function. However, they fail to elicit replicative senescence signalling which should be triggered when telomeres are critically short. Loss of checkpoint proteins p53/p21 and pRb is common in cancer cells and this may permit the accumulation of t-stumps to occur. Unlike embryonic stem cells in which telomerase preferentially elongates only the shortest telomeres (Liu et al., 2002), the ability of the cancer cell to extend the shortest telomeres seems compromised, keeping them just long enough to be stable. This promotes a telomeric situation unique to cancer cells. Failure in extending t-stumps may elicit chromosome fusions, which promote chromosome instability.

Telomerase transcriptional regulation

The expression and activity of telomerase is strictly controlled at various levels, from telomerase biogenesis (transcription, assembly, and post-translational modifications) and through recruitment, activation, and processivity at the telomeres. The transcription of the *TERT* gene is thought to be a major mechanism of telomerase regulation in human cells. This can be mediated by epigenetic modification of the *TERT* promoter. DNA methylation at specific sites in the *TERT* promoter results in the recruitment of the transcription factor GABPA/B1 (Stern et al., 2015). Many other factors can also regulate *TERT* transcription. The oncogene C-myc promotes cell growth and proliferation and can activate *TERT* transcription through interacting with regions in the promoter termed 'E-boxes' which have the sequence 5'-CACGTG (Wu et al., 1999). Conversely the tumour suppressor protein p53 can negatively regulate *TERT* expression. At the transcriptional level there are therefore many factors that influence the level of *TERT* expression and these are tightly regulated to meet the needs of the cell.

Much remains unknown about the reactivation of telomerase in cancer. However, several hotspots have been identified in the *TERT* promoter, which become mutated and lead to the inappropriate expression of telomerase. The frequency of TERT promoter mutations in cancer was studied in 741 tumour samples from patients with different cancers, including thyroid, kidney, bladder, and melanoma. Nineteen per cent of these had mutations in the TERT promoter, and most of these occurred at the –124 position. The –146 position was also mutated in a number of cases. These mutations are associated with increased *TERT* mRNA expression which sustains telomerase activity, compared to normal cells where telomerase is not expressed (Vinagre et al., 2013).

Shelterin mutations in cancer

In some rare cases of familial melanoma, patients with germline mutations to shelterin components have been identified. Firstly, mutations in POT1 were found, which are predicted to interrupt the interaction of POT1 with the telomere. Such patients have long, fragile telomeres suggesting the mutations have perturbed normal maintenance and impacted on telomere protection. This could lead to genetic instability which would promote further oncogenic mutations (Robles-Espinoza et al., 2014; Shi et al., 2014). In addition, heterozygous mutations have been identified in the gene encoding for RAP1 in four families and TPP1 in six families. These occur at evolutionarily conserved amino acid residues; however, their functional impact is unknown. The TPP1 mutations are predicted to impact on the heterodimerization of TPP1 and POT1 (Auode et al., 2014). These mutations emphasize the importance of the shelterin complex in protecting and regulating telomeres, as its perturbations can result in cancer.

Alternative lengthening of telomeres

Some cancers are telomerase-negative and a different pathway operates in order to sustain telomeres and allow proliferation to continue. This is called the alternative lengthening of telomeres pathway, or ALT, and is observed particularly frequently in osteosarcomas, soft tissue sarcomas, and astrocytic brain tumours (Bryan et al., 1997). ALT cells have very heterogeneous telomere lengths, from less than 3 kb to more than 50 kb. This differs from telomerase-positive cancers, which generally have relatively homogenous telomere lengths of less than 10 kb. However, studies have shown that the prognosis of ALT-positive cancers is no more favourable than telomerase-positive cancers (Ulaner et al., 2003).

ALT cells maintain telomeres using homologous recombination, whereby DNA sequence is copied from telomere to telomere. ALT is promoted by strand invasion of a telomere 3' G-rich overhang to any place along the double-stranded telomeric DNA. The invading 3' end is extended through the d-loop formation. The elongated telomere tail is displaced and the complementary strand is filled (**Fig. 15.5A**). The template DNA for replication can become

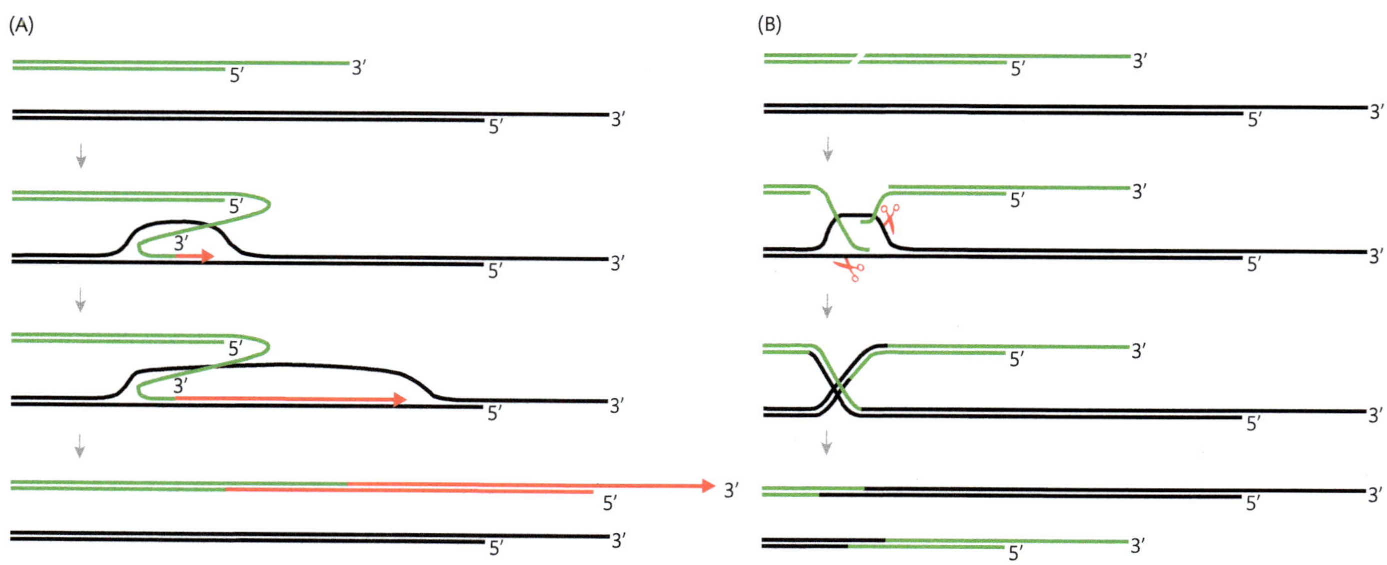

Fig. 15.5 Alternative lengthening of telomeres. Example models for possible telomere recombination/replication. (A) Pirate telomere copy model—telomere synthesis is dependent on a strand annealing pathway. Any region of the double-strand telomere repeats can be strand-invaded by the 3' end overhang. The invading 3' end can extend by unwinding the double strand. After dissolution, lagging strand is filled. (B) Robbing telomeric repeat model—homologous recombination cross-over repair-based sister chromatid exchange. The damaged telomere pairs with any region of sister chromatid telomere. The strand invasion of broken telomere establishes double-holiday junctions. A nuclease resolves the junctions and the resulting product exchanges their telomeres. Dependent on where telomere breaks and strand invasion occurred, resulting telomere length can be heterogeneous.

internal by utilizing existing d-loop formation within a t-loop, *or intertelomeric*, from sister chromatids or homologous chromosomes. Replication forks which stall at telomeres can also initiate strand invasion to other telomeric repeats at different positions or the switching of telomeres between sister chromatids (later is so-called T-SCE, telomere-sister chromatid exchanges; see **Fig. 15.5B**). Dunham et al. cloned a DNA tag within telomeric sequence of ALT cells, which could be visualized by fluorescence *in situ* hybridization. They demonstrated that the number of tags increased as cells divided, indicating that telomeric DNA was being copied to previously untagged telomeres (Dunham et al., 2000). Alternative telomere lengthening is thought to take place in structures called promyelocytic leukaemia (PML) nuclear bodies (Henson et al., 2002). Specifically, these are termed ALT-associated PML bodies, or APBs. PMLs are suborganelles found predominantly in the nucleus. The PML complexes dramatically change with the cell cycle and stress responses, and accommodate p53, DNA damage response machineries, and the histone chaperon DAXX.

The genetic changes that result in ALT are not well known, however loss of the protein ATRX, or its binding partner DAXX, has been observed in many ALT cancer cell lines and different tumour types. In many cases this is due to point mutations or large deletions within the *ATRX* and/or the *DAXX* genes (Heaphy et al., 2011; Lovejoy et al., 2012). ATRX and DAXX are parts of a chromatin remodelling complex, involved in the resolution of sister telomere cohesion which occurs prior to mitosis. However, loss of ATRX in ALT cells suppresses this resolution and telomeres remain persistently cohered. In this state, a high rate of T-SCEs occurs (Ramamoorthy and Smith, 2015). However, like reactivation of telomerase, loss of ATRX alone is not sufficient to cause an ALT phenotype. Rather it facilitates the transformation process and is thus considered a hallmark of ALT cancer.

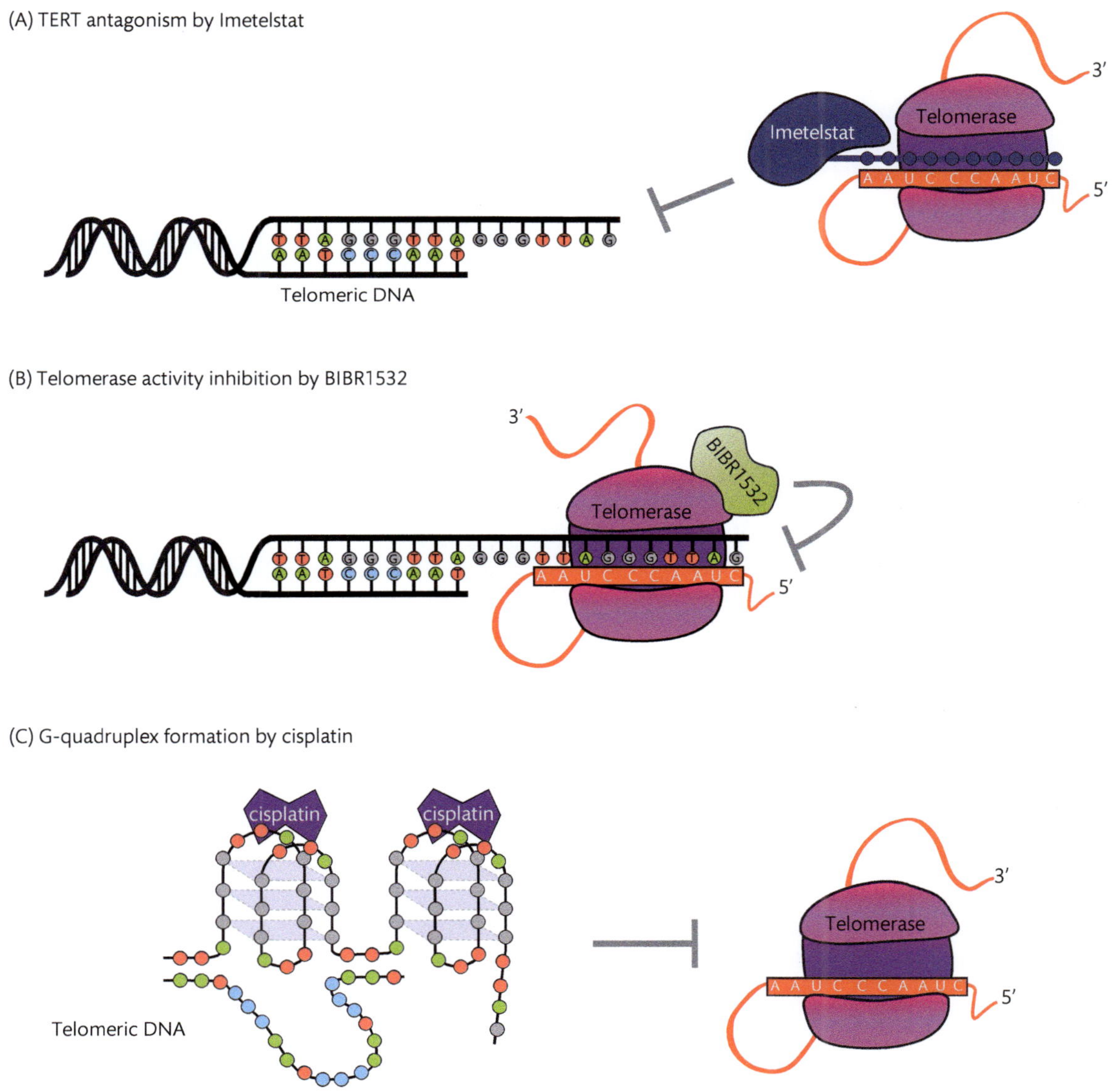

Fig. 15.6 Telomerase inhibitors for cancer therapeutics. (A) Imetelstat is a lipid-conjugated oligonucleotide that competes with telomeric DNA to hybridize telomerase RNA template. As an off-target, Imetelstat affects cytoskeleton formation and impairs cell adhesion capacity, which can increase efficiency of cancer treatment. (B) BIBR1532 binds to one of the TERC interaction domains of TERT and may cause a conformation change of telomerase, impairing its ability to function normally, leading to reduced telomerase processivity. (C) Cisplatin stabilizes G-quadruplex generated through replication of G-rich repeats. Formation of a G-quadruplex impairs interaction of telomeric DNA binding proteins at the double-strand and single-strand regions.

Another hallmark of ALT cells is an increased response to double-stranded DNA breaks at the telomeres. This triggers the telomeres to move in search for homologous DNA with which to recombine and repair the damage. This is akin to the events of meiosis, where double-stranded breaks in DNA trigger homologous recombination between non-sister chromatids (Cho et al., 2014). It is not well known how this is initiated in ALT cells, but could be a recapitulation of meiotic events.

Telomerase and telomere as target in cancer therapeutics

Due to the prominent role of telomeres and telomerase in most cancers, this pathway is an attractive target for anticancer therapies (**Fig. 15.6**). For example, the drug Imetelstat (previously named GRN163L) is currently being studied as it acts as a telomerase antagonist. It is a small oligonucleotide that binds to the TERC template with high affinity, competing with TERT to reduce telomerase activity (**Fig. 15.6A**). *In vitro* studies have demonstrated that Imetelstat causes telomere shortening over time and eventual cell death (Roth et al., 2009). Conversely, the telomerase inhibitor BIBR1532 is the small molecule which specifically targets the TERT active site and prevents its normal activity (**Fig. 15.6B**) (Pascolo et al., 2002).

In addition to telomerase, telomeres themselves are also an attractive target for anticancer therapies. For example, the drug cisplatin, currently used to treat a number of solid tumours, causes G-quadruplex formation at the telomeres (**Fig. 15.6C**). It targets the G-rich telomeric sequences, and causes them to fold from a linear formation into an intramolecular G-quadruplex structure. This prevents telomerase accessing telomeres for further extension and eventually results in cell death (Ishibashi and Lippard, 1998).

The shelterin complex can be an important target for anticancer drugs and new options are being investigated. Maria Blasco's group identified a small compound, ETP-47037, which inhibits TRF1 binding to the telomeres (Garcia-Beccaria, et al., 2015). TRF1 inhibition led to telomere replication catastrophe, genetic instability, proliferation defects and apoptosis (Sfeir et al., 2009; Garcia-Beccaria et al., 2015). Using lung carcinomas in mice as a model system, Blasco's group demonstrated that ETP-47037 was able to block tumour growth with minimum effect to normal tissues. Thus, altering shelterin function presents another attractive target in the development of anticancer therapies.

TAKE-HOME MESSAGE

Cancer is an age-related disease, with 83% of cases occurring in patients over the age of 55 (UK statistics, Cancer Research UK). Although telomere shortening and replicative senescence serve as an antitumour mechanism, the human life expectancy of 80+ years means cells have time to acquire the oncogenic mutations necessary to evade the normal 'mitotic clock' set. Although telomerase does not drive the development of cancer, it allows for the sustained growth of cells and the acquisition of oncogenic mutations.

Protection and maintenance of chromosome ends are essential even in cancer cells. However, mechanism of telomere maintenance by telomerase seems to differ in cancer cells compared to normal cells with active telomerase. Notably, telomerase processivity and the ability to target the shortest telomeres are likely to be impaired, promoting the maintenance of critically short telomeres (t-stumps).

As telomerase is reactivated in ~85% of cancer and their telomeres are shorter than in somatic cells, it is an attractive target for anticancer therapies. However, drugs that target telomerase are ineffective against telomere extension *via* the ALT pathway. Much is still unknown about the molecular events leading to telomerase activation in cancer. It therefore remains imperative to further our understanding of telomeres and telomerase in normal and cancer cells in order to identify new therapeutic targets.

OPEN QUESTIONS

- Explain the mechanism of telomerase activation and processivity at the telomere.
- Explain the difference in telomerase regulation between normal and cancer cells.
- What are the telomerase activation drivers of carcinogenesis?
- Describe the initiation and choice of alternative telomere lengthening.
- What mechanism senses replicative senescence?
- Explain the function of the telomere-led chromosome breakage-fusion bridge and aneuploidy.

FURTHER READING

de Lange, T. (2005). Shelterin: the protein complex that shapes and safeguards human telomeres. *Genes Dev*, **19**, 2100–10.

Ramamoorthy, M. & Smith, S. (2015). Loss of ATRX suppresses resolution of telomere cohesion to control recombination in ALT cancer cells. *Cancer Cell*, **28**, 57–369.

Roth, A., Harley, C., & Baerlocher, G. (2009). Imetelstat (GRN163L)—telomerase-based cancer therapy. *Small Molecules in Oncology*, **184**, 221–34.

Schmidt, J. & Cech, T. (2015). Human telomerase: biogenesis, trafficking, recruitment, and activation. *Genes Dev*, **29**, 1095–105.

Shay, J. & Wright, W. (2011). Role of telomeres and telomerase in cancer. *Seminars in Cancer Biology*, **21**(6), 349–53.

REFERENCES

Ancelin, K., Brunori, M., Bauwens, S., et al. (2002). Targeting assay to study the cis functions of human telomeric proteins: evidence for inhibition of telomerase by TRF1 and for activation of telomere degradation by TRF2. *Mol Cell Biol*, **22**(10), 3474–87.

Armstrong, C., Pearson, S., Amelina, H., Moiseeva, V., & Tomita, K. (2014). Telomerase activation after recruitment in fission yeast. *Curr Biol*, **24**(17), 2006–11.

Auode, L., Pritchard, A., Robles-Espinoza, C. D., et al. (2014). Nonsense mutations in the shelterin complex genes ACD and TERF2IP in familial melanoma. *J Natl Cancer Inst*, **107**(2), dju408.

Bianchi, A., Smith, S., Chong, L., Elias, P., & de Lange, T. (1997). TRF1 is a dimer and bends telomeric DNA. *The EMBO Journal*, **16**(7), 1785–94.

Blackburn, E. H. & Gall, J. G. (1978). A tandemly repeated sequence at the termini of the extrachromosomal ribosomal RNA genes in Tetrahymena. *J Mol Biol*, **120**(1), 33–53.

Bodnar, A., Ouellette, M., Frolkis, M., et al. (1998). Extension of life-span by introduction of telomerase into normal human cells. *Science*, **279**(5349), 349–52.

Bryan, T., Englezou, A., Dalla-Pozza, L., Melissa, D., & Reddel, R. (1997). Evidence for an alternative mechanism for maintaining telomere length in human tumors and tumor-derived cell lines. *Nature Medicine*, **3**, 1271–4.

Cho, N. W., Dilley, R., Lampson, M., & Greenberg, R. (2014). Interchromosomal homology searches drive directional ALT telomere movement and synapsis. *Cell*, **159**(1), 108–21.

Cong, Y.-S., Wright, W., & Shay, J. (2002). Human telomerase and its regulation. *Microbiol Mol Biol Rev*, **66**(3), 407–25.

de Lange, T. (2005). Shelterin: the protein complex that shapes and safeguards human telomeres. *Genes Dev*, **19**, 2100–10.

Dehe, P.-M., Rog, O., Ferreira, M., Greenwood, J., & Cooper, J. (2012). Taz1 enforces cell-cycle regulation of telomere synthesis. *Mol Cell*, **46**(6), 797–808.

Denchi, E. & de Lange, T. (2007). Protection of telomeres through independent control of ATM and ATR by TRF2 and POT1. *Nature*, **448**, 1068–71.

Doksanî, Y., Wu, J., de Lange, T., & Zhuang, X. (2013). Super-resolution fluorescence imaging of telomeres reveals TRF2-dependent T-loop formation. *Cell*, **155**(2), 345–56.

Dunham, M., Neumann, A., Fasching, C., & Reddel, R. (2000). Telomere maintenance by recombination in human cells. *Nat Genet*, **26**, 447–50.

Feuerhahn, S., Chen, L.-Y., Luke, B., & Porro, A. (2015). No DDRama at chromosome ends: TRF2 takes centre stage. *Trends Biochem Sci*, **40**(5), 275–85.

Garcia-Beccaria, M., Martinez, P., Mendez-Pertuz, M., et al. (2015). Therapeutic inhibition of TRF1 impairs the growth of p53-deficient K-RasG12V-induced lung cancer by induction of telomeric DNA damage. *EMBO Mol Med*, **7**(7), 930–50.

Greider, C. (1991). Telomerase is processive. *Mol Cell Biol*, **11**(9), 4572–80.

Greider, C. W. & Blackburn, (1985). Identification of a specific telomere terminal transferase activity in tetrahymena extracts. *Cell*, **43**(2), 405–13.

Griffith, J., Bianchi, A., & de Lange, T. (1998). TRF1 promotes parallel pairing of telomeric tracts. *J Mol Biol*, **278**, 79–88.

Griffith, J., Comeau, L., Rosenfield, S., et al. (1999). Mammalian telomeres end in a large duplex loop. *Cell*, **97**, 503–14.

Hammond, P. & Cech, T. (1997). dGTP-dependent processivity and possible template switching of euplotes telomerase. *Nucleic Acids Res*, **25**(18), 3698–704.

Harley, C., Futcher, B., & Greider, C. (1990). Telomeres shorten during ageing of human fibroblasts. *Nature*, **345**, 458–60.

Hayflick, L. & Moorhead, P. (1961). The serial cultivation of human diploid cell strains. *Experimental Cell Research*, **25**(3), 585–621.

Heaphy, C., de Wilde, R., Jiao, Y., Klein, A., Edil, B., & Meeker, A. (2011). Altered telomeres in tumors with ATRX and DAXX mutations. *Science*, **333**(6041), 425.

Henson, J., Neumann, A., Yeager, T., & Reddel, R. (2002). Alternative lengthening of telomeres in mammalian cells. *Oncogene*, **21**, 598–610.

Hug, N. & Lingner, J. (2006). Telomere length homeostasis. *Chromosoma*, **115**, 413–25.

Ishibashi, T. & Lippard, S. (1998). Telomere loss in cells treated with cisplatin. *PNAS*, **95**(8), 4219–23.

Karlseder, J., Broccoli, D., Dai, Y., Hardy, S., & de Lange, T. (1999). p53- and ATM-dependent apoptosis induced by telomeres lacking TRF2. *Science*, **283**(5406), 13211325.

Kim, N. W., Piatyszek, M. A., Prowse, K. R., et al. (1994). Specific association of human telomerase activity with immortal cells and cancer. *Science*, **266**(5193), 2011–15.

Latrick, C. & Cech, T. (2010). POT1-TPP1 enhances telomerase processivity by slowing primer dissociation and aiding translocation. *EMBO J*, **29**, 924–33.

Lendvay, T., Morris, D., Sah, J., Balasubramanian, B., & Lundblad, V. (1996). Senescence mutants of Saccharomyces cerevisiae with a defect in telomere replication identify three additional EST genes. *Genetics*, **144**(4), 1399–412.

Lingner, J., Promisel Cooper, J., & Cech, T. (1995). Telomerase and DNA end replication: no longer a lagging strand problem? *Science*, **269**, 1533–4.

Liu, Y., Kha, H., Ungrin, M., Robinson, M., & Harrington, L. (2002). Preferential maintenance of critically short telomeres in mammalian cells heterozygous for mTert. *PNAS*, **99**(6), 3597–602.

Lovejoy, C., Li, W., Reisenweber, S., et al. (2012). Loss of ATRX, genome instability, and an altered DNA damage response are hallmarks of the alternative lengthening of telomeres pathway. *PLoS Genetics*, **8**(7), e1002772.

Marcand, S., Gilson, E., & Shore, D. (1997). A protein-counting mechanism for telomere length regulation in yeast. *Science*, **275**(5302), 986–90.

McClintock, B. (1952). Chromosome organization and genetic expression. *Cold Spring Harb Symp Quant Biol*, **16**, 13–47.

Muller, H. (1938). The remaking of chromosomes. *Collecting Net-Woods Hole*, **13**, 181–98.

Nakamura, T., Morin, G., Chapman, K., et al. (1997). Telomerase catalytic subunit homologs from fission yeast and human. *Science*, **277**(5328), 955–9.

Nandakumar, J., Bell, C., Weidenfeld, I., Zaug, A., Leinwand, L., & Cech, T. (2012). The TEL patch of telomere protein TPP1 mediates telomerase recruitment and processivity. *Nature Letters*, **492**, 285–9.

Olovnikov, A. M. (1996). Telomeres, telomerase, and aging: origin of the theory. *Experimental Gerontology*, **31**(4), 443–8.

Palm, W. & de Lange, T. (2008). How shelterin protects mammalian telomeres. *Ann Rev Genet*, **42**(301), 301–34.

Pascolo, E., Wenz, C., Joachim, L., et al. (2002). Mechanism of human telomerase inhibition by BIBR1532, a synthetic, non-nucleosidic drug candidate. *J Biol Chem*, **277**, 15566–72.

Ramamoorthy, M. & Smith, S. (2015). Loss of ATRX suppresses resolution of telomere cohesion to control recombination in ALT cancer cells. *Cancer Cell*, **28**, 57–369.

Robles-Espinoza, C., Harland, M., Ramsay, A., et al. (2014). POT1 loss-of-function variants predispose to familial melanoma. *Nat Genet*, **46**(5), 478–81.

Roth, A., Harley, C., & Baerlocher, G. (2009). Imetelstat (GRN163L)—telomerase-based cancer therapy. *Small Molecules in Oncology*, **184**, 221–34.

Sarthy, J., Bae, N., Scrafford, J., & Baumann, P. (2009). Human RAP1 inhibits non-homologous end joining at telomeres. *EMBO J*, **28**(21), 3390–9.

Schmidt, J. & Cech, T. (2015). Human telomerase: biogenesis, trafficking, recruitment, and activation. *Genes Dev*, **29**, 1095–105.

Sfeir, A., Kosiyatrakul, S., Hockemeyer, D., et al. (2009). Mammalian telomeres resemble fragile sites and require TRF1 for efficient replication. *Cell*, **138**(1), 90–103.

Shay, J. W. & Bacchetti, S. (1997). A survey of telomerase activity in human cancer. *Eur J Cancer*, **33**(5), 787–91.

Shay, J. & Wright, W. (2011). Role of telomeres and telomerase in cancer. *Semin Cancer Biol*, **21**(6), 349–53.

Shi, J., Yang, X., Ballew, B., et al. (2014). Rare missense variants in POT1 predispose to familial cutaneous malignant melanoma. *Nat Genet*, **46**(5), 482–6.

Singer, M. & Gottschling, D. (1994). TLC1: template RNA component of Saccharomyces cerevisiae telomerase. *Science*, **266**(5184), 404–9.

Steczkiewicz, K., Zimmermann, M., Kurcinski, M., et al. (2011). Human telomerase model shows the role of the TEN domain in advancing the double helix for the next polymerization step. *PNAS*, **108**(23), 9443–8.

Stern, J., Theodorescu, D., Vogelstein, B., Papadopoulos, N., & Cech, T. (2015). Mutation of the TERT promoter, switch to active chromatin, and monoallelic TERT expression in multiple cancers. *Genes Dev*, **29**(21), 2219–24.

Szostak, J. & Blackburn, E. (1982). Cloning yeast telomeres on linear plasmid vectors. *Cell*, **29**(1), 245–55.

Teixeira, M., Arneric, M., Sperisen, P., & Lingner, J. (2004). Telomere length homeostasis is achieved via a switch between telomerase—extendible and nonextendible states. *Cell*, **117**(3), 323–5.

Ulaner, G., Huang, H., Otero, J., et al. (2003). Absence of a telomere maintenance mechanism as a favorable prognostic factor in patients with osteosarcoma. *Cancer Research*, **15**(63), 1759–63.

van Steensel, B., Smogorzewska, A., & de Lange, T. (1998). TRF2 protects human telomeres from end-to-end fusions. *Cell*, **92**(3), 401–13.

Venteicher, A., Abreu, E., Meng, Z., et al. (2009). A human telomerase holoenzyme protein required for Cajal body localization and telomere synthesis. *Science*, **323**(5914), 644–8.

Vinagre, J., Almeida, A., Populo, H., et al. (2013). Frequency of TERT promoter mutations in human cancers. *Nature Communications*, **4**(2185), 10.1038/ncomms3185.

Wang, F., Podell, E., Zaug, A., et al. (2007). The POT1-TPP1 telomere complex is a telomerase processivity factor. *Nature*, **445**, 506–10.

Weismann, A. (1891). *Essays Upon Heredity and Kindred Biological Problems*, 2nd edition. Oxford: Clarenden Press.

Wu, K., Grandori, C., Amacker, M., et al. (1999). Direct activation of TERT transcription by c-MYC. *Nat Genet*, **21**, 220–4.

Xi, L., Schmidt, J., Zaug, A., Ascarrunz, D., & Cech, T. (2015). A novel two-step genome editing strategy with CRISPR-Cas9 provides new insights into telomerase action and TERT gene expression. *Genome Biol*, **16**(231), doi: 10.1186/s13059-015-0791-1.

Xu, L. & Blackburn, E. H. (2007). Human cancer cells harbor t-stumps, a distinct class of extremely short telomeres. *Mol Cell*, **28**(2), 315–27.

Zhao, Y., Sfeir, A., Zou, Y., et al. (2009). Telomere extension occurs at most chromosome ends and is uncoupled from fill-in in human cancer cells. *Cell*, **138**(3), 463–75.

Zhong, F., Batista, L., Freund, A., Pech, M., Venteicher, A., & Artandi, S. (2012). TPP1 OB-fold domain controls telomere maintenance by recruiting telomerase to chromosome ends. *Cell*, **150**, 481–94.

16

Cancer metabolism

Almut Schulze, Karim Bensaad, and Adrian L. Harris

Introduction

The primary purpose of metabolic reprogramming in cancer is the generation of metabolic intermediates for macromolecule biosynthesis, which is a prerequisite of cell growth and proliferation. The metabolic pathway that has received the most attention in the context of cancer is glycolysis. Early work by Otto Warburg established that, in contrast to non-malignant tissues, tumours generate large amounts of lactate despite the ample presence of oxygen (Warburg et al., 1924). This phenotype, now generally known as the Warburg effect, has dominated much of the research into metabolic alterations in cancer. However, work over the past 15 years has uncovered a much more complex picture of the different strategies by which cancer cells modify their metabolism. This revealed that oncogenic signalling pathways enhance metabolic processes required for biomass production. Moreover, metabolic reprogramming allows cancer cells to thrive under conditions of nutrient and oxygen limitation, which is frequently found in rapidly expanding tumours due to insufficient vascularization. In this chapter, we discuss some of the major principles of metabolic reprogramming in cancer and their role in the adaptation of cancer cells to oxygen deprivation. We also highlight some of the recent advances in targeting metabolic processes for cancer therapy. However, most cancers also retain Krebs cycle function, an important source of efficient ATP generation and for interconversion of metabolites, anaplerosis (Suchorolski et al., 2013; Potter et al., 2016).

Core metabolic pathways that contribute to macromolecule biosynthesis

These core metabolic pathways provide substrates for the generation of macromolecules, including proteins, nucleic acids, and membrane lipids, required for cell growth and proliferation (**Fig. 16.1**).

Glycolysis

Glucose is a major nutrient in cancer cells. In order to be retained in the cell, glucose needs to be converted to glucose-6-phosphate through the action of hexokinase. Among the four isoforms, hexokinases 1 (HK1) and 2 (HK2) hold a special place as they can translocate to the mitochondrial membrane where they use newly synthesized ATP to catalyse this reaction. Mitochondrial translocation of hexokinase is essential to prevent apoptosis in cancer cells (Gottlob et al., 2001) and genetic deletion of HK2 blocks tumour initiation and progression in cancer models, including K-Ras-dependent adenocarcinoma of the lung (Patra et al., 2013).

In addition to entering the glycolytic cascade, glucose-6-phosphate can also be committed to glycogen synthesis or enter the pentose phosphate pathway (PPP). Therefore, the first rate-limiting reaction of glycolysis is the phosphorylation of fructose-6-phosphate to fructose-1,6-bisphosphate by phosphofructokinase 1 (PFK1), an enzyme that is tightly controlled by several allosteric regulators. High levels of citrate, the substrate for lipid biosynthesis, or ATP, an indicator of cellular energy level, reduce the activity of PFK1. In contrast, fructose-2,6-bisphosphate, the product of cellular phosphofructokinase 2 activity (PFK2), is a strong allosteric activator of PFK1. The crystal structure of mammalian PFK1 confirmed that the protein represents a tetrameric complex, in which the subunits bind fructose-6-phosphate in a cooperative manner (Webb et al., 2015).

The activity of PFK1 is often increased in cancer cells, most likely due to increased expression and activation of 6-phosphofructokinase/fructose-2,6-bisphosphatases (PFKFBs). These bifunctional enzymes control the cellular amounts of fructose-2,6-bisphosphate available for allosteric activation of PFK1 (Yalcin et al., 2009). In many cancer cell types, expression of PFKFB3, one of the four mammalian PFKFB isoforms, increases glycolytic flux (Clem et al., 2013). However, there is increasing evidence that unrestricted activity of PFK1 can limit the availability of metabolites for other biosynthetic pathways, most importantly the PPP, which is essential for the regeneration of NADPH and the synthesis of riboses for nucleotide biosynthesis (Ros and Schulze, 2013). Indeed, it has been reported that glycosylation of PFK1 on serine 529 in response to hypoxia reduces the activity of the enzyme, thereby allowing the redirection of metabolites through the PPP (Yi et al., 2012). Inhibition of glycosylation increases PFK1 activity severely restricting cancer cell survival and tumour growth (Yi et al., 2012). Similarly, PFKFB4, a PFKFB isoforms with lower relative PFK2 activity, was found to be essential for maintaining antioxidant biosynthesis and survival in prostate cancer cells both *in vitro* and *in vivo* (Ros et al., 2012).

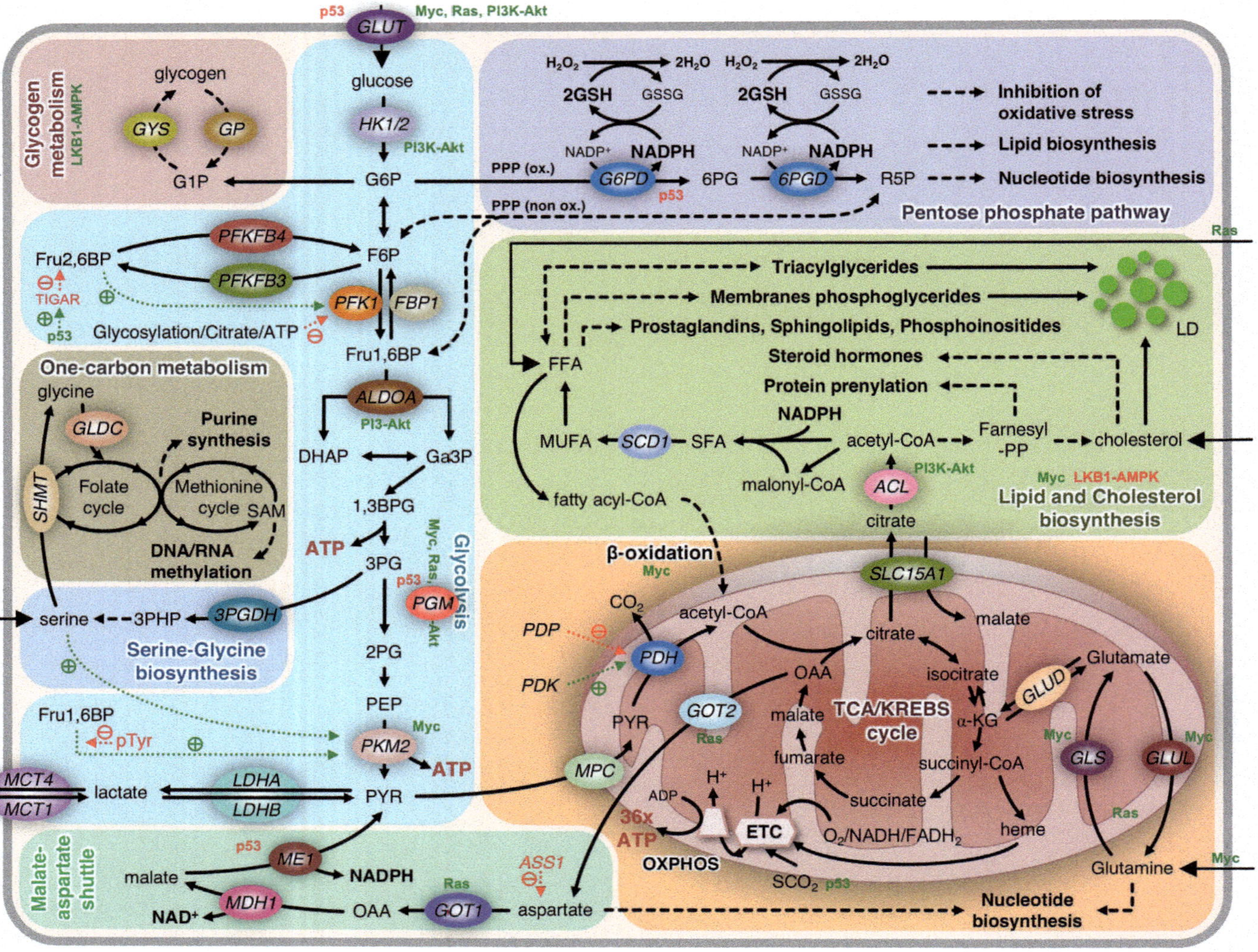

Fig. 16.1 Overview of cancer metabolism pathways and their regulation by oncogenes and tumour suppressors. Key metabolic pathways—glycolysis, glycogen metabolism, pentose phosphate pathway, one-carbon metabolism, serine-glycine biosynthesis, TCA/KREBS cycle, malate-aspartate shuttle, and lipid and cholesterol biosynthesis—are represented within coloured boxes. The main enzymes involved in these metabolic pathways are shown within coloured oval shaped boxes. Metabolic enzymes are regulated by oncogenes—MYC, RAS, and PI3K/AKT (phosphatidylinositol-3-kinase–AKT) pathways—and tumour suppressors—p53 and LKB1-AMPK (liver kinase B1 5′—adenosine monophosphate-activated protein kinase) pathways. Oncogenes and tumour suppressors are indicated near their targets when regulating proteins or metabolic pathways—in green for a positive regulation and in red for a negative regulation. A circled red plus or a circled blue minus represents a positive or negative regulation, respectively. Dashed arrows represent multiple reaction pathways. 1,3BPG, 1,3-bisphosphoglyceric acid; 2PG, 2-phosphoglyceric acid; 3PG, 3-phosphoglyceric acid; 3PGDH, 3-phosphoglycerate dehydrogenase; 3PHP, 3-phosphohydropyruvate; 6PG, 6-phosphogluconate; 6PGD, 6-phosphogluconate dehydrogenase; α-KG, α-ketoglutarate; ACL, ATP citrate lyase; ALDOA, aldolase A; ATP, adenosine-5′-triphosphate; CoA, coenzyme A; ASS1, argininosuccinate synthetase 1; DHAP, dihydroacetone phosphate; ETC, electron transport chain; FASN, fatty acid synthase; $FADH_2$, flavin adenine dinucleotide; FBP1, fructose-1, 6-bisphosphatase 1; FFA, free fatty acid; Fru1, 6BP fructose-1,6-bisphosphate; Fru2, 6BP fructose-2,6-bisphosphate; F6P, fructose-6-phosphate; Ga3P, glyceraldehyde 3-phosphate; GLDC, glycine decarboxylase; GLS, glutaminase; GLUL, glutamate-ammonia ligase; GSH, glutathione; GSSG, glutathione disulphide; G1P, glucose-1-phosphate; G6P, glucose-6-phosphate; G6PD, G6P dehydrogenase; GLUT, glucose transporter; GOT, glutamic-oxaloacetic transaminase; GP, glycogen phosphorylase; GYS, glycogen synthase; HK, hexokinase; LDH, lactate dehydrogenase; MCT, monocarboxylate transporters; MDH1, malate dehydrogenase 1; ME1, malic enzyme 1; MPC, mitochondrial pyruvate carrier; MUFA, monounsaturated fatty acid; NADH, nicotinamide adenine dinucleotide; NADPH, nicotinamide adenine dinucleotide phosphate; OAA, oxaloacetate; OXPHOS, oxidative phosphorylation; PDH, pyruvate dehydrogenase; PDK, pyruvate dehydrogenase kinase; PDP, pyruvate dehydrogenase phosphatase; PEP, phosphoenolpyruvate; PFK1, phosphofructokinase 1; PFKFB, 6-phosphofructo-2-kinase/fructose-2,6-bisphosphatase; PGM, phosphoglycerate mutase; PKM2, pyruvate kinase M2; PL, phospholipids; pTyr, phosphotyrosine; PYR, pyruvate; R5P, ribose-5-phosphate; SHMT, serine hydromethyltransferase; SFA, saturated fatty acid; SCD1, stearoyl-CoA desaturase 1; SCO_2, synthesis of cytochrome C oxidase 2; TIGAR, TP53-inducible glycolytic and apoptotic regulator; TCA, tricarboxylic acid.

Finally, some of the reported somatic mutations in the PFKP isoform are associated with reduced enzyme activity, indicating that limiting the flux through this metabolic node can provide cancer cells with a growth advantage (Webb et al., 2015).

The next step in glycolysis is the cleavage of fructose-2,6-bisphosphate into glyceraldehyde-3-phosphate (Ga3P) and dihydroxyacetone phosphate (DHAP) by aldolase. Increased expression of aldolase A (ALDOA) is part of the transcriptional response mediated by the hypoxia-inducible factors HIF-1α (HIF1A) and HIF-2α (HIF2A; see Semenza et al., 1996), to increase glycolysis in hypoxic cells. More recently, it was shown that ALDOA is activated downstream of the phosphoinositide 3-kinase (PI3K) pathway in cancer cells. This was mediated by remodelling of the actin cytoskeleton, which releases the active enzyme (Hu et al., 2016). Furthermore, treatment of mice with PI3K inhibitors caused a reduction in the generation of several glycolytic intermediates, including DHAP, in the tumour tissue (Hu et al., 2016). As one of the products of aldolase can also be used as a substrate for ribose synthesis via the non-oxidative PPP, controlling the availability of these metabolites could be important to support nucleotide biosynthesis during proliferation and DNA damage repair.

Another glycolytic enzyme that has received substantial attention in the context of metabolic reprogramming in cancer is pyruvate kinase. This enzyme catalyses the conversion of phosphoenolpyruvate (PEP) to pyruvate (PYR) in a reaction yielding ATP. The M2 splice variant of the muscle isoform of pyruvate kinase is highly expressed in cancer cells and other highly proliferative tissues (Mazurek, 2011). The inclusion of exon 10, which is specific for PKM2, is controlled by the heterogeneous ribonucleoprotein (hnRNP) proteins PTB, hnRNPA1, and hnRNPA2 (David et al., 2010). As expression of these hnRNPs is controlled by the proto-oncogene c-MYC (David et al., 2010), many cancer cells express primarily the M2 isoform. In contrast to the M1 isoform, which always forms a highly active tetramer, PKM2 can also switch to a dimeric form with lower activity (Mazurek, 2011). This allows the rerouting of glycolytic metabolites to biosynthetic pathways, including phospholipid, serine, and nucleotide synthesis (Chaneton and Gottlieb, 2012). Moreover, the M1 and M2 isoforms differ in their mode of regulation. PKM2 binds to tyrosine phosphorylated polypeptides, which causes the release of the allosteric activator fructose-1,6-bisphosphate (Christofk et al., 2008).

Another allosteric regulator of PKM2 is serine, which binds to and activates the enzyme (Chaneton et al., 2012). This ensures that PKM2 activity is blocked when cellular serine levels fall below a certain threshold and allows the fine-tuning of biosynthetic processes with the energy demands of cancer cells.

The complex role of PKM2 in the regulation of glycolysis in cancer cells was further complicated by the observation that selective deletion of exon 10 in a genetic model of breast cancer resulted in increased rather than decreased tumour growth (Israelsen et al., 2013). However, this was caused by a genetic selection for either complete loss of both isoforms of variable expression of the M1 isoform, both of which are beneficial for tumour growth. Residual expression of PKM1 was restricted to tumour areas that showed low levels of proliferation, confirming that PKM2 promotes anabolic growth. These results also suggest that the ability to switch between high and low activity provides cancer cells with the ability to adapt to the conditions of the microenvironment and that being locked in either one of these states may limit cancer growth.

The tricarboxylic acid cycle

In contrast to historically held views that cancer cells become independent of mitochondrial metabolism, it is now clear that the tricarboxylic acid (TCA) cycle represents an important hub for biosynthetic processes in cancer cells. Pyruvate is transported into the mitochondrial matrix via the recently identified mitochondrial pyruvate carriers (MPCs; see Bricker et al., 2012; Herzig et al., 2012). In the matrix, pyruvate is cleaved by pyruvate dehydrogenase (PDH) to form acetyl-CoA and CO_2. The activity of PDH, a very large multiprotein complex with up to 200 subunits, is controlled by pyruvate dehydrogenase kinases (PDKs, also known as PDHKs) and phosphatases (PDPs), which regulate the activity of the E1 catalytic subunit. Interestingly, several PDK isoforms are induced in response to hypoxia through the action of the hypoxia-inducible factor (HIF). This inhibits the activity of PDH and blocks the use of pyruvate for oxidative metabolism (Kim et al., 2006). Under these conditions, pyruvate is converted to lactate by lactate dehydrogenases (LDH) and secreted from the cell. However, as oncogenic activation of some signalling pathways leads to the activation of HIF under normoxic conditions, inhibition induction of PDHKs and inhibition of PDH results in a true 'Warburg'-type metabolism. Interestingly, targeting this mode of regulation can lead to toxicity in cancer cells as reactivation of PDH via the inhibition of PDK activity by dichloroacetate (DCA) causes apoptosis in some cancer cells (Bonnet et al., 2007).

One important function of the TCA cycle is the production of citrate, one of the precursors for the production of cytoplasmic acetyl-CoA. Citrate is shuttled out of the mitochondria via the citrate transport protein (SLC15A1), an antiporter for the exchange of mitochondrial citrate for cytoplasmic malate (Catalina-Rodriguez et al., 2012). Another important intermediate of the TCA cycle that contributes to biosynthetic pathways is α-ketoglutarate (α-KG). This metabolite is synthesized from glutamate during anaplerosis, the replenishing of TCA cycle intermediates to compensate for the efflux of metabolites into anabolic metabolism. However, α-KG can also be used to generate glutamate through transamination reactions. Glutamate can also be converted to glutamine via glutamate-ammonia ligase (GLUL), also termed glutamine synthetase. As glutamine also functions as a nitrogen donor for purine and pyrimidine synthesis, controlling glutamine levels can be important for cancer cells, particularly under conditions of low glutamine availability. Expression of GLUL is controlled by the MYC oncogene, through a mechanism involving promoter demethylation and depletion of the enzyme limited the ability of cancer cells to form tumours (Bott et al., 2015). Another TCA cycle intermediate, succinyl-CoA, can be used for the synthesis of heme groups, important electron carriers that function as prosthetic groups of cytochrome c, or complexes III and IV of the respiratory chain. Finally, oxaloacetate can be converted to aspartate by the mitochondrial glutamic-oxaloacetic transaminase 2 (GOT2). Aspartate is an important nitrogen and carbon source for purine and pyrimidine synthesis. Some cancer cells ensure sufficient aspartate supply for nucleotide biosynthesis by downregulating ASS1, the enzyme that normally feeds aspartate into the urea cycle for degradation (Rabinovich et al., 2015). In addition, aspartate participates in the malate/aspartate shuttle, which is important for the regeneration of cytoplasmic NAD+. Here, aspartate is transported out of the mitochondria, converted back to oxaloacetate by glutamic-oxaloacetic transaminase 1 (GOT1) and subsequently converted to malate by malate dehydrogenase 1 (MDH1). This reaction can also drive the regeneration of NADPH in cancer cells by allowing the synthesis of pyruvate from malate by cytoplasmic malic enzyme (ME1). Indeed, expression of oncogenic RAS in pancreatic cancer cells leads to the reprogramming of aspartate and glutamine metabolism via GOT1 and GOT2 to facilitate NADPH production and antioxidant synthesis (Son et al., 2013).

Lipid metabolism

Another biosynthetic pathway closely connected to cell growth is fatty acid and cholesterol synthesis. Fatty acids are essential intermediates for the generation of phosphoglycerides, the main building blocks of biological membranes. Together with cholesterol, they form the structural backbone of the lipid bilayer that forms the plasma membrane or surrounds membranous organelles. Specific lipid species can play particular roles in membrane homeostasis, for examples in the formation of cholesterol-rich lipid rafts, which also control cellular signalling events. Fatty acids can be converted to triacylglycerides, which are important for energy storage and which may contribute to cell survival under nutrient limited conditions. Lipids also contribute to multiple cellular processes that can affect the transformed phenotype of cancer cells. Fatty acids are used as substrates for the modification of proteins, for example, members of the WNT family of secreted signalling proteins (Nile and Hannoush, 2016). Fatty acids are also precursors for the synthesis of lipid mediators, small molecules with signalling function, such as prostaglandins, sphingolipids, and phosphoinositides (Wymann and Schneiter, 2008). The cholesterol pathway also provides isoprenoids for the prenylation of signalling proteins including RHO and RAS, as well as substrates for the synthesis of steroid hormones. Lipid metabolism therefore participates is cellular functions that can contribute to cell transformation and cancer development. It is therefore not surprising that fatty acid and cholesterol biosynthesis are frequently deregulated in human cancer (Currie et al., 2013; Mullen et al., 2016; Rohrig and Schulze, 2016).

While lipid synthesis provides essential building blocks for cellular structures, lipid degradation is an efficient source of energy. Some cancer cells depend on the update of exogenous lipids as substrates for mitochondrial beta-oxidation (Carracedo et al., 2013). The utilization of fatty acids for generating ATP through the Krebs cycle is highly efficient but requires oxygen. Although it is well recognized that most tumours have substantial zones of hypoxia, oxygen concentration around 1% is still able to maintain the beta-oxidation. Interestingly, exposure of cancer cells to hypoxia induces fatty acid uptake to provide substrates for energy generation during reoxygenation (Bensaad et al., 2014). Moreover, lipid uptake and degradation are important for metastasis formation (Nieman et al., 2011; Pascual et al., 2017).

Lipid synthesis and degradation are intricately linked to oxygen availability, as enzymes that desaturate fatty acids also require oxygen, hence the risk of accumulation of toxic saturated fatty acids arises in hypoxia. Uptake of unsaturated fatty acids from the plasma can offset this, indicating that there could be a complex balance of fatty acid synthesis, uptake, and degradation in different tumour areas, providing major avenues of therapeutic vulnerability (Carracedo et al., 2013).

Genetic changes affecting metabolism

It is now clear that oncogenic signalling pathways that are driven by the genetic chances associated with cell transformation and tumorigenesis intersect with the metabolic network in multiple ways (Fig. 16.1). In many cases, the mechanism of regulation involves modulating the expression of the enzymes involved in a specific metabolic process. However, mechanisms involving differential splicing, post-translational modifications, or direct protein-protein interactions also exist.

RAS—driving aerobic glycolysis and nutrient scavenging

The RAS proto-oncogenes (HRAS, KRAS, and NRAS) are small GTPases that are frequently activated in cancer through mutations that keep the protein in the active, GTP-bound state (Pylayeva-Gupta et al., 2011). Oncogenic RAS promotes glucose uptake and glycolysis (Vizan et al., 2005; Yun et al., 2009), thus increasing the availability of glycolytic intermediates for biosynthetic reactions to fuel rapid growth and proliferation. This is achieved by stimulating the expression of the glucose transporter GLUT1 (Chen et al., 2001) and glycolytic enzymes through activation of the hypoxia-inducible factor-1 (HIF1, discussed in more detail next). RAS also activates glutamine metabolism for anaplerosis and NADPH production, as already discussed earlier (Son et al., 2013). Remarkably, RAS-transformed cells have the ability to obtain amino acids through the uptake of whole proteins from their surroundings through a process called macropinocytosis (Commisso et al., 2013). In some cancers, in particular pancreatic cancers in which KRAS is activated by mutation in up to 95% of cases, macropinocytosis may be a major nutrient source of the cancer cells (Bryant et al., 2014). Similarly, RAS-transformed cells exhibit higher rates of lipid uptake, in particular lysophospholipids containing monounsaturated fatty acids (Kamphorst et al., 2013), suggesting that RAS drives a general programme of metabolic 'scavenging' rather than relying solely on the cell-intrinsic biosynthetic capacity.

Activation of RAS also increases the dependency of cancer cells on autophagy, a process of regulated degradation of cellular components that generates nutrients during starvation (Wilkinson and Ryan, 2010). Genetic deletion of essential components of the autophagy machinery blocks tumour development in a genetic model of pancreatic cancer (Yang et al., 2011). However, the exact role of autophagy during RAS-driven tumorigenesis remains still unclear, as the tumour-reducing effect of autophagy deletion in pancreatic cancer models was dependent on the presence of TP53 (Rosenfeldt et al., 2013).

MYC—fuelling cell proliferation

The MYC proto-oncogene also induces substantial rewiring of the cellular metabolic network to drive the growth and proliferation of cancer cells. MYC is a helix-loop-helix transcription factor that binds target sequences (E-boxes and E-box-like sequences) as a heterodimer with its partner MAX to activate the expression of genes involved in multiple cellular processes (Eilers and Eisenman, 2008). Many cancers express high levels of MYC, mainly through transcriptional activation, gene amplification, or aberrant stabilization of the protein (Dang, 2012).

In non-transformed cells, MYC drives the expression of genes involved in ribosome biogenesis and mitochondrial biogenesis (Boon et al., 2001; Arabi et al., 2005; Li et al., 2005), thus ensuring efficient protein synthesis and providing cells with the capacity to generate energy from nutrients. In cancer cells, MYC activates glucose metabolism by inducing the expression of glucose transporters and glycolytic enzymes (Osthus et al., 2000). MYC also activates the expression of the M2 isoform of pyruvate kinase through induction of three heterogeneous nuclear ribonucleoprotein (hnRNP) proteins, which bind repressively to sequences flanking exon 9 replaced by exon 10 (David et al., 2010). Expression of PKM2 limits pyruvate kinase activity and allows diversion of glycolytic intermediates into biosynthetic reactions.

In addition to enhanced glucose metabolism, MYC increases the uptake and degradation of glutamine in cancer cells. Glutaminase (GLS), the enzyme involved in mitochondrial glutaminolysis, is induced by MYC (Wise et al., 2008) through a mechanism involving suppression of the miR-23a/b microRNA (Gao et al., 2009). MYC-transformed cells are also highly dependent on glutamine availability, a phenomenon known as glutamine addiction. Metabolite analysis revealed that glutamine is mainly used to produce TCA cycle intermediates in MYC transformed cells (Yuneva et al., 2007). Unsurprisingly, targeting glutamine metabolism has been suggested as a treatment strategy for MYC-driven cancers (Hsieh et al., 2015).

Another interesting connection between the MYC proto-oncogene and cellular metabolism derives from the fact that many enzymes involved in fatty acid and cholesterol metabolism are regulated by the sterol regulatory element binding proteins, SREBP1 and SREBP2. These factors are structurally related to MYC and induce transcription through binding to sterol response elements (SREs) and E-boxes. High levels of MYC could therefore substitute for SREBP function and drive the expression of lipid biosynthesis programmes (Hsieh et al., 2015). Indeed, genetic deletion of MYC reduces the expression of genes involved in fatty acid and cholesterol biosynthesis in fibroblasts (Edmunds et al., 2014) while inhibition of MYC by 10058-F4, a compound that blocks the interaction between MYC and MAX, causes accumulation of lipid droplets in tumour cells as a consequence of mitochondrial dysfunction and inhibition of beta-oxidation (Zirath et al., 2013). Interestingly, it has been shown that cells transformed by N-MYC require the activity of two other members of the MYC superfamily, the related transcription factors MLX and Mondo A (MLXIP), to drive lipogenesis (Carroll et al., 2015). Depletion of Mondo A blocked cholesterol and phospholipid synthesis and halted proliferation and tumour growth of N-MYC transformed cells. Interestingly, cell viability was restored by supplementation of oleic acid, suggesting that this fatty acid is rate-limiting for cell growth and survival of transformed cells (Carroll et al., 2015).

Modulation of metabolic processes by TP53

TP53 tumour suppressor

In contrast to oncogenes that reprogramme cellular metabolism to drive proliferation and transformation, tumour suppressors have the opposite effect. One example is the TP53 tumour suppressor, which interferes with anabolic metabolism at multiple points. TP53 is frequently deleted or mutated in human cancers and its most established function is to induce cell cycle arrest after DNA damage (Vousden and Lane, 2007). However, TP53 also regulates cellular metabolism, mainly by counteracting anabolic processes, including glucose metabolism, while maintaining oxidative metabolism and mitochondrial energy production (Maddocks and Vousden, 2011). TP53 directly counteracts many processes that are induced by oncogenes. For example, TP53 inhibits glucose uptake by blocking the nuclear factor kappa B (NF-αB)-dependent expression of GLUT3 (Kawauchi et al., 2008). At the same time, TP53 enhances mitochondrial respiration by maintaining expression of cytochrome c oxidase 2 (SCO2) (Matoba et al., 2006), an assembly factor for cytochrome c oxidase, which functions as complex V of the electron transfer and transfers electrons to molecular oxygen.

The pentose phosphate pathway (PPP)

Another metabolic process that is regulated by TP53 is the pentose phosphate pathway. However, it seems that the regulation of this process differs between the acute activation of TP53 during DNA damage and the metabolic consequence of loss of TP53 function in cancer. It has been shown that activation of TP53 in response to DNA damage induces the expression of the TP53-induced glycolysis and apoptosis regulator (TIGAR) (Bensaad et al., 2006). This protein has structural similarities to the fructose-2,6-bisphosphatase domain of the PFKFB proteins. Induction of TIGAR reduces glycolysis through allosteric regulation of PFK1, resulting in enhanced routing of glucose-6-phosphate into the PPP for the generation of riboses and NADPH (Bensaad et al., 2006). Somewhat surprisingly, it was also shown that TP53 reduces PPP activity by inhibiting the activity of glucose-6-phosphate dehydrogenase (G6PD; Jiang et al., 2011), the first enzyme within the pathway, or by restricting the expression of phosphoglycerate mutase (PGAM1), leading to accumulation of 3-phosphoglycerate and inhibition of phosphogluconate dehydrogenase (Hitosugi et al., 2012). Moreover, TP53 also limits the expression of malic enzymes 1 and 2 (ME1, ME2), which also contribute to the production of NADPH (Jiang et al., 2013).

TP53 and amino acid metabolism

TP53 also plays an important role in the regulation of amino acid metabolism, in particular serine and cysteine. Cancer cells depend on serine for protein synthesis but also as a precursor for the synthesis of glycine and for the provision of methyl-groups for the folate cycle. As outlined here already, serine can either be taken up from extracellular sources or synthesized from glucose. Interestingly, TP53-deficient cancer cells are highly sensitive to serine deprivation (Maddocks et al., 2013). This seems to be due to their inability to arrest in response to nutrient limitation, which allows wild type cells to switch to serine synthesis (Maddocks et al., 2013). Thus, TP53 deficient cancer cells show a selective metabolic liability that could also be exploited therapeutically.

In addition to its role in the one-carbon metabolism, glycine is also an essential precursor for the synthesis of glutathione, an important antioxidant derived from cysteine, glycine, and glutamate. Reduced glutathione (GSH) is an important antioxidant and essential to counteract oxidative stress. Serine starvation reduces the levels of GSH both in wild type and TP53 deficient cancer cells (Maddocks et al., 2013). However, TP53 limits the availability of cysteine for GSH synthesis by inhibiting the expression of the cystine/glutamate antiporter SLC7A11 (also known as system X_{C-}) (Jiang et al., 2015). Inhibition of cystine uptake sensitizes cells to ferroptosis, an iron-dependent form of non-apoptotic cell death (Dixon et al., 2012). Interestingly, the ability of blocking cystine uptake was retained in an acetylation-deficient mutant of TP53 ($p53^{3KR}$), which was unable to induce cell cycle arrest, senescence, or apoptosis but maintains its functions as a tumour suppressor (Li et al., 2012; Jiang et al., 2015). These results suggest that modulating metabolic activity is an essential tumour-suppressive function.

PI3-kinase/AKT/mTORC1 and AMPK pathways

It should be mentioned that cancer stem cells require all these pathways, that are commonly upregulated or mutated in cancer compared to normal tissues, but the stems cells in normal tissues require similar pathways, suggesting that cancers are generally recapitulating normal cell biology. The PI3K/AKT/mTORC1 pathway is stimulated by the majority of growth factors via tyrosine kinase receptors (Laplante and Sabatini, 2012). Additionally, mTORC1 is activated by amino acids; the mechanisms have recently been uncovered (Chantranupong et al., 2016; Wolfson et al., 2016).

There are several different types of PI3K increasing secondary signalling phosphates which activate AKT (Thorpe et al., 2015). AKT signalling has many downstream effects (Brown and Banerji, 2016). A key step is the inhibition of the role of TSC1 and TSC2 (tuberous sclerosis 1 and 2, mutations of which cause tuberous sclerosis). Inhibition leads the activation of RHEB, which is an activator of mTOR. mTOR itself comprises two complexes, mTORC1 and mTORC2, and produces further downstream regulation of phosphorylation of 4-EBP1, which allows increased translation efficiency and S6-kinase, which activates ribosomal activity and enhances protein synthesis.

PTEN supresses PI3K activity and mutation and other methods of suppression are common in many cancer types. PIK3CA, coding a regulatory subunit P100α of the enzyme, is commonly mutated in a very wide range of tumour types including up to 30% of breast cancer, even more in endometrial cancer, colorectal cancer, and many others. Additionally, there may be gene amplification in other tumour types such as head and neck cancer and cervical cancer. PIK3CB is also commonly mutated and amplification occurs with PIK3CD.

The PI3K/AKT pathway increases glucose metabolism by upregulating glucose transporters and stimulation of hexokinase, together with activation of ATP citrate lyase, which converts citrate into acetyl-CoA. mTORC1 activated by this pathway also increases the expression of HIF 1 through increasing translational efficiency, which then can of course induce all the downstream genes by glycolysis and other pathways. There is a further feedback loop by which mTORC2, when activated, activates AKT by phosphorylation. mTORC1 also regulates the translation of nuclear encoded mitochondrial mRNAs, enhancing mitochondrial function.

So, in total, mutations in PI3K are one of the most common and broad spectrum in distribution. This has of course led to the development of many drugs to target PI3K, but it is important to recognize the aforementioned different isoforms and the need for specific targeting. The currently used inhibitors cause glucose elevation as a result of this specific targeting.

AMP-kinase is activated under low energy conditions, by a high AMP/ATP ratio (77). By directly inhibiting biosynthesis pathways such as lipid, glycogen, and ribosomal biosynthesis but by also blocking proliferation pathways, it causes a G1 arrest. AMP-kinase needs to be phosphorylated by tumour suppressor LKB1 for activity and is thought to therefore mediate the tumour supressing effects of this gene. However, reduction of AMP-kinase completely by deletion or knockout can be detrimental because it prevents the adaptation of cells to adverse environmental stress such as low glucose or nutrients. Thus, we have a situation where upregulation of AMP-kinase can slow tumour growth or suppression, which makes it very difficult to take through into clinical studies. There are other AMP-kinase subunits; some are amplified, while others are mutated, and their roles are currently poorly understood.

mTORC1 is activated by leucine, arginine, and glutamine through Rag GTPase-dependent mechanisms, which helps the anabolic reactions. In the tumour microenvironment, where amino acids may be deficient, mTORC1 activity may be reduced and since mTORC1 supresses the key kinase of autophagy ULK1/2, this may then phosphorylate glycolytic enzymes and enhance autophagy to sustain glycolysis and retain energy metabolism. ULK1 Inhibitors are being developed to block autophagy response and similarly, drugs that block both mTORC1 and mTORC2 to block the feedback loops. The complexity of mTOR signalling with 2-complexes, the multiple PI3-kinase amplifications and complexity AMP-kinase demonstrate how difficult it is to selectively target tumours in a broad approach rather than going for individual pathways and many drugs have not been designed to take into account the recent biochemical findings.

Regulation of HIF by vHL and hypoxia

The von Hippel-Lindau protein (pVHL) is a major target for mutations that generate clear cell renal carcinoma. As a result of the mutations, pVHL is no longer able to interact with HIF1α or HIF2α and modify them by ubiquitination and therefore the two transcription factors are stabilized and no longer degraded by the proteasome. HIF2α particularly contributes to renal cancer and indeed the HIF1α gene may be mutated or inactive. There is extensive metabolic reprogramming in clear cell renal carcinoma, both hereditary and sporadic, in both cases related to the vHL mutation (Robinson and Ohh, 2014). As a result of mutations upregulating the HIF pathway, there are many changes in metabolic pathways, such as glucose transporters GLUT1, GLUT3, an increase in glycolysis and an increased pentose shunt, increased lipid production. In addition, PI3-kinase, Akt, and mTOR are activated.

However, there are many proteins and epigenetic factors that interact with the vHL mutation. Examples are SETD2 and UTX, which are likely to further modify the pathways that regulate renal metabolism although their specific functions are still not fully understood (Liao et al., 2015). Besides direct mutation of vHL, other new mechanisms have been discovered that regulate vHL function and decreases activity, for example, the factor Id2 interacts with vHL, blocks its activity and stabilizes HIF2α as a result. Inhibition of the kinases that block this effect of Id2 may be a further therapeutic approach (Lee et al., 2016).

If HIF2α is a major mechanism stimulating renal cancer growth, particularly through its metabolic effects, then drugs that specifically target it should improve the growth and decrease the aggressiveness of vHL disease, and this indeed has been shown in preclinical models with specific blockade of HIF2α (Metelo et al., 2015).

A more detailed analysis of the genetic changes that occur in renal cancer and other tumour types, showed that genes involved in metabolism may be deleted and affect tumour growth. Such an enzyme is the gluconeogenic enzyme fructose-1, 6-bisphosphatase 1 (FBP1), which is commonly deleted in renal cancer (Li et al., 2014). This is partly because FBP1 reduces glycolytic flux and therefore deletion enhances flux and or the Warburg effect. A major new finding was that glycolytic enzymes can interact with HIF1α and inhibits nuclear

HIF1 function by blocking HIF1 signalling. This is an important emerging area of metabolic biology of the alternative roles of glycolytic enzymes.

Another example is enolase, which may be expressed on the cell surface of myeloid-derived suppressor cells antibodies, can modify the microenvironment and reduce the activity of MDSCs, which have otherwise potent tumours to measure activity and angiogenic effects. So, this may provide a route for linking antimetabolic therapy with immunotherapy (Cappello et al., 2016).

Additional approaches (e.g. proteomics comparing renal cancer with normal kidney tissue) have been able to identify 73 different proteins involved in glycolysis, the urea cycle, arginine, and proline metabolism. Thus, many new targets are likely to be identified from primary tissues, beyond cell line cultures (Lu et al., 2016).

Metabolic oncogenes and oncometabolites

In addition to their deregulation by oncogenic signalling pathways, metabolic enzymes can also have a more causal link to cancer development. This is exemplified by mutations of isocitrate dehydrogenases (IDH1 and IDH2) found at high prevalence in secondary glioblastoma (Parsons et al., 2008) and acute myeloid leukaemia (Mardis et al., 2009). These point mutations lead to structural changes within the catalytic centre of the enzyme and induce a neomorphic activity that catalyses the conversion of α-KG to 2-hydroxyglutarate (2HG); (Ward et al., 2010). The ®-enantiomer of 2HG alters the activity of α-KG-dependent dioxygenases, including DNA- and histone-demethylases as well as proline hydroxylases of the PHD/EGLN family (Yang et al., 2013). The resulting transcriptional and epigenetic changes are linked to the inhibition of differentiation (Sasaki et al., 2012), thus keeping cancer cells in a more stem-cell-like state.

In addition to 2HG, succinate and fumarate also have the potential to induce or promote cell transformation, for example, by inhibiting α-KG-dependent dioxygenases through product inhibition or by causing the succination of KEAP leading to activation of NRF2 (Yang et al., 2013). Similarly, as acetyl-CoA is a substrate for histone acetylation during gene regulation, changes in cellular acetyl-CoA levels can induce global changes to gene expression in cancer cells (Schug et al., 2016). Further research is likely to uncover additional mechanisms by which these and other oncometabolites affect regulatory processes linked to cell transformation and cancer development.

Specific cell types in cancer

Tumours are complex structures that contain, apart from the transformed cancer cells, numerous stromal cell types that undergo complex heterotypic interactions. These cell types have specific metabolic requirements and can participate in the local exchange of metabolites.

Cooperation between subpopulations

The role of stromal cells interacting with cancer cells and feedback between them is of critical importance in metabolism. The cancer-associated fibroblasts, for example may be important for absorbing extracellular acid because they are connected by gap junctions that allow cell-to-cell coupling and provide an acid excretion route for tumours (Hulikova et al., 2016).

Another role of tumour stromal cells has been demonstrated in pancreatic ductal adenocarcinoma where the stroma response has a major effect on drug penetration and growth, also impairing vasculature, producing a highly hypoxic tumour with poor nutrient delivery. The stellate cells undergo autophagy which is stimulated by the cancer cells and in turn releases alanine to be used in cancer cells and as a major substrate to fuel the TCA cycle (Sousa et al., 2016). This contrasts with other cell types where glucose and glutamine are more important.

Major mechanisms of symbiosis occur between cancer cells in normoxia and hypoxia; for example, the lactate released from hypoxia cells can be metabolized by lactate dehydrogenase B (Brisson et al., 2016) to convert to pyruvate and then enter the Krebs cycle in better oxygenated cells. This also results in glycolysis being less utilized by the well-oxygenated cells because of the effectiveness of the Krebs cycle, allowing more glucose to pass through to the hypoxic cells and to be metabolized by glycolysis (Sonveaux et al., 2008). The potential exchange of lactate between cancer and stromal cells has also been reported through a mechanism termed reverse Warburg effect (Lisanti et al., 2013; Fig. 16.2).

With regard to compartmentalization and lactate production, it should be remembered that lactate itself has major roles apart from being a substrate, for example, it inhibits the oxygen sensor prolyl hydroxylase 2 (PHD2); (Lee et al., 2015), thus resulting in activation of HIF1α and activates receptor tyrosine kinases such as VEGFR2, a ligand-independent mechanism. This compartmental organization can occur after antiangiogenic therapy resulting in the well-oxygenated tumour cells around remaining blood vessels supporting the growth of the more hypoxic cells and vice versa. Thus, combining inhibitions of antiangiogenesis with antimetabolism therapy maybe synergistic in the future. Similarly, the induction of different populations of cells in hypoxia expressing carbonic anhydrase 9 has been reported, with the former having more stem cell characteristics (Ledaki et al., 2015). Therefore, metabolomic heterogeneity within a tumour and even at a few cell diameters distance between cells further emphasizes the problems of targeting metabolism and the need to have combined targeting.

Endothelial cell metabolism

Each tissue and cancer will also have metabolic differences depending on the normal functions, and recently the role of endothelial metabolism has been highlighted as a therapeutic target and as having major differences from tumour cells. Endothelial cells are essentially glycolytic in normal tissues and although surprising considering they are next to oxygenated red cells, this may allow oxygen to penetrate through endothelial cells without being consumed and also primes endothelial cells to be able to respond rapidly and survive hypoxic insult and vascular damage (Harjes et al., 2016; Treps et al., 2016).

The key regulator 6-phosphofructo-2-kinase/fructose-2,6-bisphosphatase3 (PFKB3) regulates angiogenesis and controls the balance of tip versus stalk cells (De bock et al., 2013). Other endothelial cells become tip cells involved in invasion and forming of new blood vessels, as there is more glycolysis required for the sprouting. Interestingly, blockade of PFKB3 by about 50% reduces glycolysis

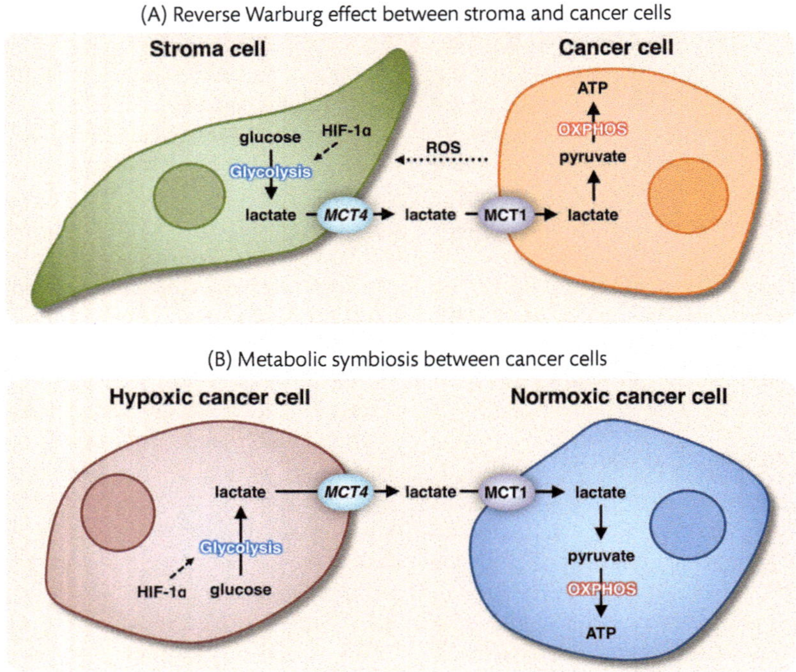

Fig. 16.2 Lactate shuttling in cancer: reverse Warburg effect and metabolic symbiosis. (A). In the reverse Warburg effect, oxidative cancer cells through ROS (reactive oxygen species) production and secretion lead to the activation of HIF-1 alpha (hypoxia-inducible factor 1 alpha) in adjacent stromal cells. This leads to an increased aerobic glycolytic flux in stromal cells and production of lactate. The lactate is then exported by stromal cells through MCT4 (monocarboxylate transporter 4) and imported by tumour cells through MCT1 (monocarboxylate transporter 1). Lactate is then converted into pyruvate which is used as a substrate for mitochondrial OXPHOS (oxidative phosphorylation). (B). In metabolic symbiosis, cancer cells in hypoxic areas of the tumour efflux lactate through MCT4. Lactate is then imported by cancer cells in normoxic areas of the tumour through MCT1, lactate is then converted into pyruvate and used by the mitochondria as a fuel for OXPHOS.

and also angiogenesis with minimal systemic cytotoxicity. So, complete inhibition is not necessary to have a clear potential therapeutic effect. In addition, access to blood vessels is likely to be better than access to the main tumour mass so it is an important consideration for antimetabolic therapy.

Oxidation of fatty acids is essential for *de novo* synthesis of nucleotides and this is an unusual aspect of endothelial cell metabolism compared to other cell types and therefore may produce a new targeting opportunity (Schoors et al., 2015).

There are other cell types involved in the stroma. The differences between various cell types including macrophages of M1 and M2 lineage, during T-cell differentiation and in cancer-associated fibroblasts, have been reviewed recently (Ghesquiere et al., 2014), but many of the pathways that are utilized by cancer cells are also important for the rapid proliferation of effector T cells such as increasing glycolysis. Blocking glycolysis will promote Treg cells. The lack of effect or modified effect of blocking metabolism in tumours may be because of detrimental effects on the T-cell population, which is important to consider in view of the recent demonstration of a board-spectrum tumour types that can benefit from specific immunotherapy. Similarly, blockade of glutaminase blocks T effector cells, and mTOR activation is important—this is obvious because rapamycin is an immunosuppressive agent. Many of these possible effects of blocking metabolism on the stromal cells are not considered in the design or potential therapeutic effects, and will not be detected in various types of immunosuppressed mice with xenografts or patient-derived xenografts.

Emerging pathways

In addition to the well-explored mechanisms described earlier, several other metabolic processes relevant in cancer cells have emerged recently.

Gluconeogenesis

Gluconeogenesis requires the enzymes phosphoenolpyruvate carboxykinase (PEPCK) and phosphoenolpyruvate carboxykinase2 (PEPCK2) to reprogramme metabolism. PEPCK2 is the mitochondrial form of the enzyme PEPCK and its activation is enhanced under low glucose conditions. Lactate under these circumstances can be converted to pyruvate. It has also been shown that cytosolic PEPCK (PCK1) in malignant melanoma promoted production of serine and glycerol 3-phosphate (Leithner et al., 2015; Montal et al., 2015; Vincent et al., 2015; Li et al., 2016). Glucose limitation also promoted the production of phosphoenol pyruvate from glutamine, which was used to provide serine and purine biosynthesis precursors. Thus, the effects of gluconeogenesis enzymes may vary from tissue type and cancer cell type, all demonstrating however an impact in low glucose conditions, which are likely to be encountered at a distance from blood vessels in a hypoxic microenvironment (Balsa-Martinez and Puigserver, 2015).

Glycogen metabolism

Another pathway that has attracted considerable interest is glycogen metabolism (**Fig. 16.3**). Gene array analysis for genes

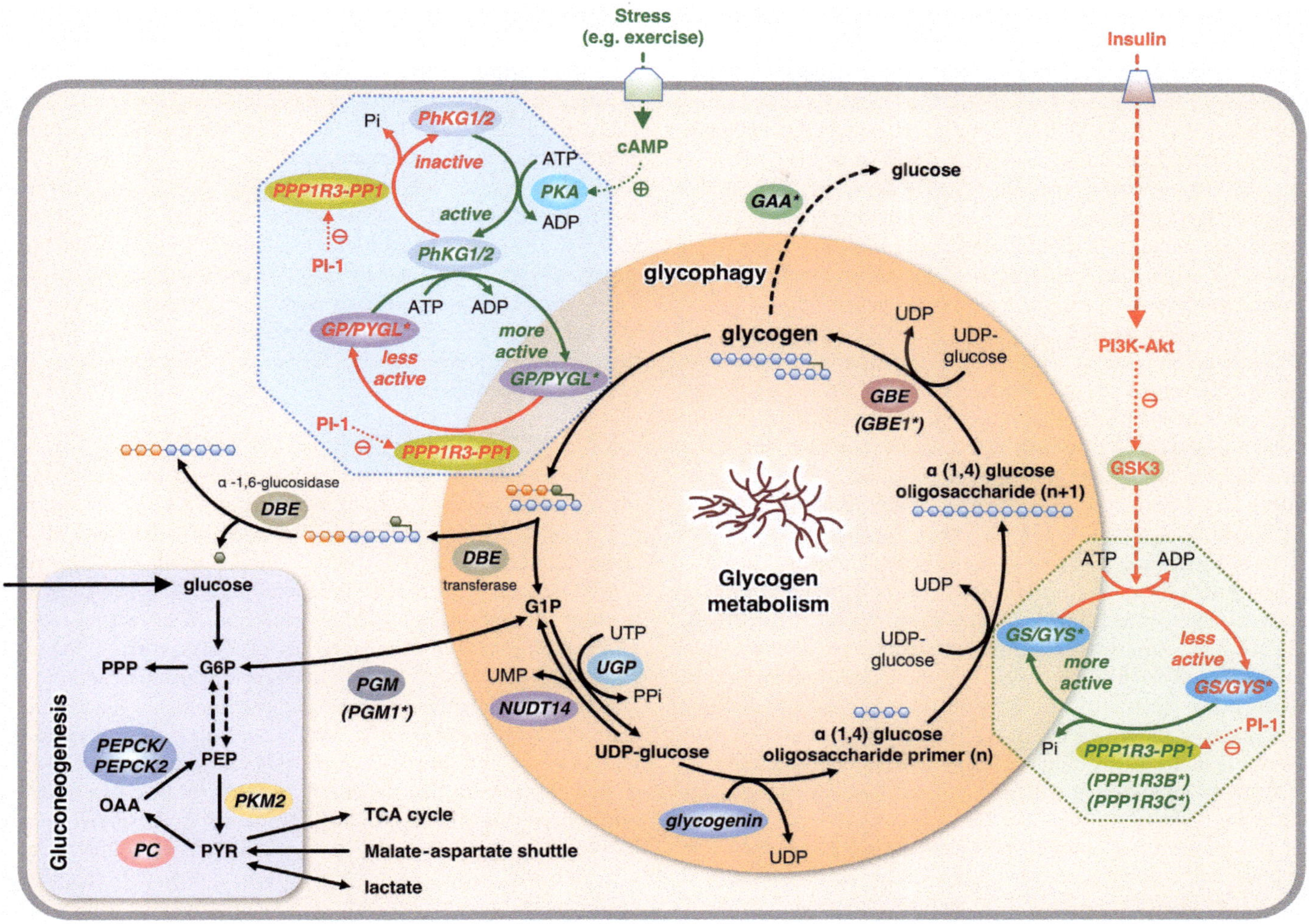

Fig. 16.3 Key pathways of glycogen metabolism and its regulation by glycogen synthase and glycogen phosphorylase. Glycogen synthesis requires formation of UDP-glucose from glucose-1-phosphate (G1P) and Uridine-5′-Phosphate (UTP) via a reaction catalysed by UDP-glucose pyrophosphorylase (UGP). The reverse reaction is catalysed by the UDP-glucose pyrophosphatase NUDT14. Glycogenin initiates the first step of glycogen synthesis by self-glycosylation of a short 8–12 glucose oligosaccharide primer. Glycogen synthase (GS) elongates the α-1,4-linked glucose oligosaccharide primer (n), which utilizes UDP-glucose as the glucosyl donor. Glycogen branching enzyme glucan 1,4-α-branching enzyme 1 (GBE) catalyses the transfer of α-1,4-linked glucose units from the outer 'non-reducing' end of a growing glycogen chain into an α-1,6 position of the same or neighbouring chain. Glycogen degradation requires the synchronous activation of the glycogen phosphorylase (GP) and the bifunctional enzyme glycogen debranching enzyme (DBE). GP catalyses the degradation of glycogen to G1P from the terminal alpha-1,4-glycosidic bond, and DBE catalyses the removal of the branches. DBE transfers three glucose blocks to another glycogen chain, and then hydrolytically cleaves the remaining glucose of the branch, generating free glucose. Phosphoglycerate mutase (PGM) catalyse the reversible interconversion between G1P and G6P. Glycogen-derived G1P and free glucose can enter glycolysis or the pentose phosphate pathway (PPP). Autophagy-dependent glycogen degradation (glycophagy) also produces free glucose via glucosidase alphas acid (GAA) activity. Stress (e.g. exercise) induces cAMP via cognate receptors. cAMP activates protein kinase A (PKA). PKA converts the inactive phosphorylase kinase catalytic subunit gamma 1 and 2 (PhKG1/2) to an active form. The active PhKG1/2 phosphorylates and activates GP. Glycogen targeting subunit (RRR1R3 family proteins) recruits protein phosphatase 1 (PP1) to glycogen to dephosphorylate the more active form of GP and the active form of PhKG1/2, this leads to the formation of less active GP and inactive PhKG1/2. RRR1R3-PP1 also converts the less active glycogen synthase (GS) to more active GS. RRR1R3-PP1 is negatively regulated by Phosphoprotein Phosphatase Inhibitor (PI-1). Glycogen synthase 3 (GSK3) activity leads to the phosphorylation of GS and reduces its activity. Insulin activates the insulin receptor tyrosine kinase, which activates the PI3K-Akt pathway and stimulates glycogen synthesis via inhibition of glycogen synthase 3 (GSK3). Pyruvate kinase isozyme M2 (PKM2) is highly expressed in cancer cells and catalyses the last step within glycolysis; the conversion of phosphoenolpyruvate (PEP) to pyruvate (PYR). PYR can be converted to lactate, can enter the TCA cycle (after being converted to acetyl-CoA) or can enter gluconeogenesis. Pyruvate carboxylase (PC) catalyses the carboxylation of PYR to form oxaloacetate OAA. Gluconeogenesis requires the enzymes phosphoenolpyruvate carboxykinase (PEPCK) and phosphoenolpyruvate carboxykinase2 (PEPCK2) to reprogramme metabolism. PEPCK/PEPCK2 drive the reaction between oxaloacetate (OAA) and PYR. Glycogen metabolism is modulated in hypoxia in tumours and enzymes marked with an asterisk are upregulated in hypoxic conditions. Green text and lines represent activation, while red text and lines represent inactivation. Dashed arrows represent multiple reaction pathways. GYS1, glycogen synthase 1; PYGL, phosphorylase, glycogen, liver; UMP, uridine monophosphate; UTP, uridine triphosphate; UDP, uridine diphosphate; PPi, inorganic pyrophosphate; Pi, inorganic phosphate.

induced by hypoxia showed that most of the enzymes involved in glycogen synthesis are induced under hypoxic conditions. This includes glycogen synthase from muscle (GYS1), UDP2-glucose-1-phosphate uridylyltransferase (UPG2), 1,4-alpha-glycan branching enzyme (GEB1). The PPP1R3B subunit is also induced in hypoxia, as is phosphoglucomutase-1, PGM1, which converts glucose 6-phosphate to glucose 1-phosphate. The latter is preferentially used to maintain the pentose phosphate shunt. Additional enzymes PYGL, glucosidase alpha acid (GAA), and glycogen branching enzyme glucan 1,4-α-branching enzyme 1 (GBE) are upregulated under hypoxic conditions. One would expect in hypoxic conditions, where the glycolysis is increased, that glucose would be maintained in the glycolytic pathway for generation of ATP, rather than accumulate glucose in glycogen stores. This suggests an adaptive response that builds up stores for utilization on either reoxygenation or during prolonged glucose deprivation in hypoxic conditions. This is indeed the case, reviewed in Zois and Harris (2016) and Massari et al. (2016). PYGL is induced 24–48 hours after glycogen synthesis is induced and it correlates with the phase or breakdown of glycogen after a previous phase of build-up in the first 48 hours.

The benefit of using glycogen as an energy store is that it is already phosphorylated and therefore when degraded can release ATP without requiring oxygen or energy sources and generates three ATPs for each glucose molecule as opposed to two via glycolysis. Also, it has a role in signalling and in neurotransmission where very rapid generation of ATP at certain sites is critical. Of interest is that inhibiting the breakdown of glycogen or blocking its synthesis or blocking glucose 6-phosphatase all have marked antitumour effects.

Glycogen synthase and phosphorylase are regulated extensively by phosphorylation and kinase cascades reviewed in Roach et al. (2012). Essentially, phosphorylation mediated via protein kinase A, activated by cAMP, increases the activity of glycogen phosphorylase, whereas glycogen synthase is inactivated by phosphorylation. These interactions are highly complex are reviewed elsewhere.

Because of the role of glycogen metabolism in maintaining glucose levels in diabetes there has been extensive investigation of glycogen phosphorylase inhibitors that either block the active site, an AMP site, or nucleoside sites, which are involved in regulation of glycogen phosphorylase (Parmenopoulou et al., 2014). A further class are dicarboxamide site inhibitors (Gaboriaud-Kolar and Skaltsounis, 2013).

Amino acids

Emerging roles of non-essential amino acids and branched amino acids has shown further controls points and targeting approaches (Geck and Toker, 2016). MYC activation increases proline biosynthesis from glutamine and also increases the enzymes required for synthesis at RNA and protein levels. Blockade of this decreases tumour growth and energy production. There is a metabolic cycle between proline and delta-1-pyrroline-5-carboxylate (P5C). The proline cycle is connected to glycolysis and the oxidative arm of the pentose shunt therefore helping maintain pyridine nucleotide levels and linking proline biosynthesis to glucose, glutamine, and pyridine nucleotide metabolism (Liu et al., 2015). The production of proline can also act as a source of glutamine, for example collagen contains large amounts of proline, which can be metabolized to serve as a reservoir for proline (Palm et al., 2015). This again emphasizes the importance of the stroma, not just in cellular mechanisms but also as an energy reservoir. Similarly, pinocytosis has a major contribution to make under certain stresses such as activation of Ras and uptake of plasma proteins (Commisso et al., 2013; Kamphorst et al., 2015). This pathway requires mTORC for activity and therefore inhibition of TORC1 may have additional affects.

Other key amino acid pathways include glutamine, glutamate, serine, glycine, arginine, and proline metabolism, all of which are shown to be important in different cell types under specific circumstances. Determination of these metabolic differences is difficult and clearly compromises selective therapy. Nevertheless, a new method that has been reported recently (Loayza-Puch et al., 2016) whereby tumour-specific proline vulnerability was uncovered using differential ribosome codon reading.

One-carbon metabolism

Another metabolic pathway that has received increased attention recently is the one-carbon metabolism. This process consists of the coupled folate and methionine cycles and generates highly reactive intermediates that provide methyl-groups for the synthesis of purine nucleotides. The methionine cycle also generates S-adenosylhomocysteine (SAM), which provides methyl-groups for the methylation of histones or DNA (Locasale, 2013). The substrates that provide one-carbon units for this process, serine (via donating a one-carbon unit during conversion to glycine by serine hydroxylmethyltransferase (SHMT)) and glycine (via cleavage into a one-carbon unit, carbon dioxide and ammonia by glycine decarboxylase (GLDC)), can either be taken up by cancer cells or synthesized from the glycolytic intermediate 3-phosphoglycerate (3PG). Phosphoglycerate dehydrogenase, the enzyme that directs 3PG into the serine/glycine biosynthesis pathway, is frequently amplified in human tumours and essential for the growth and proliferation of breast and melanoma cells (Locasale et al., 2011; Possemato et al., 2011). Moreover, enzymes of the serine and glycine metabolism were found to be upregulated in tumour initiating cells from non-small cell lung cancer patients, while overexpression of GLDC drives transformation of primary human bronchial epithelial cells (Zhang et al., 2012). More recently it was demonstrated that serine metabolism supports the synthesis of SAM through *de novo* ATP synthesis, thereby promoting DNA and RNA methylation in cancer cells (Maddocks et al., 2016). Although antifolates were among the first chemotherapeutic agents used, and inhibitors of the pathway, such as methotrexate and pemetrexed, are in clinical use today, novel aspects of the central role of one-carbon metabolism in cancer biology continue to emerge (Mattaini et al., 2016).

Imaging metabolism

Imaging metabolism will be a key development in the future, to help select patients for treatment and monitor early response to therapy. Additionally, it gives a global view of metabolism and heterogeneity in metastasis and primary tumours, which will be likely a major problem in therapeutic response. The most established and oldest technology is PET CT with 18F-FDG, which has been shown in many tumour types to be associated with poor prognosis when elevated (Fig. 16.4).

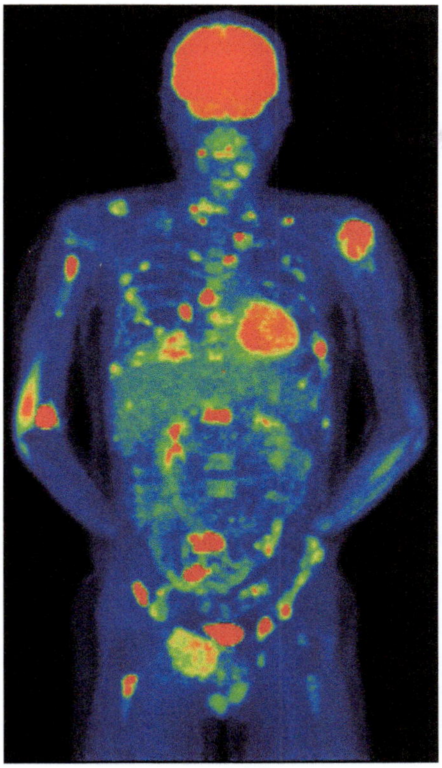

Fig. 16.4 PET ^{18}FDG scan. The patient has extensive metastases from malignant melanoma, showing very high glucose uptake.
Reproduced courtesy of Oxford University Hospitals NHS Foundation Trust.

The mechanism of elevation is through key oncogenes and pathways that target glycolysis and are responsible for the Warburg effect. Alvarez et al. (2014) showed that the oncogene pathway activation dictates FDG-PET uptake. Among these, AKT signalling seems to be predominant, but also many oncogenes such as *HER2*, P53 induce glucose uptake. It is part of routine staging for lung cancer, GIST, and lymphomas, demonstrating the importance of metabolic signatures in addition to conventional methodologies.

A more detailed evaluation correlating PET CT uptake expressed as maximal standardized uptake ratio (SUV Max) or tumour to background ratios, has shown it can distinguish different subtypes of breast cancer, that is, oestrogen-receptor negative, higher proliferation, and triple-receptor negative cancer versus other types (Miyake et al., 2014) or HER expressing gastric cancer (Chen et al., 2016).

Of more importance, however, is the detection of early response to therapies, either specifically targeting the transporters and metabolic pathways or the oncogenes that regulate them. In preclinical studies, it has been shown that response to antibodies against *HER2* can be predicted in one to two days by measuring FDG or labelled thymidine FLT uptake in correlating response three weeks later. Importantly, similar approaches have been applied in clinical studies (e.g. in predicting response to neoadjuvant chemotherapy and trastuzumab in *HER2*-positive breast cancer). The importance of this method in trial design is because it showed that randomization based on poor PET response and using another therapeutic could change the final pathological complete response (Coudert et al., 2014; Nielsen et al., 2015). Similarly, early PET responses at one month to single agent trastuzumab without the chemotherapy could predict for further response and pathways of resistance (Koukourakis et al., 2014).

There are many other imaging agents that are relevant to aspects of targeted therapy (Alam et al., 2015), for example, measuring glutamine or glutamate uptake. One that is now approved is 18F fluciclovine, a leucine analogue, which is also transported via glutamine transporters and provides better imaging of prostate cancer than the currently approved choline agent (Ulaner et al., 2016). Detection of glycogen metabolism (Witney et al., 2014) is under investigation.

Of course, it would be important if one could avoid repeated radiation exposure and additional methods are being investigated, including high-resolution magnetic resonance spectroscopy, which can be correlated with uptake of glucose in PET scanning and with metabolites measured in the tumours, such as choline, glycine and phosphocholine (Yoon et al., 2016).

Another development with MRI contrast is using chemical exchange saturation transfer (CEST) to detect a change of protons with distinct hydroxyl and amino groups in water and this can allow detection of many common metabolites and pH. This has the potential to monitor effects of drugs targeting metabolism (Chan et al., 2016). Methods using 1H MRI can detect lactate peaks in brain tumours, which are particularly suited for this type of work because of the low background (Payne et al., 2015). The ability to measure online by high-resolution magic angle spinning magnetic resonance spectroscopy (HR MAS MRS) on small tissue samples has provided the possibility of directly assessing tumour margins and in particular, metabolites such as phosphocholine, glycine, taurine, and glucose which are indicative of the difference between tumour and normal tissue. Phosphocholine compounds particularly show up on MRS, as has been known for many years, but these new techniques promise to give a much wider range of metabolites (Begley et al., 2012; Bathen et al., 2013).

Hyperpolarized metabolites give many orders of magnitude of increase in sensitivity in measuring tumour metabolism. The hyperpolarized technique has been applied for several metabolites, such as glucose, fumarate, bicarbonate, and pyruvate (Rodrigues et al., 2014; Brindle, 2015).

Thus, the potential for defining metabolic profiles in patients and their metastases before starting therapy, selecting appropriate therapies, and monitoring early response and synergistic activities of adding additional agents is now here. This is a critical approach to developing cancer therapies targeting metabolism and without such patient-specific analysis and a measurement of adaptive responses, it is likely these treatments will fail. The heterogeneity of the basic pathways plus the adaptive responses are clearly extensive and need to be taken into account with changes made early on within a few weeks of therapy.

Drugs targeting metabolism preclinically and in therapeutic trials

There have been many attempts over the past 60 years to target tumour metabolism, notably the inhibition of purine and pyrimidine metabolism by 5-FU and 6-MP, drugs that still have a key role in cancer treatment today. However, this section will concentrate only on targets other than DNA or RNA synthesis. Targeting tumour metabolism has been extensively reviewed (Vander Heiden et al., 2011; Galluzzi et al., 2013; van der Mijn et al., 2016; see also Table 16.1), and only selected pathways, which are currently under most intensive study, are discussed in detail. A common laboratory tool used to evaluate the effects on major phenotypes of metabolism, such as glycolysis, oxygen consumption, and acidification, is the 'Seahorse'—see Figure 16.5

Table 16.1 Drugs targeting metabolism and related pathways currently in clinical development. Selected compounds targeting metabolic processes currently evaluated in clinical trial

Drug	Target	Clinical stage
2DG	Glucose metabolism	Phase 2
Lonidamine	HK	Phase 2
Dichloroacetate (DCA)	PDK	Phase 3
Polyphenon E	LDH-A	Phase 1
FK866	GAPDH	Phase 1
AT-101	LDH-A	Phase 2
AZD3965	MCT1	Phase 1
PEG-ADI	Plasma arginine levels	Phase 3
L-asparaginase	Plasma asparagine levels	clinic
Indoximod	LAT1	Phase 2
Acivicin	Glutamine	Phase 2
6-Diazo-5-oxo-L-norleucine (DON)	Glutamine	Phase 2
CB-839	Glutaminase	Phase 1
Orlistat	FASN	Phase 1
TVB-2640	FASN	Phase 1
Metformin	Mitochondrial GPD2 and other targets	Phase 3
Phenformin	Mitochondrial GPD2 and other targets	Phase 1
AG-120	Mutant IDH1	Phase 1/2
AG-221	Mutant IDH2	Phase 3
NVP-BEZ235	PI3K/mTOR	Phase 2
GDC-0980 (apitolisib)	PI3K/mTOR	Phase 2
SF1126	PI3K	Phase 1
BYL719	P110a	Phase 2
MLN1117	P110a	Phase 2
AZD8186	p110β	Phase 1
GSK2636771	p110β	Phase 2
SAR260301	p110β	Phase 1
Perifosine	Akt	Phase 2
MK-2206	Akt	Phase 2
GSK690693	Akt	Phase 1
GDC-0068	Akt	Phase 2
BEZ235	PI3K/mTOR	Phase 2
XL765	PI3K/mTOR	Phase 1
GDC0890	PI3K/mTOR	Phase 2
GSK1059615	PI3K/mTOR	Phase 1
AZD8055	mTORC1/2	Phase 1
AZD2014	mTORC1/2	Phase 2

Adapted with permission from van der Mijn JC et al. 'Novel drugs that target the metabolic reprogramming in renal cell cancer', *Cancer & Metabolism*, 4:14,

Glutaminase inhibitors and amino acid depletion strategies

Because of the essential role of glutamine in cancer cells, drugs to prevent the conversion of glutamine to glutamate are already in clinical trial. CB-839 is a glutaminase inhibitor currently in phase II clinical trial. It has shown antitumour effects in xenograft models of triple-receptor negative breast cancer and also in acute myeloid leukaemia. High glutaminase levels can overcome the resistance as can α-KG, which bypasses the metabolic defect created by blocking glutamine to glutamate production and then production of α-KG from glutamate.

Inhibition of glutaminase may synergize with other targeted therapies such as *BCL2* inhibition (Jacque et al., 2015). The possibility of exploiting the high dependence of cancer cells on extracellular amino acids either by blocking their metabolism or by depleting them from the local environment by degradation is attractive but needs detailed devaluation of individual tumours. Asparaginase has been used for many decades in the treatment of childhood lymphoblastic leukaemia but recently broader metabolic dependency on arginine have been uncovered also in human adult tumours (Patil et al., 2016). This is the result of decreased expression of argininosuccinate synthetase and ornithine transcarbamylase, causing autophagy. Arginase, arginine deiminase, and arginine decarboxylase can all be used for this therapeutic strategy and also sensitize cancer cells to chemotherapy agents. These enzymes are under investigation in formulations that prolong their half-life and assays are being developed to select tumours with the aforementioned enzyme deficiencies, for therapy.

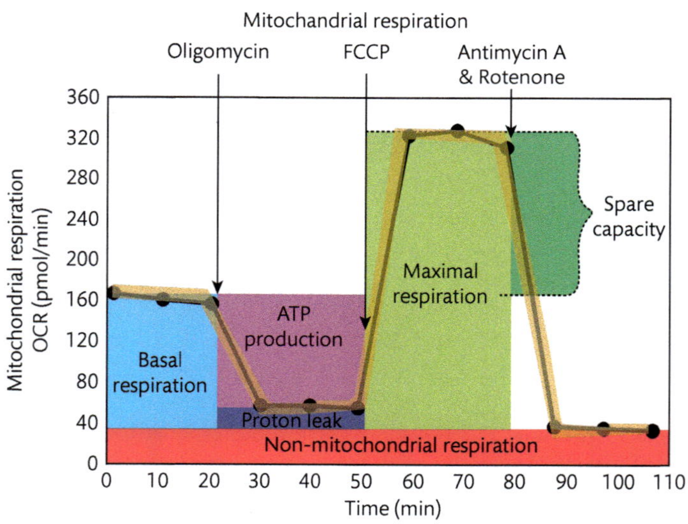

Fig. 16.5 The Seahorse apparatus. The typical readout from a Seahorse metabolic profile. Measurements of basal respiration, ATP production, maximal respiration, as well as acidification, glucose consumption, and use of fatty acids can be measured. For more detailed explanations see: http://www.agilent.com/en-us/products/cell-analysis-(seahorse)/mitochondrial-respiration-the-xf-cell-mito-stress-test
Reproduced with the permission of Agilent Technologies, Inc.

Metformin

Metformin has become one of the most widely used and investigated drugs to modify tumour metabolism. This was based initially on epidemiological evidence that insulin-requiring diabetics have an increased risk of cancer, potentially because insulin acts as tumour growth factor. However, those taking metformin as well had a reduced incidence. Many epidemiological studies have confirmed this, although the methodology has been criticized (Pernicova and Korbonits, 2014). Another influential study demonstrated increased pathological complete response in patients with breast cancer treatment with neoadjuvant therapy if they were on metformin, but this was not randomized.

Randomized trials have investigated metformin in breast cancer, and in pancreatic cancer where there is no effect; however, the results are not yet available for the former. However, one of the problems is that the *in vitro* work, evaluating the effects of metformin, uses concentrations hundreds of time higher than achievable levels in plasma.

There are two main theories about how metformin may function: one is through affecting hepatic metabolism and reducing glucose release from the liver and therefore reducing insulin response and decreasing stimulation of tumour cells indirectly.

The second proposes that because metformin can block complex-1 of the mitochondrial respiratory chain, it reduces the contribution of TCA to intermediary metabolism and ATP generation and as a result, AMP levels rise and lead to the activation of AMPK, which in turn will downregulate the mTORC pathway. However, there is little direct evidence of this in human studies so far. Phenformin is a more potent form of metformin with better tissue penetration and is now also being investigated.

The widespread use of metformin in diabetics could reveal effects on cancer incidence. While the mechanism underlying this modulation remain unclear and unproven, it has stimulated research in this area driving the development of more potent drugs targeting mitochondrial pathways. However, targeting mitochondrial metabolism has a high risk of severe side effects due to the high mitochondrial content of liver, kidney, and brain, which are likely target organs for toxicity. Certainly because of the effect of metformin on complex-1, the need for glutamine to feed into the Krebs cycle is increased and there is potential synergy between blocking of glutamine metabolism and metformin treatment (Fendt et al., 2013).

IDH inhibitors

The identification of mutations in IDH1 and IDH2 has spurred the development of compounds that selectively inhibit the neomorphic activity of the mutant enzyme. In a preclinical setting, AGI-5198, an inhibitor of mutant IDH1, efficiently blocked (R)-2HG production and induced demethylation of histone H3K9me3 as well as a gene expression programme associated with differentiation in IDH1 mutant glioma cells (Rohle et al., 2013). Furthermore, chemical inhibition or silencing of mutant IDH enzymes efficiently blocked tumour formation by human cancer cells in xenografts, suggesting that established cancers still maintain their dependence in this pathway. Two compounds, AG120 and AG221, that block the activity of mutant IDH1 and IDH2, respectively, have already entered clinical trial with the IDH2 inhibitor achieving durable responses in acute myeloid leukaemia patients (Clark et al., 2016). Further evaluation of these compounds together with efficient strategies for patient stratification will reveal the full potential of targeting IDH enzymes.

Fatty acid synthase

There is frequent reactivation of *de novo* fatty acid biosynthesis in cancer (e.g. expression of *HER2* in breast cancer is often associated with high fatty acid synthetase and the latter is associated with poor outcome). Drugs blocking FASN strongly synergize with anti-*HER2* therapy preclinically (Corominas-Faja et al., 2016).

Several compounds targeting the activity of FASN have been developed. Two of these compounds, orlistat and TVB-2640, are evaluated clinically with orlistat being investigated only in the context of obesity. An earlier compound, C75, caused anorexia and severe weight loss in mice, suggesting that targeting fatty acid synthesis could be accompanied by significant toxicity. In addition, the complex interplay between lipid uptake, synthesis, and degradation may limit the therapeutic window for FASN inhibitors. Strategies to increase the dependence of cancers on *de novo* lipogenesis, for example by combination with antiangiogenic therapies, may be required to reveal the potential of FASN inhibition. Alternatively, other enzymes within the fatty acid biosynthesis pathway, such as fatty acid desaturating enzymes, may provide promising targets.

Conclusions

Every metabolic pathway investigated has so far shown changes in the tumour microenvironment in one tumour type or another or in the relevant stromal cells. This may be driven at a molecular mutational level (e.g. by fairly commonly mutated oncogenes), but also through rarer mutations such as IDH1 and IDH2, in specific tumour types. Where the highly specific changes occur due to genetic mutations effecting a single enzyme pathway directly, a clear therapeutic target has emerged with encouraging clinical benefits already. However, many studies have been conducted in cell lines in two dimensional conditions with high ambient oxygen concentrations, high glucose concentrations, and nutrients that do not reflect the levels that are found in tumours. Often, these cell lines have adapted over many years, sometimes up to 50 years or more, so although many changes are described reproducibly from panels of cell lines and different subtypes, their direct relevance to *in vivo* measurements still remains to be demonstrated.

However, the methodology for this is complex and expensive and even now relatively few tumours have been studied directly *in vivo*, often demonstrating the maintenance of oxidative phosphorylation is important. However, the consistent changes that have been demonstrated at least at immunohistochemical level and imaging (e.g. FDG uptake), demonstrate how common the upregulation of glucose metabolism is. The measurement of proteins involved in metabolism by immunohistochemistry in tumours has generally demonstrated high levels are associated with poor outcome (e.g. for glycolysis and hypoxia induced targets, amino acid transporters).

Metabolism is the final output from the genetic changes of cancer and focused on producing an increase in cell mass, hence the switch to glycolysis which provides many intermediates required for that purpose, but the Krebs cycle also continues to run alongside as it is so efficient at generating ATP. Thus, targeting glycolysis alone may slow tumour growth but would not deprive it sufficiently to work on its own. Much of the uptake of FDG is due to activation of AKT and other signalling pathways which markedly feed into metabolism. Many of the drugs we already use (e.g. targeting *HER2*), affect metabolism as so many of the genes downstream regulate metabolic pathways to stimulate growth. So, in many respects current targeted therapies already target metabolism as a side effect, but more specific targeting and reducing it to a minimal level may have further benefit. However, many of the drugs and therapies do not reduce metabolism sufficiently because of the differential therapeutic margin is not enough, except in the case of the mutations. Nevertheless, epidemiological evidence of drugs that effect metabolism being of benefit, such as metformin, have stimulated many studies, which will be published in the near future.

HDAC inhibitors and chromatin modulators also effect metabolism substantially. It may be that through these indirect approaches we will control tumour metabolism. Antibodies targeting cell surface receptors for metabolites transporters also are encouraging in view of their specificity and lack of other off target effects. Thus, targeting tumour metabolism specifically remains challenging unless there are specific mutations in metabolic enzymes but combinations depriving tumours of the adaptation mechanism required for survival in their adverse microenvironments remain promising.

TAKE-HOME MESSAGE

- The microenvironmental factors such as hypoxia, acidosis, low glucose, and glutamine drive tumour adaptation.
- Major mutated oncogenes, such as p53, Hras, Myc, and *HER2* amplification, are key regulators of metabolism. Hence targeting existing kinase and signalling pathways for growth will also target metabolism.
- Targeting tumour metabolic pathways is an effective *in vitro* and preclinical model strategy but remains to be proven in the clinic.
- New methods have been developed to monitor metabolism *in vivo* and are essential to confirm the preceding observations.
- Tumour metabolism is not a generality in that every tumour type and subtype will have different effects depending on the tissue from which it has arisen and to which it has metastasized.
- Metabolic pathways are critical also for metastasis and invasion and are responsible resistance to drugs and radiation.
- Only where there is a mutation in a metabolic enzyme has there been an effective therapeutic development (e.g. IDH1 and 2 mutations).
- Metabolic pathways involving nucleotide metabolism in contrast to intermediary metabolism and carbon metabolism have proven effective for over six decades and continue to be developed, highlighting again the major role of metabolic pathways in increasing cell mass and division.
- However, the majority of the metabolites used by cancer cells go into protein structural components and cell mass as opposed to DNA.
- Tumour metabolism involves stromal cells, endothelial cells, and has a major effect on the immune response.
- Targeting the immunosuppression pathways mediated by tumours through metabolic blockade remains a promising technique targeting the non-tumour population.
- Heterogeneity of metabolism within tumours and from stromal cells remains to be fully evaluated but already it is clear that hypoxic areas and peripheral areas have markedly different metabolic pathways.

OPEN QUESTIONS

- How do we measure tumour metabolism in primaries tumours and metastatic disease?
- How do we classify tumour metabolism in different tumour types, because there will be multiple changes occurring?
- How do we develop drugs targeting tumour metabolism if we use cell lines?
- How do we improve differential toxicity of targeting metabolism in relevant model systems?
- What is the underlying basis of the regulation of the enzymes of metabolism, are there changes in chromatin regulation that regulate major complexes of enzymes or is each one regulated independently?
- Does every patient have a different metabolome, which will make it difficult to target, or are there enough general changes that we can pick targets that will be present for 10% of the target population (e.g. high glucose uptake)?
- Can we target endothelial metabolism and T-cell metabolism to enhance the effects of antiangiogenic and immune modulating drugs as an optimum way of exploiting metabolic changes in tumours?

FURTHER READING

Carracedo, A., Cantley, L. C., & Pandolfi, P. P. (2013). Cancer metabolism: fatty acid oxidation in the limelight. *Nat Rev Cancer*, **13**(4), 227–32.

Chaneton, B. & Gottlieb, E. (2012). Rocking cell metabolism: revised functions of the key glycolytic regulator PKM2 in cancer. *Trends Biochem Sci*, **37**(8), 309–16.

Galluzzi, L., Kepp, O., Vander Heiden, M. G., & Kroemer, G. (2013). Metabolic targets for cancer therapy. *Nat Rev Drug Disc*, **12**(11), 829–46.

Mattaini, K. R., Sullivan, M. R., & Vander Heiden, M. G. (2016). The importance of serine metabolism in cancer. *J Cell Biol*, **214**(3), 249–57.

Pernicova, I. & Korbonits, M. (2014). Metformin—mode of action and clinical implications for diabetes and cancer. *Nat Rev Endocrinol*, **10**(3), 143–56.

Roach, P. J., Depaoli-Roach, A. A., Hurley, T. D., & Tagliabracci, V. S. (2012). Glycogen and its metabolism: some new developments and old themes. *Biochem J*, **441**(3), 763–87.

Thorpe, L. M., Yuzugullu, H., & Zhao, J. J. (2015). PI3K in cancer: divergent roles of isoforms, modes of activation and therapeutic targeting. *Nat Rev Cancer*, **15**(1), 7–24.

Treps, L., Conradi, L. C., Harjes, U., & Carmeliet, P. (2016). Manipulating angiogenesis by targeting endothelial metabolism: hitting the engine rather than the drivers-a new perspective? *Pharmacol Rev*, **68**(3), 872–87.

Vander Heiden, M. G., Lunt, S. Y., Dayton, T. L., et al. (2011). Metabolic pathway alterations that support cell proliferation. *Cold Spring Harb Symp Quant Biol*, **76**, 325–34.

Warburg, O. (2015). *The Metabolism of Tumours*, updated version (Understand Cancer Book 5). Nguyen, T (ed.). EnCognitive.com: Kindle Edition

REFERENCES

Alam, I. S., Arshad, M. A., Nguyen, Q. D., & Aboagye, E. O. (2015). Radiopharmaceuticals as probes to characterize tumour tissue. *Eur J Nucl Med Mol Imaging*, **42**(4), 537–61.

Alvarez, J. V., Belka, G. K., Pan, T. C., et al. (2014). Oncogene pathway activation in mammary tumors dictates FDG-PET uptake. *Cancer Res*, **74**(24), 7583–98.

Arabi, A., Wu, S., Ridderstrale, K., et al. (2005). c-Myc associates with ribosomal DNA and activates RNA polymerase I transcription. *Nat Cell Biol*, **7**(3), 303–10.

Balsa-Martinez, E. & Puigserver, P. (2015). Cancer cells hijack gluconeogenic enzymes to fuel cell growth. *Mol Cell*, **60**(4), 509–11.

Bathen, T. F., Geurts, B., Sitter, B., et al. (2013). Feasibility of MR metabolomics for immediate analysis of resection margins during breast cancer surgery. *PloS One*, **8**(4), e61578.

Begley, J. K., Redpath, T. W., Bolan, P. J., & Gilbert, F. J. (2012). *In vivo* proton magnetic resonance spectroscopy of breast cancer: a review of the literature. Breast cancer research: *BCR*, **14**(2), 207.

Bensaad, K., Favaro, E., Lewis, C. A., et al. (2014). Fatty acid uptake and lipid storage induced by HIF-1alpha contribute to cell growth and survival after hypoxia-reoxygenation. *Cell Rep*, **9**(1), 349–65.

Bensaad, K., Tsuruta, A., Selak, M. A., et al. (2006). TIGAR, a p53-inducible regulator of glycolysis and apoptosis. *Cell*, **126**(1), 107–20.

Bonnet, S., Archer, S. L., Allalunis-Turner, J., et al. (2007). A mitochondria-k(+) channel axis is suppressed in cancer and its normalization promotes apoptosis and inhibits cancer growth. *Cancer Cell*, **11**(1), 37–51.

Boon, K., Caron, H. N., van Asperen, R., et al. (2001). N-myc enhances the expression of a large set of genes functioning in ribosome biogenesis and protein synthesis. *EMBO J*, **20**(6), 1383–93.

Bott, A. J., Peng, I. C., Fan, Y., et al. (2015). Oncogenic Myc induces expression of glutamine synthetase through promoter demethylation. *Cell Metab*, **22**(6), 1068–77.

Bricker, D. K., Taylor, E. B., Schell, J. C., et al. (2012). A mitochondrial pyruvate carrier required for pyruvate uptake in yeast, Drosophila, and humans. *Science*, **337**(6090), 96–100.

Brindle, K. M. (2015). Imaging metabolism with hyperpolarized (13) C-labeled cell substrates. *J Am Chem Soc*, **137**(20), 6418–27.

Brisson, L., Banski, P., Sboarina, M., et al. (2016). Lactate dehydrogenase B controls lysosome activity and autophagy in cancer. *Cancer Cell*, **30**(3), 418–31.

Brown, J. S. & Banerji, U. (2017). Maximising the potential of AKT inhibitors as anti-cancer treatments. *Pharmacol Ther*, **172**, 101–15.

Bryant, K. L., Mancias, J. D., Kimmelman, A. C., & Der, C. J. (2014). KRAS: feeding pancreatic cancer proliferation. *Trends Biochem Sci*, **39**(2), 91–100.

Cappello, P., Tonoli, E., Curto, R., Giordano, D., Giovarelli, M., & Novelli, F. (2016). Anti-alpha-enolase antibody limits the invasion of myeloid-derived suppressor cells and attenuates their restraining effector T cell response. *Oncoimmunology*, **5**(5), e1112940.

Carracedo, A., Cantley, L. C., & Pandolfi, P. P. (2013). Cancer metabolism: fatty acid oxidation in the limelight. *Nat Rev Cancer*, **13**(4), 227–32.

Carroll, P. A., Diolaiti, D., McFerrin, L., et al. (2015). Deregulated Myc requires MondoA/Mlx for metabolic reprogramming and tumorigenesis. *Cancer Cell*, **27**(2), 271–85.

Catalina-Rodriguez, O., Kolukula, V. K., Tomita, Y., et al. (2012). The mitochondrial citrate transporter, CIC, is essential for mitochondrial homeostasis. *Oncotarget*, **3**(10), 1220–35.

Chan, K. W., Jiang, L., Cheng, M., et al. (2016). CEST-MRI detects metabolite levels altered by breast cancer cell aggressiveness and chemotherapy response. *NMR Biomed*, **29**(6), 806–16.

Chaneton, B. & Gottlieb, E. (2012). Rocking cell metabolism: revised functions of the key glycolytic regulator PKM2 in cancer. *Trends Biochem Sci*, **37**(8), 309–16.

Chaneton, B., Hillmann, P., Zheng, L., et al. (2012). Serine is a natural ligand and allosteric activator of pyruvate kinase M2. *Nature*, **491**(7424), 458–62.

Chantranupong, L., Scaria, S. M., Saxton, R. A., et al. (2016). The CASTOR proteins are arginine sensors for the mTORC1 pathway. *Cell*, **165**(1), 153–64.

Chen, C., Pore, N., Behrooz, A., Ismail-Beigi, F., & Maity, A. (2001). Regulation of glut1 mRNA by hypoxia-inducible factor-1. Interaction between H-ras and hypoxia. J Biol Chem, **276**(12), 9519–25.

Chen, R., Zhou, X., Liu, J., & Huang, G. (2016). Relationship between 18F-FDG PET/CT findings and *HER2* expression in gastric cancer. *J Nucl Med*, **57**(7), 1040–4.

Christofk, H. R., Vander Heiden, M. G, Wu, N., Asara, J. M., & Cantley, L. C. (2008). Pyruvate kinase M2 is a phosphotyrosine-binding protein. *Nature*, **452**(7184), 181–6.

Clark, O., Yen, K., & Mellinghoff, I. K. (2016). Molecular pathways: isocitrate dehydrogenase mutations in cancer. *Clin Cancer Res*, **22**(8), 1837–42.

Clem, B. F., O'Neal, J., Tapolsky, G., et al. (2013) Targeting 6-phosphofructo-2-kinase (PFKFB3) as a therapeutic strategy against cancer. *Mol Cancer Ther*, **12**(8), 1461–70.

Commisso, C., Davidson, S. M., Soydaner-Azeloglu, R. G., et al. (2013). Macropinocytosis of protein is an amino acid supply route in Ras-transformed cells. *Nature*, **497**(7451), 633–7.

Corominas-Faja, B., Vellon, L., Cuyas, E., et al. (2017). Clinical and therapeutic relevance of the metabolic oncogene fatty acid synthase in *HER2* + breast cancer. *Histol Histopathol*, **32**(7), 687–98.

Coudert, B., Pierga, J. Y., Mouret-Reynier, M. A., et al. (2014). Use of [(18)F]-FDG PET to predict response to neoadjuvant trastuzumab and docetaxel in patients with HER2-positive breast cancer, and addition of bevacizumab to neoadjuvant trastuzumab and docetaxel in [(18)F]-FDG PET-predicted non-responders (AVATAXHER): an open-label, randomised phase 2 trial. *Lancet Oncol*, **15**(13), 1493–502.

Currie, E., Schulze, A., Zechner, R., Walther, T. C., & Farese, R. V., Jr. (2013). Cellular fatty acid metabolism and cancer. *Cell Metab*, **18**(2), 153–61.

Dang, C. V. (2012). MYC on the path to cancer. *Cell*, **149**(1), 22–35.

David, C. J., Chen, M., Assanah, M., Canoll, P., & Manley, J. L. (2010). HnRNP proteins controlled by c-Myc deregulate pyruvate kinase mRNA splicing in cancer. *Nature*, **463**(7279), 364–8.

De Bock, K., Georgiadou, M., Schoors, S., et al. (2013). Role of PFKFB3-driven glycolysis in vessel sprouting. *Cell*, **154**(3), 651–63.

Dixon, S. J., Lemberg, K. M., Lamprecht, M. R., et al. (2012). Ferroptosis: an iron-dependent form of nonapoptotic cell death. *Cell*, **149**(5), 1060–72.

Edmunds, L. R., Sharma, L., Kang A, et al. (2014). c-Myc programs fatty acid metabolism and dictates acetyl-CoA abundance and fate. *J Biol Chem*, **289**(36), 25382–92.

Eilers, M. & Eisenman, R. N. (2008). Myc's broad reach. *Genes Dev*, **22**(20), 2755–66.

Fendt, S. M., Bell, E. L., Keibler, M. A., et al. (2013). Metformin decreases glucose oxidation and increases the dependency of prostate cancer cells on reductive glutamine metabolism. *Cancer Res*, **73**(14), 4429–38.

Gaboriaud-Kolar, N. & Skaltsounis, A. L. (2013). Glycogen phosphorylase inhibitors: a patent review (2008–2012). *Expert Opin Ther Pat*, **23**(8), 1017–32.

Galluzzi, L., Kepp, O., Vander Heiden, M. G., & Kroemer, G. (2013). Metabolic targets for cancer therapy. *Nat Rev Drug Discov*, **12**(11), 829–46.

Gao, P., Tchernyshyov, I., Chang, T. C., et al. (2009). c-Myc suppression of miR-23a/b enhances mitochondrial glutaminase expression and glutamine metabolism. *Nature*, **458**(7239), 762–5.

Geck, R. C. & Toker, A. (2016). Nonessential amino acid metabolism in breast cancer. *Adv Biol Regul*, **62**, 11–17.

Ghesquiere, B., Wong, B. W., Kuchnio, A., & Carmeliet, P. (2014). Metabolism of stromal and immune cells in health and disease. *Nature*, **511**(7508), 167–76.

Gottlob, K., Majewski, N., Kennedy, S., Kandel, E., Robey, R. B., & Hay, N. (2011). Inhibition of early apoptotic events by Akt/PKB is dependent on the first committed step of glycolysis and mitochondrial hexokinase. *Genes Dev*, **15**(11), 1406–18.

Hardie, D. G., Ross, F. A., & Hawley, S. A. (2012). AMPK: a nutrient and energy sensor that maintains energy homeostasis. *Nat Rev Mol Cell Biol*, **13**(4), 251–62.

Harjes, U., Kalucka, J., & Carmeliet, P. (2016). Targeting fatty acid metabolism in cancer and endothelial cells. *Crit Rev Oncol Hematol*, **97**, 15–21.

Herzig, S., Raemy, E., Montessuit, S., et al. (2012). Identification and functional expression of the mitochondrial pyruvate carrier. *Science*, **337**(6090), 93–6.

Hitosugi, T., Zhou, L., Elf, S., et al. (2012). Phosphoglycerate mutase 1 coordinates glycolysis and biosynthesis to promote tumor growth. *Cancer Cell*, **22**(5), 585–600.

Hsieh, A. L., Walton, Z. E., Altman, B. J., Stine, Z. E., & Dang, C. V. (2015) MYC and metabolism on the path to cancer. *Semin Cell Dev Biol*, **43**, 11–21.

Hu, H., Juvekar, A., Lyssiotis, C. A., et al. (2016). Phosphoinositide 3-kinase regulates glycolysis through mobilization of aldolase from the actin cytoskeleton. *Cell*, **164**(3), 433–46.

Hulikova, A., Black, N., Hsia, L. T., Wilding, J., Bodmer, W. F., & Swietach, P. (2016). Stromal uptake and transmission of acid is a pathway for venting cancer cell-generated acid. *Proc Natl Acad Sci U S A*, **113**(36), E5344–53.

Israelsen, W. J., Dayton, T. L., Davidson, S. M., et al. (2013). PKM2 isoform-specific deletion reveals a differential requirement for pyruvate kinase in tumor cells. *Cell*, **155**(2), 397–409.

Jacque, N., Ronchetti, A. M., Larrue, C., et al. (2015). Targeting glutaminolysis has antileukemic activity in acute myeloid leukemia and synergizes with BCL-2 inhibition. *Blood*, **126**(11), 1346–56.

Jiang, L., Kon, N., Li, T., et al. (2015). Ferroptosis as a p53-mediated activity during tumour suppression. *Nature*, **520**(7545), 57–62.

Jiang, P., Du, W., Mancuso, A., Wellen, K. E., & Yang, X. (2013). Reciprocal regulation of p53 and malic enzymes modulates metabolism and senescence. *Nature*, **493**(7434), 689–93.

Jiang, P., Du, W., Wang, X., et al. (2011). p53 regulates biosynthesis through direct inactivation of glucose-6-phosphate dehydrogenase. *Nat Cell Biol*, **13**(3), 310–16.

Kamphorst, J. J., Cross, J. R., Fan, J., et al. (2013). Hypoxic and Ras-transformed cells support growth by scavenging unsaturated fatty acids from lysophospholipids. *Proc Natl Acad Sci U S A*, **110**(22), 8882–7.

Kamphorst, J. J., Nofal, M., Commisso, C., et al. (2015). Human pancreatic cancer tumors are nutrient poor and tumor cells actively scavenge extracellular protein. *Cancer Res*, **75**(3), 544–53.

Kawauchi, K., Araki, K., Tobiume, K., & Tanaka, N. (2008). p53 regulates glucose metabolism through an IKK-NF-kappaB pathway and inhibits cell transformation. *Nat Cell Biol*, **10**(5), 611–18.

Kim, J. W., Tchernyshyov, I., Semenza, G. L., & Dang, C. V. (2006). HIF-1-mediated expression of pyruvate dehydrogenase kinase: a metabolic switch required for cellular adaptation to hypoxia. *Cell Metab*, **3**(3), 177–85.

Koukourakis, M. I., Giatromanolaki, A., Bottini, A., et al. (2014). Prospective neoadjuvant analysis of PET imaging and mechanisms of resistance to Trastuzumab shows role of HIF1 and autophagy. *Br J Cancer*, **110**(9), 2209–16.

Laplante, M. & Sabatini, D. M. (2012). mTOR signaling in growth control and disease. *Cell*, **149**(2), 274–93.

Ledaki, I., McIntyre, A., Wigfield, S., et al. (2015). Carbonic anhydrase IX induction defines a heterogeneous cancer cell response to hypoxia and mediates stem cell-like properties and sensitivity to HDAC inhibition. *Oncotarget*, **6**(23), 19413–27.

Lee, D. C., Sohn, H. A., Park, Z. Y., et al. (2015). A lactate-induced response to hypoxia. *Cell*, **161**(3), 595–609.

Lee, S. B., Frattini, V., Bansal, M., et al. (2016). An ID2-dependent mechanism for VHL inactivation in cancer. *Nature*, **529**(7585), 172–7.

Leithner, K., Hrzenjak, A., Trotzmuller, M., et al. (2015). PCK2 activation mediates an adaptive response to glucose depletion in lung cancer. *Oncogene*, **34**(8), 1044–50.

Li, B., Qiu, B., Lee, D. S., et al. (2014). Fructose-1,6-bisphosphatase opposes renal carcinoma progression. *Nature*, **513**(7517), 251–5.

Li, F., Wang, Y., Zeller, K. I., et al. (2005). Myc stimulates nuclearly encoded mitochondrial genes and mitochondrial biogenesis. *Mol Cell Biol*, **25**(14), 6225–34.

Li, T., Kon, N., Jiang, L., et al. (2012). Tumor suppression in the absence of p53-mediated cell-cycle arrest, apoptosis, and senescence. *Cell*, **149**(6), 1269–83.

Li, X., Jiang, Y., Meisenhelder, J., et al. (2016). Mitochondria-translocated PGK1 functions as a protein kinase to coordinate glycolysis and the TCA cycle in tumorigenesis. *Mol Cell*, **61**(5), 705–19.

Liao, L., Testa, J. R., Yang, H. (2015). The roles of chromatin-remodelers and epigenetic modifiers in kidney cancer. *Cancer Genet*, **208**(5), 206–14.

Lisanti, M. P., Martinez-Outschoorn, U. E., & Sotgia, F. (2013). Oncogenes induce the cancer-associated fibroblast phenotype: metabolic symbiosis and 'fibroblast addiction' are new therapeutic targets for drug discovery. *Cell Cycle (Georgetown, Tex)*, **12**(17), 2723–32.

Liu, W., Hancock, C. N., Fischer, J. W., Harman, M., & Phang, J. M. (2015). Proline biosynthesis augments tumor cell growth and aerobic glycolysis: involvement of pyridine nucleotides. *Sci Rep*, **5**, 17206.

Loayza-Puch, F., Rooijers, K., Buil, L. C., et al. (2016). Tumour-specific proline vulnerability uncovered by differential ribosome codon reading. *Nature*, **530**(7591), 490–4.

Locasale, J. W., Grassian, A. R., Melman, T., et al. (2011). Phosphoglycerate dehydrogenase diverts glycolytic flux and contributes to oncogenesis. *Nature Genet*, **43**(9), 869–74.

Locasale, J. W. (2013). Serine, glycine and one-carbon units: cancer metabolism in full circle. *Nat Rev Cancer*, **13**(8), 572–83.

Lu, Z., Yao, Y., Song, Q., et al. (2016). Metabolism-related enzyme alterations identified by proteomic analysis in human renal cell carcinoma. *Onco Targets Ther*, **9**, 1327–37.

Maddocks, O. D., Berkers, C. R., Mason, S. M., et al. Serine starvation induces stress and p53-dependent metabolic remodelling in cancer cells. *Nature*, **493**(7433), 542–6.

Maddocks, O. D. & Vousden, K. H. (2011). Metabolic regulation by p53. *J Mol Med (Berl)*, **89**(3), 237–45.

Maddocks, O. D., Labuschagne, C. F., Adams, P. D., & Vousden, K. H. (2016). Serine metabolism supports the methionine cycle and DNA/RNA methylation through *de novo* ATP synthesis in cancer cells. *Mol Cell*, **61**(2), 210–21.

Mardis, E. R., Ding, L., Dooling, D. J., et al. (2009). Recurring mutations found by sequencing an acute myeloid leukemia genome. *N Engl J Med*, **361**(11), 1058–66.

Massari, F., Ciccarese, C., Santoni, M., et al. (2016). Metabolic phenotype of bladder cancer. *Cancer Treatment Rev*, **45**, 46–57.

Matoba, S., Kang, J. G., Patino, W. D., et al. (2006). p53 regulates mitochondrial respiration. *Science*, **312**(5780), 1650–3.

Mattaini, K. R., Sullivan, M. R., Vander Heiden, M. G. (2016) The importance of serine metabolism in cancer. *J Cell Biol*, **214**(3), 249–57.

Mazurek, S. (2011). Pyruvate kinase type M2: a key regulator of the metabolic budget system in tumor cells. *Int J Biochem & Cell Biol*, **43**(7), 969–80.

Metelo, A. M., Noonan, H., Iliopoulos, O. (2015). HIF2a inhibitors for the treatment of VHL disease. *Oncotarget*, **6**(27), 23036–7.

Miyake, K. K., Nakamoto, Y., Kanao, S., et al. (2014). Journal club: diagnostic value of (18)F-FDG PET/CT and MRI in predicting the clinicopathologic subtypes of invasive breast cancer. *AJR Am J Roentgenol*, **203**(2), 272–9.

Montal, E. D., Dewi, R., Bhalla, K., et al. (2015). PEPCK coordinates the regulation of central carbon metabolism to promote cancer cell growth. *Mol Cell*, **60**(4), 571–83.

Mullen, P. J., Yu, R., Longo, J., Archer, M. C., & Penn, L. Z. (2016). The interplay between cell signalling and the mevalonate pathway in cancer. *Nat Rev Cancer*, **16**(11), 718–31.

Nielsen, C. H., Jensen, M. M., Kristensen, L. K., et al. (2015). *In vivo* imaging of therapy response to a novel pan-HER antibody mixture using FDG and FLT positron emission tomography. *Oncotarget*, **6**(35), 37486–99.

Nieman, K. M., Kenny, H. A., Penicka, C. V., et al. (2011). Adipocytes promote ovarian cancer metastasis and provide energy for rapid tumor growth. *Nat Med*, **17**(11), 1498–503.

Nile, A. H. & Hannoush, R. N. (2016). Fatty acylation of Wnt proteins. *Nat Chem Biol*, **12**(2), 60–9.

Osthus, R. C., Shim, H., Kim, S., et al. (2000). Deregulation of glucose transporter 1 and glycolytic gene expression by c-Myc. *J Biol Chem*, **275**(29), 21797–800.

Palm, W., Park, Y., Wright, K., Pavlova, N. N., Tuveson, D. A., & Thompson, C. B. (2015). The utilization of extracellular proteins as nutrients is suppressed by mTORC1. *Cell*, **162**(2), 259–70.

Parmenopoulou, V., Kantsadi, A. L., Tsirkone, V. G., et al. (2014). Structure based inhibitor design targeting glycogen phosphorylase B. Virtual screening, synthesis, biochemical and biological assessment of novel N-acyl-beta-d-glucopyranosylamines. *Bioorg Med Chem*, **22**(17), 4810–25.

Parsons, D. W., Jones, S., Zhang, X., et al. (2008). An integrated genomic analysis of human glioblastoma multiforme. *Science*, **321**(5897), 1807–12.

Pascual, G., Avgustinova, A., Mejetta, S., et al. (2017). Targeting metastasis-initiating cells through the fatty acid receptor CD36. *Nature*, **541**(7635), 41–5.

Patil, M. D., Bhaumik, J., Babykutty, S., Banerjee, U. C., & Fukumura, D. (2016). Arginine dependence of tumor cells: targeting a chink in cancer's armor. *Oncogene*, **35**(38), 4957–72.

Patra, K. C., Wang, Q., Bhaskar, P. T., et al. (2013). Hexokinase 2 is required for tumor initiation and maintenance and its systemic deletion is therapeutic in mouse models of cancer. *Cancer Cell*, **24**(2), 213–28.

Payne, G. S., deSouza, N. M., Messiou, C., & Leach, M. O. (2015). Single-shot single-voxel lactate measurements using FOCI-LASER and a multiple-quantum filter. *NMR Biomed*, **28**(4), 496–504.

Pernicova, I. & Korbonits, M. (2014). Metformin—mode of action and clinical implications for diabetes and cancer. *Nat Rev Endocrinol*, **10**(3), 143–56.

Phang, J. M., Liu, W., Hancock, C. N., & Fischer, J. W. (2015). Proline metabolism and cancer: emerging links to glutamine and collagen. *Curr Opin Clin Nutr Metab Care*, **18**(1), 71–7.

Possemato, R., Marks, K. M., Shaul, Y. D., et al. (2011). Functional genomics reveal that the serine synthesis pathway is essential in breast cancer. *Nature*, **476**(7360), 346–50.

Potter, M., Newport, E., & Morten, K. J. (2016). The Warburg effect: 80 years on. *Biochem Soc Transact*, **44**(5), 1499–505.

Pylayeva-Gupta, Y., Grabocka, E., & Bar-Sagi, D. (2011). RAS oncogenes: weaving a tumorigenic web. *Nat Rev Cancer*, **11**(11), 761–74.

Rabinovich, S., Adler, L., Yizhak, K., et al. (2015). Diversion of aspartate in ASS1-deficient tumours fosters *de novo* pyrimidine synthesis. *Nature*, **527**(7578), 379–83.

Roach, P. J., Depaoli-Roach, A. A., Hurley, T. D., & Tagliabracci, V. S. (2012). Glycogen and its metabolism: some new developments and old themes. *Biochem J*, **441**(3), 763–87.

Robinson, C. M. & Ohh, M. The multifaceted von Hippel-Lindau tumour suppressor protein. *FEBS Letters*, **588**(16), 2704–11.

Rodrigues, T. B., Serrao, E. M., Kennedy, B. W., Hu, D. E., Kettunen, M. I., & Brindle, K. M. (2014). Magnetic resonance imaging of tumor glycolysis using hyperpolarized 13C-labeled glucose. *Nat Med*, **20**(1), 93–7.

Rohle, D., Popovici-Muller, J., Palaskas, N., et al. (2013). An inhibitor of mutant IDH1 delays growth and promotes differentiation of glioma cells. *Science*, **340**(6132), 626–30.

Rohrig, F. & Schulze, A. (2016). The multifaceted roles of fatty acid synthesis in cancer. *Nat Rev Cancer*, **16**(11), 732–49.

Ros, S., Santos, C. R., Moco, S., et al. (2012). Functional metabolic screen identifies 6-phosphofructo-2-kinase/fructose-2,6-biphosphatase 4 as an important regulator of prostate cancer cell survival. *Cancer Discov*, **2**(4), 328–43.

Ros, S. & Schulze, A. (2013). Balancing glycolytic flux: the role of 6-phosphofructo-2-kinase/fructose 2,6-bisphosphatases in cancer metabolism. Cancer Metab, **1**(1), 8.

Rosenfeldt, M. T., O'Prey, J., Morton, J. P., et al. (2013). p53 status determines the role of autophagy in pancreatic tumour development. *Nature*, **504**(7479), 296–300.

Sasaki, M., Knobbe, C. B., Munger, J. C., et al. (2012). IDH1(R132H) mutation increases murine haematopoietic progenitors and alters epigenetics. *Nature*, **488**(7413), 656–9.

Schoors, S., Bruning, U., Missiaen, R., et al. (2015). Fatty acid carbon is essential for dNTP synthesis in endothelial cells. *Nature*, **520**(7546), 192–7.

Schug, Z. T., Vande Voorde, J., & Gottlieb, E. (2016). The metabolic fate of acetate in cancer. *Nat Rev Cancer*, **16**(11), 708–17.

Semenza, G. L., Jiang, B. H., Leung, S. W., et al. (1996). Hypoxia response elements in the aldolase A, enolase 1, and lactate dehydrogenase A gene promoters contain essential binding sites for hypoxia-inducible factor 1. *J Biol Chem*, **271**(51), 32529–37.

Son, J., Lyssiotis, C. A., Ying, H., et al. (2013). Glutamine supports pancreatic cancer growth through a KRAS-regulated metabolic pathway. *Nature*, **496**(7443), 101–5.

Sonveaux, P., Vegran, F., Schroeder, T., et al. (2008). Targeting lactate-fueled respiration selectively kills hypoxic tumor cells in mice. *J Clin Invest*, **118**(12), 3930–42.

Sousa, C. M., Biancur, D. E., Wang, X., et al. (2016). Pancreatic stellate cells support tumour metabolism through autophagic alanine secretion. *Nature*, **536**(7617), 479–83.

Suchorolski, M. T., Paulson, T. G., Sanchez, C. A., Hockenbery, D., & Reid, B. J. (2013). Warburg and Crabtree effects in premalignant Barrett's esophagus cell lines with active mitochondria. *PloS One*, **8**(2), e56884.

Thorpe, L. M., Yuzugullu, H., & Zhao, J. J. (2015). PI3K in cancer: divergent roles of isoforms, modes of activation and therapeutic targeting. *Nat Rev Cancer*, **15**(1), 7–24.

Treps, L., Conradi, L. C., Harjes, U., & Carmeliet, P. (2016). Manipulating angiogenesis by targeting endothelial metabolism: hitting the engine rather than the drivers-a new perspective? *Pharmacol Rev*, **68**(3), 872–87.

Ulaner, G. A., Goldman, D. A., Gonen, M., et al. (2016). Initial results of a prospective clinical trial of 18f-fluciclovine PET/CT in newly diagnosed invasive ductal and invasive lobular breast cancers. *J Nucl Med*, **57**(9), 1350–6.

van der Mijn, J. C., Panka, D. J., Geissler, A. K., Verheul, H. M., & Mier, J. W. (2016). Novel drugs that target the metabolic reprogramming in renal cell cancer. *Cancer Metab*, **4**, 14.

Vander Heiden, M. G., Lunt, S. Y., Dayton, T. L., et al. (2011). Metabolic pathway alterations that support cell proliferation. *Cold Spring Harb Symp Quant Biol*, **76**, 325–34.

Vincent, E. E., Sergushichev, A., Griss, T., et al. (2015). Mitochondrial phosphoenolpyruvate carboxykinase regulates metabolic adaptation and enables glucose-independent tumor growth. *Mol Cell*, **60**(2), 195–207.

Vizan, P., Boros, L. G., Figueras, A., et al. (2005). K-ras codon-specific mutations produce distinctive metabolic phenotypes in NIH3T3 mice [corrected] fibroblasts. *Cancer Res*, **65**(13), 5512–15.

Vousden, K. H. & Lane, D. P. (2007). p53 in health and disease. *Nat Rev Mol Cell Biol*, **8**(4), 275–83.

Warburg, O., Posener, K., Negelein, E. (1924). Über den Stoffwechsel der Carcinomzelle. *Biochem Zeitschr*, **152**, 309–44.

Ward, P. S., Patel, J., Wise, D. R., et al. (2010). The common feature of leukemia-associated IDH1 and IDH2 mutations is a neomorphic enzyme activity converting alpha-ketoglutarate to 2-hydroxyglutarate. *Cancer Cell*, **17**(3), 225–34.

Webb, B. A., Forouhar, F., Szu, F. E., Seetharaman, J., Tong, L., & Barber, D. L. (2015). Structures of human phosphofructokinase-1 and atomic basis of cancer-associated mutations. *Nature*, **523**(7558), 111–14.

Wilkinson, S. & Ryan, K. M. (2010). Autophagy: an adaptable modifier of tumourigenesis. *Curr Opin Genet Dev*, **20**(1), 57–64.

Wise, D. R., DeBerardinis, R. J., Mancuso, A., et al. (2008). Myc regulates a transcriptional program that stimulates mitochondrial glutaminolysis and leads to glutamine addiction. *Proc Natl Acad Sci U S A*, **105**(48), 18782–7.

Witney, T. H., Carroll, L., Alam, I. S., et al. (2014). A novel radiotracer to image glycogen metabolism in tumors by positron emission tomography. *Cancer Res*, **74**(5), 1319–28.

Wolfson, R. L., Chantranupong, L., Saxton, R. A., et al. (2016). Sestrin2 is a leucine sensor for the mTORC1 pathway. Science (New York, NY), **351**(6268), 43–8.

Wymann, M. P. & Schneiter, R. (2008). Lipid signalling in disease. *Nat Rev Mol Cell Biol*, **9**(2), 162–76.

Yalcin, A., Telang, S., Clem, B., & Chesney, J. (2009). Regulation of glucose metabolism by 6-phosphofructo-2-kinase/fructose-2,6-bisphosphatases in cancer. *Exp Mol Pathol*, **86**(3), 174–9.

Yang, M., Soga, T., Pollard, P. J. (2013). Oncometabolites: linking altered metabolism with cancer. *J Clin Invest*, **123**(9), 3652–8.

Yang, S., Wang, X., Contino, G., et al. (2011). Pancreatic cancers require autophagy for tumor growth. *Genes Dev*, **25**(7), 717–29.

Yi, W., Clark, P. M., Mason, D. E., et al. (2012). Phosphofructokinase 1 glycosylation regulates cell growth and metabolism. *Science*, **337**(6097), 975–80.

Yoon, H., Yoon, D., Yun, M., et al. (2016). Metabolomics of breast cancer using high-resolution magic angle spinning magnetic resonance spectroscopy: correlations with 18F-FDG positron emission tomography-computed tomography, dynamic contrast-enhanced and diffusion-weighted imaging MRI. *PloS One*, **11**(7), e0159949.

Yun, J., Rago, C., Cheong, I., et al. (2009). Glucose deprivation contributes to the development of KRAS pathway mutations in tumor cells. *Science*, **325**(5947), 1555–9.

Yuneva, M., Zamboni, N., Oefner, P., Sachidanandam, R., & Lazebnik, Y. (2007). Deficiency in glutamine but not glucose induces MYC-dependent apoptosis in human cells. *J Cell Biol*, **178**(1), 93–105.

Zhang, W. C., Shyh-Chang, N., Yang, H., et al. (2012). Glycine decarboxylase activity drives non-small cell lung cancer tumor-initiating cells and tumorigenesis. *Cell*, **148**(1–2), 259–72.

Zirath, H., Frenzel, A., Oliynyk, G., et al. (2013). MYC inhibition induces metabolic changes leading to accumulation of lipid droplets in tumor cells. *Proc Natl Acad Sci U S A*, **110**(25), 10258–63.

Zois, C. E. & Harris, A. L. (2016). Glycogen metabolism has a key role in the cancer microenvironment and provides new targets for cancer therapy. *J Mol Med (Berl)*, **94**(2), 137–54.

17

Chaperones and protein quality control in the neoplastic process

Andrea Rasola

Protein quality control

Protein quality control in cell physiology

Cells are characterized by the capability of rapidly executing a variety of extremely complex and interconnected biological functions, which require the orchestrated action of several components. Most of these processes are mainly carried out by proteins, which are endowed with complex and flexible structures that can function in multimeric aggregates and in chains of related components, often confined in specific portions of the cell. Each cell must therefore maintain its proteome in a condition of perfect functionality, which depends on an exquisite control of synthesis, folding, localization, and degradation of proteins. In addition, cells must cope with fluctuations in the highly dynamic milieu in which they are located, and to this aim they incessantly exchange a flow of diverse signals, both with the surrounding environment and between different subcellular compartment, to which their proteome must rapidly respond in order to shape the correct responses. Under a biochemical point of view this constitutes a formidable challenge. As an example, protein synthesis occurs in a very short time, and new polypeptides can be released in a cytosol where steady-state protein concentration is around 300 g/l (Martin, 2004), being therefore exposed to a high probability of inappropriate or deleterious interactions and to exposure of hydrophobic surfaces (Kampinga, 2006). Similarly, improper aggregations can occur when old proteins adopt incorrect foldings, thus constituting a threat for the efficient performing of several biochemical reactions.

Molecular mechanisms of protein quality control

In order to maintain their proteome in conditions of optimal functionality, cells have evolved families of proteins termed molecular chaperones, whose role is to perform the quality control of the proteome through an integrated network of activities. These molecular machines help the correct folding of nascent polypeptides, contribute to the assembly of multimeric protein complexes, control the conformational changes that tune enzymatic activity, have a crucial role in the partition of active proteins in subcellular locations where they organize in functional pathways, and in the elimination of inactive proteins that are unfolded, misfolded or aggregated (**Fig. 17.1**). Therefore, chaperones are extremely abundant in cells, where they affect all main functions and are found in most compartments, including cytoplasm, nucleus, mitochondria, and endoplasmic reticulum (Doyle et al., 2013).

Some chaperones are called heat shock proteins (HSPs), as they were initially identified as heat-inducible proteins. Indeed, several other conditions of cellular stress, such as oxidative stress, hypoxia, nutritional deficiencies (e.g. glucose deprivation), mechanical stress, exposure to toxic radicals and carcinogens (Khalil et al., 2011), infection and a variety of pathological conditions, including neoplastic transformation, can affect protein folding and activity, and thus strongly induce chaperone expression.

Protein quality control in tumour cells

Tumour cells are exposed to a variety of stresses that can directly hit polypeptide conformation and functionality. For instance, the neoplastic process is often characterized by an unbalance in redox equilibrium of cells caused by a profound rewiring of their metabolic circuitries and by inconstant oxygen availability (Cairns et al., 2011), thus leading to a potential increase in oxidative stress and in the ensuing protein damage. Moreover, tumour cells can undergo genomic instability in a framework of relentless induction of proliferation, which leads to a high risk of synthesis of large amounts of misfolded proteins. During tumour growth, molecular chaperones are strongly induced by hypoxia, shortage of nutrients, high rate of DNA replication, RNA transcription and translation of potentially mutated proteins, and have a crucial pro-survival activity in neoplastic cells.

Chaperones: Classification, structure, and function

Mammalian HSPs are subdivided into six families (Khalil et al., 2011). Four of them are composed by high molecular weight chaperones (HSP100, HSP90, HSP70 and HSP60 families) that act in an adenosine 5'-triphosphate (ATP)-dependent way and generally require co-chaperones to modulate their conformation and ATP binding. In addition, there are small HSPs (sHSPs) that are ATP-independent chaperones and include HSP40 and chaperones with molecular weight ranging from 15 to 30 kDa. Some chaperones are targeted in specific subcellular compartments, such as endoplasmic

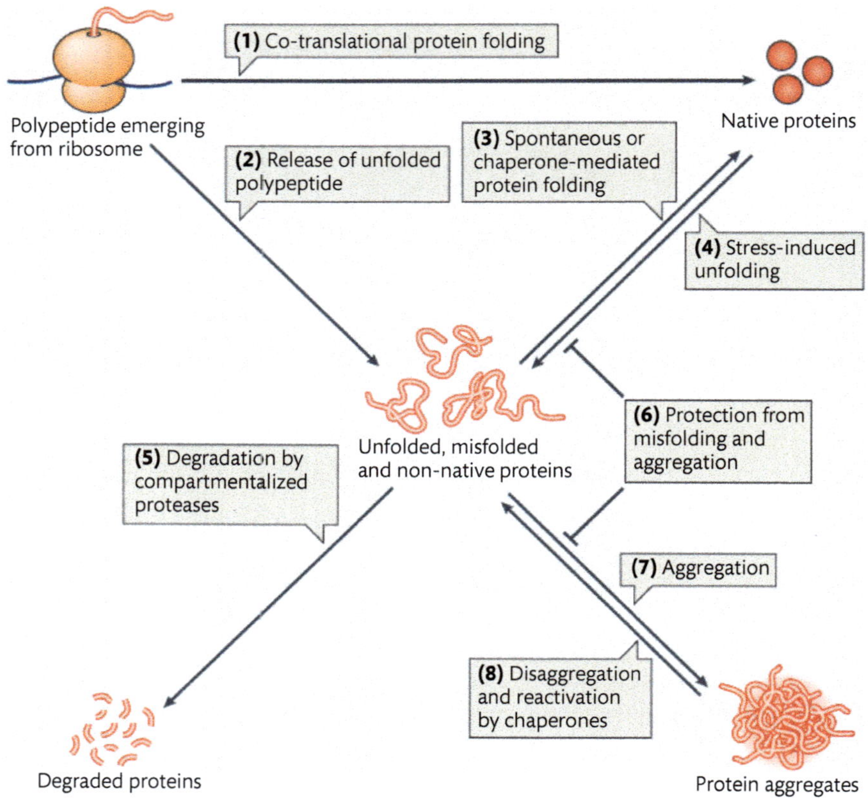

Fig. 17.1 The protein quality control network. Chaperones facilitate folding of newly synthesized proteins (1), of unfolded proteins released from ribosomes (2) and remodelling of proteins in non-native conditions caused by several types of stress (3 and 4); in the absence of proper folding, these proteins undergo chaperone-associated degradation by proteases (5). Molecular chaperones also counteract protein misfolding and aggregation (6), but if they are overwhelmed, non-native proteins may form bulky amorphous aggregates (7), from which other chaperones can remove polypeptides (8) that will be refolded (3).

reticulum (ER) and mitochondria, where they can take part in organelle-restricted protein quality control systems called unfolded protein responses (UPRs). Finally, cyclophilins are an important class of chaperones that assist protein folding and function through prolyl-isomerase reactions.

Each family of chaperones is formed both by constitutively expressed proteins, such as the HSP70 family member HSC70, and by members that are strongly induced by stressful conditions. Examples of induced chaperones include HSP70 and HSP27, whose expression levels dramatically increase following exposure to anticancer drugs, oxidative stress, or irradiation, and are therefore abundant in cancer cells, where it was proposed that they could be important prognostic factors (Garrido et al., 2006). Transcriptional induction of HSP genes is controlled by heat shock transcription factors (HSFs), whose different members can be selectively activated by different stresses (Khalil et al., 2011). The main component of this family, HSF-1, is responsible for the short-term induction of HSPs. HSPs can play important regulatory roles in gene transcription, for instance by suppressing the transcriptional activity of HSF1 in conditions of stress absence.

Chaperones in the neoplastic process

Chaperones act at several levels to block cell death and to promote survival, proliferation, or differentiation (Lanneau et al., 2008), and in general to allow the orchestrated execution of functions that can be exploited by tumour cells during the process of neoplastic progression (Fig. 17.2). In particular, the role played by chaperones in cell death inhibition was widely studied. Overexpression of either HSP27, HSP70, HSP60, or HSP90 can inhibit both the intrinsic and the extrinsic apoptotic pathways in several ways and in many different cellular models upon accumulation of misfolded proteins, reactive oxygen species (ROS) or DNA damage. These HSPs can tune the activity of important signalling pathways. For instance, HSP27 is necessary for the activity of the key survival kinase Akt or for regulating the survival transcription factor NF-κB following exposure to chemotherapeutics. HSP 70 interacts with several kinases, inhibiting those involved in stress-induced apoptosis, such as Ask1, and stabilizing Akt. Notably, both HSP70 and HSP90 can also inhibit the proapoptotic activity of p53, and HSP90 associates with a large number of kinases and transcription factors whose modulation is crucially involved in tumour growth, including, for instance, Her2 and HIF-1α.

In addition, these chaperones can interfere with cell death induction at the mitochondrial level. HSP27 can inhibit the activity of the proapoptotic Bcl2-family members Bid and Bax in the outer mitochondrial membrane and the release of the apoptogenic molecule Smac/DIABLO from mitochondria following tumour cell treatment with chemotherapeutics. HSP60, HSP90, and TRAP1, a HSP90

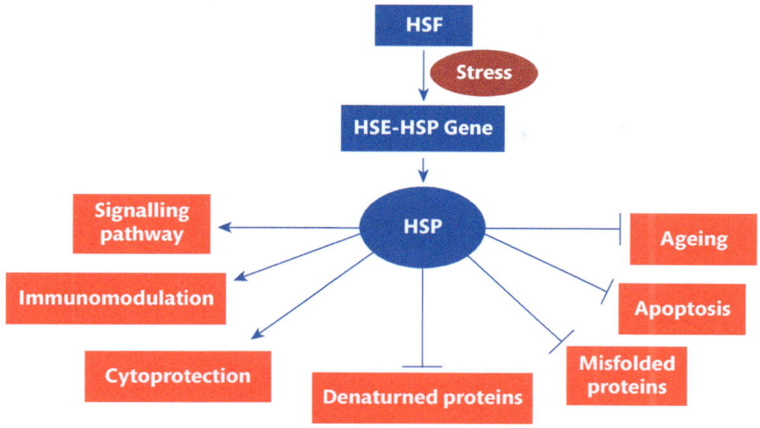

Fig. 17.2 Functions of HSPs under stress conditions. HSF, heat shock transcription factor; HSE, heat shock element in the promoter of HSP genes.
Reproduced with permission from Khalil AA et al. 'Heat shock proteins in oncology: diagnostic biomarkers or therapeutic targets?,' *Biochimica et Biophysica Acta (BBA) - Reviews on Cancer*, Volume 1816, Issue 2, pp. 89–104, https://www.sciencedirect.com/journal/biochimica-et-biophysica-acta-bba-reviews-on-cancer

family member, control cell bioenergetics and redox equilibrium in mitochondria of tumour cells, where they inhibit mitochondrial membrane permeabilization and the release of proapoptotic proteins in the cytosol. Chaperones can also act as inhibitors of the execution phases of apoptosis, directly binding and inhibiting caspase-3 (HSP27), blocking formation of the apoptosome or of the death-inducing signalling complex (HSP70 and HSP90) or the activity of the apoptosis inducing factor AIF (HSP70) (Lanneau et al., 2008).

Only when some stress overwhelms the activity of the chaperone network, accumulation of misfolded proteins can transmit a potent cell death signal. Thus, the development of chaperone inhibitors can pave the way for the synthesis of novel antineoplastic compounds that could act as strong inducers of tumour cell death (Khalil et al., 2011).

High molecular weight heat shock proteins

HSP90. The two main HSP90 isoforms, HSP90α and HSP90β, have an interactome that includes more than 400 putative clients that are involved in at least a dozen of biochemical pathways (Neckers and Workman, 2012). Therefore, HSP90s are essential for the viability of all eukaryotic cells, making up 1–2% of cytosolic proteins. Their expression can be further stimulated by stress, and in tumour cells they can reach up to the 4–7% of the expressed proteome, thus shaping all biological processes required for neoplastic progression (Fig. 17.3A). In physiological conditions, HSP90s are found as dimers in association with several intracellular proteins including actin, tubulin, kinases, receptors, and transcription factors (Lanneau et al., 2008). The stability of these proteins, called HSP90 clients, depends on the chaperone, and its inhibition induces client degradation by the proteasome. HSP90 dimers undergo a complex chaperone cycle and associate into higher-level oligomers under stress conditions. The rate of ATP hydrolysis by HSP90 is low, thus the Hsp90 cycle may be differently tuned within different tissues or subcellular compartments by complex post-translational modifications and assembly of specific co-chaperones. Hsp90 function is indeed regulated by phosphorylation, acetylation, S-nitrosylation, oxidation and ubiquitination (Mollapour and Neckers, 2012). For instance, phosphorylation at multiple sites provides a mechanism to regulate the chaperoning of different client proteins, including RAF, c-Src, and Akt kinases, whereas acetylation tends to destabilize interactions with clients, including the transcription factor HIF-1α that displays important roles in the progression to malignancy.

Co-chaperones are co-factors that regulate the rate of HSP90 ATP cycle (Lanneau et al., 2008). In addition, some co-chaperones are adaptors that deliver specific substrates to HSP90, whereas others modulate HSP90 activity via post-translational modifications. Examples include ubiquitin ligation by the E3 ubiquitin ligase CHIP that sends clients to degradation by the proteasome, or dephosphorylation by the phosphatase PP5 that inhibits HSP90-dependent stabilization of several kinases, such as RAF1. The co-chaperones Cdc37, p23, and Hop add another layer of complexity in HSP90 regulation by inhibiting its ATPase activity. Cdc37 is the co-chaperone that has the most important pro-neoplastic function, as it associates with two-thirds of the kinome, including mutant kinases that drive cancer progression (Trepel et al., 2010; Li and Buchner, 2013). Each HSP90 protomer contains three regions, a N-terminal ATP-binding domain (N-domain), a middle domain (M-domain) involved in ATP hydrolysis, and a C-terminal domain (C-domain) responsible for HSP90 dimerization and for interactions with several co-chaperones. ATP binding triggers a series of conformational changes (Fig. 17.3B) leading to the 'closed conformation' of the chaperone in which ATP hydrolysis occurs. After ATP is hydrolysed, the N-domains dissociate, release ADP as well as inorganic phosphate (Pi), and Hsp90 returns to the open (*apo*) conformation again. There is a dynamic equilibrium between the different conformations of Hsp90. This conformational plasticity is functionally important since it may allow Hsp90 to adapt to different client proteins. HSP90 might facilitate conformational rearrangements in client proteins, or it might sequester them, thus avoiding their proteasomal degradation (Scaltriti et al., 2012; Pearl, 2016). However, it remains unclear the exact effect on the clients of the ATPase activity of HSP90, and what determines whether a protein is a HSP90 client. Notably, HSP90 can also be secreted, and this occurs constitutively in a variety of tumour cells under the regulation of HIF-1α. Secreted Hsp90 inhibits TGFβ activation, decreasing the tumour-suppressing effects of TGFβ. The cell surface receptor for Hsp90 is LRP-1, which plays a critical role in receptor-stimulated ERK1/2 activation (Li et al., 2012).

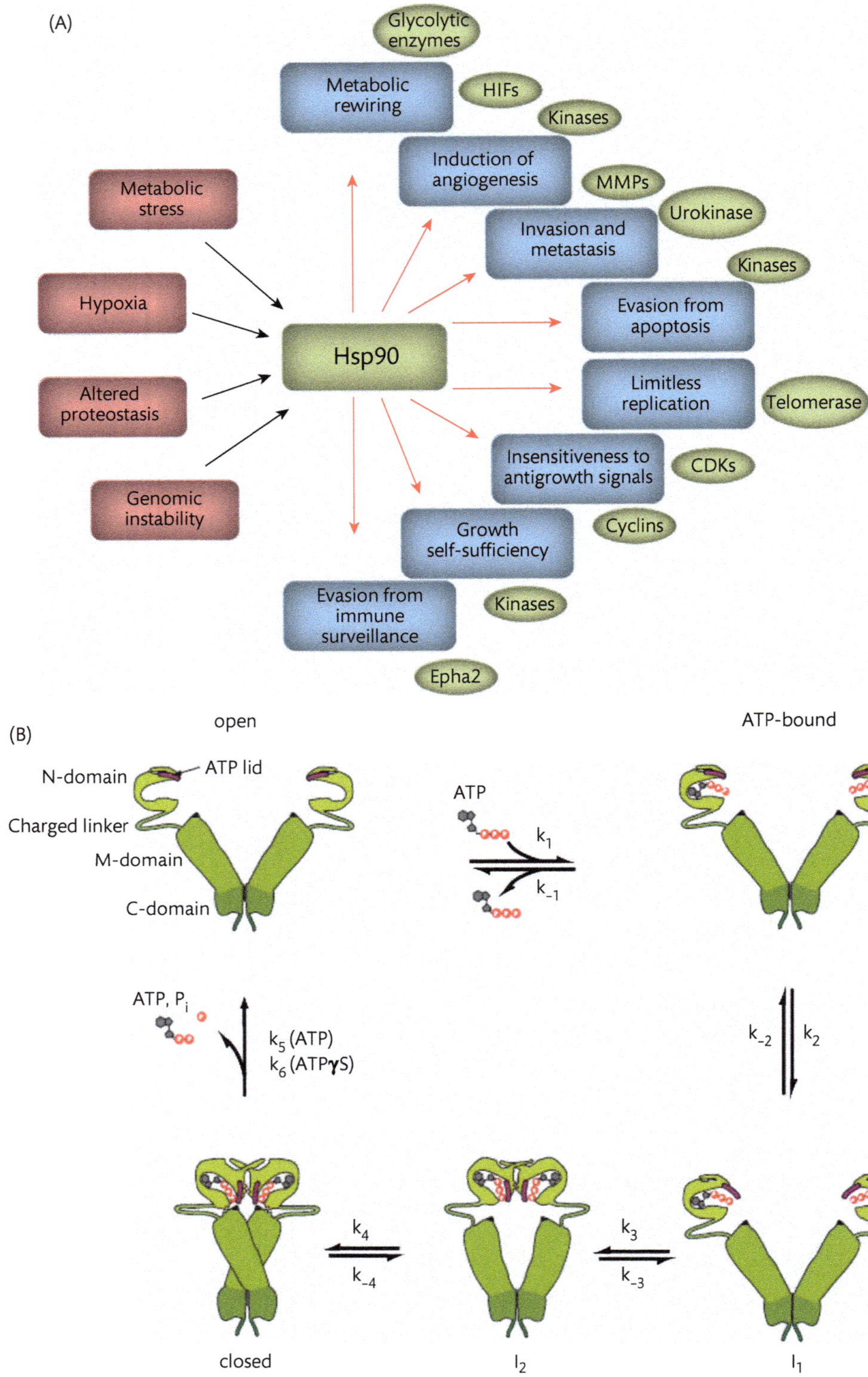

Fig. 17.3 Chaperone activity and cycle of HSP90 and HSP70. (A). In cancer cells, HSP90 regulates numerous signalling pathways and biological functions in response to a variety of environmental stresses (B). Chaperone cycle of HSP90. HSP90 rapidly binds ATP and then reaches the first intermediate state (I_1), in which the N-domains are open and the ATP lid is closed. N-terminal dimerization causes the formation of the second intermediate state (I_2): the M-domain and the N-domain interact, leading to a fully closed state in which ATP hydrolysis occurs. Then, the N-domains dissociate and HSP90 releases ADP and P_i, returning to the open conformation. (C). Mechanism of substrate remodelling action for HSP70 chaperones. ATP-bound HSP70 faintly interacts with misfolded substrates (1). A co-chaperone, the J-domain, stimulates ATP hydrolysis, thus triggering conformational changes and stabilization of substrate binding (2). Nucleotide exchange factor stimulates nucleotide exchange by HSP70 that returns to the low-affinity, ATP-bound conformation after release of the substrate (3). The substrate either refolds to the native conformation (4) or it binds again to HSP70 and the chaperone cycle is repeated.

(A) Source: data from Neckers L and Workman P, 'Hsp90 molecular chaperone inhibitors: are we there yet?', *Clinical Cancer Research*, Volume 18, Issue 1, pp. 64–76, Copyright ©2012 American Association for Cancer Research; (B) Reproduced with permission from Li J and Brenner J, 'Structure, function and regulation of the hsp90 machinery', *Biomed Journal*, Volume 36, Issue 3, pp. 106–17, Copyright © 2013; and (C) Reproduced with permission from Doyle, S et al. 'Protein rescue from aggregates by powerful molecular chaperone machines', *Nature Review of Molecular Cell Biology*, Volume 14, Issue 10, pp. 617–29, Copyright © 2013 Springer Nature.

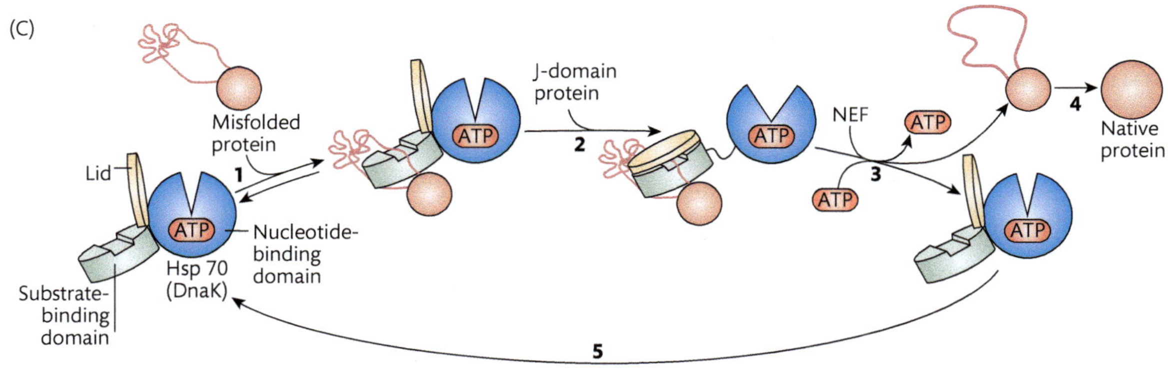

Fig. 17.3 Continued

HSP70. In humans, the HSP70 family of molecular chaperones encompasses at least 11 proteins with partially overlapping functions in the cell. These chaperones preclude the premature misfolding of nascent polypeptides and assist the assembly of multi-protein complexes and the transport of proteins across cellular membranes (Lanneau et al., 2008). HSP70 chaperones can be thought of as a potent buffering system for both extrinsic and intrinsic cellular stresses, including environmental stimuli (e.g. infective agents) or dysregulated replicative signals that prompt oncogenic transformation. Some HSP70 members display a cytosolic localization, like HSP70 itself or HSC70; others are located into mitochondria (mtHSP70) or in the endoplasmic reticulum (GRP78/Bip). Expression of most HSP70s is very low under physiological conditions but is strongly induced by different stresses. HSP70-1a and -1b, generally referred to together as HSP70-1, can account for up to 2% of total protein in a stressed cell. A range of stimuli, including anticancer agents, rapidly stimulate the synthesis of stress-inducible HSP70, whose function is critical for cell death induction. Thus, HSP70-1 chaperones are responsible for stress-induced survival, whereas HSC70 constitutively carries out the homeostatic functions of this chaperone family. Indeed, mouse embryonic cells lacking HSP70-1 are very sensitive to apoptosis induced by a wide range of lethal stimuli (Lanneau et al., 2008).

HSP70 chaperones have a peptide binding domain responsible for substrate binding/refolding and a N-terminal ATPase domain (ABD) that induces release of client proteins and accomplishes ATP hydrolysis, which is extremely slow and constitutes the rate-limiting step in the ATPase cycle of HSP70 interactions (Fig. 17.3C). Several co-chaperones modulate ATP hydrolysis. Members of the DnaJ (HSP40) family bind to ABD, increasing the rate of ATP turnover and potentiating chaperone function, whereas Bag-1, Hsp110, or HspBP1 catalyse the release of ADP, thus completing the chaperone cycle. Moreover, TPR domain co-chaperones (Hop, CHIP) bind to a C-terminal motif of both Hsp70 and Hsp90 and are essential for the assembly of complexes between these two chaperones (Khalil et al., 2011).

Several data suggest that HSP70-1 might play a causal role in cancer initiation, acting as a *bona fide* oncogene. HSP70-1 induction is sufficient to confer tumorigenicity to fibroblasts and fibrosarcoma cells and to induce T-cell lymphoma in mice (Murphy, 2013). HSP70-1 is required for transformation of mammary epithelial cells mediated by the Her2/neu receptor, where HSP70-1 accelerates cell cycle and its absence induces senescence (Meng et al., 2011). Indeed, HSP70-1 counteracts both p53-dependent and independent senescence, and HSP70-1 depletion strongly activates p53 and its downstream cell cycle inhibitor p21. HSP70-1 overexpression has been correlated with poor prognosis in different types of neoplasms including colon, bladder and breast cancer and melanoma (Lianos et al., 2015), and with therapeutic resistance, for example, to cisplatin in prostate cancer or to imatinib in chronic myeloid leukaemia (Khalil et al., 2011; Murphy, 2013). Importantly, HSP70-1 is a major component of the antiapoptotic phenotype of tumour cells, as it inhibits several components of the apoptotic machinery: assembly of death-inducing signalling complex, activation of the proapoptotic factor Bax, release of cytochrome *c*, and activation of caspases. HSP70-1 also enhances activation of several oncogenic signalling cascades, including the pro-survival PI3K/Akt transduction pathway, and has a crucial role in maintaining DNA integrity by binding to poly (ADP-ribose) polymerase-1 and regulating the base pair excision system (Lianos et al., 2015; Kumar et al., 2016). Inhibition of HSP70-1 expression with antisense RNA induces apoptosis and impairs autophagy in several cancer cell types and eradicates xenograft tumours in immunocompromised mice. Both HSP70-1 and HSC70 are involved as co-chaperones in the delivery of client proteins to HSP90, and treatment of neoplastic cells with HSP90 inhibitors usually raises the expression levels of HSP70 family members (Murphy, 2013).

HSP70 may also have antitumour functions. Several tumour cell types express HSP70 on their surface and release it in the extracellular space and in serum, where it acts as an immune-modulatory molecule that activates both innate and adaptive immune responses. HSP70 stimulates the MHC-I presentation pathway of antigen presenting cells and cytotoxic T-lymphocytes, activates natural killer cells against tumours expressing HSP70 on their cell surface, induces the release of proinflammatory cytokines from innate immune cells, primes $CD8^+$ T cells for antigen production and enhances $CD4^+$ T-cell activation via MHC-II-mediated peptide presentation (Sherman and Multhoff, 2007). Therefore, it is conceivable to use HSP70 to develop an immunotherapeutic antineoplastic approach, and a HSP70-based vaccine is in phase I clinical trials (Kumar et al., 2016).

HSP60. HSP60, also called chaperonin, is constitutively expressed, although it can slightly be induced following heat or other stresses. HSP60 is mostly contained within the mitochondrial matrix, where it participates in the folding and transport of proteins and facilitates the proteolytic degradation of misfolded or denatured proteins. These functions require the co-chaperone HSP10 that regulates both the substrate binding of the chaperone and its ATPase activity. HSP60 forms cylindrical complexes with a cavity that provides a closed, protected environment to help client folding. The efficiency of this process requires post-translational modifications such as phosphorylations, s-nitrosylations and acetylations (Khalil et al., 2011). HSP60 is up-regulated in several human cancers, for example in glioblastoma (Lianos et al., 2015), where it could contribute to several pro-neoplastic biological processes. In addition to its mitochondrial functions, HSP60 inhibits the apoptogenic activity of Bax and increases the activity of the proto-oncogene β-catenin, a transcription factor involved in proliferation and epithelial-mesenchymal transition of tumour cells (Lanneau et al., 2008).

Low molecular weight heat shock proteins

HSP40. HSP40 chaperones are also known as DnaJ chaperones and constitute a large family of co-chaperones found in different subcellular locations, encoded by at least 41 genes highly conserved throughout evolution. Most DnaJ/HSP40 proteins bind to HSP70s through a "J" domain (Fig. 17.3C), thus enhancing the ATPase activity of HSP70 family members and regulating their interactions with clients. HSP40 chaperones are therefore important in a variety of biological processes including translation, folding, unfolding, translocation and degradation of proteins, mainly via HSP70 regulation. Little is known about the exact functions of HSP40s in neoplastic transformation, where they probably contribute to the tumorigenic functions of HSP70 proteins. For instance, it was shown that HSP40/HSP70 complexes enhance the activity of the pro-survival PI3K/Akt signalling pathway (Mitra et al., 2009) and inhibit apoptosis during prostate cancer progression (Lianos et al., 2015).

HSP27. HSP27 is an ATP-independent chaperone whose main functions are protection against protein aggregation and refolding of damaged proteins. HSP27 dimers act as building blocks for the assembly of large oligomers, up to 1,000 kDa. This oligomerization is a dynamic process elicited by HSP27 dephosphorylation at three specific serine residues upon cell exposure to a wide variety of stresses such as serum deprivation, oxidants, anticancer drugs, or radiations. HSP27 is a strongly inducible chaperone, and its expression is extremely high in several cancer types, including breast, endometrial, lung and prostate cancer, where its induction correlates with increased proliferation, metastatic potential, poor response to chemotherapy, poor prognosis (Lanneau et al., 2008) and resistance to death stimuli through inhibition of several apoptotic steps, for example caspase activity (Lianos et al., 2015). In addition, HSP27 can suppress the senescence programme, as its depletion in transformed cells prompts activation of the p53/p21 pathway and expression of typical signs of cell senescence. HSP27 also takes part in EGF-mediated epithelial-mesenchymal transition via modulation of the β-catenin/Slug signalling pathway and plays a crucial role in tumour cell invasion and metastasis by eliciting cytoskeletal remodelling and activation of metalloproteases MMP2 and MMP9 (Khalil et al., 2011; Lianos et al., 2015).

Hsp22. Hsp22 is a chaperone that controls the proteasome- and autophagy-dependent degradation of unfolded proteins, which indicates a role of Hsp22 in protection of cells from insoluble protein aggregates. Hsp22 is overexpressed in melanoma, glioblastoma, and breast cancer cells, where it contributes to cell proliferation and apoptosis resistance (Acunzo et al., 2012).

Alpha B-crystallin. Alpha B-crystallin is a close Hsp27 homologue, whose overexpression was reported to transform immortalized human mammary epithelial cells, to induce anchorage-independent growth, to increase cell migration and invasion and to elicit the formation of invasive mammary carcinomas in nude mice. These biological processes would be stimulated by alpha B-crystallin-dependent induction of EGF and of Ras/ERK signalling pathway. In addition, alpha B-crystallin prevents p53 activation following oxidative stress, induces NF-κB activation and resistance to vinblastine and inhibits activation of proapoptotic Bcl-2-family proteins (Khalil et al., 2011; Acunzo et al., 2012).

HSF-1. Heat shock factor protein 1 (HSF-1) is a transcription factor that controls the expression of a variety of chaperones. HSF1 exists in an inactive cytosolic complex with Hsp40/Hsp70 and Hsp90. Upon stress, HSF1 is released from this chaperone complex and translocates into the nucleus. Deletion of HSF-1 gene does not alter basal expression of HSPs but abolishes their stress-induced expression. In mice, HSF-1 ablation inhibits tumorigenesis induced by dominant negative mutation of p53, as well as chemical-induced skin carcinogenesis associated with activating mutations of the H-Ras proto-oncogene. HSF-1 coordinates a variety of biological processes involved in tumorigenesis, such as cell proliferation, survival, ribosomal biogenesis, and glycolysis, and is involved in the activation of the RAS/MAPK and cAMP/PKA signalling pathways (Zoubeidi and Gleave, 2012).

Peptidyl-prolyl isomerases

In all peptide bonds the *trans* isomer is strongly favoured over the *cis* isomer, with the exception of peptide bonds between proline and its preceding amino acid (Xaa-Pro or prolyl bonds), where the *trans* conformation is only slightly favoured (Fig. 17.4A). Thus, peptidyl-prolyl isomerases (PPIases) can modulate protein functions that depend on the coupling between *cis/trans* isomerization at proline residues, and this regulation can be used to switch proteins between different functional states (Fig. 17.4B), or to regulate the assembly of multimeric complexes (Theuerkorn et al., 2011). These chaperones catalyse prolyl isomerizations in unfolded, partially folded, and in native state proteins during folding transitions of the polypeptide backbone. As prolyl isomerizations are slow processes, PPIases can set the rate of folding and of subsequent activation in a variety of proteins and are therefore involved in a wide spectrum of biological processes like signal transduction, cell differentiation, and apoptosis. However, at present the characterization of the biological role of prolyl isomerization reactions remains limited, and most studies are performed on isolated proteins (Schmidpeter and Schmid, 2015).

Prolyl isomerization is performed by PPIase domains of 10 to 18 kDa, but there are multidomain proteins formed by up to four PPIase domains together with accessory domains that supply additional layers of regulation and versatility to these chaperones, also contributing to their subcellular-specific targeting and to the definition of their panel of clients and interactors. There are three PPIase families (Fig. 17.4C; see Wang and Heitman, 2005). The cyclophilin family is composed of 19 members in mammals, either single-domain polypeptides or larger proteins found in virtually all organisms and subcellular compartments. The second family is formed in

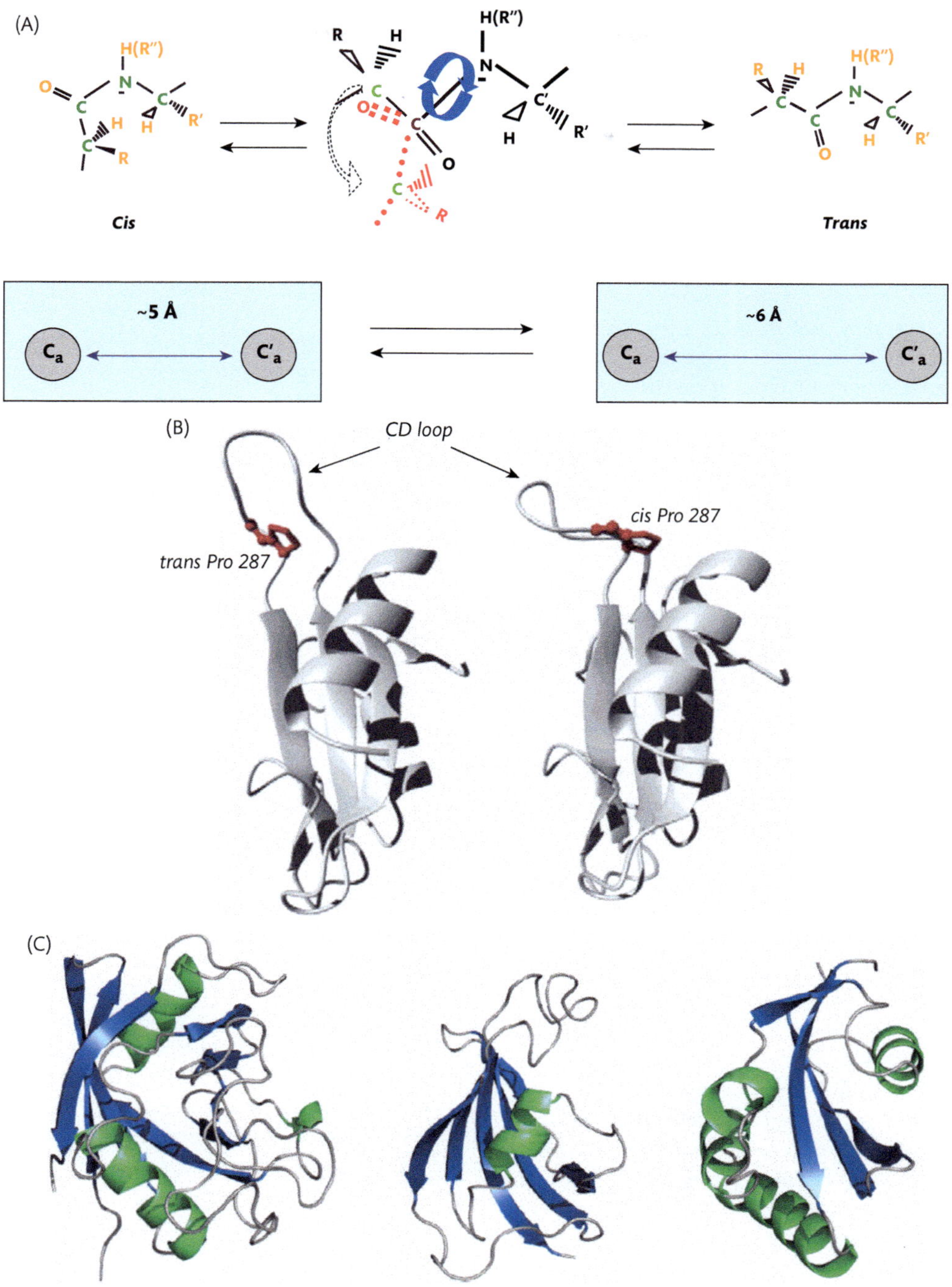

Fig. 17.4 Structure and activity of phosphatidyl-prolyl-isomerases (PPIases). (A). *cis/trans* isomerization of the secondary amide peptide bond and of proline are slow reactions, with relaxation times of 0.2 and 100 s, respectively, at room temperature. The *cis* and *trans* conformers undergo reversible isomerization by permanently rotating about the nitrogen-carbon bond. Structural changes in the protein backbone can depend on intramolecular atom displacements that are isomer-specific. In native and unfolded proteins, unfavourable steric contacts lead to a low a *cis/trans* ratio in unfolded chains, which is increased by the activity of *cis/trans* isomerases (B). NMR analyses of the *trans* (left) and *cis* (right) isomers in the SH2 domain of the kinase Itk highlight how prolyl isomerization can remodel protein structure. Pro287 in the CD loop and the preceding imide bond are shown in red. (C). Structures of representative prolyl isomerases from the three families: cyclophilins (CyP-A), FK506 binding proteins (FKBP12) and parvulins (Par10). Helices are shown in green and strands are in blue.

mammals by 16 FK506 binding proteins (FKBPs) structurally unrelated to cyclophilins. In both families PPIase activity is inhibited by immunosuppressants, cyclosporin A, and FK 506, respectively, whose immunosuppressive properties rely on inhibition of the protein phosphatase 2B (calcineurin), an interactor of both cyclophilins and FKBPs that controls cytokine transcription in T cells. Therefore, cyclophilins and FKBPs are collectively named immunophilins, whereas prolyl isomerases of the third family, termed parvulins, do not bind any immunologically active molecule (Schmidpeter and Schmid, 2015).

PPIases and cancer. Several evidences indicate that PPIases of all three families are involved in neoplastic progression. Cyclophilin A (CyP-A) is overexpressed in many types of cancers, such as pancreatic cancer, non-small cell lung cancer, hepatocellular carcinoma and breast cancer, and its expression is under the control of p53 and HIF-1α, two transcription factors with key roles in tumorigenesis. However, the exact role of CyP-A in cancer pathogenesis remains unknown, even if it was proposed that it takes part in proliferation, prevention of chemotherapeutic-induced apoptosis, and defence from oxidative stress. CyP-B and CyP-C are located primarily in the ER and form complexes with chaperones involved in the ER stress response, such as Grp78/BiP and Grp94, ensuring proper conformations of newly synthesized and secreted proteins (Lee and Kim, 2010). CyP-B shields cells from ER stress-induced death blocking the Ca^{2+} leakage from ER to cytosol, and its overexpression is associated with tumour progression. CyP-C regulates activation of the survival kinase Akt and of the pro-metastatic protease MMP-2 in murine breast cancer cells (Lee and Kim, 2010). CyP-40 and FKBP51/FKBP52 contain TPR motifs for association with the Hsp90 and Hsc70 chaperones in heterocomplexes where the PPIase domains probably target the Hsp90 client proteins and contribute to the conformational transitions required for the function of the complex. CyP-40 is overexpressed in prostate cancer where it regulates androgen-dependent cell proliferation. FKBP51 and FKBP52 could modulate neoplastic growth by regulating nuclear transport of NF-κB, microtubule stabilization, Akt activation, and apoptosis (Schiene-Fischer, 2015; Schmidpeter and Schmid, 2015).

The parvulin Pin1 catalyses the isomerization of phosphorylated Ser/Thr–Pro bonds. Proline-directed kinases are a large superfamily of proteins that includes cyclin-dependent protein kinases, mitogen-activated protein kinases (MAPKs), and glycogen synthase kinase 3 (GSK3). Isomerization of phospho-Ser/Thr–Pro motifs could define subsequent enzyme actions on the targets, such as their dephosphorylation or regulation of their stability via ubiquitination (Liou et al., 2011). Therefore, Pin1 regulates the stability, enzymatic activity, interactions, processing, post-translational modifications, or localization of a multitude of phosphorylated proteins involved in numerous biological processes, such as transcription, cell cycle progression, metabolism, apoptosis, and protein degradation (Schiene-Fischer, 2015). Several biochemical pathways undergo Pin1-dependent modulation: signalling via Ras/ERK and PI3K/AKT pathways; FAK-dependent regulation of cell migration and invasion; transcriptional programmes mastered by β-catenin, p53, and NF-κB; cell death inhibition through regulation of the expression of Bcl-2 protein family members and metabolic switch towards a pro-neoplastic Warburg phenotype (i.e. induction of aerobic glycolysis). A high expression level of Pin1 correlates with induction of a malignant phenotype and poor prognosis in several cancer models, such as hepatocarcinoma, mammary tumours and lymphomas (Theuerkorn et al., 2011; Lu and Hunter, 2014).

Protein quality control is a compartmentalized process

Quality control is mandatory for correct folding and subcellular destination of nascent proteins, interaction with biochemical partners, formation of multimeric complexes, dynamic conformational changes, and disposal of inefficient or malfolded proteins. Therefore, cells have developed an exquisitely sophisticated network of control systems, which is endowed with peculiar features in different subcellular compartments. Here we will briefly analyse the systems that are active in the ER and in mitochondria for their importance in the response to stress stimuli of transformed cells.

The unfolded protein response in the endoplasmic reticulum

ER is an extensive network of membranes that regulates Ca^{2+} homeostasis, lipid biogenesis, folding and post-translational modifications of secreted and membrane proteins (Wang and Kaufman, 2014). Cells keep under a tight control the process of protein synthesis and maturation, in order to ensure that only properly folded proteins can proceed in their maturation. In neoplastic cells, exposure to shortages of oxygen and nutrients, unbalances in Ca^{2+} and ROS levels, together with a high demand of protein synthesis can easily induce ER stress and accumulation of misfolded proteins in ER lumen.

Three major ER membrane proteins sense the presence of unfolded proteins—PERK, ATF6, and IRE—which lead to the activation of an integrated signal transduction pathway called the UPR (Fig. 17.5A; see Hetz et al., 2015). UPR attenuates protein translation and induces the activity of molecular chaperones with the aim to increase ER folding activity and to degrade malfolded proteins by the ubiquitin–proteasome machinery in a process termed ER-associated degradation (ERAD). UPR is therefore a routine that maintains ER homeostasis and cell survival. When stress is too intense or prolonged and abnormal proteins are persisting, the role of the UPR tilts towards induction of cell death. However, in cancer cells UPR activation provides a survival strategy for cells to thrive in a stressful environment (Fig. 17.5B). To adapt to the adverse conditions they encounter, neoplastic cells enhance ER folding capacity through an increased expression of chaperones, such as GRP78 and GRP94. Indeed, UPR activation has been widely observed in a variety of human cancers, such as glioblastoma, cervix, and breast carcinoma, and in several forms of lymphoma and has been associated with poor prognosis and cancer recurrence (Luo and Lee, 2013).

GRP78 and cancer. Glucose-regulated protein 78 (GRP78) belongs to the HSP70 family and is one of the best-characterized ER chaperones, but is located also in plasma membrane, cytoplasm, mitochondria and nucleus and it can be secreted. GRP78 is a central stress sensor. In normal conditions it sequesters PERK, IRE1, and ATF6, thus inhibiting UPR. Accumulation of misfolded proteins in the ER detaches GRP78 from these proteins, which results in their activation and in UPR induction (Hishiya and Takayama, 2008). GRP78 also preserves intracellular Ca^{2+} concentration by regulating both Ca^{2+} uptake through the Ca^{2+}–ATPase pump and Ca^{2+} efflux via the IP_3R channel (Zhang and Zhang, 2010). GRP78 expression increases in several tumour types, where it correlates with environmental stresses such as hypoxia, glucose deprivation, acidosis, inflammation, and exposure to chemotherapeutics. In neoplastic cells, GRP78 is implicated in proliferation, apoptosis, and therapy resistance, immune escape, metastasis, and angiogenesis both *in vitro* and in cancer mouse models. When expressed on the cell surface, GRP78 enhances PI3K/AKT signalling, increasing cell proliferation and survival. Following interaction with p53 and focal adhesion kinase, GRP78 promotes genome instability and invasion, respectively, whereas through indirect regulation of Akt1, HIF-1α and c-Myc GRP78 takes part in the metabolic rewiring of tumour cells (Li and Li, 2012).

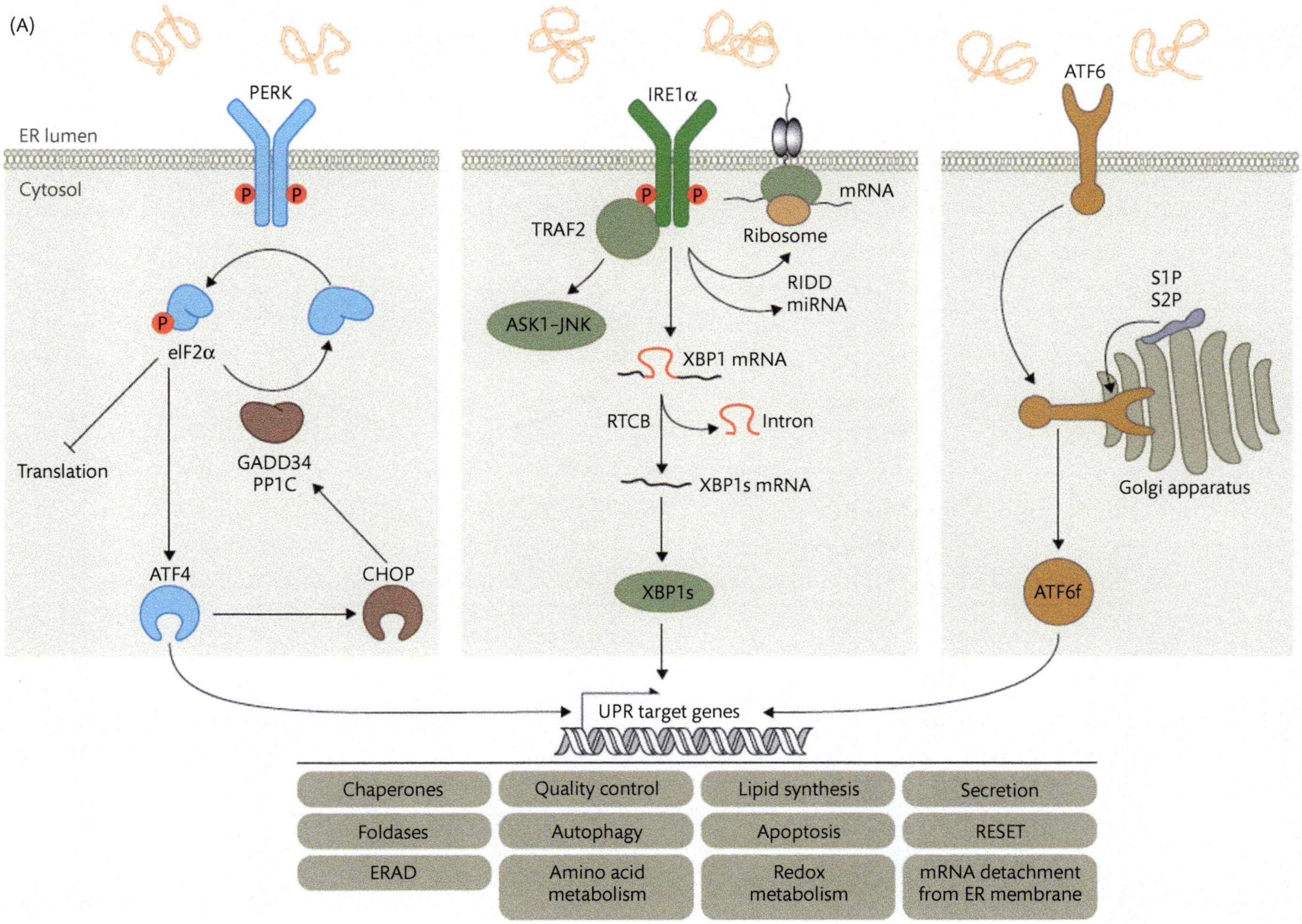

Fig. 17.5 ER stress and unfolded protein response in tumour progression. (A). The three arms of the UPR require three different ER stress sensors: PERK, IRE1α, and ATF6. Their activation induces signalling pathways that enhance protein-folding capacity of ER, tending to restore protein-folding homeostasis. PERK inhibits translation via eIF2α phosphorylation; concomitantly ATF4 promotes the transcription of selected genes with the aim of reinstating proteostasis (e.g. via a CHOP/GADD34/PP1C feedback on eIF2α itself). IRE1α activates both the TRAF2/ASK1/JNK transduction cascade and the XBP1 splicing, thus regulating gene expression both at the transcriptional and at the post-transcriptional level. ATF6 moves to the Golgi complex, where it is activated by S1P/S2P-mediated cleavage and acts as a transcription factor. Grey boxes illustrate the cell functions regulated by UPR (RESET: rapid ER stress-induced export) (B). The cross-talk between ER stress and UPR contributes to tumour progression. Tumour cells activate UPR as an adaptive response to the ER stress induced by stimulation from growth factors and environmental signals. Cancer cells under ER stress secrete angiogenic and inflammatory factors and stimulate both endothelial cells and surrounding stromal cells, mostly tumour-associated macrophages (TAMs). In return, angiogenesis favours tumour cell survival and activated TAMs secrete inflammatory cytokines that promote tumour growth, invasion, and metastasis. (C). Cross-talk between UPR and deregulation of oncogene or tumour suppressor gene products (red and green boxes, respectively) in cancer cells. Activation of UPR can promote survival, transformation, senescence, or apoptosis. Examples include: loss of BRCA1 that upregulates GRP78/BIP under chronic ER stress; loss of tuberous sclerosis complex that prompts ER stress by increasing protein synthesis in conditions of reduced expression of ATF4 and C/EBP homologous protein (CHOP), thereby causing UPR activation; loss of PTEN that induces UPR activation and increases aerobic glycolysis; Myc or Ras activation, that modulate several biological routines required for neoplastic transformation, such as apoptosis or senescence inhibition, autophagy, and metabolic rewiring.

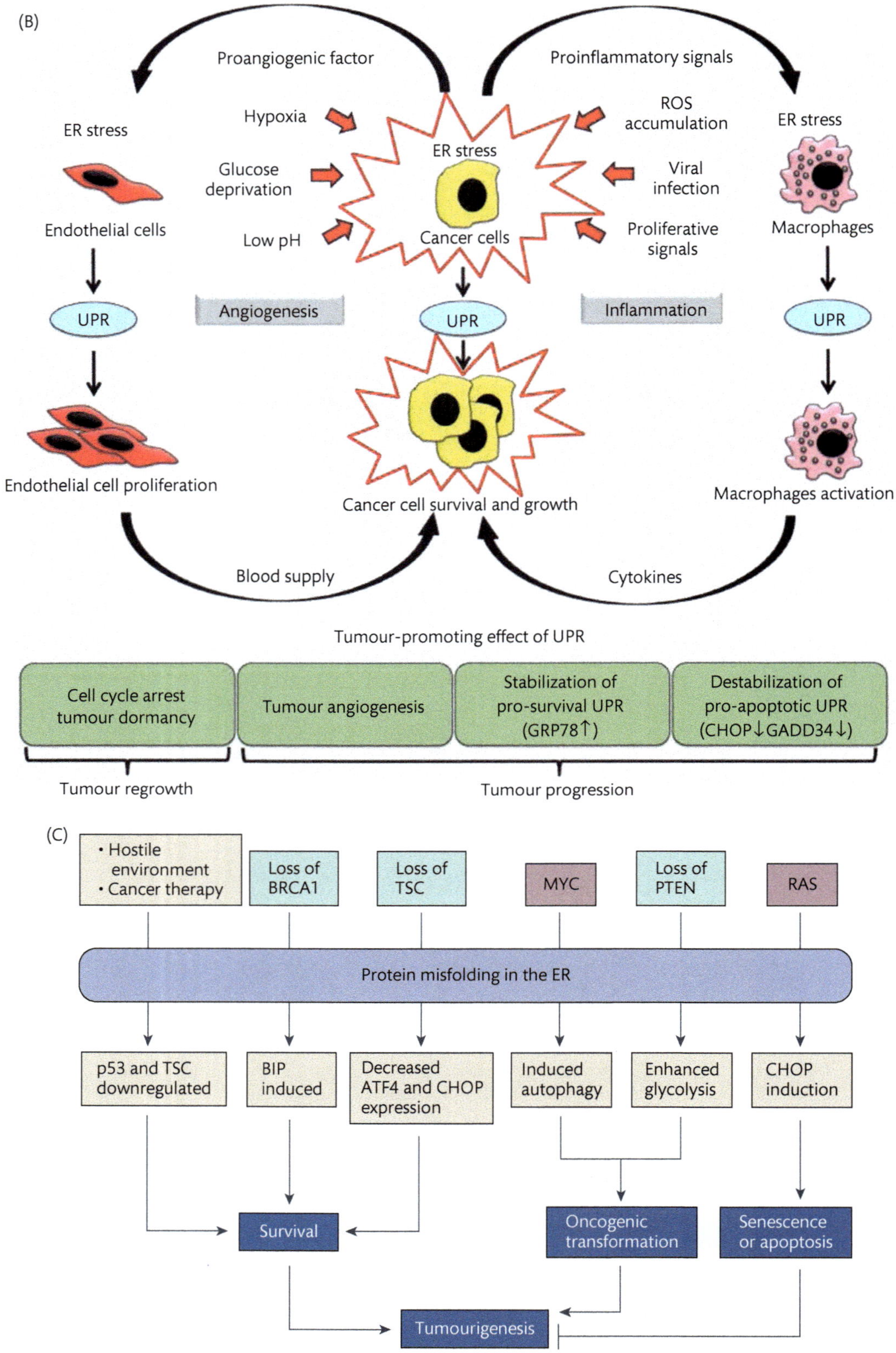

Fig. 17.5 Continued

UPR signalling network is intertwined with other signalling pathways and transcription factors (Fig. 17.5C), such as MAPK, AKT/mTOR, CREB, NRF2, HIF1α and NF-κB, whose deregulated activation is crucial for the generation of specific biological traits required for neoplastic progression (Chevet et al., 2015). For instance, hypoxia activates in an integrated way UPR in the ER, mTOR signalling and HIF-1-dependent transcription, thus adapting cell metabolism, survival, and proliferation to the paucity of oxygen (Wouters and Koritzinsky, 2008).

PERK and cancer. PERK transiently inhibits global translation and enhances the translation of selected mRNAs. Its role in tumour progression is controversial: PERK deficiency reduces tumour size and growth in transgenic mouse models, whereas in human squamous carcinoma cells PERK suppresses cell proliferation and tumour formation. PERK induces the transcription factor ATF4, which is overexpressed in many solid tumours where it prompts transcription of chaperones, antioxidants, antiapoptotic and autophagy proteins and is involved in proliferation and survival of neoplastic cells under conditions of nutrient deprivation. PERK also leads to activation of the transcription factor NRF2, which masters the response to oxidative stress together with ATF4.

ATF6 and IRE1 in cancer. ATF6 activation in tumours is poorly investigated. In a hepatocellular carcinoma model, ATF6 upregulates GRP78 as well as proteins involved in cell cycle, signal transduction, apoptosis, and adhesion (Tameire et al., 2015). IRE1 takes part in mRNA processing. One of its targets is XBP1, a major transcription factor that upregulates expression of proteins involved in protein folding, quality control, and ERAD and is required for solid tumour growth and survival under hypoxic stress.

The unfolded protein response in mitochondria

Mitochondria are dynamic organelles that house a high number of essential metabolic and bioenergetic pathways, such as the tricarboxylic acid (TCA) cycle, the oxidative phosphorylation (OXPHOS) machinery and branches of amino acid, lipid, and nucleotide metabolic pathways, and generate ATP, reducing equivalents and metabolic intermediates. Therefore, mitochondrial function contributes to numerous cellular functions, and cells have evolved mechanisms to monitor constantly mitochondrial function and to respond to mitochondrial stress. In cancer, cells incur a high level of mitochondrial damage, as they can be exposed to fluctuating levels of oxygen in the core of the tumour mass and to high levels of ROS, at least in part because of OXPHOS downregulation (Nakazawa et al., 2016; Vyas et al., 2016). As a result, mitochondrial structure and function are hampered by lack of proper protein folding and by oxidation-induced misfolding, whereas mtDNA accumulates mutations at high rates. It remains unclear whether stress-associated changes in cancer cell mitochondria provide an advantage or disadvantage to the neoplastic progression. Regardless, these observations suggest that cancer cells strongly depend on quality control and repair circuitries to preserve mitochondrial function.

Induction and molecular mechanisms of UPR in mitochondria

Accumulation of unfolded proteins within mitochondria prompts a transcriptional response termed mitochondrial unfolded protein response (UPRmt) to promote rescue mechanisms and to guarantee cell survival. UPRmt is a retrograde response, as the upstream mitochondrial stress signal initiates a nuclear transcriptional response aimed at increasing molecular chaperones and proteases in mitochondria to maintain protein homeostasis (proteostasis; see Liu and Butow, 2006). UPRmt also activates ROS scavenging machineries and mitochondrial fission and promotes a rewiring of cellular metabolism characterized by an increase in glycolysis and amino acid catabolism, with a concurrent repression of TCA cycle and OXPHOS, probably with the aim to allow cells a better withstanding of mitochondrial stress (Lin and Haynes, 2016).

Cells control the efficiency of mitochondrial function by monitoring the two systems of protein import: the translocase of the outer membrane (TOM) and the translocase of the inner membrane (TIM). These complexes recognize mitochondrial targeting sequences and allow targeting proteins encoded by nuclear genes to specific submitochondrial compartments (outer or inner mitochondrial membranes, intermembrane space, and matrix) in an exquisitely precise way. Both TOM and TIM require the chaperone activity of mtHsp70. A number of perturbations can hamper import, including OXPHOS inhibitors, oxidants and direct inhibition of mitochondrial chaperones, thus making import efficiency an accurate indicator of mitochondrial function. A key protein for UPRmt is ATFS-1 (activating transcription factor associated with Stress-1). In the absence of mitochondrial stress, ATFS-1 is imported into mitochondria and degraded by the Lon protease. When UPRmt signalling occurs, import systems become les functional and a portion of ATFS-1 cannot enter mitochondria. As ATFS-1 has also a nuclear localization sequence, it can move to the nucleus and elicit UPRmt by inducing the expression of mitochondrial chaperone genes and proteases, ROS scavenging and glycolysis proteins.

Mitophagy (i.e. mitochondrial autophagy), is also activated by mitochondrial dysfunction in order to dispose dysfunctional mitochondria through lysosome-dependent degradation. Mitochondrial inner membrane depolarization is a hallmark of organelle damage and impairs import of the Ser/Thr kinase PINK1. As a consequence, PINK1 integrates and accumulates in the outer mitochondrial membrane, where it interacts with the ubiquitin ligase Parkin that in turn recruits the autophagy machinery to degrade the defective organelle.

Mitochondrial chaperones

Once imported in mitochondria, proteins must fold and often assemble in large multimeric complexes. This difficult task is achieved with the assistance of a dedicated repertoire of localized molecular chaperones both in the intermembrane space and in the matrix, such as HSP60, mtHSP70, and HSP90 family members.

HSP60 and mtHSP70. HSP60 is a matrix chaperone composed by HSP60 and HSP10 subunits that form a barrel-shaped complex and facilitates folding of small, soluble proteins. Elevated levels of Hsp60 likely promote efficient protein folding even within a stressed folding environment. HSP60 expression is increased in several neoplasms, including tumours of the digestive, reproductive, and nervous system. RNA interference on HSP60 slows down the growth of xenografted glioblastoma cells, increasing mice survival (Siegelin, 2013). mtHSP70, also known as HSP70-9, mortalin, or Grp75, performs multiple functions in different multimeric complexes: it interacts with the TIM channel to drive polypeptides in the matrix, promotes folding and complex assembly, prevents aggregation of imported proteins, and is required for the biogenesis of iron–sulphur clusters. Several evidences have shown a tight correlation between mtHSP70 overexpression and malignancy of various cancer types (Khalil et al., 2011).

TRAP1. Expression of the HSP90 family member TRAP1 (Fig. 17.6A; see also Lavery et al., 2014) is higher in many tumours compared to surrounding non-malignant tissues, and correlates with progression, metastasis, and disease recurrence in several neoplastic models, including prostate and breast cancer, hepatocellular and colorectal carcinoma and non-small cell lung cancer (Rasola et al., 2014). TRAP1 mainly resides in mitochondria, where it provides resistance to oxidative stress and to a variety of chemotherapeutics. TRAP1 also inhibits opening of the permeability transition pore (PTP), a mitochondrial channel composed by ATP synthase dimers that can be induced by a ROS surge and whose opening leads to cell death (Bernardi et al., 2015). In tumour cell mitochondria TRAP1 could be part of a chaperone complex involving mtHSP90, HSP60, and cyclophilin D (CyP-D), a PPIase that is one of the best-characterized proteinaceous PTP inducers. Thus, TRAP1 exerts a pro-neoplastic function by counteracting ROS-induced, PTP-mediated cell death. However, as the effect of ROS on tumour growth is multifaceted, and oxidative stress can favour genetic instability and aggressiveness of tumours, mainly at advanced stages, the antioxidant effect of TRAP1 could hamper neoplastic progression in specific tumour types or stages. Accordingly, TRAP1 expression levels inversely correlate with tumour grade in cervical carcinoma, clear cell renal cell carcinoma, and high-grade ovarian cancer (Yoshida et al., 2013; Amoroso et al., 2016).

TRAP1 also contributes to the metabolic switch of tumour cells towards aerobic glycolysis (i.e. decreased OXPHOS activity paralleled by enhanced glucose utilization), by inhibiting respiration with two different molecular mechanisms. First, TRAP1 downregulates the activity of cytochrome oxidase, the complex IV of the respiratory chain (Yoshida et al., 2013); second, TRAP1 inhibits succinate dehydrogenase (SDH) (Sciacovelli et al., 2013), an enzyme that oxidizes succinate to fumarate and is at the crossroad between OXPHOS (SDH is the complex II of the respiratory chain) and the TCA cycle. TRAP1-dependent inhibition of SDH leads to an increase in intracellular succinate levels, and in turn succinate acts as an oncometabolite by stabilizing HIF1α, whose transcriptional activity enhances invasiveness, angiogenesis, and further metabolic changes in tumour cells (Semenza, 2013).

The importance of TRAP1 in neoplastic progression is further highlighted by observations that oncogenic kinase pathways directly target TRAP1. TRAP1 is both Tyr-phosphorylated in a Src-dependent way, thus favouring cytochrome oxidase inhibition, and Ser-phosphorylated by ERK1/2 (Yoshida et al., 2013; Masgras et al., 2017). ERK1/2 forms a multimeric complex with TRAP1 and SDH in mitochondria (**Fig. 17.6B**). In this complex, ERK1/2 phosphorylates TRAP1, thus stimulating TRAP1 inhibition of SDH activity and contributing to cell metabolic rewiring and tumour growth. In turn, TRAP1 stabilizes mitochondrial ERK1/2, maintaining the kinase active even under stress conditions (Masgras et al., 2017). Mitochondrial ERK1/2 can further contribute to tumorigenesis by inhibiting another chaperone, the PPIase CyP-D. In mitochondria, CyP-D activity as a PTP inducer is enhanced by the Ser/Thr kinase GSK3. ERK phosphorylation inhibits GSK3, thus down-regulating CyP-D chaperone activity. Eventually, PTP is desensitized to opening, which results in tumour cell protection from a variety of death stimuli (Rasola et al., 2010).

(A)

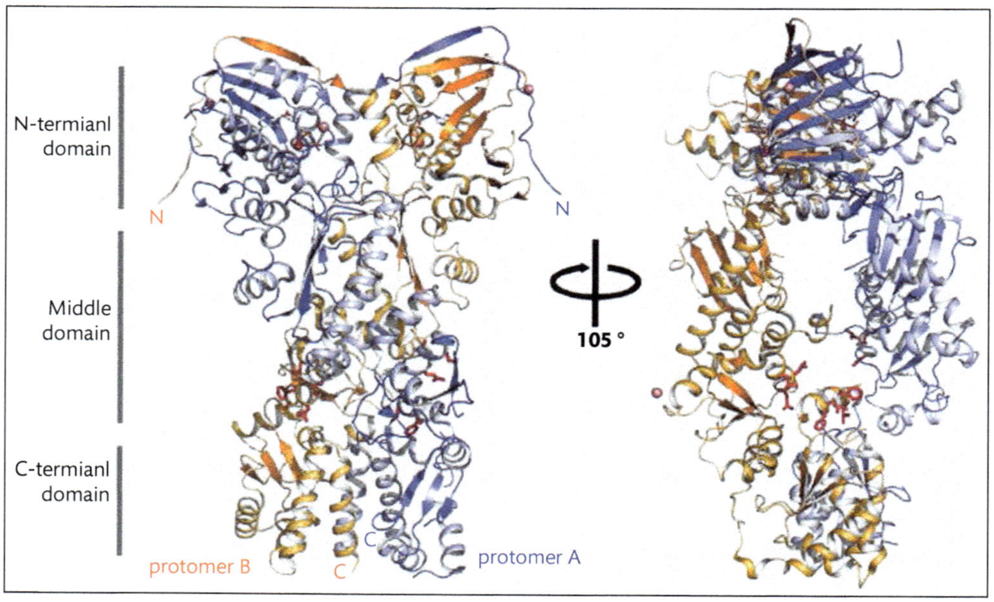

Fig. 17.6 The mitochondrial chaperone TRAP1 and its role in neoplastic progression. (A). Full-length TRAP1 homodimer from *D. rerio*. Protomer A is shown in blue and protomer B in orange. Residues known to bind clients between the middle domain and the C-terminal domain are highlighted in red. (B). TRAP1 forms a multimeric complex with SDH and ERK1/2 in mitochondria of tumour cells. Oncogenic ERK1/2 (e.g. in tumor models where Ras signalling is deregulated by loss of the Ras-GAP neurofibromin, NF1) phosphorylates TRAP1, thus enhancing TRAP1 inhibition of SDH enzymatic activity. This leads to succinate accumulation, which stabilizes HIF1α by blocking the prolyl hydroxylase (PHD) responsible for HIF1α proteasomal degradation. Eventually, HIF1α activates a pro-neoplastic transcriptional programme.

(B)

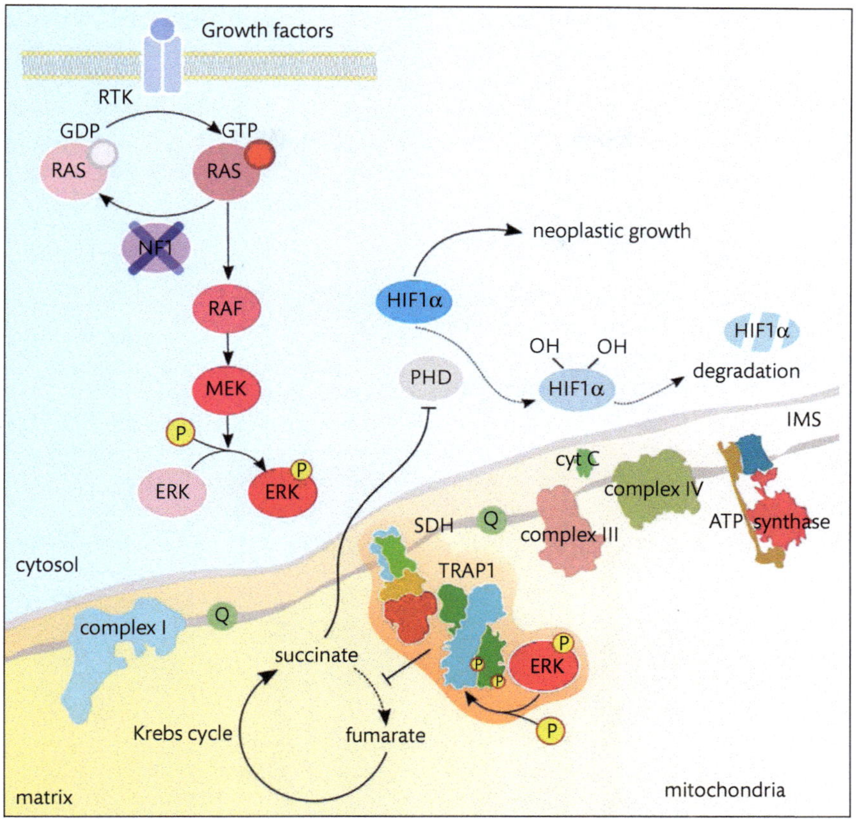

Fig. 17.6 Continued

(A) Reproduced with permission from Lavery LA et al. 'Structural Asymmetry in the Closed State of Mitochondrial Hsp90 (TRAP1) Supports a Two-Step ATP Hydrolysis Mechanism', *Molecular Cell*, Volume 53, pp. 330–43, Copyright © 2014 Elsevier Inc. All rights reserved. https://www.sciencedirect.com/science/article/pii/S1097276513009337. (B) Reproduced with permission from Masgras I et al. 'Absence of Neurofibromin Induces an Oncogenic Metabolic Switch via Mitochondrial ERK-Mediated Phosphorylation of the Chaperone TRAP1', *Cell Reports*, Volume 18, Issue 3, pp. 659–672, Copyright © 2017 The Author(s).

Chaperones as targets for antineoplastic therapies

Antineoplastic strategies aimed at targeting molecular chaperones build on the postulate that neoplastic cells need a marked induction of protein quality control machinery to cope with the variety of stress stimuli they are exposed to. Therefore, the use of HSP inhibitors in combination with other antitumour drugs (e.g. compounds delivering oxidative or proteasomal stress), could result in a better therapeutic outcome than single therapies. At present, inhibition of HSP90 and induction of ER stress constitute the most advanced antitumour approaches based on chaperone targeting.

HSP90 inhibitors. On the surface HSP90 is an unlikely target for the development of antineoplastic strategies, as it is highly expressed in all cell types. Nonetheless, HSP90 inhibitors are more toxic in most cancer cells than in their non-transformed counterparts and tend to accumulate in tumours both in animals and in humans. This selectivity could rely upon a combination of reasons. HSP90 expression is strongly induced in many tumour types, and associates with a poor overall survival of patients; post-translational modifications and interacting clients of the chaperone change following neoplastic transformation; oncogenic mutations or deregulation of many clients might induce their stable association with HSP90, creating druggable interactions; finally, HSP90 localizes to ectopic cellular or extracellular compartments in neoplasms. As an example, extracellular HSP90 assists in the activation of matrix metalloproteinase-2, an enzyme that plays an important role in tumour cell invasion. In addition, HSP90 inhibitors could prevent drug resistance in tumours because several oncoproteins are addicted to HSP90 to maintain their unstable mutant conformations. Thus, targeting HSP90 could affect sensitivity to drugs of multiple targets and signalling pathways (Barrott and Haystead, 2013).

The benzoquinone geldanamycin and the macrolide radicicol act as HSP90 inhibitors by occupying its ATP binding site, thus blocking the chaperone cycle (Mollapour and Neckers, 2012). Both compounds exhibit potent toxicity on tumour cells, but they cannot be used as drugs due to poor solubility (geldanamycin) or poor stability (radicicol; Li and Buchner, 2013). Several geldanamycin derivatives have been synthesized. 17-allylamino-17-demethoxygeldanamycin (17-AAG, KOS-953, tanespimycin) is more hydrophilic than geldanamycin and has undergone preclinical studies and phase I clinical trials. It induces a prolonged disease stabilization in various tumour types, but without any tumour regression. This limited activity could depend on partial inhibition of the target client proteins, on P-glycoprotein-dependent efflux from target cells, and on requirement for reductive metabolism to reach full activation of 17-AAG (Neckers and Workman, 2012).

A new generation of geldanamycin derivatives is currently under study, such as retaspimycin hydrochloride (IPI-504), ganetespib, or AUY922. All these molecules are well tolerated and effective in cells, with reduced liver and cardiovascular toxicity, and experiments in animal models are under way (Lianos et al., 2015). Other groups are developing strategies to target HSP90 family members in mitochondria, mainly TRAP1. These approaches are based on chemical modifications of HSP90 inhibitors such as 17-AAG to make them permeable across mitochondrial membranes. Gamitrinibs (geldanamycin mitochondrial matrix inhibitors) have shown promising results in inducing mitochondrial PTP opening, tumour cell death and efficacy in both xenografts and transgenic mouse models of prostate cancer (Siegelin, 2013).

However, pharmacological inhibition of HSP90 induces a compensatory induction in the expression of other HSPs, in particular HSP70, and might have unanticipated adverse effects in patients because of the multiplicity of biological functions that are regulated by HSP90. Moreover, in several tumour types HSP90 inhibitors elicit growth arrest, but tumours start growing again after drug removal. At present, the most promising therapeutic strategy is the combination therapy, in which HSP90 targeting molecules are associated with other chemotherapeutics, such as proteasome or HSP70 inhibitors (Khalil et al., 2011).

ER stress-inducing agents. Several ER stress-eliciting compounds are under scrutiny as potential anticancer drugs. The inhibitor of ERAD eeyarestatin and the ER stress inducer Celebrex synergize with the proteasome inhibitor bortezomib to cause cytotoxicity in several tumour types. ER stress induced by tunicamycin increases sensitivity to radiotherapy and chemotherapeutics in breast and ovarian cancer cells (Wang and Kaufman, 2014). Similarly, molecules targeting the UPR are promising chemotherapeutics. For instance, encouraging results indicate that inhibition of the IRE1α-XBP1 pathway enhances the proapoptotic effects of Bortezomib and 17-AAG in multiple myeloma. However, it is unclear whether the best strategy to inhibit the UPR is interfering with PERK, IRE1, or ATF6, or with all three arms simultaneously, and the extent of toxicity of such therapeutic approaches in patients has not been elucidated yet. Some reports have shown that ER stress induction prompts resistance to chemotherapy, suggesting that ER stress might promote both sensitivity and resistance to anticancer therapies in a context-dependent way (Nagelkerke et al., 2014).

Conclusion

Additional clues about the mode of action and regulation of chaperones are required in order to better understand their impact on the main biological routines of the cell. Data must be collected at several levels, with a particular attention to discriminate protein quality control between non-transformed and tumour cells. It is mandatory to carry out a more in-depth characterization of the mechanisms of chaperone transcriptional regulation, of their post-translational modifications, of the dynamic interactions with co-chaperones and client proteins.

At a more complex level, regulation of chaperones networks must be thoroughly dissected during orchestrated responses of the cell to a variety of noxious stimuli such as compartmentalized unfolded protein responses. These highly organized and multifaceted processes allow tumour cells to handle a wide panel of different stressors encountered during the process of neoplastic transformation.

In this scenario, a profound understanding of the biology of chaperones will be instrumental for the identification of highly selective compounds that might modulate their activity under specific conditions of stress. Combinatorial therapies could aim at simultaneously exposing tumours to specific damaging agents while blunting the activity of protecting chaperones, thus setting the stage for the definition of innovative antineoplastic strategies.

TAKE-HOME MESSAGE

- Maintenance of the proteome is vital for the physiology of the cell; this involves fine-tuning among synthesis, folding, and degradation of proteins.
- Chaperones are a class of proteins dedicated to this aim as they assist proteins in keeping their proper folding, conformational changes, and subcellular trafficking.
- As a consequence of metabolic, biochemical, and physical stress, chaperone levels are increased in malignant cells supporting their survival.
- Chaperone induction has been associated with cancer progression, resistance to chemotherapy, and poor prognosis.
- Development of chaperone-targeting drugs has emerged as a promising antineoplastic strategy.

OPEN QUESTIONS

- How is protein quality control differentially tuned in tumour vs. non-transformed cells?
- What are the molecular interactions between oncogenic signalling pathways and chaperone activity and regulation?
- What is the impact of unfolded protein responses in ER and mitochondria on biological routines deregulated in cancer?
- Is it effective to target specific chaperones as a single or combinatorial antineoplastic approach?

FURTHER READING

Harvey, R. F. & Willis, A. E. (2017). Post-transcriptional control of stress responses in cancer. *Curr Opin Genet Dev*, **48**, 30–5.

Hetz, C. & Papa, F. R. (2018). The unfolded protein response and cell fate control. *Mol Cell*, **69**, 169–81.

Ma, Y. & Hendershot, L. M. (2004). The role of the unfolded protein response in tumour development: friend or foe? *Nat Rev Cancer*, **4**, 966–77.

Obacz, J., Avril, T., Rubio-Patino, C., et al. (2017). Regulation of tumor-stroma interactions by the unfolded protein response. *FEBS J*. doi: 10.1111/febs.14359 [Epub ahead of print].

Shen, K., Johnson, D. W., Vesey, D. A., McGuckin, M. A., & Gobe, G. C. (2018). Role of the unfolded protein response in determining the fate of tumor cells and the promise of multi-targeted therapies. *Cell Stress Chaperones*, **23**, 317–34.

Vera-Ramirez, L. & Hunter, K. W. (2017). Tumor cell dormancy as an adaptive cell stress response mechanism. *F1000Res*, **6**, 2134.

REFERENCES

Acunzo, J., Katsogiannou, M., & Rocchi, P. (2012). Small heat shock proteins HSP27 (HspB1), alphaB-crystallin (HspB5) and HSP22 (HspB8) as regulators of cell death. *Int J Biochem Cell Biol*, **44**, 1622–31.

Amoroso, M. R., Matassa, D. S., Agliarulo, I., et al. (2016). TRAP1 downregulation in human ovarian cancer enhances invasion and epithelial-mesenchymal transition. *Cell Death Dis*, **7**, e2522.

Barrott, J. J. & Haystead, T. A. (2013). Hsp90, an unlikely ally in the war on cancer. *FEBS J*, **280**, 1381–96.

Bernardi, P., Rasola, A., Forte, M., & Lippe, G. (2015). The mitochondrial permeability transition pore: channel formation by F-ATP synthase, integration in signal transduction, and role in pathophysiology. *Physiol Rev*, **95**, 1111–55.

Cairns, R. A., Harris, I. S., & Mak, T. W. (2011). Regulation of cancer cell metabolism. *Nat Rev Cancer*, **11**, 85–95.

Chevet, E., Hetz, C., & Samali, A. (2015). Endoplasmic reticulum stress-activated cell reprogramming in oncogenesis. *Cancer Discov*, **5**, 586–97.

Doyle, S. M., Genest, O., & Wickner, S. (2013). Protein rescue from aggregates by powerful molecular chaperone machines. *Nat Rev Mol Cell Biol*, **14**, 617–29.

Garrido, C., Brunet, M., Didelot, C., Zermati, Y., Schmitt, E., & Kroemer, G. (2006). Heat shock proteins 27 and 70: anti-apoptotic proteins with tumorigenic properties. *Cell Cycle*, **5**, 2592–601.

Hetz, C., Chevet, E., & Oakes, S. A. (2015). Proteostasis control by the unfolded protein response. *Nat Cell Biol*, **17**, 829–38.

Hishiya, A. & Takayama, S. (2008). Molecular chaperones as regulators of cell death. *Oncogene*, **27**, 6489–506.

Kampinga, H. H. (2006). Chaperones in preventing protein denaturation in living cells and protecting against cellular stress. *Handb Exp Pharmacol*, (172), 1–42.

Khalil, A. A., Kabapy, N. F., Deraz, S. F., & Smith, C. (2011). Heat shock proteins in oncology: diagnostic biomarkers or therapeutic targets? *Biochim Biophys Acta*, **1816**, 89–104.

Kumar, S., Stokes, J., 3rd, Singh, U. P., et al. (2016). Targeting Hsp70: a possible therapy for cancer. *Cancer Lett*, **374**, 156–66.

Lanneau, D., Brunet, M., Frisan, E., Solary, E., Fontenay, M., & Garrido, C. (2008). Heat shock proteins: essential proteins for apoptosis regulation. *J Cell Mol Med*, **12**, 743–61.

Lavery, L. A., Partridge, J. R., Ramelot, T. A., et al. 2014. Structural asymmetry in the closed state of mitochondrial Hsp90 (TRAP1) supports a two-step ATP hydrolysis mechanism. *Mol Cell*, **53**, 330–43.

Lee, J. & Kim, S. S. (2010). Current implications of cyclophilins in human cancers. *J Exp Clin Cancer Res*, **29**, 97.

Li, J. & Buchner, J. (2013). Structure, function and regulation of the hsp90 machinery. *Biomed J*, **36**, 106–17.

Li, W., Sahu, D., & Tsen, F. (2012). Secreted heat shock protein-90 (Hsp90) in wound healing and cancer. *Biochim Biophys Acta*, **1823**, 730–41.

Li, Z. & Li, Z. (2012). Glucose regulated protein 78: a critical link between tumor microenvironment and cancer hallmarks. *Biochim Biophys Acta*, **1826**, 13–22.

Lianos, G. D., Alexiou, G. A., Mangano, A., et al. (2015). The role of heat shock proteins in cancer. *Cancer Lett*, **360**, 114–18.

Lin, Y. F. & Haynes, C. M. (2016). Metabolism and the UPR(mt). *Mol Cell*, **61**, 677–82.

Liou, Y. C., Zhou, X. Z., & Lu, K. P. (2011). Prolyl isomerase Pin1 as a molecular switch to determine the fate of phosphoproteins. *Trends Biochem Sci*, **36**, 501–14.

Liu, Z. & Butow, R. A. (2006). Mitochondrial retrograde signaling. *Annu Rev Genet*, **40**, 159–85.

Lu, Z. & Hunter, T. (2014). Prolyl isomerase Pin1 in cancer. *Cell Res*, **24**, 1033–49.

Luo, B. & Lee, A. S. (2013). The critical roles of endoplasmic reticulum chaperones and unfolded protein response in tumorigenesis and anticancer therapies. *Oncogene*, **32**, 805–18.

Martin, J. (2004). Chaperonin function--effects of crowding and confinement. *J Mol Recognit*, **17**, 465–72.

Masgras, I., Ciscato, F., Brunati, A. M., et al. (2017). Absence of neurofibromin induces an oncogenic metabolic switch via mitochondrial ERK-mediated phosphorylation of the chaperone TRAP1. *Cell Rep*, **18**, 659–72.

Meng, L., Hunt, C., Yaglom, J. A., Gabai, V. L., & Sherman, M. Y. (2011). Heat shock protein Hsp72 plays an essential role in Her2-induced mammary tumorigenesis. *Oncogene*, **30**, 2836–45.

Mitra, A., Shevde, L. A., & Samant, R. S. (2009). Multi-faceted role of HSP40 in cancer. *Clin Exp Metastasis*, **26**, 559–67.

Mollapour, M. & Neckers, L. (2012). Post-translational modifications of Hsp90 and their contributions to chaperone regulation. *Biochim Biophys Acta*, **1823**, 648–55.

Murphy, M. E. (2013). The HSP70 family and cancer. *Carcinogenesis*, **34**, 1181–8.

Nagelkerke, A., Bussink, J., Sweep, F. C., & Span, P. N. (2014). The unfolded protein response as a target for cancer therapy. *Biochim Biophys Acta*, **1846**, 277–84.

Nakazawa, M. S., Keith, B., & Simon, M. C. (2016). Oxygen availability and metabolic adaptations. *Nat Rev Cancer*, **16**, 663–73.

Neckers, L. & Workman, P. (2012). Hsp90 molecular chaperone inhibitors: are we there yet? *Clin Cancer Res*, **18**, 64–76.

Pearl, L. H. (2016). Review: the HSP90 molecular chaperone-an enigmatic ATPase. *Biopolymers*, **105**, 594–607.

Rasola, A., Neckers, L., & Picard, D. (2014). Mitochondrial oxidative phosphorylation TRAP(1)ped in tumor cells. *Trends Cell Biol*, **24**, 455–63.

Rasola, A., Sciacovelli, M., Chiara, F., Pantic, B., Brusilow, W. S., & Bernardi, P. (2010). Activation of mitochondrial ERK protects cancer cells from death through inhibition of the permeability transition. *Proc Natl Acad Sci U S A*, **107**, 726–31.

Scaltriti, M., Dawood, S., & Cortes, J. (2012). Molecular pathways: targeting hsp90—who benefits and who does not. *Clin Cancer Res*, **18**, 4508–13.

Schiene-Fischer, C. (2015). Multidomain peptidyl prolyl cis/trans isomerases. *Biochim Biophys Acta*, **1850**, 2005–16.

Schmidpeter, P. A. & Schmid, F. X. (2015). Prolyl isomerization and its catalysis in protein folding and protein function. *J Mol Biol*, **427**, 1609–31.

Sciacovelli, M., Guzzo, G., Morello, V., et al. (2013). The mitochondrial chaperone TRAP1 promotes neoplastic growth by inhibiting succinate dehydrogenase. *Cell Metab*, **17**, 988–99.

Semenza, G. L. (2013). HIF-1 mediates metabolic responses to intratumoral hypoxia and oncogenic mutations. *J Clin Invest*, **123**, 3664–71.

Sherman, M. & Multhoff, G. (2007). Heat shock proteins in cancer. *Ann N Y Acad Sci*, **1113**, 192–201.

Siegelin, M. D. (2013). Inhibition of the mitochondrial Hsp90 chaperone network: a novel, efficient treatment strategy for cancer? *Cancer Lett*, **333**, 133–46.

Tameire, F., Verginadis, I. I., & Koumenis, C. (2015). Cell intrinsic and extrinsic activators of the unfolded protein response in cancer: Mechanisms and targets for therapy. *Semin Cancer Biol*, **33**, 3–15.

Theuerkorn, M., Fischer, G., & Schiene-Fischer, C. (2011). Prolyl cis/trans isomerase signalling pathways in cancer. *Curr Opin Pharmacol*, **11**, 281–7.

Trepel, J., Mollapour, M., Giaccone, G., & Neckers, L. (2010). Targeting the dynamic HSP90 complex in cancer. *Nat Rev Cancer*, **10**, 537–49.

Vyas, S., Zaganjor, E., & Haigis, M. C. (2016). Mitochondria and cancer. *Cell*, **166**, 555–66.

Wang, M. & Kaufman, R. J. (2014). The impact of the endoplasmic reticulum protein-folding environment on cancer development. *Nat Rev Cancer*, **14**, 581–97.

Wang, P. & Heitman, J. (2005). The cyclophilins. *Genome Biol*, **6**, 226.

Wouters, B. G. & Koritzinsky, M. (2008). Hypoxia signalling through mTOR and the unfolded protein response in cancer. *Nat Rev Cancer*, **8**, 851–64.

Yoshida, S., Tsutsumi, S., Muhlebach, G., et al. (2013). Molecular chaperone TRAP1 regulates a metabolic switch between mitochondrial respiration and aerobic glycolysis. *Proc Natl Acad Sci U S A*, **110**, E1604–12.

Zhang, L. H. & Zhang, X. (2010). Roles of GRP78 in physiology and cancer. *J Cell Biochem*, **110**, 1299–305.

Zoubeidi, A. & Gleave, M. (2012). Small heat shock proteins in cancer therapy and prognosis. *Int J Biochem Cell Biol*, **44**, 1646–56.

18

Oxygen and cancer

The response to hypoxia

Adrian L. Harris and Margaret Ashcroft

Introduction

Oxygen is the most abundant element in the Earth's crust and the third most abundant element in the universe, after hydrogen and helium. The oxidative status of the Earth's atmosphere varied greatly over the 4.5 billion years of its history. Oxygen produced by photosynthetic organisms produced a massive increase of the oxygen levels between 2,500–5,000 million years ago. This was a major evolutionally selective mechanism and oxygen sensing evolved as part of the challenge in meeting these conditions. Recent evidence has shown that the simplest metazoan, *Trichoplax adhaerens*, has the components of the HIF signalling pathway and the mechanism is conserved in all organisms since bacteria. Maintenance of oxygenation levels became a major element of survival.

In cancer, the mechanism of adaptation to hypoxia and effects of hypoxia has a critical effect on its biology. In this review, we cover the normal roles of oxygen in physiology and the signalling pathways responsible for adaptation to hypoxia, as well as the downstream mechanisms that are mediated by hypoxia. The importance of hypoxia in prediction of outcome and response to treatments such as radiotherapy as well as the overall prognosis of cancer is discussed, along with new approaches for dealing with genes and the mechanisms of hypoxia to improve cancer therapies.

A key aspect of this is the ability to detect hypoxia in tumours and select appropriate patients for treatment as well as the multiple pathways involved and the difficulty of inhibiting them for individual patients. Currently, there are only a few countries where hypoxia-modifying drugs are licenced for use for radiotherapy but there are now many new approaches being developed and likely to come into the clinic in the next few years.

Oxygen in physiology

O_2 delivery system (vasculature, Hg)

Oxygen is essential for aerobic multicellular organisms to survive. Primary functions of the respiratory system, cardiovascular system, and blood ensure appropriate oxygenation of cells, tissues, and organs (Pittman, 2011). Oxygen is taken up from the atmosphere by the respiratory system and diffuses from alveoli in the lungs into the bloodstream via the pulmonary vasculature. Haemoglobin in red blood cells carries oxygen in the blood. Oxygenated blood is pumped by the heart to tissues and organs.

O_2 levels in tissues/body: What is normoxia vs. hypoxia?

The amount of dissolved oxygen in the blood is referred to as normal oxygen pressure or partial pressure of oxygen (PO_2). There are several different units used for partial pressure. The international pressure unit is the Pascal (Pa). Other units include: the bar (1 bar = 100 kPa), the atmosphere (1 atm = 101.325 kPa), the Torr (1 Torr =133.322 Pa), and two units most commonly used in medicine are the millimetre of mercury (1 mmHg = 133.322 Pa) and the percentage of oxygen (1% = 1.013 kPa). The PO_2 in arterial blood is ~100 mmHg, while the PO_2 in venous blood is much lower ~40 mmHg reflecting the difference in blood oxygen levels to and from tissues. The partial pressure of oxygen varies between tissues and within organs (**Table 18.1**; see Carreau et al., 2011). For example, the PO_2 within the skin increases from the surface (superficial region) with depth (**Table 18.1**).

Table 18.1 Normal PO_2 of tissues

Organ, normal PO_2 (mmHg)	% O_2
Air, 160	21.1
Trachea, 150	19.7
Air in the alveoli, 110	14.5
Arterial blood, 100	13.2
Venous blood, 40	5.3
Brain, 33.8 +/−2.6	4.4 +/−0.3
Lung, 42.8	5.6
Skin (subpapillary plexus), 35.2 +/− 8	4.6 +/−1.1
Skin (dermal papillae), 24 +/−6.4	3.2 +/−0.8
Skin (superficial region), 8 +/−3.2	1.1 +/−0.4
Intestinal tissue, 57.6 +/−2.3	7.6 +/−0.3
Liver, 40.6 +/−5.4	5.4 +/−0.7
Kidney, 72 +/−20	9.5 +/−2.6
Muscle, 29.2 +/−1.8	3.8 +/−0.2
Bone marrow, 48.9 +/−4.5	6.4 +/−0.6

The terms normoxia and hypoxia are frequently used to distinguish between physiologically normal and low levels of oxygenation in tissues, respectively. Physiological hypoxia occurs during development and can occur at high altitude or during vigorous physical exercise. When hypoxia leads to a complete deprivation of oxygen, this is called anoxia. In the context of cells cultured outside the body, normoxia is usually atmospheric oxygen pressure (~160 mmHg or 21.1% oxygen).

Mitochondria and cellular respiration.

In multicellular organisms, mitochondria are the sites of oxygen consumption when cells undergo aerobic respiration. Mitochondria are membrane-bound organelles that utilize molecular oxygen to generate chemical energy in the form of adenosine triphosphate (ATP), by a process called oxidative phosphorylation (OxPhos). OxPhos occurs with the transfer of electrons through a series of acceptor cytochromes (complexes I–IV) housed within the inner mitochondrial membrane and collectively called the electron transport chain (ETC) or respiratory chain. Electrons are donated to the ETC at complex I and complex II from the reducing agents, nicotinamide adenine dinucleotide (NADH) and flavin adenine dinucleotide ($FADH_2$), respectively. Transfer of electrons occurs sequentially from complex I or II through to complex IV, where molecular oxygen is utilized as the final electron acceptor to produce water. This serial transfer of electrons within the ETC is coupled to a flow of protons from the mitochondrial matrix into the intermembrane space (IMS) which creates an electrochemical gradient of protons across the mitochondrial inner membrane. The electrochemical gradient (or proton motive force, $\Delta\mu H+$) is used by F1FO-ATPase (complex V) to generate ATP from adenosine diphosphate (ADP) and phosphate.

Mitochondria contain their own genome or mitochondrial DNA (mtDNA). In humans, the mtDNA is a double-stranded, circular molecule of 16,569 base pairs and contains 37 genes coding for two rRNAs, 22 tRNAs, and 13 polypeptides which include components the ETC complexes (Taanman, 1999). Mitochondria and mtDNA are important for influencing the cellular response to hypoxia (Chandel et al., 1998; Agani et al., 2000; Vaux et al., 2001; Guzy et al., 2005; Mansfield et al., 2005). Hypoxia regulates the activity of the ETC in part, by regulating the availability of key complex IV subunits (Fukuda et al., 2007).

O_2 sensing machinery and cellular response pathway

Hypoxia activates a major transcriptional programme regulated by the hypoxia-inducible factor (HIF) family of basic helix-loop-helix (bHLH), PAS (Per-ARNT-Sim) domain containing dimeric transcription factors (Semenza and Wang, 1992; Wang et al., 1995; Wang and Semenza, 1995; Jiang et al., 1996). HIF-1 is the best known member of the HIF family, and was originally identified by its ability to bind to a cis-acting sequence (5'-RCGTG-3') termed the hypoxia-responsive element (HRE) within the 3'-enhancer region of the hypoxia-responsive gene, *erythropoietin EPO* (Semenza et al., 1991; Semenza and Wang, 1992). HIF-1 binding to the HRE sequence of the *EPO* gene was found to be required for *EPO* transcriptional activation in Hep3B cells, when these cells were exposed to hypoxia (Semenza et al., 1991; Semenza and Wang, 1992). Since then, HRE sequences have been identified in well over 150 hypoxia-responsive HIF target genes (Benita et al., 2009) involved in a broad range of cellular responses (Webb et al., 2009; Samanta et al., 2017) that underpin both physiological and pathological processes (Semenza, 2014, 2017).

HIF-1 comprises a tightly regulated HIF-1α subunit and constitutively expressed HIF-1β subunit (also known as aryl hydrocarbon nuclear translocator, ARNT). In addition to HIF-1α, other structurally related HIF-α subunits have also been described, including HIF-2α (also known as endothelial PAS domain protein-1, EPAS-1; see Tian et al., 1997) and HIF-3α (Gu et al., 1998). These three HIF-α subunits are highly structurally related within their bHLH and PAS domains. However, while both HIF-1α and HIF-2α contain N-terminal (N-TAD) and C-terminal (C-TAD) transactivation domains, HIF-3α has no C-TAD. Furthermore, there are several HIF-3α isoforms (Duan, 2016), some of which have been shown to provide a dominant negative function (reviewed in Duan, 2016). Like HIF-1α, HIF-2α and HIF-3α are also able to bind independently to HIF-1β, forming HIF-2 and HIF-3 transcriptional complexes, respectively.

HIF transcriptional activity is controlled primarily through regulation of HIF-α protein availability and has been best studied for HIF-1α and HIF-2α. HIF-α (HIF-1α or HIF-2α) protein levels are regulated by oxygen-dependent protein stability and non-oxygen-dependent protein synthesis mechanisms (Bardos and Ashcroft, 2005; Poon et al., 2009). In normoxia, HIF-α is synthesized and continuously undergoes ubiquitin-mediated degradation by the proteasome. HIF-α is targeted for degradation via post-translational site-specific prolyl hydroxylation by the oxygen-dependent domain (PHD) enzymes (Epstein et al., 2001; Jaakkola et al., 2001; see Fig. 18.1). PHD enzymes are Krebs cycle (tri-carboxylic acid cycle) intermediate α-ketoglutarate (2-oxoglutarate)-dependent dioxygenases, requiring non-haem iron (Fe^{2+}) and ascorbate as cofactors. PHD hydroxylation occurs in proline residues Pro402 and Pro564 in HIF-1α, and Pro405 and Pro531 in HIF-2α (Epstein et al., 2001; Jaakkola et al., 2001; Patel and Simon, 2008). Hydroxylated HIF-α is recognized by the von Hippel Lindau (pVHL) E3-ubiquitin ligase complex, which polyubiquitinates HIF-α and targets it for degradation by the 26S proteasome (Maxwell et al., 1999; Ivan et al., 2001; Fig. 18.1). HIF-1α is degraded in both the nucleus and cytoplasm and has a half-life of ~5–8 minutes in normoxia (Berra et al., 2001).

In addition to regulation of HIF-α protein stability by the PHD-VHL axis (Fig. 18.1), HIF-α transactivation is also regulated by the asparaginyl hydroxylase enzyme, factor inhibiting HIF-1 (FIH-1) in normoxia (Mahon et al., 2001; Lando et al., 2002a; Lando et al., 2002b; McNeill et al., 2002). Factor-inhibiting hypoxia-inducible (FIH)-1 is an α-ketoglutarate and (Fe^{2+}) dependent dioxygenase. Site-specific asparaginyl hydroxylation of HIF-α (Asp803 in HIF-1α and Asp851 in HIF-2α) by FIH-1 requires molecular oxygen and blocks the interaction with the coactivation p300/CBP and HIF-α which is required for HIF transcriptional activity (Lancaster et al., 2004). In this way, FIH-1 negatively regulates HIF-α transactivation in normoxia.

In response to hypoxia, HIF-α protein is rapidly stabilized, localizes to the nucleus, and binds to HIF-1β and DNA (Semenza et al., 1997; Bracken et al., 2006). Dimerization of HIF-α and HIF-1β, leads to a conformational change in HIF-α (Kallio et al., 1997) and recruitment of coactivators including p300/CBP (CREB; cAMP-response element binding protein; see Arany et al., 1996; Jiang et al., 1996) to the HIF-α TADs, which are required for HIF transactivation.

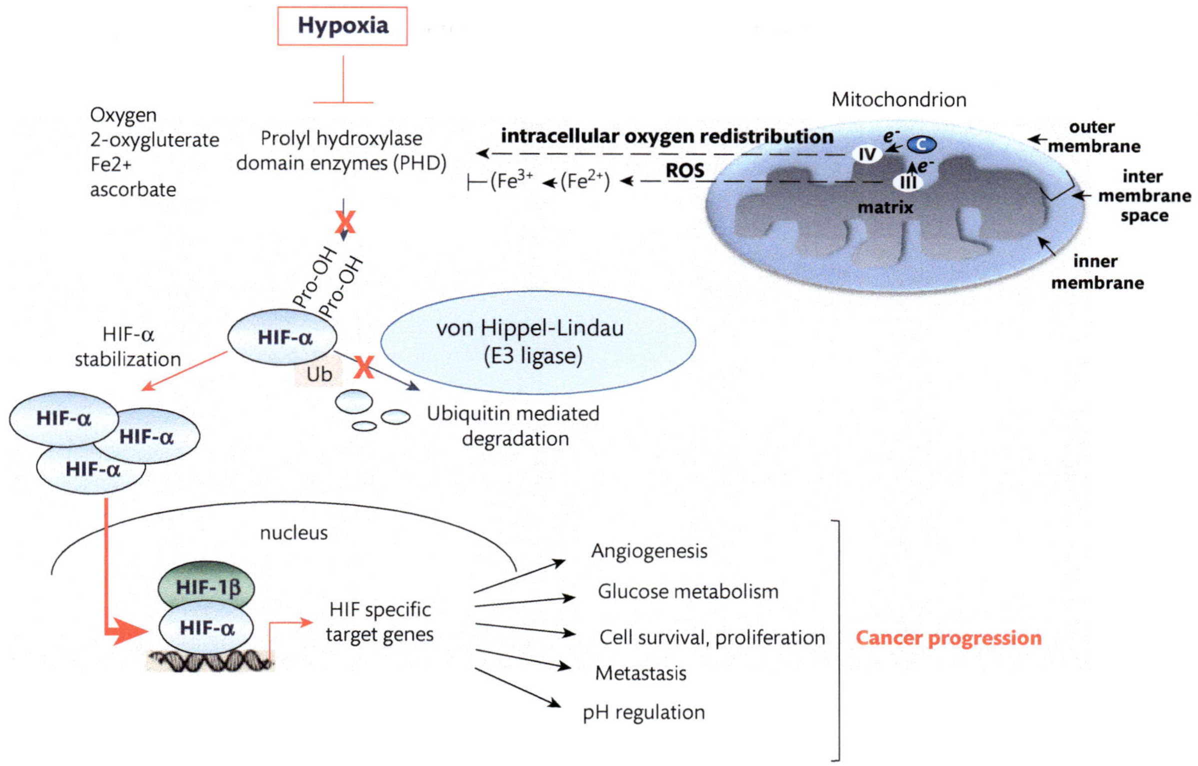

Fig. 18.1 Regulation of HIF-α protein stability.
Reproduced courtesy of Adrian L. Harris, Department of Oncology, University of Oxford, UK and Margaret Ashcroft Department of Medicine, University of Cambridge, UK.

Mitochondria provide an important role in influencing hypoxia-induced HIF activity in mammalian cells (Fig. 18.1). Conversely, HIF plays an important role in regulating mitochondrial function to optimize efficiency of respiration in hypoxic cells by modulating the expression of cytochrome oxidase subunits (Fukuda et al., 2007). Recent studies have discovered that the coiled-coil-helix-coiled-coil-helix domain containing 4 (CHCHD4) redox-sensitive mitochondrial protein regulates intracellular oxygenation and HIF activity (Yang et al., 2012; Thomas et al., 2017). CHCHD4 provides an import and oxidoreductase-mediated protein folding function along with the sulfhydryl oxidase GFER (ALR/Erv1; see Banci et al., 2009; Fischer et al., 2013) as a key part of the disulphide relay system (DRS) within the mitochondrial IMS. As such, CHCHD4 controls the import of a number of mitochondrial proteins that contain a twin-CX_9C or twin-CX_3C motif (Yang et al., 2012; Fischer et al., 2013; Modjtahedi et al., 2016). As a component of the DRS, CHCHD4 participates in electron transfer to complex IV (the oxygen-accepting complex) of the ETC (Bihlmaier et al., 2007).

Hypoxia and cancer

O_2 levels across tumours

The tumour microenvironment (TME) is a central aspect of tumour biology, contributing to tumour initiation and progression, and impacts the response to therapy. The TME is a multicellular compartment in which tumour cells exist, and includes surrounding blood vessels, fibroblasts, bone marrow-derived inflammatory cells, immune cells, lymphocytes, signalling molecules, and the extracellular matrix. Tumour cells and the surrounding microenvironment continuously interact. Tumours cells can influence other components of the microenvironment by releasing an array of extracellular factors, which mediate local cellular responses such as angiogenesis and peripheral responses such as immune cell recruitment and infiltration to the TME. Tumours of different origins and different stages of progression contain different proportions of TME components and due to genetic instability (see next) become genetically heterogeneous, exhibiting differential cellular phenotypes and morphologies.

A key feature of the TME is hypoxia. Most solid tumours contain areas of hypoxia. As abnormally proliferating tumours cells grow, they become hypoxic due to oxygen deprivation, and as such, HIF-α is upregulated. Activation of HIF leads to tumour cell survival, metabolic adaptive responses that enable tumour cells to generate ATP by glycolysis and the induction of angiogenesis which enables surrounding blood vessels from normal tissue to invade the proliferating tumour mass (Table 18.2). Hypoxia within the TME also leads to immune and inflammatory cell responses.

There are at least three known mechanisms for hypoxia in tumours. These include diffusion limited hypoxia, which occurs as a result of distance from vessels, is constant and on a scale of hundreds of microns. Perfusion-limited hypoxia, which results from perturbations in tumour vessel blood flow, and can be both transient and recurring, and diffusion-limited hypoxia occurs on a scale of hundreds of microns. In addition to diffusion- and perfusion-limited hypoxia, there is a stable gradient of oxygen that can occurs over multimillimetre distances along the length of tumour vessels, leading to hypoxia (Fig. 18.2). Additionally, red cell rouleaux may form, which reduces the surface area for oxygen exchange. Tumour

Table 18.2 Selected HIF-1 target genes whose products contribute to cancer progression

Gene product	Role in cancer progression
Angiopoietin 2	Angiogenesis, lymphangiogenesis
Angiopoietin-like 4	Metastasis
Breast cancer resistance protein (ABCG2)	Multidrug transport, stem cell maintenance
Carbonic anhydrase 9 and 12	pH regulation
C-MET	Invasion
CXCR4	Metastasis
DEC1	Genomic instability
Endothelin 1	Invasion
Fibronectin 1	Invasion
Glucose phosphate isomerase	Cell motility, glucose metabolism, immortalization
Glucose transporter 1	Glucose uptake
Hexokinase 1 and 2	Glucose phosphorylation; cell survival
Inhibitor of differentiation 2	Angiogenesis, proliferation
Insulin-like growth factor 2	Cell survival, proliferation
JARID1B	Stem cell maintenance
Kit ligand (stem cell factor)	Angiogenesis, stem cell maintenance
Lactate dehydrogenase A	Glucose metabolism
Lysyl oxidase	Metastasis
Matrix metalloproteinase 2 and 14	Invasion
NT5E (ecto-5′-nucleotidase/CD73)	Immune evasion, multidrug resistance
OCT4	Stem cell maintenance
Placental growth factor	Angiogenesis
Platelet-derived growth factor B	Cell proliferation/survival, angiogenesis
Pyruvate dehydrogenase kinase 1	Glucose metabolism
Pyruvate kinase M2	Glucose metabolism
Stromal-derived factor 1	Angiogenesis
Survivin	Cell survival
Telomerase	Immortalization
Transforming growth factor α	Cell proliferation/survival
TWIST1	Epithelial-mesenchymal transition
Urokinase plasminogen activator receptor	Invasion
Vascular endothelial growth factor	Angiogenesis
WSB1	Cell survival
ZEB1 (ZFHX1A), ZEB2 (ZFHX1B)	Epithelial-mesenchymal transition

hypoxia is associated with overexpression of HIFs, tumour progression, treatment resistance, and poor patient survival.

pH regulation

The management of tumour pH under metabolic stress is critical both for lactate and also carbon dioxide generated from the Krebs cycle, fatty acid oxidation and other pathways (McIntyre and Harris, 2016) Thus the extracellular milieu is maintained at an acid pH due to both lactic acid and carbonic acid. It is important to note however the acid pH generated in tumours comes from the hydrogen ions generated by ATP hydrolysis. In fact, the conversion of pyruvate to lactate will consume hydrogen ions, as ATP is generated.

The alkaline pH within tumours and extracellular acidic pH is maintained through several mechanisms and is a clear survival advantage to tumour cells competing with normal cells, which can survive less well at such an acid pH (Estrella et al., 2013; Damaghi et al., 2015). Lactate is well recognized as being secreted from tumour cells and there are two major lactate transporters, MCT1 and MCT4, involved in its transport and uptake between cells. MCT4 is induced by hypoxia and involved in the secretion of lactate and many cells will express both MCT4 and other components such as carbonic anhydrase IX involved in pH regulation (Counillon et al., 2016). Conversely, MCT1 is involved in uptake of lactate from the medium and this provides the possibility of a synergistic route of lactate being produced by hypoxic cells helping provide lactate for oxidative phosphorylation in more oxygenated cells. So, lactate needs to be considered a major metabolite of survival benefit besides a method of reducing acid accumulation in tumour cells. Inhibitors of MCT1 and 4 are now available and MCT1 inhibitors are in clinical studies to see if blocking the uptake of lactate can increase extracellular acidity or reduce the survival of MCT1 positive cells.

Additionally, uptake of bicarbonate from the environment provides another major route for maintaining intracellular alkaline conditions and there are several bicarbonate transporters, such as SLC4A4 and SLC4A7 (Parks and Pouyssegur, 2015; McIntyre et al., 2016), that are induced by HIF and their inhibition can greatly reduce tumour survival *in vivo* and *in vitro*. There are no drugs that inhibit these exclusively that are currently available.

Carbonic anhydrase IX is a transmembrane protein that acts as a dimer with the active face on the extracellular surface (Robertson et al., 2004). It requires intracellular phosphorylation for function and greatly facilitates the transfer of CO_2 across the cell membrane by converting it to bicarbonate and hydrogen ions and allowing the hydrogen ions to rapidly diffuse in the extracellular environment (Hulikova et al., 2012). This also provides a source of bicarbonate for recycling to maintain the intracellular pH at a high level. Intracellular carbonic anhydrases will convert hydrogen ions plus bicarbonate to CO_2, which will then transfer across the cell membrane (Hulikova et al., 2012).

CAIX is relatively specific to cancer cells, very rarely seen in normal tissues, but is present in crypt cells in the colon and in normal stem cells. However, it is markedly upregulated in epithelial tumours compared to normal tissues and is one of the most responsive HIF genes (Robertson et al., 2004). It is highly specifically regulated by HIF1 versus HIF2. Because of its long plasma cell membrane half-life of 2–3 days once induced, it has been a useful marker for tissue studies of previous hypoxia or concurrent hypoxia as one of the most widely used endogenous markers of hypoxia because of this (van Kuijk et al., 2016). There are many clinical studies now produced in most epithelial tumour types showing that the high level of CA9 is an independent predictor of poor prognosis and as a result of the extensive biology investing CA9, demonstrating it is not down *in vitro* and *in vivo* as antitumour extracellular or in combination with antiangiogenic drugs and

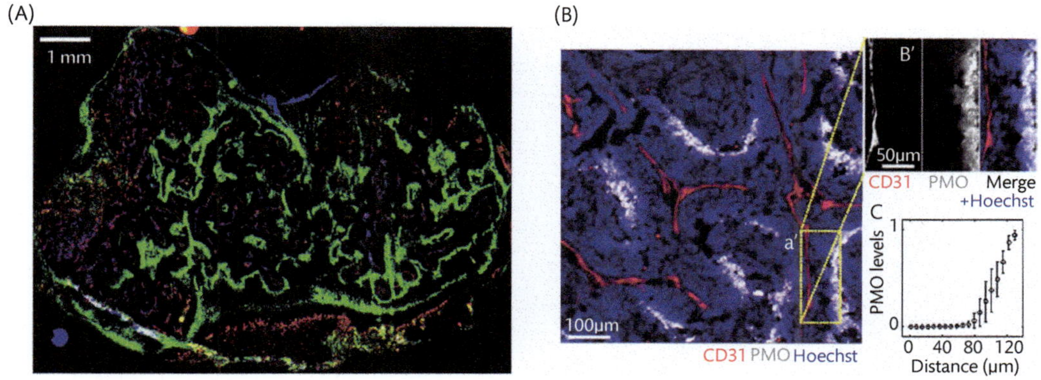

Fig. 18.2 Hypoxia across tumours.
(A) Reproduced with permission from Poon E et al. 'Targeting the hypoxia-inducible factor (HIF) pathway in cancer', *Expert Reviews in Molecular Medicine*, Volume 11, e22, Copyright © Cambridge University Press 2009; and (B) Reproduced with permission from Carmona-Fintaine C et al. 'Metabolic origins of spatial organization in the tumor microenvironment', *PNAS*, Volume 114, pp. 2934–9, Copyright © 2017.

radiotherapy (Dubois et al., 2011) there are now major efforts to develop inhibitors.

By targeting the excretion of lactate and carbon dioxide through lactate transport inhibitors or carbonic anhydrase inhibitors or indeed the combination, selective effects on tumour metabolism may be produced, affecting the intracellular pH, and reducing the more alkaline milieu inside the tumours to acidic, which will decrease signalling by mTORC1 and other pathways, as well as being directly toxic (Parks et al., 2011).

Activation of HIF genes–targets

It has been estimated that up to 1.5% of the tumour genome is transcriptionally responsive to hypoxia and many studies have identified genes that are induced by hypoxia (Ratcliffe, 2013). All the hallmarks of cancer include genes induced by hypoxia (Wigerup et al., 2016; see Fig. 18.3) and indeed hypoxia-induced factors have a major role to play in nearly every aspect of cancer biology. Thus, the HIF targets are involved in proliferation and survival (e.g. IGF2, IGF binding protein 3, TGFα, erythropoietin, cyclin D1); metabolism; invasion, metastasis, angiogenesis; pH regulation: stem cell maintenance (Semenza, 2016a); immune escape; inflammation; and drug resistance.

Hypoxia at the level that induces HIF1α is associated with and a major determinant of radiotherapy resistance. However, interestingly, many DNA repair genes are downregulated by hypoxia and this reflects the reduction in proliferation and cell cycle progression. Therefore, if one could overcome the hypoxia-mediated resistance to radiotherapy, for example, radiosensitizers, may also be worthwhile using additional DNA inhibitors to potentiate these effects, such as ataxia telangiectasia mutated (ATM) and ATR kinase inhibitors, PARP inhibitors, CHK1 and CHK2 inhibitors.

However, it should be apparent that trying to block just one of these many genes may be insufficient, because of the great heterogeneity of response, some patients may show much greater induction of one pathway that another. This makes the case for the detailed analysis by gene expression profiling and trying to match relevant treatments. But clearly approaches blocking HIF signalling itself may be better than going for the individual targets. Many approaches are being tried and the drugs summarized in Table 18.3 (Masoud and Li, 2015).

Hypoxia and metabolism

The pathways involved in metabolism are extensively covered elsewhere but it is clear that hypoxia is one of the major mediators of the metabolic response to environmental stress in cancer and synergizes, via the complex regulation of HIF1α with many oncogenes (Semenza, 2013).

Apart from loss of VHL, the signalling pathway of PI3 kinase/Akt/mTOR, which is downstream of many oncogenes, regulates HIF1α and other transcription factors that regulate metabolism, such as sterol regulatory element-binding protein (SREBP). One of the clearest and earliest effects discovered for HIF1 induction in hypoxia was the upregulation of the enzymes of the glycolytic pathway plus the downregulation of the Krebs cycle (Semenza, 2016b). Isozymes of every step are hypoxia-regulated compared to others that are basally regulated independently of hypoxia. The regulation of metabolism in hypoxia is covered in the section of metabolism but suffice to say that the following pathways are induction of glycolysis, inhibition of the Krebs cycle, increase of the pentose shunt, enhanced lipid synthesis, and increase of the metabolic flux through glycolysis providing other essential metabolites such as serine. Additionally, in hypoxia, reduced mitochondrial function triggers reactive oxygen species (ROS) induction and there is a switch in the subunits of cytochrome C oxidase to allow increased efficiency of electron transport.

One of the major mechanisms by which the Krebs cycle has reduced activity and oxygen consumption is via induction of pyruvate dehydrogenase isozymes 1. This phosphorylates and activates pyruvate dehydrogenase and thus blocks the very first entrance step to the Krebs cycle of pyruvate conversion to acetyl CoA. The pyruvate is also selectively converted to lactate by lactate dehydrogenase A. A further change on the enzyme metabolism in the mitochondria is the result of the subunit switch of COX4I1 to COX4I2 in cytochrome C oxidases that increase the efficiency of electron transfer and allows continued progress without increasing ROS levels. HIF1 induces activate miR210, which is induced in hypoxic cells and switches off the expression of the *ISCU* gene, which is essential for assembly of the mitochondrial electron transport complex 1. This is a further way of reducing electron transport through the mitochondria. Finally, mitochondria damaged by the free radicals need to

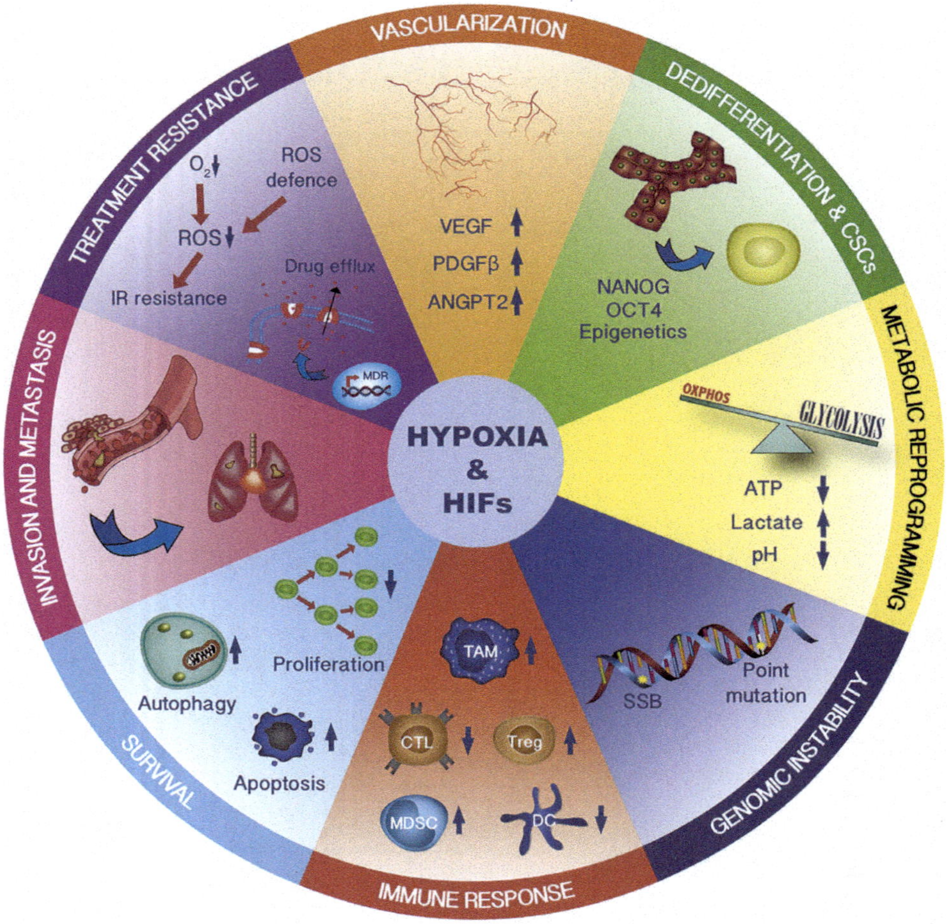

Fig. 18.3 Hallmarks of cancer regulated by hypoxia and HIFs. Hypoxia and HIFs regulate multiple cancer phenotypes. Targeting hypoxia/HIFs will thus inhibit several traits of tumour progression, metastasis, and treatment resistance. Abbreviations: VEGF, vascular endothelial growth factor; PDGF-β, platelet-derived growth factor-β; ANGPT2, angiopoietin 2; oxphos, oxidative phosphorylation; SSB, single-strand break; TAM, tumour-associated macrophage; CTL, cytotoxic T lymphocyte; Treg, regulatory T-cell; MDSC, myeloid-derived suppressor cell; DC, dendritic cell; MDR, multidrug resistance; ROS, reactive oxygen species; IR, ionizing radiation.

be cleared by mitophagy and hypoxia induces BNIP3 and BNIP3L, both of which synergized to signal mitochondrial autophagy.

Hypoxia and non-coding ribonucleic acids

There are many types of non-coding ribonucleic acids (RNAs) such as microRNAs, snRNAs, transfer RNAS and other long non-coding RNAs (LNC RNAs) which have more than 200 nucleotides, the former groups being less than 200 nucleotides. The long non-coding RNAs include many antisense transcripts. They have many functions regulating transcription, posttranslational regulation, and epigenetics. However, a substantial number of these are regulated by hypoxia and have an important role in the regulating the response to hypoxia (Choudhry et al., 2014; Choudhry et al., 2016).

Analysis of the role of HIF1α and HIF2α show that many microRNAs have HIF binding sites and one of the most upregulated is mir-210 (Favaro et al., 2010). Many microRNAs induced by hypoxia are now known and signatures of hypoxia mRNAs have been reported. Mir-210 is prognostic and generates information on poor prognosis and has been included in hypoxia gene profiles. The potential key roles of microRNAs are demonstrated in the analysis of just one mir, mir-210. It has many validated targets, which include enzymes involved in DNA repair, homeobox genes, metabolic enzymes, and regulation of mitochondrial complex assembly and enzymes such as succinate dehydrogenase and cytochrome c oxidase. MicroRNAs can be detected in the circulation either free, or bound to exosomes or other proteins, and these may be another useful way in the future to classify hypoxic tumours.

Furthermore, precursors of microRNA, pri-miRNAs and mature microRNAs can be identified and there is substantial regulation of the microRNA processing machinery in hypoxia. Dicer expression is reduced in hypoxic cancers (van den Beucken et al., 2014) and compensatory changes occur with AGO4 being upregulated. Other processing genes such as AGO1, AGO2, TRBP are also downregulated in hypoxia. The result of this is to reduce processing of microRNAs and the repression of genes such as ZEB1, which can activate the epithelial-mesenchymal transition (EMT). DROSHA is downregulated, reducing microRNA biogenesis. Interestingly,

Table 18.3 Classification of molecules intervening HIF-1α pathway according to their putative mechanisms of action

Mechanism of inhibition	Inhibitor	Target
HIF-1α mRNA expression	EZN-2698 Aminoflavone	HIF-1α mRNA
HIF-1α translation	CPTs: Topotecan, EZN-2208, SN38, Irinotecan Temsirolimus Everolimus Sirolimus LY294002 Wortmannin	Topoisomerase I mTOR
	Cardiac glycosides: Digoxin, ouabain, proscillaridin	HIF-1α protein/HIF-2α mRNA
	2ME2's: 2ME2, ENDM-1198, ENMD-1200, ENMD-1237	Microtubules
HIF-1α stabilization	*GAs:* Radicicol, SCH66336, GA, KF58333, Apigenin 17-AAG 17AG 17-DMAG	Hsp90
	Romidepsin (KF228)	
	Trichostatin	
	LW6	
HIF-1α dimerization	Acriflavin	
HIF-1/DNA binding	Echinomycin	5^0-CGTG-3^0
	Anthracyclines: doxorubicin, daunorubicin	HRE
HIF-1 transcriptional activity	Chetomin	CH1 domain of p300
	Bortezomib	C-TAD of HIF-1α and Asn^{803} of FIH
HIF-1α at multiple levels	YC-1	HIF-1&2 protein/FIH
	PX-478	HIF-1&2 protein

EGFR suppresses the maturation of microRNAs by phosphorylating AGO2, which reduces association with Dicer and the effect is enhanced in hypoxia (Shen et al., 2013).

HIF itself is regulated by microRNAs and suppressed by mir-119A, mir-105 for HIF2, and mir-3383P, mir145, and others for HIF1. Additionally, vHL may be supressed by microRNAs and other components of the complex and microRNAs may block FIH again allowing increased function of HIF1 and HIF2. Thus, the microRNA network regulating the hypoxia signalling pathway is a complex variable. They target many other genes in the hallmarks of cancer and will synergize or antagonize the HIF transcriptional and translational programme. Mirs induced by hypoxia have a major impact on invasion, metastasis (mir-210) (mir-191), autophagy (mir-155), metabolism (mir-143 targeting hexokinase 2).

Hypoxia also alters a number of long non-coding RNAs including NEAT1, H19 LincRNA-P21, and UCA1. LincRNA-P21 has been shown to bind HIF1α and VHL disrupting their interaction and allowing HIF1α accumulation and also promoted glycolysis and the Warburg effect. Again, this emphasizes the potential importance of being able to block the HIF1 pathway itself directly and indirectly rather than individual downstream products.

Role of hypoxia in tumorigenesis

Sustaining proliferative signals, evading growth suppressors, and resisting cell death

Hypoxia and HIF activation leads to the upregulation of the expression of a variety of cell survival and proliferative/growth factors and their receptors including IGF-1 and IGFR and EGFR. Hypoxia activates the PI-3K-AKT/PKB pathway, a key cellular signalling pathway involved in promoting survival. Hypoxia and other TME stimuli such as growth factors can cooperate via the PI-3K-AKT/PKB pathway to enhance HIF activation and drive tumour cell survival and metastasis.

Oncogenes such as c-MYC can be activated in hypoxia or in response to HIF activation to promote cell cycle progression. Other oncogenes such as HDM2 (also known as MDM2) can be induced by growth factors and other oncogenic signals in tumours to enhance HIF activation in hypoxia and drive tumour progression. In addition, MET has been shown to mediate hypoxia-driven metastasis induced in tumours when there is a reduction in pericyte cell coverage of tumour blood vessels in response the therapy (Cooke et al 2012).

Hypoxia can lead to tumour progression by assisting the survival of tumour cells harbouring mutations in tumour suppressor genes such as p53. The p53 tumour suppressor protein is a transcription factor crucially involved in inducing cell cycle arrest and/or apoptosis in response to a variety of cellular stressors. Loss of p53 function occurs in most cancers due to mutation and/or deregulation of the p53 pathway. Tumour cell apoptosis is reduced in hypoxic tumour cells exhibiting loss of p53 function, and is proposed to be a mechanism that enables these tumour cells to survive, proliferate, and metastasize.

Genome instability and enabling replicative immortality

DNA damage-associated checkpoint control mechanisms enable cells to be executed when their DNA is damaged and unable to be repaired. The canonical DNA damage response pathway involves the activation of the anoxia-mediated intra-S phase arrest mediated by the ATM and ataxia telangiectasia and RAD3-related (ATR) kinases. Activation of ATM and ATR leads to the phosphorylation and activation of checkpoint kinases 1 and 2 (CHK1 and CHK2) respectively, which in turn leads to the activation of key cell cycle regulators p53 and CDC25. This DNA damage pathway prevents the accumulation of cells with damaged DNA and protects against carcinogenesis.

Hypoxia within the TME can lead to altered cell cycle checkpoint activation and/or the sensing and repair of damaged DNA, and thus can result in tumour cells accumulating damaged DNA. However, the extent to which tumour hypoxia mediates DNA damage-associated cell cycle checkpoint activation and/or genetic instability depends on whether the hypoxia is acute and cycling or chronic (Bristow and Hill, 2008; Luoto et al., 2013).

In an experimental setting, the canonical DNA damage-associated response pathways involving ATM and ATR kinases are generally not activated above 0% oxygen (i.e. anoxia). However, hypoxia has been shown to increase the rate of spontaneous DNA mutations. When hypoxia (0.2–1% O_2) is acute and cycling, this leads to sporadic re-oxygenation. Re-oxygenation induces a burst of mitochondrial ROS. ROS are a highly reactive chemical species containing oxygen, which reacts with DNA causing DNA damage and genetic instability (Cooke et al., 2003). Genetic instability (both microsatellite instability (MSI) and chromosomal instability) is associated with hypoxia, anoxia, and HIF activation in tumours (Luoto et al., 2013).

The enzyme human telomerase reverse transcriptase (hTERT) is the major mechanism for cancer-specific activation of telomerase, which is activated in over 90% of tumours and important for genomic stabilization after immortalization. Hypoxia via HIF1 is one of the main transcription factors that regulate the promoter (Wu et al., 2013).

Inducing angiogenesis, activating invasion and metastasis

Angiogenesis

VEGF was one of the first major secreted proteins discovered to be regulated by HIF1α in hypoxia but many other angiogenic factors are also induced, including stromal-derived factor CDF1 (CXCL12), stem cell factor, adrenomedullin, and other growth factors that also affect vessels such as IGF1 and platelet-derived growth factor and additional hypoxia pathways controlling vascular mimicry (Li et al., 2016). Because of the outstanding evidence of the role of VEGF in tumour angiogenesis (see Section III, 'How the cancer cell works') it has become a key target for therapeutic agents, which have wide usage in many tumour types. This was one of the first examples of a specific targeting agent against a HIF-targeted genes

Role of HIF in invasion and metastasis

Most cancer deaths are associated with metastasis (see Chapter 19) and hypoxia controls many of the key steps from local invasion to establishment of distant sites in preprogrammed niches (Gilkes and Semenza, 2013). The EMT is induced by HIF in many cell types and HIF induces key EMT factors ZEB1, Snail, and Twist. It also induces many cell signalling pathways such as TGFβ, ILK, integrin-linked kinase. Additionally, many genes involved in local invasion are activated including the urokinase receptor, lysyl oxidases, and RIO-kinase 3. A combined EMT switch plus production of MMP2 and other matrix metalloproteinases (MMPs) facilitates local invasion, which is also needed at distant site for metastasis.

Lysyl oxidases are a large gene family but LOXL2 and LOXL4 are the key ones induced by hypoxia and induce collagen cross-linking. The secreted LOX also helps establish the metastatic niche within the lung and remodel existing collagen. Thus, not only does invasion and degradation occur but also remodelling and laying down of new collagen as HIF regulates expression of PLOD1 and PLOD2, PLOD1 being procollagen-lysine, 2-oxoglutarate 5-dioxygenase, and PLOD2 and PLOD3 also contribute to the remodelling process.

Prolyl 4-hydroxylases, P4HA1, P4HA2, and P4HA3 are essential for collagen synthesis. The collagenase type IV collagen degrading enzymes MMP2 and MMP9 are induced by hypoxia and MMP1 (MT1-MMP also known as MMP14) is upregulated by HIF. Cancer cells are attracted to and migrate towards lymphatic endothelium which secretes chemoattractants such SDF1, which binds CXCR4 on cancer cells, which is induced by HIF. The stromal cells also produce hypoxia-induced cytokines that can sustain tumour growth (Allaoui et al., 2016).

In the case of colon cancer cells metastatic to the liver, they release the enzyme creatinine-kinase brain type, which produces phosphocreatine. In hypoxia the tumour cells use the phosphocreatine to generate ATP. This helps them survive the hypoxic zones in liver metastasis (Kazak et al., 2015).

Of course, angiogenesis is another key factor involved in metastasis, both in making vessels more permeable at the primary site and essential for continuous growth of the metastases.

Inflammation (AH) and evading immune destruction

Hypoxia and inflammation

Cancers have been described as 'wounds that do not heal'. Inflammation is one of the hallmarks of cancer, which includes inflammatory infiltrates and production of large numbers of cytokines, chemokines, and other proinflammatory molecules such as prostaglandins and ATP. There is considerable overlap between the genes involved in angiogenesis and inflammation, and the inflammation pathways involve interleukin-1, and COX2. However, rather than HIF regulating inflammation, the hypoxic regulation of the inflammatory pathways is often via NFkB (Taylor et al., 2016). In normoxia, the PHDs hydroxylate beta-subunit of the IkB-kinase complex (IKKβ) and inactive it. The IkBα keeps NFkB inactive in the cytosol. However, hypoxia reduces PHD activity, which allows IKKβ to become active leading to phosphorylation-dependent degradation of IKβα and releasing of NFkB. NFkB is translocated to the nucleus and induces IL6, COX-2, TNFα, ICAM, VCAM, IL8, CCL5, inducible nitric oxide synthesize and IL1β. Additionally, HIF1α can be induced by NFkB and IkBα antagonizes FIH. Therefore, the cross talk between NFkB and HIF1α may be important for future targeting with combined inhibition of both pathways.

Evading immune destruction

It is well known that tumours are antigenic, and it is clear that hypoxia within the TME can mediate local and peripheral inflammatory and immune responses. Tumour cells express a diverse array of antigens that can be differentially represented due to heterogeneity within the tumour mass. Incredibly, tumour cells are able to evade destruction by the immune system at least in part due to their incapacity to mediate appropriate and sustained immune responses (Stewart and Abrams, 2008). Tumour cells that dislodge from the primary tumour into the circulation during metastasis are able to evade immune destruction in the circulation and thus are able to lodge themselves in distant sites.

Mechanisms underlying precisely how tumour cells evade immune destruction are still emerging (de Visser et al., 2006; Vinay et al., 2015). It is evident that tumour cell subpopulations can avoid detection and evade destruction by both the innate and adaptive actions of the immune system through their differential expression of antigens and cell-surface molecules (markers). Hypoxia induces the expression of markers (e.g. PD-L1, and CTLA-4) on tumour cells increasing their resistance to immunity (Barsoum et al., 2014). In addition, hypoxia-mediated TGF-beta signalling, can enable tumour cells to evade destruction by the immune system (Park et al., 2013). Re-oxygenation has been shown to be a mechanism of loss of immune resistance in some cancer types (Ai et al., 2015), and has important implications for immunotherapy approaches in the treatment of cancer.

Clinical impact of hypoxia and therapeutic agents

Detection of hypoxia in tumours

Historically, the gold standard for measuring oxygen was using an oxygen electrode, which is a probe passing through the tumour and measuring oxygen tension as it passes, pioneered by (Vaupel et al., 1991). However, the technology is not widely available and is prone to artefacts and has never been routinely utilized. In addition, there are often variable results between PET and oxygen electrode measurements. Many methods are being developed, reviewed in (Dubois et al., 2015).

Imaging

The definition of what is a hypoxic tumour and what degree of hypoxia is relevant to resistance is extremely difficult. There is no obvious cut point in any of the analyses since it is a continuous variable and very heterogeneous, even within a tumour. For example, around necrotic areas there will be hypoxia, but also intermittent shut down of vessels would cause hypoxia. As already discussed, the heterogeneity response to hypoxia is substantial in both different areas of the tumour as well as between cancers.

The method most extensively investigated for imaging is positron emission tomography with agents that become reactive under hypoxic conditions. Nitroimidazoles such as misonidazole (FMISO) diffuse passively into cells and then become activated in hypoxia, whereas in normoxic conditions they become reoxidized to the parent compound and can diffuse out again (**Fig. 18.4**). The degree of trapping is proportional to the PO_2. It does however require enzymes for this, but the enzymes are not HIF targets. Additionally, if there is no enzyme activity (i.e. necrotic hypoxic cells), these areas will not be detected FMISO.

Besides FMISO, FAZA and EF5 are the agents most widely investigated or used. The main measures are the SUV max (standardized uptake value; see Fleming et al., 2015) and tumour to muscle ratio. They correlate partly with each other, although because there have been small patient numbers in many studies and different tumour types investigated. The advantage of imaging is in investigating heterogeneity in tumours, primaries and multiple secondaries and it has been shown to be reproducible and reported to have association with poor survival and be responsive to chemotherapy. Key data required for its wider spread use is evidence that it predicts response rate and outcome to radiotherapy in a wide range of tumours. Correlations with immunohistochemistry markers and imaging vary from study to study but in general there is a correlation between carbonic anhydrase IX and HIF1α, although not strong.

Other imaging modalities include MRI and BOLD (blood oxygen level determination) but this is measuring oxygenation in the blood rather than in the tumour. Although this should be widely available, there are problems in reproducibility between centres in contrast to PET, which is more reliable and standardized.

Gene expression and immunochemistry

Based on the gene expression described, immunohistochemistry of formaldehyde fixed-paraffin embedded (FFPE) sections is a highly desirable approach to score hypoxia and this may be through giving patients pimonidazole before surgery and then staining for pimonidazole or using endogenous hypoxia-induced genes such as carbonic anhydrase IX and HIF1α. The problems here are reproducibility of staining of sections and samples and thresholds for positive and negative.

Another methodology is using gene expression arrays and developing expression profiles for key genes in the hypoxia pathway, which then represents an average across the whole tumour, but includes multiple genes (Winter et al., 2007). So, on one hand there is a wider sample, but it brings down the average level of expression. Nevertheless, over 20 different gene expression arrays have been published and linked to clinical outcome.

It interesting how little overlap there is between the genes and the different profiles, even though they all have hypoxia-induced genes within their lists (Harris et al., 2015). It needs to be considered that the hypoxia profiles and genes expressed may differ from tissue to tissue and a recent study has shown, using a hypoxia profile derived from head and neck cancer could predict the benefit

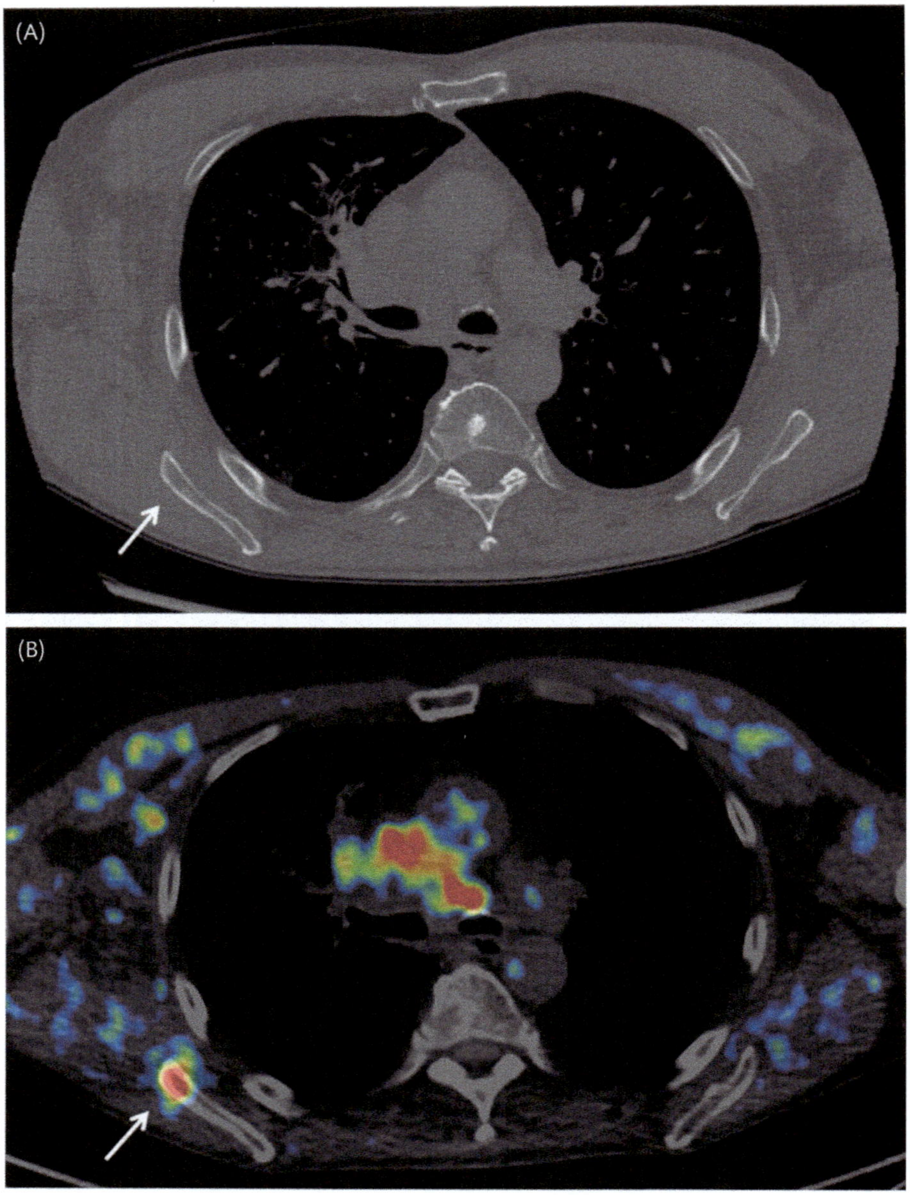

Fig. 18.4 CT contrast enhanced image reformatted using 1.25 mm axial thickness for (A) a bone reconstruction showing that the bone metastasis visible on FMISO PET is not visible on CT. (B) FMISO PET image fused with the corresponding CT displayed on a tumour-to-blood ratio (TBR) colour scale. Red regions depict a TBR greater than 1.4, indicating hypoxia, and no visible PET depicts a TBR below 1, indicating normoxia. This axial image shows the unexpected hypoxic bone metastasis indicated by the white arrow in addition to hypoxic nodal disease in the mediastinum.

Reproduced with permission from McGowan DR et at. '^{18}F-Misonidazole PET-CT scan detection of occult bone metastasis', *Thorax*, Volume 71, Issue 1, p. 97, Copyright © 2016 BMJ Publishing Group Ltd and the British Thoracic Society.

of hypoxia-modifying therapy in a randomized trial in laryngeal cancer (ARCON). However, this could not predict the outcome of a similar hypoxia-modifying therapy in bladder cancer (BACON; see Eustace et al., 2013b).

Bioreductive drugs

Drugs that are hypoxically activated (Hunter et al., 2016) have been validated in randomized trials. These include tirapazamine (Rischin et al., 2006) and nimorazole (Overgaard et al., 1998), the latter is licenced as a hypoxic sensitizer with radiotherapy for head and neck cancer.

A wide variety of hypoxic modifiers have been utilized and meta-analysis showed in 32 randomized trials involving 4,805 patients, that using carbogen breathing, hyperbaric oxygen, hypoxic radiosensitizers overall had a clear effect on local regional control and disease-specific survival (Overgaard, 2011). It is not surprising that these are the strongest indicators because radiotherapy is a local treatment and many patients with head and neck cancer die from local recurrence. Additionally, there was an effect on overall survival, as many patients will die from the local recurrence, but not a significant effect on metastasis at distant sites or radiotherapy complications. These results are encouraging and make the case for

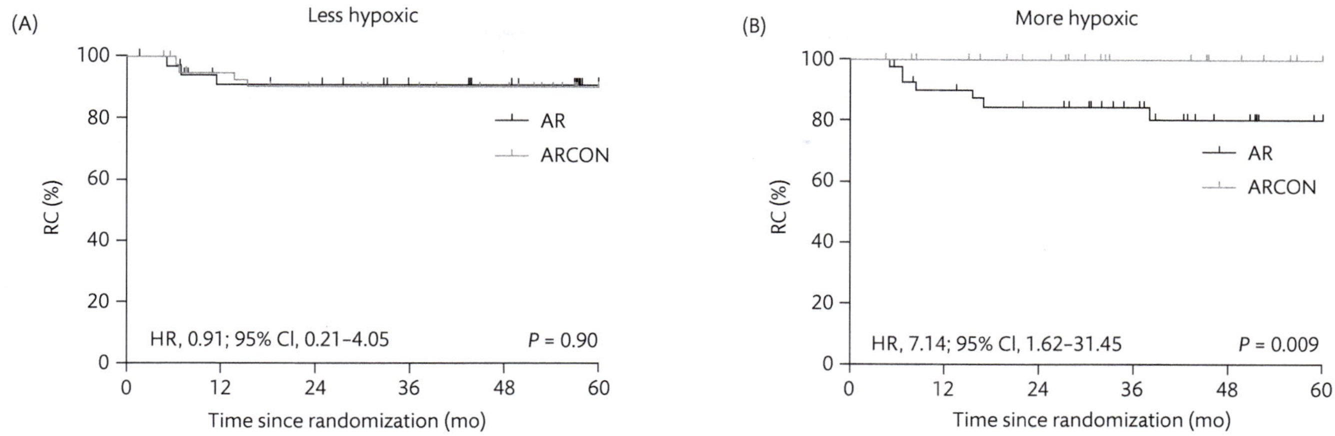

AR	36	29	28	22	20	14
ARCON	43	38	34	32	29	19

AR	41	36	32	25	15	10
ARCON	37	35	31	23	20	15

Fig. 18.5 (A,B): Hypoxia gene profile and patient prognosis.

routine utilization of this approach for head and neck cancer, which is one of the most hypoxic of tumours. It is the tumour type in which tumour cords with hypoxic areas distant from blood vessels, was first described by Tomlinson and Gray. However, there has not been such intensive research in other tumour types.

Hypoxia modification trials and resistance to radiotherapy

Because of the importance of hypoxia in head and neck cancer, several studies have evaluated biomarkers that could detect those that are most likely to benefit from nimorazole (i.e. those that are most hypoxic, and a series of trials run in Denmark), starting with DAHANCA 5, have been very influential in this area (Toustrup et al., 2012). They have shown that gene expression profiles of 15 genes can predict the patients who are going to benefit, the first study to use prediction of this type of array. In their study of 323 HNSEC patients, classification of patients of more hypoxic based upon expression of a 15-gene hypoxia signature predictive benefit from the hypoxic radiosensitizer nimorazole.

Additionally, using misonidazole PET scans they were able to classify patients into those who are most hypoxic and again, only those that had the most hypoxia benefitted. This is a clear demonstration of the value of predictive assessment by gene expression or imaging for the worse subgroup of head and neck cancer patients and avoids using hypoxia modifiers in those that have a better prognosis. Hypoxia scans also predict prognosis (Mortensen et al., 2012).

It should be pointed out that much simpler measures associated with hypoxia such as necrosis of tumours have been associated with response to hypoxia-modifying therapy, for example, in the randomized phase III trial of BACON (Eustace et al., 2013a; Fig. 18.5). Overall, however, it is clear that there is a range of hypoxia resistance modification therapies available that are not currently utilized and form whom we can predict a subgroup of patients who do very poorly without the utilization of these agents.

The most recent drug is THP-302, evofosfamide, which has improved pharmacology compared to many predecessors, being reduced by a one electron oxygen reductase to a potent mustard causing DNA cross-linking and wide activity against many tumour types *in vitro*. It could also diffuse once activated to have a bystander effect in other cell types. However, such a strategy would not target the oxygenated part of the tumour and combination with drugs that are activate in oxygenated in cancer cells such as doxorubicin or epirubicin, or radiotherapy, would be important.

The drug failed its phase III trial. This may have been for several reasons. Tumour types were studied that show notorious resistance to chemotherapy, metastatic pancreatic carcinoma, and soft tissue sarcoma. Treatments were given in combination with chemotherapy rather than radiotherapy because of the extensive metastatic disease. Potentially the problem is of trial design in that patient stratification is critical. We do not know what proportion of the patients had hypoxic tumours or what proportion of the tumours were hypoxic. It would also have been extremely useful to have had FMISO scans before treatment and stratify patients as well as an assessment of HIF1α and other markers and potentially a gene signature, which can be obtained from paraffin sections.

Reduction of oxygen consumption

There are essentially two ways that hypoxia could be targeted, one by utilizing hypoxia to activate prodrugs and synthetic lethality approaches and the other is to reduce the hypoxia by reducing oxygen consumption. This latter approach is being followed by drugs that block the Krebs cycle such as metformin or drugs that block kinase signalling pathways such as PI3-kinase, EGFR, which reduce mitochondrial activity and oxygen consumption (Storozhuk et al., 2013) and more recently the antimalarial atovaquone (Ashton et al., 2016). The results indicating a successful approach in decreasing oxygen as opposed to harnessing the bioreductive effects of drugs, but as stated earlier, many of these have problems due to poor trial design.

Resistance to angiogenesis (covered partly in metabolism, partly in angiogenesis)

Clearly one of the major effects of inhibition of angiogenesis is reducing the blood supply to tumours and thus inducing hypoxia. Whether the mechanism of action is complex, for example induction of vascular normalization as well, nevertheless hypoxia is commonly reported as being induced in cancers on therapy in preclinical models and human cancers.

In view of the aforementioned biology, hypoxia will switch tumours to survival pathways that can grow in the absence or low levels of oxygen, thus may be major mediators of resistance through metabolic adaptation (McIntyre and Harris, 2015). Indeed, several studies have shown blocking the adaptation pathways such as carbonic anhydrase IX or GLUT1 and HIF1α itself can produce synergistic effects or add to the effects in preclinical models, but none of these have yet been translated through into clinical studies. However, just as above with the hypoxia sensitizer, it will be critical to ensure biomarkers are used to see which patients have developed hypoxic tumours and upregulation of the target pathways for this approach to synthetic lethality, blocking a pathway and also the adaptive response required to maintain survival.

Additionally, lipid uptake is critical under hypoxic conditions, and drugs that target fatty acid synthesis or desaturation have been shown to help prevent regrowth of cancers after antiangiogenic therapy has ceased.

Overall, therefore, to significantly target hypoxia and utilize many of these aforementioned approaches, the coordinated assessment of the patient populations, suitable drugs, classification, and early response assessment is critical to develop this area. So far this has not been achieved, apart from the use of radiotherapy and FMISO imaging.

TAKE-HOME MESSAGE

- Hypoxia occurs in tumours because of an inadequate supply of oxygen due to suboptimal blood supply.
- The occurrence of hypoxia is a key feature of most solid tumours as enhances some cancer features like proliferation, motility, and survival.
- HIFs are a highly evolutionarily conserved family of dimeric transcription factors central to the cellular response to hypoxia.
- HIFs regulate the expression of a diverse array of pathways.
- Hypoxia-induced pathways offer some promising targets for treatment.

OPEN QUESTIONS

- The definition of what is a hypoxic tumour and what degree of hypoxia is relevant to resistance is extremely difficult. There is no obvious cut-off point in any of the analyses, since it is a continuous variable and very heterogeneous, even within a tumour.
- Some drugs that are hypoxically activated are licenced as a hypoxic sensitizer with radiotherapy for head and neck cancer but not in other tumour types, which need to be properly investigated.
- It will be critical to ensure that biomarkers, suitable for use in clinical setting, are developed and used to see which patients have developed hypoxic tumours and upregulation of the target pathways.
- To significantly target hypoxia in cancer, the coordinated assessment of the patient populations, suitable drugs, classification, and early response assessment, needs to be developed.

FURTHER READING

Choudhry, H., & Harris, A. L. (2018). Advances in hypoxia-inducible factor biology. *Cell Metab*, **27**(2), 281–98.

Nakazawa, M. S., Keith, B., & Simon, M. C. (2016). Oxygen availability and metabolic adaptations. *Nat Rev Cancer*, **16**(10), 663–73.

Pugh, C. W., & Ratcliffe, P. J. (2017). New horizons in hypoxia signaling pathways. *Exp Cell Res*, **356**(2), 116–21.

Schito, L., & Semenza, G. L. (2016). Hypoxia-inducible factors: master regulators of cancer progression. *Trends Cancer*, **2**(12), 758–70.

Rankin, E. B., Nam, J. M., & Giaccia, A. J. (2016). Hypoxia: signaling the metastatic cascade. *Trends Cancer*, **2**(6), 295–304.

REFERENCES

Agani, F. H., Pichiule, P., Chavez, J. C., & LaManna, J. C. (2000). The role of mitochondria in the regulation of hypoxia-inducible factor 1 expression during hypoxia. *J Biol Chem*, **275**, 35863–7.

Ai, M., Budhani, P., Sheng, J, et al. (2015). Tumor hypoxia drives immune suppression and immunotherapy resistance. *J Immunother Cancer*, **3**(Suppl 2), P392.

Allaoui, R., Bergenfelz, C., Mohlin, S., et al. (2016). Cancer-associated fibroblast-secreted CXCL16 attracts monocytes to promote stroma activation in triple-negative breast cancers. *Nature Communications*, **7**, 13050.

Arany, Z., Huang, L. E., Eckner, R., et al. (1996). An essential role for p300/CBP in the cellular response to hypoxia. *Proc Natl Acad Sci U S A*, **93**, 12969–73.

Ashton, T. M., Fokas, E., Kunz-Schughart, L. A., et al. (2016). The antimalarial atovaquone increases radiosensitivity by alleviating tumour hypoxia. *Nature Communications*, **7**, 12308.

Banci, L., Bertini, I., Cefaro, C., et al. (2009). MIA40 is an oxidoreductase that catalyzes oxidative protein folding in mitochondria. *Nat Struct Mol Biol*, **16**, 198–206.

Bardos, J. I. & Ashcroft, M. (2005). Negative and positive regulation of HIF-1: a complex network. *Biochim Biophys Acta*, **1755**, 107–20.

Barsoum, I. B., Smallwood, C. A., Siemens, D. R., & Graham, C. H. (2014). A mechanism of hypoxia-mediated escape from adaptive immunity in cancer cells. *Cancer Res*, **74**, 665–74.

Benita, Y., Kikuchi, H., Smith, A. D., Zhang, M. Q., Chung, D. C., & Xavier, R. J. (2009). An integrative genomics approach identifies hypoxia inducible factor-1 (HIF-1)-target genes that form the core response to hypoxia. *Nucleic Acids Res*, **37**, 4587–602.

Berra, E., Roux, D., Richard, D. E., & Pouyssegur, J. (2001). Hypoxia-inducible factor-1 alpha (HIF-1 alpha) escapes O(2)-driven proteasomal degradation irrespective of its subcellular localization: nucleus or cytoplasm. *EMBO Rep*, **2**, 615–20.

Bihlmaier, K., Mesecke, N., Terziyska, N., Bien, M., Hell, K., & Herrmann, J. M. (2007). The disulfide relay system of mitochondria is connected to the respiratory chain. *J Cell Biol*, **179**, 389–95.

Bracken, C. P., Fedele, A. O., Linke, S., et al. (2006). Cell-specific regulation of hypoxia-inducible factor (HIF)-1alpha and HIF-2alpha stabilization and transactivation in a graded oxygen environment. *J Biol Chem*, **281**, 22575–85.

Bristow, R. G. & Hill, R. P. (2008). Hypoxia and metabolism. Hypoxia, DNA repair and genetic instability. *Nat Rev Cancer*, **8**(3), 180–92.

Carreau, A., El Hafny-Rahbi, B., Matejuk, A., Grillon, C., & Kieda, C. (2011). Why is the partial oxygen pressure of human tissues a crucial parameter? Small molecules and hypoxia. *J Cell Mol Med*, **15**, 1239–53.

Chandel, N. S., Maltepe, E., Goldwasser, E., Mathieu, C. E., Simon, M. C., & Schumacker, P. T. (1998). Mitochondrial reactive oxygen species trigger hypoxia-induced transcription. *Proc Natl Acad Sci U S A*, **95**, 11715–20.

Choudhry, H., Harris, A. L., & McIntyre, A. (2016). The tumour hypoxia induced non-coding transcriptome. *Mol Aspects Med*, **47–8**, 35–53.

Choudhry, H., Schodel, J., Oikonomopoulos, S., et al. (2014). Extensive regulation of the non-coding transcriptome by hypoxia: role of HIF in releasing paused RNApol2. *EMBO Rep*, **15**, 70–6.

Cooke, M. S., Evans, M. D., Dizdaroglu, M., & Lunec, J. (2003). Oxidative DNA damage: mechanisms, mutation, and disease. *FASEB J*, **17**(10), 1195–214.

Cooke, V. G., LeBleu, V. S., Keskin, D., et al. (2012). Pericyte depletion results in hypoxia-associated epithelial-to-mesenchymal transition and metastasis mediated by met signaling pathway. *Cancer Cell*, **21**(1), 66–81.

Counillon, L., Bouret, Y., Marchiq, I., & Pouyssegur, J. (2016). Na(+)/H(+) antiporter (NHE1) and lactate/H(+) symporters (MCTs) in pH homeostasis and cancer metabolism. *Biochim Biophysi Acta*, **1863**, 2465–80.

Damaghi, M., Tafreshi, N. K., Lloyd, M. C., et al. (2015). Chronic acidosis in the tumour microenvironment selects for overexpression of LAMP2 in the plasma membrane. *Nature Communications*, **6**, 8752.

Duan, C. (2016). Hypoxia-inducible factor 3 biology: complexities and emerging themes. *Am J Physiol Cell Physiol*, **310**, C260–9.

Dubois, L., Peeters, S., Lieuwes, N. G., et al. (2011). Specific inhibition of carbonic anhydrase IX activity enhances the *in vivo* therapeutic effect of tumor irradiation. *Radiother Oncol*, **99**, 424–31.

Dubois, L. J., Niemans, R., van Kuijk, S. J., et al. (2015). New ways to image and target tumour hypoxia and its molecular responses. *Radiother Oncol*, **116**, 352–7.

Epstein, A. C., Gleadle, J. M., McNeill, L. A., et al. (2001). C. elegans EGL-9 and mammalian homologs define a family of dioxygenases that regulate HIF by prolyl hydroxylation. *Cell*, **107**, 43–54.

Estrella, V., Chen, T., Lloyd, M., et al. (2013). Acidity generated by the tumor microenvironment drives local invasion. *Cancer Res*, **73**, 1524–35.

Eustace, A., Irlam, J. J., Taylor, J., et al. (2013a). Necrosis predicts benefit from hypoxia-modifying therapy in patients with high risk bladder cancer enrolled in a phase III randomised trial. *Radiother Oncol*, **108**, 40–7.

Eustace, A., Mani, N., Span, P. N., et al. (2013b). A 26-gene hypoxia signature predicts benefit from hypoxia-modifying therapy in laryngeal cancer but not bladder cancer. *Clin Cancer Res*, **19**, 4879–88.

Favaro, E., Ramachandran, A., McCormick, R., et al. (2010). MicroRNA-210 regulates mitochondrial free radical response to hypoxia and krebs cycle in cancer cells by targeting iron sulfur cluster protein ISCU. *PloS One*, **5**, e10345.

Fischer, M., Horn, S., Belkacemi, A., et al. (2013). Protein import and oxidative folding in the mitochondrial intermembrane space of intact mammalian cells. *Mol Biol Cell*, **24**, 2160–70.

Fleming, I. N., Manavaki, R., Blower, P. J., et al. (2015). Imaging tumour hypoxia with positron emission tomography. *Br J Cancer*, **112**, 238–50.

Fukuda, R., Zhang, H., Kim, J. W., Shimoda, L., Dang, C. V., & Semenza, G. L. (2007). HIF-1 regulates cytochrome oxidase subunits to optimize efficiency of respiration in hypoxic cells. *Cell*, **129**, 111–22.

Gilkes, D. M., & Semenza, G. L. (2013). Role of hypoxia-inducible factors in breast cancer metastasis. *Future Oncol*, **9**, 1623–36.

Gu, Y. Z., Moran, S. M., Hogenesch, J. B., Wartman, L., & Bradfield, C. A. (1998). Molecular characterization and chromosomal localization of a third alpha-class hypoxia inducible factor subunit, HIF3alpha. *Gene Expr*, **7**, 205–13.

Guzy, R. D., Hoyos, B., Robin, E., et al. (2005). Mitochondrial complex III is required for hypoxia-induced ROS production and cellular oxygen sensing. *Cell Metab*, **1**, 401–8.

Harris, B. H., Barberis, A., West, C. M., & Buffa, F. M. (2015). Gene expression signatures as biomarkers of tumour hypoxia. *Clinical Oncol*, **27**, 547–60.

Hulikova, A., Harris, A. L., Vaughan-Jones, R. D., & Swietach, P. (2012). Acid-extrusion from tissue: the interplay between membrane transporters and pH buffers. *Curr Pharma Des*, **18**, 1331–7.

Hunter, F. W., Wouters, B. G., & Wilson, W. R. (2016). Hypoxia-activated prodrugs: paths forward in the era of personalised medicine. *Br J Cancer*, **114**, 1071–7.

Ivan, M., Kondo, K., Yang, H., et al. (2001). HIFalpha targeted for VHL-mediated destruction by proline hydroxylation: implications for O_2 sensing. *Science*, **292**, 464–8.

Jaakkola, P., Mole, D. R., Tian, Y. M., et al. (2001). Targeting of HIF-alpha to the von Hippel-Lindau ubiquitylation complex by O2-regulated prolyl hydroxylation. *Science*, **292**, 468–72.

Jiang, B. H., Rue, E., Wang, G. L., Roe, R., & Semenza, G. L. (1996). Dimerization, DNA binding, and transactivation properties of hypoxia-inducible factor 1. *J Biol Chem*, **271**, 17771–8.

Kallio, P. J., Pongratz, I., Gradin, K., McGuire, J., & Poellinger, L. (1997). Activation of hypoxia-inducible factor 1alpha: posttranscriptional regulation and conformational change by recruitment of the Arnt transcription factor. *Proc Natl Acad Sci U S A*, **94**, 5667–72.

Kazak, L., Chouchani, E. T., Jedrychowski, M. P., et al. (2015). A creatine-driven substrate cycle enhances energy expenditure and thermogenesis in beige fat. *Cell*, **163**, 643–55.

Lancaster, D. E., McNeill, L. A., McDonough, M. A., et al. (2004). Disruption of dimerization and substrate phosphorylation inhibit factor inhibiting hypoxia-inducible factor (FIH) activity. *Biochem J*, **383**, 429–37.

Lando, D., Peet, D. J., Gorman, J. J., Whelan, D. A., Whitelaw, M. L., & Bruick, R. K. (2002a). FIH-1 is an asparaginyl hydroxylase enzyme that regulates the transcriptional activity of hypoxia-inducible factor. *Genes Dev*, **16**, 1466–71.

Lando, D., Peet, D. J., Whelan, D. A., Gorman, J. J., & Whitelaw, M. L. (2002b). Asparagine hydroxylation of the HIF transactivation domain a hypoxic switch. *Science*, **295**, 858–61.

Li, S., Meng, W., Guan, Z., Guo, Y., & Han, X. (2016). The hypoxia-related signaling pathways of vasculogenic mimicry in tumor treatment. *Biomed Pharmacother*, **80**, 127–35.

Luoto, K. R., Kumareswaran, R., & Bristow, R. G. (2013). Tumor hypoxia as a driving force in genetic instability. *Genome Integr*, **4**(1), 5.

Mahon, P. C., Hirota, K., & Semenza, G. L. (2001). FIH-1: a novel protein that interacts with HIF-1alpha and VHL to mediate repression of HIF-1 transcriptional activity. *Genes Dev*, **15**, 2675–86.

Mansfield, K. D., Guzy, R. D., Pan, Y., et al. (2005). Mitochondrial dysfunction resulting from loss of cytochrome c impairs cellular oxygen sensing and hypoxic HIF-alpha activation. *Cell Metab*, **1**, 393–9.

Masoud, G. N., & Li, W. (2015). HIF-1alpha pathway: role, regulation and intervention for cancer therapy. *Acta Pharm Sin B*, **5**, 378–89.

Maxwell, P. H., Wiesener, M. S., Chang, G. W., et al. (1999). The tumour suppressor protein VHL targets hypoxia-inducible factors for oxygen-dependent proteolysis. *Nature*, **399**, 271–5.

McIntyre, A. & Harris, A. L. (2015). Metabolic and hypoxic adaptation to anti-angiogenic therapy: a target for induced essentiality. *EMBO Mol Med*, **7**, 368–79.

McIntyre, A. & Harris, A. L. (2016). The role of pH regulation in cancer progression. Recent results in cancer research. *Fortschritte der Krebsforschung Progres Dans les Recherches sur le Cancer*, **207**, 93–134.

McIntyre, A., Hulikova, A., Ledaki, I., et al. (2016). Disrupting hypoxia-induced bicarbonate transport acidifies tumor cells and suppresses tumor growth. *Cancer Res*, **76**, 3744–55.

McNeill, L. A., Hewitson, K. S., Claridge, T. D., Seibel, J. F., Horsfall, L. E., & Schofield, C. J. (2002). Hypoxia-inducible factor asparaginyl hydroxylase (FIH-1) catalyses hydroxylation at the beta-carbon of asparagine-803. *Biochem J*, **367**, 571–5.

Modjtahedi, N., Tokatlidis, K., Dessen, P., & Kroemer, G. (2016). Mitochondrial proteins containing coiled-coil-helix-coiled-coil-helix (CHCH) domains in health and disease. *Trends Biochem Sci*, **41**, 245–60.

Mortensen, L. S., Johansen, J., Kallehauge, J., et al. (2012). FAZA PET/CT hypoxia imaging in patients with squamous cell carcinoma of the head and neck treated with radiotherapy: results from the DAHANCA 24 trial. *Radiother Oncol*, **105**, 14–20.

Overgaard, J. (2011). Hypoxic modification of radiotherapy in squamous cell carcinoma of the head and neck--a systematic review and meta-analysis. *Radiother Oncol*, **100**, 22–32.

Overgaard, J., Hansen, H. S., Overgaard, M., et al. (1998). A randomized double-blind phase III study of nimorazole as a hypoxic radiosensitizer of primary radiotherapy in supraglottic larynx and pharynx carcinoma. Results of the Danish Head and Neck Cancer Study (DAHANCA) Protocol 5-85. *Radiother Oncol*, **46**, 135–46.

Park, J., Lee, J., Kang, W., Chang, S., Shin, E. C., & Choi, C. (2013). TGF-beta1 and hypoxia-dependent expression of MKP-1 leads tumor resistance to death receptor-mediated cell death. *Cell Death Dis*, **4**, e521.

Parks, S. K., Chiche, J., & Pouyssegur, J. (2011). pH control mechanisms of tumor survival and growth. *J Cell Physiol*, **226**, 299–308.

Parks, S. K., & Pouyssegur, J. (2015). The Na(+)/HCO_3(-) co-transporter SLC4A4 Plays a role in growth and migration of colon and breast cancer cells. *J Cell Physiol*, **230**, 1954–63.

Patel, S. A., & Simon, M. C. (2008). Biology of hypoxia-inducible factor-2alpha in development and disease. *Cell Death Differ*, **15**, 628–34.

Pittman, R. N. (2011). *Regulation of Tissue Oxygenation*. San Rafael, CA: Morgan & Claypool Life Sciences.

Poon, E., Harris, A. L., & Ashcroft, M. (2009). Targeting the hypoxia-inducible factor (HIF) pathway in cancer. *Expert Rev Mol Med*, **11**, e26.

Ratcliffe, P. J. (2013). Oxygen sensing and hypoxia signalling pathways in animals: the implications of physiology for cancer. *J Physiol*, **591**, 2027–42.

Rischin, D., Hicks, R. J., Fisher, R., et al. (2006). Prognostic significance of [18F]-misonidazole positron emission tomography-detected tumor hypoxia in patients with advanced head and neck cancer randomly assigned to chemoradiation with or without tirapazamine: a substudy of Trans-Tasman Radiation Oncology Group Study 98.02. *J Clin Oncol*, **24**, 2098–104.

Robertson, N., Potter, C., & Harris, A. L. (2004). Role of carbonic anhydrase IX in human tumor cell growth, survival, and invasion. *Cancer Res*, **64**, 6160–5.

Samanta, D., Prabhakar, N. R., & Semenza, G. L. (2017). Systems biology of oxygen homeostasis. *Wiley Interdiscip Rev Syst Biol Med*, **9**(4). doi: 10.1002/wsbm.1382.

Semenza, G. L. (2013). HIF-1 mediates metabolic responses to intratumoral hypoxia and oncogenic mutations. *J Clin Invest*, **123**, 3664–71.

Semenza, G. L. (2014). Oxygen sensing, hypoxia-inducible factors, and disease pathophysiology. *Annu Rev Pathol*, **9**, 47–71.

Semenza, G. L. (2016a). Dynamic regulation of stem cell specification and maintenance by hypoxia-inducible factors. *Mol Aspects Med*, **47–8**, 15–23.

Semenza, G. L. (2016b). Hypoxia-inducible factors: coupling glucose metabolism and redox regulation with induction of the breast cancer stem cell phenotype. *EMBO J*, **36**(3), 252–9.

Semenza, G. L. (2017). A compendium of proteins that interact with HIF-1alpha. *Exp Cell Res*, **356**, 128–35.

Semenza, G. L., Agani, F., Booth, G., et al. (1997). Structural and functional analysis of hypoxia-inducible factor 1. *Kidney Int*, **51**, 553–5.

Semenza, G. L., Nejfelt, M. K., Chi, S. M., & Antonarakis, S. E. (1991). Hypoxia-inducible nuclear factors bind to an enhancer element located 3' to the human erythropoietin gene. *Proc Natl Acad Sci U S A*, **88**, 5680–4.

Semenza, G. L. & Wang, G. L. (1992). A nuclear factor induced by hypoxia via de novo protein synthesis binds to the human erythropoietin gene enhancer at a site required for transcriptional activation. *Mol Cell Biol*, **12**, 5447–54.

Shen, J., Xia, W., Khotskaya, Y. B., et al. (2013). EGFR modulates microRNA maturation in response to hypoxia through phosphorylation of AGO2. *Nature*, **497**, 383–7.

Stewart, T. J. & Abrams, S. I. (2008). How tumours escape mass destruction. *Oncogene*, **27**, 5894–903.

Storozhuk, Y., Hopmans, S. N., Sanli, T., et al. (2013). Metformin inhibits growth and enhances radiation response of non-small cell lung cancer (NSCLC) through ATM and AMPK. *Br J Cancer*, **108**, 2021–32.

Taanman, J. W. (1999). The mitochondrial genome: structure, transcription, translation and replication. *Biochim Biophys Acta*, **1410**, 103–23.

Taylor, C. T., Doherty, G., Fallon, P. G., & Cummins, E. P. (2016). Hypoxia-dependent regulation of inflammatory pathways in immune cells. *J Clin Invest*, **126**, 3716–24.

Thomas, L. W., Staples, O., Turmaine, M., & Ashcroft, M. (2017). CHCHD4 regulates intracellular oxygenation and perinuclear distribution of mitochondria. *Front Oncol*, **7**, 71.

Tian, H., McKnight, S. L., & Russell, D. W. (1997). Endothelial PAS domain protein 1 (EPAS1), a transcription factor selectively expressed in endothelial cells. *Genes Dev*, **11**, 72–82.

Toustrup, K., Sorensen, B. S., Lassen, P., Wiuf, C., Alsner, J., & Overgaard, J. (2012). Gene expression classifier predicts for hypoxic modification of radiotherapy with nimorazole in squamous cell carcinomas of the head and neck. *Radiother Oncol*, **102**, 122–9.

van den Beucken, T., Koch, E., Chu, K., et al. (2014). Hypoxia promotes stem cell phenotypes and poor prognosis through epigenetic regulation of DICER. *Nat Commun*, **5**, 5203.

van Kuijk, S. J., Yaromina, A., Houben, R., Niemans, R., Lambin, P., & Dubois, L. J. (2016). Prognostic significance of carbonic anhydrase IX expression in cancer patients: a meta-analysis. *Front Oncol*, **6**, 69.

Vaupel, P., Schlenger, K., Knoop, C., & Hockel, M. (1991). Oxygenation of human tumors: evaluation of tissue oxygen distribution in breast cancers by computerized O_2 tension measurements. *Cancer Res*, **51**, 3316–22.

Vaux, E. C., Metzen, E., Yeates, K. M., & Ratcliffe, P. J. (2001). Regulation of hypoxia-inducible factor is preserved in the absence of a functioning mitochondrial respiratory chain. *Blood*, **98**, 296–302.

Vinay, D. S., Ryan, E. P., Pawelec, G., et al. (2015). Immune evasion in cancer: Mechanistic basis and therapeutic strategies. *Semin Cancer Biol*, **35** Suppl, S185-S198.

de Visser, K. E., Eichten, A., & Coussens, L. M. (2006). Paradoxical roles of the immune system during cancer development. *Nat Rev Cancer*, **6**, 24–37.

Wang, G. L., Jiang, B. H., Rue, E. A., & Semenza, G. L. (1995). Hypoxia-inducible factor 1 is a basic-helix-loop-helix-PAS heterodimer regulated by cellular O2 tension. *Proc Natl Acad Sci U S A*, **92**, 5510–4.

Wang, G. L., & Semenza, G. L. (1995). Purification and characterization of hypoxia-inducible factor 1. *J Biol Chem*, **270**, 1230–7.

Webb, J. D., Coleman, M. L., & Pugh, C. W. (2009). Hypoxia, hypoxia-inducible factors (HIF), HIF hydroxylases and oxygen sensing. *Cell Mol Life Sci*, **66**, 3539–54.

Wigerup, C., Pahlman, S., & Bexell, D. (2016). Therapeutic targeting of hypoxia and hypoxia-inducible factors in cancer. *Pharmacol Ther*, **164**, 152–69.

Winter, S. C., Buffa, F. M., Silva, P., et al. (2007). Relation of a hypoxia metagene derived from head and neck cancer to prognosis of multiple cancers. *Cancer Res*, **67**, 3441–9.

Wu, X. Q., Huang, C., He, X., et al. (2013). Feedback regulation of telomerase reverse transcriptase: new insight into the evolving field of telomerase in cancer. *Cell Signal*, **25**, 2462–8.

Yang, J., Staples, O., Thomas, L. W., et al. (2012). Human CHCHD4 mitochondrial proteins regulate cellular oxygen consumption rate and metabolism and provide a critical role in hypoxia signaling and tumor progression. *J Clin Invest*, **122**, 600–11.

19

Invasion, metastasis, and tumour dormancy

Andrey Ugolkov and Andrew P. Mazar

Introduction

Tumours (neoplasms) can be either malignant or benign. Malignant (from Latin *malignus* = born to be bad) tumour or cancer has several primary characteristics: hyperplasia (from ancient Greek *huper* = over + *plasis* = formation, increased number of cells), anaplasia (from ancient Greek *ana* = backward + *plasis* = formation, poor cellular differentiation), invasiveness and ability to metastasize. (the ability to disseminate and spread beyond its site of origin). In contrast to anaplastic malignant tumours, benign tumours are slow growing, encapsulated, and non-invasive and are comprised of differentiated cells resembling mature tissue that are not capable of metastasis (Cotran and Robbins, 2009). Usually, benign lesions or neoplasms (e.g. PanINs) as well as normal-looking cells, glands, or ducts are adjacent to or part of a malignant tumour. If a tissue component does not look completely normal (as is the case for most tissues adjacent to malignant tumours), a pathologist would classify it as 'benign' rather than 'normal' tissue. In this chapter, we use 'benign' as the equivalent of any 'non-malignant' tissue (i.e. also pathological but not malignant tissue). Clinically, benign tumours can often be surgically removed and do not usually lead to death from the tumour. Malignant tumour growth is characteristically invasive, extending from the tissue of origin into adjacent organs. Metastasis (from Greek *meta* = beyond; *stasis* = a standing still) is defined as the spread of cancer cells from a primary malignant tumour to a distal anatomical location followed by growth of the disseminated cancer cells and formation of the metastatic lesion. Although most malignant tumours can metastasize, there are two exceptions: glioma and basal cell carcinoma rarely form metastases (Cotran and Robbins, 2009). Metastatic growth represents a major clinical challenge in the treatment of cancer and most patients with metastatic disease have a poor clinical prognosis in terms of long-term survival (Havrilesky et al., 2015).

Cancer invasion

Malignant tumours grow by progressive invasion and destruction of adjacent benign tissue. Proteases that are expressed or recruited by cancer cells such as matrix metalloproteinases (MMPs), cathepsins, plasmin, and urokinase plasminogen activator can breakdown and remodel extracellular matrix (ECM) adjacent to tumour cells as well as the basement membrane that surrounds blood vessels, thus promoting invasion (Mason and Joyce, 2011; Yue et al., 2012). In addition, the transformed metabolism of cancer cells supports destruction of adjacent normal cells further facilitating the invasive process (McCawley and Matrisian, 2000; Gatenby and Gawlinski, 2003). In contrast to normal cells, many malignant cells rely on anaerobic glycolysis even in the presence of abundant oxygen (Warburg effect; see Warburg, 1930; Warburg, 1956). Because anaerobic metabolism is energy inefficient, it is compensated by increased glucose uptake by cancer cells (Garber, 2004). This high consumption of glucose by cancer cells (up to 50-times higher than normal cells; see Warburg, 1930; Warburg, 1956) was used as a basis to establish the positron emission tomography imaging technique to detect primary and metastatic cancer (Czernin and Phelps, 2002), which utilizes the positron-emitting radionuclide glucose analogue fluorodeoxyglucose (^{18}FdG). This inefficiency in ATP production might be thought to represent a significant competitive disadvantage for cancer cell survival. However, mathematical modelling of the tumour–host interface proposes that anaerobic glycolysis actually supports tumour cell invasion (Gottfried et al., 2012). In cancer cells, glycolysis leads to increased excretion of protons that decreases the extracellular pH around the tumour, leading to p53 mediated apoptosis of normal cells that are adjacent to cancer cells, impaired immune cell response, and surveillance and the loss of intercellular gap junctions. Cancer cells often survive this low pH milieu since they are p53-mutant or p53-null and thus circumvent p53-mediated apoptosis.

Mechanisms of cancer invasion: epithelial-mesenchymal transition (EMT). EMT is defined as a process of cell transdifferentiation from an epithelial to a mesenchymal phenotype characterized by the decreased cell adhesion, which could potentially lead to cancer cell invasion by migration and metastasis (Lee et al., 2006). The molecular mechanisms that drive the process of cancer cell invasion through benign tissue and into lymphatic and blood vessels remain poorly understood. Several decades of research have modelled the various steps of invasion and metastasis in a general way, but many gaps still exist in understanding the continuum of events that lead to invasion and metastasis and how they integrate. Further, mechanisms of invasion and metastasis may be context dependent and influenced not only be the genotype and phenotype of the tumour cell but also by its microenvironment (Clark and Vignjevic, 2015; Yaacoub et al., 2016). The predominant existing dogma suggests that cancer cells can invade from the primary tumour through the migration of individual cancer cells, using amoeboid- or mesenchymal-type movement, or as collective migration of cell sheets, strands and clusters (Friedl and Wolf, 2003). Can cancer cells migrate and invade independently within benign human tissue? Despite many years of investigation into cancer cell invasion and metastasis, this simple but critical question remains to be addressed. *In vitro* migration and invasion assays (e.g. Matrigel, wound healing) indirectly support the theory of cancer cell migration and EMT (Crnic and Christofori, 2004; Lee et al., 2006). However, these *in vitro* systems are extremely artificial in that they do not recapitulate the process of cancer cell invasion faithfully. For example, primary tumour invasion goes through layers of muscle in the case of colon or gastric cancer. For individual cancer cells, it is hardly possible to migrate through muscle layer in the absence of other processes because this well-organized tissue contains tightly connected individual muscle fibres and a network of supporting collagenous tissue between fasciculi.

Although many complex systems were developed by researchers to investigate the various cell-cell and cell-ECM interactions within a tumour, many of the conclusions drawn based on older migration and invasion *in vitro* results are being called into question with the advent of more sophisticated approaches to modelling invasion. For example, EMT is characterized by the loss of E-cadherin (a tumour suppressor) expression and decreased cell adhesion, leading to cancer cell invasion through increased migration (Lee et al., 2006). However, it has been shown that, when two sublines of the hepatocellular carcinoma cell line Li7, one positive and another negative for E-cadherin expression, were injected into mouse spleens, only E-cadherin positive cancer cells formed liver metastasis (Osada et al., 1996). Moreover, E-cadherin expression localized to the membrane was observed in primary gastric carcinoma in 13 of 16 cases of gastric cancer with liver metastasis, and all 16 metastatic liver metastases also showed membranous expression of E-cadherin (Ougolkov et al., 2003). These studies support a critical

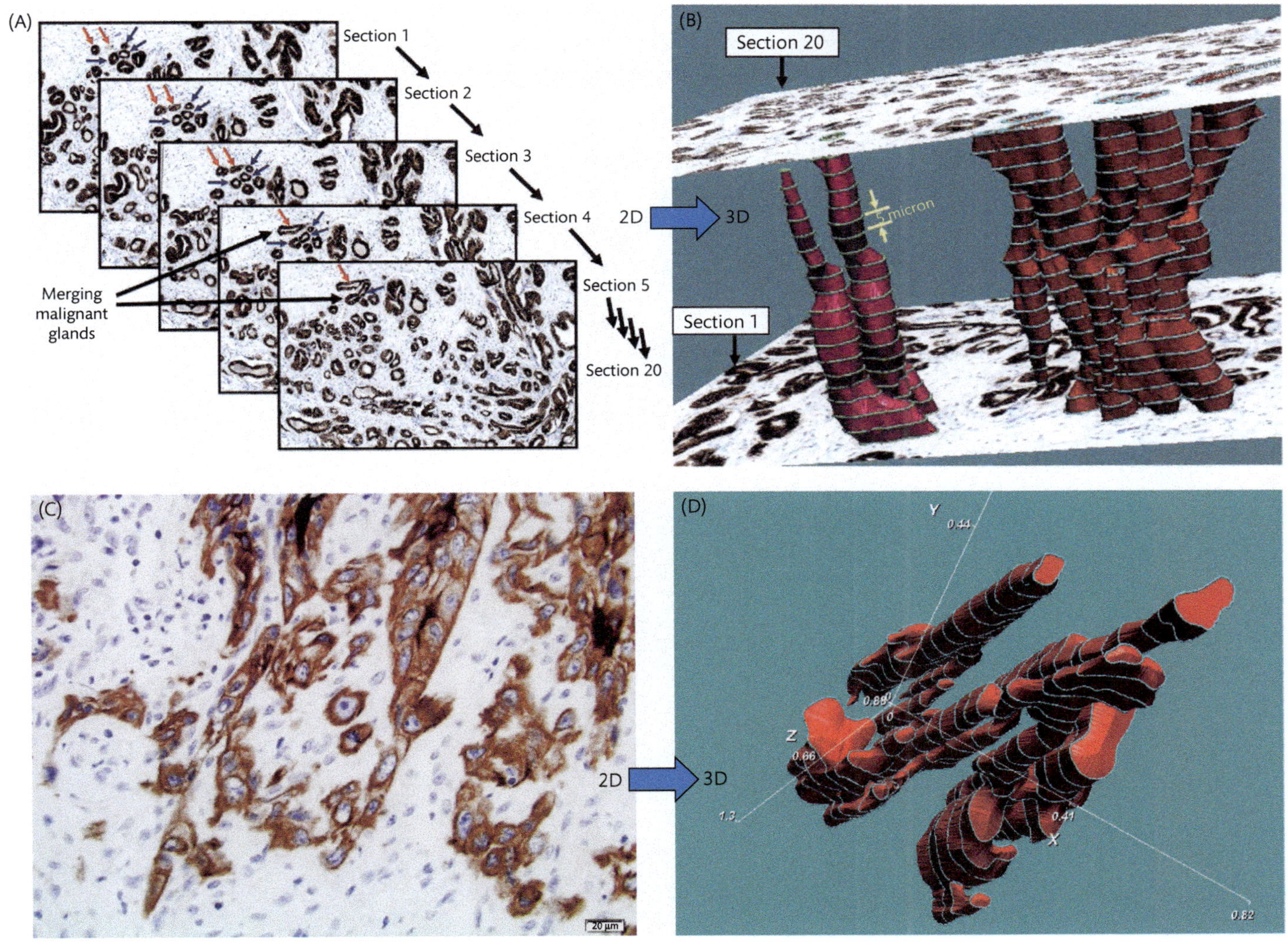

Fig. 19.1 Microscopic 3D reconstruction of prostate and head and neck carcinomas. (A) Stack of registered E-cadherin stained serial sections of prostate adenocarcinoma. (B) Partial microscopic 3D reconstruction of prostate cancer cell network from a stack of 20 serial 5 mm section stained for E-cadherin as shown in A. Cancer cell network from two different microscopic areas (pink and red) of prostate adenocarcinoma was reconstructed as 3D using BioVis3D software. (C) Cytokeratin-stained section of head and neck carcinoma. (D) Partial microscopic 3D reconstruction of cancer cell network from a stack of 15 serial 5 mm sections of head and neck carcinoma stained for cytokeratin as shown in (C).

Cancer cell network was reconstructed as 3D Fig. 19.1 using BioVis3D software.

role of E-cadherin expression in formation of liver metastasis. The hypothesis of EMT reversibility has been proposed based on the obvious similarity in E-cadherin expression pattern in primary and metastatic tumours although EMT reversibility has not been directly demonstrated experimentally (Thompson et al., 2005; Christiansen and Rajasekaran, 2006).

The concept of EMT is based on developmental phenomenon in embryogenesis although none of the cells during embryogenesis are defined as epithelial or mesenchymal at that stage and thus the extrapolation of what is observed during embryogenesis to tumour invasion is tenuous at best (Tarin et al., 2005). The experimental approach to identifying cancer cells *in vitro* as epithelial or mesenchymal is based on cell morphology and the expression of a few putative markers of EMT, which is highly subjective. The EMT hypothesis in cancer, derived from comparisons of *in vitro* results with embryologic data, is based on microscopic features of human tumours. Specifically, the major mechanistic support for EMT occurring in the first place is based on microscopic images of 2D (2-Dimensional) tumour sections that provides the impression that small separated clumps of 'migrating' cancer cells invade adjacent benign tissue (Fig. 19.1). However, recent experimental results using more sophisticated systems that reconstruct a tumour in 3D (rather than just looking at a tumour section in a single plane) indicate that all cancer cells within a tumour specimen are interconnected, forming a 3D (3-Dimensional) network of cancer cell invading roots resembling a 'hair ball' in which the space between microscopic invading roots of cancer cells is filled with benign cells and stromal tissue (Fig. 19.1 and 19.2). These observations suggest that cancer cell proliferation rather than migration may be a driving force of cancer invasion (Fig. 19.1 and 19.2). 3D modelling of tumour sections suggests that invading roots might penetrate venules at multiple points, budding and releasing proliferating cancer cells into the blood circulation leading to metastasis (Fig. 19.2). Thus, cancer cell invasion and seeding into blood circulation (metastasis) may be driven by continuous proliferation of cancer cells and may not require EMT (loss of E-cadherin) for migration of certain cells from primary tumour and MET (mesenchymal-epithelial transition or reversed EMT, E-cadherin expression) at the metastatic site (Figs. 19.2 and 19.3).

Tumour cell proliferation as a driver of invasion. The hypothesis of tumour invasion driven by proliferation of cancer cells is further supported by the fact that most proliferating cancer cells are typically located at the invasive front of a tumour. First step of metastatic process is intravasation of cancer cells. Thin-walled venules are

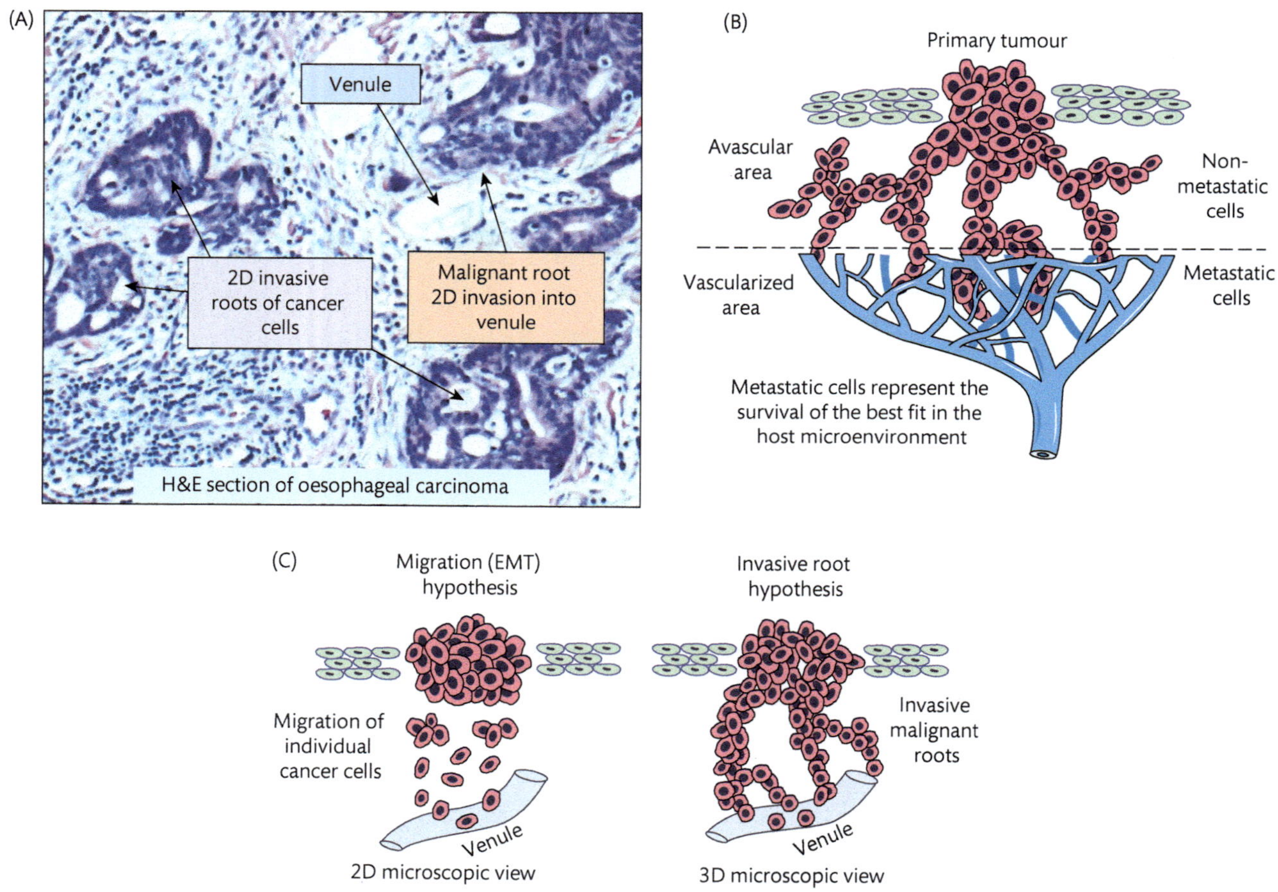

Fig. 19.2 Invasion and intravasation of malignant tumour. (A) Representative H&E image of oesophageal carcinoma showing intravasation of invasive tumour root into venule (2D microscopic view). (B) Schematic representation of tumour invasion and intravasation showing proliferating metastatic cancer cells forming invasive roots in tumour invasion front (vascular area). (C) Illustrations of 2D migration (EMT) and 3D invasive root hypotheses.

penetrated by cancer cells and provide the most common pathways for cancer cells to enter into the blood circulation (**Fig. 19.2**). It has been demonstrated that less than 0.1% of intravasated cancer cells survive to produce metastases after introduction of radiolabelled mouse B16 melanoma cells into mouse blood circulation. Because the vast majority (99.9%) of circulating cancer cells are destroyed quickly by the number of host factors, only continuous massive proliferation of cancer cells into venules but not migration of individual cells can provide a critically large number of cancer cells with few surviving cells nesting in distant organs to form metastasis (**Fig. 19.2**). The 'invading roots' hypothesis can explain the clonal selection of metastatic cancer cells (located at invasive ends of roots and penetrating blood vessels in benign tissue). These metastatic cells at the edge of the growing tumour, selected by their capacity to survive and proliferate in this harsh environment, would have evolved over thousands of generations of proliferation from the originating cancer cells (**Fig. 19.2**).

Invasion by continuous proliferation and genomic instability of cancer cells support the accumulation of beneficial genetic and epigenetic molecular alterations in cancer cells that result in tumour cell survival and lead to the formation of metastatic cells at the invasive front of a tumour (**Fig. 19.2B**). Cancer cells at the invasive front may represent metastasis initiating cells and they should be considered as potential target cells for systemic therapy of metastasis. The accumulation of genetic alterations and the development of fully metastatic clones takes time and is linked to primary tumour size. In breast cancer, mutations of the tumour suppressor gene *TP53* (p53) are significantly more frequent in T3 stage (>5 cm) than in T1 stage (<2 cm). For example, clonal selection of p53 mutated cancer cells occurs when breast carcinoma grows beyond 2 cm in size (Olivier et al., 2006), likely due to the survival advantage conferred in cells expressing p53 mutations.

Other mechanisms of invasion. New model coculture systems that better mimic the microenvironment as well as the heterogenous assortment of cells present in a tumour are beginning to reveal new mechanisms of invasion that illustrate the opportunistic nature of an invading tumour. In areas of sparse or absent matrix or non-tumour cells, invading cancer cells may still migrate stochastically or in response to chemotactic gradients. Proliferation and mechanisms where proliferating tumour can push existing structures or proliferate into channels formed by these structures and both single cell and collective modes (cell jamming) of invasion have been described (Haeger et al., 2014). In reality, an invading tumour can potentially utilize all of these mechanisms to spread and these have been recently reviewed in great detail (Friedl and Alexander, 2011; Haeger et al., 2015).

Cancer metastasis

The process of metastasis requires a series of sequential events leading to tumour cell dissemination and seeding at site distal from the primary tumour (**Fig. 19.3A**). When tumour invasive roots reach and penetrate blood vessels in adjacent benign tissue and/or tumour is vascularized, cancer cells intravasate into the lumina of blood vessels (**Fig. 19.3A**). Cancer cells that survive in blood circulation are frequently trapped in the first microvascular bed that they encounter. Once cancer cell circulation is arrested, it is followed by extravasation and proliferation of cancer cells within the organ leading to formation of metastatic tumour (**Fig. 19.3A**). In many cases, metastatic lesions can grow at this first site of arrest but in some tumours, growth only occurs when the site that has been seeded by the disseminated tumour cell (DTC) is permissive to its growth. Permissive sites that support metastatic growth are termed the 'pre-metastatic niche'. A complex programme of molecular and cellular events is invoked to create this niche as described briefly as follows and in detail in several recent reviews (Chin and Wang, 2016; Tüting and de Visser, 2016). Metastatic tumours can also invade host organ tissue, penetrate into blood vessels, and enter the circulation to produce secondary metastases. Cancer cell dissemination and formation of metastasis occurs in several ways: (1) hematogenous (through the blood) dissemination; (2) lymphatic (lymphogenous) dissemination; (3) direct dissemination to cavities and surfaces of the human body.

Hematogenous metastasis. Hematogenous spreading of cancer cells is typical for sarcomas but is also observed in most carcinomas (Cotran and Robbins, 2009). After intravasation of cancer cells from the primary tumour into venous flow, disseminated cancer cells arrest in the microvasculature of target organ and extravasate to form metastases (**Fig. 19.3**). Two hypotheses explain the process of hematogenous metastasis: 'mechanistic' (blood flow) and 'seed and soil' (organ microenvironment) attributed to Paget in 1989 and recently reviewed (Fidler, 2003; Wirtz et al., 2011). These hypotheses complement each other to explain the development of metastasis. Mechanistically, the prevalence of certain metastatic sites depends on where the primary tumour is localized in the body (**Fig. 19.3B**). For example, liver is the first organ through which venous blood flow from colon, gastric, and pancreatic cancer passes, therefore these types of cancer frequently metastasize to the liver (**Fig. 19.3B**). Liver is not an initial target organ for disseminated metastatic breast cancer cells delivered by venous flow to the right heart which next delivers blood to the lungs, a typical site for breast cancer metastasis (**Fig. 19.3**B). However, patients with metastatic breast cancer do also develop liver metastases, possibly as secondary metastases or through arrest of primary metastatic cells. Metastasis to brain and bone are formed by cancer cells delivered with arterial blood coming from the lungs (**Fig. 19.3B**). In this case, metastatic cancer cells (from the primary tumour) pass through the pulmonary vascular system or lung metastatic lesions release cancer cells to seed secondary metastasis in brain and bone as seen for breast and lung cancer (**Fig. 19.3B**).

The 'seed and soil' hypothesis suggests that seeding and growth of cancer cells is regulated by the biologically specific microenvironment of the target organ ('soil') (Fidler, 2003). Indeed, it has been demonstrated *in vivo* that certain kinds of cancer prefer to metastasize to specific organs regardless of blood flow or vascular anatomy (Fidler, 2003). However, the 'seed and soil' hypothesis is also compatible with the mechanistic hypothesis of metastasis since tumour cells must reach target organs by flow and be trapped there at which time the 'soil' determines whether those trapped cells will progress to true metastases. The development of hematogenous metastasis *in vivo* was first demonstrated by Wood et al. (1967) when rabbit carcinoma cells were injected into rabbit ear arteries where a chamber was inserted to observe the ear capillary network. Cancer cells were rapidly trapped in the capillaries

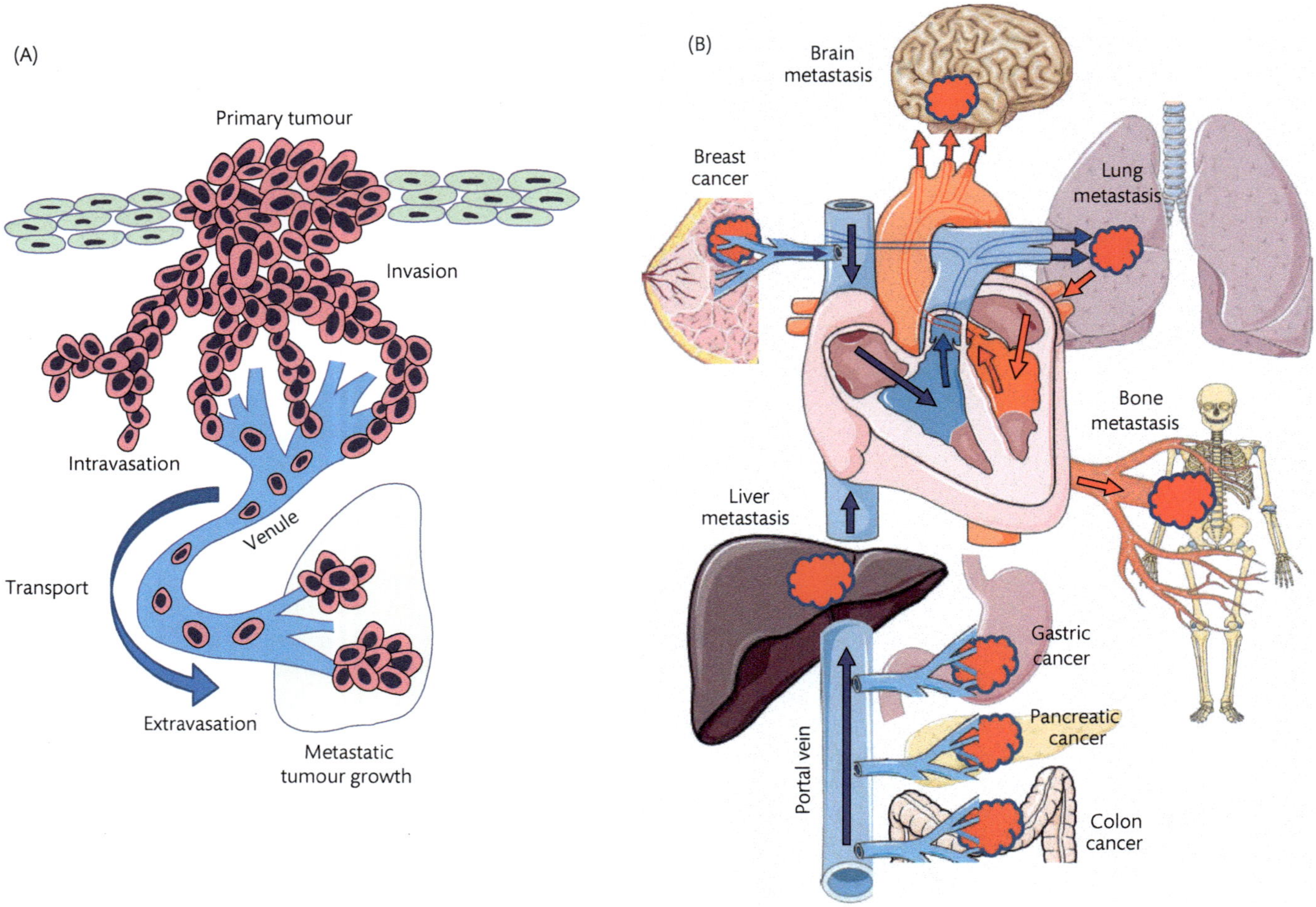

Fig. 19.3 The process of metastasis formation. (A) Sequential steps of metastatic process. (B) Scheme of blood circulation and metastatic dissemination of different types of tumour in certain organs depending on primary tumour localization.

and a thrombus formed within 30 min of tumour cell injection. In 24 hours post-injection, cancer cells started to divide and extravasate through the endothelium. A metastatic lesion was established 72 hours post-injection. This study was the first to illustrate that the process of hematogenous metastasis consisted of a series of sequential steps that involve dissociated single DTC (**Fig. 19.3A**). This was also the first study to suggest that metastasis is an inefficient process, later confirmed by Fidler and coworkers (Fidler, 1970) and later others, since the vast majority (99.9%) of circulating cancer cells were destroyed quickly in those studies. Since potential metastatic cells are detached from the ECM and are essentially anchorage-independent until they seed, many of these cells likely undergo anoikis, a form of programmed cell death, upon detachment at this stage of metastasis (20).

A variety of host factors (e.g. blood viscosity and turbulence, platelets, T cells, natural killer cells, and macrophages) also contribute to the rapid death of circulating cancer cells (Wirtz et al., 2011). Moreover, the passage of disseminated tumour cells through capillaries leads to cancer cell lysis by shear forces (Wirtz et al., 2011). Because only a few cancer cells end up seeding a metastatic site, a recent hypothesis that micrometastases (defined as metastatic tumours that are not clinically detectable) arise from cancer stem cells has been proposed (Steeg, 2006; Diehn and Majeti, 2010). In addition, recent data that contradicts the single cells metastasis models described here provides support for some tumours metastasizing as clumps of tumour cells, thereby retaining the survival advantages of cell-cell interactions that contribute to promoting metastasis (Cheung and Ewald, 2016). It is likely that all of these mechanisms play a role in metastasis that differs based on tumour type.

The most common sites of distant hematogenous metastasis include lung, liver, brain, and bones (Cotran and Robbins, 2009).

Lung metastasis. Metastasis to the lung is common in sarcoma, breast, kidney, bladder, endometrial, and other types of carcinoma. Sarcoma metastasizes almost exclusively to the lungs (Cotran and Robbins, 2009). In many cases, there are no lung-related symptoms when lung metastases are detected. However, symptoms of lung metastases can include chest pain, bloody sputum, cough, and shortness of breath. Complications of lung metastases include pneumonia, bleeding, respiratory failure, collapsed lung, and formation of effusions.

Liver metastasis. The colonization of cancer cells in the liver leads to progressive liver damage. Liver metastases can block bile ducts leading to jaundice and intoxication. Liver metastasis is generally not curable (Lygidakis and Pearl, 1997), although surgery and local and systemic treatment of metastatic lesions might improve clinical outcomes. Life-threatening complications of liver metastasis include hepatic failure and bleeding.

Brain metastasis. Metastases in the brain are the most frequent intracranial tumours (Eichler et al., 2011). Clinical symptoms of brain metastasis develop in up to 10% of cancer patients (Eichler et al., 2011). These include seizures, headache, and neurological deficit. Brain metastases originate from lung cancer (50%), melanoma (20%) and breast cancer (25%; see Eichler et al., 2011). Brain metastases are typically located in the cerebral hemispheres (80%), cerebellum (15%), and brainstem (5%). Multiple brain metastatic lesions are detected in most of patients at the time of diagnosis of brain metastasis. The development of novel therapeutic approaches for the treatment of brain metastases remains a challenging task because the blood–brain barrier is often not penetrated by systemically delivered drugs and patients with brain metastases are typically excluded from clinical trials testing novel anticancer agents. Brain metastases are also not curable although palliation with radiation therapy is utilized to improve quality of life.

Bone metastasis. Bone metastases are detectable in over 50% of patients who die of cancer (Mundy, 2002). There are two types of bone metastases: osteoclastic (breast cancer, lung cancer, multiple myeloma, some prostate cancer) and osteoblastic (prostate cancer). Reciprocal interactions are observed between tumour and bone cells in osteoclastic (osteolytic) lesions where cancer cells secrete factors that activate osteoclasts leading to degradation of the bone matrix, which releases growth factors to stimulate cancer cells to secrete more of these factors (Steeg, 2006). Osteoblastic metastases activate osteoblasts leading to abnormal bone formation. Clinical signs of bone metastasis include severe pain, bone fractures, anaemia, affected mobility, and spinal cord compression. Osteolytic metastasis can also lead to life-threatening hypercalcaemia.

Lymphogenous metastasis. The most common route of cancer cell dissemination for many carcinomas is the lymphatics (Cotran and Robbins, 2009). Lymph nodes metastasis follows the routes of lymphatic drainage. In many types of cancer, regional lymph nodes represent barriers for further spreading of cancer cells. There are numerous interconnections between the lymphatic and vascular systems suggesting the possible involvement of lymphatic spreading also in hematogenous metastasis (Cotran and Robbins, 2009).

Direct dissemination to body cavities and surfaces. Direct seeding and dissemination of cancer cells into the body cavity occurs when cancer grows through the target organ and into a body cavity (e.g. penetration of colon serosal layer by colon carcinoma leads to peritoneal dissemination). In addition, surgical resection of gastrointestinal tumours (e.g. liver, gastric, and pancreatic cancer, cholangiocarcinoma) can also lead to direct dissemination of tumour cells through peritoneal carcinomatosis. Pleural effusions and peritoneal dissemination are the most common metastases to represent direct spreading of cancer cells into a body cavity (Cotran and Robbins, 2009). The peritoneum is a common site for surface metastasis in ovarian and gastrointestinal cancers with median survival of less than 6 months once they occur (Nissan et al., 2009). Peritoneal dissemination may not produce any unusual symptoms in its early stages. Ascites is a typical feature of progressing peritoneal dissemination and may cause abdominal pain, breathing problems, loss of appetite, nausea, constipation. In many cases, cytological analysis of ascitic fluid is performed to confirm peritoneal dissemination.

Clinical implication of metastasis

The kinetics of metastasis development varies in different tumours. For example, breast cancer metastasis can arise 10–15 years after primary tumour resection (Karrison et al., 1999), leading some investigators to hypothesize that metastatic cells in breast cancer lie dormant as discussed next. In contrast, pancreatic cancer often demonstrates a very short time (<1 year) between primary tumour detection and clinical manifestation of metastasis. In general, tumour size, depth of cancer cell invasion and in some cancer types, necrosis (which correlates with tumour size) are considered as valuable clinical predictors of increased risk for metastasis. The larger and more invasive the primary tumour, the higher the probability of metastasis. For example, gastric carcinoma with invasion to mucosal, submucosal, muscle, and serosal layer have an incidence of lymph node metastasis in 5%, 23%, 52%, and 82% cases, respectively (Akagi et al., 2011).

Malignant primary tumours in human vital organs (brain, lung, liver) can lead to patient death because of primary tumour growth, however, the majority of cancer deaths (>90%) occur in patients with metastatic disease (Fidler, 2003). Many of these patients do not die from their tumour burden but rather from complications associated with metastatic disease. For example, approximately 30% of cancer deaths are caused by cachexia (Tisdale, 1997), a wasting syndrome, which develops in advanced stages of cancer and is therefore associated with metastasis. Cachexia is characterized by anorexia, progressive loss of body fat and muscles, weakness, and anaemia (Tisdale, 1997). Patients with cachexia have a decreased survival due to the loss of total body protein leading to significant impairment of respiratory and heart muscle function (Tisdale, 1997). In contrast to starvation, providing extra calories does not reverse the loss of fat and muscle in cancer patients with cachexia and the mechanism of cachexia is not well understood. In other cases, metastasis leads to cancer-related death due to the failure of organs affected by progressive metastatic disease (lungs, liver, and brain).

Modelling of metastasis

Animal models remain the best option for cancer researchers to study metastasis. Animal tumour models include cancer cell line xenografts, patient-derived xenograft (PDX) models, syngeneic tumour models and genetically engineered mouse (GEM).

GEM tumour models. In GEM models, mice have been genetically engineered for over-expression ('knock in') or knock-down ('knock out') of an oncogene or a tumour suppressor gene that leads to formation of a tumour in that GEM model. Generally, single hit models where only one oncogene is knocked in or out are most accurate in recapitulating human disease when the target of interest is a driver mutation for that disease (e.g. the EML4-ALK fusions protein in lung cancer). In other single hit models where the transgene is not a driver mutation, certain steps of tumour formation may be observed but these models rarely metastasize, and addition of other transgenes is required to recapitulate metastatic disease. GEM models generate in this way can reproduce multiple steps of carcinogenesis on the way of development of primary tumour and metastasis replicating human tumour initiation and progression. GEM models generally show a low incidence of distant metastases which develop asynchronically after long periods of latency making GEM models difficult to use as a reliable model of metastasis (Francia et al., 2011). Several popular GEMs that are exceptions to this low incidence of metastasis include MMTV-PyMT and MMTV-neu models of breast cancer, which form metastasis in mouse lung in more than 70% of animals in 3 months (Khanna and Hunter, 2005). GEM models are useful in that they retain an intact normal mouse immune system and can be used to study immune checkpoint and other immunotherapeutic agents, but the prolonged latency period of these models hinders their more frequent use in new drug development.

Syngeneic tumour models. Syngeneic transplantation models also known as allograft mouse tumour systems, consist of mouse tumour tissues derived from the same genetic background as a given mouse strain. Syngeneic mouse tumour cell lines can be derived from transgene or carcinogen-induced mouse tumours or spontaneously developed tumours (Khanna and Hunter, 2005). Several popular syngeneic models including B16 mouse melanoma, 4T1 breast cancer, and 3LL lung cancer models form rapid metastases from subcutaneous tumours with high penetration (Bobek et al., 2010) and also be used to study experimental metastasis (or lung colonization) where tumour cells are injected intravenously in the tail vein. The drawback of these models is that they progress very rapidly (much more quickly than human metastatic disease) and their aggressiveness makes them difficult to use for new drug screening. Further, genomic analysis of syngeneic tumour cell lines shows that many characteristics differ from human tumours from the same histology (Xi et al., 2008). Syngeneic mouse models utilize immunocompetent mice and these models serve as surrogates for human patients and allow to study how cancer therapies perform in animals with metastasis and a functional immune system (Khanna and Hunter, 2005).

Xenograft tumour models. Xenograft models involve the transplantation of human cancer cells or human tumour in immunodeficient mouse. The lack of a functional mouse immune system is required to prevent a rejection of the human graft. Due to convenience, utilization of human tumour, and ease of use, subcutaneous xenograft tumour models have become the prevalent animal tumour model. However, the vast majority of subcutaneous human solid xenograft tumours do not metastasize in immunodeficient mice showing localized expansive growth without obvious invasion (Hoffman, 2015). Subcutaneous engraftment of human tumour does not allow the cancer cells to grow in an appropriate microenvironment provided by tissue of tumour origin. Advances in xenograft models that have overcome some of these limitations include the development of orthotopic xenograft tumour models that attempt to mimic the patient tumour microenvironment, vascularization, and the development of metastatic disease (Cespedes et al., 2006). The spontaneous metastasis models utilize the ability of cancer cells to disseminate from an orthotopic xenograft tumour established by the transplantation of cancer cells or human tumour into a mouse tissue reflecting the organ or tissue of origin of the human cancer. For example, the human breast cancer model MDA-MB-231 is an orthotopic xenograft model when inoculated in the mammary fat pad with a capacity to form spontaneous metastasis in mouse lung (Munoz et al., 2006; Fig. 19.4A). The resection of the primary MDA-MB-231 orthotopic tumour is required to allow the formation of metastasis. Although the MDA-MB-231 orthotopic xenograft tumour is defined as a 'primary' in the model, it is really a metastatic breast tumour grown in the mouse mammary fat pad since the MDA-MB-231 cell line was established from a metastatic pleural effusion obtained from breast cancer patient. This cell line can also be used for an experimental model of lung metastasis induced by injection of cancer cells in mouse tail vein (Fig. 19.4A). Orthotopic xenograft models of breast and other types of cancer frequently use cell lines developed from metastatic lesions and are useful for studying metastasis in mouse models, especially in a translational setting. One drawback of these models is that the metastatic lesions tend to be microscopic and require reporter constructs to track and quantitate *in vivo*. The development of new imaging modalities and the detection of bioluminescent signal from cancer cells expressing firefly luciferase, has revolutionized non-invasive monitoring of orthotopic xenograft tumour growth and metastasis (Contag et al., 2000).

Because the role of the immune system in metastasis development cannot be assessed in immunodeficient mice, new models of humanized mice have been developed for xenograft transplantation studies. By coengrafting mice with human tumours and human haematopoietic cells, a new platform to study the complex relationship between the human immune system and tumour progression was created. Using this approach, immuno-oncology researchers can test the efficacy of novel compounds to mediate human immune cell responses in a tumour model that represents a human patient. For example, humanized mice can be engrafted with human cord blood-derived haematopoietic stem cells homing in mouse bone marrow and this model exhibits multilineage engraftment of human immune cell populations including T-cell maturation and T-cell dependent inflammatory responses (Ito et al., 2012). These humanized mouse models are still being optimized to more closely reflect human immune cell behaviour (Morton et al., 2016).

PDX models. PDX tumour models have many advantages over xenografts established from cancer cell lines. PDX tumours are established by transplantation of surgically resected tumour specimens in immunodeficient mice. These tumours are not first cultured *in vitro* and avoid selection bias associated with ability to grow under cell culture conditions. PDX tumours closely resemble corresponding malignant tumours from patients showing consistency in tumour growth, histological appearance, mutations, and gene expression profile and maintaining cellular heterogeneity of original human tumour (Tentler et al., 2012). Orthotopic PDX models of mesothelioma, colon, pancreatic, breast, ovarian, lung and stomach cancer have been established (Hoffman, 2015). These PDX models showed a similarity in the metastatic tumour growth observed in the patients. It has also been demonstrated that orthotopic PDX tumours (established from human primary gastric cancer) formed liver metastasis in five cases representing all five gastric cancer patients which had clinical liver metastases whereas orthotopic PDX growth of clinically non-metastatic tumours did not produce liver metastasis (Furukawa

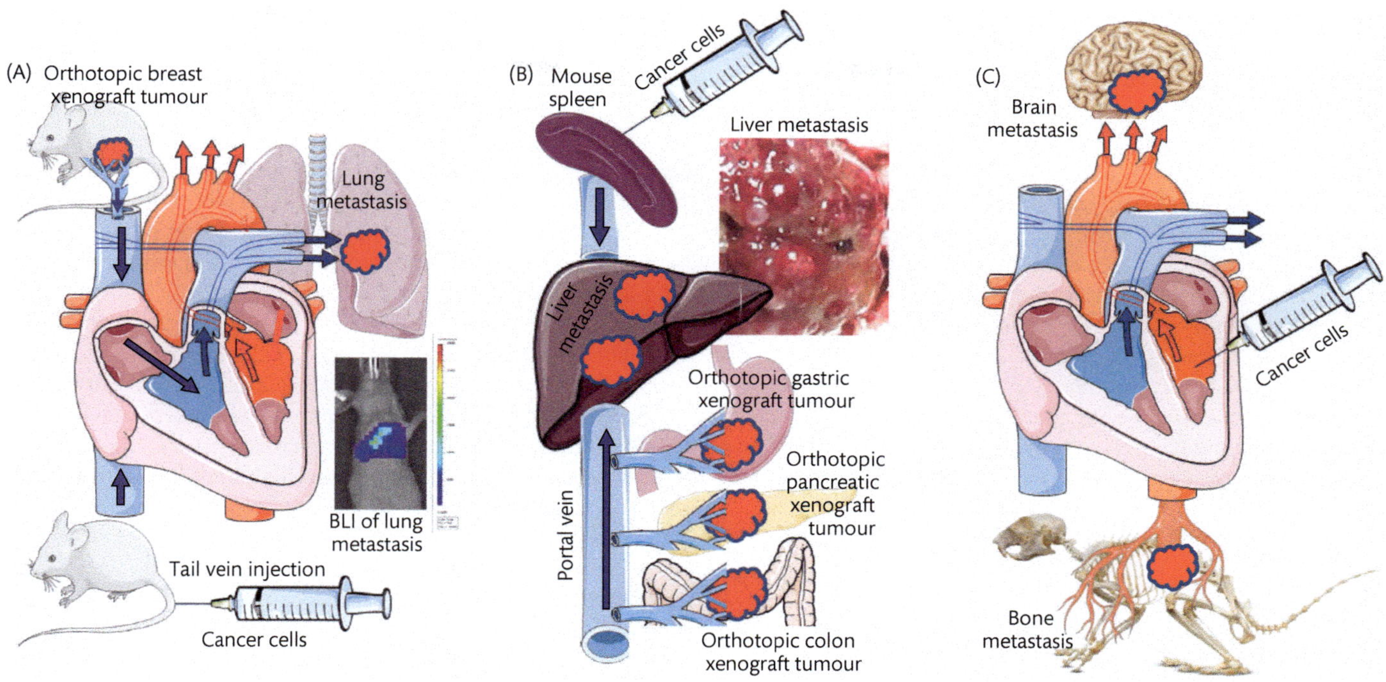

Fig. 19.4 Spontaneous and experimental animal models of metastasis. (A) Breast cancer lung metastasis could be established by either the growth of orthotopic breast xenograft tumour (spontaneous model) or intravenous injection of breast cancer cells (experimental model). Bioluminescent image (BLI) of mouse with lung metastasis is shown. Mouse was injected into tail vein with 5×10^5 MDA-MB-231/LM2-4 breast cancer cells expressing luciferase 4 weeks before the BLI image was acquired. (B) Liver metastasis can be developed from either orthotopic xenograft tumour (spontaneous model) or intrasplenic injection of cancer cells (experimental model). The image of numerous metastases in mouse liver is shown. The picture of mouse liver was taken in 4 weeks after intrasplenic injection of 5×10^5 SK-N-BE(2) neuroblastoma cells. (C) Intracardial injection of cancer cells in left heart ventricle (experimental model) results in the development of brain and bone metastasis.

et al., 1993). Orthotopic xenograft tumours of stomach, pancreas and colon are now used in many labs to model spontaneous liver metastasis (Fig. 19.4B) and many new PDX models are being described weekly that more closely resemble human metastatic disease. Thus, there is significant enthusiasm surrounding the use of metastatic PDX models as tools to advance the study of metastasis and the development of new drugs to treat advanced cancer.

Experimental metastasis models. Most tumour cell lines including orthotopic and syngeneic can be adapted for experimental metastasis models. Suspensions of tumour cells can be inoculated virtually anywhere, and their growth studied. Although these models do not recapitulate the full metastatic process, they are representative of the frequent presentation of advanced human disease in the clinic in which patient often present with metastatic lesions already disseminated.

New drug development for metastatic cancer assumes that metastases are present when the patients begins treatment and thus, experimental metastasis models have tremendous value in assessing drugs to be used in this clinical setting. PDX derived experimental metastasis models are also possible if the PDX tumour can be prepared as a homogeneous suspension of cells after dissociation of the tumour. Intravenous metastatic models are initiated by the injection of a cancer cell suspension into the mouse tail vein. This model allows the study of circulating cancer cells, extravasation, and arrest of cancer cells in target organ and growth of metastasis. The lung is the most common site of metastasis in this model, also referred to as lung colonization (Fig. 19.4A). Using cell lines established from prostate tumour and lung metastasis, it has been demonstrated that only metastasis derived cell lines could produce new spontaneous lung metastasis in an orthotopic model whereas all cell lines developed lung metastasis after intravenous injection in an experimental metastasis model (Hall and Thompson, 1997). In intracardial model, cancer cells are delivered to arterial blood by injection into the left ventricle of anaesthetized mouse (Fig. 19.4C). Typically, arterial blood flow delivers cancer cells to mouse bone and brain where metastases are formed (Fig. 19.4C). These animal models are useful to study breast, lung, melanoma, and prostate cancer metastases. A variation of this model involves inoculation of tumour cells in the tibia and allows the study of bone metastasis (Abdelaziz et al., 2015). There is also a model of liver metastasis induced by intrasplenic injection of cancer cells. Cancer cells flow from spleen with venous blood through splenic vein and entering the liver through portal vein. This resembles the delivery of metastatic cells through portal vein and formation of liver metastasis in case of gastric, colon, and pancreatic cancer (Fig. 19.4B). It has been shown that in 10 min after intrasplenic injection, 90% of radiolabelled colon cancer cells arrest in the liver (Price et al., 1989).

Remodelling the microenvironment in metastasis

DTCs interact with ECM and the microenvironment within a tumour. The heterogeneity in a tumour can be manifested as islands of tumour cells surrounded by non-malignant cells (Fukumura and Jain, 2007). Tumour cell cross talk with stromal cells, endothelial cells and endothelial progenitor cells (EPCs) during angiogenesis, and interactions (or suppression of interactions) with immune system cells contribute to the survival and dissemination of the tumour (Miles and Sikes, 2014). These interactions also contribute to the escape of micrometastases from dormancy, leading to cancer progression (Kareva, 2016). For example, genetic alterations in a tumour cell induce changes in stromal gene expression in models of invasion (Reddy et al., 2012) and these could contribute to the progression of metastasis by altering gene expression in the tumour through epigenetic or mechanical alterations that influence tumour cell phenotype. In tumour dormancy, the opposite may be true, and the failure of tumour cells to interact with the tumour microenvironment or to engage in a different way than during disease progression leads to senescence (Ordóñez-Morán and Huelsken, 2014).

3D cell culture. Molecular and cellular evidence for these conclusions comes from 3D cell culture systems comprised of basement membranes depleted of growth factors that model the transitions of tumour cells to or from dormancy and demonstrate the ability of ECM components to mediate signalling pathways that can lead to inhibition of tumour cells growth and senescence. Dormant tumour cells show a unique cytoskeletal organization with minimal adhesive interactions with the ECM in these models. The switch to proliferation of the tumour cell is driven by cytoskeletal rearrangements that form stress fibres and lead to new interactions with the tumour microenvironment. These cytoskeletal stress fibres are formed mainly through α5β1 integrin outside-in signalling leading to myosin light chain phosphorylation. The upregulation of fibronectin (FN) expression by tumour cells is an early event that mediates the interaction with α5β1 (Allgayer and Aguirre-Ghiso, 2008; Barkan et al., 2008), inducing the switch to malignant growth and proliferation. In this model system, dormant cell lines were induced to switch to aggressive behaviour by the simple addition of FN to the cell culture system.

Head and neck cancer dormancy study. In a separate study of head and neck cancer dormancy, upregulation of urokinase plasminogen activator receptor (uPAR) expression led to increased tumour cell proliferation through FN fibril formation and the binding of uPAR to α5β1 integrin. Downregulation of urokinase receptor (uPAR) had the opposite effect and suppressed α5β1 signalling leading to dormancy (Allgayer and Aguirre-Ghiso, 2008). Assembly of an FN matrix through uPAR-α5β1 binding led to the activation of the RAS/ERK pathway, a major driver of tumour cell proliferation (Allgayer and Aguirre-Ghiso, 2008).

In general, normal tissue architecture and intact ECM suppress cancer cell growth and dissemination. Myoepithelial cells, fibroblasts, and immune system cells such as macrophages likewise initially resist tumour progression. However, these cells exhibit plasticity and can be commandeered in the tumour stromal milieu to create a tumour-permissive or enhancing environment. Further, the cross talk of the tumour with other immune cells and components such as T cells and myeloid-derived suppressor cells (MDSC) leads to evasion of immune surveillance and contributes to metastasis (Peranzoni et al., 2013).

The premetastatic niche. Overwhelming evidence supports the paradigm that tumours are highly interactive systems with stromal cells (fibroblasts), endothelial cells and immune cells acting in concert to promote cancer progression and metastasis (Gao and Mittal, 2009). Non-immune cells that contribute to the progression of metastatic disease include resident stromal cells (cancer-associated fibroblasts or CAFs) and endothelial cells that comprise tumour vasculature. Immune cells, which have gained a more prominent role in how cancer is treated, are derived from bone marrow precursors and include macrophage, neutrophils, mast cells, MDSC, and progenitor cells (Ch'ng et al., 2006; Gao and Mittal, 2009) as well as different T-cell populations that may infiltrate a tumour. These immune cells allow tumour cells to evade immune surveillance leading to dissemination and survival of metastatic cancer cells (Mohme et al., 2017). In many tumours, the microenvironment is characterized by a chronic inflammatory state with prominent infiltration of bone marrow derived cells, including neutrophils, lymphocytes, and macrophage. Over time, this chronic inflammatory state undergoes cross talk with the tumour cells to promote the transformation of normal stromal cells to tumour-associated cells that further contribute to metastatic progression (Mantovani et al., 2008; Shalapour and Karin, 2015). Cross talk between chemokines and cytokines, proteases, cell-surface adhesive proteins, and the tumour ECM contributes to this process.

The immune system. The immune system has a dual role in the development of cancer and metastasis and the recent advances of the cells and pathways involved in these processes has led to the development of several new therapeutics to treat advanced solid cancer (de Visser et al., 2006; La Beck et al., 2015). Most immune cells are normally tumour suppressive. However, in the tumour microenvironment, immune cells can be co-opted by the tumour to promote local invasion, angiogenesis, and metastasis. Tumour-associated macrophages (TAM) are a key player in this process and a growing body of evidence supports the role of TAM in tumour invasion, tumour growth, tumour angiogenesis, and metastasis by producing chemokines and growth factors, secreting proteases, and directly interacting with tumour cells (Giuliano et al., 2011).

The role of TAM. Further, the role of TAM may depend on its local environment, with individual subpopulations of TAM promoting angiogenesis or invasion (Lawrence, 2011; Lin et al., 2016). In advanced disease, immune suppression allows a tumour to evade immune surveillance and may have a principal role in the development and promotion of metastasis (Shalapour and Karin, 2015; Vinay et al., 2015). Several immune cells interact to mediate evasion of tumour immune surveillance. MDSC that are present in tumours, peripheral blood, lymph nodes, and bone marrow of cancer patients and preclinical cancer models (Ostrand-Rosenberg and Sinha, 2009). MDSC are CD11b + immature bone marrow derived cells that suppress CD8 + T cells, increase regulatory T-cells levels and inhibit natural killer cells. MDSC are heterogeneous populations of cells with a number of CD11b + subpopulations identified and thus, eliciting their exact role has been challenging (Pan et al., 2015). Peripheral blood MDSC have been identified in many solid tumours and increased peripheral blood MDSC is associated with advanced stage and certain chemotherapy regimens (Diaz-Montero et al., 2009).

The presence of the CD11b + MDSC characterizes the pre-metastatic niche that allows metastatic tumour cells to seed and proliferate through the suppression of the immune system. In addition, tumours evade immune surveillance by suppressing cytotoxic T-cell infiltration into the tumour and favouring the regulatory T suppressor cell populations (Vinay et al., 2015; Muenst et al., 2016). Tumours express factors such as PD-L1 that regulate T-cell differentiation and expansion and an entire spectrum of immune cell regulating factors that may influence tumour metastasis are a major topic of current investigation. Detailed discussion of these factors is beyond the scope of this chapter. In addition, recent data demonstrate the interaction between immune cells in the tumour, such as MDSC and T cells, that also may also mediate tumour evasion of immune surveillance (Umansky et al., 2017).

In addition to TAM, T cells, and MDSC, other cellular components that regulate metastasis include various bone marrow-derived progenitor cells. The bone marrow is a reservoir of haematopoietic progenitor cells. When tumour outgrowth begins at the metastatic site, tumour cells begin to secrete soluble growth factors which mobilize progenitor cells and recruit them to the metastatic niche via chemo-attractant gradients. Soluble actors such as vascular endothelial growth factor (VEGF) are mobilized by local remodelling of the microenvironment in the metastatic niche and stimulate progenitor cells recruitment. VEGF promotes recruitment of EPCs to sites of tumour neovascularization (Rabbany et al., 2003; Li et al., 2006). EPC have been proposed to be required for metastatic outgrowth (Mancuso et al., 2012; Moschetta et al., 2014). EPC merge with the walls of existing blood vessels, sprout neovessels and then differentiate into endothelial cells. EPC may be less critical for established tumours but contribute to the growth of small tumour lesions such as metastases (Bertolini et al., 2012). Further, EPC-driven neovascularization may be responsible for stimulating dormant tumours to grow (Chen et al., 2010).

Tumour dormancy and metastasis

Micrometastatic disease is often established at sites of tumour dissemination even after patients have gone through curative surgery or other treatments with curative intent. Emerging data has shown that even patients with early stage disease can experience disease recurrence and progression leading to poor clinical outcomes. Recent approaches to molecular profiling of tumours are beginning to identify biomarker patterns that characterize these patients (Muratori et al., 2016; Tarhoni et al., 2016). Many tumours such as colorectal and pancreatic cancer, can disseminate early whereas others such as renal cell cancer disseminate late (Macintyre et al., 2017).

Micrometastases that disseminate early are already established at various distant sites may become permissive for outgrowth. Molecular data shows that early metastases may share mutations with the primary sites, but that evolution of metastatic disease may be complex. Parallel progression of the DTC predicts for an independent programme of genetic and epigenetic changes although how many of these are actual drivers of tumour progression remains to be elucidated. The development of extensive heterogeneity among DTCs would be predicted even if they start our as single clones arising from the primary tumour. In addition, recent studies show evidence for polyclonal seeding of metastases where a lesion of disseminated tumour may have multiple clones that allow each other to subsist and collectively drive metastasis (Macintyre et al., 2017).

The examination of patients and their metastatic samples in relation to primary tumour stage is consistent with the parallel progression hypothesis in cases of long-term dormancy (Klein, 2009). Unfortunately, despite intensive study, the mechanisms responsible for tumour cell dormancy and ultimate outgrowth are still not well understood although the understanding of metastatic evolution has increased to reveal multiple mechanism of tumour dissemination (Macintyre et al., 2017). Based on both experimental and clinical information, two general mechanisms of tumour dormancy have been postulated. In the first case, tumour cells that survive adjuvant treatment exist in a quiescent or senescent solitary state because they are resistant to therapies targeting rapid cell proliferation (Naumov et al., 2003). The acquisition of additional mutations or alterations in the microenvironment provide a permissive environment for tumour growth. In the second case, a balance between tumour growth and apoptosis exists that gives no net increase in tumour size. These dormant tumours only begin to progress when they experience additional genetic changes that drive progression (Klein and Holzel, 2006). Control of tumour growth and the switch to progression once dormant tumour cells are activated is regulated through the interaction of micrometastatic tumour cells with their microenvironment and the immune system, ('soil'). The same characteristics that support survival of these cells may also enable their outgrowth, and the metastatic seeds that lead to this outgrowth are characterized by being more malignant and resistant to treatment (Aguirre-Ghiso, 2007; Touil et al., 2014). Given that these metastatic lesions are the most likely to lead to the demise of the patient, a deeper understanding of tumour dormancy and metastatic progression remains a major focus in the development of novel therapies to treat advanced cancer.

Conclusion

Cancer metastasis remains the major therapeutic challenge in cancer. Although the understanding of metastatic disease and its modelling have advanced tremendously over the past 30 years, very few treatments for patients with metastatic disease have had a significant impact on clinical outcomes such as overall survival. However, the advent of immunotherapy, such as immune checkpoint inhibition, and the increase in molecular knowledge that is driving the understanding of how metastatic disease occurs are opening new insights into therapeutic targeting. The exponential increase in both therapeutic and investigational tools promises to alter and improve the course of research in the area of cancer invasion, dormancy, and metastasis.

Acknowledgement

Chapter adapted with permission from Kerr D et al., 'Invasion and metastasis' in Kerr D et al. (Eds), *Oxford Textbook of Oncology, Third Edition,* Oxford University Press, Oxford, UK, Copyright © 2016.

TAKE-HOME MESSAGE

- Despite the recent advances in understanding the biology of metastases, very little advancement has been made in the ability to treat them.
- Invasion and metastasis models are identifying new paradigms for how DTC spread from primary to metastatic sites.
- The study of the immune system in metastasis is revealing new pathways for metastatic spread and new target for therapy.
- New analytical approaches including genomic and expression analysis, as well as functional analytics that look at protein expression and metabolism are leading to new insights in the understanding of invasion and metastasis.
- The role of the tumour microenvironment and host immune system is being revealed through improved *in vitro* and *in vivo* models and will identify new points of therapeutic intervention.

OPEN QUESTIONS

- To translate our knowledge of the biology of the metastases into effective ways to treat and prevent them.
- To understand and unravel the biology of dormancy.

FURTHER READING

Lambert, A. W., Pattabiraman, D. R., & Weinberg, R. A. (2017). Emerging biological principles of metastasis. *Cell*, **168**, 670–91.

Sethi, N. & Kang, Y. (2011). Unravelling the complexity of metastasis—molecular understanding and targeted therapies. *Nat Rev Cancer*, **11**, 735–48.

Stuelten, C. H., Parent, C. A., & Montell, D. J. (2018). Cell motility in cancer invasion and metastasis: insights from simple model organisms. *Nat Rev Cancer*, **18**, 296–312.

Valiente, M., Ahluwalia, S., Boire, A., et al. (2018). The evolving landscape of brain metastasis. *Trends in Cancer*, **4**, 176–96.

Winkler, F. (2015). The brain metastatic niche. *J Mol*, **93**, 1213–20.

REFERENCES

Abdelaziz, D. M., Stone, L. S., & Komarova, S. V. (2015). Localized experimental bone metastasis drives osteolysis and sensory hypersensitivity at distant non-tumor-bearing sites. *Breast Cancer Res Treat*, **153**, 9–20.

Aguirre-Ghiso, J. A. (2007). Models, mechanisms and clinical evidence for cancer dormancy. *Nat Rev Cancer*, **7**, 834–46.

Akagi, T., Shiraishi, N., & Kitano, S. (2011). Lymph node metastasis of gastric cancer. *Cancers (Basel)*, **3**, 2141–59.

Allgayer, H. & Aguirre-Ghiso, J. A. (2008). The urokinase receptor (u-PAR)—a link between tumor cell dormancy and minimal residual disease in bone marrow? *Apmis*, **116**, 602–14.

Barkan, D., Kleinman, H., Simmons, J. L., et al. (2008). Inhibition of metastatic outgrowth from single dormant tumor cells by targeting the cytoskeleton. *Cancer Res*, **68**, 6241–50.

Bertolini, F., Mancuso, P., Benayoun, L., Gingis-Velitski, S., & Shaked, Y. (2012). Evaluation of circulating endothelial precursor cells in cancer patients. *Methods Mol Biol*, **904**, 165–72.

Bobek, V., Kolostova, K., Pinterova, D., et al. (2010). A clinically relevant, syngeneic model of spontaneous, highly metastatic B16 mouse melanoma. *Anticancer Res*, **30**, 4799–803.

Cespedes, M. V., Casanova, I., Parreno, M., & Mangues, R. (2006). Mouse models in oncogenesis and cancer therapy. *Clin Transl Oncol*, **8**, 318–29.

Chen, J. A., Shi, M., Li, J. Q., & Qian, C. N. (2010). Angiogenesis: multiple masks in hepatocellular carcinoma and liver regeneration. *Hepatol Int*, **4**, 537–47.

Cheung, K. J. & Ewald, A. J. (2016). A collective route to metastasis: seeding by tumor cell clusters. *Science*, **352**, 167–9.

Chin, A. R. & Wang, S. E. (2016). Cancer tills the premetastatic field: mechanistic basis and clinical implications. *Clin Cancer Res*, **22**, 3725–33.

Ch'ng, S., Wallis, R. A., Yuan, L., Davis, P. F., & Tan, S. T. (2006). Mast cells and cutaneous malignancies. *Mod Pathol*, **19**, 149–59.

Christiansen, J. J. & Rajasekaran, A. K. (2006). Reassessing epithelial to mesenchymal transition as a prerequisite for carcinoma invasion and metastasis. *Cancer Res*, **66**, 8319–26.

Clark, A. G. & Vignjevic, D. M. (2015). Modes of cancer cell invasion and the role of the microenvironment. *Curr Opin Cell Biol*, **36**, 13–22.

Contag, C. H., Jenkins, D., Contag, P. R., & Negrin, R. S. (2000). Use of reporter genes for optical measurements of neoplastic disease *in vivo*. *Neoplasia*, **2**, 41–52.

Cotran RS, K. V., Robbins SL (2009). *Robbins Pathologic Basis of Disease*, Philadelphia, PA: W. B. Saunders Company.

Crnic, I. & Christofori, G. (2004). Novel technologies and recent advances in metastasis research. *Int J Dev Biol*, **48**, 573–81.

Czernin, J. & Phelps, M. E. (2002). Positron emission tomography scanning: current and future applications. *Annu Rev Med*, **53**, 89–112.

De Visser, K. E., Eichten, A., & Coussens, L. M. (2006). Paradoxical roles of the immune system during cancer development. *Nat Rev Cancer*, **6**, 24–37.

Diaz-Montero, C. M., Salem, M. L., Nishimura, M. I., Garrett-Mayer, E., Cole, D. J., & Montero, A. J. (2009). Increased circulating myeloid-derived suppressor cells correlate with clinical cancer stage, metastatic tumor burden, and doxorubicin-cyclophosphamide chemotherapy. *Cancer Immunol Immunother*, **58**, 49–59.

Diehn, M. & Majeti, R. (2010). Metastatic cancer stem cells: an opportunity for improving cancer treatment? *Cell Stem Cell*, **6**, 502–3.

Eichler, A. F., Chung, E., Kodack, D. P., Loeffler, J. S., Fukumura, D., & Jain, R. K. 2011. The biology of brain metastases-translation to new therapies. *Nat Rev Clin Oncol*, **8**, 344–56.

Fidler, I. J. (1970). Metastasis: quantitative analysis of distribution and fate of tumor emboli labeled with 125 I-5-iodo-2'-deoxyuridine. *J Natl Cancer Inst*, **45**, 773–82.

Fidler, I. J. (2003). The pathogenesis of cancer metastasis: the 'seed and soil' hypothesis revisited. *Nat Rev Cancer*, **3**, 453–8.

Francia, G., Cruz-Munoz, W., Man, S., Xu, P., & Kerbel, R. S. (2011). Mouse models of advanced spontaneous metastasis for experimental therapeutics. *Nat Rev Cancer*, **11**, 135–41.

Friedl, P. & Wolf, K. (2003). Tumour-cell invasion and migration: diversity and escape mechanisms. *Nat Rev Cancer*, **3**, 362–74.

Friedl, P. & Alexander, S. (2011). Cancer invasion and the microenvironment: plasticity and reciprocity. *Cell*, **147**, 992–1009.

Fukumura, D. & Jain, R. K. (2007). Tumor microenvironment abnormalities: causes, consequences, and strategies to normalize. *J Cell Biochem*, **101**, 937–49.

Furukawa, T., Kubota, T., Watanabe, M., Kitajima, M., & Hoffman, R. M. (1993). Orthotopic transplantation of histologically intact clinical specimens of stomach cancer to nude mice: correlation of metastatic sites in mouse and individual patient donors. *Int J Cancer*, **53**, 608–12.

Gao, D. & Mittal, V. (2009). The role of bone-marrow-derived cells in tumor growth, metastasis initiation and progression. *Trends Mol Med*, **15**, 333–43.

Garber, K. (2004). Energy boost: the Warburg effect returns in a new theory of cancer. *J Natl Cancer Inst*, **96**, 1805–6.

Gatenby, R. A. & Gawlinski, E. T. (2003). The glycolytic phenotype in carcinogenesis and tumor invasion: insights through mathematical models. *Cancer Res*, **63**, 3847–54.

Giuliano, A. E., Hunt, K. K., Ballman, K. V., et al. (2011). Axillary dissection vs no axillary dissection in women with invasive breast cancer and sentinel node metastasis: a randomized clinical trial. *Jama*, **305**, 569–75.

Gottfried, E., Kreutz, M., & Mackensen, A. (2012). Tumor metabolism as modulator of immune response and tumor progression. *Semin Cancer Biol*, **22**, 335–41.

Haeger, A., Krause, M., Wolf, K., & Friedl, P. (2014). Cell jamming: collective invasion of mesenchymal tumor cells imposed by tissue confinement. *Biochim Biophys Acta*, **1840**, 2386–95.

Haeger, A., Wolf, K., Zegers, M. M., & Friedl, P. (2015). Collective cell migration: guidance principles and hierarchies. *Trends Cell Biol*, **25**, 556–66.

Hall, S. J. & Thompson, T. C. (1997). Spontaneous but not experimental metastatic activities differentiate primary tumor-derived vs metastasis-derived mouse prostate cancer cell lines. *Clin Exp Metastasis*, **15**, 630–8.

Havrilesky, L. J., Reiner, M., Morrow, P. K., Watson, H., & Crawford, J. (2015). A review of relative dose intensity and survival in patients with metastatic solid tumors. *Crit Rev Oncol Hematol*, **93**, 203–10.

Hoffman, R. M. (2015). Patient-derived orthotopic xenografts: better mimic of metastasis than subcutaneous xenografts. *Nat Rev Cancer*, **15**, 451–2.

Ito, R., Takahashi, T., Katano, I., & Ito, M. (2012). Current advances in humanized mouse models. *Cell Mol Immunol*, **9**, 208–14.

Kareva, I. (2016). Escape from tumor dormancy and time to angiogenic switch as mitigated by tumor-induced stimulation of stroma. *J Theor Biol*, **395**, 11–22.

Karrison, T. G., Ferguson, D. J., & Meier, P. (1999). Dormancy of mammary carcinoma after mastectomy. *J Natl Cancer Inst*, **91**, 80–5.

Khanna, C. & Hunter, K. (2005). Modeling metastasis *in vivo*. *Carcinogenesis*, **26**, 513–23.

Klein, C. A. (2009). Parallel progression of primary tumours and metastases. *Nat Rev Cancer*, **9**, 302–12.

Klein, C. A. & Holzel, D. (2006). Systemic cancer progression and tumor dormancy: mathematical models meet single cell genomics. *Cell Cycle*, **5**, 1788–98.

La Beck, N. M., Jean, G. W., Huynh, C., Alzghari, S. K., & Lowe, D. B. (2015). Immune checkpoint inhibitors: new insights and current place in cancer therapy. *Pharmacotherapy*, **35**, 963–76.

Lawrence, T. (2011). Macrophages and NF-κB in cancer. *Curr Top Microbiol Immunol*, **349**, 171–84.

Lee, J. M., Dedhar, S., Kalluri, R., & Thompson, E. W. (2006). The epithelial-mesenchymal transition: new insights in signaling, development, and disease. *J Cell Biol*, **172**, 973–81.

LI, B., Sharpe, E. E., Maupin, A. B., et al. (2006). VEGF and PlGF promote adult vasculogenesis by enhancing EPC recruitment and vessel formation at the site of tumor neovascularization. *Faseb J*, **20**, 1495–7.

Lin, E. W., Karakasheva, T. A., Hicks, P. D., Bass, A. J., & Rustgi, A. K. (2016). The tumor microenvironment in esophageal cancer. *Oncogene*, **35**, 5337–49.

Lygidakis, N. J. & Pearl, A. (1997). Metastatic liver disease—a review. *Hepatogastroenterology*, **44**, 1484–7.

Macintyre, G., Van Loo, P., Corcoran, N. M., Wedge, D. C., Markowetz, F., & Hovens, C. M. (2017). How subclonal modeling is changing the metastatic paradigm. *Clin Cancer Res*, **23**, 630–5.

Mancuso, P., Calleri, A., & Bertolini, F. (2012). Circulating endothelial cells and circulating endothelial progenitors. *Recent Results Cancer Res*, **195**, 163–70.

Mantovani, A., Allavena, P., Sica, A., & Balkwill, F. (2008). Cancer-related inflammation. *Nature*, **454**, 436–44.

Mason, S. D. & Joyce, J. A. (2011). Proteolytic networks in cancer. *Trends Cell Biol*, **21**, 228–37.

Mccawley, L. J. & Matrisian, L. M. (2000). Matrix metalloproteinases: multifunctional contributors to tumor progression. *Mol Med Today*, **6**, 149–56.

Miles, F. L. & Sikes, R. A. (2014). Insidious changes in stromal matrix fuel cancer progression. *Mol Cancer Res*, **12**, 297–312.

Mohme, M., Riethdorf, S., & Pantel, K. (2017). Circulating and disseminated tumour cells—mechanisms of immune surveillance and escape. Nat Rev Clin Oncol, **14**(3), 155–67.

Morton, J. J., Bird, G., Refaeli, Y., & Jimeno, A. (2016). Humanized mouse xenograft models: narrowing the tumor-microenvironment gap. *Cancer Res*, **76**, 6153–8.

Moschetta, M., Mishima, Y., Sahin, I., et al. (2014). Role of endothelial progenitor cells in cancer progression. *Biochim Biophys Acta*, **1846**, 26–39.

Muenst, S., Läubli, H., Soysal, S. D., Zippelius, A., Tzankov, A., & Hoeller, S. (2016). The immune system and cancer evasion strategies: therapeutic concepts. *J Intern Med*, **279**(6), 541–62.

Mundy, G. R. (2002). Metastasis to bone: causes, consequences and therapeutic opportunities. *Nat Rev Cancer*, **2**, 584–93.

Munoz, R., Man, S., Shaked, Y., et al. (2006). Highly efficacious nontoxic preclinical treatment for advanced metastatic breast cancer using combination oral UFT-cyclophosphamide metronomic chemotherapy. *Cancer Res*, **66**, 3386–91.

Muratori, L., Petroni, G., Antonuzzo, L., et al. (2016). hERG1 positivity and Glut-1 negativity identifies high-risk TNM stage I and II colorectal cancer patients, regardless of adjuvant chemotherapy. *Onco Targets Ther*, **9**, 6325–332.

Naumov, G. N., Townson, J. L., Macdonald, I. C., et al. (2003). Ineffectiveness of doxorubicin treatment on solitary dormant mammary carcinoma cells or late-developing metastases. *Breast Cancer Res Treat*, **82**, 199–206.

Nissan, A., Stojadinovic, A., Garofalo, A., Esquivel, J., & Piso, P. (2009). Evidence-based medicine in the treatment of peritoneal carcinomatosis: past, present, and future. *J Surg Oncol*, **100**, 335–44.

Olivier, M., Langerod, A., Carrieri, P., et al. (2006). The clinical value of somatic TP53 gene mutations in 1,794 patients with breast cancer. *Clin Cancer Res*, **12**, 1157–67.

Ordóñez-Morán, P. & Huelsken, J. (2014). Complex metastatic niches: already a target for therapy? *Curr Opin Cell Biol*, **31**, 29–38.

Osada, T., Sakamoto, M., Ino, Y., et al. (1996). E-cadherin is involved in the intrahepatic metastasis of hepatocellular carcinoma. *Hepatology*, **24**, 1460–7.

Ostrand-Rosenberg, S. & Sinha, P. (2009). Myeloid-derived suppressor cells: linking inflammation and cancer. *J Immunol*, **182**, 4499–506.

Ougolkov, A., Yamashita, K., Bilim, V., Takahashi, Y., Mai, M., & Minamoto, T. (2003). Abnormal expression of E-cadherin, beta-catenin, and c-erbB-2 in advanced gastric cancer: its association with liver metastasis. *Int J Colorectal Dis*, **18**, 160–6.

Pan, W., Sun, Q., Wang, Y., Wang, J., Cao, S., & Ren, X. (2015). Highlights on mechanisms of drugs targeting MDSCs: providing a novel perspective on cancer treatment. *Tumour Biol*, **36**, 3159–69.

Peranzoni, E., Rivas-Caicedo, A., Bougherara, H., Salmon, H., & Donnadieu, E. (2013). Positive and negative influence of the matrix architecture on antitumor immune surveillance. *Cell Mol Life Sci*, **70**, 4431–48.

Price, J. E., Daniels, L. M., Campbell, D. E., & Giavazzi, R. (1989). Organ distribution of experimental metastases of a human colorectal carcinoma injected in nude mice. *Clin Exp Metastasis*, **7**, 55–68.

Rabbany, S. Y., Heissig, B., Hattori, K., & Rafii, S. (2003). Molecular pathways regulating mobilization of marrow-derived stem cells for tissue revascularization. *Trends Mol Med*, **9**, 109–17.

Reddy, B. Y., Lim, P. K., Silverio, K., Patel, S. A., Won, B. W., & Rameshwar, P. (2012). The microenvironmental effect in the progression, metastasis, and dormancy of breast cancer: a model system within bone marrow. *Int J Breast Cancer*, **2012**, 721659.

Shalapour, S. & Karin, M. (2015). Immunity, inflammation, and cancer: an eternal fight between good and evil. *J Clin Invest*, **125**, 3347–55.

Steeg, P. S. (2006). Tumor metastasis: mechanistic insights and clinical challenges. *Nat Med*, **12**, 895–904.

Tarhoni, I., Fhied, C., Pithadia, R., et al. (2016). Circulating angiogenesis biomarkers are predictive for early stage non-small cell lung cancer progression and clinical outcome: topic: pathology. *J Thorac Oncol*, **11**, S259–60.

Tarin, D., Thompson, E. W., & Newgreen, D. F. (2005). The fallacy of epithelial mesenchymal transition in neoplasia. *Cancer Res*, **65**, 5996–6000; discussion 6000–1.

Tentler, J. J., Tan, A. C., Weekes, C. D., et al. (2012). Patient-derived tumour xenografts as models for oncology drug development. *Nat Rev Clin Oncol*, **9**, 338–50.

Thompson, E. W., Newgreen, D. F., & Tarin, D. (2005). Carcinoma invasion and metastasis: a role for epithelial-mesenchymal transition? *Cancer Res*, **65**, 5991–5; discussion 5995.

Tisdale, M. J. (1997). Biology of cachexia. *J Natl Cancer Inst*, **89**, 1763–73.

Touil, Y., Igoudjil, W., Corvaisier, M., et al. (2014). Colon cancer cells escape 5FU chemotherapy-induced cell death by entering stemness and quiescence associated with the c-Yes/YAP axis. *Clin Cancer Res*, **20**, 837–46.

Tüting, T. & De Visser, K. E. (2016). CANCER. How neutrophils promote metastasis. *Science*, **352**, 145–6.

Umansky, V., Blattner, C., Fleming, V., et al. (2017). Myeloid-derived suppressor cells and tumor escape from immune surveillance. *Semin Immunopathol*, **39**(3), 295–305.

Vinay, D. S., Ryan, E. P., Pawelec, G., et al. (2015). Immune evasion in cancer: Mechanistic basis and therapeutic strategies. *Semin Cancer Biol*, **35**, Suppl, S185–98.

Warburg, O. (1930). *The Metabolism of Tumors*. London: Constable Press.

Warburg, O. (1956). On the origin of cancer cells. *Science*, **123**, 309–14.

Wirtz, D., Konstantopoulos, K., & Searson, P. C. (2011). The physics of cancer: the role of physical interactions and mechanical forces in metastasis. *Nat Rev Cancer*, **11**, 512–22.

Wood S, J., Robinson, R. R., & Marzocchi, B. (1967). Factors influencing the spread of cancer: locomotion of normal and malignant cells *in vivo*. In: Wissler, R. W., Dao, T. L., & Wood, S., Jr. (eds) *Endogenous Factors Influencing Host–Tumor Balance*. Chicago: University of Chicago Press, pp. 223–7.

XI, Y., Riker, A., Shevde-Samant, L., et al. (2008). Global comparative gene expression analysis of melanoma patient samples, derived cell lines and corresponding tumor xenografts. *Cancer Genomics Proteomics*, **5**, 1–35.

Yaacoub, K., Pedeux, R., Tarte, K., & Guillaudeux, T. (2016). Role of the tumor microenvironment in regulating apoptosis and cancer progression. *Cancer Lett*, **378**, 150–9.

Yue, J., Zhang, K., & Chen, J. (2012). Role of integrins in regulating proteases to mediate extracellular matrix remodeling. *Cancer Microenviron*, **5**, 275–83.

20

Cancer stem cells

Connor Sweeney, Lynn Quek, Betty Gration, and Paresh Vyas

Introduction

Conventional anticancer chemotherapy targets the enhanced rate of neoplastic proliferation but despite therapeutic advances, treatment resistance and relapse remain a major cause of mortality. Despite their clonal origin, individual cells within a cancer exhibit considerable heterogeneity in terms of genetic mutations, morphology, cell surface markers and function (Dick, 2008). In addition to malignant cells, infiltrating cells such as stromal and endothelial cells contribute to the tumour microenvironment and influence the function of cancer cells. Increasing evidence indicates that this heterogeneity contributes to treatment failure.

In at least some cancers, cells are organized hierarchically, with the cancer stem cell (CSC) at the apex of the hierarchy maintaining growth of the cancer. CSCs act as caricatures of the normal tissue stem cells that produce differentiated cells in embryogenesis and normal tissue renewal. Instead of giving rise to normal tissues, disordered growth from CSCs leads to dysfunctional tissues and disease. CSCs are predicted to be relatively quiescent and this propensity for slow growth may cause CSCs to be resistant to conventional chemotherapy, leading to relapse after treatment (Ishikawa et al., 2007). Indeed, evidence from experimental models, and to a lesser extent from clinical studies, indicates that CSCs survive many commonly employed cancer therapeutics (Kreso and Dick, 2014a). Therefore, one hypothesis for long-term disease management is that new treatments should be developed to specifically eliminate these cellular reservoirs of cancer cell growth.

This chapter will examine the CSC model and other models of tumorigenesis, summarizing the evidence for the CSC model. Genetic and functional models of cancer evolution from premalignant stem cells to fully transformed CSCs will be discussed. The genetic, epigenetic, and microenvironmental factors that influence tumour heterogeneity both between individuals and within a tumour are explored. The chapter concludes by discussing novel therapeutic strategies targeting CSCs.

The stem cell concept

Regenerating tissues including the haematopoietic system, skin, and gastrointestinal mucosa exhibit a cellular hierarchy of differentiation, which is maintained by stem cells. These cells are functionally defined by their capacity for long-term self-renewal and multipotential differentiation and are responsible for maintaining tissues during steady state and replenishing cells lost during episodes of stress, such as injury or chemotherapy.

The paradigm of the multipotent stem cell in haematopoiesis was first expounded by Till and McCullock (1961) when they demonstrated that bone marrow cells injected into lethally irradiated mice gave rise to clonal colonies of myeloid/erythroid cells in the spleen and bone marrow. The haematopoietic stem cells (HSCs) responsible for repopulation are rare, highly specialized cells enriched in fetal liver, umbilical cord blood, and adult bone marrow and have the capacity to generate multilineage haematopoiesis while retaining the unique stem cell property of indefinite self-renewal.

Self-renewal is the essential process whereby a stem cell undergoing cell division produces either one further stem cell with the capacity for self-renewal and a more differentiated daughter cell (asymmetric division) or two daughter cells with self-renewal ability (symmetric division). These processes are likely to coexist and ensure that the parent stem cell clone is preserved. A further form of symmetric cell division involves the production of two differentiated daughter cells, which lack long-term self-renewal capacity and if this were to occur alone, would lead to stem cell exhaustion. Studies using haematopoietic tissue have identified multipotential cells capable of repopulation but lacking self-renewal potential. Therefore, although a capacity for multipotential differentiation is often considered a stem property, indefinite self-renewal is the key feature of a stem cell.

Based on an understanding of the role of stem cells in normal adult tissues, it was hypothesized that as cancer cells have long-term self-renewal ability, they too have stem cells. The main difference between a CSC and a normal multipotent stem cell is that the CSC harbours oncogenic mutations that drive dysfunctional growth of the tumour.

Tumour heterogeneity

For over 70 years, cancer therapeutics have targeted the enhanced rate of neoplastic proliferation. Clinicians have used cytotoxic drugs to kill rapidly dividing malignant cells, while minimizing toxicity to normal tissues. Current therapeutic strategies are based on a deepening understanding of the molecular basis of cancer cell growth, incorporating genetic and epigenetic factors as well as the microenvironmental factors that support this growth. This has led to rational design of disease-specific drugs that inhibit the molecular drivers of cancer (e.g. transtuzumab—Her2, imatinib—BCR-ABL) or interfere with extrinsic factors that support growth (e.g. vascular endothelial growth factor). Implicit in both approaches is the perspective of cancer as a homogenous entity. Indeed, laboratory approaches to understanding tumour biology have historically used large numbers of unfractionated cells from cancer tissue culture to perform genetic or biochemical studies, as the technology was unable to operate with low cell numbers. However, despite using this research approach, it is well established that tumours are heterogeneous.

Morphologic heterogeneity within a tumour was observed over a century ago (Heppner, 1984), but evidence for a hierarchy defined by a maturation process was provided by Pierce and colleagues (1960), who observed that malignant teratocarcinoma cells spontaneously differentiate into mature cells. Non-differentiated cells were able to form tumours in secondary recipients, in contrast to their differentiated progeny. In addition to histology, technological advances have enabled further characterization of the heterogeneity both between tumours and within an individual tumour, by assessing growth kinetics, cell surface markers, and genetic mutations.

In vivo experiments in the 1960s and 1970s using human acute myeloid leukaemia (AML) and acute lymphoblastic leukaemia (ALL) samples used radiolabelling with tritiated thymidine to study leukaemia blast proliferation kinetics. Three groups of leukaemic blast cells were demonstrated: the majority were postmitotic and continuously replenished from a small fraction (5%) of rapidly dividing cells. Careful analysis revealed a smaller fraction of slow-cycling cells that remained dormant for weeks to months. Due to their cytokinetic similarities to normal HSCs these were postulated to represent a leukaemic stem cell (LSC) population (Clarkson et al., 1970).

Two models have been proposed to explain how tumour heterogeneity arises (Fig. 20.1). According to the stochastic model, tumours

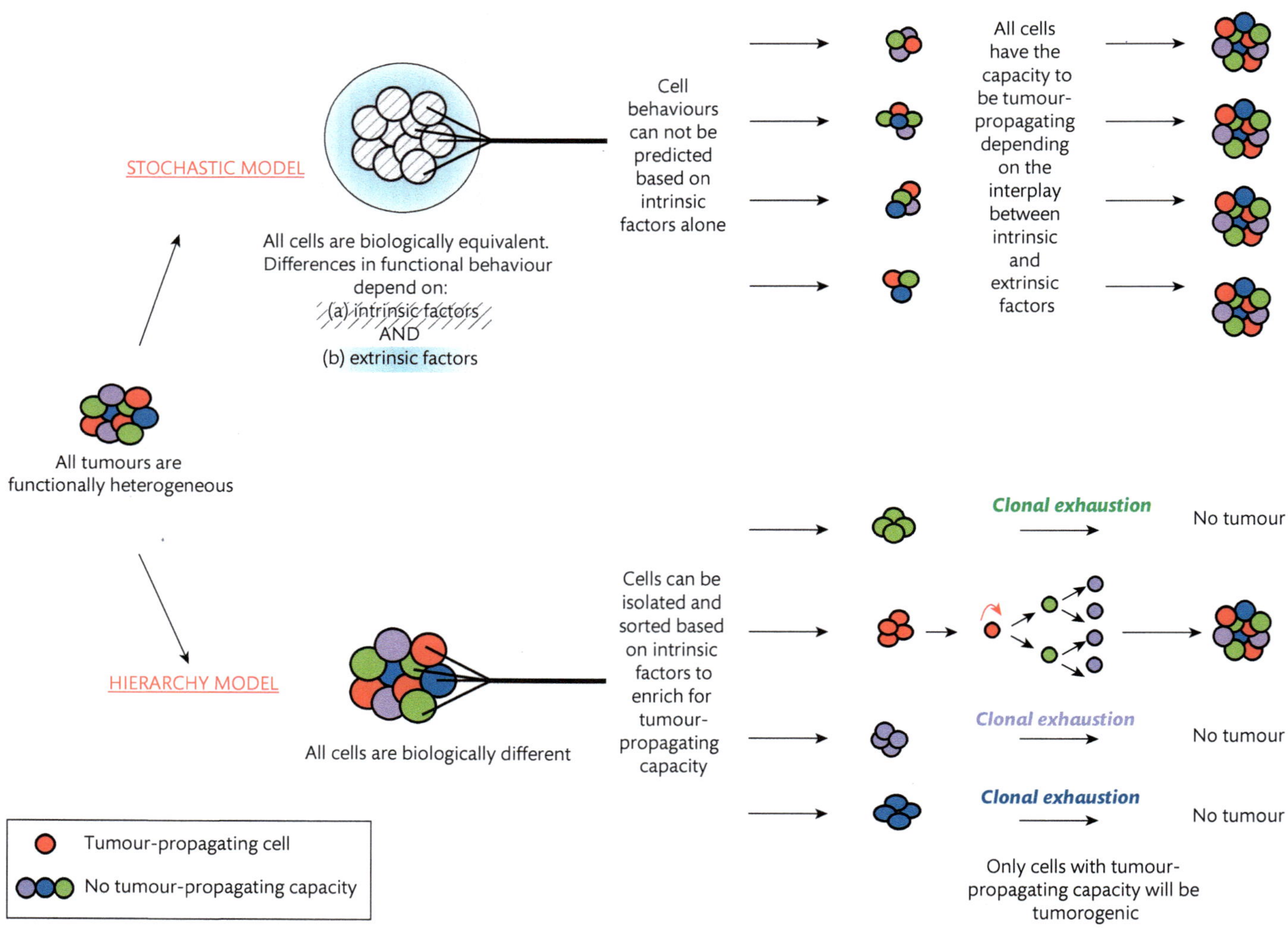

Fig. 20.1 Stochastic and hierarchy models of tumour heterogeneity. Tumours are composed of cells that are phenotypically and functionally distinct. Two models have been proposed to explain this heterogeneity. According to the stochastic model, tumours are biologically homogenous but cells are subjected to intrinsic and extrinsic factors that influence their behaviour. All cells have the potential to propagate tumours, depending on the interplay between intrinsic and extrinsic factors. In contrast, the hierarchy model describes biologically distinct cells with differing phenotypes and functions. Only a subset of cells can propagate tumour growth and these cancer stem cells can be fractionated based on intrinsic characteristics.

are biologically homogenous, but cells are subjected to intrinsic and extrinsic factors that influence their behaviour, thereby making cells functionally heterogeneous. In this model, tumour propagation in xenotransplantation experiments can occur with any cell given the appropriate genetic, epigenetic, or microenvironmental influences. In contrast, the hierarchy model describes distinct, biologically heterogeneous subpopulations of cells. Only a subset of cells, the CSCs, possesses indefinite self-renewal and produces non-propagating progeny that constitute the bulk of the cancer.

The distinction between these models is important because the developmental rigidity inherent in the hierarchy model predicts that CSCs can be fractionated from the non-tumorigenic population according to intrinsic properties. In contrast, if the stochastic model is correct, each tumour cell has the potential to propagate tumours and so cell fractionation according to intrinsic characteristics would not enhance our understanding of the tumour-propagating cell fraction. In addition to the inherent heterogeneity of the malignant clone, other infiltrating cells such as endothelial, stromal, and haematopoietic cells may also influence the function of tumour cells.

Stem cell assays

The key stem cell property of long-term self-renewal can be assessed *in vitro* using serial replating assays, whereby samples that contain stem cells can regenerate tissues in perpetuity. However, stem cells can be difficult to maintain *in vitro*, most likely due to suboptimal microenvironmental conditions. *In vivo* modelling can overcome this challenge and also provides a more realistic system to study stem cell function. If a sample exhibits serial engraftment in transplantation recipients, it must contain a cell that possesses long-term self-renewal capacity—that is, a stem cell (Table 20.1).

Engraftment of a human tumour in serial immunodeficient mouse xenotransplantation experiments is used to assess whether a particular tumour fraction contains a stem cell. Immunodeficient mouse models have defects in adaptive immunity (severe combined immunodeficiency (SCID) mice) and innate immunity, permitting engraftment of human cells across species barriers. These tumour-initiating cell (T-IC) assays assess self-renewal, as demonstrated by

Table 20.1 Characteristics of cancer stem cells and bulk tumour cells. Tumours are composed of functionally heterogeneous cells. Cancer stem cells form a key subpopulation with important properties that distinguish them from bulk tumour cells

Cancer stem cell	Bulk tumour cell
Indefinite self-renewal	Limited or absent self-renewal
Early stage of differentiation	Variable degree of differentiation
Can enter quiescent state	Proliferative
Maintains long-term growth of malignant clone	In absence of stem cell, malignant clone will eventually exhaust
Can propagate tumour in serial xenotransplantation recipients	Unable to propagate tumour in serial xenotransplantation recipients
Implicated in treatment resistance and relapse	Targeted by cytotoxic chemotherapy

long-term repopulation after serial transplantation. As well as giving rise to further self-renewing cells, T-ICs also give rise to daughter cells that proliferate but are unable to maintain the tumour on serial passage (Fig. 20.2). T-ICs so defined (i.e. initiating the tumour in an animal model) must be distinguished from cells in a patient, which contain the first genetic or stable epigenetic change. In a patient, these cells are usually not fully transformed and will not usually engraft in an immunodeficient mouse model. It is for this reason that cells that initiate a tumour in an animal are probably better termed 'tumour-propagating cells'.

Despite being the current gold standard *in vivo* system to assess CSC activity, there are several limitations of xenotransplantation assays. Functional CSCs in a human may not act as tumour-propagating cells in an animal model. These mice lack an adaptive immune system and the tumour microenvironment and growth factors are of mouse origin. Therefore, the selective pressure is likely to differ from that in a human. This may produce a cancer model that is not representative of the native tumour. In addition, isolation of tumour cells for the assessment of tumour-propagating activity may affect their growth. This includes digestion of solid tumours, treatment with antibodies used in fluorescence-activated cell sorting (FACS) separation and growth in non-physiological nutrient conditions (Kreso and Dick, 2014a).

The frequency of CSCs in a tumour is estimated using limit dilution analysis, whereby a series of dilutions of the tumour cells are assessed for engraftment. The number of mice negative for engraftment in each cell dose is measured and the frequency of stem cells is estimated using Poisson statistics (Purton and Scadden, 2007). Several experimental parameters influence the efficiency of tumour engraftment and therefore the estimation of CSC frequency. These include:

1. **The immunodeficient animal model.** Immunodeficient mouse models have been refined from the SCID background to produce increasingly permissive conditions for CSC engraftment. The first SCID model, exhibiting functional defects in T-cell and B-cell receptors, was discovered in 1983 (Bosma et al., 1983). Genetic crossings enabled the generation of models with additional deficiency in NK cell activity, NOD/SCID/$\beta 2m^{null}$ mice. Subsequently, the immunodeficient NSG (NOD/SCID/γ^{null}) mouse was produced, lacking the IL-2 gamma chain, which is more permissive for the engraftment of human haematopoietic cells (Ito et al., 2002). Compared with the NOD/SCID model, which will permit engraftment of 49% of human AML samples (Pearce et al., 2006), the NSG mouse will engraft up to two-thirds of samples (Sanchez et al., 2009). Recent refinements of immunodeficient models include the generation of NBSGW mice, which permit engraftment of human haematopoietic tissue without the requirement for pretransplant irradiation (McIntosh et al., 2015). Humanized mouse models have been used to enhance engraftment by secreting human cytokines (MITRG/MISTRG mice), or by providing a humanized bone marrow microenvironment through ossicle xenotransplantation (Rongvaux et al., 2014; Reinisch et al., 2016).
2. **Type of injection used for transplantation.** Intrafemoral transplantation can increase the sensitivity of AML stem cell detection when compared with intravenous injection (Eppert et al., 2011).

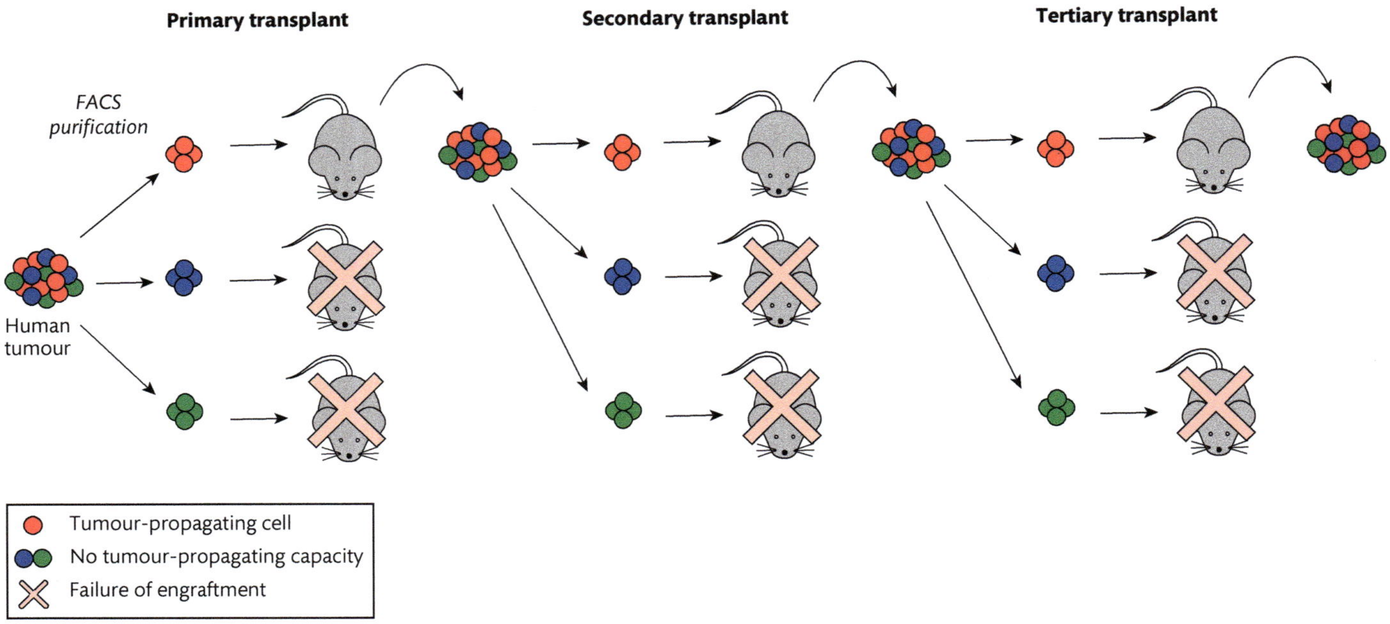

Fig. 20.2 *In vivo* assessment of human cancer stem cells (CSCs) using mouse xenotransplantation assays. The current 'gold standard' experimental system to assess CSC activity is an immunodeficient animal model. Only a cell with capacity for long-term self-renewal (i.e. a stem cell) can propagate tumour in serial transplant recipients.

3. Primary tumour type. Experimental estimations of CSC frequency show marked variation between different patient samples and between different tumour types. Biomarkers have been used to enrich for CSCs in different tumours (Table 20.2).

Given the significant variation in CSC frequency in different tumour types with different animal models, the true frequency in the human patient is unknown. While some of the experimental variation may due to sampling bias and the permissiveness of the xenotransplantation model used, it is likely that some variation in CSC frequency truly reflects tumour biology. Pearce et al. (2006) examined samples from AML patients for their ability to initiate leukaemia in NOD/SCID mice and showed that engraftment was independent of homing to the bone marrow or the cell dose administered but was directly related to prognosis. AML samples with poor prognostic karyotypes were significantly and independently associated with engraftment.

Experimental evidence for normal haematopoietic stem cells and cancer stem cells

Pioneering studies on normal and malignant haematological tissues led the way to experiments characterizing CSCs in solid tumours. To demonstrate the hierarchical organization of normal or malignant tissue that is maintained by stem cells, it is essential to detect, purify, and assay functionally distinct cells. In 1988, Spangrude and colleagues were first to purify normal murine multipotential HSCs, using FACS with monoclonal antibodies directed against cell surface antigens. These cells were identified by a lack of expression of mature lineage-affiliated markers and presence of cell surface Sca-1 and c-Kit. However, these Lin$^-$Sca-1$^+$c-Kit$^+$ (LSK) cells were mostly capable of only short-term multipotential repopulation in murine transplantation experiments. In 2005, Morrison's group identified that within the LSK fraction, true long-term HSCs can be distinguished from multipotent progenitors using cell surface markers from the signalling lymphocyte activation molecule (SLAM) family. Long-term HSCs are in the CD150$^+$CD244$^-$CD48$^-$ fraction, whereas multipotent progenitors with short-term repopulation capacity are in the CD150$^-$CD244$^+$CD48$^-$ fraction (Kiel et al., 2005).

Parallel studies used FACS purification of human HSC and LSCs, followed by xenotransplantation into immunodeficient mice to characterize human HSC and LSC. Studies from the early 1980s had already established that AML cells could engraft in the SCID mouse system. Quantitative xenotransplantation assays indicated that the initiating cell in AML had an average frequency of 1 per 1 x 10^6 cells (Dick, 2008). To purify the T-IC population, Lapidot and colleagues (1994) fractionated AML cells by FACS using cell surface expression of CD34 and CD38 and subsequently transplanted cell fractions into SCID mice. They showed that the CD34$^+$CD38$^-$ fraction was the only fraction capable of leukaemic engraftment, generating large numbers of more mature CD34$^+$CD38$^+$ cells and mature blasts in the mice. The tumour-propagating cell frequency within the CD34$^+$CD38$^-$ fraction was estimated at between 1 in 1 × 10^4 and 5 × 10^6 cells. The hierarchical nature of AML was demonstrated with LSC-enriched populations being able to generate both LSCs and non-LSCs, but non-LSCs were unable to produce LSCs. A parallel study purifying HSCs showed that they also lie within the CD34$^+$CD38$^-$ fraction, suggesting that LSCs may arise from HSCs (Pflumio et al., 1996).

Although antibodies used for FACS purification are treated as being neutral, anti-CD38 antibodies have been shown to have an inhibitory effect on engraftment, due to antibody-mediated clearance by recipient innate immune cells (Taussig et al., 2008). This may account for the lack of tumour propagation with CD38$^+$ cells in early studies using NOD/

Table 20.2 Estimation of cancer stem cell (CSC) frequency. Estimation of frequency is influenced by tumour type as well as experimental variables and there is significant variation in results between studies. Biomarkers have been used to enrich for CSC populations

Experimental variables				Results			References
Cancer	Animal model	Type of transplantation	Biomarkers	Estimated frequency of CSCs in bulk tumour (%)	Biomarker-positive cells in cancer (%)	Estimated frequency of CSCs in biomarker-positive cells (%)	
AML	NOD/SCID	Intravenous	$CD34^+/CD38^-$	0.00002–0.01	0.02–2	0.002	Bonnet & Dick, 1997;
	NOD/SCID, NOD/SCID/β2m^{null}, NSG	Intravenous and intraosseous	$CD34^+/CD38^-$ and $CD34^+/CD38^+$	?	0.076	0.0001–0.06 0.0004–0.02	Taussig et al., 2008;
	NOD/SCID	Intrafemoral	$CD34^+/CD38^-$ and $CD34^+/CD38^+$	? ?	? ?	0.00009–0.01 0.0004–0.01	Eppert et al., 2011;
	NOD/SCID, NSG	Intravenous	$CD34^+$/CD38-/$CD45RA^+$ $CD34^+/CD38^+/CD123^{+/low}$/$CD110^-/CD45RA^+$	? ?	2.82–60.1 16–82	0.009–0.09 0.001–0.01	Goardon et al., 2011;
	NSG, NRG, NSGS	Intratibial	$CD34^-/CD244^+/CD117^+$ $CD34^-/CD244^-/CD117^+$	0.0003–0.1 0–0.03	38.13–89.57 0.38–19.65	0.0004–0.2 0.0003–0.2	Quek et al., 2016
Brain	NOD/SCID	Intracranial	$CD133^+$	?	62–9	1	Singh et al., 2004
Breast	NOD/SCID	Mammary fat pad	$CD44^+/CD24^{-/low}$ ESA^+	0.01	11–35	1	Al-Hajj et al., 2003
Colorectal	NOD/SCID	Renal capsule	$CD133^+$	0.002	1.8–24.5	0.4	O'Brien et al., 2007
Head and Neck SCC	NOD/SCID Rag2γDKO	Subcutaneous	$CD44^+$	0.0001	<10	0.02	Prince et al., 2007
Lung	SCID	Subcutaneous	$CD133^+$	?	0.32–3.5	0.01	Eramo et al., 2008
Melanoma	NSG NOD/SCID	Subcutaneous Subcutaneous	NA $ABCB5^+$	100 0.0001	NA 1.6–20.4	NA 0.001	Quintana et al., 2010 Schatton et al., 2008
Pancreas	NUDE	Intrapancreatic	$CD133^+$	0.0001	3.6 cells per high-power field	0.2	Hermann et al., 2007
Ovarian	NOD/SCID	Subcutaneous	$CD133^+$	0.01	0.3–35	1	Curley et al., 2009

SCID mice and can be at least partially be overcome by using NSG mouse models. Subsequent xenotransplantation studies using NSG mice have enabled the identification of functional human AML LSCs not only in $CD34^+CD38^-$ cells, but within multiple immunophenotypically defined compartments, with considerable interpatient heterogeneity (Goardon et al., 2011; Eppert et al., 2011; Sarry et al., 2011; Ho et al., 2016; Quek et al., 2016). In approximately 25% of AML, >98% of cells are $CD34^-$ and within these cancers, tumour-propagating LSCs are found in both $CD34^+$ and $CD34^-$ populations with similar frequencies (Taussig et al., 2010; Quek et al., 2016). Cell surface marker expression can be aberrant in AML and therefore may not reflect the maturity of normal haematopoietic cells that share an immunophenotype. Nevertheless, evidence of a hierarchy persists, as there are almost always cell fractions that lack tumour-initiating ability.

FACS combined with global gene expression profiling has shown that functional LSCs in AML more closely resemble normal haematopoietic progenitors and precursors than primitive stem cells. However, compared to normal progenitors, the gene expression of LSCs is enriched for a stem cell signature (Martelli et al., 2010; Goardon et al., 2011; Quek et al., 2016). This suggests that either LSCs arise from progenitors and then aberrantly acquire stem cell properties that convey long-term self-renewal, or they arise from stem cells and subsequently acquire progenitor traits such as enhanced proliferative ability.

Using a similar strategy to that employed in AML, CSCs in solid tumours have been identified and characterized using FACS separation and serial transplantation in immunodeficient mouse models. Serial transplantation of the fractionated CSC population reproduces a histologically heterogeneous tumour with a hierarchy of differentiation. This is demonstrated by further tumour-propagating cells being produced in xenografts but most daughter cells in the bulk tumour lacking tumour-propagating capacity. In pioneering work in breast cancer, the $CD44^+/CD24^{-/low}/Lineage^-$ fraction was found to have tumour-propagating ability and is proposed to contain CSCs (Al-Hajj et al., 2003). Similarly, CSCs have been identified in many other cancers including brain, colon, prostate, lung, pancreatic and head and neck squamous cell carcinoma (Singh et al., 2004; Collins et al., 2005; Hermann et al., 2007; Prince et al., 2007; O'Brien et al., 2007; Ricci-Vitiani et al., 2007; Eramo et al., 2008). In many cases, the tumour-propagating fraction expresses the cell surface glycoprotein CD133.

CSCs may not exist in all tumours

Although the presence of a differentiation hierarchy with a CSC at its apex is well established for AML, studies in other tumours have often yielded discordant results. In some studies, the majority of cancer cells have been found to have CSC function, which may be either due to a very shallow hierarchy or the absence of a hierarchical tumour cell organization. In melanoma, Quintana found that most cells can act as stem cells in xenotransplantation assays. Within each cancer, diverse tumorigenic cells were found that undergo reversible phenotypic changes *in vivo* and this phenotypic heterogeneity was not associated with loss of tumorigenic potential (Quintana et al., 2010). In contrast, other studies have found CSCs in melanoma to be an uncommon population that sustains a developmental hierarchy, in keeping with the CSC model (Schatton et al., 2008; Boiko et al., 2010).

In B-ALL, xenotransplantation studies using fractionated cells of different maturity have revealed that blasts at different stages of immunophenotypic maturation exhibit stem cell properties, providing evidence against a differentiation hierarchy (Le Viseur et al., 2008; Rehe et al., 2013). Le Viseur and colleagues (2008) sorted paediatric leukaemic bone marrow cells according to the B-cell precursor differentiation markers CD34 and CD19 and these cells were transplanted into SCID mice by intrafemoral injection. Purified populations of $CD34^{+}CD19^{-}$, $CD34^{+}CD19^{+}$ and $CD34^{-}CD19^{+}$, which represent immunophenotypes of increasing maturational stage, all demonstrated leukaemia-initiating ability in primary and serial grafts up to quaternary recipients. The original immunophenotype and clinical disease were reproduced in all transplantations, as demonstrated by flow cytometry and findings of splenomegaly and lymphoblast infiltration of bone marrow. Interestingly, even the cell populations predicted to be less mature than the engrafting cells were reconstituted. A model was proposed whereby leukaemic cells can move back and forth within their stages of differentiation, adopting biology that mirrors the relevant normal populations. A caveat to this interpretation is that as leukaemia cells can aberrantly alter expression of surface markers, the immunophenotype may not accurately reflect the stage of maturation.

Clinical significance of cancer stem cells

Despite the limitations of xenotransplantation assays highlighted previously, several lines of evidence indicate that CSCs are clinically important and not simply an artefact of experimental models. CSCs have been hypothesized to be responsible for chemotherapy resistance and relapse and a proposed mechanism is by entering slow-cycling states or quiescence. Most conventional chemotherapy targets proliferating cells and therefore quiescence may provide CSCs with a means to survive. Ishikawa et al. (2007) demonstrated that human AML stem cells are relatively resistant to chemotherapy. Using the NSG mouse xenograft model, they showed that although bulk leukaemia cells were killed by administration of cytarabine, LSCs were spared. Using cell cycle analysis, they showed that LSCs were quiescent, suggesting that this may be a mechanism underlying resistance to cytotoxic chemotherapy. Support for the role of quiescence or slow-cycling states of CSCs in therapy resistance has been found using clonal tracking techniques in colorectal carcinoma. Within single genetic clones, there were significant differences in cellular behaviour. Although some cells were detectable at each transplant, there was evidence of a dormant population, which was detectable only on serial transplantation. Furthermore, chemotherapy was shown to promote dominance of these previously minor, slow-cycling cells within a single genetic clone (Kreso et al., 2013).

Although targeted chemotherapy can induce clinical remission, CSCs may survive. In most cases of CML, bulk leukaemia cells are exquisitely sensitive to tyrosine kinase inhibitors and treatment often leads to undetectable BCR-ABL transcripts by RT-PCR. However, the LSCs are resistant and accordingly, on cessation of treatment many patients relapse (Graham et al., 2002). Similarly, in myelodysplasia with 5q deletion, treatment with lenalidomide can lead to clinical remission but most patients relapse with expansion of CSCs with the 5q deletion (Tehranchi et al., 2010). Normal stem cells spend much of their time quiescent but re-enter the cell cycle to regenerate tissue following an insult. It has been hypothesized that CSCs can be reactivated in a similar manner to cause disease relapse and this may occur years after initial treatment (Wilson et al., 2008; Essers et al., 2009). If CSCs are responsible for relapse, their frequency would be expected to rise at relapse when compared with diagnosis. Indeed, in paired human AML samples the frequency of LSCs increases up to 90-fold, with LSCs functionally defined as capable of repopulating the tumour in serial mouse xenotransplantation experiments (Ho et al., 2016).

The expression of 'stem' genes influences clinical outcome in human AML (Gentles et al., 2010). By identifying the core transcriptional programme shared by functionally defined LSCs and normal HSCs, Eppert et al. (2011) showed that a stem cell gene expression signature was a significant predictor of poor survival. This group have recently produced a score based on the expression of 17 'stem cell genes' that can predict prognosis across a range of genetic subtypes of AML (Ng et al., 2016).

Intratumoural genetic diversity leads to clonal evolution

Cancer is initiated by the sequential acquisition of genetic mutations that leads to expansion of a malignant clone (Knudson, 1971; Nowell, 1976). Seminal work in colorectal cancer established that multiple genetics 'hits' are acquired that cause phenotypic changes representing the different stages of tumorigenesis from adenoma to carcinoma (Vogelstein et al., 1988). If genetic mutations in a cell lead to a survival advantage, the clone will expand. Further mutation acquisition in cells within an individual cancer leads to heterogeneous, genetically distinct subclones that compete for survival. According to Darwinian principles, selection of cells with a survival advantage produces branching evolution. Sequencing studies have provided support for this model in AML, ALL, renal, lung and breast cancer (Anderson et al., 2011; Notta et al., 2011; Jan et al., 2012; Gerlinger et al., 2012; Govindan et al., 2012; Nik-Zainal et al., 2012; Quek et al., 2016).

Yachida and colleagues (2010) elegantly demonstrated the branching evolution of tumorigenesis through exome sequencing of seven cases of metastatic pancreatic cancer. The primary tumours consisted of multiple, genetically distinct subclones with different metastatic potential. Chronological analysis of the genetic evolution revealed that more than ten years elapsed between the initiating mutation that leads to intraepithelial neoplasia, and subsequent mutations that produce an infiltrating carcinoma. A further five years separated this stage from the acquisition of mutations that enable metastasis.

A further body of evidence in support of the multihit model of tumorigenesis is found in transgenic mouse models, where expression of a single oncogene is rarely sufficient to cause cancer. Mice with aberrant expression of a single gene frequently found to be mutated in human AML, such as *Dnmt3a, Tet2, Idh2, Npm1*, or *Flt3*, do not develop AML. Instead, many of these models manifest abnormal haematopoiesis with features of myeloproliferative or myelodysplastic disease. Crossing these single-mutant mice with mice harbouring a second mutation (most commonly in *Flt3* or *Nras*) is required to induce AML (Li et al., 2011; Greenblatt et al., 2012; Kats et al., 2014; Sportoletti et al., 2015; Poitras et al., 2016).

Premalignancy

Premalignancy describes a state in which a tissue or organ retains adequate normal function despite detectable expansion of an abnormal clone, at risk of transforming to cancer. Examples of such precursor states in haematological malignancies include monoclonal gammopathy of undetermined significance, monoclonal B-cell lymphocytosis, and myelodysplasia. In solid tumours, the premalignant stage can be a non-invasive stage of tumour development (such as carcinoma *in situ*) or a morphological description of cellular dysplasia (such as Barrett's oesophagus), both of which are at increased risk of transformation to invasive disease. Early work by Vogelstein et al. (1988) demonstrated that certain mutations found commonly in colorectal carcinoma are more likely to support a premalignant state whereas others can cause frank carcinoma. More recently, sputum and bronchial lavage samples from 25 chronic smokers and 51 confirmed lung cancer patients revealed that $p16^{INK4a}$ promoter hypermethylation and *TP53* mutations occur as precancerous insults in chronic smokers before evidence of malignancy. By contrast, *K-RAS* mutations occurred exclusively in the presence of clinically evident neoplasia (Kersting et al., 2000).

Age-related clonal haematopoiesis

In 2014, three landmark papers (Jaiswal et al., 2014; Genovese et al., 2014; Xie et al., 2014) reported that mutations frequently identified in AML, such as in *DNMT3A, TET2, ASXL1*, and *IDH2*, can be found in the peripheral blood of individuals with no clinically apparent haematological disease. The incidence increased sharply with age such that evidence of clonal haematopoiesis was found in 5.6% of 60–69-year-olds compared with 18.4% of persons aged over 90 years. Longitudinal follow-up of these patients demonstrated a positive correlation between clonal haematopoiesis and development of haematological malignancy, suggesting a premalignant state within healthy individuals. These mutations were found in multiple mature cell types in patients including T-lymphocytes, suggesting that the cell of origin of the putative preleukaemic clone is one capable of generating multilineage progeny and is most likely to be an HSC. An expanded population with clonal advantage is formed but differentiation is permitted, leaving the peripheral blood counts unperturbed.

The term clonal haematopoiesis of indeterminate potential (CHIP) has been used to define this precursor state in clinical practice as individuals with at least 2% variant allele frequency of somatic mutations known to be associated with haematopoietic neoplasia without fulfilling diagnostic criteria for the malignancy itself (Steensma et al., 2015). Whether CHIP may be used as a means of screening healthy individuals for future haematological malignancies is currently unclear, due to a low risk of progression to malignancy (0.5–1% per year) and the absence of therapy to target the precursor state. Interestingly, a correlation has also been reported between clonal haematopoiesis and non-cancer mortality. In a multivariate analysis, the presence of clonal haematopoiesis was found to be an independent risk factor for coronary artery disease (hazard ratio, HR 2.0) and ischaemic stroke (HR 2.6), which may be due to a pro-inflammatory environment (Fuster et al., 2017). Interestingly, the risk imparted by clonal haematopoiesis was greater than the contribution of traditional atherosclerosis risk factors (Genovese et al., 2014; Jaiswal et al., 2014). Although there may be unknown confounding variables, these findings open an interesting area of research with significant relevance to public health.

Malignant transformation

Malignant transformation from premalignancy is associated with acquisition of additional 'driver' mutations. In AML samples, immunophenotypic HSC can give rise to normal, mature B-lymphocytes and myelomonocytic cells in NSG mice (Jan et al., 2012). These mature cells have some, but not all, of the mutations present in the patient's AML. Imputed clonal structures of AML suggests that cells with preserved normal HSC function coexist with LSCs. Whereas HSCs belong to clones that have preleukaemic mutations acquired early in leukaemogenesis, LSCs have the full complement of both preleukaemic and driver mutations.

Capacity for self-renewal is a prerequisite for a premalignant clone to survive and progress to become cancerous. If an oncogenic mutation occurs in a cell that is incapable of self-renewal, the clone will become exhausted through differentiation. Therefore, the precancerous mutation must either be attained in a normal stem cell, which already exhibits indefinite self-renewal, or in a more differentiated cell that subsequently acquires self-renewal capacity through the acquisition of additional genetic or epigenetic abnormalities (Fig. 20.3).

The initiating genetic event in CML, the t(9;22) translocation that generates the BCR-ABL fusion protein, most probably occurs in a haematopoietic stem cell (Takahashi et al., 1998). Despite the success of treatment with targeted tyrosine kinase inhibitors (TKIs), most patients do not eradicate all malignant cells. Copland et al. (2006) investigated the sensitivities of CML cells of different maturity to TKIs and found that the primitive $CD34^{+}CD38^{-}$ fraction was resistant to TKIs, despite having higher levels of BCR-ABL transcripts than more mature CML cells. This cellular fraction is likely to contain CML stem cells, whose quiescent nature may protect them from the cell cycle-dependent apoptotic machinery. If the TKI is removed, the disease will frequently re-establish itself with relapse (Mahon et al., 2010). Thus, the CML stem cell provides a refuge for the t(9;22) translocation, while the TKI eliminates its proliferating progeny. More recently, it has been observed that CML patients who achieve deep molecular remission (<0.001% BCR-ABL transcripts) for prolonged periods may have eliminated the CML stem cell and be cured. Whether TKI treatment can be stopped in these patients is being addressed in ongoing clinical trials. Progression to advanced-stage blast crisis CML is associated with increased genetic instability that causes proliferation to proceed independently of BCR-ABL. At this stage, functional LSCs reside in multiple populations with the immunophenotype of normal progenitors as well as stem cells (Jamieson et al., 2004; Kinstrie et al., 2016).

Mutations differ in their capacity to transform a cell. 'Driver' mutations promote tumorigenesis, whereas 'passenger' mutations do

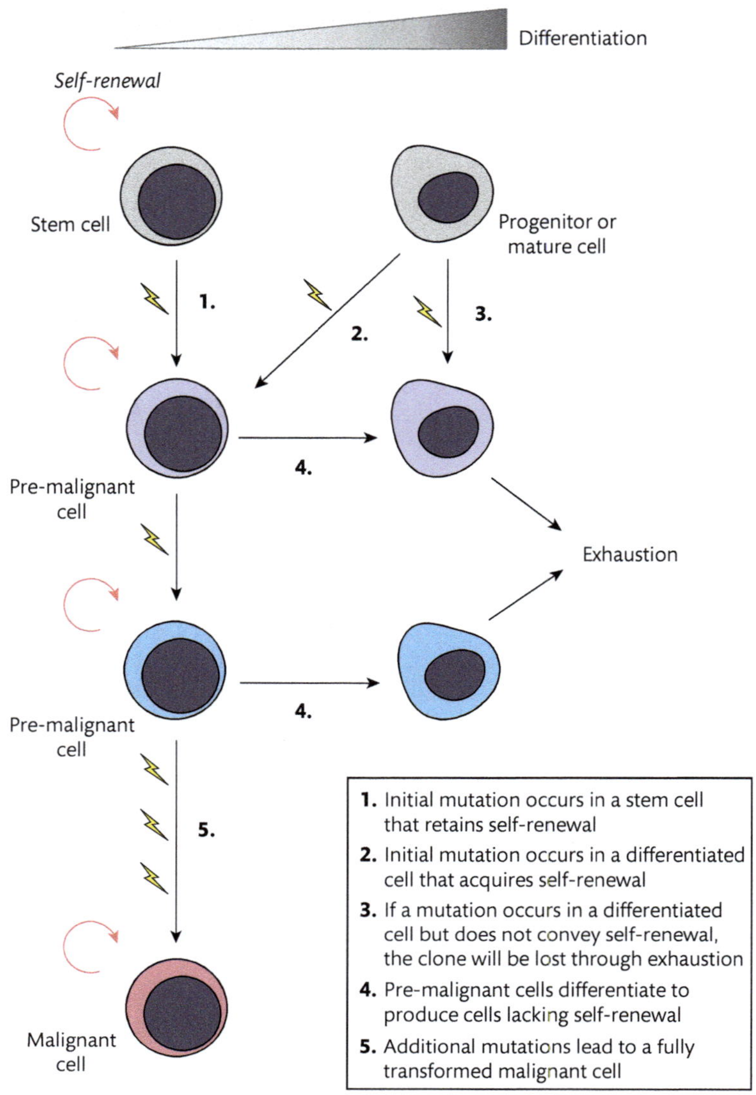

Fig. 20.3 Model for clonal evolution driven by sequential acquisition of genetic mutations. Malignant transformation is the result of sequential acquisition of genetic mutations. Capacity for self-renewal is essential for a premalignant clone to survive and progress to become cancerous. If an oncogenic mutation occurs in a cell that is incapable of self-renewal, the clone will become exhausted.

not confer a selective growth advantage. Melanomas and lung cancers contain up to 200 mutations per tumour, whereas AML has an average of eight mutations. Paediatric tumours also generally contain fewer driver mutations than adult cancers (Vogelstein et al., 2013). Within AML, samples with cytogenetic anomalies involving balanced translocations typically have fewer cooccurring mutations. In contrast, AML that has developed from preceding myelodysplastic syndrome typically has more karyotypic abnormalities including aneuploidies and additional mutations (Papaemmanuil et al., 2016). It is often unclear whether a single clone requires one 'driver' mutation or combinations of mutations for malignant transformation and to understand this process, we must understand clonal structures.

Clonal dynamics

If a mutation is detected in a bulk tumour, there must be a CSC with that particular mutation that propagates the clone. We can gain insight into the behaviour of CSC clones in patients by tracking them during the course of their disease. This is most frequently performed in the context of remission following therapy, and at relapse. There is evidence that the genetic composition of cancers changes under the selection pressure exerted by chemo-, radio-, or immune-based therapies. Relapse can be caused by persistence of the same CSC clones as present at diagnosis. There may be a process of clonal selection, whereby a clone gains a survival advantage over other clones that may have been more dominant prior to therapy. Alternatively, clonal evolution may occur, whereby a pre-existing clone acquires a new mutation. In a Darwinian system, previously uncompetitive clones can gain dominance as larger clones succumb to therapy, if they exhibit self-renewal and proliferative mechanisms that are independent of those targeted by therapy (Fig. 20.4).

In some cases of AML, self-renewing, premalignant clones that harbour early mutations, but not the full complement that produce a cancerous clone, may survive therapy (Jan et al., 2012; Shlush et al.,

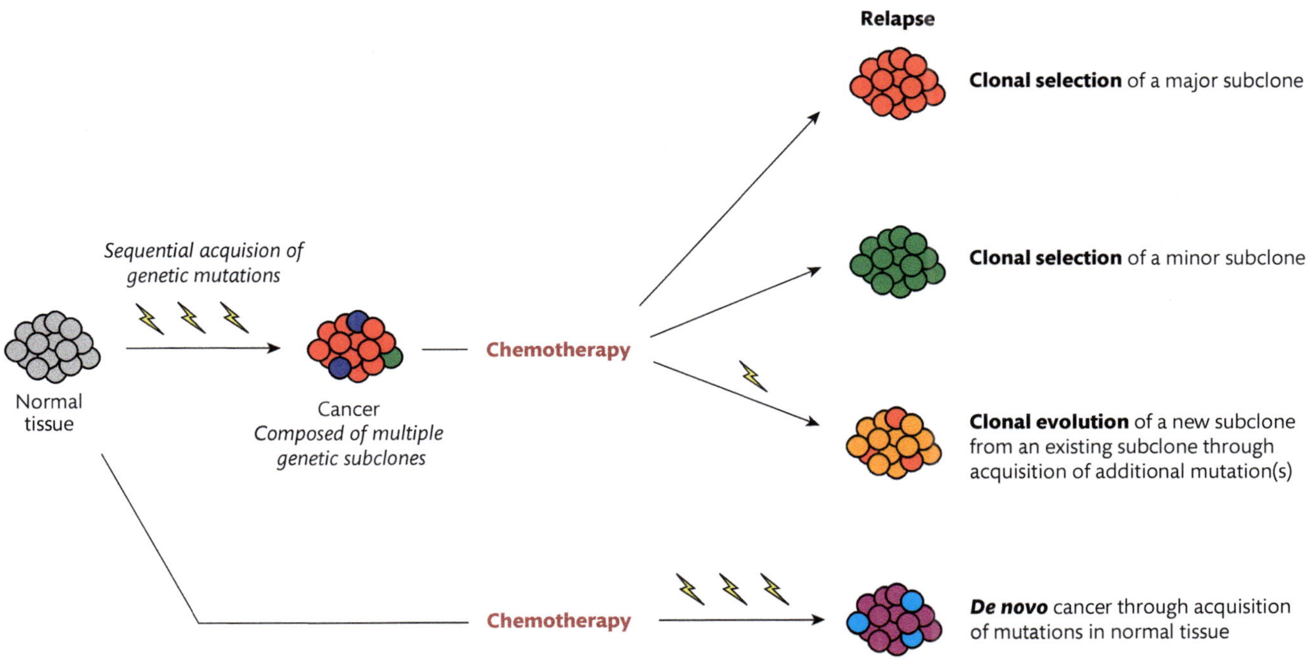

Fig. 20.4 Subclone selection and evolution of a cancer following treatment. The genetic composition of cancers changes under the selection pressure exerted by treatment. Relapse can be caused by clonal selection of a CSC clone that was present at diagnosis, which may have previously been a minor subclone. Alternatively, clonal evolution may occur, whereby a pre-existing clone acquires a new mutation and a survival advantage. Lastly, mutations may occur in normal tissue cells following treatment that produce *de novo* cancerous clones.

2014). These antecedent clones could therefore act as seeds for genetically related but divergent relapsed disease. Indeed, selective pressure imparted by therapy may lead to clonal expansion of the preleukaemic clones through acquisition of new mutations. Sequencing studies using samples from AML patients have shown that in some patients, the only mutations shared between clones at diagnosis and relapse are those present in the premalignant HSCs. This suggests that mutations acquired by the leukaemic clone later in its evolution have been targeted by therapy and do not contribute at relapse (Corces-Zimmerman et al., 2014). These observations emphasize the importance of therapeutically targeting cells with self-renewal capacity, as only these have the ability to survive and re-initiate the cancer. Whereas mutations in certain genes have a high likelihood of persisting between diagnosis and relapse (e.g. *DNMT3A, NPM1, IDH1*), some are readily lost (e.g. kinase domain mutations in *FLT3*), and others are frequently gained (e.g. *FLT3* internal tandem duplication or *NRAS* mutations). This likely reflects the order of acquisition of mutations; those acquired early are present in parental clones, whereas those acquired late (e.g. *FLT3, NRAS*) are typically subclonal and more susceptible to processes of clonal selection.

Minimal residual disease tracking of CSCs

Our increasing understanding of the genetics and functional properties of CSC has profound implications not only for diagnosis, but also the way in which tumour burden is monitored. Minimal residual disease (MRD) techniques measure CSCs that persist during and after therapy. Assessment methods depend on the malignancy but include high-resolution flow cytometry or tracking of genetic mutations using molecular techniques (Grimwade et al., 2009; Freeman et al., 2013; Ivey et al., 2016). Quantitation of AML LSCs using flow cytometry following treatment provides a biomarker for disease response and detection of MRD confers an increased risk of relapse and poor prognosis. In patients who achieve a clinical remission, later expansion of LSCs predicts relapse (Gerber et al., 2012; Craddock et al., 2013; Terwijn et al., 2014).

Many patients relapse despite being MRD negative. Therefore, another way of considering MRD is 'measurable' residual disease, which implies that we may not be detecting residual disease either due to limited sensitivity of the assay or because we are not using the correct marker to track disease. A minority of patients with detectable MRD do not relapse, which raises the interesting possibility that premalignant stem cells can be compatible with long-term disease remission. If MRD were to be performed using an early mutation, then it would incorporate the precancerous population as well as the frankly cancerous clone and therefore could overestimate MRD burden. This may explain the success of NPM1 and PML-RARA molecular MRD in AML, as these mutations are not found in preleukaemic HSCs.

Epigenetic determinants of CSC biology

It has been known for several decades that heritable changes in gene expression exist that are not encoded by the genomic DNA sequence. These epigenetic changes increase the diversity within a tumour and include:

1. Covalent modifications of DNA (such as methylation and hydroxymethylation) and histone proteins (such as acetylation, methylation, and ubiquitination)
2. Expression of non-coding ribonucleic acids (RNAs), such as microRNAs

Epigenetic changes are likely to contribute to the heterogeneity within cancers and increasing evidence is emerging for their role in maintaining self-renewal of CSCs. Within a cancer, cells of the same genotype can exist in different epigenetic states that influence their function. Although changes in the epigenome, such as hypermethylation of tumour suppressors and hypomethylation of oncogenes, has been described in cancers, the mechanisms by which such changes alter gene expression to drive tumorigenesis are not well understood.

It is becoming increasingly apparent that epigenetic changes are often downstream of genetic abnormalities (Baylin and Jones, 2011). Mutations in genes that encode proteins involved in epigenetic regulation are common in cancer and are found in over 50% of all cases of AML. Mutations in epigenetic regulators are often early events in tumorigenesis and are common drivers of age-related clonal haematopoiesis. Mutations in the DNA methyltransferase *DNMT3A* are found in 22% of AML and this gene is critical for self-renewal of normal haematopoietic stem cells (Tadokoro et al., 2007; Ley et al., 2010; Shah and Licht, 2011). Other recurrent mutations found in AML encode epigenetic regulators that have also been implicated in stem cell function, including *TET2, IDH1*, and *IDH2*.

The polycomb group of proteins are transcriptional repressors that influence normal adult stem cell function and are likely to be key regulators of cell fate in CSCs. The Polycomb Repressive Complex 2 (PRC2) trimethylates histones at the H3K27 position (Lund et al., 2014). PRC1 is recruited to the H3K27 trimethylation mark left by PRC2 and catalyses monoubiquitination of histone H2A. This leads to transcriptional repression by compaction of chromatin, which prevents transcription factors accessing DNA. EZH2 is the catalytic component of PRC2 and aberrant *EZH2* expression directs malignant cells towards a stem cell state by switching off pro-differentiation genes (Kamminga et al., 2006). BMI1 and RING1B are core subunits of PRC1. RING1B catalyses the ubiquitination of histone H2A and this reaction is stabilized by the binding of BMI1 to RING1B (Li et al., 2006). BMI1 also represses $p16^{INK4A}$ and $p14^{ARF}$, which normally inhibit cell cycle progression, and contributes to the cellular response to DNA damage (Ginjala et al., 2011). Additionally, BMI1 plays an essential role in self-renewal of normal adult stem cells and has also been implicated in CSC function (Lessard and Sauvageau, 2003; Iwama et al., 2004). Expression levels of BMI1 correlate with poor prognosis in human myeloid leukaemias and it is more highly expressed in LSCs, suggesting this transcription factor may represent a novel therapeutic target (Chowdhury et al., 2007; Saudy et al., 2014; Quek et al., 2016).

Epigenetic modifications may provide a mechanism for therapy resistance. Sharma and colleagues demonstrated that following treatment of cancer cell lines with chemotherapy, a small proportion of cancer stem cell-like cells survived. Their drug-resistant phenotype was associated with global chromatin changes and elevated expression of the Jumonji histone demethylase, JARID1A, which removes the activating transcription mark, H3K4me3. Interestingly, multiple histone modifying drugs were shown to reverse the therapeutic resistance of these cells (Sharma et al., 2010).

The *HOX* genes are a family of highly conserved homeodomain-containing transcription factors that specify cell identity in embryonic development and subsequently in haemopoiesis, where their expression is restricted to stem and early progenitor cells (Alharabi et al., 2013). As well as playing a key role in transcriptional regulation in normal haematopoiesis, *HOX* genes and their cofactor *MEIS1* are highly expressed in AML and convey stem cell properties in AML LSCs, where their expression is regulated by epigenetic mechanisms (Wong et al., 2007; Argiropoulos et al., 2007; Spencer et al., 2015). *HOX* genes are expressed in 80% of AML samples and expression is highly correlated with recurrent genetic abnormalities. Upstream epigenetic regulators of *HOX* and *MEIS1* expression include MLL (a histone methyltransferase) and DNMT3A. Mutations in these genes are found in AML and may lead to upregulation of *HOX* and *MEIS1* expression by disruption of BMI1-mediated transcriptional repression (Tan et al., 2016).

Signalling pathways influence CSC function

The tumour microenvironment is composed of a multitude of cells, including haematopoietic, stromal, and endothelial cells, as well as extracellular components. These factors influence the function of malignant cells and are likely to contribute to a 'stem cell niche' that maintains the CSC (Hanahan and Coussens, 2012). Through interactions between neighbouring cells and also between secreted ligands and transmembrane receptors on the CSC surface, extracellular signals are transduced to activate intracellular pathways. Many of the pathways that are critical for stem cell function during embryogenesis and normal tissue self-renewal are also employed by CSCs and include the Notch, Wnt/β-catenin, and hedgehog pathways (Takebe et al., 2011).

Notch pathway

Coordinated interactions between CSCs and other cells in the niche lead to the activation of stem cell intracellular signalling pathways. The pairing of a membrane-bound Notch ligand with a transmembrane Notch receptor leads to proteolytic cleavage and the release of the active Notch intracellular domain (NICD) into the cytoplasm (Gordon et al., 2007). Following translocation to the nucleus, transcription of target genes is activated, including *MYC*, a known oncogene. Gain-of-function *NOTCH1* mutations are found in over 50% of human T-cell acute lymphoblastic leukaemias (T-ALLs) and lead to constitutive activation of the signalling pathway, resulting in upregulation of *MYC*, which drives malignant proliferation (Weng et al., 2004). Additionally, 20% of T-ALLs harbour *FBXW7* mutations, which disrupts its normal role in directing NICD for ubiquitination and proteasomal degradation. Therefore, two mechanisms contribute to aberrant NOTCH1 activation in T-ALL: ligand-independent receptor activation and stabilization of NICD (Girardi et al., 2017; Sanchez-Martin and Ferrando, 2017).

Wnt/β-catenin pathway

Interactions between soluble ligands and receptors allow communication between the microenvironment and the CSC. Wnt proteins serve as extracellular ligands for the Frizzled transmembrane receptor and play a key role in directing cell fate in embryogenesis and in normal regenerative adult tissues. Following Wnt binding, β-catenin is released from a cytoplasmic multiprotein complex and translocates to the nucleus, where it influences gene expression. Aberrant Wnt signalling has been found in numerous tumours including Wilms tumour, hepatocellular, and gastric carcinomas (de La Coste et al., 1998; Koesters et al., 1999; Kim et al., 2009).

Additionally, the signalling pathway has been specifically implicated in CSC activity in AML and skin cancers. A comparison of gene expression between human LSCs and HSCs revealed dysregulation of the Wnt/β-catenin pathway in LSCs (Majeti et al., 2009). In skin cancer, $CD34^+$ CSCs preserve the hierarchical tumour organization on transplantation, but removal of β-catenin signalling results in loss of CSC activity. Interestingly, this pathway is not essential for normal epidermal homeostasis and therefore it may represent a therapeutic target (Malanchi et al., 2008).

Hedgehog pathway

The hedgehog (Hh) family of secreted proteins that were first described in embryonic development of Drosophila but have subsequently been implicated in tumorigenesis. Hh proteins act as ligands for the Ptch transmembrane receptor, leading to activation of the Smo receptor and release of Gli1/2. Gli1/2 then translocates to the nucleus, resulting in upregulation of Hh-associated genes. The role of this pathway in tumorigenesis was highlighted by the discovery that germline mutations in the human *PTCH1* gene are found in Gorlin's syndrome, which is associated with the development of basal cell carcinomas and other tumours (Hahn et al., 1996; Johnson et al., 1996). Hh signalling has also been shown to be important for CSC function in CML and breast cancer (Liu et al., 2006; Zhao et al., 2009). Interestingly the polycomb protein BMI-1, which is a key transcriptional regulator of normal and cancer stem cell function, is activated downstream target of Hh signalling in breast CSCs.

Therapeutic targeting of cancer stem cells

Much of the clinical importance of the CSC hypothesis arises from the prospect that it could provide an explanation for the poor correlation between bulk tumour regression following therapy and increased patient survival. Volume-dependent measures of tumour control do not take into account that the survival of a single CSC can lead to relapse. This holds further impetus if CSCs are truly more resistant to current therapies (Bao, 2006; Ishikawa et al., 2007; Diehn, 2009; Saito et al., 2010; Viale et al., 2010) as a reduction in tumour mass will not necessarily bear equivalence to permanent tumour eradication.

How best to target the elusive CSC population is a major focus of novel therapeutic design. Approaches have included sensitization of CSCs to conventional therapeutics and also directly targeting factors that contribute to 'stemness'. The latter is challenging due to the susceptibility of normal stem cells to toxicity from drugs targeting CSCs and therefore CSC-directed therapies should target unique vulnerabilities.

Sensitization of CSCs to current therapeutics

CSCs have been shown to express high levels antioxidants, which is of interest as reactive oxygen species (ROS) act as a point of convergence in mediating the DNA damage effects of both chemotherapy and radiotherapy (Trachootham et al., 2009). Diehn (2009) demonstrated that relative to non-CSC populations, the CSCs of human head and neck and breast cancers expressed higher levels of free radical scavengers, reduced ROS, and lower levels of DNA damage after irradiation. Correspondingly, pharmacological inhibition of free radical scavengers reduced *in vitro* colony-forming capacity and radiosensitized the CSCs. Therefore, sensitizing CSCs to ROS-mediated toxicity of conventional therapeutics may enhance the therapeutic impact. However, translation of this preclinical rationale into successful therapeutic strategies remains a challenge, as demonstrated by disappointing Phase III trial results in advanced melanoma with elesclomol, a drug that acts by inducing oxidative stress (O'Day et al., 2013).

Another CSC resistance mechanism that represents a potential therapeutic target is the increased expression of drug efflux transporters (Frank et al., 2005; Wang et al., 2010; Cheung et al., 2011; Tsai et al., 2011). Unfortunately, as these transporters serve essential cellular functions beyond that of CSC survival, inhibiting their function may lead to unacceptable side effects.

Targeting self-renewal

Capacity for indefinite self-renewal is the cardinal feature of a stem cell. It therefore follows that inhibiting self-renewal is a key aim of CSC-directed therapy. Identification of novel agents with CSC-specific toxicity has proven difficult when applying standard high-throughput cell viability assays to the entire tumour because the CSC population is small, and propagation of enriched CSC populations *in vitro* can prove difficult. However, induction of epithelial-mesenchymal transition (EMT) in normal or neoplastic mammary epithelial cell populations leads to enrichment of cells bearing stem-like properties (Gupta et al., 2009; Clevers, 2011). It has therefore been postulated that via high-throughput screening, this EMT+ population could be used to identify compounds with EMT-specific toxicity that might also exhibit CSC-specific toxicity. Using this methodology, they report that sensitivity of breast cancer cells to the compound salinomycin, a potassium ionophore antibiotic, correlated with the proportion of breast CSCs in the population. Furthermore, this compound reduced mammosphere expression of CSC-associated genes that had been previously associated with poor patient survival. Anti-CSC activity of salinomycin has also been shown in a lung cancer cell line (Wang, 2011). However, both studies used genetically induced tumour cells rather than primary, patient tumour cells and later reports showed that the concentrations required to exert such effects on CSCs *in vivo* would have associated neuronal toxicity (Boehmerle and Endres, 2011). The precise mechanism of salinomycin's toxicity is still to be elucidated but it was recently reported to act as a potent inhibitor of Wnt/β-catenin signalling (Lu et al., 2011). Signalling pathways that drive self-renewal of CSCs, including the Wnt/β-catenin pathway, appear to be promising targets. However, these pathways are shared with normal tissue stem cells and avoiding toxicity represents a major obstacle.

One promising development for targeted CSC therapy was the identification that the NF-κB survival pathway is active in LSCs but not normal HSCs (Guzman et al., 2007a; Guzman et al., 2007b). Correspondingly, an orally bioavailable parthenolide analogue inhibited engraftment potential of primary AML cells by up to 98.2% through inhibition of the NF-κB pathway but did not influence growth or engraftment of normal stem cells. Although poor bioavailability of the compound was found, this has been circumvented by incorporation of parthenolide into nanoparticles and the subsequent use of a multistage vector system to penetrate the protective CSC niche. This has resulted in reduced AML burden and dramatically impaired CSC function *in vivo* (Zong et al., 2016).

BMI1, a polycomb complex protein, is overexpressed in multiple malignancies including lung, breast, colorectal, skin, neuroblastoma and AML and is essential for CSC function (Vonlanthen et al., 2001; Kim et al., 2004a; Kim et al., 2004b; Cui et al., 2007; Chowdhury et al., 2007; Lee et al., 2008; Vrzalikova et al., 2008; Saudy et al., 2014). Pharmacological inhibition of BMI1 leads to long-term, irreversible control of tumour burden in murine xenograft models of human colorectal cancer and lung adenocarcinoma (Kreso et al., 2014b; Yong et al., 2016). BMI1 also plays a key role in stem cell maintenance in many normal tissues and therefore off-target effects are a concern. Promisingly, peripheral blood counts and the numbers of haematopoietic stem cells were preserved in these models, but whether this translates to therapeutic doses in humans remains to be answered. Indeed, in studies of BMI1 inhibition in AML, the small molecule inhibitor was effective at inducing apoptosis of AML cells but also showed toxicity to normal HSCs (Nishida et al., 2015). BMI1 inhibition is currently being evaluated in an early phase clinical trial in solid tumours.

Stimulating differentiation

Another method of interrupting CSC survival mechanisms is to stimulate differentiation. By induction of symmetrical cell division that generates only more mature daughter cells, the progeny will lack self-renewing capacity and be purged by differentiation.

This approach has been extremely successful in treating acute promyelocytic leukaemia with all-transretinoic acid or arsenic trioxide. Therapeutics to induce differentiation in other malignancies remain experimental but are showing potential. In glioblastoma, bone morphogenetic protein signalling leads to a reduction in proliferation and increased expression of markers of neural differentiation. Exploitation of this through cotransplantation of BMP4-saturated polyacrylic beads with glioblastomas cells reduced tumour growth and prevented mortality in animal models. Together with *in vitro* data demonstrating a reduction in $CD133^+$ CSCs after BMP4 application, this shows promise for future treatments of this aggressive tumour (Piccirillo et al., 2006). BMP4 has also been shown to promote terminal differentiation, apoptosis and chemosensitization of CSCs in a murine model of colorectal carcinoma (Lombardo et al., 2011).

New therapies that induce differentiation using small molecular inhibitors of mutant IDH1/2 enzymes are undergoing early phase clinical trials. Mutations in IDH have been described in AML, B-cell lymphoma, glioma, thyroid, breast, colorectal and prostate cancers and result in a neomorphic enzyme that produces an oncometabolite, 2-hydroxyglutarate (Kang et al., 2009; Watanabe et al., 2009; Yan et al., 2009; Abbas et al., 2010; Green and Beer, 2010; Ho et al., 2010; Murugan et al., 2010; Ichimura, 2012; Fathi et al., 2014). This inhibits the function of several key enzymes involved in epigenetic regulation including TET2 and Jumonji histone demethylases, resulting in a differentiation block. Pharmacological inhibition of these mutants therefore aims to induce differentiation of tumour cells, reduce tumour bulk, and restore normal cellular function. However, early data has suggested that suboptimal elimination of IDH mutant CSCs could underlie disease relapse. Nevertheless, induction of differentiation may render CSCs more susceptible to other drugs, a hypothesis that will be tested in clinical trials of IDH inhibitors in combination with other modalities such as cytotoxic chemotherapy.

Stimulating phagocytosis

CD47 is a cell surface protein that serves as a ligand for signal regulatory protein-α (SIRPα), which is expressed on phagocytic cells and this interaction produces an inhibitory signal for phagocytosis. It has been found to be expressed by CSCs in a range of malignancies and is expressed at a significantly higher level by AML LSCs relative to normal HSCs (Majeti, 2011). CD47 is therefore proposed to represent a means by which CSCs evade cell death and correspondingly, its expression is associated with adverse clinical outcome. A novel monoclonal antibody, Hu5F9-G4, blocks the interaction between CD47 and SIRPα and has been shown to promote phagocytosis of AML cells *in vitro* and in xenograft models. It also been shown to eradicate multiple other cancers, including lung cancer, glioblastoma, B-cell lymphoma, and acute lymphoblastic leukaemia (Chao et al., 2009; Chao et al., 2011; Weiskopf et al., 2016; Zhang et al., 2016). Anti-CD47 treatment has entered early phase clinical trials for several malignancies.

Targeting the tumour microenvironment

There is an increasing appreciation of the importance of the tumour microenvironment (TME), as drug resistance appears to be influenced by interactions between cancer cells and their surroundings (Hanahan and Weinberg, 2011). Discrepancies between *in vitro* and *in vivo* sensitivities of CSCs to experimental therapeutics suggests this may be of clinical relevance (Junttila and de Sauvage, 2013).

Extracellular factors have been shown to confer self-renewal to cancer cells. For example, hypoxia-inducible factors, expressed in CSC populations under hypoxic conditions, have been shown to play instrumental roles in maintenance and promotion of 'stemness' (Heddleston et al., 2010). HIF2α has been shown to enhance expression of the embryonic stem cell markers OCT4 and NANOG and promote Notch signalling, which supports an undifferentiated state (Ma et al., 2011; Bhaskara et al., 2012). Correspondingly, local tumour control correlates inversely with tumour hypoxia and hypoxia confers radioresistance in murine models of medulloblastoma (Woodward et al., 2009). Reduction of hypoxia is therefore an obvious strategy for sensitization of tumours to therapeutics. Indeed, use of erythropoietin to increase radiosensitivity was explored in a large clinical trial of head and neck cancer (Henke et al., 2003). Unexpectedly, the presence of erythropoietin was associated with reduced overall survival that was accompanied by an increase in the number of CSCs.

Direct targeting of the interactions within the TME has also been explored. For example, CD44, a mediator of cell–cell and cell–extracellular matrix interactions, is highly expressed on AML cells and its glycosaminoglycan ligand hyaluronan is enriched in human bone marrow endosteal regions that are proposed to act as stem cell niches (Bendall et al., 2000). A monoclonal antibody to CD44v6, an isoform of CD44 unique to CSCs, therefore looked a promising means to specifically interrupt CSC interactions with their supportive TME. Unfortunately, its development was halted in Phase I trials due to skin toxicity (Riechelmann et al., 2008).

Conclusion

The CSC model explains the heterogeneity and hierarchy of cellular differentiation that is observed in many cancers. Curative therapies must eradicate the CSC, as this cell is essential for the survival of the cancer and is likely to contribute to treatment resistance and relapse. The CSC model provides a framework to consider novel therapeutic strategies. Although promising CSC-directed treatments are undergoing investigation, these come with caveats as even with good preclinical rationale, translation of bench-side work to the bedside is a challenge.

TAKE-HOME MESSAGE

- Regenerative tissues exhibit a cellular hierarchy of differentiation, maintained by stem cells, which exhibit the key property of long-term self-renewal.
- Cancer stem cells maintain the growth of some cancers.
- For a genetic clone to survive, oncogenic mutations must occur in a self-renewing cell.
- CSCs have been implicated in treatment resistance and relapse.
- As CSCs are the only cells able to propagate the cancer, curative treatment strategies must effectively target these cells.

OPEN QUESTIONS

- Do cancer stem cells exist in all cancers?
- What is the true CSC frequency in different type of human cancers?
- How can minimal residual disease tracking of CSCs be used to guide treatment?
- Can CSC-directed therapy reduce relapse?

FURTHER READING

Blanpain, C. & Fuchs, E. (2014). Stem cell plasticity. Plasticity of epithelial stem cells in tissue regeneration. *Science*, **344**, 1242281.

Chaffer, C. L. & Weinberg, R. A. (2010). Cancer cell of origin: spotlight on luminal progenitors. *Cell Stem Cell*, **7**(3), 271–2.

Clevers, H. (2011). The cancer stem cell: premises, promises and challenges. *Nat Med*, **17**(3), 313–9.

Colom, B. & Jones, P. H. (2016). Clonal analysis of stem cells in differentiation and disease. *Curr Opin Cell Biol*, **43**, 14–21.

Douer, D. & Tallman, M. S. (2005). Arsenic trioxide: new clinical experience with an old medication in hematologic malignancies. *J Clin Oncol*, **23**, 2396–410.

Hogan, B. L. M., Barkauskas, C. E., Chapman, H. A., et al. (2014). Repair and regeneration of the respiratory system: complexity, plasticity, and mechanisms of lung stem cell function. *Cell Stem Cell*, **15**(2), 123–38.

Nassar, D. & Blanpain, C. (2016). Cancer stem cells: basic concepts and therapeutic implications. *Annu Rev Pathol*, **11**, 47–76.

Pollyea, D. A. & Jordan, C. T. (2017). Therapeutic targeting of acute myeloid leukemia stem cells. *Blood*, **129**(12), 1627–35.

Stange, D. E. & Clevers, H. (2013). Concise review: the yin and yang of intestinal (cancer) stem cells and their progenitors. *Stem Cells*, **31**(11), 2287–95.

Thomas, D. & Majeti, R. (2017). Biology and relevance of human acute myeloid leukemia stem cells. *Blood*, **129**(12), 1577–85.

Zhang, M., Lee, A. V., & Rosen, J. M. (2017). The cellular origin and evolution of breast cancer. *Cold Spring Harb Perspect Med*, **7**(3), pii: a027128.

REFERENCES

Abbas, S., Lugthart, S., Kavelaars, F. G., et al. (2010). Acquired mutations in the genes encoding IDH1 and IDH2 both are recurrent aberrations in acute myeloid leukemia: prevalence and prognostic value. *Blood*, **116**(12), 2122–6.

Al-Hajj, M., Wicha, M. S., Benito-Hernandez, A., Morrison, S. J., & Clarke, M. F. (2003). Prospective identification of tumorigenic breast cancer cells. *Proc Natl Acad Sci USA*, **100**, 3983–8.

Alharabi, R. A., Pettengell, R., Pandha, H. S., & Morgan, R. (2013). The role of HOX genes in normal hematopoiesis and acute leukemia. *Leukemia*, **27**, 1000–8.

Anderson, K., Lutz, C., van Delft, F. W., et al. (2011). Genetic variegation of clonal architecture and propagating cells in leukaemia. *Nature*, **469**, 356–61.

Argiropoulos, B., Yung, E., Humphries, R. K. (2007). Unravelling the crucial roles of Meis1 in leukemogenesis and normal hematopoiesis. *Genes Dev*, **21**(22), 2845–9.

Bao, S. (2006). Glioma stem cells promote radioresistance by preferential activation of the DNA damage response. *Nature*, **444**, 756–60.

Baylin, S. B. & Jones, P. A. (2011). A decade of exploring the cancer epigenome—biological and translational implications. *Nat Rev Cancer*, **11**(10), 726–34.

Bendall, L. J., Bradstock, K. F., & Gottlieb, D. J. (2000). Expression of CD44 variant exons in acute myeloid leukemia is more common and more complex than that observed in normal blood, bone marrow or CD34 + cells. *Leukemia*, **14**, 1239–46.

Bhaskara, V. K., Mohanam, I., Rao, J. S., & Mohanam, S. (2012). Intermittent hypoxia regulates stem-like characteristics and differentiation of neuroblastoma cells. *PLoS One*, **7**, e30905.

Boehmerle, W. & Endres, M. (2011). Salinomycin induces calpain and cytochrome c-mediated neuronal cell death. *Cell Death Dis*, **2**, e168.

Boiko, A. D., Razorenova, O. V., van de Rijn, M., et al. (2010). Human melanoma-initiating cells express neural crest nerve growth factor receptor CD271. *Nature*, **466**, 133–7.

Bonnet, D. & Dick, J. E. (1997). Human acute myeloid leukemia is organized as a hierarchy that originates from a primitive hematopoietic cell. *Nat Med*, **3**(7), 730–7.

Bosma, G. C., Custer, R. P., & Bosma, M. J. (1983). A severe combined immunodeficiency mutation in the mouse. *Nature*, **301**, 527–30.

Chao, M. P., Alizadeh, A. A., Tang, C. Z., et al. (2009). Therapeutic antibody targeting of CD47 synergizes with rituximab to completely eradicate human B-cell lymphoma xenografts. *Blood*, **114**, 1063–4.

Chao, M. P., Alizadeh, A. A., Tang, C. Z., et al. (2011). Therapeutic antibody targeting of CD47 eliminates human acute lymphoblastic leukemia. *Cancer Res*, **71**, 1374–84.

Cheung, S. T., Cheung, P. F., Cheng, C. K., Wong, N. C., & Fan, S. T. (2011). Granulin-epithelin precursor and ATP-dependent binding cassette (ABC)B5 regulate liver cancer cell chemoresistance. *Gastroenterology*, **140**, 344–55.

Chowdhury, M., Mihara, K., Yasunaga, S., et al. (2007). Expression of polycomb-group (PcG) protein BMI-1 predicts prognosis in patients with acute myeloid leukemia. *Leukemia*, **21**, 1116–22.

Clarkson, B., Fried, J., Strife, A., et al. (1970). Studies of cellular proliferation in human leukemia. III. Behavior of leukemic cells in three adults with acute leukemia given continuous infusions of 3H-thymidine for 8 or 10 days. *Cancer*, **25**, 1237–60.

Clevers, H. (2011). The cancer stem cell: premises, promises and challenges. *Nat Med*, **17**, 313–19.

Collins, A. T., Berry, P. A., Hyde, C., Stower, M. J., & Maitland, N. J. (2005). Prospective identification of tumorigenic prostate cancer stem cells. *Cancer Res*, **65**, 10946–51.

Copland, M., Hamilton, A., Elrick, L., et al. (2006). Dasatinib (BMS-354825) targets an earlier progenitor population than imatinib in primary CML but does not eliminate the quiescent fraction. *Blood*, **107**(11), 4532–9.

Corces-Zimmerman, M. R., Hong, W. J., Weissman, I. L., Medeiros, B. C., & Majeti, R. (2014). Preleukemic mutations in human acute myeloid leukemia affect epigenetic regulators and persist in remission. *Proc Natl Acad Sci USA*, **111**(7), 2548–53.

Craddock, C., Quek, L., Goardon, N., et al. (2013). Azacitidine fails to eradicate leukemic stem/progenitor cell populations in patients with acute myeloid leukemia and myelodysplasia. *Leukemia*, **27**(5), 1028–36.

Cui, H., Hu, B., Li, T., et al. (2007). Bmi-1 is essential for the tumorigenicity of neuroblastoma cells. *Am J Pathol*, **170**, 1370–8.

Curley, M. D., Therrien, V. A., Cummings, C. L., et al. (2009). CD133 expression defines a tumor initiating cell population in primary human ovarian cancer. *Stem Cells*, **27**, 2875–83.

de La Coste, A., Romagnolo, B., Billuart, P., et al. (1998). Somatic mutations of the β-catenin gene are frequent in mouse and human hepatocellular carcinomas. *Proc Natl Acad Sci USA*, **95**, 8847–51.

Dick, J. E. (2008). Stem cell concepts renew cancer research. *Blood*, **112**, 4793–807.

Diehn, M. (2009). Association of reactive oxygen species levels and radioresistance in cancer stem cells. *Nature*, **458**, 780–3.

Eppert, K., Takenaka, K., Lechman, E. R., et al. (2011). Stem cell gene expression programs influence clinical outcome in human leukaemia. *Nat Med*, **19**(9), 1086–93.

Eramo, A., Lotti, F., Sette, G., et al. (2008). Identification and expansion of the tumorigenic lung cancer stem cell population. *Cell Death Differ*, **15**, 504–14.

Essers, M. A. G., Offner, S., Blanco-Rose, W. E., et al. (2009). IFNα activates dormant haematopoietic stem cells *in vivo*. *Nature*, **458**, 904–9.

Fathi, A. T., Sadrzadeh, H., Comander, A. H., et al. (2014). Isocitrate dehydrogenase 1 (IDH1) mutation in breast adenocarcinoma is associated with elevated levels of serum and urine 2-hydroxyglutarate. *Oncologist*, **19**(6), 602–7.

Frank, N. Y., Margaryan, A., Huang, Y., et al. (2005). ABCB5-mediated doxorubicin transport and chemoresistance in human malignant melanoma. *Cancer Res*, **65**, 4320–33.

Freeman, S. D., Virgo, P., Couzens, S., et al. (2013). Prognostic relevance of treatment response measured by flow cytometric residual disease detection in older patients with acute myeloid leukemia. *J Clin Oncol*, **31**(32), 4123–31.

Fuster, J. J., MacLauchlan, S, Zuriaga, M. A., et al. (2017). Clonal hematopoiesis associated with TET2 deficiency accelerates atherosclerosis development in mice. *Science*, **355**, 842–7.

Genovese, G., Kähler, A. K., Handsaker, R. E., et al. (2014). Clonal hematopoiesis and blood-cancer risk inferred from blood DNA sequence. *N Engl J Med*, **371**(26), 2477–87.

Gentles, A. J., Plevritis, S. K., Majeti, R., et al. (2010). Association of a leukemic stem cell gene expression signature with clinical outcomes in acute myeloid leukemia. *JAMA*, **304**(24), 2706–15.

Gerber, J. M., Smith, B. D., Ngwang, B., et al. (2012). A clinically relevant population of leukemic CD34(+)CD38(-) cells in acute myeloid leukemia. *Blood*, **119**(15), 3571–7.

Gerlinger, M., Rowan, A. J., Horswell, S., et al. (2012). Intratumor heterogeneity and branched evolution revealed by multiregion sequencing. *N Eng J Med*, **366**(10), 883–92.

Ginjala, V., Nacerddine, K., Kulkarni, A., et al. (2011). BMI1 is recruited to DNA breaks and contributes to DNA damage-induced H2A ubiquitination and repair. *Mol Cell Biol*, **31**(10), 1972–82.

Girardi, T., Vicente, C., Cools, J., De Keersmaecker, K. (2017). The genetics and molecular biology of T-ALL. *Blood*, **129**, 1113–23.

Goardon, N., Marchi, E., Atzberger, A., et al. (2011). Coexistence of LMPP-like and GMP-like leukemia stem cells in acute myeloid leukemia. *Cancer Cell*, **19**, 138–52.

Gordon, W. R., Vardar-Ulu, D., Histen, G., et al. (2007). Structural basis for autoinhibition of Notch. *Nat Struct Mol Biol*, **14**(4), 295–300.

Govindan, R., Ding, L., Griffith, M., et al. (2012). Genomic landscape of non-small cell lung cancer in smokers and never-smokers. *Cell*, **150**, 1121–34.

Graham, S. M., Jørgensen, H. G., Allan, E., et al. (2002). Primitive, quiescent, Philadelphia-positive stem cells from patients with chronic myeloid leukemia are insensitive to STI571 *in vitro*. *Blood*, **99**(1), 319–25.

Green, A. & Beer, P. (2010). Somatic mutations of IDH1 and IDH2 in the leukemic transformation of myeloproliferative neoplasms. *NEJM*, **362**(4), 369–70.

Greenblatt, S., Li, L., Slape, C., et al. (2012). Knock-in of a FLT3/ITD mutation cooperates with a NUP98-HOXD13 fusion to generate acute myeloid leukemia in a mouse model. *Blood*, **119**(12), 2883–94.

Grimwade, D., Jovanovic, J. V., Hills, R. K., et al. (2009). Prospective minimal residual disease monitoring to predict relapse of acute promyelocytic leukemia and to direct pre-emptive arsenic trioxide therapy. *J Clin Oncol*, **27**(22), 3650–8.

Gupta, P. B., Onder, T. T., Jiang, G., et al. (2009). Identification of selective inhibitors of cancer stem cells by high-throughput screening. *Cell*, **138**, 645–59.

Guzman, M. L., Rossi, R. M., Neelakantan, S., et al. (2007a). An orally bioavailable parthenolide analog selectively eradicates acute myelogenous leukemia stem and progenitor cells. *Blood*, **110**, 4427–35.

Guzman, M. L., Li, X., Corbett, C. A., et al. (2007b). Rapid and selective death of leukemia stem and progenitor cells induced by the compound 4-benzyl, 2-methyl, 1,2,4-thiadiazolidine, 3,5 dione (TDZD-8). *Blood*, **110**, 4436–44.

Hahn, H., Wicking, C., Zaphiropoulos, P. G., et al. (1996). Mutations of the human homolog of Drosophila *patched* in the nevoid basal cell carcinoma syndrome. *Cell*, **85**, 841–51.

Hanahan, D. & Weinberg, R. A. (2011). Hallmarks of cancer: the next generation. *Cell*, **144**, 646–74.

Hanahan, D. & Coussens, L. M. (2012). Accessories to the crime, functions of cells recruited to the tumor microenvironment. *Cancer Cell*, **21**, 309–22.

Heddleston, J. M., Li, Z., Lathia, J. D., et al. (2010). Hypoxia inducible factors in cancer stem cells. *Br J Cancer*, **102**, 789–95.

Henke, M., Laszig, R, Rübe, C., et al. (2003). Erythropoietin to treat head and neck cancer patients with anaemia undergoing radiotherapy: randomised, double-blind, placebo-controlled trial. *Lancet*, **362**, 1255–60.

Heppner, G. H. (1984). Tumor heterogeneity. *Cancer Res*, **44**, 2259–65.

Hermann, P. C., Huber, S. L., Herrler, T., et al. (2007). Distinct populations of cancer stem cells determine tumor growth and metastatic activity in human pancreatic cancer. *Cell Stem Cell*, **1**, 313–23.

Ho, P. A., Alonzo, T. A., Kopecky, K. J., et al. (2010). Molecular alterations of the IDH1 gene in AML: a children's oncology group and southwest oncology group study. *Leukemia*, **24**(5), 909–13.

Ho, T-C., LaMere, M., Stevens, B. M., et al. (2016). Evolution of acute myelogenous leukemia stem cell properties after treatment and progression. *Blood*, **128**(13), 1671–8.

Ichimura, K. (2012). Molecular pathogenesis of IDH mutations in gliomas. *Brain Tumor Pathol*, **29**(3), 131–9.

Ishikawa, F., Yoshida, S., Saito, Y., et al. (2007). Chemotherapy-resistant human AML stem cells home to and engraft within the bone-marrow endosteal region. *Nat Biotechnol*, **25**(11), 1315–21.

Ito, M., Hiramatsu, H., Kobayashi, K., et al. (2002). NOD/SCID/γ^{null} mouse: an excellent recipient mouse model for engraftment of human cells. *Blood*, **100**(9), 3175–82.

Ivey, A., Hills, R. K., Simpson, M. A., et al. (2016). Assessment of minimal residual disease in standard-risk AML. *N Engl J Med*, **374**(5), 422–33.

Iwama, A., Oguro, H., Negishi, M., et al. (2004). Enhanced self-renewal of hematopoietic stem cells mediated by the polycomb gene product Bmi-1. *Immunity*, **21**, 843–51.

Jaiswal, S., Fontanillas, P., Flannick, J., et al. (2014). Age-related clonal hematopoiesis associated with adverse outcomes. *N Engl J Med*, **371**(26), 2488–98.

Jamieson, C. H. M., Ailles, L. E., Dylla, S. J., et al. (2004). Granulocyte-macrophage progenitors as candidate leukemic stem cells in blast-crisis CML. *N Engl J Med*, **351**(7), 657–67.

Jan, M., Snyder, T. M., Corces-Zimmerman, M. R., et al. (2012). Clonal evolution of preleukemic hematopoietic stem cells precedes human acute myeloid leukemia. *Sci Transl Med*, **4**, 149ra118.

Johnson, R. L., Rothman, A. L., Xie, J., et al. (1996). Human homolog of *patched*, a candidate gene for the basal cell nevus syndrome. *Science*, **272**, 1668–71.

Junttila, M. R. & de Sauvage, F. J. (2013). Influence of tumor microenvironment heterogeneity on therapeutic response. *Nature*, **501**, 346–54.

Kamminga, L. M., Bystrykh, L. V., De Boer, A., et al. (2006). The polycomb group gene *Ezh2* prevents hematopoietic stem cell exhaustion. *Blood*, **107**(5), 2170–9.

Kang, M. R., Kim, M. S., Oh, J. E., et al. (2009). Mutational analysis of IDH1 codon 132 in glioblastomas and other common cancers. *Int J Cancer*, **125**(2), 353–5.

Kats, L. M., Reschke, M., Taulli, R., et al. (2014). Proto-oncogenic role of mutant *IDH2* in leukemia initiation and maintenance. *Cell Stem Cell*, **14**, 329–41.

Kersting, M., Friedl, C., Kraus, A., et al. (2000). Differential frequencies of $p16^{INK4a}$ promoter hypermethylation, *p53* mutation and K-*ras* mutation in exfoliative material mark the development of lung cancer in symptomatic chronic smokers. *J Clin Oncol*, **18**, 3221–9.

Kiel, M. J., Yilmaz, Ö. H., Iwashita, T., et al. (2005). SLAM family receptors distinguish hematopoietic stem and progenitor cells and reveal endothelial niches for stem cells. *Cell*, **121**, 1109–21.

Kim, J. H., Yoon, S. Y., Kim, C. N., et al. (2004a). The Bmi-1 oncoprotein is overexpressed in human colorectal cancer and correlates with the reduced p16INK4a/p14ARF proteins. *Cancer Lett*, **203**, 217–24.

Kim, J. H., Yoon, S. Y., Jeong, S. Y., et al. (2004b). Overexpression of Bmi-1 oncoprotein correlates with axillary lymph node metastases in invasive ductal breast cancer. *Breast*, **13**, 383–8.

Kim, M. S., Kim, S. S., Ahn, C. H., Yoo, N. J., & Lee, S. H. (2009). Frameshift mutations of Wnt pathway genes *AXIN2* and *TCF7L2* in gastric carcinomas with high microsatellite instability. *Hum Pathol*, **40**(1), 58–64.

Kinstrie, R., Karamitros, D., Goardon, N., et al. (2016). Heterogeneous leukemia stem cells in myeloid blast phase chronic myeloid leukemia. *Blood Advances*, **1**, 160–9.

Knudson, A. G. (1971). Mutation and cancer: statistical study of retinoblastoma. *Proc Natl Acad Sci U S A*, **68**(4), 820–3.

Koesters, R., Ridder, R., Kopp-Schneider, A., et al. (1999). Mutational activation of the β-catenin proto-oncogene is a common event in the development of Wilms' tumors. *Cancer Res*, **59**, 3880–2.

Kreso, A., O'Brien, C. A., Van Galen, P., et al. (2013). Variable clonal repopulation dynamics influence chemotherapy response in colorectal cancer. *Science*, **339**, 543–8.

Kreso, A. & Dick, J. E. (2014a). Evolution of the cancer stem cell model. *Cell Stem Cell*, **14**(3), 275–91.

Kreso, A., van Galen, P., Pedley, N. M., et al. (2014b). Self-renewal as a therapeutic target in human colorectal cancer. *Nat Med*, **20**, 29–36.

Lapidot, T., Sirard, C., Vormoor, J., et al. (1994). A cell initiating human acute myeloid leukaemia after transplantation into SCID mice. *Nature*, **367**, 645–8.

Le Viseur, C., Hotfilder, M., Bomken, S., et al. (2008). In childhood acute lymphoblastic leukemia, blasts at different stages of immunophenotypic maturation have stem cell properties. *Cancer Cell*, **14**, 47–58.

Lee, K., Adhikary, G., Balasubramanian, S., et al. (2008). Expression of Bmi-1 in epidermis enhances cell survival by altering cell cycle regulatory protein expression and inhibiting apoptosis. *J Invest Dermatol*, **128**, 9–17.

Lessard, J. & Sauvageau, G. (2003). *Bmi-1* determines the proliferative capacity of normal and leukaemic stem cells. *Nature*, **423**, 255–60.

Ley, T. J., Ding, L., Walter, M. J., et al. (2010). *DNMT3A* mutations in acute myeloid leukemia. *N Engl J Med*, **363**(25), 2424–33.

Li, Z., Cai, X., Cai, C. L., et al. (2011). Deletion of Tet2 in mice leads to dysregulated hematopoietic stem cells and subsequent development of myeloid malignancies. *Blood*, **118**(17), 4509–18.

Li, Z., Cao, R., Wang, M., et al. (2006). Structure of a Bmi1-Ring1B polycomb group ubiquitin ligase complex. *J Biol Chem*, **281**(29), 20643–9.

Liu, S., Dontu, G., Mantle, I. D., et al. (2006). Hedgehog signaling and Bmi-1 regulate self-renewal of normal and malignant human mammary stem cells. *Cancer Res*, **66**(12), 6063–71.

Lombardo, Y., Scopelliti, A., Cammareri, P., et al. (2011). Bone morphogenetic protein 4 induces differentiation of colorectal cancer stem cells and increases their response to chemotherapy in mice. *Gastroenterology*, **140**, 297–309.

Lu, D., Choi, M. Y., Yu, J., et al. (2011). Salinomycin inhibits Wnt signaling and selectively induces apoptosis in chronic lymphocytic leukemia cells. *Proc Natl Acad Sci U S A*, **108**, 13253–7.

Lund, K., Adams, P. D., & Copland, M. (2014). EZH2 in normal and malignant hematopoiesis. *Leukemia*, **28**, 44–9.

Ma, Y., Liang, D., Liu, J., et al. (2011). Prostate cancer cell lines under hypoxia exhibit greater stem-like properties. *PLoS One*, **6**, e29170.

Mahon, F. X., Ré, D., Guilhot, F., et al. (2010). Discontinuation of imatinib in patients with chronic myeloid leukaemia who have maintained complete molecular remission for at least 2 years: the prospective, multicenter Stop Imatinib (STIM) trial. *Lancet Oncol*, **11**(11), 1029–35.

Majeti, R., Becker, M. W., Tian, Q., et al. (2009). Dysregulated gene expression networks in human acute myelogenous leukemia stem cells. *Proc Natl Acad Sci U S A*, **106**(9), 3396–401.

Majeti, R. (2011). Monoclonal antibody therapy directed against human acute myeloid leukemia stem cells. *Oncogene*, **30**(9), 1009–19.

Malanchi, I., Peinado, H., Kassen, D., et al. (2008). Cutaneous cancer stem cell maintenance is dependent on β-catenin signalling. *Nature*, **452**, 650–4.

Martelli, M. P., Pettirossi, V., Thiede, C., et al. (2010). CD34 + cells from AML with mutated NPM1 harbor cytoplasmic mutated nucleophosmin and generate leukemia in immunocompromised mice. *Blood*, **116**(19), 3907–22.

McIntosh, B. E., Brown, M. E., Duffin, B. M., et al. (2015). Nonirradiated NOD, B6. SCIDIl2rγ-/-Kit(W41/W41) (NBSGW) mice support multilineage engraftment of human hematopoietic cells. *Stem Cell Reports*, **4**(2), 171–80.

Murugan, A. K., Bojdani, E., & Xing, M. (2010). Identification and functional characterization of isocitrate dehydrogenase 1 (IDH1) mutations in thyroid cancer. *Biochem Biophys Res Commun*, **393**(3), 555–9.

Ng, S. W. K., Mitchell, A., Kennedy, J. A., et al. (2016). A 17-gene stemness score for rapid determination of risk in acute leukaemia. *Nature*, **540**, 433–7.

Nik-Zainal, S., Van Loo, P., Wedge, D. C., et al. (2012). The life history of 21 breast cancers. *Cell*, **149**, 994–1007.

Nishida, Y., Maeda, A., Chachad, D., et al. (2015). Preclinical activity of the novel B-cell-specific Moloney murine leukemia virus integration site 1 inhibitor PTC-209 in acute myeloid leukemia: implications for leukemia therapy *Cancer Sci*, **106**(12), 1705–13.

Notta, F., Mullighan, C. G., Wang, J. C., et al. (2011). Evolution of human BCR-ABL1 lymphoblastic leukaemia-initiating cells. *Nature*, **469**, 362–7.

Nowell, P. C. (1976). The clonal evolution of tumor cell populations. *Science*, **194**, 23–28.

O'Brien, C. A., Pollett, A., Gallinger, S., et al. (2007). A human colon cancer cell capable of initiating tumour growth in immunodeficient mice. *Nature*, **445**, 106–10.

O'Day, S. J., Eggermont, A. M., Chiarion-Sileni, V., et al. (2013). Final results of phase III SYMMETRY study: randomized, double-blind trial of elesclomol plus paclitaxel versus paclitaxel alone as treatment for chemotherapy-naive patients with advanced melanoma. *J Clin Oncol*, **31**(9), 1211–8.

Papaemmanuil, E., Gerstung, M., Bullinger, L., et al. (2016). Genomic classification and prognosis in acute myeloid leukemia. *N Engl J Med*, **374**, 2209–21.

Pearce, D. J., Taussig, D., Zibara, K., et al. (2006). AML engraftment in the NOD/SCID assay reflects the outcome of AML: implications for our understanding of the heterogeneity of AML. *Blood*, **107**(3), 1166–73.

Pflumio, F., Izac, B., Katz, A., et al. (1996). Phenotype and function of human hematopoietic cells engrafting immune-deficient CB17-severe combined immunodeficiency mice after transplantation of human cord blood mononuclear cells. *Blood*, **8**(10), 3731–40.

Piccirillo, S. G., Reynolds, B. A., Zanetti, N., et al. (2006). Bone morphogenetic proteins inhibit the tumorigenic potential of human brain tumour-initiating cells. *Nature*, **444**, 761–5.

Pierce, G. B., Dixon, F. J., & Verney, E. L. (1960). Teratocarcinogenic and tissue-forming potentials of the cell types comprising neoplastic embryoid bodies. *Lab Invest*, **9**, 583–602.

Poitras, J. L., Heiser, D., Li, L., et al. (2016). *Dnmt3a* deletion cooperates with the *Flt3/ITD* mutation to drive leukemogenesis in a murine model. *Oncotarget*, **7**(43), 69124–35.

Prince, M. E., Sivanandan, R., Kaczorowski, A., et al. (2007). Identification of a subpopulation of cells with cancer stem cell properties in head and neck squamous cell carcinoma. *Proc Natl Acad Sci U S A*, **104**, 973–8.

Purton, L. E. & Scadden, D. T. (2007). Limiting factors in murine hematopoietic stem cell assays. *Cell Stem Cell*, **1**(3), 263–70.

Quek, L., Otto, G. W., Garnett, C., et al. (2016). Genetically distinct leukemic stem cells in human CD34—acute myeloid leukemia are arrested at a hematopoietic precursor-like stage. *J Exp Med*, **213**(8), 1513–35.

Quintana, E., Shackleton, M., Foster, H. R., et al. (2010). Phenotypic heterogeneity among tumorigenic melanoma cells from patients that is reversible and not hierarchically organized. *Cancer Cell*, **18**, 510–23.

Rehe, K., Wilson, K., Bomken, S., et al. (2013). Acute B lymphoblastic leukaemia-propagating cells are present at high frequency in diverse lymphoblast populations. *EMBO Mol Med*, **5**(1), 38–51.

Reinisch, A., Thomas, D., Corces, M. R., et al. (2016). A humanized bone marrow ossicle xenotransplantation model enables improved engraftment of healthy and leukemic human hematopoietic cells. *Nat Med*, **22**(7), 812–21.

Ricci-Vitiani, L., Lombardi, D. G., Pilozzi, E., et al. (2007). Identification and expansion of human colon-cancer-initiating cells. *Nature*, **445**, 111–15.

Riechelmann, H., Sauter, A., Golze, W., et al. (2008). Phase I trial with the CD44v6-targeting immunoconjugate bivatuzumab mertansine in head and neck squamous cell carcinoma. *Oral Oncol*, **44**, 823–9.

Rongvaux, A., Willinger, T., Martinek, J., et al. (2014). Development and function of human innate immune cells in a humanized mouse model. *Nat Biotechnol*, **32**(4), 364–72.

Saito, Y., Uchida, N., Tanaka, S., et al. (2010). Induction of cell cycle entry eliminates human leukemia stem cells in a mouse model of AML. *Nat Biotechnol*, **28**, 275–80.

Sanchez, P. V., Perry, R. L., Sarry, J. E., et al. (2009). A robust xenotransplantation model for acute myeloid leukemia. *Leukemia*, **23**(11), 2109–17.

Sanchez-Martin, M. & Ferrando, A. (2017). The NOTCH1-MYC highway towards T-cell acute lymphoblastic leukemia. *Blood*, **129**, 1124–33.

Sarry, J. E., Murphy, K., Perry, R., et al. (2011). Human acute myelogenous leukemia stem cells are rare and heterogeneous when assayed in NOD/SCID/IL2Rγc-deficient mice. *J Clin Invest*, **121**(1), 384–95.

Saudy, N. S., Fawzy, I. M., Azmy, E., et al. (2014). BMI1 gene expression in myeloid leukemias and its impact on prognosis. *Blood Cells Mol Dis*, **53**, 194–8.

Schatton, T., Murphy, G. F., Frank, N. Y., et al. (2008). Identification of cells initiating human melanomas. *Nature*, **451**, 345–9.

Shah, M. Y. & Licht, J. D. (2011). *DNMT3A* mutations in acute myeloid leukemia. *Nat Genet*, **43**(4), 289–90.

Sharma, S. V., Lee, D. Y., Li, D., et al. (2010). A chromatin-mediated reversible drug-tolerant state in cancer cell subpopulations. *Cell*, **141**, 69–80.

Shlush, L. I., Zandi, S., Mitchell, A., et al. (2014). Identification of pre-leukaemic haematopoietic stem cells in acute leukaemia. *Nature*, **506**, 328–33.

Singh, S. K., Hawkins, C., Clarke, I. D., et al. (2004). Identification of human brain tumour initiating cells. *Nature*, **432**, 396–401.

Spangrude, G. J., Heimfeld, S., Weissman, I. L. (1988). Purification and characterisation of mouse haematopoietic stem cells. *Science*, **241**, 58–62.

Spencer, D. H., Young, M. A., Lamprecht, T. L., et al. (2015). Epigenomic analysis of the HOX gene loci reveals mechanisms that may control canonical expression patterns in AML and normal hematopoietic cells. *Leukemia*, **29**, 1279–89.

Sportoletti, P., Varasano, E., Rossi, R., et al. (2015). Mouse models of NPM1-mutated acute myeloid leukemia, biological and clinical implications. *Leukemia*, **29**, 269–78.

Steensma, D. P., Bejar, R., Jaiswal, S., et al. (2015). Clonal hematopoiesis of indeterminate potential and distinction from myelodysplastic syndromes. *Blood*, **126**, 9–16.

Tadokoro, Y., Ema, H., Okano, M., Li, E., & Nakauchi, H. (2007). *De novo* DNA methyltransferase is essential for self-renewal, but not for differentiation, in hematopoietic stem cells. *J Exp Med*, **204**(4), 715–22.

Takahashi, N., Miura, I., Saitoh, K., & Miura, A. B. (1998). Lineage involvement of stem cells bearing the Philadelphia chromosome in chronic myeloid leukemia in the chronic phase as shown by a

combination of fluorescence-activated cell sorting and fluorescence *in situ* hybridization. *Blood*, **92**(12), 4758–63.
Takebe, N., Harris, P. J., Warren, R. Q., & Ivy, S. P. (2011). Targeting cancer stem cells by inhibiting Wnt, Notch and Hedgehog pathways. *Nat Rev Clin Oncol*, **8**(2), 97–106.
Tan, Y. T., Sun, Y., Zhu, S. H., et al. (2016). Deregulation of HOX genes by DNMT3A and MLL mutations converges on BMI1. *Leukemia*, **30**, 1609–12.
Taussig, D. C., Miraki-Moud, F., Anjos-Afonso, F., et al. (2008). Anti-CD38 antibody-mediated clearance of human repopulating cells masks the heterogeneity of leukemia-initiating cells. *Blood*, **112**(3), 568–75.
Taussig, D. C., Vargaftig, J., Miraki-Moud, F., et al. (2010). Leukemia-initiating cells from some acute myeloid leukemia patients with mutated nucleophosmin reside in the CD34(-) fraction. *Blood*, **115**(10), 1976–84.
Tehranchi, R., Woll, P. S., Anderson, K., et al. (2010). Persistent malignant stem cells in del(5q) myelodysplasia in remission. *N Engl J Med*, **363**(11), 1025–37.
Terwijn, M., Zeijlemaker, W., Kelder, A., et al. (2014). Leukemic stem cell frequency: a strong biomarker for clinical outcome in acute myeloid leukemia. *PLoS One*, **9**(9), e107587.
Till, J. E. & McCullock, E. A. (1961). A direct measurement of the radiation sensitivity of normal mouse bone marrow cells. *Radiat Res*, **14**(2), 213–22.
Trachootham, D., Alexandre, J., & Huang, P. (2009). Targeting cancer cells by ROS-mediated mechanisms: a radical therapeutic approach? *Nat Rev Drug Discov*, **8**, 579–91.
Tsai, L. L., Yu, C. C., Chang, Y. C., et al. (2011). Markedly increased Oct4 and Nanog expression correlates with cisplatin resistance in oral squamous cell carcinoma. *J Oral Pathol Med*, **40**, 621–8.
Viale, A., De Franco, F., Orleth, A., et al. (2010). Cell-cycle restriction limits DNA damage and maintains self-renewal of leukaemia stem cells. *Nature*, **457**, 51–6.
Vogelstein, B., Fearon, E. R., Hamilton, S. R., et al. (1988). Genetic alterations during colorectal-tumor development. *N Engl J Med*, **319**(9), 525–32.
Vogelstein, B., Papadopoulos, N., Velculescu, V. E., et al. (2013). Cancer genome landscapes. *Science*, **339**, 1546–58.
Vonlanthen, S., Heighway, J., Altermatt, H. J., et al. (2001). The bmi-1 oncoprotein is differentially expressed in non-small cell lung cancer and correlates with INK4A-ARF locus expression. *Br J Cancer*, **84**, 1372–6.
Vrzalikova, K., Skarda, J., Ehrmann, J., et al. (2008). Prognostic value of Bmi-1 oncoprotein expression in NSCLC patients: a tissue microarray study. *J Cancer Res Clin Oncol*, **134**, 1037–42.
Wang, X. Q., Ongkeko, W. M., Chen, L., et al. (2010). Octamer 4 (Oct4) mediates chemotherapeutic drug resistance in liver cancer cells through a potential Oct4-AKT-ATP-binding cassette G2 pathway. *Hepatology*, **52**, 528–53.
Wang, Y. (2011). Effects of salinomycin on cancer stem cell in human lung adenocarcinoma A549 cells. *Med Chem*, **7**, 106–11.
Watanabe, T., Nobusawa, S., Kleihues, P., & Ohgaki, H. (2009). IDH1 mutations are early events in the development of astrocytomas and oligodendrogliomas. *Am J Pathol*, **174**(4), 1149–53.
Weiskopf, K., Jahchan, N. S., Schnorr, P. J., et al. (2016). CD47-blocking immunotherapies stimulate macrophage-mediated destruction of small-cell lung cancer. *J Clin Invest*, **126**(7), 2610–20.
Weng, A. P., Ferrando, A. A., Lee, W., et al. (2004). Activating mutations of NOTCH1 in human T cell acute lymphoblastic leukemia. *Science*, **306**, 269–71.
Wilson, A., Laurenti, E., Oser, G., et al. (2008). Hematopoietic stem cells reversibly switch from dormancy to self-renewal during homeostasis and repair. *Cell*, **135**, 1118–29.
Wong, P., Iwasaki, M., Somervaille, T. C., So, C. W., & Cleary, M. L. (2007). Meis1 is an essential and rate-limiting regulator of MLL leukemia stem cell potential. *Genes Dev*, **21**, 2762–74.
Woodward, W. A., Bristow, R. G., Clarke, M. F., et al. (2009). Radiation therapy oncology group translational research program stem cell symposium: incorporating stem cell hypotheses into clinical trials. *Int J Radiat Oncol Biol Phys*, **74**, 1580–91.
Xie, M., Lu, C., Wang, J., et al. (2014). Age-related mutations associated with clonal hematopoietic expansion and malignancies. *Nat Med*, **20**, 1472–78.
Yachida, S., Jones, S., Bozic, I., et al. (2010). Distant metastasis occurs late during the genetic evolution of pancreatic cancer. *Nature*, **467**, 1114–17.
Yan, H., Parsons, D. W., Jin, G., et al. (2009). IDH1 and IDH2 mutations in gliomas. *N Engl J Med*, **360**(8), 765–73.
Yong, K. J., Basseres, D. S., Welner, R. S., et al. (2016). Targeted BMI1 inhibition impairs tumor growth in lung adenocarcinomas with low CEBPα expression. *Sci Transl Med*, **8**(350), 350ra104.
Zhang, M., Hutter, G., Kahn, S. A., et al. (2016). Anti-CD47 treatment stimulates phagocytosis of glioblastoma by M1 and M2 polarized macrophages and promotes M1 polarized macrophages *in vivo*. *PLoS*, **11**(4), e0153550.
Zhao, C., Chen, A., Jamieson, C. H., et al. (2009). Hedgehog signalling is essential for maintenance of cancer stem cells in myeloid leukaemia. *Nature*, **458**, 776–80.
Zong, H., Sen, S., Zhang, G., et al. (2016). *In vivo* targeting of leukemia stem cells by directing parthenolide-loaded nanoparticles to the bone marrow niche. *Leukemia*, **30**(7), 1582–6.

SECTION IV

Cancer microenvironment

21

Cancer-associated stroma

Wilma Mesker and Rob Tollenaar

Introduction to cancer-associated stroma

For many years, tumour stromal formation or desmoplasia was considered to be a passive bystander of tumorigenesis and tumour progression. However, during the last 10 years, attention has shifted towards the tumour microenvironment and it is now well established that tumour stroma plays an important role in cancer initiation and progression. The stroma interacts with non-malignant cells as well as with malignant cells at different stages of tumorigenesis ranging from tumour onset to invasion and metastasis (Park et al., 2000). This complex crosstalk between tumour cells and microenvironmental cells leads to tumour migration, invasion, angiogenesis, and metastasis (Hanahan and Weinberg, 2011). See also **Figure 21.1**. The tumour stroma consists of different cell types, including cancer-associated fibroblasts (CAFs), extracellular matrix (ECM), immune cells and blood vessel cells, and is characterized by a 'reactive' phenotype. Activated fibroblasts are called CAFs in tumour stroma and myofibroblasts in healthy tissue. CAFs play an important role in tumour progression by modifying the ECM so that tumour cells can evade the primary tumour and metastasize to distant organs (Mueller and Fusenig, 2004; Prud'homme, 2007). TGF-beta and PDGF are the two main pathways, which are involved in stromal conversion. Tumour cells induced by TGF-beta loosen their intimate cell-cell contacts and acquire mesenchymal properties. During wound healing, fibroblasts are activated through the same TGF-beta process suggesting that epithelial cells can also convert into fibroblast-like cells (Siegel and Massague, 2003). See also **Figure 21.2**.

Apart from tumour promoting properties, the tumour stroma is also able to inhibit tumour growth. The heterogeneity of the stroma and particularly of CAFs may be related to the contradictory functions of the tumour stroma.

In addition, tumour stroma is relevant in the clinic. It has been confirmed that a high tumour stroma level was associated with a bad patient prognosis in different cancer types.

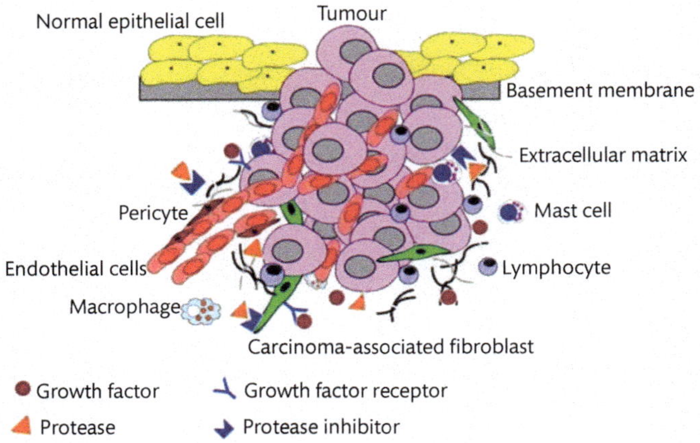

Fig. 21.1 The tumour microenvironment (TME). TME comprises different stromal cells in addition to tumour cells. These include vascular or lymphatic endothelial cells, supporting pericytes, fibroblasts, and both innate and adaptive infiltrating immune cells. Moreover, TME contains non-cellular components, including extracellular matrixes, growth factors, proteases, protease inhibitors, and other signalling molecules that play important roles in stromal reactions in TME.

Fibroblasts and cancer-associated fibroblasts

Fibroblasts present in healthy tissue have a spindle-like morphology and once they become myofibroblasts, they play a crucial role during development, wound healing, and inflammation by contributing to the synthesis and the degradation of ECM components (Marsh et al., 2013). Fibroblasts found in tumours, called CAFs are different from myofibroblasts in healthy tissue; CAFs show an increased proliferation, release more growth factors and ECM components, and have a tumour promoting effect (Ohlund et al., 2014). Although CAFs have a crucial effect on tumorigenesis, they have not yet been accurately defined. No exclusive marker characterizes CAFs solely suggesting that CAFs are a heterogeneous cell population. Cortez et al. (2014) divided CAFs into four different subpopulations based on the following markers: fibroblast activation protein (FAP), fibroblast specific protein (FSP1/S100A4) and platelet-derived growth factor receptor-α and -β (PDGFR-α and PDGFR-β: see Cortez et al., 2014). See also **Figure 21.3**.

Another important marker of CAFs as well as myofibroblasts is α-smooth muscle actin (α-SMA). Fibroblasts residing in the tumour and healthy tissue are stimulated by the growth factor

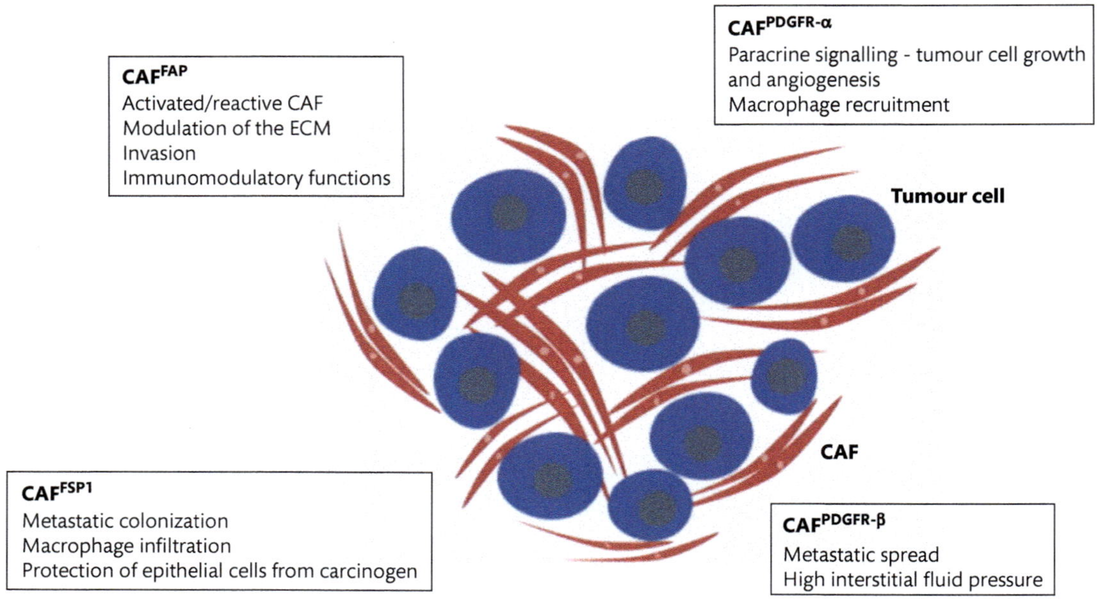

Fig. 21.2 Functional subsets of cancer-associated fibroblasts (CAF). Several studies show the existence of different subpopulations of CAF in the tumour microenvironment based on marker expression and functional analysis using different mouse models of cancer and human tumour material. The importance of CAF for tumour development and metastatic dissemination has been widely investigated. Here we summarize the current understanding on CAF subsets and their role in tumour development.

TGF-beta to activate the TGF-beta/Smad signalling pathway and to transdifferentiate into a α-SMA expressing fibroblasts (Hawinkels et al., 2014). CAFs are also denoted by the expression of many other markers like vimentin, thrombospondin-1 (TSP-1), tenascin-C, periostin, osteonectin, paladin, podoplanin, and stromelysin.

Resident fibroblasts and fibroblast precursors stimulated by TGF-β are generally considered to be the prominent source of myofibroblasts. However, CAFs are likely also originating from other sources such as smooth muscle cells and differentiation of adipocytes (Karnoub et al., 2007; Bochet et al., 2013). Most studies

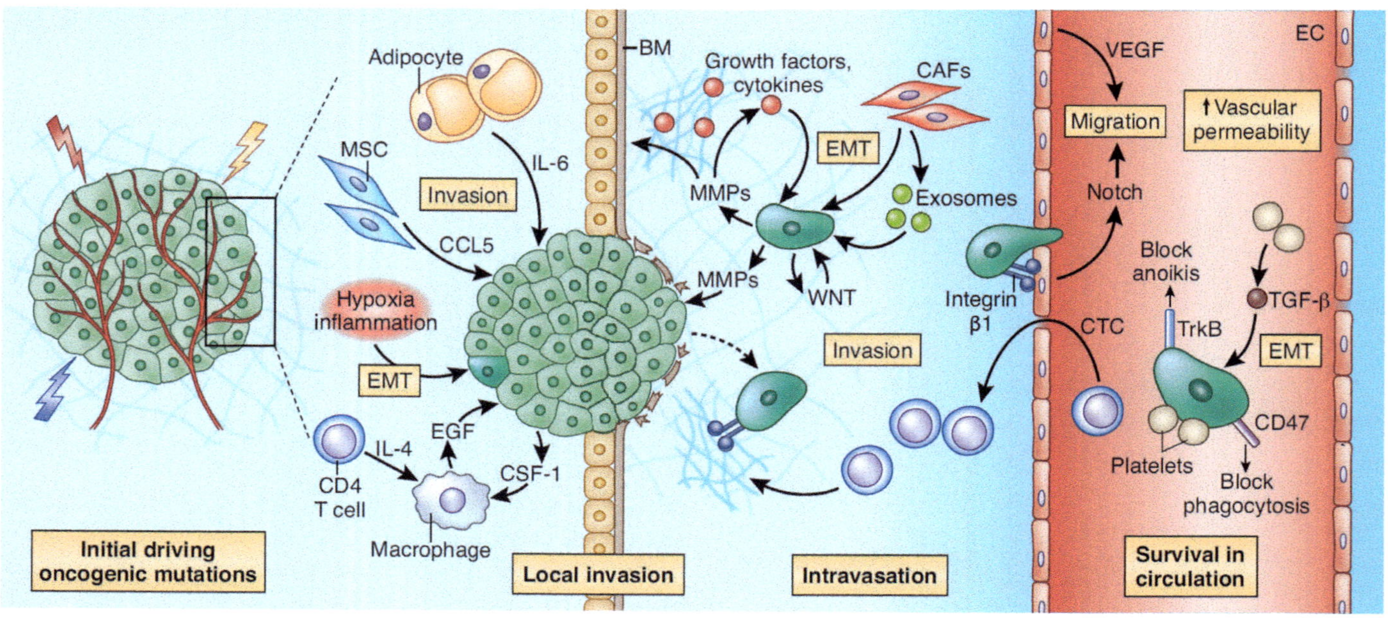

Fig. 21.3 The early steps of metastasis: tumour invasion, dissemination, and survival in the circulation. To disseminate, carcinoma cells have to acquire the capabilities to break down the basement membrane, invade into the stroma (local invasion), enter the blood circulation (intravasation) and manage to survive before they can arrest at a distant organ and grow into clinically detectable metastases. In addition to cell-autonomous mechanisms, carcinoma cells enlist a myriad of stromal cells to aid in each step during this invasion-dissemination cascade. Selective examples of tumour-stroma crosstalk are illustrated, with detailed molecular mechanisms described in the text. EC, endothelial cells.

therefore refer to CAFs as activated fibroblasts originating from different sources. Intriguingly, epithelial cancer cells themselves can undergo epithelial-mesenchymal transition (EMT) and become fibroblasts due to their mesenchymal phenotype. Mesenchymal stem cells (MSCs) have also been shown to be a precursor of CAFs. MSCs are non-haematopoietic self-renewing multipotent cells that are able to differentiate into lineage of bone, cartilage, and fat tissues (Pittenger et al., 1999). The function of MSCs in the tumour is likely to be similar to its function in wound healing. In a wound, MSCs differentiate into myofibroblasts and pericytes and secrete cytokines to recruit other cell types. Human bone marrow-derived MSCs (hMSCs) cocultured with human breast cancer cells acquired *in vitro* a CAF-like myofibroblastic phenotype. Gene expression profiles confirmed these large similarities between hMSCs and CAFs (Mishra et al., 2008). Moreover, in an elegant *in vivo* experiment, MSCs were injected into the stroma of an orthotopic mouse model of colon cancer. The tumour stroma in the mouse showed expression of α-smooth muscle actin and platelet-derived growth factor receptor-β markers that correspond to CAF markers, suggesting that the injected MSCs differentiated into CAFs (Shinagawa et al., 2010).

CAFs orchestrate and educate other cells in the tumour area through the secretion of a cascade of growth factors, metalloproteases, cytokines, and proteins. For instance, CAFs contribute to angiogenesis by releasing desmin-1, fibroblast growth factor-2 (FGF-2), Vascular endothelial growth factor (VEGF) and angiopoietin-1 (Ang-1), and support tumour proliferation and progression by releasing epidermal growth factor (EGF), hepatocyte growth factor/scatter factor (HGF/SF), IL-6, FGF-2. CAFs express matrix-metalloproteinases-1 and -3 (MMP-1, MMP-3), produce collagens, and release cytokines (IL-6, IL-8) exerting tumour promoting activities and CAF-derived factors such as chemokines (CXCL12, CXCL14), recruit bone marrow-derived MSCs, macrophages, and other immune cells into the growing tumour (Orimo et al., 2005; Ostman and Augsten, 2009; Mishra et al., 2011).

The effect of cancer-associated fibroblasts on tumour progression

Interestingly, the function of CAFs in tumorigenesis remains contradictory. Some studies report that CAFs contribute to protumour properties while other studies show that CAFs have antitumorigenic activity (Klopp et al., 2011). The reason for this discrepancy remains unknown, but these differences might be attributed to the heterogeneous population of CAFs originating from different cell populations (see 'Fibroblasts and cancer-associated fibroblasts') and to the difference in models used in experiments (see 'Stromal models in cancer').

CAFs affect tumour cells directly or indirectly, through various different and complex mechanisms. The following mechanisms will be discussed: EMT (a); tumour cell proliferation and survival (b); immune system modulation (c); maintenance of stemness (d); angiogenesis (e); ECM remodelling (f); and stimulation to metastasize (g) (Harper and Sainson, 2014).

Epithelial-mesenchymal transition

Epithelial-mesenchymal transition (EMT) is the process in which epithelial cells lose their defining characteristics as motility and cell polarity and gain mesenchymal-like cell migration and invasion which promotes the metastatic cascade (Thiery, 2002). EMT has a central role in tumour progression such as reduced cell-cell and cell-matrix adhesion, reorganization of the cytoskeleton for migration, and degradation and remodelling of the ECM through matrix-degrading proteases (Guarino, 2007). By signals from the microenvironment, cells at the invasive front of the tumour undergo EMT, bud into the stroma and initiate metastatic lesion formation (Jiang et al., 2013). Dynamic interactions between tumour cells and constituents of the tumour stroma influence EMT induction with subsequent cancer progression and metastases.

Tumour cell proliferation and survival

As discussed previously, the tumour microenvironment contributes to tumour initiation. In a rat experiment, neoplastic transformation of epithelial cells induced mammary gland cancer only when the stroma was exposed to the chemical carcinogen N-nitrosomethylurea, regardless of whether or not the epithelial cells were exposed to the carcinogen (Maffini et al., 2004). CAFs stimulate tumour growth by secreting a cascade of chemokines and growth factors. The main chemokines are CXCL1, CXCL2, CXCL12, IL-6, IL-8 and TNF-beta and the main growth factors are HGF, insulin-like growth factor (IGF), and EGF. Inactivating the TGFBR II in FSP-1-expressing fibroblasts in a mouse model resulted in neoplastic epithelial cells of the prostate confirming the importance of the epithelial-mesenchymal interaction in tumour progression (Bhowmick et al., 2004). Moreover, CAFs can inhibit tumour cell apoptosis. This was observed in the experiment by Shinagawa et al. who injected colon cancer cells with and without MSCs and observed that the apoptosis level was higher in the colon cancer cells injected alone suggesting that MSCs contribute to tumour cell survival (Shinagawa et al., 2010). See also section on 'Stimulation to metastasize'.

Immune system modulation

The inflammatory immune environment can underlie tumour initiation and progression. Colitis-associated cancer, also called Crohn's disease, is a good example that illustrate inflammatory-induced cancer. Inflammatory cells influence cancer initiation via the release of cytokines, growth factors, and chemokines. Chronic inflammation cause for instance alterations in the cytokine level of the microenvironment, which contributes to the progression of cancer via MMPs.

Furthermore, immune cells and CAFs are in close crosstalk and participate in both an antitumour immune response as well as a protumour immune response. An important inflammatory cytokine produced by CAFs is IL-6 that promotes tumour cell survival and stimulates the production of VEGF.

At the moment, successful immunotherapies against cancer are being developed and launched. Targeting immune checkpoints might be an effective manner to boost the immune system of the patient. For instance, PD-L1 is expressed by some cancer cells to, when bound to PD-1 on immune cells, ward off the immune response (Herbst et al., 2016).

Maintenance of stemness

Normal stem cells (SCs) are self-renewing and have multilineage differentiation properties to maintain and repair adult tissue.

A tumour is thought to have a similar hierarchy as healthy tissue including cancer stem cells (CSCs) at the top. CSCs sustain tumour growth and result in relapse after therapy by giving rise to a new tumour. Tumours with high levels of CSCs are associated with bad prognosis (Bao et al., 2006). Like SCs residing in a 'SC niche', the tumour microenvironment ensures that CSCs remain in a stem-like state. The Wnt pathway and its main oncoprotein beta-catenin are major players in sustaining CSCs. In colon cancer, high activity of the Wnt pathway is observed in tumour cells, which are located in close proximity to stromal myofibroblasts, indicating that Wnt activity and cancer stemness is regulated by extrinsic cues (Vermeulen et al., 2010). Intriguingly, the microenvironment, in particular the growth factor HGF secreted by myofibroblasts, has the ability to dedifferentiate cancer cells into CSCs by reactivating the Wnt pathway (Biswas et al., 2015).

ECM remodelling

CAFs pro-tumorigenic properties can function through the remodelling of the ECM. CAFs together with other stromal cells secrete enzymes such as plasminogen activators and MMPs that degrade the ECM resulting in the release of growth factors (e.g. HGF, VEGF, and EGF). These growth factors released due to structural or mechanical modifications of the ECM lead to an indirect effect of CAFs on tumour growth, invasion, and dissemination of tumour cells. ECM modifications may also lead to a shift of non-malignant cells to malignant cells. Overexpression of ECM degrading enzymes in stroma (proteinase MMP3/stromelysin-1) of transgenic mice indeed induced the transformation of normal breast epithelium into cancerous tissue (Sternlicht et al., 1999).

MMP inhibitors (MMPi) have been developed to inhibit the function of MMP and eventually counteract tumour cell metastasize (Gialeli et al., 2011). However, clinical trials testing MMPi have not achieved the expected results.

Angiogenesis

MSCs support tumour vasculature directly by differentiating into pericytes and endothelial cells (Rajantie et al., 2004; Bexell et al., 2009). CAFs promote new blood vessel formation indirectly by secreting proangiogenic factors such as VEGF, IL-6, TGF-beta, and SDF-1 that promote endothelial smooth muscle migration at the tumour site. Although one might think that an increase in blood vessels correlates with an increase in immune infiltration in the tumour stroma, tumour blood vessels are poorly permeable.

Stimulation to metastasize

The metastasis cascade consists of tumour cells that invade the adjacent ECM to exit the primary tumour, migrate to a distant organ through the blood circulation and eventually colonize the organ. To initiate metastasis, tumour cells first develop invasive and migratory abilities by undergoing EMT, an important determinant of tumour progression and metastases (Thiery, 2002; Kalluri and Zeisberg, 2006; Wan et al., 2013). See also Figure 21.1. EMT is often induced by stromal signals. This is observed by the fact that tumour cells at the invasive edge are in contact with stromal cells and undergo EMT while tumour cells positioned at the centre of the tumour do not undergo EMT (Thiery, 2002). The reactive tumour stroma cells modify the ECM so that the mesenchymal-like tumour cells travel through the ECM (Mueller and Fusenig, 2004; Prud'homme, 2007). Different growth factors related to cell proliferation stimulate tumour cells to activate a series of EMT-transcription factors such as Snail, Slug and TWIST (Cano et al., 2000; Wang et al., 2012). MSCs promote invasion and influence different steps of the metastatic cascade, and the formation of the metastatic niche (Karnoub et al., 2007).

Altogether, genomic instability may be considered to be the driving force for cancer initiation and progression, but the contribution of the microenvironment cannot be neglected for onset and neoplastic progression. The tumour stroma does play an initiating and supporting role in malignant epithelial tumour growth.

Stromal models in cancer

The rapidly growing field of stroma-cancer interaction has been advanced by *in vitro* and *in vivo* studies. In *in vitro* studies, stromal cells isolated from human carcinomas are cocultured with cancer cells to measure the effect of the stroma on tumour growth. Isolated stromal fibroblasts from prostatic cancer significantly increased growth of prostate epithelium. This tumour supporting effect was however not seen when epithelium cells were cocultured with healthy fibroblasts (Olumi et al., 1999). The drawback of *in vitro* experiments is that 2D petri dishes are used that lack many components of the *in vivo* tumour microenvironment interaction. An alternative is the three-dimensional transwell, where the ECM can be added. This simplified model consists of type I collagen gel with cancer cells put on a transwell filter and placed in a larger recipient where CAFs are present. Soluble factors secreted by CAFs now reach cancer cells through the filter in a gradient (De Vlieghere et al., 2015). The transwell confirms that CAFs act in a paracrine manner instead of a direct cell-cell contact and the soluble factors can be measured.

The effect of stromal cells on cancer cells can also be observed in *in vivo* experiments. MCF7-ras BC tumour cells were co-injected in an immunodeficient mouse with CAFs or with fibroblasts originating from healthy tissue. CAFs indeed provided better support for MCF7 growth (Orimo et al., 2005). Another experiment showed that senescent fibroblasts promote cancer in certain circumstances. Senescent human stromal cells stimulated premalignant and malignant epithelial cells to proliferate in culture and formed tumours when injected in mice (Krtolica et al., 2001). The role of fibroblasts was further confirmed in a study conditionally inactivating the TGF-type II receptor gene in fibroblasts. This resulted in prostatic intraepithelial neoplasia and invasive squamous cell carcinoma of the forestomach suggesting that the TGF-signalling pathway in fibroblasts can modulate the oncogenic potential of adjacent epithelium (Bhowmick et al., 2004; West and van de Rijn, 2007; Isella et al., 2015).

The identification of distinct gene expression patterns in tumour cells and the adjacent fibroblasts can explain activated cell communication pathways. Studies using laser capture microscopy have measured gene expression of stromal cells using microarray gene analysis and next-generation sequencing (Gregg et al., 2010). Molecular profiling revealed a specific CAF transcriptional set of 11 genes expressed in CAFs isolated from NSCLC patients and involved in the MAPK signalling pathway and focal adhesion, which correlate with patient survival (Navab et al., 2011). A so-called wound response signature, consisting of genes induced in the 'CAF serum-response programme', of the response of fibroblasts to serum, was able to predict for breast, gastric and lung cancer patients with an increased risk to metastases (Chang et al., 2004).

By measuring the gene expression of cancer cells, four colorectal cancer (CRC) subtypes have emerged based on a meta-analysis of independent research groups (Guinney et al., 2015). All groups identified one subtype associated with poor prognosis, and more importantly, this subtype was recently observed to be associated with a high stromal content (Melo et al., 2013; Isella et al., 2015).

Clinical implications

Tumour–stroma interactions in routine pathology

The proportion and the composition of the tumour stroma differ between tumours and is distinct from normal tissue stroma. The association of the presence of stroma within the primary tumour with patient outcome has gained interest in the clinic.

Morphological changes in the stromal compartment of tumours include fibroblast proliferation and dense fibrosis (or desmoplasia). Compared to their normal counterparts, CAFs are morphologically different, including increased cell size and spindle-cell shaped appearance (Orimo et al., 2005; Soon et al., 2013). Tumour stroma can be classified into desmoplastic, sclerotic, normal-like, or inflammatory type based on histological findings (Park et al., 2015). The most frequently reported marker associated with CAFs and stromal activation is α-SMA. α-SMA is also expressed in other cell types, which hinders the use of this protein as a specific marker for activated fibroblasts. α-SMA also lacks sufficient sensitivity to identify all CAFs (Sugimoto et al., 2006). Although other biomarkers have been reported that are thought to identify CAFs, it was shown that CAF-related protein expression differed according to stromal phenotype (Park et al., 2015).

Scoring the tumour–stroma ratio (TSR) in simple haematoxylin-eosin (H&E) stained tumour slides contains good prognostic information. The scoring of the TSR is performed at routine pathology analysis of the primary tumour with high reproducibility (K >0.80). Morphological evaluation of H&E stained sections for TSR showed that tumours from patients with a bad prognosis have a high proportion of stroma and few tumour cells. The tumours from patients with a good prognosis showed the opposite: abundant tumour and less stroma. See also **Figure 21.4**. This phenomenon has led to the prognostic evaluation of this parameter in large patient series and has been shown to be prognostic in several types of cancers including colon (Mesker et al., 2007; Mesker et al., 2009; West et al., 2010; Huijbers et al., 2013; Park et al., 2014), breast (de Kruijf et al., 2011; Moorman et al., 2012; Dekker et al., 2013; Gujam et al., 2014), oesophagus (Courrech Staal et al., 2010; Courrech Staal et al., 2011; Wang et al., 2012), NSCLC (Zhang et al., 2015), ovary (Chen et al., 2015), HCC (Lv et al., 2015), nasopharynx (Zhang et al., 2014), and cervical cancer (Liu et al., 2014). See also **Table 21.1**.

The prognostic information assessed by scoring the TSR has been validated in the VICTOR (Vioxx in Colorectal Cancer Therapy: Definition of Optimal Regime) clinical trial of 710 stage II and III patients. When adding the TSR to the ASCO criteria the rate of correctly classified patients increased with an additional 14% (Huijbers et al., 2013). Although the TSR is assessed based on only a small part of the total tumour mass, a stroma-high level can be linked to an unfavourable patient prognosis, independent of other prognostic parameters. Possibly, this particular part of the tumour has obtained the capability to orchestrate its direct environment to facilitate its invasive and metastatic behaviour.

Although the reproducibility between pathologists is high, using an automated scoring system may make the TSR technique more objective. Therefore, an automated scoring of TSR was investigated using the microscope Aperio system that counts tumour cells on virtual sections. For stroma-high tumours, significantly lower cancer-specific survival rates were found (West et al., 2010; Park et al., 2014). The use of image features based on visual perception for discriminating epithelium and stroma of histologically verified, well-defined images of epithelium and stroma of digitized tissue micro-arrays of colorectal cancer were used to train classifiers based on support vector machines (SVM), nearest neighbour rule (1-NN), and the Naïve Bayes rule (NB). The experiments demonstrated that the proposed features can correctly discriminate epithelium from stroma with state-of-the-art accuracy (Bianconi et al., 2015).

Changes in the directionality of the collagen bundles that compose the ECM have been shown to promote tumour cell dissemination (Provenzano et al., 2006). Under the influence of tumour cells, the stromal tissue reorganizes into straight, aligned bundles (referred to as TACS3), which can be used to predict breast cancer disease-specific survival (Conklin et al., 2011). Also, a relationship was observed between stroma consisting of organized collagen

(A) (B)

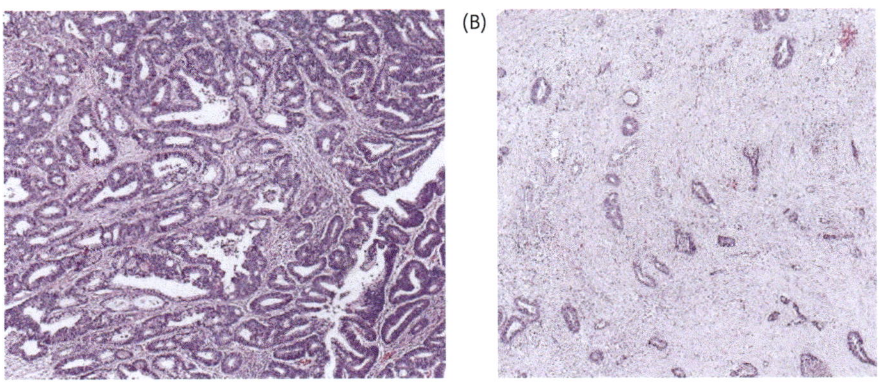

Fig. 21.4 Haematoxylin and eosin (H&E) stained 5 μm paraffin sections examined of the most invasive part of primary colon tumours. (A) Stroma low (20%); (B) stroma high (80%).

Table 21.1 Overview of studies evaluating the prognostic capacity of the tumour-stroma ratio (TSR) for different cancer types

Study	Cancer type	Univariate		Multivariate	
		OS	DFS	OS	DFS
Mesker et al., 2007	Colon ca. Stage I–III	*P* <0.0001 HR 3.74	*P* <0.0001 HR 4.18	*P* <0.001 HR 0.39	*P* <0.001 HR 0.34
Mesker et al., 2009	Colon ca. Stage I–II	*P* <0.001 HR 2.70	*P* <0.001 HR 2.40	n.a.	n.a.
West et al., 2010	Colorectal ca. Stage I–IV Colonca Stage I–IV		CSS [a] *P* = 0.024 HR 2.087 *P* = 0.019 HR 2.47		*P* = 0.017 *P* = 0.018 HR 2.70
Huijbers et al., 2013 VICTOR trial	Colon ca. II high-risk–III	*P* <0.0001 HR 1.96	*P* <0.0001 HR 2.15	*P* = 0.002 HR 1.70	*P* <0.001 HR 1.90
de Kruijf et al., 2011	Breast ca. Stage I–IV triple neg.	HR 1.29 *P* = 0.025 HR 1.60 *P* = 0.088	RFP [b] HR 1.62 *P* = 0.001 HR 3.91 *P* = 0.003	HR 1.50 *P* = 0.001 HR 1.87 *P* = 0.028	RFP [b] HR 1.97 *P* = 0.001 HR 2.92 *P* = 0.006
Courech Staal et al., 2010	Oesophageal adenoca. Stage I–IIa III–IV	*P* = 0.008 HR 2.80 Not significant	*P* = 0.019 HR 2.50 Not significant	n.a.	n.a.
Wang et al., 2012	Oesophageal squamous ca. Stage I–III	*P* = 0.001 HR 3.55	*P* = 0.001 HR 3.45	*P* = 0.001 HR 2.99	*P* = 0.001 HR 2.99
Moorman et al., 2012	Breast ca. triple neg. N0-N3	RFP [b] *P* = 0.004 HR 2.93	*P* = 0.035 HR 2.56		*P* = 0.034
Dekker et al., 2013	Breast ca. Premenstrual N0	*P* = 0.001 HR 1.68	*P* = 0.024 HR 1.55	*P* <0.001 HR 1.85	*P* = 0.050 HR 1.60
Park et al., 2014	Colorectal ca. Stage I–III		CSS [a] *P* = 0.009 HR 1.84		
Gujam et al., 2014	Ductal breast ca. TNM unknown		CSS [a] *P* <0.001 HR 2.12		*P* <0.001 HR 2.12
Liu et al., 2014	Cervical ca. FIGO I-II	*P* = 0.001 HR 3.86	*P* = 0.001 HR 4.16	*P* = 0.005 HR 3.12	*P* = 0.003 HR 3.46
Zhang et al., 2015	NSCLC Stage I–III	*P* <0.001 HR 1.82	*P* <0.001 HR 1.72	*P* <0.05 HR 1.75	*P* <0.05 HR 1.57
Zhang et al., 2014	NPC Stage I–IVa	*P* <0.022 HR 1.98	*P* <0.015 HR 2.07	*P* = 0.030 HR 1.99	*P* = 0.042 HR 1.93
Chen et al., 2015	Ovarian ca. FIGO I–IV	*P* <0.001 HR 1.17	*P* <0.001 HR 0.73	*P* = 0.076	*P* = 0.001
Lv et al., 2015	HCC Stage I–IV Liver resection Liver transplant	*P* <0.001 HR 4.35 *P* = 0.001 HR 2.92	*P* = 0.002 HR 2.63	*P* = 0.001 HR 2.55 *P* = 0.001 HR 2.67	*P* = 0.042 HR 1.93

[a] CSS = cancer-specific survival

[b] RFP = relapse-free period

and pathological response to neoadjuvant chemotherapy showing intratumoural stromal organization determined based on pretreatment breast cancer biopsies was related to pathological response to chemotherapy (Dekker et al., 2015).

Prognostic stromal biomarkers

Although current prognostication to predict cancer progression and patient survival has benefited many patients, improvement is still required. Treatment guidelines have included the stratification of patients in high-risk, chemotherapy-eligible groups, yet low-risk patients also experience recurrence of disease. In order to reduce under- and overtreatment, improved strategies for risk stratification are needed. No stromal markers are currently incorporated for patient prognostication. Retrospective and prospective studies have investigated specific individual components of the microenvironment for prognostic information.

For example, in patients affected by colorectal, oesophageal, or breast cancers the amount of CAFs expressing α-SMA, FAP, and FSP1 correlated with a poorer prognosis (Shi et al., 2012; Yamashita et al., 2012; Wikberg et al., 2013). In fibroblasts caveolin-1 downregulation has been shown to be predictive of worse survival in invasive breast cancer. Interestingly, differential caveolin-1 expression has not shown to be of any prognostic value in epithelial cells (Witkiewicz et al., 2010). The expression of the PDGFR by fibroblasts has been shown in preclinical models to contribute to the recruitment and activation of tumour-associated stroma. Presence of the PDGF-β receptor in the stromal compartment of breast tumours has been linked to an adverse effect on the disease-free survival period (Paulsson et al., 2009). The p53 tumour suppressor gene has been linked to a multitude of malignancies. While traditionally studied for its role in tumour epithelium, loss of p53 expression in the stromal compartment of p53 mutant fibroblasts have been shown to promote metastases and p53 protein accumulation in fibroblasts has been linked to prognosis (Hasebe et al., 2009; Addadi et al., 2010).

The presence and extent of the inflammatory infiltrate in breast tumours has been shown to hold prognostic information. For instance, immuno-histochemical assessment of the amount of CD8+, CD4 + FOXP3 + cells, γδ T-cells, and macrophages within the stromal compartment has been found to result in prognostic information (Bianchini et al., 2010; DeNardo et al., 2011; Liu et al., 2012; Mahmoud et al., 2012; Hidalgo et al., 2014).

It is shown that direct contact between lymphocytes and tumour cells is crucial to allow a successful T-cell immune response. Structures such as an intact basal membrane correlate with less infiltrate and a worse survival. This might suggest that a thorough extracellular matrix can encapsulate tumour cells from infiltrating lymphocytes, and that large amounts of tumour-associated stroma might function alike (Hewitt et al., 1991; Disis, 2010; Jiang et al., 2013). If such mechanisms caused by stroma are of significant influence, it is to be expected that tumours with much stroma effectively restrain lymphocytic infiltrate and have a negative effect on patient survival.

Angiogenesis is the formation of novel blood vessels induced by the presence of tumour epithelial cells. VEGF is considered as the major contributor to this process and is expressed by both tumour epithelial and stromal cells (Folkman and Shing, 1992). The stromal environment contributes to tumour angiogenesis, which supplies the oxygen and nutrients needed for tumour growth and progression (Hanahan and Coussens, 2012; Peiris-Pages et al., 2015). Antiangiogenic therapy, for example bevacizumab, a monoclonal antibody against VEGF, can therefore play an important role in treating patients with 'increased' tumour angiogenesis. Therapy targeting the TME could make a difference in survival, especially for the stroma-high group. This patient group shows a worse survival compared to stroma-low patients and recent literature indicates the resistance of stroma-high patients to current standard chemotherapy regimens (Meads et al., 2009; Liu et al., 2012).

Predictive stromal biomarkers

The tumour stroma plays an important role in tumour growth and can be a key issue for targeted therapy. Moreover, it may determine the success of the current therapy regimens. According to recent literature stroma fulfils several important roles in the response and resistance of malignant tumours to anticancer therapy. The physical properties of the tumour stroma contribute to hypoxia and acidity induced response and resistance to chemotherapy (Brown, 2002; Tredan et al., 2007; Cukierman and Khan, 2010). Moreover, they are key determinants of effective drug delivery (Minchinton et al., 1997; Tannock et al., 2002; Cukierman and Khan, 2010). Cellular adhesions by means of integrins between neoplastic cells and the ECM or stromal cells mediate drug response and resistance to apoptosis (Desgrosellier and Cheresh, 2010; Cukierman and Bassi, 2012). Stroma-related paracrine signalling makes neoplastic cells less responsive or even resistant to chemotherapy and targeted therapy (Castells et al., 2012; Straussman et al., 2012; Hale et al., 2013; Paraiso and Smalley, 2013). Many efforts are undertaken to elucidate the complex stroma–tumour interactions into the stromal biology. Most of the effort in the development of cancer therapeutics has been focused on targeting epithelial cancer cells.

Studies performed in pancreatic cancer *in vitro* and in transplanted immunodeficient *in vivo* models showed the complex composition of the microenvironment. No approved antistromal therapy has yet entered the clinical practice and most of antistromal therapies failed in clinical phase II or III. Recent experimental evidence has shown that stromal depletion approaches may favour tumour aggressiveness and spread indicating that the mechanisms and functional consequences of the tumour–stroma crosstalk are more complex than previously anticipated (Neesse et al., 2015).

The organization of the stromal matrix formation can also be an important factor for prediction of therapy response. Efficient organization of this matrix might increase the effective path of molecules towards the target cells (Ramanujan et al., 2002). This might influence drug diffusion and treatment efficacy. A recent study confirms this hypothesis by describing tumours (breast cancer patients) whose stroma consisted of organized collagen showing a higher benefit from neoadjuvant chemotherapy compared to tumours with disorganized tumour stroma (Dekker et al., 2013).

Considering the major role contributed to the tumour stroma in drug sensitivity and resistance, high stroma producing tumours may respond differently to anticancer therapy compared to low stroma producing tumours.

Few studies have directly investigated the response of chemotherapy with regards to expression of stromal markers. Farmer et al. demonstrated that high expression of a stromal metagene signature was related to resistance to chemotherapy (Farmer et al., 2009). This gene signature was also related to the amount of reactive stroma present in the tumour biopsy. The composition of this gene signature did reveal that several factors (e.g. PDGF-βR and MMP2) that were found to be prognostic stromal biomarkers were also factors involved in response to chemotherapy, suggesting that significant overlap might exist between these two mechanisms. In human CRC, CAFs and pericytes of the tumour vasculature express high levels of PDGF-R, while cancer cells express PDGF-A and -B. The blockade of PDGF-R signalling pathways in tumour-associated stromal cells using drugs such as imatinib inhibits growth and metastasis of colon cancer cells (Kitadai et al., 2006). Stromal myofibroblasts surrounding colon adenocarcinomas are an important source of COX-2, suggesting that myofibroblasts are an important target cells for NSAIDs and selective COX-2 inhibitors in the chemoprevention of CRC (Vandoros et al., 2006).

New insights into antitumour therapy reported that bone marrow-derived MSCs contribute to formation of tumour promoting stromal cells. MSCs travel to tumour stroma, where they differentiate into CAF-like cells (Shinagawa et al., 2013). The tumour-homing property of MSCs allows targeted delivery of therapeutic genes into the tumour microenvironment and allows a more focused expression within primary tumours, as the adoptively transferred MSC develop CAF-like characteristics (Knoop et al., 2015).

Conclusion

In conclusion, recent data have provided novel insights into the role of the tumour microenvironment in cancer development, progression, and therapeutic resistance. An increasing amount of prognostic and predictive data has emerged from the tumour-associated stroma in recent years. These data vary from parameters derived from morphological information to individual biomarkers and gene expression profiles. Evidence now suggests that CAFs, with their effects on antitumour immunity, tumour growth, and malignant dissemination provide fertile ground for drug development. The communication interface between stromal cells and host cells can be used as for personalized therapy and as target for the development of new cancer therapies. The differentiation of stroma-high and stroma-low tumours is useful in clarifying these pathophysiological mechanisms.

Perturbations of cancer cell-stromal interactions might attenuate early stages of the metastatic cascade as the EMT and cell invasion, leading to decreased metastasis

One might imagine a scenario where chemotherapy that targets proliferating cells might be deleterious for tumours with a high mitotic rate in fibroblasts. Thus, stromal density and proliferation should be carefully considered for both prognostic and therapeutic classifications as the biological behaviour, differentiation, and tumour vasculature might be profoundly different.

TAKE-HOME MESSAGE

- The non-neoplastic cells making the tumour-associated stroma affect cancer growth and progression.
- The stroma interacts with malignant cells at different stages of carcinogenesis ranging from tumour onset to invasion and metastasis.
- Tumour-associated stroma can have a tumour-inhibiting as well as a tumour promoting effect.
- The composition of tumour-associated stroma changes according to the type of malignancy.
- Tumour stroma can affect the response to treatment, its potential to influence therapy response must therefore be taken into account.

OPEN QUESTIONS

- Fibroblast in tumour-associated stroma, or cancer-associated fibroblasts (CAF) are likely to make a heterogeneous population of cells, however their classification into subgroups is still at its early phases.
- The biology and also the cell of origin of theses CAF is still not clear, with evidences pointing at various types of cells like mesenchymal stem cells, smooth muscles, and epithelium.
- Why in some cases stroma can promote tumour growth and in other suppress it, is still poorly understood.
- More studies are required to clarify how biomarkers of the stroma can help to provide both prognostic and predictive information.

FURTHER READING

Gai, P., Varinska, L., Faber, L., et al. (2017). How signaling molecules regulate tumor microenvironment: parallels to wound repair. *Molecules*, **22**, pii: E1818.

Huang, W., Lu, S., Burgess, R., Yi, Y.-H., Huang, G. F., & Huang, R.-P. (2018). New insights into the tumor microenvironment utilizing protein array technology. *Int J Mol Sci*, **19**, 559–71.

Joyce, J. A. & Pollard, J. W. (2009). Microenvironmental regulation of metastasis. *Nat Rev Cancer*, **9**, 239–52.

Maman, S. & Witz, I. P. (2018). A history of exploring cancer in context. *Nat Rev Cancer*, **18**(6), 359–76.

Santi, A., Kugeratski, F. G., & Zaniva, S. (2018). Cancer associated fibroblasts: the architects of stroma remodeling. *Proteomics*, **18**, e1700167.

Valkenburg, K. C., de Groot, A. E., & Pienta, K. J. (2018). Targeting the tumour stroma to improve cancer therapy. *Nat Rev Clin Oncol*, **15**(6), 366–81.

REFERENCES

Addadi, Y., Moskovits, N., Granot, D., et al. (2010). p53 Status in stromal fibroblasts modulates tumor growth in an SDF1-dependent manner. *Cancer Res*, **70**, 9650–8.

Bao, S., Wu, Q., Mclendon, R. E., et al. (2006). Glioma stem cells promote radioresistance by preferential activation of the DNA damage response. *Nature*, **444**, 756–60.

Bexell, D., Gunnarsson, S., Tormin, A., et al. (2009). Bone marrow multipotent mesenchymal stroma cells act as pericyte-like migratory vehicles in experimental gliomas. *Mol Ther*, **17**, 183–90.

Bhowmick, N. A., Chytil, A., Plieth, D., et al. (2004). TGF-beta signaling in fibroblasts modulates the oncogenic potential of adjacent epithelia. *Science*, **303**, 848–51.

Bianchini, G., QI, Y., Alvarez, R. H., Iwamoto, T., et al. (2010). Molecular anatomy of breast cancer stroma and its prognostic value in estrogen receptor-positive and -negative cancers. *J Clin Oncol*, **28**, 4316–23.

Bianconi, F., Alvarez-Larran, A., & Fernandez, A. (2015). Discrimination between tumour epithelium and stroma via perception-based features. *Neurocomputing*, **154**, 119–26.

Biswas, S., Davis, H., Irshad, S., Sandberg, T., Worthley, D., & Leedham, S. (2015). Microenvironmental control of stem cell fate in intestinal homeostasis and disease. *J Pathol*, **237**, 135–45.

Bochet, L., Lehuede, C., Dauvillier, S., et al. (2013). Adipocyte-derived fibroblasts promote tumor progression and contribute to the desmoplastic reaction in breast cancer. *Cancer Res*, **73**, 5657–68.

Brown, J. M. (2002). Tumor microenvironment and the response to anticancer therapy. *Cancer Biol Ther*, **1**, 453–8.

Cano, A., Perez-Moreno, M. A., Rodrigo, I., et al. (2000). The transcription factor snail controls epithelial-mesenchymal transitions by repressing E-cadherin expression. *Nat Cell Biol*, **2**, 76–83.

Castells, M., Thibault, B., Delord, J. P., & Couderc, B. (2012). Implication of tumor microenvironment in chemoresistance: tumor-associated stromal cells protect tumor cells from cell death. *Int J Mol Sci*, **13**, 9545–71.

Chang, H. Y., Sneddon, J. B., Alizadeh, A. A., et al. (2004). Gene expression signature of fibroblast serum response predicts human cancer progression: similarities between tumors and wounds. *PLoS. Biol*, **2**, E7.

Chen, Y., Zhang, L., Liu, W., & Liu, X. (2015). Prognostic significance of the tumor-stroma ratio in epithelial ovarian cancer. *Biomed Res Int*, **2015**, 589301.

Conklin, M. W., Eickhoff, J. C., Riching, K. M., et al. (2011). Aligned collagen is a prognostic signature for survival in human breast carcinoma. *Am J Pathol*, **178**, 1221–32.

Cortez, E., Roswall, P., & Pietras, K. (2014). Functional subsets of mesenchymal cell types in the tumor microenvironment. *Semin Cancer Biol*, **25**, 3–9.

Courrech Staal, E. F., Smit, V. T., Van Velthuysen, M. L., et al. (2011). Reproducibility and validation of tumour stroma ratio scoring on oesophageal adenocarcinoma biopsies. *Eur J Cancer*, **47**, 375–82.

Courrech Staal, E. F., Wouters, M. W., Van Sandick, J. W., et al. (2010). The stromal part of adenocarcinomas of the oesophagus: does it conceal targets for therapy? *Eur J Cancer*, **46**, 720–8.

Cukierman, E. & Bassi, D. E. (2012). The mesenchymal tumor microenvironment A drug-resistant niche. *Cell Adhesion & Migration*, **6**, 285–96.

Cukierman, E. & Khan, D. R. (2010). The benefits and challenges associated with the use of drug delivery systems in cancer therapy. *Biochem Pharmacol*, **80**, 762–70.

De Kruijf, E. M., Van Nes, J. G., Van De Velde, C. J., et al. (2011). Tumor-stroma ratio in the primary tumor is a prognostic factor in early breast cancer patients, especially in triple-negative carcinoma patients. *Breast Cancer Res Treat*, **125**, 687–96.

De Vlieghere, E., Gremonprez, F., Verset, L., et al. (2015). Tumor-environment biomimetics delay peritoneal metastasis formation by deceiving and redirecting disseminated cancer cells. *Biomaterials*, **54**, 148–57.

Dekker, T. J., Charehbili, A., Smit, V. T., et al. (2015). Disorganised stroma determined on pre-treatment breast cancer biopsies is associated with poor response to neoadjuvant chemotherapy: results from the NEOZOTAC trial. *Mol Oncol*, **9**, 1120–8.

Dekker, T. J., Van De Velde, C. J., et al. (2013). Prognostic significance of the tumor-stroma ratio: validation study in node-negative premenopausal breast cancer patients from the EORTC perioperative chemotherapy (POP) trial (10854). *Breast Cancer Res Treat*, **139**, 371–9.

Denardo, D. G., Brennan, D. J., Rexhepaj, E., et al. (2011). Leukocyte complexity predicts breast cancer survival and functionally regulates response to chemotherapy. *Cancer Discov*, **1**, 54–67.

Desgrosellier, J. S. & Cheresh, D. A. (2010). Integrins in cancer: biological implications and therapeutic opportunities. *Nat Rev Cancer*, **10**, 9–22.

Disis, M. L. (2010). Immune regulation of cancer. *J Clin Oncol*, **28**, 4531–8.

Farmer, P., Bonnefoi, H., Anderle, P., et al. (2009). A stroma-related gene signature predicts resistance to neoadjuvant chemotherapy in breast cancer. *Nat Med*, **15**, 68–74.

Folkman, J. & Shing, Y. (1992). Angiogenesis. *J Biol Chem*, **267**, 10931–4.

Gialeli, C., Theocharis, A. D., & Karamanos, N. K. (2011). Roles of matrix metalloproteinases in cancer progression and their pharmacological targeting. *FEBS J*, **278**, 16–27.

Gregg, J. L., Brown, K. E., Mintz, E. M., Piontkivska, H., & Fraizer, G. C. (2010). Analysis of gene expression in prostate cancer epithelial and interstitial stromal cells using laser capture microdissection. *BMC Cancer*, **10**, 165.

Guarino, M. (2007). Epithelial-mesenchymal transition and tumour invasion. *Int J Biochem Cell Biol*, **39**, 2153–60.

Guinney, J., Dienstmann, R., Wang, X., et al. (2015). The consensus molecular subtypes of colorectal cancer. *Nat Med*, **21**, 1350–6.

Gujam, F. J. A., Edwards, J., Mohammed, Z. M. A., Going, J. J., & Mcmillan, D. C. (2014). The relationship between the tumour stroma percentage, clinicopathological characteristics and outcome in patients with operable ductal breast cancer. *Br J Cancer*, **111**, 157–65.

Hale, M. D., Hayden, J. D., & Grabsch, H. I. (2013). Tumour-microenvironment interactions: role of tumour stroma and proteins produced by cancer-associated fibroblasts in chemotherapy response. *Cell Oncol*, **36**, 95–112.

Hanahan, D. & Coussens, L. M. (2012). Accessories to the crime: functions of cells recruited to the tumor microenvironment. *Cancer Cell*, **21**, 309–22.

Hanahan, D. & Weinberg, R. A. (2011). Hallmarks of cancer: the next generation. *Cell*, **144**, 646–74.

Harper, J. & Sainson, R. C. (2014). Regulation of the anti-tumour immune response by cancer-associated fibroblasts. *Semin Cancer Biol*, **25**, 69–77.

Hasebe, T., Okada, N., Tamura, N., et al. (2009). p53 expression in tumor stromal fibroblasts is associated with the outcome of patients with invasive ductal carcinoma of the breast. *Cancer Science*, **100**, 2101–8.

Hawinkels, L. J., Paauwe, M., Verspaget, H. W., et al. (2014). Interaction with colon cancer cells hyperactivates TGF-beta signaling in cancer-associated fibroblasts. *Oncogene*, **33**, 97–107.

Herbst, R. S., Baas, P., Kim, D. W., et al. (2016). Pembrolizumab versus docetaxel for previously treated, PD-L1-positive, advanced non-small-cell lung cancer (Keynote-010): a randomised controlled trial. *Lancet*, **387**, 1540–50.

Hewitt, R. E., Powe, D. G., Griffin, N. R., & Turner, D. R. (1991). Relationships between epithelial basement membrane staining patterns in primary colorectal carcinomas and the extent of tumour spread. *Int J Cancer*, **48**, 855–60.

Hidalgo, J. V., Bronsert, P., Orlowska-Volk, M., et al. (2014). Histological analysis of gammadelta T lymphocytes infiltrating human triple-negative breast carcinomas. *Front Immunol*, **5**, 632.

Huijbers, A., Tollenaar, R. A., Pelt, G. W., et al. (2013). The proportion of tumor-stroma as a strong prognosticator for stage II and III colon cancer patients: validation in the VICTOR trial. *Ann Oncol*, **24**, 179–85.

Isella, C., Terrasi, A., Bellomo, S. E., et al. (2015). Stromal contribution to the colorectal cancer transcriptome. *Nat Genet*, **47**, 312–19.

Jiang, B., Mason, J., Jewett, A., et al. (2013). Tumor-infiltrating immune cells: triggers for tumor capsule disruption and tumor progression? *Int J Med Sci*, **10**, 475–97.

Kalluri, R. & Zeisberg, M. (2006). Fibroblasts in cancer. *Nat Rev Cancer*, **6**, 392–401.

Karnoub, A. E., Dash, A. B., Vo, A. P., et al. (2007). Mesenchymal stem cells within tumour stroma promote breast cancer metastasis. *Nature*, **449**, 557–63.

Kitadai, Y., Sasaki, T., Kuwai, T., et al. (2006). Expression of activated platelet-derived growth factor receptor in stromal cells of human colon carcinomas is associated with metastatic potential. *Int J Cancer*, **119**, 2567–74.

Klopp, A. H., Gupta, A., Spaeth, E., Andreeff, M., & Marini, F., 3rd (2011). Concise review: dissecting a discrepancy in the literature: do mesenchymal stem cells support or suppress tumor growth? *Stem Cells*, **29**, 11–19.

Knoop, K., Schwenk, N., Schmohl, K., et al. (2015). Mesenchymal stem cell-mediated, tumor stroma-targeted radioiodine therapy of metastatic colon cancer using the sodium iodide symporter as theranostic gene. *J Nucl Med*, **56**, 600–6.

Krtolica, A., Parrinello, S., Lockett, S., Desprez, P. Y., & Campisi, J. (2001). Senescent fibroblasts promote epithelial cell growth and tumorigenesis: a link between cancer and aging. *Proc Natl Acad Sci U S A*, **98**, 12072–7.

Liu, J., Liao, S., Diop-Frimpong, B., et al. (2012). TGF-beta blockade improves the distribution and efficacy of therapeutics in breast carcinoma by normalizing the tumor stroma. *Proc Natl Acad Sci U S A*, **109**, 16618–23.

Liu, J., Liu, J., Li, J. S., et al. (2014). Tumor-stroma ratio is an independent predictor for survival in early cervical carcinoma. *Gynecol Oncol*, **132**, 81–6.

Lv, Z., Cai, X., Weng, X., et al. (2015). Tumor-stroma ratio is a prognostic factor for survival in hepatocellular carcinoma patients after liver resection or transplantation. *Surgery*, **158**, 142–50.

Maffini, M. V., Soto, A. M., Calabro, J. M., Ucci, A. A., & Sonnenschein, C. (2004). The stroma as a crucial target in rat mammary gland carcinogenesis. *J Cell Sci*, **117**, 1495–502.

Mahmoud, S. M., Lee, A. H., Paish, E. C., Macmillan, R. D., Ellis, I. O., & Green, A. R. (2012). Tumour-infiltrating macrophages and clinical outcome in breast cancer. *J Clin Pathol*, **65**, 159–63.

Marsh, T., Pietras, K., & McAllister, S. S. (2013). Fibroblasts as architects of cancer pathogenesis. *Biochim Biophys Acta*, **1832**, 1070–8.

Meads, M. B., Gatenby, R. A., & Dalton, W. S. (2009). Environment-mediated drug resistance: a major contributor to minimal residual disease. *Nat Rev Cancer*, **9**, 665–74.

Melo, D. S. E., Wang, X., Jansen, M., et al. (2013). Poor-prognosis colon cancer is defined by a molecularly distinct subtype and develops from serrated precursor lesions. *Nat. Med*, **19**, 614–8.

Mesker, W. E., Junggeburt, J. M., Szuhai, K., et al. (2007). The carcinoma-stromal ratio of colon carcinoma is an independent factor for survival compared to lymph node status and tumor stage. *Cell Oncol*, **29**, 387–98.

Mesker, W. E., Liefers, G. J., Junggeburt, J. M., et al. (2009). Presence of a high amount of stroma and downregulation of SMAD4 predict for worse survival for stage I-II colon cancer patients. *Cell Oncol*, **31**, 169–78.

Minchinton, A. I., Wendt, K. R., Clow, K. A., & Fryer, K. H. (1997). Multilayers of cells growing on a permeable support—an *in vitro* tumour model. *Acta Oncologica*, **36**, 13–16.

Mishra, P., Banerjee, D., & Ben-Baruch, A. (2011). Chemokines at the crossroads of tumor-fibroblast interactions that promote malignancy. *J Leukoc Biol*, **89**, 31–9.

Mishra, P. J., Mishra, P. J., Humeniuk, R., et al. (2008). Carcinoma-associated fibroblast-like differentiation of human mesenchymal stem cells. *Cancer Res*, **68**, 4331–9.

Moorman, A. M., Vink, R., Heijmans, H. J., Van Der Palen, J., & Kouwenhoven, E. A. (2012). The prognostic value of tumour-stroma ratio in triple negative breast cancer. *Eur J Surg Oncol*, **38**(4), 307–13.

Mueller, M. M. & Fusenig, N. E. (2004). Friends or foes—bipolar effects of the tumour stroma in cancer. *Nat Rev Cancer*, **4**, 839–49.

Navab, R., Strumpf, D., Bandarchi, B., et al. (2011). Prognostic gene-expression signature of carcinoma-associated fibroblasts in non-small cell lung cancer. *Proc Natl Acad Sci U S A*, **108**, 7160–5.

Neesse, A., Algul, H., Tuveson, D. A., & Gress, T. M. (2015). Stromal biology and therapy in pancreatic cancer: a changing paradigm. *Gut*, **64**, 1476–84.

Ohlund, D., Elyada, E., & Tuveson, D. (2014). Fibroblast heterogeneity in the cancer wound. *J Exp Med*, **211**, 1503–23.

Olumi, A. F., Grossfeld, G. D., Hayward, S. W., Carroll, P. R., Tlsty, T. D., & Cunha, G. R. (1999). Carcinoma-associated fibroblasts direct tumor progression of initiated human prostatic epithelium. *Cancer Res*, **59**, 5002–11.

Orimo, A., Gupta, P. B., & Sgroi, D. C., et al. (2005). Stromal fibroblasts present in invasive human breast carcinomas promote tumor growth and angiogenesis through elevated SDF-1/CXCL12 secretion. *Cell*, **121**, 335–48.

Ostman, A. & Augsten, M. (2009). Cancer-associated fibroblasts and tumor growth--bystanders turning into key players. *Curr Opin Genet Dev*, **19**, 67–73.

Paraiso, K. H. T. & Smalley, K. S. M. (2013). Fibroblast-mediated drug resistance in cancer. *Biochem Pharmacol*, **85**, 1033–41.

Park, C. C., Bissell, M. J., & Barcellos-Hoff, M. H. (2000). The influence of the microenvironment on the malignant phenotype. *Mol Med Today*, **6**, 324–9.

Park, J. H., Richards, C. H., McMillan, D. C., Horgan, P. G., & Roxburgh, C. S. (2014). The relationship between tumour stroma percentage, the tumour microenvironment and survival in patients with primary operable colorectal cancer. *Ann Oncol*, **25**, 644–51.

Park, S. Y., Kim, H. M., & Koo, J. S. (2015). Differential expression of cancer-associated fibroblast-related proteins according to molecular subtype and stromal histology in breast cancer. *Breast Cancer Res Treat*, **149**, 727–41.

Paulsson, J., Sjoblom, T., Micke, P., et al. (2009). Prognostic significance of stromal platelet-derived growth factor beta-receptor expression in human breast cancer. *Am J Pathol*, **175**, 334–41.

Peiris-Pages, M., Smith, D. L., Gyorffy, B., Sotgia, F., & Lisanti, M. P. (2015). Proteomic identification of prognostic tumour biomarkers, using chemotherapy-induced cancer-associated fibroblasts. *Aging (Albany. NY)*, **7**, 816–38.

Pittenger, M. F., Mackay, A. M., Beck, S. C., et al. (1999). Multilineage potential of adult human mesenchymal stem cells. *Science*, **284**, 143–7.

Provenzano, P. P., Eliceiri, K. W., Campbell, J. M., Inman, D. R., White, J. G., & Keely, P. J. (2006). Collagen reorganization at the tumor-stromal interface facilitates local invasion. *BMC Med*, **4**, 38.

Prud'Homme, G. J. (2007). Pathobiology of transforming growth factor beta in cancer, fibrosis and immunologic disease, and therapeutic considerations. *Lab Invest*, **87**, 1077–91.

Rajantie, I., Ilmonen, M., Alminaite, A., Ozerdem, U., Alitalo, K., & Salven, P. (2004). Adult bone marrow-derived cells recruited during angiogenesis comprise precursors for periendothelial vascular mural cells. *Blood*, **104**, 2084–6.

Ramanujan, S., Pluen, A., Mckee, T. D., Brown, E. B., Boucher, Y., & Jain, R. K. (2002). Diffusion and convection in collagen gels: implications for transport in the tumor interstitium. *Biophys J*, **83**, 1650–60.

Shi, M., Yu, D. H., Chen, Y., et al. (2012). Expression of fibroblast activation protein in human pancreatic adenocarcinoma and its clinicopathological significance. *World J Gastroenterol*, **18**, 840–6.

Shinagawa, K., Kitadai, Y., Tanaka, M., et al. (2010). Mesenchymal stem cells enhance growth and metastasis of colon cancer. *Int J Cancer*, **127**, 2323–33.

Shinagawa, K., Kitadai, Y., Tanaka, M., et al. (2013). Stroma-directed imatinib therapy impairs the tumor-promoting effect of bone marrow-derived mesenchymal stem cells in an orthotopic transplantation model of colon cancer. *Int J Cancer*, **132**, 813–23.

Siegel, P. M. & Massague, J. (2003). Cytostatic and apoptotic actions of TGF-beta in homeostasis and cancer. *Nat Rev Cancer*, **3**, 807–21.

Soon, P. S., Kim, E., Pon, C. K., et al. (2013). Breast cancer-associated fibroblasts induce epithelial-to-mesenchymal transition in breast cancer cells. *Endocr Relat Cancer*, **20**, 1–12.

Sternlicht, M. D., Lochter, A., Sympson, C. J., et al. (1999). The stromal proteinase Mmp3/stromelysin-1 promotes mammary carcinogenesis. *Cell*, **98**, 137–46.

Straussman, R., Morikawa, T., Shee, K., et al. (2012). Tumour microenvironment elicits innate resistance to RAF inhibitors through HGF secretion. *Nature*, **487**, 500–U118.

Sugimoto, H., Mundel, T. M., Kieran, M. W., & Kalluri, R. (2006). Identification of fibroblast heterogeneity in the tumor microenvironment. *Cancer Biol Ther*, **5**, 1640–6.

Tannock, I. F., Lee, C. M., Tunggal, J. K., Cowan, D. S. M., & Egorin, M. J. (2002). Limited penetration of anticancer drugs through tumor tissue: a potential cause of resistance of solid tumors to chemotherapy. *Clinical Cancer Research*, **8**, 878–84.

Thiery, J. P. (2002). Epithelial-mesenchymal transitions in tumour progression. *Nat Rev Cancer*, **2**, 442–54.

Tredan, O., Galmarini, C. M., Patel, K., & Tannock, I. F. (2007). Drug resistance and the solid tumor microenvironment. *J Natl Cancer Inst*, **99**, 1441–54.

Vandoros, G. P., Konstantinopoulos, P. A., Sotiropoulou-Bonikou, G., et al. (2006). PPAR-gamma is expressed and NF-kB pathway is activated and correlates positively with COX-2 expression in stromal myofibroblasts surrounding colon adenocarcinomas. *J Cancer Res Clin Oncol*, **132**, 76–84.

Vermeulen, L., Melo, D. S. E., Van Der Heijden, M., et al. (2010). Wnt activity defines colon cancer stem cells and is regulated by the microenvironment. *Nat Cell Biol*, **12**, 468–76.

Wan, L., Pantel, K., & Kang, Y. (2013). Tumor metastasis: moving new biological insights into the clinic. *Nat Med*, **19**, 1450–64.

Wang, K., MA, W., Wang, J. B., et al. (2012). Tumor-stroma ratio is an independent predictor for survival in esophageal squamous cell carcinoma. *J Thorac Oncol*, **7**, 1457–61.

West, N. P., Dattani, M., Mcshane, P., et al. (2010). The proportion of tumour cells is an independent predictor for survival in colorectal cancer patients. *Br J Cancer*, **102**, 1519–23.

West, R. B. & Van De Rijn, M. (2007). Experimental approaches to the study of cancer-stroma interactions: recent findings suggest a pivotal role for stroma in carcinogenesis. *Lab Invest*, **87**, 967–70.

Wikberg, M. L., Edin, S., Lundberg, I. V., et al. (2013). High intratumoral expression of fibroblast activation protein (FAP) in colon cancer is associated with poorer patient prognosis. *Tumour Biol*, **34**, 1013–20.

Witkiewicz, A. K., Dasgupta, A., Sammons, S., et al. (2010). Loss of stromal caveolin-1 expression predicts poor clinical outcome in triple negative and basal-like breast cancers. *Cancer Biology & Therapy*, **10**, 135–43.

Yamashita, M., Ogawa, T., Zhang, X., et al. (2012). Role of stromal myofibroblasts in invasive breast cancer: stromal expression of alpha-smooth muscle actin correlates with worse clinical outcome. *Breast Cancer*, **19**, 170–6.

Zhang, T., Xu, J., Shen, H., Dong, W., Ni, Y., & Du, J. (2015). Tumor-stroma ratio is an independent predictor for survival in NSCLC. *Int J Clin Exp Pathol*, **8**, 11348–55.

Zhang, X. L., Jiang, C., Zhang, Z. X., Liu, F., Zhang, F., & Cheng, Y. F. (2014). The tumor-stroma ratio is an independent predictor for survival in nasopharyngeal cancer. *Oncol Res Treat*, **37**, 480–4.

22

Blood vessels and cancer

Francesco Pezzella and Robert Kerbel

Cancer and vessels: A complex relationship

Blood vessels in the human body can be formed by two main processes: vasculogenesis and angiogenesis. Vasculogenesis is predominant during embryonal development and consists of the haemangioblasts differentiating into endothelial cells and blood vessels. Angiogenesis instead is the formation of sprouting new blood vessels from the pre-existing ones.

The role played by the blood vessels in tumours has been the object of studies for centuries with the introduction of the concept of angiogenesis dated far back as 1787 (Hall, 2005), although the word 'angiogenesis' was coined much later (Natale et al., 2016). In 1939, Ida et al. described that tumour implants in the ear of rabbits were accompanied by the formation of new capillaries (Ide et al., 1939) leading, later on, to the idea that angiogenesis is necessary to support tumours. In the same period several histopathologists maintained that both pre-existing and newly formed vessels coexisted in tumours (Willis, 1934; Ritchie, 1962). However, it was not until over 40 years ago that the subject started to be systematically studied by Judah Folkman (Zetter, 2008) following the path indicated by Ide.

In 1971 Folkman published his classic hypothesis paper (Folkman, 1971) in which the idea that 'the growth of a solid neoplasm is always accompanied by neo-vascularization' was put forward. This hypothesis relied mainly on '*in vitro*' and animal models (Folkman, 1990) mostly with experiments conducted implanting tumours in avascular sites, like the cornea of rabbits (Gimbrone et al., 1972) regarded as a classic proof of concept. Subsequent work on mice has not only confirmed the need for angiogenesis but also showed that its induction is an early event in tumour development (Hanahan and Folkman, 1996). Immunohistochemical studies of human '*in situ*' breast (Guidi et al., 1994) and cervical (Guidi et al., 1995) carcinomas have demonstrated the presence of microvessels in the underlying basal membranes suggesting that angiogenesis may represent an intermediate phase between *in situ* and infiltrating carcinomas (Hanahan and Folkman, 1996). The direct correlation between microvessel density and clinical outcome (Weidner et al., 1992) further strengthened the idea of a link between angiogenesis and tumour growth although the validity of such a correlation has been subsequently strongly questioned (Trivella et al., 2007). Furthermore, in the last 20 years, evidence has been accumulating showing that tumours can also grow in the absence of angiogenesis by exploiting pre-existing vessels (Pezzella et al., 1996) or, less commonly, by vasculogenic mimicry (Maniotis et al., 1999).

Tumour growth with formation of new vessels

As described in Chapter 18 on hypoxia, cancer cells exposed to diminished concentrations of oxygen activate a series of pathways leading to the expression of a number of proteins which can induce formation of new vessels. Adaptive angiogenesis therefore follows the hypoxia suffered by cancer cells in the tumour microenvironment, but this does not happen in all tumours. Some hypoxic malignancies will still remain non-angiogenic and grow, while many neoplastic lesions will contain both angiogenic and non-angiogenic areas. When angiogenic cancer cells are present, different types of angiogenesis can be triggered which can coexist, such as sprouting angiogenesis, intussusceptive microvascular growth (IMG), and glomeruloid microvascular proliferation.

Classic angiogenesis: Vascular sprouting

Vascular sprouting is by far the commonest type of angiogenesis occurring in cancers (Fig. 22.1). Vessels in normal conditions are quiescent and their homeostasis is mainly controlled by the Angiopoietin family (ANG1, ANG2 and ANG4) and their receptors (TIE-1 and TIE-2; see Augustin et al., 2009). Sprouting angiogenesis is largely affected /induced by the vascular endothelial growth factor (VEGF) family of proteins (Fig. 22.2). There are four VEGF proteins, VEGF-A, B, C, and D, and three receptors, VEGF1R/flt-1, VEGFR2/kdr, and VEGFR3/flt4. As far as angiogenesis is concerned, the circulating VGFA isoforms 121 and 165 plus the VGFR-2, a type III receptor tyrosine kinase, are the most relevant. Another membrane protein, neuropilin, has been shown to link to VEGFA165 and to produce an angiogenic effect similar to that observed following activation of VEGFR-2 (Kerbel, 2008).

In presence of a hypoxic signal leading to the secretion of angiogenic factors endothelial cells loosen their junctions and detach by the basal membrane through the action of matrix metalloproteases (MMPs) and, following increasing levels of VEGF-A, the vessels become leaky (Dvorak et al., 1995). Endothelial cells break through the basal membrane and start to migrate towards the angiogenic signal. The stimulated endothelial cells also begin

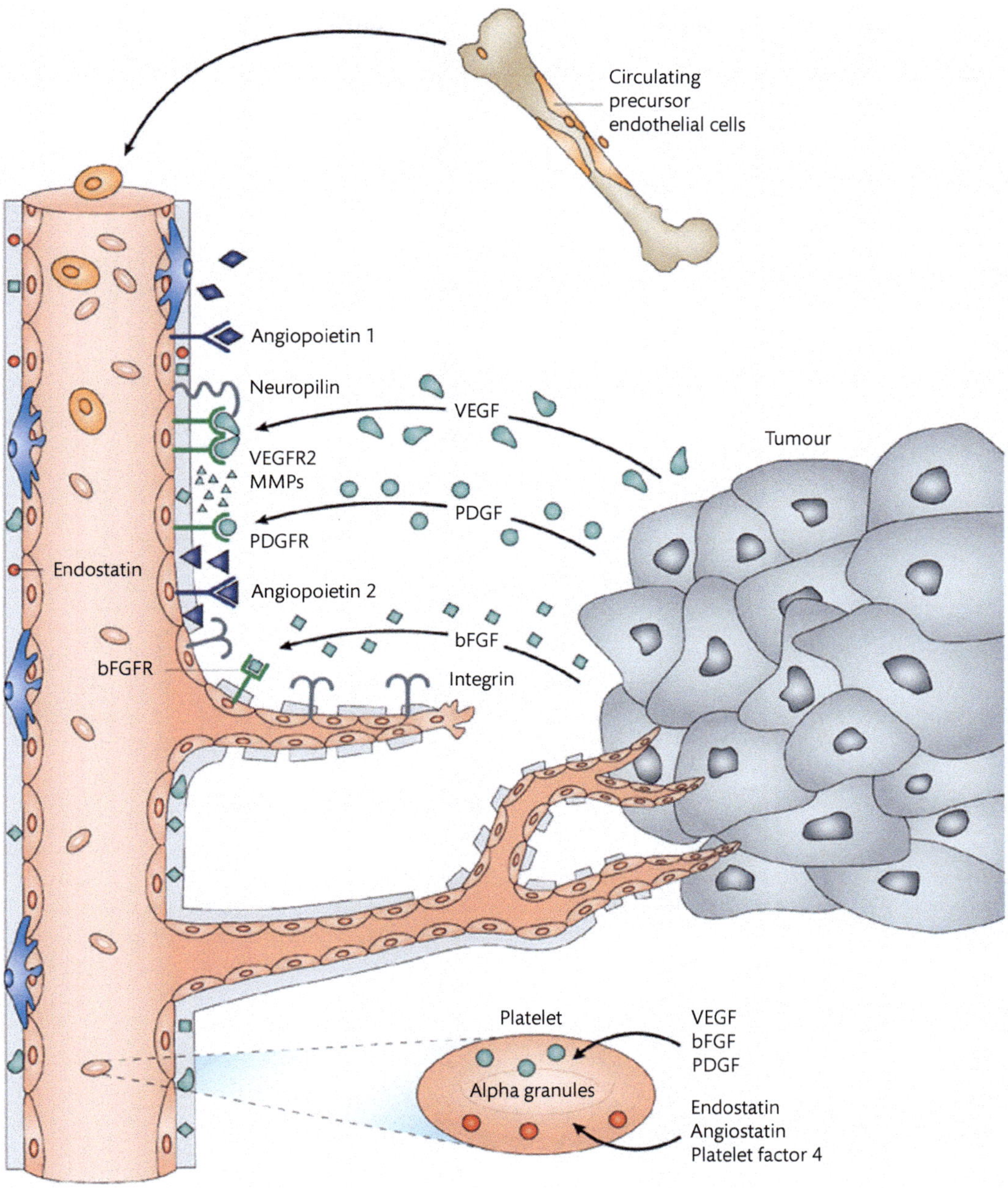

Fig. 22.1 Sprouting angiogenesis. In normal circumstances, angiopoietin-1 acts as an agonist, and maintains the endothelial cell quiescence, binding to receptor Tie2. Secretion of growth factor from the neoplastic cells (e.g. VEGF, PDGF, and bFGF), induces endothelial cell proliferation. Matrix metalloproteinases (MMPs), released by the tumour and/or the endothelial cells following VEGF stimulation, break the basal membrane. Tumours cells also secrete angiopoietin-2 which competes with the agonistic function of angiopoietin-1, further destabilizing the vessels. Some of the newly formed endothelial cells may also derive from bone marrow-derived endothelial cell progenitor cell population.

Reproduced with permission from Folkman J, 'Opinion: Angiogenesis: an organizing principle for drug discovery?', *Nature Reviews Drug Discovery*, Volume 6, pp. 273,

to secrete ANG2 which competes with ANG1 for the TIE2 receptor, further inducing endothelial cell detachment, vascular permeability, and endothelial proliferation and sprouting (Augustin et al., 2009).

Subsequently endothelial cells start to divide (Dvorak et al., 1995) following the docking of VEGF-A to VEGFR-2 triggering activation of the PLC gamma/PKC/RAF MEK/MAPK pathway leading to cell proliferation and of the PI3K/AKT, supporting cell survival (Kerbel, 2008). As the endothelial cells proliferate and accumulate, after crossing the basal membrane, they form a cord of cells divided into two types: the Stalk cells, forming most of the growing capillaries, and the most advanced cell, the Tip cell, leading the migration. The differentiation into either Stalk or Tip cells is regulated by VEGF-A through the NOTCH/WNT signalling pathway (Fig. 22.3). NOTCH (1, 2, 3, and 4) are receptors expressed on different cell types; their ligands (Jag 1 and 2 plus DLL3 and 4) are also present on the cell membrane. This pathway regulates mainly proliferation and differentiation.

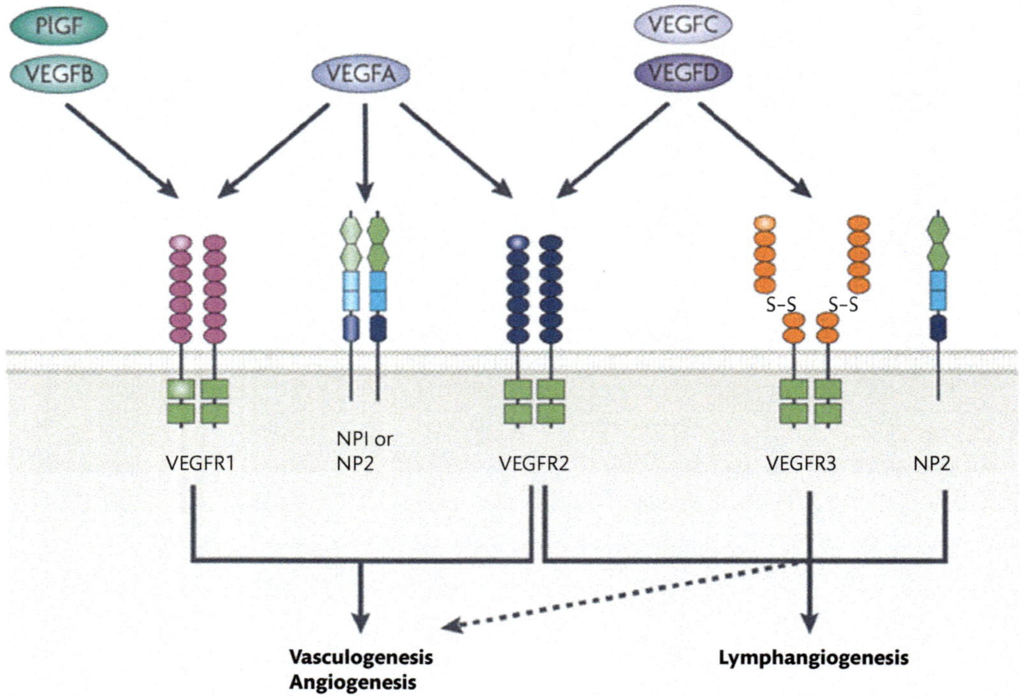

Fig. 22.2 The VEGF (vascular endothelial growth factor) family. This family is made up by five VEGF proteins: VEGF-A, VEGF-B, VEGF-C, VEGF-D, and placental growth factor (PlGF). There are five receptors: VEGFR-1 (Flt1), VEGFR-2 (Kdr), VEGFR-3 (Flt4), neuropilin 1, and neuropilin 2.

All the endothelial cells express both NOTCH and NOTCH ligands (Jag 1, 2 DLL1, 3 and 4; see Fig. 22.4). Following VEGFR2 activation by VEGF-A, the endothelial cells start to increase the transcription of DLL4. The cells which are most advanced towards the hypoxic area are, among the endothelial sprout cells, therefore subject to a higher VEGF-A concentration. These tip cells will accumulate more DLL4 leading to a higher DLL4:NOTCH ratio compared to the endothelial cells found further back along the vascular sprout.

In these Tip cells therefore, the higher levels of DLL4 will link to the extracellular domain of NOTCH and will import it into

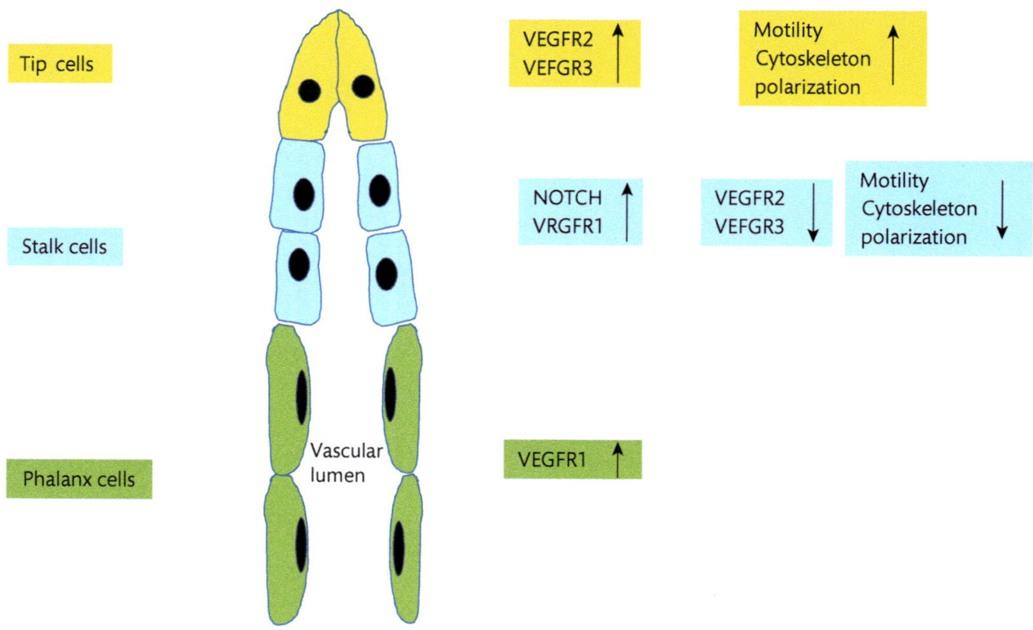

Fig. 22.3 The sprouting new vessel. The sprouting new vessel is divided into the Tip cells, at the advancing end, the matured phalanx cells and, in between, the stalk cells. VEGFR2 and VEGFR3 are highly expressed on the advancing Tip. The cells behind in the stalk start to mature and stabilize and express NOTCH and VEGFR1, while VEGFR2 and VEGFR3 expression diminishes. As cells completely stabilize, the phalanx, VEGFR2, and VEGFR3 are no longer expressed.

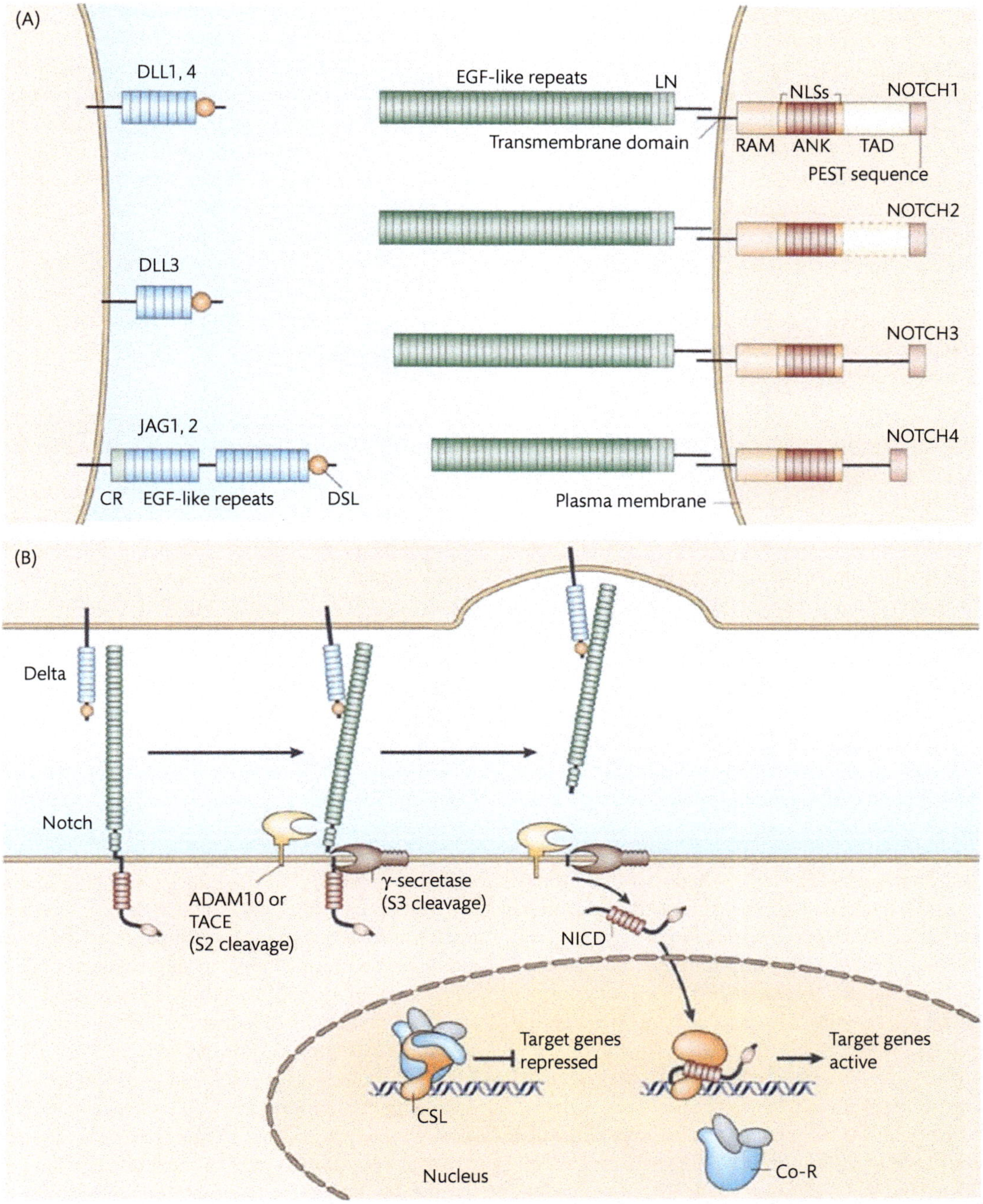

Fig. 22.4 (A,B) The NOTCH family. There are four membrane associated NOTCH receptors. Their ligands, also tethered to the cell membrane rather than being secreted, are DLL4, DLL3, JAG1, and JAG2. Upon ligation with the ligand, the intracellular component of the receptors is cleaved, transported to the nucleus where it links to DNA, inducing transcription of its downstream target genes. In addition, the extracellular component is also cleaved and remains attached to the ligand.

the cytoplasm. This complex though is not a transcription factor. Instead the cells with more NOTCH, having NOTCH lost the extra-cellular domain following linkage with DLL4, will be importing into the cytoplasm the intracellular domain of NOTCH which has transcription factor activity, and can trigger the induction of several downstream pathways. NOTCH-activated pathways will then reduce the proliferative effect of VEGF, contributing to the maturation process of the newly formed vessel. Therefore, the most advanced cells of the endothelium, having more DLL4, will acquire a phenotype characterized by low proliferation, lack of vascular lumen formation but VEGF-A dependent production of filopodia rich in actin, resulting in increased motility. The stalk cells will have instead a phenotype characterized by proliferation which is progressively switched off as NOTCH activation increases leading to vascular lumen formation, establishment of interendothelial junctions, and deposition of basal membrane.

As a lumen is established, the arrival of oxygenated blood starts to inhibit VEGF-A further decreasing the proliferation. Eventually new pericytes are recruited and a mature vessel is ready (Phng and Gerhardt, 2009).

During the angiogenic process, therefore, there will be new vessel 'formation' made up by endothelial cells with a 'gradient of phenotype': highly mobile but not proliferating at the Tip, progressively less mobile going down the new vessels, and with higher proliferation in the middle section but re-establishment of the lumen, oxygenation, and quiescence at the beginning. Eventually the newly formed branches will fuse with other vessels allowing the blood to flow and all the vessels, well oxygenated, will acquire a quiescent phenotype.

Intussusceptive microvascular growth or splitting angiogenesis

In intussusceptive microvascular growth (IMG), there is an increase in number of vessels by the longitudinal splitting in two of the pre-existing vessels. The initial step is the growth of endothelial cells into forming a transluminal endothelial septum with reorganization of the endothelial cells which cover the newly formed channels. Once this endothelial septum is in place, connective tissue moves into it, completing the division into two (Fig. 22.5). A slightly different type also occurs: in some vessels, rather than starting with the creation of the endothelial bridge, the process begins with the two sides of the vessel getting into contact. Subsequently some fenestration occurs in the opposite endothelial cells getting into touch and through these holes the stroma proliferates, creating a septum ultimately completely covered by new endothelial cells (Makanaya et al., 2009). During the whole process blood flow is maintained and indeed these flow patterns can influence IMG (Patan et al., 1996).

IMG occurs physiologically both during the embryogenesis and the postnatal growth: it was first described in the lung microvasculature of the rabbit embryo (Short, 1950) and then in the postnatal development of the microvasculature of the rat lung (Caduff et al., 1986) and 10 years later also in cancer, being detected in colonic adenocarcinoma in mouse xenografts (Patan et al., 1996). The biology underlying IMG is still not very well known but some data have been published. On chicken embryo chorioallantoic membranes, treatment with the VEGF121 isoform or with recombinant human erythropoietin (Crivellato et al., 2004) induces predominantly IMG rather than sprouting angiogenesis (Wilting et al., 1996). TIE1 and TIE2 are also involved: in mouse embryos TIE2 positively regulates endothelial cell stretching, motility, and adhesion to the extracellular matrix while TIE1 inhibits the endothelial cell motility. Therefore, if the system made up by TIE2, its ligand angiopoietin-1 and TIE1 are disrupted IMG is severely disrupted. In the absence of TIE2 the endothelial cells fail to migrate properly or to adhere to the extracellular matrix, thus failing in the formation of the intussusceptive septa. On the other hand, deficiency of TIE1 leads to excessive thinning of the endothelial septa and increased numbers of leaky endothelium on the septa and outside, impairing again the final intussusceptive process (Patan, 1998).

Glomeruloid microvascular proliferation

Glomeruloid bodies, also known as glomeruloid microvascular proliferation, are small vascular structures reminiscent in their appearance to the kidney glomeruli. They are very common in glioblastomas but have been detected also in other malignancies (Dome et al., 2003). Their anatomical structure has been described in glioblastoma and they are made up by small vessels lined by high endothelial cells. Around them there is a discontinuous layer of pericytes. Each capillary loop is then surrounded by a basal membrane. In the centre of the glomeruloid there are a very few capillaries with flat endothelium, possibly the vessels from which the glomeruli originate (Wesseling et al., 1993; Rojiani and Dorovini-Zis, 1996).

There are two models of how glomeruloid bodies form, one called 'active' and the second called 'passive'. The two are not mutually exclusive as there could well be glomeruloid bodies with different origins (Dome et al., 2003). In the first (so-called active) there is formation of new microvessels that form the glomeruloid bodies. In a mouse model, subcutaneous injections of a viral vector carrying VEGF-A results in the formation of glomeruloid-like bodies. In melanomas where such bodies have been observed, the endothelial cells are proliferating and are positive for VEGFR2, FLT1, Neuropilin and VEGF-A, further supporting the idea that their formation is VEGF-A dependent (Sundberg et al., 2001; Straume and Akslen, 2003). The second type (passive) has been described in another mouse model where neoplastic cells were injected into the carotid artery in order to generate brain metastases. Glomeruloid bodies were observed in

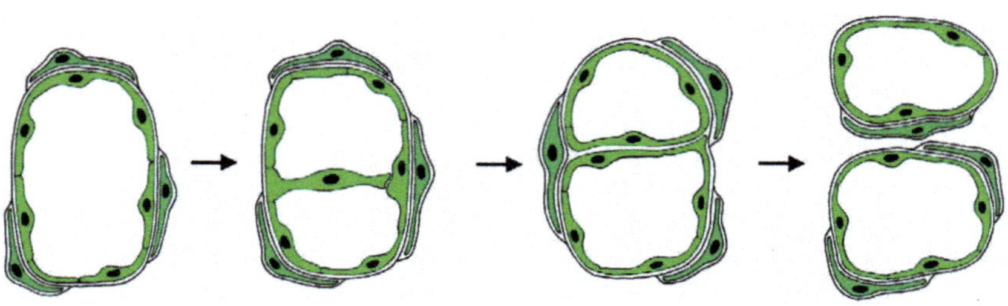

Fig. 22.5 Intussusceptive (splitting), vascular growth. Transversal section. First the endothelial cells proliferate and forms a bridge inside the lumen, followed by the realignment of these endothelial cells. Finally, the division is completed by the migration inside the bridge of connective tissue. Endothelial cells: red; pericytes: green.

these experimental metastases but the proliferating rate in the endothelial cells was low.

By studying the lesions as they happened with serial tissue collection and three-dimensional reconstructions, Dome et al described as the first step the extravasation of single tumour cells. Metastatic cells are then found to locate on the exterior of the capillaries adhering on their basal membrane. By increasing in number and the contractile activity of their cytoskeleton, the neoplastic cells pull the capillaries and entrap them into the tumour nodules, looping and coiling them up. Eventually the trapped capillaries acquire the appearance of florid glomeruloid bodies, this time made up by pre-existing normal vessels, with a superficial resemblance to renal glomeruli. As the phenomenon proceeds, occasional rupture of capillaries is observed (Dome et al., 2003).

Postnatal vasculogenesis: Bone marrow-derived endothelial progenitor cells

During angiogenesis however, the endothelial cells involved in tumour vascularization can be also derived from circulating endothelial progenitor cells (EPCs; see Nolan et al., 2007). These cells are mainly found in the bone marrow but are also isolated from peripheral blood, fetal liver and umbilical cord blood (Peichev et al., 2000), and their phenotype is variable according to the organ of provenience but always includes CD34 and VEGFR2 (de la Puente et al., 2013). When a tumour develops both the secretion of cytokines and hypoxia related proteins, can attract EPCs to the neoplasm where it is assumed that they can play two main roles: further increase the angiogenic stimulus by secreting proangiogenic factors and to be recruited inside the newly establish vascular structure where the EPCs differentiate into mature endothelial cells (de la Puente et al., 2013). However, the biology of the intratumour vasculogenesis is more complex than the aforementioned model. There is a number of different bone marrow-derived cells, other than the classic EPCs, these are candidates for the origin for endothelial cells (e.g. Tie-2 positive monocytes, CD11b positive myeloid cells, VEGFR1 positive hemangiocytes, VE cadherine positive leukocytes, lymphocytes stimulated by pleiotrophin, and intratumour macrophages). How each of these cells contributes to making of intratumour endothelium is uncertain. Furthermore, there is disagreement on whether these cells can form endothelium or remain just closely adjacent to the vessels and influence the vessel formation by a paracrine effect. It also not clear how much of a contribution the marrow-derived stem cells make compared to those present in the other tissues of the body (Bussolati et al., 2011).

Tumour growth without angiogenesis

An increasing body of evidences from histopathology studies first, and later from preclinical animal model studies and clinical trials, has uncovered an added layer of complexity: the possibility that malignant tumours in some circumstances are able to grow in the absence of neoangiogenesis by exploiting the pre-existing vasculature. Such 'non-angiogenic' cancers, both primary tumours and metastatic lesions, have so far been described in lung, liver, brain, and lymph nodes.

Evidence that tumours can be non-angiogenic

In lung, both primary and secondary non-angiogenic tumours can grow by filling the alveolar space (Pezzella et al., 1996; Pezzella et al., 1997). The only vessels, highlighted by immunostains for endothelial cells, evident in such tumours, arise from the alveolar septa, and highlight the alveolar network of the lung entrapped by neoplastic cells. The fact that the vessels observed are the (pre-existing) alveolar vessels is suggested not only by their arrangement according to the normal lung architecture but also by the presence in the surrounding stroma tissue of anthracitic pigment. The amount of anthracitic pigment is highly variable from case to case but within the same lung appears to be comparable inside the tumour and in the surrounding uninvolved parenchyma. The tumour grows in a solid fashion filling the alveolar spaces. Neither endothelial cells nor tumour-associated stroma are present among the neoplastic cell masses within the alveolar sacs. Tumours having both angiogenic and non-angiogenic areas are also commonly seen (Pezzella et al., 1997).

Serial sections of angiogenic lung cancers, non-angiogenic lung cancers and normal lung were cut and immunostained both for cytokeratin, present in the malignant epithelial cells and normal pneumocytes, and CD34 expressed by the endothelial cells. Eventually epithelium was labelled in green and endothelium in red. Images were than captured using an image handling software. The colours of individual images were superimposed and consecutive slide images were then stacked on top of each other for 3D reconstruction/restoration of the spatial orientation of section images. The stacked images were then imported a further software for the three-dimensional reconstruction.

When looking only at the red signal (endothelium) the three-dimensional reconstruction of normal tissue vessels was indistinguishable from that of the normal tissue whereas a disorganized network of vessels was seen in the angiogenic tumours (Adighibe et al., 2006; see Fig. 22.6).

Animal studies using both immunofluorescence techniques and electron microscopy models have revealed more details of how the metastatic cells interact with the normal lung in these tumours. First the metastatic cells extravasate and form small aggregates within the alveolar wall between the epithelial cells and the basal membrane of the alveolar vessels. Subsequently the neoplastic cells move inside the alveolar space and progressively fill it, moving then to nearby space through the alveolar pores. Once the alveolar space is filled, the neoplastic cells pass again between the alveolar epithelial cells and infiltrate again the basement membrane dissecting the alveolar from the endothelial cell/alveolar wall. Eventually the alveolar epithelial cells, now separated from their basal membrane, detach, and die and the neoplastic cells lie in the alveolar vessels directly (Szabo et al., 2015; see Fig. 22.7).

Phenotype characterization of human intratumour vessels has shown that in normal lung the alveolar vessels have the basal membrane containing LH39, an antigen expressed in the basal membrane of mature but not newly formed vessels (Almeida et al., 1992) while the integrin $\alpha_v\beta_3$ is known to be expressed, albeit usually only weakly, and in a patchy way. Vessels in the non-angiogenic tumours had the same phenotype while most of the intratumour vessels in angiogenic carcinoma were $\alpha_v\beta_3$ strongly positive and/ or LH39 negative (Passalidou et al., 2002) with only occasional proliferating cells (Sardari Nia et al., 2008).

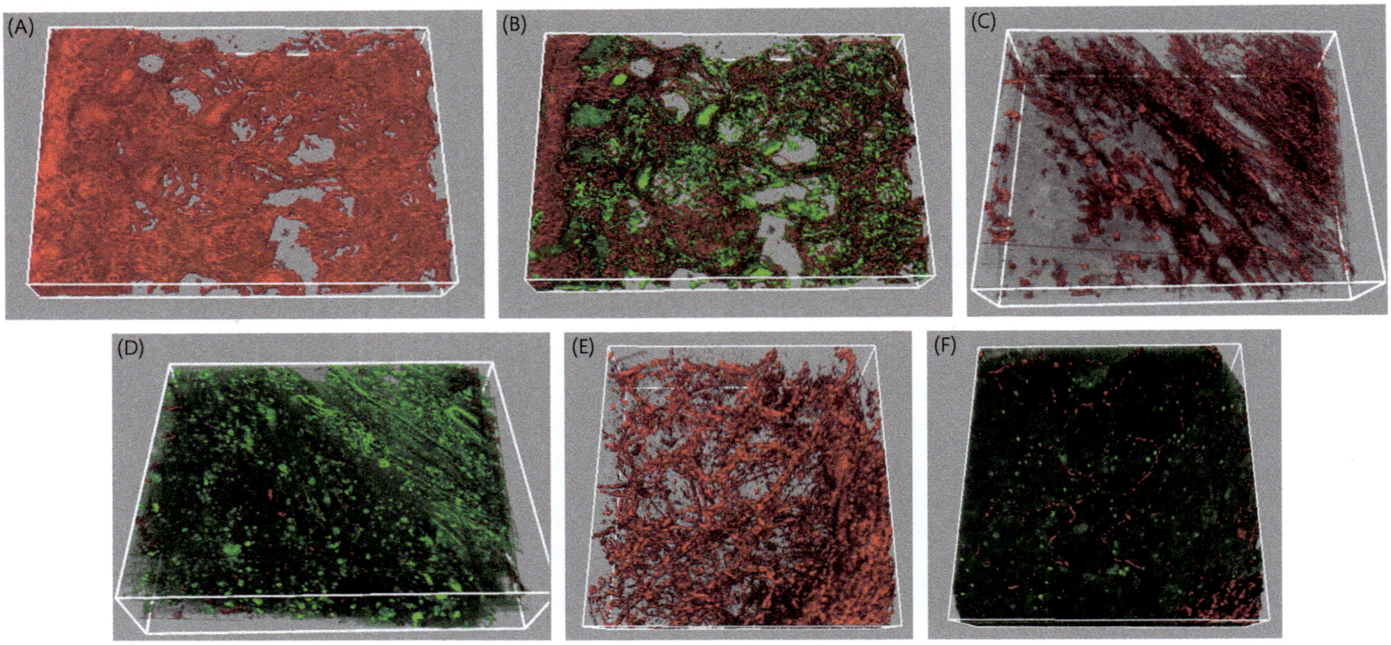

Fig. 22.6 Three-dimensional reconstruction and double staining of (A, B) normal lung, (C, D) angiogenic, and (E, F) non-angiogenic primary non-small cell lung carcinoma. Endothelial cells were stained with anti-CD34 mouse antibody QBEnd/10 (DAKO, UK) and epithelial cells with pan-cytokeratin rabbit polyclonal antiserum (Novacastra, UK). Following the primary antibodies, immunostaining was completed for blood vessels by incubation with Alexa Fluor 568 (red staining) goat anti-mouse (Molecular Probes, USA) and for epithelial cells with Alexa Fluor 488 (green staining) goat anti-rabbit (Molecular Probes, USA). (A) Normal lung vasculature (red) highlights the lung alveolar structure; (B) when the staining for epithelial cells is also shown, the pneumocytes (green) lining the alveoli can be seen. (C) Vasculature (red) in an angiogenic tumour, in the upper right corner squashed normal vessels can be seen, in the lower left corner the normal vascular structure is no longer present, only irregularly scattered vessels are seen. (D) When staining for epithelial cells (green) is added the lower left corner is filled by cancer cells. In the residual upper left normal tissue some empty alveolar spaces can be seen. (E) Vessels in a non-angiogenic tumour are stained in red: the same pattern seen in normal lung is present. However, this time when showing the epithelial cells (red) (F) the alveolar cavities are filled completely by carcinoma cells.

In the liver, the non-angiogenic tumours grow with a pattern called 'replacement' as the neoplastic cells infiltrate the liver parenchyma without any disturbance of the pre-existing tissue architecture/structure. The reticulin framework of the normal liver is preserved and there is no fibrosis and only minimal inflammation. As the tumour cells colonize the liver trabecula, they replace the hepatocyte and the frontal neoplastic cells and have intimate cell–cell contact with the facing hepatocyte. The tumour cells grow by co-opting the stroma with the sinusoidal blood vessels (Vermeulen et al., 2001). Alternatively, in a smaller number of cases, the neoplastic cells grow inside the sinusoids (Terayama et al., 1996).

The possibility that some human brain tumours, namely glioblastoma multiforme, may not be entirely angiogenesis dependent was implicated both in clinical (Wesseling et al., 1994) and animal model studies (Holash et al., 1999; Rubenstein et al., 2000; Kunkel et al., 2001). This includes human glioblastomas surgically removed before any other treatment as well as animal models in which glioblastoma was observed progressing after anti-VEGF treatment, where it was reported that in several tumour histopathological fields which were analysed, the number of vessels observed was in the same range of the number in the normal cerebral white matter. On this basis, as well as on the morphological patterns of the intratumour vessels, it was proposed that few new 'sprouting' blood vessels had formed in such areas.

The occurrence of vascular co-option in human glioblastoma after antiangiogenic treatment has been further supported by immunohistochemical studies (di Tomaso et al., 2011). While the newly formed vessels in the primary tumours have a thick membrane positive for Collagen IV with the endothelial cells strongly expressing PDGFR beta, in post-mortem samples the co-opted vessels have, like the normal vessels of the brain, a thin Collagen IV positive basal membrane, PDGFR beta-positive pericytes while the actual endothelial cells are PDGFR beta-negative. The ability of glioblastoma cells to grow along pre-existing brain vessels has been confirmed in mouse models using *in vivo* multiphoton laser scanning microscopy which allows visually monitoring of the movement and proliferation of neoplastic cells in close relationship to the brain vessels (Winkler et al., 2009).

Not only brain primary tumours but brain metastases as well can be non-angiogenic and grow by co-opting pre-existing normal vessels. This has been observed both in human tumours (Berghoff et al., 2013) and has also been described in animal models (Leenders et al., 2004; Bugyik et al., 2011; Simonsen et al., 2015).

Solid tumours can also grow in a non-angiogenic manner in lymph nodes as shown using a mouse model in which a chronic lymph node window is installed. Tetramethylrhodamine isothiocyanate–Dextran (TRITC-Dextran) angiography was performed to visualize the lymph node vessels by intravital

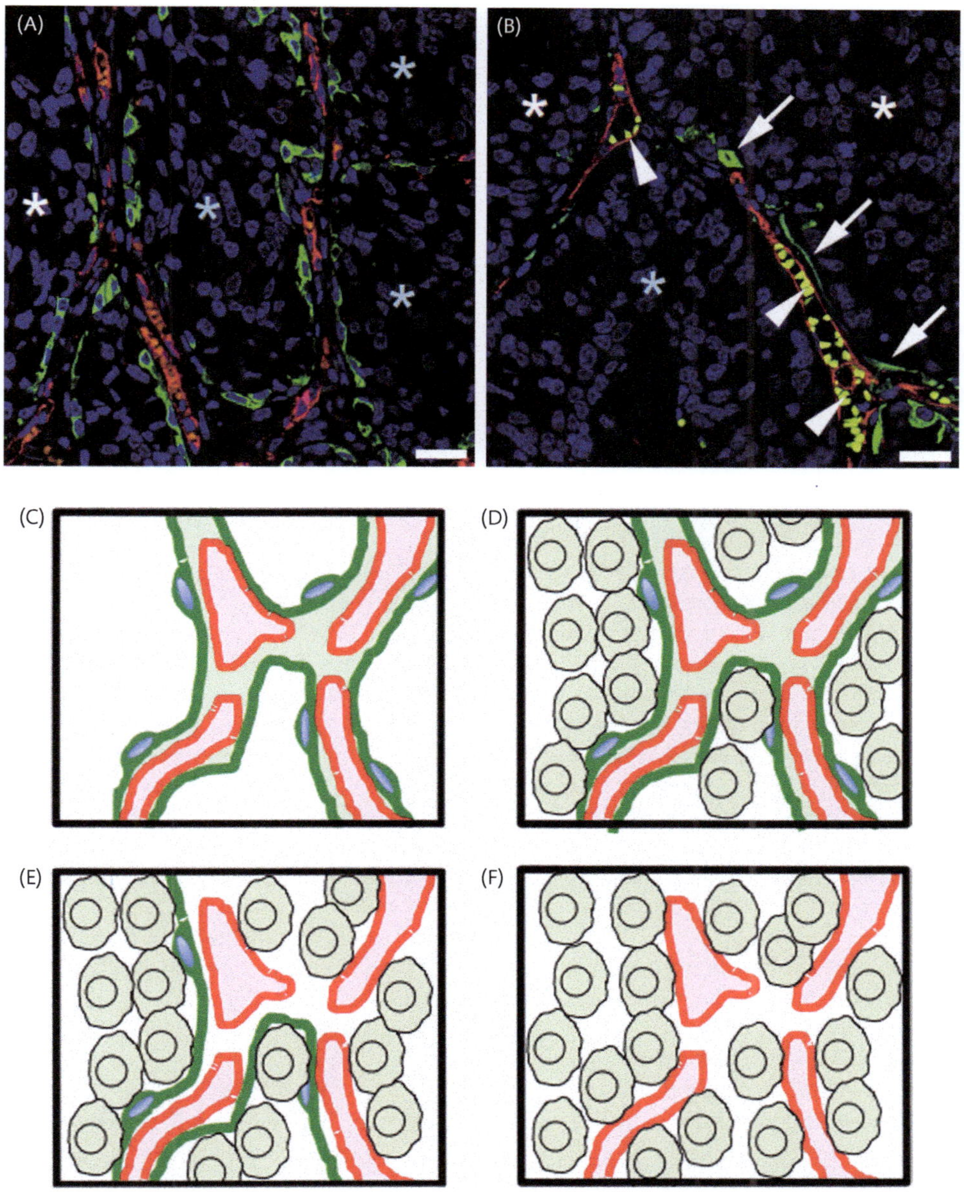

Fig. 22.7 Interaction between neoplastic cells and lung alveolar wall in non-angiogenic metastatic carcinoma to the lung. Endothelial cells were stained with anti-CD31 antibody (red), the normal pneumocytes of the lung with an anticytokeratin 7 (green) which does not stain the metastatic cells. The metastatic cells are identified by blue counterstaining of their nuclei. After extravasation, the metastatic cells (nuclei in blue) (A) are filling the alveolar cavity (asterisk). The normal alveolar walls are still preserved, pneumocytes in green and endothelial cells in red. Later on (B) the pneumocytes (green) start to disappear and the metastatic cells are in contact with the basal membrane of the pre-existing lung vessels (triangle). Residual pneumocytes are identified by arrows. The sequence is schematically represented in (C)-(F): vessels in red, pneumocytes dark green, metastatic cells in light green.

multiphoton microscopy whereas the metastatic carcinoma cells were detectable since they were transfected with a gene-encoding green fluorescent protein (SCCVII-GFP; see Jeong et al., 2015). The first neoplastic cells are detected in the subcapsular sinuses, initially as single cells than as small aggregates after which they invade deeper in the lymph node cortex where they start to dispose themselves along the vessels when they reach a depth of 50–100 microns. Monitoring by angiography of the vessels did not reveal any change or evidence of angiogenic blood vessel sprouting while the metastatic cells colonized lymph nodes over a period of 40 days (Jeong et al., 2015).

The vascular counts showed consistently that the reactive nodes have the same (Breast Cancer Progression Working Party, 2000) or even a higher (Naresh et al., 2001; Jeong et al., 2015) vascular density than the nodal metastases. The intrametastatic vessels have also the same (Naresh et al., 2001) or even lower (Jeong et al.,

2015) proliferating cell fractions than those in the reactive nodes as detected using immunohistochemistry for expression of Ki67 in the endothelial cells.

Non-angiogenic tumour growth: Why does it happen?

When comparing angiogenic and non-angiogenic lung carcinomas, only minor differences were found as far as necrosis is concerned whereas chronic inflammation and fibrosis characterized the angiogenic tumours, but no differences with respect to microvessel density nor apoptosis (Ferguson, 2008). An immunohistochemical study failed to demonstrate any major difference in the expression of markers of angiogenesis and hypoxia, as also observed in vasculogenic mimicry (Paulis et al., 2010), the only exception being differences in stromal thrombospondin, almost absent in non-angiogenic tumours but widely present in the angiogenic ones (Ferguson, 2008; Adighibe, 2012; Jeong et al., 2015). This study however was done using total tumour tissue including stroma where vascular remodelling occurs. Thrombospondin, which is antiangiogenic, may be highly expressed in angiogenesis to exert its antiangiogenic action within the remodelling activity.

mRNA expression profiling by microarray studies confirmed that no differences in classic hypoxia/angiogenesis pathways could be found with the exception, again, of thrombospondin-1. An unexpected finding was instead the increased expression in non-angiogenic tumours of a set of genes linked to oxidative phosphorylation, suggesting the possibility of metabolic reprogramming in the non-angiogenic tumours. The second surprising finding was the decreased level, in the same tumours, of a set of adhesion molecule genes, raising the hypothesis that diminished cell to cell contact could, again, be associated with failure to develop a vascular infrastructure (Hu et al., 2005).

A third finding was the association between expression of cytoplasmic p53 and non-angiogenic tumours. A pilot study on a small number of these cases demonstrated a higher incidence of P53 mutations in these cases (Adighibe, 2012). If confirmed, these data would be consistent with the report that, in animal models, inactivation of P53 leads to resistance to antiangiogenic drugs by increasing the ability of the cells to survive in hypoxia (Yu et al., 2002).

Less data are available as far as the biological characterization of the non-angiogenic liver metastases is concerned. However the different type of spreading compared to the lung (replacement of normal cells versus filling empty spaces) and the finding so far that very scanty CA9 expression at the edge of the replacement pattern in the liver indicate that these metastases are not very hypoxic (Van den Eynden et al., 2013), which suggests that liver and lung non-angiogenic tumours could have different underlying mechanisms of development and biology (Jeong et al., 2015).

Vasculogenic mimicry

Alongside exploitation of normal vessels, vasculogenic mimicry (VM), that is, the ability of tumour cells to form vessel-like networks, is another way for tumours to receive oxygen and nutrients in the absence of sprouting neoangiogenesis (Maniotis et al., 1999). First described in uveal melanoma, it has now been reported to occur in several types of tumour (Hendrix et al., 2003; Seftor et al., 2012). It differs from vascular co-optation because functional channels are created but they are not comprised of newly formed vessels lined by endothelial cells, but rather by the very same tumour cells. VM appears to recapitulate vasculogenesis network as neoplastic cells possibly reverse to an embryonic-like phenotype therefore becoming competent to mimic endothelial cells (Maniotis et al., 1999).

The main signalling pathway regulating VM is ignited by upregulation of VE-cadherin through HIF2. VE-cadherin colocalizes with, and phosphorylates EphA2 which in turn activates PI3K both directly and throughout the FAK/ERK1/2 pathway. PIK3 induces the cleavage of pro-MT1-MMP and pro-MMP2 into their corresponding active forms which promote the cleavage of laminin5γ2 to the pro-migratory fragments 5γ2′ and 5γ2x. These two latter protein fragments induce and guide the arrangement of the neoplastic cells into channel structures. This pathway can be on one side inhibited by cAMP, inhibiting ERK1/2 function. However, cAMP can also sustain VM by positively regulating VE-cadherin expression throughout the nodal/notch 1/4 pathway with Gal-3 having a similar positive effect on VM (Hendrix et al., 2003; Paulis et al., 2010).

Vascular co-option by neoplastic cells

Vascular co-option is a circumstance in which cells come into contact with the abluminal side of the blood vessels in order to thrive. This is a physiological phenomenon (e.g. plasma cells can be seen to co-opt vessels in bone marrow). In tumours, so far, co-option has been identified as a mechanism by which non-angiogenic neoplastic cells 'parasitize' normal vessels to obtain oxygen and nutrients. However, as co-option is an active mechanism that allows the cancer cell to exploit the presence of vessels, it is likely that co-option is also used by angiogenic cancer cells to interact with and exploit the newly formed vessels. Co-option therefore is likely to be relevant to both angiogenic and non-angiogenic tumours.

Only a few studies are so far available addressing the mechanism of vascular co-option by neoplastic cells and they are mainly focused on vascular co-option of normal vessels in the brain.

Primary brain tumours

As far as the primary brain tumours are concerned, the first question is how the neoplastic cells are attracted towards the vessel (Montana and Sontheimer, 2011). There is evidence to show that this is mediated by bradykinin signalling pathways: the glioma cells express bradykinin 2 receptor (B2R) which is activated by the bradykinin secreted by the endothelial cells. Once activated B2R induces intracellular Ca^{2+} oscillations triggering the migration of the neoplastic cells towards the vessel along the gradient of secreted bradykinin. In another study, Caspani and colleagues (Caspani et al., 2014; **Fig. 22.8A**) studied the interaction between glioblastoma multiforme (GBM) cells and pericytes associated with the brain blood vessels, both *in vivo* animal and *in vitro* models. They reported that GMB cells produce cytoplasmic extensions, denominated flectopodia, which rely on CDC42, a GTPAse which regulates the actin-dependent cytoplasm extension. These flectopodia then adhere to the pericytes through the adhesion molecule CD44. Following the co-option, the pericytes contract inducing changes of the vascular structure from linear to convoluted. Fusion between the cytoplasm of the GMB cells and the pericytes occurs, with some hybrid cells formed '*in vitro*'. Inhibition of CDC42 results in impaired vascular co-option and differentiation of some pericytes into macrophage like cells with antitumour activity. The possibility

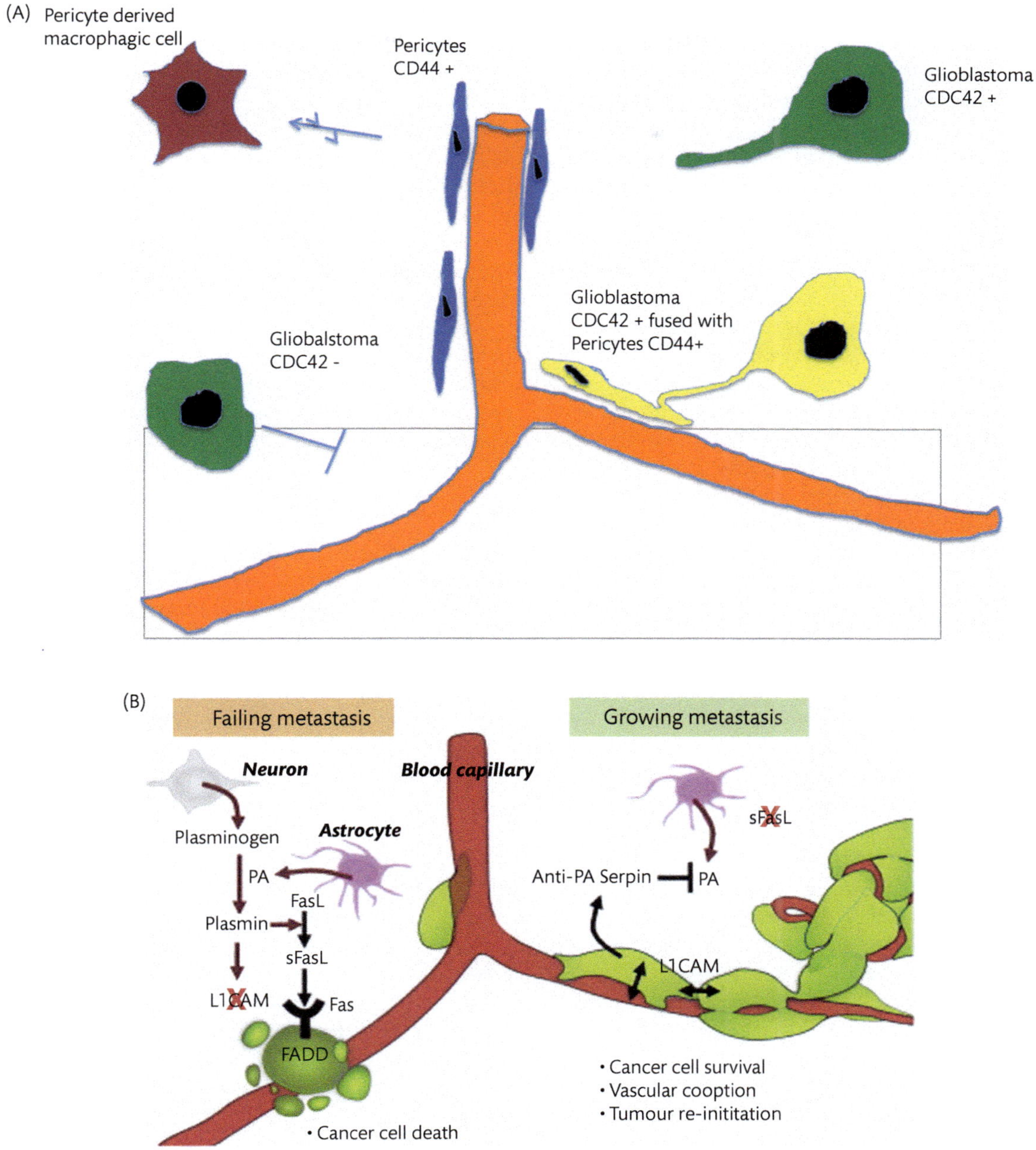

Fig. 22.8 Models for mechanism of brain vascular co-option of pre-existing normal vessels. (A) Co-option by a primary brain tumour, Glioblastoma Multiforme (GBM). On the right side a CDC42 positive GBM cell (green) is extending a flectopodia in direction of the pericytes and when in contact, the cell membrane of the two cells fuse and a hybrid cell is created (yellow). On the left side a GBM cell has had the expression of CD42 inhibited and is unable to get in contact with the pericytes. Some pericytes instead differentiate into cells with macrophagic antitumour activity. (B) Co-option by metastatic carcinoma cells. On the right, the extravasated metastatic cells expressing Serpin and L1CAM, co-opt the vessel. Serpin is released and inactivates the plasminogen activator (PA) produced by the astrocytes. This leads to prevention of L1CAM destruction and sFasL-induced apoptosis of the cancer cells. When Serpin is blocked, on the left side, PA cleaves plasminogen into active plasmin which, on one side, inactivates L1CAM blocking adhesion and co-option, and on the other releases sFasL which induces apoptotic death of the metastatic cells.

Source: data from Caspani E et al. 'Glioblastoma: A Pathogenic Crosstalk between Tumor Cells and Pericytes', *PLoS ONE*, Volume 9, Issue 7, e101402, Copyright © 2014 Caspani et al. This is an open-access article distributed under the terms of the Creative Commons Attribution License, which permits unrestricted use, distribution, and reproduction in any medium, provided the original author and source are credited. (B) Reproduced with permission from Valiente M et al. 'Serpins promote cancer cell survival and vascular co-option in brain metastasis', *Cell*, Volume 156, pp. 1002–16, Copyright © 2014 Elsevier Inc. All rights reserved. https://www.sciencedirect.com/science/article/pii/S0092867414000920.

that this process could actually occur in humans is supported by immunohistochemical stains demonstrating increased levels of CDC42 and CD44 expression in perivascular GBM cells as assessed by GBM histological sections.

Both in a murine orthotopic and in a chorioallantoic membrane model of glioblastoma using the cell line U87, the IRE1 gene, which encodes a stress sensing protein that triggers the unfolded protein response, is a key regulator for angiogenesis. Wild type U87 glioma cells expressing this gene grow as a discrete, highly angiogenic mass, while the U87 cells, in which the expression of IRE1 has been blocked, grow as non-angiogenic and more infiltrative tumours co-opting pre-existing vessels. These non-angiogenic cells have a lower expression of proangiogenic genes encoding proteins such as IL1beta, IL6, IL8, BEFGA, MMP1, MMP9, and PLAU while there is upregulation of SPARC (Auf et al., 2010).

In brain metastases

With respect to brain metastases by other types of tumour, Valiente and colleagues (Valiente et al., 2014) studied characteristics of tumour cells co-opting brain vessels (Fig. 22.8B). Through analyses of human tumours and also data obtained from mouse models, they showed how metastatic cells resist apoptosis and co-opt brain vessels by expressing the protein neuroserpin, which blocks the generation of plasmin. Plasmin protects the brain from metastasis by promoting the apoptosis of cancer cells and inhibiting their spread along the vasculature—a defensive reaction of the brain tissue to malignant cells entering the parenchyma. Linking these two events (the ingress of metastatic tumour cells and the production of plasmin) is the astrocyte which, in response to inflammation, parenchymal damage, and the passage of metastatic cells across the brain barrier, expresses high levels of two proteins: Fas ligand (FasL) and plasminogen activator (PA). Increased levels of PA lead to cleavage of plasminogen (which is secreted by neurons) into the active form, plasmin, which can then act on several other proteins, including FasL and the adhesion molecule L1CAM. FasL is bound to the membrane of astrocytes, but plasmin is able to cleave and release it as soluble form, sFasL, which in turn can link to its receptor, Fas, on the tumour cell and trigger apoptosis. However, in the presence of cancer cells expressing high levels of neuroserpin, the astrocyte's production of PA is inhibited, thereby blocking the release of plasmin from plasminogen, and suppressing the secretion of sFasL and thus apoptosis of tumour cells.

Once the increased survival of the cancer cells is achieved there is the problem of how they will adhere to the co-opted vessels; a number of studies have been undertaken to examine the relationship between neoplastic cells and different components of the vessels: endothelium, basal membrane, and pericytes. One mechanism follows again neuroserpin expression: by inhibiting generation of plasmin, the L1CAM molecule expressed by the tumour cell is not cleaved and remains intact, thus allowing the metastatic cells to adhere to the outbound surface of the vessels (Valiente et al., 2014).

Models of metastatic breast cancer and melanoma vascular co-option *in vivo* and cancer cell-endothelium adhesion *in vitro* are reported to be dependent upon the induction of Connexin 26 (Cx26) and Connexin 43 (Cx43) expression by the 'twist' transcription factor. These two connexins induce the formation of functional gap junctions between cancer cells and endothelium. Silencing of these connexins in both breast adenocarcinoma and melanoma cells results in loss of the ability not only to adhere at the endothelial cells *in vitro* but also to co-opt vessels *in vivo* in zebra fish, chicken embryos, and mouse models (Stoletov et al., 2013), while presence of the beta1 integrin mediates the adhesion between neoplastic cells and the vascular basal membrane (Carbonell et al., 2009).

But what are the effects on the vessels of being co-opted?

In models of glioma, lung carcinoma, and breast adenocarcinoma, Holash and coworkers (Holash et al., 1999) have observed vascular co-option, both in the brain and the lung, but only as a transient phenomenon in the early stages of tumour growth, before the triggering of angiogenesis. Adult mouse blood vessels express high levels of angiopoietin-1 which antagonizes the action of angiopoietin-2 with the Tie2 receptor. This maintains the stability of mature vessels by inducing anchorage to the basal membrane and protection from apoptosis. The first event observed following co-option was an increase in the levels of angiopoietin-2 in the pre-existing vessels surrounded by tumour cells. Higher levels of angiopoietin-2, which compete for Tie2, without increases of VEGF expression, induced vascular regression by detachment of the endothelium from the basal membrane. However, in this model VEGF expression was induced on the edge of the tumour, triggering neoangiogenesis sprouting from the angiopoietin-2 regressing vessels. It is therefore likely that a different response occurs in long term non-angiogenic cancer.

Vascular co-option and extravascular migratory metastasis

It is widely accepted that metastatic cells, in order to spread, need to enter the vessels and subsequently extravasate. However, alongside the role of vascular co-option in sustaining the growth of a neoplastic lesion, another role has also been proposed: as a mechanism for metastatic spread. Melanoma cells have been demonstrated by histological observations to dispose themselves along blood vessels, so-called angiotropism. Neural Crest cells can migrate on the abluminal surface of vessels in chick embryo models and melanoma cells, implanted in the same model, 'travelling' in a comparable way, possibly by reviving some embryonic properties. As the cancer cells replace the pericytes on the abluminal surfaces, their ability to migrate and avoid intravasation has been called pericyte mimicry (Lugassy et al., 2014). *In vivo* experiments in a mouse model have also demonstrated pericyte mimicry. GFP-positive human melanoma cells were injected into immune-compromised Nu/J mice brain whose vessels had been injected with fluorescent lectin. Three-dimensional reconstruction showed the tumour cells progressing outside the vessels, not inside them. Strong serpin B2 expression was associated with angiotropic cells and presence of laminin with the basal membrane of the co-opted vessels, alongside laminin subunit gamma 2 (LAMC2). Laminin subunit alpha 4 (LAMA4) and integrin subunit beta 3 (ITGB3) were found upregulated in angiotropic melanoma cells compared to non-angiotropic cells. After 4 weeks, cells were found to have migrated, on average, for 581 micron meters (Lugassy et al., 2014)

These findings suggest that vascular co-option and angiotropism/pericytic mimicry are closely related, if not identical processes in animal models. Whether extravascular migratory metastasis plays a role in cancer metastasis in human tumours needs remains to be fully established.

Blood vessels and antiangiogenic treatment

One of the early theoretical incentives and rationale for undertaking basic and clinical research of antiangiogenic therapy was the hypothesis that such agents would not be susceptible to acquired drug resistance (Kerbel, 1991). The thinking behind this hypothesis was that the endothelial cells lining newly forming blood vessels being normal host cells, would lack the extensive genetic abnormalities associated with tumour cells—and which are known to be a major factor driving the evolutionary development of acquired resistance to virtually every type and form of cancer therapy (Kerbel, 1991). Some early preclinical studies of antiangiogenic drugs such as endostatin or TNP-470 did appear to provide some preliminary evidence in support of this hypothesis (Boehm et al., 1997). However, extensive clinical trial results and experience in real world practice with patients, as well as many follow up preclinical studies, have indicated that both intrinsic and acquired resistance are a common (and unfortunate) occurrence with antiangiogenic therapies. Moreover, the mechanisms that have been reported or hypothesized to account for resistance to antiangiogenic drugs, especially VEGF pathway targeting agents, are both numerous and diverse, and have been discussed in numerous reviews (Bergers and Hanahan, 2008; Rini and Atkins, 2009; Bottsford-Miller et al., 2012; Kuczynski et al., 2015). These proposed mechanisms mostly come from preclinical studies. Figure 22.9 and Figure 22.10 list several of these mechanisms. A few of some better-known mechanisms are discussed next.

Redundancy of proangiogenic pathways

Perhaps the most studied and generally accepted hypothesis for acquired resistance to a targeted antiangiogenic drug is based on redundancy of proangiogenic pathways (Casanovas et al., 2005). Simply put, there are multiple proangiogenic factors such that targeting one may lead to the induction or upregulation of another through a bypass compensatory mechanism. Thus targeting VEGF functions using antibodies to VEGF or VEGFR-2 or using small molecule TKIs that target VEGF receptors, may initially provide an antiangiogenic and hence concurrent antitumour effect (when angiogenesis is present in the tumours being treated) which is lost over time as a result of induction or upregulation of an alternative proangiogenic factor/pathway such as interleukin-8, basic FGF (Casanovas et al., 2005) angiopoietin-2 (Rigamonti et al., 2014), among a number of others including inflammatory cytokines. Hence concurrently or sequentially targeting the upregulated bypass pathway may prolong the benefits of an antiangiogenic therapy, for example, targeting both the angiopoietin-2/tie-2 and VEGF/VEGFR-2 pathways or both bFGF/bFGF receptor/VEGF/VEGFR-2 (Casanovas et al., 2005; Rigamonti et al., 2014). While the body of work supporting this theory is substantial, it is almost entirely based on preclinical studies and as such, has not yet been definitively validated in the clinic. The hypothesis for the compensatory/bypass redundancy mechanism is based largely on the notion that an antiangiogenic therapy/drug, by reducing blood vessel numbers, and functions including flow and perfusion, would inevitably lead to a drop in tumour oxygenation (i.e. an increase in tumour hypoxia). This would trigger the upregulation of transcription factors such as hypoxia inducible factor 1α (HIF-1) or hypoxia inducible factor 2α(HIF-2α) which would then induce the expression or upregulation of other proangiogenic factors, for example, bFGF, but also including VEGF itself or other forms of VEGF in addition to VEGF-A (Casanovas et al., 2005). Hence, another way of delaying or preventing, at least in theory, this form of resistance to antiangiogenic drugs would be to target the expression or function of HIF-1α (or HIF-2α). While there are no specific validated inhibitors of these transcription factors, there are a diverse range of drugs that have, as a secondary effect, inhibition of HIF-1α, for example, topoisomerase-2 inhibitors such as topotecan (Rapisarda and Melillo, 2012) or an investigational nanoparticle camptothecin drug conjugate called CRLX101 (Pham et al., 2015; Pham et al., 2016). Combination of such drugs with an agent such as bevacizumab, therefore, should prevent or suppress the induced upregulation of HIF-1α by bevacizumab, and as such, would constitute a seemingly ideal combination treatment partner to increase the benefits of the antiangiogenic drug (Rapisarda and Melillo, 2012). There is evidence to support this hypothesis, based mainly, once again, on preclinical studies (Rapisarda et al., 2009; Pham et al., 2015; Pham et al., 2016).

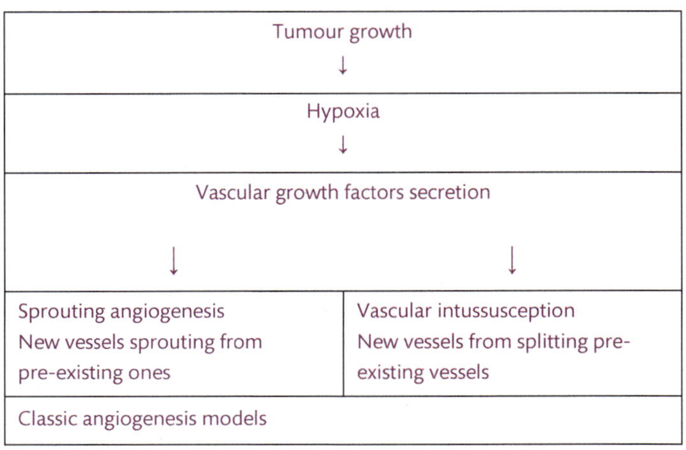

Fig. 22.9 Schematic events occurring during cancer growth cancer according to the classic theory of angiogenesis.

Intratumour blood vessels heterogeneity

Another mechanism of resistance is related to the heterogeneous nature of the tumour vasculature, including the angiogenic vasculature populating tumours, and this especially applies to spontaneous tumours that have existed for a considerable period of time—perhaps years—in patients, as opposed to rapidly growing transplanted tumours in mice.

The work of Dvorak and colleagues has revealed evidence for such extensive tumour blood vessel heterogeneity and the fact that certain types of abnormal tumour blood vessels are not necessarily VEGF-dependent (Sitohy et al., 2011). Moreover, antiangiogenic drug treatments may alter such morphologic and functional tumour blood vessel heterogeneity, and indeed, 'vessel normalization' in tumours as a consequence of antiangiogenic therapy may be an example of this (Goel et al., 2012). The nature of this heterogeneity may be strongly influenced by metastatic spread, and this may help explain why in some preclinical studies distant lung metastases are unresponsive to a drug such as sunitinib, in contrast the same tumour cell line when treated as a primary tumour mass (Guerin et al., 2013). Such results highlight one of the deficiencies of most types of conventional mouse tumour therapy models, which rarely involve treatment of established visceral metastatic disease (Francia et al., 2011).

Genetic mutations

Some other mechanisms that have been hypothesized to account for resistance to antiangiogenic drugs, interestingly, involve tumour cell associated genetic mutations. For example, mutation or loss of

Tumour growth ↓					
↓ Hypoxia					
Vascular growth factors secretion					
?	?	?	↓	↓	↓
Non-angiogenic tumours	Vasculogenic mimicry	Vasculogenic mimicry	Vasculogenesis	Angiogenesis	Angiogenesis
No new vessels or channels formed	Channels formed from stem cells	Vessels formed from cancer stem cells	Vessels formed from mobilized normal marrow stem cells	Sprouting (classic) angiogenesis New vessesls sprouting from pre-existing one	Vascular intussusception New vessesls from splitting of pre-existing one
Co-option of pre-existing vessels		? Co-option of new vessels	? Co-option of new vessels	? Co-option of new vessels	? Co-option of new vessels
All these patterns can co-exist within a single neoplastic lesion. Synchronous lesions: each lesion can have a different pattern Patterns can change in any order during disease progression					

Fig. 22.10 Actual events occurring during cancer growth according to the present status of knowledge. Induction of classic angiogenesis is not a necessary event.

the p53 suppressor gene can lead to relative resistance to increased hypoxia (Yu et al., 2002; Panka et al., 2013). This could allow tumours being treated with an antiangiogenic drug to survive and possibly continue to grow under highly hypoxic conditions. There is some limited evidence in support of this hypothesis; for example, colorectal cancer cells or renal cancer cells which are deficient in p53 functions appear less responsive to antiangiogenic therapies targeting the VEGF pathway, at least in preclinical studies (Yu et al., 2002).

Vascular co-option

As stated earlier, there are many proposed mechanisms to account for resistance to antiangiogenic therapies, too numerous to review here. However, a number of detailed reviews have been published on this topic, which can be consulted for more details (Bergers and Hanahan, 2008; Rini and Atkins, 2009; Bottsford-Miller et al., 2012; Kuczynski et al., 2015). These published reviews have either neglected to discuss or have minimized the impact of vessel co-option as possibly the single most important clinical factor in explaining intrinsic resistance to antiangiogenic drugs (Donnem et al., 2013). Moreover, evidence is beginning to emerge that vessel co-option may also be a driver of acquired resistance as well (Rubenstein et al., 2000; Grunewald et al., 2006; Frentzas et al., 2016; Kuczynski et al., 2016). Considering that vessel co-option is likely a common event in some of the most common sites of primary and secondary malignancy, for example, the lungs, liver, brain, and lymph nodes (Pezzella et al., 1996; Pezzella et al., 1997; Naresh et al., 2001; Vermeulen et al., 2001; Donnem et al., 2013; Jeong et al., 2015), it is highly likely that it is a major determinant for explaining the limited clinical benefits as well as failed randomized clinical trials evaluating antiangiogenic drugs. These failed trials now include more than a dozen randomized phase III trials of adjuvant therapy of early stage micro-metastatic breast, colorectal, lung, renal, and liver cancers. Vessel co-option and lack of tumour angiogenesis may have contributed to these disappointing results as well. Hence, the need to investigate the molecular, biochemical, metabolic, and physiologic characteristics of co-opted tumour vessels represents an important new area of investigation in 'antivascular' therapy. If co-opted vessels can indeed be targeted with a reasonable therapeutic index, it would open up an entirely new research area of experimental therapeutics—one that could have ramifications for other types of therapy as well. Thus, studies evaluating the possible vulnerability of co-opted vessels to immune therapies, including vascular disrupting agents (i.e. 'VDAs'), oncolytic viral therapy, immune checkpoint inhibitors, chemotherapy regimens, including metronomic chemotherapy, radiation therapy, among many others, would appear to be a promising area of research in the vascular biology of tumours. This could apply to other types of therapy as well. To cite one example, recent studies showed that either sorafenib treatment of preclinical orthotopic liver tumours (Kuczynski et al., 2016) or colorectal cancer patients treated with bevacizumab treatment of human liver metastases from breast cancer (Frentzas et al., 2016) led to an initial antiangiogenic and antitumour effect which is however lost as resistance developed to the therapy as it always does in hepatocellular carcinoma patients treated with sorafenib (Kuczynski et al., 2015) or bevacizumab in other indications (Frentzas et al., 2016). In the mouse model the development of resistance was preceded by an increase in invasiveness of the HCC cells and a transition to an epithelial to mesenchyme transition (EMT) phenotype (Kuczynski et al.,

2015). Thus, this process or switch to relying on vessel co-option from sprouting neoangiogenesis, induced by the sorafenib treatment, may be delayed, or prevented by using drugs which prevent tumour cell invasion/migration or the switch to an EMT phenotype. Although this would not necessarily target co-opted vessels per se, it could prevent the actual switch to co-option and thus prolong the benefit of therapy.

Conclusion

In summary, the ways in which tumours can develop resistance, both intrinsic or acquired to antiangiogenic drugs—especially those that target the VEGF pathway—are surprisingly numerous and varied in nature. Among them, and quite possibly the most important and underestimated, is the impact of vessel co-option. Understanding these mechanisms and especially the impact of vessel co-option, may open up promising new ways to improve the therapeutic impact of targeting the tumour vasculature.

TAKE-HOME MESSAGE

- The blood supply of a tumour can be provided not only by neovascularization, but also by the ability of tumours to exploit the pre-existing host vasculature in various organs and expanding into non-angiogenic primary and metastatic tumours.
- Alongside pure non-angiogenic tumours, a mixed pattern of pre-existing and newly formed vessels is also commonly seen. Therefore, contrary to the theory of Folkman (Folkman, 1990), still regarded as one of the hallmarks of cancer (Hanahan and Weinberg, 2011) it is now well established that some tumours can grow and metastasize in the absence of angiogenesis (Pezzella et al., 1996). The reason for the delay in reaching a more global view of the role of blood vessels in cancer can likely be traced to a gap between bedside and benchtop research (Ebos and Kerbel, 2011)
- One biological implication is that the triggering of hypoxia related pathways does not necessarily lead to angiogenesis, and that targeting the tumour blood supply using existing antiangiogenic drugs, may fail in some circumstances because of vessel co-option. Following the initial mostly modest results obtained so far with antiangiogenic drugs (Ebos and Kerbel, 2011) understanding the mechanisms driving this behaviour is likely to generate new antivascular therapy approaches for tumours that are intrinsically resistant to conventional antiangiogenic drugs or which acquire resistance over time.

OPEN QUESTIONS

- How common is vessel co-option in human cancer?
- Are co-opted vessels in organs such as the liver, lungs, brain, and lymph nodes targetable using new or existing 'antivascular' therapies?
- Is vessel co-option the most important factor in explaining why antiangiogenic therapies tend to show much better therapeutic results in animal studies, compared to patients?

FURTHER READING

BMC. A collection of papers on cancer and blood vessels from the *Chinese Journal of Cancer*. Available at: http://www.biomedcentral.com/collections/tumorangiogenesis

Hendrix, M. J. C. (2015). Cancer: an extravascular route for tumour cells. *Nature*, **520**, 300–2.

Nature Publishing Group. A collection of classic angiogenesis papers. Available at: http://www.nature.com/focus/angiogenesis/classics/establish.html

Pezzella, F. & Gatter, K. C. (2016). Evidence showing that tumors can grow without angiogenesis and can switch between angiogenic and nonangiogenic phenotypes. *J Natl Cancer Inst*, **108**(8), djw032.

Sadler, T. W. (2014). *Langman's medical embryology*, 13th edition. Wolter Kluwer, Philadelphia.

Zetter, B. R. (2008). The scientific contributions of M. Judah Folkman to cancer research. *Nat Rev Cancer*, **8**, 647–54.

REFERENCES

Breast Cancer Progression Working Party (2000). Evidence for novel non-angiogenic pathway in breast-cancer metastasis. *Lancet*, **355**, 1787–8.

Adighibe, O. (2012). *Loss of chaperone protein in human cancer.* DPhil, University of Oxford.

Adighibe, O., Micklem, K., Campo, L., et al. (2006). Is nonangiogenesis a novel pathway for cancer progression? A study using 3-dimensional tumour reconstructions. *Br J Cancer*, **94**, 1176–9.

Almeida, B. M., Challacombe, S. J., Eveson, J. W., Smith, C. G., & Leigh, I. M. (1992). A novel lamina lucida component of epithelial and endothelial basement membranes detected by LH39 monoclonal antibody. *J Pathol*, **166**, 243–53.

Auf, G., Jabouille, A., Guerit, S., et al. (2010). Inositol-requiring enzyme 1alpha is a key regulator of angiogenesis and invasion in malignant glioma. *Proc Natl Acad Sci U S A*, **107**, 15553–8.

Augustin, H. G., Koh, G. Y., Thurston, G., & Alitalo, K. (2009). Control of vascular morphogenesis and homeostasis through the angiopoietin-Tie system. *Nat Rev Mol Cell Biol*, **10**, 165–77.

Bergers, G. & Hanahan, D. (2008). Modes of resistance to anti-angiogenic therapy. *Nat Rev Cancer*, **8**, 592–603.

Berghoff, A. S., Rajky, O., Winkler, F., et al. (2013). Invasion patterns in brain metastases of solid cancers. *Neuro Oncol*, **15**, 1664–72.

Boehm, T., Folkman, J., Browder, T., & O'Reilly, M. S. (1997). Antiangiogenic therapy of experimental cancer does not induce acquired drug resistance. *Nature*, **390**, 404–7.

Bottsford-Miller, J. N., Coleman, R. L., & Sood, A. K. (2012). Resistance and escape from antiangiogenesis therapy: clinical implications and future strategies. *J Clin Oncol*, **30**, 4026–34.

Bugyik, E., Dezso, K., Reiniger, L., et al. (2011). Lack of angiogenesis in experimental brain metastases. *J Neuropathol Exp Neurol*, **70**, 979–91.

Bussolati, B., Grange, C., & Camussi, G. (2011). Tumor exploits alternative strategies to achieve vascularization. *FASEB J*, **25**, 2874–82.

Caduff, J. H., Fischer, L. C., & Burri, P. H. (1986). Scanning electron microscope study of the developing microvasculature in the postnatal rat lung. *The Anatamical Record*, **216**, 154–64.

Carbonell, W. S., Ansorge, O., Sibson, N., & Muschel, R. (2009). The vascular basement membrane as 'soil' in brain metastasis. *PLoS One*, **4**, e5857.

Casanovas, O., Hicklin, D. J., Bergers, G., & Hanahan, D. (2005). Drug resistance by evasion of antiangiogenic targeting of VEGF signaling in late-stage pancreatic islet tumors. *Cancer Cell*, **8**, 299–309.

Caspani, E. M., Crossley, P. H., Redondo-Garcia, C., & Martinez, S. (2014). Glioblastoma: a pathogenic crosstalk between tumor cells and pericytes. *PLoS One*, **9**, e101402.

Crivellato, E., Nico, B., Vacca, A., Djonov, V., Presta, M., & Ribatti, D. (2004). Recombinant human erythropoietin induces intussusceptive microvascular growth *in vivo*. *Leukemia*, **18**, 331–6.

De La Puente, P., Muz, B., Azab, F., & Azab, A. K. (2013). Cell trafficking of endothelial progenitor cells in tumor progression. *Clin Cancer Res*, **19**, 3360–8.

Di Tomaso, E., Snuderl, M., Kamoun, W. S., et al. (2011). Glioblastoma recurrence after cediranib therapy in patients: lack of 'rebound' revascularization as mode of escape. *Cancer Res*, **71**, 19–28.

Dome, B., Timar, J., & Paku, S. (2003). A novel concept of glomeruloid body formation in experimental cerebral metastases. *J Neuropathol Exp Neurol*, **62**, 655–61.

Donnem, T., Hu, J., Ferguson, M., et al. (2013). Vessel co-option in primary human tumors and metastases: an obstacle to effective anti-angiogenic treatment? *Cancer Med*, **2**, 427–36.

Dvorak, H. F., Brown, L. F., Detmar, M., & Dvorak, A. M. (1995). Vascular permeability factor/vascular endothelial growth factor, microvascular hyperpermeability, and angiogenesis. *Am J Pathol*, **146**, 1029–39.

Ebos, J. M. & Kerbel, R. S. (2011). Antiangiogenic therapy: impact on invasion, disease progression, and metastasis. *Nat Rev Clin Oncol*, **8**, 210–21.

Ferguson, M. (2008). *Angiogenesis in Human Lung Tumours*. DPhil, University of Oxford.

Folkman, J. (1971). Tumor angiogenesis: therapeutic implications. *N Engl J Med*, **285**, 1182–6.

Folkman, J. (1990). What is the evidence that tumors are angiogenesis dependent? *J Natl Cancer Inst*, **82**, 4–6.

Francia, G., Cruz-Munoz, W., Man, S., Xu, P., & Kerbel, R. S. (2011). Mouse models of advanced spontaneous metastasis for experimental therapeutics. *Nat Rev Cancer*, **11**, 135–41.

Frentzas, S., Simoneau, E., Bridgeman, V. L., et al. (2016). Vessel co-option mediates resistance to anti-angiogenic therapy in liver metastases. *Nat Med*, **22**(11), 1294–302.

Gimbrone, M. A., JR., Leapman, S. B., Cotran, R. S., & Folkman, J. (1972). Tumor dormancy *in vivo* by prevention of neovascularization. *J Exp Med*, **136**, 261–76.

Goel, S., Wong, A. H., & Jain, R. K. (2012). Vascular normalization as a therapeutic strategy for malignant and nonmalignant disease. *Cold Spring Harb Perspect Med*, **2**, a006486.

Grunewald, M., Avraham, I., Dor, Y., et al. (2006). VEGF-induced adult neovascularization: recruitment, retention, and role of accessory cells. *Cell*, **124**, 175–89.

Guerin, E., Man, S., Xu, P., & Kerbel, R. S. (2013). A model of postsurgical advanced metastatic breast cancer more accurately replicates the clinical efficacy of antiangiogenic drugs. *Cancer Res*, **73**, 2743–8.

Guidi, A. J., Abu-Jawdeh, G., Berse, B., et al. (1995). Vascular permeability factor (vascular endothelial growth factor) expression and angiogenesis in cervical neoplasia. *J Natl Cancer Inst*, **87**, 1237–45.

Guidi, A. J., Fischer, L., Harris, J. R., & Schnitt, S. J. (1994). Microvessel density and distribution in ductal carcinoma *in situ* of the breast. *J Natl Cancer Inst*, **86**, 614–19.

Hall, A. P. (2005). The role of angiogenesis in cancer. *Comp Clin Path*, **13**, 95–9.

Hanahan, D. & Folkman, J. (1996). Patterns and emerging mechanisms of the angiogenic switch during tumorigenesis. *Cell*, **86**, 353–64.

Hanahan, D. & Weinberg, R. A. (2011). Hallmarks of cancer: the next generation. *Cell*, **144**, 646–74.

Hendrix, M. J., Seftor, E. A., Hess, A. R., & Seftor, R. E. (2003). Vasculogenic mimicry and tumour-cell plasticity: lessons from melanoma. *Nat Rev Cancer*, **3**, 411–21.

Holash, J., Maisonpierre, P. C., Compton, D., et al. (1999). Vessel cooption, regression, and growth in tumors mediated by angiopoietins and VEGF. *Science*, **284**, 1994–8.

Hu, J., Bianchi, F., Ferguson, M., et al. (2005). Gene expression signature for angiogenic and nonangiogenic non-small-cell lung cancer. *Oncogene*, **24**, 1212–19.

Ide, A. G., Baker, N. H., & Warren, S. L. (1939). Vascularization of the Brown Pearce rabbit epithelioma transplant as seen in the transparent ear chamber. *Am J Roentgenol*, **42**, 891–9.

Jeong, H. S., Jones, D., Liao, S., et al. (2015). Investigation of the lack of angiogenesis in the formation of lymph node metastases. *J Natl Cancer Inst*, **107**(9), pii: djv155.

Kerbel, R. S. (1991). Inhibition of tumor angiogenesis as a strategy to circumvent acquired resistance to anti-cancer therapeutic agents. *Bioessays*, **13**, 31–6.

Kerbel, R. S. (2008). Tumor angiogenesis. *N Engl J Med*, **358**, 2039–49.

Kuczynski, E. A., Lee, C. R., Man, S., Chen, E., & Kerbel, R. S. (2015). Effects of sorafenib dose on acquired reversible resistance and toxicity in hepatocellular carcinoma. *Cancer Res*, **75**, 2510–19.

Kuczynski, E. A., Yin, M., Bar-Zion, A., et al. (2016). Co-option of liver vessels and not sprouting angiogenesis drives acquired sorafenib resistance in hepatocellular carcinoma. *J Natl Cancer Inst*, **108**(8). doi: 10.1093/jnci/djw030.

Kunkel, P., Ulbricht, U., Bohlen, P., et al. (2001). Inhibition of glioma angiogenesis and growth *in vivo* by systemic treatment with a monoclonal antibody against vascular endothelial growth factor receptor-2. *Cancer Res*, **61**, 6624–8.

Leenders, W. P., Kusters, B., Verrijp, K., et al. (2004). Antiangiogenic therapy of cerebral melanoma metastases results in sustained tumor progression via vessel co-option. *Clin Cancer Res*, **10**, 6222–30.

Lugassy, C., Zadran, S., Bentolila, L. A., et al. (2014). Angiotropism, pericytic mimicry and extravascular migratory metastasis in melanoma: an alternative to intravascular cancer dissemination. *Cancer Microenviron*, **7**, 139–52.

Makanaya, A. N., Hlushchuk, R., & Djonov, V. G. (2009). Intussusceptive angiogenesis and its role in vascular morphogenesis, patterning, and remodeling. *Angiogenesis*, **12**, 113–23.

Maniotis, A. J., Folberg, R., Hess, A., et al. (1999). Vascular channel formation by human melanoma cells *in vivo* and *in vitro*: vasculogenic mimicry. *Am J Pathol*, **155**, 739–52.

Montana, V. & Sontheimer, H. (2011). Bradykinin promotes the chemotactic invasion of primary brain tumors. *J Neurosci*, **31**, 4858–67.

Naresh, K. N., Nerurkar, A. Y., & Borges, A. M. (2001). Angiogenesis is redundant for tumour growth in lymph node metastases. *Histopathology*, **38**, 466–70.

Natale, G., Bocci, G., & Lenzi, P. (2016). Looking for the word 'angiogenesis' in the history of health sciences: from ancient times to the first decades of the twentieth century. *World J Surg*, **41**(6). doi: 10.1007/s00268-016-3680-1.

Nolan, D. J., Ciarrocchi, A., Mellick, A. S., et al. (2007). Bone marrow-derived endothelial progenitor cells are a major determinant of nascent tumor neovascularization. *Genes Dev*, **21**, 1546–58.

Panka, D. J., Liu, Q., Geissler, A. K., & Mier, J. W. (2013). Effects of HDM2 antagonism on sunitinib resistance, p53 activation, SDF-1 induction, and tumor infiltration by CD11b+/Gr-1+ myeloid derived suppressor cells. *Mol Cancer*, **12**, 17.

Passalidou, E., Trivella, M., Singh, N., et al. (2002). Vascular phenotype in angiogenic and non-angiogenic lung non-small cell carcinomas. *Br J Cancer*, **86**, 244–9.

Patan, S. (1998). TIE1 and TIE2 receptor tyrosine kinases inversely regulate embryonic angiogenesis by the mechanism of intussusceptive microvascular growth. *Microvasc Res*, **56**, 1–21.

Patan, S., Munn, L. M., & Jain, R. K. (1996). Intussusceptive microvascular growth in a human colon adenocarcinoma xenograft: a novel mechanism of tumor angiogenesis. *Microvasc Res*, **51**, 260–72.

Paulis, Y. W., Soetekouw, P. M., Verheul, H. M., Tjan-Heijnen, V. C., & Griffioen, A. W. (2010). Signalling pathways in vasculogenic mimicry. *Biochim Biophys Acta*, **1806**, 18–28.

Peichev, M., Naiyer, A. J., Pereira, D., et al. (2000). Expression of VEGFR-2 and AC133 by circulating human CD34(+) cells identifies a population of functional endothelial precursors. *Blood*, **95**, 952–8.

Pezzella, F., Di Bacco, A., Andreola, S., Nicholson, A. G., Pastorino, U., & Harris, A. L. (1996). Angiogenesis in primary lung cancer and lung secondaries. *Eur J Cancer*, **32A**, 2494–500.

Pezzella, F., Pastorino, U., Tagliabue, E., et al. (1997). Non-small-cell lung carcinoma tumor growth without morphological evidence of neo-angiogenesis. *Am J Pathol*, **151**, 1417–23.

Pham, E., Birrer, M. J., Eliasof, S., et al. (2015). Translational impact of nanoparticle-drug conjugate CRLX101 with or without bevacizumab in advanced ovarian cancer. *Clin Cancer Res*, **21**, 808–18.

Pham, E., Yin, M., Peters, C. G., et al. (2016). Preclinical efficacy of bevacizumab with CRLX101, an investigational nanoparticle-drug conjugate, in treatment of metastatic triple-negative breast cancer. *Cancer Res*, **76**, 4493–503.

Phng, L. K. & Gerhardt, H. (2009). Angiogenesis: a team effort coordinated by notch. *Dev Cell*, **16**, 196–208.

Rapisarda, A., Hollingshead, M., Uranchimeg, B., et al. (2009). Increased antitumor activity of bevacizumab in combination with hypoxia inducible factor-1 inhibition. *Mol Cancer Ther*, **8**, 1867–77.

Rapisarda, A. & Melillo, G. (2012). Overcoming disappointing results with antiangiogenic therapy by targeting hypoxia. *Nat Rev Clin Oncol*, **9**, 378–90.

Rigamonti, N., Kadioglu, E., Keklikoglou, I., Wyser Rmili, C., Leow, C. C., & De Palma, M. (2014). Role of angiopoietin-2 in adaptive tumor resistance to VEGF signaling blockade. *Cell Rep*, **8**, 696–706.

Rini, B. I. & Atkins, M. B. (2009). Resistance to targeted therapy in renal-cell carcinoma. *Lancet Oncol*, **10**, 992–1000.

Ritchie, A. C. (1962). The classification, morphology, and behaviour of tumours. In: Florey, H. (ed.) *General Pathology*, 3rd edition. London: Lloyd-Luke (Medical Books) Ltd, Chapter 22.

Rojiani, A. M. & Dorovini-Zis, K. (1996). Glomeruloid vascular structures in glioblastoma multiforme: an immunohistochemical and ultrastructural study. *J Neurosurg*, **85**, 1078–84.

Rubenstein, J. L., Kim, J., Ozawa, T., et al. (2000). Anti-VEGF antibody treatment of glioblastoma prolongs survival but results in increased vascular cooption. *Neoplasia*, **2**, 306–14.

Sardari Nia, P., Colpaert, C., Vermeulen, P., et al. (2008). Different growth patterns of non-small cell lung cancer represent distinct biologic subtypes. *Ann Thorac Surg*, **85**, 395–405.

Seftor, R. E., Hess, A. R., Seftor, E. A., et al. (2012). Tumor cell vasculogenic mimicry: from controversy to therapeutic promise. *Am J Pathol*, **181**, 1115–25.

Short, R. D. H. (1950). Alveolar epithelium in relation to growth of the lung. *Philos Trans R Soc Lond B Biol Sci*, **235**, 35–87.

Simonsen, T. G., Gaustad, J. V., & Rofstad, E. K. (2015). Intertumor heterogeneity in vascularity and invasiveness of artificial melanoma brain metastases. *J Exp Clin Cancer Res*, **34**, 150.

Sitohy, B., Nagy, J. A., Jaminet, S. C., & Dvorak, H. F. (2011). Tumor-surrogate blood vessel subtypes exhibit differential susceptibility to anti-VEGF therapy. *Cancer Res*, **71**, 7021–8.

Stoletov, K., Strnadel, J., Zardouzian, E., et al. (2013). Role of connexins in metastatic breast cancer and melanoma brain colonization. *J Cell Sci*, **126**, 904–13.

Straume, O. & Akslen, L. A. (2003). Increased expression of VEGF-receptors (FLT-1, KDR, NRP-1) and thrombospondin-1 is associated with glomeruloid microvascular proliferation, an aggressive angiogenic phenotype, in malignant melanoma. *Angiogenesis*, **6**, 295–301.

Sundberg, C., Nagy, J. A., Brown, L. F., et al. (2001). Glomeruloid microvascular proliferation follows adenoviral vascular permeability factor/vascular endothelial growth factor-164 gene delivery. *Am J Pathol*, **158**, 1145–60.

Szabo, V., Bugyik, E., Dezso, K., et al. (2015). Mechanism of tumour vascularization in experimental lung metastases. *J Pathol*, **235**, 384–96.

Terayama, N., Terada, T., & Nakanuma, Y. (1996). Histologic growth patterns of metastatic carcinomas of the liver. *Jpn J Clin Oncol*, **26**, 24–9.

Trivella, M., Pezzella, F., Pastorino, U., Harris, A. L., & Altman, D. G. (2007). Microvessel density as a prognostic factor in non-small-cell lung carcinoma: a meta-analysis of individual patient data. *Lancet Oncol*, **8**, 488–99.

Valiente, M., Obenauf, A. C., Jin, X., et al. (2014). Serpins promote cancer cell survival and vascular co-option in brain metastasis. *Cell*, **156**, 1002–16.

Van Den Eynden, G. G., Majeed, A. W., Illemann, M., et al. (2013). The multifaceted role of the microenvironment in liver metastasis: biology and clinical implications. *Cancer Res*, **73**, 2031–43.

Vermeulen, P. B., Colpaert, C., Salgado, R., et al. (2001). Liver metastases from colorectal adenocarcinomas grow in three patterns with different angiogenesis and desmoplasia. *J Pathol*, **195**, 336–42.

Weidner, N., Folkman, J., Pozza, F., et al. (1992). Tumor angiogenesis: a new significant and independent prognostic indicator in early-stage breast carcinoma. *J Natl Cancer Inst*, **84**, 1875–87.

Wesseling, P., Van Der Laak, J. A., De Leeuw, H., Ruiter, D. J., & Burger, P. C. (1994). Quantitative immunohistological analysis of the microvasculature in untreated human glioblastoma multiforme. Computer-assisted image analysis of whole-tumor sections. *J Neurosurg*, **81**, 902–9.

Wesseling, P., Vandersteenhoven, J. J., Downey, B. T., Ruiter, D. J., & Burger, P. C. (1993). Cellular components of microvascular proliferation in human glial and metastatic brain neoplasms. A light microscopic and immunohistochemical study of formalin-fixed, routinely processed material. *Acta Neuropathologica*, **85**, 508–14.

Willis, R. A. (1934). *The Spread of Tumours in the Human Body*. London: Churchill.

Wilting, J., Birkenhager, R., A., E., Kurz, H., et al. (1996). VEGF121 induces proliferation of vascular endothelial cells and expression of FLK-1 without affecting lymphatic vessels of the chorioallantoic membrane. *Dev Biol*, **176**, 76–85.

Winkler, F., Kienast, Y., Fuhrmann, M., et al. (2009). Imaging glioma cell invasion *in vivo* reveals mechanisms of dissemination and peritumoral angiogenesis. *Glia*, **57**, 1306–15.

Yu, J. L., Rak, J. W., Coomber, B. L., Hicklin, D. J., & Kerbel, R. S. (2002). Effect of p53 status on tumor response to antiangiogenic therapy. *Science*, **295**, 1526–8.

Zetter, B. R. (2008). The scientific contributions of M. Judah Folkman to cancer research. *Nat Rev Cancer*, **8**, 647–54.

23

Cancer immunology

Herman Waldmann

Why cancer immunology?

The reader may wonder why cancer immunology should be considered a topic distinct from immunology in general, or from transplantation immunology from which many fundamental concepts of the immune system have evolved. In confronting pathogens or transplants the immune system is often rapidly exposed to large doses of non-self-antigens in the context of molecular adjuvant-creating moieties (so-called molecular patterns) that elicit multiple cascades of immunity (Matzinger, 1994; Medzhitov and Janeway, 1997). In contrast, it is not easy to imagine how a 'self' cell which has recently developed sufficient mutations to become malignant, can provide adequate adjuvant-creating signals, let alone non-self-antigens, to evoke equivalent destructive immune responses. As the malignancy expands what cues, if any, can it give the immune system that it should respond?

Early conjectures were that growing 'self' tumour cells could somehow 'sneak through' the various pro-immunity checkpoints, so bypassing immune reactivity.

The issue of tumour immunogenicity has been, and continues to be, a dilemma for immunologists. Leaving aside virally induced tumours, would we not be tolerant of the protein's tumour expressed in our own tumours? Even if we were not naturally tolerant of some tumour antigens, would these not be shared with normal tissues? Presumably then, tumour immunity might also risk autoimmunity?

Although answers to these questions are still incomplete, the explosion in fundamental immunology has, over the past 25 years or so, provided the groundwork for a better understanding of the relationships between tumours and the immune system. The advent of monoclonal antibodies, both as diagnostics and as probes for molecular function, have been important. The identification of dendritic cells as a major intermediary between the antigen source and T-cell activation has been crucial (Banchereau and Steinman, 1998). Major advances in molecular biology and the creation of mutant mice lacking defined gene products, has pinpointed key molecules influencing immune function. Finally, many translational efforts in vaccination, autoimmune disease, and transplantation have enabled identification of hitherto undervalued mechanisms that the immune system uses to regulate itself. Historically, these had been poorly studied because it was easier to quantitate immune responses than the lack of them (i.e. self-tolerance). Moreover, as self-tolerance was largely seen as a fait accompli from birth, research into mechanisms of self-tolerance has always been particularly demanding.

Against the backdrop of advances in fundamental immunology, there have been several clues that the immune system could attack cancers, promoting their rejection as if they were 'foreign' transplants. From all these contributions, the notion that tumours simply sneak through the immune system has changed. We now accept that some tumours can evoke and/or provide sufficiently adjuvant-like signals which, if adequately sensed, can promote immune reactivity in the host. With this accepted a new hallmark of cancer can be defined as the cross-talk between tumours, their supportive stroma, and the immune system, that impedes immune reactivity using diverse inhibitory mechanisms acting throughout the evolution of that tumour (Hanahan and Weinberg, 2011). Such suppressive mechanisms, formerly studied solely in relation to carcinogenesis, are now becoming increasingly relevant to immunoregulation more generally (Zitvogel et al., 2013; Belvin and Mellman, 2015). At a practical level, an improved understanding of these mechanisms has suddenly created a major growth industry in practical efforts to give the immune system the upper hand. These range from conventional small molecule drugs, antibodies, cytokines, and cellular therapies, as well as a recognition that what were formerly thought of as chemotherapeutic drugs, are, in retrospect, agents that can also liberate host antitumour immune function in diverse ways. In short, our concepts of tumour immunology and our application of them are developing fast, being both informed by, but also informing, fundamental immunology.

The immune system and protection against pathogens

The vertebrate immune system has evolved to protect against microbial infection while minimizing damage to self. To achieve this, it uses innate mechanisms mediated by plasma proteins and diverse cells to contain infection long enough for the adaptive system of antigen-specific lymphocytes to engage, and to generate responses, best able to clear the microbe. The innate mechanisms have no specificity for the foreign antigen, but reflect recognition by preformed receptors of one or more forms of perturbation, including microbially-encoded molecular patterns, so-called pathogen-associated molecular patterns or PAMPS (Medzhitov and Janeway, 1997; Beutler and Rehli, 2002), comprising proteins, lipids, carbohydrates, and/or nucleic acids, and self-encoded molecular beacons of 'cell stress' or danger. The consequent innate effector mechanisms include activation of complement, enhancement of phagocytosis, and exocytosis by neutrophils, eosinophils, basophils, macrophages, natural killer (NK), γδ T cells, and other cells, all aimed at elimination of the pathogen.

In contrast to the rapidly-activated, generic innate response, the adaptive response is highly-specific but delayed. It depends upon the expansion and differentiation of clones of (thymic-derived) T cells and (bursa-equivalent) B cells that are selected by the binding of antigen to clonally distributed receptors produced by seemingly-random somatic gene rearrangement within T-cell and B-cell progenitors (Kurosawa and Tonegawa, 1982). Following their antigen-driven selection, the activated B cells and T cells generate differentiated progeny with effector function, and additionally commit cells to reservoirs that retain memory for the eliciting antigens. Whereas T-cell receptors of all progeny stay true to the parental cell, B-cell (immunoglobulin) receptors undergo somatic mutation of their heavy and light variable (V-) regions to derive better fitting (higher affinity) receptors, culminating in best-fit antibody molecules being released into the circulation. Antibodies of identical specificities can be made using distinct Heavy (H-) chains (classes and subclasses or isotypes) responsible for mediating different functions, some of these being critical to elimination of the organism or neutralization of its key attachment and virulence molecules, through localizing and amplifying innate mechanisms. Conversely, T-cell receptor structures are constant, but can be expressed by T cells of different effector potentials, thereby suiting the host response to the nature of the challenge, as is elaborated on next.

To a first approximation, the adaptive system uses antibodies to target pathogens that exist outside the body's cells but exploits T cells to deal with microbes that reside within cells. In most vertebrates studied in any detail, the great majority of circulating T-cells bear receptors (TCR) made up of so-called α and β chains, although this is not the case in, for example, chickens and in cattle during the first year of life, in which very many T cells bear TCRs composed of γ and δ chains. Moreover, seemingly in all vertebrates, T cells that reside within non-lymphoid tissues (e.g. skin and gut), are enriched in γδ T cells distinct from those in the circulation.

For αβ T cells to recognize intracellular protein antigens, it is necessary for the antigens to be processed into shorter peptide fragments which are loaded intracellularly into the clefts of major histocompatibility complex (MHC) Class I or MHC Class II molecules, thereby stabilizing those MHC molecules and their expression at the cell surface. MHC Class I-associated peptides provide the target for CD8 + T-cell-mediated cytotoxicity, while MHC Class II-associated peptides provide the target for CD4 + T cells.

In order for T cells to be activated, it is essential that they see their peptide antigen on activated 'antigen-presenting cells', particularly dendritic cells (DC) (Banchereau and Steinman, 1998; Mellman, 2013), which are specialized forms of innate myelomonocytic cells. Initial and necessary activation or 'licensing' of DC largely occurs through the previously mentioned innate-sensing receptors. Among these, Toll-like receptors (TLRs) are the best studied (Medzhitov and Janeway, 1997; Beutler and Rehli, 2002). Suffice it to say, the sensing of pathogens by engagement of TLRs and/or other equivalent receptors is crucial to the licensing of DC that is, in turn, critical for their capacity to activate naïve T cells. Licensing is manifest in an altered DC phenotype involving, in particular, upregulation of sets of costimulatory molecules such as CD40, CD80, and CD86. These interact with corresponding receptors on T cells to provide so-called 'second signals' that enable antigen to drive T cell activation. In short, it is the second signal that connects innate immunity to adaptive immunity. Without such licensing DC, antigen presentation will not only fail to immunize T cells but may even end up tolerating them. In fact, DC can be 'decommissioned' away from this licensing function in many ways, including exposure to diverse agents such as IL-10, TGFβ, Vitamin D, and interaction with regulatory T cells (Steinman et al., 2003; Hammer and Ma, 2013).

As was introduced earlier, antigen-specific CD4+ T cells can differentiate into subsets with different immune functions. Th1 cells making IFNγ, Th2 cells making IL-4, Th17-cells making IL-17, and T cells expressing Foxp3 and able to regulate other immune cells. These different functional subsets are responsible for orchestrating and sculpting the behaviour of diverse cellular elements through their interaction with MHC Class II-expressing myelomonocytic cells of which DC are key players. Not only can DC be licensed by diverse sensors of perturbation, but they can also be activated by CD4+ T cells, as a route to providing help to enable further cohorts of naïve CD4+ or CD8+ T cells to be recruited into immune responses (Ridge et al., 1998). Among these, CD8 + T cells are best known for their cytotoxic functions, but they too can secrete influential cytokines affecting their local milieu.

For an adaptive immune response to occur, antigens must gain entry to DC undergoing activation-induced enhancement of antigen processing and presentation. To facilitate their interactions with (their so-called 'priming of') naïve T cells, licensed DC migrate from the tissues, within which they were exposed to pathogens, moving to draining lymphoid tissues, wherein they promote the clonal proliferation and differentiation of those T cells with TCRs specific for the antigens presented by the licensed DC Thereafter, primed T cells re-enter the circulation, with some making their way to the peripheral sites where the initiating antigens are located. The clues that direct them to the correct locations arise from the inflammation evoked by the infection. Local inflammation induces the neighbouring post-capillary venules to upregulate distinct patterns of adhesion molecules (as if barcodes) which can divert the recently activated T cells and other leukocytes out of the circulation at that site and no other. Leukocytes with different functions perceive different barcodes, so determining the functional composition of white cells entering the tissue. The nature of the infective agent is indeed influential in determining what barcodes are exhibited on the endothelium. Once in the tissues, both CD4+ and CD8+ T cells can liberate influential cytokines, interact with tissue-resident and other tissue-infiltrating immune cells, and deliver protective functions. Altruistic self-sacrifice of virally infected tissue cells that can be easily replenished provides an effective way to eliminate pathogens.

While a proportion of T cells are recruited to peripheral tissues, others may be retained in the lymphoid tissues, as in the case of some CD4 helper T cells that localize to germinal centres where they provide cytokines to help antigen-specific B cells generate antibodies. Other CD4 + and CD8 + T cells may differentiate as memory cells, using their own particular barcodes to locate to sites appropriate for their role.

These steps to achieve beneficial pathogen-directed immunity can be considered as necessary stages for virtually all forms of adaptive immunity associated with αβ T cells.

As might be expected, the immune system has inbuilt checks and balances to prevent inflammation-generating immunity getting out of hand. In part, this is achieved by a wide range of co-inhibitory molecules that counter secondary signals induced by licensed DC (Chambers and Allison, 1999). The best known of these are CTLA4 and PDL1, molecules which have been extensively studied in their capacity as 'checkpoint inhibitors'. Understanding the circumstances where co-inhibition dominates costimulation offers a major theme in cancer immunology.

Whereas most αβ T cells provide the means to monitor specific intracellular microbes by their response to processed peptide antigens presented by MHC, other αβ T cells can sense microbial lipids presented by a set of MHC-related molecules, known collectively as CD1. Moreover, the repertoire of moieties sensed by T cells is increased further by γδ T cells whose recognition is not limited to either MHC or CD1-associated antigens. Furthermore, activation of γδ T cells does not necessarily require elaborate antigen processing, and presentation, and consequent clonal outgrowths. Consequently, γδ T cells are not limited to the delayed adaptive response mode of αβ T cells. Indeed γδ T-cell recognition of self-encoded beacons of cell and tissue dysregulation (Willcox et al., 2012) can provide a rapid mechanism for the immune system to eliminate stressed cells in tissues, such as those infected by pathogens, or those subject to carcinogens (Willcox et al., 2012; Vantourout and Hayday, 2013). Thus, γδ T cells illustrate how some lymphocytes can contribute both to the early innate phase and to the delayed adaptive phase of an immune response. The same may be true of some subsets of B cells. Importantly, we currently lack an understanding of how the responses of such innate-like lymphocytes resident within tissues impacts upon the temporal progression of the immune response through its innate and adaptive phases, and the development of antigen-specific memory.

Relevance of antipathogen immunity to cancer immunology

Given the necessary stages required for T cells to become activated, to deliver effector functions, and to generate challenge-specific memory, what would be needed for tumour antigens to achieve the same outcome?

The necessary elements would be:

1. antigens to which T cells with cognate TCRs are available in the circulating immune repertoire;
2. appropriate molecular signals of cell/tissue perturbation to license dendritic cells;
3. absence of tumour-associated elements that might interfere with licensing, or even decommission DC;
4. adequate bar-coding of primed T cells to permit their homing to sites of tumour growth and to sites of memory reservoirs;
5. receptiveness of the tumour and its vasculature to entry of primed T cells;
6. a tumour-associated environment (TAM) that enables the effector T cells to express their full effector potentials.

Failures in routinely delivering destructive immunity to cancers suggests constraints at one or more of these levels. A major theme of modern cancer is to identify routes to bypass any, or all, of these constraints.

Self-tolerance and its relationship to cancer

The immune system adopts three general strategies to minimize reactivity to 'self'

The first involves tolerising self-reactive lymphocytes as they develop in the primary lymphoid organs. For T cells this happens in the thymus, and for B cells largely in the bone marrow. Remarkably the thymic medullary epithelial cells, through the function of the **Autoi**mmune **Re**gulator transcription factor AIRE, promiscuously express a vast array of self-proteins and their processed peptides, enabling a massive purging of the antiself T cells that engage them at this staging post (Anderson and Su, 2016; Richards et al., 2016).

Simultaneously, the actions of another transcription factor (Foxp3) expressed in some CD4 + T cells can rescue T cells with antiself reactivity while endowing them with a protective regulatory function (Hori et al., 2003; Ramsdell and Ziegler, 2014). These natural or thymic regulatory T cells (tTreg) seem critical for self-tolerance, as their absence in humans and mice can result in severe multisystem autoimmune disease. Clearly, promiscuous gene expression through AIRE does not ensure sufficient clonal deletion to protect against all autoimmunity. This suggests that any residual self-reactive T cells spared from clonal deletion are likely held in check by the action of Treg.

The antigen-dose-sensitivity of B cells to clonal deletion appears less than that of T cells (Chiller et al., 1971; Goodnow and Basten, 1989). This means that B cells against abundant, ubiquitous antigens are deleted or inactivated during development, whereas those against sparse self-antigens may be spared. As B cells need T-cell help to generate antibody responses (Mitchell and Miller, 1968), and as any potential antiself helper cells are normally deleted, T-cell-independent B-cell autoimmunity is unlikely. Whenever IgG autoantibodies arise, there must always have been an antecedent CD4 + helper T-cell response to an antigen (Fig. 23.1)

The second category of tolerance processes, known as 'peripheral tolerance' embraces several mechanisms (Mueller, 2010). For example, a T cell seeing its antigen in the absence of other triggering signals may be inactivated or die, or even become outnumbered by lymphocytes that regulate immune function. The best studied regulatory lymphocytes are CD4 + Treg, some of which (tTreg) arise in the thymus, as previously mentioned, while other Treg are induced *de-novo* in tissue microenvironments rich in TGFβ and/or lacking pro inflammatory mediators, or in circumstances of mTOR inhibition (Chen and Konkel, 2010; Waldmann et al., 2016). Intimately involved in dictating these pathways are diverse tissue-infiltrating myelomonocytic cells-including monocytes, granulocytes, and the heterogenous population of myeloid suppressor cells (MSDC; see Gabrilovich et al., 2012; Marvel and Gabrilovich, 2015); tumour-induced endothelial cells and other stromal elements (Quail and Joyce, 2013; Joyce and Fearon, 2015); and innate-like tissue-resident lymphocytes (Hayday and Tigelaar, 2003).

Third, and rarely discussed, are the range of 'silent' events in normal or healing tissues that discourage or curtail immune

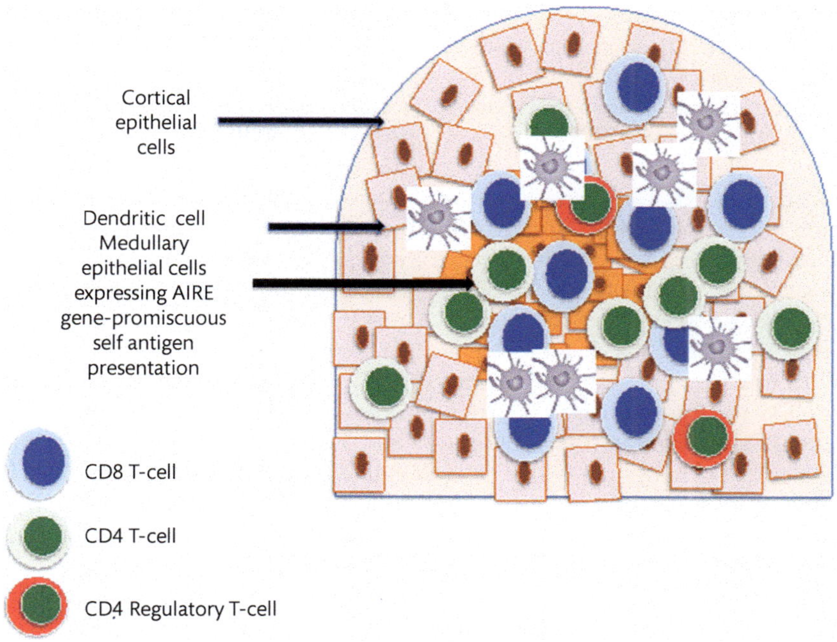

Fig. 23.1 T cells become tolerant to most self-antigens during their development in the thymus. Expression of the AIRE gene in thymic medullary epithelial cells results in promiscuous expression of the majority of self-proteins with each medullary cell expressing a small proportion of such antigens, while the whole population provide coverage for all. Whether these antigens are expressed on the epithelium or reprocessed by dendritic cells, the outcome is the deletion of most self-reactive T cells. Some self-MHC restricted CD4 T cells, which come to express the transcription factor FoxP3, are not deleted, and emerge from the thymus as regulatory T cells, that provide an additional failsafe against autoimmune disease.

reactivity. It is quite likely, given the different physiological demands on different tissues, that each might use different mechanisms to ensure that immune responses to eliminate microbes do not get out of hand and cause irreparable tissue damage. At the extreme end of these are tissues, identified from transplantation research, wherein immune responses are vigorously constrained. These have been referred to 'immunologically privileged sites', of which the placenta and anterior chamber of the eye, and to a lesser extent, brain, testis, and ovaries are the best known (Stein-Streilein and Streilein, 2002; Simpson, 2006). Almost certainly all tissues will, through a coordinated activity of their diverse cellular components, create molecular environments that limit immune activity within them. Equally tissues in the process of healing and remodelling provide many negative feedback signals, of which local active TGFβ inducing immunosuppressive adenosine-generating ectonucleotidases on many tissue cells provides one good example (Regateiro et al., 2011).

Tolerogenic processes can also be promoted in circumstances where some immune activation has already occurred. These have been highlighted in the analysis of experimental animals, where foreign tissues have been accepted long term (Baas et al., 2016; Waldmann et al., 2016), and in reversal of organ-specific autoimmune diseases, after short term therapeutic intervention (Chatenoud et al., 1997). Studies of the tissue components in these situations have emphasized a role for newly established microenvironments within such tissues, whereby cross talk between innate, adaptive immune cells as well as tissue cells renders the tissue less permissive for immune attack.

Collectively, these examples of the tissue 'fighting back' can be considered as forms of acquired immunological privilege.

Research into mechanisms of natural and acquired privilege provides information on the diversity of protective tissue reactions that can operate. On the one hand, these may reflect the directions any given tissue can follow to sculpt the immune infiltrates within it to ensure damage limitation. On the other hand, they may give clues on the ingredients that tumours use to achieve the same outcome. In short, the study of immune interactions with tumours has lots to learn from the study of immune interactions with normal tissues, and *vice versa*.

In fact, the analysis of how the tumour microenvironment allows tumours to escape T-cell damage is now offering the most rapidly advancing evidence for privilege mechanisms in general. Undoubtedly, its understanding will have broad ramifications, not only in the treatment of cancer, but also in reversal of autoimmune diseases, the prevention of allogeneic cell and organ rejection, and the limitation of major inflammatory diseases.

The combined knowledge emerging from all these directions, and especially from cancer studies, challenges the view of immunity and tolerance as binary discrete processes. Elements of each are always contributing to the way that the immune system reacts to antigen encounter. What we call an immune response, is one that emerges despite many constraining influences, whereas when we speak of tolerance to tumours, we are recognizing that tolerance mechanisms have simply prevailed over immune ones that were present but that passed unnoticed.

Cells hitherto thought of as mediators of immunity, can, in particular contexts, also behave as mediators of tolerance and *vice versa* (Wang et al., 2010). The mere enumeration of cell types in biopsy cannot tell us whether the immune infiltrate is an indicator of imminent destruction or of protection. More detailed parameters of cell subpopulations, their relative proportions, their micro-anatomical locations, frequencies of discrete cell-cell interactions, and indicators of their functional capacity, are all relevant in what has come to be called the analysis of cell 'contexture' as a guide to cancer prognosis and treatment (Fridman et al., 2012; Galon et al., 2013; Gentles et al., 2015; Galon et al., 2016).

The acceptance of the notion, that most of our well-studied cells of the innate and adaptive immune system, can be a force for tumour rejection as well as for tumour acceptance, is key to therapeutic intervention in a way appropriate to the patient's tumour. Despite, the possible diversity of tolerance-related mechanisms operating we need, to establish priority lists of the most generic and tractable of these that can be targeted, rather than aiming at treatments that become too personalized. Ultimately, our chosen drugs will need to offer a commercial incentive to their manufacturer, and fiduciary practicality to healthcare providers.

Why do we think that tumours may have antigens the immune system can target?

Evidence for immunosurveillance of cancer

It has long been known that some cancer patients undergo spontaneous remissions of their disease. Patients receiving 'immunosuppressive' drugs to prevent graft rejection are more susceptible to diverse cancers. Some cancers are clearly caused by oncogenic viruses (e.g. human papilloma virus (HPV), EBV, or hepatitis), but others seem not to be. The possibility that immune mechanisms might have been involved in control of cancer was recognized some 60 years ago, when it was shown that chemically induced tumours could be rejected in inbred mouse strains. These and related experimental findings spawned the notion of cancer immunosurveillance proposed by Burnet (Dunn et al., 2004). It was initially somewhat discouraging that T-cell deficient mice did not show a higher susceptibility to chemically–induced cancers compared to their normal counterparts, although their increased susceptibility to tumours caused by polyoma virus was consistent with immunosurveillance (Simpson and Nehlsen, 1971). However, later evidence that neutralization of IFNγ by antibodies enhanced tumour growth, and that IFNγ deficient mice were more susceptible to tumours, restored optimism to the concept of immunosurveillance, further boosted by clear evidence that mice deficient specifically in γδ T cells have heightened sensitivity to environmental carcinogens (Girardi et al., 2001). Likewise, tumour incidence is higher in mice in which elements of the T-cell killing machinery and members of the TNF family were compromised.

Later work showed that tumours developing in immunocompromised mice demonstrated greater immunogenicity when transferred into fresh immunocompetent hosts, than those developing in controls. This observation led to the immunoediting hypothesis (Dunn et al., 2004), a modification of the original immunosurveillance theory. Here it was envisaged that the interaction of the immune system with cancers operates in three phases: elimination, equilibrium, and escape. Although no experiment has definitely established that this hypothesis satisfactorily reflects a general facet of cancer biology, it provides a useful framework for considering how the immune system may interact with tumours, and hence how the balance may be swung toward tumour eradication and patient benefit. 'Elimination' reflects the capacity of innate and adaptive mechanisms to clear early tumour cells. Should this fail, then the 'equilibrium' phase reflects the capacity of T cells to keep the tumour under control. When that eventually fails for whatever reason, tumours enter the 'escape' phase. The proposed equilibrium phase offers a period during which the immune system exerts pressure on tumours to develop escape mutants. Many tumours at time of presentation are likely to be in this proposed escape period. Identifying the reasons for tumour escape may offer therapeutic routes to its prevention.

As mentioned, evidence supporting the immunoediting idea has been provided from chemically induced and spontaneous tumours in mice, based on identifying changes in immunogenic antigens. In recent years, abundant data from human tumours has been consistent with a role for the immune system in shaping tumour evolution. Analysis of the quality, location, and intensity of immune infiltrates in tumours has provided prognostic information for tumour outcomes with a series of Th1 related indicators proving a more favourable prognosis for many tumours. Surprisingly, 'signatures' reflecting Th2-like, Th17, or regulatory T-cell signatures have not been so clear cut. Perhaps consistent with the mouse data mentioned earlier, a γδ T-cell signature has proved to be the strongest correlate of overall survival in an unprecedentedly large survey of 18,000 patients (Gentles et al., 2015). Understandably, many attempts are currently being made to take many of the immune parameters into account in providing an 'immunoscore' to predict the potential tumour outcome (Galon et al., 2016; Mlecnik et al., 2016).

The nature of tumour antigens

Early evidence for immunogenic tumour-associated antigens came from serology of individuals bearing tumours, and in paraneoplastic immune disorders with antitumour antibodies detecting shared antigens on neuronal cells. Although indicative of adaptive immunity, little evidence existed on whether the response was driven by self-antigens, or neoantigens, derived from altered self-proteins encoded by genes mutated by genotoxic carcinogens. Spontaneous antitumour T-cell responses (both CD4+ and CD8+ T cells) have been identified in some patients with melanoma, and as mentioned earlier, these examples give indications that the host may be able to mount response to self-antigens in tumours, where perhaps clonal deletion had not occurred.

Virally induced tumours (e.g. EBV, HPV, hepatitis virus) all offer potential tumour-associated antigens that are 'foreign', to which the host will not be naturally tolerant. The natural CD8 T-cell mediated immune-surveillance that prohibits malignant B-cell accumulation in most EBV infected people is a good indicator that antiviral immunity enables a long-term equilibrium between host and cells harbouring the oncogenic virus.

However, the most compelling evidence that neoantigens emerge in cancers, that these are subject to immune attack, and that they change over the time course of tumour development, comes from advances using modern DNA recombinant technologies involving exome sequencing of tumours and T cells infiltrating them, in-silico extrapolations of whether peptides predicted to derive from mutated proteins are potential immunogens able to bind to MHC molecules, and technologies to show that T cells can bind these and become

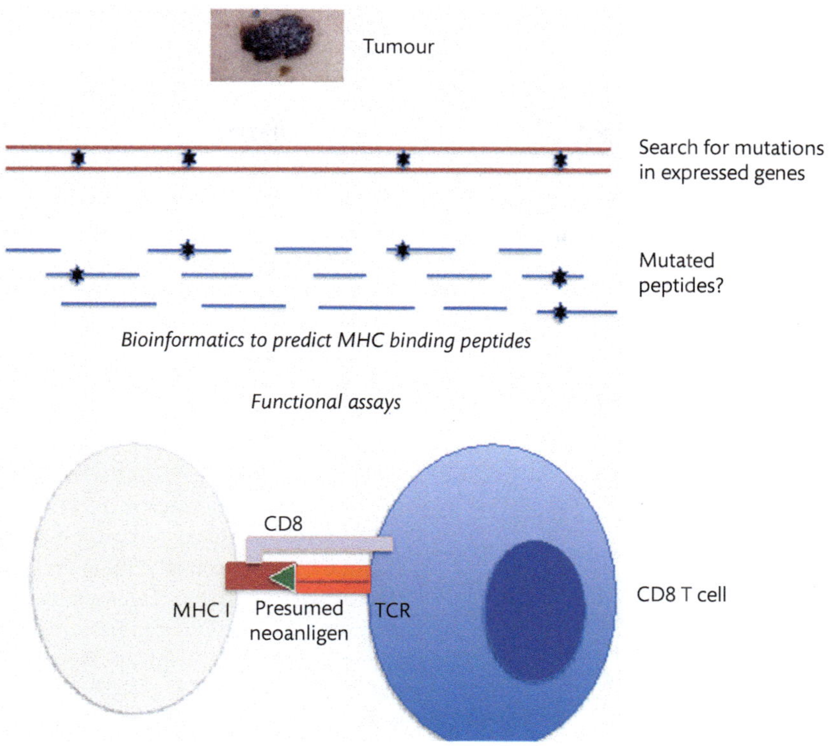

Fig. 23.2 Identification of cancer neoantigens. Ribonucleic acid (RNA) sequencing of tumour extracts RNA sequencing data enables detection of mutations in expressed genes. Likely mutated peptides are catalogued, and those likely to bind to MHC Class I are predicted and tested for possible immunogenicity in *in vitro* functional assays. Similar approaches are used to detect MHC Class II binding neoantigens that stimulate CD4 + T cells.
Source: data from Schumacher TN and Schreiber RD, 'Neoantigens in cancer immunotherapy', *Science*, Volume 348, Issue 6230, pp. 69–74, Copyright © 2015 American Association for the Advancement of Science.

activated (Linnemann et al., 2014; Gubin et al., 2015; McGranahan et al., 2016; Schumacher and Schreiber, 2015; Ward et al., 2016). After successful studies in mouse cancer models, such technology is now being applied to human cancers, albeit with a propensity of studies largely, thus far, confined to melanoma (Fig. 23.2).

Many neoantigens have now been identified that bind to both MHC Class I and II molecules. It is also clear that some tumours exhibit a high mutational load, and others less so. The expectation is that those with a higher mutational load will generate more neoantigens. There is now the opportunity to pool such information with that of the contexture scores, as well as therapeutic outcomes to assess the impact of neoantigenic load on outcomes of therapy (Rizvi et al., 2015). As expected, given the stochastic nature of mutational neoantigens, it may be difficult to find broadly-shared antigens within tumour types. That, together with MHC polymorphisms may make it difficult to identify worthwhile tumour-vaccine candidates, other than on a personalized basis. In the context of personalized therapies (if such were commercially viable), knowledge of an individual's neoantigen repertoire may enable selection and expansion of antigen-specific cells from those infiltrating the tumour, or engineering of T cells to express TCRs recognizing such antigens for adoptive immunotherapy.

A word of caution though! These aspirations may be confounded by limitations imposed by the patient's available TCR repertoire, as well as our lack of knowledge of how both qualitative and quantitative MHC/peptide epitope expression at the cell surface determines immunogenicity.

The demonstration of tumour-specific T cells in tumours

The strategy for studying T cells infiltrating tumours owes much to early studies of adoptive T-cell transferred from mice immunized to tumour antigens. Poor protection was initially observed until transfers were attempted into lymphocyte-depleted recipients. Pioneering studies by Rosenberg using extracted human melanoma infiltrating lymphocytes expanded in culture with IL-2, and transferred into lymphocyte-depleted patients provided evidence that these cells could attack the cancer and persist in the circulation (Rosenberg and Restifo, 2015; Topalian et al., 2015). The necessity for prior host lymphocyte depletion for efficacy could reflect a need for homeostatic cytokines and 'space' to enable transfused lymphocytes to expand further and differentiate, or relate to, ablation of other immune inhibitory pathways operating in the lymphocyte-replete host.

These studies with TILs have confirmed that melanomas contain T cells that recognize both 'self' as well as mutated antigens. Studies with epithelial cancers, using exomic sequencing to identify possible neoantigens, have provide evidence that TILs include T cells responsive to some of these neoantigens. Once expanded *in vitro*, some such TILs could achieve dramatic tumour regression on transfer (Hinrichs and Rosenberg, 2014; Rosenberg and Restifo, 2015). Another approach to identify reactive T cells has been to engineer their receptors (TCR) so that they can be transfected into naïve T cells to give them target specificity (Rosenberg and Restifo, 2015). In some examples using TCRs to non-mutated antigens, where cancer regressions were observed, damage to other normal tissues could also be seen, indicating broader display of those antigens beyond the tumour itself. Somehow, the host from which such TILs had been derived, had not exhibited such off-target damage-perhaps pointing to protective regulatory processes operating to limit damage. Nonetheless, it is the case that some melanoma patients with immune responses to their tumours present clinically with vitiligo in normal areas of skin.

How do tumours provide the innate signalling mechanisms in the way that microbial pathogens do?

It is now widely acknowledged that there exist a large range of mechanisms for sensing cell stress, damage, and death, which can provide adjuvanticity for tumour antigens, in the way that PAMPS do for microbial antigens. These damage-associated molecular patterns (DAMPs) are distinct entities but can engage in much cross talk with PAMP signalling mechanisms. A detailed understanding of DAMPS and their signalling pathways will be an important avenue of future research aimed at maximizing the immune response to tumour antigens (Schaefer, 2014; Woo et al., 2015).

The tumour-associated microenvironment limits immune activity against tumour antigens

The accumulating evidence, that some tumours contain T cells with specificity to tumour neoantigens, raises the question of why these are not eradicating the tumour. A compelling answer to this question has recently been provided by the massive and effective immune response to some tumours unleashed by blockade of known co-inhibitory molecules such as PD1 or CTLA4 expressed on T cells. These findings, together with previously mentioned prognostic studies on the nature of tumour infiltrates, have clearly indicated that the tumour-associated microenvironment, shaped both by the tumour and by immune components, determines the extent to which T cells can exert their antitumour potential. As each contributor is defined, then additional novel targets for checkpoint blockade become available.

Cells infiltrating tumours may have the potential to contribute to tumour damage, but this may be outweighed by other contributions to contribute tumour growth, and to immune privilege (Hanahan and Coussens, 2012; Hagerling et al., 2015; Joyce and Fearon, 2015; Marvel and Gabrilovich, 2015; Quail and Joyce, 2013). Growth promotion may be mediated by diverse growth factors, and proteolytic enzymes that might alter the extracellular matrix may permit tumour cell spread. Cell sources contributing such influence are not limited to leukocytes but may also include tumour-associated fibroblasts and blood vessels, for example.

As indicated earlier, T-cell priming requires licensing and migration of antigen-presenting cell (APC). There is much evidence to indicate that DC isolated from human tumours are functionally impaired, if not decommissioned. Various products in the TAM have been implicated, including cytokines such as TGFβ, IL-10, and vascular endothelial growth factor (VEGF), macrophage colony-stimulating factor (MCSF), and even hypoxia, in some cases associated with enhanced expression of the co-inhibitory molecule PDL1.

Even if T cells have been primed, they need to gain access to tumour cells. This requires that they gain entry through the tumour vasculature and migrate along chemotactic gradients to their targets. Failure of tumour vasculature endothelial cells to properly coordinate expression of key adhesion molecules is one element to tumour protection. VEGF, endothelins, and nitric oxide have been implicated as key mediators at this checkpoint, in addition to a number of well established immunosuppressive molecules.

Beyond the endothelium there are many other cell types that can prevent immune engagement, including Treg, myeloid-derived suppressor cells (Gabrilovich et al., 2012; Marvel and Gabrilovich, 2015), and monocytes that are 'alternatively activated' as if exposed to Th2 T-cell products (Hanahan and Coussens, 2012). Some of these effects on effector T cells can be mediated by enzymes catabolizing among others, tryptophan (Munn and Mellor), and L-arginine (Molon et al., 2011) creating a nutrient-deprived local microenvironment, and generating range of anti-inflammatory moieties, such as kynurenines, active radicals of oxygen and nitric oxide, and peroxynitrite. Peroxynitration and inactivation (by nitration/nitrosylation) of chemokines and cancer neoantigens is one proposed mechanism by which products of arginine metabolism can impede immune functions. All these factors influence the balance of cell types that accumulate in tumours, and the quality of immune privilege that results.

Analysis of the components of the tumour microenvironments indicate widespread cross talk between cancer cells, their stroma, and the immune system (Fig. 23.3). Evidence is accumulating that diverse signalling pathways influencing tissue development and carcinogenesis (e.g. Wnt; see Yamabuki et al., 2007; Sato et al., 2010), as well as tumour suppression (e.g. p53, PTEN, RB1, and ARF; Munoz-Fontela et al., 2016), have profound local immune subversive functions in addition to their cell intrinsic functions.

It is possible that, in the not-distant future, tumour biopsies can be described by a multidisciplinary read-out of cell composition, interactions, and molecules present, that can be interpreted for prognostic outcome and selection of optimal treatment protocols to promote immunological control (Bindea et al., 2013). At this stage, however, the goal should be to identify the more critical parameters lending themselves to therapeutic decisions.

To this end, the identification of cell expressing co-inhibitory molecules and their ligands in tumour and infiltrating cells (e.g. T cells expressing PD1 and Treg expressing CTLA4) and diverse cells expressing PDL1, offers some indication as to which tumours might be responsive to checkpoint inhibition with anti-CTLA4 and anti-PD1 antibodies (Mahoney et al., 2015; Smyth et al., 2016). In this arena, there is much need for improved reagents and standardized protocols.

In the same vein, there may be other less well-documented interactions that may provide biomarker clues for intervention. These could include indicators for costimulation with agonist antibodies such as TNFR family members, antibodies to Wnt ligand members (Sato et al., 2010), drugs influencing TGFβ signalling, and those targeting (immunosuppressive) adenosine-generating ectonucleotidases CD39, CD73 in diverse immune and epithelial cells (Regateiro et al., 2011; Young et al., 2016), and the upregulation

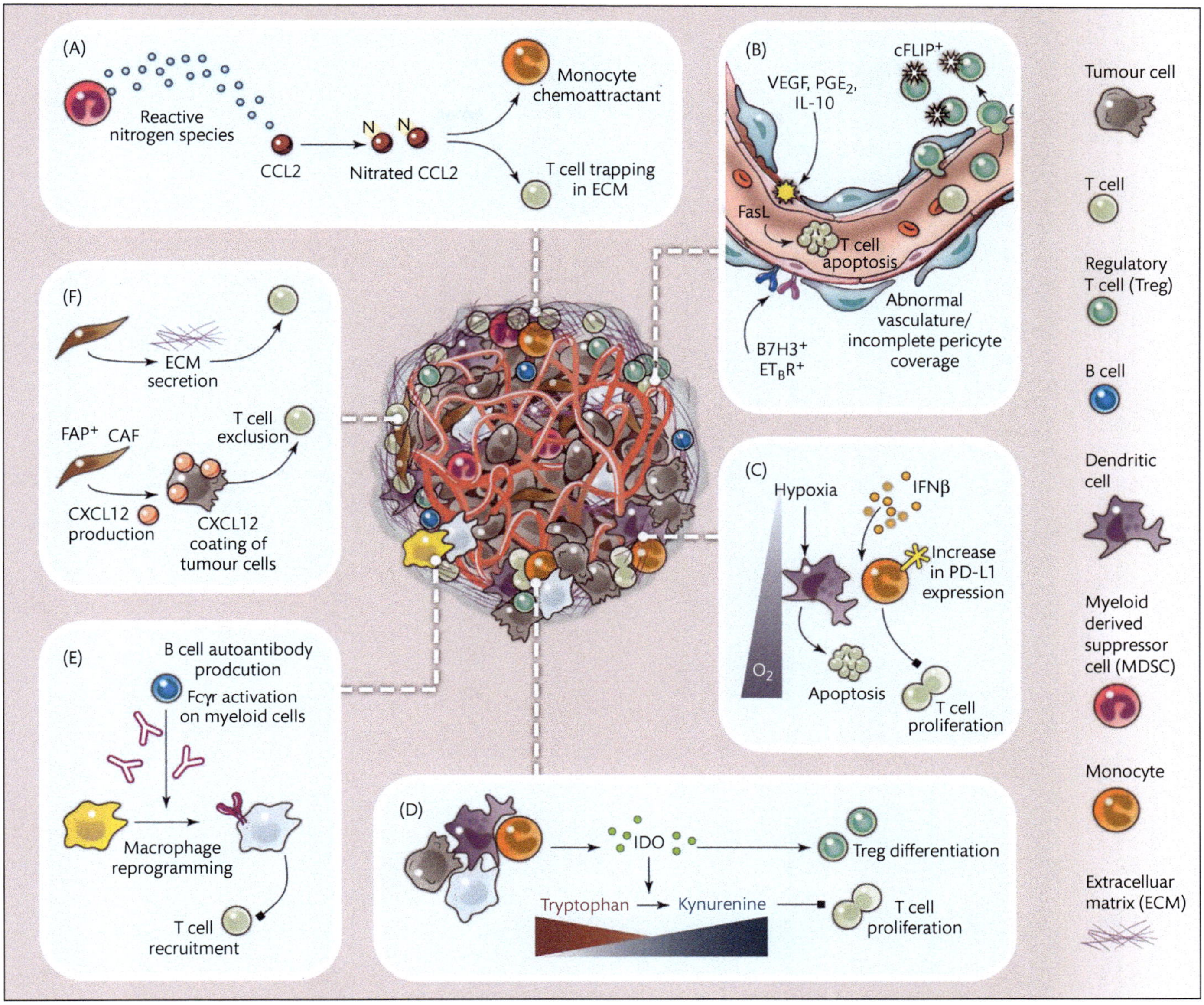

Fig. 23.3 (A–F) Mechanisms of immune suppression within the tumour microenvironment. Diverse cellular components in the tumour microenvironment regulate the entry, accumulation, activation, and expansion of T cells in tumours exemplified in the representative frames shown.

of these by TGFβ, offer further opportunities for another type of checkpoint blockade.

The overall principles underlying tumour immunity and its natural resistance are summarized in Figure 23.4 (Chen and Mellman, 2013).

Overriding tumour privilege: Immune intervention in cancer

Introductory remarks

It is becoming increasingly clear that many of the currently adopted chemotherapeutic drugs for cancer affect the immune system. It is not straightforward to determine how much of the therapeutic benefit derives from direct drug kill, rather than from effects on immunoregulatory mechanisms (Zitvogel et al., 2013; Belvin and Mellman, 2015), but it may be appropriate to consider both effects when optimizing drug dosing. Clearly, the pleiotropy of chemotherapeutics offers potentials for combinatorial synergies with emerging immunotherapeutic modalities.

In considering strategies involving vaccination or adoptive cell therapy the question arises as to which antigens could be targeted? In principle these could be self-antigens shared by normal tissues with attendant risks of debilitating collateral damage; self-antigens shared by normal cells whose loss would not be a health risk (e.g. melanocytes); self-antigens expressed in ontogeny but not in the mature individual (so-called oncofoetal antigens); and, of course,

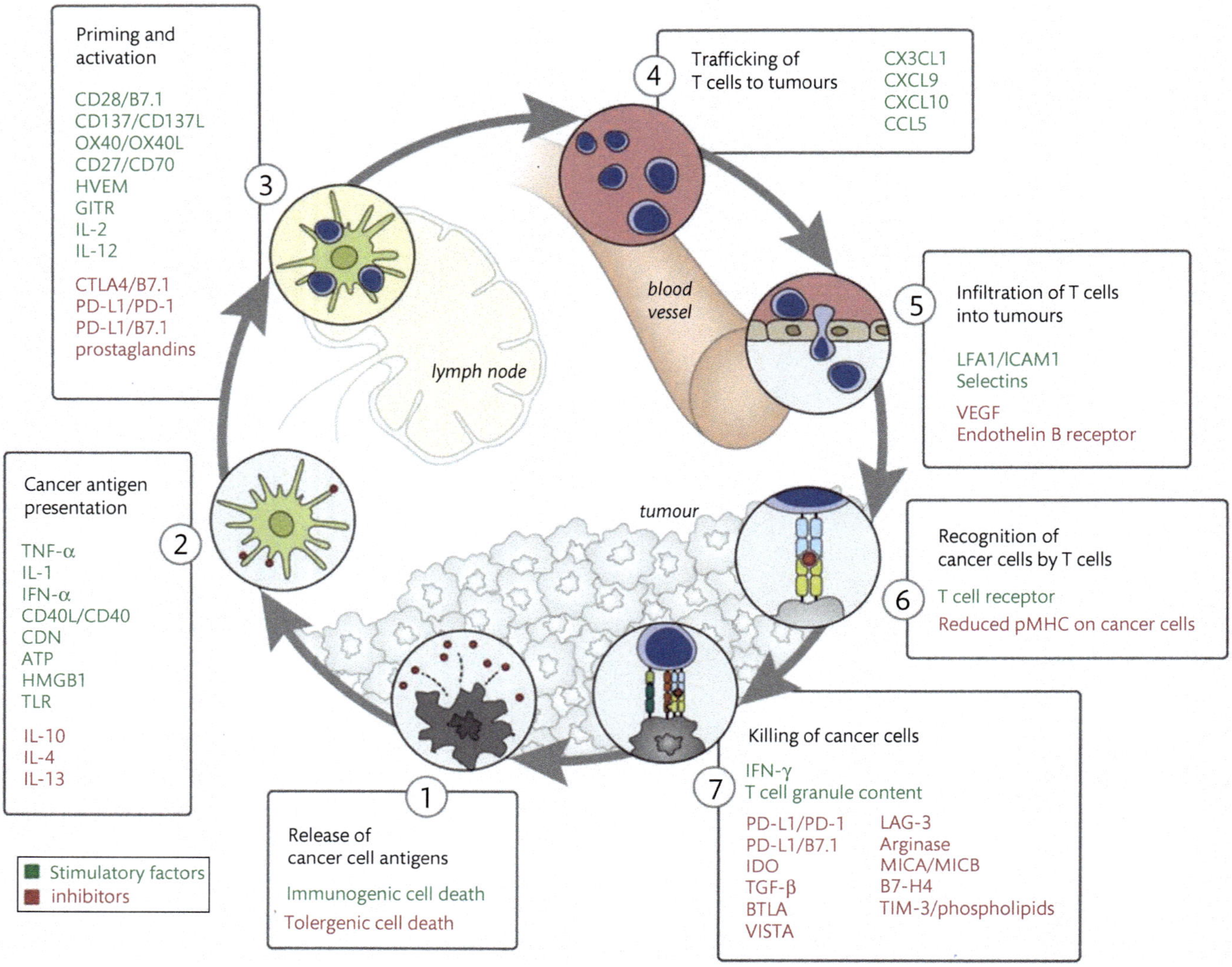

Fig. 23.4 Stimulatory and inhibitory factors in the cancer-immunity cycle. The pathway used by the immune system to generate antitumour immunity (in green). Released tumour antigens are processed by dendritic cells that are licensed by 'danger' signals to leave the tissue and make their way to secondary lymphoid tissues. There they filter out antigen reactive lymphocytes from the circulation and immunize them to the tumour antigens. These 'selected' T cells migrate out of the lymphoid tissue into the circulation, some providing memory cells for future encounters, while others make their way into the tumour after traversing the local vascular endothelium. If the microenvironment does not impede their activity, they will then go on to damage tumour cells, release more tumour antigens, and enable the cycle to be perpetuated. For all stages in the cycle there are a myriad of potentially inhibitory mechanisms (examples in red) that, in the modern era, constitute potential targets for checkpoint blockade.

Reproduced with permission from Chen DS and Mellman I, 'Oncology meets immunology: the cancer-immunity cycle', *Immunity*, Volume 39, pp. 1–10, Copyright © 2013 Elsevier Inc. https://www.sciencedirect.com/science/article/pii/S1074761313002963

tumour neoantigens. Exploitation of the latter may require personalized therapies with their attendant commercial limitations. Of the various self-antigens, one category currently attracting attention for this reason are the numerous cancer testis antigens (Djureinovic et al., 2016; Li et al., 2016; Park et al., 2016), normally expressed on male germ cells, but aberrantly expressed on some tumours. Given the potentially tolerable collateral damage some may prove useful targets, especially for evolving generic strategies to enhance tumour kill.

Vaccination against non-viral cancers

The notion that one might have vaccines against defined self-antigens is predicated on the fact that central tolerance is not perfect, and that the autoantigens concerned may not be expressed on vital cells, or cells that cannot easily be renewed. Any therapeutic vaccine strategy needs to incorporate components that enable the licensing of dendritic cells, and numerous strategies for doing this have emerged (Palucka and Banchereau, 2013; Romagnoli et al., 2014; Shimizu et al., 2016). Although there are examples of some therapeutic benefit from vaccination to 'self' tumour antigens, no really effective therapeutic vaccine strategy has yet emerged (Melief et al., 2015; van der Burg et al., 2016).

By contrast, the more likely targets for vaccine development should be tumour neoantigens. However, these would, more often, need to be patient-specific and hence potentially expensive to develop. In the context of adoptive cell therapy, there may be a route to identify vaccine-peptides to select and expand TILs from individual patients for adoptive T-cell therapy.

Vaccination against viral cancers

Some 20% of human cancers are known to have an infectious origin. Prophylactic vaccines have emerged for only two of these (hepatitis B and HPV). Persistent infections with HPV offer a challenge for the development of a therapeutic vaccine, as the persistent virus needs to be eradicated before cancer develops. EBV vaccines would be highly desirable given the number of malignancies arising, but no useful vaccine has yet been demonstrated. Even in the case of HPV best efficacy data are provided in patients with premalignant lesions.

In short, our marked inability to overcome many infectious diseases through vaccination, highlights the huge challenges in developing efficacious prophylactic vaccines. Those challenges are likely to be greater for therapeutic vaccines, even where viral non-self-antigens are concerned. On this basis, it will be no easy matter to immunize cancer patients to eradicate their tumours. The presumption is that even if antigen-specific responses are induced, the ability of the host to mount efficacious effector responses will be limited by peripheral tolerance mechanisms of the kind discussed earlier. From our knowledge of the adaptive immune response, the best therapeutic immunization outcomes will most likely come where the vaccines promote the responses of both antigen-specific CD4+ T cells as well as CD8+ cells, given in such a way as to provide strong and sustained licensing of dendritic cells. Moreover, to ensure access of effector cells to tumour cells we will need to break open the strangleholds of the tumour microenvironment. This will undoubtedly require combinatorial therapy based on the emerging knowledge of its organization.

Therapeutic antibodies

The application of preformed antibodies as magic bullets became a reality when Kohler and Milstein (Kohler and Milstein, 1975) described the monoclonal antibody technology (Fig. 23.5). The initial optimism was not immediately realized for many reasons. Many of the initial rodent antibodies (Mabs) were poorly lytic even for circulating blood cells. Very few were able to utilize the human complement system for killing, while others were directed to antigens which were simply not disposed to effective lysis by cell mediated Fc dependent killing mechanisms. It also became clear that rodent IgG isotypes vary enormously in recruiting natural lytic mechanisms, with mouse IgG2a and rat IgG2b subclasses being the most effective. Some antigens were easily modulated off the cell surface, and other were simply too sparse. Rodent antibodies proved immunogenic for man, meaning that they could only be used for a short time. This early pessimism led to early studies with antibody-toxin conjugates, aiming to achieve adequate kill from a short pulse of therapy, but these generated their own problems of unwanted toxicities and additional immunogenicity.

Extensive knowledge of the structure of immunoglobulin genes, and ideas on how antibody diversity was generated, rapidly enabled the application of genetic engineering technologies to create improved immunoglobulins that nature had itself never made.

Initially, chimeric antibodies having rodent variable regions and human constant regions allowed for selection of appropriate IgG isotypes (e.g. IgG1 and IgG3 for most efficient lysis; see Bruggemann et al., 1987), and diminished their degree of foreignness to man. Shortly after, Winter and his colleagues (Jones et al., 1986) demonstrated an even greater degree of humanization by creating antibodies where only the complementary determining regions remained rodent, so reducing the risk of immunogenicity even further.

These initial observations spawned a revolution in the biotechnology sector, resulting in methods to generate antigen-specific human antibodies wholly in silico, or even replacing the rodent immunoglobulin genes with human (Bruggemann et al., 1989), so that mice could be immunized to generate human antibodies from the outset.

Even with this technology, the attempts to directly kill cancer cells have been no simple matter, except perhaps for more easily accessible cancers of the blood system. More recently toxins for conjugation have been mutated to eliminate potential T and B-cell epitopes, so enabling further efforts to exploit antibody-toxin conjugates (Mazor et al., 2016). Antibodies have also been engineered to be able to deliver chemotherapeutic drugs more selectively to the tumour site.

Overall though, failure of naked antibodies to achieve an adequate kill of solid tissue malignancies comes from limitations in tumour penetration, and failure of the Mabs to recruit a sufficient innate effector mechanism (for Fc-mediated lysis). Off-target toxicities are also real issues sometimes related to the expression of the target antigen on other body cells.

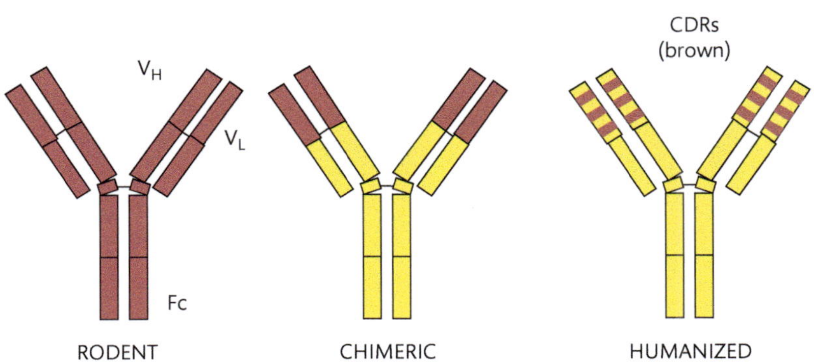

Fig. 23.5 Genetic engineering has been successful in transforming rodent antibodies into human forms that are far less immunogenic. The technologies that evolved to achieve this have provided a platform which has catalysed the therapeutic exploitation of many facets of immune function as novel cell-bound, and also, cell-free fusion proteins.

Where human monoclonal antibodies have fared better has been where they have been used to block inhibitory immune function as checkpoint inhibitors (Ribas and Wolchok, 2018) or alter cell signalling (Herceptin and Erbitux; see Katsumata et al., 2006), in addition to any immune-mediated mechanisms.

Although combination therapy with other Mabs or therapeutic modalities can give improved outcomes, there remains a commercial incentive to engineer antibody forms that may be more potent by focussing multiple effector moieties, within the one antibody construct to the tumour target. This can be achieved by incorporating multiple non-immunogenic 'self' effector modules in fusion proteins expressing the appropriate antibody variable regions to target the tumour. Perhaps the best early examples of such engineering are bispecific antibodies having one arm directed to the cancer antigen and the other to the CD3 epsilon chain of the TCR (Staerz and Bevan, 1989). This approach theoretically allows any T cell, regardless of its own antigen specificity, to be recruited and activated to the site of the tumour with demonstrable efficacy (Staerz and Bevan, 1989; Riethmuller, 2012). Generation of intracellular antibody fragments to critical functional sites on oncogenes is proving an exciting route to the generation of novel chemical drugs blocking interactions of oncogenes with other protein partners in malignant cells (Quevedo, 2018, Cruz-Migoni.2019).

Adoptive lymphocyte therapy

The previously mentioned work of Rosenberg and colleagues using *in vitro* antigen expanded TILs, has demonstrated the potential of adoptive immunotherapy (Rosenberg and Restifo, 2015). Host lymphocyte depletion was important in allowing the expanded cells to mediate tumour kill.

The unpredictability of tumour antigens coupled with human MHC polymorphisms has rendered such T-cell therapy very personalized, and not ideal for commercial development, even allowing for advances that might reduce the resistance of the tumour microenvironment.

The ideal adoptive immune cell therapy would be one where the effector cells recognized and could be activated and expanded by some endogenous ligand that is reproducibly expressed on tumour cells, and few normal cells. It is this notion that is driving interest in the possible use of γδ T cells and NKT cells for cancer immunotherapy (Silva-Santos et al., 2015). The natural propensity for γδ T cells to be resident in tissues may allow them to access tumours and to perform more effectively than αβ T cells. Consistent with this view, is the claim that a γδ T-cell signature local to the tumour is the strongest correlate of overall survival (Gentles et al., 2015).

Engineering of T cells

As is discussed in other chapters of this book, there has been extensive efforts to enhance the therapeutic power of adoptively transferred T cells. There are two general categories. The first involves use of cloned antigen-specific TCR genes introduced into human T cells (Johnson et al., 2009; Perro et al., 2010). Preclinical mouse models may allow selection of the optimal TCRs for therapeutic use (Obenaus et al., 2015). By their nature these will be MHC restricted and limited to targeting 'self' antigens whose off-target expression runs little risk, or to neoantigens that will likely limit utility to the individual patient. The best use of such cells will likely involve sets of gene manipulations that increase their sensitivity to antigen stimulation, limit their ability to be tolerated by their cognate antigen, endow them with adhesion and chemotactic receptors that offer optimal migration to tumour tissue, enhanced lytic machinery to kill their target, and provide them with the where-with-all to be 'suicided', if need be. It remains unclear, at this stage then, of what sort of 'off-the-shelf therapies' will offer the necessary commercial incentive to developers.

The introduction of antibody variable regions into chimeric TCR-like signalling molecules in T cells (CAR-T cells; Kalos and June, 2013; Fesnak et al., 2016; Gross and Eshhar, 2016) offers a more general approach to target cell surface antigens shared by many individuals. Once again, the most useful target antigens for these are membrane-exposed differentiation antigens on cells that are accessible and easily replenished by the host.

Perhaps not surprising therefore that malignancies of the haemopoietic system are those that have proven most targetable, thus far.

A major challenge in using third party CAR-T cells is that the host will inevitably reject them. Host preconditioning of recipients to temporarily prevent rejection might provide a sufficient window for the cells to expand, deliver, and their kill, so potentially rendering CAR-T cells as off-the-shelf-therapies. Another major challenge in CAR-T and in checkpoint blockade approaches is to avoid a cytokine storm and rampant immunopathology initiations where large numbers of de-repressed, primed T cells flood the body with an excess of immune effector molecules.

Conclusion

Cancer immunology is in a challenging phase and the major requirements to building on the therapeutic successes thus far are reported in the 'Open Questions'. Overall, there remain many challenges. Yet, the field is in far better shape than it was 20 years ago. Then, cancer immunology was deemed to be a potential graveyard for the young immunologist. How things have changed!

Acknowledgements

I am indebted to Adrian Hayday and Elizabeth Simpson for their expert comments and many helpful suggestions on the manuscript.

TAKE-HOME MESSAGE

- In recent years, advancement in the field of tumour immunology have started to produce a range of new therapeutic strategies.
- The notion that tumours can simply 'escape' the immune system has changed. It is now established that some tumours can evoke and/or provide sufficiently adjuvant-like signals which, if adequately sensed, can promote immune reactivity in the host.

- Cross talk between tumours, stroma, and immune system however can result into an inhibition of the immune response.
- Hence one of the aims of treatment is to override this inhibition of the immune response.

OPEN QUESTIONS

- Establishing safer and more effective ways to license dendritic cells to achieve high quality antigen presentation. Put another way, immunologists have still not cracked the goals of achieving super-adjuvants that are safe.
- To break open diverse types of tumour microenvironments that limit access of T cells and antibodies that could be far more tumoricidal were more of them to get to the right place.
- The need to identify ways of stimulating and empowering host T cells and other innate cells without depending on identifying neoantigens, as discussed in relation to γδ T cells and stress surveillance (Hayday, 2009).
- Inevitably, the application of combinatorial therapies will make in-roads (Mahoney et al., 2015; Sharma and Allison, 2015; Khalil et al., 2016), but it would be valuable to have some sort of biomarker-based algorithms that might predict the best agents to put together for individual patients.
- For therapies directed to self-antigens, it will be imperative to establish procedures to minimize autoimmune damage to normal tissues.

FURTHER READING

Baumeister, S. H., Freeman, G. J., Dranoff, G., & Sharpe, A. H. (2016). Coinhibitory pathways in immunotherapy for cancer. *Annu Rev Immunol*, **34**, 539–73.

Chen, D. S. & Mellman, I. (2013). Oncology meets immunology: the cancer-immunity cycle. *Immunity*, **39**, 1–10.

Cruz-Migoni, A., Canning, P., Quevedo, C.E., et al. C. 2019. Structure-based development of new RAS-effector inhibitors from a combination of active and inactive RAS-binding compounds. *Proc Natl Acad Sci*, U S A.

Fredriksen, T., Lafontaine, L., Berger, A., et al. (2013). Spatiotemporal dynamics of intratumoral immune cells reveal the immune landscape in human cancer. *Immunity*, **39**, 782–95.

Joyce, J. A. & Fearon, D. T. (2015). T cell exclusion, immune privilege, and the tumor microenvironment. *Science*, **348**, 74–80.

Palucka, K. & Banchereau, J. (2013). Dendritic-cell-based therapeutic cancer vaccines. *Immunity*, **39**, 38–48.

Prendergast, G. & Jaffee, E. (eds) (2013). *Cancer Immunotherapy*. Cambridge, MA: Elsevier Inc.

Quevedo, C.E., Cruz-Migoni, A., Bedry, N., et al. 2018. Small molecule inhibitors of RAS-effector protein interactions derived using an intracellular antibody fragment. *Nat Commun*, 9, 3169.

Rabbitts, T.H., and Stocks, M.R. 2003. Chromosomal translocation products engender new intracellular therapeutic technologies. *Nat Med*, 9, 383–6.

Rosenberg, S. A. & Restifo, N. P. (2015). Adoptive cell transfer as personalized immunotherapy for human cancer. *Science*, **348**, 62–8.

Schumacher, T. N. & Schreiber, R. D. (2015). Neoantigens in cancer immunotherapy. *Science*, **348**, 69–74.

Sharma, P. & Allison, J. P. (2015). Immune checkpoint targeting in cancer therapy: toward combination strategies with curative potential. *Cell*, **161**, 205–14.

Silva-Santos, B., Serre, K., & Norell, H. (2015). Gammadelta T cells in cancer. *Nat Rev Immunol*, **15**, 683–91.

Zitvogel, L., Galluzzi, L., Smyth, M. J., & Kroemer, G. (2013). Mechanism of action of conventional and targeted anticancer therapies: reinstating immunosurveillance. *Immunity*, **39**, 74–88.

REFERENCES

Anderson, M. S. & Su, M. A. (2016). AIRE expands: new roles in immune tolerance and beyond. *Nat Rev Immunol*, **16**, 247–58.

Baas, M., Besancon, A., Goncalves, T., et al. (2016). TGFbeta-dependent expression of PD-1 and PD-L1 controls CD8(+) T cell anergy in transplant tolerance. *Elife*, **5**, e08133.

Banchereau, J. & Steinman, R. M. (1998). Dendritic cells and the control of immunity. *Nature*, **392**, 245–52.

Belvin, M. & Mellman, I. (2015). Is all cancer therapy immunotherapy? *Sci Transl Med*, **7**, 315fs348.

Beutler, B. & Rehli, M. (2002). Evolution of the TIR, tolls and TLRs: functional inferences from computational biology. *Curr Top Microbiol Immunol*, **270**, 1–21.

Bindea, G., Mlecnik, B., Tosolini, M., et al. (2013). Spatiotemporal dynamics of intratumoral immune cells reveal the immune landscape in human cancer. *Immunity*, **39**, 782–95.

Bruggemann, M., Caskey, H. M., Teale, C., et al. (1989). A repertoire of monoclonal antibodies with human heavy chains from transgenic mice. *Proc Natl Acad Sci U S A*, **86**, 6709–13.

Bruggemann, M., Williams, G. T., Bindon, C. I., et al. (1987). Comparison of the effector functions of human immunoglobulins using a matched set of chimeric antibodies. *J Exp Med*, **166**, 1351–61.

Chambers, C. A. & Allison, J. P. (1999). Costimulatory regulation of T cell function. *Curr Opin Cell Biol*, **11**, 203–10.

Chatenoud, L., Primo, J., & Bach, J. F. (1997). CD3 antibody-induced dominant self tolerance in overtly diabetic NOD mice. *J Immunol*, **158**, 2947–54.

Chen, D. S. & Mellman, I. (2013). Oncology meets immunology: the cancer-immunity cycle. *Immunity*, **39**, 1–10.

Chen, W. & Konkel, J. E. (2010). TGF-beta and 'adaptive' Foxp3(+) regulatory T cells. *J Mol Cell Biol*, **2**, 30–6.

Chiller, J. M., Habicht, G. S., & Weigle, W. O. (1971). Kinetic differences in unresponsiveness of thymus and bone marrow cells. *Science*, **171**, 813–15.

Djureinovic, D., Hallstrom, B. M., Horie, M., et al. (2016). Profiling cancer testis antigens in non-small-cell lung cancer. *JCI Insight*, **1**, e86837.

Dunn, G. P., Old, L. J., & Schreiber, R. D. (2004). The immunobiology of cancer immunosurveillance and immunoediting. *Immunity*, **21**, 137–48.

Fesnak, A. D., June, C. H., & Levine, B. L. (2016). Engineered T cells: the promise and challenges of cancer immunotherapy. *Nat Rev Cancer*, **16**, 566–81.

Fridman, W. H., Pages, F., Sautes-Fridman, C., & Galon, J. (2012). The immune contexture in human tumours: impact on clinical outcome. *Nat Rev Cancer*, **12**, 298–306.

Gabrilovich, D. I., Ostrand-Rosenberg, S., & Bronte, V. (2012). Coordinated regulation of myeloid cells by tumours. *Nat Rev Immunol*, **12**, 253–68.

Galon, J., Angell, H. K., Bedognetti, D., & Marincola, F. M. (2013). The continuum of cancer immunosurveillance: prognostic, predictive, and mechanistic signatures. *Immunity*, **39**, 11–26.

Galon, J., Fox, B. A., Bifulco, C. B., et al. (2016). Immunoscore and immunoprofiling in cancer: an update from the melanoma and immunotherapy bridge 2015. *J Transl Med*, **14**, 273.

Gentles, A. J., Newman, A. M., Liu, C. L., et al. (2015). The prognostic landscape of genes and infiltrating immune cells across human cancers. *Nat Med*, **21**, 938–45.

Girardi, M., Oppenheom, D. E., Steele, C. R., et al (2001). Regulation of cutaneous malignancy by gammadelta T cells. *Science*, **294**, 605–9.

Goodnow, C. C. & Basten, A. (1989). Self-tolerance in B lymphocytes. *Semin Immunol*, **1**, 125–35.

Gross, G. & Eshhar, Z. (2016). Therapeutic potential of T cell chimeric antigen receptors (CARs) in cancer treatment: counteracting off-tumor toxicities for safe CAR T cell therapy. *Annu Rev Pharmacol Toxicol*, **56**, 59–83.

Gubin, M. M., Artyomov, M. N., Mardis, E. R., & Schreiber, R. D. (2015). Tumor neoantigens: building a framework for personalized cancer immunotherapy. *J Clin Invest*, **125**, 3413–21.

Hagerling, C., Casbon, A. J., & Werb, Z. (2015). Balancing the innate immune system in tumor development. *Trends Cell Biol*, **25**, 214–20.

Hammer, G. E. & Ma, A. (2013). Molecular control of steady-state dendritic cell maturation and immune homeostasis. *Annu Rev Immunol*, **31**, 743–91.

Hanahan, D. & Coussens, L. M. (2012). Accessories to the crime: functions of cells recruited to the tumor microenvironment. *Cancer Cell*, **21**, 309–22.

Hanahan, D. & Weinberg, R. A. (2011). Hallmarks of cancer: the next generation. *Cell*, **144**, 646–74.

Hayday, A. & Tigelaar, R. (2003). Immunoregulation by gammadelta cells. *Nat Rev Immunol*, **3**, 233–42.

Hayday, A. C. (2009). Gammadelta T cells and the lymphoid stress-surveillance response. *Immunity*, **31**, 184–96.

Hinrichs, C. S., & Rosenberg, S. A. (2014). Exploiting the curative potential of adoptive T-cell therapy for cancer. *Immunol Rev*, **257**, 56–71.

Hori, S., Nomura, T., & Sakaguchi, S. (2003). Control of regulatory T cell development by the transcription factor Foxp3. *Science*, **299**, 1057–61.

Johnson, L. A., Morgan, R. A., Dudley, M. E., et al. (2009). Gene therapy with human and mouse T-cell receptors mediates cancer regression and targets normal tissues expressing cognate antigen. *Blood*, **114**, 535–46.

Jones, P. T., Dear, P. H., Foote, J., Neuberger, M. S., & Winter, G. (1986). Replacing the complementarity-determining regions in a human antibody with those from a mouse. *Nature*, **321**, 522–5.

Joyce, J. A. & Fearon, D. T. (2015). T cell exclusion, immune privilege, and the tumor microenvironment. *Science*, **348**, 74–80.

Kalos, M. & June, C. H. (2013). Adoptive T cell transfer for cancer immunotherapy in the era of synthetic biology. *Immunity*, **39**, 49–60.

Katsumata, M., Drebin, J. A., & Greene, M. I. (2006). Trastuzumab in breast cancer. *N Engl J Med*, **354**, 640–4; author reply 640–4.

Khalil, D. N., Smith, E. L., Brentjens, R. J., & Wolchok, J. D. (2016). The future of cancer treatment: immunomodulation, CARs and combination immunotherapy. *Nat Rev Clin Oncol*, **13**, 394.

Kohler, G. & Milstein, C. (1975). Continuous cultures of fused cells secreting antibody of predefined specificity. *Nature*, **256**, 495–7.

Kurosawa, Y. & Tonegawa, S. (1982). Organization, structure, and assembly of immunoglobulin heavy chain diversity DNA segments. *J Exp Med*, **155**, 201–18.

Li, B., Li, T., Pignon, J. C., et al. (2016). Landscape of tumor-infiltrating T cell repertoire of human cancers. *Nat Genet*, **48**, 725–32.

Linnemann, C., Mezzadra, R., & Schumacher, T. N. (2014). TCR repertoires of intratumoral T-cell subsets. *Immunol Rev*, **257**, 72–82.

Mahoney, K. M., Rennert, P. D., & Freeman, G. J. (2015). Combination cancer immunotherapy and new immunomodulatory targets. *Nat Rev Drug Discov*, **14**, 561–84.

Marvel, D. & Gabrilovich, D. I. (2015). Myeloid-derived suppressor cells in the tumor microenvironment: expect the unexpected. *J Clin Invest*, **125**, 3356–64.

Matzinger, P. (1994). Tolerance, danger, and the extended family. *Annu Rev Immunol*, **12**, 991–1045.

Mazor, R., Onda, M., & Pastan, I. (2016). Immunogenicity of therapeutic recombinant immunotoxins. *Immunol Rev*, **270**, 152–64.

McGranahan, N., Furness, A. J., Rosenthal, R., et al. (2016). Clonal neoantigens elicit T cell immunoreactivity and sensitivity to immune checkpoint blockade. *Science*, **351**, 1463–9.

Medzhitov, R. & Janeway, C. A., Jr. (1997). Innate immunity: impact on the adaptive immune response. *Curr Opin Immunol*, **9**, 4–9.

Melief, C. J., van Hall, T., Arens, R., Ossendorp, F., & van der Burg, S. H. (2015). Therapeutic cancer vaccines. *J Clin Invest*, **125**, 3401–12.

Mellman, I. (2013). Dendritic cells: master regulators of the immune response. *Cancer Immunol Res*, **1**, 145–9.

Mitchell, G. F. & Miller, J. F. (1968). Cell to cell interaction in the immune response. II. The source of hemolysin-forming cells in irradiated mice given bone marrow and thymus or thoracic duct lymphocytes. *J Exp Med*, **128**, 821–37.

Mlecnik, B., Bindea, G., Kirilovsky, A., et al. (2016). The tumor microenvironment and Immunoscore are critical determinants of dissemination to distant metastasis. *Sci Transl Med*, **8**, 327ra326.

Molon, B., Ugel, S., Del Pozzo, F., et al. (2011). Chemokine nitration prevents intratumoral infiltration of antigen-specific T cells. *J Exp Med*, **208**, 1949–62.

Mueller, D. L. (2010). Mechanisms maintaining peripheral tolerance. *Nat Immunol*, **11**, 21–7.

Munoz-Fontela, C., Mandinova, A., Aaronson, S. A., & Lee, S. W. (2016). Emerging roles of p53 and other tumour-suppressor genes in immune regulation. *Nat Rev Immunol*, **16**, 741–50.

Obenaus, M., Leitao, C., Leisegang, M., et al. (2015). Identification of human T-cell receptors with optimal affinity to cancer antigens using antigen-negative humanized mice. *Nat Biotechnol*, **33**, 402–7.

Palucka, K. & Banchereau, J. (2013). Dendritic-cell-based therapeutic cancer vaccines. *Immunity*, **39**, 38–48.

Park, T. S., Groh, E. M., Patel, K., Kerkar, S. P., Lee, C. C., & Rosenberg, S. A. (2016). Expression of MAGE-A and NY-ESO-1 in primary and metastatic cancers. *J Immunother*, **39**, 1–7.

Perro, M., Tsang, J., Xue, S. A., et al. (2010). Generation of multi-functional antigen-specific human T-cells by lentiviral TCR gene transfer. *Gene Ther*, **17**, 721–32.

Quail, D. F. & Joyce, J. A. (2013). Microenvironmental regulation of tumor progression and metastasis. *Nat Med*, **19**, 1423–37.

Ramsdell, F. & Ziegler, S. F. (2014). FOXP3 and scurfy: how it all began. *Nat Rev Immunol*, **14**, 343–9.

Regateiro, F. S., Howie, D., Nolan, K. F., et al. (2011). Generation of anti-inflammatory adenosine by leukocytes is regulated by TGF-beta. *Eur J Immunol*, **41**, 2955–65.

Ribas, A. & Wolchok, J. D (2018). Cancer immunotherapy using checkpoint blockade. *Science*, **359**, 1350–5.

Richards, D. M., Kyewski, B., & Feuerer, M. (2016). Re-examining the nature and function of self-reactive T cells. *Trends Immunol*, **37**, 114–25.

Ridge, J. P., Di Rosa, F., & Matzinger, P. (1998). A conditioned dendritic cell can be a temporal bridge between a CD4 + T-helper and a T-killer cell. *Nature*, **393**, 474–8.

Riethmuller, G. (2012). Symmetry breaking: bispecific antibodies, the beginnings, and 50 years on. *Cancer Immun*, **12**, 12.

Rizvi, N. A., Hellmann, M. D., Snyder, A., et al. (2015). Cancer immunology. Mutational landscape determines sensitivity to PD-1 blockade in non-small cell lung cancer. *Science*, **348**, 124–8.

Romagnoli, G. G., Zelante, B. B., Toniolo, P. A., Migliori, I. K., & Barbuto, J. A. (2014). Dendritic cell-derived exosomes may be a tool for cancer immunotherapy by converting tumor cells into immunogenic targets. *Front Immunol*, **5**, 692.

Rosenberg, S. A. & Restifo, N. P. (2015). Adoptive cell transfer as personalized immunotherapy for human cancer. *Science*, **348**, 62–8.

Sato, N., Yamabuki, T., Takano, A., et al. (2010). Wnt inhibitor Dickkopf-1 as a target for passive cancer immunotherapy. *Cancer Res*, **70**, 5326–36.

Schaefer, L. (2014). Complexity of danger: the diverse nature of damage-associated molecular patterns. *J Biol Chem*, **289**, 35237–45.

Schumacher, T. N. & Schreiber, R. D. (2015). Neoantigens in cancer immunotherapy. *Science*, **348**, 69–74.

Sharma, P. & Allison, J. P. (2015). Immune checkpoint targeting in cancer therapy: toward combination strategies with curative potential. *Cell*, **161**, 205–14.

Shimizu, K., Yamasaki, S., Shinga, J., et al. (2016). Systemic DC activation modulates the tumor microenvironment and shapes the long-lived tumor-specific memory mediated by CD8 + T cells. *Cancer Res*, **76**, 3756–66.

Silva-Santos, B., Serre, K., & Norell, H. (2015). Gammadelta T cells in cancer. *Nat Rev Immunol*, **15**, 683–91.

Simpson, E. (2006). A historical perspective on immunological privilege. *Immunol Rev*, **213**, 12–22.

Simpson, E. & Nehlsen, S. L. (1971). Prolonged administration of antithymocyte serum in mice. II. Histopathological investigation. *Clin Exp Immunol*, **9**, 79–98.

Smyth, M. J., Ngiow, S. F., Ribas, A., & Teng, M. W. (2016). Combination cancer immunotherapies tailored to the tumour microenvironment. *Nat Rev Clin Oncol*, **13**, 143–58.

Staerz, U. D. & Bevan, M. J. (1989). Redirecting the cellular immune response. *Int Rev Immunol*, **4**, 159–73.

Stein-Streilein, J. & Streilein, J. W. (2002). Anterior chamber associated immune deviation (ACAID): regulation, biological relevance, and implications for therapy. *Int Rev Immunol*, **21**, 123–52.

Steinman, R. M., Hawiger, D., & Nussenzweig, M. C. (2003). Tolerogenic dendritic cells. *Annu Rev Immunol*, **21**, 685–711.

Topalian, S. L., Wolchok, J. D., Chan, T. A., et al. (2015). Immunotherapy: The path to win the war on cancer? *Cell*, **161**, 185–6.

van der Burg, S. H., Arens, R., Ossendorp, F., van Hall, T., & Melief, C. J. (2016). Vaccines for established cancer: overcoming the challenges posed by immune evasion. *Nat Rev Cancer*, **16**, 219–33.

Vantourout, P. & Hayday, A. (2013). Six-of-the-best: unique contributions of gammadelta T cells to immunology. *Nat Rev Immunol*, **13**, 88–100.

Waldmann, H., Howie, D., & Cobbold, S. (2016). Induction of immunological tolerance as a therapeutic procedure. *Microbiol Spectr*, **4**(4). doi: 10.1128/microbiolspec.MCHD-0019-2015.

Wang, L., Yi, T., Zhang, W., Pardoll, D. M., & Yu, H. (2010). IL-17 enhances tumor development in carcinogen-induced skin cancer. *Cancer Res*, **70**, 10112–20.

Ward, J. P., Gubin, M. M., & Schreiber, R. D. (2016). The role of neoantigens in naturally occurring and therapeutically induced immune responses to cancer. *Adv Immunol*, **130**, 25–74.

Willcox, C. R., Pitard, V., Netzer, S., et al. (2012). Cytomegalovirus and tumor stress surveillance by binding of a human gammadelta T cell antigen receptor to endothelial protein C receptor. *Nat Immunol*, **13**, 872–9.

Woo, S. R., Corrales, L., & Gajewski, T. F. (2015). Innate immune recognition of cancer. *Annu Rev Immunol*, **33**, 445–74.

Yamabuki, T., Takano, A., Hayama, S., et al. (2007). Dikkopf-1 as a novel serologic and prognostic biomarker for lung and esophageal carcinomas. *Cancer Res*, **67**, 2517–25.

Young, A., Ngiow, S. F., Barkauskas, D. S., et al. (2016). Co-inhibition of CD73 and A2AR adenosine signaling improves anti-tumor immune responses. *Cancer Cell*, **30**, 391–403.

Zitvogel, L., Galluzzi, L., Smyth, M. J., & Kroemer, G. (2013). Mechanism of action of conventional and targeted anticancer therapies: reinstating immunosurveillance. *Immunity*, **39**, 74–88.

SECTION V

Global vision of cancer

24

Molecular profiling in cancer research and personalized medicine

Pieter-Jan van Dam and Steven Van Laere

Molecular profiling of breast cancer: The state-of-the-art

Breast cancer is a heterogeneous disease with respect to molecular alterations, cellular composition, and clinical outcome (**Fig. 24.1**). This diversity creates a challenge in developing tumour classifications that are clinically useful with respect to prognosis or prediction (Parker et al., 2009). Currently, a wide array of clinical and pathological characteristics, such as tumour size, histological grade, number of affected axillary lymph nodes, oestrogen receptor (ER) and progesterone receptor (PR) expression levels, and presence of HER2 amplification, are used in routine diagnostics to guide patient management and perform breast tumour classifications. Hence, subgroups of patients can be identified that are either more or less sensitive to specific therapies, or that have an elevated risk of metastatic recurrence and therefore should be treated more aggressively. In the past two decades, efforts towards a more rationalized breast cancer classification system based on genome- and transcriptome-wide molecular alterations have been undertaken, the majority of which spearheaded by The Cancer Genome Atlas (TCGA) and the International Cancer Genome Consortium (ICGC). In the following sections, the contributions of these studies to understanding breast cancer biology are summarized.

Gene expression

Historically, genome-wide gene expression profiling has contributed significantly to the establishment of novel insights into breast cancer, both from a biological and clinical perspective. Already at the beginning of this century, Perou and colleagues reported the existence of different, gene expression-based breast cancer subtypes. In the following years, these subtypes have been repeatedly observed in multiple breast cancer data sets (Smid et al., 2016). Importantly, they recapitulated the classical breast cancer classification scheme. The luminal subtypes (i.e. luminal A and luminal B), characterized by the expression of genes typical for the luminal epithelial cells of the mammary gland, largely coincided with the subgroup of breast cancers expressing ER and PR. The HER2-enriched subtypes, characterized by the expression of genes associated with HER2-signalling, were enriched for breast cancer samples with documented HER2 protein overexpression and/or gene amplification. The basal-like and the normal-like subtypes, characterized respectively by the expression of genes typical for myoepithelial cells and for normal breast epithelium, represent the fraction of so-called triple-negative breast cancer (TNBC) samples that have no notable hormone receptor or HER2 protein expression. In 2011, the group of TNBCs was further classified in to six molecular subtypes: two distinct basal-like subtypes; an immunomodulatory subtype; a mesenchymal subtype; a mesenchymal stem-like subtype; and a luminal androgen receptor subtype (Lehmann et al., 2011).

More importantly, numerous studies have shown that these molecular classification schemes have clinical relevance. In multivariate analysis, the molecular breast cancer subtypes are independent predictors of prognosis in a group of therapy-naïve patients with lymph node-negative disease. In addition, in patients treated with neo-adjuvant chemotherapy, the intrinsic subtypes are able to predict therapeutic efficacy (Parker et al., 2009). These observations paved the way for the development of the Prosigna assay, which measures 50 genes that are related to the molecular subtypes to assess a patient's risk of distant recurrence at 10 years. Similarly, also the TNBC-subtypes are significantly associated with the rate of pathological complete response to neo-adjuvant chemotherapy (Masuda et al., 2013). Other examples of the successful application of gene expression profiling in breast cancer from a clinical perspective include the 21-gene OncotypeDX assay and the Genomic Grade assay, which can be used to risk-stratify early-stage ER-positive breast cancer and the 70-gene MammaPrint assay that has documented prognostic significance in early-stage node-negative breast cancer, regardless of ER status. Altogether, gene expression profiling demonstrated that breast cancer is a heterogeneous disease at the molecular level and that molecular profiling can be used guide patient management. This statement is further corroborated when considering molecular profiles other than those based on gene expression data.

Copy number variations

In 2013, Ciriello and colleagues reported an analysis across 12 different cancer types and showed that the landscape of somatic genomic variations in cancer is either dominated by mutations (i.e. M-class) or recurrent copy number variations (i.e. C-class). As such, it was found that breast cancer is primarily CNV-driven, except for

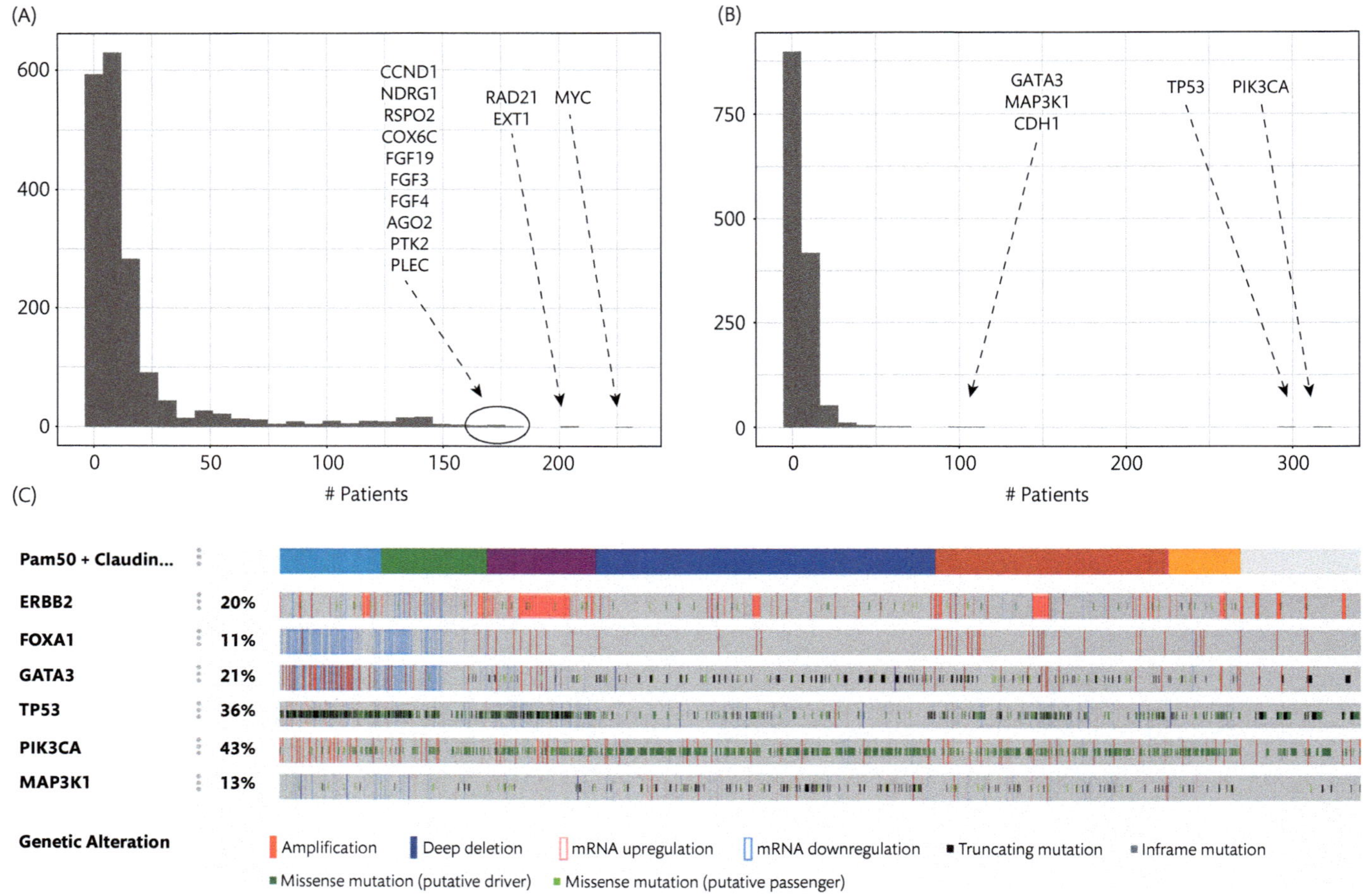

Fig. 24.1 Overview of the mutational landscape and the copy number alteration profile of breast cancer. Frequency histograms for copy number variations (A) and point mutations (B) show that the majority of genes harbour DNA alterations in just a few samples. It can also be appreciated that copy number variations are also more prevalent in breast cancer as compared to base substitutions. In the presented data set (TCGA, 1,105 samples), only three genes exhibit point mutations in at least 100 samples, whereas 87 genes exhibit copy number in the same frequency range. Panel C represents an oncoprint image showing the molecular landscape of six recurrently altered genes in 2,509 breast cancer patients (METABRIC). On top, the classifications of these samples according to the PAM50 classification (blue = basal, green = claudin-low, purple = Her2 enriched, blue = luminal A, red = luminal B, orange = normal-like) is shown. The legend for the different molecular alteration types is represented underneath the oncoprint. It can be observed that ERBB2 amplifications are enriched in the Her2 enriched subtype and also to some extend in the luminal B subtype. Further, TP53 and PIK3CA are the most frequently altered genes with truncating mutations affecting particularly TP53 and missense mutations being more prevalent in PIK3CA. In addition, the mutational profile of both genes is mutually exclusive. Finally, FOXA1 is frequently deleted in basal-like and Claudin-low samples and truncating MAP3K1 and GATA3 mutations are more prevalent in luminal-type breast cancer samples.

Source: data from *Breast Cancer* (METABRIC, Nature 2012, and Nat Commun 2016), cBioPortal for Cancer Genomics, http://www.cbioportal.org/study?id=brca_metabric#summary

the luminal A subtype in which somatic mutations are more frequent (Ciriello et al., 2013). This observation is corroborated by the fact that the mutational load in breast cancer (i.e. ~2 mutations per tumour) is considerably lower as compared to other cancer types (Kandoth et al., 2014).

Curtis and colleagues investigated the genomic and transcriptomic architecture of 2,000 breast cancer samples (Curtis et al., 2013) and revealed that somatic copy number variations (CNVs) have a dramatic impact on the expression levels of genes, both in *cis* (i.e. effecting genes on the same chromosome) and in *trans* (i.e. effecting genes on a different chromosome). Ten clinically relevant breast cancer subtypes (i.e. iClusters) that were loosely associated with the more classical intrinsic subtypes were identified. Particularly the ER-positive luminal tumours were further stratified into subgroups based on the genomic information. For example, an ER-positive subgroup composed of 11q13/14 cis-acting luminal tumours was noted that exhibited a steep mortality trajectory with elevated hazard ratios. A second group of luminal tumours was characterized by low genomic instability and was enriched in histotypes that typically have good prognosis. The classical basal-like and HER2-enriched breast cancers were predominantly allocated in specific clusters (Curtis et al., 2013). Importantly, CNV profiles described by Curtis and colleagues have also been described in a series of 773 breast cancer patient samples reported by TCGA, including the characteristic loss of 5q and gain of 10p in basal-like cancers and gain of 1q and/or 16q loss in luminal tumours (Koboldt et al., 2012). Breast cancer genes most frequently affected by CNVs include *CNND1, CDKN2A, EGFR, ERBB2, FGFR1, FOXA1, MAP2K4, MDM2, MLL3, MYC, PIK3CA, PTEN, RB1*, and *ZNF217* (Koboldt et al., 2012; Stephens et al., 2013).

Somatic mutations

In 2005, Stephens and colleagues described the detection of mutations in a panel of 518 protein kinases using targeted amplicon sequencing (Stephens et al., 2005). The results of this study can be summarized according to three major themes that have since dominated the discussions related to the mutational characteristics of breast cancer. First, no commonly mutated protein kinase was identified across the series of 25 breast cancer samples, suggesting significant heterogeneity in the landscape of somatic mutations across breast cancer. This observation has been repeatedly observed in other breast cancer profiling studies (Koboldt et al., 2012; Stephens et al., 2013). Second, the majority of the observed mutations were considered as 'passengers', and therefore are not subjected to natural selection and thus did not contribute to cancer development or progression. Finally, specific mutations (e.g. C>G transversions) were not evenly distributed within the genome, but are actually characterized by a specific genomic context, determined by the sequence immediately at the 5' and the 3' site of the mutation. This observation has led to the concept of mutational signatures, in which mutations are classified according to their genomic context and which betray the processes that are fundamental to the development of the mutational landscape.

In a recent analysis of 560 whole breast cancer genomes, 12 mutational signatures were identified in breast cancer (Nik-Zainal et al., 2016). Five of these were reported in earlier breast cancer studies and are associated with age-related deamination of methylated cytosine, BRCA-deficiency, and APOBEC-activity (Alexandrov et al., 2013). The remaining seven mutational signatures, of which three are linked to DNA mismatch repair deficiency, are either novel or have been previously reported in other cancer types (Alexandrov et al., 2013). Importantly, the fact that mutational signatures can be related to BRCA-deficiency and defective DNA mismatch repair indicates that mechanisms involved in DNA maintenance (i.e. BRCA is involved in repair of double strand DNA breaks by homologous recombination) contribute significantly to the mutational burden in breast cancer. This is even further corroborated when inspecting the genomic context of indel mutations, which may be caused by DNA break repair by microhomology-mediated end joining (Nik-Zainal et al., 2012; Nik-Zainal et al., 2016). Other mechanisms such as APOBEC-activity, which is part of the cellular antiretroviral defence mechanism, are involved in establishing somatic mutations in breast cancer. APOBEC-enzymes can deaminate cytosine to uracil, preferentially in single-stranded RNA molecules like the retroviral genome. However, single-stranded genomic DNA, for example generated during replication of the lagging DNA-strand, can also be targeted leading to C>T transitions. Importantly, cancer samples that exhibit a mutational profile related to APOBEC-activity are also characterized by immune response-specific gene expression patterns, as well as increased number of tumour-infiltrating lymphocytes, suggesting that APOBEC-activity is associated with a more effective immune response (Lin et al., 2015; Smid et al., 2016).

Sequencing efforts in breast cancer, among other those led by TCGA, have identified lists of genes that are frequently affected by mutations, among which *AFF2, AKT1, CBFB, CCND3, CDH1, CDKN1B, FOXP1, GATA3, MAP3K1, MED23, MLL3, MLLT4, NF1, PIK3CA, PIK3R1, PTEN, PTPN22, PTPRD, RB1, RUNX1, SF3B1, TBX3, TP53, XBP1*, and *ZFP36L1* are the most recurrently mutated (Koboldt et al., 2012; Nik-Zainal et al., 2016). Interestingly, a significant pattern of mutual exclusion among *AKT1, PIK3CA, PIK3R1*, and *PTEN* was observed, eluding at the importance of the PI3K/AKT pathway in breast cancer. In addition, several of the recurrently mutated genes show a molecular subtype-specific mutation pattern. In basal-like breast cancer, TP53 is the most dominantly mutated gene (i.e. 80% of the cases) whereas PIK3CA-mutations are most frequent in luminal A breast cancer samples (i.e. 45% of the cases). Luminal B and HER2-enriched breast cancer samples have a more hybrid pattern with relative high frequencies for both TP53 and PIK3CA.

Structural variations

Structural variations or genomic rearrangements include tandem duplications, inversions, deletions, or interchromosomal translocations. In agreement with the mutational signatures, six rearrangement signatures have been described in a series 560 complete breast cancer genomes. Two rearrangement signatures were characterized by the presence of tandem duplications, one of which was attributable to BRCA1 deficiency, and thus defective repair of double strand DNA breaks by homologous recombination. Importantly, the other tandem duplication signature also revealed evidence of defective homologous recombination but no mutations or promotor hypermethylation for the BRCA genes was detected, suggesting that other mechanisms besides BRCA contribute to DNA repair by homologous recombination. Both tandem duplication signatures were characteristic for basal-like or triple-negative breast cancers. The remaining signatures were characterized to a variable extend by clustered and non-clustered deletions, inversions, and intra- and interchromosomal translocations. Importantly, all rearrangement signatures were associated with different patterns of microhomology, indicating that different end-joining mediated DNA repair mechanisms may operate with different rearrangement processes (Nik-Zainal et al., 2016).

Structural genomic variations may contribute to the establishment of fusion genes, which can result in inappropriate gene expressions and functions and thus drive cancer biology. Stephens and colleagues described 29 rearrangements that generated in-frame gene fusions involving candidate cancer genes like the ETS family of transcription factors. Importantly, none of the in-frame fusions genes were recurrent (Stephens et al., 2009). Also in the series of 560 complete breast cancer genomes, recurrent in-frame fusion was detected only infrequently whereas the majority of the in-frame fusions were unique (Nik-Zainal et al., 2016). Comparing in-frame fusion events between the molecular subtypes, an important difference in the number of rearrangements was found, with luminal A samples reporting the lowest frequency and HER2-enriched and basal-like samples reporting the highest frequencies (Banerji et al., 2013).

MicroRNA and epigenome profiling

Dvinge and colleagues profiled a subset of 2,000 breast tumours described by Curtis et al. for microRNA expression and revealed that the variability of microRNA expression is generally associated with the presence of the molecular subtypes or the iCluster subtypes (Dvinge et al., 2014). Similar observations were reported by TCGA (Koboldt et al., 2012). However, microRNA expression is generally not associated with CNVs or gene expression changes that affect their host genes, suggesting that specific mechanisms regulating the expression of microRNAs exist. More detailed analysis revealed 10 microRNA-mRNA interaction groups in breast cancer, with the majority of these interactions being stimulatory. These interaction groups represented several key processes associated with breast cancer such as cell proliferation, cell adhesion, sex-hormone activity, and immunity. An innovative screen for mRNA-modulating microRNA-activity revealed that a subgroup of copy number neutral breast tumours exhibit evidence for a pathogenic role of microRNAs possibly by affecting the immune response that characterizes these tumours.

DNA methylation was assessed in a series of 802 breast cancer patients reported by TCGA, in which five distinct DNA methylation groups were described. One group, enriched for luminal B samples, revealed significant levels of DNA hypermethylation, whereas DNA hypomethylation was more often associated with the basal-like subtype. Relating patterns of differential methylation to patterns of differential expression showed that the Wnt signalling pathway is often affected by DNA hypermethylation in the luminal B breast cancer subtype (Koboldt et al., 2012).

Protein expression profiling

Finally, microarray-based technology can also be used to evaluate patterns of protein expression. Using reverse phase protein arrays focussed on 171 cancer-related proteins and phospho-proteins, seven subtypes were described in a series of 403 breast cancer samples. These protein-based subtypes were again closely related to the mRNA-based subtypes indicating that changes observed at the mRNA level are representative of changes on a more functional level. Importantly, two potentially novel protein-defined subgroups were identified: reactive I consisted primarily of a subset of luminal A tumours, whereas reactive II consisted of a mixture of mRNA subtypes. These groups are termed 'reactive' because many of the characteristic proteins are probably produced by the microenvironment and/or cancer-activated fibroblasts including fibronectin, caveolin 1, and collagen VI (Koboldt et al., 2012). These findings were confirmed on an independent patient series using mass spectrometry as an alternative profiling technique (Mertins et al., 2016).

The study by Mertins and colleagues also evaluated how variation at the genomic and transcriptomic level relates to variation in phosphoprotein expression in breast cancer (Mertins et al., 2016). Overall, gene and protein expression data were well correlated, but the influence of genomic variation on phosphoprotein expression levels is less evident. First, about 30% of the proteins and 20% of the phosphoproteins have expression levels that are positively correlated to somatic copy number changes. These associations were primarily limited to CNVs affecting oncogenes or tumour suppressor genes. In addition, copy number gains for some genes (e.g. GATA3) were associated with decreased protein expression levels. Second, the impact of point mutations on protein expression levels is versatile and varies according to the mutation type and the affected gene. Frameshift mutations in TP53 decrease protein expression, whereas frameshift mutations in GATA3 have no effect on protein expression. Notably, the authors provide strong evidence that the functional impact of a somatic mutation varies according to the affected protein domain. For example, mutations in the helical domain of PIK3CA are more likely to active the PI3K pathway as compared to mutations in the kinase domains. Finally, the number of genomic (and transcriptomic) variants that were confirmed by peptide sequence changes was low. Together, these data reveal that protein profiling provides an alternative and complimentary perspective on the molecular landscape of breast cancer. Further, integration of genomic, transcriptomic, and proteomic information will undoubtedly ease the identification of crucial events that contribute to the development of breast cancer.

Towards an integrated molecular landscape

Based on the complete landscape of somatic mutations, copy number variations, and recurrent rearrangements in a series of 560 complete breast cancer genomes, the catalogue of 93 breast cancer genes representing 1,628 likely driver changes was established (Nik-Zainal et al., 2016). At least one of these driver genes was compromised in 95% of the evaluated breast cancer samples, indicating that the reported catalogue is near complete. The 10 most frequently mutated genes were *TP53, PIK3CA, MYC, CCND1, PTEN, ERBB2, ZNF703/FGFR1* locus, *GATA3, RB1*, and *MAP3K1*, which accounted for 62% of drivers.

The molecular characteristics of the molecular subtypes based on an integrated analysis of gene mutation, methylation and expression data, CNVs, and protein expression data is reported in Table 24.1. From these data, it is evident that the PI3K-, TP53-, and RB1-pathways are often implicated in breast cancer, albeit with significant variation across the different molecular subtypes.

Profiling of intrapatient heterogeneity

Tumours consist of a complex patchwork of genetically related and competing subpopulations. These cell populations arise through the continued actions of mutational processes during which 'hallmark' cellular processes are cumulatively co-opted or ablated, leading to a growth benefit for some but not all cancer cell clones that will thus become the dominant cancer cell populations. It is obvious that, as selective pressures shift (e.g. due to treatment), other cancer cell clones may become better adapted. Therefore, the composition of a tumour in terms of its cell populations is dynamic and thus cancer heterogeneity is a possible contender to explain therapy resistance.

Several studies have documented that breast tumours are indeed composed of heterogeneous cell populations. Nik-Zainal and colleagues reported the life history of 21 breast cancer. In this study, different cancer cell subclones were evident in all evaluated breast cancer samples (Jiao et al., 2011; Koboldt et al., 2012; Nik-Zainal et al., 2012). In another study, distinct breast cancer subclones were only detected in 35% of the evaluated samples (Kandoth et al., 2014). However, it should be mentioned that in this study only somatic point mutations were assessed, and thus the dominant patterns of heterogeneity may have been missed as breast cancer is primarily CNV-driven. Using single nucleus sequencing of hormone receptor positive and TNBC, it was even suggested that every cancer cell is characterized by a unique mutational profile (Wang et al., 2015).

The clinical relevance of breast cancer heterogeneity is documented in a study by Yates and colleagues. Comparing pre- and post-treatment samples identified a set of mutations present in the post-treatment samples, some of which were shared with the pre-treatment samples. These mutations possibly reflect therapy resistance mechanisms. In addition, comparing metastases to the primary tumour revealed that some subclones are responsible for

Table 24.1 The molecular characteristics of the molecular subtypes based on an integrated analysis of gene mutation, methylation, and expression data, CNVs, and protein expression data

	Luminal A	Luminal B	Basal	HER2E
TP53 pathway	TP53 mut (12%); gain of MDM2 (14%)	TP53 mut (32%); gain of MDM2 (31%)	TP53 mut (84%); gain of MDM2 (14%)	TP53 mut (75%); gain of MDM2 (30%)
PIK3CA/PTEN pathway	PIK3CA mut (49%); PTEN mut/loss (13%); INPP4B loss (9%)	PIK3CA mut (32%) PTEN mut/loss (24%) INPP4B loss (16%)	PIK3CA mut (7%); PTEN mut/loss (35%); INPP4B loss (30%)	PIK3CA mut (42%); PTEN mut/loss (19%); INPP4B loss (30%)
RB1 pathway	Cyclin D1 amp (29%); CDK4 gain (14%); low expression of CDKN2C; high expression of RB1	Cyclin D1 amp (58%); CDK4 gain (25%)	RB1 mut/loss (20%); cyclin E1 amp (9%); high expression of CDKN2A; low expression of RB1	Cyclin D1 amp (38%); CDK4 gain (24%)
mRNA expression	High ER cluster; low proliferation	Lower ER cluster; high proliferation	Basal signature; high proliferation	HER2 amplicon signature; high proliferation
Copy number	Most diploid; many with quiet genomes; 1q, 8q, 8p11 gain; 8p, 16q loss; 11q13.3 amp (24%)	Most aneuploid; many with focal amp; 1q, 8q, 8p11 gain; 8p, 16q loss; 11q13.3 amp (51%); 8p11.23 amp (28%)	Most aneuploid; high genomic instability; 1q, 10p gain; 8p, 5q loss; MYC focal gain (40%)	Most aneuploid; high genomic instability; 1q, 8q gain; 8p loss; 17q12 focal ERRB2 amp (71%)
DNA mutations	PIK3CA (49%); TP53 (12%); GATA3 (14%); MAP3K1 (14%)	TP53 (32%); PIK3CA (32%); MAP3K1 (5%)	TP53 (84%); PIK3CA (7%)	TP53 (75%); PIK3CA (42%); PIK3R1 (8%)
DNA methylation	-	Hypermethylated phenotype for subset	Hypomethylated	-
Protein expression	High oestrogen signalling; high MYB; RPPA reactive subtypes	Less oestrogen signalling; high FOXM1 and MYC; RPPA reactive subtypes	High expression of DNA repair proteins, PTEN and INPP4B loss signature (pAKT)	High protein and phospho-protein expression of EGFR and HER2
Molecular subtypes	M-class; intClust2-3-6-7-8-9; ReactiveI&II	C-class; intClust1-2-6-9; ReactiveII	C-class; intClust10; ReactiveII	C-class; intClust5; ReactiveII
Mutational signatures	Low frequency of Signature8	-	Signature1-3-8-13	Signature2-5
Rearrangement signatures	Signature2	Signature2	Signature 3(BRCA1)-1(HRD)-5(BRCA1)	-

the formation of the metastases and that these subclones can be present early during breast cancer development, even well before the onset of subclonal variation (Yates et al., 2015).

Molecular profiling technologies

Molecular profiling is defined as the identification and documentation of the complete or partial set of DNA, RNA, or proteins molecules that characterize a condition of interest. With the development of high-throughput technologies such as microarrays (MA) and next-generation sequencing (NGS), molecular cancer profiles have been generated on an unprecedented scale, as exemplified for breast cancer in the previous section. In this section, we aim to provide a synopsis of these two popular profiling techniques. It should be mentioned that other high-throughput profiling techniques exist, such as mass spectrometry for proteomics and chromosome confirmation capture assays. The latter aims to understand the complex organization of the genome into a three-dimensional structure including chromatin loops and bridges that bring distant element of the genome in close proximity and allow coregulation of genes on different chromosomes but occupying the same spatial localization in the nucleus (i.e. transcription factories). However, both mass spectrometry and chromosome confirmation capture assays will not be further discussed within the context of this chapter.

Microarray technology

Microarray technology appeared in 1995 and can be considered as one of the major biotechnological revolutions of the last 15 years, thanks to improvements in chemistry, physics, optics, robotics, software engineering, and molecular biology. The MA, also known as (bio)chip, provides a miniaturized sensor tool allowing the interrogation of genome-wide events on a surface smaller than two square centimetres. Originally, MAs emerged in the field of transcriptomics, but the technology was rapidly applied within the context of alternative 'omics' approaches. As a result, a large variety of MA techniques have been developed for various applications, although the overall protocol relies on similar characteristics.

MA technology involves tethering probes to a solid support (i.e. the chip) such as glass, plastic, or silica. The probes are deposited on a microscopic area of the chip called a spot or feature and a typical MA contains thousands or millions of spots making it a powerful tool for genome-wide screening. Alternatively, probes can be attached to silica beads, which self-assemble in the micro-wells (i.e. bead arrays). A key concept is that probes are chosen specifically to report the quantification of their expected target. For example, in case of cDNA or oligonucleotide probes, specificity is guaranteed by the choice of a unique base-paired complementarily between the probe and the target sequence. A variety of MAs exist that can be classified according to the contained probe types and their coverage profiles. Possible probe types include long cDNA clones, short oligonucleotides, antibodies, and sample analytes (e.g. reverse phase protein arrays, tissue MA). The first two probe types are dedicated for nucleic acid-based analyses (e.g. profiling of CNVs and gene expression), whereas the latter two are used to quantify protein expression. The development of oligonucleotide arrays led to an ever-increasing number of probes per chip and thus a denser coverage profile. By consequence, the number of target sequences that can be interrogated expanded, allowing a more detailed analysis of both genomic and transcriptomic landscapes to be performed. Figure 24.2 illustrates the probe coverage profile for different MA classes.

Applying MA technology starts by extracting DNA, RNA, or proteins from tissue samples or cells. The extracted biomolecules are then hybridized onto the MA, during which the probes capture their target molecules, if they are present in the sample. In general, one sample is hybridized per MA, but two samples can be hybridized competitively. Subsequently, the amount of target captured by the probes is measured using confocal laser scanners. For this purpose, fluorescent labels are used that are attached to either the target molecules during library preparation in case of DNA or RNA analyses or to a secondary reporter antibody in case or protein analyses. The intensity of the emitted fluorescent signal, captured as an image, is directly related to the amount of target present in the sample allowing it to be quantified during image analysis.

Next-generation sequencing

In 2004, the first sequence that nearly completely covered the complete human genome was published by the International Human Genome Sequencing Consortium. This effort, established using classical Sanger sequencing (a.k.a. first-generation sequencing), took 13 years, involved about 3,000 scientists worldwide and cost about 2.7 billion dollars. Despite improvements, Sanger sequencing was unable to deliver complete human sequences at a reasonable cost or in a reasonable time frame. To overcome these limitations, second-generation sequencing (a.k.a. next-generation sequencing) was developed. The new methodology, applied for the first time in 2004, relies on parallel sequencing of a massive amount of DNA-fragments and provides a dramatic increase in throughput capacity at a lower cost. Using NGS, James Watson's complete genome was sequenced in 4.5 months, involved 30 scientists and cost less than 1.5 million dollars. In 2008, third-generation sequencing, based on single molecule analysis, was established allowing the complete genome to be sequenced in several days, at a cost of about 50,000 dollars.

NGS relies on high-level technologies including enzymology, chemistry, high-resolution optics, hardware, and software engineering. Although different platforms exist, a generally applicable workflow can be discerned. First, polymerase chain reaction (PCR)-technology is used to prepare a sequencing library that contains small fragments of template. During this library preparation step, adapters are typically attached to the templates. These adapters allow binding of the templates to a solid support and provide a start site for the sequencing enzymes. In a second step, library templates are clonally amplified (e.g. bridge amplification or emulsion PCR) and immobilized onto solid support (e.g. a flow cell or beads). Notably, single molecule sequencing does not involve an additional amplification step. Finally, the library is sequenced, allowing a fixed number of bases for each template to be read in the forward-only direction (i.e. single-end sequencing) or in both the forward and reverse direction (i.e. paired-end sequencing). This process generates signals from which the analysed sequence can be deduced. Most sequencing systems depend on nucleotides (i.e. A, C, G, or T) labelled with fluorochromes, one for each type of nucleotide, allowing the emission of a fluorescent signal as the nucleotides are incorporated into a growing stretch of DNA. The fluorescent signal is captured and image analysis allows the sequence of the template to be obtained. From an analytical perspective, the identified sequences, termed reads, are on the same level as the probes of a MA. Alternative technologies exist, in which the optical detection system is replaced by semiconductor technology that detects a change in impedance each time a nucleotide is incorporated.

Low-end data analysis

Low-end data analysis involves all statistical procedures applied to the raw signal intensities, obtained after image analysis, to extract a biological profile (e.g. expression, mutation or CNV profile) for each sample in the analysis. It is required for two main reasons: A. to minimize the influence of technical artefacts (i.e. spatial artefacts, GC-bias, batch- and dye-effects) and B. to summarize the signals obtained after image analysis according to meaningful genomic entities (e.g. genes and transcripts). In contrast, high-end data analysis (a.k.a. data mining; *vide infra*) relates to the application of statistical methods in order to organize complex data sets, composed of many individual profiles, into novel and insightful biological concepts. Without appropriate application of low-end data analysis, data mining techniques will generate non-sensical and/or biased output. Many different approaches for low-end data analysis have been described, each with their own benefits and drawbacks. It is not the purpose of this section to provide an in-depth tutorial. Interested readers are advised to explore the BioConductor repository and particularly the software packages tailored to low-end data analysis. Instead, a general workflow (Fig. 24.3) that encompasses both MA and NGS data analysis is presented.

In general, five distinct phases can be discerned: A. image capture and analysis; B. feature-level quality control; C. read alignment; D. summarization; and E. normalization. During image capture and analysis, the fluorescent signals are acquired and analysed to generate feature-level data. These represent either probe-level fluorescence intensities in case of MA analysis or reads with a per base quality score (a.k.a. Phred score) in case of NGS analysis. Next, feature-level quality control is performed to detect sources of technical bias for which correction is required or that necessitates the exclusion of specific features or ultimately of complete samples in order to improve overall comparability. The third phase in the low-end data analysis workflow is read alignment, which is specific for NGS data analysis. During read alignment, the individual reads are mapped to the (human) reference genome, allowing the genomic origin of the individual reads to be traced. By consequence, reads are associated with specific genomic entities such as genes, promoters, or intergenic regulatory sequences. Of note, reads may align to the reference genome in a non-continuous fashion (i.e. split reads) indicating the presence of structural rearrangements, fusion genes, and/or splice events. For MA data analysis, read alignment is irrelevant since the probes are annotated beforehand, and thus the fluorescent signals can be readily interpreted in their appropriate genomic context. To read the alignment of NGS data is essential to allow data summarization, during which the feature-level data are

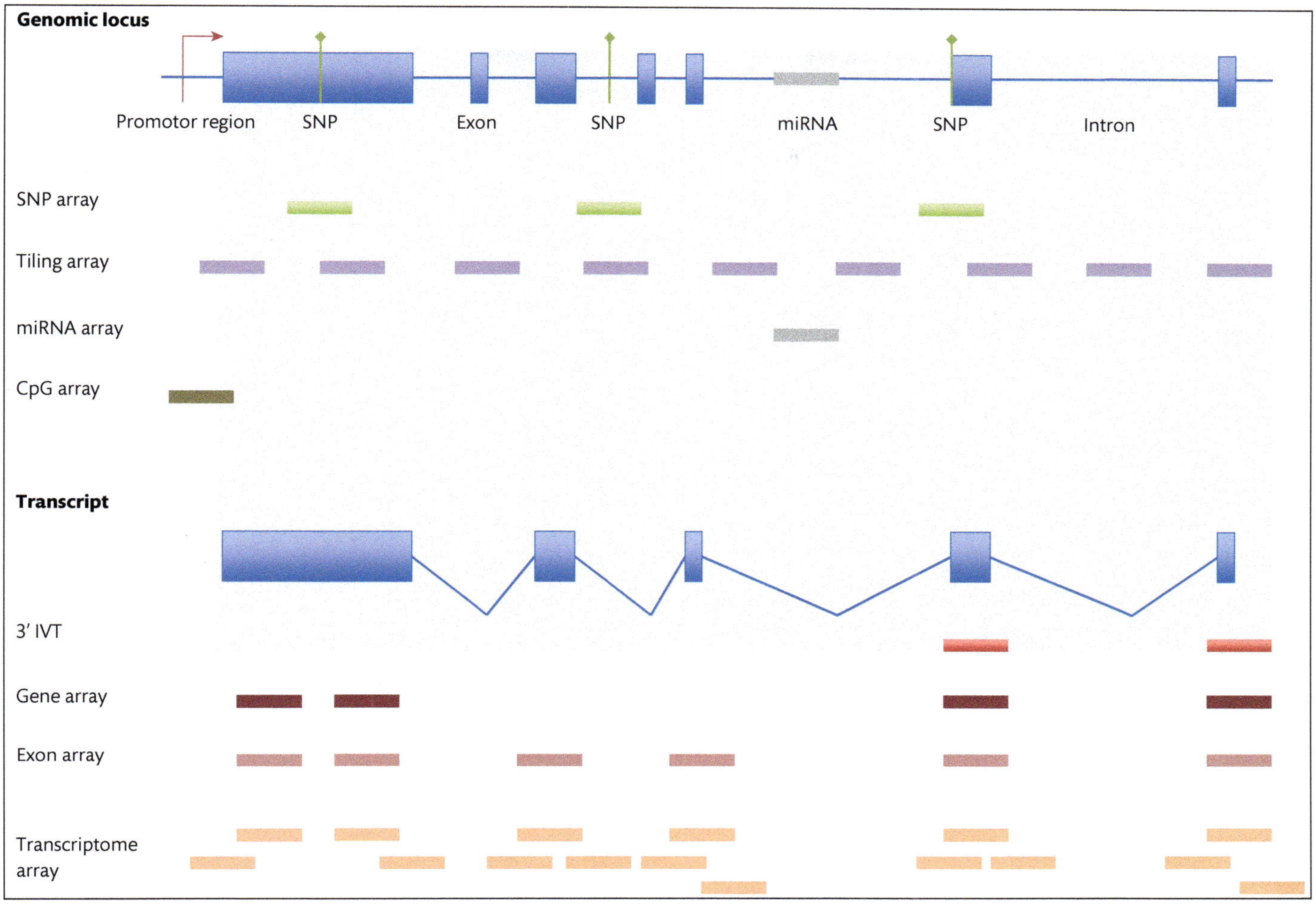

Fig. 24.2 Microarray probes. The type of data that can be obtained from a microarray experiment is largely determined by the nature of the probes. In this figure, a hypothetical structure of a genetic locus is depicted, including the promotor region (red arrow), single nucleotide polymorphism (SNPs) (green spike), exons (blue box), and a miRNA (grey box). A transcript originating from this genomic locus with two spliced exons is also shown. Probes directed at specific features of the genomic locus or the transcript allow different types of information to be obtained. Using SNP arrays with probes directed at specific SNPs, CNV profiles can be established, loss-of-heterozygosity events can be detected, and genotyping can be performed to identify cancer susceptibility genes. Genome-wide DNA-protein interactions, including histone modifications, can be evaluated using tiling arrays, in which probes cover the entire genome, either contiguously or partially overlapping. In addition, tiling arrays can also be used to analyse genome-wide methylation patters. For both applications, DNA should be isolated using chromatin immunoprecipitation using an antibody against a protein of interest or against methylated DNA. Genome-wide methylation patterns can also be investigated using CpG arrays, to which bisulfite-treated DNA is hybridized. Bisulfite-treatment results in a sequence change (i.e. unmethylated cytosines are converted into uracils, methylated cytosines are not) and different probes inspecting either the methylated or the unmethylated sequence will be used to evaluate methylation differences. Finally, transcript abundance can be measured using probes located in the 3′ *in vitro* transcribed region of the mRNA molecules (i.e. 3′IVT arrays) or more equally distributed and overlapping most or all exons (i.e. exon or gene arrays respectively). In recent designs, probes additionally overlap two consecutive exons (i.e. transcriptome array), allowing for the analysis of expression patterns at the level of genes and transcripts as well as the identification of alternative splicing events. Finally, probes may also be designed to detect non-coding RNA expression changes.

collapsed according to specific genomic characteristics. For example, all probes or reads corresponding to one gene or one genomic location (e.g. transcription factor binding site) are combined to generate a composite result. During the summarization phase, the actual molecular profile is generated. Particularly for NGS data, a plethora of summarization methods exist, particularly depending on nature of the molecular profile that is being investigated (e.g. transcriptome analysis *vs.* mutational analysis). The final step involved in low-end data analysis is data normalization, which is performed to remove systematic differences between the samples that are due to noise rather than true biological variability (e.g. differences in hybridization conditions, amounts of input material, probe, and read characteristics like the GC content). It should be emphasized that the specific procedures for low-end data analysis are highly dependent on the applied technology, as indicated in Figure 24.3. Adding to that, software packages (e.g. robust multiarray average; RMA) may adopt workflows with a slightly different order.

Microarray technology or NGS: some considerations

Over the past few years, the application of NGS has gained enormous momentum, topping more than 1,000 publications in 2011 according to PubMed. Nevertheless, in the same year, more than 6,000

	Microarrays	Next generation sequencing
1. Image capture & analysis	• Recording of **fluorescence intensities** per probe across the array	• Recording of **fluorescence intensities** per sequencing cycle • Calculation of the **base quality** scores
2. Feature-level quality control	• **Pseudo-image** plots to assess spatially distributed artifacts • **MA-plots** comparing the log transformed fold changes (M-value; Y-axis) to the average log transformed intensities (A-value; X-axis) to visualize within and between array biases • **Probe-level analysis**: mRNA degradation, GC-dependent fluorescence • **Background correction**: adjusts signal for background fluorescence intensities	• **Assessment of quality**: per base and overall per sequence • **Per base sequence and GC content**: in a random library, little difference with respect to the different bases and the GC content is expected between reads • **Library statistics**: uniform sequence length distribution, limited sequence duplication, presence of K-mers revealing non-uniform sequence distributions • **Sequence manipulation**: removal of sequencing adapters, filtering of low quality reads, filtering of short reads
3. Alignment	• **Not Applicable**	• **Reference:** genome or transcriptome, different builds (e.g. Hg19 *vs.* Hg38) • **Splice-awareness**: split or non-split reads (e.g. RNA-sequencing *vs.*DNA-sequencing) • **Base recalibration and indel realignment**: to optimize mutation calling • **Genome/transcriptome assembly**: In case a reference genome/transcriptome is not available or to analyze unmapped reads
4. Summarization	• **Collapse fluorescent signals** from probes directed at identical genomic features (i.e. genes, exons, genomic regions, miRNAs), log2 transformation • **Expression:** Correction for GC content • **CNVs:** Additional data segment the data • **DNA/protein interactions:** Identify genomic regions enriched for probe signals • **Methylation:** Calculate beta-values that reflect the methylation status of the probed region	• **Expression:** counting the number of reads per genomic feature (i.e. exons, transcripts, genes, splice-junctions, miRNAs) • **Mutations**: compare the observed sequence to a reference to identify differences, filter SNPs and technical artifacts • **CNVs, DNA/protein interactions and methylated DNA analysis**: evaluate fluctuations in the coverage profile along chromosomes (i.e. peaks for DNA/protein interactions), segmentation to identify regions of uniform coverage • **Structural variations:** evaluate unexpected patterns of read mapping
5. Normalization	• Correct for **technical variation** between samples (**between-array** normalization) • **Within-array normalization** to correct for dye-bias when operating on two-color arrays • Normalization for specific **probe effects**	• Correct for differences **in library sequencing depth** • Correct for differences in **GC content** and **gene lengths** • Correct for **dispersion**, being specifc over-or underrepresentations of specific genomic features due to biological variability (e.g. gene length, abundance) • Correct **technical variation** between runs and samples

Fig. 24.3 Low-end data analysis. Overview of the five phases of low-end data analysis, including image analysis (blue), feature-level quality control (green), alignment (purple), summarization (red) and normalization (orange).

publications reported MA data and this number has not dropped drastically ever since. These data suggest that the popularity of MAs has not been blunted by the recent success of NGS and thus both techniques offer mutually exclusive benefits. In this section, some considerations regarding the use of gene chip technology or NGS are put forward.

First, MA technology is the oldest and, up to date, most widely applied profiling technique. Therefore, the technological and analytical aspects are well characterized. In particular, the low-end MA data analysis is simpler and more straightforward, whereas many aspects of low-end NGS data analysis are still subjected to intense research and novel algorithms are constantly being developed. A good example relates the read alignment procedure, which is not applied in MA data analysis but is quintessential for NGS data analysis. Unfortunately, a generally applicable alignment strategy currently does not exist and the optimal algorithm is largely application dependent. Second, both MA technology and NGS have a broad application spectrum (*vide supra*). However, due to the fact that NGS offers data at single base resolution, detection of somatic mutations and more accurate identification of splice sites and genomic breakpoints is only feasible with NGS. However, it should be realized that NGS data is error prone, particularly at the level of individual bases. Many potential error sources exist including PCR artefacts, sequencing artefacts, and erroneous base calls, all of which generate false positive findings. Adequate filtering techniques to remove spurious results can enhance specificity but will never completely resolve the issue and thus care must be taken when interpreting the data. Third, compared to MA technology, NGS is generally considered to be a more sensitive technology with an enhanced dynamic range. This is particularly relevant in the context of expression studies. The superior sensitivity of NGS relates to the sequencing depth, which indicates how many times a genomic or transcriptomic feature of interest is analysed and roughly associates with the number of reads in a library. It should be noted that sequencing depth, and thus sensitivity, directly translates into sequencing cost. From this perspective, MAs may represent a more cost-effective alternative, especially when considering the fact that MAs can be more sensitive and generate more detailed profiles when compared to NGS with shallow sequencing depth. Last but not least, NGS data are chip-independent meaning that the information that can be obtained is not restrained by the probe content of a chip, thus allowing the assessment of an unbiased molecular profile. On the other hand, the availability of probes on a MA chip guarantees that data with respect to the probe-specific target will be obtained, even when this target is not abundantly present in analyte (e.g. poorly expressed genes). In fact, the reason why MAs can be more sensitive than NGS with shallow sequencing depth resides in its probe dependent design.

High-end analysis and data mining

High-end data mining techniques can be categorized as either unsupervised (Fig. 24.4) or supervised (Fig. 24.5). Unsupervised methods are applied without considering prior sample knowledge. Put in other words, sample annotations are not allowed to guide the analysis, which is fundamental when the aim is to identify novel cancer classes based on the molecular data. Therefore, unsupervised techniques are often used for class discovery. In contrast, supervised techniques are applied when the aim is to define a molecular profile or signature for a specific subgroup of samples (e.g. patients that respond well to therapy). Under these conditions, prior knowledge that determines the sample classification is mandatory, allowing for the different classes to be compared (i.e. class comparison) to identify, for example, differentially expressed or mutated genes as well as to establish predictive models (i.e. class prediction). In the following paragraphs, methods for class discovery, comparison, and prediction will be discussed with particular focus on how they can be applied in cancer research. Readers interested in the mathematical foundations of these techniques and how they are implemented algorithmically are advised to explore the BioConductor repository and particularly the collection of software packages dedicated to data mining.

Unsupervised analysis using clustering techniques

Capturing and understanding the diversity of cancers from profiling data can be achieved using a variety of different methods. In general, these methods aim at organizing the data into more of less uniform subgroups that, in addition, are also sufficiently distinct from each other. Finding such subgroups paves the way for new taxonomies that may parallel or refine classical classification schemes. The most commonly applied method for subgroup identification is clustering. Importantly, clustering can also be performed to identify groups of genes that behave similarly across samples. Inspecting such gene clusters may identify novel molecular factors, such as the activation of a particular signal transduction pathway or the deletion of a specific genomic region, that are fundamental to the new taxonomy, and thereby help interpreting the classification and shed new light onto the molecular biology of cancer. In addition, the guilt-by-association principle dictates that genes with similar behaviour across samples may be involved in the same biological process, allowing their functional characterization.

Cluster analysis can be applied using a variety of different methods. By far the most popular method is hierarchical clustering (Fig. 24.4), the result of which is represented using dendrograms. These tree-like structures provide an overview of the hierarchical organization of the data into nested clusters with individual samples at terminal branches, and clusters of increasing size closer to the root. Cutting the tree at a given depth defines a clustering of the data into a finite number of groups. Hierarchical algorithms can be either agglomerative or divisive. Agglomerative methods iteratively join the most similar groups into larger ones, whereas divisive methods start from a single group that is iteratively split into two subgroups that are maximally dissimilar.

An alternative to hierarchical clustering is data partitioning (Fig. 24.4). Partitioning methods (e.g. K-means, partitioning around medoids, self-organizing maps) try to optimize a partitioning of all samples into a finite number of groups such that the similarity between samples within each group is large but small between samples from different groups. A general drawback of partitioning methods is that the number of clusters in which the data should be partitioned has to be specified in advance. This requires prior knowledge of the data structure that is generally not available. In addition, the solution presented by partitioning methods is highly dependent on how the algorithms were initialized (e.g. number of clusters for data partitioning, number of iterations allowed for the algorithm to find a solution), and a unique partitioning generally does not exist. Therefore, it is common practice to run data partitioning methods multiple times with different initializations and keep the best solution (*vide infra*). An advantage of partitioning methods over hierarchical methods is that they do not impose a hierarchical structure onto the data.

To organize samples or features into a smaller number of groups, the relative similarity between samples or genes first needs to be quantified, which is done using a distance of dissimilarity measure (e.g. correlation coefficient, Euclidean distance). The choice of a distance measure has a strong influence on the clustering result but no rules to guide its choice exist. In a second phase, the collection of all pair-wise distances between samples or features is used to define clusters. At this stage, algorithms for hierarchical clustering and data partitioning function differently, but the overarching goal is to both minimize the within-group distance and maximize the between-group distance.

A generic problem with clustering methods is choosing the adequate number of clusters that best represents the data. In this context, cluster robustness is an important concept that reflects the extent to which clusters are reliably identified in a data set. Robust clusters are very likely observed in other, independent data sets and therefore potentially reflect existing subtypes. Various indicators of cluster robustness have been described (i.e. silhouette score, Hartigan's index, within-group, and between-group sum of squares) which can be monitored as a data set is being divided in an increasing number of clusters (Fig. 24.4). Unfortunately, this strategy may not always be successful in finding the optimal cluster number. Alternative strategies, based on repetitive sampling with replacement (i.e. bootstrapping), can be applied to identify clusters that are repetitively observed in a data set, and therefore can be considered robust.

Unsupervised analysis using matrix decomposition

Classifying cancer into several distinct subgroups is useful for human understanding but it may be too limited to capture the intrinsic nature of cancer heterogeneity. Consider the example of biological process (e.g. cell motility) that is being progressively activated across a set of cancer samples. In such a context, the assumption that cancer samples can be organized into well-separated subgroups based on

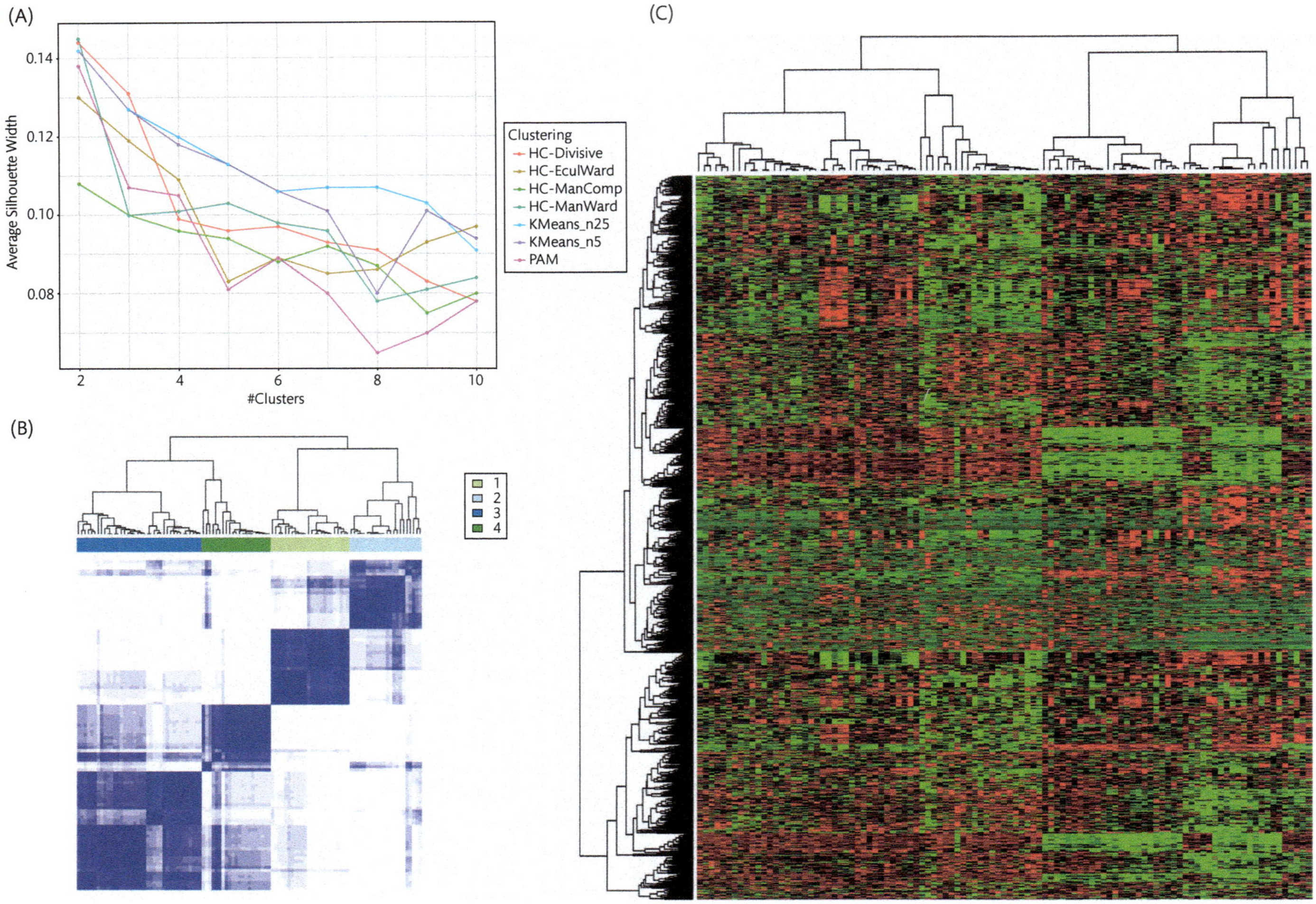

Fig. 24.4 (A–C) Unsupervised analysis. A collection 105 breast cancer samples was subjected to various unsupervised clustering approaches, including both agglomerative and divisive (i.e. divisive hierarchical clustering, K-means, and partitioning around medoids) clustering methods. For the agglomerative methods, hierarchical clustering was performed using various distance (i.e. Euclidean and Manhattan) and linkage (i.e. complete linkage and Ward linkage) methods. The resulting clustering was subjected to a robustness analysis using the silhouette score, which measures how well each sample represents the entire cluster. The results are shown in panel A. Irrespective of the clustering method used, the robustness is an inversely correlated to the number of clusters the data is divided into. It can be observed that overall, the divisive methods perform better in finding robust clusters as compared to the agglomerative methods. One problem with divisive methods is that the number of clusters the data should be partitioned into needs to be defined beforehand. This can be solved using a consensus clustering approach, the result of which is shown in panel B. Consensus clustering repeatedly partitions the complete data set by randomly selecting samples and features. Each partition is then subjected to clustering and a consensus block-diagonal matrix is constructed that demonstrates how often two samples have been clustered together (dark blue = high consensus score; white indicates a low consensus score). From this consensus matrix, it can be observed that four clusters can be robustly identified using K-means clustering, resembling the presence of the distinct molecular subtypes in the data set (1 = luminal A, 2 = luminal B, 3 = basal-like, and 4 = HER2 enriched). Using the result from panel B, a heatmap was constructed, which is shown in panel C. Here, molecular profiles across many cancer samples are structured in a matrix-format, with rows and columns indicating respectively features and samples. A matrix value then represents a molecular characteristic measured for a specific feature in a specific sample, in this case gene expression. Heatmaps can be reordered so that samples and features with similar characteristics are placed next to each other, allowing for a visual identification of features with uniform characteristics in a group of samples. This visual identification is made possible thanks to the fact that matrix-values are colour-coded according to signal strength with red and green indicating over- and underexpression, respectively.

molecular data may be not very accurate. In addition, it is likely that samples in different clusters may share common features (e.g. increased turnover of actin filaments). Therefore, it may be more pertinent to describe a sample not by a cluster assignment but as a superposition of a few well-defined molecular properties. Detection and quantification of such properties from high-throughput molecular profiling data can be performed using matrix decomposition techniques like matrix factorization, principal component analysis (PCA) and independent component analysis (ICA).

Understanding matrix decomposition requires a different perspective of high-throughput molecular profiling data. Briefly, in such data sets, each cancer sample can be considered as a single point in a multidimensional space, the coordinates of which are determined by the measurements made for each of the features in the data set. Dealing with such high-dimensional data spaces is intractable (e.g. consider the constraints for graphical data representation that do not allow more than three dimensions). Therefore, special analytical methods using matrix algebra can be applied to decompose data

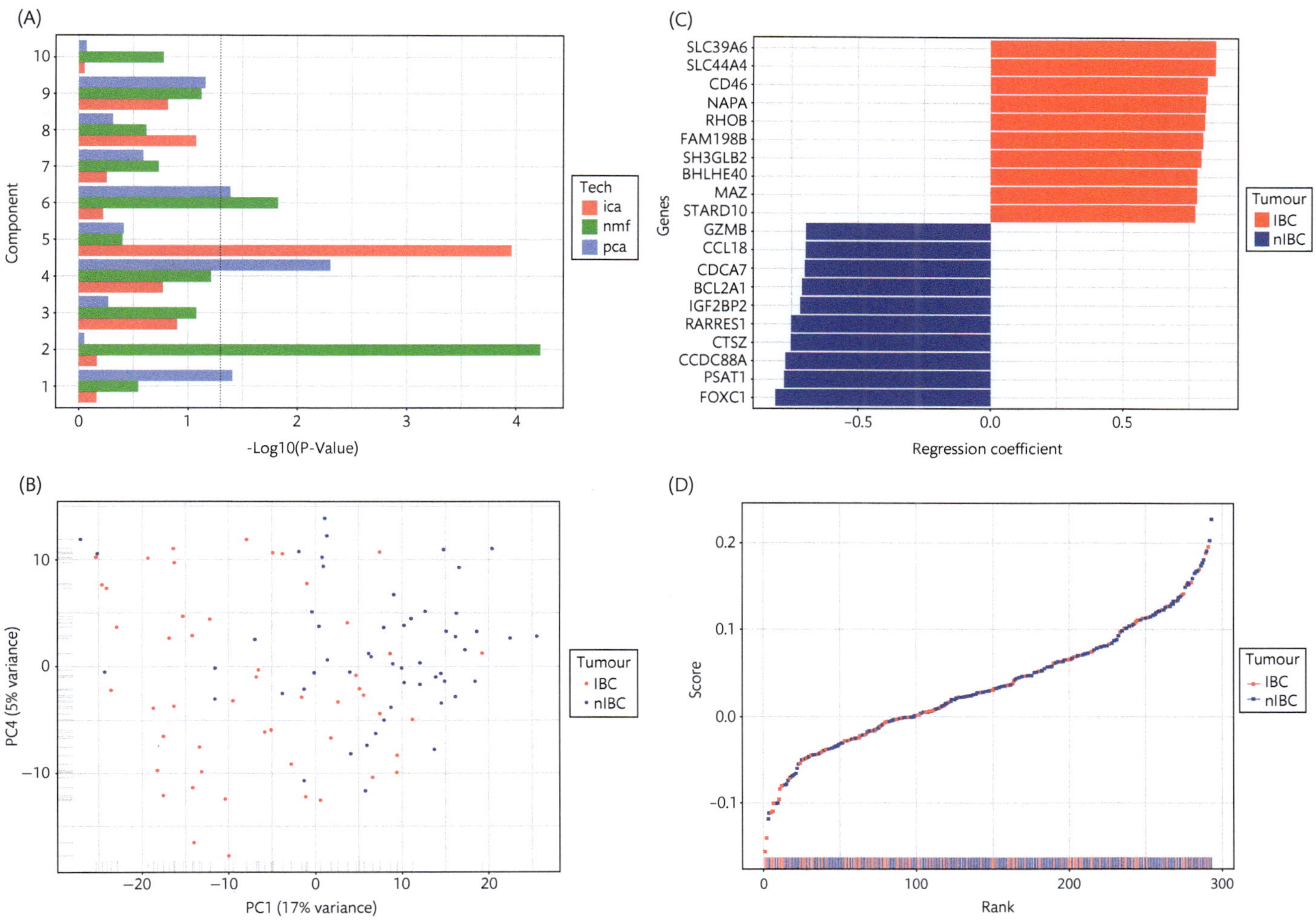

Fig. 24.5 (A–D) Matrix decomposition and supervised analysis. Three different matrix decomposition algorithms (i.e. ICA, NMF, and PCA) were applied to a gene expression data set of 105 breast cancer samples, each identifying 10 components that capture dominant expression themes. These expression themes were compared between samples with and without inflammatory breast cancer (IBC). The results are shown in panel A, which demonstrates that all matrix decomposition techniques identify components (i.e. colour-coded bars) that are significantly different between the two conditions (i.e. values exceed the threshold for significance that is indicated by the dashed line). The first and fourth component identified by PCA was selected to generate a two-dimensional scatter plot representation of the data (panel B). Ticks along the X- and Y-axes indicate the positions of the IBC samples, from which it can be observed that the red dots are in general scattered towards the left side of the plot, indicating that PC1 (X-axis) captures expression differences between IBC and nIBC samples. The regression coefficients of the 20 genes most strongly associated to PC1 are shown in panel C, with positive regression coefficients indicating overexpression in IBC and vice versa. All components in panel A are associated with such a profile and thus it can be understood that every component represents a biological process that contributes to defining the difference between the profiled samples. The gene profile in panel C can be applied as a correlation-based classifier on an independent data set, the results of which are shown in panel D. The samples are ordered according to increasing correlation coefficient and colour-coded ticks along the X-axis specify the position of the IBC and nIBC samples. From these data it is clear that the PC1 expression profile is unable to discriminate between IBC and nIBC samples in an independent test set, and thus further model optimization is needed.

sets such that new and composite dimensions (a.k.a. metagenes) are defined, along which the original data show significant variability. Put in other words, high-dimensional data structures are projected onto a low-dimensional data space that preserves the most relevant themes.

After matrix decomposition, the metagenes constitute linear combination of all available features in a data set and represent basic processes that are responsible for the biological variation observed in the data set. In addition, each cancer sample can be described as a superposition of those metagenes and their associated biological processes. Therefore, matrix decomposition techniques are particularly suited to describe cancer biology for two reasons. First, decomposing a molecular profile into a linear combination of metagenes provides a mathematical foundation for the fact that cancer development and progression are complex phenomena in which many biological processes cooperate. More importantly, these biological processes can be identified and quantified. Second, the formulation of a metagene as a linear combination of all genes in the data set recognizes the fact that biological processes are coordinated by many genes, albeit with variable importance. Thus, metagenes reflect the fact that perturbations of biological processes in cancer are rarely caused by the isolated actions of single genes.

Existing matrix decomposition techniques are computationally versatile and therefore provide alternative decompositions of the same data set, each with their own benefits. For example, PCA uses single value decomposition (SVD) to determine the metagenes. Although very effective and widely applied, the biological interpretation of PCA metagenes is not always easy, mainly because SVD identifies metagenes that are orthogonal to each other. Unfortunately, independent biological processes are not necessarily orthogonal (e.g. consider two pathways that share some genes but are activated independent from each other). PCA will fail to identify these individual processes and instead forces them into a single metagene, complicating the biological interpretation. ICA, which is a variant of PCA, overcomes the limitation of orthogonal metagenes by decomposing the data into a series of statistically independent processes.

Class comparison analysis

Class comparison analysis aims at identifying molecular features associated with a condition of interest. This generally involves performing a statistical comparison of measurements between the condition(s) of interest and a comparator (e.g. normal tissue). An important issue in class comparison analysis relates to the false discovery rate (FDR), which is particularly relevant when performing multiple tests. To understand the concept of false discovery, imagine the expression of 1,000, of which 100 genes are differentially expressed at a nominal *P* value of 5% between two conditions. At this significance level, 50 genes (i.e. 5% of 1,000) are expected to be significant by chance and thus the FDR equals 50% (i.e. 50/100). Put in other words, 50% of the statistically significant genes are probably false positives.

A number of approaches have been developed to overcome the problems related to multiple testing, all of which attempt to assign an adjusted *P* value to each test. Many traditional techniques such as the Bonferroni correction are too conservative in the sense that they aim to reduce the probability of having false positive results, and in doing so, they also reduce the number of true discoveries. Competing approaches accept that false positive results will be present and aims at controlling the number of false discoveries at a fixed level (e.g. 10%). Because of this, these approaches are less conservative that the Bonferroni approach and has greater power to find truly significant results. The latter is the method of preference when dealing with high-throughput data, because it least compromises the statistical power that is already limited by the small sample sizes that are often associated with molecular profiling experiments.

Class prediction analysis

Class prediction analysis aims at identifying models that are able to distinguish between two or more conditions. Such discriminative models are based on the molecular characteristics (e.g. expression or mutational profiles) of the investigated cancer samples. Although discriminative models can be built from observations related to single features, it is general practice to establish such models by combining measurements made on several features. The idea is that the combined information offers greater discriminative power, which alludes to the fact that the molecular aspects of cancer are heterogeneous across many patients. The set of condition-specific molecular features is called a molecular signature, and these can be used as biomarkers in clinical practice.

Traditionally, discriminative models have been established from three distinct perspectives: molecular diagnosis, prognostication, and therapy prediction. A diagnostic model is established to allow the identification of different cancer types based on molecular characteristics (e.g. PAM50). A prognostic model aims at estimating the risk of future tumour relapse, either local or distant, and can be used to assess the need for adjuvant therapy. Finally, a predictive model aims at assessing the probability that a cancer will react to a specific treatment and thus can be used to guide treatment-decisions.

The development of a discriminative model starts with the selection of the features that constitute the molecular profile. This is a crucial step for several reasons. First, inferring a classification model with many features from only a limited number of observations is difficult and leads to poor models, among others due to model overfitting which occurs when a model is excessively complex. Second, reducing the number of features in a model allows an easier design of dedicated devices for cheaper and faster prediction. Finally, predictive models based on only a few features may simplify biological interpretation and thus lead to better understanding of the molecular basis of the investigated phenotype. Performing feature selection solely based on discriminative power is not always good practice. Rather, due to the complex and correlated structure of biological data, complementarity should also be considered. Indeed, once a first feature is selected to enter a predictive model, adding a second feature strongly correlated to the first one can be less interesting than adding a third one that adds complementary information to the first one, even though its discriminative power is lower than that of the second feature. Therefore, it seems more interesting to attempt to select a subset of features which together establish a good model. This can be achieved using stepwise selection strategies, in which features that maximally optimize a model are iteratively added (i.e. forward stepwise selection) or removed (i.e. backward stepwise selection).

A second important aspect in the development of a discriminatory model is the selection of the appropriate statistical model. Many different algorithms exist (e.g. regression-based methods, discriminant analysis, naive Bayes, k-nearest neighbours, decision trees, support vector machines), none of which outperforms all others. Their relative performance depends on many factors, including the nature of the data and the number of different parameters that can be used to optimize the model. Therefore, choosing the appropriate algorithm is not straightforward. However, Occam's razor implies that any given complex function is a priori less probable than any given simple function. Therefore, simple statistical models should be preferred over complex models. Indeed, complex statistical models have a greater tendency for overfitting because they will adjust to very specific random fluctuations in the training data.

A final aspect of class prediction analysis relates to validation, which is primarily involved in assessing the performance of the model. Ideally, this is achieved by splitting a set of samples randomly into a training and a test set. The training set is used for model construction, whereas the test set is used only to evaluate a model's discriminating power in terms of its accuracy, error rate, sensitivity, specificity, positive predictive value, and negative predictive value. Unfortunately, in case of limited sample sizes, it is impracticable to establish an independent test set. Under these conditions, cross-validation strategies can be applied, which essentially evaluates the model using the data it was trained on. The basic principle for all cross-validation procedures is to split the data into a set of equal parts. A model is then constructed on all parts but one, which is left out for validation. The model construction and testing cycle is repeated using all possible combinations of the data partitions. Importantly, cross-validates estimates of model performance are generally too optimistic, and therefore a distinction is made between training error (i.e. cross-validated error rate on a training set) and test error (i.e. error rate on an independent test set).

Molecular cancer profiling: Challenges and perspectives

Long tails of infrequently mutated cancer genes

For many cancer types, a handful of cancer genes are mutated at high frequency, but many more cancer-related genes are found mutated at much lower frequencies (Garraway and Lander, 2013). For example, a recent genomic study of breast cancer reported more than 90 genes that were mutated at statistically significant rates (Nik-Zainal et al., 2016). The 10 most frequently mutated genes accounted for 62% of the driver genes, meaning that remaining driver mutation were dispersed across more than 80 genes, exemplifying the long tail distribution. Some of the genes found mutated at low frequencies in some cancers are more commonly mutated in other cancers (e.g. the SWI/SNF complex genes), suggesting that the discovery of new driver genes is approaching a plateau. On the other hand, the fact that so many driver genes occur at lower frequencies raises the possibility that many such genes may yet remain undiscovered. The problem is due, in part, to the fact that most studies to date have been insufficiently powered—lacking adequate sample size to detect low-frequency events and/or adequate depth of sequence coverage to overcome impurity due to stroma contamination (Garraway and Lander 2013). This statement is corroborated by the power analysis reported by Lawrence and colleagues (Lawrence et al., 2015), which demonstrated that in order to create a reasonably comprehensive catalogue of cancer genes mutated in more than 2% of the patients are required approximately between 650 and 5,300 samples, depending on the background mutation rate. Their analysis also showed that for most cancer types, the numbers of currently sequenced samples allows identifying cancer genes with a mutation frequency between 5% and 10% (Lawrence et al., 2015).

The dark matter of non-coding mutations and structural variations

In contrast to point mutations in coding regions, the ability to discover and understand other types of driver mutations is still distressingly limited. Many more important cancer drivers may be lurking in the places that we cannot currently interpret. For example, gains and losses spanning whole chromosome arms occur commonly in most types of cancer, but it is difficult to pinpoint the key genes for which the presence of a few extra copies contributes to cancer. Even for focal amplifications or deletions, finding the target genes can be difficult. A study of somatic cDNAs across cancer types found that proven cancer genes were known for less than half of recurrent focal amplifications and an even smaller proportion of recurrent focal deletions (Zack et al., 2013). Along the same lines, chromosomal rearrangements that are pervasive in many cancers only rarely implicate the emergence of new cancer genes, possibly owing to the challenges of interpreting such events. However, the great uncharted frontier is the more than 98% of the human genome that does not encode proteins, partly due to the fact that adequate analytical techniques to recognize recurrent mutations in non-genic territory are lacking (Garraway and Lander 2013). The presence of non-coding mutations, recurrent across many cancer types, is demonstrated by a recent study in which the upstream regulatory elements of TERT, PLEKHS1, WDR74, and SDHD are frequently affected (Weinhold et al., 2014). However, a systematic analysis of non-coding mutations and gene expression across 14 tumour types revealed that mutations in TERT regulatory DNA were exceptional in showing an association with altered promoter activity (Fredriksson et al., 2014). Other recurrent non-coding mutations exhibit clear evidence for selection (e.g. DPH3 promoter mutations) but cannot be explained with the current level of knowledge.

Cancer heredity

Many of the genetic factors that drive a cancer are acquired through somatic mutation, but some are inherited at birth. Genomics has revealed many genes that influence predisposition to cancer, although the picture remains far from complete. Cancer genes responsible for highly penetrant cancer syndromes (e.g. *BRCA1*) account together for less than 5% of the heritability of cancer. Genome-wide association studies have identified cancer genes that confer a more modest risk and revealed that common cancer alleles (i.e. population frequency of >1%) include regulatory variants and confer a lower increased risk (i.e. <0.3-fold increase), whereas the rare cancer alleles (i.e. population frequency of <1%) affect coding regions of known cancer genes (e.g. *ATM, BRIP1, CHEK2, PALB2*, and *RAD51C*) and tend to be associated with higher risk (i.e. 2- to 3-fold increase). Importantly, the risk factors identified to date explain only a fraction of the heritability of cancer and genomic studies with much larger samples will be needed to obtain a fuller picture of the inherited basis of cancer risk. Of note, understanding cancer heritability may help to explain disparities among ethnic groups (e.g. higher risk of prostate cancer in African Americans and other men of African descent; see Garraway and Lander, 2013).

Intrapatient heterogeneity

As already discussed earlier in this chapter, tumours show extensive and clinically relevant cellular and molecular heterogeneity. Tumour heterogeneity could have important implications for 'precision' cancer medicine. Studies that seek to stratify patients for clinical trials of targeted agents based on specific 'actionable' mutations may be confounded if a biopsy sample is not representative of the whole tumour. On the other hand, the ability to identify driver or resistance mutations within subclonal populations may allow improved prediction of clinical outcomes. The growing understanding on the matter may inform the design of clinical studies to account for heterogeneity and circumvent its subversive effects (Garraway and Lander, 2013). A critical question that needs to be addressed relates to the extent of the intratumor heterogeneity that should be considered as clinically relevant.

The focus in this chapter was primarily on primary tumours. However, the development of metastatic cancer is a crucial event in a patient's life history. Studies have shown that the molecular characteristics of metastases may differ from those of the primary tumour and as such metastases can contribute to the total intrapatient heterogeneity. Moreover, the development of metastatic disease usually delineates the start of a new line of therapy. Obtaining metastatic tissue biopsies for diagnostic testing can be a risky and painful procedure. In this context, a liquid biopsy, in which tumour material is obtained using a simple blood draw offers an elegant alternative. An added benefit of liquid biopsies resides in its potential for repeated sampling, thereby providing the means for high-resolution monitoring of therapeutic responses or emerging resistance mechanisms and facilitating detection of aggressive tumour subclones. Liquid biopsies can be performed using circulating tumour cells (CTCs), which are actively metastasizing tumour cells that are circulating in the bloodstream. Recent advances in both the sequencing technology and the analysis pipelines now allow detailed whole-genome and whole-transcriptome analysis down to single cell resolution. In order to apply molecular testing on a CTC-based biopsy, a detailed analysis of the heterogeneity within the CTC population is required in order to establish the number of CTCs that need to be analysed for the biopsy to be representative. Most likely, the answer to this question will also vary on a per patient basis. In addition to CTCs, DNA and RNA molecules that are either floating freely in the bloodstream or that are encapsulated by exosomes can be isolated from plasma and can serve as an alternative means to perform blood borne testing.

Charting and understanding the full spectrum of molecular alterations in cancer

Cancer profiling has largely focussed on documenting molecular changes in the primary tumours. However, the focus should be expanded to gather systematic information regarding a wider range of biological and clinical questions. First, the complete catalogue of molecular driver alterations in cancer including those from infrequent cancer types should be established, which is of interest from both a clinical (i.e. for therapeutic decision-making) and a scientific (i.e. for comprehensive knowledge of tumour biology) point of view. This requires analysing many tumour/normal pairs, including those from less frequent cancer types and possibly starting from formalin-fixed and paraffin embedded tissue material. Second, the catalogue of molecular driver alterations should expand beyond primary tumours, to allow studying the natural history of human cancer, as well as the homology to cancer biology in key model systems. To this end, a wide array of cancer samples should be profiled including preneoplastic lesions, metastases from various organ sites, CTCs, tumours with different types of responses to therapies, and animal models including naturally occurring cancers in large animals. Third, a functional encyclopaedia of altered pathways and acquired vulnerabilities should be created, which is the domain of systems biology that is extensively captured in one of the following chapters. Briefly, genomic, epigenomic, transcriptomic and proteomic data should be integrated into holistic models that can be interrogated and from which predictions on drug sensitivity can be made by confronting the models with sensitivities to anticancer drugs, RNAi-, or CRISPR-based gene inactivation and microenvironmental interactions, that have been obtained on cell line data (Barretina et al., 2013; Garnett et al., 2013). Finally, and crucial to successfully accomplishing the three points mentioned earlier, cancer profiling information should be shared. The shared information would not only speed up the identification of cancer genes, but also the correlations between therapeutic responses and specific tumour genotypes. Therefore, the computational infrastructure should be created and a culture of data sharing be promoted (Garraway and Lander 2013).

Molecular profiling and personalized cancer medicine

The ultimate goal of molecular profiling in cancer resides in the improvement of diagnostics and therapeutics. As discussed in the first section of this chapter, several molecular tests, both genome and transcriptome based, are progressively being incorporated in routine molecular diagnostics. Early results from amplicon sequencing suggest that in 40% to 60% of the cancer patients, at least one molecular alteration that might influence therapeutic decision-making or might suggest enrolment in a particular clinical trial can be identified (Garraway and Lander, 2013). Exploring MyCancerGenome identifies more than 20 cancer types for which at least some kind of evidence that molecular alterations in specific genes affects therapeutic sensitivity is available. The Precision Oncology Knowledge Base reports 82 variants in 12 genes that are recognized by the FDA as being predictive biomarkers of response to and FDA-approved drug in a specific indication (i.e. level 1 evidence). Another set of about 130 variants in more than 30 different genes are reported as being either a standard of care predictive biomarker of response to an FDA-approved drug (i.e. level 2 evidence; 85 variants in 11 genes) or a predictive biomarker of response to a drug that is supported by compelling clinical evidence (i.e. level 3 evidence; 55 variants in 25 genes). In addition, more than 70 FDA-approved drugs are available for the treatment of patients according to the presence of level 1 or level 2 molecular alterations that represent approximately 20 genes and many more drugs are currently being developed or tested. When extrapolating these numbers to the level of the full human genome/transcriptome, the potential of targeted therapies cannot be underestimated. Along the same lines, the importance of molecular profiling in cancer medicine, also from an economical point of view when the costs of these drugs are considered, is self-evident.

In order to unlock the full potential of molecular profiling in cancer medicine, techniques allowing the analysis of full human genomes and transcriptomes should gradually be introduced in routine diagnostic settings. Nevertheless, the challenges facing such an introduction and the consequences are many-fold. First, an efficient validation strategy should be developed. Data obtained with molecular profiling techniques are fraught with uncertainty, resulting from various sources of sampling and technical bias. In a research setting, such uncertainties can be alleviated by analysing an appropriately sized population and by including samples of superior quality. In a diagnostic setting, these luxuries cannot be afforded and data should be accurate on a per sample basis, requiring stringent analysis settings. With the current testing procedures that focus on specific genes or limited gene sets, it is possible to identify these analysis settings for each target separately. However, with full genome or transcriptome data, such a validation strategy is impractical. Therefore, for each technique, a detailed mapping of the various potential error sources and their error profiles is required such that their influences can be mitigated. Second, a major challenge resides in filtering and annotating a staggering amount of molecular alterations, particularly when realizing that the landscape of scientific knowledge is swiftly changing. This is particularly relevant for the infrequent alterations that are in general less profoundly investigated. Adding to that, the clinical impact of many molecular alterations with respect to specific treatments, is currently unknown. To accelerate this knowledge, the American Association for Cancer Research (AACR) has established the GENIE (Genomics, Evidence, Neoplasia, Information, Exchange) project, which fulfils an unmet need in oncology by providing the statistical power necessary to improve clinical decision-making, to identify novel therapeutic targets, to understand of patient response to therapy, and to design new biomarker-driven clinical trials. Third, acquiring complete sets of molecular alterations at various levels (i.e. genomic, transcriptomic) in addition to histological and clinical information requires careful consideration of how such data should be (jointly) analysed. In such analyses, the focus might shift from individual genes to molecular subtypes, pathways, and gene networks. Ultimately, this should enable the identification of key signalling genes on per patient basis, ease patient stratification for basket trials and/or combination therapy and ameliorate the prediction of therapeutic efficacy.

TAKE-HOME MESSAGE

- The application of high-throughput molecular profiling techniques, which allow an unbiased whole-genome analysis of cancer samples at various molecular levels (i.e. genome, epigenome, transcriptome, and proteome), has greatly expanded our knowledge of the molecular foundations of human cancer.
- From a biological perspective, current profiling experiments demonstrated that cancer is a complex disease characterized by extreme molecular heterogeneity, both between and within the classical, tissue-defined cancer types. In addition, the full spectrum of mutational processes in cancer is gradually being revealed and the first near complete catalogues of cancer genes are being published.
- From a clinical perspective, the extreme molecular heterogeneity between cancer samples warrants a paradigm shift in patient management, away from generalized therapy schemes and towards more personalized treatments. Thanks to the application of molecular profiling techniques, the choice of a therapeutic drug can be rationalized. In addition, data mining techniques allow the development of more accurate diagnostic models based on joint characteristics of many genes.
- From a scientific perspective, high-throughput molecular profiling techniques have paved the way for translational cancer research to become a mature, independent, and hypothesis-generating research field. The staggering amount of data provides significant opportunities to many investigators around the world to study human cancer biology in its most appropriate context.

OPEN QUESTIONS

- The ability to discover and understand driver mutations, other than base substitutions in coding exons is limited. In addition, the repertoire of cancer predisposition genes is incompletely charted and understanding their inheritance patterns at a population level is restricted. For a detailed understanding of cancer biology, efforts should be taken to resolve these issues.
- A critical question that needs to be addressed relates to the extent of the intratumor heterogeneity that should be considered as clinically relevant.
- The molecular landscape of rare primary tumours, metastatic lesions, and preclinical models should be analysed in more detail. In addition, there is urgent need to translate the available data into functional models of activated signalling pathways.
- Integration of molecular and clinical data is of utmost importance to allow more accurate prognostication or prediction of therapy response is mandatory. To this end, rigid analysis pipelines need to be established.
- To allow accurate judgement of the molecular profiling results in routine diagnostics, detailed assessment of the applied technology, the applied analysis pipelines, and the matrix on which the analysis is performed need to be established.

FURTHER READING

Ciriello, G., Miller, M. L., Aksoy, B. A., Senbabaoglu, Y., Schultz, N., & Sander, C. (2013). Emerging landscape of oncogenic signatures across human cancers. *Nat Genet*, **45**(10), 1127–33.

Fredriksson, N. J., Ny, L., Nilsson, J. A., & Larsson, E. (2014). Systematic analysis of noncoding somatic mutations and gene expression alterations across 14 tumor types. *Nat Genet*, **46**(12), 1258–63.

Garraway, L. A. & Lander, E. S. (2013). Lessons from the cancer genome. *Cell*, **153**, 17–37.

Koboldt, D. C., Fulton, R. S., McLellan, M. D., et al. (2012). Comprehensive molecular portraits of human breast tumours. *Nature*, **490**, 61–70.

Lawrence, M. S., Stojanov, P., Mermel, C. H., et al. (2015). Discovery and saturation analysis of cancer genes across 21 tumour types. *Nature*, **505**, 495–501.

Mertins, P., Mani, D. R., Ruggles, K. V., et al. (2016). Proteogenomics connects somatic mutations to signalling in breast cancer. *Nature*, **534**(7605), 55–62.

Nik-Zainal, S., Davies, H., Staaf, J., et al. (2016). Landscape of somatic mutations in 560 breast cancer whole-genome sequences. *Nature*, **534**(7605), 47–54.

Nik-Zainal, S., Van Loo, P., Wedge, D. C., et al. (2012). The life history of 21 breast cancers. *Cell*, **149**, 994–1007.

Zack, T. I., Schumacher, S. E., Carter, S. L., et al. (2013). Pan-cancer patterns of somatic copy number alteration. *Nat Genet*, **45**(10), 1134–40.

REFERENCES

Alexandrov, L. B., Nik-Zainal, S., Wedge, D. C., et al. (2013). Signatures of mutational processes in human cancer. *Nature*, **500**(7463), 415–21.

Banerji, S., Cibulskis, K., Rangel-Escareno, C., et al. (2013). Sequence analysis of mutations and translocations across breast cancer subtypes. *Nature*, **486**(7403), 405–9.

Barretina, J., Caponigro, G., Stransky, N., et al. (2013). The cancer cell line encyclopedia enables predictive modelling of anticancer drug sensitivity. *Nature*, **483**(7391), 603–7.

BioConductor repository. Available at: https://www.bioconductor.org

Ciriello, G., Miller, M. L., Aksoy, B. A., Senbabaoglu, Y., Schultz, N., & Sander, C. (2013). Emerging landscape of oncogenic signatures across human cancers. *Nat Genet*, **45**(10), 1127–33.

Curtis, C., Shah, S. P., Chin, S. F., et al. (2013). The genomic and transcriptomic architecture of 2,000 breast tumours reveals novel subgroups. *Nature*, **486**(7403), 346–52.

Dvinge, H., Git, A., Gräf, S., et al. (2014). The shaping and functional consequences of the microRNA landscape in breast cancer. *Nature*, **497**(7449), 378–82.

Fredriksson, N. J., Ny, L., Nilsson, J. A., & Larsson, E. (2014). Systematic analysis of noncoding somatic mutations and gene expression alterations across 14 tumor types. *Nat Genet*, **46**(12), 1258–63.

Garnett, M. J., Edelman, E. J., Heidorn, S. J., et al. (2013). Systematic identification of genomic markers of drug sensitivity in cancer cells. *Nature*, **483**(7391), 570–5.

Garraway, L. A. & Lander, E. S. (2013). Lessons from the cancer genome. *Cell*, **153**(1), 17–37.

Jiao, Y., Lawler, K., Patel, G. S., et al. (2011). DART: denoising algorithm based on relevancenetwork topology improves molecular pathway activity inference. *BMC Bioinformatics*, **12**(1), 403.

Kandoth, C., McLellan, M. D., Vandin, F., et al. (2014). Mutational landscape and significance across 12 major cancer types. *Nature*, **502**(7471), 333–9.

Koboldt, D. C., Fulton, R. S., McLellan, M. D., et al. (2012). Comprehensive molecular portraits of human breast tumours. *Nature*, **490**(7418), 61–70.

Lawrence, M. S., Stojanov, P., Mermel, C. H., et al. (2015). Discovery and saturation analysis of cancer genes across 21 tumour types. *Nature*, **505**(7484), 495–501.

Lehmann, B. D., Bauer, J. A., Chen, X., et al. (2011). Identification of human triple-negative breast cancer subtypes and preclinical models for selection of targeted therapies. *J Clin Invest*, **121**(7), 2750–67.

Lin, I.-H., Chen, D. T., Chang, Y. F., et al. (2015). Hierarchical clustering of breast cancer methylomes revealed differentially methylated and expressed breast cancer genes O. El-Maarri, ed. *PLoS One*, **10**(2), p.e0118453.

Masuda, H., Baggerly, K. A., Wang, Y., et al. (2013). Comparison of molecular subtype distribution in triple-negative inflammatory and non-inflammatory breast cancers. *Breast Cancer Res*, **15**(6), R112.

Mertins, P., Mani, D. R., Ruggles, K. V., et al. (2016). Proteogenomics connects somatic mutations to signalling in breast cancer. *Nature*, **534**(7605), 55–62.

MyCancerGenome. Available at: https://www.mycancergenome.org

Nik-Zainal, S., Alexandrov, L. B., Wedge, D. C., et al. (2012). Mutational processes molding the genomes of 21 breast cancers. *Cell*, **149**(5), 979–93.

Nik-Zainal, S., Van Loo, P., Wedge, D. C., et al. (2012). The life history of 21 breast cancers. *Cell*, **149**(5), 994–1007.

Nik-Zainal, S., Davies, H., Staaf, J., et al. (2016). Landscape of somatic mutations in 560 breast cancer whole-genome sequences. *Nature*, **534**(7605), 47–54.

Parker, J. S., Mullins, M., Cheang, M. C., et al. (2009). Supervised risk predictor of breast cancer based on intrinsic subtypes. *J Clin Oncol*, **27**(8), 1160–7.

Precision Oncology Knowledge Base. Available at: http://www.oncokb.org

Smid, M., Rodríguez-González, F. G., Sieuwerts, A. M., et al. (2016). Breast cancer genome and transcriptome integration implicates specific mutational signatures with immune cell infiltration. *Nat Commun*, **7**, 12910.

Stephens, P., Edkins, S., Davies, H., et al. (2005). A screen of the complete protein kinase gene family identifies diverse patterns of somatic mutations in human breast cancer. *Nat Genet*, **37**(6), 590–2.

Stephens, P. J. et al. (2009). Complex landscapes of somatic rearrangement in human breast cancergenomes. *Nature*, **462**(7276), 1005–10.

Stephens, P. J., McBride, D. J., Lin, M. L., et al. (2013). The landscape of cancer genes and mutational processes in breast cancer. *Nature*, **486**(7403), 400–4.

Wang, Y., Waters, J., Leung, M. L., et al. (2015). Clonal evolution in breast cancer revealed by single nucleus genome sequencing. *Nature*, **512**(7513), 155–60.

Weinhold, N., Jacobsen, A., Schultz, N., Sander, C., & Lee, W. (2014). Genome-wide analysis of noncoding regulatory mutations in cancer. *Nat Genet*, **46**(11), 1160–5.

Yates, L. R., Gerstung, M., Knappskog, S., et al. (2015). Subclonal diversification of primary breast cancer revealed by multiregion sequencing. *Nat Med*, **21**(7), 751–9.

Zack, T. I., Schumacher, S. E., Carter, S. L., et al. (2013). Pan-cancer patterns of somatic copy number alteration. *Nat Genet*, **45**(10), 1134–40.

25

Proteomics and metabolomics applications in cancer biology

Pedro Cutillas and Benedikt M. Kessler

Delving into the cancer -ome through advances in mass spectrometry technology

Biological heterogeneity and diversity is one of the main challenges to understand the causes of cancer. This has caused major obstacles for experimentalists to classify cancer types and the understanding of molecular processes the underlie oncogenesis.

Technical developments and breakthroughs in analytical methodologies often promote new insights into biological details in normal physiology and pathological processes underlying human disease. More specifically, molecular insights into cancer biology have heavily benefited from analytical -omics technologies that allow the profiling of thousands of molecular entities in a single experiment. Current state-of-the-art mass spectrometry is capable of profiling deep cellular proteomes covering ~10,000 proteins, and metabolomics profiling includes the measurement of more than 1,000 human and non-human metabolites per analysis (Aebersold and Mann, 2016). This has been made possible through advancements in mass spectrometry technology, mainly at the level of improved sensitivity, acquisition speed, and more sophisticated data analysis tools. For protein analysis, referred to as proteomics, the most relevant experimental approach is based on a shotgun approach where proteins are proteolytically digested into peptides of ~8–30 amino acids in length, for which mass spectrometers have greatest sensitivity, and for which sequencing information can be obtained directly via tandem mass spectrometry (MS/MS, Fig. 25.1). This in combination with refined sample preparation protocols capable of handling small cellular tissue or body fluid samples including prefractionation have resulted in wider proteome coverage.

Based on these advances, comprehensive 'molecular snapshots' are now becoming available for many cancer types and are also deposited in publicly available databases and resources, such as Oncomine, COSMIC, The Cancer Genome Atlas, Progenetix, Human Protein Atlas, Human Proteome DB, Clinical proteomic Tumor Analysis Consortium (CPTAC), Human Proteome Map, and the Human Metabolome (Table 25.1).

Next to genomics and transcriptomics data sets becoming readily available for a variety of cancer types, global protein and metabolite profiles are now also being collected, predominantly through mass spectrometry-based proteomics and metabolomic studies.

Applications in cancer biology—limitations and challenges

Biological mass spectrometry has been applied for more than a decade (Aebersold and Mann, 2003). Despite substantial advances, there are some limitations in this approach. For instance, it is not currently possible to cover entire proteomes in single experiments, which is in part due to suboptimal sample processing, detection limits in mass spectrometers and data processing. In addition, it is still under debate to what extent the full proteome is characterized. Many proteins exist as different isoforms, contain mutations or deletions, and carry a range of posttranslational modifications. Often, it is a subset of these protein forms that is highly relevant for cancer disease progression.

In the sections to follow, we will discuss the impact that -omics approaches have made on the advancement of cancer biomarker discovery and our general understanding of aspect of cancer biology at a molecular level.

Translational applications in cancer—biomarker discovery

One of the most desirable applications of -omics in cancer has been the discovery of molecular markers (biomarkers) that are capable of diagnosing tumours at early stages, predicting disease outcome, or estimating success rates of particular treatment regimes. In recent years, considerable efforts were made using mass spectrometry, proteomics and metabolomics to identify markers correlating with cancer progression (Issaq et al., 2011; Crutchfield et al., 2016; Borrebaeck, 2017), in tumours such as lung cancer (Planque et al., 2009), liver cancer (Kimhofer et al., 2015; Nie et al., 2016), breast cancer (Kulasingam and Diamandis, 2007), prostate cancer (Petricoin et al., 2002), bladder cancer (Frantzi and Vlahou, 2017), oesophageal adenocarcinoma (O'Neill et al., 2017), colon cancer (Ward et al., 2006), and non-small lung cancers (Sandfeld-Paulsen et al., 2016). However, despite these considerable efforts, where many molecules have been claimed to full-fill these criteria, it has been challenging to take them forward towards validation on larger patient cohorts and develop them into robust clinical tools.

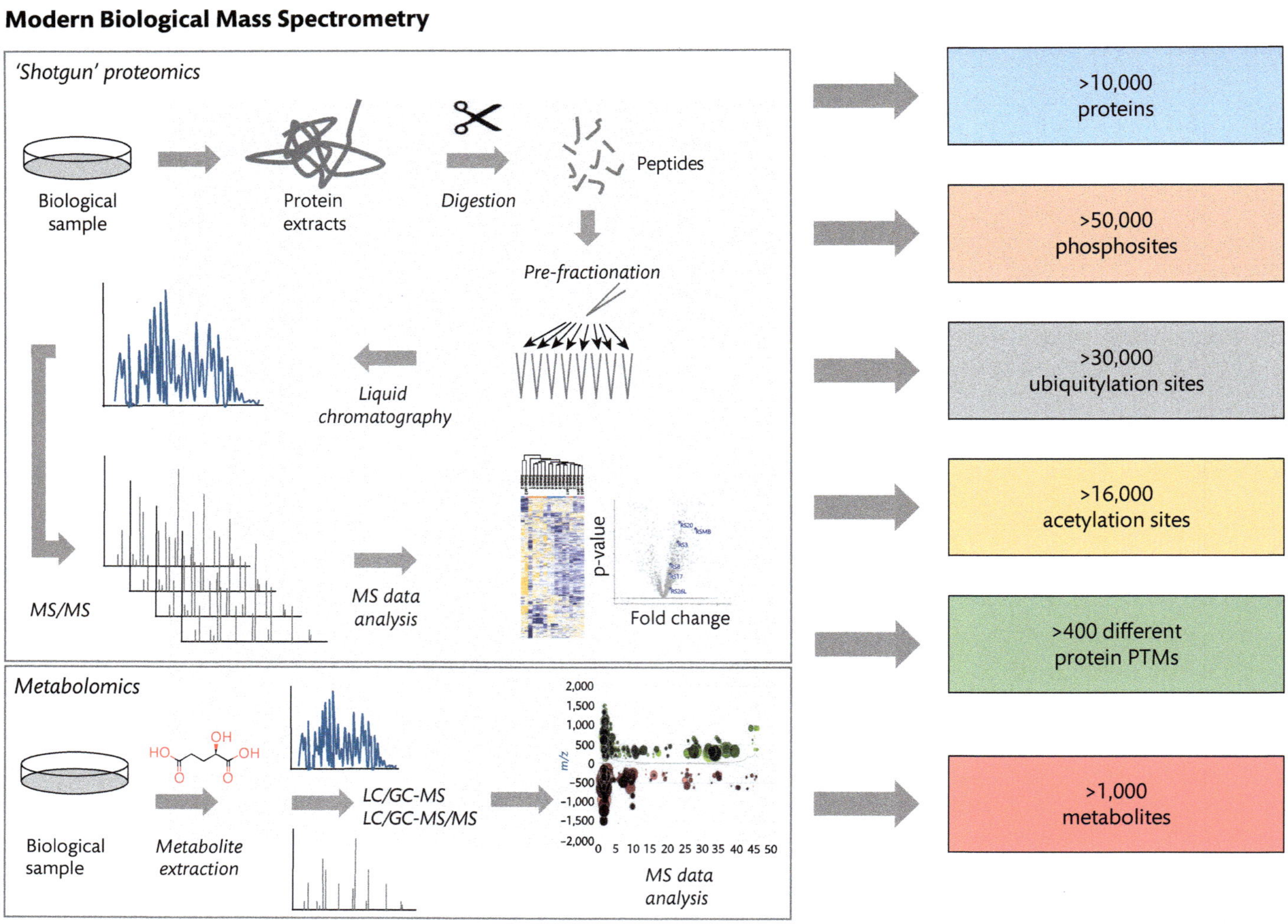

Fig. 25.1 Advances in mass spectrometry technology for deeper coverage of cellular proteomes, protein modifications, and metabolites. Workflows are shown for 'shotgun proteomics' and metabolomics. MS, mass spectrometry; MS/MS, tandem mass spectrometry; LC, liquid chromatography; GC, gas chromatography.

Even prominent molecules such as sarcosine, a glycine metabolic derivative, claimed to be a marker for prostate cancer progression (Sreekumar et al., 2009), have later been questioned in their clinical usefulness (Issaq and Veenstra, 2011). One of the major reasons for this is the fact that most of these molecules (irrespective of these being proteins, nucleic acids or small molecule metabolites) are part of normal physiological processes and often fluctuate due to either normal homeostasis or other 'confounding' diseases, predominantly inflammation, infection, injury, food poisoning, and ageing. In addition, preanalytical variability due to inconsistent clinical sample handling remains a problem. For instance, collection of blood samples and handling at 4°C leads to platelet activation and promotes blood clotting (Kaisar et al., 2016a; Kaisar et al., 2016b), and many blood markers may simply relate to these common events irrespective of the patient's disease status. Consequently, a very small number of candidate molecules have made the cut and were taken forward towards the clinic (Swami, 2010; Henry and Hayes, 2012).

Nevertheless, promising attempts include the combination of multiple proteins in 'biomarker panels', such as the 34-marker cancer panel (Bio-Plex Pro, Biorad), the 21-gene signature for breast cancer (AACC, 2013), or the TruSeq Amplicon Cancer panel (Illumina) detecting hundreds of known critical cancer mutations. Future developments may go towards combining orthogonal molecular markers, and perhaps also different molecule species, such as proteins, lipids, and other metabolites. Promising developments towards this goal include proteogenomic and integrative -omics approaches, as discussed in the next section.

Integrative -omics boosts cancer molecular profiles

Deep insights into cancer biology, in general, and biomarker discovery in particular, have more recently benefited from efforts to integrate -omics data sets, such as transcriptomics, genomics, proteomics, and metabolomics data (Casado-Vela et al., 2011). Such 'precision analysis' approaches have gained popularity as more accessible bioinformatic tools are becoming available and are easier to use also for experimentalists with less specialized bioinformatics backgrounds (Nagaraj et al., 2015; Proietti et al.,

Table 25.1 Public databases available for -omic cancer research

Name	Organization	Organisms	Type of date	Web site link
Oncomine	Compendia Bioscience Inc	Human	Gene exp.	https://www.oncomine.org/resource/login.html
COSMIC	Wellcome Trust, Sanger Centre	Human	Mutation	http://cancer.sanger.ac.uk/cosmic
ClinicalTrials.gov	NIH	Human	Gene exp., mRNA, mut.	https://clinicaltrials.gov/
cBio Cancer Genomics Portal	Memorial Sloan Kettering	Human	Gene exp., mRNA	http://www.cbioportal.org/
The Cancer Genome Atlas (TCGA)	NCI	Human	Gene exp., mRNA, mut.	https://cancergenome.nih.gov/
Progenetix	University of Zurich	Human	Copy nr.	http://progenetix.org
Network of Cancer Genes	King's College	Human	Mut.	http://ncg.kcl.ac.uk/
Human Proteome Map	Pandey Lab, USA	Human	Proteome	http://humanproteomemap.org/
Human Proteome DB	TUM, Germany	Human	Proteome	http://www.proteomicsdb.org/
Human Protein Atlas	Wallenbery Foundation (Swe)	Human	Proteome, transcriptome	http://www.proteinatlas.org/
Human Metabolome	Alberta Canada	Human	Metabolome	http://www.hmdb.ca/
Clinical Proteomic CPTAC Portal	Tumor Analysis Consortium	Human	Cancer proteome data	http://proteomics.cancer.gov/programs/cptacnetwork
Genomics of Drug Sensibility in Cancer (GDSC)	Sanger/Wellcome Trust	Human	1,000 cell lines, drug panel	http://www.cancerrxgene.org/
Division of Cancer Treatment and Diagnosis (DCTD)	NCI	Human	60 cell lines, drug panel	http://dtp.cancer.gov/databases_tools/default.htm
Achilles Database of Cancer Cell Lines	Broad Institute	Human	Cancer cell genome / transcriptome	https://portals.broadinstitute.org/achilles

2016) see also Table 25.1). Proteogenomics approaches make use of available transcriptomics and genomics data that are then combined with cognate proteomics information. Many studies matched mRNA and protein expression levels finding poor a correlation (R2~0.4) although there are conflicting reports (Schwanhausser et al., 2011). In addition, transcriptomics and proteomics data collected from the same biological sample has the potential to reveal much more 'individualized' information such as isoforms, deletions, insertions, and single mutations that are not readily present in public data repositories such as Uniprot, trEMBL, and NCBInr. Multi-omics approaches benefit from measurements at different time-points or physiological interventions and can reveal more than the classical flow from transcription, translation, protein degradation, enzymatic activity, and metabolite turnover. For instance, allosteric interactions between metabolites and enzymes, post-translational modifications of proteins that affect their function and stability of epigenetic regulations can often be revealed by the integration of -omics data, revealing more than their sum (Fig. 25.2; see Buescher and Driggers, 2016). Spurred by the availability of -omics tools and analysis platforms, the number of studies in which integrative approaches are used to study cancer biology is increasing (Possemato et al., 2011; Shaw et al., 2013; Bansal et al., 2015; Valli et al., 2015). A particular example is the classification of breast cancer into four distinct subtypes using additional differentially expressed kinases and q5 chromosome loss, demonstrating the power of phosphoproteomics, proteomics and in-depth RNA seq on 77 samples (Mertins et al., 2016). Proteogenomic approaches have also been applied to liver cancer (Kimhofer et al., 2015; Nie et al., 2016), bladder cancer (Bansal et al., 2015), and colorectal cancer (Ma et al., 2012). One of the concepts emerging from these studies is the dependence of tumour growth, proliferation, and its metastatic properties on nutrition availability through the tumour microenvironment (Possemato et al., 2011; Maddocks et al., 2017) that subsequently may affect therapeutic resistance by orchestrating changes in gene expression at the level of translation (Falletta et al., 2017).

Molecular biology of cancer—contributions by -omics studies

Proteomics and metabolomics have made numerous contributions to our understanding of cancer biology, and these methodologies are also starting to be used in translational research. Instead of providing an exhaustive review of all contributions that proteomics and metabolomics have made to the field of cancer research, we will illustrate the methodologies with examples of how proteomics and metabolomics are being used to advance our understanding of cancer biology (Fig. 25.2).

Cancer cell signalling and survival mechanisms

Omics technology is contributing to our understanding of the mechanisms that govern intracellular cell signalling and cell-to-cell communication in different tumour types. In particular, phosphoproteomics methodologies are providing unprecedented insights into the complexity of cell signalling regulation. Since protein phosphorylation is the consequence of net protein kinase activity, the quantification of phosphorylation offers the opportunity to quantify cell signalling comprehensively and without bias. Recent studies have uncovered the existence of tens of thousands of phosphorylation events and it is estimated that cells contain more than 200,000

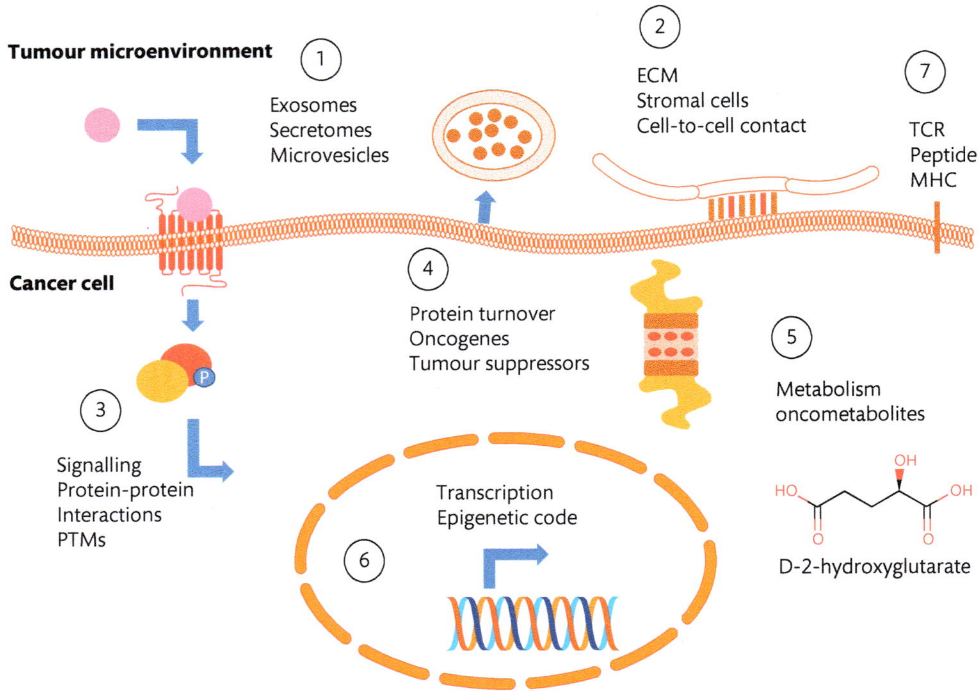

Fig. 25.2 Biological processes in cancer cells where -omics is making an impact. Proteomics and metabolomics studies can reveal molecular details of cancer cells and their interaction with the tumour microenvironment. 1—Exosomes, microvesicles, secretomes; 2—Extracellular matrix (ECM), stromal cells, cell-to-cell contact; 3—Cell signalling, protein-protein interactions, posttranslational modifications (PTMs); 4—Protein turnover by the ubiquitin-proteasome system, stability of tumour suppressor and oncogenes; 5—Metabolism, oncometabolites; 6—Transcriptional and epigenetic control of gene expression; 7—Tumour antigen presentation: major histocompatibility complex (MHC) class I presentation of antigenic peptides to T-lymphocytes for immune surveillance.

different phosphorylation sites (Vlastaridis et al., 2017). Proteomics technologies now allow cataloguing phosphorylation with high depth (Ballif et al., 2004; Nita-Lazar et al., 2008; Macek et al., 2009; Bodenmiller et al., 2010; Yu et al., 2011; de Graaf et al., 2014), but as high coverage comes at the expense of prohibitive analysis time, the challenge is to design experiments that can compare sufficient number of conditions and replicates to provide statistically and biologically meaningful results. The investigation of signalling events across tissues and cell types, or as a consequence of genetic or environmental challenges, requires the implementation of experimental workflows that allow comparing large number of samples and conditions with technical and biological replicates. Thus, methodologies for direct and medium throughput phosphoproteomics are preferred when the aim is to compare large number of samples (Casado and Cutillas, 2011; Alcolea et al., 2012; de Graaf et al., 2014; Humphrey et al., 2015; Lawrence et al., 2016; Cutillas, 2017).

The different cell types within tissues communicate with each other in paracrine signalling loops that regulate cell behaviour. This form of cell-to-cell communication is disrupted in cancer cells at different levels. On the one hand, genes with roles in regulating intracellular signalling pathways are frequently mutated or otherwise dysregulated in tumour cells. On the other hand, tumour cells secrete factors that corrupt 'normal' untransformed cells present in their vicinity. These non-cancerous tumour cells include immune cells, fibroblasts, and adipocytes, which together create a microenvironment that supports tumour progression (Joyce and Pollard, 2009; Klemm and Joyce, 2015).

Cancer mutations

Large-scale sequencing of genomic DNA has identified recurrent mutations in essentially all cancer types. One lesson derived from these studies is that essentially all cancers deregulate genes with roles in regulating intracellular signalling (Greenman et al., 2007; Forbes et al., 2015). Further refinements on essential cancer genes were provided by the "Cancer dependency map" from the Achilles project (Cowley et al., 2014; Tsherniak et al., 2017). Examples of frequently mutated signalling genes include *NRAS* in colorectal cancer, *PIK3CA* in breast cancer (Bachman et al., 2004) and *BRAF* in melanoma (Davies et al., 2002); these mutations are known to produce constitutive overactive enzymes. Proteomics studies, some of which are reviewed next, are now beginning to shed light into how these mutations drive cell biochemistry to effect cell transformation and other phenotypic characteristics of cancer cells (Samuels et al., 2005).

Phosphoproteomics and proteogenomics

In a recent study, Mertins et al. profiled the proteomes and phosphoproteomes of 105 breast tumours using iTRAQ labelling methods at a depth of 12,553 proteins and 33,239 phosphorylation sites (Mertins et al., 2016). Because these tumours had previously been characterized for genomic mutations and gene expression by The Cancer Genome Atlas (TCGA) breast cancer study, this study allowed the authors to assess the extent by which genetic mutations correlate with gene expression and biochemical pathway activation. The authors obtained putative signatures of pathway activation by correlating genetic mutations with the phosphorylation sites quantified in the experiment (across the 77 cases for which good quality data was obtained). An interesting, albeit perhaps counterintuitive, finding was that signatures of pathway activation were not always associated with activating mutations on genes that regulate such pathways. For example, only 15 of the 26 of the tumours having activating mutations on *PIK3CA* (which codes for the alpha isoform of PI3K) exhibited a signature of active PI3K signalling (Mertins et al., 2016), suggesting that other forms of regulation, independent of genetic mutations, fine tune PI3K signalling pathway activation in breast cancer cells. These conclusions are consistent with clinical data showing that *PIK3CA* mutations do not predictive the efficacy of PI3K-targeting drugs (Krop et al., 2016; Schmid et al., 2016).

In addition to genetic mutations, the activity of kinase-driven signalling pathways may be modulated by signals emanating from the tumour microenvironment and by the expression of receptors and adaptors that act upstream of signalling cascades. For example, by acting on c-Met, the hepatocyte growth factor (HGF), which is present in the tumour microenvironment of melanomas, activates intracellular signalling pathways including MAPK and PI3K in the absence of mutations (Straussman et al., 2012). Such activation confers melanoma cells with innate resistance to treatments with RAF inhibitors. Thus, activation of pathways parallel to those being targeted may contribute to resistance to targeted compounds. In this regard, using a proteomic approach consisting of measuring the phosphoproteomes of cells treated with BRAF inhibitors, Parker et al. found that the activity of the serine/threonine casein kinase 2 (CK2) increased as a result of RAF/MAPK pharmacological blockade. Cotreatment with BRAF and CK2 inhibitors synergized in reducing melanoma cell viability and proliferation, suggesting that CK2 acts as a compensatory mechanism that results from MAPK pathway inhibition (Parker et al., 2014). Similarly, phosphoproteomic studies of AML showed that cells resistant to PI3K inhibitors activated the MAPK and PKC pathways, which act in parallel to PI3K to promote cell viability (Alcolea et al., 2012; Casado et al., 2013a; Casado et al., 2013b). Collectively these and other studies highlight that the combination of target and parallel pathways determine the extent by which cancer cells respond to signalling inhibitors.

Given the existence of a multitude of mechanism by which kinase signalling pathways may be aberrantly activated in cancer, inferring kinase pathway activation using proxies such as genetic mutations or protein expression is likely to result in erroneous conclusions. As an example of this, studies have shown that clinical responses of cancer patients treated with PI3K inhibitors were not associated with *PIK3CA* mutations in their tumour cells (Janku et al., 2012; Krop et al., 2016; Schmid et al., 2016). Consequently, there is a strong interest in developing proteomics methodologies that may be used to infer network circuitry from proteomics data (Kothari et al., 2013; Workman et al., 2013; Cutillas, 2015; von Stechow and Olsen, 2017; Labots et al., 2017), with the view that such approaches may give more faithfully readouts of pathway activity.

Computational approaches to derive information on cell signalling from omics data

Computational approaches that use mass spectrometry proteomics to derive signalling pathway activity and circuitry from the phosphoproteomics data fall into two main groups: one consists of overlying phosphoproteomics data onto networks obtained from compiling data from the literature; the other involves using phosphoproteomics data as a means to infer kinase activity, thus providing a direct measure of pathway activities, also referred to as *activitomics* (**Fig. 25.3**). In this, measuring the product of given biochemical reactions may reveal the extent of activation of enzymes and the biochemical pathways that they regulate (**Fig. 25.3A**). For signalling and metabolic pathways, this entails measuring protein phosphorylation and metabolites, respectively, as these are the products of kinases and metabolic enzymes that regulate the flow of information and pathway fluxes (**Fig. 25.3B,C**). Thus, in general, interpretation of LC-MS/MS data using specialized computational methods may be used to quantify pathway activities in a broadly manner and without preconceptions on the molecules or biochemical pathway that may play a role in the system under investigation (**Fig. 25.3D**).

Different approaches to infer protein kinase activation from phosphoproteomics data have been developed (Schwartz and Gygi, 2005; Linding et al., 2007; Bakal et al., 2008; Miller et al., 2008; Casado et al., 2013b; Creixell et al., 2015; Drake et al., 2016; Ochoa et al., 2016). A common feature of these methods is that sites of phosphorylation are linked to the kinases that are thought to phosphorylate such sites. Statistics are then used to infer enrichment of kinase activities between the samples that are being compared. Phosphorylation sites may be the substrates of different kinases and therefore approaches that use only one site as a marker of kinase activity may give erroneous results. In contrast, by calculating enrichment using several substrates per kinase, the contribution of outliers may be diluted so that statistical tests may reveal true activities from the data. Recently, Ochoa et al. investigated the performance of the different algorithms used to calculate kinase substrate enrichment in phosphoproteomics data (Hernandez-Armenta et al., 2017). The authors reported that the Z-test (as reported in (Casado et al., 2013b)) and the statistics of the GSEA (as reported by Subramanian et al., 2005) provided similar levels of accuracy and outperformed methods based on more complex statistics such as those based on multiple linear regression and machine learning (Hernandez-Armenta et al., 2017).

The application of KSEA to rationalize why some AML cases were sensitive to PI3K inhibitors whereas others were resistant to the same treatments (Casado et al., 2013b) showed that while there were no significant differences in PI3K pathway activation between resistant and sensitive cells, resistant cells activated pathways, such as those driven by ERK and PKCD, that act in parallel to PI3K to sustain cell viability and proliferation. Treatment of cells with PKC inhibitors synergized with PI3K in inhibiting AML cell viability (Alcolea et al., 2012; Casado et al., 2013a). In a separate study Ochoa et al. used KSEA to compile an atlas of kinase activation across different tissues and to identify new hypotheses on kinase regulation (Ochoa et al., 2016).

In addition to KSEA, phosphoproteomics data may also be mined by overlying such data onto signalling networks derived from the literature (Huang and White, 2008; Mitsos et al., 2009; Terfve and Saez-Rodriguez, 2012). These approaches involve analysing the phosphoproteomes of cells perturbed with either agonists or antagonists of cell signalling, such as growth factors or pharmacological inhibitors. Signalling circuitry is then determined from the perturbed phosphoproteomes (Saez-Rodriguez

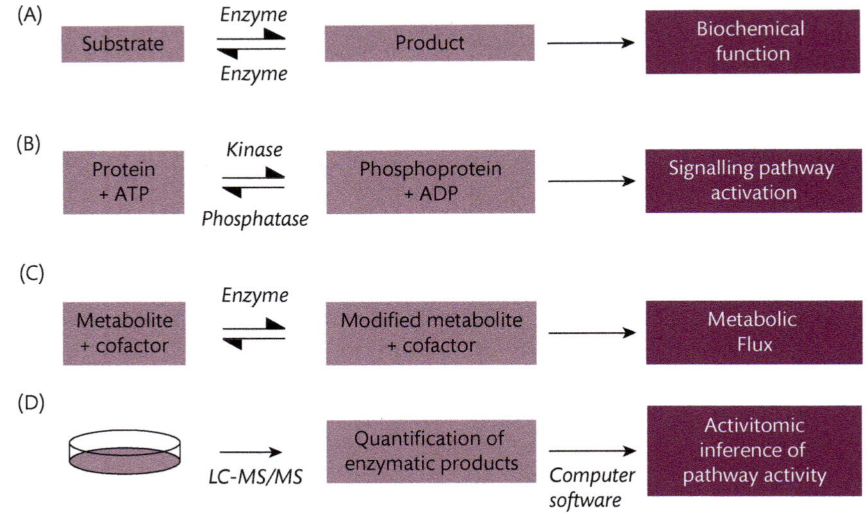

Fig. 25.3 Activitomics strategies to infer biochemical pathway activity from proteomics/metabolomics experiments. Advanced interrogation strategies of metabolomic and proteomic data sets can provide insights into enzyme function using knowledge of substrates for enzymes such as kinases and metabolic activities and how these link to pathway activation. (A) Profiling enzyme function by MS based quantitation of substrate and product levels. (B) Protein phosphorylation and kinase substrate discovery using algorithm such as KSEA (C) Metabolic flux studies using isotope-labelled metabolite precursors such as $^{13}C_6$-glucose or $^{15}N/^{13}C$-labelled glutamine. (D) In general, specialized software may be used to derive knowledge on pathway activity from enzyme products detected and quantified by LC-MS/MS (see also Table 25.1).

et al., 2009). One such approach, named PhoNemes, identified that specific CDK isoforms were regulated by mTORC1 (Terfve et al., 2015).

As an alternative to models based on networks reconstructed from the literature, Wilkes et al. showed that signalling circuitry may also be determined from drug-perturbed phosphoproteomes without reference to networks constructed from literature searches (Wilkes et al., 2015). In this study the authors assembled groups of phosphorylation sites commonly regulated by a battery of 22 kinase inhibitors; networks based on such sites allowed the authors to explore the relationships between the targets of the kinase inhibitors. The application of the approach to the MCF7 cancer cell line showed that PI3K/mTOR/S6K and MEK/ERK relationships were covered in the results, while allowing the authors to uncover unexpected circuitry such as the existence of PI3K/mTOR signalling independent of AKT. Interestingly, it has recently been found that non-canonical AKT-independent PI3K/mTOR signalling seems to contribute to cancer drug resistance (Elkabets et al., 2015; Dermit et al., 2017). The advantage of this approach is that it is not based on previous knowledge and therefore it can identify unexpected cell-specific circuitry.

Cell-to-cell communication

The intracellular signal transduction pathways discussed here are activated because of cues present in the extracellular environment. Most of these extracellular signals are derived from other cells present in adjacent tissues or in the tumour microenvironment. Indeed, cells do not work in isolation *in vivo* and instead communicate with each other using autocrine and paracrine loops that allow tissues and organs to organize coordinate responses to physiological challenges. In tumours, cell-to-cell communication allows cancer cells to recruit stromal cells to their vicinity which in turn provide an environment where cancer cells can proliferate and ultimately invade adjacent and distant tissue. This form of heterotypic communication is mediated by secreted factors, which may be soluble or embedded in vesicles, and by direct cell-to-cell contacts. Proteomics approaches are being fundamental to our understanding of the protein composition of the tumour microenvironment, including those present in the soluble form (i.e. the secretome) and those that are present in vesicles (exosomes). Proteomic approaches are also contributing to the understanding of how cells to communicate by direct contact with each other.

Cancer secretomes

The protein composition of cancer secretomes has been extensively investigated in order to identify biomarkers of disease (Gronborg et al., 2006; Pavlou and Diamandis, 2010). In addition, several studies have investigated how secreted proteins influence the biology of adjacent tumour cells. For example, proteomics analysis identified the composition of a complex secretome induced because of oncogene-induced senescence. These secreted proteins, which include interleukins, VEGF, CCL2 and CCL20, induce inflammation and impact senescence *in vivo* (Acosta et al., 2013). Interestingly, KSEA identified the kinases mTORC1 and MAPK2K2 as being involved in the regulation of the oncogene-induced secretome (Herranz et al., 2015) thus providing a potential rationale for the observed beneficial effects of mTOR inhibitors in preventing ageing and cancer (Zhuo et al., 2009; Hasty, 2010).

Exosomes

Exosomes are vesicles secreted by most cell types with a role in cell–cell communication. Proteomic studies have identified the protein composition of cancer exosomes. For example, Welton et al. found that the HT1376 cell line which is derived from bladder cancer secreted exosomes containing several antigens involved in immune modulation, including β1 and α6 integrins, CD36 and CD44 among others (Welton et al., 2010). These proteins are cell surface receptors with roles in promoting downstream signalling events (Chen et al., 2009; Grass et al., 2013).

Heterotypic signalling

In addition to responding to soluble factors and exosomes, cancer cells interact with other cells present in the microenvironment by direct cell–cell contacts. Proteomics methodologies have been used to understand how such form of cell communication affects intracellular signalling networks. Jorgensen used Stable Isotope Labelling with Amino acids in Cell culture (SILAC) quantitative phosphoproteomics to investigate reciprocal intercellular communication mediated by cell–cell contacts (Jorgensen et al., 2009). The strategy consisted of labelling cells lines (engineered to express EphB2, a transmembrane receptor tyrosine kinase) with heavy arginine and lysine amino acids, while cells expressing ephrin-B1, an EphB2 ligand, were labelled with the light version of the same amino acids. Global phosphoproteomics approaches were then used to compare signalling events between the cells growing in isolation with those that were induced after 10 minutes in coculture. The labelling made possible the quantification of signalling in a cell line-specific manner, which allowed the authors to discover asymmetric responses to stimuli, cell-specific signalling events, and to assemble cell-specific models of signals initiated by cell–cell contacts.

Although the SILAC strategy provided an elegant way of investigating rapid changes in phosphorylation as a result of direct cell–cell contacts, this approach does not allow investigating how long periods of coculturing affects the biochemistry of distinct cell populations in heterocellular systems, as SILAC protein labelling is quickly lost as a result of protein turnover. Thus, in order to investigate the long-term effects of cell–cell communication on the *in vivo* biochemistry of cancer cells, Rajeeve et al. designed a cross-species proteomics approach which involved comparing the proteomes and phosphoproteomes of cancer cells grown in isolation *in vitro* with those of the same cells grown in mouse xenografts. The study focused on the differential effects of two different PI3K inhibitors on the DLD-1 colorectal cancer cell line grown in culture or as xenografts. As expected, growing conditions had a profound effect on the activation of cell signalling pathways downstream of growth factors and those that regulate hypoxia. The authors discovered specific effects of the inhibitors in cancer and stromal cells. Interestingly, a PI3Kδ-specific inhibitor induced apoptosis in cells grown *in vivo* (as xenografts) but not when these cells were cultured in the absence of stroma (i.e. *in vitro* cell culture), suggesting that microenvironmental factors determined the extent by which cells responded to the compound (Rajeeve et al., 2014).

Recently, Gauthier et al. reported an elegant cell-selective labelling approach to investigate cell–cell communication in human to human cells at the proteomic level (Gauthier et al., 2013). The methodology, termed cell type-specific labelling using amino acid precursors (CTAP), exploits the well-known inability of animal cells to synthesize essential amino acids, such as lysine. Transgenic expression of amino acid synthesizing enzymes (from fungi or plants) in human or mouse cells allows these to incorporate labelled amino acid precursors into proteins. Expression of distinct amino acid synthesizing enzymes in different cell lines permits these to stably express labelled proteins with amino acids differing in mass, thus enabling the determination of relative abundance of proteins from different cell types growing in coculture (Gauthier et al., 2013).

Tape et al. employed the CTAP approach to investigate heterotypic communication between pancreatic ductal adenocarcinoma (PDA) cells harbouring the $KRAS^{G12D}$ mutation and pancreatic stellate cells (PSC), a myofibroblast-like cell type that interacts with PDA cells in tumours (Tape et al., 2016). As expected, mutated *KRAS* signalled to the classical MEK/ERK cascade in a cell autonomous manner. More interestingly, phosphoproteomic analysis of CTAP-labelled cells showed that mutations on *KRAS* also led to the activation of AKT through a reciprocal signalling axis; *KRAS* mutated PDA secrete of sonic hedgehog, which by acting on PSC cells activates a reciprocal signalling axis that promote AKT activity in PDA cells. This example illustrates the utility of CTAP and proteomics to dissect molecular events that regulate paracrine signalling in complex heterocellular systems such as solid tumours. However, a limitation of CTAP is that, as cells need to be cultured with labelled amino acid precursors, it cannot easily be applied in *in vivo* studies.

Alterations in metabolism

Metabolomics and proteomics technologies have made substantial contributions to our understanding of cancer metabolism. While other chapters of this volume cover this field in more detail, here we use selected studies to exemplify how proteo-metabolomic technology has been used, and can be used in the future, to investigate cancer metabolism.

The reliance on anaerobic glycolytic metabolism as a source of energy and enhanced lactate production relative to untransformed cells are well understood traits of cancer cells. This phenomenon, known as the Warburg effect, is consistent with the results of early proteomic studies, which showed that cancer cells alter the expression of enzymes involved in several metabolic processes, including those that regulate bioenergetic metabolism (Gharbi et al., 2002, Unwin et al., 2003; Zhang et al., 2005; Bi et al., 2006). Changes in the expression of glycolytic enzymes probably contribute to the observed increase of glycolytic flux of cancer cells.

In addition to changes in gene expression, posttranslational modifications on metabolic enzymes contribute to the Warburg effect. For example, tyrosine phosphorylation regulates the pyruvate kinase (PK) isozyme PNK2, which is predominantly expressed in fetal tissue and cancer cells (Christofk et al., 2008; Hitosugi et al., 2009). PK catalyses the production of pyruvate and ATP from phosphoenolpyruvate and ADP, a rate limiting step in glycolysis. A phosphoproteomics study found that PKM2 is phosphorylated at several tyrosine residues by the fibroblast growth factor receptor oncogenic form. When the tyrosine in 105 position is phosphorylated PKM2 is inhibited. (Hitosugi et al., 2009). More recently, using a phosphoproteomics approach, Chae et al. found that the serine/threonine kinase AKT regulates the Warburg effect in the prostate adenocarcinoma PC3 cell line by direct phosphorylation of pyruvate dehydrogenase kinase (PDK1), which, by direct phosphorylation of the pyruvate dehydrogenase complex, shuts down oxidative phosphorylation and diverts energy production towards glycolysis (Chae et al., 2016).

Proteomic and metabolic studies are also contributing to delineating the mechanisms by which onco-metabolites contribute to cancer onset and progression. For example, proteomics identified a role of fumarate in eliciting epithelial-to-mesenchymal-transition, which is a phenotype associated with oncogenesis and malignancy (Sciacovelli et al., 2016), and a metabolomics study found a role of this metabolite in senescence (Zheng et al., 2015). A combination of proteomic, metabolomic, and phosphoproteomic approaches identified a feedback loop in which the polyamine and PI3K pathways regulate each other in the regulation of colorectal cancer cell proliferation (Rajeeve et al., 2013). Intense research in this area is resulting in the development of increasingly sophisticated approaches to investigate the intricacies of metabolic processes, their fluxes, and how these contribute to cancer initiation and progression (Vizan et al., 2005; Mackay et al., 2015; Liu et al., 2016; Tumanov et al., 2016).

Onco-immunology—proteomics on major histocompatibility complex molecules

Major histocompatibility complex (MHC) peptide sequencing has been an interest for the discovery of cancer-specific antigens. Initially this was focused on proteins that are uniquely expressed in cancer tissue as compared to non-pathological counterparts, but this has proven to be problematic as these proteins are usually expressed in some normal tissue types. The most remarkable case are the MAGE antigens, but all of them are also expressed in normal tissue, such as testis (Bassani-Sternberg and Coukos, 2016; Desrichard et al., 2016; Schumacher and Hacohen, 2016). Recent efforts have been focused on the discovery of neoantigens based on somatic or genomic mutations. State-of-the-art mass spectrometry has now become sufficiently powerful to sequence more than 20,000 MHC peptide antigens from cellular material (Abelin et al., 2017; Khodadoust et al., 2017).

Angiogenesis and metastasis—proteomics and metabolomics studies

Angiogenesis is one of the way cancer interact with blood vessels, alongside non-angiogenic growth and vascular co-option. Considerable efforts have thus been placed in developing inhibitors that block vessel growth and proteo-metabolomic studies have contributed in the elucidation of the mode of action of angiogenic inhibitors. One such study found that the angiogenic inhibitor SU6668 inhibited Aurora kinases and TANK-binding kinase 1 protein kinases, suggesting a role of these kinases in angiogenesis (Godl et al., 2005). Park et al. performed a proteomics characterization of the A431 cell line in hypoxic conditions and found that hypoxia promote the secretion of proteins and exosomes with potential roles in angiogenesis and metastasis (Park et al., 2010). The role of hypoxia in mediating the release of exosomes with roles in invasion was later confirmed to occur in breast cancer and glioma cells (King et al., 2012; Kucharzewska et al., 2013). A phosphoproteomic study of signalling downstream of CXCR4, a G-coupled receptor with roles in breast cancer metastasis, identified several kinase-driven pathways that may be implicated in promoting invasion and which could represent new targets for therapy (Yi et al., 2014). This receptor is activated by CXCL12, a cytokine released by cancer-associated fibroblast (Izumi et al., 2016). Similarly, a phosphoproteomic study of integrin signalling identified several potential kinases involved in metastasis, of which integrin-regulated kinases, DBF4, PAK2, and GRK6, were validated by siRNA to have a role in cell adhesion and migration possibly through their regulation of actin cytoskeleton dynamics (Chen et al., 2009).

TAKE-HOME MESSAGE

- The maturation of mass spectrometry has contributed to the development of proteomics and metabolomics approaches to investigate cancer biology from different standpoints, and LC-MS/MS methodologies are now indispensable in molecular biology research and oncology.
- Identification and quantification of thousands of proteins, metabolites, and posttranslational modifications (PTMs) are now routinely carried out in many mass spectrometry laboratories. Although the illustrative studies discussed earlier have focused on well-established approaches, such as phosphoproteomics as these are more mature than the analysis of other forms of PTMs, most modifications and endogenous analytes are amenable to be analysed by LC-MS/MS in a global manner.
- The challenge now is to develop computational approaches capable of integrating the large volumes of data obtained with state-of-the-art mass spectrometry and with those from other -omic data sets, such as those obtained by whole genome sequence and RNAseq gene expression analyses.
- In this regard, considerable advances have recently been made in the development of proteogenomic approaches that integrate different omics data sets to provide a systems level of cancer biology. We expect that application of these integrative methodologies in the near future will provide an even deeper understanding of cancer biology, from which approaches for personalized therapies may be designed and applied in the clinic.

OPEN QUESTIONS

- Cancer biomarkers—what are the real contributions of -omics technologies that are translatable?
- What coverage is required so that information derived from proteomics and metabolomics studies provide faithful readouts of biochemical pathway activity?
- Are there synergistic effects of integrated -omics that can reveal more valuable information about complex biology networks in cancer?
- How can we translate the vast biological information obtained with proteo-metabolomic approaches into novel cancer therapies and biomarkers for early identification of cancer?

FURTHER READING

Aebersold, R. & Mann, M. (2016). Mass-spectrometric exploration of proteome structure and function. *Nature*, **537**, 347–55.

Creixell, P., Schoof, E. M., Simpson, C. D., et al. (2015). Kinome-wide decoding of network-attacking mutations rewiring cancer signaling. *Cell*, **163**, 202–17.

Drake, J. M., Paull, E. O., Graham, N. A., et al. (2016). Phosphoproteome integration reveals patient-specific networks in prostate cancer. *Cell*, **166**, 1041–54.

Khodadoust, M. S., Olsson, N., Wagar, L. E., et al. (2017). Antigen presentation profiling reveals recognition of lymphoma immunoglobulin neoantigens. *Nature*, **543**, 723–7.

Mertins, P., Mani, D. R., Ruggles, K. V., et al. (2016). Proteogenomics connects somatic mutations to signalling in breast cancer. *Nature*, **534**, 55–62.

Patti, G. J., Yanes, O., & Siuzdak, G. (2012). Innovation: metabolomics: the apogee of the omics trilogy. *Nat Rev Mol Cell Biol*, **13**(4), 263–9.

Sciacovelli, M., Goncalves, E., Johnson, T. I., et al. (2016). Fumarate is an epigenetic modifier that elicits epithelial-to-mesenchymal transition. *Nature*, **537**, 544–7.

Tape, C. J., Ling, S., Dimitriadi, M., et al. (2016). Oncogenic KRAS regulates tumor cell signaling via stromal reciprocation. *Cell*, **165**, 910–20.

REFERENCES

AACC (2013). Tumor markers. Available at: https://labtestsonline.org/understanding/analytes/tumor-markers)

Abelin, J. G., Keskin, D. B., Sarkizova, S., et al. (2017). Mass spectrometry profiling of HLA-associated peptidomes in mono-allelic cells enables more accurate epitope prediction. *Immunity*, **46**, 315–26.

Acosta, J. C., Banito, A., Wuestefeld, T., et al. (2013). A complex secretory program orchestrated by the inflammasome controls paracrine senescence. *Nat Cell Biol*, **15**, 978–90.

Aebersold, R. & Mann, M. (2003). Mass spectrometry-based proteomics. *Nature*, **422**, 198–207.

Aebersold, R. & Mann, M. (2016). Mass-spectrometric exploration of proteome structure and function. *Nature*, **537**, 347–55.

Alcolea, M. P., Casado, P., Rodriguez-Prados, J. C., Vanhaesebroeck, B., & Cutillas, P. R. (2012). Phosphoproteomic analysis of leukemia cells under basal and drug-treated conditions identifies markers of kinase pathway activation and mechanisms of resistance. *Mol Cell Proteomics*, **11**, 453–66.

Bachman, K. E., Argani, P., Samuels, Y., et al. (2004). The *PIK3CA* gene is mutated with high frequency in human breast cancers. *Cancer Biol Ther*, **3**, 772–5.

Bakal, C., Linding, R., Llense, F., et al. (2008). Phosphorylation networks regulating JNK activity in diverse genetic backgrounds. *Science*, **322**, 453–6.

Ballif, B. A., Villen, J., Beausoleil, S. A., Schwartz, D., & Gygi, S. P. (2004). Phosphoproteomic analysis of the developing mouse brain. *Mol Cell Proteomics*, **3**, 1093–101.

Bansal, N., Gupta, A., & Sankhwar, S. N. (2015). Proteometabolomics of bladder cancer: current and future prospects. *Cancer Biomark*, **15**, 339–48.

Bassani-Sternberg, M. & Coukos, G. (2016). Mass spectrometry-based antigen discovery for cancer immunotherapy. *Curr Opin Immunol*, **41**, 9–17.

Bi, X., Lin, Q., Foo, T. W., et al. (2006). Proteomic analysis of colorectal cancer reveals alterations in metabolic pathways: mechanism of tumorigenesis. *Mol Cell Proteomics*, **5**, 1119–30.

Bodenmiller, B., Wanka, S., Kraft, C., et al. (2010). Phosphoproteomic analysis reveals interconnected system-wide responses to perturbations of kinases and phosphatases in yeast. *Science Signaling*, **3**(153), rs4.

Borrebaeck, C. A. (2017). Precision diagnostics: moving towards protein biomarker signatures of clinical utility in cancer. *Nat Rev Cancer*, **17**, 199–204.

Buescher, J. M. & Driggers, E. M. (2016). Integration of omics: more than the sum of its parts. *Cancer Metab*, **4**, 4.

Casado, P. & Cutillas, P. R. (2011). A self-validating quantitative mass spectrometry method for assessing the accuracy of high-content phosphoproteomic experiments. *Mol Cell Proteomics*, **10**, M110.003079.

Casado, P., Alcolea, M. P., Iorio, F., et al. (2013a). Phosphoproteomics data classify hematological cancer cell lines according to tumor type and sensitivity to kinase inhibitors. *Genome Biol*, **14**, R37.

Casado, P., Rodriguez-Prados, J. C., Cosulich, S. C., et al. (2013b). Kinase-substrate enrichment analysis provides insights into the heterogeneity of signaling pathway activation in leukemia cells. *Sci Signal*, **6**, rs6.

Casado-Vela, J., Cebrian, A., Gomez Del Pulgar, M. T., & Lacal, J. C. (2011). Approaches for the study of cancer: towards the integration of genomics, proteomics and metabolomics. *Clin Transl Oncol*, **13**, 617–28.

Chae, Y. C., Vaira, V., Caino, M. C., et al. (2016). Mitochondrial Akt regulation of hypoxic tumor reprogramming. *Cancer Cell*, **30**, 257–72.

Chen, Y., Lu, B., Yang, Q., Fearns, C., Yates, J. R., 3rd, & Lee, J. D. (2009). Combined integrin phosphoproteomic analyses and small interfering RNA-based functional screening identify key regulators for cancer cell adhesion and migration. *Cancer Res*, **69**, 3713–20.

Christofk, H. R., Vander Heiden, M. G., Harris, M. H., et al. (2008). The M2 splice isoform of pyruvate kinase is important for cancer metabolism and tumour growth. *Nature*, **452**, 230–3.

Cowley, G. S., Weir, B. A., Vazquez, F., et al. (2014). Parallel genome-scale loss of function screens in 216 cancer cell lines for the identification of context-specific genetic dependencies. *Nat Sci Data*, **1**, Article number: 140035. September 30, 2014.

Creixell, P., Schoof, E. M., Simpson, C. D., et al. (2015). Kinome-wide decoding of network-attacking mutations rewiring cancer signaling. *Cell*, **163**, 202–17.

Crutchfield, C. A., Thomas, S. N., Sokoll, L. J., & Chan, D. W. (2016). Advances in mass spectrometry-based clinical biomarker discovery. *Clin Proteomics*, **13**, 1.

Cutillas, P. R. (2015). Role of phosphoproteomics in the development of personalized cancer therapies. *Proteomics Clin Appl*, **9**, 383–95.

Cutillas, P. R. (2017). Targeted in-depth quantification of signaling using label-free mass spectrometry. *Methods Enzymol*, **585**, 245–68.

Davies, H., Bignell, G. R., Cox, C., et al. (2002). Mutations of the *BRAF* gene in human cancer. *Nature*, **417**, 949–54.

De Graaf, E. L., Giansanti, P., Altelaar, A. F., & Heck, A. J. (2014). Single step enrichment by Ti4+-IMAC and label free quantitation enables in-depth monitoring of phosphorylation dynamics with high reproducibility and temporal resolution. *Mol Cell Proteomics*, **13**, 2426–34.

Dermit, M., Casado, P., Rajeeve, V., et al. (2017). Oxidative stress downstream of mTORC1 but not AKT causes a proliferative defect in cancer cells resistant to PI3K inhibition. *Oncogene*, **36**(19), 2762–74.

Desrichard, A., Snyder, A., & Chan, T. A. (2016). Cancer neoantigens and applications for immunotherapy. *Clin Cancer Res*, **22**, 807–12.

Drake, J. M., Paull, E. O., Graham, N. A., et al. (2016). Phosphoproteome integration reveals patient-specific networks in prostate cancer. *Cell*, **166**, 1041–54.

Elkabets, M., Pazarentzos, E., Juric, D., et al. (2015). AXL mediates resistance to PI3Kalpha inhibition by activating the EGFR/PKC/mTOR axis in head and neck and esophageal squamous cell carcinomas. *Cancer Cell*, **27**, 533–46.

Falletta, P., Sanchez-Del-Campo, L., Chauhan, J., et al. (2017). Translation reprogramming is an evolutionarily conserved driver of phenotypic plasticity and therapeutic resistance in melanoma. *Genes Dev*, **31**, 18–33.

Forbes, S. A., Beare, D., Gunasekaran, P., et al. (2015). COSMIC: exploring the world's knowledge of somatic mutations in human cancer. *Nucleic Acids Res*, **43**, D805–11.

Frantzi, M. & Vlahou, A. (2017). Ten years of proteomics in bladder cancer: progress and future directions. *Bladder Cancer*, **3**, 1–18.

Gauthier, N. P., Soufi, B., Walkowicz, W. E., et al. (2013). Cell-selective labeling using amino acid precursors for proteomic studies of multicellular environments. *Nat Methods*, **10**, 768–73.

Gharbi, S., Gaffney, P., Yang, A., et al. (2002). Evaluation of two-dimensional differential gel electrophoresis for proteomic expression analysis of a model breast cancer cell system. *Mol Cell Proteomics*, **1**, 91–8.

Godl, K., Gruss, O. J., Eickhoff, J., et al. (2005). Proteomic characterization of the angiogenesis inhibitor SU6668 reveals multiple impacts on cellular kinase signaling. *Cancer Res*, **65**, 6919–26.

Grass, G. D., Tolliver, L. B., Bratoeva, M., & Toole, B. P. (2013). CD147, CD44, and the epidermal growth factor receptor (EGFR) signaling pathway cooperate to regulate breast epithelial cell invasiveness. *J Biol Chem*, **288**, 26089–104.

Greenman, C., Stephens, P., Smith, R., et al. (2007). Patterns of somatic mutation in human cancer genomes. *Nature*, **446**, 153–8.

Gronborg, M., Kristiansen, T. Z., Iwahori, A., et al. (2006). Biomarker discovery from pancreatic cancer secretome using a differential proteomic approach. *Mol Cell Proteomics*, **5**, 157–71.

Hasty, P. (2010). Rapamycin: the cure for all that ails. *J Mol Cell Biol*, **2**, 17–19.

Henry, N. L. & Hayes, D. F. (2012). Cancer biomarkers. *Mol Oncol*, **6**, 140–6.

Hernandez-Armenta, C., Ochoa, D., Goncalves, E., Saez-Rodriguez, J., & Beltrao, P. (2017). Benchmarking substrate-based kinase activity inference using phosphoproteomic data. *Bioinformatics*, **33**(12), 1845–51.

Herranz, N., Gallage, S., Mellone, M., et al. (2015). mTOR regulates Mapkapk2 translation to control the senescence-associated secretory phenotype. *Nat Cell Biol*, **17**, 1205–17.

Hitosugi, T., Kang, S., Vander Heiden, M. G., et al. (2009). Tyrosine phosphorylation inhibits PKM2 to promote the Warburg effect and tumor growth. *Sci Signal*, **2**, ra73.

Huang, P. H. & White, F. M. (2008). Phosphoproteomics: unraveling the signaling web. *Mol Cell*, **31**, 777–81.

Humphrey, S. J., Azimifar, S. B., & Mann, M. (2015). High-throughput phosphoproteomics reveals *in vivo* insulin signaling dynamics. *Nat Biotechnol*, **33**, 990–5.

Issaq, H. J., Fox, S. D., Chan, K. C., & Veenstra, T. D. (2011). Global proteomics and metabolomics in cancer biomarker discovery. *J Sep Sci*, **34**, 3484–92.

Issaq, H. J. & Veenstra, T. D. (2011). Is sarcosine a biomarker for prostate cancer? *J Sep Sci*, **34**, 3619–21.

Izumi, D., Ishimoto, T., Miyake, K., et al. (2016). CXCL12/CXCR4 activation by cancer-associated fibroblasts promotes integrin beta1 clustering and invasiveness in gastric cancer. *Int J Cancer*, **138**, 1207–19.

Janku, F., Wheler, J. J., Westin, S. N., et al. (2012). PI3K/AKT/mTOR inhibitors in patients with breast and gynecologic malignancies harboring Pik3CA mutations. *J Clin Oncol*, **30**, 777–82.

Jorgensen, C., Sherman, A., Chen, G. I., et al. (2009). Cell-specific information processing in segregating populations of Eph receptor ephrin-expressing cells. *Science*, **326**, 1502–9.

Joyce, J. A. & Pollard, J. W. (2009). Microenvironmental regulation of metastasis. *Nat Rev Cancer*, **9**, 239–52.

Kaisar, M., Van Dullemen, L. F., Thezenas, M. L., Charles, P. D., Ploeg, R. J., & Kessler, B. M. (2016a). Plasma biomarker profile alterations during variable blood storage. *Clin Chem*, **62**, 1272–4.

Kaisar, M., Van Dullemen, L. F., Thezenas, M. L., et al. (2016b). Plasma degradome affected by variable storage of human blood. *Clin Proteomics*, **13**, 26.

Khodadoust, M. S., Olsson, N., Wagar, L. E., et al. (2017). Antigen presentation profiling reveals recognition of lymphoma immunoglobulin neoantigens. *Nature*, **543**, 723–7.

Kimhofer, T., Fye, H., Taylor-Robinson, S., Thursz, M., & Holmes, E. (2015). Proteomic and metabonomic biomarkers for hepatocellular carcinoma: a comprehensive review. *Br J Cancer*, **112**, 1141–56.

King, H. W., Michael, M. Z., & Gleadle, J. M. (2012). Hypoxic enhancement of exosome release by breast cancer cells. *BMC Cancer*, **12**, 421.

Klemm, F. & Joyce, J. A. (2015). Microenvironmental regulation of therapeutic response in cancer. *Trends Cell Biol*, **25**, 198–213.

Kothari, V., Wei, I., Shankar, S., et al. (2013). Outlier kinase expression by RNA sequencing as targets for precision therapy. *Cancer Discov*, **3**, 280–93.

Krop, I. E., Mayer, I. A., Ganju, V., et al. (2016). Pictilisib for oestrogen receptor-positive, aromatase inhibitor-resistant, advanced or metastatic breast cancer (FERGI): a randomised, double-blind, placebo-controlled, phase 2 trial. *Lancet Oncology*, **17**, 811–21.

Kucharzewska, P., Christianson, H. C., Welch, J. E., et al. (2013). Exosomes reflect the hypoxic status of glioma cells and mediate hypoxia-dependent activation of vascular cells during tumor development. *Proc Natl Acad Sci U S A*, **110**, 7312–17.

Kulasingam, V. & Diamandis, E. P. (2007). Proteomics analysis of conditioned media from three breast cancer cell lines: a mine for biomarkers and therapeutic targets. *Mol Cell Proteomics*, **6**, 1997–2011.

Labots, M., Van Der Mijn, J. C., Beekhof, R., et al. (2017). Phosphotyrosine-based-phosphoproteomics scaled-down to biopsy level for analysis of individual tumor biology and treatment selection. *J Proteomics*, **162**, 99–107.

Lawrence, R. T., Searle, B. C., Llovet, A., & Villen, J. (2016). Plug-and-play analysis of the human phosphoproteome by targeted high-resolution mass spectrometry. *Nat Methods*, **13**, 431–4.

Linding, R., Jensen, L. J., Ostheimer, G. J., et al. (2007). Systematic discovery of *in vivo* phosphorylation networks. *Cell*, **129**, 1415–26.

Liu, X., Romero, I. L., Litchfield, L. M., Lengyel, E., & Locasale, J. W. (2016). Metformin targets central carbon metabolism and reveals mitochondrial requirements in human cancers. *Cell Metab*, **24**, 728–39.

Ma, Y., Zhang, P., Wang, F., Liu, W., Yang, J., & Qin, H. (2012). An integrated proteomics and metabolomics approach for defining oncofetal biomarkers in the colorectal cancer. *Ann Surg*, **255**, 720–30.

Macek, B., Mann, M., & Olsen, J. V. (2009). Global and site-specific quantitative phosphoproteomics: principles and applications. *Annu Rev Pharmacol Toxicol*, **49**, 199–221.

Mackay, G. M., Zheng, L., Van Den Broek, N. J., & Gottlieb, E. (2015). Analysis of cell metabolism using LC-MS and isotope tracers. *Methods Enzymol*, **561**, 171–96.

Maddocks, O. D. K., Athineos, D., Cheung, E. C., et al. (2017). Modulating the therapeutic response of tumours to dietary serine and glycine starvation. *Nature*, **544**, 372–6.

Mertins, P., Mani, D. R., Ruggles, K. V., et al. (2016). Proteogenomics connects somatic mutations to signalling in breast cancer. *Nature*, **534**, 55–62.

Miller, M. L., Jensen, L. J., Diella, F., et al. (2008). Linear motif atlas for phosphorylation-dependent signaling. *Sci Signal*, **1**, ra2.

Mitsos, A., Melas, I. N., Siminelakis, P., Chairakaki, A. D., Saez-Rodriguez, J., & Alexopoulos, L. G. (2009). Identifying drug effects via pathway alterations using an integer linear programming optimization formulation on phosphoproteomic data. *PLoS Comput Biol*, **5**, e1000591.

Nagaraj, S. H., Waddell, N., Madugundu, A. K., et al. (2015). PGTools: a software suite for proteogenomic data analysis and visualization. *J Proteome Res*, **14**, 2255–66.

Nie, W., Yan, L., Lee, Y. H., Guha, C., Kurland, I. J., & Lu, H. (2016). Advanced mass spectrometry-based multi-omics technologies for exploring the pathogenesis of hepatocellular carcinoma. *Mass Spectrom Rev*, **35**, 331–49.

Nita-Lazar, A., Saito-Benz, H., & White, F. M. (2008). Quantitative phosphoproteomics by mass spectrometry: past, present, and future. *Proteomics*, **8**, 4433–43.

O'Neill, J. R., Pak, H. S., Pairo-Castineira, E., et al. (2017). Quantitative shotgun proteomics unveils candidate novel oesophageal adenocarcinoma-specific proteins. *Mol Cell Proteomics*, **16**(6), 1138–50.

Ochoa, D., Jonikas, M., Lawrence, R. T., et al. (2016). An atlas of human kinase regulation. *Mol Syst Biol*, **12**, 888.

Park, J. E., Tan, H. S., Datta, A., et al. (2010). Hypoxic tumor cell modulates its microenvironment to enhance angiogenic and metastatic potential by secretion of proteins and exosomes. *Mol Cell Proteomics*, **9**, 1085–99.

Parker, R., Clifton-Bligh, R., & Molloy, M. P. (2014). Phosphoproteomics of MAPK inhibition in BRAF-mutated cells and a role for the lethal synergism of dual BRAF and CK2 inhibition. *Mol Cancer Ther*, **13**, 1894–906.

Pavlou, M. P. & Diamandis, E. P. (2010). The cancer cell secretome: a good source for discovering biomarkers? *J Proteomics*, **73**, 1896–906.

Petricoin, E. F., 3rd, Ornstein, D. K., Paweletz, C. P., et al. (2002). Serum proteomic patterns for detection of prostate cancer. *J Natl Cancer Inst*, **94**, 1576–8.

Planque, C., Kulasingam, V., Smith, C. R., Reckamp, K., Goodglick, L., & Diamandis, E. P. (2009). Identification of five candidate lung cancer biomarkers by proteomics analysis of conditioned media of four lung cancer cell lines. *Mol Cell Proteomics*, **8**, 2746–58.

Possemato, R., Marks, K. M., Shaul, Y. D., et al. (2011). Functional genomics reveal that the serine synthesis pathway is essential in breast cancer. *Nature*, **476**, 346–50.

Proietti, C., Zakrzewski, M., Watkins, T. S., et al. (2016). Mining, visualizing and comparing multidimensional biomolecular data using the Genomics Data Miner (GMine) Web-Server. *Sci Rep*, **6**, 38178.

Rajeeve, V., Pearce, W., Cascante, M., Vanhaesebroeck, B., & Cutillas, P. R. (2013). Polyamine production is downstream and upstream of oncogenic PI3K signalling and contributes to tumour cell growth. *Biochem J*, **450**, 619–28.

Rajeeve, V., Vendrell, I., Wilkes, E., Torbett, N., & Cutillas, P. R. (2014). Cross-species proteomics reveals specific modulation of signaling in cancer and stromal cells by phosphoinositide 3-kinase (PI3K) inhibitors. *Mol Cell Proteomics*, **13**, 1457–70.

Saez-Rodriguez, J., Alexopoulos, L. G., Epperlein, J., et al. (2009). Discrete logic modelling as a means to link protein signalling networks with functional analysis of mammalian signal transduction. *Mol Syst Biol*, **5**, 331.

Samuels, Y., Diaz, L. A., JR., Schmidt-Kittler, O., et al. (2005). Mutant PIK3CA promotes cell growth and invasion of human cancer cells. *Cancer Cell*, **7**, 561–73.

Sandfeld-Paulsen, B., Aggerholm-Pedersen, N., Baek, R., et al. (2016). Exosomal proteins as prognostic biomarkers in non-small cell lung cancer. *Mol Oncol*, **10**, 1595–602.

Schmid, P., Pinder, S. E., Wheatley, D., et al. (2016). Phase II randomized preoperative window-of-opportunity study of the PI3K inhibitor pictilisib plus anastrozole compared with anastrozole alone in patients with estrogen receptor-positive breast cancer. *Journal of Clinical Oncology*, **34**, 1987–94.

Schumacher, T. N. & Hacohen, N. (2016). Editorial overview: cancer immunology: genomics and biomarkers: cancer immunity through the prism of genomics and proteomics. *Curr Opin Immunol*, **41**, ix–x.

Schwanhausser, B., Busse, D., Li, N., et al. (2011). Global quantification of mammalian gene expression control. *Nature*, **473**, 337–42.

Schwartz, D. & Gygi, S. P. (2005). An iterative statistical approach to the identification of protein phosphorylation motifs from large-scale data sets. *Nat Biotechnol*, **23**, 1391–8.

Sciacovelli, M., Goncalves, E., Johnson, T. I., et al. (2016). Fumarate is an epigenetic modifier that elicits epithelial-to-mesenchymal transition. *Nature*, **537**, 544–7.

Shaw, P. G., Chaerkady, R., Wang, T., et al. (2013). Integrated proteomic and metabolic analysis of breast cancer progression. *PLoS One*, **8**, e76220.

Sreekumar, A., Poisson, L. M., Rajendiran, T. M., et al. (2009). Metabolomic profiles delineate potential role for sarcosine in prostate cancer progression. *Nature*, **457**, 910–4.

Straussman, R., Morikawa, T., Shee, K., et al. (2012). Tumour micro-environment elicits innate resistance to RAF inhibitors through HGF secretion. *Nature*, **487**, 500–4.

Subramanian, A., Tamayo, P., Mootha, V. K., et al. (2005). Gene set enrichment analysis: a knowledge-based approach for interpreting genome-wide expression profiles. *Proc Natl Acad Sci U S A*, **102**, 15545–50.

Swami, M. (2010). Proteomics: a discovery strategy for novel cancer biomarkers. *Nat Rev Cancer*, **10**, 597.

Tape, C. J., Ling, S., Dimitriadi, M., et al. (2016). Oncogenic KRAS regulates tumor cell signaling via stromal reciprocation. *Cell*, **165**, 910–20.

Terfve, C. & Saez-Rodriguez, J. (2012). Modeling signaling networks using high-throughput phospho-proteomics. *Adv Exp Med Biol*, **736**, 19–57.

Terfve, C. D., Wilkes, E. H., Casado, P., Cutillas, P. R., & Saez-Rodriguez, J. (2015). Large-scale models of signal propagation in human cells derived from discovery phosphoproteomic data. *Nat Commun*, **6**, 8033.

Tsherniak, A., Vazquez, F., Phil G. Montgomery, et al. (2017). Defining a cancer Dependency Map. *Cell*. doi: j.cell.2017.06.010.

Tumanov, S., Bulusu, V., Gottlieb, E., & Kamphorst, J. J. (2016). A rapid method for quantifying free and bound acetate based on alkylation and GC-MS analysis. *Cancer Metab*, **4**, 17.

Unwin, R. D., Craven, R. A., Harnden, P., et al. (2003). Proteomic changes in renal cancer and co-ordinate demonstration of both the glycolytic and mitochondrial aspects of the Warburg effect. *Proteomics*, **3**, 1620–32.

Valli, A., Rodriguez, M., Moutsianas, L., et al. (2015). Hypoxia induces a lipogenic cancer cell phenotype via HIF1alpha-dependent and -independent pathways. *Oncotarget*, **6**, 1920–41.

Vizan, P., Boros, L. G., Figueras, A., et al. (2005). K-ras codon-specific mutations produce distinctive metabolic phenotypes in NIH3T3 mice [corrected] fibroblasts. *Cancer Res*, **65**, 5512–15.

Vlastaridis, P., Kyriakidou, P., Chaliotis, A., et al. (2017). Estimating the total number of phosphoproteins and phosphorylation sites in eukaryotic proteomes. *Gigascience*, **6**, 1–11.

Von Stechow, L. & Olsen, J. V. (2017). Proteomics insights into DNA damage response and translating this knowledge to clinical strategies. *Proteomics*, **17**(3–4). doi: 10.1002/pmic.201600018.

Ward, D. G., Suggett, N., Cheng, Y., et al. (2006). Identification of serum biomarkers for colon cancer by proteomic analysis. *Br J Cancer*, **94**, 1898–905.

Welton, J. L., Khanna, S., Giles, P. J., et al. (2010). Proteomics analysis of bladder cancer exosomes. *Mol Cell Proteomics*, **9**, 1324–38.

Wilkes, E. H., Terfve, C., Gribben, J. G., Saez-Rodriguez, J., & Cutillas, P. R. (2015). Empirical inference of circuitry and plasticity in a kinase signaling network. *Proc Natl Acad Sci U S A*, **112**, 7719–24.

Workman, P., AL-Lazikani, B., & Clarke, P. A. (2013). Genome-based cancer therapeutics: targets, kinase drug resistance and future strategies for precision oncology. *Curr Opin Pharmacol*, **13**, 486–96.

Yi, T., Zhai, B., Yu, Y., et al. (2014). Quantitative phosphoproteomic analysis reveals system-wide signaling pathways downstream of SDF-1/CXCR4 in breast cancer stem cells. *Proc Natl Acad Sci U S A*, **111**, E2182–90.

Yu, Y., Yoon, S. O., Poulogiannis, G., et al. (2011). Phosphoproteomic analysis identifies Grb10 as an mTORC1 substrate that negatively regulates insulin signaling. *Science*, **332**, 1322–6.

Zhang, D., Tai, L. K., Wong, L. L., Chiu, L. L., Sethi, S. K., & Koay, E. S. (2005). Proteomic study reveals that proteins involved in metabolic and detoxification pathways are highly expressed in HER-2/neu-positive breast cancer. *Mol Cell Proteomics*, **4**, 1686–96.

Zheng, L., Cardaci, S., Jerby, L., et al. (2015). Fumarate induces redox-dependent senescence by modifying glutathione metabolism. *Nat Commun*, **6**, 6001.

Zhuo, L., Cai, G., Liu, F., et al. (2009). Expression and mechanism of mammalian target of rapamycin in age-related renal cell senescence and organ aging. *Mech Ageing Dev*, **130**, 700–8.

26

Cancer systems biology: From molecular profiles to pathways, signalling networks, and therapeutic vulnerabilities

Lieven Verbeke and Steven Van Laere

Introduction

Unravelling the biology of human tumours, either primary or metastatic, is daily practice for many researchers worldwide. The advent of high-throughput technologies, like microarrays and next-generation sequencing, has greatly accelerated our understanding of the complex molecular underpinnings of cancer development and progression. Big sequencing consortia like The Cancer Genome Atlas (TCGA) and the International Cancer Genome Consortium (ICGC) provide the research community an unprecedented wealth of genomic, epigenomic, transcriptomic, and proteomic details of various tumour and cell types. Recent landscape papers represent only the initial analyses, and answers to many other important questions may well be buried deep inside these data. It is the primary purpose of systems biology to provide the keys to unlock the secrets underpinning the molecular complexity of cancer biology. In this chapter, an overview will be provided of several important concepts in systems biology, how gained knowledge has reshaped the field and what opportunities still lie ahead.

The systems biology paradigm shift: From pathways to networks

A common goal of many cancer profiling experiments is to identify the signal transduction mechanisms and processes that characterize tumour biology, with the ultimate aim of revealing novel targets for treatment. The current mass production of molecular cancer profiles has spurred the development of tools particularly designed for this purpose. The majority of these tools build on the concept of interpreting profiling experiments in the context of reference genes sets that are based on prior knowledge gained through decades of basic and translational research. These gene sets can be found in various publicly available databases (a.k.a. knowledge bases), among others: Gene Ontology (GO), Kyoto Encyclopedia of Gene and Genomes (KEGG), and BioCarta. Up to date, the interpretation of molecular data in the context of pathway related gene sets has been a successful approach. Table 26.1 provides an overview of pathways and processes significantly altered in 12 common cancer types according to TCGA (Kandoth et al., 2014).

With the acquisition of novel biological insights, the classical representation of processes and particularly pathways as hierarchically structured and mostly linear diagrams of protein–protein interactions (PPIs) that are sharply and precisely delineated from the broader cellular transduction network is increasingly being questioned. Instead, the now prevailing vision in systems biology, is to think of pathways and processes as warm and fuzzy clouds: warm since their representation is close to the truth but not necessarily exact; fuzzy since the membership of components in a pathway is graded and dynamic and therefore not all components of a pathway are equally important and might vary; and a cloud since the boundaries of a pathway are not sharply defined, among others because many of the pathways and processes connect to form a network (Amit et al., 2007; Caron et al., 2010; Kleensang et al., 2014).

This paradigm shift, which is represented in Figure 26.1, offers two crucial advantages. First, a holistic view of cancer cell signalling can be obtained, which is more appropriate for describing the complexity inherent to cancer biology as opposed to the reductionist, gene-by-gene approach that is currently still prevailing. The advantage of a holistic approach can be understood from the concept of mutually exclusive gene mutations. Imagine that a tumour's development is dependent of the disturbance of a particular pathway, then an individual patient will almost never exhibit more than one driver mutation in the genes belonging to this pathway, since extra mutations in this pathway will not give an extra selective advantage to the tumour cell. Consequently, different patients in which the same pathway is disturbed can all have mutations in different genes of this pathway, complicating the study of tumour biology based on a gene-by-gene approach. Hence, it is clear that cancer genes should be analysed in relation to the pathways they are involved in, rather than as isolated events. A second notable advantage relates to the fact that important cancer cell characteristics like robustness and plasticity can only be explained as emerging or network-level properties, which is explained in the next paragraph.

Fitness, robustness, and evolvability

Systems biology has provided a new perspective on signal transduction with particular implications for cancer biology. First, it is

Table 26.1 Overview of pathways and processes significantly altered in 12 common cancer types according to TCGA

#	Biological concept	Genes	Cancer type
1	*Transcriptional regulation*	*VHL, GATA3, TSHZ3, EP300, CTCF, TAF1, TSHZ2, RUNX1, MECOM, TBX3, SIN3A, WT1, EIF4A2, FOXA1, PHF6, CBFB, SOX9, ELF3, VEZF1, CEBPA, FOXA2*	BLCA, BRCA, HNSC, KIRC, LUAD, AML, LUSC, UCEC
2	*Histone modifier*	*MLL2, MLL3, ARID1A, PBRM1, SETD2, NSD1, SETBP1, KDM5C, KDM6A, MLL4, ARID5B, ASXL1, EZH2, HIST1H1C, H3F3C, HIST1H2BD*	BLCA, BRCA, COAD/READ, HNSC, KIRC, LUAD, LUSC, UCEC
3	*Genome integrity*	*TP53, ATM, ATRX, BRCA2, ATR, STAG2, BAP1, BRCA1, SMC1A, SMC3, CHEK2, RAD21, ERCC2*	BLCA, BRCA, COAD/READ, GBM, HNSC, KIRC, AML, LUAD, LUSC, OV, UCEC
4	*RTK signalling*	*EGFR, FLT3, EPHA3, ERBB4, PDGFRA, EPHB6, FGFR2, KIT, FGFR3*	BLCA, GBM, AML, LUAD, LUSC, UCEC
5	*Cell cycle*	*CDKN2A, RB1, CDK12, CDKN1B, CCND1, CDKN1A, CDKN2C*	BLCA, GBM, LUAD, LUSC, UCEC
6	*MAPK signalling*	*KRAS, NF1, MAP3K1, BRAF, NRAS, MAP2K4, MAPK8IP1*	BLCA, BRCA, COAD/READ, GBM, AML, LUAD, LUSC, UCEC
7	*PI3K signalling*	*PIK3CA, PTEN, PIK3R1, TLR4, PIK3CG, AKT1*	BLCA, BRCA, COAD/READ, GBM, HNSC, LUAD, LUSC, UCEC
8	*TGFβ signalling*	*SMAD4, TGFBR2, ACVR1B, SMAD2, ACVR2A*	COAD/READ
9	*WNT/β-catenin signalling*	*APC, CTNNB1, AXIN2, TBL1XR1, SOX17*	COAD/READ, LUAD, UCEC
10	*Proteolysis*	*FBXW7, KEAP1, SPOP*	BLCA, COAD/READ, HNSC, LUAD, LUSC, UCEC

BLCA, bladder urothelial carcinoma; BRCA, breast adenocarcinoma; COAD; colon adenocarcinoma; READ, rectal adenocarcinoma; HNSC, head and neck squamous cell carcinoma; GBM, glioblastoma multiforme; KIRC, kidney renal clear cell carcinoma; LUAD, lung adenocarcinoma; LUSC, lung squamous cell carcinoma; AML, acute myeloid leukaemia; OV, ovarian serous carcinoma; UCEC, uterine corpus endometrial carcinoma.

Source: data from Kandoth C et al. Mutational landscape and significance across 12 major cancer types, *Nature*, Volume 502, Issue 7471, pp. 333–9, Copyright © 2013 Springer Nature. DOI: 10.1038/nature12634.

important to realize that the eukaryotic signalling network evolved over time from simple, linear, and sparsely connected cascades into layered and richly interconnected structures. This evolutionary process has conveyed two key concepts of eukaryotic signalling that are also crucial for cancer biology: evolvability and robustness. Robustness is inherently associated with a network topology enriched for highly connected proteins. These so-called hub proteins and their interconnected neighbourhoods constitute diverse signalling modules that are dominated by multiple network motifs and feedback loops, stabilizing the flow of signals through the hub proteins allowing output reproducibility despite input variation and stochastic signal processing. In addition, signalling modules are redundant or degenerate, enabling other signalling modules to compensate for their eventual loss of function (i.e. dynamic switching), thus allowing cell plasticity. The consequence of a robust signal transduction network resides in the fact that intrinsic (e.g. gene mutations) and extrinsic (e.g. therapeutic intervention) perturbations can be tolerated, and thus the functionality of the network will not be substantially altered. Moreover, the fact that network perturbations can be tolerated is fundamental to the concept of evolvability, which is defined as the ability of a cell population to generate phenotypic variability. Indeed, due to a networks' tolerance towards molecular alterations, genetic alterations are allowed to accumulate in a cell population. Thus, robustness augments evolvability. The resulting phenotypic variability is a crucial feature that allows the survival of cell populations through adaptation to changing micro-environments.

When applying these principles to cancer biology, it is important to realize that cancer cells are characterized by extremely rewired signal transduction networks due to multiple molecular alterations (Leiserson et al., 2014), including mutational activation or inhibition of a small group of signalling components and gene expression changes that strengthen positive feedback loops and curtail negative feedback loops. Combined, these events amount to powerful and sustained signalling by locking modules in an active state, which guarantees a sustained flow of information. As discussed earlier, such robust signal transduction networks, and particularly those characterized by many redundant or degenerate signalling modules (i.e. high network entropy), are intrinsically highly resistant to therapeutic interventions. In addition, the continued action of mutational processes will lead to the acquisition of novel and tolerated gene mutations, thus constituting the basis for phenotypic heterogeneity on which natural selection can act allowing the outgrowth of therapy resistant or metastasis-prone cancer cell clones. Given the fact that cancer cells rely on robust signal transduction networks, the key towards successful cancer treatment resides in identifying a networks' fragile sites. As signal transduction networks have been trained by evolution to tolerate single perturbations, the key towards successful cancer therapy may reside in targeting a networks' fragile sites simultaneously by either using combination therapies or by exploiting synthetic lethality (Amit et al., 2007).

Understanding inflammatory breast cancer: A textbook systems biology example

Throughout this chapter, a practical example on how systems biology approaches can be applied to unravel the molecular underpinnings of tumour biology will be provided. To this end, we will utilize gene expression data obtained on a series of samples from patients with and without inflammatory breast cancer (IBC), an aggressive and highly metastatic type of breast cancer, the molecular underpinnings of which are only poorly characterized. By consequence, options for targeted treatment are limited. Hence, we apply several systems biology approaches to refine the current knowledge, detect relevant signal transduction processes and identify putative targets for therapy.

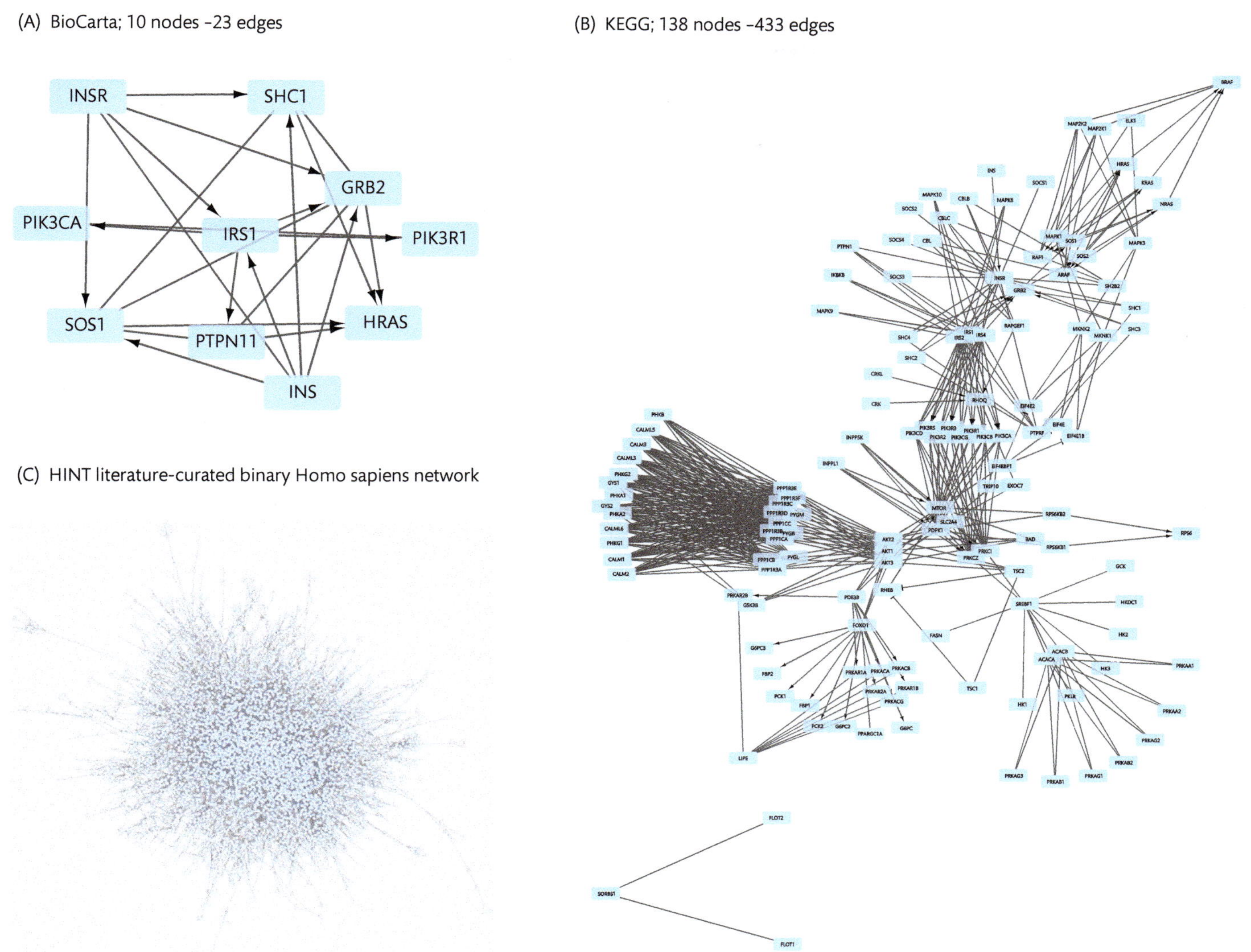

Fig. 26.1 Comparison of classical insulin signalling pathway representations (A and B) to a global signal transduction network (C), exemplifies the pathway paradigm shift away from linear diagrams and towards warm and fuzzy clouds (i.e. typical hairball structure). Comparison of the insulin signalling pathway diagrams extracted from two different knowledge bases (A: BioCarta; and B: KEGG) reveals that the number of pathway nodes and edges as well as the pathway topologies are different. For example, the BioCarta diagram is composed of one connected component whereas the KEGG diagram reveals two connected components. Also, the neighbourhood connectivity is more heterogeneous in the KEGG diagram as compared to the BioCarta diagram.

Source: data from Kandoth C et al. Mutational landscape and significance across 12 major cancer types, Nature, Volume 502, Issue 7471, pp. 333–9, Copyright © 2013 Springer Nature. DOI: 10.1038/nature12634.

Reference gene set libraries

An important aspect of systems biology relates to the concept of reference gene sets, which are lists of genes with a common biological characteristic. Up to date, many gene sets are available and most of them are organized in distinct knowledge bases that are accessible in the public domain. A comprehensive overview is given by the PathGuide website (https://www.pathguide.org). These knowledge bases, also described in literature as gene set libraries, can be classified according to different molecular concepts. In this chapter, a slightly modified version of the classification system described by Chen and colleagues (Chen et al., 2013) is adopted and includes following concept classes: transcription, signal transduction, gene ontology, disease and drugs, and cell types. A restrictive overview of available knowledge bases is given in Table 26.2.

In the transcription category, genes are grouped based on the transcription factors, non-coding RNAs, or histone modifications that regulate their transcription. Gene set assembly depends on either theoretical predictions (i.e. target gene prediction) or experimental evidence (e.g. sequencing of immunoprecipitated chromatin or perturbation experiments). In the signal transduction category, gene sets represent pathways, signalling complexes, and kinase substrates obtained from knowledge bases or experimental observations. In addition, gene sets can be constructed using protein–protein

Table 26.2 Classification of different knowledge bases according to different molecular concepts

Category	Gene set library	Description	URL
Transcription	Chip-X database	Manually extracted transcription factor and target gene interactions	http://amp.pharm.mssm.edu/lib/chea.jsp
	TRANScription FACTor database (TRANSFAC)	Eukaryotic transcription factors, their binding sites, and binding profiles	http://www.gene-regulation.com/
	JASPAR (alternative: UCSC Genome Browser)	Transcription factor binding profiles	http://jaspar.genereg.net
	ENCyclopedia of DNA Elements (ENCODE)	Functional elements at DNA, RNA, and protein level	http://genome.ucsc.edu/ENCODE
	Epigenomics Roadmap	Human epigenomic data including DNA methylation, histone modifications, and chromatin accessibility	http://www.roadmapepigenomics.org/
	TargetScan (alternative: PicTar, MicroCosm, Miranda)	Predicted microRNA target genes	http://www.targetscan.org
Signal Transduction	Kyoto Encyclopedia of Genes and Genomes (KEGG)	Manually drawn pathway maps representing experimental knowledge on metabolism and signal transduction	http://www.kegg.jp/kegg/download/
	WikiPathways	Pathway diagrams curated by the scientific community	http://www.wikipathways.org
	Reactome	Experimentally validated reactions between nucleic acids, proteins, complexes, and small molecules	http://www.reactome.org/download
	NCI-Nature Pathway Interaction Database (PID)	Molecular interactions and reactions, with a particular focus on cancer research and treatment	http://pid.nci.nih.gov/
	Nuclear Receptor Signaling Atlas (NURSA)	Complexes of nuclear receptor coregulators	https://www.nursa.org/
	Comprehensive Resource of Mammalian Protein Complex (CORUM)	Manually annotated protein complexes from mammalian organisms	http://mips.helmholtz-muenchen.de/corum/
	Human Dephosphorylation Database (DEPOD)	Manually curated interactions between phosphatases and their substrates	http://www.koehn.embl.de/depod/
	Kinase Enrichment Analysis (KEA)	Manually curated interactions between kinases and their substrates	http://amp.pharm.mssm.edu/lib/kea.jsp
Gene Ontology	Gene Ontology–Biological Process	Genes annotated according to their species-independent associated biological processes	http://www.geneontology.org
	Gene Ontology–Molecular Function	Genes annotated according to their species-independent associated molecular functions	http://www.geneontology.org
	Gene Ontology–Cellular Component	Genes annotated according to their species-independent associated cellular components	http://www.geneontology.org
Diseases and Drugs	GeneSignature Database	Gene sets curated from publications in PubMed	http://www.genesigdb.org/
	Online Mendelian Inheritance in Man (OMIM)	Human genes and their associated genetic phenotypes	https://www.ncbi.nlm.nih.gov/omim
	Achilles Fitness	Gene sets related to genetic vulnerabilities	http://www.broadinstitute.org/achilles
	VirusMINT	Interactions between human and viral proteins	http://mint.bio.uniroma2.it
	Library of Integrated Cellular Signatures (LINCS), Connectivity Map (CMAP)	Transcriptional expression data from cultured human cells treated with bioactive small moleucles	http://www.lincscloud.org/
	Human Phenotype Ontology	Genes annotated according to their associated phenotypic abnormalities	http://www.human-phenotype- ontology.org/
Cell Types	Human Gene Atlas	Tissue specific expression profiles	http://biogps.org/
	Allen Brain Atlas	Brain specific expression profiles	http://www.brain-map.org/
	Tissue sample gene expression profiles (GTex)	Relationships between eQTLs and gene expression per tissue type	http://www.gtexportal.org/
	Embryonic Stem Cell Atlas and Pluripotency Evidence (ESCAPE)	Protein-DNA interaction in embryonic stem and pluripotent cells	http://www.maayanlab.net/ESCAPE/

Source: data from Chen EY et al. Enrichr: interactive and collaborative HTML5 gene list enrichment analysis tool, *BMC Bioinformatics*, Volume 14: 128, 201314:128, Copyright © Chen et al.; licensee BioMed Central Ltd. 2013. https://doi.org/10.1186/1471-2105-14-128

interaction (PPI) data (e.g. genes that interact with transcription factors or key signalling/hub proteins), allowing dominant signalling components to be identified. The ontology category classifies genes based on their biological function. A well-known example is gene ontology (GO) that describes genes in terms of their associated biological processes, cellular components, and molecular functions. In the disease and drugs category, genes are organized according to their relation to drug responses, diseases, or phenotypical abnormalities. In the cell type category, gene sets are constructed based on cell type specific expression patterns. Gene sets not included in these categories classify genes based on chromosomal location, presence of structural protein domains and metabolites interaction patterns.

Gene set libraries can be used to predict novel gene functions or to build functional association networks allowing the discovery of distal relationships between biological processes (*vide infra*; see Clark et al., 2012; Dannenfelser et al., 2012; Chen et al., 2013). However, the major application of these reference gene sets is to translate lists of genes that characterize a certain condition (e.g. breast cancer) into biological meaningful data. The underlying principle is to evaluate whether or not a gene list of interest significantly overlaps (i.e. the intersection is significantly greater than expected by chance) with any of the reference gene sets present in a knowledge base. Significant overlaps indicate that the molecular concept reflected by the reference gene set may be relevant to the condition under investigation. Importantly, by analysing a gene list from the perspective of different gene set library categories (*vide supra)*, a multifaceted picture of the underlying biology can be shaped. For example, gene ontology analysis may reveal cell proliferation as a central theme in breast cancer biology, whereas target gene enrichment analysis may identify the driving transcription factor. Therefore, it is advisable to evaluate a gene list according to multiple gene set categories.

However, this recommendation also comes with a drawback. The strategy of interpreting lists of relevant genes in the context of reference gene sets is only useful when gene lists can be projected onto a lower dimensional structure that is determined by a limited number of gene groupings, allowing a dimensionality reduction. However, the amount of gene sets that is currently available in different knowledge bases far exceeds the number of genes present in the human genome, which complicates biological interpretation. This complexity is further elaborated by the fact molecular concepts may be represented by distinct gene sets in different knowledge bases (Doderer et al., 2012). As an example, one could compare the insulin signalling pathways from the KEGG and BioCarta knowledge bases (**Fig. 26.1**). Differences with respect to the sizes of the gene sets and the network topologies are clearly visible. The BioCarta pathway includes fewer nodes and emphasizes on PI3K-signalling, whereas the KEGG pathway includes additional proteins that are involved in, among others, MAPK- and PTP1B-signalling.

A final aspect that should be mentioned relates to the fact that many gene sets reveal significant overlaps (**Fig. 26.2**), which can be considered both a blessing and a curse. On the one hand, these interactions provide a theoretical framework for gene set consolidation, allowing a significant dimensionality reduction. For example, the gene sets termed 'Synthesis and Degradation of Ketone Bodies' from WikiPathways and 'Ketone body metabolism' from Pathway Commons contain exactly the same five genes and hence should be merged into one pathway (Doderer et al., 2012). In addition, consolidation of distinct gene sets associated with a similar molecular concept has the supplemental benefit that differences across multiple knowledge bases can be circumvented, leading to more robust information. On the other hand, overlapping gene sets have the tendency of generating confounded results. Indeed, lists of differentially expressed genes can be characterized by multiple overlapping gene sets, a subset of which is truly relevant. Therefore, it can be argued that interacting gene sets severely impede specificity and thus data interpretability. In order to enhance specificity, gene set consolidation is again recommended. Thus, gene set consolidation can assist both in reducing the number of evaluable gene sets, thereby enhancing statistical power and sensitivity, and in finding true positive results leading ameliorated specificity. Distinct methods for gene set consolidation are proposed by Doderer and colleagues (Doderer et al., 2012).

Network theory

The far most important limitation of representing pathways and processes as reference gene sets relates to the fact that these gene sets are considered to be applicable in all possible conditions. However, this assumption is flawed, since pathways are increasingly considered as modular structures, implying that certain signalling components can be replaced without affecting pathway functionality (Amit et al., 2007). Hence, gene-to-pathway membership is graded and depends on the molecular background (e.g. different cancer cell lines) against which a pathway operates. Unfortunately, due to the fact that this specific biological knowledge is limited, pathway dynamics are often not fully recapitulated in the gene set libraries, and thus gene sets may not fully represent a certain molecular concept. Moreover, when dealing with diseases such as cancer that are characterized by extremely rewired signal transduction networks (Hofree et al., 2013), the content of gene sets should even be increasingly questioned. Another drawback of testing gene sets is that the pathways and processes they represent are considered as isolated from the global signal transduction network, whereas in fact biology is often shaped by cooperation between different molecular systems.

In order to compensate for these limitations, an alternative approach for extracting biological knowledge from high-throughput '-omics' studies that is now progressively being adopted in systems biology relies on gene networks. A gene network can be regarded as a web composed of nodes and edges, where nodes represent a biological entity (i.e. generally genes or gene products) and edges represent either physical, functional or influence interactions (*vide infra*). In fact, many pathway representations in various knowledge bases (e.g. KEGG) can be considered as small physical interaction networks conditional on a predefined set of genes. However, a defining feature of the network analysis strategies discussed in this section resides in the fact that all possible interactions between genes of interest are considered. This may result in the discovery of network modules, defined as a subset of genes that regulate each other through multiple (indirect) interactions, but with few connections to other genes outside the module. Such network modules represent functional signalling entities and are often related to relevant signal transduction pathways. As network modules constitute the central entities of robust and evolvable signal transduction networks, their identification and characterization are of extreme biological importance (Ma'ayan, 2011; Doncheva et al., 2012). In addition, network modules are important to reveal previously unrecognized associations between genes, between

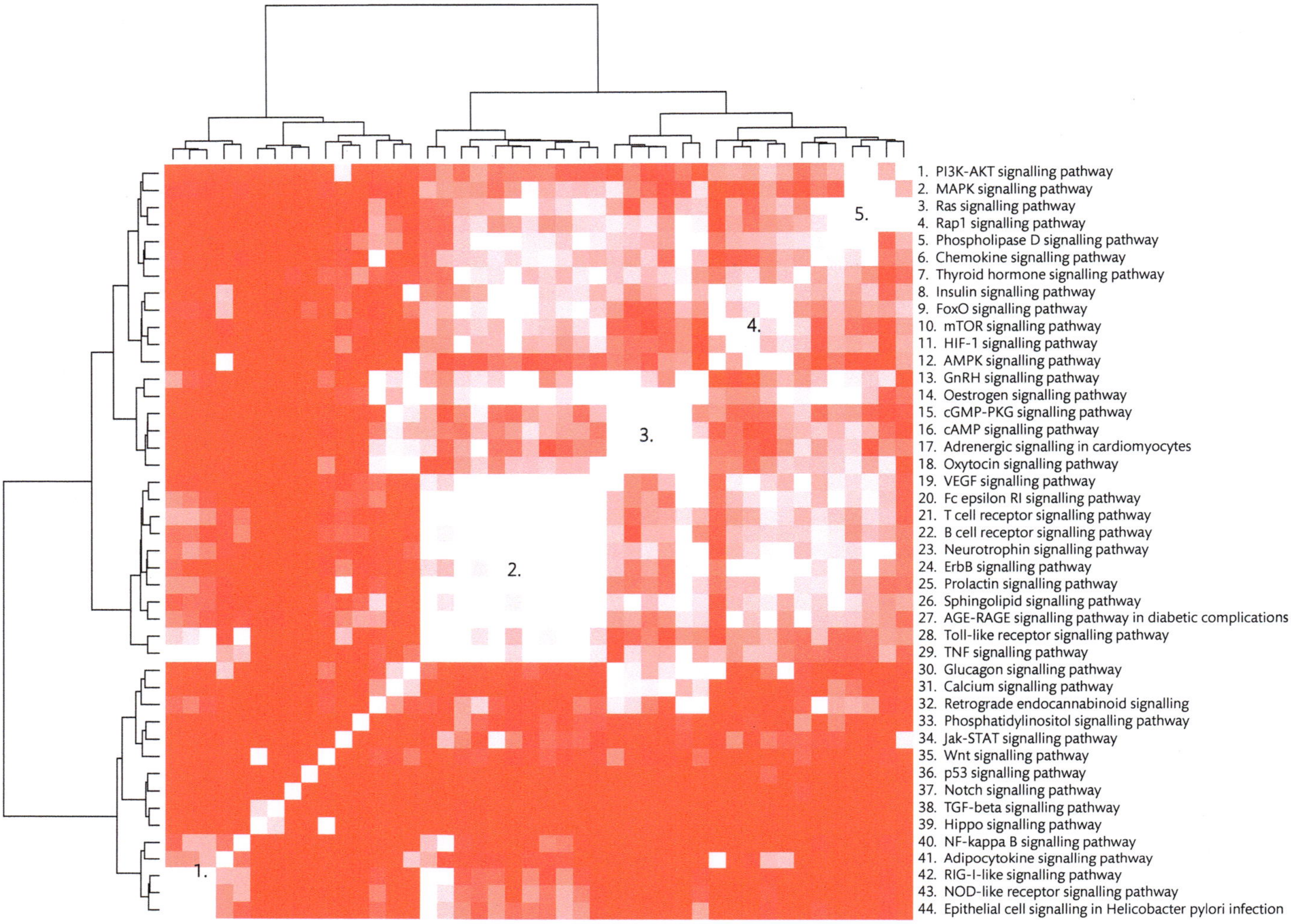

Fig. 26.2 Overlaps between multiple gene sets related to KEGG signalling pathways are represented in heatmap format. The Jaccard index, which is defined as the ratio of the number of overlapping genes (i.e. intersection) and the total number of genes in two gene sets (i.e. union), was calculated to quantify gene set overlaps. Unsupervised hierarchical clustering of the resulting dissimilarity matrix was performed using Ward linkage. The Jaccard index is colour-coded using a red-to-white colour scale, with red indicating small gene set overlaps and white indicating large gene set overlaps. Inspection of the cluster dendrogram and the heatmap reveals five clusters with strongly overlapping gene sets. Gene set analysis should take into account these gene set overlaps to reduce the chance of reporting false-positive results.

Source: data from Chen EY et al. Enrichr: interactive and collaborative HTML5 gene list enrichment analysis tool, BMC Bioinformatics, Volume 14: 128, 201314:128, Copyright © Chen et al.; licensee BioMed Central Ltd. 2013. https://doi.org/10.1186/1471-2105-14-128.

genes and pathways, and between pathways (i.e. cross-talk). Thus, instead of interpreting biological mechanisms solely in the context of gene sets and the limited information they represent, network analysis provides a more holistic view to signal transduction mechanisms.

Network representation: Graph models and the adjacency matrix

Networks can be presented in a variety of ways (Table 26.3). Most signal transduction pathways (e.g. KEGG and BioCarta), metabolic pathways and PPI networks adopt the 'single-type' graph model for pathway representation, in which nodes represent a single biological component (e.g. genes, proteins). Special consideration should be given to metabolic networks, which can be represented by two single-type graph models: A. the chemical networks in which nodes are metabolites and edges are enzymes that catalyse the chemical reactions; or B. the protein networks in which the representation is reversed. In contrast to the single-type graph model, the 'multitype' model allows multiple nodes types. Multitype graph models are more complex, but often capture more pathway characteristics. Bipartite graphs are a particular case of multitype models, which contain two types of nodes and allow connection only between nodes of different types. Bipartite graphs are adopted by Reactome and NCI knowledge bases, in which biological component nodes are connected to interaction nodes that define the type of interaction between the connected component nodes. The edges specify the component's type of contribution to the interaction (e.g. positive or negative regulation; Mitrea et al., 2013). Importantly, network graph models can either be undirected or directed (i.e. when the flow of information in the network is indicated). Directed acyclic graphs (DAGs) constitute a special case of directed networks in which no sequence of edges will eventually loop back to its starting position. Hence, DAGs contain no cycles and therefore are said to be acyclic.

In addition to the classical network representations, a network can be formulated in a more mathematical sense. Each network can

Table 26.3 Networks representations: graphs, adjacency matrices, and node degree distributions

		Undirected	Directed
$G\{V, E\}$	A set of vertices V connected by edges E.	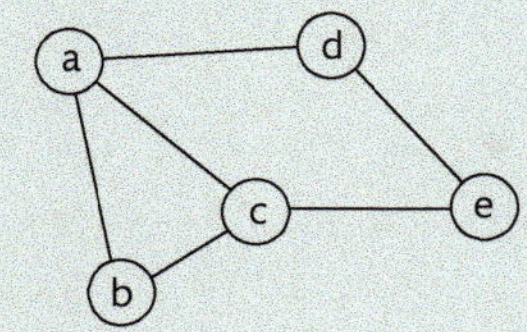	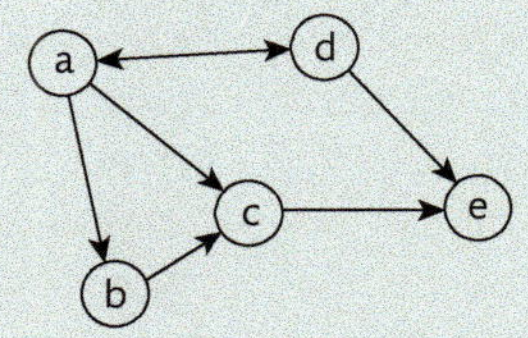
A	The adjacency matrix representation of a graph. Rows and columns are vertices, and a 1 indicates a connection between two vertices.	$\begin{bmatrix} 0 & 1 & 1 & 1 & 0 \\ 1 & 0 & 1 & 0 & 0 \\ 1 & 1 & 0 & 0 & 1 \\ 1 & 0 & 0 & 0 & 1 \\ 0 & 0 & 1 & 1 & 0 \end{bmatrix}$	$\begin{bmatrix} 0 & 1 & 1 & 1 & 0 \\ 0 & 0 & 1 & 0 & 0 \\ 0 & 0 & 0 & 0 & 1 \\ 1 & 0 & 0 & 0 & 1 \\ 0 & 0 & 0 & 0 & 0 \end{bmatrix}$
D, D_{IN}, D_{OUT}	Degree, indegree and out-degree. Respectively the total number, the number of incoming and the number of outgoing connections of a vertex.	$D = \begin{bmatrix} 3 \\ 2 \\ 3 \\ 2 \\ 2 \end{bmatrix} D_{IN} = \begin{bmatrix} 3 \\ 2 \\ 3 \\ 2 \\ 2 \end{bmatrix} D_{OUT} = \begin{bmatrix} 3 \\ 2 \\ 3 \\ 2 \\ 2 \end{bmatrix}$	$D = \begin{bmatrix} 3 \\ 2 \\ 3 \\ 2 \\ 2 \end{bmatrix} D_{IN} = \begin{bmatrix} 3 \\ 2 \\ 3 \\ 2 \\ 2 \end{bmatrix} D_{OUT} = \begin{bmatrix} 3 \\ 1 \\ 1 \\ 1 \\ 0 \end{bmatrix}$

be expressed in terms of an adjacency matrix, which is essentially a symmetric matrix where each matrix element represents the connection strength between two nodes. The connection strength can be represented by a binary variable, indicating whether nodes are connected or not (i.e. respectively 1 or 0). This is often the case when dealing with physical or functional interaction networks in which the connection is based on prior knowledge and thus its strength cannot be easily quantified. Alternatively, the connection strength can be expressed as a continuous variable (e.g. correlation coefficient) that reflects the extent to which two genes behave similarly. Under these conditions, a network is weighted. Influence networks, in which the connection strength is directly inferred from the data and represents the mutual influence between two genes on a continuous scale, are examples of weighted networks. Of note, continuous variables can be dichotomized, transforming a weighted network into an unweighted network.

Network topology

The complete collection of all nodes and edges in a network defines the network topology, which can provide insight into the organizational principles of the network. Network topology can be described using a series of indices (**Table 26.3** and **Table 26.4**), classified as

Table 26.4 A short definition and example calculation of frequently-used centrality measures

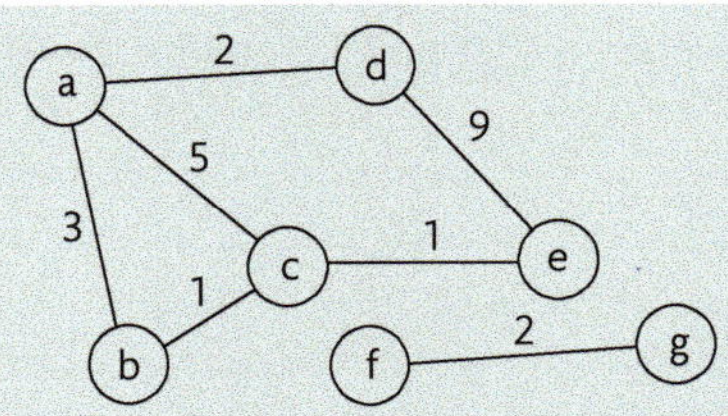

Connected component	A subgraph of an undirected graph wherein each vertex can be reached from any other vertex in the subgraph, and wherein no vertex that is not a member of the subgraph	This graph contains to connected components, *{a, b, c, d e}* and *{f, g}*
Closeness	The average length of the shortest path between a vertex and all other vertices in the graph	The closeness for node c is calculated as (4+1+6+1)/4=3
Betweenness	The fraction of the shortest paths between two other vertices in the graph (the connected component) that pass through a vertex of interest.	The betweenness of node *c* = 0.5, indicating that half of the shortest paths not starting or ending in vertex *c* pass through vertext *c*
Local clustering coefficient	The number of triangles a vertex v is part of divided by the number of triplets (all combinations of two other nodes that are connected to *v*) vertex v is part of.	The local clustering coefficient for node *c* = 1/3 (one triangle *bca* and three triplets *bca*, and *ace*)
Global clustering coefficient	3x the number of triangles in the graph divided by the total number of open triplets	for the largest connected component of the graph, the global clustering coefficient = 3/9

either node-, neighbourhood-, or edge-specific. Node-specific parameters reflect the importance of a node relative to all other nodes in a network. For example, the node degree or connectivity amounts to the number of connections a node has in a network (Table 26.3). Nodes with many connections are often referred to as hubs, conditional on the fact that a network exhibits a scale-free topology. The latter is the case when a small number of nodes have many connections whereas the remaining nodes are sparsely connected, a condition regularly encountered in mammalian signal transduction networks. Another important network topology measure is the path length, which is defined as the number of edges connecting two nodes. The path length of two directly connected nodes in a network is essentially one and the shortest path or distance between two nodes is the smallest number of edges connecting two nodes. The importance of the shortest path length between two nodes is reflected by its incorporation in the definition of several centrality measures (Table 26.4) that reflect the amount of control exerted by a node on a network. Betweenness centrality relates to the amount of control a node exerts over the interactions between other nodes in the network. Closeness centrality reflects how quickly information from a given node in the network spreads to other reachable nodes. Path length can be used to define edge-specific measures, such as the edge betweenness centrality, which is the number of shortest paths that go through an edge among all possible shortest paths between all the pairs of nodes. Neighbourhood-specific parameters reflect how well the direct interaction partners of a node are connected. For example, the global and local clustering coefficients of a node are defined in function of the number of triangles (i.e. revolving path among three nodes) and the number of triples (i.e. non-revolving path among three nodes) the node is involved in (Table 26.4). A node with high clustering coefficient is involved in a higher number of triangles and thus its neighbourhood is more densely connected. Nodes with densely connected neighbourhoods are likely part of a network module.

Identification of biologically relevant network modules can be performed using various clustering techniques. Therefore, a dissimilarity-transformed version of the adjacency matrix (i.e. every matrix element is subtracted from 1) can be used as input. Network clustering algorithms can be applied on both nodes (e.g. unsupervised hierarchical clustering, density peak clustering) and edges (i.e. link community clustering, spectral clustering). The advantage of clustering edges resides in the fact that it allows nodes to be part of multiple modules, revealing overlapping and nested network structures, which is inherent to densely connected networks (Newman, 2006; Kalinka and Tomancak, 2011).

Biological networks in systems biology

Physical interaction networks

Networks can represent different types of physical interactions that are contained in various publicly available repositories: gene regulatory interactions, metabolic interactions, protein-compound interactions, and PPIs. For an exhaustive overview, the website https://www.pathguide.org can be explored. Gene regulatory interactions relate to interactions involving nucleic acids that affect the expression of genes or gene products, including binding of transcription factors or non-coding RNAs to their target sequences. Metabolic interactions contain all interactions that take part in cellular metabolism, including interactions with cofactors. Protein-compound interactions involve interactions between gene products and any other compounds exogenous to the system under investigation (e.g. interactions between proteins and drugs).

However, the most generally investigated interaction type in systems biology involves PPIs. Many major biological processes such as immunity (e.g. antigen–antibody interaction) and signal transduction (e.g. ligand-receptor interaction), are mediated through PPIs. A PPI is defined as a physical interaction that is intentional (i.e. the result of specific events) and non-generic (i.e. the interaction should have a specific purpose distinct from totally generic functions such as protein production or degradation). It should be noted that physical and functional interactions (*vide infra*), are not to be confused. Proteins in a complex often exhibit functional interactions without physical interactions. The latter is the only interaction type that is considered when dealing with PPIs (Las Rivas and Fontanillo, 2010). Physical interactions between proteins can be inferred using either experimental approaches (e.g. yeast2hybird systems, co-immunoprecipitation, tandem affinity purification coupled with mass spectrometry) or computational approaches. Methods for computationally predicting PPIs often involve tracing evolutionary conserved patterns in protein sequences, protein structures or genome structures (Shoemaker and Panchenko, 2007). Although PPIs presently constitute an essential component in systems biology, some limitations have to be acknowledged. First, the conditions under which PPI are evaluated do not reflect normal cellular physiology. Hence, experimentally observed PPIs may in fact be false-positive results. Therefore, more reliable PPI are needed, possibly by incorporating 3D-structural information about the interacting proteins. A second issue that needs to be noted relates to the low coverage of PPIs across individual databases. For example, the data on human PPIs coming from six different primary databases show a small overlap. In contrast, the number of PPIs exclusively reported by each database is large. This observation reflects the low specificity in predicting PPIs and it implies that results obtained through the analysis of PPIs will likely be dependent on the evaluated database. A final aspect that needs to be taken into account when dealing with PPIs is the biological context. Not all possible interactions will occur in any cell at any time. Instead, interactions depend on cell type, cell cycle phase, developmental stage, environmental conditions, protein modifications, presence of cofactors, and presence of other binding partners. Therefore, development of analysis strategies that take the biological context, in the broadest possible sense, into consideration is mandatory (Las Rivas and Fontanillo, 2010).

Influence networks

An alternative approach for constructing gene networks involves inferring or reverse engineering signalling mechanisms directly from the data. In these data-driven networks, interactions between nodes not necessarily imply a physical interaction, but can also refer to indirect regulations via proteins, metabolites, or non-coding RNAs that have not been measured directly. Thus, connected genes are influenced by each other's behaviour, and therefore, the networks are called influence networks. Since influence networks are constructed directly from the data and do not rely on prior knowledge, network inference algorithms are mandatory. These can be classified as deterministic or probabilistic, indicating whether or not they rely on the estimation of conditional probabilities.

Deterministic approaches include clustering algorithms in which genes are grouped based on similarities in their expression profiles. Coexpressed or clustered genes have a good probability of being functionally related. In a classical network representation, clustered genes are all connected and recover an undirected graph. Alternatively, gene interactions can be estimated from the data using ordinary differential equations (ODEs) that relate expression changes to gene-gene interactions and external perturbations (e.g. treatment, disease state). Using ODEs, a directed graph model is inferred from the data for each gene separately. An important advantage of ODEs is that once the interaction parameters for all genes are known, the equation can be used to predict the behaviour of the network in different conditions.

Probabilistic methods for reverse engineering gene networks include Bayesian network modelling and information theoretic approaches. In a Bayesian network, the relationships between genes are encoded in the structure of a DAG and represent conditional probabilities. In other words, genes will be connected when the observations made with respect to one gene have a high (posterior) probability given the observations made on its direct regulators. Many network models can be constructed and finding the optimal model that best describes the gene expression data is essential. With theoretic information approaches (e.g. ARACNE), the mutual information (MI) criterion is calculated which is a probability function that represents the extent to which the behaviours of two genes are non-randomly associated to each other. Edges between genes are defined by comparing the MI values to a predefined threshold and the resulting graphs are undirected. Alternative measures that can be used to assess the probability of non-random association between two genes include correlation and covariance coefficients.

Functional association networks

Functional association networks (FANs) are networks where edges indicate whether or not two genes have a shared functionality. Functionally can be defined in various ways, among others by evaluating membership patterns of genes across gene ontology terms or pathways (Dannenfelser et al., 2012). Alternatives to constructing FANs involve evaluating co-citation patterns or genetic interactions. Co-citation evaluates if two or more genes or proteins are reported in the same publication. A limitation of network construction based on co-citation is that literature is dominated by studies looking into a limited number of genes. Hence, novel associations will not be easily recovered. Nevertheless, co-citation networks are an attractive way of quickly creating an overview of available knowledge related to genes of interest. Genetic interactions, which are fundamental to the concept of epistasis, are defined by evaluating the phenotypic effect of combined gene perturbations. The phenotypic effect can be either more or less extreme than expected. In case of more extreme effects, the genetic interactions are termed synergistic or synthetic, with synthetic lethality being the ultimate example. Diminished effects can be caused by the perturbation of genes that operate in the same pathway. Alleviating effects may be due to gene permutations impairing a whole process, thereby masking additional permutations in the same process. Mutual exclusivity and cooccurrence of gene mutations are examples of genetic interactions that are often reported in cancer profiling studies.

Gene set analysis in systems biology

Extracting biological knowledge from lists of differentially expressed genes by comparing them to gene sets requires specialized algorithms, called gene set analysis algorithms. These consist of three generations (Khatri et al., 2012) that reflect gradual improvements with respect to their statistics and computational methods: overrepresentation (OR) analyses, functional class scoring (FCS) analyses and pathway topology-based (PT) analyses. Although gene set analysis methods are diverse in their statistical approaches and performed computations, a general workflow along which most of these algorithms operate can be defined (Table 26.5). Regardless of the method of choice, the input always consists of a two gene lists, containing either the genes of interest or a reference list containing genes associated with a molecular concept (e.g. NOTCH pathway). In both gene lists, gene identifiers (e.g. gene symbols or EntrezIDs) must be similar. Next, a statistical test is performed to evaluate the probability that the reference gene list is represented by the genes of interest. A key step in the setup of the analysis is the selection of null hypothesis, which can be either competitive (i.e. the genes in a pathway are differentially expressed as often as the rest of the genes) or self-contained (i.e. no genes in a pathway are differentially expressed). Several arguments have been raised in favour of self-contained tests: A. they represent an immediate generalization of single-gene tests; B. their null hypothesis has a clear biological interpretation; C. they make sense even if all genes on a chip are considered simultaneously, whereas a competitive test does not; and D. a competitive test assumes independence of genes, which we know is not the case (Emmert-Streib and Glazko, 2011; Bayerlová et al., 2015). On the other hand, self-contained tests have been criticized for being too powerful and thus yielding too many significant gene sets (Bayerlová et al., 2015). It should be noted that the primary reason for discrepant results obtained by alternative gene set analysis methods relates mainly to the type and formulation of the null hypothesis that is being tested (Glazko and Emmert-Streib, 2009). However, different null hypotheses may also project on distinct, but complementary aspects of the same data, meaning that discrepant results may in fact offer complementary insights. Therefore, using a combination of different tests is advisable. Following statistical assessment, the results are presented as a ranked list, ordered by confidence values such as false discovery rate corrected p-values (García-Campos et al., 2015). Consolidation methods can be applied to exclude confounded, false-positive results.

A key methodological difference that divides all gene set analysis methods is the use of univariate or multivariate statistics. Univariate analyses account for one variable at a time, while multivariate analyses consider multiple variables simultaneously. Intuitively it is expected that multivariate analyses, testing the joint distribution of genes and accounting for interdependencies, are of greater statistical power than univariate analyses. However, benchmarking performed on simulated and biological data showed that both approaches have similar statistical power at less conservative significance cut-offs, whereas univariate approaches are in fact more powerful at more stringent significance levels (Glazko and Emmert-Streib, 2009).

Table 26.5 A practical example of gene set enrichment analysis

You performed an analysis that resulted in the selection of nine genes. You are curious if these genes are participating in similar biological processes, especially NOTCH1 and TGF-β signalling. In a public repository (MSigDB), you can find curated sets of genes that have previously been linked to the biological functions you are interested in. The question that needs to be answered now is: Does your gene set contain more genes participating in these processes than you would expect purely by chance?

Set of genes found by your analysis	*DTXI, ID1, CCND1, SAP30, NOTCH1, NOTCH2, PRKCA, WNT2, RBX1*	#9
Set of genes that are known to participate in NOTGH signalling	*APH1A, ARRB1, CCND1, CUL1, DLL1, DTX1, DTX2, DTX4, FBXW11, FZD1, FZD5, FZD7, HES1, HEYL, JAG1, KAT2A, LFNG, MAML2, NOTCH1, NOTGH2, NOTGH3, PPARD, PRKCA, PSEN2, PSENEN, RBX1, SAP30, SKP1, ST3GAL6, TGF7L2, WNT2, WNT5A*	#32
Set of genes that are known to participate in TGF-β signalling	*ACVR1, APC, ARID4B, BCAR3, BMP2, BMPR1A, BMPR2, CDH1, CDX9, CDKNIC, CTNNB1, ENG, FKBP1A, FNTA, FURIN, HDAC1, HIPK2, ID1, ID2, ID3, IFNGR2, JUNB, KLF10, LEFTY2, LTBP2, MAP3K7, NCOR2, NOG, PMEPA1, PPM1A, PPP1CA, PPP1R15A, RAB31, RHOA, SERPINE1, SKI, SKIL, SLC20A1, SMAD1, SMAD3, SMAD6, SMAD7, SMURF1, SMURF2, SPTBN1, TGFB1, TGFBR1, TGIF1, THBS1, TJP1, TRIM33, UBE2D3, WWTR1, XIAP*	#54

Is our gene set enriched for genes that are involved in NOTCH1 or TGF-β signalling? You find that eight of the genes in your gene set are found in NOTCH1 signalling, and one to TGF-*beta* signalling. Additionally, in total 86 different genes were annotated. Let's have a look at the relative frequency of genes belonging to a particular signalling pathway:

	Fraction of gene set	Fraction of universe
NOTCH1	8/9 = 0.89	32/86 = 0.37
TGF-β	1/9 = 0.11	54/86 = 0.63

Our gene set appears to be enriched for NOTCH1 signalling. You can calculate the probability of finding 8 NOTCH1-related genes out of 9 when 32 genes out 86 were annotated as related to NOTCH1 using the hypergeometric distribution. More specifically, you want to obtain the probability that you would find at least 8 genes related to NOTCH1 when you randomly select genes from the annotated universe. Using a statistical package or an online tool (e.g. http://stattrek.com/online-calculator/hypergeometric.aspx), you find

	p
NOTCH1	0.0013
TGF-β	0.999

suggesting that our gene set is indeed highly enriched for NOTCH1 signalling-related genes!

Overrepresentation analysis

Approaches for OR analysis were developed due to an immediate need for functional analysis of microarray gene expression data. OR analysis tools evaluate whether the overlap between a gene list of interest (e.g. differentially expressed or mutated genes) and a reference gene set is greater (or smaller) than expected by chance (**Table 26.5**). OR analysis always takes into an account the complete gene universe from which a gene list of interest is generated (e.g. all genes in the genome or all genes on a microarray chip) and therefore, the null hypothesis is essentially competitive. Statistical significance is assessed using the hypergeometric, chi-square, or binomial distribution. The major limitations associated with OR analysis-based methods are threefold. First, the different statistics used by OR analysis are independent of the measured changes, and thus all genes are treated equally. However, quantitative data like expression levels or fold changes can be useful in assigning different weights to input genes, as well as to the pathways they are involved in. Second, OR analysis typically uses only the most significant genes and discard the others. Marginally less significant genes are missed, leading to loss of potentially relevant information (e.g. when the majority of the marginally less significant genes are involved in one signal transduction pathway). Third, by treating all genes equally, OR analysis assumes independence between genes and ignores the correlation structure that constitutes the complex web of interactions between genes, leading to biased significance estimates.

Functional class scoring analysis

FCS approaches build on the hypothesis that weak but coordinated changes in sets of functionally related genes might be biologically relevant. FCS analysis adopts a general framework that consists of following three steps: A. a gene-level statistic (e.g. z-score, fold change) is computed using the molecular measurements from an experiment; B. the gene-level statistics for all genes in the reference gene set are aggregated into a single statistic, which can be uni- or multivariate, respectively ignoring or accounting for gene interdependence; and C. significance of the aggregated statistic is assessed by performing random gene permutations (i.e. self-contained null hypothesis) or random sample permutations (i.e. competitive null hypothesis), which both reflect the possibility that a molecular concept is represented by randomly generated data. Importantly, FCS methods address all three limitations of OR analysis-based approaches as they do not require an arbitrary threshold for classifying genes as relevant, they do consider all available molecular measurements to detect coordinated changes and they are accountable for gene interdependence. A limitation inherent to some FCS approaches is that during the gene ranking procedure, relative differences between the test statistics of all genes in a gene set are discarded (e.g. a gene changing expression by 20-fold is treated similarly to a gene changing expression by 2-fold).

Pathway topology-based analysis

The third generation of gene set analysis algorithms incorporates topological information about gene interaction patterns, which is the key discriminant with respect to earlier generation tests that consider only the number of relevant genes (i.e. OR analysis) or quantifiable changes (i.e. FCS). Topological knowledge is primarily relevant when dealing with signal transduction pathways, and therefore these methods are referred to as pathway topology (PT)-based approaches. For the majority of the PT-based tests, the computational aspects are similar to those used in FCS methods, with the exception that pathway topology is used to compute either gene- or pathway-level statistics. Calculation of the gene-level statistics is versatile and incorporates a number of factors such as expression levels, expression covariance measures, pathway centrality measures, or perturbation estimates. Perturbation estimates take expression changes, gene functions, interaction types and directions, and the efficiencies by which perturbations diffuse to downstream genes into account for all genes in the pathway. Gene-level statistics are aggregated into estimates of pathway activity, using linear functions (e.g. summation or averaging) or non-linear functions. Finally, statistical significance is evaluated using either competitive tests (i.e. random gene permutations) or self-contained tests (i.e. random sample permutations). A detailed review of possible methods for aggregation and statistical assessment is provided in Mitrea et al. (2013). It should be noted though that not all PT-based approaches adopt the scheme described earlier, and some return the node-statistics only (e.g. PARADIGM) whereas others provide pathway statistics without performing statistical inference (e.g. TopoGSA, EnrichNet).

The advantages of PT-based gene set analysis reside in the fact that both biologically relevant differences between components of a pathway (e.g. activators and inhibitors) and causal interactions within a pathway, either direct or indirect, are taken into consideration. In addition, by taking into account the fact that pathway topology may depend on the biological condition under investigation, a more precise estimation of pathway activity can be obtained. However, this implies that the true pathway topology in all cell types and under all conditions is known, which is rarely the case. A recent effort by Greene and colleagues to understand the molecular underpinnings of individual cell lineages and complex tissues exemplifies the crucial steps that need to be taken to unlock the full power of PT-based algorithms. Other limitations of PT-based methods include the limited ability to model spatiotemporal dynamics of a system (e.g. cellular compartmentalization and pathway kinetics) as well as the inability to consider interactions between pathways. The latter may be circumvented using network-based analyses, which will be covered in the next section. An additional drawback of PT-based approaches relates to the fact that consolidation methods that consider gene set overlap are more difficult to apply. This is explained by the fact that simply counting overlaps between gene sets denies the fact that genes play different roles in different pathways resulting in altered topological characteristics. Nevertheless, gene set overlap is an issue when performing PT-based gene set analyses, since two independent studies have shown that gene set overlaps have a detrimental effect particularly on the performance of PT-based methods (Donato et al., 2013).

Gene set analysis on inflammatory breast cancer expression data

The aforementioned gene set analysis strategies were applied onto a gene expression data set of samples from patients with IBC (N = 50) and with non-IBC breast tumour (N = 55). Using generalized linear models, we identified 192 differentially expressed genes related to the contrast of interest (i.e. IBC *vs.* non-IBC). In addition, fold changes were defined for 11,435 unique genes. Gene expression data were translated into molecular concepts defined according to the Reactome database. Prior to analysis, gene sets containing less than 10 genes were excluded, resulting in 816 evaluable pathways and processes. The 15 most significant molecular concepts identified using OR-, FCS- and PT-approaches are shown in **Figure 26.3**. Inspection of the data reveals major differences between the results. The OR analysis identifies concepts related to extracellular matrix, G-protein signalling, and biological oxidations, whereas the FCS-based analysis underscores the role of cell proliferation and immunity-related concepts in the biology of IBC. The primary reason for the observed discordance relates to the computational differences between both methods, as exemplified by the PD-1 signalling pathway (**Fig. 26.4**). The majority of the evaluated genes (i.e. 93%) related to this molecular concept are overexpressed in IBC, albeit not significantly, and with small fold changes. Nevertheless, the consistent pattern of expression changes leads to a highly significant FCS test result (P <0.001). On the other hand, significant expression changes were observed for only 3/15 PD-1 signalling genes (i.e. 20%) resulting in an insignificant OR test (P = 0.564). Put in other words, 3/15 significant genes provide insufficient evidence to reject the null hypothesis that transcription of PD-1 signalling genes is not altered in IBC. Of note, a consolidated OR analysis and FCS analysis led to a fivefold reduction in the number of molecular concepts called significant and removed gene set overlap-based false-positive results.

The result of the PT-based analysis (**Fig. 26.3**) demonstrates the added value topology information for pathway analysis. The 15 most significant molecular concepts are either enriched for differentially expressed genes, have significantly altered signalling activities or both. Compared to the output of the OR analysis, the role of extracellular matrix remodelling and G-protein signalling is confirmed. However, other concepts that have not been identified using the OR analysis (e.g. hypoxia-related signalling) seem to be relevant based on perturbation scores, which can only be assessed by taking pathway topology into account. Of note, some molecular concepts are characterized by both enrichment for differentially expressed genes and perturbed signalling activities (i.e. assembly of collagen fibrils and other multimeric structures), providing significant evidence for role in the biology of IBC.

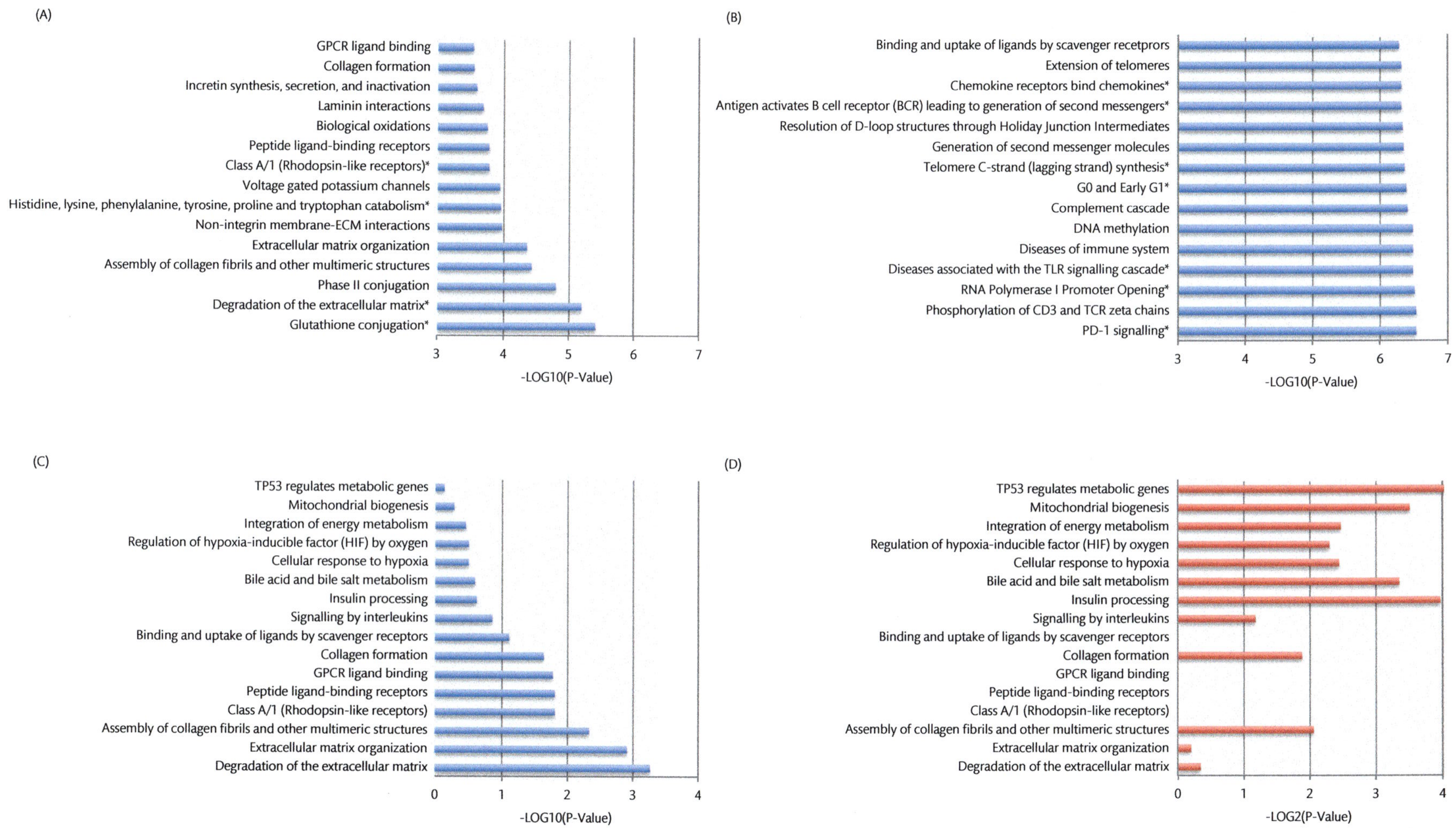

Fig. 26.3 The results of different gene set analyses performed on the IBC/nIBC data set are summarized in bar plot format (A: ORA; B: FCS; and C&D: PWA). For each analysis, the 15 most significant results are displayed. The X-axis represent the −log10 of the *P* value of the gene set analysis test results and is proportionate the number of digits behind the decimal. In panel D, the −log2 of the *P* value is provided. For the ORA and FCS analysis, the consolidated results are indicated using a star. Of note, panels C and D both represent data obtained using the same test (i.e. SPIA) and the pathways are ordered in the same fashion based on the significance of the general *P* value. In panel C, the result of a classical ORA analysis, which is part of the SPIA-routine, is shown. In panel D, the result of the perturbation analysis based on the pathway topology is shown. When comparing panels C and D, it can be appreciated that some pathways have significant values for the pathway perturbation analysis but not for the enrichment analysis (e.g. 'TP53 regulates metabolic genes') and vice versa. This underscores the added value of incorporating pathway topology information for translating gene expression profiles into biological concepts.

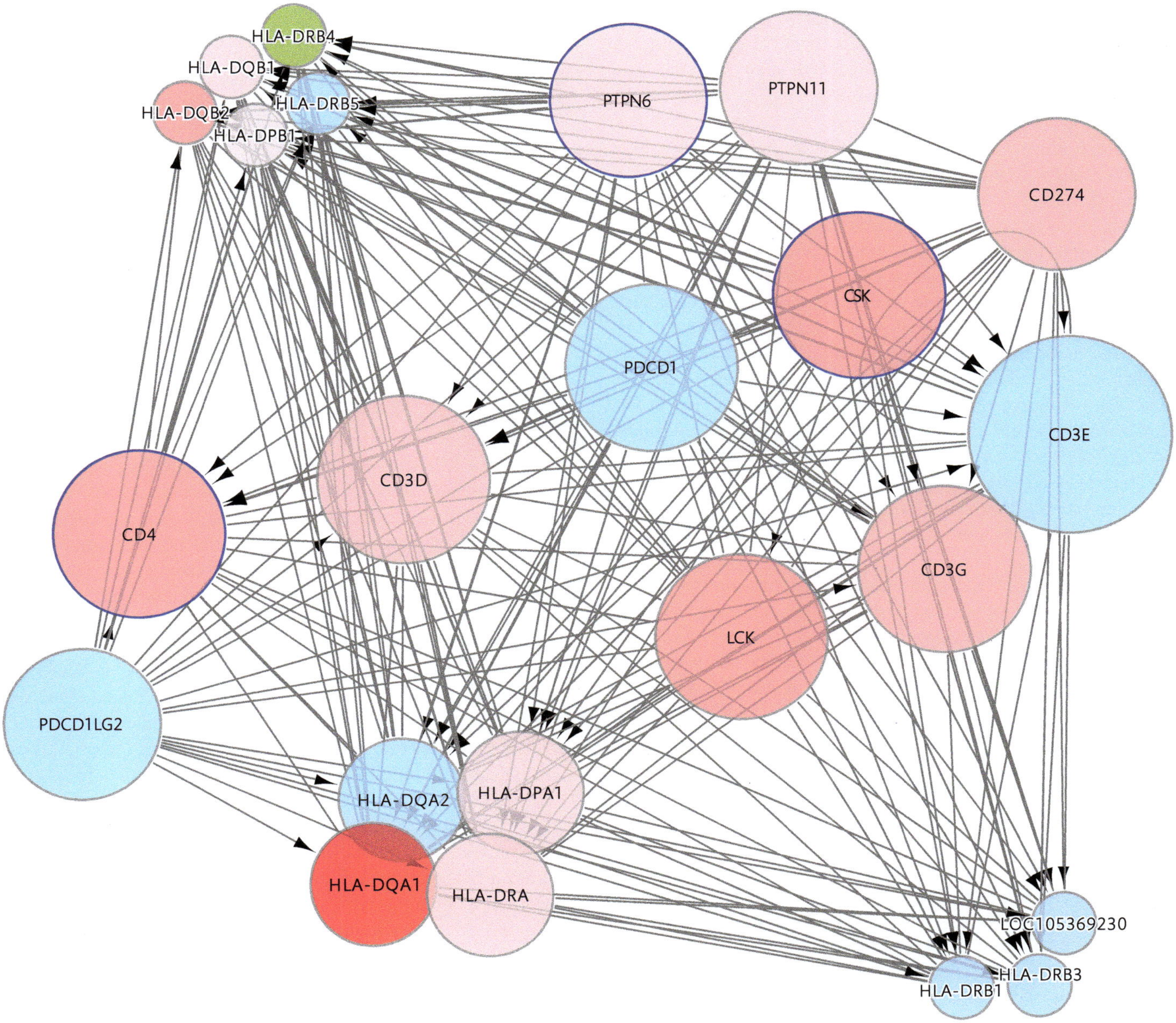

Fig. 26.4 A graph showing the PD-1 signalling network in IBC. The size of the nodes related to the node degree. Blue nodes denote genes with no measured expression data, green nodes are genes overexpressed in nIBC and red nodes are genes overexpressed in IBC. The colour saturation indicates the fold change (i.e. light: small fold change; dark: high fold change). Nodes with a blue border denote genes with a significant (i.e. $P < 0.05$) expression difference. It can be appreciated that only three genes are significantly different between IBC and nIBC. Therefore, ORA will not identify this pathway as significant. However, 14 out of 15 measured genes are overexpressed in IBC. These small but concordant expression changes will result in a significant test results when applying FCS.

Gene network analysis in cancer systems biology

Gene network analysis strategies

To investigate tumour biology using network-based approaches, two major strategies can be discerned. First, a gene list of interest can be organized into a network using network inference algorithms or using knowledge contained in various interaction databases. In this context, the method developed by Chen and colleagues (Chen et al., 2011) termed Expression2Kinases (E2K) merits specific attention. The algorithm is designed to operate on expression data, and it treats these data as a read-out of activated transcription factors. Therefore, the first step is to perform a gene set analysis using transcription factor target knowledge bases as a reference. The identified transcription factors are then organized into a PPI network. In a final step, the PPI network is analysed for enrichment of kinase-specific substrates. Thus, E2K performs an integrated analysis using various knowledge bases to establish a gene interaction network from which signalling pathways and biological processes that drive gene expression data can be identified. The resulting PPI network can be regarded as the collection of all signal transduction pathways that can theoretically explain the observed gene expression profile. Because the algorithm treats the expression data as the output layer (i.e. expression data) and networks are constructed using the causal transcription factors, this strategy is considered a 'bottom-up' approach, which is different from the 'top-down' approaches where the relevant genes are considered as the input layer that is laid out in a network context. In a sense, the 'bottom-up' approach aims at identifying the cause of the gene expression data, whereas the 'top-down' approach evaluates the consequence. Importantly, E2K demonstrates that information contained in various knowledge bases is complementary and can be used to obtain an integrated picture of the biological condition under study.

An alternative strategy to gain a network-level understanding of molecular tumour biology, is to analyse relevant genes in the context of a global background model, which can be constructed to optimally represent the appropriate context. The rationale for using a global background network is that genes with significant expression changes not necessarily interact with each other but may in fact interact with genes of much greater importance. Imagine a hub gene in the global background model that is not differentially expressed, but interacts with 20% of the genes deemed relevant by a differential expression screen. Only by analysing expression profiles in the context of a global background model, the importance of the hub gene will be revealed.

Having established the network models, the next question then is how to extract meaningful data. First, network genes may be prioritized using topological statistics or by applying machine learning approaches such as heat kernel diffusion ranking and random walking. These methods simulate the flow of information through a network starting from a set of seed nodes, which can be differentially expressed genes. In the course of this simulation, the nodes most frequently affected by the flow of information originating from the seed nodes are recorded, leading to gene prioritization. Expression differences can be incorporated by modelling the strength of the simulated signal in function of gene expression levels (Nitsch et al., 2010; Han et al., 2014). As such, important network genes may be identified. Using prioritized gene lists, a weighted and consolidated gene set analysis can be performed to identify relevant signal transduction pathways. Second, the network topology can be exploited to identify important network structures like motifs, modules, and subgraphs. Network motifs, which constitute recurrent regulatory patterns are of vital importance in the organization of robust signal transduction networks. They can be identified using various algorithms (e.g. G-Tries algorithm) and offer clues towards important network perturbation points that can be targeted therapeutically. Modules, which reflect biologically relevant network entities, can be identified using clustering techniques as explained before. Key module genes can be identified by evaluating module-specific network topologies (Dezső et al., 2009) and can be translated into pathways allowing network-level pathway interactions to be studied and novel gene-pathway associations to be made. Finally, relevant subgraphs may be identified by mapping a gene list of interest (e.g. differentially expressed or genes with a mutually exclusive mutation pattern) onto the network and identifying enriched network regions. It should be stressed that the computational properties of many network analysis algorithms do not allow for a unique solution to be found (i.e. these algorithms identify local minima instead of global minima) and repeated analyses of the same data may provide divergent results. Therefore, it is advisable to perform multiple analysis run and present the most common results.

Gene network analysis applied to inflammatory breast cancer expression data

Using the list of 195 differentially expressed genes between tumour samples from patients with and without IBC, gene interaction networks were generated. To that end we applied a network inference strategy by calculating Spearman correlation coefficients and mutual information between all 192 differentially expressed genes. Thresholding the data at specified levels resulted in networks presented in **Fig. 26.5**. An additional third gene interaction network was generated using E2K (**Fig. 26.6**). The presented networks exhibit striking differences in network topology. The Spearman correlation and mutual information network are more sparsely connected with multiple independent subnetworks (i.e. connected components). The E2K network on the other hand is composed of a single and densely connected component. It should be stressed that the topological differences rather reflect the parameter settings used during network construction and thus should not be considered intrinsic to the network inference methodology. For example, thresholding the Spearman correlation coefficient at 0.8 rather than 0.9 as was done in **Figure 26.5A** would result in a more densely connected network.

Topology weighted gene set analysis of the E2K network identified significantly enriched gene sets and revealed a central role for TGFβ signalling, which included the four most connected network genes (i.e. *MAPK1, EP300, CREBBP*, and *SMAD3*). Spectral clustering identified five network modules that are partially redundant. For example, two modules were enriched for genes involved in TFGβ-dependent *SMAD2/3* signalling. A third module is involved in NFκB signalling, a previously identified molecular hallmark of IBC. The E2K network thus also suggest intricate cross-talk between TFGβ- and NFκB-signalling pathways mediated by among others AKT1 and HDAC-proteins. The influence networks can be analysed to assess the functional consequences of the identified gene expression differences. In these networks, processes related to the immune system and cell cycle take centre stage, which agrees with the result obtained using FCS-based gene set analysis.

Last, but not least, we mapped the list of 192 differentially expressed genes onto a backbone network composed of Biogrid pathways and Encode transcription factors. By diffusing the gene expression data through the backbone network, key molecules with perturbed signalling activities were identified and translated into pathways. Hence, we learned that the observed gene expression differences in IBC have a major impact on NOTCH1-signalling. Interestingly, NOTCH1-signalling events were also enriched in the E2K network, with YY1 and MYC as crucial signalling mediators.

Caveats in systems biology

When analysing high-throughput '-omics' profiles of tumour samples using systems biology approaches, several caveats need to be taken into account. Deciphering tumour biology from molecular profiles critically depends on the quality of the tissue sampling procedure. From this perspective, the 'garbage in, garbage out' principle known from computer science, and referring to the fact that computers will process unintended input data to produce nonsensical output, is also applicable in this setting. Several technical aspects related to tissue sampling (i.e. ischaemia and fixation time, used fixatives and sampling bias) may have deleterious effects on data quality

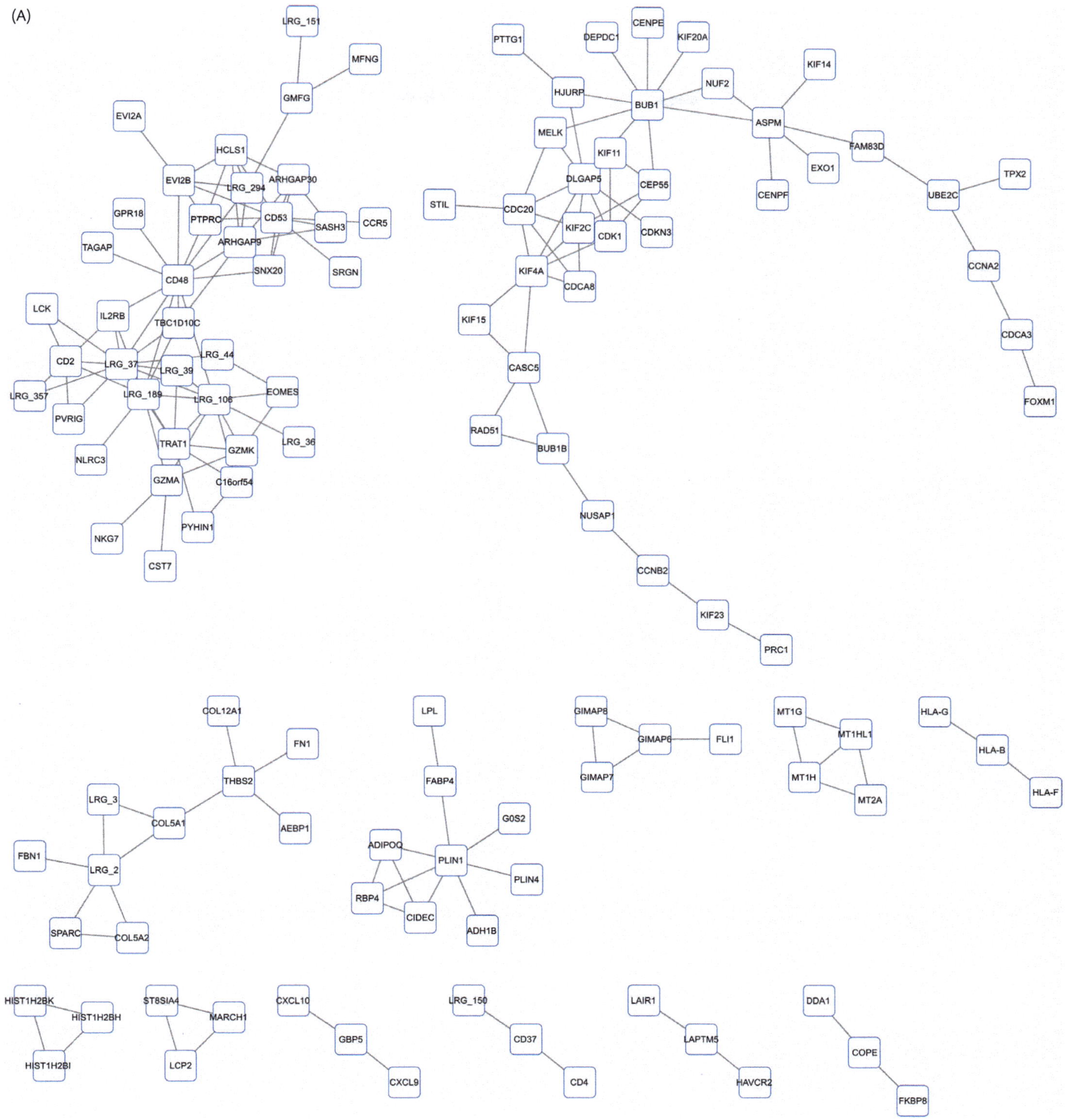

Fig. 26.5 Influence networks constructed using the Spearman correlation coefficients (A) and mutual information (B). First, Spearman correlations and mutual information between all pairs of gene expression signals were calculated. Then, the absolute values were thresholded at 0.9 for the Spearman correlations and at 0.6 for the mutual information resulting in the presented gene networks. The resulting correlation network contains 179 genes, and 201 correlation-based edges. The resulting mutual information network contains fewer genes (114), but far more connections (566). Both networks are enriched for particular tumour-, immune system- and signalling-related functions.

and should be considered. No matter how sophisticated the analysis pipeline, such quality issues cannot be resolved. Thus, establishing a rigid tissue sampling protocol should the primary focus for cancer profiling studies.

Regardless of tissue quality, it should be stressed that molecular profiles are unlikely to recapitulate the full complexity of a disease for two reasons. First, it is increasingly being recognized that tumours consist of heterogeneous cell populations. Intratumor

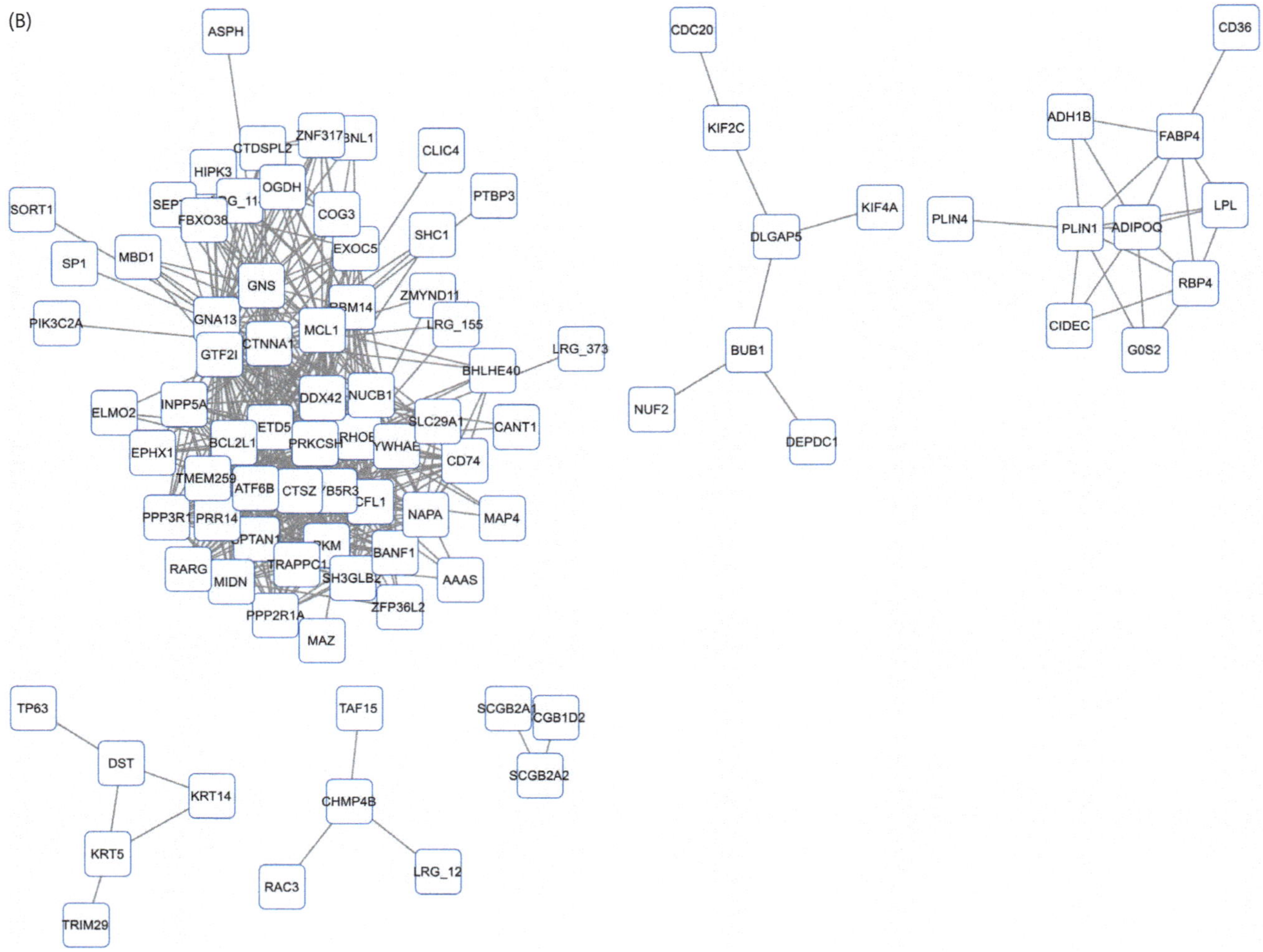

Fig. 26.5 Continued

heterogeneity implies that a single biopsy most likely contains phenotypically distinct cancer cells, which are additionally surrounded by different populations of normal cells that constitute the tumour stroma. When generating molecular profiles, information from these different cell populations is simultaneously extracted and interrogated with no means to trace back the cellular origin of a signal. Therefore, signal transduction networks inferred from such profiles should be interpreted with care. For example, genes connected in a network due to coordinated expression changes may in fact never interact physically because their concerted behaviours take place in distinct cell types. In fact, recent evidence demonstrated that these so-called non-cell autonomous interactions are crucial in tumour biology as they drive tumour growth and further support subclonal heterogeneity (Marusyk et al., 2015). Hence, in order to fully comprehend the complexity of tumour biology, it will become mandatory to develop strategies that allow taking cellular diversity into account. A second argument that dictates careful interpretation of molecular profiles resides in the fact that these profiles only represent a snapshot of a dynamic process. In the course of time, cancer cells will be exposed to different stimuli that will each contribute differently to tumour development or progression. Unfortunately, these dynamic properties are not reflected in molecular profiles obtained on a single time point, nor can they be easily inferred.

A final item that needs to be considered relates to the fact that most biological processes and signal transduction mechanisms involve interaction at the protein level. On the other hand, the majority of the present-day cancer profiling studies have generated genomic and/or transcriptomic data. However, changes at DNA or RNA level not necessarily translate into proteomic differences. For example, mutations in unexpressed genes will have little influence on the cellular protein interaction network. Similarly, non-coding RNAs might inhibit the translation of overexpressed transcripts. Regardless of these impediments, it is common practice to interpret genomic and transcriptomic profiles from a functional, protein-driven perspective, particularly when performing topology-based pathway analyses or when studying PPI networks. Some recommendations can be made to alleviate these issues to some extent. First, gene expression data can be translated into protein-centric data, for example by performing target gene enrichment analysis (e.g. E2K) or by identifying transcription factors or signalling hubs that have many established interactions with a set of genes of interest. Once protein-centric data are obtained, these could be subjected to further analyses. Second, molecular profiles that capture different aspects of a molecular landscape may provide complementary data that can be integrated to obtain an even more holistic view. For example, integration of gene and micro-RNA expression data will allow a more precise judgement of the functional consequences

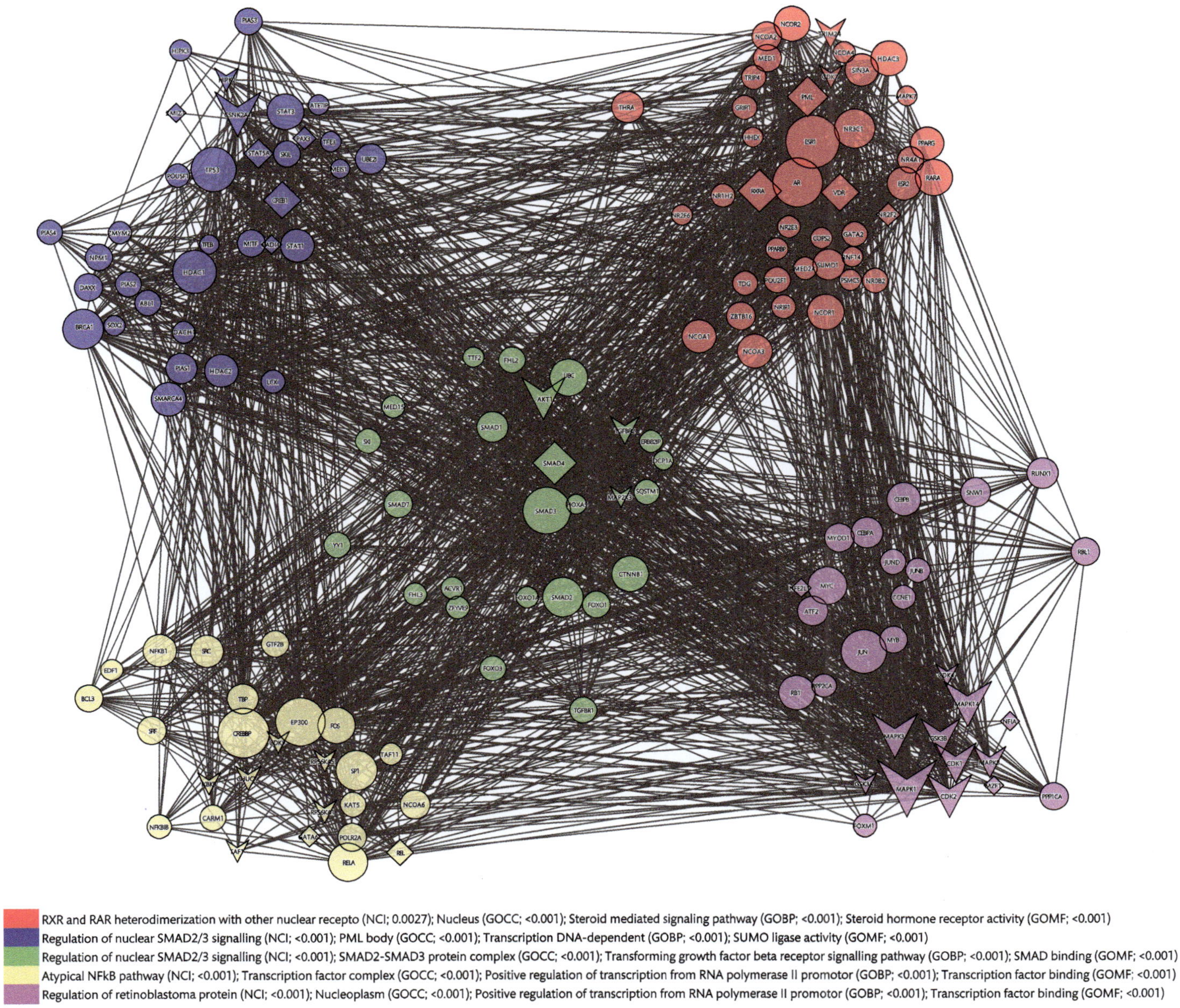

Fig. 26.6 E2K was applied on the list of 192 genes differentially expressed between IBC and nIBC. The resulting PPI network was clustered and five different modules were identified (yellow, green, magenta, red, and blue). The size of the nodes is proportionate to the node degree. The shape of the node indicates functionality: transcription factors (diamonds), kinases (arrow heads) and signalling proteins (circles). The most relevant processes each of the modules are involved in are summarized underneath the network.

of gene expression changes. Similarly, integrating mutational profiles with gene expression data, will shed light onto the relevance of each single mutation. Finally, it should be mentioned that not all analysis procedures are equally sensitive to the issues discussed in this paragraph. For example, FANs and influence networks do not rely on physical interactions.

Challenges in systems biology: Annotations and methodology

Annotations play a vital role in systems biology since they provide the necessary information to translate lists of individual genes into biological concepts. However, the development of new approaches in systems biology will require improvement of the existing annotations. The reason for this is many-fold. First, current gene annotations are very much gene-centric, whereas evidence indicates that many genes generate alternative transcripts that may have related, distinct, or even opposing functions. Second, the information contained in various knowledge bases is incomplete and inaccurate. More than half of the gene symbols used by the HUGO Gene Nomenclature Committee have no allocation in any pathway reported in Pathway Commons (https://www.pathwaycommons.org), one of the biggest composite pathway databases available. In addition to incomplete gene coverage, other relevant aspects of signal transduction such as cellular compartmentalization and second messengers are often completely ignored. When evaluating the last release of GO, about 25% of the annotations are inferred from indirect evidence, which is considered to be of low quality and therefore potentially incorrect. Third, many associations reported in various knowledge bases represent context-specific interactions, which contributes to the ambiguous representations of pathways and processes across various databases. Unfortunately, these data are generally not recorded, but ideally, they should be considered during data analysis. Fourth, integrating data available in different resources will be crucial to unlock the full potential of systems biology. However, data formats often differ, and efforts are currently undertaken to develop a standard language for systems biology that will allow substantial unification. This will have the additional benefit that interconnections between pathways can be established, which is a prerequisite to obtain integrative biological models. It is self-evident that solving these annotation issues will be crucial to achieve more accurate interpretations of biological data.

In addition, from a methodological perspective systems biology deals with some unmet challenges. First, in the past few years, the number of systems biology algorithms has increased exponentially. The performance of each method relative to the others is poorly assessed, among others due to absent benchmark data sets. To compare sensitivity and specificity, simulated data sets can be used. However, these lack the biological complexity, such as confounding variables, presence of outliers, and technical bias. Therefore, it is desirable to use real biological data. The downside of using real data sets is that the actual biology is never fully known. Second, the presently available methods are limited from the perspective analysing data as a single dynamic system. A typical approach for analysing dynamic responses at the pathway level is to measure changes at multiple time points individually and evaluate which pathways are significant at each time point. However, this approach implicitly assumes that pathways at different time points are independent of each other. Topology-based analyses can accommodate this problem to a certain extent, but these methods also assume that expression levels of all genes are constant. This assumption almost never holds, as there are positive and negative feedback loops that continuously regulate gene expression. Third, biological systems are complex and their outcome is determined by the concerted action of a large numbers of components that act following dynamic rules. A central task of systems biology is to develop testable models that reflect this complexity, which can be accomplished by integrating different levels of biological information including different types of molecular data (e.g. mutations, transcript expressions) and condition-specific and spatiotemporal signalling properties. Special attention should be given to genomic alterations outside the protein-coding domain of the human genome.

TAKE-HOME MESSAGE

Computational methods to translate molecular profiles into biological concepts have contributed significantly towards gaining novel insights in cancer biology. Historically, these methods focus on analysing genes of interest in the context of predefined gene sets that are associated with pathways and processes. Recently, this paradigm has shifted towards applying analysis methods based on networks that reflect molecular interactions. By consequence, the output of molecular profiling experiment is no longer limited to a descriptive list of relevant genes. Rather, by taking molecular interactions between genes into account, a more mechanistical interpretation of the data is feasible and individual genes can be related to their appropriate molecular contexts, which is encoded in the topology of a network. More importantly, a gene interaction network can be considered as an independent system, which is characterized by a series of network-level properties that cannot be predicted from the individual network components. These emergent properties provide insights into the organizational principles of cancer cell robustness, plasticity, and evolvability. Exploiting this knowledge should allow identifying fragile network points for therapeutic intervention. Therefore, cancer systems biology holds great promise from both a research and a clinical perspective. However, it should be noted that cancer systems biology is still a relatively young discipline, prone to active developments with respect to both the computational and strategic issues (see open questions section). These developments should best be performed in concert with the stakeholders in order to accelerate a full-blown deployment of cancer systems biology in both research and diagnostic settings.

OPEN QUESTIONS

- Different approaches to apply systems biology are currently available. However, clear evidence towards the optimal model is missing. Head-to-head comparisons of different methods with respect to different research questions need to be performed, preferentially in combination with functional validation studies.
- Molecular interactions are dependent on the genetic background in which they occur. Workflows to integrate cell and tissue specific molecular backgrounds during network construction need to be established.
- Computational methods for network analysis rarely identify one optimal result. Rather, repeated analysis will generate many alternative near-optimal solutions. Methods need to be defined on how such ambiguous results should be interpreted.
- Cancer samples are often a mixture of heterogenous cell types, including both cancer cells with different molecular profiles and normal stromal cells. Methods should be developed to deal with this cellular heterogeneity, allowing the identification of cancer cell intrinsic signal transduction patterns.
- Identifying fragile network points is one of the goals of cancer systems biology, allowing the identification of potential targets for therapy. To that end, clinical data should be integrated in systems biology approaches.

FURTHER READING

Bansal, M., Belcastro, V., Ambesi-Impiombato, A., & di Bernardo, D. (2007). How to infer gene networks from expression profiles. *Mol Syst Biol*, **3**, 78.

Bartocci, E. & Lió, P. (2016). Computational modeling, formal analysis, and tools for systems biology. *PLoS Comput Biol*, **12**, e1004591.

Donato, M., Xu, Z., Tomoiaga, A., et al. (2013). Analysis and correction of crosstalk effects in pathway analysis. *Genome Res*, **23**, 1885–93.

Doncheva, N. T., Assenov, Y., Domingues, F. S., & Albrecht, M. (2012). Topological analysis and interactive visualization of biological networks and protein structures. *Nature Protocols*, **7**, 670–85.

Ghosh, S., Matsuoka, Y., Asai, Y., Hsin, K.-Y., & Kitano, H. (2011). Software for systems biology: from tools to integrated platforms. *Nat Rev Genet*, **12**, 821–32.

Khatri, P., Sirota, M. & Butte, A. J. (2012). Ten years of pathway analysis: current approaches and outstanding challenges. *PLoS Comput Biol*, **8**, e1002375.

Ryan, C. J., Cimermančič, P., Szpiech, Z. A., et al. (2013). High-resolution network biology: connecting sequence with function. *Nat Rev Genet*, **14**, 865–79.

Tarca, A. L., Draghici, S., Khatri, P., et al. (2008). A novel signaling pathway impact analysis. *Bioinformatics* **25**, 75–82.

Werner, H. M. J., Mills, G. B., & Ram, P. T. (2014). Cancer systems biology: a peek into the future of patient care?. *Nat Rev Clin Oncol*, **11**, 167–76.

REFERENCES

Amit, I., Wides, R., & Yarden, Y. (2007). Evolvable signaling networks of receptor tyrosine kinases: relevance of robustness to malignancy and to cancer therapy. *Mol Syst Biol*, **3**, 151.

Bayerlová, M., Jung, K., Kramer, F., Klemm, F., Bleckmann, A., & Beißbarth, T. (2015). Comparative study on gene set and pathway topology-based enrichment methods. *BMC Bioinformatics*, **16**, 334.

Caron, E., Ghosh, S., Matsuoka, Y., et al. (2010). A comprehensive map of the mTOR signaling network. *Mol Syst Biol*, **6**, 453.

Chen, E. Y. Tan, C. M., Kou, Y., et al. (2013). Enrichr: interactive and collaborative HTML5 gene list enrichment analysis tool. *BMC Bioinformatics*, **14**, 128.

Chen, E. Y., Xu, H., Gordonov, S. et al. (2011). Expression2Kinases: mRNA profiling linked to multiple upstream regulatory layers. *Bioinformatics*, **28**(1), 105–11.

Clark, N. R., Dannenfelser, R., Tan, C. M., Komosinski, M. E., & Ma'ayan, A. (2012). Sets2Networks: network inference from repeated observations of sets. *BMC Systems Biology*, **6**(1), 1–1.

Dannenfelser, R., Clark, N. R., & Ma'ayan, A. (2012). Genes2FANs: connecting genes through functional association networks. *BMC Bioinformatics*, **13**, 156.

Dezső, Z., Nikolsky, Y., Nikolskaya, T., et al. (2009). Identifying disease-specific genes based on their topological significance in protein networks. *BMC Syst Biol*, **3**(1), 36.

Doderer, M. S., Anguiano Z, Suresh U, et al. (2012). Pathway Distiller—multisource biological pathway consolidation. *BMC Genomics*, **13**(Suppl 6), S18.

Donato, M., Xu, Z., Tomoiaga, A., et al. (2013). Analysis and correction of crosstalk effects in pathway analysis. *Genome Res*, **23**(11), 1885–93.

Doncheva, N. T., Assenov, Y., Domingues, F. S., & Albrecht, M. (2012). Topological analysis and interactive visualization of biological networks and protein structures. *Nat Protoc*, **7**(4), 670–85.

Mitrea, C., Taghavi, Z., Bokanizad, B., et al. (2013). Methods and approaches in the topology-based analysis of biological pathways. *Front Physiol*, **4**, 278.

Emmert-Streib, F. & Glazko, G. V. (2011). Pathway analysis of expression data: deciphering functional building blocks of complex diseases. *PLoS Comput Biology*, **7**(5), e1002053.

García-Campos, M. A., Espinal-Enríquez, J., & Hernández-Lemus, E. (2015). Pathway analysis: state of the art. *Front Physiol*, **6**(278), 47.

Glazko, G. V. & Emmert-Streib, F. (2009). Unite and conquer: univariate and multivariate approaches for finding differentially expressed gene sets. *Bioinformatics*, **25**(18), 2348–54.

Han, J., Li, C., Yang, H., et al. (2014). A novel dysregulated pathway-identification analysis based on global influence of within-pathway effects and crosstalk between pathways. *Journal of The Royal Society Interface*, **12**(102), 20140937.

Hofree, M., Shen, J. P., Carter, H., Gross, A., & Ideker T (2013). Network-based stratification of tumor mutations. *Nat Methods*, **10**(11), 1108–15.

Kalinka, A. T. & Tomancak, P. (2011). Linkcomm: an R package for the generation, visualization, and analysis of link communities in networks of arbitrary size and type. *Bioinformatics*, **27**(14), 2011–12.

Kandoth, C., McLellan, M. D., Vandin, F., et al., (2014). Mutational landscape and significance across 12 major cancer types. *Nature*, **502**(7471), 333–9.

Khatri, P., Sirota, M., & Butte, A. J. (2012). Ten years of pathway analysis: current approaches and outstanding challenges. *PLoS Comput Biol*, **8**(2), e1002375.

Kleensang, A., Maertens, A., Rosenberg, M., et al., (2014). t4 workshop report: pathways of toxicity. *ALTEX*, **31**(1), 53–61.

Las Rivas, De, J. & Fontanillo, C. (2010). Protein–protein interactions essentials: key concepts to building and analyzing interactome networks. *PLoS Comput Biol*, **6**(6), e1000807.

Leiserson, M. D. M., Vandin, F., Wu, H. T., et al., (2014). Pan-cancer network analysis identifies combinations of rare somatic mutations across pathways and protein complexes. *Nat Genet*, **47**(2), 106–14.

Ma'ayan, A. (2011). Introduction to network analysis in systems biology. *Sci Signal*, **4**(190), tr5.

Marusyk, A., Tabassum, D. P., Altrock, P. M., Almendro, V., Michor, F., & Polyak, K. (2015). Non-cell-autonomous driving of tumour growth supports sub-clonal heterogeneity. *Nature*, **514**(7520), 54–8.

Newman, M. E. J. (2006). Modularity and community structure in networks. *Proc Natl Acad Sci U S A*, **103**(23), 8577–82.

Nitsch, D., Gonçalves, J. P., Ojeda, F., de Moor, B., & Moreau, Y. (2010). Candidate gene prioritization by network analysis of differential expression using machine learning approaches. *BMC Bioinformatics*, **11**(1), 460.

Shoemaker, B. A. & Panchenko, A. R. (2007). Deciphering protein–protein interactions. Part II. Computational methods to predict protein and domain interaction partners. *PLoS Comput Biol*, **3**(4), e43.

27

Cancer biology through immunohistology

Karen Pulford and Kevin Gatter

Introduction to cancer biology through immunohistology

Cancer is a heterogeneous and dynamic disease arising from different cell types and each cancer is associated with different biological activities. Accurate identification of the origin of the tumour cell is, therefore, essential not only for diagnostic purposes but also provides invaluable information about the biology underlying tumour development. Furthermore, such knowledge is fundamental for decisions regarding therapy and the development of future, more effective therapeutic approaches for patients.

Prior to the availability of immunohistological techniques, pathologists relied upon cell morphology for diagnosis. There were, however, occasions when it was difficult to distinguish the normal from the tumour cells. The 'antibody revolution', with the availability of specific antibodies and the development of improved staining techniques, led to the evolution of the science of immunohistology. This enabled pathologists and researchers to distinguish between cell types and normal and abnormal cells (Soilleux and Gatter, 2006), but also provided an invaluable opportunity to study the tumour within its tissue microenvironment.

Immunohistology has resulted in the establishment of improved tumour classification systems with accompanying advances in diagnosis. Moreover, immunohistology has performed key roles in unravelling cellular pathways involved in the biology of the cancer cell. This increased understanding of oncogenetic pathways has enabled the development of improved treatment options. The latter includes the identification of new therapeutic targets and the increased likelihood of 'off the shelf' personalized treatments.

This chapter describes the use of immunohistology, its roles in the classification of tumours, the investigation of mechanisms underlying tumour development, the provision of prognostic and predictive information, and in the development of effective innovative therapies. A detailed example illustrating the importance of immunohistology in the discovery and the unravelling of the role of the oncogenic anaplastic lymphoma kinase (ALK) protein is provided. Reference is also made to future directions taken by new generation immunohistology and its relevance to other platforms used in the study of cancer.

What is immunohistology?

Immunohistology is the microscopic study of tissues using specific antibodies that bind to individual molecules expressed by the cellular and non-cellular components of the tissues. The combination of highly specific antibodies, effective antigen retrieval methods, and staining techniques provides important information about cell lineage, the subcellular distribution of the relevant molecules, diagnosis, metastasis, and the tumour microenvironment. The science and art of immunohistology advanced in multiple areas as described next.

Increasing availability of antibodies

The availability of antibodies raised against a huge variety of antigens, including mutated proteins, revolutionized the study of cancer biology. Monoclonal antibody technology by Kohler and Milstein (1975), with its ability to produce large quantities of highly specific antibodies, was pivotal in the development of immunohistochemistry. The popularity of the technology (outlined in **Fig. 27.1**) led to the production of highly specific reagents that could be used to study the immunophenotype of tumour cells and the functional properties of the tumour cells. Their availability in unlimited quantities provided excellent opportunities for performing immunohistological studies in centres worldwide. Although the majority of monoclonal antibodies are of mouse origin, technical advances permitted monoclonal antibodies to be obtained from a variety of other animals (e.g. rats and rabbits), as well as being manufactured from technologies harnessing phage and yeast display libraries (Dubel et al., 2010). The increased knowledge in monoclonal and antibody production is mirrored in the presence of more than 300 antibody suppliers providing more than half a million antibodies (Bird, 2012).

Developments in staining technologies

Immunohistochemistry was originally performed using antibodies directly conjugated with fluorescent tags (**Fig. 27.2**). The subsequent development of indirect staining techniques, in which the Fc region of a secondary polyclonal antispecies specific antibody was labelled with enzymes such as horseradish peroxidase (HRP), enabled the use of unlabelled antibodies as a first step in staining procedure (**Figs. 27.2 and 27.3A**). This had the advantages of increasing flexibility and the sensitivity of the labelling. Methods employing

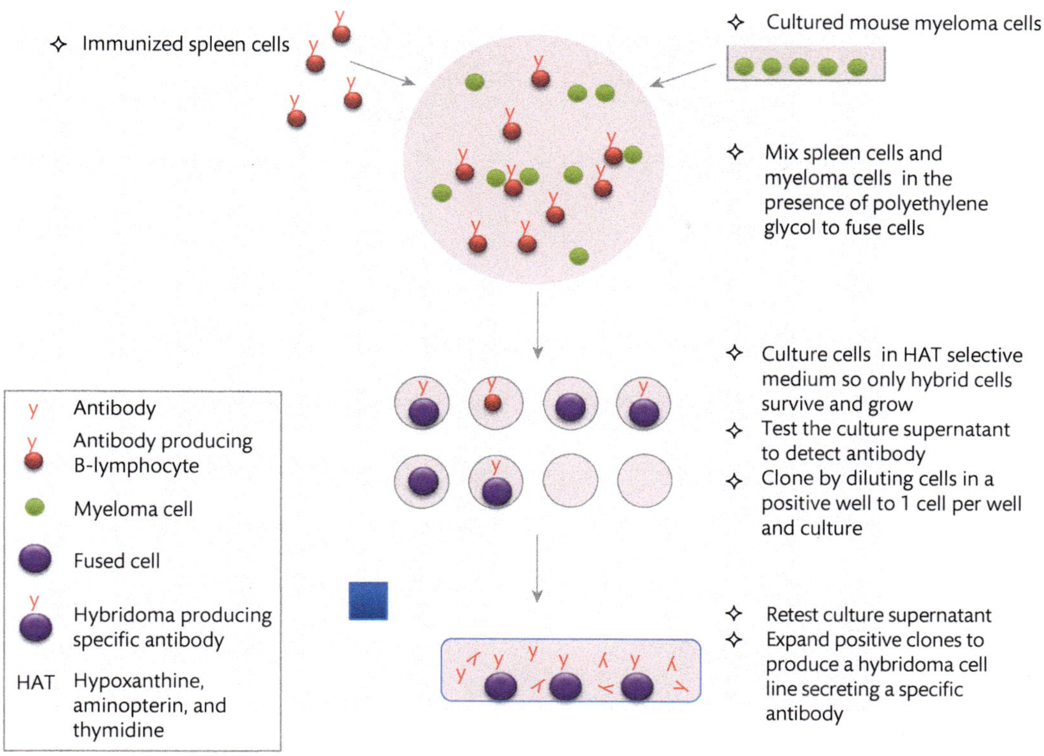

Fig. 27.1 Schematic outline of the production of monoclonal antibodies. These highly specific reagents are produced initially from a single immune mouse B-cell that is immortalized to produce an antibody secreting cell line that can be maintained in culture indefinitely.

additional steps, such as the alkaline phosphatase:antialkaline phosphatase (APAAP; see Cordell et al., 1984) technique (Figs. 27.2 and 27.3B), and subsequent avidin-biotin and polymer-based systems further increased the sensitivity of the staining procedures.

Combining immunolabelling techniques on a cell or tissue preparation permitted multiplex labelling studies to be performed (Mason et al., 2000). These techniques, using immunoenzymatics, immunofluorescence, or a combination of the two, enabled the

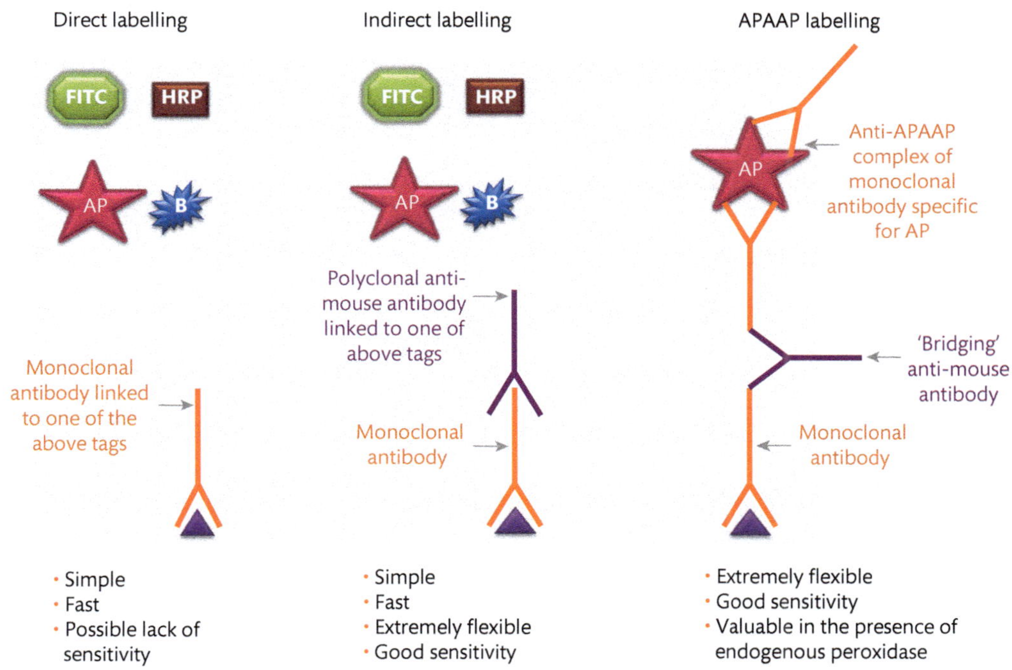

Fig. 27.2 Examples of immunolabelling. An enzyme or fluorescent tag is directly attached to the primary antibody in direct labelling. For indirect labelling, the tag can be conjugated to the secondary (usually polyclonal) antibody. A polyclonal bridging antibody recognizes an antialkaline phosphatase antibody and alkaline phosphatase immune complex thus providing greater sensitivity to the technique.

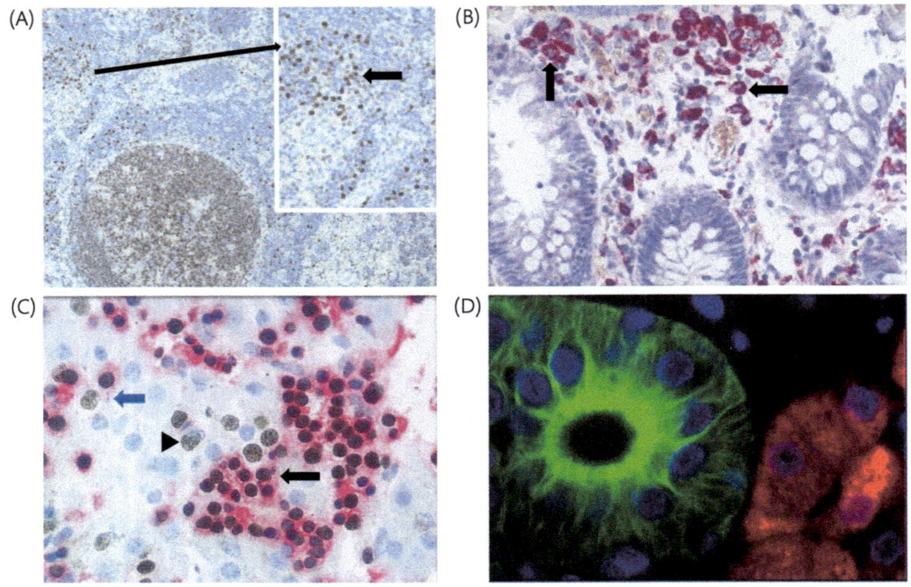

Fig. 27.3 (A) Immunoperoxidase labelling of tonsil sections showing the distribution of BCL11A$_{XL}$ protein (brown). The nuclei of the cells are counterstained with haematoxylin. The inset shows the labelled cells in greater detail. (B) APAAP labelling of intestine showing the distribution of CD68 macrophages (red, arrowed). (C) Double immunoenzymatic labelling of foetal liver using immunoperoxidase and APAAP techniques of foetal liver to study the expression of the TAL-1 transcription factor in foetal haematopoietic cells. Note the presence of the transcription factor TAL-1 (brown nuclei) in the erythroid precursors expressing glycophorin A or CD235a (red, arrowed). The TAL-1-positive nuclei (grey, arrowhead) are haematopoietic cells and occasional hepatocytes. (D) Double immunofluorescent labelling of the salivary gland to show the position of epithelium (red) and lysozyme (green).

simultaneous study of more than one antigen in the same cell type (Fig. 27.3C) or in different cell populations in the same slide preparation (Fig. 27.3D). Fluorescent tags were particularly useful for the detection of target proteins at the same or similar cellular locations where the physical size of the chromogenic substrate would limit binding in subsequent staining steps. With appropriate software it was also possible to perform double immunofluorescence using immunoperoxidase tags when it was difficult to find appropriate primary antibodies of different isotypes and species (Pulford et al., 2006).

Combinations of immunolabelling techniques and fluorescence *in situ* hybridization (FISH) have also provided excellent opportunities to combine investigations on protein expression associated with genetic changes (Gellrich et al., 2004) and, more recently, microRNA. The development and use of quantum dots that could be directly conjugated to antibodies, cDNA, or RNA, has further opened up the possibility of performing even more complex multiplex staining (Mansfield, 2014).

Cells and tissues, particularly clinical samples, require fixation in order to maintain their morphology and pretreatment to permit the reliable detection of target molecules. Many antibodies were initially unable to recognize their target in routinely formalin fixed material, even following proteolytic pretreatment. This problem has now been largely overcome by the innovative use of different pH buffers and heat-based antigen retrieval methods (Leong and Leong, 2006).

The availability of high throughput techniques, such as gene expression profiling and Next Generation Sequencing, has resulted in the identification of thousands of biomarkers that are of potential importance in cancer biology. The development of tissue microarrays (TMAs), in which cores of tissue or cell preparations (as small as 1 mm) can be geometrically arranged on a single slide, enabled an approach to investigate the expression of these new biomarkers in a more efficient and cost-effective way than using a single section on a slide (Kallioniemi et al., 2001). TMAs, when combined with new automated staining machines, provide an excellent platform for performing rapid immunohistochemical investigations.

Advance of imaging equipment

Microscopes have evolved alongside immunohistochemistry. Scanning laser confocal microscopy has enabled the precise subcellular position of targeted molecules and genes to be visualized (Blazquez et al., 2006). Furthermore, the use of thick tissue sections and *in vitro* models of tissues, such as tumour spheroids, provided the opportunity to perform three-dimensional studies of tumour cells and their tissue microenvironment (Matsumura et al., 2000). The development of digital imaging and image analysis software, combined with improvements in equipment, has resulted in dramatic leaps forward in multispectral imaging. Furthermore, there are now opportunities for 'virtual microscopy' in which investigators in multiple geographical sites worldwide can examine the same stained tissue/cell preparation.

Immunophenotyping cancers

Information on the biology of the cancer cell starts with the correct diagnosis. In the past, a frequent problem concerned the differential diagnosis of poorly differentiated tumours, such as cases of lymphoma and carcinoma (Soilleux and Gatter, 2006). Diagnostic accuracy was improved by the use of immunohistochemistry to assign a 'fingerprint' or immunophenotype to tumours. Table 27.1 shows antigens that are invaluable for differentiating between the

Table 27.1 Examples of antigens used to identify major tumour subtypes of carcinoma and lymphoma

Tumour	Antigen	Cellular localization
Non-haematopoietic malignancies		
Breast	Low molecular weight cytokeratins	Cytoplasm
Breast	Oestrogen receptors	Nuclei
Breast	Progesterone receptor	Nuclei
Breast	Gata binding protein 3 (GATA-3)	Nuclei
Colorectal and gastrointestinal	Villin	Membrane
Colorectal	High molecular weight cytokeratins)	Cytoplasm
Colorectal	Homeobox protein CDX-2	Nuclei
Gynaecological cancers	Paired box gene 8 (PAX8)	Nuclei
Hepatocellular	Arginase	Granules in cytoplasm
Lung adenocarcinoma	Napsin A	Cytoplasm
Ovarian	Wilms tumour protein 1	Nuclei
Prostate	Prostate specific antigen	Cytoplasm
Prostate	Homeobox protein NKX3.1	Nuclei
Haematopoietic malignancies		
Lymphoma	CD45	Cytoplasm
B-cell lymphoma	CD20	Cytoplasm
B-cell lymphoma	CD79a	Cytoplasm
Follicular lymphoma	CD10	Cytoplasm
Follicular lymphoma	B-cell lymphoma 2 (BCL-2)	Nuclei
Burkitt's lymphoma	Epstein-Barr virus	Nuclei
Chronic lymphocytic leukaemia	Zeta-chain-associated protein kinase 70 (ZAP-70)	Cytoplasm
T-cell lymphoma	CD2	Membrane and cytoplasm
T-cell lymphoma	CD3	Membrane and cytoplasm
Hodgkin's lymphoma	CD15	Cytoplasm (may be focal)
Hodgkin's lymphoma	CD30	Golgi apparatus and cell membrane
Anaplastic large cell lymphoma	Anaplastic lymphoma kinase (ALK)	Distribution dependent upon ALK protein
Myeloma	Transformation related protein 63	Cytoplasm
Myeloma	CD38	Cytoplasm and membrane

Source: data from Soilleux E. and Gatter KC, ' The Antibody Revolution: How 'Immuno' changed pathology', *Pathological Society Annual Meeting and Centenary Meeting*, pp. 173–84, Copyright © 2006 Pathological Society; and Kandalaft PL and Gown AM, 'Practical Applications in Immunohistochemistry: Carcinomas of Unknown Primary Site', *Archives of Pathology and Laboratory Medicine*, Volume 140, Issue 6, pp. 508–23, Copyright © 2016 College of American Pathologists.

major tumour types. Excellent examples demonstrating the value of immunohistology in immunophenotyping tumours include the World Health Organization classification of haematological malignancies and the subtyping of breast cancer. In the case of breast cancer, then immunohistochemistry was able to diagnose subgroups previously identified from gene expression analyses (Goldhirsch et al., 2013; Matsumoto et al., 2016).

Investigating mechanisms of cancer biology

Although genomic studies provide invaluable information on potential mechanisms underlying tumour development, the presence of RNA and DNA does not always mean that the encoded protein is transcribed. Immunohistology, in contrast, provides direct confirmation that the target protein is actually expressed, its precise subcellular distribution, the amount present and whether it is mutated and/or activated (Falini and Mason, 2002; Chan et al., 2013). Some relevant examples are summarized in Table 27.2.

Important offshoots of the direct visualization of protein include studies on the biochemical and molecular interactions of the target protein. A specific example of the study of the detection and functional role of a tumour antigen initiated by immunohistochemistry is described in more detail in the section, 'Example of the potential of immunohistochemistry'.

Multiple studies have highlighted the importance of the microenvironment in tumour growth. One area of special interest is the relationship between the cancer and the immune response of the patient. It is, therefore, crucial to be able to distinguish the cancer cells from the normal cell population. Immunohistochemistry has and continues to play an invaluable role in unravelling the information obtained from these studies. One example concerns the CD25 + FOXP3 + regulatory T cells (Tregs) that possess major roles in maintaining immune homeostasis and in tumour immunity. High

Table 27.2 Examples of antigens used to identify genetic changes in haematological malignancies

Molecular alterations and tumour type	Antigen	Mechanism	Change in staining pattern
Translocation			
t(2;5)(p23;q35) Anaplastic large cell lymphoma	ALK	Expression of the resulting fusion protein	Nuclear and cytoplasmic NPM-ALK
t(14;18)(q32;q21) Follicular lymphoma	BCL-2	Overexpression of one of the involved genes	Aberrant BCL-2
t(11;14)(q13;q32) Mantle cell lymphoma	Cyclin D1	BCL-1 and IGH genes leading to deregulation cyclin D1	Aberrant expression of cyclin-D1
Gene mutations			
Myeloid leukaemia	NPM	Aberrant localization	Cytoplasmic NPM
Various tumour types	p53	Stabilization of protein	Overexpression of p53
Chronic lymphocytic leukaemia	ZAP-70	Lack of mutation of IGH gene	Associated with overexpression of ZAP-70
Melanoma	[a] BRAF mutation	Altered protein	Presence of mutated protein
Gene deletion			
Breast	E-cadherin	Loss of expression of encoded protein	No protein detected
Gene amplification			
Breast	HER2	Increase in copy number	Overexpression of encoded protein
Virus			
Burkitt's lymphoma	EBV	Viral infection	Overexpression of EBV proteins
Cervical cancer	p16	Surrogate marker associated with papillomavirus infection	Aberrant expression of p16

[a] V600E is the most common mutation; ALK, anaplastic lymphoma kinase; BCL-2, B-cell lymphoma 2; EBV, Epstein-Barr virus; HER2, human epidermal growth factor receptor 2; NPM, nucleophosmin; ZAP-70, Zeta-chain-associated protein kinase 70

Source: data from Chan JK et al., 'The utility of immunohistochemistry for providing genetic information on tumors', *International Journal of Surgical Pathology*, Volume 21, Number 5, pp. 455–75, Copyright © 2013 by SAGE Publications; and Falini B and Mason DY, 'Proteins encoded by genes involved in chromosomal alterations in lymphoma and leukemia: clinical value of their detection by immunocytochemistry', *Blood*, Volume 99, pp. 409–26, Copyright © 2002 by The American Society of Hematology.

numbers of these cells have been associated a good prognosis in follicular lymphoma (Carreras et al., 2006) but with a poor prognosis in breast cancer (Bates et al., 2006). One possible explanation for the difference is the relationship of the CD8 + T cells, the Tregs, and their position within the tumour (Ling et al., 2014). The importance of the location of immune cells for prognosis has resulted in the introduction of the term Immunoscore that uses immunohistochemistry to identify the immune cells in and around the tumour site (Galon et al., 2013).

Cancer stem cells have been implicated in tumour development and spread in both haematological and solid tumours. The ability to identify these cells within tissue sections has, therefore, produced invaluable information with regard to their origin, renewal, their interaction within the tumour microenvironment and their response to therapies (Wang et al., 2015).

Provision of prognostic and predictive information

The quality of prognostic information improved when immunohistochemical results were included in staging systems such as the Tumour Node Metastasis (TNM), Ann Arbor, and the International Prognostic Index classifications. A list of antigens providing prognostic information is summarized in Table 27.3. One example is a panel of antibodies used to identify high-risk patients with breast cancer (Goldhirsch et al., 2013; Matsumoto et al., 2016; see Table 27.2). In the case of DLBCL, antibodies recognizing CD10, BCL-6, MUM-1, and FOXP1 can be used to determine whether the lymphoma was of germinal centre (good prognosis) or activated B-cell (poor prognosis) origin (Choi et al., 2009), a categorization that closely agrees with gene expression profiling.

Initial studies have shown that a combination of the Immunoscore with the TNM identified CD8 + T cells and CD45RO + T cells as candidate prognostic markers in certain cancers, including colorectal tumours. Reduced numbers of lymphatic vessels and the Immunoscore (cytotoxic T-cell signature), rather than genetic instability, were also associated with metastasis (Mlecnik et al., 2016).

Immunohistology performs an essential step in determining treatment options and is, perhaps, the first step in directing personalized therapy. Its use, confirming expression of the relevant target antigen(s), provides key information given the heterogeneity of protein expression found in certain cancers and also reduces the risk of patients being given inappropriate therapies.

Development of therapeutic targets

Immunohistochemistry has been instrumental in the development of therapies directed against antigens expressed by the tumour cells. These include therapeutic antibodies, cellular therapies, and the use of small molecule inhibitors (Table 27.4).

Table 27.3 Examples of antigens used for prognostic and predictive information

Antigen	Tumour	Cell type	Comments
NPM-ALK fusion protein	Anaplastic large cell lymphoma	Tumour	Good prognosis
ALK	Neuroblastoma	Tumour	Poor prognosis
FOXP3	Follicular lymphoma	Tregs	High numbers FOXP3 + cells associated with good prognosis
FOXP3	Breast	Tregs	High numbers FOXP3 + cells associated with bad prognosis
HER2	Breast	Tumour	Overexpression linked to poor prognosis
ER and PR	Breast	Tumour	Both positive then linked to better survival than ER+/PR−, who have better survival than ER−/PR−
ER, PR, and HER2	Breast	Tumour	Triple positive has improved prognosis compared to triple negative
FOXP1	Diffuse large B-cell lymphoma	Tumour	High expression linked to poor prognosis
CD10 and BCL6	Diffuse large B-cell lymphoma	Tumour	Good prognosis
BCL6 and MUM1/IRF4	Diffuse large B-cell lymphoma	Tumour	Poor prognosis
Ki67	Variety of malignancies	Tumour	High expression and poor prognosis in haematological and breast cancers
CD3, CD8, CD45RO	Variety of malignancies	T cells	Cytotoxic Immunoscore
ZAP70, CD38	Chronic lymphocytic leukaemia	Tumour	Poor prognosis
p53	Colon	Tumour	Poor prognosis
Trk-A	Neuroblastoma	Tumour	Good prognosis

ALK, anaplastic lymphoma kinase; BCL-6, B-cell lymphoma 6; ER, oestrogen receptor; FOXP, Forkhead box protein; HER2, human epidermal growth factor receptor 2; IRF-4, interferon regulatory factor 4; NPM, nucleophosmin; PR, progesterone receptor; p53, tumour protein 53; TrkA, tropomyosin receptor kinase A; ZAP-70, Zeta-chain-associated protein kinase 70.

Ideal therapeutic targets are those molecules expressed only by the cancer cells and that also play a significant role in the underlying cancer biology. A multitude of new tumour-associated markers and potential therapeutic targets have been identified using a range of high throughput methodologies, including the serological identification of antigens by recombinant expression cloning (SEREX) technique. Immunohistochemistry continues to play an invaluable role identifying which of these antigens are of potential therapeutic value. An example of this is the identification of a new cancer testis antigen, per ARNT SIM (PAS) domain containing (PASD1) protein. Immunohistochemical studies showed limited PASD1 protein in normal tissues (present in the immune-privileged testis) but its presence in a variety of lymphomas and multiple myeloma. These results highlighted the PASD1 cancer testis antigen to be a promising therapeutic target for a range of haematological malignancies (Cooper et al., 2006).

Table 27.4 Targeted therapies to known antigens confirmed using immunohistochemistry

Cancer	Target molecule	Treatment
Breast	ER, PR hormone receptors	Oestrogen receptor modulators and aromatase inhibitors (e.g. tamoxifen and analogues)
	HER2 receptor tyrosine kinase	Trastuzumab/Herceptin antibody (Gown, 2008) and TKIs such as lapatinib, pertuzumab, and antibody-drug conjugate T-DMI –efficacy being studied (Matsumoto et al., 2016)
Breast and colon	EGFR receptor tyrosine kinase	Cetuximab (Erbitux)
Liver, kidney, and colon	VEGFR, FGFR receptor tyrosine kinases	Sorafenib, sunitinib check these
Gastrointestinal stromal tumour	PDGFR receptor tyrosine kinase	Imatinib mesylate/STI571/Glivec
Chronic myeloid leukaemia	BCR-ABL—tyrosine kinase	Glivec
B-cell lymphoma	CD20	Rituximab
Anaplastic large cell lymphoma	ALK—receptor tyrosine kinase	Crizotinib
Multiple myeloma	CD117—receptor tyrosine kinase	Imatinib mesylate/STI571?Glivec
Non-small cell lung cancer	EGFR—receptor tyrosine kinase	Erlotinib and gefitinib
Non-small cell lung cancer	EML4-ALK fusion protein	Crizotinib and others
Colorectal cancer	Mutated KRAS	Cetuximab

ALK, anaplastic lymphoma kinase; BCR-ABL, breakpoint cluster region-Abelson; EGFR, epithelial growth factor receptor; EML4, Echinoderm microtubule-associated protein-like 4; ER, oestrogen receptor; FGFR, fibroblast growth factor receptor; FOXP, Forkhead box protein; HER2, human epidermal growth factor receptor 2; IRF-4, Interferon regulatory factor 4; KRAS, -Ki-ras2 Kirsten rat sarcoma viral oncogene homologue; NPM, nucleophosmin; PR, progesterone receptor; p53, tumour protein 53; TrkA, tropomyosin receptor kinase A; VEGFR, vascular endothelial growth factor receptor; ZAP-70, Zeta-chain-associated protein kinase 70.

Example of the potential of immunohistochemistry

The next few sections illustrate an example of the crucial role of immunohistology in the study of cancer biology in linking genetic abnormalities with the functions of a molecule, ALK, possessing a pivotal role(s) in the development of a variety of cancers.

Identification of ALK and its use in diagnosis

The diagnosis of anaplastic large cell lymphoma (ALCL), a CD30-positive tumour, was complicated by heterogeneity in clinical features, the appearance of the tumour cells, and by the presence of CD30 in other lymphoma types. ALCL was, however, associated with the (2;5)(p23;q35) chromosomal translocation involving the genes encoding ALK at 2p23 and nucleophosmin (NPM) at 5q35 (Morris et al., 1994). The resulting *NPM-ALK* gene was predicted to encode NPM-ALK, a fusion protein consisting of the amino region of NPM and the carboxyl terminal of ALK (Fig. 27.4A).

A monoclonal antibody, ALK1, recognizing both the NPM-ALK and the full length ALK proteins, was produced using a recombinant protein of the intracytoplasmic region of ALK (Fig. 27.4A). Immunolabelling studies confirmed the absence of ALK protein in normal tissues (with the exception of scattered cells in the brain) but showed high levels of NPM-ALK in ALCL exhibiting the t(2;5)(p23;q35) (Pulford et al., 1997). Importantly, the NPM-ALK protein exhibited a distinctive distribution, being present in the nuclei, nucleoli, and cytoplasm of the tumour cells (Fig. 27.4B). The sensitivity and specificity of the antibody meant that even individual tumour cells could be identified (arrowed). The subsequent use of the ALK1 monoclonal antibody was instrumental in identifying ALK-positive ALCL as a new tumour entity (Delsol et al., 2008).

Subsequent immunohistochemical studies showed varying patterns of labelling in some cases of ALCL suggesting that ALK proteins other than NPM-ALK were present (Fig. 27.4C). Further immunohistological studies, combined with molecular biological techniques, identified ALK fusion proteins in a subset of DLBCL and also in non-haematological tumours, such NSCLC, GIST, and breast cancer. The majority of the ALK fusion proteins contain the final 563 C-terminal aminoacid residues of ALK. Aberrant expression of the full length ALK protein has also been identified in rhabdomyosarcoma, neuroblastomas, and other neural tumours as summarized in Table 27.5 (Pulford, 2013). ALK is now considered to be a major familial neuroblastoma predisposition gene (Mosse et al., 2008) and a

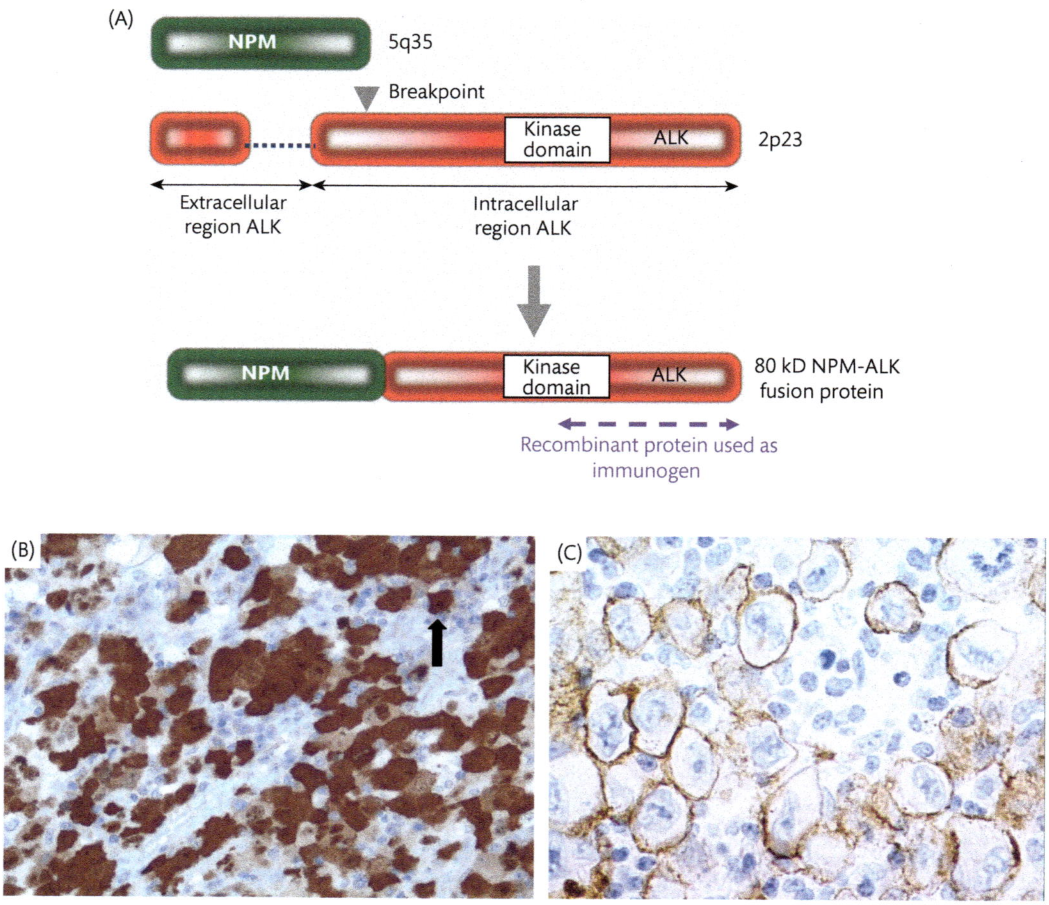

Fig. 27.4 The value of immunohistochemistry in the diagnosis of ALK-positive ALCL. (A) Diagram illustrating the t(2;5)(p23;q35) during which the first 117 aminoacids of the amino region of NPM are fused to the intracellular aminoacids 1058 to 1620 of ALK to form the 80 kD NPM-ALK fusion protein. Antibody ALK1 was produced using a 680 aminoacid recombinant protein (dotted line) from the intracytoplasmic region of ALK. (B) Immunoperoxidase labelling to show the nuclear and cytoplasmic distribution of NPM-ALK in a case of t(2;5)-positive ALCL. Even scattered tumour cells (arrowed) are easily visible. (C) Immunoperoxidase labelling shows the membrane expression of MSN-ALK (brown).

Table 27.5 Tumour expression patterns of the most common aberrant ALK fusion proteins

Translocation	ALK partner protein	Fusion protein	Subcellular distribution	Tumour
t(2;5)(p23;q35)	*Nucleophosmin (NPM)	NPM-ALK	Nucleus, nucleolus, and cytoplasm	ALCL, ALK+DLBCL
t(1;2)(p25;p23)	*Tropomyosin 3 (TPM3)	TPM3-ALK	Cytoplasm	ALCL, IMT
t(2;3)(p23;q21)	*TRK-fused gene (TFG)	TFG-ALK_S	Cytoplasm	ALCL
inv(2)(p23q35)	5-aminoimidazole-4-carboxamide ribonucleotide formyltransferase/IMP cyclohydrolase (ATIC)	ATIC-ALK	Cytoplasm	ALCL
t(2;17)(p23;q23)	*Clathrin heavy chain (CLTC)	CLTC-ALK	Cytoplasmic granules	ALCL, ALK+DLBCL, IMT
t(2;X)(p23;q11-12)	Moesin (MSN)	MSN-ALK Variants also described	Internal cell membrane	ALCL
t(2;19)(p23;p13.1)	Tropomyosin 4 (TPM3)	TPM4-ALK	Cytoplasm	ALCL, IMT, oesophageal cancer
t(2;2)(p23;q13) or inv(2) (p23q11-13)	RAN binding protein 2 (RANBP2)	RANBP2-ALK	Nuclear periphery	IMT
t(2;22)(p23;q11.2)	Non-muscle myosin heavy chain (MYH9)	MYH9-ALK	Cytoplasm	ALCL
t(2;11;2)(p23;p15;q31)	Cysteinyl-tRNA synthetase (CARS)	CARS-ALK	Cytoplasm (predicted)	IMT
t(2;4)(p23;q21)	SEC31 homologue A *(Saccharomyces cerevisiae)* (SEC31L1)	SEC31L1-ALK	Cytoplasm	IMT, ALK+DLBCL
inv(2)(p21;q21)	Echinoderm microtubule-associated protein-like4 (EML4)	EML4-ALK (variant 1—there are potentially multiple variants multiple variants)	Cytoplasm	NSCLC, breast, and colorectal cancer
t(2;10)(p23;q22.1)	Kinesin family member 5B (KIF5B)	KIF5B-ALK	Cytoplasm	NSCLC

target for genetic abnormalities in both sporadic and familial neuroblastoma (Janoueix-Lerosey et al., 2008).

The majority of the ALK fusion proteins contained the final 563 aminoacids of ALK linked to the N-terminals of the partner protein (Table 27.5) containing domains permitting dimerization. The subcellular distribution of these variant ALK fusion proteins was dependent upon the identity of the partner protein (Fig. 27.5). These results emphasized the value of simple immunolabelling screening studies to detect new ALK proteins in ALCL.

Immunostaining for ALK remains an essential part of the tool box for diagnosing ALK-positive ALCL. At present, FISH and molecular-based techniques are primarily used to detect *ALK* abnormalities in NSCLC, where ALK fusion proteins are present in low quantities. The combination of new antibodies able to detect ALK proteins with greater sensitivity, and automated staining platforms have demonstrated that immunohistochemistry can now perform a major role in identifying ALK-positive lung and colorectal tumours (Lee et al., 2015; Jiang et al., 2016) thus opening up new therapeutic opportunities for patients.

Information on the oncogenic role of ALK fusion proteins and full length ALK

The dysregulation of ALK, like other receptor tyrosine kinases, has been shown to play a major role in oncogenesis through the constitutive activation of its tyrosine kinase domain. Labelling for phosphotyrosine on tissue sections ALK-positive ALCL confirmed the presence of high levels of phosphorylation in the tumour cells. The pattern of immunolabelling was similar to that obtained with the ALK1 antibody suggesting that the ALK fusion proteins were constitutively phosphorylated (Fig. 27.6A), a result confirmed through Western blotting studies (Fig. 27.6B and C; Pulford et al., 1999). The oligomerization motif present in the majority of the partner proteins permitted dimerization of the ALK fusion proteins leading to the constitutive activation of the kinase domains (Pulford, 2013).

In vitro and *in vivo* studies using the NPM-ALK fusion protein have provided the majority of information on the pivotal biological roles of ALK proteins in ALK-positive tumours (Pulford, 2013; Giuriato and Turner, 2015). Importantly, these studies were underpinned by immunohistochemical studies. An example using a combination of quantitative proteomic analysis, immunohistochemical, biochemical and DNA-based techniques to study the activity of constitutively activated NPM-ALK in cancer development is outlined in Fig. 27.7 (Lim et al., 2009). An overview to show some of the wide-ranging roles of NPM-ALK in cell proliferation, survival, invasion, and metastasis is shown in Figure 27.8.

Immunohistochemistry has been used in preclinical models for the investigation of the role of point mutations in ALK in neural tumours. These include the preparation of ALK mutants found in neuroblastoma and tested in drosophila (Hallberg and Palmer, 2016).

The presence of NPM-ALK in the nucleus, nucleoli, and cytoplasm initially suggested that its transforming effect was exerted at all these sites. However, studies on transfected cells in a TPR-ALK murine model demonstrated that it was the NPM-ALK protein in the cytoplasm, rather than in the nucleoli, that exhibited the transformative activity (Mason et al., 1998). This raises that important caveat with regard to immunohistology that care should be taken not to assume that the distribution of the molecules involved in oncogenesis does not necessarily always reflect their activity at that site. Further functional studies should also be performed to further support the immunohistochemical results.

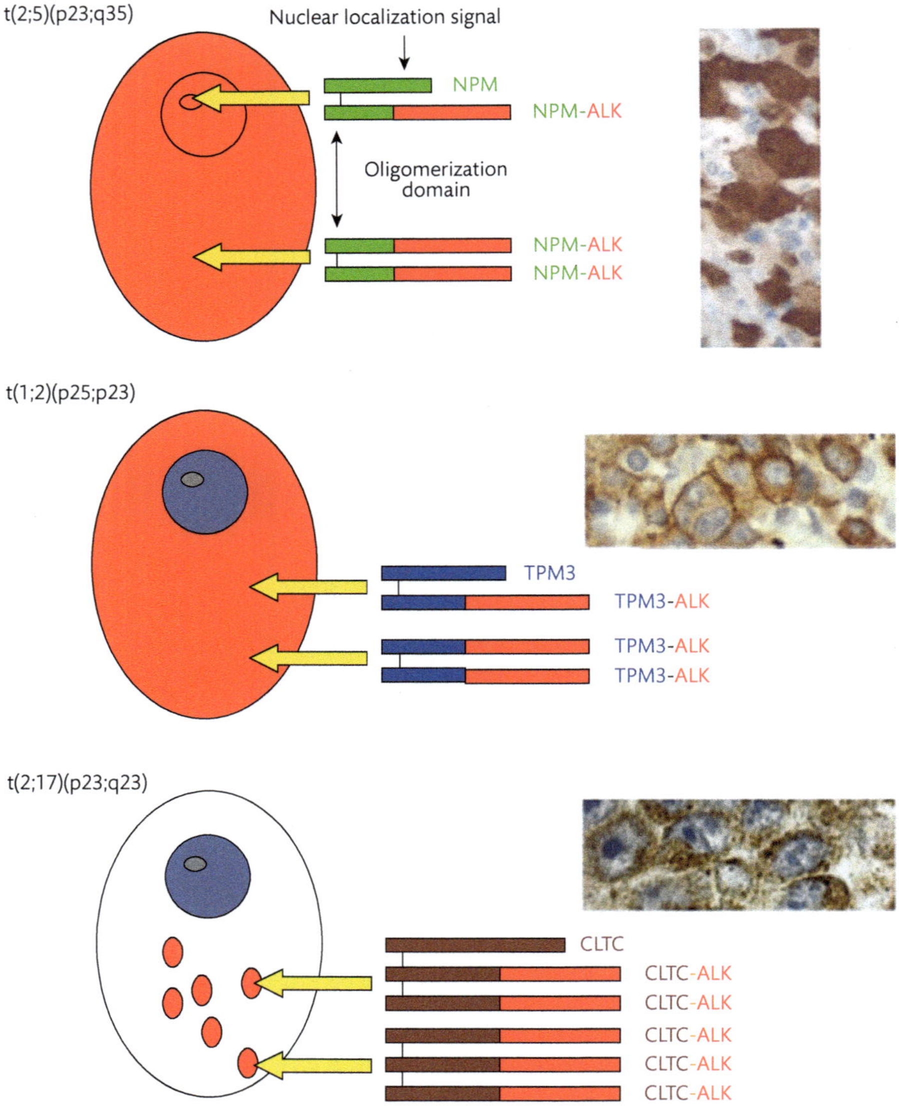

Fig. 27.5 Examples of mechanisms for the distribution of ALK fusion proteins in tumour cells. Immunoperoxidase labelling of the different ALK proteins are shown on the right. Representative diagrams of the variant ALK proteins are shown on the left. The N-termini of all the three partner proteins, NPM, TPM3, and CLTC, contain oligomerization domains permitting the formation of both homo- and heterodimers of the fusion proteins. NPM-ALK and the t(2;5)(p23;q35): NPM-ALK homodimers are present only in the cytoplasm. The nuclear localization motifs present in the N-terminus of NPM allows the entry of NPM/NPM-ALK heterodimers to the nucleolus and nucleus. TPM3-ALK and t(1;2)(p25;p23): The TPM3/TPM3-ALK/ hetero- and TPM3-ALK/TPM3-ALK homodimers are limited to the cytoplasm to give a diffuse labelling pattern of the TPM3-ALK-positive ALCL. Stronger labelling towards the cell membrane is also present. CLTC-ALK and t(2;17)(p23;q23): the CLTC/CLTC-ALK hetero- and CLTC-ALK/CLTC-ALK homodimers are present within the clathrin-coated vesicles of the ALCL pattern, visible in the NPM-ALK-positive ALCL.

Prognostic role of ALK

The prognostic significance of ALK protein expression in tumours appears to be tumour dependent. ALK expression is a powerful independent marker associated with good prognosis in ALK-positive ALCL with 80% to 90% of children achieving complete remission with chemotherapy (Delsol et al., 2008). In contrast, ALK expression in cases of breast cancer is associated with *ALK* gene amplification and with aggressive tumours (Siraj et al., 2015). EML4-ALK is associated with distinct clinicopathological features and a poorer prognosis in NSCLC and rhabdomyosarcoma (Shaw et al., 2009; Gasparini et al., 2016).

Predictive role of ALK

Alternative treatments are still being pursued for those patients with ALK-positive tumours who fail to respond to conventional chemotherapy. Small molecule inhibitors that inhibit the kinase activity of ALK represent a promising therapeutic approach. Indeed, crizotinib a tyrosine kinase inhibitor (TKI), has been shown to improve survival in treatment refractory ALK-positive ALCL (Gambacorti-Passerini et al., 2014) as well as neuroblastoma (Mosse et al., 2013). The presence of ALK in subsets of lung and breast cancer (which represent the two most common types of cancer; see Word Cancer Research Fund, 'References') also opens up new therapeutic options

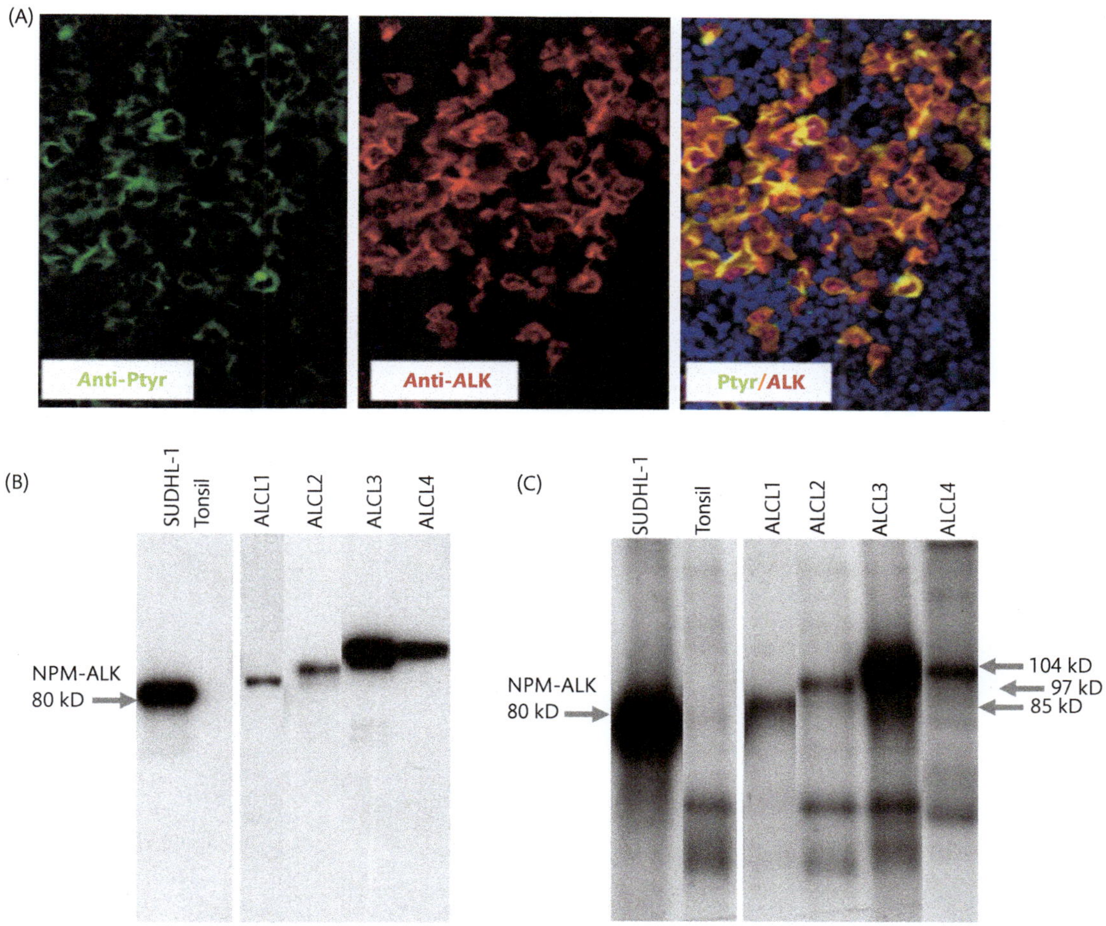

Fig. 27.6 Phosphorylation status of ALK proteins. (A) Immunofluorescent labelling studies showing (a) tyrosine phosphorylation and (b) NPM-ALK expression of cells in a case t(2;5) ALK-positive ALCL. The distribution of phosphotyrosine (green) is similar to that of NPM-ALK (red). Combining the two images shows areas of colocalization (yellow) of the two antigens in the cytoplasm. (B) Western blotting results obtained from four different cases of ALK-positive ALCL showing the presence of 85 kD (ALCL2), 97 kD (ALCL2) and 104 kD (ALCL3 and ALCL4) ALK proteins. The position of 80 kD NPM-ALK is shown in the SUDHL-1 lysates while tonsil is a negative control. (C) Results of a tyrosine kinase assay confirm the activation of the different ALK proteins. The diagram illustrates an example in which immunohistochemistry is used to validate data obtained from mass spectroscopy studies to identify functional roles for NPM-ALK.

Fig. 27.6B and C reprinted from *The American Journal of Pathology*, Volume 154, Issue 6, Karen Pulford et al. 'Biochemical Detection of Novel Anaplastic Lymphoma Kinase Proteins in Tissue Sections of Anaplastic Large Cell Lymphoma', pp. 1657–63,

using TKIs in these diseases. Indeed crizotinib received US Food and Drug Administration approval in 2011 for use in patients with NSCLC.

An important aspect of more effective targeting of treatment is the identification of high-risk patients. In the case of ALCL, immunohistochemistry has played an important role in identifying high-risk patients on the basis of testing for the presence of autoantibodies to ALK in transfectant cells (Ait-Tahar et al., 2010). There is also increasing evidence to support the utility of immunohistochemistry to perform an initial screen to detect ALK in those patients with lung cancer who would benefit from molecularly based treatments such as ALK inhibitors (Takeuchi et al., 2016).

ALK and the development of new therapies

ALK is an excellent example of an ideal therapeutic target having extremely limited expression in normal tissues but expressed in a wide range of tumours and possessing functional roles pivotal to the development of the cancer (Pulford, 2013). The latter attribute means that ALK expression, rather than being lost due to the process of immunoediting, remains available for therapeutic targeting. Understanding the biology of ALK should, therefore, provide important information for the development of new treatments.

Resistance to crizotinib occurs possibly through secondary mutations in ALK and/or through the activation of the multiple different signalling pathways involving ALK (Shaw et al., 2016). Second-generation TKIs, such as ceritinib and alectinib, are currently under investigation for use in those patients who become resistant to first-line crizotinib. Ceritinib inhibits the L1196M and G1269A ALK point mutations and has 20 times the potency of crizotinib, while alectinib inhibits EML4-ALK more effectively than crizotinib. Lorlatinib is promising as a third-line TKI for patients with mutated ALK. Other possible approaches include HSP90 inhibitors which

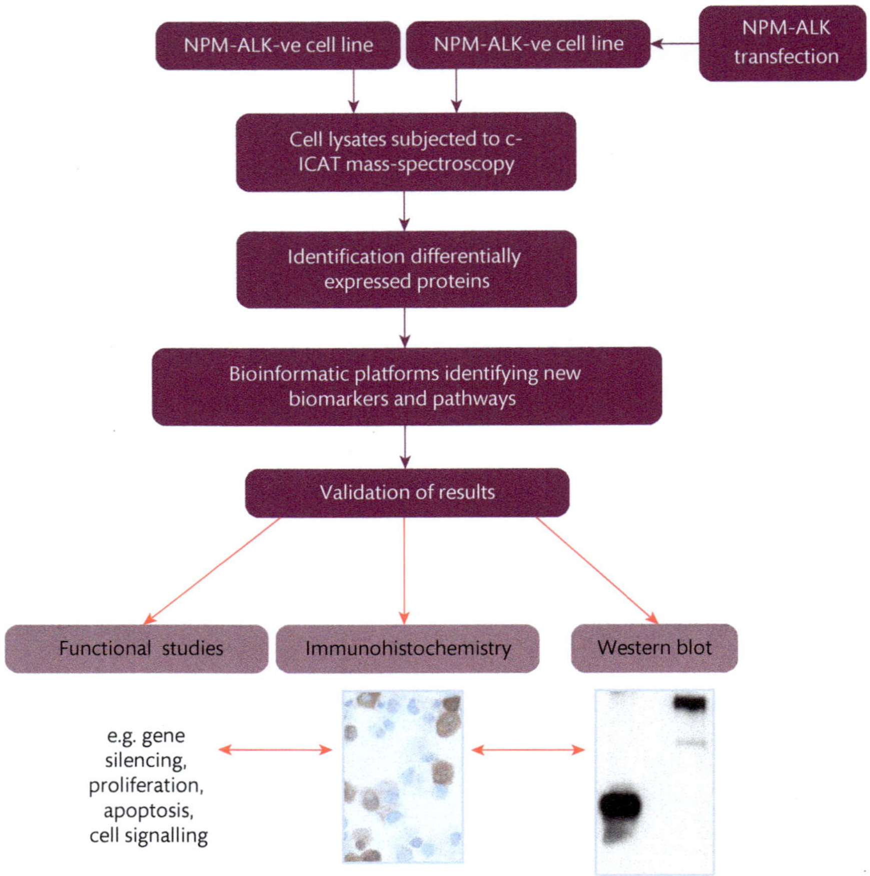

Fig. 27.7 An example showing how immunohistochemistry can occupy a central role in validating data obtained from mass spectroscopy studies to identify functional roles for NPM-ALK

Source: data from Lim, MS et al., 'The proteomic signature of NPM/ALK reveals deregulation of multiple cellular pathways', *Blood*, Volume 114, Issue 8, pp. 1585–95, Copyright © 2009 by American Society of Hematology.

would increase the proteasomal degradation of ALK (Farina and Gambacorti-Passerini, 2016).

While NPM-ALK can elicit both a B- and T-cell immune response in patients with ALK-positive ALCL (Pulford, 2013), ALK is also immunogenic in patients with other ALK-positive tumours (Damm-Welk et al., 2016). Such immunogenicity opens up the possibility of the development of vaccines targeting ALK in ALK-positive tumours (Voena et al., 2015) to be used alone, or in combination with other therapy types.

Next-generation immunohistology in cancer biology

Immunohistochemistry holds a key position linking myriad findings from the genomic and proteomic based studies to the confirmation and investigation of functional molecules with important roles in cancer biology. The presence of more than 300 million human tissue specimens located in clinical biorepositories in the United States only 15 years ago (Baker, 2012) opens up the possibility of performing multiple retrospective studies resulting in the confirmation of ever more important information. There are, however, practical challenges given the speed at which new information is becoming available.

Improvements in antibody production and validation

Immunohistochemistry is absolutely dependent upon the availability of properly characterized fully validated antibodies that are specific for the target antigen. It is highly probable that thousands of currently available antibodies are insufficiently characterized. Journals are requesting that information on the provenance of antibodies be provided and that only properly characterized antibodies of known specificity are used in clinical and research contexts (Bradbury and Pluckthun, 2015). This is accompanied by a demand for both providers and users of antibodies to make all relevant technical information available to readers and to follow prescribed protocols to ensure the correct validation of antibodies (Roncador et al., 2016). EuroMAbNet, a European network of laboratories specializing in the production, validation, and use of antibodies, is one example of a movement that has been established to meet this need (see EuroMAbNet, 'References').

Improvements in staining and imaging technologies

The combination of developments in microscopes, improvements in the range and combinations of chromogenic and fluorescent tags, and nanotechnology has enabled imaging to be performed at the level of a single molecule. Stimulated emission depletion (STED)

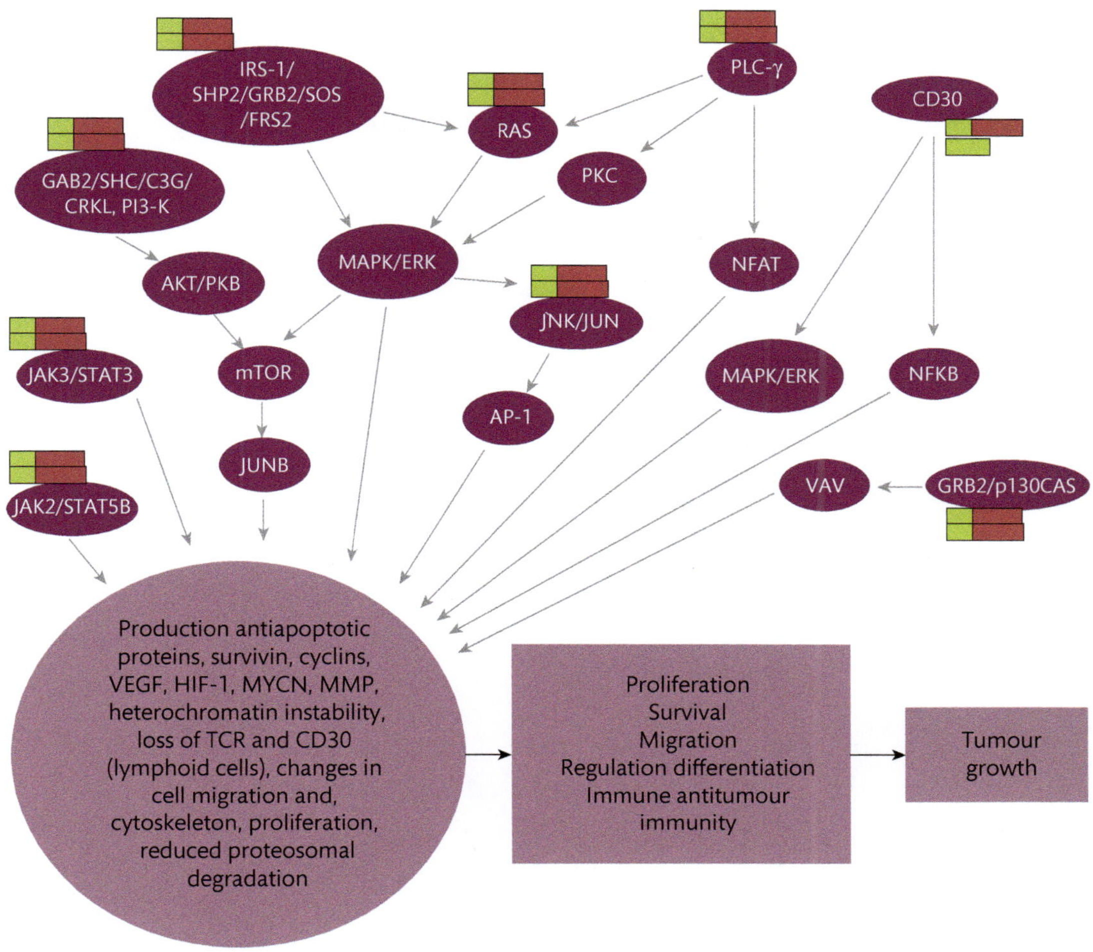

Fig. 27.8 A summary to illustrate some of the multiple pathways linking NPM-ALK to tumour development.
Adapted from Karen Pulford, 'ALK: Anaplastic lymphoma kinase', in Edward P. Gelmann et al. (Eds.), *Molecular Oncology: Causes of Cancer and Targets for Treatment*, Cambridge University Press, Cambridge, UK, Copyright © 2013. Reprinted with permission.

microscopy is a good example, being well suited to the use of immunohistochemistry both in tissue sections and live cells (Blom and Brismar, 2014).

Two approaches to performing multiplexed staining analyses in cells and tissue sections combined immunohistochemistry with matrix-assisted laser desorption/ionization (MALDI)-TOF time of flight (TOF) imaging mass cytometry (Giesen et al., 2014) and multiplexed ion beam imaging (Angelo et al., 2014). Antibodies were directly linked to isotopically pure non-biological elemental metal reporters or tags and opened up the possibility of screening a single section with up to 100 antibodies from a single tissue section (Angelo et al., 2014). These advances enabled changes in the profiles of molecules and their cellular distribution occurring in cancer, in addition to changes occurring as a result of drug treatment, to be studied at the molecular level (Fig. 27.9B). The introduction of laser capture microdissection, a procedure enabling specific groups of cells to be removed for study, should further complement these approaches.

Analysis of images

The ability to generate and stain high-quality TMAs is an essential step in the analysis of the thousands of results obtained from molecular studies. However, while the staining of the TMAs and image capturing is rapid and largely automated, the analysis of the results represents a 'bottleneck' in the whole process. Although the presence or absence of a target molecule in a cell is easy to confirm, whether the cells or tissues under study are considered 'positive' for diagnostic and/or prognostic purposes requires a more thorough analysis. Various options of scoring, such as the Quickscore, have been proposed to meet this demand (reviewed by Amaral et al., 2013). A number of image analysis software systems have been proposed for scoring the more demanding multiplexed stained images (Stack et al., 2014).

Another problem that needs to be addressed is the accurate quantitation of the amount of target molecule present. The use of quantum dots offers one approach to provide quantitative information concerning both amount of target protein present but also its dynamics at a nanoscale level (Vu et al., 2015).

Integration of 'omics' platforms and technology

Individual technologies have given rise to the evolution of a variety of different knowledge platforms. Examples include the proteome, the transcriptome, immunome, interactome, and the metabolome,

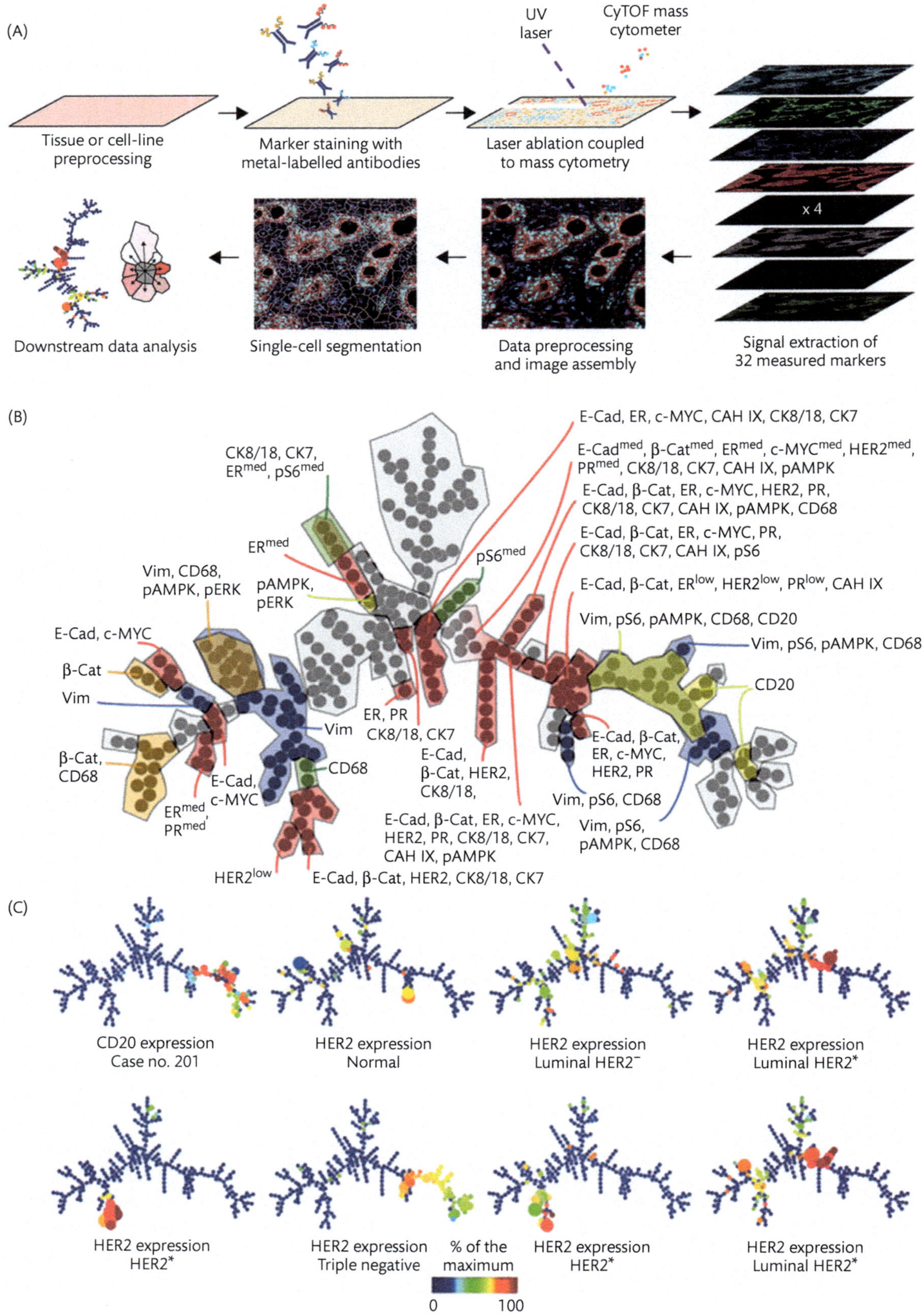

Fig. 27.9 Combination of mass spectroscopy and immunohistochemistry in the study of breast cancer. (A) An outline of the methodology used in which 32-plex imaging mass cytometry was performed on a TMA of samples from 20 breast cancer patients. (B) A SPADE (spanning-tree progression analysis of density-normalized events) tree illustrating differences in the cell populations. (C) Cluster trees showing the heterogeneity of HER2 expression in different cases of breast cancer subtypes.

each specific to the particular discipline under study. The integration of these platforms is essential to obtain the optimal benefit on cancer and its underlying biology. The Human Protein Atlas and BioGPS gene portal system (see 'References') represent two examples of integration between the immunohistochemical, transcriptomics, and metabolomics platforms.

TAKE-HOME MESSAGE

Immunohistology is an invaluable science and art that is not limited to the clinic. It works 'hand in hand' with other techniques to provide a filter directing the clinician or researcher to those molecules with the most relevance in cancer biology. The ability to visualize and study the expression of molecules *in situ* has provided unrivalled opportunities to investigate the influence of genomic and molecular changes at both the cellular level and within the context of the tumour environment. Compared to many molecular-based technologies, immunohistochemistry employs relatively quick and simple techniques requiring a limited amount of specialist equipment. Its use also reduces the risk of problems associated with the contamination of tumour samples by the presence of other cell types.

The insight gained from immunohistochemistry has revolutionized the diagnosis of cancer and improved therapeutic options for patients. Information obtained from investigating oncogenetic mechanisms and the effects of therapies at the cellular and molecular levels should continue to improve patient outcomes. Indeed, the accumulation of knowledge of the immunophenotype of the tumour cell and the functions of the target molecules is a major advance in the bid towards personalized medicine.

The global future of immunohistology is assured. The quest for increasing the availability of excellent antibodies is ongoing. This development, combined with advances in staining methodologies and the integration of other research platforms, should ensure that immunohistochemistry becomes even more important in the future.

OPEN QUESTIONS

- Will the production of specific antibodies keep up with the number of molecules identified through other the high throughput technologies?
- Who will monitor the quality of these antibodies?
- How will the integration of different technology platforms be facilitated?
- How will further knowledge of cancer biology be able to better define (a) patients who benefit, or (b) become resistant to specific treatments?
- How will further knowledge of cancer biology facilitate the development of new therapeutic approaches?
- How near are 'off the shelf' therapies?

FURTHER READING

Belizario, J. E., Sangiuliano, B. A., Perez-Sosa, M., Santos, B. V., & Machado-Santelli, G. (2015). Advances in the integration of optical and mass spectrometry molecular imaging technologies: from omics data to molecular signature discovery. *Discov Med*, **20**, 393–401.

Galon, J., Mlecnik, B., Bindea, G., et al. (2014). Towards the introduction of the 'Immunoscore' in the classification of malignant tumours. *J Pathol*, **232**, 199–209.

Krogan, N. J., Lippman, S., Agard, D. A., Ashworth, A., & Ideker, T. (2015). The cancer cell map initiative: defining the hallmark networks of cancer. *Mol Cell*, **58**, 690–8.

McCourt, C. M., Boyle, D., James, J., & Salto-Tellez, M. (2013). Immunohistochemistry in the era of personalised medicine. *J Clin Pathol*, **66**, 58–61.

Meller, S., Meyer, H. A., Bethan, B., et al. (2015). Integration of tissue metabolomics, transcriptomics and immunohistochemistry reveals ERG- and Gleason score-specific metabolomic alterations in prostate cancer. *Oncotarget*, **7**, 1421–38.

Muenst, S., Laubli, H., Soysal, S. D., Zippelius, A., Tzankov, A., & Hoeller, S. (2016). The immune system and cancer evasion strategies: therapeutic concepts. *J Intern Med*, **279**(6), 541–62.

Regad, T. (2015). Targeting RTK signaling pathways in cancer. *Cancers (Basel)*, **7**, 1758–84.

Uhlen, M., Bandrowski, A., Carr, S., et al. (2016). A proposal for validation of antibodies. *Nat Methods*, **13**(10), 823–7.

Wang, T., Shigdar, S., Gantier, M. P., et al. (2015). Cancer stem cell targeted therapy: progress amid controversies. *Oncotarget*, **6**, 44191–206.

REFERENCES

Ait-Tahar, K., Damm-Welk, C., Burkhardt, B., et al. (2010). Correlation of the autoantibody response to the ALK oncoantigen in pediatric anaplastic lymphoma kinase-positive anaplastic large cell lymphoma with tumor dissemination and relapse risk. *Blood*, **115**, 3314–9.

Amaral, T., Mckenna, S. J., Robertson, K., & Thompson, A. (2013). Classification and immunohistochemical scoring of breast tissue microarray spots. *IEEE Trans Biomed Eng*, **60**, 2806–14.

Angelo, M., Bendall, S. C., Finck, R., et al. (2014). Multiplexed ion beam imaging of human breast tumors. *Nat Med*, **20**, 436–42.

Baker, M. (2012). Biorepositories: building better biobanks. *Nature*, **486**, 141–6.

Bates, G. J., Fox, S. B., Han, C., et al. (2006). Quantification of regulatory T cells enables the identification of high-risk breast cancer patients and those at risk of late relapse. *J Clin Oncol*, **24**, 5373–80.

BioGPS. A free extensible and customizable gene annotation portal, a complete resource for learning about gene and protein function. Available at: http://biogps.org

Bird, C. (2012). Antibodies User Survey: findings show researchers value quality over low price. *The Scientist*. Available at: https://www.the-scientist.com/lab-tools/antibodies-user-survey-41051

Blazquez, C., Cook, N., Micklem, K., Harris, A. L., Gatter, K. C., & Pezzella, F. (2006). Phosphorylated KDR can be located in the nucleus of neoplastic cells. *Cell Res*, **16**, 93–8.

Blom, H. & Brismar, H. (2014). STED microscopy: increased resolution for medical research? *J Intern Med*, **276**, 560–78.

Bradbury, A. & Pluckthun, A. (2015). Reproducibility: standardize antibodies used in research. *Nature*, **518**, 27–9.

Carreras, J., Lopez-Guillermo, A., Fox, B. C., et al. (2006). High numbers of tumor-infiltrating FOXP3-positive regulatory T cells are associated with improved overall survival in follicular lymphoma. *Blood*, **108**, 2957–64.

Chan, J. K., Ip, Y. T., & Cheuk, W. (2013). The utility of immunohistochemistry for providing genetic information on tumors. *Int J Surg Pathol*, **21**, 455–75.

Choi, W. W., Weisenburger, D. D., Greiner, T. C., et al. (2009). A new immunostain algorithm classifies diffuse large B-cell lymphoma into molecular subtypes with high accuracy. *Clin Cancer Res*, **15**, 5494–502.

Cooper, C. D., Liggins, A. P., Ait-Tahar, K., Roncador, G., Banham, A. H., & Pulford, K. (2006). PASD1, a DLBCL-associated cancer testis antigen and candidate for lymphoma immunotherapy. *Leukemia*, **20**, 2172–4.

Cordell, J. L., Falini, B., Erber, W. N., et al. (1984). Immunoenzymatic labeling of monoclonal antibodies using immune complexes of alkaline phosphatase and monoclonal anti-alkaline phosphatase (APAAP complexes). *J Histochem Cytochem*, **32**, 219–29.

Damm-Welk, C., Siddiqi, F., Fischer, M., et al. (2016). Anti-ALK antibodies in patients with ALK-positive malignancies not expressing NPM-ALK. *J Cancer*, **7**, 1383–7.

Delsol, G., Jaffe, E., Falini, B., et al. (2008). *Anaplastic large cell lymphoma (ALCL), ALK-positive*. Lyon, France: International Agency for Research on Cancer.

Dubel, S., Stoevesandt, O., Taussig, M. J., & Hust, M. (2010). Generating recombinant antibodies to the complete human proteome. *Trends Biotechnol*, **28**, 333–9.

EuroMabNet. European Monoclonal Antibodies Network. Available at: https://www.euromabnet.com/

Falini, B. & Mason, D. Y. (2002). Proteins encoded by genes involved in chromosomal alterations in lymphoma and leukemia: clinical value of their detection by immunocytochemistry. *Blood*, **99**, 409–26.

Farina, F. & Gambacorti-Passerini, C. (2016). ALK inhibitors for clinical use in cancer therapy. *Front Biosci (Elite Ed)*, **8**, 46–60.

Galon, J., Angell, H. K., Bedognetti, D., & Marincola, F. M. (2013). The continuum of cancer immunosurveillance: prognostic, predictive, and mechanistic signatures. *Immunity*, **39**, 11–26.

Gambacorti-Passerini, C., Farina, F., Stasia, A., et al. (2014). Crizotinib in advanced, chemoresistant anaplastic lymphoma kinase-positive lymphoma patients. *J Natl Cancer Inst*, **106**, djt378.

Gasparini, P., Casanova, M., Villa, R., et al. (2016). Anaplastic lymphoma kinase aberrations correlate with metastatic features in pediatric rhabdomyosarcoma. *Oncotarget*, **7**(37), 58903–14.

Gellrich, S., Ventura, R., Jones, M., Tan, S. Y., & Mason, D. Y. (2004). Immunofluorescent and FISH analysis of skin biopsies. *Am J Dermatopathol*, **26**, 242–7.

Giesen, C., Wang, H. A., Schapiro, D., et al. (2014). Highly multiplexed imaging of tumor tissues with subcellular resolution by mass cytometry. *Nat Methods*, **11**, 417–22.

Giuriato, S. & Turner, S. D. (2015). Twenty years of modelling NPM-ALK-induced lymphomagenesis. *Front Biosci (Schol Ed)*, **7**, 236–47.

Goldhirsch, A., Winer, E. P., Coates, A. S., et al. (2013). Personalizing the treatment of women with early breast cancer: highlights of the St Gallen International Expert Consensus on the Primary Therapy of Early Breast Cancer 2013. *Ann Oncol*, **24**, 2206–23.

Gown, A. M. (2008). Current issues in ER and HER2 testing by IHC in breast cancer. *Mod Pathol*, **21** Suppl 2, S8–S15.

Hallberg, B. & Palmer, R. H. (2016). The role of the ALK receptor in cancer biology. *Ann Oncol*, **27** Suppl 3, iii4–iii15.

Janoueix-Lerosey, I., Lequin, D., Brugieres, L., et al. (2008). Somatic and germline activating mutations of the ALK kinase receptor in neuroblastoma. *Nature*, **455**, 967–70.

Jiang, L., Yang, H., HE, P., Liang, W., et al. (2016). Improving selection criteria for ALK inhibitor therapy in non-small cell lung cancer: a pooled-data analysis on diagnostic operating characteristics of immunohistochemistry. *Am J Surg Pathol*, **40**, 697–703.

Kallioniemi, O. P., Wagner, U., Kononen, J., & Sauter, G. (2001). Tissue microarray technology for high-throughput molecular profiling of cancer. *Hum Mol Genet*, **10**, 657–62.

Kandalaft, P. L. & Gown, A. M. (2015). Practical applications in immunohistochemistry: carcinomas of unknown primary site. *Arch Pathol Lab Med*, **140**(6), 508–23.

Kohler, G. & Milstein, C. (1975). Continuous cultures of fused cells secreting antibody of predefined specificity. *Nature*, **256**, 495–7.

Lee, J., Kim, H. C., Hong, J. Y., et al. (2015). Detection of novel and potentially actionable anaplastic lymphoma kinase (ALK) rearrangement in colorectal adenocarcinoma by immunohistochemistry screening. *Oncotarget*, **6**, 24320–32.

Leong, A. S. & Leong, T. Y. (2006). Newer developments in immunohistology. *J Clin Pathol*, **59**, 1117–26.

Lim, M. S., Carlson, M. L., Crockett, D. K., et al. (2009). The proteomic signature of NPM/ALK reveals deregulation of multiple cellular pathways. *Blood*, **114**, 1585–95.

Ling, A., Edin, S., Wikberg, M. L., Oberg, A., & Palmqvist, R. (2014). The intratumoural subsite and relation of CD8(+) and FOXP3(+) T lymphocytes in colorectal cancer provide important prognostic clues. *Br J Cancer*, **110**, 2551–9.

Mansfield, J. R. (2014). Multispectral imaging: a review of its technical aspects and applications in anatomic pathology. *Vet Pathol*, **51**, 185–210.

Mason, D. Y., Micklem, K., & Jones, M. (2000). Double immunofluorescence labelling of routinely processed paraffin sections. *J Pathol*, **191**, 452–61.

Mason, D. Y., Pulford, K. A., Bischof, D., et al. (1998). Nucleolar localization of the nucleophosmin-anaplastic lymphoma kinase is not required for malignant transformation. *Cancer Res*, **58**, 1057–62.

Matsumoto, A., Jinno, H., Ando, T., et al. (2016). Biological markers of invasive breast cancer. *Jpn J Clin Oncol*, **46**, 99–105.

Matsumura, H., Ohnishi, T., Kanemura, Y., Maruno, M., & Yoshimine, T. (2000). Quantitative analysis of glioma cell invasion by confocal laser scanning microscopy in a novel brain slice model. *Biochem Biophys Res Commun*, **269**, 513–20.

Mlecnik, B., Bindea, G., Kirilovsky, A., et al. (2016). The tumor microenvironment and Immunoscore are critical determinants of dissemination to distant metastasis. *Sci Transl Med*, **8**, 327ra26.

Morris, S. W., Kirstein, M. N., Valentine, M. B., et al. (1994). Fusion of a kinase gene, ALK, to a nucleolar protein gene, NPM, in non-Hodgkin's lymphoma. *Science*, **263**, 1281–4.

Mosse, Y. P., Laudenslager, M., Longo, L., et al. (2008). Identification of ALK as a major familial neuroblastoma predisposition gene. *Nature*, **455**, 930–5.

Mosse, Y. P., Lim, M. S., Voss, S. D., et al. (2013). Safety and activity of crizotinib for paediatric patients with refractory solid tumours or anaplastic large-cell lymphoma: a Children's Oncology Group phase 1 consortium study. *Lancet Oncol*, **14**, 472–80.

Pulford, K. (2013). ALK: Anaplastic lymphoma kinase. In: Gelmann E. P., Sawyers C. L., & Rauscher F. J. (eds), *Molecular Oncology: Causes of Cancer and Targets for Treatment*. Cambridge: Cambridge University Press.

Pulford, K., Banham, A. H., Lyne, L., et al. (2006). The BCL11AXL transcription factor: its distribution in normal and malignant tissues and use as a marker for plasmacytoid dendritic cells. *Leukemia*, **20**, 1439–41.

Pulford, K., Falini, B., Cordell, J., et al. (1999). Biochemical detection of novel anaplastic lymphoma kinase proteins in tissue sections of anaplastic large cell lymphoma. *Am J Pathol*, **154**, 1657–63.

Pulford, K., Lamant, L., Espinos, E., et al. (2004). The emerging normal and disease-related roles of anaplastic lymphoma kinase. *Cell Mol Life Sci*, **61**, 2939–53.

Pulford, K., Lamant, L., Morris, S. W., et al. (1997). Detection of anaplastic lymphoma kinase (ALK) and nucleolar protein nucleophosmin (NPM)-ALK proteins in normal and neoplastic cells with the monoclonal antibody ALK1. *Blood*, **89**, 1394–404.

Roncador, G., Engel, P., Maestre, L., et al. (2016). The European antibody network's practical guide to finding and validating suitable antibodies for research. *MAbs*, **8**, 27–36.

Shaw, A. T., Gandhi, L., Gadgeel, S., et al. (2016). Alectinib in ALK-positive, crizotinib-resistant, non-small-cell lung cancer: a single-group, multicentre, phase 2 trial. *Lancet Oncol*, **17**, 234–42.

Shaw, A. T., Yeap, B. Y., Mino-Kenudson, M., et al. (2009). Clinical features and outcome of patients with non-small-cell lung cancer who harbor EML4-ALK. *J Clin Oncol*, **27**, 4247–53.

Siraj, A. K., Beg, S., Jehan, Z., et al. (2015). ALK alteration is a frequent event in aggressive breast cancers. *Breast Cancer Res*, **17**, 127.

Soilleux, E. & Gatter, K. C. (2006). The antibody revolution: how 'immuno' changed pathology. Pathological Society Annual Meeting & Centenary Meeting, 2006 Manchester, UK: Wiley, pp. 173–84.

Stack, E. C., Wang, C., Roman, K. A., & Hoyt, C. C. (2014). Multiplexed immunohistochemistry, imaging, and quantitation: a review, with an assessment of tyramide signal amplification, multispectral imaging and multiplex analysis. *Methods*, **70**, 46–58.

Takeuchi, K., Togashi, Y., Kamihara, Y., et al. (2016). Prospective and clinical validation of ALK immunohistochemistry: results from the phase I/II study of alectinib for ALK-positive lung cancer (AF-001JP study). *Ann Oncol*, **27**, 185–92.

The Human Protein Atlas. Available at: https://www.proteinatlas.org/

Voena, C., Menotti, M., Mastini, C., et al. (2015). Efficacy of a cancer vaccine against ALK-rearranged lung tumors. *Cancer Immunol Res*, **3**, 1333–43.

Vu, T. Q., Lam, W. Y., Hatch, E. W., & Lidke, D. S. (2015). Quantum dots for quantitative imaging: from single molecules to tissue. *Cell Tissue Res*, **360**, 71–86.

Wang, T., Shigdar, S., Gantier, M. P., et al. (2015). Cancer stem cell targeted therapy: progress amid controversies. *Oncotarget*, **6**, 44191–206.

World Cancer Research Fund. Cancer trends. Available at: https://www.wcrf.org/dietandcancer/cancer-trends

SECTION VI
The biology of cancer treatment

28

Principles of chemotherapy

David J. Kerr, Daniel Haller, and Jaap Verweij

Introduction

The treatment of cancer is one of the best settings for a multidisciplinary approach to treatment in medicine. Surgery and radiotherapy are frequently still the primary choice of treatment for patients with malignant tumours. However, since 60–70% of patients with cancer will develop metastatic disease during their lifetime despite local control of their cancer, for most patients cancer may be considered as a systemic disease, requiring systemic treatment. In addition, given with radiation and surgery, systemic therapy is frequently given as part of primary treatments with curative intent. The development of systemic therapy over the last few decades has therefore created an important role for medical oncologists in the care of patients with cancer. One of the dominant reasons for the emergence of medical oncology as a subspecialty was the significance of the toxicity associated with conventional cytotoxic drugs. Classically, these agents have steep dose–response curves and narrow therapeutic ratios, and therefore a small increment in dose can lead to a large increase in toxicity. Famously, the French philosopher Rene Descartes declared, 'Cogito ergo sum', I think therefore I am, and we may be able to modify this statement to justify the emergence of medical oncology in the early days to 'Veneno, ergo sum'.

The types of systemic treatment available to the medical oncologist are continually expanding with newly emerging pharmacological and biological therapies that have clinical activity. We can predict that the role of medical oncology will become increasingly important in the near future.

Every medical oncologist must be aware of the scientific rationale for choosing specific drugs, combinations of drugs, or combinations of different types of treatment. This chapter summarizes the basis of chemotherapy and addresses important issues in the development of new approaches using molecular targets in a more sophisticated way to try to obtain tumour cell kill, dormancy, or enhanced immune rejection of the cancer.

Principles of chemotherapy

As indicated earlier, in the majority of patients with cancer chemotherapy will be considered for use at some time during the course of their illness, either aiming at cure, prolongation of life, or palliation, depending on tumour type, stage, and the relative fitness of the patient.

In an ideal world, the design of chemotherapy regimens should be based on specific knowledge about cell cycle kinetics, pharmacokinetics, biochemical–pharmacological factors, and bioinformatic analysis of the consequences of inhibition of signal transduction pathways. However, this is often still trumped by empirical knowledge of the responsiveness of specific tumours to specific drugs, and a more traditional disease-oriented approach to anticancer drug development. Although much has been made of the remarkable insights that cell and molecular biology has given the oncology community and a belief that it would be possible to treat cell targets and pathways rather than specific tumour types, it has become clear that a 'driver mutation' in, for example, lung cancer may not operate in the same way in colorectal or breast cancer, given the differing mutational landscape in these diverse tumour types. This lends greater weight to the computational statisticians' attempts to model what the consequences of inhibition of pathway X are on pathway Y, and how different resistance escape mechanisms might operate depending on the constellation of background mutations (Domingo et al., 2013).

Mechanism of action of commonly available drugs

Anthracyclines (doxorubicin, daunomycin, epirubicin, daunorubicin) are antitumour antibiotics used in cancer chemotherapy derived from the *Streptomyces* bacteria and have three mechanisms of action:

- Inhibits DNA and RNA synthesis by intercalating between base pairs of the DNA/RNA strand, thus preventing the replication of rapidly-growing cancer cells.
- Inhibits topoisomerase II enzyme, preventing the relaxing of super-coiled DNA and thus blocking DNA transcription and replication.
- Creates iron-mediated free oxygen radicals that damage the DNA and the lipid domain of cell membranes.

Vinca alkaloids (vincristine, vinorelbine, vinblastine, vindesine) are antimitotic and antimicrotubule agents that were originally derived from the Periwinkle plant *Catharanthus roseus*. The principal mechanisms of cytotoxicity relate to their interactions with tubulin and disruption of microtubule function, particularly of microtubules comprising the mitotic spindle apparatus, leading to metaphase arrest.

Taxanes (docetaxel, paclitaxel, nab-paclitaxel) are diterpenes produced by the plants of the genus Taxus (yews). The principal mechanism of action of the taxane class of drugs is the disruption of microtubule function. Microtubules are essential to cell division, and taxanes stabilize GDP-bound tubulin in the microtubule, thereby inhibiting the process of cell division—a 'frozen mitosis'. Thus, in essence, taxanes are mitotic inhibitors. In contrast to the taxanes, the vinca alkaloids destroy mitotic spindles. Both taxanes and vinca alkaloids are together named spindle poisons

Alkylating agents (nitrogen mustards—cyclophosphamide, ifosfamide, melphalan, chlorambucil; nitrosoureas—carmustine, lomustine, streptozotocin; alkyl sulphonate—busulfan).

Their principal mechanism of action is to alkylate the N7 residue of the guanine which can cross-link nucleobases in DNA double-helix strands. This makes the strands unable to uncoil and separate leading ultimately to apoptotic cell death. The antitumour antibiotic, mitomycin C, isolated from *Streptomyces caespitosus* can also be considered an alkylating agent as it is activated to produce a species which crosslinks guanine residues in the sequence 5'-CpG-3'.

Bleomycin is a glycopeptide antibiotic produced by the bacterium *Streptomyces verticillus* which induces DNA strand breaks by generating superoxide free radicals.

Platinum-based agents, including cisplatin, carboplatin, oxaliplatin, also bind N7 guanine and can cause intra- and interstrand DNA crosslinks, inhibiting DNA synthesis and inducing programmed cell death.

Antimetabolites (purines, pyrimidines, antifolates) are chemicals that inhibit the use of a naturally occurring metabolite, which is essential for the cell's normal economy (e.g. DNA and protein synthesis). Such drugs are often similar in structure to the metabolite with which they interfere. Purine analogues (mercaptopurine, thioguanine, fludarabine, pentostatin, and cladribine) are antimetabolites that mimic the structure of metabolic purines and which therefore inhibit DNA synthesis. Pyrimidine analogues (5-fluorouracil, gemcitabine, floxuridine, cytosine arabinoside) are antimetabolites which mimic the structure of metabolic pyrimidines and inhibit DNA and RNA synthesis. Antifolate analogues (methotrexate, pemetrexed) are drugs that impair the function of folic acids. A well-known example is methotrexate. This is a folic acid analogue which, owing to structural similarity, binds and inhibits the enzyme dihydrofolate reductase (DHFR) and thus prevents the formation of tetrahydrofolate. Because tetrahydrofolate is essential for purine and pyrimidine synthesis, methotrexate inhibits production of DNA, RNA, and proteins (as tetrahydrofolate is also involved in the synthesis of amino acids serine and methionine).

Topoisomerase inhibitors(topo 1—irinotecan, topotecan; topo 2—etoposide, amsacrine, teniposide) inhibit the two enzymes that regulate the overwinding or underwinding of DNA and lead to single and double strand DNA breaks that can induce apoptosis.

Hormonal agents: Hormonal therapy involves the manipulation of the endocrine system through exogenous administration of specific hormones, particularly steroid hormones, or drugs which inhibit the production or activity of such hormones (hormone antagonists). Because steroid hormones are powerful drivers of gene expression in certain cancer cells, changing the levels or activity of certain hormones can cause cytostasis, or cell death.

Aromatase inhibitors: At menopause, oestrogen production in the ovaries ceases, but other tissues continue to produce oestrogen through the action of the enzyme aromatase on androgens produced by the adrenal glands. Aromatase blockade reduces oestrogen levels in postmenopausal women, causing growth arrest and/or apoptosis of hormone-responsive cancer cells. Letrozole and anastrozole are aromatase inhibitors which have been shown to be superior to tamoxifen for the first-line treatment of breast cancer in postmenopausal women, while exemestane is an irreversible 'aromatase inactivator' which is superior to megestrol for treatment of tamoxifen-refractory metastatic breast cancer. Analogues of gonadotropin-releasing hormone (GnRH) can induce chemical castration, complete suppression of the production of oestrogen, progesterone, and testosterone from the reproductive organs via the negative feedback effect of continuous stimulation of the pituitary gland by these hormones. Leuprolide and goserelin are GnRH analogues which are used primarily for the treatment of hormone-responsive prostate cancer.

Oestrogen receptor antagonists (SERM): Tamoxifen is a partial agonist, which can actually increase oestrogen receptor signalling in some tissues, such as the endometrium. Raloxifene is another partial agonist SERM which does not seem to promote endometrial cancer and is used primarily for chemoprevention of breast cancer in high-risk individuals. Toremifene and fulvestrant are SERMs with little or no agonist activity.

Antiandrogens are a class of drug which bind and inhibit the androgen receptor, blocking the growth- and survival-promoting effects of testosterone on certain prostate cancers. There are steroidal antiandrogens and 'pure' antiandrogens. The steroidal antiandrogens include megestrol. The 'pure' or non-steroidal antiandrogens include bicalutamide, flutamide, and nilutamide.

Cell signalling inhibitors: This broad classification underpins the remarkable insights that cell and molecular biology have yielded over the past two decades that have resulted in druggable targets, several of which are used as biomarkers to select chemosensitive patient subpopulations.

Antiangiogenic inhibitors (bevacizumab, aflibercept, sorafenib, sunitinib, pazopanib, and everolimus) Angiogenesis requires the binding of signalling molecules, such as vascular endothelial growth factor (VEGF), to receptors on the surface of normal endothelial cells. When VEGF and other endothelial growth factors bind to their receptors on endothelial cells, signals within these cells are initiated that promote the growth and survival of new blood vessels. Angiogenesis inhibitors interfere with various steps in this process (for example bevacizumab is a monoclonal antibody that specifically recognizes and binds to VEGF, preventing it from activating its receptor). Other angiogenesis inhibitors, including sorafenib, regorafenib, and sunitinib, bind to receptors on the surface of endothelial cells or to other proteins in the downstream signalling pathways, blocking their activities. Ramucirumab is a fully human monoclonal antibody (IgG1) being developed for the treatment of solid tumours. It is directed against the vascular endothelial growth factor receptor 2 (VEGFR2). By binding to VEGFR2 it works as a receptor antagonist blocking the binding of VEGF to VEGFR2. VEGFR2 is known to mediate the majority of the downstream effects of VEGF in angiogenesis.

Growth factor receptor inhibitors

Aberrant expression of the epidermal growth factor receptor (EGFR) system has been reported in a wide range of epithelial cancers. In some studies, this has also been associated with a poor prognosis and resistance to the conventional forms of therapies. These discoveries have led to the strategic development of several kinds of EGFR inhibitors, which comprise the anti-EGFR monoclonal antibodies cetuximab and panitumumab and the small molecule EGFR tyrosine kinase inhibitors gefitinib and erlotinib. Human EGFR 2 (HER2) is the target of the monoclonal antibody trastuzumab, which is effective only in cancers where HER2 is overexpressed.

One of the difficulties in classifying signal transduction inhibitors is that they often have multiple targets and it may be impossible to be precise about the dominant mode of action. Some of the main pharmacologic types are:

- Multitargeted tyrosine kinase inhibitors which include erlotinib, imatinib, gefitinib, dasatinib, sunitinib, nilotinib, lapatinib, sorafenib, crizotinib, regorafenib
- Proteasome inhibitors (e.g. bortezomib)
- mTOR inhibitors (e.g. temsirolimus, everolimus)
- Histone deacetylase inhibitors (e.g. vorinostat (SAHA))
- Hedgehog pathway inhibitors (e.g. vismodegib)

Cellular principles of chemotherapy

For cytotoxic treatment, the following characteristics of tumour growth are important in determining outcome:

- cell cycle time
- tumour doubling time
- growth fraction
- tumour size, or the number of cells in the population.

Decades ago, Skipper (1979) found that the doubling time of proliferating murine cancer cells is constant, forming a straight line on a semilog plot. Death of the animals resulted when the malignant cells reached a critical fraction of body weight. Since other experiments (Furth and Kahn, 1937) had shown that a single surviving cell leads to treatment failure, and there is still no evidence that normal levels of host defence are capable of eliminating a few remaining tumour cells, for a given amount of chemotherapy survival was related to tumour size at the time of diagnosis (Fig. 28.1).

These studies were performed in model systems, often using murine leukaemic cell lines, showing logarithmic (exponential) growth. All of the cells were in cycle and dividing, with no cells in a resting phase, and the cell number doubling at a tumour-specific rate. While knowledge based on these model characteristics is important, the rules only apply to the cells in the proliferation compartment. Unfortunately, only a few human cancers have a large proportion of such responsive proliferating cells. In most tumours there is a large non-proliferating compartment; therefore, these model systems, though important in their time, are not truly representative of the kinetics of human solid tumours (Mendelsohn, 1960).

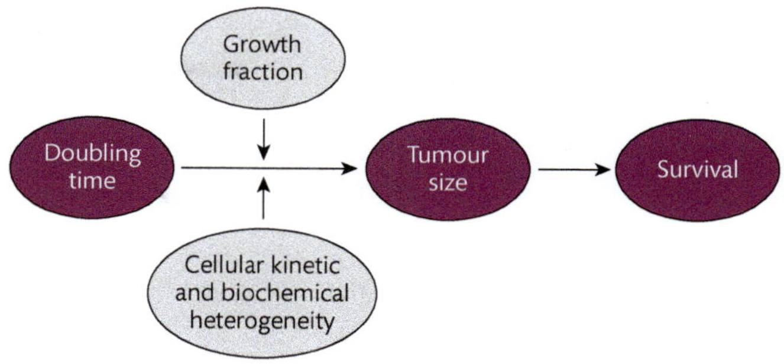

Fig. 28.1 Cancer dynamics.

Most human tumours are diagnosed when they are relatively large and kinetically and genetically heterogeneous. Due to a variety of factors such as poor vascularity, hypoxia (Tannock, 1968), and competition for nutrients, they exhibit decelerating growth at this stage. Larger tumours contain a high fraction of slowly- or non-dividing cells (termed G0 cells) and as a consequence the growth fraction is low. Therefore, in treating human tumours, the fractional cell kill hypothesis probably does not apply as well as in animal tumour models. Non-proliferating cells are less sensitive to antineoplastic agents, particularly because they have time to repair the damaged DNA. As many antineoplastic agents are most effective against rapidly dividing cells, the cell-kinetic situation at tumour diagnosis is unfavourable for treatment with most drugs.

Unlike the tumour models used by Skipper and related to the fact that the proliferating cell population is distinct from the non-proliferating population, human tumours are thought to follow a different growth pattern. Attempts have been undertaken to describe human tumour growth by mathematical models. Two available models are the so-called Gompertzian growth model and the exponential growth model. The primary distinction between the two models is that in the Gompertzian growth kinetics the growth fraction of the tumour is not constant but decreases exponentially with time. Exponential growth implies that the time taken for a tumour to double its volume is constant. A significant problem is that most tumours only become clinically manifest by approximately 10^9 tumour cells (equating approximately to a size of 1 cm^3, representing the last part of the tumour's growth curve). Thus, estimating growth curves of human primary tumours based on multiple time points of tumour volume appeared to be difficult, if not impossible. Overall, the available data suggest that the Gompertzian growth model (sigmoid in shape on a log scale) is the most probable model (it is interesting to reflect that Gompertz was an economist and developed his models around how industrialized economies behaved).

Apart from cell-kinetic heterogeneity, genetic and biochemical heterogeneity of human tumours may also reduce the likelihood of cure. Although most human tumours evolve from a single clone of malignant cells (Fialkow, 1976), more recent studies have shown that this homogeneity does not persist during further stages of tumour growth, presumably as a result of further somatic mutations of the original tumour line. When non-homogeneous tumour cells are exposed to drugs, sensitive tumour cells will be destroyed while resistant cells will survive and proliferate (Curt and Chabner, 1984).

As a result, tumour cell kill tends to decrease with subsequent courses of treatment, as resistant cells are selected. Paradoxically, normal tissues never change their level of sensitivity to chemotherapy, emphasizing an important difference in relative genetic stability.

Pharmacological principles of chemotherapy

The scheduling of drug treatments is based on both practical and theoretical considerations. Intermittent cycles of treatment are used to allow periods of recovery of normal tissues. This strategy aims at retreatment with full therapeutic doses as frequently as possible in keeping with the fractional cell kill hypothesis, but also allowing the normal tissues to recover from the unintended effects of cytotoxic treatment. The outcome of chemotherapy will obviously largely depend on the overall intrinsic sensitivity of the treated tumour.

Fractional cell kill means that a given drug concentration applied for a defined time period will kill a constant fraction of the total cell population, independent of the absolute number of cells. In other words, each treatment cycle kills a specific fraction (percentage) of the remaining cells. Since this fraction is never 100%, a single drug administration will never be sufficient to eradicate a tumour completely. Therefore, treatment results will also be a direct function of the drug concentration and exposure time and the frequency of repeating treatment. Drug concentration and exposure time will be dependent on pharmacokinetic (PK) factors such as drug absorption, metabolism, and elimination (Fig. 28.2). These will have to be considered in general in determining the dose, schedule, and route of drug administration. In addition, interpatient variations in pharmacokinetic parameters are usually large and this may be one reason for the inconsistency in responses of 'sensitive' tumours.

The bioavailability of a drug describes the proportion of a dose of a drug that enters the systemic circulation (e.g. for intravenous 5-FU this would be 100% compared to 10–25% for oral 5-FU). For drugs taken orally that are intended for systemic action, a significant proportion of a given dose may not even enter the systemic circulation. This may be due to poor absorption from the gastrointestinal (GI) tract, or metabolism in the gut wall or liver (called first-pass metabolism).

Various processes are involved in drug elimination, although hepatic and renal processes are the most important.

The liver is the main organ of drug metabolism. There are generally two types of reaction (Phase I and Phase II) that have two important effects:

- Make the drug more water soluble—to aid excretion by the kidneys.
- Inactivate the drug—in most cases, the metabolite is less active than the parent drug, although in some cases the metabolite can be as active, or more so, than the parent (e.g. irinotecan, tamoxifen).

Phase I metabolism involves oxidation, reduction, or hydrolysis reactions. Oxidation reactions are most common and are catalysed by cytochrome P450 isoenzymes located primarily in the liver. Phase II metabolism involves conjugation reactions, such as glucuronidation (such as UGT1a1) or sulphation, which produce more water-soluble compounds, enabling rapid elimination.

The main route of excretion of drugs is the kidney. Renal elimination is dependent on multiple factors that include:

- Glomerular filtration rate (GFR)
- Active tubular secretion (may involve P-gp)
- Passive tubular secretion

If a drug is metabolized to mainly inactive compounds (e.g. 5-fluorouracil), renal function will not greatly affect the elimination. If, however, the drug is excreted largely unchanged (e.g. carboplatin), or an active metabolite is excreted via the kidney (e.g. morphine), changes in renal function will influence the elimination and dose adjustments may be necessary.

Effect of hepatic and renal impairment

Impaired liver function can affect the pharmacokinetics and pharmacodynamics of many anticancer drugs (e.g. anthracyclines, taxanes, vinca alkaloids, and 5-FU). Reduction in hepatic blood flow and a potential fall in the number and the activity of hepatocytes can alter liver function and impact on drug clearance. A reduced synthesis of albumin can result in reduced drug–protein binding, thereby affecting the volume of distribution. Cholestasis can affect the biliary excretion of drugs and metabolites. Patients with impaired hepatic function may also develop a degree of renal impairment due to decreased renal plasma flow and GFR.

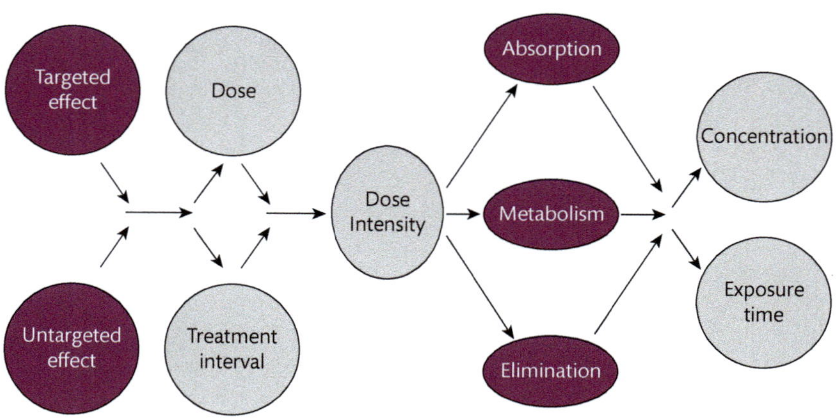

Fig. 28.2 Pharmacokinetic (PK) factors.

Unlike impaired renal function, there is no simple test that can determine the impact of liver disease on drug handling. A combination of factors needs to be considered before such impact can be assessed, which include liver function tests, diagnosis, and physical symptoms.

In general, the metabolism of drugs is unlikely to be affected unless the patient has severe liver disease. Most problems are seen in patients with jaundice, ascites, and hepatic encephalopathy. As such, doses of drugs should be reviewed in the following situations:

- Hepatically metabolized drug with narrow therapeutic index.
- There is a significant involvement of the cytochrome P450 system (CYP3A4/5 is highly susceptible to liver disease, while CYP2D6 appears relatively refractory).
- International normalized ratio (INR) more than 1.2.
- Bilirubin more than 100 micromol/L.
- Albumin less than 30 g/L.
- Signs of ascites and/or encephalopathy.

The elimination of several cytotoxic drugs and their metabolites is dependent upon renal function (capecitabine, carboplatin, cisplatin, methotrexate). Impaired renal function, coupled with rising urea plasma concentrations, induces changes in drug pharmacokinetics and pharmacodynamics. Implications for drug therapy include:

- Increased risk of undesirable effects and toxicity through reduced excretion of the drug and/or metabolite(s);
- Increased sensitivity to drug effects, irrespective of route of elimination, (e.g. antipsychotics);
- Increased risk of further renal impairment (e.g. NSAIDs).

Many of these problems can be avoided by simple adjustment of daily dose or frequency of administration. The dose nomogram for carboplatin is an excellent example of this. In other situations, however, an alternative drug may need to be chosen.

Pharmacogenetics

'If it were not for the great variability among individuals, medicine might as well be a science and not an art.'

Sir William Osler, *The Principles and Practice of Medicine*, Appleton, NY, 1892

Pharmacogenetics is the study of how variation in an individual gene affects the response to drugs which can lead to adverse drug reactions, drug toxicity, therapeutic failure, and drug interactions.

Genetic variability can affect an individual's response to drug treatment by influencing pharmacokinetic and pharmacodynamic processes (e.g. variations in genes that encode cytochrome P450 isoenzymes, drug receptors, or transport proteins can determine clinical response). Pharmacogenetics can aid in the optimization of drug therapy through the identification of individuals who are likely to respond to treatment, or those who are most likely at risk of an adverse drug reaction. For example, with tamoxifen as the active metabolite, endoxifen is produced by a reaction involving CYP2D6. Patients with a PM phenotype are at risk of therapeutic failure with tamoxifen. Drugs that inhibit CYP2D6 will also mimic the PM phenotype and should be avoided. Warfarin bleeding effects are more common with CYP 2D9 polymorphisms; patients with mutations in UGT1a1 reduce metabolism of irinotecan and are associated with more toxicity.

Dose intensity

Most cancer chemotherapeutic agents *in vitro* and *in vivo* exhibit a steep dose–response curve. Consequently, it is considered desirable to administer them in humans at the highest possible dose intensity, since in theory even small reductions in dose would lead to substantial reductions of tumour cell kill. The importance of dose intensity in tumour responsiveness in humans was first suggested in heavily criticized retrospective analyses performed by Levin and Hryniuk (1987) and others (Longo et al., 1987). These investigators suggested dose–response correlations for 5-fluorouracil in colon cancer, cisplatin in ovarian cancer, doxorubicin in breast cancer, and vincristine in Hodgkin's disease. Subsequently, for some of these, prospective randomized trials of relatively small sample size have supported or refuted the concept.

It is generally accepted that the most important measure of drug exposure would be the area under the curve in a plot of local tumour drug concentration against time.

Obviously, there is a hierarchy ranging from simple plasma levels of unbound and activated drug (where appropriate) to activated drug concentrations within the target tumour cell. Unfortunately, there are scarcely any data from humans to suggest that plasma drug levels do correlate with tumour levels and response.

Loco-regional drug administration

Regional perfusion with antineoplastic drugs can generate higher than expected drug concentrations in the body compartment which harbours the dominant tumour burden.

a. Direct intrathecal administration of drugs such as methotrexate or cytarabine enables a higher local dose in a sanctuary site.
b. Hepatic artery or portal vein infusion in theory allows an increased drug concentration to liver primary or secondary tumours (Kemeny and Schneider, 1989).
c. Intra-arterial perfusion can be used for metastatic and primary liver cancer, limb melanomas, and soft tissue sarcomas.
d. Intraperitoneal chemotherapy is effective for ovarian cancer (Howell, 1988; Alberts et al., 1995).

Dose intensification using local administration in uncontrolled studies has suggested improved response rates. However, with the exception of intrathecal therapy in acute lymphocytic leukaemia, hepatic arterial infusion for metastatic colorectal cancer, and intraperitoneal therapy for ovarian cancer, none of the regional methods has been reported in randomized trials to produce a longer survival than conventional systemic methods of drug administration.

An increase of the dose per administration

The dose per administration can be increased without changing the intervals between administrations. One such method to increase tumour cell kill is to use doses of chemotherapy that cause prolonged bone marrow suppression and to rescue the host either with autologous bone marrow harvested before treatment or with allogenic marrow from a histocompatible donor.

With the use of high-dose chemotherapy, new dose-limiting toxicities to organs other than bone marrow emerge, such as nitrosourea toxicity to lung, kidneys, or liver (Philips et al., 1983; Tchekmedyian et al., 1986). For this reason, its use is still limited to certain classes of drugs. To date, marrow-ablative chemotherapy in humans has only been shown effective in leukaemias and lymphomas, and in tumour types with high growth fractions and intrinsic sensitivity to chemotherapy.

Shortening treatment intervals

This method has been used, especially in solid tumours, as another way of increasing dose intensity. The introduction of drugs such as the 5-hydroxytryptamine-3 antagonist antiemetics and the haematopoietic growth factors has greatly facilitated this approach. Although weekly (Planting et al., 1993) or biweekly administration of relatively high doses of drugs previously given at 3- to 4-week intervals seems feasible and yields interesting results in uncontrolled trials, it is too early to conclude if the resulting increases in dose intensity produce an increase in cure rate (an exception may be in the use of dose-dense adjuvant chemotherapy for breast cancer).

Principles of combination chemotherapy

As a consequence of somatic mutations, tumour cell kill tends to decrease with subsequent courses of treatment, as genetically resistant cell types are selected out.

For this reason, single-agent treatment is rarely curative. Therefore, and for a variety of other reasons discussed here, cancer chemotherapy is most frequently given as a combination of different drugs. Favourable and unfavourable interactions between drugs must be considered in developing such combination regimens. These interactions may be pharmacokinetic, cytokinetic, or biochemical, and may influence the effectiveness of the components of the combination. The theoretical and sometimes proven superiority of combination chemotherapy over single-agent treatment is derived from the principles listed in Box 28.1.

When any drugs are administered in combination, three outcomes are possible: additive, subtractive, or synergistic effects. With additive effects, the drugs act completely independently of each other, presumably through non-overlapping biochemical pathways and have no pharmacokinetic interactions which alter the quantum of drug reaching its active site. For subtractive, or negative synergistic effects, one drug interferes with the other to reduce efficacy. This could be competition at the active site, altered pharmacokinetics inducing one of the enzyme systems responsible for the other drug's metabolism, or the unexpected consequence that inhibition of a biochemical pathway might have on up/down regulation of the target kinase of the companion drug. Synergy is defined as an interaction between drugs where the effects are stronger than their mere sum and may be driven by both pharmacokinetic and pharmacodynamic interactions. The best way of demonstrating true synergy is to use the Chou–Talalay method for drug combination. This is based on the median-effect equation, derived from the mass-action law principle, which is the unified theory that provides the common link between single entity and multiple entities, and first order and higher order dynamics. It is possible, therefore, to apply stringent equations in the preclinical setting, both *in vitro* and *in vivo* to empirical cytotoxic drug combinations to determine whether true synergy can be documented and therefore used as a piece of supporting evidence to take specific anticancer drug combinations into the clinic.

Box 28.1 Principles of combination chemotherapy

- Use drugs active as a single agent
- Use drugs with different mechanisms of action
- Use drugs with different mechanisms of resistance
- Use drugs with different side effects
- Be aware of drug-drug interactions

Activity as a single agent

Drugs with at least activity as a single agent should be selected. Because of primary resistance (see later), which is frequent for any single agent even in the most responsive tumours, complete response rates of single agents rarely exceed 20%.

Different mechanisms of action

Drugs with different mechanisms of action (and toxicity profiles) should be combined. The various anticancer agent classes have different targets in the cell. Thus, even if a certain target cannot be exploited in a given tumour cell, another target might. Even if tumours are initially sensitive, they usually rapidly acquire resistance after drug exposure. This is probably due to selection of pre-existing resistant tumour cells in the biochemically heterogeneous tumour cell population. In other words, the chemotherapy destroys the sensitive cell population, but is less effective against the non-sensitive population of cells that subsequently is able to continue expanding. In addition, cytotoxic drugs themselves appear to increase the rate of mutation to resistance, at least in tumour models (Rice at el., 1986). The use of multiple agents with different mechanisms of action enables independent cell killing by each agent. Cells that are resistant to one agent might still be sensitive to the other drug(s) in the regimen and might thus still be killed. Known patterns of cross-resistance must be taken into consideration in the design of drug combinations.

Different mechanisms of resistance

Drugs with different mechanisms of resistance should be combined. Resistance to many agents may be the result of mutational changes unique to those agents.

However, in other circumstances a single mutational change may lead to resistance to a variety of different drugs. The number of potential mechanisms of resistance is continually increasing and is partly drug dependent.

The most investigated form of multidrug resistance is the one mediated by increased expression of the P170 membrane glycoprotein (PGP). Primary (intrinsic) overexpression of this protein has been identified in tumours derived from healthy tissues in which the protein also occurs (Fojo et al., 1987). Secondary (acquired) overexpression has also been found following chemotherapy in various diseases. PGP

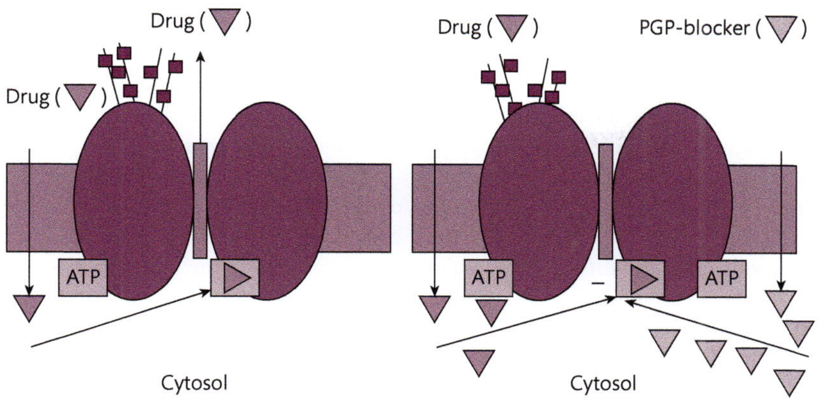

Fig. 28.3 P-glycoprotein: action and inhibition. The blocking agent binds to the cytoplasmic binding site of the drug, inhibiting its efflux.
Reproduced with permission from Kerr D et al. 'Principles of chemotherapy' in Kerr D et al. (Eds), *Oxford Textbook of Oncology, Third Edition*, Oxford University Press, Oxford, UK, Copyright © 2016.

mediates the efflux of drugs such as anthracyclines, vinca alkaloids, epipodophyllotoxins, and various others, all derived from natural resources. PGP-mediated active efflux results in decreased intracellular drug levels resulting in decreased cell kill. Agents that compete for the protein such as calcium-channel blockers, cyclosporin, and various synthetically produced drugs can, at least in models, reverse the effect of PGP (Fig. 28.3). The clinical relevance of this type of resistance and its reversal is still unclear.

For several of the classic alkylating agents (cisplatin, cyclophosphamide, melphalan) resistance appears to be related to enhanced repair of drug-induced DNA damage.

There are many other mechanisms of resistance, including alterations in target proteins (e.g. dihydrofolate reductase with altered affinity for methotrexate) and carrier-mediated drug uptake (e.g. reduced folate). In the heterogeneous human tumours, there are frequently various mechanisms of resistance that play a role.

Reversing only one of these is unlikely to yield a major impact. Resistance reversal is therefore pursued in only limited numbers of tumours. Because of the presence of drug-resistant mutants at the time of clinical diagnosis, the earliest possible use of drugs that are not cross-resistant is recommended to avoid the selection of double mutants by sequential chemotherapy. Adequate cytotoxic doses of drugs have to be administered as frequently as possible to achieve maximal kill of both sensitive and moderately resistant cells. Less desirable alternatives are the use of different regimens in alternating cycles of therapy or the use of multiple cycles of one regimen given to the point of maximal response, followed by a second regimen.

Different dose-limiting toxicities

If possible, drugs with different dose-limiting toxicities should be combined. In the case of non-overlapping toxicity, it is more likely that each of the drugs can be used at full dose and thus the effectiveness of each agent will be maintained in the combination. Unfortunately, for many cytotoxic agents the side effects frequently involve myelosuppression. If there is overlapping myelotoxicity, arbitrary scales of dose adjustment according to bone marrow toxicity can be used, or haematopoietic growth factors can be added to reduce the risk related to myelosuppression. Clearly, drugs that have low bone marrow toxicity, such as vincristine, cisplatin, and bleomycin, are favoured to combine with myelosuppressive agents.

Drugs with renal or hepatic side effects in theory can alter the elimination of other agents through these routes and therefore must be used with caution in combinations. For example, cisplatin causes renal toxicity and is known to alter the pharmacokinetics of other agents (such as methotrexate or bleomycin) that depend on renal elimination as their primary mechanism of excretion.

Cell cycle-related and biochemical interactions

Cell cycle-related and biochemical interactions between drugs can also be used to design combinations. Examples of drug interactions are listed in Box 28.2. Provided that the drugs used are active in a particular disease, knowledge of cell kinetics can be used to consider initiation of therapy with agents that are not specific for cell cycle (the alkylating agents and nitrosoureas), first to reduce tumour bulk and second to recruit slowly dividing cells into active DNA synthesis. Once the latter is achieved, therapy can be continued within the same cycle of treatment with agents specific for cell cycle phase (such as methotrexate or the fluoropyrimidines) that mainly affect

Box 28.2 Drug interactions in combination chemotherapy

Antagonism of antitumour effect

- L-Asparaginase prior to methotrexate
- 5-Fluorouracil prior to methotrexate

Enhancement of antitumour effect

- Nitroimidazoles enhance alkylating agent activity
- Leucovorin increases 5-fluorouracil inhibition of thymidylate synthase
- Inhibitors of pyrimidine synthesis enhance 5-fluorouracil incorporation into RNA

Reversal of drug resistance

- Calcium-channel blocker inhibits efflux by P-glycoprotein

Prevention or reversal of toxicity

- Allopurinol blocks 5-fluorouracil activation by normal tissues
- Leucovorin prevents methotrexate toxicity
- Deoxycytidine prevents toxicity of cytarabine

Reproduced with permission from Kerr D et al. 'Principles of chemotherapy' in Kerr D et al. (Eds), *Oxford Textbook of Oncology, Third Edition*, Oxford University Press, Oxford, UK, Copyright © 2016.

cells during periods of DNA synthesis. Further, repeated courses with S-phase-specific drugs, such as cytosine arabinoside and methotrexate, that block cells during the period of DNA synthesis, are most effective if they are administered during the rebound rapid recovery of DNA synthesis that follows the period of suppression of DNA synthesis (Vaughan et al., 1984). For cytosine arabinoside, for instance, this occurs approximately 10 days after the first treatment with this agent.

An example of biochemical interaction suggesting a rational combination of drugs is shown by the example of leucovorin (5-formyltetrahydrofolate) enhancing 5-fluorouracil, by ternary stabilization of its metabolite (fluoro deoxy uridine monophosphate) with thymidylate synthase (Grem et al., 1987).

Adjuvant and neoadjuvant systemic therapy

Drug therapy is also used to benefit patients who exhibit no evidence of residual disease after initial therapy but are at high risk of relapse (adjuvant therapy), or those with bulky primary disease to reduce this bulk preceding local therapy (neoadjuvant therapy).

Adjuvant therapy

There are two major reasons why adjuvant chemotherapy might be considered: (i) the high rate of recurrence after surgery for some, apparently localized tumours; and (ii) the failure of systemic therapy or combined modality treatment to cure patients with clinically apparent metastatic disease. These issues may in turn be related to the following three points.

First, once the tumour bulk is reduced to clinically undetectable levels with local therapy, the number of cells yet to be destroyed by subsequent chemotherapy is relatively small, in contrast to the large numbers of cells in the case of clinical metastatic disease. According to the principles of log cell kill in the first situation, the likelihood of completely eradicating the remaining tumour cells is much higher than in the latter (**Fig. 28.4**).

Second, as mentioned earlier, there is evidence from tumour models that supports the hypothesis that tumours are most sensitive to chemotherapy at their earliest stages of growth. This is thought to be related to their high growth fraction (most cells are in active progression through the cell cycle), shorter cell cycle times, and therefore greater fractional cell kill for a given dose of drug (Salmon, 1979). Once tumours progress to clinical detectability, their growth fraction falls, the cell cycle time lengthens, and they become much less sensitive to treatment.

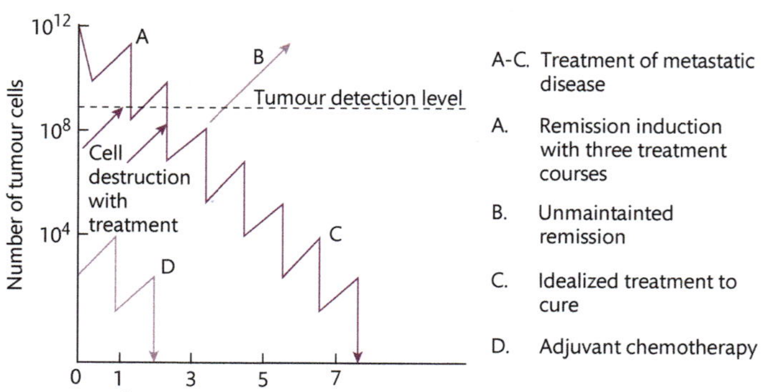

Fig. 28.4 The principle of adjuvant chemotherapy as opposed to treatment of measurable disease.

Third, there is a relationship between tumour bulk and tumour cell resistance. The probability of the occurrence of resistant cells in a tumour population is a function of the total number of cells present and mutation rates. Therefore, subclinical (occult) tumours rather than clinically detectable tumours are more likely to be cured by chemotherapy.

On the other hand, there are obviously potential disadvantages of adjuvant chemotherapy that have to be taken into account. They relate to immediate, short-, and long-term side effects of such treatment. Since a significant fraction of patients receiving adjuvant treatment is already cured by the primary surgical procedure, they would therefore experience needless toxicity and risks if treated with adjuvant therapy.

The immediate side effects obviously relate to potentially lethal infectious complications from neutropenia, bleeding, and less hazardous—but very inconvenient—side effects such as alopecia and nausea and vomiting.

Late complications such as carcinogenicity and permanent sterility assume greater importance, but neither risk has been adequately quantified for all drugs and combinations. The risks of other late effects such as bone marrow hypoplasia from alkylating agents and nitrosoureas, the cardiotoxicity of doxorubicin, neurotoxicity of platins and taxanes, and pulmonary toxicity of bleomycin and the nitrosoureas should obviously not outweigh the potential benefits. In balancing these risks with the outcome as far as tumour eradication is concerned, adjuvant chemotherapy has become a standard of care in many cancers of childhood, the breast, the colon, and the ovary.

Neoadjuvant therapy

If chemotherapy is used as initial treatment, preceding a form of local therapy, it is called 'neoadjuvant therapy' or 'induction therapy'. Such neoadjuvant chemotherapy has attracted increasing attention in the treatment of some adult solid tumours. Factors related to the response to neoadjuvant chemotherapy are similar to those that may affect adjuvant chemotherapy, namely growth rate, presence of drug resistance, and tumour mass. The potential benefits of neoadjuvant chemotherapy involve the control of the primary tumour as well as of potential micrometastases.

There are several potential advantages for control of the primary tumour (Goldie and Coldman, 1979; Frei et al., 1986). First, local reduction of the tumour may facilitate the use of more conservative surgery and/or radiotherapy. Second, administering chemotherapy prior to the local therapy avoids the potential of poor distribution and penetration of drug at the tumour site due to the compromised vascularity that may result from surgery and/or radiation therapy. Third, cytotoxic drugs can be combined concurrently with radiation therapy to increase the local control rate. Fourth, post-chemotherapy surgery offers a unique opportunity to assess the correlation between clinical tumour response measurements and actual pathological changes. Fifth, the response of the primary tumour to chemotherapy may reflect the response of micrometastases and therefore influence further patient management. The disadvantages of neoadjuvant chemotherapy are similar to those mentioned for adjuvant chemotherapy. Additional disadvantages may be:

a. Selection of drug-resistant clones;
b. An increase in toxicity from subsequent therapies;
c. A failure of cytotoxic agents to reduce the primary tumour significantly, thereby possibly allowing further subclinical progression of disease; and
d. A loss of the advantage of attacking micrometastases after surgical resection of the primary tumour mass, when they may exhibit more favourable cell kinetics.

Conclusion

Despite the insights in tumour biology gained by modern molecular genetics, the broad principles governing medical oncology and the delivery of systemic anticancer therapy have remained intact over the past few decades. It is all about delivering sufficient quantities of the drug to its site of action and a molecular target which is differentially or uniquely expressed relative to normal tissue. We can respond to—but not control—the genetic background of the tumour, utilizing predictive and prognostic biomarkers, but we can control drug dose and this must remain the cornerstone of effective medical oncology.

Acknowledgements

This chapter has been adapted with permission from Kerr, D. et al., 'Principles of chemotherapy', in Kerr, D. et al. (eds), *Oxford Textbook of Oncology*, 3rd edition, Oxford University Press, Oxford, UK, copyright © 2016.

TAKE-HOME MESSAGE

- Conventional chemotherapy based on DNA synthesis inhibition is here to stay.
- Targeted chemotherapy takes advantage of cell biological insights into growth factor-receptor signalling pathways, vascular biology, and epigenesis.
- Molecular immunology has led to the development of clinically effective immune checkpoint inhibitors and genetically engineered T cells for adoptive therapy.
- The hypothesis that next-generation sequencing will reveal clinically useful actionable mutations is yet to be proven.

OPEN QUESTIONS

- 'Veneno, ergo sum': still many anticancer agents have steep dose–response curves and narrow therapeutic ratios.
- Precision medicine aims to have treatment guided by the genome alterations present; however, it has become clear that a 'driver mutation' in, for example, lung cancer may not operate as it does in colorectal or breast cancer, given the differing mutational landscape in these diverse tumour types.
- Cell-kinetic, genetic, and biochemical heterogeneity of human tumours is still an obstacle to effective treatment in many patients.

FURTHER READING

La Thangue, N. B. & Kerr, D. J. (2011). Predictive biomarkers: a shift towards personalised cancer medicine. *Nat Rev Clin Oncol*, **8**, 587–96.

Midgley, R., Middleton, M. R., Dickman, A., & Kerr, D. (eds) (2013). *Drugs in Cancer Care*. Oxford: Oxford University Press.

REFERENCES

Alberts, D. S., et al. (1995). Phase III study of intraperitoneal cisplatin (CDDP)/intravenous cyclophosphamide (CPA) vs i.v. CDDP/i.v. CPA in patients with optimal disease stage III ovarian cancer: a SWOG-GOG-ECOG Intergroup study. *Proc Am Soc Clin Oncol*, **14**, 273.

Curt, G. A. & Chabner, B. A. (1984). Gene amplification in drug resistance of mice and men. *J Clin Oncol*, **2**, 62–4.

Domingo, E., Church, D. N., Sieber, O., et al. (2013). Evaluation of *PIK3CA* mutation as a predictor of benefit from nonsteroidal anti-inflammatory drug therapy in colorectal cancer. *J Clin Oncol*, **31**(34), 4297–305.

Fialkow, P. J. (1976). Clonal origin of human tumors. *Biochimica et Biophysica Acta*, **458**, 283–321.

Fojo AT, Ueda, K., Slamon, D. J., Poplack, D. G., Gottesman, M. M., & Pastan, I. (1987). Expression of a multidrug-resistance gene in human tumors and tissues. *Proc Natl Acad Sci U S A*, **84**, 265–9.

Frei, E., Miller, D., Clark, J. R., Fallon, B. G., & Ervin, T. J. (1986). Clinical and scientific consideration in preoperative (neoadjuvant) chemotherapy. *Recent Results in Cancer Research*, **103**, 1–5.

Furth, J. & Kahn, M. C. (1937). The transmission of leukemia of mice with a single cell. *Am J Cancer*, **31**, 276–82.

Goldie, J. H. & Coldman, A. J. (1979). A mathematic model for relating the drug sensitivity of tumors to their spontaneous mutation rate. *Cancer Treat Rep*, **63**, 1727–33.

Grem, J. L., Hoth, D. F., Hamilton, J. M., King, S. A., & Leyland-Jones, B. (1987). Overview of current status and future directions of clinical trials with fluorouracil and folinic acid. *Cancer Treat Rep*, **71**, 1249–64.

Howell, S. B. (1988). Intraperitoneal chemotherapy for ovarian carcinoma. *J Clin Oncol*, **6**, 1673–5.

Kemeny, N. & Schneider, A. (1989). Regional treatment of hepatic metastases and hepatocellular carcinoma. *Current Problems in Cancer*, **13**, 197–284.

Levin, L. & Hryniuk, W. M. (1987). Dose intensity analysis of chemotherapy regimens in ovarian carcinoma. *J Clin Oncol*, **5**, 756–67.

Longo, D. L., Young, R. C., Wesley, M., et al. (1987) Twenty years of MOPP therapy for Hodgkin's disease. *J Clin Oncol*, **5**, 756–67.

Mendelsohn, M. L. (1960). The growth fraction: a new concept applied to tumors. *Science*, **132**, 1496.

Philips, G. L., Fay, J. W., Herzig, G. P., et al. (1983). Intensive 1.3-bis(2-chloroethyl)-1 nitrosourea (BCNU) (NSC-4366650) and cryopreserved autologous marrow transplantation for refractory cancer. A phase I–II study. *Cancer*, **52**, 1792–802.

Planting, A. S., van der Burg, M. E., De Boer-Dennert, M., Stoter, G., & Verweij J (1993). Phase I/II study of a short course of weekly cisplatin in patients with advanced solid tumors. *Br J Cancer*, **68**, 789–92.

Rice, G. C., Hoy, C., & Schimke, R. T. (1986). Transient hypoxia enhances the frequency of DHFR gene amplification in Chinese hamster ovary cells. *Proc Natl Acad Sci U S A*, **83**, 5978–82.

Salmon, S. E. (1979). Kinetics of minimal residual disease. *Recent Results in Cancer Research*, **67**, 5–15.

Skipper, H. E. (1979). Historic milestones in cancer biology: a few that are important to cancer treatment (revisited). *Semin Oncol*, **6**, 506–14.

Tannock, I. F. (1968). The relationship between proliferation and the vascular system in a transplanted mouse mammary tumor. *Br J Cancer*, **22**, 258–73.

Tchekmedyian, N. S., Tait, N., Van Echo, D., & Aisner, J. (1986). High-dose chemotherapy without autologous bone marrow transplantation in melanoma. *J Clin Oncol*, **4**, 1811–18.

Vaughan, W. P., Karp, J. E., & Burke, P. J. (1984). Two-cycle-times sequential chemotherapy for adult acute nonlymphocytic leukemia. *Blood*, **64**, 975–80.

29

Immunotherapy and tumour resistance to immune-mediated control and elimination

Gwennaëlle C. Monnot and Pedro Romero

Introduction

It is now well understood that the core characteristic of a cancerous cell is its mutated DNA. The overexpression of certain pro-tumorigenic proteins, as well as the deletion of important proliferation checkpoints is sometimes sufficient to start tumour formation. However, it is usually not a single mutation but rather an accumulation of mutations that can unleash a fully aggressive cancer form.

Indeed, a tumoural lesion has to possess several characteristics in order to be able to develop into a full cancer. These characteristics, which were named *hallmarks of cancer* firstly by Hanahan and Weinberg in 2000 (Hanahan and Weinberg, 2000), encompass many aspects of the tumour. Some hallmarks can be features that are linked only to the tumour cells themselves and their gene expression. For example, tumour cells should express survival and proliferation signals, and have activated pathways to resist apoptosis. They should also be able to withstand many rounds of replication by blocking telomere shortening. In contrast, there are other hallmarks that take into account the relationship of the tumour with its environment, such as its ability to resist to growth suppressors, as well as to induce the creation of new blood and lymphatic vessels. Finally, another hallmark of cancer is its ability to go through epithelial-mesenchymal transition and invade new areas of the body. More recently, another critical tumour ability was added to the list, as an 'emerging hallmark': immune escape (Hanahan and Weinberg, 2011; also see Fig. 29.1). Indeed, controlling the formation of an immune response is necessary for the survival of malignant tumours, as strong immune activation together with infiltration of immune cells can be fatal to the tumour.

The first observation that a strong immune response could induce tumour regression was actually made in the nineteenth century. At this time, W. B. Coley noticed that the development at the site of the sarcoma of erysipelas, an acute skin infection, increased the likelihood of complete tumour regression and long-term remission. Following this observation, he managed to cure patients' sarcoma by injecting it with cultures of streptococcus (Coley, 1893). This shows that the strong inflammation and immune activation resulting from the infection induced antitumour effect as a bystander effect. Later on, the ability of the immune system to mediate tumour regression was further demonstrated via the injection of mice with tumour lysate from a congenic strain. Once vaccinated, the mice were resistant to future tumour cells injections and did not develop cancers (Old and Boyse, 1964). Lastly, in cancer patients, most studies could show a beneficial prognosis for patients whose tumours were highly infiltrated with cells of the adaptive immune system, namely cytotoxic CD8+ T lymphocytes (Gooden et al., 2011). Other cells from the innate immune system have been shown to be potent at inducing tumour lysis, such as natural killer (NK) cells. They are from the innate immune system and are able to recognize and destroy cells from the 'missing' or 'altered' self. This includes cells that are from a different organism or mutated cancer cells, as well as infected cells that have downregulated their major histocompatibility complex (MHC) class I to hide from the adaptive immune system.

However, the immune system presents itself as a double-edged sword. Indeed, another one of the new hallmarks is the ability of the tumour to induce a special type of inflammation, which is actually tumour-promoting.

The immune system is composed of a lot of different cell types, some of which can have the role of dampening the immune response. This is important to preserve the balance of the organism at homeostasis. It ensures local inflammations can be resolved, helps to avoid autoimmunity as well as disproportionate reactions to innocuous substance. There are several cellular subsets of the immune system whose main function is to manage inflammation and to help with tissue remodelling and wound healing. These cell subtypes are classically present in tumours. Indeed, cancerous cells are tragically expert in attracting immune cells that will in turn promote tumorigenesis and protect the tumour cells from the cytotoxic cells.

For example, macrophages are present at high number within the tumour microenvironment. Although they were firstly identified as tumoricidal from *in vitro* experiments, it was soon shown that *in vivo*, macrophages can acquire a differentiation phenotype called M2, which renders them actively tumour-promoting.

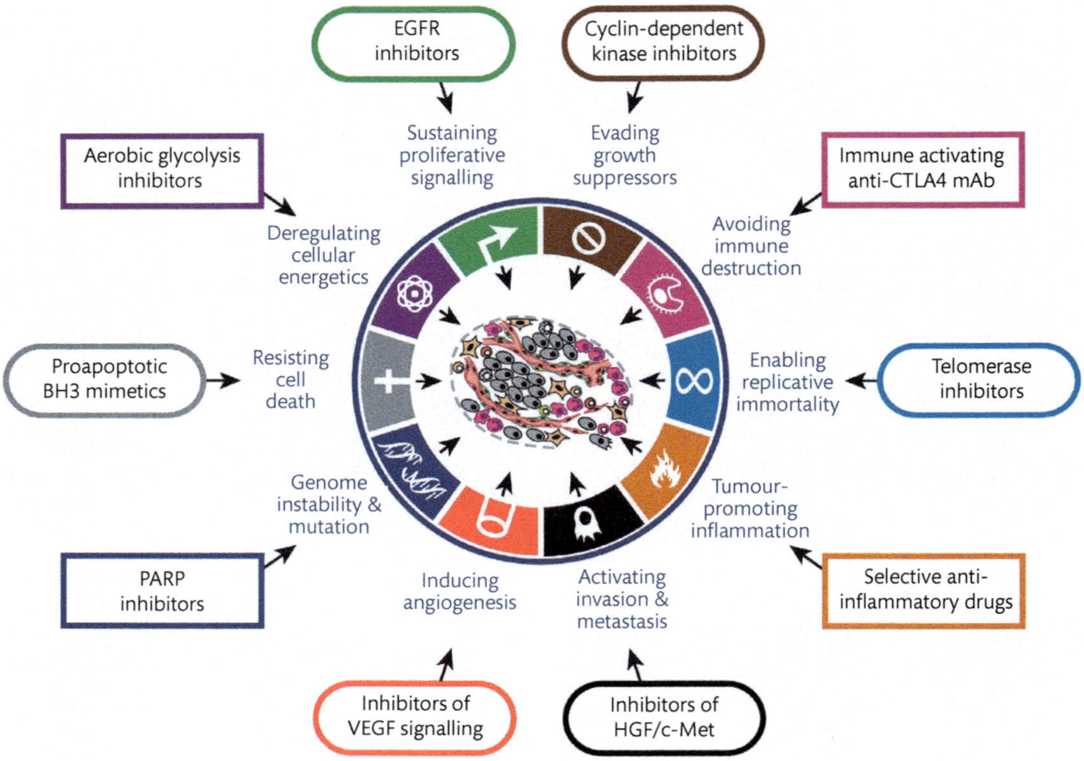

Fig. 29.1 The hallmarks of cancer and drugs interfering with them.

Reproduced with permission from Hanahan D and Weinberg RA, 'Hallmarks of cancer: The next generation', *Cell*, Volume 144, Issue 5, pp. 646–74, Copyright © 2011 Elsevier Inc. https://www.sciencedirect.com/science/article/pii/S0092867411001279

M2-macrophages can provide help at the tumour-initiation stage, by creating a favourable inflammatory environment. M2-macrophages can promote angiogenesis and mediate suppression of cytotoxic T cells (Ruffell and Coussens, 2015). Additionally, cancerous cells are able to recruit and subvert more macrophages as they grow, by producing attracting chemokines, such as CCL2, and cytokines such as colony-stimulating factor-1 (CSF-1) and transforming growth factor beta 1 (TGF-β1). The importance of these molecules was underlined recently by a study using an inhibitor of CSF-1R in a mouse model of glioblastoma, which resulted in decreased M2-associated gene expression on the tumour-associated macrophages (TAMs) as well as increased survival (Pyonteck et al., 2013). Other immune cell types have been identified as tumour-promoting, such as FoxP3+ CD4+ T lymphocytes (or Tregs). Tregs mediate immune suppression via the depletion of activatory signal and the production of immunosuppressive cytokines (Yamaguchi et al., 2011).

Overall, the interaction of tumour cells with the immune system is rather complex. Immune cells can be both tumour-destructive and tumour-promoting, depending on their subtype and their differentiation status. As a consequence, the immune system is a crucial element that can lead to either the tumour development or its elimination. This explains the excitement of modulating the immune system in the fight against cancer.

The central role of the immune system in tumour protection was nicely illustrated by studies showing that several of the widely used chemo- and radiotherapies were inefficient in immunocompromised animals. It was believed that the effect of these therapies was almost exclusively the consequence of a direct killing of the tumour cells. However, these therapies also have immunostimulatory properties, and can both elicit immunogenic cell death or directly activate immune effectors. Firstly, it was shown that doxorubicin could elicit immunogenic tumour cell death and thus increase dendritic cells antigen presentation, which in turn would produce a more potent adaptive immune response (Casares et al., 2005). Indeed, the adaptive immune response is dependent on the recognition of a specific molecular target—an antigen—and induces a subsequent specific response against that one target. Additionally, immunogenic cell death was crucial for the efficiency of both doxorubicin and X-ray treatment. Indeed, these treatments failed to induce tumour protection in mice that had a genetic deletion of Toll-like receptor 4 (TLR4), an activatory receptor that binds the lipopolysaccharide (LPS) on the surface of Gram-negative bacteria. This was due to the inability of dendritic cells to efficiently cross-present tumour debris to cytotoxic T cells and hence mount a potent cytotoxic immune response (Apetoh et al., 2007). This pathway was also found to be relevant in the treatment of human cancer. Indeed, breast cancer patients carrying a TLR4 mutation preventing the response to HMBG1, an alarmin protein released by dying tumour cells, had decreased metastasis-free survival after treatment, compared to patients without the mutation (Apetoh et al., 2007). Moreover, a multidrug chemotherapy containing doxorubicin was shown to repolarize TAMs, which possess the immune inhibitory phenotype M2, into M1 effector macrophages (Buhtoiarov et al., 2011). These evidences highlight the importance of the immune system in mediating tumour regression upon chemo- or radiotherapy.

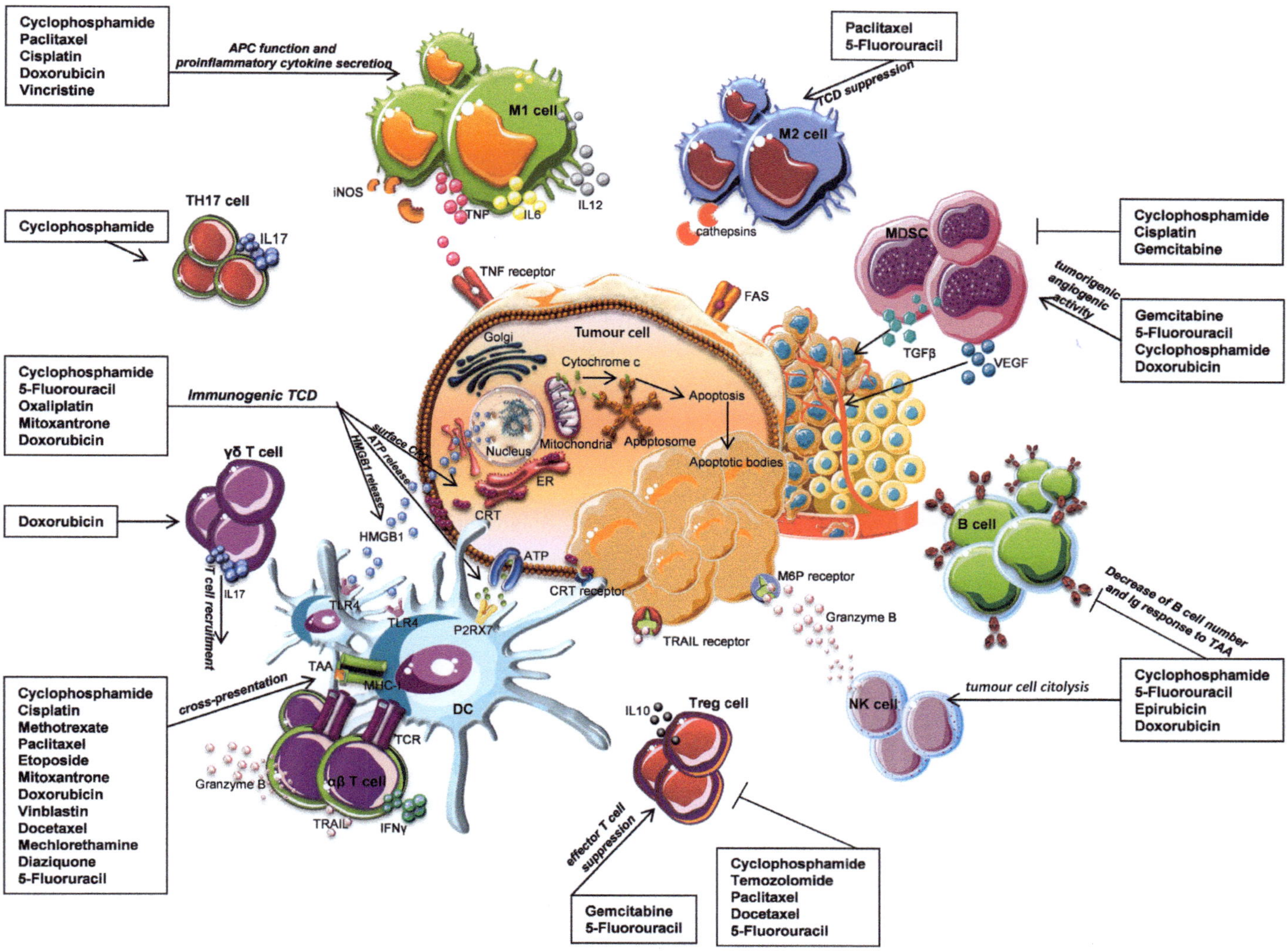

Fig. 29.2 The influence of conventional cytotoxic drugs on the immune system and the tumour microenvironment.

Reproduced with permission from Bracci L et al. 'Immune-based mechanisms of cytotoxic chemotherapy: implications for the design of novel and rationale-based combined treatments against cancer', *Cell Death and Differentiation*, Volume 21, Issue 1, pp. 15–25. Copyright © 2014 Macmillan Publishers Limited.

Additionally, it was shown that some specific cancer treatments also have a broader effect and act on the immune system. For example, the multitargeted tyrosine kinase inhibitor *sunitinib*, was shown to induce direct apoptosis on a renal cell carcinoma cell line by inhibition of p-STAT3. However, several types of tumour-promoting myeloid cells also rely on STAT3 signalling. Consequently, inhibition of STAT3 could also induce a significant reduction of myeloid-derived suppressor cells (MDSCs) as well as Tregs (Xin et al., 2009), hence hindering the tumour-derived immunosuppression. These data show that a potent immune response is crucial not only for spontaneous tumour regression, but also for an efficient therapeutic effect of most current treatments (Fig. 29.2).

Immunosurveillance

The immune system is important in mediating the regression of existing tumours. It also has a critical role in patrolling the body and preventing the rise of new carcinomas. Intrinsic safety mechanisms do exist, such as DNA repair and cell apoptosis. Their function is to make sure abnormal cells die swiftly, and do not acquire further mutations. However, some cells manage to escape these safety barriers and survive with their mutations. In that case, the immune system has the ability to recognize those cells and eliminate them, before they can cause harm to the organism. This ability of the immune system to prevent tumour formation without external intervention is known as cancer immunosurveillance.

Immunocompromised animals develop tumours at higher frequency

The first attempts to prove this hypothesis were unfruitful, as athymic nude mice and wildtype controls did not differ in their tumour development rate upon subcutaneous injection of 3-methylcholanthrene (MCA), a highly carcinogenic substance (Stutman, 1974). Both wildtype and nude MCA-treated cohorts developed subcutaneous sarcomas at similar rates—less than 20%—120 days after treatment. Moreover, lung adenomas were detected in all cohorts, with an increase for mice that had received MCA but no difference relative to the mouse strain. From this study, Stutman and colleagues concluded that tumour immunosurveillance did not occur in their model. However, as it was later discovered, athymic nude mice do retain some immune cells and the ability to mount limited but useful immune responses (Hunig, 1983). Despite their lack of thymus, these mice possess some residual functional T cells, as well as an intact repertoire of many innate immune players such as NK cells. These elements might have been enough to compensate for the absence of thymus-maturated T cells and mask the effect of immunosurveillance in Stutman's aggressive tumour model.

Later on, the ability of the immune system to prevent tumour development was demonstrated using RAG2$^{-/-}$ mice. The recombination-activating gene-2 (*RAG2*) is necessary for the rearrangement of the T and B-cell receptor during the lymphocyte development, and mice deficient in this protein lack T, B, and NKT cells. In 2001, Shankaran and colleagues showed that RAG2$^{-/-}$ mice developed tumours earlier and with greater frequency than their wildtype counterparts. They also tested mouse strains knockout (KO) for the IFN-γ receptor gene (*IFNGR*) or *STAT1*, which is necessary to the signalling cascade in response to IFN-γ. Both of these strains turned out to be more sensitive to MCA treatment compared to wild type animals. Interestingly, double KO for RAG2 and STAT1 did not do worse that single KO in terms of tumour control, which shows a high level of overlap between the adaptive immune system and the IFN-γ response pathway. From this study, they concluded that the immune system, and in particular lymphocytes, were playing an important role in the control of tumour development (Shankaran et al., 2001).

The existence of immunosurveillance in humans was tricky to demonstrate because of confounding factors, as immunosuppressed individuals are more susceptible to infections, and it is known that chronic inflammation increases the risk of developing cancer (Coussens and Werb, 2002). However, some studies indicate that immunosurveillance is also relevant for humans. Indeed, immunosuppressed transplant recipients generally display an increased incidence of non-virally induced cancers, in addition to the known virally induced cancers (Dunn et al., 2004). For example, in Sweden, patients were found to have a 100-fold increase in the risk of melanoma after transplantation (Lindelöf et al., 2000). Moreover, the incidence of both virally and non-virally induced cancer is increased in individuals that suffer from genetic immunodeficiencies (Shapiro, 2011).

The increase in incidence of cancer and metastatic lesions in immunosuppressed mice and humans demonstrates that the presence of an efficient immune system does have a protective effect against tumour progression. Several pathways are involved in this defence mechanism. Firstly, the presence of functioning lymphocytes seems to be critical, but another important factor is the IFN-γ pathway.

The need for IFN-γ in tumour regression

Cytokines are molecules that are used to signal and give instructions to immune cells in a paracrine and sometimes autocrine fashion. Cytokines can have stimulatory effects, such as IL-2, which instructs lymphocytes to proliferate and survive, or inhibitory effects, such as TGF-β. IFN-γ is a mediator of inflammation. It is produced mainly by activated T lymphocytes, and the innate NK cells and it promotes the activation and maturation of innate cells such as macrophages. In an infection setting, IFN-γ instructs macrophages to destroy the phagocytized pathogens. In a cancer setting, the cytokine IFN-γ was shown multiple times to be a very important immune mediator of protection against tumours. However, in some cases it can also contribute to tumour progression (Zaidi and Merlino, 2011).

Dighe et al. demonstrated that the neutralization of IFN-γ by the injection of monoclonal antibodies completely abrogated the LPS-induced rejection of Meth A tumours (Dighe et al., 1994). IFN-γ can directly increase tumour cells' immunogenicity. Indeed, it was shown in 1988 by Weber and Rosenberg that surface expression of the MHC class I was increased in tumour cells upon IFN-α or IFN-γ treatment, both *in vitro* and *in vivo* (Weber and Rosenberg, 1988). The surface expression of MHC class I is crucial to the CD8+ T lymphocytes recognition and lysis of the tumour cells. Hence, this upregulation, as well as the presence of a functional immune system, were critical components of the observed IFN-γ-mediated tumour rejection (Dighe et al., 1994). To confirm these findings, it was shown that a tumour cell line engineered to express an MHC class I incompatible with the host's immune effector cells could not be rejected (Weber and Rosenberg 1988; Shankaran et al., 2001). IFN-γ-mediated tumour rejection does not only act directly on the tumour, but also provides an appropriate inflammatory environment for the immune system to be efficiently activated. Indeed, it was shown that mice lacking STAT1 signalling, which is necessary for the immune response to IFN-γ, could not reject highly immunogenic tumours. This was due to deficiency in both cytokine production and cytolytic activity of T and NK cells (Fallarino and Gajewski, 1999). It was later observed that the absence of STAT1 signalling was not intrinsically deleterious to the T cells effector functions but was necessary in CD8α+ dendritic cells to mediate efficient antigen cross-presentation and priming of antitumour T lymphocytes (Fuertes et al., 2011).

However, IFN-γ signalling can also make the tumour more resistant to the immune system. Indeed, mouse melanoma cells were shown to upregulate PD-L1 after IFN-γ treatment, inducing immunosuppression of tumour-specific CD8+ T cells (Blank et al., 2004). In humans, a clinical trial evaluating the potential of IFN-γ for the immunotherapeutic treatment of ovarian cancer showed increased mortality in the treated group and had to be interrupted (Alberts et al., 2008). It was later demonstrated that human ovarian cancer cells expressed the IFN-γ receptor and could upregulate upon IFN-γ treatment (Abiko et al., 2015).

The observation of the IFN-γ pathways shows the intricate and dynamic relationship between tumour cells and the immune system. Indeed, without an immune system there is no tumour protection. However, the tumour cells are able to adapt to certain inflammatory signal and strengthen their immunosuppressive pathways in reaction, which will in turn make them more aggressive. Therefore, both increasing the inflammatory response, and simultaneously blocking the immunosuppressive pathways of the tumour should be considered in the design of immunotherapeutic treatments against cancer.

Immunoediting

An apparent paradox in the immunosurveillance hypothesis is that not only immunocompromised but also immunocompetent host develop tumours. Why is that so, if the immune system is so good at preventing tumour growth? To more accurately represent the complex interaction between tumours and their environment, the cancer immunosurveillance idea was replaced by the more refined immunoediting concept (**Fig. 29.3**).

Cancer immunoediting occurs in three phases. This has already been explained in Chapter 23 and can be briefly recapitulated as follows. First, arising mutant cells are detected and removed by the host. This is the elimination phase. Secondly, there is the equilibrium phase, in which the host's immune system is unable to completely get rid of all malignant cells but manages to control their growth. During this time, the immune system will exert a strong pressure on the tumour cells and only the least immunogenic mutants will survive. Depending on their tumorigenicity, these cells will go on to the third phase, the escape. In that final step, low immunogenic and highly malignant tumour cells are no longer controlled by the immune system and develop into a cancer (Dunn et al., 2004).

Tumour escape

Some elements that allow cancer cells to avoid immune control and increase their malignancy have been recently identified. To evade elimination by effectors from the adaptive branch of the immune system, tumours can mutate key antigens as well as downregulate their expression of MHC and/or of various components of the antigen processing machinery. Since recognition by T lymphocytes involves the binding of their antigen receptor to the peptide-MHC complexes, cells that express low or no MHC will be poorly recognized and lysed by cytotoxic T cells. In human tumours, complete MHC class I loss is frequent, and occurs in many types of cancer (Khong and Restifo, 2002; Garrido et al., 2016).

Cells in the body losing expression of MHC class I molecules can be detected by NK cells, which are innate lymphoid cells with strong

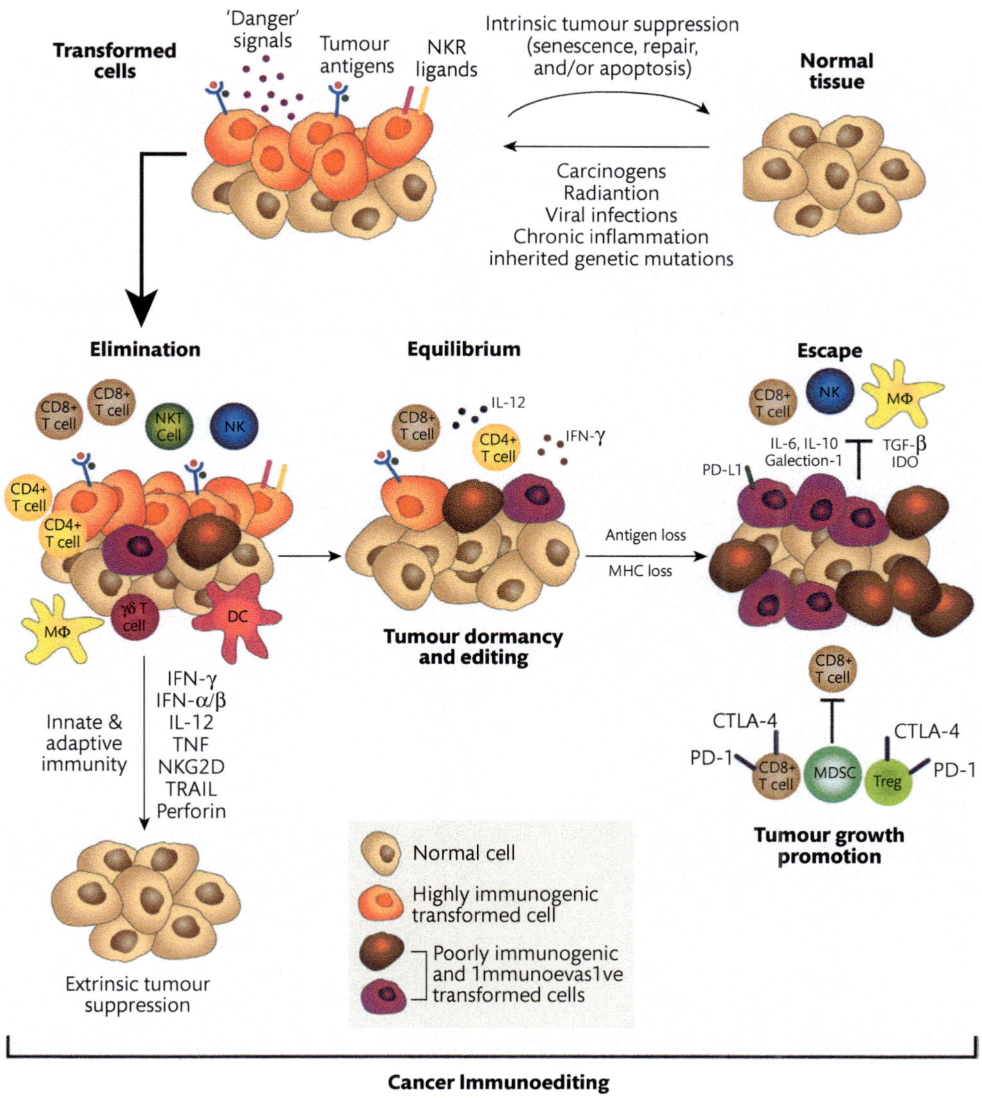

Fig. 29.3 Elimination, equilibrium, and escape are three steps of interaction between the immune system and cancerous cells. When malignant cells escape the immune system, a full cancer can develop.

cytolytic and cytokine secretion function. Thus, the loss of MHC class I would render tumour cells targets for lysing by NK cells. However, frequently this does not seem to be the case. Two mechanisms might protect the malignant cells from NK cell recognition. Firstly, the pro-tumoural environment is usually low on IL-12 and immunosuppressive, which is not adequate for the proper activation of NK. Secondly, NK cells also require binding to the stress-induced ligands MICA/B via their NKG2D receptor for their activation. It has been observed that many tumours lose the expression of these ligands, or even shed them in soluble form, to induce a downregulation of NKG2D at the surface of NK cells before they reach the tumour cells (Groh et al., 2002; Lanier, 2015). These mechanisms protect the tumour cells from NK cell cytotoxicity.

Moreover, tumours may also evade antitumour immunity through the secretion of immunosuppressive cytokines and factors, which deactivate the immune response and deplete the tumour microenvironment of crucial nutrients. Immunosuppressive cytokines typically overexpressed by tumours are transforming growth factor β (TGF-β) and IL-10. TGF-β inhibits the efficient activity of cytotoxic T lymphocytes *in vivo*. Additionally, TGF-β can convert CD4+ T cells into inducible T regulatory cells (iTregs), which are in turn able to suppress the activity of other lymphocytes (Li and Flavell, 2008). Moreover, this cytokine was also shown to induce the differentiation of myeloid cells into terminally differentiated myeloid mononuclear cells, which express CD39 and CD73. The CD39/CD73 enzyme pair degrades ATP into adenosine, which has further immunosuppressive effects (Ryzhov et al., 2014). In humans, TGF-β expression was found in several types of cancer and was associated with disease progression in breast cancer patients (Gorsch et al., 1992).

A further way in which tumours are able to deactivate the cytotoxic T-cell response is by promoting the degradation the crucial amino acid tryptophan through the overexpression of indoleamine-2,3-dioxygenase (IDO). This enzyme, as well as the closely related tryptophan 2, 3-dioxygenase, catalyses the conversion of tryptophan to kynurenine. The depletion of tryptophan and the production of kynurenine have been shown as inducers of Th1 cell apoptosis (Fallarino et al., 2002). Moreover, absence of tryptophan induces the activation of the stress-response kinase GCN2 in CD4+ T cells and drives their differentiation into Tregs (Platten et al., 2012; Munn and Mellor, 2016). In CD8+ T cells, IDO-induced activation of GCN2 leads to cell cycle arrest and a state of functional anergy (Munn et al., 2005; Trotta et al., 2013). A large range of human tumours, such as prostatic, colorectal, pancreatic, cervical, gastric, ovarian, head, lung, and others, overexpress human IDO (Uyttenhove et al., 2003; Munn and Mellor, 2016).

Another major pathway to localized immunosuppression follows the ability of tumours to recruit myeloid cells such as monocytes and immature precursors from the bone marrow which are transformed into polarized M2 macrophages and MDSCs. Tumour cells can subvert myeloid cells via the expression of several cytokines and chemokines such as CSF1, VEGFA, SEMA3A, CCL2, and CXCL12 (Kitamura et al., 2015). The MDSCs in turn express inducible nitric oxide synthase (iNOS) as well as arginase 1, that both are deleterious to efficient T-cell function through the production of reactive oxygen species and depletion of arginine (Rodríguez and Ochoa, 2008). Finally, tumours are able to express immunosuppressive ligands, such as PD-L1, which when bound to their receptor on the activated T cells, renders them tolerogenic and inactive (see **Fig. 29.4**).

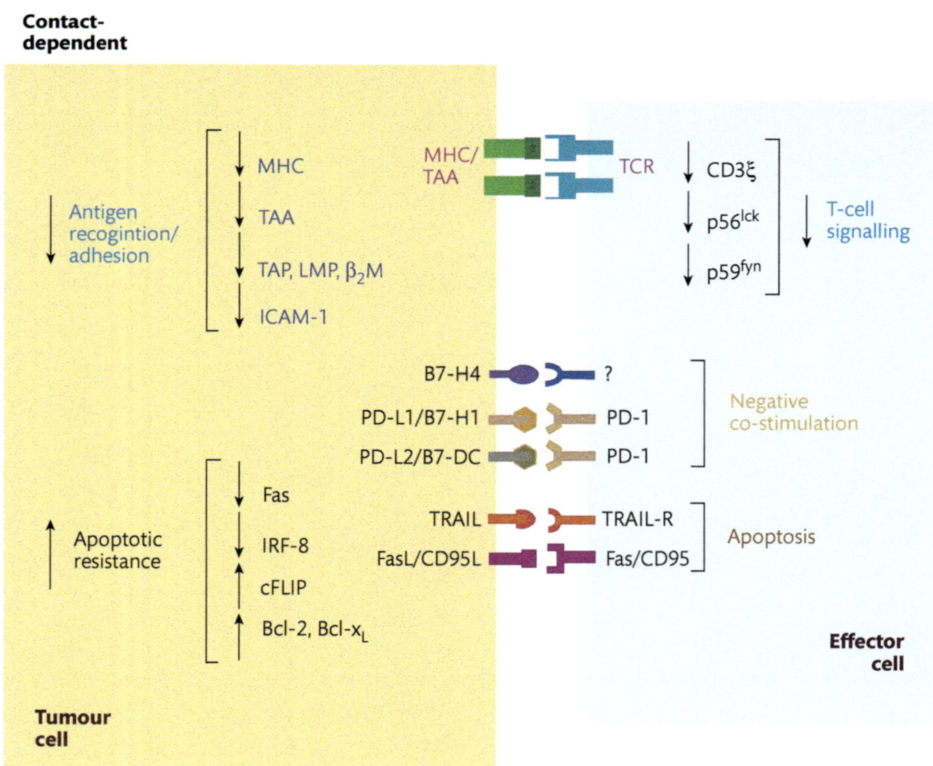

Fig. 29.4 Ligands expressed at the surface of the tumour cell which can decrease the functionality of effector T cells.

In summary, in addition to being able to hide from the immune system by downregulating MHC class I, tumours can directly deactivate it by creating a local immunosuppressive environment. They do so via the secretion of immunosuppressive factors, as well as the surface expression of inhibitory receptors and the recruitment of immunosuppressive cells. All these immunosuppressive mechanisms have to be carefully considered for the design of efficient immunotherapies.

Cancer immunotherapy

The strong evidence of the importance of the immune system in mediating tumour protection promoted the development of antitumour immunotherapies. Given the fact that lymphocytes have the ability to kill tumour cells, and that CD8+ T-cell infiltration correlates with good prognosis (Gooden et al., 2011), the main aims of immunotherapies against cancer are both to increase the lymphocyte numbers on site, as well as to reactivate them. Indeed, lymphocytes present in the tumour microenvironment are usually of the exhausted phenotype and have often lost most of their effector functions and tumour lysing abilities (Jiang et al., 2015).

Different types of immunotherapies have been tested to boost the immune system against tumours: antibodies can either be used to target tumours epitopes and induce the activation of immune cells and lysis of the tumour cell as a consequence; or be used to block immunosuppressive pathways. The latter therapeutic application has been coined in recent years as immune checkpoint blockade (discussed next). It would also involve small molecules specifically interfering with immune checkpoints that are in active development and may reach the clinic in the near future. Treatments with cytokines aim at systemically boosting the immune system. Cancer vaccines are made to elicit a specific response against one or several relevant tumour antigens. Additionally, adoptive cell transfer can be performed; by isolating and expanding a patient's own lymphocytes before transferring them back into the circulation.

Antibodies

Using antibodies for tumour targeting has been of strong interest for several decades, and the biggest hurdles have been (i) to engineer antibodies with minimal immunogenicity allowing for repeated administration over long periods of time; and (ii) to identify target antigens that are specific to the tumour cells. An appropriate target for antibody immunotherapy needs to be expressed both homogeneously and at a high enough level on the surface of the tumour. Once a target that fulfils those criteria has been identified, different types of antibodies can be produced against it.

Monoclonal antibodies against tumour surface antigen. Several monoclonal antibodies targeting different kind of tumour antigens have shown success in clinical trials. Consequently, they have become commercially available for cancer treatment (Scott et al 2012). For example, Cetuximab is a monoclonal antibody targeting the epidermal growth factor receptor (EGFR). It was found to be an efficient therapy for colorectal cancer patients, if they did not have a KRAS mutation. For the patients with wild type KRAS, treatment with Cetuximab improved the overall survival median from 4.8 months to 9.5 months (Karapetis et al., 2008). This led in 2012 to the approval of Cetuximab, together with chemotherapy, for first-line care of EGFR+ KRAS WT colorectal cancer patients (Pazdur, n.d.). Trastuzumab, is another FDA-approved monoclonal antibody used for the treatment of HER2+ breast cancer (Vogel et al., 2002). It showed clinical benefit in 48% of the patients whose tumours where highly positive for HER2. The latter is approximately one-third of breast carcinomas. The efficacy of both of these treatments were linked to the inhibition of signalling, as well as to antibody-dependant cell-mediated cytotoxicity (Fig. 29.5).

Bispecific antibodies. Bispecific antibodies have the advantage that they can either simultaneously target two antigens and thus increase the specificity of the antibody to the tumour and reduce the non-specific toxicity, or alternatively specifically attract some immune cells to the tumour site. An example from this second category is catumaxomab. This non-humanized mouse bispecific antibody simultaneously targets EPCAM and CD3 and has a functional Fc domain to activate innate immune cells. It is used for the treatment of malignant ascites together with paracentesis (removal of fluid from the peritoneal cavity). Overall survival, as well as median puncture-free survival, were significantly increased for patients treated with catumaxomab (Heiss et al., 2010). Another bispecific antibody used in the clinic is blinatumomab. It does not have a functional Fc domain but it is composed of two single chain antibody fragments (scFv) one targeting CD19, and the other one targeting CD3. This allows attracting cytotoxic lymphocytes, which express CD3, and put them in contact with the malignant B cells, which express the marker CD19. Indeed, the FDA-approved blinatumomab as treatment for refractory B-ALL, Philadelphia chromosome negative. Treatment with blinatumomab increased the complete response in these patients to 30% compared to 5–12% with previous chemotherapeutic regiments (Przepiorka et al., 2015).

Other bispecific molecules have been investigated, such as ImmTACs, which comprise a TCR-like antibody linked to a T-cell receptor (TCR). Such a molecule, IMCgp100, was made by linking an anti-CD3 scFv to a TCR specific for the gp100 peptide:MHC complex. It is currently in phase II trial after encouraging reports from the phase I trial, which results are not yet published (Immunocore, n.d.). Similar to TCR-transgenic T cells, such a construct creates the challenge of having HLA-matching between the patients and the drug, as well as a high enough MHC expression on the surface of the tumour.

Additional bispecific antibodies have sought to redirect NK cells, for example by linking the Fv domains of an anti-CD16A and an anti-CD30. Such a molecule proved efficient *in vitro* to mediate lysis of human lymphoma cells (Reusch et al., 2014). Another investigated method to reactivate iNKT cells against tumours was to take advantage of the CD1d invariant molecule loaded with alpha-galactosylceramide (αGC), which is capable of strongly activating iNKT cells. However, iNKT cells become anergic after one stimulation with this molecule. A new approach was then investigated by fusing the αGC-loaded CD1d with an scFv against the cancer antigens Her2 or CEA. This method proved capable of initiating a potent and specific antitumour effect in a mouse model of adenocarcinoma (Corgnac et al., 2013).

Immunomodulatory antibodies—the emerging class of immune checkpoint blockade agents. Lastly, immunomodulatory antibodies function by either blocking the interaction of important

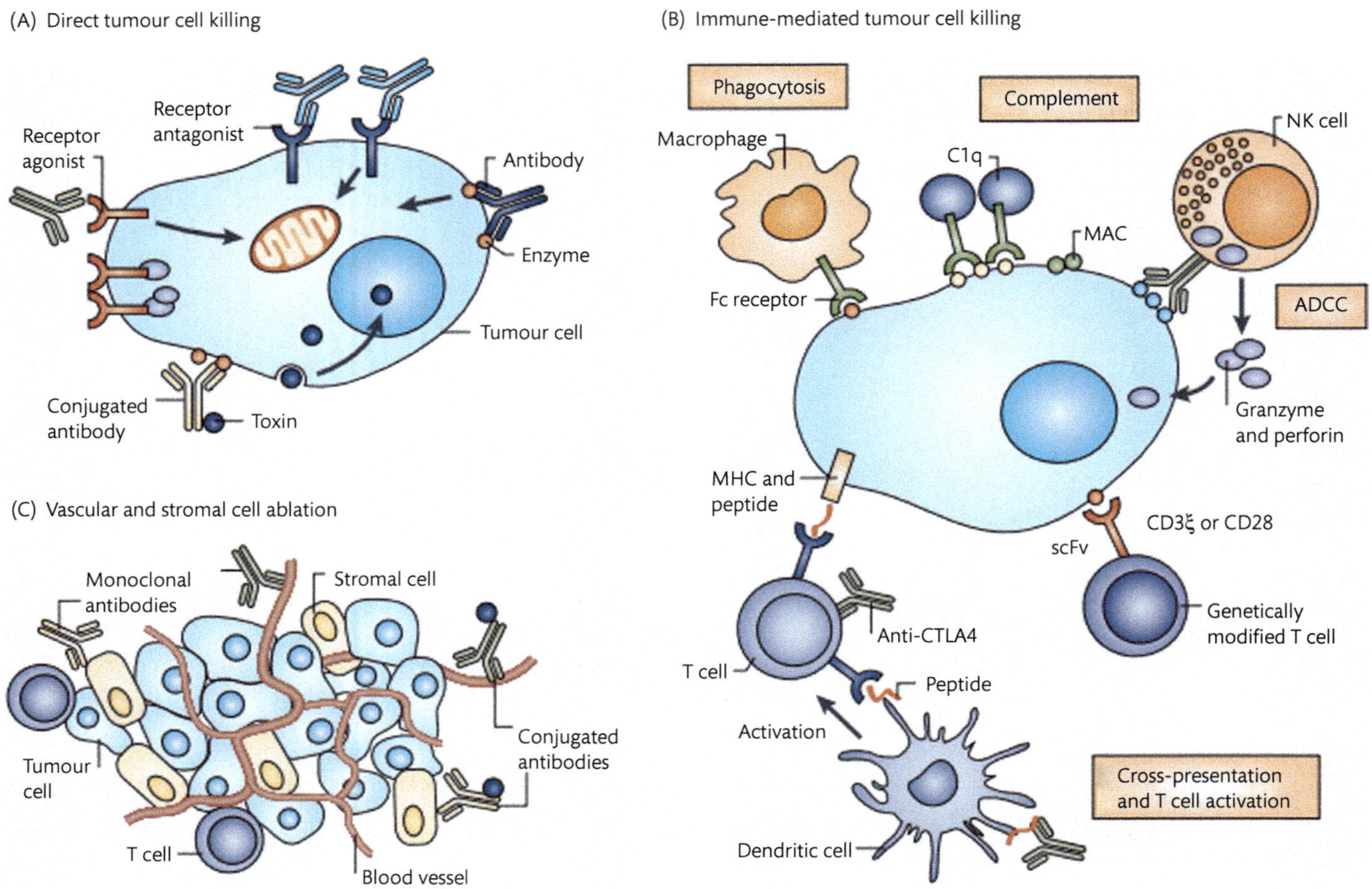

Fig. 29.5 (A–C) Mechanisms of tumour targeting by antibodies.
Reproduced with permission from Scott AM et al. 'Antibody therapy of cancer', *Nature Review Cancer*, Volume 12, Issue 4, pp. 278–87, Copyright © 2012 Springer Nature.

immunosuppressive receptors with their ligands or by ligating and promoting the activity of costimulatory receptors expressed on recently activated T cells. In both cases, the net effect is the rekindling of tumour-infiltrating lymphocytes which exert direct potent antitumour activity and may also redirect the local immune response towards inflammation and a type I polarity of other infiltrating immune cells.

Immunosuppressive receptors are usually upregulated on the surface of activated T cells and provide a negative feedback, hence preventing an uncontrolled immune response and fine-tuning the effector activity output of differentiating T cells. For this reason, they have been named *immune checkpoints*. They are necessary for the normal function of the immune system and the restoration of homeostasis after the resolution of an infection. As a consequence, the absence of immune checkpoints can drive fatal immunopathology, for example, in CTLA-4$^{-/-}$ mice (Tivol et al., 1995), or in PD-1L$^{-/-}$ mice upon chronic lymphocytic choriomeningitis virus (LCMV) infection (Barber et al., 2006).

CTLA-4 has two immunosuppressive functions. On one hand it binds costimulatory ligands B7.1 and B7.2 and hence prevents the proper activation of the naive T lymphocyte via the activatory coreceptor CD28. On the other hand, it delivers a cascade of immunosuppressive signalling in the T cell. Moreover, CTLA-4 is constitutively expressed in Tregs, and blocking of CTLA-4 via a monoclonal antibody abrogates their suppressive function (Takahashi et al., 2000).

PD-1 is another major coinhibitory receptor upregulated upon T-cell activation (Agata et al., 1996). PD-1 expression levels are further increased in T cells during chronic infection or in tumour-infiltrating lymphocytes (Ahmadzadeh et al., 2009) due to the repeated engagement of antigen receptors. These PD-1 high T cells are usually impaired in their effector functions and are hence named 'exhausted' (Zajac et al., 1998). Several tumour cell lines were shown to either endogenously express the PD-1 ligand PD-L1, or to upregulate it upon IFN-γ treatment. Subsequent binding of PD-L1 to its receptor on T cells induced their apoptosis (Dong et al., 2002). Indeed, tumours are able to hijack mechanisms designed to control the immune response and use them to suppress the function of CTLs. Hence, researchers have sought to block the interaction of those immunosuppressive receptors with their ligands via the injection of blocking antibodies. This line of treatment is called *immune checkpoint blockade*.

In 1996, Leach et al. first demonstrated that anti-CTLA-4 could block the interaction of CTLA-4 with its ligand and hence prevents immunosuppressive signalling as well as restores the availability of CD80/CD86 for CD28 costimulation. Treatment with this antibody induced antitumour immunity and protected the mice against a further tumour challenge (Leach et al., 1996). A fully humanized anti-CTLA-4 called ipilimumab was subsequently produced and used in a phase III trial. The trial demonstrated that treatment with ipilimumab improved both overall survival and progression-free survival in metastatic melanoma patients. Moreover, 18% of the patients treated with ipilimumab survived beyond 2 years, compared to only 5% of the patients receiving a peptide vaccine alone (Hodi et al., 2010).

Another common target of immunomodulatory antibodies is PD-1. Nivolumab and pembrolizumab both target this receptor and have been found to have an antitumour effect in advanced melanoma. Nivolumab improved both overall survival as well as progression-free survival in metastatic melanoma without a BRAF mutation compared to dacarbazine chemotherapy. Interestingly, in both treatment arms, the expression of PD-L1 on the tumour was correlated with better overall survival (Robert et al., 2015). Pembrolizumab is currently used in 16 phase III trials for many types of cancer (breast, melanoma, lymphoma, lung, and others; see NIH, n.d.). In 2015, pembrolizumab was proved to be more efficient than ipilimumab in a phase III clinical trial for the treatment of advanced melanoma, as well as to induce less high-grade adverse events (Robert et al., 2015). Additionally, pembrolizumab was shown to be more efficient than chemotherapy in patients that had ipilimumab-refractory melanoma (Ribas et al., 2015). Indeed, it had been shown that CTLA-4 and PD-1 are not redundant in their functionality and inhibit T-cell function by distinct pathways (Parry et al., 2005). Consequently, combination therapies, which aim at using different immunotherapies together to increase the response and prevent possible resistance mechanisms, were experimented. Postow et al. measured an improved response rate in advanced melanoma patients treated with both nivolumab and ipilimumab compared to ipilimumab administered as a monotherapy (Postow et al., 2015).

The unprecedented efficiency of immune checkpoint blockade prompted FDA approval of these therapies for a myriad of cancer namely non-small-cell lung cancer (Garon et al., 2015; Lim et al., 2016), kidney cancer (Motzer et al., 2015), Hodgkin lymphoma (U.S. Food and Drug, n.d.), and bladder cancer (Rosenberg et al., 2016). Immune checkpoint blockade antibodies have generated unprecedented responses and are hence being incorporated in many trials across several types of cancer. Moreover, the possibility of blocking other immunosuppressive receptors is under research. Finally, combining immune checkpoint blockade with other immunostimulatory treatments, such as cancer vaccines, is of high interest. Indeed, since the tumours can respond to inflammation by increasing their immunosuppressive ligands, blocking this pathway during a vaccination response would make sense.

Cytokines as immunotherapeutic agents

It was in the late seventies that the ability of cytokines to mediate lymphocyte survival and proliferation was discovered. This new development allowed researchers to grow T and B as well as innate immune cells in culture for *in vitro* studies and greatly improved the understanding of lymphocytes and immune mechanisms. Following this discovery, and with the ability to produce and purify recombinant cytokines came the interest to use them in anticancer treatment.

Two cytokines have mainly been tried in systemic cancer treatment. IL-2 is a cytokine promoting lymphocyte survival and proliferation which explains the rationale behind its use in cancer therapy. However, IL-2 promotes the proliferation of all lymphocytes, including the immunosuppressive Tregs. It is hence a very non-specific form of stimulation. Firstly, IL-2 was injected in patients with metastatic melanoma and renal cell cancers, which induced complete and durable responses in 6.6% and 9.3% of cases, respectively (Turcotte and Rosenberg, 2011). Systemic administration of IL-2 is a non-specific treatment, aiming at boosting the proliferation of NK and T cells. Hence, adverse toxicities were important at first, with 92% of patients experimenting grade 3/4 diarrhoea, and 81% experimenting grade 3/4 hypotension (Kammula et al., 1998). However, once appropriate dosing and administration of IL-2 were experimentally determined, adverse events following the treatment could be lowered from 2 to 4% fatality to less than 1%, and other non-fatal symptoms could be resolved by appropriate additional care (Rosenberg, 2014).

Another cytokine that drew interest for immunotherapy was IL-12. IL-12 is produced by activated dendritic cells and induces the activation of NK cells as well as helps the differentiation of helper T cells into their Th1 pro-inflammatory form. However, in contrast to IL-2, treatment with IL-12 was found to have strong adverse effects that could not be controlled. Despite encouraging results using IL-12 therapies in animals (either via systemic injection of purified IL-12 or via local production of IL-12 in the tumour), severe toxicities were found in human trials using IL-12, which led to the immediate halt of those trials. In general, cytokines are powerful small molecules which upon *in vivo* systemic administration display both pleiotropic immune and inflammatory effects and an extraordinarily narrow therapeutic window. Clearly, IL-2, IL-15, and IL-12 almost completely overlap therapeutic and highly toxic doses. This imposes serious constraints in their use in the clinic and await innovative technologies of delivery to the target tumour sites without systemic spill over. This is supported by more recent studies which, using local IL-12 expression in the tumour microenvironment, in combination with other therapies, gave encouraging results and are reviving the interest in this cytokine for cancer immunotherapy (Lasek et al., 2014).

Cancer vaccines

Vaccines are usually developed for their ability to mount an immune response against a known pathogen and prevent vaccinated individuals from further contracting a disease. Their prophylactic activity is generally mediated by the induction of stable, long lived high levels of neutralizing IgG or IgA antibody responses. Such prophylactic vaccines cannot be applied to most cancers, with the exception of virally induced cancers. For example, the human papilloma virus (HPV) is a known trigger for cervical neoplasia. Vaccines against HPV have been developed and could successfully prevent HPV infection as well as neoplasia development (Paavonen et al., 2009).

For most cancers however, research has focused on designing vaccines, which aim to set off an effective antitumour immune response once the cancer is already established. In contrast to prophylactic vaccines, cancer vaccines may protect the host from tumour growth not by neutralizing antibodies but through the induction of tumour antigen-specific T-cell responses, both those of CD4+ T helper and especially CD8+ cytotoxic T-cells. While prophylactic vaccines have been refined over many decades of research and technological advances, therapeutic cancer vaccines are still in their infancy. Their development relies on research on the mechanisms of induction of specific T cells, of their expansion and differentiation, on their life cycles, migration through various tissues, their residency, and even fates such as activation induced cell death, self-renewal, survival, and senescence. In addition, while neutralizing antibodies can be accurately measured by simple biochemical assays in body fluids using robust and high throughput assays, antigen-specific T cells are difficult to monitor and to measure in a quantitative and sensitive fashion. The assays to measure T-cell responses, and hence vaccine efficacy, are still in their first infancy.

Such therapeutic vaccines can be formulated in many ways, which will affect the type of the response, whether the inflammation is more local or systemic, or if the cells are activated against only one antigen or several of them.

Peptide-based cancer vaccines. Peptide and adjuvant-based vaccines aim at igniting an adaptive immune response against a chosen target tumour antigen. However, most antigens expressed by the tumour are of low affinity and are subject to peripheral tolerance since T cells recognize them as 'self'. Indeed, as we previously discussed, the immune system has already taken care of weeding out high immunogenic cancerous cells at the early stage of tumour development during tumour immunoediting. Consequently, the antigen of choice for a therapeutic vaccine has to be carefully considered. Antigens of highest interest for therapeutic vaccines include mutated antigens (neoantigens), cancer testis antigens—which are expressed only in male germ cells but not adult somatic tissues with the exception of tumours—and antigens that are overexpressed in the tumour compared to normal tissue. Finally, tissue-specific antigens can be considered when the tissue happens to be dispensable for the survival of the patient, for example the prostate specific membrane antigen PSMA, or the B-cell antigen CD19 (Melief et al., 2015).

Naked DNA and recombinant vector-based vaccines. The delivery of tumour antigen via its genetic sequence has also been studied. Bacterial plasmids modified to include the gene of the antigen of interest can be directly injected into the muscles of patients, and some cells, such as dendritic cells, will uptake the genetic material and start expressing the protein. It can then be presented to T cells via the classical MHC class I pathway of antigen processing and presentation. The efficacy of DNA incorporation into muscle cells can be enhanced by *in vivo* electroporation and dedicated devices have been developed to achieve reproducible rates of cell transduction with the plasmid DNA vaccine (Aurisicchio et al., 2013). Vaccination with naked DNA is generally efficient at priming *de novo* antigen-specific CD8+ T-cell responses, which is the aim of therapeutic cancer vaccines. However, these responses are weak and cannot be boosted by repeated DNA vaccination.

To optimize this approach, packaging vectors for DNA delivery have been engineered. For example, bacterial or viral vector can be used. They possess the advantage of infecting the target cells and delivering the protein of interest more effectively. An additional advantage of this approach is the fact that pathogens carry pathogen-associated molecules, such as LPS, which can trigger an inflammatory immune response. In a recent clinical trial for the treatment of prostate cancer, Kantoff and colleagues administered subcutaneously a recombinant vaccinia virus expressing prostate-specific antigen (PSA) and three immune costimulatory molecules, namely B7.1, ICAM-1, and LFA3, together with recombinant GM-CSF. At three years post-treatment, 30.5% of the vaccinated patients were still alive, compared to 17.5% in the control arm (Kantoff et al., 2010b). The diversity of recombinant vectors, including naked DNA plasmids, offers plenty of possibilities for alternating these vehicles for priming and boosting. Priming with plasmid DNA and boosting with a viral vector seems to be particularly efficient (Hanke, 2014).

Of note, oncolytic viral vectors have also been used. Their advantage is that they are both directly harmful to the tumour, and at the same time inducing immunogenic cell death, which improves the quality of the subsequent immune response. A recent clinical trial using a herpes simplex oncolytic virus expressing GM-CSF (T-VEC) to attract antigen-presenting cells to the tumour site showed encouraging results. Specifically, patients treated with T-VEC had an overall survival of 23.3 months compared to 18.9 months in the GM-CSF control arm (Andtbacka et al., 2015).

Whole tumour cell vaccines. To circumvent the hurdle of the antigen's identification and selection, researchers have explored the use of whole tumour cell vaccines. Lysates of autologous tumours provide the whole spectrum of tumour antigen and neoepitopes thus allowing to skip the identification of target antigens. Unfortunately, however, the success of this approach in clinical trials has been rather limited albeit slightly higher than that of peptide-based vaccination. Indeed, a comparison of objective clinical response rates recorded in 173 small trials, using either whole cell-based approach or peptide-based approach as immunotherapies for a wide range of cancers, indicated that whole cell vaccines had an 8.1% objective clinical response, compared to 3.6% for molecularly defined antigens. A similar result was found when the analysis was restricted to advanced metastatic melanoma (12.7% for whole cell versus 6.7% for peptides; Neller et al., 2008). From a practical standpoint, the use of well-defined allogeneic tumour cell lines would be advantageous. Moreover, these could be engineered to express additional immunoattractant or immunostimulatory molecules. One such modification entails expression of GM-CSF, which would increase further the efficiency of the vaccine by providing maturation signals to antigen-presenting cells. Unfortunately, most clinical trials using this approach so far did not show clinical benefits (Srivatsan et al., 2014).

Dendritic cells vaccines. A more controlled way of delivering antigen is to extract cells from the patient and pulse them with the peptide or peptides of choice. Moreover, the best type of antigen-presenting cell (APC) can be selectively isolated. Since they are the best-known cells for antigen presentation, dendritic cells (DCs) have been the cells of choice for this type of immunotherapy. Indeed, after maturation, DCs are the most potent APCs, capable of triggering robust immune response and to provide all the signals to T lymphocytes for their expansion and activation. DCs can be purified from the blood of cancer patients, maturated and pulsed with antigen or whole tumour lysate *in vitro*. Several clinical trials have been started and have shown marginal improvement in terms of tumour responses (Guo et al., 2013). Phase III clinical trials completed in 2010 showed that Sipuleucel T, a vaccine comprising activated monocytes pulsed with a GM-CSF-prostate antigen fusion protein, improved overall survival in castration-resistant prostate cancer patients, a result that led to FDA approval (Kantoff et al., 2010a).

Adoptive cell transfer

A fourth modality of cancer immunotherapy is the adoptive cell transfer with autologous immune cells into the cancer patient, after *ex vivo* expansion, treatment, or genetic modification. While a preclinical study showed that the combination of radiation therapy and transfer of preactivated NK cells could mediate tumour rejection in mice (Ni et al., 2012), and, additionally, several studies showed CAR-modified NK cells for the immunotherapy of cancer (Chu et al., 2014; Guillerey et al., 2016), the vast majority of the adoptive cell transfer therapy research has focused on expanding or genetically modifying T lymphocytes.

Tumour-infiltrating lymphocytes. As it was previously mentioned, tumours are often infiltrated with T lymphocytes (TILs). However, due to the chronic antigen exposure as well as the immunosuppressive environment, these T cells generally express high levels of inhibitory receptors such as PD-1 and are functionally impaired. Researchers sought to isolate those TILs and expand them *in vitro*, in the presence of high IL-2 levels, aiming to artificially increase their total number and to reactivate them away from the deleterious cancer environment. This allows the T cells to recover their functionality. After *in vitro* expansion, the lymphocytes can be re-infused into the circulation via an intravenous injection, usually after the patient has been pretreated with chemotherapy or irradiation to induce non-myeloablative lymphodepletion. This technique proved very successful for metastatic melanoma, where 22% of patients receiving TILs after *ex vivo* expansion, in combination with lymphodepletion, reached complete response (Rosenberg et al., 2011).

Unfortunately, this method can only be applied for the kind of tumours that are reasonably well infiltrated by TILs. Additionally, some tumours are hardly accessible to surgical resection. Moreover, as previously discussed, tumours have been edited to express antigen of low immunogenicity, which means that the retrieved TILs might not have an optimal affinity to the tumour antigens. Hence, researchers aimed to genetically modify T cells to redirect them against the tumour with high affinity antigen recognition.

TCR-transgenic (engineered) T cells. It was sought to use lymphocytes from peripheral blood, as they are more easily accessible. These cells can be genetically modified to enforce the expression of a tumour antigen-specific TCR. Such receptors are isolated either from patients who happen to have a high affinity CD8+ T-cell clone against a shared antigen, or sometimes from mouse model. Retro- or lentiviruses are often used as a way to insert genetic material into the cells and induce the expression of the new receptor (Restifo et al., 2012). There are drawbacks in this method. Firstly, different human leukocyte antigen (HLA) types exist among the population, and each TCR is compatible only with a certain HLA type. Secondly, TCRs are composed of two chains, and hence mispairing between one of the newly introduced TCR chains and the complementary endogenous TCR chain can occur and prevent the correct expression of the tumour-specific TCR. Thirdly, like naturally occurring T cells, TCR-engineered T cells depend on the expression of the peptide-MHC complex at the surface of the tumour for their activity.

Chimeric antigen receptors. A way to overcome these issues is to construct chimeric antigen receptors (CARs). These receptors do not occur naturally and are composed of two parts. The extracellular domain comprises the scFv of an antibody specific for a tumour antigen exposed on the tumour cell surface. It is fused to the intracellular CD3ζ signalling domain, as well as one or more costimulatory domains depending on the CAR generation (see **Fig. 29.6**). It was observed that the addition of the costimulatory domain could efficiently compensate the fact that tumour cells generally do not express the appropriate costimulatory ligands (Krause et al., 1998). The expression of a CAR on the surface of a lymphocyte allows for direct recognition of a non-processed antigen on the surface of the tumour, thus avoiding the issue of MHC class I expression. Moreover, proteins as well as carbohydrates and glycolipids can be recognized by CARs, which expands the list of possible targets (Sadelain et al., 2013).

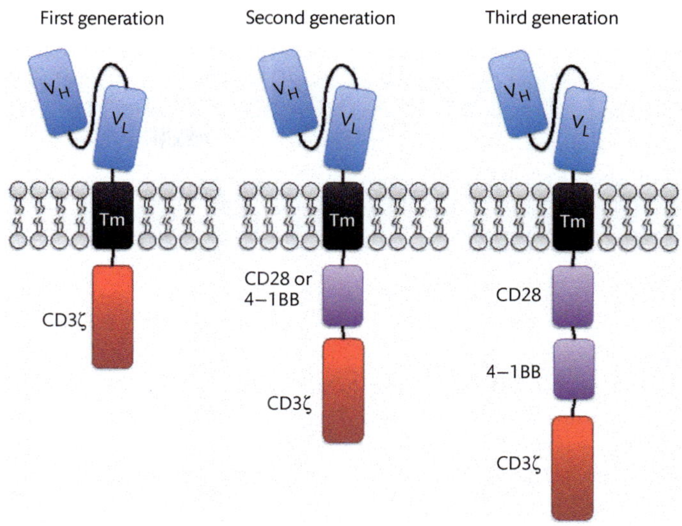

Fig. 29.6 First, second and third CAR generations incorporate respectively none, one or two costimulatory domains in their intracellular moiety. V_h, variable heavy chain; V_L, variable light chain; Tm, transmembrane domain.
Reproduced courtesy of GC Monnot.

The excitement around CARs grew strong when the results of a pilot clinical trial conducted by Carl June were published in 2011. Out of three advanced leukaemia patients, two had a complete response and the last one a partial response (Kalos et al., 2011). Since then, more than 20 clinical trials have been launched and published, and shown great promise for the treatment of leukaemia and other haematologic malignancies (Dai et al., 2016). The adoptively transferred cells were targeting the antigen CD19, which is expressed by leukaemic cells but also by normal B cells. A successful anti-CD19 CAR treatment is thus always accompanied by a long-term B-cell aplasia.

Targeting solid tumours via CAR therapy proved more challenging and did not yet yield such outstanding results as for leukaemias (Dai et al., 2016). A hurdle in the development of CAR T cells against solid tumours was the selection of the appropriate antigen to avoid on-target off-tumour toxicity. A publication from 2010 reported that a metastatic colon cancer patient succumbed after treatment with CAR T cells redirected against HER2 using the scFv of the monoclonal antibody trastuzumab. CAR T cells were able to recognize HER2 expressed at low levels in the lung and induced a lethal pulmonary oedema (Morgan et al., 2010).

A way to better manage the sensitivity of CAR T cells to tumour antigens, and avoid the recognition of low antigen expression, might be to modulate and lower the affinity of the scFv to its ligand. In their recent publication, Liu et al. showed that reducing the anti-HER2 scFv affinity prevented recognition of tumours expressing physiologic levels of the target without affecting the ability of CAR T cells to lyse tumours overexpressing HER2 (Liu et al., 2015).

Combination immunotherapy

All the previously mentioned approaches function via different pathways and effector cells to redirect the immune system against the tumour. To increase further the efficiency of treatments, current approaches are seeking to combine different treatments and hence augment the overall efficiency. For example, the efficiency of a whole cell cancer vaccine in preclinical model of melanoma was improved to 50% rejection using coinhibition of both aCTLA-4 and aPD-1, compared to 10% and 25% when immune checkpoints were administered individually. In a similar manner, combining a vaccine comprising GM-CSF-transduced whole tumour cells with coblockade of CTLA-4 and PD-1 proved the most efficient for two different mouse tumour models (Duraiswamy et al., 2013). Similarly, combining immunotherapies and tumour-targeted drugs is an approach that shows great promises, as both methods can synergistically improve the efficiency of each other (Vanneman and Dranoff, 2012).

Conclusions

In conclusion, modern tumour immunology is approaching a deep level of understanding of the complex cross-talk between immune cells, tumour cells, and stromal cells forming the tumour microenvironment. Recent progress in systems biology is providing the next generation with a set of insights into these interactions. Importantly, key mechanisms responsible for maintaining tumour immunosuppression have been successfully targeted and the first therapies harnessing the immune system have been approved for clinical use. This is a revolution in oncology and it is becoming clear that immunotherapy is a new powerful therapeutic option possibly at a pair with chemotherapy, radiotherapy, and targeted therapies. Going forward new combinatorial therapies together with an increasing number of immunotherapeutic agents showing clinical efficacy may contribute to render cancer a manageable chronic disease in the worst-case scenario or even achieve durable cures of cancer.

TAKE-HOME MESSAGE

- The immune system can recognize tumours early during carcinogenesis and may either eliminate some of those tumours or be coopted by tumours to their advantage in promoting tumour growth, invasion, and metastasis.
- The immune system may eliminate tumours, even of large size. The major effectors in tumour elimination are tumour antigen-specific CD8+ T lymphocytes recognizing complexes of self-MHC class I molecules and short tumour-derived peptides.
- The tumour microenvironment in patients with advanced cancer is generally highly immunosuppressive. Both tumour intrinsic and extrinsic mechanisms signal paralysis of immune cells or promote functions such as the release of type 2 cytokines that favour tumour growth.
- Three major immunotherapeutic approaches seek to harness the immune system for the treatment of cancer. These include (i) therapeutic vaccines, (ii) adoptive cell transfer therapy, and (iii) immunomodulatory agents, notably immune checkpoint blocking monoclonal antibodies. The latter have met major successes and some have been approved in recent years for the treatment of a growing number of malignancies.
- Basic research is needed to gain additional insights into the interactions between immune system and cancer. In turn, these results will illuminate the improvement of current immunotherapies and bring about novel and more effective therapies based on the exploitation of the sophisticated cell and molecular systems of tissue defence evolved by the immune system.

OPEN QUESTIONS

- What makes tumours immunogenic? How to consistently turn 'immune desert' tumours into immunogenic tumours heavily infiltrated by immune effector cells and harbouring an inflamed microenvironment?
- Which are the best antigens for tumour rejection? Is it only the so-called neoantigens or are there meaningful numbers of common tumour rejection antigens?
- What are the useful biomarkers predictive of response to immunotherapy?
- Can we use cancer vaccines as a means to prevent cancer rather than to treat advanced tumours?
- Can systems immunology be used to profile immune fitness in cancer patients as well as to identify predictive or prognostic biomarkers?

FURTHER READING

Fridman, W. H., Pages, F., Suates-Fridman, C., & Galon, J. (2012). The immune contexture in human tumors: impact on clinical outcome. *Nat Rev Cancer*, **12**, 298–306.

Hall, S. (1997). *A Commotion in the Blood: Life, Death, and the Immune System*. New York: Henry Holt and Company.

Hanahan, D. & Weinberg, R. A. (2011). Hallmarks of cancer: the next generation. *Cell*, **144**, 646–74.

Lim, W. A. & June, C. H. (2017). The principles of engineering immune cells to treat cancer. *Cell*, **168**, 724–40.

Prendergast, G. & Jaffee, E. M. (2013). *Cancer Immunotherapy: Immune Suppression and Tumor Growth*, 2nd edition. Elsevier Inc.

Rosenberg, S. A. (2014). IL-2: the first effective immunotherapy for human cancer. *J Immunol*, **192**, 5451–8.

Schreiber, R. D., Old, L. J., & Smyth, M. J. (2011). Cancer immunoediting: integrating immunity's role in cancer suppression and promotion. *Science*, **331**, 1565–70.

Topalian, S. L., Drake, C. G., & Pardoll, D. M. (2015). Immune checkpoint blockade: a common denominator approach to cancer therapy. *Cancer Cell*, **27**, 450–61.

Tran, E., Robbins, P. F., & Rosenberg, S. A. (2017). 'Final common pathway' of human cancer immunotherapy: targeting random somatic mutations. *Nat Immunol*, **18**, 255–62.

REFERENCES

Abiko, K., Matsumura, N., Hamanishi, J., et al. (2015). IFN-γ from lymphocytes induces PD-L1 expression and promotes progression of ovarian cancer. *Br J Cancer*, **112**(9), 1501–9.

Agata, Y., Kawasaki, A., Nishimura, H., et al. (1996). Expression of the PD-1 antigen on the surface of stimulated mouse T and B lymphocytes. *Int Immunol*, **8**(5), 765–72.

Ahmadzadeh, M., Johnson, L., Heemskerk, B., et al. (2009). Tumor antigen-specific CD8 T cells infiltrating the tumor express high levels of PD-1 and are functionally impaired. *Blood*, **114**(8), 1537–44.

Alberts, D. S., Marth, C., Alvarez, R. D., et al. (2008). Randomized phase 3 trial of interferon gamma-1b plus standard carboplatin/paclitaxel versus carboplatin/paclitaxel alone for first-line treatment of advanced ovarian and primary peritoneal carcinomas: results from a prospectively designed analysis of prog. *Gynecol Oncol*, **109**(2), 174–81.

Andtbacka, R. H. I., Kaufman, H. L., Collichio, F., et al. (2015). Talimogene laherparepvec improves durable response rate in patients with advanced melanoma. *J Clin Oncol*, **33**(25), 2780–8.

Apetoh, L., Ghiringhelli, F., Tesniere, A., et al. (2007). Toll-like receptor 4-dependent contribution of the immune system to anticancer chemotherapy and radiotherapy. *Nat Med*, **13**(9), 1050–9.

Aurisicchio, L., Mancini, R., & Ciliberto, G. (2013). Cancer vaccination by electro-gene-transfer. *Expert Rev Vaccines*, **11**(10), 1–11.

Barber, D. L., Wherry, E. J., Masopust, D., et al. (2006). Restoring function in exhausted CD8 T cells during chronic viral infection. *Nature*, **439**(7077), 682–7.

Blank, C., Brown, I., Peterson, A. C., et al. (2004). PD-L1/B7H-1 inhibits the effector phase of tumor rejection by T cell receptor (TCR) transgenic CD8+ T cells PD-L1/B7H-1 inhibits the effector phase of tumor rejection by T cell receptor. *Cancer Res*, **64**(3), 1140–5.

Bracci, L., Schiavoni, G., Sistigu, A., & Belardelli, F. (2014). Immune-based mechanisms of cytotoxic chemotherapy: implications for the design of novel and rationale-based combined treatments against cancer. *Cell Death and Differ*, **21**(1), 15–25.

Buhtoiarov, I. N., Sondel, P. M., Wigginton, J. M., et al. (2011). Anti-tumour synergy of cytotoxic chemotherapy and anti-CD40 plus CpG-ODN immunotherapy through repolarization of tumour-associated macrophages. *Immunology*, **132**(2), 226–39.

Casares, N., Pequignot, M. O., Tesniere, A., et al. (2005). Caspase-dependent immunogenicity of doxorubicin-induced tumor cell death. *J Experimental Med*, **202**(12), 1691–701.

Chu, J., Deng, Y., Benson, D. M., et al. (2014). CS1-specific chimeric antigen receptor (CAR)-engineered natural killer cells enhance *in vitro* and *in vivo* antitumor activity against human multiple myeloma. *Leukemia*, **28**(4), 917–27.

Coley, W. B. (1893). The treatment of malignant tumors by repeated inoculations of erysipelas. *Am J Med* Sci, **105**(5), 487–510.

Corgnac, S., Perret, R., Derré, L., et al. (2013). CD1d-antibody fusion proteins target iNKT cells to the tumor and trigger long-term therapeutic responses. *Cancer Immunol Immunother*, **62**(4), 747–60.

Coussens, L. M. & Werb, Z. (2002). Inflammation and cancer. *Nature*, **420**(6917), 860–7.

Dai, H., Wang, Y., Lu, X., & Han, W. (2016). Chimeric antigen receptors modified T-cells for cancer therapy. *J Natl Cancer Institute*, **108**(7), 1–15.

Dighe, A. S., Richards, E., Old, L. J., & Schreiber, R. D. (1994). Enhanced *in vivo* growth and resistance to rejection of tumor cells expressing dominant negative IFN gamma receptors. *Immunity*, **1**(6), 447–56.

Dong, H., Strome, S. E., Salomao, D. R., et al. (2002). Tumor-associated B7-H1 promotes T-cell apoptosis: a potential mechanism of immune evasion. *Nature Med*, **8**(8), 793–800.

Dunn, G. P., Old, L. J., & Schreiber, R. D. (2004). The three Es of cancer immunoediting. *Ann Rev Immunol*, **22**(4), 329–60.

Duraiswamy, J., Kaluza, K. M., Freeman, G. J. & Coukos, G. (2013). Dual blockade of PD-1 and CTLA-4 combined with tumor vaccine effectively restores T-cell rejection function in tumors. *Cancer Res*, **73**(12), 3591–603.

Fallarino, F. & Gajewski, T. F. (1999). Cutting edge: differentiation of antitumor CTL *in vivo* requires host expression of Stat1. *J Immunol*, **163**(8), 4109–13.

Fallarino, F., Grohmann, U., Vacca, C., et al. (2002). T cell apoptosis by tryptophan catabolism. *Cell Death Differ*, **9**(10), 1069–77.

Fuertes, M. B., Kacha, A. K., Kline, J., et al. (2011). Host type I IFN signals are required for antitumor CD8+ T cell responses through CD8+ dendritic cells. *J Exp Med*, **208**(10), 2005–16.

Garon, E. B., Rizvi, N. A., Hui, R., et al., & KEYNOTE-001 investigators (2015). Pembrolizumab for the treatment of non-small-cell lung cancer. *N Engl J Med*, **372**(21), 2018–28.

Garrido, F., Aptsiauri, N., Doorduijn, E. M., Garcia Lora, A. M., & van Hall, T. (2016). The urgent need to recover MHC class I in cancers for effective immunotherapy. *Curr Opin Immunol*, **39**, 44–51.

Gooden, M. J. M., de Bock, G. H., Leffers, N., Daemen, T., & Nijman, H. W. (2011). The prognostic influence of tumour-infiltrating lymphocytes in cancer: a systematic review with meta-analysis. *Br J Cancer*, **105**(1), 93–103.

Gorsch, S. M., Memoli, V. A., Stukel, T. A., Gold, L. I., & Arrick, B. A. (1992). Immunohistochemical staining for transforming growth factor beta 1 associates with disease progression in human breast cancer. *Cancer Res*, **52**(24), 6949–52.

Groh, V., Wu, J., Yee, C., & Spies, T. (2002). Tumour-derived soluble MIC ligands impair expression of NKG2D and T-cell activation. *Nature*, **419**(6908), 734–8.

Guillerey, C., Huntington, N. D., & Smyth, M. J. (2016). Targeting natural killer cells in cancer immunotherapy. *Nat Immunol*, **17**(9), 1025–36.

Guo, C., Manjili, M. H., Subjeck, J. R., Sarkar, D., Fisher, P. B., & Wang, X.-.Y. (2013). Therapeutic cancer vaccines: past, present, and future. *Adv Cancer Res*, **119**(9), 421–75.

Hanahan, D. & Weinberg, R. A. (2000). The hallmarks of cancer. *Cell*, **100**(1), 57–70.

Hanahan, D. & Weinberg, R. A. (2011). Hallmarks of cancer: the next generation. *Cell*, **144**(5), 646–74.

Hanke, T. (2014). Expert opinion on biological therapy conserved immunogens in prime-boost strategies for the next-generation HIV-1 vaccines. *Expert Opin Biol Ther*, **14**(5), 601–16.

Heiss, M. M., Murawa, P., Koralewski, P., et al. (2010). The trifunctional antibody catumaxomab for the treatment of malignant ascites due to epithelial cancer: Results of a prospective randomized phase II/III trial. *Int J Cancer*, **127**(9), 2209–21.

Hodi, F. S., O'Day, S. J., McDermott, D. F., et al. (2010). Improved survival with ipilimumab in patients with metastatic melanoma. *New Engl J Med*, **363**(8), 711–23.

Hunig, T. (1983). T-cell function and specificity in athymic mice. *Immunology Today*, **4**(3), 84–7.

Immunocore (n.d.). *Immunocore Presents Positive IMCgp100 Phase I Data at the 2016 ASCO Annual Meeting*. Available at: http://www.immunocore.com/news/2016/06/immunocore-presents-positive-imcgp100-phase-i-data-at-the-2016-asco-annual-meeting

Jiang, Y., Li, Y., & Zhu, B. (2015). T-cell exhaustion in the tumor microenvironment. *Cell Death Dis*, **6**, e1792.

Kalos, M., Levine, B. L., Porter, D. L., Katz, S., Grupp, S. A., Bagg, A., & June, C. H. (2011). T cells with chimeric antigen receptors have potent antitumor effects and can establish memory in patients with advanced leukemia. *Sci Trans Med*, **3**(95), 95ra73.

Kammula, U. S., White, D. E., & Rosenberg, S. A. (1998). Trends in the safety of high dose bolus interleukin-2 administration in patients with metastatic cancer. *Cancer* **83**(4), 797–805.

Kantoff, P. W., Higano, C. S., Shore, N. D., et al. (2010a). Sipuleucel-T immunotherapy for castration-resistant prostate cancer. *New Engl J Med*, **363**(5), 411–22.

Kantoff, P. W., Schuetz, T. J., Blumenstein, B. A., et al. (2010b). Overall survival analysis of a phase II randomized controlled trial of a poxviral-based PSA-targeted immunotherapy in metastatic castration-resistant prostate cancer. *J Clin Oncol*, **28**(7), 1099–105.

Karapetis, C. S., Khambata-Ford, S., Jonker, D. J., et al. (2008). K-ras mutations and benefit from cetuximab in advanced colorectal cancer. *New Engl J Med*, **359**(17), 1757–65.

Khong, H. T. & Restifo, N. P. (2002). Natural selection of tumor variants in the generation of 'tumor escape' phenotypes. *Nat Immunol*, **3**(11), 999–1005.

Kitamura, T., Qian, B.-Z., & Pollard, J. W. (2015). Immune cell promotion of metastasis. *Nat Rev Immunol*, **15**(2), 73–86.

Krause, A., Guo, H. F., Latouche, J. B., Tan, C., Cheung, N. K., & Sadelain, M. (1998). Antigen-dependent CD28 signaling selectively enhances survival and proliferation in genetically modified activated human primary T lymphocytes. *J Exp Med*, **188**(4), 619–26.

Lanier, L. L. (2015). NKG2D receptor and its ligands in host defense. *Cancer Immunol Res*, **3**(6), 575–82.

Lasek, W., Zagożdżon, R., & Jakobisiak, M. (2014). Interleukin 12: still a promising candidate for tumor immunotherapy? *Cancer Immunol Immunother*, **63**(5), 419–35.

Leach, D. R., Krummel, M. F., & Allison, J. P. (1996). Enhancement of antitumor immunity by CTLA-4 blockade. *Science*, **271**(5256), 1734–6.

Li, M. O. & Flavell, R. A. (2008). TGF-beta: a master of all T cell trades. *Cell*, **134**(3), 392–404.

Lim, S. H., Sun, J. -M., Lee, S. -H., Ahn, J. S., Park, K., & Ahn, M. -J. (2016). Pembrolizumab for the treatment of non-small cell lung cancer. *Exp Opin Biol Ther*, **16**(3), 397–406.

Lindelöf, B., Sigurgeirsson, B., Gäbel, H., & Stern, R. S. (2000). Incidence of skin cancer in 5356 patients following organ transplantation. *Br J Dermatol*, **143**(3), 513–19.

Liu, X., Jiang, S., Fang, C., et al. (2015). Affinity-tuned ErbB2 or EGFR chimeric antigen receptor T cells exhibit an increased therapeutic index against tumors in mice. *Cancer Res*, **75**(17), 3596–607.

Melief, C. J. M., van Hall, T., Arens, R., Ossendorp, F., & van der Burg, S. H. (2015). Therapeutic cancer vaccines. *J Clin Invest*, **125**(9), 3401–12.

Morgan, R. A., Yang, J. C., Kitano, M., Dudley, M. E., Laurencot, C. M., & Rosenberg, S. A. (2010). Case report of a serious adverse event following the administration of T cells transduced with a chimeric antigen receptor recognizing ERBB2. *Mol Ther*, **18**(4), 843–51.

Motzer, R. J., Escudier, B., McDermott, D. F., et al., & CheckMate 025 Investigators (2015). Nivolumab versus everolimus in advanced renal-cell carcinoma. *New Engl J Med*, 1803–13.

Munn, D. H. & Mellor, A. L. (2016). IDO in the tumor microenvironment: inflammation, counter-regulation, and tolerance. *Trends Immunol*, **37**(3), 193–207.

Munn, D. H., Sharma, M. D., Baban, B., et al. (2005). GCN2 kinase in T cells mediates proliferative arrest and anergy induction in response to indoleamine 2,3-dioxygenase. *Immunity*, **22**(5), 633–42.

Neller, M. A., López, J. A., & Schmidt, C. W. (2008). Antigens for cancer immunotherapy. *Semin Immunol*, **20**(5), 286–95.

Ni, J., Miller, M., Stojanovic, A., Garbi, N., & Cerwenka, A. (2012). Sustained effector function of IL-12/15/18-preactivated NK cells against established tumors. *J Exp Med*, **209**(13), 2351–65.

NIH (n.d.). *Ongoing phase III clinical trials using pembrolizumab.* Available at: http://www.cancer.gov/about-cancer/treatment/clinical-trials/search/results?protocolsearchid=15276083

Old, L. J. & Boyse, E. A. (1964). Immunology of experimental tumors. *Ann Rev Med*, **15**(0014–1755), 167–86.

Paavonen, J., Naud, P., Salmerón, J., et al. (2009). Efficacy of human papillomavirus (HPV)-16/18 AS04-adjuvanted vaccine against cervical infection and precancer caused by oncogenic HPV types (PATRICIA): final analysis of a double-blind, randomised study in young women. *Lancet*, **374**(9686), 301–14.

Parry, R. V., Chemnitz, J. M., Frauwirth, K. A., et al. (2005). CTLA-4 and PD-1 receptors inhibit T-cell activation by distinct mechanisms. *Mol Cell Biol*, **25**(21), 9543–53.

Pazdur, R. (n.d.). *FDA approval for cetuximab.* Available at: http://www.cancer.gov/about-cancer/treatment/drugs/fda-cetuximab

Platten, M., Wick, W., & Van Den Eynde, B. J. (2012). Tryptophan catabolism in cancer: Beyond IDO and tryptophan depletion. *Cancer Res*, **72**(21), 5435–40.

Postow, M. A., Chesney, J., Pavlick, A. C., et al. (2015). Nivolumab and ipilimumab versus ipilimumab in untreated melanoma. *New Engl J Med*, **372**(21), 2006–17.

Przepiorka, D., Ko, C. W., Deisseroth, A., et al. (2015). FDA approval: blinatumomab. *Clin Cancer Res*, **21**(18), 4035–9.

Pyonteck, S. M., Akkari, L., Schuhmacher, A. J., et al. (2013). CSF-1R inhibition alters macrophage polarization and blocks glioma progression. *Nat Med*, **19**(10), 1264–72.

Restifo, N., Dudley, M., & Rosenberg, S. (2012). Adoptive immunotherapy for cancer: harnessing the T cell response. *Nat Rev Immunol*, **12**, 269–81.

Reusch, U., Burkhardt, C., Fucek, I., et al. (2014). A novel tetravalent bispecific TandAb (CD30/CD16A) efficiently recruits NK cells for the lysis of CD30 + tumor cells. MAbs, **6**(3), 728–39.

Ribas, A., Puzanov, I., Dummer, R., et al. (2015). Pembrolizumab versus investigator-choice chemotherapy for ipilimumab-refractory melanoma (KEYNOTE-002): a randomised, controlled, phase 2 trial. *Lancet Oncol*, **16**(8), 908–18.

Robert, C., Long, G. V., Brady, B., et al. (2015). Nivolumab in previously untreated melanoma without BRAF mutation. *New Engl J Med*, **372**(4), 320–30.

Robert, C., Schachter, J., Long, G. V., et al., & KEYNOTE-006 investigators (2015). Pembrolizumab versus ipilimumab in advanced melanoma. *New Engl J Med*, **372**(26), 2521–32.

Rodríguez, P. C. & Ochoa, A. C. (2008). Arginine regulation by myeloid derived suppressor cells and tolerance in cancer: mechanisms and therapeutic perspectives. *Immunol Rev*, **222**(1), 180–91.

Rosenberg, J. E., Hoffman-Censits, J., Powles, T., et al. (2016). Atezolizumab in patients with locally advanced and metastatic urothelial carcinoma who have progressed following treatment with platinum-based chemotherapy: a single-arm, multicentre, phase 2 trial. *Lancet*, **387**(10031), 1909–20.

Rosenberg, S. A. (2014). IL-2: the first effective immunotherapy for human cancer. *J Immunol*, **192**(12), 5451–8.

Rosenberg, S. A., Yang, J. C., Sherry, R. M., et al. (2011). Durable complete responses in heavily pretreated patients with metastatic melanoma using T-cell transfer immunotherapy. *Clin Cancer Res*, **17**(13), 4550–7.

Ruffell, B. & Coussens, L. M. (2015). Macrophages and therapeutic resistance in cancer. *Cancer Cell*, **27**(4), 462–72.

Ryzhov, S. V., Pickup, M. W., Chytil, A., et al. (2014). Role of TGF-β signaling in generation of CD39+CD73 + myeloid cells in tumors. *J Immunol*, **193**(6), 3155–64.

Sadelain, M., Brentjens, R., & Riviere, I. (2013). The basic principles of chimeric antigen receptor design. *Cancer Discov*, **3**(4), 388–98.

Schreiber, R. D., Old, L. J., & Smyth, M. J. (2011). Cancer immunoediting: integrating immunity's roles in cancer suppression and promotion. *Science*, **331**(6024), 1565–70.

Scott, A. M., Wolchok, J. D., & Old, L. J. (2012). Antibody therapy of cancer. *Nat Rev Cancer*, **12**(4), 278–87.

Shankaran, V., Ikeda, H., Bruce, A. T., et al. (2001). IFNγ and lymphocytes prevent primary tumour development and shape tumour immunogenicity. *Nature*, **410**(6832), 1107–11.

Shapiro, R. S. (2011). Malignancies in the setting of primary immunodeficiency: implications for hematologists/oncologists. *Am J Hematol*, **86**(1), 48–55.

Srivatsan, S., Patel, J. M., Bozeman, E. N., et al. (2014). Allogeneic tumor cell vaccines: the promise and limitations in clinical trials. *Human Vacc Immunother*, **10**(1), 52–63.

Stewart, T. J. & Abrams, S. I. (2008). How tumours escape mass destruction. *Oncogene*, **27**(45), 5894–903.

Stutman, O. (1974). Tumor development after 3-methylcholanthrene in immunologically deficient athymic-nude mice. *Science*, **183**(4124), 534–6.

Takahashi, T., Tagami, T., Yamazaki, S., et al. (2000). Immunologic self-tolerance maintained by CD25(+)CD4(+) regulatory T cells constitutively expressing cytotoxic T lymphocyte-associated antigen 4. *J Exp Med*, **192**(2), 303–10.

Tivol, E. A., Borriello, F., Schweitzer, A. N., Lynch, W. P., Bluestone, J. A., & Sharpe, A. H. (1995). Loss of CTLA-4 leads to massive lymphoproliferation and fatal multiorgan tissue destruction, revealing a critical negative regulatory role of CTLA-4. *Immunity*, **3**(5), 541–7.

Trotta, R., Chen, L., Costinean, S., et al. (2013). Overexpression of miR-155 causes expansion, arrest in terminal differentiation and functional activation of mouse natural killer cells. *Blood*, **121**(16), 3126–34.

Turcotte, S. & Rosenberg, S. A. (2011). Immunotherapy for metastatic solid cancers. *Adv Surg*, **45**, 341–60.

U.S. Food and Drug Administration (n.d.). *Nivolumab (Opdivo) for Hodgkin lymphoma*. Available at: http://www.fda.gov/Drugs/InformationOnDrugs/ApprovedDrugs/ucm501412.htm

Uyttenhove, C., Pilotte, L., Théate, I., et al. (2003). Evidence for a tumoral immune resistance mechanism based on tryptophan degradation by indoleamine 2,3-dioxygenase. *Nat Med*, **9**(10), 1269–74.

Vanneman, M. & Dranoff, G. (2012). Combining immunotherapy and targeted therapies in cancer treatment. *Nat Rev Cancer*, **12**(4), 237–51.

Vogel, C. L., Cobleigh, M. A., Tripathy, D., et al. (2002). Efficacy and safety of trastuzumab as a single agent in first-line treatment of HER2-overexpressing metastatic breast cancer. *J Clin Oncol*, **20**(3), 719–26.

Weber, J. S. & Rosenberg, S. A. (1988). Modulation of murine tumor major histocompatibility antigens by cytokines *in vivo* and *in vitro* modulation of murine tumor major histocompatibility antigens by cytokines *in vivo* and *in vitro*. *Cancer Res*, **48**(20), 5818–24.

Xin, H., Zhang, C., Herrmann, A., Du, Y., Figlin, R., & Yu, H. (2009). Sunitinib inhibition of Stat3 induces renal cell carcinoma tumor cell apoptosis and reduces immunosuppressive cells. *Cancer Res*, **69**(6), 2506–13.

Yamaguchi, T., Wing, J. B., & Sakaguchi, S. (2011). Two modes of immune suppression by Foxp3 + regulatory T cells under inflammatory or non-inflammatory conditions. *Semin Immunol*, **23**(6), 424–30.

Zaidi, M. R. & Merlino, G. (2011). The two faces of interferon-γ in cancer. *Clin Cancer Res*, **17**(19), 6118–24.

Zajac, A. J., Blattman, J. N., Murali-Krishna, K., et al. (1998). Viral immune evasion due to persistence of activated T cells without effector function. *J Exp Med*, **188**(12), 2205–13.

30

Biological effect of radiotherapy on cancer cells

Anna Dubrovska, Mechthild Krause, and Michael Baumann

Introduction to biological effect of radiotherapy on cancer cells

Brief historical background

Radiation therapy, or radiotherapy, remains one of the main modalities for cancer treatment alone or in combination with surgery or chemotherapy. Today, it is applied in about 50–60% of all cancer patients during the course of treatment (Atun et al., 2015; Baumann et al., 2016). Radiation can be defined as propagation of energy in the form of electromagnetic waves or particles in matter or space. Radiotherapy may eradicate tumour cells through energy transfer that induces tissue ionization and leads to a chain of chemical reactions, from damage of chemical molecules to cell death and tissue damage.

Despite the indispensable role of radiation for cancer treatment, its ability to damage living tissues was not initially recognized. In 1895 the German physicist Wilhelm Conrad Röntgen was the first who discovered X-rays while experimenting with a cathode-ray tube. In these experiments, cathode rays, which are electrons, collided with a target leading to the disturbance in the atoms of matter and to release of photons that was seen by Röntgen as the glow of fluorescence. X-rays are highly energetic electromagnetic radiation capable to penetrate most solid objects, but does not or only very slightly pass through bones or metals (Lederman, 1981). The therapeutic potential of X-rays was first discovered by Emile Grubbe, a German emigrant living in Chicago, who assembled the X-ray machine and used it for treatment of the first cancer patient with a breast tumour in 1896. The first well documented treatment, in this case of a hairy naevus, was performed by Leopold Freund (Kogelnik, 1997), and the first proven successful treatment of tumour by radiation was reported by Swedish doctor Thor Stenbeck. In 1899, he applied X-rays in a course of small daily doses over a period of several months for treatment of a basal cell carcinoma of the nose. Notably, the patient was healthy 30 years later (Panizzon and Seegenschmiedt, 2015).

However, it was difficult to use X-rays in medicine before 1913 since the high-voltage power source required for adequate images was not available. A high vacuum tube invented by William D Coolidge called Coolidge tube was a more powerful and stable source of X-ray, and together with development of powerful high-voltage sources it laid the ground for the routine use of X-rays for medicine and industry (Reed, 2011). In parallel with the discovery of X-rays by Roentgen, Henri Becquerel, and Marie Curie discovered the natural sources of penetrating rays which they called radioactivity. The British scientist Ernest Rutherford performed many experiments aiming to characterize the nature of this radiation which he called alpha, beta, and gamma rays. In 1899 Rutherford demonstrated that these rays can be distinguished by their trajectories in a magnetic field: the alpha rays are positively charged; the beta rays carry a negative charge, and gamma rays have no charge and do not deflect in the magnetic field. Further investigations demonstrated that alpha rays are helium nuclei, beta rays are electrons, and gamma rays are electromagnetic waves (Blaufox, 1996; Heilbron, 1968).

Since the first cancer patients treated by X-rays, many additional types of ionizing radiation have been discovered and employed for therapeutic applications including gamma rays, electrons, heavy ions (carbon, silicon, neon, helium), and hadrons (negative pimesons, neutrons, protons).

Major types of radiation used to treat cancer

The curative rate of radiotherapy depends on the quality of radiation and its controllability within the anatomic site. One of the most important physical measures of radiation quality is the amount of energy transferred to the tissues per unit of irradiation track called linear energy transfer, LET (Hein et al., 2014). Since radiation transfers energy to matter in the form of ionization, LET can be explained as a pattern of ionizing density along the length of the radiation track. The amount of energy transferred to a medium by ionizing radiation is defined as absorbed dose and measured in units of Gray, Gy, or earlier used rad, where 1 Gy = 100 rads = 1 J/kg. A high LET irradiation, transfers more energy per unit of track in tissues of the body and might cause a higher biological effect compared to a low LET irradiation such as X-rays or gamma rays (Goodhead, 1999). The relative biological effectiveness, RBE of the different types of irradiation is traditionally measured against X-ray as standard using a variety of endpoints (e.g. cell death, chromosomal aberrations, DNA damage, and others) and can be defined as: $RBE = D_{250}/D_r$, where D_{250} and D_r are the doses of 250 keV X-rays and tested radiation, respectively, required for the same biological effect. High LET radiation has RBE values higher than 1 (reviewed in Jakel, 2008; Baumann et al., 2016). High ionizing density might be associated with a higher RBE due to an increased probability of creating irreversible damage in a small target volume of interest such as an individual tumour cell.

To date, X-rays (photons) remain the most available type of radiation therapy due to low production and service cost (Slater, 2006). A host of preclinical and clinical studies demonstrated that particle therapy such as hydrogen ions (protons) and carbon ions have more precise dose delivery and thereby may achieve a higher therapeutic efficacy compared to X-rays. Particle radiation has a specific tissue ionization pattern called Bragg peak allowing distributing maximum energy into the tumour site and therefore minimizing injury of the surrounding normal tissues that makes this therapy especially beneficial for eradication of tumours located near radiosensitive organs. Despite a currently low level of evidence in most tumour entities, proton and carbon ion radiotherapy have a high potential for the treatment of different types of cancer including base of skull tumours, retroperitoneal and pelvic tumours, subgroups of head and neck squamous cell carcinoma (HNSCC), brain, and paediatric cancers (Uhl et al., 2014; Wisenbaugh et al., 2014; Wang, 2015). The major obstacles for the broader adoption of particle therapy are the current scarcity of prospective clinical trials comparing particle beams versus state-of-the-art photon therapy, and the high equipment, facility, and service costs that are expected to be potentially reduced by future technological improvements (reviewed in Loeffler and Durante, 2013; Baumann et al., 2016). Furthermore, economical evaluation of photon and proton therapies is still in its infancy, although (e.g. for paediatric patients) such evaluation suggested that proton therapy might in specific cases be even more cost-effective with respect to its low side effects (Mailhot Vega et al., 2015).

Types of radiation-cellular interaction

Direct effect of irradiation

The integrity of the cellular genome is fundamental for cell survival. Throughout the cell lifespan, genome integrity is challenged by different exogenous and endogenous factors leading to arising of many thousands of DNA lesions in each human cell every day (Polo and Jackson, 2011). A dose of 2 Gy of gamma rays yields in about 80 DNA double-strand breaks (DSB), 2,000 single-strand breaks (SSB), and about 6,000 nucleotide bases lesions per single cell (Ward, 1994). The principle difference of radiation-induced DNA lesions from endogenously induced DNA damages consist in their clustering which is defined by the pattern of deposited energy. Low LET radiation has lower ionizing density compared to high LET radiation and therefore induces less complex DNA damage clusters (Lomax et al., 2013). Genome integrity is preserved by DNA damage signalling and DNA repair mechanisms. However, if severe DNA damage cannot be repaired, it might lead to cell death.

Indirect effect of irradiation

The curative rate of radiation depends not only on its ability to induce irreparable DNA damage in tumour cells by direct ionization, but also on the radiolysis of water and production of chemically reactive oxygen species, ROS along with production of the arginine-derived reactive nitrogen species, RNS (Mikkelsen and Wardman, 2003; Azzam et al., 2012).

Endogenous ROS are formed as by-products during different metabolic processes including mitochondrial electron transport or as intermediates of the metal catalysed oxidation reactions. Under physiological conditions, these metabolites present in the cells at very low concentrations and its physiological level is maintained by the scavenging molecules neutralizing ROS including glutathione, dismutase, peroxidase, thioredoxin and catalase (Trachootham et al., 2009). Cancer cells have activated metabolic pathways that drive their proliferation and survival and therefore generate elevated levels of ROS. To deal with this, many tumour cells have reprogrammed glycolysis to utilize glucose 6-phosphate for production of an adequate amount of reduced nicotinamide adenine dinucleotide phosphate, NADPH which fuel thioredoxin antioxidant system (Schumacker, 2015).

In addition to water, another key source of the ROS production is atomic oxygen which is susceptible to radical formation upon irradiation. Oxygen supply is indispensable for cellular respiration to convert biochemical energy into ATP. However, in solid tumours oxygen distribution is not homogenous since tumours often have a chaotic vascular network. During tumour growth, some of its regions can outgrow their blood supplies and therefore have a low oxygen tension, or hypoxia. Cancer cells residing in the hypoxic areas can be more shielded from the radiation-induced DNA damage through less ROS generation and activation of the pro-survival signalling pathways such as hypoxia-inducible factor (HIF) signalling (Begg et al., 2011; Peitzsch et al., 2014). To meet their energetic demands, tumour cells can adapt their metabolism to low oxygen tension by switching glucose metabolism to anaerobic glycolysis. This metabolic shift from aerobic to anaerobic respiration that occurs in tumour cells even under normoxic conditions was first discovered by Otto Warburg nearly a century ago and called 'Warburg effect' (Hanahan and Weinberg, 2011).

Biological effects and type of radiation

Radiotherapy is usually given in a course of repeated small doses of radiation (i.e. fractionated radiotherapy) of around 2 Gy at five days per week that are less damaging to normal tissues than an equivalent single dose. Reoxygenation between radiation fractions is generally believed to improve the efficacy of radiotherapy (reviewed in Peitzsch et al., 2014). The experiments of Louis H. Gray in 1955 demonstrated that hypoxic cells need two to three times higher doses to be inactivated compared to cells which are well oxygenated, although cell radiosensitivity is little affected by oxygenation for high LET radiation such as neutrons (Gray et al., 1953). These experiments demonstrated that the proportion of direct or indirect ionization depends on the type of radiation. Low LET radiation deposits its energy mostly in water, which is consistent to the fact that human cells and tissues contain about 70% of water. DNA damage by high LET radiation is less dependent on ROS production due to the high probability of direct DNA ionization and therefore can more efficiently treat hypoxic tumours (reviewed in Baumann et al., 2016). The direct and indirect effects of irradiation initiate a set of intracellular biochemical signals that may repair the damage or lead to permanent genomic changes (mutations) or to cell death if mutations are too severe and not compatible with cell survival.

Activation of the DNA damage response mechanisms and its dependence on the phase of cell cycle

The decision of a cell between life and death is highly influenced by the activation of the DNA damage response (DDR) mechanisms. Among the different kinds of DNA lesions produced in response to irradiation, that is, SSB, DSBs, damaged nucleotide bases, or abasic sites, DSB is one of the most toxic and mutagenic lesions because this type of damage does not leave an intact complimentary DNA strain that can be used as a template for repair (Polo and Jackson, 2011). Both normal and tumour cells can repair DSBs by either error-free homology-directed recombination, HR, which is accomplished by using a homologous sequence of the sister chromatid as template, or error-prone non-homologous end joining, NHEJ, which works in a template-independent manner (Helleday et al., 2007; Morgan and Lawrence, 2015; see **Figs. 30.1–30.3**).

Both of these mechanisms are acting in a complementary and competitive manner, and the balance between HR and NHEJ depends on various factors including the stage of the cell cycle and availability of a sister chromatid for homologous recombination

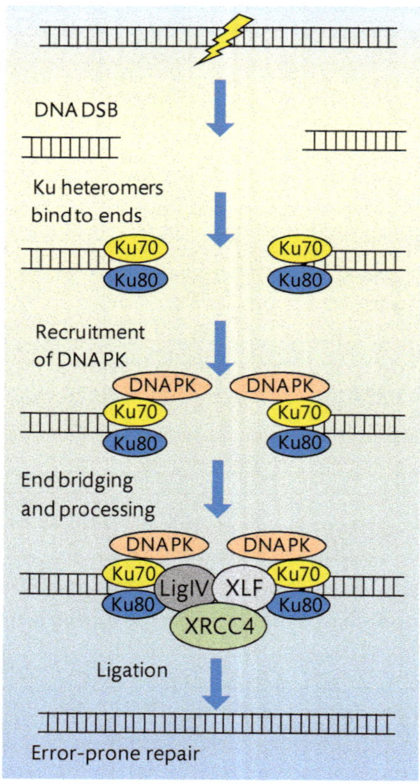

Fig. 30.1 Non-homologous end joining (NHEJ) mechanism. NHEJ is a major and simpler mechanism for DSB repair, which in mammalian cells proceeds in a few consequent steps. First, DNA ends are bound by the Ku70/80 heterodimer, which forms a complex with DNA protein kinase, DNA-PK. In turn, DNA-PK attracts the ligase IV complex, which includes ligase IV that seals the DNA ends and X-ray cross-complementing protein 4, XRCC4/XRCC4-like factor, XLF protein dimers which are bridging broken DNA ends prior to repair and stimulate DNA ligation (Helleday et al., 2007; Roy et al., 2015). NHEJ is an error-prone DNA repair mechanism.

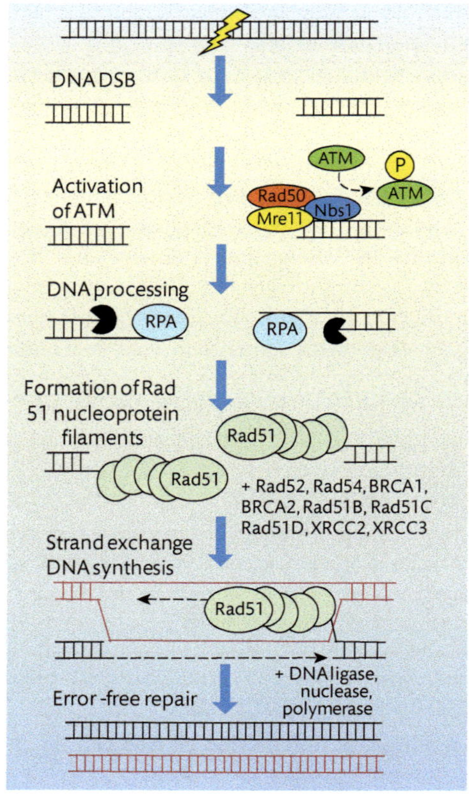

Fig. 30.2 Homology-directed DNA double-strand break (DSB) repair pathways. The initial step of homology-directed DNA repair pathways includes an extensive 5′ to 3′ processing of DSBs by nucleases and helicases to produce 3′ single-strand (ss) DNA tails that is regulated by MRN complex consisting of Mre11/Rad50/Nbs1 proteins. The resulting 3′ tails of ssDNA at each end of break are occupied by replication protein A, RPA protein which has a protective function and removes secondary DNA structures. Next, Rad51 protein is assembling onto the ssDNA with the aid of other associated proteins such as Rad52, Rad54, Rad54B, BRCA1, BRCA2, Rad51B, Rad51C, Rad51D, XRCC2, and XRCC3. Rad51 plays a key role in the homology-directed repair by forming nucleoprotein filaments which facilitate searching of a homologous sequence and invasion of the ssDNA into the dsDNA of the sister chromatid. Strand invasion into the homologous DNA duplex allows DNA polymerases to re-synthesize missing DNA sequences using the intact template and to form two identical sister chromatids (Forget and Kowalczykowski, 2010).

(S and G2 phases; see Helleday et al., 2007; Shrivastav et al., 2008). The requirement for a template for the high-fidelity HR DSBs repair can explain uniformly high cell radioresistance in the latter part of S phase and increased radiosensitivity for cells irradiated during mitosis and to a less extent just before beginning of DNA replication at G1/S interphase (Tamulevicius et al., 2007). Tumour cells are asynchronous, therefore exhibit differential sensitivity to the DNA damage at the time of irradiation. Cancer cells are dividing more rapidly than normal cells and accelerated tumour repopulation can be one of the common reasons for radiotherapy failure (reviewed in Ng et al., 2013).

Repopulation can explain so-called 'time factor' of fractionated radiotherapy (i.e. the impairment of local tumour control probability with prolongation of the overall treatment time between the first and the last radiation fraction). This time factor has been shown by

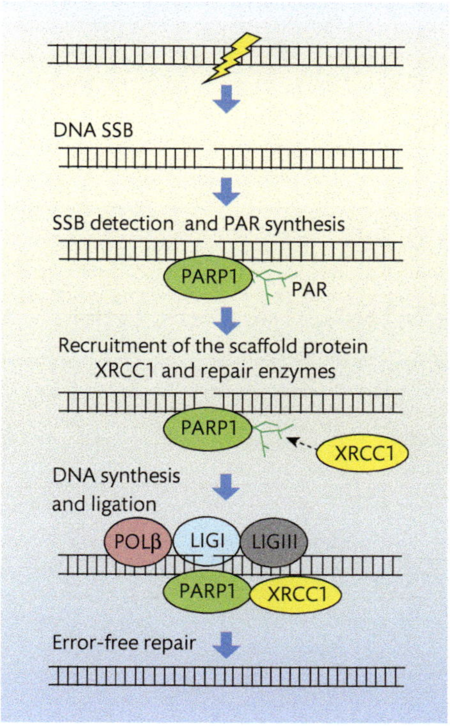

Fig. 30.3 DNA single-strand break (SSB) repair pathways. In contrast to DNA DSBs, single-strand breaks, SSBs are detected by poly (ADP-ribose) polymerase, PARP1 which synthesize the chains of poly-ADP-ribose, PAR to facilitate the recruitment of other DNA repair proteins such as X-ray cross-complementing protein 1 (XRCC1), DNA polymerase Polβ and DNA ligase I and III (Caldecott, 2014).

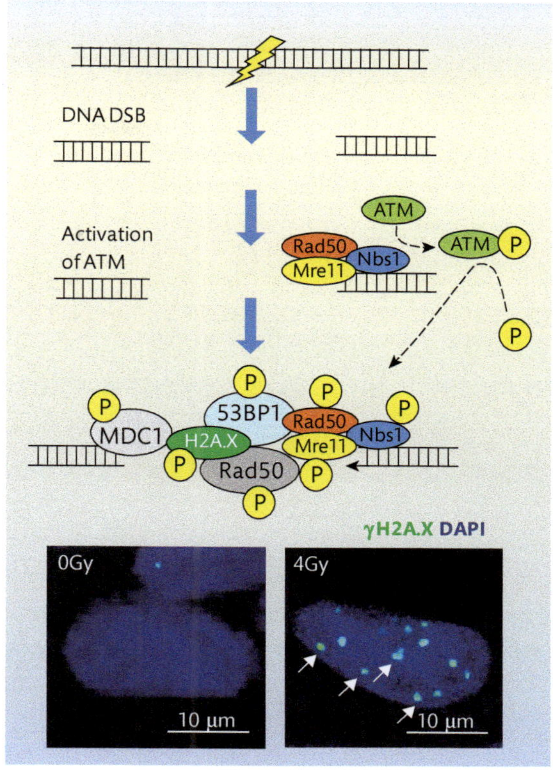

Fig. 30.4 DNA damage-induced ataxia telangiectasia mutated (ATM)-dependent protein phosphorylation. Activation of ATM kinase mediated by its interaction with MRN proteins is one of the initial molecular responses to DNA DSBs. Activation of ATM by autophosphorylation, acetylation and recruitment at the sites of DSBs called 'radiation-induced foci' initiates phosphorylation of multiple proteins, including histone H2A.X. Phosphorylation of H2A.X by ATM is required for the assembly of DNA repair proteins at the DNA lesion sites and for activation of the checkpoint proteins, which arrest cell cycle progression. The residual nuclear foci of the phosphorylated H2A.X (γ-H2A.X) 24 h after irradiation are indicative of unrepaired DNA DSBs and therefore may be used as predictive test for radiotherapy. To this end, number of the residual γ-H2A.X was correlated with intrinsic tumour cell radiosensitivity in tumour cell lines and then has been employed as a clinical marker of patient radioresponsiveness (Banath et al., 2004; Menegakis et al., 2015). In addition to H2A.X, ATM phosphorylates other proteins involved in DNA damage repair and checkpoint signalling including NBS1, BRCA1, MDC1, p53BP1, Chk2 and also acts on many other nuclear and cytoplasmic substrates (Smith et al., 2010).

clinical studies for different tumour entities and seems to be most pronounced in squamous cell carcinoma, for example, in the head and neck region or in the lung (reviewed in Bütof and Baumann, 2013).

Because surviving tumour cells undergo cell cycle progression between radiation fractions, dose fractionation results in redistribution of the radioresistant tumour cells in S phase to a more radiosensitive state (Withers, 1975b). Variation of cell radiosensitivity depending on the phase of cell cycle suggests that efficacy of the above-described DNA repair mechanisms is a critical factor influencing tumour radiocurability.

DNA damage-induced biological responses regulating cell cycle check points

In addition to the activation of the DNA repair process, DNA damage triggers a chain of biological responses regulating cell cycle checkpoints and cell death. Activation of ataxia telangiectasia mutated (ATM) kinase mediated by its interaction with MRN proteins is the earliest cell response to DNA DSBs (Paull, 2015; Fig. 30.4).

Phosphorylation and activation of checkpoint kinase 2 (Chk2) by ATM kinase leads to phosphorylation of multiple Chk2 substrates involved in regulation of cell cycle progression, gene expression and cell death including tumour suppressors p53 and BRCA1, transcription factors FOXM1 and E2F1, regulators of G1/S checkpoint Rb and CDC25A, regulators of G2/M checkpoint CDC25C and TTK and many other targets including proteins with unknown function (Smith et al., 2010; Zannini et al., 2014).

In addition to ATM, another kinase called ATM-Rad3-related, ATR is involved in the initial cell response to different types of DNA lesions and replication problems, and contributes to cell cycle checkpoint response (Flynn and Zou, 2011). The key structure triggering ATR activation is ssDNA. ATR-interacting protein, ATRIP directly binds to the ssDNA associated RPA protein and recruits ATR to the sites of DNA damage. To activate DNA damage-induced checkpoint, ATR requires additional regulators such as RAD17-RFC complex, RAD9, RAD1, and HUS1, which stimulate ATR kinase activity and DNA topoisomerases TOP1 and TOP2, which induce DNA replication and checkpoint activation and are also a substrate for ATM and ATR. Checkpoint kinase 1 (Chk1) is the best characterized ATR

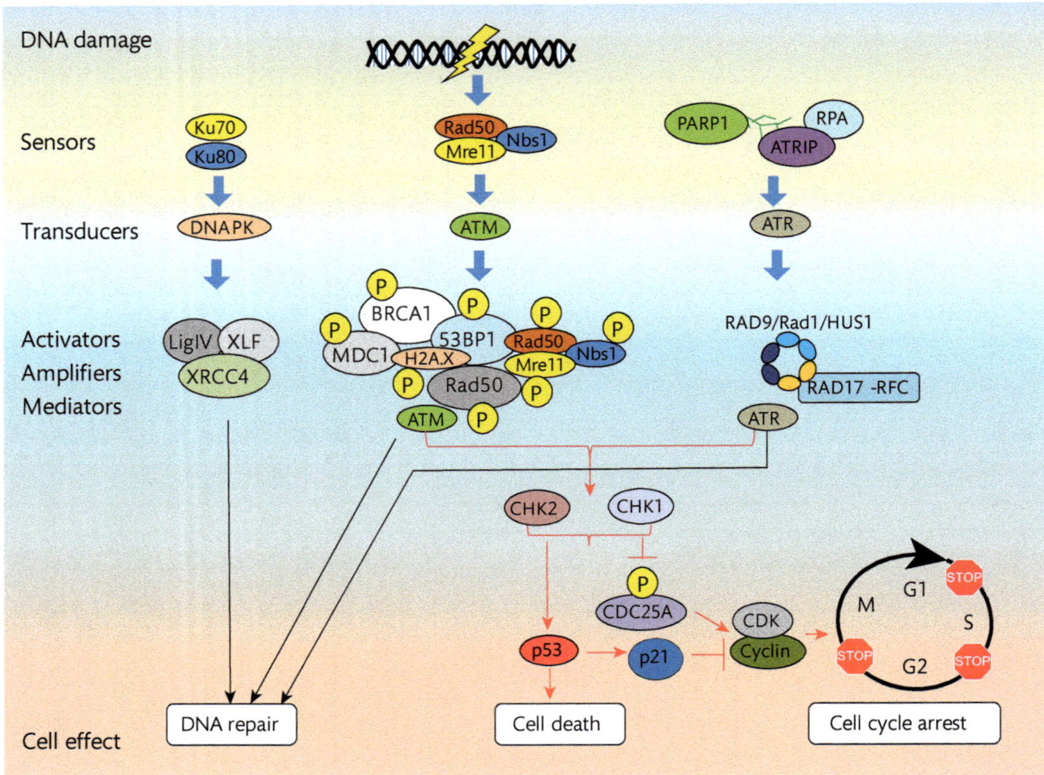

Fig. 30.5 DNA damage response. After DNA damage, different cellular proteins detect, transduce, and amplify the signal leading to the distinct cell effects such as DNA repair, cell cycle arrest, or cell death.

substrate in the DNA damage-induced checkpoint response (Flynn and Zou, 2011). Both, ATM-Chk2 and ATR-Chk1 signalling pathways delay cell cycle progression to allow proper DNA reparation and to prevent propagation of cells with damaged DNA (Morgan and Lawrence, 2015; see Fig. 30.5).

DNA damage induces checkpoints at the G1/S transition of cell cycle (G1 checkpoint), G2/M transition (G2 checkpoint) and during S phase of cell cycle (S checkpoint). G1 arrest is induced by AMT/Chk2 dependent phosphorylation of p53 and activation of its transcriptional activity leading to increase in the expression of p21, which inhibits kinase activity of the different classes of cyclin-CDK complexes (Abbas and Dutta, 2009). Irradiation of the cells in S phase results in a transient decrease in the rate of DNA synthesis mediated by two distinct mechanisms. The first ATM/NBS1/SMC1 pathway is based on direct NBS1 phosphorylation by ATM which is required for formation of MRN complexes. Although the exact mechanism how MRN complex inhibit DNA synthesis is unknown, there is evidence that MRN complex regulates S-phase arrest through interaction with PRA since abolishing this interaction leads to pronounced DNA synthesis in the irradiated cells (Olson et al., 2007). One of the downstream effectors of ATM/MRN is the SMC1 protein which is phosphorylated by ATM and regulates cohesion of the sister chromatid after replication and possibly can regulate the replication elongation process. However, the exact role of MRN and SMC1 in the S2 checkpoint warrants further investigation (Woo and Poon, 2003; Morgan and Lawrence, 2015). Another mechanism of the S-phase checkpoint is based on the ATM/Chk2 mediated phosphorylation and degradation of CDC25A phosphatase. This leads to inhibition of CDK2, which controls the efficacy of DNA replication in association with cyclin E and cyclin A (Woo and Poon, 2003). Activation of the radiation-induced G2 checkpoint prevents entry of cells with damaged DNA into mitosis and is initiated by ATM- and ATR-mediated phosphorylation of Chk1 and Chk2. Similar to the S checkpoint arrest, G2 arrest is mediated by inhibition of CDK2 and CDK1 kinases by degradation of CDC25A and inhibition of Polo kinase 1 (PLK1). In addition, ATM/Chk2 mediated phosphorylation and transcriptional activation of p53 induces expression of its target genes p21, *GADD45*, and 14-3-3σ, which inhibit cyclin-dependent kinases and therefore block entry of the cells into mitosis (Taylor and Stark, 2001).

Radiation-induced cell death

Molecular mechanisms of radiation-induced cell death

By definition, cell cycle checkpoints are transient states from which cells can escape after DNA repair is completed. DNA lesions beyond the repair capacity of a cell leads to cellular death or loss of proliferative capacity that is mediated by proteins of DDR signalling through activation of different mechanisms such as senescence, which is an irreversible cell cycle arrest, apoptosis, which is programmed cell death, or mitotic catastrophe, which is a type of cell death occurring during mitosis (Castedo et al., 2004; D'Adda and di Fagagna, 2008; Eriksson and Stigbrand, 2010; Matt and Hofmann, 2016). Therefore, efficacy of the DNA repair machinery activated by the radiation-induced DNA damage is critical to decide between cell death and repair. Radiation-induced cell death is orchestrated in a p53-dependent fashion, and different DNA damage levels define different levels of p53-dependent transcription of pro-apoptotic genes. This process is controlled by the dynamic balance between p53 activating signalling pathways involved in DDR response such as ATM/Chk2/p53 and ATR/Chk1/p53 and negative regulation by E3 ligase Mdm2 mediating p53 degradation (Aylon and Oren, 2007; Roos et al., 2016; **Fig. 30.5**).

In addition to DNA damage-induced death of the cells directly targeted by irradiation, adjacent cells can be also affected by stress signal factors released from irradiated cells (including ROS, RNS, transforming growth factor α (TGFα), TGFβ, tumour necrosis factor α (TNFα) and cytokines such as IL-1). In response to these secreted factors, bystander cells activate multiple signalling pathways such as mitogen-activated protein kinase (MAPK) and NF-κB, and exhibit signs of radiation exposure such as oxidative stress, chromosomal aberrations, genomic instability, micronucleation, and cell death. This effect is called bystander effect (Jalal et al., 2014; Wang et al., 2015). Although the radiation-induced genotoxic stress in non-irradiated cells was described almost 25 years ago, the exact molecular mechanisms underlying the bystander effect are not completely understood (Nagasawa and Little, 1992).

A distinctive hallmark of cancer cells is activation of the pro-survival signalling pathways including phosphatidylinositol-3 kinases (PI3K/AKT), nuclear factor κB (NF κB), and MAPK that increase their tolerance to DNA damage induced cell death and reduce efficacy of radiotherapy (Begg et al., 2011; Hanahan and Weinberg, 2011; Hein et al., 2014). Cells with a defective activation of p53-dependent DNA damage—induced checkpoints can progress through cell cycle even after their exposure to genotypic stress that might lead to mitotic catastrophe or accumulation of genetic alterations and tumour development. Although DSBs are among the most lethal lesions induced by irradiation, they are a likely source of genomic mutations if cells escape DNA damage-induced death (Halazonetis et al., 2008; Eriksson and Stigbrand, 2010).

Dose-response relationship

According to the classical view, the endpoint of radiotherapy is reproductive cell death as result of unrepaired DNA damage leading to removal of cells from the clonogenic pool. Philip Marcus and Theodore Puck were the first who developed a method for measurement of reproductive potential of mammalian cells by plotting clonogenic survival of HeLa cells as logarithm of the surviving fraction versus X-ray dose (**Fig. 30.6**). They noticed that dose dependence is not a straight line and hypothesized that existence of the shoulder in the cell survival curve could be an evidence for multiple hit killing mechanisms, where 'hit' is an ionizing event leading to unrepaired DNA damage and reproductive cell death (Puck and Marcus, 1956). Further radiobiological data obtained by a number of investigators with a variety of different cell types have been explained on the same ground, and existence of the shoulder in the survival curves has been interpreted as a multitarget effect of irradiation (Chadwick and Leenhouts, 1973). According to the target theory of cell survival (i) cell inactivation requires at least two sublethal lesions leading to DNA DSB, which can be produced by one or two radiation tracks, wherein probability of cell survival after one track action can be described as $P = \alpha d$, and probability of survival after two track action can be defined as $P = \beta d^2$, where d is the dose, α is a constant representing the direct killing of cells, and β represents the impact of cell killing from 'double hits'; (ii) reproductive cell death can result from a single lethal event or from accumulation of sublethal lesions; (iii) cells are able to repair sublethal DNA lesions even after multiple irradiations; (iv) low LET radiation can sterilize the cells as result

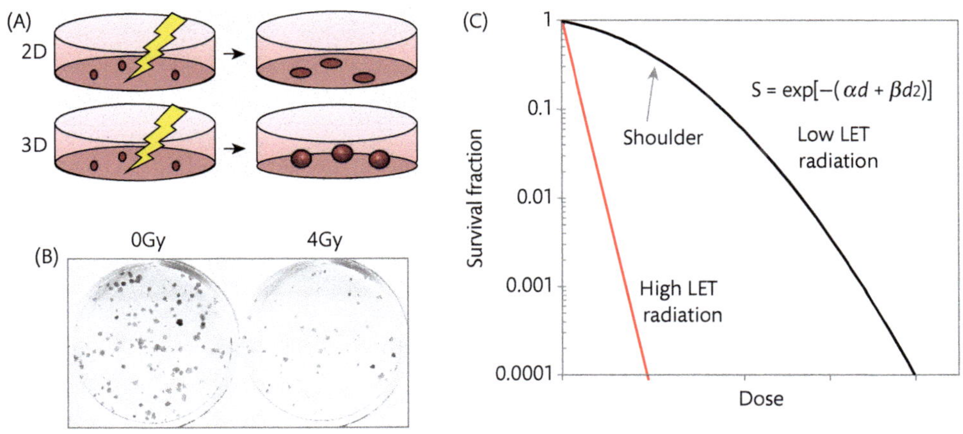

Fig. 30.6 Clonogenic cell survival analysis. (A). The colony formation assay is applied for the analysis of clonogenic cell survival after irradiation. Cells are seeded in single cell suspension on plastic plates (2D conditions) or embedded in Matrigel which mimic the *in vivo* tumour microenvironment (3D conditions) and exposed to the different doses of radiation or sham-irradiated. The plates are incubated for 7–14 days and results of the clonogenic cell survival analysis are calculated as plating efficacy: $PE = \frac{\text{counted colonies}}{\text{seeded cells}} \times 100$ and survival fraction: $SF = \frac{\textit{counted colonies}}{(\textit{seeded cells} \times \textit{PE})} \times 100$.
(B) Images of the multicellular 2D colonies formed by head and neck squamous cell carcinoma (HNSCC) FaDu cells which were exposed to 4Gy of X-rays or sham-irradiated. (C) Schematic survival curves for the cells exposed to high LET irradiation (such as particle irradiation: protons, carbon ions, neutrons) or low LET irradiation (such as X-rays). The survival curve is defined by the two components: linear component which is proportional to dose (αd) and quadratic component which is proportional to the square of the dose (βd^2). (Hall and Giaccia, 2012).

of accumulation of DNA damage proportionally to the dose. In contrast, cells are less able to repair DNA lesions produced by high LET (Chadwick and Leenhouts, 1973; Coleman, 1993). These formulations were used to develop the linear-quadratic (LQ) model for description of dose/radiation fractionation dependencies in radiotherapy (Thames, 1985; Jones et al., 2001; Brenner, 2008). This model was developed and modified by a number of investigators and can mathematically be described as follows: $S = \exp[-(\alpha d + \beta d^2)]$, where S is the fraction of surviving cells, D is the dose, α and β are tissue-dependent constants. The ratio α/β is often used to define sensitivity of tissue to fraction size. A low α/β suggests that the linear component is less important compared to the quadratic component that is more typical for late-responding normal tissues. A low α/β is indicative of a high sensitivity to fraction size and dose and a large repair capacity. In contract, a high α/β is typical for a situation where the linear component is more important and, consequently, there is less dependence on the fraction size. High α/β values are usually found in tumours and fast dividing normal tissues. This model has now evolved into a more broader view since some tumours, such as prostate or breast carcinomas, have a lower α/β ratio in the range or even lower than the one in the surrounding organs and therefore would be treated most efficiently with hypofractionated treatment schedules (Yarnold et al., 2011; Hegemann et al., 2014).

The LQ model can be employed to include the effect of fractionation and time factor as such: $S = \exp[-n(\alpha d + \beta d^2) + \ln 2 \frac{T - T_k}{T_p}$ where n is the number of fractions, d is the dose per fraction, T is the total treatment time, T_k is the 'kick off time' from start of treatment to the beginning of accelerated proliferation, and T_p is the potential doubling time (Jones et al., 2001). The LQ model is now in a widespread use in analysis of *in vitro* cell survival and for clinical dose–response data as well as for designing new treatment regimens for radiation oncology.

Factors affecting local tumour control in light of the cancer stem cell concept

The 5Rs of radiation biology

Radiotherapy failures can be attributed to the treatment-related, clinical, and biological factors. Treatment-related factors include, for example, treatment inaccuracy issues attributed to the uncertainties in target identification and to the tumour motion, or conflicting objective issues when precise radiation delivery is challenged by the vital normal tissues surrounding the target volume (Baumann et al., 2016). Local control probability is also reduced for large tumours that are treated with conventional radiation doses. However, many radiotherapy failures remain unclear based only on clinical information (i.e. tumour size, grade, stage, treatment regimen, and given dose). To this end, in addition to the clinical factors, success of the fractionated therapy has been explained by the radiobiological concept of '4R' parameters that was first summarized by Rodney Withers in 1975. These parameters include (i) DNA repair, (ii) cell repopulation, (iii) redistribution of the cells in cell cycle and (iv) reoxygenation (Withers, 1975a) that are described in the previous sections of this chapter. Later, intrinsic tumour cell radiosensitivity was added as fifth R of radiobiology suggesting that the standard 4R are often insufficient to explain different tumour radiocurability at the mechanistic level (Good and Harrington, 2013; see Fig. 30.7). For example, genetic and epigenetic differences between tumours of the same type with similar clinical characteristics might underlay different responses to ionizing radiation through modulation the DNA damage response and cell cycle checkpoints (e.g. mutations in p53) or by upregulation of the antiapoptotic signalling pathways (e.g. epidermal growth factor (EGFR) mediated activation of PI3K/AKT pathway or NF-κB and MAPK pathway activation; Bussink et al., 2008; Begg et al., 2011; Skinner et al., 2012). Therefore,

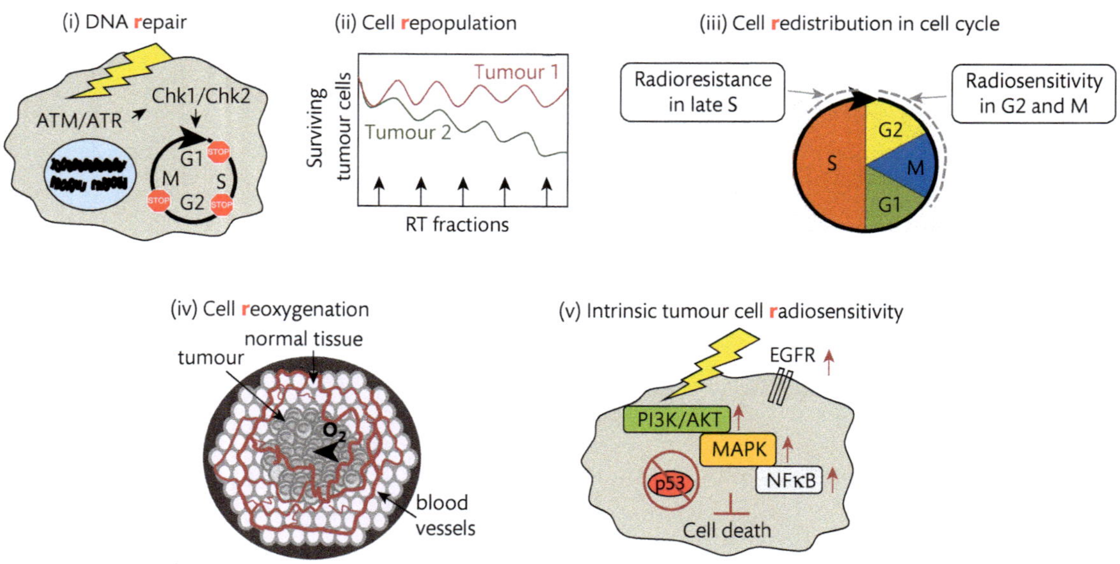

Fig. 30.7 The 5Rs of radiation biology. In addition to the clinical factors, tumour radiocurability has been explained by the radiobiological parameters which include (i) DNA repair, (ii) cell repopulation, (iii) redistribution of the cells in cell cycle, (iv) reoxygenation, and (v) intrinsic tumour cell radiosensitivity.

characterization of the deregulated pathways in the individual tumours and modulation of the response to radiotherapy through its combination with pathway-specific inhibitors could lead to improvement of therapy outcome.

Tumour heterogeneity and stem cell concept of cancer

In addition to the intertumour heterogeneity, intratumour cell diversity plays an important role in tumour radiocurability. German pathologist Rudolf Virchow was the first who described the morphological diversity of cancer cells within individual tumours (Brown and Fee, 2006). Later, a growing body of genetic evidence demonstrated that tumour cells are heterogeneous, and their heterogeneity fuels the Darwinian evolution within the tumours during tumour development and treatment, and correlates with therapeutic resistance (Gerlinger and Swanton, 2010).

In addition to the genetic diversity within each individual tumour, epigenetically regulated properties are playing a critical role to define cell fate during tumour growth and therapy. For many types of tumours, only distinct cell populations called cancer stem cells are tumorigenic. The stemness state of tumour cells depends on the different genetic and epigenetic mechanisms and is regulated by tumour microenvironment (Kreso and Dick, 2014). The cancer stem cell model proposes the hierarchical tumour organization where cancer stem cells (CSCs) are able to self-renew and to generate all other non-CSC tumour cells during asymmetric division. Analogous to the normal tissues, where stem cells can switch from asymmetric to symmetric division to increase stem cell pool and its regenerative potential following tissue injury, CSCs predominantly undergo symmetric division following irradiation (Tang, 2012). The changes of the microenvironmental conditions surrounding CSCs such as increase of oxygen tension following tumour irradiation trigger activation of the developmental signalling pathways, for example, wingless-type MMTV integration site family (Wnt), Notch and Sonic hedgehog (Shh) that leads to switch to symmetric type of cell division and allows CSCs to replenish tumour cells. In contrast to the non-CSC cell populations, CSCs are immortal and have unlimited proliferative potential. Therefore, CSCs contribute to the tumour initiation, growth, maintenance, and recurrence after therapy. The discovery of CSCs has important clinical application for tumour diagnostics and treatment and suggests that tumour cure requires eradication of entire CSC population.

The impact of the absolute number of CSCs and their intrinsic radiosensitivity on local tumour control

The cancer stem cell hypothesis explains the discordance between the volume-related endpoint parameters (e.g. tumour growth delay and tumour regression) and local tumour control probability (TCP). The probability of permanent control tumour control is described by a sigmoid function where TCP increases with increasing irradiation dose. A threshold indicates a dose which is almost never enough to inactivate all CSCs, whereas TCP might approach 100% if dose is high enough to inactivate all CSCs. The sigmoid shape of the dose–response relationship reflects a Poisson distribution of the surviving CSC and suggests that TCP depends on the density of CSCs within a given tumour. The impact of the absolute number of CSCs and their intrinsic radiosensitivity on local tumour control was first experimentally demonstrated more than 30 years ago in a study of Hill and Milas by using isogenic mouse models when they first demonstrated a negative correlation between TCD_{50} (tumour control dose 50%) and TD_{50} (median transplantation dose). These findings were confirmed in a study of Baumann and Suit by using human squamous carcinoma xenografts (Hill and Milas, 1989; Baumann et al., 1990; Baumann et al., 2008). These studies inspired the idea to use tumour biopsies for the development of predictive clonogenic assays for radiotherapy, although many of these studies revealed a lack of correlation between cell survival *in vitro* and tumour response *in vivo* that might be due to the differences in microenvironmental factors such as oxygen pressure, extracellular matrix components, interaction with stroma cells, and so on (West et al., 1993; Kurth et al., 2014). Discovery of CSC-specific markers in acute myeloid leukaemia by Dick and colleagues in 1994 and identification of CSC markers in solid tumours opened a new era for experimental investigation of CSCs and development of CSC-based predictive tests (Lapidot et al., 1994; Peitzsch et al., 2013; Linge et al., 2016). Since then, many retrospective clinical studies for different types of cancer demonstrated that analysis of CSC number in pretreatment tumour biopsies can be used for prediction of the patients' outcome after radiotherapy and for individualized treatment selection. The details for some of these studies—including analysed markers, clinical parameters used for analysis, prognostic correlation, and shortcomings of the study design—are reviewed by Bütof et al. (Bütof et al., 2013).

The mechanisms of CSC radioresistance

In addition, compelling experimental data suggest that some CSC populations might possess a higher resistance to radiation compared to non-CSCs. Therapy resistance and accelerated repopulation of CSC during and after radiotherapy has been attributed to the activation of the antiapoptotic signalling pathways such as PI3K/AKT, Wnt or Notch (Cojoc et al., 2015; Krause et al., 2016). Furthermore, a high activation of the DNA repair machinery has been described for CSCs in the different tumour entities (Krause et al., 2016). A highly efficient DNA repair mechanism in CSCs is attributed to the activation of the ATR-Chk1 and ATM-Chk2 signalling pathways. Additional intrinsic mechanism providing a high resistance of CSC populations to genotoxic stress is a highly efficient ROS scavenging system (Diehn et al., 2009; Bensimon et al., 2016). Beside the above-described intrinsic mechanisms of therapy resistance, CSCs residing in the hypoxic tumour areas might be shielded from radiation-induced oxidative stress through less ROS generation and by activation of the HIF pro-survival signalling pathway which is also associated with accelerated repopulation of CSCs during radiotherapy (Kurth et al., 2014; Krause et al., 2016).

Taken together, these findings on CSCs suggest that factors affecting local tumour control such as repair of DNA damage, repopulation, redistribution of cancer cells in the cell cycle, reoxygenation and intrinsic tumour cell radiosensitivity need to be revisited (Pajonk et al., 2010). The radiobiological concept of '5R' parameters can explain the differences in tumour curability only if attributed to the tumorigenic (i.e. CSC populations). Therefore, analysis of CSC density and inherent radioresistance might be indicative of tumour radiocurability and can be used for the development of predictive tests for radiotherapy. Novel treatment strategies based on combination of radiotherapy and CSC-targeting therapies are expected to improve therapy outcome (Krause et al., 2016).

Nevertheless, there are still some pitfalls that need to be taken into consideration. First of all, a high inter- and intratumoral genetic diversity of CSC cells challenges the clinical translation of CSC-specific predictive tests (reviewed in Peitzsch et al., 2013). Further studies are expected to integrate the analysis of marker expression and clinical data for the patients treated with radiotherapy with studies of tumour genetic heterogeneity.

In addition to the spatial heterogeneity of CSCs that might be attributed to the intratumoral diversity of the microenvironmental conditions (e.g. oxygen tension, pH, nutrient and growth factor availability), CSCs might undergo temporal changes during tumour development and treatment. Recent studies demonstrated that CSC-specific markers can be dynamically regulated during the course of irradiation. Furthermore, analogous to the induced reprogramming of the differentiated somatic cells to pluripotent cells, non-CSC population can be reprogrammed to CSCs under induced conditions such as microenvironmental factors (e.g. hypoxia, inflammation), oncogene expression or by irradiation (reviewed in Friedmann-Morvinski and Verma, 2014; Kuhlmann et al., 2015). These observations demonstrate an importance of the continuous biomarker evaluation during the course of treatment and also suggest that inhibition of the epigenetic reprograming in combination with radiotherapy could be a promising therapeutic approach (Peitzsch et al., 2016).

The fate of each individual tumour cell after irradiation depends on the activation status of DNA repair machinery and nature of the DNA lesions. Cells may repair DNA damage and escape from the cell cycle checkpoint. Alternatively, the outcome of DNA injury can be cell death in case of intolerable damage or mutagenesis of surviving cells in case of inaccurate DNA repair, which in turn might result in genomic instability. If these mutations occur in CSCs or induce cell reprogramming to CSC, they could lead to the tumour development in the long-term perspective (Khanna, 2015). X-rays and gamma-radiation were classified as Group 1 carcinogens by the IARC monograph, based on the associated increased risk of different types of cancer including leukaemia, breast, thyroid, non-melanoma skin, stomach, colon, and lung cancer (2000). The tumourigenic effect of radiation is discussed in detail in Chapter 8 of this book, 'Radiation as a carcinogen'.

TAKE-HOME MESSAGE

- Recent advances in dose delivery techniques and new treatment modalities such as proton therapy are leading to improving clinical outcomes and reducing toxicity for normal tissues. However, tumour radioresistance and radiotherapy-associated side effects which limit the maximal radiation dose that can be safely delivered to target volume still remain a serious challenge.
- The cure rate of radiation depends on its ability to induce cellular death or loss of proliferative capacity through inducing of non-repairable DNA damage. In addition to the treatment conformity and clinical factors, success of the fractionated therapy has been explained by the radiobiological concept of '5R' parameters which include **R**epair or DNA damage, **R**epopulation, **R**edistribution of cancer cells in the cell cycle, **R**eoxygenation, and intrinsic tumour cell **R**adiosensitivity.
- Compelling evidence suggests that cancer cells within each individual tumour are heterogeneous in their tumorigenic properties and only distinct cell population called cancer stem cells, capable of tumour initiation, maintenance, and regrowth after treatment. Taken into account the stem cell model of cancer development, radiobiological concept of '5R' parameters can explain the differences in tumour radiosensitivity only if applicable to the CSC populations.
- In addition, a higher resistance to radiation for some of CSCs compared to non-CSCs creates a serious challenge for current cancer treatment. Seminal clinical studies demonstrated that analysis of CSC density in pretreatment biopsies can be correlated with tumour radiocurability and therefore can be used for the development of predictive tests for radiotherapy. Novel treatment strategies inducing CSC differentiation and apoptosis, targeting protective CSC niche, quiescent state, and preventing cancer cell reprogramming bring a promise to improve efficacy of radiotherapy.

OPEN QUESTIONS

- Why can some non-CSC-specific markers (e.g. EGFR expression, HPV status, osteopontin blood level) effectively work as predictive or prognostic markers for radiotherapy if CSC express higher radioresistance and thus a different response to radiation?
- What is the mechanism of radiation-induced cell reprogramming?
- How to achieve effective delivery of the radiosensitizing drugs to more radioresistant hypoxic cells which are distant from the blood supplies?
- How to explain that duration of hypoxia is a critical factor regulating cell radiosensitivity wherein short-term hypoxia (e.g. 4 –24 h) increases cell radioresistance and long-term chronic hypoxia makes cells more radiosensitive?
- What are the mechanisms of radiation-induced non-targeted effect such as bystander effect and genomic instability and how are they related?
- Can targeted inhibition of the cell cycle checkpoint machinery be used for tumour radiosensitization with tolerable side effect?
- How relevant are commonly acceptable laboratory endpoints of radiobiological assays (e. g. tumour volume, colony forming ability, marker expression or cell death) for the inactivation of CSCs and tumour cure?
- How reflective are currently used tumour volume-dependent experimental endpoints for CSC populations which often constitute the minority of tumour cells?

FURTHER READING

Baumann, M., Overgaard, J., Debus, J., et al. (2016). Radiation oncology in the era of precision medicine. *Nat Rev Cancer*, **16**(4), 234–49.

Begg, A. C., Stewart, F. A., & Vens, C. (2011). Strategies to improve radiotherapy with targeted drugs. *Nat Rev Cancer*, **11**, 239–53.

Bütof, R., Dubrovska, A., & Baumann, M. (2013). Clinical perspectives of cancer stem cell research in radiation oncology. *Radiother Oncol*, **108**, 388–96.

Eriksson, D. & Stigbrand, T. (2010). Radiation-induced cell death mechanisms. *Tumour Biol*, **31**, 363–72.

Good, J. S. & Harrington, K. J. (2013). The hallmarks of cancer and the radiation oncologist: updating the 5Rs of radiobiology. *Clin Oncol (R Coll Radiol)*, **25**, 569–77.

Helleday, T., Lo, J., Van Gent, D. C., & Engelward, B. P. (2007). DNA double-strand break repair: from mechanistic understanding to cancer treatment. *DNA Repair (Amst)*, **6**, 923–35.

Krause, M., Dubrovska, A., Linge, A., & Baumann, M. (2016). Cancer stem cells: radioresistance, prediction of radiotherapy outcome and specific targets for combined treatments. *Adv Drug Deliv Rev*, **109**, 63–73.

Kurth, I., Baumann, M., & Dubrovska, A. (2014). The role of cancer stem cells in tumor radioresistance. In: Rajasekhar, V. K. (ed.) *Cancer Stem Cells*, Chapter 35. Hoboken, NJ: John Wiley and Sons.

Morgan, M. A. & Lawrence, T. S. (2015). Molecular pathways: overcoming radiation resistance by targeting DNA damage response pathways. *Clin Cancer Res*, **21**, 2898–904.

Hall, E. J. & Giaccia, A. J. (2012). *Radiobiology for the Radiologist*, 7th edition. Philadelphia: Lippincott Williams and Wilkins, Wolters Kluwer.

REFERENCES

[No authors listed] (2000). IARC Working group on the evaluation of carcinogenic risks to humans: ionizing radiation, part I: X- and gamma-radiation and neutrons. Lyon, France, 26 May–2 June 1999. *IARC Monogr Eval Carcinog Risks Hum*, **75** Pt 1, 1–448.

Abbas, T. & Dutta, A. (2009). p21 in cancer: intricate networks and multiple activities. *Nat Rev Cancer*, **9**, 400–14.

Atun, R., Jaffray, D. A., Barton, M. B., et al. (2015). Expanding global access to radiotherapy. *Lancet Oncol*, **16**, 1153–86.

Aylon, Y. & Oren, M. (2007). Living with p53, dying of p53. *Cell*, **130**, 597–600.

Azzam, E. I., Jay-Gerin, J. P., & Pain, D. (2012). Ionizing radiation-induced metabolic oxidative stress and prolonged cell injury. *Cancer Lett*, **327**, 48–60.

Banath, J. P., Macphail, S. H., & Olive, P. L. (2004). Radiation sensitivity, H2AX phosphorylation, and kinetics of repair of DNA strand breaks in irradiated cervical cancer cell lines. *Cancer Res*, **64**, 7144–9.

Baumann, M., Dubois, W., & Suit, H. D. (1990). Response of human squamous cell carcinoma xenografts of different sizes to irradiation: relationship of clonogenic cells, cellular radiation sensitivity *in vivo*, and tumor rescuing units. *Radiat Res*, **123**, 325–30.

Baumann, M., Krause, M., Overgaard, J., et al. (2016). Radiation oncology in the era of precision medicine. *Nat Rev Cancer*, **16**(4), 2234–49.

Baumann, M., Krause, M., & Hill, R. (2008). Exploring the role of cancer stem cells in radioresistance. *Nat Rev Cancer*, **8**, 545–54.

Begg, A. C., Stewart, F. A., & Vens, C. (2011). Strategies to improve radiotherapy with targeted drugs. *Nat Rev Cancer*, **11**, 239–53.

Bensimon, J., Biard, D., Paget, V., et al. (2016). Forced extinction of CD24 stem-like breast cancer marker alone promotes radiation resistance through the control of oxidative stress. *Mol Carcinog*, **55**(3), 245–54.

Blaufox, M. D. (1996). Becquerel and the discovery of radioactivity: early concepts. *Semin Nucl Med*, **26**(3), 145–54.

Brenner, D. J. (2008). The linear-quadratic model is an appropriate methodology for determining isoeffective doses at large doses per fraction. *Semin Radiat Oncol*, **18**, 234–9.

Brown, T. M. & Fee, E. (2006). Rudolf Carl Virchow: medical scientist, social reformer, role model. *Am J Public Health*, **96**, 2104–5.

Bussink, J., Van Der Kogel, A. J., & Kaanders, J. H. (2008). Activation of the PI3-K/AKT pathway and implications for radioresistance mechanisms in head and neck cancer. *Lancet Oncol*, **9**, 288–96.

Bütof, R. & Baumann, M. (2013). Time in radiation oncology—keep it short! *Radiother Oncol*, **106**, 271–5.

Bütof, R., Dubrovska, A., & Baumann, M. (2013). Clinical perspectives of cancer stem cell research in radiation oncology. *Radiother Oncol*, **108**, 388–96.

Caldecott, K. W. (2014). DNA single-strand break repair. *Exp Cell Res*, **329**, 2–8.

Castedo, M., Perfettini, J. L., Roumier, T., Andreau, K., Medema, R., & Kroemer, G. (2004). Cell death by mitotic catastrophe: a molecular definition. *Oncogene*, **23**, 2825–37.

Chadwick, K. H. & Leenhouts, H. P. (1973). A molecular theory of cell survival. *Phys Med Biol*, **18**, 78–87.

Cojoc, M., Peitzsch, C., Kurth, I., et al. (2015). Aldehyde dehydrogenase is regulated by beta-catenin/TCF and promotes radioresistance in prostate cancer progenitor cells. *Cancer Res*, **75**, 1482–94.

Coleman, C. N. (1993). Beneficial liaisons: radiobiology meets cellular and molecular biology. *Radiother Oncol*, **28**, 1–15.

D'Adda, D. I. & di Fagagna, F. (2008). Living on a break: cellular senescence as a DNA-damage response. *Nat Rev Cancer*, **8**, 512–22.

Diehn, M., Cho, R. W., Lobo, N. A., et al. (2009). Association of reactive oxygen species levels and radioresistance in cancer stem cells. *Nature*, **458**, 780–3.

Eriksson, D. & Stigbrand, T. (2010). Radiation-induced cell death mechanisms. *Tumour Biol*, **31**, 363–72.

Flynn, R. L. & Zou, L. (2011). ATR: a master conductor of cellular responses to DNA replication stress. *Trends Biochem Sci*, **36**, 133–40.

Forget, A. L. & Kowalczykowski, S. C. (2010). Single-molecule imaging brings Rad51 nucleoprotein filaments into focus. *Trends Cell Biol*, **20**, 269–76.

Friedmann-Morvinski, D. & Verma, I. M. (2014). Dedifferentiation and reprogramming: origins of cancer stem cells. *EMBO Rep*, **15**, 244–53.

Gerlinger, M. & Swanton, C. (2010). How Darwinian models inform therapeutic failure initiated by clonal heterogeneity in cancer medicine. *Br J Cancer*, **103**, 1139–43.

Good, J. S. & Harrington, K. J. (2013). The hallmarks of cancer and the radiation oncologist: updating the 5Rs of radiobiology. *Clin Oncol (R Coll Radiol)*, **25**, 569–77.

Goodhead, D. T. (1999). Mechanisms for the biological effectiveness of high-LET radiations. *J Radiat Res*, **40** Suppl, 1–13.

Gray, L. H., Conger, A. D., Ebert, M., Hornsey, S., & Scott, O. C. (1953). The concentration of oxygen dissolved in tissues at the time of irradiation as a factor in radiotherapy. *Br J Radiol*, **26**, 638–48.

Halazonetis, T. D., Gorgoulis, V. G., & Bartek, J. (2008). An oncogene-induced DNA damage model for cancer development. *Science*, **319**, 1352–5.

Hall, E. J. & Giaccia, A. J. (2012). *Radiobiology for the Radiologist*, 7th edition. Philadelphia: Lippincott Williams and Wilkins, Wolters Kluwer.

Hanahan, D. & Weinberg, R. A. (2011). Hallmarks of cancer: the next generation. *Cell*, **144**, 646–74.

Hegemann, N. S., Guckenberger, M., Belka, C., Ganswindt, U., Manapov, F., & Li, M. (2014). Hypofractionated radiotherapy for prostate cancer. *Radiat Oncol*, **9**, 275.

Heilbron, J. L. (1968). The scattering of α and ß particles and Rutherford's atom. *Arch Hist Exact Sci*, **4**, 247–307.

Hein, A. L., Ouellette, M. M., & Yan, Y. (2014). Radiation-induced signaling pathways that promote cancer cell survival (review). *Int J Oncol*, **45**, 1813–19.

Helleday, T., LO, J., Van Gent, D. C., & Engelward, B. P. (2007). DNA double-strand break repair: from mechanistic understanding to cancer treatment. *DNA Repair (Amst)*, **6**, 923–35.

Hill, R. P. & Milas, L. (1989). The proportion of stem cells in murine tumors. *Int J Radiat Oncol Biol Phys*, **16**, 513–18.

Jakel, O. (2008). The relative biological effectiveness of proton and ion beams. *Z Med Phys*, **18**, 276–85.

Jalal, N., Haq, S., Anwar, N., Nazeer, S., & Saeed, U. (2014). Radiation induced bystander effect and DNA damage. *J Cancer Res Ther*, **10**, 819–33.

Jones, L., Hoban, P., & Metcalfe, P. (2001). The use of the linear quadratic model in radiotherapy: a review. *Australas Phys Eng Sci Med*, **24**, 132–46.

Khanna, A. (2015). DNA damage in cancer therapeutics: a boon or a curse? *Cancer Res*, **75**, 2133–8.

Kogelnik, H. D. (1997). Inauguration of radiotherapy as a new scientific speciality by Leopold Freund 100 years ago. *Radiother Oncol*, **42**, 203–11.

Krause, M., Dubrovska, A., Linge, A., & Baumann, M. (2016). Cancer stem cells: radioresistance, prediction of radiotherapy outcome and specific targets for combined treatments. *Adv Drug Deliv Rev*, **15**(109), 63–73.

Kreso, A. & Dick, J. E. (2014). Evolution of the cancer stem cell model. *Cell Stem Cell*, **14**, 275–91.

Kuhlmann, J. D., Hein, L., Kurth, I., Wimberger, P., & Dubrovska, A. (2015). Targeting cancer stem cells: promises and challenges. *Anticancer Agents Med Chem*, **16**, 38–58.

Kurth, I., Peitzsche, C., Baumann, M., & Dubrovska, A. (2014). The role of cancer stem cells in tumor radioresistance. In: Rajasekhar, V. K. (ed.) *Cancer Stem Cells*, Chapter 35. Hoboken, NJ: John Wiley and Sons.

Lapidot, T., Sirard, C., Vormoor, J., et al. (1994). A cell initiating human acute myeloid leukaemia after transplantation into SCID mice. *Nature*, **367**, 645–8.

Lederman, M. (1981). The early history of radiotherapy: 1895–1939. *Int J Radiat Oncol Biol Phys*, **7**, 639–48.

Linge, A., Lock, S., Gudziol, V., et al. (2016). Low CSC marker expression and low hypoxia identify good prognosis subgroups in HPV(-) HNSCC after postoperative radiochemotherapy: a multicenter study of the DKTK-ROG. *Clin Cancer Res*, **22**(11), 2639–49.

Loeffler, J. S., Durante, M. (2013). Charged particle therapy—optimization, challenges and future directions. *Nat Rev Clin Oncol*, **10**(7), 411–24.

Lomax, M. E., Folkes, L. K., & O'Neill, P. (2013). Biological consequences of radiation-induced DNA damage: relevance to radiotherapy. *Clin Oncol (R Coll Radiol)*, **25**, 578–85.

Mailhot Vega, R., Kim, J., Hollander, A., et al. (2015). Cost effectiveness of proton versus photon radiation therapy with respect to the risk of growth hormone deficiency in children. *Cancer*, **121**, 1694–702.

Matt, S. & Hofmann, T. G. (2016). The DNA damage-induced cell death response: a roadmap to kill cancer cells. *Cell Mol Life Sci*, **73**(15), 2829–50.

Menegakis, A., Von Neubeck, C., Yaromina, A., et al. (2015). gammaH2AX assay in ex vivo irradiated tumour specimens: a novel method to determine tumour radiation sensitivity in patient-derived material. *Radiother Oncol*, **116**, 473–9.

Mikkelsen, R. B. & Wardman, P. (2003). Biological chemistry of reactive oxygen and nitrogen and radiation-induced signal transduction mechanisms. *Oncogene*, **22**, 5734–54.

Morgan, M. A. & Lawrence, T. S. (2015). Molecular pathways: overcoming radiation resistance by targeting DNA damage response pathways. *Clin Cancer Res*, **21**, 2898–904.

Nagasawa, H. & Little, J. B. (1992). Induction of sister chromatid exchanges by extremely low doses of alpha-particles. *Cancer Res*, **52**, 6394–6.

Ng, W.-L., Huang, Q., Liu, X., Zimmerman, M., Li, F., & Li, C. Y. (2013). Molecular mechanisms involved in tumor repopulation after radiotherapy. *Transl Cancer Res*, **2**, 442–8.

Olson, E., Nievera, C. J., Liu, E., Lee, A. Y., Chen, L., & Wu, X. (2007). The Mre11 complex mediates the S-phase checkpoint through an interaction with replication protein A. *Mol Cell Biol*, **27**, 6053–67.

Pajonk, F., Vlashi, E., & McBride, W. H. (2010). Radiation resistance of cancer stem cells: the 4 R's of radiobiology revisited. *Stem Cells*, **28**, 639–48.

Panizzon, R. G. & Seegenschmiedt, M. H. (2015). *Radiation Treatment and Radiation Reactions in Dermatology*. Berlin: Springer.

Paull, T. T. (2015). Mechanisms of ATM activation. *Annu Rev Biochem*, **84**, 711–38.

Peitzsch C, C. M., Hein L, Kurth I, et al. (2016). An epigenetic reprogramming strategy to re-sensitize radioresistant prostate cancer cells. *Cancer Res*, **76**(9), 2637–51.

Peitzsch, C., Kurth, I., Kunz-Schughart, L., Baumann, M., & Dubrovska, A. (2013). Discovery of the cancer stem cell related determinants of radioresistance. *Radiother Oncol*, **108**, 378–87.

Peitzsch, C., Perrin, R., Hill, R. P., Dubrovska, A., & Kurth, I. (2014). Hypoxia as a biomarker for radioresistant cancer stem cells. *Int J Radiat Biol*, **90**, 636–52.

Polo, S. E. & Jackson, S. P. (2011). Dynamics of DNA damage response proteins at DNA breaks: a focus on protein modifications. *Genes Dev*, **25**, 409–33.

Puck, T. T. & Marcus, P. I. (1956). Action of X-rays on mammalian cells. *J Exp Med*, **103**, 653–66.

Reed, A. B. (2011). The history of radiation use in medicine. *J Vasc Surg*, **53**, 3S–5S.

Roos, W. P., Thomas, A. D., & Kaina, B. (2016). DNA damage and the balance between survival and death in cancer biology. *Nat Rev Cancer*, **16**, 20–33.

Roy, S., De Melo, A. J., Xu, Y., et al. (2015). XRCC4/XLF interaction is variably required for DNA repair and is not required for ligase IV stimulation. *Mol Cell Biol*, **35**, 3017–28.

Schumacker, P. T. (2015). Reactive oxygen species in cancer: a dance with the devil. *Cancer Cell*, **27**, 156–7.

Shrivastav, M., De Haro, L. P., & Nickoloff, J. A. (2008). Regulation of DNA double-strand break repair pathway choice. *Cell Res*, **18**, 134–47.

Skinner, H. D., Sandulache, V. C., Ow, T. J., et al. (2012). TP53 disruptive mutations lead to head and neck cancer treatment failure through inhibition of radiation-induced senescence. *Clin Cancer Res*, **18**, 290–300.

Slater, J. M. (2006). Considerations in identifying optimal particles for radiation medicine. *Technol Cancer Res Treat*, **5**, 73–9.

Smith, J., Tho, L. M., Xu, N., & Gillespie, D. A. 2010. The ATM-Chk2 and ATR-Chk1 pathways in DNA damage signaling and cancer. *Adv Cancer Res*, **108**, 73–112.

Tamulevicius, P., Wang, M., & Iliakis, G. (2007). Homology-directed repair is required for the development of radioresistance during S phase: interplay between double-strand break repair and checkpoint response. *Radiat Res*, **167**, 1–11.

Tang, D. G. (2012). Understanding cancer stem cell heterogeneity and plasticity. *Cell Res*, **22**, 457–72.

Taylor, W. R. & Stark, G. R. (2001). Regulation of the G2/M transition by p53. *Oncogene*, **20**, 1803–15.

Thames, H. D. (1985). An 'incomplete-repair' model for survival after fractionated and continuous irradiations. *Int J Radiat Biol Relat Stud Phys Chem Med*, **47**, 319–39.

Trachootham, D., Alexandre, J., & Huang, P. (2009). Targeting cancer cells by ROS-mediated mechanisms: a radical therapeutic approach? *Nat Rev Drug Discov*, **8**, 579–91.

Uhl, M., Herfarth, K., & Debus, J. (2014). Comparing the use of protons and carbon ions for treatment. *Cancer J*, **20**, 433–9.

Wang, H., Yu, K. N., Hou, J., Liu, Q., & Han, W. (2015). Radiation-induced bystander effect: early process and rapid assessment. *Cancer Lett*, **356**, 137–44.

Ward, J. F. (1994). DNA damage as the cause of ionizing radiation-induced gene activation. *Radiat Res*, **138**, S85–8.

West, C. M., Davidson, S. E., Roberts, S. A., & Hunter, R. D. (1993). Intrinsic radiosensitivity and prediction of patient response to radiotherapy for carcinoma of the cervix. *Br J Cancer*, **68**, 819–23.

Wisenbaugh, E. S., Andrews, P. E., Ferrigni, R. G., et al. (2014). Proton beam therapy for localized prostate cancer 101: basics, controversies, and facts. *Rev Urol*, **16**, 67–75.

Withers, H. (1975a). *The Four R's of Radiotherapy.*, New York: Academic Press.

Withers, H. R. (1975b). Cell cycle redistribution as a factor in multifraction irradiation. *Radiology*, **114**, 199–202.

Woo, R. A. & Poon, R. Y. (2003). Cyclin-dependent kinases and S phase control in mammalian cells. *Cell Cycle*, **2**, 316–24.

Yarnold, J., Bentzen, S. M., Coles, C., & Haviland, J. (2011). Hypofractionated whole-breast radiotherapy for women with early breast cancer: myths and realities. *Int J Radiat Oncol Biol Phys*, **79**, 1–9.

Zannini, L., Delia, D., & Buscemi, G. (2014). CHK2 kinase in the DNA damage response and beyond. *J Mol Cell Biol*, **6**, 442–57.

SECTION VII

Conclusions

31

Benign tumours: The forgotten neoplasms

Francesco Pezzella, Adrian L. Harris, and Mahvash Tavassoli

Introduction: The forgotten tumours

As already mentioned in this book, alongside the malignant tumours there are also benign tumours. Benign tumours are localized, clonal neoplasms, which outnumber the occurrence of malignant lesions (Guray and Sahin, 2006; Marino-Enriquez and Fletcher, 2014) although reliable statistics are not available (RightDiagnosis.com). Benign tumours are rarely fatal because they do not metastasize and only a minority are at risk of malignant transformation (Marino-Enriquez and Fletcher, 2014). Because of their non-aggressive 'benign' nature and, in many cases, easy, straightforward possibility of surgical treatment, the biology of these tumours has not received much attention compared to that of the malignant neoplasms. However, several studies recently have started to unveil the involvement of mechanisms overlapping with those of the malignant lesions, raising the issue that, by ignoring the biology of benign tumours, we are perhaps missing information vital for understanding malignancies (Marino-Enriquez and Fletcher, 2014).

Similar to their malignant counterpart, benign tumours can arise from any type of tissue. In Figure 31.1 is a schematic representation of a colonic adenoma, arising from the gut epithelium. Colon adenoma are among the commonest benign tumours.

It is not surprising that the types of benign tumours which have received more attention are the one for which the treatment options are, for various reasons, not yet fully satisfactory: three types are representative of this limited group. One first type is those of the pituitary adenomas: in this case the difficulty in treating them is mostly due to the anatomical location, which makes the surgical treatment complex and a medical option would be favoured, if possible. The second type is the uterine leiomyoma (so called uterine fibroma): is extremely common but, although treatable by surgery, in many cases it would be desirable to manage it by medical treatment in order to avoid invasive surgery and, in some patients, total hysterectomy. This is especially important in young patients where the issue of infertility arises. The third type is the colorectal adenoma which are amongst the ones at highest risk of progression towards cancer, as discussed later within this chapter.

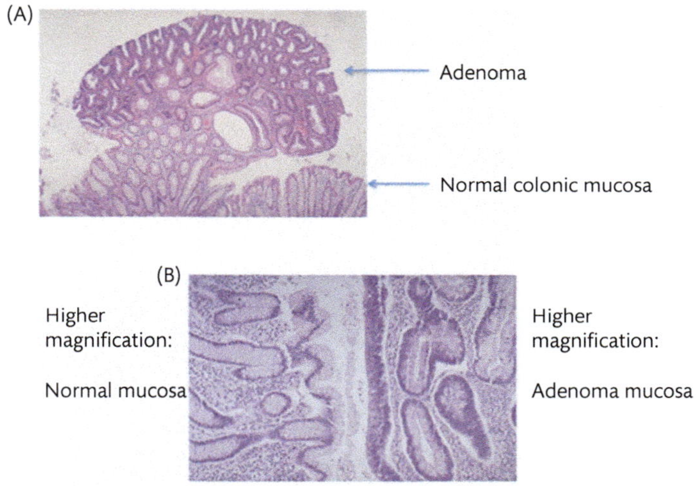

Fig. 31.1 (A and B) Microscopic image of a colonic adenoma. (B) (high magnification). In (A) is normal colonic mucosa, in (B) is mucosa from and adenoma. Although the overall architecture is preserved, glands are a bit more irregularly shaped in the adenoma and the nuclei are enlarged giving to the cells a darker appearance. Still each cell is polarized with the nucleus at the bottom and the formation of mucus in the portion of the cell facing the lumen of the gland.

Definition and comparison between benign and malignant tumours

Benign tumours are defined as neoplasms growing locally, usually but not always compressing rather than infiltrating the surrounding tissue and, by definition, never originating metastases. As they do not shed neoplastic cells able to metastasize, their surgical excision invariably results in successful treatment. They are usually well differentiated as their architecture and morphology is very similar to that of the corresponding normal tissue (Fig. 31.1A). Also, the cytological features are usually very close to normal. There is generally average, or slightly increased, nuclear: cytoplasmic ratio with only very mild atypias (Fig. 31.1B). The symptoms they cause are due to the local physical effects such as compression and /or burden caused by the neoplastic mass. There is one notable exception: the benign secreting tumours. These are endocrine benign tumours which retain the ability to secrete their hormones but as they escape the regulatory feedback mechanisms, they secrete hormones in a quantity larger than the matching normal tissue. Therefore, according to the hormone released, there will be symptoms caused by its excess, one of the commonest types is a 'secreting' or 'hot' thyroid adenoma appearing with the signs and symptoms of hyperthyroidism.

Several other features are also shared with the malignant neoplasms as benign lesions also have increased proliferation, loss of suppressor activity, escape from apoptosis, escape from immune surveillance, and metabolic changes.

Basic alterations in benign tumours

Genetic damage

Sequencing the genome of benign neoplasms has revealed a wealth of mutations occurring which are mostly the same as observed in cancer, both with regard to activating changes in oncogenes and inactivation of tumour suppressor genes. Some of the commonest of these mutations are listed in Table 31.1: only three of these genes have not been reported mutated in malignant tumours so far, although they are involved in other molecular alterations.

Table 31.1 Main types of gene mutations in benign tumours

	Benign tumour carrying point mutations	Mutation reported in malignant tumours?
PRKACA	Adrenocortical adenoma cortisol producing	No *
KCNJ5	Adrenocortical adenoma aldosterone producing	No*
AIP	Pituitary adenoma	No*
HRAS	Nevus sebaceous	Yes
KRAS	Nevus sebaceous	Yes
BRAF V600E	Papillary craniopharyngioma Cutaneous naevi	Yes
MED12	Uterine leiomyoma Breast fibroadenoma	Yes
SMARCB1	Schwannomatosis	Yes
IDH1	Spindle cell haemangioma	Yes
IDH1	Spindle cell haemangioma	Yes
GNAQ	Blue naevi	Yes
NF2	Meningioma Shwannoma	Yes
TRAF7	Meningioma (non-NF2) Perineurioma	Yes
KLF4	Meningioma (non-NF2)	Yes
AKT1	Meningioma (non-NF2)	Yes
SMO	Meningioma (non-NF2)	Yes
PTEN	Hamartoma	Yes

* Other molecular alterations found in malignant tumours (e.g. amplification, translocation).

Adapted with permission from Springer Nature, *Nature Reviews Cancer*, 'Shouldn't we care about the biology of benign tumours?' Marino-Enriquez A and Fletcher CD, Volume 14, pp. 701–2, Copyright © 2014 Macmillan Publishers Limited.

The reason for which no risk of malignant growth is increased in most benign tumours compared to normal tissue, is because the genetic changes by themselves are believed to be insufficient to trigger it (Marino-Enriquez and Fletcher, 2014). One example is meningioma, a benign tumour of the meninges, the layer of tissue covering the brain. Two main types occur: those associated with mutation or deletion of NF2 gene and those in which this gene is normal (Meningioma non-NF2; see Benmaamar, 2013). Genome sequencing has revealed that PI3K/AKT pathway, which is commonly altered in human cancers (Mayer and Arteaga, 2016) is affected in non-NF2 meningioma. In these tumours the AKT gene can be mutated alongside several other genes like TRAF7, KLF4, and SMO. (Brastianos et al., 2013; Clark et al., 2013). Another common type of genetic alteration found in benign tumours is chromosomal imbalance as detected by cytogenetic techniques. In a study looking at *Phyllodes* tumours, a rare group of fibro-epithelial benign tumours of the breast, both gain and loss of large chromosomal areas were found including amplification at 12q24 the MDM2 locus, and at 8p24, involving the MYC oncogene (Lae et al., 2007)

Genetic instability

As discussed in Chapter 4 on genetic instability, the integrity of the genome is continuously under threat and a number of mechanisms are in place to preserve its integrity. However, some genetic alterations, once established, inactivate such a mechanism diminishing the possibility of maintaining the genome integrity. These events lead to yet more alterations occurring, some further inactivating the remaining safety pathways therefore producing an even faster accumulation of additional damages. As the defences of the genomic stability are eroded, the genetic instability increases. In a minority of benign tumours, the genetic instability increases and these are the ones with a high risk of developing malignant transformations, as in the adenoma-carcinoma sequence of the colon (see later on in this chapter, 'Benign tumours progression to cancer and the borderline tumours'). However, most of the benign tumours have genetic alterations, but do not progress towards malignancy: this suggests that the ability to maintain the genoma stable is preserved. The question, formally raised in 1988 by Volpe, is therefore why most benign tumours are not associated with an increased risk of cancer. He formulated the hypothesis that in benign tumours genes affecting genomic stability are not, or very little involved (Volpe, 1988). Up to now, not much progress has been made to answer this question as the knowledge of the benign molecular pathology of benign tumours remains limited. For example, a study in breast benign lesions has shown that increased p53 protein expression was associated with increased risk of transformation but the occurrence of a point mutation was not (Rohan et al., 2006) while Med12 gain of function causes an increase of genomic instability, which is however self-limiting, in leiomyomas (Mittal et al., 2015).

Epigenetic changes

Epigenetics is the study of the changes in gene expression in absence of any genetic alteration but rather due to interaction between DNA and other molecules. One of the most common mechanism in cancer is DNA methylation, a process by which methyl groups link to DNA segments changing their behaviour but maintaining their normal sequence. The most common methylation in cancer is that of a promoter sequence which leads to repression of gene transcription. Diminished methylation, so called hypomethylation, can instead lead to increased transcription (Kulis and Esteller, 2010). Altered methylation has been reported in various types of benign lesions including fibroadenoma of the breast (Billard et al., 2002), benign ovarian epithelial tumours (Takehara et al., 2009), and benign adrenocortical adenomas (Rechache et al., 2012).

Stem cells and benign tumours

The role of cancer stem cells in benign tumours has started to emerge only recently but so far it is unclear as what role stem cells play and, if so, how it differs from the role played by normal and cancer stem cells, respectively. First evidence of the presence of 'benign tumour stem cells' comes from teratomas, benign tumours in which many different types of tissues are represented in just one neoplastic lesion. These tumours derive from embryonal-like stem cells and in a single teratoma different types of tissues are present such as gut epithelium, bone, connective, and endocrine tissue (Qin et al., 2016). Subsequently, stem cells have been isolated form a neuronal benign tumour, the pituitary adenomas (Xu et al., 2009). Xu and colleague proved the presence of stem cell in these adenomas as they isolated cells that:

- were able to grow as neurospheres, similar to the stem cells from brain malignant glioblastoma;
- cells from these neurospheres generated new neurospheres;
- neural stem cells markers nestin and CD133, present on malignant neuronal stem cells, were also expressed on these 'benign' stem cells;
- early differentiation neuronal markers, GFPA and S100, were positive but differentiation markers (e.g. growth hormone), were negative;
- other stem cells markers, CD90, Oct4, Musashi1, Notch4, Jag 2, and DLL1, were positive;
- antiapoptotic proteins were present;
- if implanted in mice these stem cells developed benign lesions from which similar stem cells could be isolated.

Using a similar approach neoplastic stem cells have been identified in a variety of benign tumours (Table 31.2). These benign stem cells seem to be subject to a variety of signals which depend upon the anatomical micro-environment, or niche, in which they are located (Qin et al., 2016; see Fig. 31.2).

Table 31.2 Stem cells in benign tumours

Benign tumour	Phenotype	Functional test
Palmar fibrosis	CD73, CD90, CD105	*In vitro* differentiation; Transplantation
Giant cell tumour	CD73, CD105, CD34, FGFR3	*In vitro* differentiation
Keloid	CD105, CD146, STRO-1	*In vitro* differentiation; Serial transplantation;
Lipoma	---	*In vitro* differentiation
Meningioma	CD44, CD166	Transplantation
Haemangioma	CD133	Serial transplantation
Osteolipoma	---	*In vitro* differentiation
Pituitary adenoma	CD133, Nestin, CD90, Oct4, Musashi1, Notch4, Jag 2, DLL1	Sphere formation; Serial transplantation
Uterine leiomyoma	---	Transplantation
Endometriosis	---	Transplantation
Neurofibroma	EGFR, P75, CD34	*In vitro* differentiation; Transplantation
Intestinal adenoma	Lgr5, CD133, CD44	Lineage tracing

Adapted with permission from *Tumor Biology*, 'The role of stem cells in benign tumors', Qin H et al., Volume 37, Issue 12, pp. 15349–57, Copyright © 2016 Springer Nature.

The multicellularity hallmarks and the benign tumours

We have previously discussed how the biological functions, which are the markers of multicellularity, are affected in malignancy (Chapter 1). Next we will discuss these in the context of benign tumours.

Regulation and control of cell cycle

As a general rule, benign tumours grow more slowly than most of the malignant ones and have a very low proliferating rate (Rosai, 2011). However genetic alterations can cause escape from normal cell cycle regulation. In one of the best studied benign tumour types, the pituitary adenoma, sonic hedgehog (Shh) is not expressed: as a consequence, the Smoothened (Smo)/Glioblastoma1 (Gli1) pathway remains active and induces cellular proliferation (Yavropoulou et al., 2015).

Apoptosis and telomerases: Control of lifespan

In the majority of tumours investigated, telomerase activity was present in benign lesions although at much lower levels than in the malignant counterparts. The active full-length hTERT transcript was present in thyroid adenomas, although at levels lower than the inactive splice variants, while the opposite was observed in malignant thyroid tumours (Wang et al., 2008). The same abnormal pattern of hTERT expression has been described in benign soft tissue tumours (Umehara et al., 2004), pheochromocytomas (Boltze et al., 2003), breast fibroadenomas (Yano et al., 2002) and colon adenomas (Abe et al., 2001). Conversely, no expression of telomerase reverse transcriptase (hTERT) was detected in parathyroid adenoma, while the carcinomas were positive (Osawa et al., 2009).

There is also indication that pathological inhibition of apoptosis plays a role in benign tumours.

In seborrheic keratosis, a benign epithelial tumour of the skin, and in vestibular schwannoma, a benign neoplasm of the peripheral nervous system, protection form apoptosis is maintained by the activation of AKT pathway that, by phosphorylating its downstream targets, leads to inhibition of apoptosis (Agnihotri et al., 2014; Neel et al., 2016). In benign breast lesions, such as fibroadenoma, instead protection from apoptosis is mediated by the loss of the expression of PKGIα, PKGIβ, and PKGII: these cyclic GMP-dependent protein kinases not only play a role inhibiting cell cycle but also in induction of apoptosis (Karami-Tehrani et al., 2012). In uterine leiomyoma, fewer cells expressed caspase 3 compared to normal myometrium, matched by menopausal status. However, while in

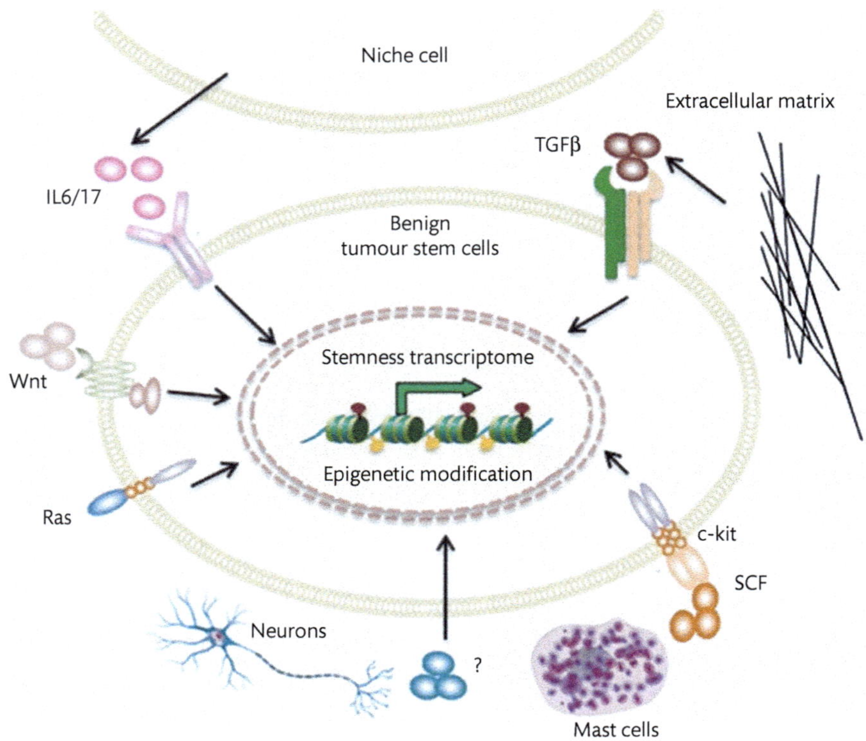

Fig. 31.2 Regulation of stem cells in a benign tumour. Stem cells are influenced by several different factors which eventually influence the phenotype of the tumour.

women of reproductive age both loss of caspase3 and some proliferating activity was present, in postmenopausal women no proliferation was detected in uterine leiomyomas while less expression of caspase3 was present—suggesting an even more important role for loss of apoptosis in these older patients (Plewka et al., 2011). COX2 high levels are linked to tumorigenesis in a wide number of organs. Growth of xenografts of meningioma cells in mice also has been found to be in part supported by reduced apoptosis, as abnormal high levels of COX2 were present and COX2 inhibition by Celecoxib led to increase in the number of apoptotic cells (Ragel et al., 2006). In a series of different benign tumours of the central nervous system, meningioma, schwannoma, low-grade ependymoma, pilocytic astrocytoma and pituitary adenoma, instead high levels of survivin (Hassounah et al., 2005) mediate the increased protection form apoptosis.

The interaction with extracellular environment and cell-cell and cell-matrix adhesion

One of the morphological characteristics of benign tumours is that their histology is very similar to that of the correspondent normal tissue, including the extracellular environment (Rosai, 2011). Nevertheless, a few studies have reported abnormalities in their extracellular matrix. Tenascin and fibronectin are up regulated in breast cancers, but also in breast fibroadenoma increased expression of these two proteins has been detected although at lower levels and not in the epithelial cells but only in the extracellular stromal matrix of the adenoma. This is of interest, as coexpression of these two proteins in the stroma and epithelium, is associated with migration which is absent in benign epithelial tumours (Yoshida et al., 1997).

Alterations in the proteins involved in remodelling and invasion are present in benign tumour cells but the patterns that has been seen so far differs in benign from malignant lesions. Matrix metalloproteases are expressed in pituitary adenomas (Paez-Pereda et al., 2005) and, as they switch towards an aggressive phenotype, their expression tends to increase. Instead cathepsin D is highly expressed in malignant tumours while absent in fibroadenomas (Zheng et al., 1999). Despite the high levels of MMP expression in the extracellular matrix in the pituitary adenomas no invasion occurs: this is thought to be due to the increased levels of tissue inhibitors of metalloproteases (TIMPs) in adenomas protecting them from the metalloproteases action. TIMPs levels decrease with progression towards malignancy suggesting that a more aggressive phenotype of the extracellular matrix is due not only to the increase in the levels of MMP but also loss of TIMPs. Pituitary adenomas have an abnormal extracellular matrix but are associated with a phenotype which blocks invasion (Paez-Pereda et al., 2005; Gurlek et al., 2007).

The sarcoglycan complex is a multimember transmembrane system involved in anchoring the epithelial cells to the extracellular matrix by providing a mechano-signalling (i.e. a mechanic event able to cause an alteration on proteins which activate an intracellular pathway, that is, muscular exercise leads to enhanced transcription of muscle proteins) connection from the cytoskeleton to the extracellular matrix. A severe reduction of this family of proteins is present in breast fibroadenomas (Taylor et al., 1996; Arco et al., 2012) theoretically suggesting a reduction in adherence of the fibroadenoma epithelial cells to the extracellular matrix although eventually the adenoma cells remains firmly anchored to the stroma.

Most integrins, catenins, and N and P cadherins are expressed equally in fibroadenomas, uterine leiomyoma and in their correspondent normal tissues are down regulated only in malignant lesions (Damjanovich et al., 1997). In leiomyomas, integrin beta 1 plays an important role: uterine leiomyoma increase progressively in size and integrin beta 1 is more expressed than in normal myometrium together with laminins 5 alpha, 5 beta, and 5 gamma. Laminins are the substrate on which the leiomyoma cells grow. These tumours have also an increased amount of ECM compared to normal uterine wall, while inhibition of integrin beta1 leads to less motility of the cells (although the adhesion among them is not altered) and to inhibition of cyclinD1 with consequent slowing of proliferation by inhibition of ERK pathway (Malik et al., 2012; Lin et al., 2013).

E-cadherin expression is instead increased in uterine leiomyoma (Tai et al., 2003). In an opposite fashion, both catenins and E-cadherin expression are down regulated in pituitary adenomas (Sano et al., 2004) and meningiomas (Brunner et al., 2006), suggesting that their final mechanisms of action depend on the type of tissue.

Specialization of cell types and division of work

Growth factors and hormones

Growth factors are widely involved in benign tumour formation like breast fibroadenomas and *Phyllodes* tumours. The mRNA levels for EGFR, its ligand RGF, and IGFR1 protein expression were found increased in both fibroadenomas and carcinomas. The mRNA for insulin-like growth factor 2, platelet-derived growth factor A chain, and B chain (c-sis) were instead highly expressed in benign tumours but not so much in carcinomas (Travers et al., 1988; Happerfield et al., 1997). Transforming growth factor β2 was also present, although the levels were lower than in carcinomas (Zheng et al., 2014). An increased risk of breast fibroadenoma formation has been described in carriers of the Val allele, which have a polymorphism at codon 655 of the human epithelial growth factor receptor 2 (HER-2; Zubor et al., 2008). Overexpression of both HER-2 and transforming growth factor alpha induces development of mammary fibroadenomas and papillary adenomas in rats (Davies et al., 1999) further highlighting the role of growth factors in benign tumours.

As in carcinomas, in benign tumours of breast and reproductive organs, hormones play an important role. In fact, the commonest type of benign tumours in women is the uterine leiomyoma, commonly called fibroids, which occur in approximately 70–80% of women, with 30% needing treatment. Uterine leiomyomas start to appear when women become fertile and tend to shrink in both physiological and pharmacological menopause, suggesting a dependence from oestrogen and/or progesterone. A higher number of receptors for both oestrogen and progesterone are present in leiomyoma cells compared to normal myometrium. Activation by both oestradiol and progesterone lead to rapid activation of MAPK and other signalling pathways causing expression of several different growth factors (Fujii et al., 2004; Kim and Sefton, 2012).

The secreting benign tumours

The physiological function of endocrine cells is to produce hormones which are then released in the blood flow to be delivered to the target organs, inducing their effect. Similar to other benign neoplasms, the endocrine cells are very well differentiated and not surprisingly are frequently able to maintain their hormone secretion. However, these tumours escape such regulation by negative feedback mechanisms, resulting in the continuous secretion of an excess of hormones. This leads to a series of symptoms changing from one type of endocrine adenoma to another according to the type of hormone secreted. The prerogative of secreting active hormones is not only of the benign endocrine tumours: malignant carcinomas of endocrine origin can also secrete active hormones, although less commonly. One group of tumours which exemplify the biological and clinical problems raised by these lesions are the gastroenteropancreatic neuroendocrine tumours (Weaver et al., 2016), which can be either benign or malignant, with the benign more likely to be actively secreting hormones. The benign ones can be very small and hence difficult to identify; however, their release of hormone can induce significant and potentially life-threatening syndromes. In the Zollinger–Ellison syndrome abnormally high levels of gastrin are released causing recurrent peptic ulcers, at high risk of perforating, plus causing diarrhoea, steatorrhea (loos of lipids) and malabsorption.

Resources transport and allocation: Blood supply, hypoxia, and metabolism

Necrosis has been found to be associated with malignant tumours and although expression of hypoxia surrogate markers has been detected in benign tumours, these markers (e.g. CAIX and HIF1) have a lower level of expression in benign tumours compared to carcinomas (Couvelard et al., 2005; Couvelard et al., 2008; Madej et al., 2013). However, this is not a general rule: a study on uterine leiomyoma suggests that although not much HIF1 expression is observed, still high levels of hypoxia are present as suggested by oxygen sensor electrodes applied during surgery (Mayer et al., 2008).

Inversely, pseudo hypoxic response (i.e. activation of hypoxic pathways in presence of normal levels of oxygen), has been reported in two benign neuronal tumours: pheochromocytoma, arising from adrenal gland, and paraganglioma, arising from extra-adrenal sympathetic- or parasympathetic-derived chromaffin tissue (Jochmanova et al., 2013). A group of 10 genes, affected by mutations, have been found variably involved in 50% of cases. These genes are grouped into two clusters: the *Pseudohypoxia* cluster (**Table 31.3**), and the kinase receptor signalling and protein translation pathways cluster (Gimenez-Roqueplo et al., 2012). A further pattern is found in the colorectal adenomas, where high levels of HIF1 and hypoxia have been demonstrated: this is probably due to the upper layer of the normal colorectal mucosa being physiologically hypoxic, as demonstrated by HIF1 expression (Greijer et al., 2008). VEGFA and VEGFR2 have been reported to be expressed in different types of benign tumours at higher levels than the correspondent normal tissues as in the female genital tract, where uterine leiomyoma tumour (or fibroids) tissue and the intratumour endothelial cells have been found to carry genetic polymorphism of these two genes (Couvelard et al., 2005; Swelam et al., 2005; Segiet et al., 2015).

No specific connection between benign/malignant lesions and number of intratumour vessels has been observed. Pituitary adenomas and small uterine leiomyomas are less vascularized than their normal tissue counterparts, but as the size of uterine fibroma increases they become more vascularized than the normal myometrium (Fleischer et al., 2008; Di Ieva et al., 2013). The vessels are a mixture of mature and immature (Balinisteanu et al., 2014), and are

Table 31.3 The *Pseudohypoxia* cluster of genes carrying point mutations in *Pheochromocytomas* and *Paragangliomas*

Gene	Function	Consequences of inactivation
Pseudohypoxia subcluster VHL		
VHL	Essential component of the E3 ubiquitin ligases complex which causes proteosomal degradation of hydroxylated HIF1alpha	Hif1alpha remains stable and active even in normoxia
Pseudohypoxia subcluster SDH family		
SDHA	SDHs couple the oxidation of succinate to fumarate in the Krebs cycle. This results in reduction of ubiquitin in the electron transport chain.	Succinate accumulation inhibiting HIF1alpha hydroxylation by prolyl hydroxylase enzyme. HIF1alpha remain therefore stable in presence of oxygen as it is not degraded by E3 ubiquitin ligases complex
SDHB		
SDHC		
SDHD		
SDHAF2	Flavination of a succinate dehydrogenase complex subunit required for activity of the complex	

Source: data from Jochmanova I et al. 'Hypoxia-inducible factor signaling in pheochromocytoma: turning the rudder in the right direction', *Journal of the National Cancer Institute*, Volume 105, Issue 17, pp. 1270–83, Copyright © 2013 Oxford University Press; and Gimenez-Roqueplo AP et al. 'An update on the genetics of paraganglioma, pheochromocytoma, and associated hereditary syndromes', Hormone Metabolic Research, Volume 44, Issue 5, pp. 328–33, Copyright © 2012 Georg Thieme Verlag KG Stuttgart.

also organized in an ordered/'non-random' fashion as established using the Microvascular structure entropy technique, which distinguishes between 'chaotic' and 'non-random' organization of micro anatomic structures (Di Ieva et al., 2013). In an opposing fashion, the benign pancreatic tumours have a higher microvessel density then the malignant ones (Couvelard et al., 2005).

Signalling and gene regulation

The RAS network is one of the most commonly involved in cancer and it includes three main pathways: the RAS/RAF/MEK/ERK—(MAP kinases); the PI3K/AKT/PKB; and the Ral-GEF. In the limited number of studies, the first two are the most commonly involved in benign tumours.

The RAS/RAF/MEK/ ERK–(MAP kinases) pathway

This pathway is mostly involved in controlling cell cycle progression and transcription through Myc, Mad, Max, Fos, and Jun. Is under the control of a series of membrane receptors and in uterine leiomyoma, is regulated by oestrogen and progesterone receptors (Islam et al., 2013). This pathway is also activated in pituitary adenomas (Dworakowska and Grossman, 2009; Kim and Sefton, 2012; Suojun et al., 2012; Borahay et al., 2015), melanocytic nevi (Michaloglou et al., 2008), colon adenoma (Goldstein, 2006), and neurofibroma (Harrisingh and Lloyd, 2004; Harrisingh et al., 2004).

It is not known what triggering the activation of this pathway in most of the pituitary adenomas. Ras mutations (Dworakowska and Grossman, 2009) have been detected in opituitary adenoma while in some colon adenomas BRAF mutations have been described (Goldstein, 2006). In benign nevi a range of mutations for BRAF, HRAS, and NRAS have been detected but no clear-cut difference between malignant and benign lesions has been found. While any Ras-Raf mutation accounted for only 9% of benign Spitz naevi, it accounted for 84% of common benign nevi and 40% of malignant melanoma. raising the hypothesis that other alterations could be occurring dictating the benign versus malignant development of the lesions (Indsto et al., 2007; Michaloglou et al., 2008).

The PI3K/AKT/PKB

This pathway controls a broad range of activities ranging from cell cycle, transcription, cell survival, growth and migration. It is activated in uterine leiomyoma by progesterone and oestrogen (Kim and Sefton, 2012; Islam et al., 2013; Borahay et al., 2015). In pituitary adenomas this pathway is also active but the mechanisms are unknown (Dworakowska and Grossman, 2009), while in neurofibroma its activation appears to be due to aberrant activation of G protein signalling (Woods et al., 2002; Suojun et al., 2012).

Finally, no involvement of the Ral/GEF pathway has so far been described in benign tumours.

Other signalling pathways involved in benign tumours include Jak/STAT, the WNT/beta-catenin, the delta/notch, the hedgehog, and the TGF-beta/SMAD.

The Jak-STAT pathway is active in hepatocellular adenoma and is associated with an inflammatory phenotype and with mutations in IL6ST, STAT3, and GNAS (Nault et al., 2013; Calderaro et al., 2016). In colorectal adenomas instead, it is associated with a high level of Leptin phosphorylated at Tyr 1141 which activates this pathway (Uchiyama et al., 2011).

The WNT beta-catenin pathway is also active in different types of benign neoplasms. In uterine fibroids mutations of MED12 leads to its increased activity (Borahay et al., 2015). In colon adenoma the presence of a truncated adenomatosis polyposis coli (APC) protein leads to abnormal activity of the WNT/beta-catenin pathway with increased proliferation and diminished apoptosis (Goldstein, 2006) while in hepatocellular adenoma its activity is due to beta-catenin mutations (Nault et al., 2013). Finally, in patients affected by tuberous sclerosis mutation in either *TSC1* or *TSC2* genes can lead to the development of astrocytoma with activated WNT/beta-catenin pathways (Jozwiak et al., 2007).

The hedgehog pathway is centred on three genes: sonic hedgehog, desert hedgehog, and indian hedgehog. Its involvement in human tumours leads to increase in the expression of cyclins D1 and B1 plus an antiapoptotic activity. Not widely involved in malignancies, it is surprising therefore that its activation is relatively common in benign tumours. In colon adenoma sonic hedgehog is overexpressed alongside the antiapoptotic protein survivin (Parfitt and Driman, 2007) while in an adenoma mouse model indian hedgehog is required for colonic adenoma formation (Buller et al., 2015). It is also involved in pituitary adenomas (Yavropoulou et al., 2015), adrenal adenoma (Werminghaus et al., 2014), and meningiomas (Laurendeau et al., 2010).

Finally, the delta/notch pathway is involved in pituitary (Yavropoulou et al., 2015) and ovarian adenoma (Hopfer et al., 2005), while TGFbeta/SMAD is active in uterine and hepatocellular adenoma (Nault et al., 2013).

Benign tumours progression to cancer and the borderline tumours

Although the majority of the benign tumours do not have any particular risk of malignant progression, some carry a high risk of progression: the paradigmatic type of tumour progressing from benign to malignant is the colon adenoma: in these lesions the progression sequence has been well characterized with the discovery of the accumulation of genetic damages leading to the malignant transformation. Described formally by Morson (Morson, 1974) in respect to progressive morphological changes, subsequently the underlying basic genetic changes were described by Fearon and Vogelstein (Fearon and Vogelstein, 1990).

Three are the main types of colonic adenomas described by Morson: tubular adenoma (75% of all the adenomas), tubulovillous adenoma (15%), and villous adenoma (10%). The latter has the greater incidence of progression into adenocarcinoma: 41% of the cases versus 5% of the tubular and 22% of the tubular villous (Morson, 1974).

Progresses in molecular biology have subsequently allowed the understanding of the underlying genetic defects accumulating with the time and driving these benign lesions into malignant phenotype, described in the classic paper by Fearon and Vogelstein (Fearon and Vogelstein, 1990; see Fig. 31.3). The colorectal neoplasia proved to be an excellent model for a number of reasons: the progression from benign to malignant is common in this organ surgery is widely practised at all the stages of tumour progression. In this model a step-wise accumulation of alterations, by either activation of an oncogene or inactivation of a tumour suppressor occurs and they broadly match the progressive morphological changes. The most relevant alterations identified in the Fearon–Vogelstein model are five genetic changes.

The first one is the loss or inactivation of the APC gene on chromosome 5 leading to hyperproliferative epithelium. The second one is DNA hypomethylation, contributing to genetic instability found in about one-third of the DNA regions investigated in adenomas. The third is Ras gene mutations which confer oncogenic properties to the cells. This mutation has been detected in 50% of colorectal carcinoma and in the same proportion of adenomas with a diameter larger than 1 cm but only in 10% of smaller adenomas. It is therefore considered as pivotal mutation in the progression from an earlier low risk to an intermediate higher risk adenoma. The development into late adenoma is instead associated with loss or inactivation of the tumour suppressor gene DCC (Deleted in Colon Cancer) encoding a netrin1 receptor located on chromosome 18. Fifth and last one, situated at the transition between very high-risk adenoma and cancer is mutation or loss of p53 on chromosome 17. Eventually a number of further damages will cause metastatic disease to develop.

The authors of this model stresses that this is obviously an oversimplification of the events and the progression from normal mucosa to adenoma and eventually cancer is a continuous process. In addition, although the different discrete genetic changes are positioned at different steps of this progression, this is according to the likelihood of occurrence, but this order of changes does not necessarily occur in all the patients. It is the progressive accumulation of changes, rather than their orders that are more important for the development of colorectal cancer highlighting the importance of the number rather than the order of alterations which is always higher in cancer compared to precancerous steps (Fearon and Vogelstein, 1990).

Conclusion

Our knowledge of benign tumour biology is very limited, still they appear to share with the malignant counterpart much more pathogenetic mechanisms than previously thought but intriguingly most of the benign tumours do not bring an increased risk of malignant progression. It is therefore likely that a better knowledge of these neoplastic processes, together with a system biology approach looking at interaction between all the genetic and epigenetic lesions present, rather than to examining alterations one by one, will shed light on key differences between self-limiting benign lesions and malignant tumours.

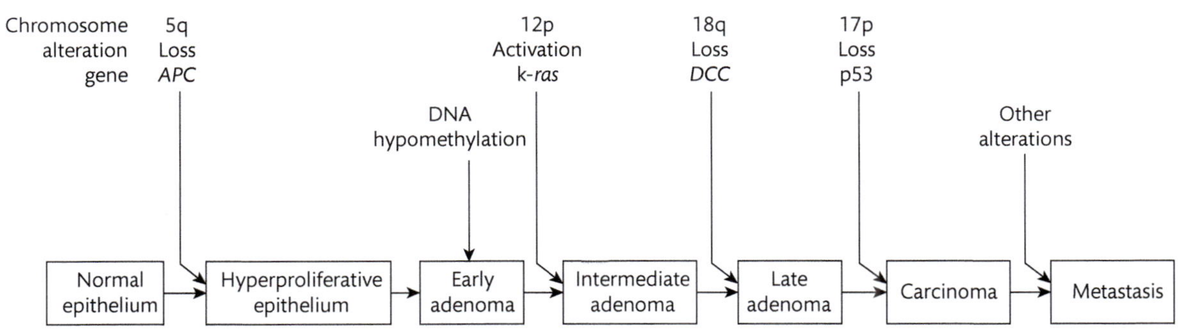

Fig. 31.3 The Fearon–Vogelstein model for colorectal tumorigenesis. The succession of the genetic and epigenetic damages leading from normal mucosa to benign adenoma and eventually carcinoma is represented.

TAKE-HOME MESSAGE

- Although the morphological features of the being tumours are very well described, very little is known about their biology.
- Despite having genetic mutations no genetic instability is present in most of the benign tumours.

OPEN QUESTIONS

- Only a very few and incomplete aspects of benign tumours biology are known. It is still an almost unexplored field.
- Despite sharing with malignant tumours some basic genetic defects, still most of the benign tumours do not have a risk to progress to malignancy higher than that of normal tissue. The reason for this is unknown.

FURTHER READING

Islam, M. S., Protic, O., Stortoni, P., et al. (2013). Complex networks of multiple factors in the pathogenesis of uterine leiomyoma. *Fertility Sterility*, **100**(1), 178–93.

Qin, H., Bao, D., Tong, X., Hu, Q., Sun, Q., & Huang, X. (2016). The role of stem cells in benign tumours. *Tumour Biol*, **37**(12), 15349–57.

REFERENCES

Abe, N., Watanabe, T., Nakashima, M., et al. (2001). Quantitative analysis of telomerase activity: a potential diagnostic tool for colorectal carcinoma. *Hepatogastroenterology*, **48**, 692–5.

Agnihotri, S., Gugel, I., Remke, M., et al. (2014). Gene-expression profiling elucidates molecular signaling networks that can be therapeutically targeted in vestibular schwannoma. *J Neurosurg*, **121**, 1434–45.

Arco, A., Favaloro, A., Gioffre, M., et al. (2012). Sarcoglycans in the normal and pathological breast tissue of humans: an immunohistochemical and molecular study. *Cells Tissues Organs*, **195**, 550–62.

Balinisteanu, B., Cimpean, A. M., Melnic, E., Coculescu, M., Ceausu, R. A., & Raica, M. (2014). Crosstalk between tumor blood vessels heterogeneity and hormonal profile of pituitary adenomas: evidence and controversies. *Anticancer Res*, **34**, 5413–20.

Benmaamar, R. (2013). Non-NF2 mutations in meningioma. *Lancet Oncol*, **14**, e91.

Billard, L. M., Magdinier, F., Lenoir, G. M., Frappart, L., & Dante, R. (2002). MeCP2 and MBD2 expression during normal and pathological growth of the human mammary gland. *Oncogene*, **21**, 2704–12.

Boltze, C., Mundschenk, J., Unger, N., et al. (2003). Expression profile of the telomeric complex discriminates between benign and malignant pheochromocytoma. *J Clin Endocrinol Metab*, **88**, 4280–6.

Borahay, M. A., AL-Hendy, A., Kilic, G. S., & Boehning, D. (2015). Signaling pathways in leiomyoma: understanding pathobiology and implications for therapy. *Mol Med*, **21**, 242–56.

Brastianos, P. K., Horowitz, P. M., Santagata, S., et al. (2013). Genomic sequencing of meningiomas identifies oncogenic SMO and AKT1 mutations. *Nat Genet*, **45**, 285–9.

Brunner, E. C., Romeike, B. F., Jung, M., Comtesse, N., & Meese, E. (2006). Altered expression of beta-catenin/E-cadherin in meningiomas. *Histopathology*, **49**, 178–87.

Buller, N. V., Rosekrans, S. L., Metcalfe, C., et al. (2015). Stromal Indian hedgehog signaling is required for intestinal adenoma formation in mice. *Gastroenterology*, **148**, 170–80 e6.

Calderaro, J., Nault, J. C., Balabaud, C., et al. (2016). Inflammatory hepatocellular adenomas developed in the setting of chronic liver disease and cirrhosis. *Mod Pathol*, **29**, 43–50.

Clark, V. E., Erson-Omay, E. Z., Serin, A., et al. (2013). Genomic analysis of non-NF2 meningiomas reveals mutations in TRAF7, KLF4, AKT1, and SMO. *Science*, **339**, 1077–80.

Couvelard, A., Deschamps, L., Rebours, V., et al. (2008). Overexpression of the oxygen sensors PHD-1, PHD-2, PHD-3, and FIH is associated with tumor aggressiveness in pancreatic endocrine tumors. *Clin Cancer Res*, **14**, 6634–9.

Couvelard, A., O'Toole, D., Turley, H., et al. (2005). Microvascular density and hypoxia-inducible factor pathway in pancreatic endocrine tumours: negative correlation of microvascular density and VEGF expression with tumour progression. *Br J Cancer*, **92**, 94–101.

Damjanovich, L., Fulop, B., Adany, R., & Nemes, Z. (1997). Integrin expression on normal and neoplastic human breast epithelium. *Acta Chir Hung*, **36**, 69–71.

Davies, B. R., Platt-Higgins, A. M., Schmidt, G., & Rudland, P. S. (1999). Development of hyperplasias, preneoplasias, and mammary tumors in MMTV-c-erbB-2 and MMTV-TGFalpha transgenic rats. *Am J Pathol*, **155**, 303–14.

Di Ieva, A., Weckman, A., Di Michele, J., et al. (2013). Microvascular morphometrics of the hypophysis and pituitary tumors: from bench to operating theatre. *Microvasc Res*, **89**, 7–14.

Dworakowska, D. & Grossman, A. B. (2009). The pathophysiology of pituitary adenomas. *Best Pract Res Clin Endocrinol Metab*, **23**, 525–41.

Fearon, E. R. & Vogelstein, B. (1990). A genetic model for colorectal tumorigenesis. *Cell*, **61**, 759–67.

Fleischer, R., Weston, G. C., Vollenhoven, B. J., & Rogers, P. A. (2008). Pathophysiology of fibroid disease: angiogenesis and regulation of smooth muscle proliferation. *Best Pract Res Clin Obstet Gynaecol*, **22**, 603–14.

Fujii, S., Suzuki, A., Matsumura, N., et al. (2004). Fibroids: basic science and etiology. *International Congress Series*, **1266**, 183–90.

Gimenez-Roqueplo, A. P., Dahia, P. L., & Robledo, M. (2012). An update on the genetics of paraganglioma, pheochromocytoma, and associated hereditary syndromes. *Horm Metab Res*, **44**, 328–33.

Goldstein, N. S. 2006. Serrated pathway and APC (conventional)-type colorectal polyps: molecular-morphologic correlations, genetic pathways, and implications for classification. *Am J Clin Pathol*, **125**, 146–53.

Greijer, A. E., Delis-Van Diemen, P. M., Fijneman, R. J., et al. (2008). Presence of HIF-1 and related genes in normal mucosa, adenomas and carcinomas of the colorectum. *Virchows Arch*, **452**, 535–44.

Guray, M. & Sahin, A. A. (2006). Benign breast diseases: classification, diagnosis, and management. *Oncologist*, **11**, 435–49.

Gurlek, A., Karavitaki, N., Ansorge, O., & Wass, J. A. (2007). What are the markers of aggressiveness in prolactinomas? Changes in cell biology, extracellular matrix components, angiogenesis and genetics. *Eur J Endocrinol*, **156**, 143–53.

Happerfield, L. C., Miles, D. W., Barnes, D. M., Thomsen, L. L., Smith, P., & Hanby, A. (1997). The localization of the insulin-like growth factor receptor 1 (IGFR-1) in benign and malignant breast tissue. *J Pathol*, **183**, 412–17.

Harrisingh, M. C. & Lloyd, A. C. (2004). Ras/Raf/ERK signalling and NF1. *Cell Cycle*, **3**, 1255–8.

Harrisingh, M. C., Perez-Nadales, E., Parkinson, D. B., Malcolm, D. S., Mudge, A. W., & Lloyd, A. C. (2004). The Ras/Raf/ERK signalling pathway drives Schwann cell dedifferentiation. *Embo J*, **23**, 3061–71.

Hassounah, M., Lach, B., Allam, A., et al. (2005). Benign tumors from the human nervous system express high levels of survivin and are resistant to spontaneous and radiation-induced apoptosis. *J Neurooncol*, **72**, 203–8.

Hopfer, O., Zwahlen, D., Fey, M. F., & Aebi, S. (2005). The Notch pathway in ovarian carcinomas and adenomas. *Br J Cancer*, **93**, 709–18.

Indsto, J. O., Kumar, S., Wang, L., Crotty, K. A., Arbuckle, S. M., & Mann, G. J. (2007). Low prevalence of RAS-RAF-activating mutations in Spitz melanocytic nevi compared with other melanocytic lesions. *J Cutan Pathol*, **34**, 448–55.

Islam, M. S., Protic, O., Stortoni, P., et al. (2013). Complex networks of multiple factors in the pathogenesis of uterine leiomyoma. *Fertil Steril*, **100**, 178–93.

Jochmanova, I., Yang, C., Zhuang, Z., & Pacak, K. (2013). Hypoxia-inducible factor signaling in pheochromocytoma: turning the rudder in the right direction. *J Natl Cancer Inst*, **105**, 1270–83.

Jozwiak, J., Kotulska, K., Grajkowska, W., et al. (2007). Upregulation of the WNT pathway in tuberous sclerosis-associated subependymal giant cell astrocytomas. *Brain Dev*, **29**, 273–80.

Karami-Tehrani, F., Fallahian, F., & Atri, M. (2012). Expression of cGMP-dependent protein kinase, PKGIalpha, PKGIbeta, and PKGII in malignant and benign breast tumors. *Tumour Biol*, **33**, 1927–32.

Kim, J. J. & Sefton, E. C. (2012). The role of progesterone signaling in the pathogenesis of uterine leiomyoma. *Mol Cell Endocrinol*, **358**, 223–31.

Kulis, M. & Esteller, M. (2010). DNA methylation and cancer. *Adv Genet*, **70**, 27–56.

Lae, M., Vincent-Salomon, A., Savignoni, A., et al. (2007). Phyllodes tumors of the breast segregate in two groups according to genetic criteria. *Mod Pathol*, **20**, 435–44.

Laurendeau, I., Ferrer, M., Garrido, D., et al. (2010). Gene expression profiling of the hedgehog signaling pathway in human meningiomas. *Mol Med*, **16**, 262–70.

Lin, T. H., Liu, H. H., Tsai, T. H., et al. (2013). CCL2 increases alphavbeta3 integrin expression and subsequently promotes prostate cancer migration. *Biochim Biophys Acta*, **1830**, 4917–27.

Madej, J. A., Madej, J. P., Dziegiel, P., Pula, B., & Nowak, M. (2013). Expression of hypoxia-inducible factor-1alpha and vascular density in mammary adenomas and adenocarcinomas in bitches. *Acta Vet Scand*, **55**, 73.

Malik, M., Segars, J., & Catherino, W. H. (2012). Integrin beta1 regulates leiomyoma cytoskeletal integrity and growth. *Matrix Biol*, **31**, 389–97.

Marino-Enriquez, A. & Fletcher, C. D. (2014). Shouldn't we care about the biology of benign tumours? *Nat Rev Cancer*, **14**, 701–2.

Mayer, A., Hockel, M., Wree, A., Leo, C., Horn, L. C., & Vaupel, P. (2008). Lack of hypoxic response in uterine leiomyomas despite severe tissue hypoxia. *Cancer Res*, **68**, 4719–26.

Mayer, I. A. & Arteaga, C. L. (2016). The PI3K/AKT pathway as a target for cancer treatment. *Annu Rev Med*, **67**, 11–28.

Michaloglou, C., Vredeveld, L. C., Mooi, W. J., & Peeper, D. S. (2008). BRAF(E600) in benign and malignant human tumours. *Oncogene*, **27**, 877–95.

Mittal, P., Shin, Y. H., Yatsenko, S. A., Castro, C. A., Surti, U., & Rajkovic, A. (2015). Med12 gain-of-function mutation causes leiomyomas and genomic instability. *J Clin Invest*, **125**, 3280–4.

Morson, B. (1974). President's address. The polyp-cancer sequence in the large bowel. *Proc R Soc Med*, **67**, 451–7.

Nault, J. C., Bioulac-Sage, P., & Zucman-Rossi, J. (2013). Hepatocellular benign tumors—from molecular classification to personalized clinical care. *Gastroenterology*, **144**, 888–902.

Neel, V. A., Todorova, K., Wang, J., et al. (2016). Sustained Akt activity is required to maintain cell viability in seborrheic keratosis, a benign epithelial tumor. *J Invest Dermatol*, **136**, 696–705.

Osawa, N., Onoda, N., Kawajiri, H., et al. (2009). Diagnosis of parathyroid carcinoma using immunohistochemical staining against hTERT. *Int J Mol Med*, **24**, 733–41.

Paez-Pereda, M., Kuchenbauer, F., Arzt, E., & Stalla, G. K. (2005). Regulation of pituitary hormones and cell proliferation by components of the extracellular matrix. *Braz J Med Biol Res*, **38**, 1487–94.

Parfitt, J. R. & Driman, D. K. (2007). Survivin and hedgehog protein expression in serrated colorectal polyps: an immunohistochemical study. *Hum Pathol*, **38**, 710–7.

Plewka, A., Plewka, D., Madej, P., et al. 2011). Processes of apoptosis and cell proliferation in uterine myomas originating from reproductive and perimenopausal women. *Folia Histochem Cytobiol*, **49**, 398–404.

Qin, H., Bao, D., Tong, X., Hu, Q., Sun, G., & Huang, X. (2016). The role of stem cells in benign tumors. *Tumour Biol*, **37**, 15349–57.

Ragel, B. T., Jensen, R. L., Gillespie, D. L., Prescott, S. M., & Couldwell, W. T. (2006). Celecoxib inhibits meningioma tumor growth in a mouse xenograft model. *Cancer*, **109**, 588–97.

Rechache, N. S., Wang, Y., Stevenson, H. S., et al. (2012). DNA methylation profiling identifies global methylation differences and markers of adrenocortical tumors. *J Clin Endocrinol Metab*, **97**, E1004–13.

RightDiagnosis.com. *Prevalence Statistics for Types of Benign Tumor*. Available at: http://www.rightdiagnosis.com/b/benign/prevalence-types.htm

Rohan, T. E., LI, S. Q., Hartwick, R., & Kandel, R. A. (2006). P53 Alterations and protein accumulation in benign breast tissue and breast cancer risk: a cohort study. *Cancer Epidemiol Biomarkers Prev*, **15**, 1316–23.

Rosai, J. (2011). *Rosai and Ackerman's Surgical Pathology*. Edinburgh: Mosby.

Sano, T., Rong, Q. Z., Kagawa, N., & Yamada, S. (2004). Down-regulation of E-cadherin and catenins in human pituitary growth hormone-producing adenomas. *Front Horm Res*, **32**, 127–32.

Segiet, O. A., Michalski, M., Brzozowa-Zasada, M., et al. (2015). Angiogenesis in primary hyperparathyroidism. *Ann Diagn Pathol*, **19**, 91–8.

Suojun, Z., Feng, W., Dongsheng, G., & Ting, L. (2012). Targeting Raf/MEK/ERK pathway in pituitary adenomas. *Eur J Cancer*, **48**, 389–95.

Swelam, W., Ida-Yonemochi, H., Maruyama, S., Ohshiro, K., Cheng, J., & Saku, T. (2005). Vascular endothelial growth factor in salivary pleomorphic adenomas: one of the reasons for their poorly vascularized stroma. *Virchows Arch*, **446**, 653–62.

Tai, C. T., Lin, W. C., Chang, W. C., Chiu, T. H., & Chen, G. T. (2003). Classical cadherin and catenin expression in normal myometrial tissues and uterine leiomyomas. *Mol Reprod Dev*, **64**, 172–8.

Takehara, K., Miyamoto, K., Kawakami, Y., et al. (2009). Epigenetic alteration of *BRCA1* in human ovarian tumors. *J Clin Oncol*, **27**, e16532.

Taylor, C. V., Letarte, M., & Lye, S. J. (1996). The expression of integrins and cadherins in normal human uterus and uterine leiomyomas. *Am J Obstet Gynecol*, **175**, 411–19.

Travers, M. T., Barrett-Lee, P. J., Berger, U., et al. (1988). Growth factor expression in normal, benign, and malignant breast tissue. *Br Med J (Clin Res Ed)*, **296**, 1621–4.

Uchiyama, T., Takahashi, H., Sugiyama, M., et al. (2011). Leptin receptor is involved in STAT3 activation in human colorectal adenoma. *Cancer Sci*, **102**, 367–72.

Umehara, N., Ozaki, T., Sugihara, S., et al. (2004). Influence of telomerase activity on bone and soft tissue tumors. *J Cancer Res Clin Oncol*, **130**, 411–16.

Volpe, J. P. (1988). Genetic instability of cancer. Why a metastatic tumor is unstable and a benign tumor is stable. *Cancer Genet Cytogenet*, **34**, 125–34.

Wang, Y., Kowalski, J., Tsai, H. L., et al. (2008). Differentiating alternative splice variant patterns of human telomerase reverse transcriptase in thyroid neoplasms. *Thyroid*, **18**, 1055–63.

Weaver, A., Weetman, A. P., Grimm, O., et al. (2016). Endocrine cancers. In: Kerr, D. J., Haller, D. G., Van De Velde, C. J. H., & Baumann, M. (eds) *Oxford Textbook of Oncology*, Chapter 58. Oxford: Oxford University Press.

Werminghaus, P., Haase, M., Hornsby, P. J., et al. (2014). Hedgehog-signaling is upregulated in non-producing human adrenal adenomas and antagonism of hedgehog-signaling inhibits proliferation of NCI-H295R cells and an immortalized primary human adrenal cell line. *J Steroid Biochem Mol Biol*, **139**, 7–15.

Woods, S. A., Marmor, E., Feldkamp, M., et al. (2002). Aberrant G protein signaling in nervous system tumors. *J Neurosurg*, **97**, 627–42.

Xu, Q., Yuan, X., Tunici, P., et al. (2009). Isolation of tumour stem-like cells from benign tumours. *Br J Cancer*, **101**, 303–11.

Yano, Y., Yoshida, K., Osaki, A., et al. (2002). Expression and distribution of human telomerase catalytic component, hTERT, in human breast tissues. *Anticancer Res*, **22**, 4101–7.

Yavropoulou, M. P., Maladaki, A., & Yovos, J. G. (2015). The role of Notch and Hedgehog signaling pathways in pituitary development and pathogenesis of pituitary adenomas. *Hormones (Athens)*, **14**, 5–18.

Yoshida, T., Matsumoto, E., Hanamura, N., et al. (1997). Co-expression of tenascin and fibronectin in epithelial and stromal cells of benign lesions and ductal carcinomas in the human breast. *J Pathol*, **182**, 421–8.

Zheng, R., Wang, J., Wu, Q., et al. (2014). Expression of ALDH1 and TGFbeta2 in benign and malignant breast tumors and their prognostic implications. *Int J Clin Exp Pathol*, **7**, 4173–83.

Zheng, W. Q., Looi, L. M., & Cheah, P. L. (1999). A comparison of the pattern of cathepsin-D expression in fibroadenoma, fibrocystic disease, preinvasive and invasive ductal breast carcinoma. *Pathology*, **31**, 247–51.

Zubor, P., Kajo, K., Stanclova, A., et al. (2008). Human epithelial growth factor receptor 2[Ile655Val] polymorphism and risk of breast fibroadenoma. *Eur J Cancer Prev*, **17**, 33–8.

32

Conclusions: Cancer biology, a moveable feast

David J. Kerr, Francesco Pezzella, and Mahvash Tavassoli

Introduction

Tumours include many of different types of abnormal growths. The main separating between benign and malignant, is based on the ability of a neoplastic lesion to produce metastases during the natural course of the disease. Malignant tumours are a heterogeneous group of lesions, which are brought together under the generic name of cancer and its synonymous. Although we recognize a tumour as malignant intuitively, to identify the basic characteristics common to all the types of cancers is challenging and may not be possible. Delivering the 1925 Erasmus Wilson Lectures (Nicholson, 1926). G. W. Nicholson, Professor of Pathology at the Guy's Hospital in London, stated '. . . that is impossible to define a tumour. Wherever we look we see that tumours exhibit no differences in kind but only differences in degree-and these often the slightest-from the other tissues of the body. I have tried for years to formulate a definition, but have failed. Others have been bolder. I will not weary you with their definitions, every one of which breaks down at one or more points' (quoted in Willis, 1960).

The hallmarks of cancer

Still, it has been recognized that try to identify and describe the main biological characteristics which define what is cancer, is not a sterile exercise but it is very helpful. These features on the one hand describe in a concise way the 'state of the art' of our knowledge, on the other hand are an excellent tool for teaching and learning about cancer, allowing to easily access the basic notion necessary to start to understand such a complex subject. It is here the importance of the work of Hannahan and Weinberg. In the essay published in 2000 (Hanahan and Weinberg, 2000), which immediately acquired the status of classic, and in the second update text published in 2011 (Hanahan and Weinberg, 2011), the two authors have framed the main features of cancer biology into a series of hallmarks, or characteristics, providing the kind of tool discussed earlier.

Keeping in mind that knowledge in this field issue is continuosly growing, and changes and updates are necessary, we have followed their guidelines for outlining which are the basic biological features of cancer. Other authors in the past have defined what they identified as cancer-typical features; some of these are reported in Table 32.1.

There are threads going through the older and the contemporary authors which highlights the importance of this type of exercise in identifying the fundamental aspects of cancer biology through the years, for example, the relevance of energy metabolism (Table 32.1 and Table 32.2). Understandably the model of Hannahan and Weinberg comes under scrutiny with pros and cons debated (Lazebnik, 2010; Floor et al., 2012; Sonnenschein and Soto, 2013; Horne et al., 2015), although the discussion is more about the contents, which are bound to evolve and change—and the form used to represent them, rather than the concept of synthesis. In Table 32.2 we present an update 'state of the art' list.

Cancer biology and cancer care

Malignant tumours are a heterogeneous collection of many different types of disease. As such doctors treating cancer patients have been relying on the knowledge of cancer biology in order to identify, as accurately as possible, discrete types of tumours each amenable of its own therapeutic approach (Song et al., 2015). The first steps we take when seeing a new a patient, are: to establish from which organ the lesion is from and what are the histopathological features, including from which tissue, within the organ, the tumour is coming from and whether is a primary or rather a metastatic lesion. As in every organ there are always many types of cells, different tumours can raise from the same organ: histopathology allow us, by what the cells looks like, to establish which, among the of the possible tumours arising form that organ, we are dealing with. For example, a tumour from the breast can be proven by histology to originate from the stroma of the breast (i.e. a sarcoma), rather than from the glandular epithelium, which would have been a carcinoma. The first problem encountered by this approach is that some tumours can present a rather homogeneous type of morphology although being quite different entity and even arising from different tissues: this can occur because tumours can become very poorly differentiated and therefore a carcinoma from the lung can have very poorly differentiated aspect indistinguishable from that of an aggressive lymphoma originating form B lymphocytes.

Table 32.1 Some of the biological characteristics of cancer described through the years before 'the hallmarks of cancer'

Nicholson 1925	Warburg 1930	Willis 1947, 1960	Berenblum I. 1962 (Florey 1962)
Impossible to define			
	Reprogramming energy metabolism High glycolysis in presence of oxygen		Reprogramming energy metabolism High glycolysis in presence of oxygen
		An abnormal growth of tissue	Growth is progressive (i.e. without limits)
			Cancer stem cells maintain the tumour growing
		Growth in excess of and uncoordinated with that of normal tissue	Tumour growth is autonomous
		Excessive growth persisting after cessation of the stimuli which evoked the change	
		Indefinitely progressing	
			Alteration of the metabolic homeostasis
			Neoplastic changes are irreversible
			Malignant cells produce metastases
			Cancer cells can be dormant

Source: data from Nicholson GW *'The nature of tumour formation'*, Erasmus Wilson Lectures, Heller, Cambridge, UK, Copyright © 1926; Warberg O. *'The metabolism of tumours'*. Nguyen, U. B. D. B. T., Ed. EnCognitive, Copyright © 1930; Willis RA *'Pathology of tumours'*, p. 992, Butterworths, London, UK, Copyright © 1947; Willis RA, *'Pathology of Tumours'*, Butterworths, London, UK, Copyright © 1960; and Berenblum I 'The nature of tumour growth' pp 528–50 in Florey H. (Ed.) *Lloyd-Luke, General pathology*, Medical Books Lt, London, UK, Copyright © 1962.

As the depth of information which we can obtain from cancer biology increases, two remaining important questions for the clinician and the patient which are the prognostic and predictive markers present. The prognosis inform us at the point of the pathological tissue analysis what are the chances of survival, while, more importantly, the predictive factors tell us which type of treatment is 'predicted' to be the best for that particular patient.

A problem in achieving this goal is that, although cancer is caused by genetic damages, its behaviour is dictated by the phenotype of the cancer cells. But between the genoma and the final phenotypes other steps are present that can strongly affect the latter. Already in 1909 it was observed that inbred isogenic bean plants growing in a tightly controlled environment produced bean pod of different dimensions (Johannsen, 1909; Vidal et al., 2011). This observation has been widely confirmed: in complex organisms, as differences have been found among human genetically identical twins, and in very simple organisms such as bacterial or yeast cells, with identical genoma, and growing in comparable environment, can still express

Table 32.2 Update of the definition of characteristics of the cancer since 2000

Hallmarks of cancer 2000 Hanahan and Weinberg (2000)	Hallmarks of cancer update 2011 Hanahan and Weinberg (2011)	Update 2017 Pezzella and Winkler (2018)
Characteristics		
Self-sufficiency in signal growth*	Sustaining proliferative signal#	
Insensitivity to antigrowth signals*	Evading growth suppressors#	
Evading apoptosis*	Resisting cell death#	
Limitless replicative potential*	Enabling replicative immortality#	
Sustained angiogenesis*	Inducing angiogenesis*	Exploitation of blood vessels, pre-existing and/or newly formed#
Tissue invasion and metastasis*	Activating invasion and metastases#	
	Reprogramming energy metabolism#	
	Evading immune destruction#	
Enabling characteristics		
	Genome instability and mutations#	
	Tumour promoting inflammation#	

*out of date capabilities, #update capabilities.

Source: data from Hanahan D and Weinberg RA, 'The hallmarks of cancer', *Cell*, Volume 100, Issue 1, pp. 57–70, Copyright © 2000; Hanahan D and Weinberg RA, 'Hallmarks of cancer: the next generation', *Cell*, 2011; Volume 144, Issue 5, pp. 646–74, Copyright © 2011; and Pezzella F and Winkler F, 'Tumors and Blood Vessel Interactions: A Changing Hallmark of Cancer', in Boffetta P et al. (Eds.), *Encyclopedia of Cancer*, Third Edition, Elsevier, Copyright © 2018.

different subsets of transcripts and gene products at any given moment (Vidal et al., 2011).

We summarize below the best-known among the factors that are involved in the formation of a phenotype from a genotype (**Fig. 32.1**).

Genome. This is the complete set of DNA strands coding for an organism. Genomic medicine aims to utilize information concerning the status of the genoma sequence to make a correct diagnosis and/or a treatment plan. One example is the presence of mutations in the sequence of the epidermal growth factor receptor in lung cancer patients or the occurrence of the translocation between chromosomes 9 and 22 in chronic myeloid leukaemia. In the latter case one of the two abnormal chromosomes was the first genetic alteration described in tumours, in 1960, and is best known as 'Philadelphia chromosome' (Nowell and Hungerford, 1960).

Epigenome. This term was coined in 1947 by C. H. Waddington. It is defined as the branch of biology which deals with the modification other than genetics of the genoma, that lead to changes in the final phenotype (Hickman et al., 2004). It includes different chemical modifications at DNA levels affecting the transcription but leaving the DNA sequence unchanged.

Transcriptomics-I mRNAs. One of the original tenets of molecular biology is that one DNA sequence will produce one RNA transcript. However, post-translational modification can lead to one sequence of mRNA being modified into final different mRNAs each coding for a different variant protein. Therefore, the analysis of which mRNAs are present, and to what amount, allow us a further insight in the status of a cell as two cells with the same genome can have some qualitative differences in their transcriptomes, alongside the quantitative ones. These differences are usually subtle in normal cells but can be more evident in dysregulated cancer cells. Application of transcriptome to cancer has allowed so far to divide tumours into clinically meaningful subtype like, for example, large B-cell lymphomas are currently classified as of 'germinal centre' or 'non-germinal centre' types based on their transcriptional profile (Alizadeh et al., 2000).

Transcriptomics-II, the non-coding RNAs. In recent years it has been well established that approximately 80% of the transcribed genome codes for RNAs that do not translate into protein. The number of newly discovered non-coding RNAs is fast growing and they appears to be fundamental in regulating transcription and translation of the coding mRNAs. Some, like the long non-coding RNA, can also affect the behaviour of proteins and DNA. Therefore they add a third layer of complexity in dictating the final cancer phenotype.

Proteomics. This techniques maps which proteins are present in a cell or in the extracellular tissue. It represents the total content of proteins of a tissue. Understanding proteins is the ultimate goal as they build up organisms and carry on all the functions. Proteomics alone however is not enough: we need also to know where, in the cell, the proteins are, as the same protein can have different functions in different cell compartments. Immunohistochemistry, or the more complex technique allowing to perform proteomics on subcellular fraction allow us to answer these question.

Further changes affecting proteins (further epigenetics). The official definition of epigenetics is that of 'epigenetic' changes to the DNA, leading to changes in the phenotype. However, many types of interaction can still be affected by protein changes outside the realm of genetics. Such biochemical changes can affect how a protein will perform, the best-known example is phosphorylation status, which deeply affects the behaviour of a protein. As any protein can shape into different conformational status or link with a variety of other

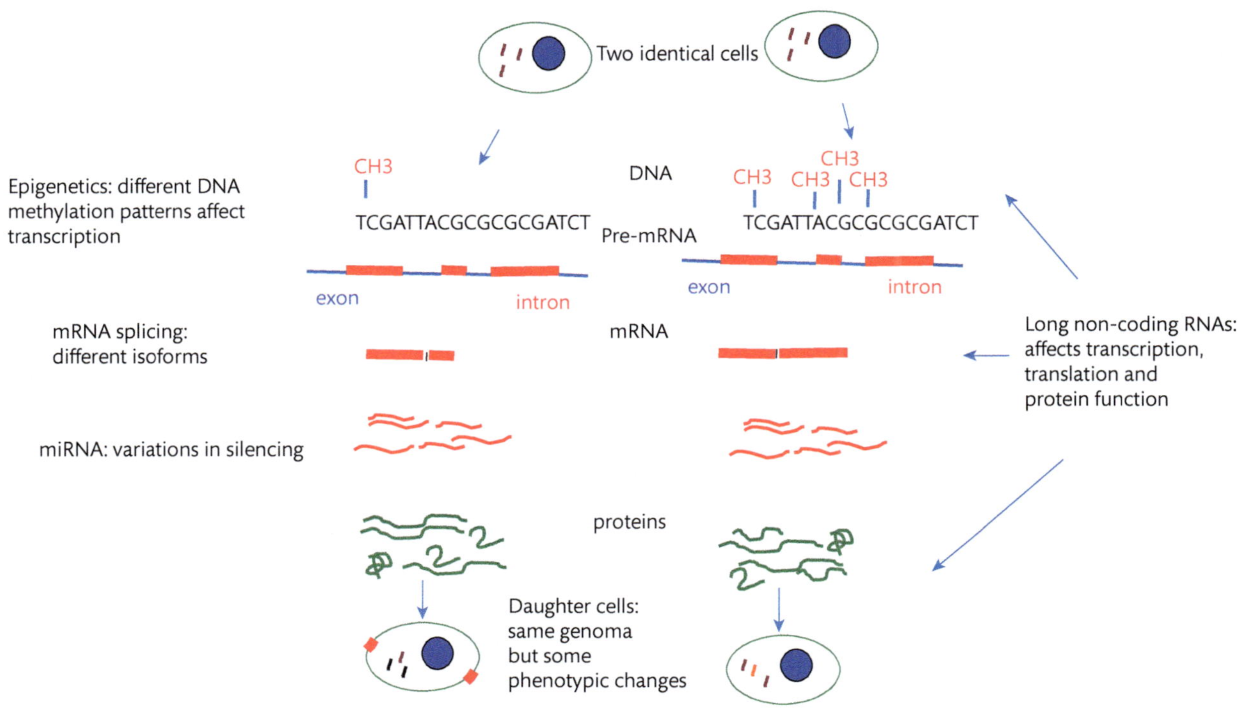

Fig. 32.1 A schematic representation of how identical cells, with the same genoma, can achieve differences in their phenotypes.

proteins, lipids, sugars, nucleic acids, or other chemicals, this implies that the simple mapping of the protein presents in a tissue still does not provide a great deal of the functional data

How shall we use these data?

In order to understand cell physiology, cell pathology, and eventually, the patient's disease, we have therefore to make sense of increasingly large, complex, and qualitative data sets. For the clinician, the goal is to better understand the patient's disease in order to offer effective treatment (TAGCN, 2012; Song et al., 2015).

The interactome is (so far) the last frontier: it defines the huge number of ever-changing interactions of the component macromolecules, both in the cell and the extracellular matrix) forming living organisms and it is meant to describe, eventually, all the interactions between all the molecules present in a cell (Vidal et al., 2011; Aitchison and Rout, 2015) or, in a more restrictive definition, all the protein-protein interactions (Cusick et al., 2005). The final product of the events described by the interactome is meant to be the cell phenotype; that is, the cell itself and its behaviour (Aitchison and Rout, 2015). Although we can now have the genetic map of an organism quickly and accurately, we still have a poor understanding of the interactome and we know almost nothing about the dynamic of the macromolecule involved (Aitchison and Rout, 2015). While highly complex, the interactome is still a simplification based on representing each macromolecule as a node and the interactions as edges and will be as informative as the quality of the input data allow. Of course, this method does not need to be applied to all cell scenarios, but it is currently used to study mostly simpler situations (e.g. the metabolic or the gene regulatory interactome, the cell cycle interactome; Vidal et al., 2011). The proteins are the central hubs in interactome and, as a cell can contain thousands of proteins, there is an estimate of approximately 50,000 pairs of interactions in a human cell, notwithstanding the interaction between proteins and other molecules (Aitchison and Rout, 2015).

Instability: From bench to bedside

In order to increase in complexity, biological systems had to become increasingly stable; however, instability is always present at variable degrees in different systems. A simple example is the variability of the blood flow pulse (Rew, 1999). In a linear system, a change will always produce the same predictable response but when the stimulation is not proportional to the effect, and when the same stimulus can produce variable outcome, then we have a non-linear system (Rew, 1999; Higgins, 2002). Chaos is the branch of mathematics that explore the disproportionate changes originating form a given stimulus (Rew, 1999). Chaos is defined by three characteristics: (1) it is an oscillatory, irregular process; (2) chaotic phenomenon resemble, but are not, random events; and (3) a chaotic system is highly dependent on initial conditions. This means that even small changes, like the changes of abnormal transcript of a gene following a chromosomal translocation, can lead to profoundly different outcome (Rew, 1999) making cancer subject to chaotic behaviour (Michor et al., 2011). As this chaotic behaviour affects cancer at all levels, from the biology of a single cells, to the development of metastases and response to treatment, integration of mathematical and biological approaches is a promising way to develop further cancer studies (Michor et al., 2011).

Conclusions

Recent breakthrough in biology have significantly improved our knowledge of cancer and the possibility to treat it more effectively. However, these very same discoveries have unveiled an increasingly complex picture of this disease—so we can conclude that our present level of knowledge '. . . is not the end. It is not even the beginning of the end. But it is, perhaps, the end of the beginning' (Winston Churchill, on the British victory at El Alamein, London, 10 November 1942).

TAKE-HOME MESSAGE

- It's complicated!

OPEN QUESTIONS

- Increasing!

FURTHER READING

Bray, D. (2009). *Wetware: A Computer in Every Living Cell.* New Haven and London: Yale University Press.

Carey, N. (2012). *The Epigenetic Revolution: How Modern Biology is Rewriting Our Understanding of Genetics, Disease and Inheritance.* London: Icons books.

DeVita, V. T., Lawrence, T. S., & Rosenberg, S. A. (2015). *Primer of the Molecular Biology of Cancer*, 2nd edition. Philadelphia, Wolters Kluwer.

Goldacre, B. (2014). *I Think You'll Find It's A Bit More Complicated Than That.* London: Fourth Estate.

Mishra, N. (2010). Introduction to proteomics. Principles and applications. Hoboken, NJ: Wiley.

Monod, J. (1997). Chance and Necessity. London: Penguin Books.

Pezzella, F. & Winkler, F. (2018). Tumors and blood vessel interactions: a changing hallmark of cancer. In: Boffetta, P., Bosman, F., Colditz, G. A., et al. (eds) *Encyclopedia of Cancer*, 3rd edition. Oxford: Elsevier.

Robinson, C. V., Ali, A., Baumeister, W. (2007). The molecular sociology of the cell. *Nature*, **450**, 973–82.

Smith, L. (2007). *Chaos: A Very Short Introduction.* Oxford: Oxford University Press.

REFERENCES

Aitchison, J. D. & Rout, M. P. (2015). The interactome challenge. *J Cell Biol*, **211**(4), 729–32.

Alizadeh, A. A., Eisen, M. B., Davis, R. E., et al. (2000). Distinct types of diffuse large B-cell lymphoma identified by gene expression profiling. *Nature*, **403**(6769), 503–11.

Cusick, M. E., Klitgord, N., Vidal, M., & Hill, D. E. (2005). Interactome: gateway into systems biology. *Hum Mol Genet*, **14**(2), R171–81.

Floor, S. L., Dumont, J. E., Maenhaut, C., & Raspe, E. (2012). Hallmarks of cancer: of all cancer cells, all the time? *Trends Mol Med*, **18**(9), 509–15.

Hanahan, D. & Weinberg, R. A. (2000). The hallmarks of cancer. *Cell*, **100**(1), 57–70.

Hanahan, D. & Weinberg, R. A. (2011). Hallmarks of cancer: the next generation. *Cell*, **144**(5), 646–74.

Hickman, M., Thain, M., Turvey, R., et al (2004). *Dictionary of Biology*. London: Penguin Books.

Higgins, J. P. (2002). Nonlinear systems in medicine. *Yale J Biol Med*, **75**(5–6), 247–60.

Horne, S. D., Pollick, S. A., & Heng, H. H. (2015). Evolutionary mechanism unifies the hallmarks of cancer. *Int J Cancer*, **136**(9), 2012–21.

Johannsen, W. (1909). *Elemente der exakten Erblichkeitslehre*. Jena: Gustav Fischer.

Lazebnik, Y. (2010). What are the hallmarks of cancer? *Nat Rev Cancer*, **10**(4), 232–3.

Michor, F., Liphardt, J., Ferrari, M., & Widom, J. (2011). What does physics have to do with cancer? *Nat Rev Cancer*, **11**, 657–70.

Nicholson, G. W. (1926). *The Nature of Tumour Formation*. Erasmus Wilson Lectures. Cambridge: Heller.

Nowell, P. C. & Hungerford, D. A. (1960). A minute chromosome in human chronic granulocytic leukaemia. *Science*, **132**(3438), 1497.

Rew, D. A. (1999). Tumour biology, chaos and non-linear dynamics. *Eur J Surg Oncol*, **25**(1), 86–9.

Song, Q., Merajver, S. D., & Li, J. Z. (2015). Cancer classification in the genomic era: five contemporary problems. *Hum Genomics*, **9**, 27.

Sonnenschein, C. & Soto, A. M. (2013). The aging of the 2000 and 2011 Hallmarks of Cancer reviews: a critique. *J Biosci*, **38**(3), 651–63.

TAGCN (2012). Comprehensive molecular portraits of human breast tumours. *Nature*, **490**(7418), 61–70.

Vidal, M., Cusick, M. E., & Barabasi, A. L. (2011). Interactome networks and human disease. *Cell*, **144**(6), 986–98.

Willis, R. A. (1960). *Pathology of Tumours*. London: Butterworths.

Index

Notes: Tables, figures and boxes are indicated by an italic *t*, *f* and *b* following the page number.